Chapter 4 Systems of Equations and Inequalities

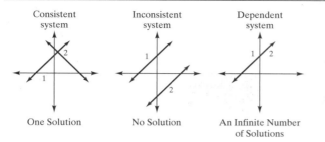

Consistent system	Inconsistent system	Dependent system
One Solution	No Solution	An Infinite Number of Solutions

Cramer's Rule:

Given a system of equations of the form

$$a_1 x + b_1 y = c_1 \atop a_2 x + b_2 y = c_2 \quad \text{then } x = \frac{\begin{vmatrix} c_1 & b_1 \\ c_2 & b_2 \end{vmatrix}}{\begin{vmatrix} a_1 & b_1 \\ a_2 & b_2 \end{vmatrix}} \text{ and } y = \frac{\begin{vmatrix} a_1 & c_1 \\ a_2 & c_2 \end{vmatrix}}{\begin{vmatrix} a_1 & b_1 \\ a_2 & b_2 \end{vmatrix}}$$

A system of linear equations may be solved: (a) graphically, (b) by the substitution method, (c) by the addition or elimination method, (d) by matrices or (e) by determinants.

$$\begin{vmatrix} a_1 & b_1 \\ a_2 & b_2 \end{vmatrix} = a_1 b_2 - a_2 b_1$$

Chapter 5 Polynomials and Polynomial Functions

FOIL method to multiply two binomials

$$(a + b)(c + d) = a \cdot c + a \cdot d + b \cdot c + b \cdot d$$
$$\qquad\qquad\qquad\quad F \qquad O \qquad I \qquad L$$

Pythagorean Theorem:

$$\text{leg}^2 + \text{leg}^2 = \text{hyp}^2 \quad \text{or} \quad a^2 + b^2 = c^2$$

Square of a binomial:

$(a + b)^2 = a^2 + 2ab + b^2.$
$(a - b)^2 = a^2 - 2ab + b^2$

Product of the sum and difference of the same two terms (also called the difference of two squares):

$(a + b)(a - b) = a^2 - b^2$

Perfect square trinomials:

$a^2 + 2ab + b^2 = (a + b)^2, \quad a^2 - 2ab + b^2 = (a - b)^2$

Sum of two cubes:

$a^3 + b^3 = (a + b)(a^2 - ab + b^2)$

Difference of two cubes:

$a^3 - b^3 = (a - b)(a^2 + ab + b^2)$

Standard form of a quadratic equation:

$ax^2 + bx + c = 0, a \neq 0$

Zero-factor property: If $a \cdot b = 0$, then either $a = 0$ or $b = 0$, or both a and $b = 0$.

Chapter 6 Rational Expressions and Equations

To Multiply Rational Expressions:

1. Factor all numerators and denominators.
2. Divide out any common factors.
3. Multiply numerators and multiply denominators.
4. Simplify the answer when possible.

To Divide Fractional Expressions:

Invert the divisor and then multiply the resulting rational expressions.

To Add or Subtract Rational Expressions:

1. Write each fraction with a common denominator.
2. Add or subtract the numerators while keeping the common denominator.
3. When possible factor the numerator and simplify the fraction.

Similar Figures: Corresponding angles and corresponding sides are in proportion.

Proportion: If $\dfrac{a}{b} = \dfrac{c}{d}$ then $ad = bc$

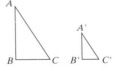

Variation: direct, $y = kx$; inverse, $y = \dfrac{k}{x}$; joint, $y = kxz$

Algebra
For College
Students

Algebra
For College
Students

Second Edition

Allen R. Angel
Monroe Community College

with assistance from
Richard Semmler
Northern Virginia Community College

PEARSON EDUCATION, Inc.
Upper Saddle River, New Jersey 07458

Library of Congress Cataloging-in-Publication Data

Algebra for college students.—2nd ed. / Allen R. Angel, with assistance from Richard Semmler.
    p. cm.
Includes bibliographical references and index.
ISBN 0-13-141085-7 (student ed.) — ISBN 0-13-141092-X (annotated instructor's ed.)
1. Algebra. I. Semmler, Richard. II. Title.

QA154.3 .A53 2004
512.9—dc21

2003193065

Senior Acquisitions Editor: Paul Murphy
Editor in Chief: Christine Hoag
Senior Managing Editor: Linda Mihatov Behrens
Production Management: Elm Street Publishing Service, Inc.
Executive Managing Editor: Kathleen Schiaparelli
Vice President/Director of Production and Manufacturing: David W. Riccardi
Production Assistant: Nancy Bauer
Assistant Managing Editor, Math Media Production: John Matthews
Assistant Manufacturing Manager/Buyer: Michael Bell
Manufacturing Manager: Trudy Pisciotti
Executive Marketing Manager: Eilish Collins Main
Marketing Project Manager: Barbara Herbst
Marketing Assistant: Annett Uebel
Development Editor: Don Gecewicz
Editor in Chief, Development: Carol Trueheart
Media Project Manager, Development Math: Audra J. Walsh
Editorial Assistant/Print Supplements Editor: Kerri-Ann O'Donnell
Art Director/Cover Designer: John Christiana
Interior Designer: Jonathan Boylan
Art Editor: Thomas Benfatti
Creative Director: Carole Anson
Director of Creative Services: Paul Belfanti
Director, Image Resource Center: Melinda Reo
Manager, Rights and Permissions: Zina Arabia
Interior Image Specialist: Beth Brenzel
Image Permission Coordinator: Charles Morris
Photo Researcher: Sheila Norman
Cover Photo: ©Royalty-Free/CORBIS
Cover Image Specialist: Karen Sanatar
Art Studio: Scientific Illustrators
Compositor: Preparé, Inc.

© 2004, 2000 by Pearson Education, Inc.
Pearson Education, Inc.
Upper Saddle River, New Jersey 07458

Printed in the United States of America
10 9 8 7 6 5 4 3 2 1

ISBN 0-13-141085-7

Pearson Education LTD., *London*
Pearson Education Australia PTY, Limited, *Sydney*
Pearson Education Singapore, Pte. Ltd
Pearson Education North Asia Ltd, *Hong Kong*
Pearson Education Canada, Ltd., *Toronto*
Pearson Educación de Mexico, S.A. de C.V.
Pearson Education — Japan, *Tokyo*
Pearson Education Malaysia, Pte. Ltd

To my mother,
Sylvia Angel-Baumgarten

And to my step-father,
Lenny Baumgarten

And to my mother-in-law,
Ruth Pollinger

Contents

Preface   **xii**

To the Student   **xxv**

1   Basic Concepts   1

1.1   Study Skills for Success in Mathematics,
and Using a Calculator   **2**
1.2   Sets and Other Basic Concepts   **6**
1.3   Properties of and Operations with Real Numbers   **18**
1.4   Order of Operations   **29**
1.5   Exponents   **42**
1.6   Scientific Notation   **52**
Chapter Summary   **60**
Chapter Review Exercises   **62**
Chapter Practice Test   **64**

2   Equations and Inequalities   66

2.1   Solving Linear Equations   **67**
2.2   Problem Solving and Using Formulas   **79**
2.3   Applications of Algebra   **90**
2.4   Additional Application Problems   **104**
2.5   Solving Linear Inequalities   **115**
2.6   Solving Equations and Inequalities Containing
Absolute Values   **129**
Chapter Summary   **140**
Chapter Review Exercises   **141**
Chapter Practice Test   **144**
Cumulative Review Test   **145**

3   Graphs and Functions   147

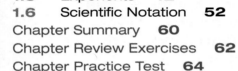

3.1   Graphs   **148**
3.2   Functions   **163**
3.3   Linear Functions: Graphs and Applications   **179**
3.4   The Slope–Intercept Form of a Linear Equation   **191**
3.5   The Point–Slope Form of a Linear Equation   **205**
3.6   The Algebra of Functions   **215**
3.7   Graphing Linear Inequalities   **224**

Chapter Summary **228**
Chapter Review Exercises **229**
Chapter Practice Test **233**
Cumulative Review Test **234**

4   Systems of Equations and Inequalities   236

4.1   Solving Systems of Linear Equations in Two Variables   **237**
4.2   Solving Systems of Linear Equations in Three Variables   **249**
4.3   Systems of Linear Equations: Applications and Problem Solving   **256**
4.4   Solving Systems of Equations Using Matrices   **270**
4.5   Solving Systems of Equations Using Determinants and Cramer's Rule   **279**
4.6   Solving Systems of Linear Inequalities   **287**
Chapter Summary **292**
Chapter Review Exercises **294**
Chapter Practice Test **296**
Cumulative Review Test **297**

5   Polynomials and Polynomial Functions   299

5.1   Addition and Subtraction of Polynomials   **300**
5.2   Multiplication of Polynomials   **311**
5.3   Division of Polynomials and Synthetic Division   **321**
5.4   Factoring a Monomial from a Polynomial and Factoring by Grouping   **332**
5.5   Factoring Trinomials   **340**
5.6   Special Factoring Formulas   **351**
5.7   A General Review of Factoring   **359**
5.8   Polynomial Equations   **364**
Chapter Summary **377**
Chapter Review Exercises **378**
Chapter Practice Test **383**
Cumulative Review Test **384**

6   Rational Expressions and Equations   385

6.1   The Domains of Rational Functions and Multiplication and Division of Rational Expressions   **386**
6.2   Addition and Subtraction of Rational Expressions   **396**
6.3   Complex Fractions   **408**
6.4   Solving Rational Equations   **414**
6.5   Rational Equations: Applications and Problem Solving   **427**
6.6   Variation   **438**

Chapter Summary   447
Chapter Review Exercises   448
Chapter Practice Test   451
Cumulative Review Test   452

7 Roots, Radicals, and Complex Numbers   454

7.1   Roots and Radicals   455
7.2   Rational Exponents   464
7.3   Simplifying Radicals   473
7.4   Adding, Subtracting, and Multiplying Radicals   481
7.5   Dividing Radicals   488
7.6   Solving Radical Equations   497
7.7   Complex Numbers   509
Chapter Summary   518
Chapter Review Exercises   519
Chapter Practice Test   523
Cumulative Review Test   524

8 Quadratic Functions   525

8.1   Solving Quadratic Equations by Completing the Square   526
8.2   Solving Quadratic Equations by the Quadratic Formula   536
8.3   Quadratic Equations: Applications and Problem Solving   549
8.4   Writing Equations in Quadratic Form   559
8.5   Graphing Quadratic Functions   566
8.6   Quadratic and Other Inequalities in One Variable   585
Chapter Summary   596
Chapter Review Exercises   596
Chapter Practice Test   600
Cumulative Review Test   601

9 Exponential and Logarithmic Functions   603

9.1   Composite and Inverse Functions   604
9.2   Exponential Functions   616
9.3   Logarithmic Functions   626
9.4   Properties of Logarithms   633
9.5   Common Logarithms   639
9.6   Exponential and Logarithmic Equations   646
9.7   Natural Exponential and Natural Logarithmic Functions   652

Chapter Summary **664**
Chapter Review Exercises **665**
Chapter Practice Test **668**
Cumulative Review Test **669**

10   Characteristics of Functions and Their Graphs   671

10.1 Continuous and Discontinuous Functions **672**
10.2 Characteristics of Graphs of Polynomial Functions **680**
10.3 Finding the Rational Zeros of a Polynomial Function **691**
10.4 Finding the Real Zeros of a Polynomial Function **704**
10.5 Characteristics of Graphs of Rational Functions **713**
Chapter Summary **727**
Chapter Review Exercises **728**
Chapter Practice Test **731**
Cumulative Review Test **732**

11   Conic Sections   734

11.1 The Parabola and the Circle **735**
11.2 The Ellipse **746**
11.3 The Hyperbola **753**
11.4 Nonlinear Systems of Equations
and Their Applications **761**
11.5 Nonlinear Systems of Inequalities **770**
Chapter Summary **775**
Chapter Review Exercises **775**
Chapter Practice Test **777**
Cumulative Review Test **778**

12   Sequences, Series, and Probability   780

12.1 Sequences and Series **781**
12.2 Arithmetic Sequences and Series **789**
12.3 Geometric Sequences and Series **796**
12.4 Mathematical Induction **808**
12.5 The Binomial Theorem **815**
12.6 The Counting Principle, Permutations,
and Combinations **820**
12.7 Probability **832**
Chapter Summary **841**
Chapter Review Exercises **843**
Chapter Practice Test **846**
Cumulative Review Test **847**

Appendix

Geometric Formulas   **849**

Answers   **A1**

Index   **I1**

Photo Credits   **P1**

Preface

This book was written for college students who have successfully completed a first course in elementary algebra. My primary goal was to write a book that students can read, understand, and enjoy. To achieve this goal I have used short sentences, clear explanations, and many detailed, worked-out examples. I have tried to make the book relevant to college students by using practical applications of algebra throughout the text.

Features of the Text

Full-Color Format   Color is used pedagogically in the following ways:

- Important definitions and procedures are color screened.

- Color screening or color type is used to make other important items stand out.

- Artwork is enhanced and clarified with use of multiple colors.

- The full-color format allows for easy identification of important features by students.

- The full-color format makes the text more appealing and interesting to students.

Readability   One of the most important features of the text is its readability. The book is very readable, even for those with weak reading skills. Short, clear sentences are used and more easily recognized, and easy-to-understand language is used whenever possible.

Accuracy   Accuracy in a mathematics text is essential. To ensure accuracy in this book, mathematicians from around the country have read the pages carefully for typographical errors and have checked all the answers.

Connections   Many of our students do not thoroughly grasp new concepts the first time they are presented. In this text, we encourage students to make connections. That is, we introduce a concept, then later in the text briefly reintroduce it and build upon it.

Often an important concept is used in many sections of the text. Students are reminded where the material was seen before, or where it will be used again. This also serves to emphasize the importance of the concept. Important concepts are also reinforced throughout the text in the Cumulative Review Exercises and Cumulative Review Tests.

Chapter Opening Application   Each chapter begins with a real-life application related to the material covered in the chapter. By the time students complete the chapter, they should have the knowledge to work the problem.

A Look Ahead   This feature at the beginning of each chapter gives students a preview of the chapter. This feature also indicates where this material will be used again in other chapters of the book. This material helps students see the connections between various topics in the book and the connection to real-world situations.

The Use of Icons   At the beginning of each chapter and of each section, a variety of icons are illustrated. These icons are provided to tell students where they may be able to get extra help if needed. There are icons for the *Student's Solution Manual*, ▮ ; the *Student's Study Guide*, ▯ ; *CDs and videotapes*, ▤ ; *Math Pro 4/5 Software*, ◉ ; the *Prentice Hall Tutor Center*, ▯ ; and the *Angel Website*, 🖳 . Each of these items will be discussed shortly.

Keyed Section Objectives   Each section opens with a list of skills that the student should learn in that section. The objectives are then keyed to the appropriate portions of the sections with red numbers such as **1**.

Problem Solving   George Polya's five-step problem-solving procedure is discussed in Section 1.2. Throughout the book problem solving and Polya's problem-solving procedure are emphasized.

Practical Applications   Practical applications of algebra are stressed throughout the text. Students need to learn how to translate application problems into algebraic symbols. The problem-solving approach used throughout this text gives students ample practice in setting up and solving application problems. The use of practical applications motivates students.

Detailed, Worked-Out Examples   A wealth of examples have been worked out in a step-by-step, detailed manner. Important steps are highlighted in color, and no steps are omitted until after the student has seen a sufficient number of similar examples.

Now Try Exercise   In each section, students are asked to work exercises that parallel the examples given in the text. These Now Try Exercises make the students *active*, rather than passive, learners and they reinforce the concepts as students work the exercises. Through these exercises students have the opportunity to immediately apply what they have learned. Now Try Exercises are indicated in green type, such as **35**, in the exercise sets.

Study Skills Section   Many students taking this course have poor study skills in mathematics. Section 1.1, the first section of this text, discusses the study skills needed to be successful in mathematics. This section should be very beneficial for your students and should help them to achieve success in mathematics.

Helpful Hints   The Helpful Hint boxes offer useful suggestions for problem solving and other varied topics. They are set off in a special manner so that students will be sure to read them.

Helpful Hints—Study Tips   This is a new feature. These Helpful Hint–Study Tips boxes offer valuable information on items related to studying and learning the material.

Avoiding Common Errors   Errors that students often make are illustrated. The reasons why certain procedures are wrong are explained, and the correct procedure for working the problem is illustrated. These Avoiding Common Errors boxes will help prevent your students from making those errors we see so often.

Mathematics in Action   This new feature stresses the need for and the uses of mathematics in real-life situations. Examples of the use of mathematics in many professions, and how we use mathematics daily without ever giving it much thought, are given. This can be a motivational feature for your students and can give them a better appreciation of mathematics.

Using Your Calculator   The Using Your Calculator boxes, placed at appropriate locations in the text, are written to reinforce the algebraic topics presented in the section and to give the student pertinent information on using a scientific calculator to solve algebraic problems.

Using Your Graphing Calculator   Using Your Graphing Calculator boxes are placed at appropriate locations throughout the text. They reinforce the algebraic topics taught and sometimes offer alternate methods of working problems. This book is designed to give the instructor the option of using or not using a graphing calculator in their course. Many Using Your Graphing Calculator boxes contain graphing calculator exercises, whose answers appear in the answer section of the book. The illustrations shown in the Using Your Graphing Calculator boxes are from a Texas Instruments 83 Plus calculator. The Using Your Graphing Calculator boxes are written assuming that the student has no prior graphing calculator experience.

Exercise Sets

The exercise sets are broken into three main categories: Concept/Writing Exercises, Practice the Skills, and Problem Solving. Many exercise sets also contain Challenge Problems and/or Group Activities. Each exercise set is graded in difficulty. The early problems help develop the student's confidence, and then students are eased gradually into the more difficult problems. A sufficient number and variety of examples are given in each section for the student to successfully complete even the more difficult exercises. The number of exercises in each section is more than ample for student assignments and practice.

Concept/Writing Exercises   Most exercise sets include exercises that require students to write out the answers in words. These exercises improve students' understanding and comprehension of the material. Many of these exercises involve problem solving and conceptualization and help develop better reasoning and critical thinking skills. Writing exercises are indicated by the symbol .

Problem Solving Exercises   These exercises help students become better thinkers and problem solvers. Many of these exercises involve real-life applications of algebra. It is important for students to be able to apply what they learn to real-life situations. Many problem solving exercises help with this.

Challenge Problems   These exercises, which are part of many exercise sets, provide a variety of problems. Many were written to stimulate student thinking. Others provide additional applications of algebra or present material from future sections of the book so that students can see and learn the material on their own before it is covered in class. Others are more challenging than those in the regular exercise set.

Cumulative Review Exercises   All exercise sets (after the first two) contain questions from previous sections in the chapter and from previous chapters. These cumulative review exercises reinforce topics that were previously covered and help students retain the earlier material, while they are learning the new material. For the students' benefit the Cumulative Review Exercises are keyed to the section where the material is covered using brackets, such as *[3.4]*.

Video Icon Exercises   The exercises that are worked out in detail on the videotapes are marked with the video icon, ⬛. This will prove helpful to your students.

Group Activities   Many exercise sets have group activity exercises that lead to interesting group discussions. Many students learn well in a cooperative learning atmosphere, and these exercises will get students talking mathematics to one another.

Chapter Summary   At the end of each chapter is a chapter summary that includes a glossary and important chapter facts.

Chapter Review Exercises   At the end of each chapter are review exercises that cover all types of exercises presented in the chapter. The review exercises are keyed, using color numbers and brackets, to the sections where the material was first introduced.

Chapter Practice Tests   The comprehensive end-of-chapter practice test will enable the students to see how well they are prepared for the actual class test. The Instructor's Test Manual includes several forms of each chapter test that are similar to the student's practice test. Multiple choice tests are also included in the Instructor's Test Manual.

Cumulative Review Tests   These tests, which appear at the end of each chapter after the first, test the students' knowledge of material from the beginning of the book to the end of that chapter. Students can use these tests for review, as well as for preparation for the final exam. These exams, like the cumulative review exercises, will serve to reinforce topics taught earlier.

The answers to the Cumulative Review Test questions directly follow the test so that students can quickly check their work. After each answer, the section and objective numbers where that material was covered are given using brackets, such as [Sec. 4.2, Obj. 5].

Answers   The *odd answers* are provided for the exercise sets. *All answers* are provided for the Using Your Graphing Calculator Exercises, Cumulative Review Exercises, Review Exercises, Practice Tests, and the Cumulative Review Tests. *Answers* are not provided for the Group Activity exercises since we want students to reach agreement by themselves on the answers to these exercises.

National Standards

Recommendations of the *Curriculum and Evaluation Standards for School Mathematics*, prepared by the National Council of Teachers of Mathematics (NCTM), and *Crossroads in Mathematics: Standards for Introductory College Mathematics Before Calculus*, prepared by the American Mathematical Association of Two Year Colleges (AMATYC), are incorporated into this edition.

Prerequisite

The prerequisite for this course is a working knowledge of elementary algebra. Although some elementary algebra topics are briefly reviewed in the text, students should have a basic understanding of elementary algebra before taking this course.

Modes of Instruction

The format and readability of this book lends itself to many different modes of instruction. The constant reinforcement of concepts will result in greater understanding and retention of the material by your students.

The features of the text and the large variety of supplements available make this text suitable for many types of instructional modes including:

- lecture
- distance learning
- self-paced instruction
- modified lecture
- cooperative or group study
- learning laboratory

Changes in the Second Edition

When I wrote the second edition, I considered many letters and reviews I got from students and faculty alike. I would like to thank all of you who made suggestions for improving the second edition. I would also like to thank the many instructors and students who wrote to inform me of how much they enjoyed, appreciated, and learned from the text.

Some of the changes made in the second edition of the text include:

- The chapter on Roots, Radicals, and Complex Numbers has been reorganized and rewritten for greater clarity. The chapter now has a better flow of material.

- More emphasis has been placed on geometry. There are more geometry-related examples and exercises throughout the book.

- A greater variety of exercises has been added to exercise sets throughout the book. In general, the exercise sets have been greatly enhanced.

- In selected sections, more difficult exercises have been added at the end of the exercise sets.

- More varied, real-life situations involving intermediate algebra material have been added to the examples and exercises throughout the text.

- Complementary and supplementary angles have been moved from the exercises into the body of the text.

- Selected material in the Graphs and Functions chapter has been reorganized and rewritten for better understanding by students.

- More concept/writing exercises have been added where appropriate.

- In the Polynomials and Polynomial Functions chapter, the pertinent rules of exponents are now reviewed, and examples of the rules are now provided, before the appropriate material on multiplication and division of polynomials is discussed.

- More material on multiplication of radicals is provided.

- The discussion of complex fractions more clearly indicates to students when it is advantageous to use each of the methods to simplify complex fractions.

- Selected material in the Exponential and Logarithmic Functions chapter has been rewritten for greater clarity.

- The basic characteristics of both exponential and logarithmic graphs are now discussed more completely.

- The number of both examples and exercises in the Exponential and Logarithmic Functions chapter has been increased.

- Material on finding the area of an ellipse has been added to the Conic Sections chapter.

- Certain definitions in the Sequences, Series, and the Binomial Theorem chapter have been rewritten for greater clarity and for better understanding by students.

- The book has a new design with the purpose of making the exercise sets flow more smoothly so that exercises will be easy to spot and identify.

- The Cumulative Review Tests now have the answers directly following the test so that students can get immediate feedback. In addition, the section and objective numbers where the material was discussed are given after the answer.

- A Look Ahead has replaced the Preview and Perspective feature. The information provided gives students an overview of the chapter and how it relates to other material in the book and to real-life situations.

- A new feature called Mathematics in Action has been added. This feature stresses the need for and the importance of mathematics in real life. This may be motivational for your students.

- More and new Helpful Hints and Avoiding Common Errors boxes have been added where appropriate.

- Helpful Hint—Study Tips boxes have been added. These reinforce and expand upon the Study Skills for Success in Mathematics covered in Section 1.1.

- Application problems throughout the book have been updated and made more interesting.

- Using Your Graphing Calculator boxes now show keystrokes and screens from a Texas Instruments 83 Plus calculator.

- When factoring by grouping, the common factor is now placed on the left for consistency with other factoring problems.

- More photos have been added to the text to make the text more attractive and interesting for students.

- The basic colors used in the text have been softened to make the text easier to read.

Supplements to the Second Edition

For this edition of the book the author has personally coordinated the development of the *Student's Solution Manual* and the *Instructor's Solution Manual*.

Experienced mathematics professors who have prior experience in writing supplements, and whose works have been of superior quality, have been carefully selected for authoring the supplements.

For Instructors

Printed Supplements

Annotated Instructor's Edition (0-13-141092-X)

- Contains all of the content found in the current edition.
- Answers to exercises are printed on the same text page (graphed answers are in a special graphing section at the back of the text).
- Teaching Tips throughout the text are placed at key points in the margin.

NEW! Instructor's Examples Manual (0-13-141772-X)

- Provides similar examples for every example presented in the text. Instructors will find this very convenient when they wish to work additional examples on the chalkboard.

Instructor's Solutions Manual (0-13-141091-1)

- Solutions to even-numbered section exercises.
- Solutions to every exercise found in the Chapter Reviews, Chapter Tests, and Cumulative Review Tests.

Instructor's Test Manual (0-13-141090-3)

- Two free-response Pretests per chapter.
- Eight Chapter Tests per chapter (3 multiple choice, 5 free response).
- Two Cumulative Review Tests (one multiple choice, one free response) every two chapters.
- Eight Final Exams (3 multiple choice, 5 free response).
- Twenty additional exercises per section for added test exercises if needed.

Media Supplements

NEW! TestGen with QuizMaster CD-ROM (Windows/Macintosh) (0-13-141089-X)

- Algorithmically driven, text-specific testing program.
- Networkable for administering tests and capturing grades online.
- Edit and add your own questions to create a nearly unlimited number of tests and worksheets.
- Use the new "Function Plotter" to create graphs.
- Tests can be easily exported to HTML so they can be posted to the Web for student practice.
- Includes an e-mail function for network users, enabling instructors to send a message to a specific student or an entire group.
- Network-based reports and summaries for a class or student and for cumulative or selected scores available.

 NEW! MathPro5 (Instructor Version) (0-13-141096-2)

- Online customizable tutorial, diagnostic, and assessment program.
- Text-specific at the learning objective level.
- Diagnostic option identifies student skills and provides individual learning plan and tutorial reinforcement.
- Integration of TestGen allows for testing to operate within the tutorial environment.
- Course management tracking of tutorial and testing activity.

MathPro Explorer 4.0 (Network Version IBM/Mac) (0-13-141093-8)

- Enables instructors to create either customized or algorithmically generated practice quizzes from any section of a chapter.
- Includes e-mail function for network users enabling instructors to send a message to a specific student or to an entire group.
- Network-based reports and summaries for a class or student and for cumulative or selected scores.

Companion Website (*www.prenhall.com/angel*)

- Create a customized online syllabus with Syllabus Manager.
- Assign quizzes or monitor student self-quizzes by having students e-mail results.
- Destination links provide additional opportunities to explore related sites.

For Students

Printed Supplements

 Student's Solutions Manual
SSM (0-13-141769-X)

- Solutions to odd-numbered section exercises.

- Solutions to every (even and odd) exercise found in the Chapter Reviews, Chapter Tests, and Cumulative Review Tests.

Student's Study Guide
Study Guide (0-13-141771-1)

- Includes additional worked-out examples and additional exercises, practice tests, and answers.

- Emphasizes important concepts and includes information to help students study and succeed in mathematics.

Media Supplements

NEW! MathPro5 (Student Version)
MathPro 4/5 (0-13-142484-X)

- Online customizable tutorial, diagnostic, and assessment program.

- Text-specific at the learning objective level.

- Algorithmically driven for a virtually unlimited number of practice problems with immediate feedback.

- Includes "Watch" screen videoclips.

- Step-by-step solutions.

- Summary of progress.

MathPro 4.0 Explorer Student Version
MathPro 4/5 (IBM/Mac) (0-13-141095-4)

- Available on CD-ROM for stand-alone use or can be networked in the school laboratory.

- Text-specific tutorial exercises and instructions at the objective level.

- Algorithmically generated Practice Problems.

- Includes "Watch" screen videoclips.

 Videotape Series
CD/Video (0-13-141088-1)

- Keyed to each section of the text.

- Step-by-step solutions to exercises from each section of the text. Exercises from the text that are worked in the videos are marked with a video icon.

 NEW! Digitized Lecture Videos on CD-ROM
CD/Video (0-13-141087-3)

- The entire set of Angel, *Algebra For College Students*, Second Edition, lecture videotapes in digital form.

- Convenient access anytime to video tutorial support from a computer at home or on campus.

- Available shrinkwrapped with the text or stand-alone.

NEW! Prentice Hall Mathematics
PH Math
Tutor Center **Tutor Center**

- Staffed with developmental math instructors and open 5 days a week, 7 hours per day.

- Obtain help for examples and exercises in Angel, *Algebra For College Students*, Second Edition, via toll-free telephone, fax, or e-mail.

- The Prentice Hall Mathematics Tutor Center is accessed through a registration number that may be bundled with a new text or purchased separately with a used book. Visit *http://www.prenhall.com/ tutorcenter* to learn more.

Companion Website
prenhall.com/Angel (*www.prenhall.com/angel*)

- Practice problems and quizzes with instant feedback.

- Graphing calculator keystroke instructions.

- Destination links provide additional opportunities to explore related sites.

Acknowledgments

Writing a textbook is a long and time-consuming project. Many people deserve thanks for encouraging and assisting me with this project. Most importantly I would like to thank my wife, Kathy and sons, Robert and Steven. Without their constant encouragement and understanding, this project would not have become a reality. I would also like to thank my daughter-in-law, Kathy, for her support.

I would like to thank Richard Semmler and Larry Clar of Northern Virginia Community College and Dennis Runde of Manatee Community College for assisting me with many aspects of the book. Special thanks to Richard, who also worked with me throughout the entire production process.

I would like to thank Aimee Calhoun and Larry Clar of Monroe Community College and Lauri Semarne for reading the pages and checking the answers in the book. Mitchel Levy of Broward Community College also deserves my thanks for helping with the Cumulative Review Tests.

I would like to thank my editors at Prentice Hall, Paul Murphy and Ann Heath, my developmental editor, Don Gecewicz, and my project editor, Phyllis Crittenden, of Elm Street Publishing Services, Inc., for their valuable suggestions and conscientiousness with this project.

I would like to thank the following reviewers and proofreaders for their thoughtful comments and suggestions:

Laura Adkins, *Missouri Southern State College*
Arthur Altshiller, *Los Angeles Valley College*
Peter Arvanites, *State University of New York—Rockland Community College*
Jon Becker, *Indiana University*
Paul Boisvert, *Oakton Community College (IL)*
Beverly Broomell, *Suffolk County Community College (NY)*
Lavon Burton, *Abilene Christian University (TX)*
Marc Campbell, *Daytona Beach Community College (FL)*
Mitzi Chaffer, *Central Michigan University*
Terry Cheng, *Irvine Valley College (CA)*
Ted Corley, *Arizona State University and Glendale Community College*
Charles Curtis, *Missouri Southern State College*
Gary Egan, *Monroe Community College (NY)*
Elizabeth Farber, *Bucks County Community College (PA)*
Warrene Ferry, *Jones County Junior College (MS)*

Gary Glaze, *Spokane Falls Community College (WA)*
James Griffiths, *San Jacinto College (TX)*
Kathy Gross, *Cayuga Community College (NY)*
Abdollah Hajikandi, *State University of New York—Buffalo*
Mary Beth Headlee, *Manatee Community College (FL)*
Kelly Jahns, *Spokane Community College (WA)*
Judy Kasabian, *El Camino College (CA)*
Maryanne Kirkpatrick, *Laramie County Community College (WY)*
Marcia Kleinz, *Atlantic Cape Community College (NJ)*
Shannon Lavey, *Cayuga Community College (NY)*
Shywanda Moore, *Meridian Community College (MS)*
Catherine Moushon, *Elgin Community College (IL)*
Kathy Nickell, *College of DuPage (IL)*
Shelle Patterson, *Moberly Area Community College (MO)*
Dennis Reissig, *Suffolk County Community College (NY)*
Linda Retterath, *Mission College (CA)*
Dale Rohm, *University of Wisconsin—Stevens Point*
Troy Rux, *Spokane Falls Community College (WA)*
Hassan Saffari, *Prestonburg Community College (KY)*
Rick Silvey, *St. Mary College (KS)*
Julia Simms, *Southern Illinois University—Edwardsville*
Jed Soifer, *Atlantic Cape Community College (NJ)*
Elizabeth Suco, *Miami-Dade Community College (FL)*
Harold Tanner, *Orangeburg-Calhoun Technological College (SC)*
Dale Thielker, *Ranken Technological College (MO)*
Patrick Ward, *Illinois Central College*
Cindy Wilson, *Henderson State University (AZ)*

Focus on Pedagogy

The Angel series is well-known and widely respected for its realistic, practical approach to algebra. The Angel approach offers thoughtful pedagogy, integrated, up-to-date real-world examples and data, and superior exercise sets.

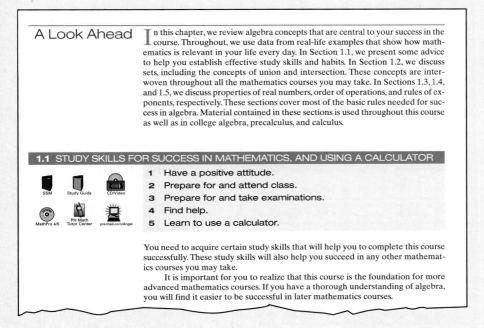

A Look Ahead

In this chapter, we review algebra concepts that are central to your success in the course. Throughout, we use data from real-life examples that show how mathematics is relevant in your life every day. In Section 1.1, we present some advice to help you establish effective study skills and habits. In Section 1.2, we discuss sets, including the concepts of union and intersection. These concepts are interwoven throughout all the mathematics courses you may take. In Sections 1.3, 1.4, and 1.5, we discuss properties of real numbers, order of operations, and rules of exponents, respectively. These sections cover most of the basic rules needed for success in algebra. Material contained in these sections is used throughout this course as well as in college algebra, precalculus, and calculus.

1.1 STUDY SKILLS FOR SUCCESS IN MATHEMATICS, AND USING A CALCULATOR

1. Have a positive attitude.
2. Prepare for and attend class.
3. Prepare for and take examinations.
4. Find help.
5. Learn to use a calculator.

You need to acquire certain study skills that will help you to complete this course successfully. These study skills will also help you succeed in any other mathematics courses you may take.

It is important for you to realize that this course is the foundation for more advanced mathematics courses. If you have a thorough understanding of algebra, you will find it easier to be successful in later mathematics courses.

▲ Page 2

A Look Ahead

Every chapter begins with **A Look Ahead** to give students an overview of the chapter and how it relates to other material in the book and to real-life situations.

Study Skills for Success In Mathematics (Section 1.1)

Following the study skills presented in this section greatly increases a student's chance for success in this and all other mathematics courses.

In-text Examples

A wealth of **in-text examples** illustrate the concept being presented and provide a step-by-step annotated solution.

Now Try Exercises

Now Try Exercises appear after selected examples to reinforce important concepts. Now Try Exercises provide students with immediate practice and make the student an active learner.

EXAMPLE 2 Evaluate using the definition of absolute value. **a)** $-|5|$ **b)** $-|-6.43|$

Solution **a)** We are finding the opposite of the absolute value of 5. Since the absolute value of 5 is positive, its opposite must be negative.
$$-|5| = -(5) = -5$$
b) We are finding the opposite of the absolute value of -6.43. Since the absolute value of -6.43 is positive, its opposite must be negative.
$$-|-6.43| = -(6.43) = -6.43$$

EXAMPLE 3 Insert $<$, $>$, or $=$ in the shaded area between the two values to make each statement true.
a) $|8|$ ▨ $-|-8|$ **b)** $|-1|$ ▨ $-|-3|$

Solution **a)** Since both $|8|$ and $-|-8|$ equal 8, we have $|8| = |-8|$.

NOW TRY EXERCISE 39   **b)** Since $|-1| = 1$ and $-|-3| = -3$, we have $|-1| > -|-3|$.

▲ Page 19

The Angel series is designed to help students see the important information that they need to learn concepts and topics.

Definitions, Procedures and Important Facts

Definitions, **Procedures** and **Important Facts** are presented in boxes throughout the text to make it easy for students to focus on this material when studying or preparing for quizzes and tests.

> **DEFINITION**
>
> **Additive Inverse**
> For any real number a, its additive inverse is $-a$.
>
> Consider the number -5. Its additive inverse is $-(-5)$. Since we know this number must be positive, this implies that $-(-5) = 5$. This is an example of the double negative property.
>
> **Double Negative Property**
> For any real number a, $-(-a) = a$.
>
> By the double negative property, $-(-7.4) = 7.4$ and $-\left(-\frac{12}{5}\right) = \frac{12}{5}$.

Page 18 ▲

Helpful Hints

Helpful Hints offer useful suggestions for problem solving and various other topics.

> **HELPFUL HINT**
>
> There are various ways to write the solution to an inequality problem. Be sure to indicate the solution to an inequality problem in the form requested by your professor. Examples of various forms follow.

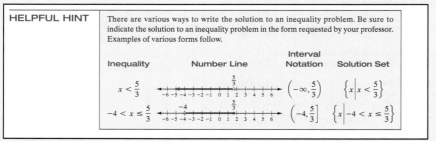

▲ Page 125

Helpful Hints— Study Tips

Helpful Hints—Study Tips reinforce and expand upon the Study Skills for success in Mathematics covered in Section 1.1

> **HELPFUL HINT**
>
> *STUDY TIP*
>
> Here are some suggestions if you are having some difficulty with application problems.
>
> 1. Instructor—Make an appointment to see your instructor. Make sure you have read the material in the book and attempted all the homework problems. Go with specific questions for your instructor.
> 2. Videotapes—Find out if the videotapes that accompany this book are available at your college. If so, view the videotapes that go with this chapter. Using the pause control, you can watch the videotapes at your own pace.
> 3. Student's Study Guide—If a copy of the Student's Study Guide is available, you may wish to read the material related to this chapter.
> 4. Tutoring—If your college learning center offers free tutoring, as many colleges do, you may wish to take advantage of tutoring.
> 5. Study Group—Form a study group with classmates. Exchange phone numbers and e-mail addresses. You may be able to help one another.
> 6. Student's Solutions Manual—If you get stuck on an exercise, you may want to use the Student's Solutions Manual to help you understand a problem. Do not use the Solutions Manual in place of working the exercises. In general, the Solutions Manual should be used only to check your work.

▲ Page 99

Avoiding Common Errors

Avoiding Common Errors boxes illustrate common mistakes, explain why certain procedures are wrong, and show the correct methods for working the problem.

> **AVOIDING COMMON ERRORS**
>
> Remember the middle term when squaring a binomial.
>
> | CORRECT | INCORRECT |
> |---|---|
> | $(x + 2)^2 = (x + 2)(x + 2)$ | $(x + 2)^2 = x^2 + 4$ |
> | $\quad = x^2 + 4x + 4$ | |
> | $(x - 3)^2 = (x - 3)(x - 3)$ | $(x - 3)^2 = x^2 + 9$ |
> | $\quad = x^2 - 6x + 9$ | |

▲ Page 315

Focus on Problem Solving

The revisions of the Angel series continue to place a strong focus on problem solving. Angel's exemplary approach to Problem Solving helps students learn to solve problems with confidence. In the process, the Angel texts help students understand *why* they are working on a certain operation while teaching them *how* to perform it. Problem solving is introduced early and incorporated as a theme throughout the texts.

Five-Step Problem-Solving Procedure

The in-text examples demonstrate how to solve each application exercise based on Polya's five-step problem-solving procedure: **Understand, Translate, Carry Out, Check,** and **State Answer.**

Guidelines for Problem Solving

1. **Understand the problem.**
   - Read the problem **carefully** at least twice. In the first reading, get a general overview of the problem. In the second reading, determine a) exactly what you are being asked to find and b) what information the problem provides.
   - If possible, make a sketch to illustrate the problem. Label the information given.
   - List the information in a table if it will help in solving the problem.
2. **Translate the problem into mathematical language.**
   - This will generally involve expressing the problem algebraically.
   - Sometimes this involves selecting a particular formula to use, whereas other times it is a matter of generating your own equation. It may be necessary to check other sources for the appropriate formula to use.
3. **Carry out the mathematical calculations necessary to solve the problem.**
4. **Check the answer obtained in step 3.**
   - Ask yourself: "Does the answer make sense?" "Is the answer reasonable?" If the answer is not reasonable, recheck your method for solving the problem and your calculations.
   - Check the solution in the original problem if possible.
5. **Answer the question**. Make sure you have answered the question asked. State the answer clearly.

▲ Page 80

Problem Solving

93. **Submarine Depth** A submarine is 160 feet below sea level. It has rock formations above and below it, and should not change its depth by more than 28 feet. Its distance below sea level, d, can be described by the inequality $|d - 160| \le 28$.
   a) Solve this inequality for d. Write your answer in interval notation.
   b) Between what vertical distances, measured from sea level, may the submarine move?

94. **A Bouncing Spring** A spring hanging from a ceiling is bouncing up and down so that its distance, d, above the ground satisfies the inequality $|d - 4| \le \frac{1}{2}$ ft (see the figure).
   a) Solve this inequality for d. Write your answer in interval notation.
   b) Between what distances, measured from the ground, will the spring oscillate?

95. **Glass Thickness** Certain types of glass manufactured by PPG Industries ideally will have a thickness of 0.089 inches. However, due to limitations in the manufacturing process, the thickness is allowed to vary from the ideal thickness by up to 0.004 inch. If t represents the actual thickness of the glass, then the allowable range of thicknesses can be represented using the inequality

96. **Plywood Guarantee** Certain plywood manufactured by Lafor International is guaranteed to be $\frac{5}{8}$ inch thick with a tolerance of plus or minus $\frac{1}{56}$ of an inch. If t represents the actual thickness of the plywood, then the allowable range of thicknesses can be represented using the inequality $|t - \frac{5}{8}| \le \frac{1}{56}$. *Source:* www.sticktrade.com

Problem-Solving Exercises

Problem Solving exercises are designed to help students become better thinkers.

▲ Page 139

Focus on Real-World Applications & Data

Each chapter begins with an illustrated, real-world application to motivate students and encourage them to see algebra as an important part of their daily lives. Problems based on real data from a broad range of subjects appear throughout the text, in the end-of-chapter material, and in the exercise sets.

Real-World Applications

An abundance of wonderfully updated, **real-world applications** give students needed practice with practical applications of algebra. **Real data** is used and real-world situations emphasize the relevance of the material being covered to students' everyday lives.

◀ Page 222

Chapter-Opening Applications

New chapter-opening applications emphasize the role of mathematics in everyday life, and in the workplace, giving students an applied, real-world introduction to the chapter material.

Page 66 ▶

Mathematics In Action

Mathematics In Action emphasizes the need for and the importance of mathematics in real-life.

◀ Page 13

Focus on Exercises

Exercise sets throughout the text have been greatly enhanced. Each exercise set progresses in difficulty to help students gain confidence and succeed with more difficult exercises. The end of each exercise set also includes a set of Challenge Problems.

▼Page 57

Concept/Writing Exercises

Concept/Writing exercises encourage students to analyze and write about the concepts they are learning.

Exercise Set 1.6

Concept/Writing Exercises

1. What is the form of a number in scientific notation?
2. Can 1×10^n ever be a negative number for any positive integer n? Explain.
3. Which is greater, 1×10^{-2} or 1×10^{-3}? Explain.
4. Can 1×10^{-n} ever be a negative number for any positive integer n? Explain.

Practice the Skills

Express each number in scientific notation.

5. 4700
6. 560
7. 0.031
8. 0.000000718
9. 360,000
10. 5,260,000,000

Practice the Skills Exercises

Practice the Skills exercises cover all types of exercises presented in the chapter.

Practice the Skills

Evaluate each expression.

13. $2^3 \cdot 2^2$
14. $3^2 \cdot 3^3$
15. $\dfrac{3^7}{3^5}$
16. $\dfrac{7^4}{7^3}$

17. 6^{-2}
18. 4^{-3}
19. $\dfrac{1}{2^{-3}}$
20. $\dfrac{1}{3^{-2}}$

21. 13^0
22. 17^0
23. $(2^3)^2$
24. $(3^2)^2$

25. $(2 \cdot 3)^2$
26. $(3 \cdot 5)^2$
27. $\left(\dfrac{2}{3}\right)^4$
28. $\left(\dfrac{3}{5}\right)^2$

▲Page 49

Problem Solving

93. **Submarine Depth** A submarine is 160 feet below sea level. It has rock formations above and below it, and should not change its depth by more than 28 feet. Its distance below sea level, d, can be described by the inequality $|d - 160| \le 28$.
   a) Solve this inequality for d. Write your answer in interval notation.
   b) Between what vertical distances, measured from sea level, may the submarine move?

94. **A Bouncing Spring** A spring hanging from a ceiling is bouncing up and down so that its distance, d, above the ground satisfies the inequality $|d - 4| \le \frac{1}{2}$ ft (see the figure).
   a) Solve this inequality for d. Write your answer in interval notation.
   b) Between what distances, measured from the ground, will the spring oscillate?

160 ft
28 ft
28 ft

$\frac{1}{2}$ ft
$\frac{1}{2}$ ft
4 ft

Problem-Solving Exercises

Problem-Solving exercises are designed to help students become better thinkers.

▲Page 139

Challenge Problems

Challenge Problems stimulate student interest with exercises that are conceptually and computationally more demanding.

Group Activity

Group Activities provide students with opportunities for collaborative learning.

Cumulative Review Exercises

Cumulative Review Exercises reinforce previously covered topics. These exercises are keyed to sections where the material is explained.

Challenge Problems

We will learn in Section 7.2 that the rules of exponents given in this section also apply when the exponents are rational numbers. Using this information and the rules of exponents, evaluate each expression.

138. $\left(\dfrac{x^{1/2}}{x^{-1}}\right)^{3/2}$

139. $\left(\dfrac{x^{5/8}}{x^{1/4}}\right)^{3}$

140. $\left(\dfrac{x^{4}}{x^{-1/2}}\right)^{-1}$

141. $\dfrac{x^{1/2}y^{-3/2}}{x^{5}y^{5/3}}$

142. $\left(\dfrac{x^{1/2}y^{4}}{x^{-3}y^{5/2}}\right)^{2}$

Group Activity

Discuss and answer Exercise 143 as a group.

143. *Doubling a Penny* On day 1 you are given a penny. On each following day, you are given double the amount you were given on the previous day.
   a) Write down the amounts you would be given on each of the first 6 days.
   b) Express each of these numbers as an exponential expression with a base of 2.
   c) By looking at the pattern, determine an exponential expression for the number of cents you will receive on day 10.
   d) Write a general exponential expression for the number of cents you will receive on day n.
   e) Write an exponential expression for the number of cents you will receive on day 30.
   f) Calculate the value of the expression in part **e)**. Use a calculator if one is available.
   g) Determine the amount found in part **f)** in dollars.
   h) Write a general exponential expression for the number of dollars you will receive on day n.

Cumulative Review Exercises

[1.2]**144.** If $A = \{3, 4, 6\}$ and $B = \{1, 2, 5, 8\}$, find
   a) $A \cup B$ and **b)** $A \cap B$.

145. Illustrate the following set on the number line: $\{x | -3 \leq x < 2\}$.

[1.4]**146.** Evaluate $6 + |12| \div |-3| - 4 \cdot 2^2$.

147. Evaluate $\sqrt[3]{-125}$.

1.6 SCIENTIFIC NOTATION

1 Write numbers in scientific notation.
2 Change numbers in scientific notation to decimal form.
3 Use scientific notation in problem solving.

SSM   Study Guide   CD/Video
MathPro 4/5   PH Math Tutor Center   prenhall.com/Angel

1 Write Numbers in Scientific Notation

Scientists and engineers often deal with very large and very small numbers. For example, the frequency of an FM radio signal may be 14,200,000,000 hertz (or cycles per second) and the diameter of a hydrogen atom is about 0.0000000001 meter. Because it is difficult to work with many zeros, scientists often express such numbers with exponents. For example, the number 14,200,000,000 might be written as 1.42×10^{10}, and 0.0000000001 as 1×10^{-10}. Numbers such as 1.42×10^{10} and 1×10^{-10} are in a form called **scientific notation**. In scientific notation, numbers

To the Student

Algebra is a course that cannot be learned by observation. To learn algebra you must become an active participant. You must read the text, pay attention in class, and, most importantly, you must work the exercises. The more exercises you work, the better.

The text was written with you in mind. Short, clear sentences are used, and many examples are given to illustrate specific points. The text stresses useful applications of algebra. Hopefully, as you progress through the course, you will come to realize that algebra is not just another math course that you are required to take, but a course that offers a wealth of useful information and applications.

This text makes full use of color. The different colors are used to highlight important information. Important procedures, definitions, and formulas are placed within colored boxes.

The boxes marked **Helpful Hints** should be studied carefully, for they stress important information. The boxes marked **Avoiding Common Errors** should also be studied carefully. These boxes point out errors that students commonly make, and provide the correct procedures for doing these problems.

Ask your professor early in the course to explain the policy on when the calculator may be used. Pay particular attention to the **Using Your Calculator** boxes. You should also read the **Using Your Graphing Calculator** boxes even if you are not using a graphing calculator in class. You may find the information presented here helps you better understand the algebraic concepts.

Other questions you should ask your professor early in the course include: What supplements are available for use? Where can help be obtained when the professor is not available? Supplements that may be available, and the icons used to represent them in the text, include: Student's Solution Manual, ▉; Student's Study Guide, ▉; Math Pro tutorial software, ◉; and CDs and videotapes, 🔒 (including a videotape on study skills needed for success in mathematics). Other sources of help include the Prentice Hall Tutor Cen-

ter, 📱; and the Angel Website, 💻. These items are discussed under the heading of Supplements in Section 1.1.

You may wish to form a study group with other students in your class. Many students find that working in small groups provides an excellent way to learn the material. By discussing and explaining the concepts and exercises to one another you reinforce your own understanding. Once guidelines and procedures are determined by your group, make sure to follow them.

One of the first things you should do is to read Section 1.1, Study Skills for Success in Mathematics. Read this section slowly and carefully, and pay particular attention to the advice and information given. Occasionally, refer back to this section. This could be the most important section of the book. Carefully read the material on doing your homework and on attending class.

At the end of all Exercise Sets (after the first two) are **Cumulative Review Exercises**. You should work these problems on a regular basis, even if they are not assigned. These problems are from earlier sections and chapters of the text, and they will refresh your memory and reinforce those topics. If you have a problem when working these exercises, read the appropriate section of the text or study your notes that correspond to that material. The section of the text where the Cumulative Review Exercise was introduced is indicated in brackets, [ ], to the left of the exercise. After reviewing the material, if you still have a problem, make an appointment to see your professor. Working the Cumulative Review Exercises throughout the semester will also help prepare you to take your final exam.

At the end of each chapter are a **Chapter Summary, Chapter Review Exercises**, and a **Chapter Practice Test**. Before each examination, you should review this material carefully and take the Practice Test. If you do well on the Practice Test, you should do well on the class test. The questions in the Review Exercises are marked to indicate the section in which that material was first introduced. If you have a problem with a Review Exercise question, reread the section indicated. You may also wish to take the **Cumulative Review Test** that appears at the end of every chapter.

In the back of the text there is an **Answer Section** which contains the answers to the *odd-numbered* exercises, including the Challenge Problems. Answers to *all* Using Your Graphing Calculator Exercises, Cumulative Review Exercises, Chapter Review Exercises, and Chapter Practice Tests are provided. Answers to the Group Activity exercises are not provided, for we wish students to reach agreement by themselves on answers to these exercises. The answers should be used only to check your work. The answers to the Cumulative Review Test exercises are provided directly after the test for immediate feedback. After each answer, the section and objective number where that type of exercise was covered is provided.

I have tried to make this text as clear and error free as possible. No text is perfect, however. If you find an error in the text, or an example or section that you believe can be improved, I would greatly appreciate hearing from you. If you enjoy the text, I would also appreciate hearing from you. You can contact me at *www.prenhall.com/angel*.

Allen R. Angel

Chapter 1

Basic Concepts

1.1 Study Skills for Success in Mathematics, and Using a Calculator

1.2 Sets and Other Basic Concepts

1.3 Properties of and Operations with Real Numbers

1.4 Order of Operations

1.5 Exponents

1.6 Scientific Notation

Chapter Summary
Chapter Review Exercises
Chapter Practice Test

Have you ever asked yourself, "When am I ever going to use mathematics?" In this chapter we will see many areas in which we can use algebra to study everyday life situations. These situations will range from the use of alternate medical remedies (see Example 10, page 35) to the increase in carbon dioxide emissions (see Exercise 116, page 40). We will find that mathematics can be used in virtually every area of our lives.

SSM

Study Guide

CD/Video

MathPro 4/5

PH Math
Tutor Center

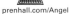
prenhall.com/Angel

A Look Ahead

In this chapter, we review algebra concepts that are central to your success in the course. Throughout, we use data from real-life examples that show how mathematics is relevant in your life every day. In Section 1.1, we present some advice to help you establish effective study skills and habits. In Section 1.2, we discuss sets, including the concepts of union and intersection. These concepts are interwoven throughout all the mathematics courses you may take. In Sections 1.3, 1.4, and 1.5, we discuss properties of real numbers, order of operations, and rules of exponents, respectively. These sections cover most of the basic rules needed for success in algebra. Material contained in these sections is used throughout this course as well as in college algebra, precalculus, and calculus.

1.1 STUDY SKILLS FOR SUCCESS IN MATHEMATICS, AND USING A CALCULATOR

SSM   Study Guide   CD/Video
MathPro 4/5   PH Math Tutor Center   prenhall.com/Angel

1 Have a positive attitude.
2 Prepare for and attend class.
3 Prepare for and take examinations.
4 Find help.
5 Learn to use a calculator.

You need to acquire certain study skills that will help you to complete this course successfully. These study skills will also help you succeed in any other mathematics courses you may take.

It is important for you to realize that this course is the foundation for more advanced mathematics courses. If you have a thorough understanding of algebra, you will find it easier to be successful in later mathematics courses.

1 Have a Positive Attitude

You may be thinking to yourself, "I hate math" or "I wish I did not have to take this class." You may have heard the term *math anxiety* and feel that you fall into this category. The first thing you need to do to be successful in this course is to change your attitude to a more positive one. You must be willing to give this course and yourself a fair chance.

Based on past experiences in mathematics, you may feel this will be difficult. However, mathematics is something you need to work at. Many of you taking this course are more mature now than when you took previous mathematics courses. Your maturity and your desire to learn are extremely important and can make a tremendous difference in your ability to succeed in mathematics. I believe you can be successful in this course, but you also need to believe it.

2 Prepare for and Attend Class

Preview the Material

Before class, you should spend a few minutes previewing any new material in the textbook. You do not have to understand everything you read yet. Just get a feeling for the definitions and concepts that will be discussed. This quick preview will help you to understand what your instructor is explaining during class. After the material is explained in class, read the corresponding sections of the text slowly and carefully, word by word.

Read the Text

A mathematics text is not a novel. Mathematics textbooks should be read slowly and carefully. If you do not understand what you are reading, reread the material. When you come across a new concept or definition, you may wish to underline or highlight it so that it stands out. This way, when you look for it later, it will be easier to find. When you come across a worked-out example, read and follow the example carefully. Do not just skim it. Try working out the example yourself on another sheet of paper. Also, work the **Now Try Exercises** that appear in the text next to many examples. The Now Try Exercises are designed so that you have the opportunity to immediately apply new ideas. Make notes of anything that you do not understand to ask your instructor.

Do the Homework

Two very important commitments that you must make to be successful in this course are to attend class and do your homework regularly. Your assignments must be worked conscientiously and completely. Mathematics cannot be learned by observation. You need to practice what you have heard in class. By doing homework you truly learn the material.

Don't forget to check the answers to your homework assignments. Answers to the odd-numbered exercises are in the back of this book. In addition, the answers to all the Cumulative Review Exercises, Chapter Review Exercises, and Chapter Practice Tests, are in the back of the book. The answers to the Cumulative Review Test questions follow the tests themselves. In addition, the section and objective number where the material was first introduced is provided in brackets after each answer. Answers to the Group Activity Exercises are not provided for we want you to arrive at the answers as a group.

If you have difficulty with some of the exercises, mark them and do not hesitate to ask questions about them in class. You should not feel comfortable until you understand all the concepts needed to work every assigned problem.

When you do your homework, make sure that you write it neatly and carefully. Pay particular attention to copying signs and exponents correctly. Do your homework in a step-by-step manner. This way you can refer back to it later and still understand what was written.

Attend and Participate in Class

You should attend every class. Most instructors will agree that there is an inverse relationship between absences and grades: The more absences you have, the lower your grade will be. Every time you miss a class, you miss important information. If you must miss a class, contact your instructor ahead of time and get the reading assignment and homework.

While in class, pay attention to what your instructor is saying. If you do not understand something, ask your instructor to repeat or explain the material. If you have read the upcoming material before class and have questions that have not been answered, ask your instructor. If you do not ask questions, your instructor will not know that you have a problem understanding the material.

In class, take careful notes. Write numbers and letters clearly so that you can read them later. It is not necessary to write down every word your instructor says. Copy down the major points and the examples that do not appear in the text. You should not be taking notes so frantically that you lose track of what your instructor is saying. It is a mistake to believe that you can copy down material in class without understanding it and then figure it out later.

Study

Study in the proper atmosphere. Study in an area where you are not constantly disturbed so that your attention can be devoted to what you are reading. The area where you study should be well ventilated and well lit. You should have sufficient desk space to spread out all your materials. Your chair should be comfortable. You should try to minimize distractions while you are studying. You should not study for hours on end. Short study breaks are a good idea.

When studying, you should not only understand how to work a problem, you should also know why you follow the specific steps you do to work the problem. If you do not have an understanding of why you follow the specific process, you will not be able to solve similar problems.

Time Management

It is recommended that students spend at least 2 hours studying and doing homework for every hour of class time. Some students require more time than others. Finding the necessary time to study is not always easy. The following are some suggestions that you may find helpful.

1. Plan ahead. Determine when you will have time to study and do your homework. Do not schedule other activities for these time periods. Try to space these periods evenly over the week.
2. Be organized so that you will not have to waste time looking for your books, pen, calculator, or notes.
3. Use a calculator to perform tedious calculations.
4. When you stop studying, clearly mark where you stopped in the text.
5. Try not to take on added responsibilities. You must set your priorities. If your education is a top priority, as it should be, you may have to cut the time spent on other activities.
6. If time is a problem, do not overburden yourself with too many courses. Consider taking fewer credits. If you do not have sufficient time to study, your understanding and your grades in all of your courses may suffer.

3 Prepare for and Take Examinations

Study for an Exam

If you do some studying each day, you should not need to cram the night before an exam. If you wait until the last minute, you will not have time to seek the help you may need. To review for an exam:

1. Read your class notes.
2. Review your homework assignments.
3. Study the formulas, definitions, and procedures you will need for the exam.
4. Read the Avoiding Common Errors boxes and Helpful Hint boxes carefully.
5. Read the summary at the end of each chapter.
6. Work the review exercises at the end of each chapter. If you have difficulties, restudy those sections. If you still have trouble, seek help.
7. Work the chapter practice test.
8. Rework quizzes previously given if the material covered in the quizzes will be included on the test.
9. Work the Cumulative Review Test if material from earlier chapters will be included on the test.

Take an Exam

Make sure that you get a good night's sleep the day before the test. If you studied properly, you should not have to stay up late the night before to prepare for the test. Arrive at the exam site early so that you have a few minutes to relax before the exam. If you need to rush to get to the exam, you will start out nervous and anxious. After you receive the exam, do the following:

1. Carefully write down any formulas or ideas that you want to remember.
2. Look over the entire exam quickly to get an idea of its length and to make sure that no pages are missing. You will need to pace yourself to make sure that you complete the entire exam. Be prepared to spend more time on problems worth more points.
3. Read the test directions carefully.
4. Read each problem carefully. Answer each question completely and make sure that you have answered the specific question asked.
5. Starting with number 1, work each question in order. If you come across a question that you are not sure of, do not spend too much time on it. Continue working the questions that you understand. After completing all other questions, go back and finish those questions you were not sure of. Do not spend too much time on any one question.
6. Attempt each problem. You may be able to earn at least partial credit.
7. Work carefully and write clearly so that your instructor can read your work. Also, it is easy to make mistakes when your writing is unclear.
8. Check your work and your answers if you have time.
9. Do not be concerned if others finish the test before you. Do not be disturbed if you are the last to finish. Use all your extra time to check your work.

4 Find Help

Use the Supplements

This text comes with many supplements. Find out from your instructor early in the semester which of these supplements are available and which supplements might be beneficial for you to use. Reading supplements should not replace reading the text. Instead supplements should enhance your understanding of the material. If you miss a class, you may want to review the videotape on the topic you missed before attending the next class. The supplements are listed on each chapter opening page, and on each page with a section head. The icons used for the supplements are:

for the Student's Solution Manual (SSM); for the Student's Study Guide;

for CD's and videotapes (which show about 20 minutes of lecture per section

and includes the worked out solutions to the exercises marked with this icon);

for Math Pro (tutorial and practice test) software; for the Prentice Hall Mathematics Tutor (The tutor center offers individual tutoring on examples, exercises, and problems contained in the text. Students can access the tutor center by phone, fax, or email once your instructor has initiated the process with the publisher.);

and for the Allen R. Angel textbook website at Prentice Hall. The website includes additional exercises, practice quizzes that can be graded, graphing calculator instructions for all brands of graphing calculators, and chapter projects.

Seek Help

One thing I stress with my own students is to *get help as soon as you need it!* Do not wait! In mathematics, one day's material is usually based on the previous day's material. So if you don't understand the material today, you may not be able to understand the material tomorrow.

Where should you seek help? There are often a number of places to obtain help on campus. You should try to make a friend in the class with whom you can study. Often you can help one another. You may wish to form a study group with other students in your class. Discussing the concepts and homework with your peers will reinforce your own understanding of the material.

You should not hesitate to visit your instructor when you are having problems with the material. Be sure you read the assigned material and attempt the homework before meeting with your instructor. Come prepared with specific questions to ask.

Often other sources of help are available. Many colleges have a mathematics laboratory or a mathematics learning center where tutors are available to help students. Ask your instructor early in the semester if any tutors are available, and find out where the tutors are located. Then use these tutors as needed.

5 Learn to Use a Calculator

Many instructors require their students to purchase and to use a calculator in class. You should find out as soon as possible which calculator, if any, your instructor expects you to use. If you plan on taking additional mathematics courses, you should determine which calculator will be required in those courses and consider purchasing that calculator for use in this course if its use is permitted by your instructor. Many instructors require a scientific calculator and many others require a graphing calculator.

In this book we provide information about both types of calculators. Always read and save the user's manual for whatever calculator you purchase.

Exercise Set 1.1

Do you know all of the following information? If not, ask your instructor as soon as possible.

1. What is your instructor's name?
2. What are your instructor's office hours?
3. Where is your instructor's office located?
4. How can you best reach your instructor?
5. Where can you obtain help if your instructor is not available?
6. What supplements are available to assist you in learning?
7. Does your instructor recommend or require a specific calculator? If so, which one?
8. When can you use a calculator? Can it be used in class, on homework, on tests?
9. What is your instructor's attendance policy?
10. Why is it important that you attend every class possible?
11. Do you know the name and phone number of a friend in class?
12. For each hour of class time, how many hours outside class are recommended for homework and studying?

13. List what you should do to be properly prepared for each class.
14. Explain how a mathematics textbook should be read.
15. Write a summary of the steps you should follow when taking an exam.
16. Having a positive attitude is very important for success in this course. Are you beginning this course with a positive attitude? It is important that you do!
17. You need to make a commitment to spend the time necessary to learn the material, to do the homework, and to attend class regularly. Explain why you believe this commitment is necessary to be successful in this course.
18. What are your reasons for taking this course?
19. What are your goals for this course?
20. Have you given any thought to studying with a friend or a group of friends? Can you see any advantages in doing so? Can you see any disadvantages in doing so?

1.2 SETS AND OTHER BASIC CONCEPTS

SSM   Study Guide   CD/Video

MathPro 4/5   PH Math Tutor Center   prenhall.com/Angel

1 Identify sets.
2 Identify and use inequalities.
3 Use set builder notation.
4 Find the union and intersection of sets.
5 Identify important sets of numbers.

Let's start with some important definitions. When a letter is used to represent various numbers it is called a **variable**. For instance, if t = the time, in hours, that a car is traveling, then t is a variable since the time is constantly changing as the car is traveling. We often use the letters x, y, z, and t to represent variables. However, other letters may be used. When presenting properties or rules, the letters a, b, and c are often used as variables.

If a letter represents one particular value it is called a **constant**. For example, if s = the number of seconds in a minute, then s represents a constant because there are always 60 seconds in a minute. The number of seconds in a minute does not vary. In this book, letters representing both variables and constants are italicized.

The term **algebraic expression**, or simply **expression**, will be used often in the text. An expression is any combination of numbers, variables, exponents, mathematical symbols (other than equal signs), and mathematical operations.

1 Identify Sets

Sets are used in many areas of mathematics, so an understanding of sets and set notation is important. A **set** is a collection of objects. The objects in a set are called **elements** of the set. Sets are indicated by means of braces, { }, and are often named with capital letters. When the elements of a set are listed within the braces, as illustrated below, the set is said to be in **roster form**.

$$A = \{a, b, c\}$$
$$B = \{\text{yellow, green, blue, red}\}$$
$$C = \{1, 2, 3, 4, 5\}$$

Set A has three elements, set B has four elements, and set C has five elements. The symbol $\in$ is used to indicate that an item is an element of a set. Since 2 is an element of set C, we may write $2 \in C$; this is read "2 is an element of set C."

A set may be finite or infinite. Sets A, B, and C each have a finite number of elements and are therefore *finite sets*. In some sets it is impossible to list all the elements. These are *infinite sets*. The following set, called the set of **natural numbers** or **counting numbers**, is an example of an infinite set.

$$N = \{1, 2, 3, 4, 5, \ldots\}$$

The three dots after the last comma, called an *ellipsis*, indicate that the set continues in the same manner.

Another important infinite set is the integers. The set of **integers** follows.

$$I = \{\ldots, -4, -3, -2, -1, 0, 1, 2, 3, 4, \ldots\}$$

Notice that the set of integers includes both positive and negative integers and the number 0.

If we write

$$D = \{1, 2, 3, 4, 5, \ldots, 280\}$$

we mean that the set continues in the same manner until the number 280. Set D is the set of the first 280 natural numbers. D is therefore a finite set.

A special set that contains no elements is called the **null set**, or **empty set**, written { } or $\emptyset$. For example, the set of students in your class over the age of 150 is the null or empty set.

2 Identify and Use Inequalities

Before we introduce a second method of writing a set, called *set builder notation*, we will introduce the inequality symbols.

Inequality Symbols

> $>$ is read "is greater than."
>
> $\geq$ is read "is greater than or equal to."
>
> $<$ is read "is less than."
>
> $\leq$ is read "is less than or equal to."
>
> $\neq$ is read "is not equal to."

Inequalities can be explained using the real number line (Fig. 1.1).

FIGURE 1.1

The number a is greater than the number b, $a > b$, when a is to the right of b on the number line (Fig. 1.2). We can also state that the number b is less than a, $b < a$, when b is to the left of a on the number line. The inequality $a \neq b$ means either $a < b$ or $a > b$.

FIGURE 1.2

EXAMPLE 1 Insert either $>$ or $<$ in the shaded area between the numbers to make each statement true.

a) 6 ▨ 2   **b)** -7 ▨ 1   **c)** -4 ▨ -5

Solution Draw a number line illustrating the location of all the given values (Fig. 1.3).

FIGURE 1.3

a) $6 > 2$ Note that 6 is to the right of 2 on the number line.

b) $-7 < 1$ Note that -7 is to the left of 1 on the number line.

NOW TRY EXERCISE 29   **c)** $-4 > -5$ Note that -4 is to the right of -5 on the number line.   ✳

Remember that the symbol used in an inequality, if it is true, always points to the smaller of the two numbers.

We use the notation $x > 2$, read "x is greater than 2," to represent *all* real numbers greater than 2. We use the notation $x \leq -3$, read "x is less than or equal to -3," to represent all real numbers that are less than or equal to -3. The notation $-4 \leq x < 3$ means all real numbers that are greater than or equal to -4 and also less than 3. In the inequalities $x > 2$ and $x \leq -3$, the 2 and the -3 are called **endpoints**. In the inequality $-4 \leq x < 3$, the -4 and 3 are the endpoints. The solutions to inequalities that use either $<$ or $>$ do not include the endpoints, but the solutions to inequalities that use either $\leq$ or $\geq$ do include the endpoints. When inequalities are illustrated on the number line, a solid circle is used to show that the endpoint is included in the answer, and an open circle is used to show that the endpoint is not included. On the top of the next page are some illustrations of how certain inequalities are indicated on the number line.

Inequality	Inequality Indicated on the Number Line
$x > 2$	
$x \le -3$	
$-4 \le x < 3$	

Some students misunderstand the word *between*. The word *between* indicates that the endpoints are not included in the answer. For example, the set of natural numbers between 2 and 6 is $\{3, 4, 5\}$. If we wish to include the endpoints, we can use the word *inclusive*. For example, the set of natural numbers between 2 and 6 inclusive is $\{2, 3, 4, 5, 6\}$.

3 Use Set Builder Notation

Now that we have introduced the inequality symbols we will discuss another method of indicating a set, called **set builder notation**. An example of set builder notation is

$$E = \{x | x \text{ is a natural number greater than } 6\}$$

This is read "Set E is the set of all elements x, such that x is a natural number greater than 6." In roster form, this set is written

$$E = \{7, 8, 9, 10, 11, \ldots\}$$

The general form of set builder notation is

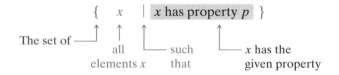

We often will use the variable x when using set builder notation, although any variable can be used.

Two condensed ways of writing set $E = \{x | x \text{ is a natural number greater than } 6\}$ in set builder notation follow.

$$E = \{x | x > 6 \text{ and } x \in N\} \quad \text{or} \quad E = \{x | x \ge 7 \text{ and } x \in N\}$$

The set $A = \{x | -3 < x \le 4 \text{ and } x \in I\}$ is the set of integers greater than -3 and less than or equal to 4. The set written in roster form is $\{-2, -1, 0, 1, 2, 3, 4\}$. Notice that the endpoint -3 is not included in the set but the endpoint 4 is included.

How do the sets $B = \{x | x > 2 \text{ and } x \in N\}$ and $C = \{x | x > 2\}$ differ? Can you write each set in roster form? Can you illustrate both sets on the number line? Set B contains only the natural numbers greater than 2, that is, $\{3, 4, 5, 6, \ldots\}$. Set C contains not only the natural numbers greater than 2 but also fractions and decimal numbers greater than 2. If you attempted to write set C in roster form, where would you begin? What is the smallest number greater than 2? Is it 2.1 or 2.01 or 2.001? Since there is no smallest number greater than 2, this set cannot be written in roster form. On the top of the next page we illustrate these two sets on the number line. We have also illustrated two other sets.

Set	Set Indicated on the Number Line
$\{x \mid x > 2 \text{ and } x \in N\}$	
$\{x \mid x > 2\}$	
$\{x \mid -1 \le x < 4 \text{ and } x \in I\}$	
$\{x \mid -1 \le x < 4\}$	

NOW TRY EXERCISE 75   Another method of indicating inequalities, called *interval notation*, will be discussed in Section 2.5.

4 Find the Union and Intersection of Sets

Just as *operations* such as addition and multiplication are performed on numbers, operations can be performed on sets. Two set operations are *union* and *intersection*.

DEFINITION | The **union** of set A and set B, written $A \cup B$, is the set of elements that belong to either set A *or* set B.

Because the word *or*, as used in this context, means belonging to set A or set B or both sets, the union is formed by combining, or joining together, the elements in set A with those in set B. If an item is an element in either set A, or set B, or in both sets, then it is an element in the union of the sets.

Examples of Union of Sets

$$A = \{1, 2, 3, 4, 5\}, \quad B = \{3, 4, 5, 6, 7\}, \quad A \cup B = \{1, 2, 3, 4, 5, 6, 7\}$$
$$A = \{a, b, c, d, e\}, \quad B = \{x, y, z\}, \quad A \cup B = \{a, b, c, d, e, x, y, z\}$$

In set builder notation we can express $A \cup B$ as

$$A \cup B = \{x \mid x \in A \text{ or } x \in B\}$$

DEFINITION | The **intersection** of set A and set B, written $A \cap B$, is the set of all elements that are common to both set A *and* set B.

Because the word *and*, as used in this context, means belonging to *both* set A and set B, the intersection is formed by using only those elements that are in both set A and set B. If an item is an element in only one of the two sets, then it is not an element in the intersection of the sets.

Examples of Intersection of Sets

$$A = \{1, 2, 3, 4, 5\}, \quad B = \{3, 4, 5, 6, 7\}, \quad A \cap B = \{3, 4, 5\}$$
$$A = \{a, b, c, d, e\}, \quad B = \{x, y, z\}, \quad A \cap B = \{\ \}$$

Note that in the last example, sets A and B have no elements in common. Therefore, their intersection is the empty set. In set builder notation we can express $A \cap B$ as

NOW TRY EXERCISE 57

$$A \cap B = \{x \mid x \in A \text{ and } x \in B\}$$

5 Identify Important Sets of Numbers

At this point we have all the necessary information to discuss important sets of real numbers. In the box below, we describe these sets and provide letters that are often used to represent these sets of numbers.

Important Sets of Real Numbers

Real numbers	$\mathbb{R} = \{x \mid x \text{ is a point on the number line}\}$
Natural or counting numbers	$N = \{1, 2, 3, 4, 5, \ldots\}$
Whole numbers	$W = \{0, 1, 2, 3, 4, 5, \ldots\}$
Integers	$I = \{\ldots, -3, -2, -1, 0, 1, 2, 3, \ldots\}$
Rational numbers	$Q = \left\{\dfrac{p}{q} \middle\mid p \text{ and } q \text{ are integers}, q \neq 0\right\}$
Irrational numbers	$H = \{x \mid x \text{ is a real number that is not rational}\}$

Let us briefly look at the rational, irrational, and real numbers. A **rational number** is any number that can be represented as a quotient of two integers, with the denominator not 0.

Examples of Rational Numbers

$$\frac{3}{5}, \quad \frac{-2}{3}, \quad 0, \quad 1.63, \quad 7, \quad -12, \quad \sqrt{4}$$

Notice that 0, or any other integer, is also a rational number since it can be written as a fraction with a denominator of 1. For example, $0 = \frac{0}{1}$ and $7 = \frac{7}{1}$.

The number 1.63 can be written $\frac{163}{100}$ and is thus a quotient of two integers. Since $\sqrt{4} = 2$ and 2 is an integer, $\sqrt{4}$ is a rational number. *Every rational number when written as a decimal number will be either a repeating or a terminating decimal number.*

Examples of Repeating Decimals	Examples of Terminating Decimals
$\frac{2}{3} = 0.6666\ldots$ The 6 repeats.	$\frac{1}{2} = 0.5$
$\frac{1}{7} = 0.142857142857\ldots$ The block 142857 repeats.	$\frac{7}{4} = 1.75$

To show that a digit or group of digits repeat, we can place a bar above the digit or group of digits that repeat. For example, we may write

$$\frac{2}{3} = 0.\overline{6} \quad \text{and} \quad \frac{1}{7} = 0.\overline{142857}$$

Although $\sqrt{4}$ is a rational number, the square roots of most integers are not. Most square roots will be neither terminating nor repeating decimals when expressed as decimal numbers and are **irrational numbers**. Some irrational numbers are $\sqrt{2}$, $\sqrt{3}$, $\sqrt{5}$, and $\sqrt{6}$. Another irrational number is pi, π. When we give a decimal value for an irrational number, we are giving only an *approximation* of the value of the irrational number. The symbol $\approx$ means "is approximately equal to."

$$\pi \approx 3.14 \qquad \sqrt{2} \approx 1.41$$

The **real numbers** are formed by taking the *union* of the rational numbers and the irrational numbers. Therefore, any real number must be either a rational number or an irrational number. The symbol $\mathbb{R}$ is often used to represent the set of real numbers. Figure 1.4 illustrates various real numbers on the number line.

FIGURE 1.4

$$-\sqrt{23} \quad -3.62 \quad -\tfrac{8}{5} \quad 0 \quad \tfrac{1}{2} \quad \sqrt{2} \quad 2 \quad \tfrac{20}{7}\pi \quad 4.3$$

$$\begin{array}{ccccccccccccc} & & & & & & & & & & & & \\ -6 & -5 & -4 & -3 & -2 & -1 & 0 & 1 & 2 & 3 & 4 & 5 & 6 \end{array}$$

A first set is a **subset** of a second set when every element of the first set is also an element of the second set. For example, the set of natural numbers, $\{1, 2, 3, 4, \ldots\}$, is a subset of the set of whole numbers, $\{0, 1, 2, 3, 4, \ldots\}$, because every element in the set of natural numbers is also an element in the set of whole numbers. Figure 1.5 illustrates the relationships between the various subsets of the real numbers. In Figure 1.5a, you see that the set of natural numbers is a subset of the set of whole numbers, of the set of integers, and of the set of rational numbers. Therefore, every natural number must also be a whole number, an integer, and a rational number. Using the same reasoning, we can see that the set of whole numbers is a subset of the set of integers and of the set of rational numbers, and that the set of integers is a subset of the set of rational numbers.

Looking at Figure 1.5b we see that the positive integers, 0, and the negative integers form the integers, that the integers and noninteger rational numbers form the rational numbers, and so on.

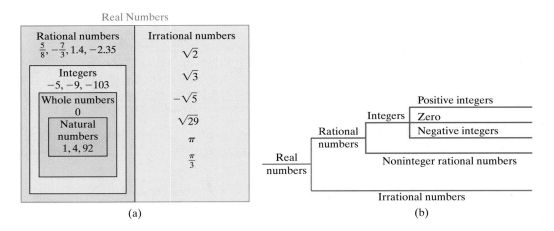

FIGURE 1.5   (a)   (b)

EXAMPLE 2   Consider the following set:

$$\left\{ -3, 0, \frac{5}{7}, 12.25, \sqrt{7}, -\sqrt{11}, \frac{22}{7}, 5, 7.1, -17, \pi \right\}$$

List the elements of the set that are

a) natural numbers.   **b)** whole numbers.   **c)** integers.

d) rational numbers.   **e)** irrational numbers.   **f)** real numbers.

Solution   **a)** Natural numbers: 5   **b)** Whole numbers: 0, 5   **c)** Integers: $-3, 0, 5, -17$

d) Rational numbers can be written in the form $p/q, q \neq 0$. Each of the following can be written in this form and is a rational number.

$$-3, 0, \frac{5}{7}, 12.25, \frac{22}{7}, 5, 7.1, -17$$

e) Irrational numbers are real numbers that are not rational. The following numbers are irrational.

$$\sqrt{7}, -\sqrt{11}, \pi$$

f) All of the numbers in the set are real numbers. The union of the rational numbers and the irrational numbers forms the real numbers.

$$-3, 0, \frac{5}{7}, 12.25, \sqrt{7}, -\sqrt{11}, \frac{22}{7}, 5, 7.1, -17, \pi$$

NOW TRY EXERCISE 49

Not all numbers are real numbers. Some numbers that we discuss later in the text that are not real numbers are complex numbers and imaginary numbers.

Mathematics in Action

Crisp and Fuzzy Sets

The type of sets you are learning about in this chapter are often referred to as *crisp sets*. An item either *is* or *is not* a member of a crisp set. *Q* clearly is an element of the set $\{T, R, Q, F\}$ and *B* is not. Reaching outside of mathematics, we can recognize that the statement, "Abraham Lincoln is an element of the set of all presidents of the United States," reflects the notion of the crisp set.

Now, consider the set of all men with a full head of hair, and let's say a full head of hair is defined as 100,000 or more hairs. Fred wakes up one morning with 100,000 hairs on his head, so he belongs to the set. He takes a shower and five hairs from his head go down the drain. Now he doesn't belong to the set.

Well, that's correct, silly as it sounds. Rigid categorization like this would not be very helpful to a company gathering data for marketing a line of hair care products.

In 1965, the notion of the *fuzzy set* was first mentioned. Simply stated, an element in a fuzzy set has associated with it the degree of its membership in the set. To get back to Fred, if a fuzzy set were defined as "all men with a full head of hair," then when Fred got up in the morning he would have been a member of that set with a value of 1 (meaning 100%). After his shower, he would still be in that set but with a value of, say, 0.999.

The power of fuzzy sets, governed by the rules of fuzzy logic, drives computer programs dealing with pattern recognition, such as speech recognition. The pronunciation of a rancher from Butte, Montana, differs considerably from that of a stockbroker from London, England. But the program will apply a large number of fuzzy logic tests to determine the *degree* to which a sound matches up with a sound in the program's fuzzy-set database. By not insisting on a perfect match for any single letter or syllable—by being fuzzy—the program can "understand" human speech with amazing accuracy.

Exercise Set 1.2

Concept/Writing Exercises

1. What is a variable?
2. What is an algebraic expression?
3. What is a set?
4. What do we call the objects in a set?
5. What is the null or empty set?

6. Is the set of natural or counting numbers a finite or infinite set? Explain.
7. List the five inequality symbols and write down how each is read.
8. Give an example of a set that is empty.

9. List the set of integers *between* 2 and 7.

10. List the set of integers *between* −1 and 3 *inclusive*.

11. Explain why every integer is also a rational number.

12. Describe the counting numbers, whole numbers, integers, rational numbers, irrational numbers, and real numbers. Explain the relationships among the sets of numbers.

In Exercises 13–22, indicate whether each statement is true or false.

13. Every natural number is a whole number.

14. Some rational numbers are integers.

15. Every whole number is a natural number.

16. Every rational number is an integer.

17. Every integer is a rational number.

18. The union of the set of rational numbers with the set of irrational numbers forms the set of real numbers.

19. The intersection of the set of rational numbers and the set of irrational numbers is the empty set.

20. The set of natural numbers is a finite set.

21. The set of integers between π and 4 is the null set.

22. The set of rational numbers between 3 and π is an infinite set.

Practice the Skills

Insert either $<$ or $>$ in the shaded area to make each statement true.

23. 3 ▦ 4

24. −1 ▦ 5

25. −3 ▦ 0

26. −1 ▦ 1

27. −1 ▦ −1.01

28. 4 ▦ −3

29. −5 ▦ −3

30. −6 ▦ −1

31. −14.98 ▦ −14.99

32. −3.6 ▦ −3.2

33. 1.1 ▦ 1.9

34. −1.1 ▦ −1.9

35. $-\pi$ ▦ −4

36. −780 ▦ −655

37. $-\dfrac{7}{8}$ ▦ $-\dfrac{8}{9}$

38. $-\dfrac{4}{7}$ ▦ $-\dfrac{5}{9}$

In Exercises 39–48, list each set in roster form.

39. $A = \{x|-1 < x < 1 \text{ and } x \in I\}$

40. $B = \{y|y \text{ is an odd natural number less than 8}\}$

41. $C = \{z|z \text{ is an even integer greater than 16 and less than or equal to 20}\}$

42. $D = \{x|x \geq -3 \text{ and } x \in I\}$

43. $E = \{x|x < 3 \text{ and } x \in W\}$

44. $F = \left\{x|-\dfrac{6}{5} \leq x < \dfrac{15}{4} \text{ and } x \in N\right\}$

45. $H = \{x|x \text{ is a whole number multiple of 7}\}$

46. $L = \{x|x \text{ is an integer greater than } -5\}$

47. $J = \{x|x > 0 \text{ and } x \in I\}$

48. $K = \{x|x \text{ is a whole number between 3 and 4}\}$

49. Consider the set $\left\{-3, 4, \frac{1}{2}, \frac{5}{9}, 0, \sqrt{2}, \sqrt{8}, -1.23, \frac{99}{100}\right\}$. List the elements that are:
a) natural numbers.
b) whole numbers.
c) integers.
d) rational numbers.
e) irrational numbers.
f) real numbers.

50. Consider the set $\left\{2, 4, -5.33, \frac{9}{2}, \sqrt{7}, \sqrt{2}, -100, -7, 4.7\right\}$. List the elements that are:
a) whole numbers.
b) natural numbers.
c) rational numbers.
d) integers.
e) irrational numbers.
f) real numbers.

Find $A \cup B$ and $A \cap B$ for each set A and B.

51. $A = \{1, 2, 3, 4, 5\}, B = \{2, 4, 6, 8\}$

52. $A = \{1, 2, 3\}, B = \{4, 5, 6\}$

53. $A = \{-3, -2, -1, 0\}, B = \{-1, 0, 1, 2\}$

54. $A = \{-3, -1, 1, 3\}, B = \{-4, -3, -2, -1, 0\}$

55. $A = \{\ \}, B = \{2, 4, 6, 8, 10\}$

56. $A = \{2, 4, 6\}, B = \{2, 4, 6, 8, \ldots\}$

57. $A = \{0, 10, 20, 30\}, B = \{5, 15, 25\}$

58. $A = \{1, 3, 5\}, B = \{1, 3, 5, 7, \ldots\}$

59. $A = \{-1, 0, 1, e, i, \pi\}, B = \{-1, 0, 1\}$

60. $A = \left\{1, \dfrac{1}{2}, \dfrac{1}{4}, \dfrac{1}{6}, \ldots\right\}, B = \left\{\dfrac{1}{4}, \dfrac{1}{6}, \dfrac{1}{8}\right\}$

Describe each set.

61. $A = \{1, 2, 3, 4, \ldots\}$

62. $B = \{2, 4, 6, 8, \ldots\}$

63. $C = \{0, 3, 6, 9, \ldots\}$

64. $A = \{a, b, c, d, \ldots, z\}$

65. $B = \{\ldots, -5, -3, -1, 1, 3, 5, \ldots\}$

66. $C = \{\text{Alabama, Alaska}, \ldots, \text{Wyoming}\}$

*In Exercises 67 and 68, **a)** write out how you would read each set; **b)** write the set in roster form.*

67. $A = \{x | x < 8 \text{ and } x \in N\}$

68. $B = \{x | x \text{ is one of the last five capital letters in the English alphabet}\}$

Illustrate each set on a number line.

69. $\{x | x \geq 0\}$

70. $\{y | y < 4\}$

71. $\{z | z \leq 3\}$

72. $\{w | w > -5\}$

73. $\{p | -4 \leq p < 3\}$

74. $\{x | -1.67 \leq x < 5.02\}$

75. $\{q | q > -2 \text{ and } q \in N\}$

76. $\{x | -1.90 \leq x \leq 2.1 \text{ and } x \in I\}$

77. $\{r | r \leq \pi \text{ and } r \in W\}$

78. $\left\{ x \left| \dfrac{1}{2} < x \leq \dfrac{5}{9} \text{ and } x \in N \right. \right\}$

Express in set builder notation each set of numbers that are indicated on the number line.

79.

80.

81.

82.

83.

84.

85.

86.

87.

88.

Refer to the box on page 11 for the meanings of $\mathbb{R}$, N, W, I, Q, and H. Then determine whether the first set is a subset of the second set for each pair of sets.

89. N, W

90. W, Q

91. I, Q

92. W, N

93. Q, H

94. $Q, \mathbb{R}$

95. $H, \mathbb{R}$

96. Q, I

Problem Solving

97. Construct a set that contains five rational numbers between 1 and 2.

98. Construct a set that contains five rational numbers between 0 and 1.

99. Determine two sets A and B such that $A \cup B = \{2, 4, 5, 6, 8, 9\}$ and $A \cap B = \{4, 5, 9\}$.

100. Determine two sets A and B such that $A \cup B = \{3, 5, 7, 8, 9\}$ and $A \cap B = \{5, 7\}$.

101. *Teenage Clothing* The following table shows the top teenage girls' and boys' clothing brand preferences based on the American Express Retail Index on a back-to-school shopping survey taken in summer 2001.

a) Find the set of brands that were listed in *either* the boys' or the girls' category.

b) Does part **a)** represent the union or intersection of the brand names? Explain.

c) Find the set of brands that were listed in *both* the boys' and the girls' categories.

d) Does part **b)** represent the union or intersection of the brand names? Explain.

Most Popular Teenage Clothing Brands 2001	
Girls'	**Boys'**
1. Old Navy	1. Tommy Hilfiger
2. The Gap	2. Nike
3. Tommy Hilfiger	3. Adidas
4. Abercrombie & Fitch	4. Polo/Ralph Lauren
5. Express/The Limited	5. FUBU
Source: American Express and M. Booth and Associates	

102. *Advertising Dollars* The following table shows the top five broadcast television advertisers and the top five cable television advertisers for April–June 2001.

a) Find the set of companies that were in *either* category during this period.

b) Does part **a)** represent the union or the intersection of the companies? Explain.

c) Find the set of companies that were in *both* categories during this period.

d) Does part **c)** represent the union or the intersection of the companies? Explain.

Broadcast TV	**Cable TV**
1. General Motors Corp.	1. Philip Morris Cos., Inc.
2. Philip Morris Cos., Inc.	2. AOL Time Warner, Inc.
3. Johnson & Johnson	3. General Motors Corp.
4. Pfizer, Inc.	4. Procter & Gamble Co.
5. Pepsico, Inc.	5. Johnson & Johnson
Source: Nielson Media Research	

103. *Cub Scouts* The Cub Scouts in Pack 108 must complete four achievements to earn their Wolf Badge. Doug Wedding, their den leader, has the following table in his record book. A *Yes* indicates that the Cub Scout has completed that achievement.

Let A = the set of scouts who have completed Achievement 1: *Feats of Skill*.

Let B = the set of scouts who have completed Achievement 2: *Your Flag*.

Let C = the set of scouts who have completed Achievement 3: *Cooking and Eating*.

Let D = the set of scouts who have completed Achievement 4: *Making Choices*.

a) Give each of the sets A, B, C, and D using the roster method.

b) Determine the set $A \cap B \cap C \cap D$, that is, find the set of elements that are common to all four sets.

c) Which scouts have met all the requirements to receive their Wolf Badge?

	Achievement			
Scout	**1**	**2**	**3**	**4**
Alex	Yes	Yes	Yes	Yes
James	Yes	Yes	No	No
George	No	Yes	No	Yes
Connor	No	Yes	No	Yes
Stephen	No	No	Yes	No

104. *Goods and Services* The following graph shows the percentage weight given to different goods and services in the consumer price index.

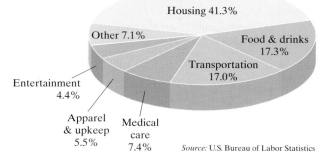

Source: U.S. Bureau of Labor Statistics

a) List the set of goods and services that have a weight of 17% or greater.

b) List the set of goods and services that have a weight of less than 6%.

105. The following diagram is called a *Venn diagram.* From the diagram determine the following sets:

a) A,

b) B,

c) $A \cup B$,

d) $A \cap B$.

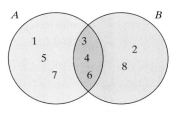

106. Use the following Venn diagram to determine the following sets:

a) A,

b) B,

c) $A \cup B$,

d) $A \cap B$.

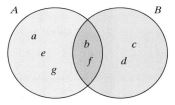

107. a) Explain the difference between the following sets of numbers: $\{x|x > 1 \text{ and } x \in N\}$ and $\{x|x > 1\}$.

b) Write the first set given in roster form.

c) Can you write the second set in roster form? Explain your answer.

108. Repeat Exercise 107 for the sets $\{x|2 < x < 6 \text{ and } x \in N\}$ and $\{x|2 < x < 6\}$.

Challenge Problems

109. a) Write the decimal numbers equivalent to $\frac{1}{9}, \frac{2}{9}$, and $\frac{3}{9}$.

b) Write the fractions equivalent to $0.\overline{4}, 0.\overline{5}$, and $0.\overline{6}$.

c) What is $0.\overline{9}$ equal to? Explain how you determined your answer.

Group Activity

110. *Newspaper Preferences* The Venn diagram that follows shows the results of a survey given to 45 people. The diagram shows the number of people in the survey who read the *New York Post*, the *New York Daily News*, and *The Wall Street Journal*.

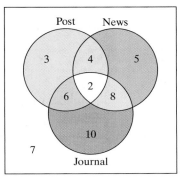

a) Group member 1: Determine the number surveyed who read *both* the *News* and the *Post*, that is, *News* ∩ *Post*.

b) Group member 2: Determine the number who read both the *Post* and the *Journal*, that is, *Post* ∩ *Journal*.

c) Group member 3: Determine the number who read both the *News* and the *Journal*, that is, *News* ∩ *Journal*.

d) Share your answer with the other members of the group and see if the group agrees with your answer.

e) As a group, determine the number of people who read all three papers.

f) As a group, determine the number of people who do not read any of the three papers.

1.3 PROPERTIES OF AND OPERATIONS WITH REAL NUMBERS

SSM   Study Guide   CD/Video
MathPro 4/5   PH Math Tutor Center   prenhall.com/Angel

1 Evaluate absolute values.
2 Add real numbers.
3 Subtract real numbers.
4 Multiply real numbers.
5 Divide real numbers.
6 Use the properties of real numbers.

To succeed in algebra, you must understand how to add, subtract, multiply, and divide real numbers. Before we can explain addition and subtraction of real numbers, we need to discuss absolute value.

Two numbers that are the same distance from 0 on the number line but in opposite directions are called **additive inverses**, or **opposites**, of each other. For example, 3 is the additive inverse of -3, and -3 is the additive inverse of 3. The number 0 is its own additive inverse. The sum of a number and its additive inverse is 0. What are the additive inverses of -56.3 and $\frac{76}{5}$? Their additive inverses are 56.3 and $-\frac{76}{5}$, respectively. Notice that the additive inverse of a positive number is a negative number and the additive inverse of a negative number is a positive number.

DEFINITION

Additive Inverse
For any real number a, its additive inverse is $-a$.

Consider the number -5. Its additive inverse is $-(-5)$. Since we know this number must be positive, this implies that $-(-5) = 5$. This is an example of the double negative property.

Double Negative Property

For any real number a, $-(-a) = a$.

By the double negative property, $-(-7.4) = 7.4$ and $-\left(-\frac{12}{5}\right) = \frac{12}{5}$.

1 Evaluate Absolute Values

The **absolute value** of a number is its distance from the number 0 on a number line. The symbol $|\ |$ is used to indicate absolute value.

FIGURE 1.6

Consider the numbers 3 and -3 (Fig. 1.6). Both numbers are 3 units from 0 on the number line. Thus

$$|3| = 3 \quad \text{and} \quad |-3| = 3$$

EXAMPLE 1   Evaluate.   **a)** $|7|$ **b)** $|-6.1|$ **c)** $|0|$

Solution   **a)** $|7| = 7$, since 7 is 7 units from 0 on the number line.

b) $|-6.1| = 6.1$, since -6.1 is 6.1 units from 0 on the number line.

c) $|0| = 0$.

The absolute value of any nonzero number will always be a positive number, and the absolute value of 0 is 0.

To find the absolute value of a real number without using a number line, use the following definition.   ✳

DEFINITION

Absolute Value

If a represents any real number, then

$$|a| = \begin{cases} a & \text{if } a \geq 0 \\ -a & \text{if } a < 0 \end{cases}$$

The definition of absolute value indicates that the absolute value of any nonnegative number is the number itself, and the absolute value of any negative number is the additive inverse (or opposite) of the number. The absolute value of a number can be found by using the definition, as illustrated below.

$|8.4| = 8.4$ Since 8.4 is greater than or equal to 0, its absolute value is 8.4.

$|0| = 0$ Since 0 is greater than or equal to 0, its absolute value is 0.

$|-12| = -(-12) = 12$ Since -12 is less than 0, its absolute value is $-(-12)$ or 12.

EXAMPLE 2   Evaluate using the definition of absolute value.   **a)** $-|5|$ **b)** $-|-6.43|$

Solution   **a)** We are finding the opposite of the absolute value of 5. Since the absolute value of 5 is positive, its opposite must be negative.

$$-|5| = -(5) = -5$$

b) We are finding the opposite of the absolute value of -6.43. Since the absolute value of -6.43 is positive, its opposite must be negative.

$$-|-6.43| = -(6.43) = -6.43$$ ✳

EXAMPLE 3   Insert $<$, $>$, or $=$ in the shaded area between the two values to make each statement true.

a) $|8| \blacksquare |-8|$ **b)** $|-1| \blacksquare -|-3|$

Solution   **a)** Since both $|8|$ and $|-8|$ equal 8, we have $|8| = |-8|$.

NOW TRY EXERCISE 39   **b)** Since $|-1| = 1$ and $-|-3| = -3$, we have $|-1| > -|-3|$.   ✳

2 Add Real Numbers

We first discuss how to add two numbers with the same sign, either both positive or both negative, and then we will discuss how to add two numbers with different signs, one positive and the other negative.

To Add Two Numbers with the Same Sign (Both Positive or Both Negative)

Add their absolute values and place the common sign before the sum.

The sum of two positive numbers will be a positive number, and the sum of two negative numbers will be a negative number.

EXAMPLE 4   Evaluate $-4 + (-7)$.

Solution   Since both numbers being added are negative, the sum will be negative. To find the sum, add the absolute values of these numbers and then place a negative sign before the value.

$$|-4| = 4 \qquad |-7| = 7$$

Now add the absolute values.

$$|-4| + |-7| = 4 + 7 = 11$$

Since both numbers are negative, the sum must be negative. Thus,

$$-4 + (-7) = -11$$

To Add Two Numbers with Different Signs (One Positive and the Other Negative)

Subtract the smaller absolute from the larger absolute value. The answer has the sign of the number with the larger absolute value.

The sum of a positive number and a negative number may be either positive, negative, or zero. The sign of the answer will be the same as the sign of the number with the larger absolute value.

EXAMPLE 5   Evaluate $5 + (-9)$.

Solution   Since the numbers being added are of opposite signs, we subtract the smaller absolute value from the larger absolute value. First we take each absolute value.

$$|5| = 5 \qquad |-9| = 9$$

Now we find the difference, $9 - 5 = 4$. The number -9 has a larger absolute value than the number 5, so their sum is negative.

$$5 + (-9) = -4$$

EXAMPLE 6   Evaluate.   **a)** $1.3 + (-2.7)$ **b)** $-\dfrac{7}{8} + \dfrac{5}{6}$

Solution   **a)** $1.3 + (-2.7) = -1.4$

b) Begin by writing both fractions with the least common denominator, 24.

$$-\frac{7}{8} + \frac{5}{6} = -\frac{21}{24} + \frac{20}{24} = \frac{(-21) + 20}{24} = \frac{-1}{24} = -\frac{1}{24}$$

EXAMPLE 7   **Depth of Ocean Trenches**   The Palau Trench in the Pacific Ocean lies 26,424 feet below sea level. The deepest ocean trench, the Mariana Trench, is 9416 feet deeper than the Palau Trench (See Fig. 1.7 on the next page). Find the depth of the Mariana Trench.

Solution   Consider distance below sea level to be negative. Therefore, the total depth is

$$-26,424 + (-9416) = -35,840 \text{ feet}$$

NOW TRY EXERCISE 135   or 35,840 feet below sea level.

3   Subtract Real Numbers

Depth below sea level

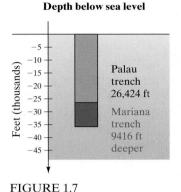

Feet (thousands)

Palau trench 26,424 ft

Mariana trench 9416 ft deeper

FIGURE 1.7

Every subtraction problem can be expressed as an addition problem using the following rule.

Subtraction of Real Numbers

$$a - b = a + (-b)$$

To subtract b from a, add the opposite (or additive inverse) of b to a.
    For example, $5 - 7$ means $5 - (+7)$. To subtract $5 - 7$, add the opposite of $+7$, which is -7, to 5.

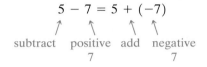

$$5 - 7 = 5 + (-7)$$

subtract   positive   add   negative
            7   7

Since $5 + (-7) = -2$, then $5 - 7 = -2$.

EXAMPLE 8   Evaluate.   **a)** $3 - 8$ **b)** $-6 - 4$

Solution   **a)** $3 - 8 = 3 + (-8) = -5$ **b)** $-6 - 4 = -6 + (-4) = -10$ ✳

EXAMPLE 9   Evaluate $8 - (-10)$.

Solution   In this problem, we are subtracting a negative number. The procedure to subtract remains the same.

$$8 - (-10) = 8 + 10 = 18$$

subtract   negative   add   positive
            10   10

Thus, $8 - (-10) = 18$.   ✳

By studying Example 9 and similar problems, we can see that for any real numbers a and b,

$$a - (-b) = a + b$$

We can use this principle to evaluate problems such as $8 - (-10)$ and other problems where we *subtract a negative quantity*.

EXAMPLE 10   Evaluate $-4 - (-12)$.

Solution $-4 - (-12) = -4 + 12 = 8$ ✳

EXAMPLE 11   **a)** Subtract 35 from -42.   **b)** Subtract $-\dfrac{3}{5}$ from $-\dfrac{5}{9}$.

Solution   **a)** $-42 - 35 = -77$

b) $-\dfrac{5}{9} - \left(-\dfrac{3}{5}\right) = -\dfrac{5}{9} + \dfrac{3}{5} = -\dfrac{25}{45} + \dfrac{27}{45} = \dfrac{2}{45}$ ✳

NOW TRY EXERCISE 57

EXAMPLE 12   **Extreme Temperatures**   The hottest temperature ever recorded in the United States was 134°F, which occurred at Greenland Ranch, California, in Death Valley on July 10, 1913. The coldest temperature ever recorded in the United States

FIGURE 1.8

was −79.8°F, which occurred at Prospect Creek Camp, Alaska, in the Endicott Mountains on January 23, 1971 (see Fig 1.8). Determine the difference between these two temperatures. *Source:* Learning Network Internet Web site

Solution   To find the difference, we subtract.

$$134° − (−79.8°) = 134° + 79.8° = 213.8°$$ ✳

Addition and subtraction are often combined in the same problem, as in the following examples. Unless parentheses are present, if the expression involves only addition and subtraction, we add and subtract from left to right. When parentheses are used, we add and subtract within the parentheses first. Then we add and subtract from left to right.

EXAMPLE 13   Evaluate $−15 + (−37) − (5 − 8)$.

Solution
$$\begin{aligned}
−15 + (−37) − (5 − 8) &= −15 + (−37) − (−3) \\
&= −15 − 37 + 3 \\
&= −52 + 3 = −49
\end{aligned}$$ ✳

EXAMPLE 14   Evaluate $2 − |−3| + 4 − (6 − |−7|)$.

Solution   Begin by replacing the numbers in absolute value signs with their numerical equivalents; then evaluate.

$$\begin{aligned}
2 − |−3| + 4 − (6 − |−7|) &= 2 − 3 + 4 − (6 − 7) \\
&= 2 − 3 + 4 − (−1) \\
&= 2 − 3 + 4 + 1 \\
&= −1 + 4 + 1 \\
&= 3 + 1 = 4
\end{aligned}$$ ✳

NOW TRY EXERCISE 69

4 Multiply Real Numbers

The following rules are used in determining the product when two numbers are multiplied.

> **Multiply Two Real Numbers**
>
> 1. To multiply two numbers with **like signs**, either both positive or both negative, multiply their absolute values. The answer is **positive**.
>
> 2. To multiply two numbers with **unlike signs**, one positive and the other negative, multiply their absolute values. The answer is **negative**.

EXAMPLE 15   Evaluate.   **a)** $(4.2)(−1.6)$ **b)** $(−16)\left(−\dfrac{1}{2}\right)$.

Solution   **a)** $(4.2)(−1.6) = −6.72$ *The numbers have unlike signs.*

b) $(−16)\left(−\dfrac{1}{2}\right) = 8$ *The numbers have like signs, both negative.* ✳

EXAMPLE 16   Evaluate $4(−2)(−3)(1)$.

Solution $4(−2)(−3)(1) = (−8)(−3)(1) = 24(1) = 24$ ✳

When multiplying more than two numbers, the product will be *negative* when there is an *odd* number of negative numbers. The product will be *positive* when there is an *even* number of negative numbers.

The multiplicative property of zero indicates that the product of 0 and any number is 0.

Multiplicative Property of Zero

For any number a,

$$a \cdot 0 = 0 \cdot a = 0$$

By the multiplicative property of zero, $5(0) = 0$ and $(-7.3)(0) = 0$.

EXAMPLE 17   Evaluate $9(5)(-2.63)(0)(4)$.

Solution   If one or more of the factors is 0, the product is 0. Thus, $9(5)(-2.63)(0)(4) = 0$. Can you explain why the product of any number of factors will be 0 if any factor is 0? ☀

5   Divide Real Numbers

The rules for the division of real numbers are similar to those for multiplication of real numbers.

Divide Two Real Numbers

1. To divide two numbers with **like signs**, either both positive or both negative, divide their absolute values. The answer is **positive**.

2. To divide two numbers with **unlike signs**, one positive and the other negative, divide their absolute values. The answer is **negative**.

EXAMPLE 18   Evaluate.   **a)** $-24 \div 6$ **b)** $-6.45 \div (-0.4)$

Solution   **a)** $\dfrac{-24}{6} = -4$ *The numbers have unlike signs.*

      b) $\dfrac{-6.45}{-0.4} = 16.125$ *The numbers have like signs.*   ☀

EXAMPLE 19   Evaluate $\dfrac{-3}{8} \div \left| \dfrac{-2}{5} \right|$.

Solution   Since $\left| \dfrac{-2}{5} \right|$ is equal to $\dfrac{2}{5}$, we write

$$\frac{-3}{8} \div \left| \frac{-2}{5} \right| = \frac{-3}{8} \div \frac{2}{5}$$

Now invert the divisor and proceed as in multiplication.

$$\frac{-3}{8} \div \frac{2}{5} = \frac{-3}{8} \cdot \frac{5}{2} = \frac{-3 \cdot 5}{8 \cdot 2} = -\frac{15}{16}$$

NOW TRY EXERCISE 85   ☀

When the denominator of a fraction is a negative number, we usually rewrite the fraction with a positive denominator. To do this, we use the following fact.

Sign of a Fraction

For any number a and any nonzero number b,

$$\frac{a}{-b} = \frac{-a}{b} = -\frac{a}{b}$$

Thus, when we have a quotient of $\dfrac{1}{-2}$, we rewrite it as either $\dfrac{-1}{2}$ or $-\dfrac{1}{2}$.

6 Use the Properties of Real Numbers

We have already discussed the double negative property and the multiplicative property of zero. Table 1.1 lists other basic properties for the operations of addition and multiplication on the real numbers.

TABLE 1.1		
For real numbers a, b, and c	**Addition**	**Multiplication**
Commutative property	$a + b = b + a$	$ab = ba$
Associative property	$(a + b) + c = a + (b + c)$	$(ab)c = a(bc)$
Identity property	$a + 0 = 0 + a = a$ (0 is called the **additive identity element.**)	$a \cdot 1 = 1 \cdot a = a$ (1 is called the **multiplicative identity element.**)
Inverse property	$a + (-a) = (-a) + a = 0$ ($-a$ is called the **additive inverse** or **opposite** of a.)	$a \cdot \dfrac{1}{a} = \dfrac{1}{a} \cdot a = 1$ ($1/a$ is called the **multiplicative inverse** or **reciprocal** of a, $a \neq 0$.)
Distributive property (of multiplication over addition)	$a(b + c) = ab + ac$	

Note that the commutative property involves a change in *order*, and the associative property involves a change in *grouping*.

The distributive property applies when there are more than two numbers within the parentheses.

$$a(b + c + d + \cdots + n) = ab + ac + ad + \cdots + an$$

This expanded form of the distributive property is often called the *extended distributive property*. However, when using the extended distributive property, we will just refer to it as the distributive property.

EXAMPLE 20 Name each property illustrated.

a) $7 \cdot m = m \cdot 7$ **b)** $(a + 4) + 2b = a + (4 + 2b)$

c) $4s + 5t = 5t + 4s$ **d)** $2v(w + 3) = 2v \cdot w + 2v \cdot 3$

Solution **a)** Commutative property of multiplication: change of order, $7 \cdot m = m \cdot 7$

b) Associative property of addition: change of grouping,
$(a + 4) + 2b = a + (4 + 2b)$

c) Commutative property of addition: change of order, $4s + 5t = 5t + 4s$

d) Distributive property: $2v(w + 3) = 2v \cdot w + 2v \cdot 3$

In Example 20 **d)** the expression $2v \cdot w + 2v \cdot 3$ can be simplified to $2vw + 6v$ using the properties of the real numbers. Can you explain why?

NOW TRY EXERCISE 119

EXAMPLE 21 Name each property illustrated.

a) $4 \cdot 1 = 4$ **b)** $x + 0 = x$

c) $4 + (-4) = 0$ **d)** $1(x + y) = x + y$

Solution
a) Identity property of multiplication

b) Identity property of addition

c) Inverse property of addition

d) Identity property of multiplication

EXAMPLE 22 Write the additive inverse (or opposite) and multiplicative inverse (or reciprocal) of each of the following.

a) -3 **b)** $\dfrac{2}{3}$

Solution
a) The additive inverse is 3. The multiplicative inverse is $\dfrac{1}{-3} = -\dfrac{1}{3}$.

b) The additive inverse is $-\dfrac{2}{3}$. The multiplicative inverse is $\dfrac{1}{\frac{2}{3}} = \dfrac{3}{2}$.

Exercise Set 1.3

Concept/Writing Exercises

1. What are additive inverses or opposites?

2. Give an example of the double negative property.

3. Will the absolute value of every real number be a positive number? Explain.

4. Give the definition of absolute value.

Find the unknown number(s). Explain how you determined your answer.

5. All numbers a such that $|a| = |-a|$

6. All numbers a such that $|a| = a$

7. All numbers a such that $|a| = 5$

8. All numbers a such that $|a| = -a$

9. All numbers a such that $|a| = -3$

10. All numbers x such that $|x - 3| = |3 - x|$

11. In your own words, explain how to add two numbers with the same sign.

12. Explain in your own words how to add two numbers with different signs.

13. In your own words, explain how to subtract real numbers.

14. In your own words, explain how the rules for multiplication and division of real numbers are similar.

15. List two other ways that the fraction $\dfrac{a}{-b}$ may be written.

16. a) Write the associative property of multiplication.
b) In your own words, explain the property.

17. a) Write the commutative property of addition.
b) In your own words, explain the property.

18. a) Write the distributive property of multiplication over addition.
b) In your own words, explain the property.

19. Using an example, explain why addition is not distributive over multiplication. That is, explain why $a + (b \cdot c) \neq (a + b) \cdot (a + c)$.

20. Give an example of the extended distributive property.

Practice the Skills

Evaluate each absolute value expression.

21. $|5|$

22. $|-4|$

23. $|-7|$

24. $|1.3|$

25. $\left|-\dfrac{7}{8}\right|$

26. $|-9.34|$

27. $|0|$

28. $-|1|$

29. $-|-7|$

30. $-|-\pi|$

31. $-\left|\dfrac{5}{9}\right|$

32. $-\left|-\dfrac{7}{8}\right|$

Insert <, >, or = in the shaded area to make each statement true.

33. $|-7|$ ▨ $|7|$

34. $|-4|$ ▨ $|5|$

35. $|-5|$ ▨ -5

36. $|-10|$ ▨ -5

37. $|-\pi|$ ▨ -3

38. $-|-1|$ ▨ -1

39. $|-7|$ ▨ $-|3|$

40. $-|9|$ ▨ $-|11|$

41. $-(-3)$ ▨ $-|-3|$

42. $|-(-4)|$ ▨ -4

43. $|19|$ ▨ $|-25|$

44. $-|-1|$ ▨ $|-6|$

List the values from smallest to largest.

45. $-1, -2, |-3|, 4, -|5|$

46. $\pi, -\pi, |-3|, -|-3|, -2, |-2|$

47. $-32, |-7|, 15, -|4|, 4$

48. $-8, -12, -|9|, -|20|, -|-18|$

49. $-2.1, -2, -2.4, |-2.8|, -|2.9|$

50. $-6.1, |-6.3|, -|-6.5|, 6.8, |6.4|$

51. $\dfrac{1}{3}, \left|-\dfrac{1}{2}\right|, -2, \left|\dfrac{3}{5}\right|, -\left|\dfrac{3}{4}\right|$

52. $\left|-\dfrac{5}{2}\right|, \dfrac{3}{5}, |-3|, -\left|\dfrac{5}{3}\right|, -\left|\dfrac{2}{3}\right|$

Evaluate each addition and subtraction problem.

53. $7 + (-6)$

54. $-2 + 5$

55. $-11 + 10$

56. $-12 - (-4)$

57. $-9 - (-5)$

58. $-2.18 - 3.14$

59. $-14.21 - (-13.22)$

60. $\dfrac{4}{5} - \dfrac{6}{7}$

61. $-\dfrac{5}{12} - \left(-\dfrac{7}{8}\right)$

62. $-1 - \dfrac{9}{16}$

63. $9 - (-2.31) + (-4.39)$

64. $-|7.31| - (-3.28) + 4.76$

65. $8.9 - |8.5| - |17.6|$

66. $|9 - 4| - 6$

67. $|5 - 12| - |3|$

68. $|12 - 5| - |5 - 12|$

69. $-|-3| - |7| + (6 + |-2|)$

70. $|-4| - |-4| - |-4 - 4|$

71. $\left(\dfrac{3}{5} + \dfrac{3}{4}\right) - \dfrac{1}{2}$

72. $\dfrac{4}{5} - \left(\dfrac{3}{4} - \dfrac{2}{3}\right)$

Evaluate each multiplication and division problem.

73. $-5 \cdot 7$

74. $(-9)(-3)$

75. $-4\left(-\dfrac{5}{16}\right)$

76. $-4\left(-\dfrac{3}{4}\right)\left(-\dfrac{1}{2}\right)$

77. $(-1)(-2)(-1)(2)(-3)$

78. $(-2.1)(-7.8)(-9.1)$

79. $(-1.1)(3.4)(8.3)(-7.6)$

80. $-16 \div 8$

81. $-55 \div (-11)$

82. $-4 \div \left(-\dfrac{1}{4}\right)$

83. $-\dfrac{5}{9} \div \dfrac{-5}{9}$

84. $\left|-\dfrac{1}{2}\right| \cdot \left|\dfrac{-3}{4}\right|$

85. $\left(-\dfrac{3}{4}\right) \div |-8|$

86. $\left|\dfrac{3}{8}\right| \div (-2)$

87. $\left|\dfrac{-5}{6}\right| \div \left|\dfrac{-1}{2}\right|$

88. $\dfrac{-5}{9} \div |-5|$

Evaluate.

89. $10 - 14$

90. $-12 - 15$

91. $-56 \div 7$

92. $-\dfrac{1}{8} + \left(-\dfrac{1}{16}\right)$

93. $3\left(-\dfrac{2}{3}\right)\left(-\dfrac{5}{2}\right)$

94. $(1.32 - 2.76) - (-3.85 + 4.28)$

95. $(-3.2)(4.9)(-2.73)$

96. $(4.2)(-1)(-9.6)(3.8)$

97. $-16.4 - (-9.6) - 14.8$

98. $9 - (4 - 3) - (-2 - 1)$ **99.** $-|8| \cdot \left|\dfrac{-1}{2}\right|$ **100.** $-\left|\dfrac{-12}{5}\right| \cdot \left|\dfrac{3}{4}\right|$

101. $\left|\dfrac{-9}{4}\right| \div \left|\dfrac{-4}{9}\right|$ **102.** $(-|3| + |5|) - (1 - |-9|)$ 🔒**103.** $5 - |-7| + 3 - |-2|$

104. $\left(\dfrac{3}{8} - \dfrac{4}{7}\right) - \left(-\dfrac{1}{2}\right)$ **105.** $\left(-\dfrac{3}{5} - \dfrac{4}{9}\right) - \left(-\dfrac{2}{3}\right)$ **106.** $(|-4| - 3) - (3 \cdot |-5|)$

107. $(25 - |32|)(-6 - 5)$ **108.** $\left[(-2)\left|-\dfrac{1}{2}\right|\right] \div \left|-\dfrac{1}{4}\right|$

Name each property illustrated.

109. $c + d = d + c$
110. $5(v + w) = 5v + 5w$
111. $b \cdot 0 = 0$
112. $c \cdot d = d \cdot c$
113. $(x + 3) + 6 = x + (3 + 6)$
114. $x + 0 = x$
115. $x = 1 \cdot x$
116. $x(y + z) = xy + xz$
117. $5(xy) = (5x)y$
118. $(2x \cdot 3y) \cdot 4y = 2x \cdot (3y \cdot 4y)$
119. $4(x + y + 2) = 4x + 4y + 8$
120. $-(-1) = 1$
121. $5 + 0 = 5$
122. $4 \cdot \dfrac{1}{4} = 1$
123. $3 + (-3) = 0$
124. $6 \cdot 0 = 0$
125. $x \cdot \dfrac{1}{x} = 1$
126. $(x + y) = 1(x + y)$
🔒**127.** $-(-x) = x$
128. $x + (-x) = 0$

List both the additive inverse and the multiplicative inverse for each problem.

129. 6 **130.** -11 **131.** $-\dfrac{22}{7}$ **132.** $-\dfrac{3}{7}$

Problem Solving

133. ***The Film* Gold** During the production of the documentary film *Gold*, the film crew experienced severe changes in temperature. In a South African gold mine 3 miles below the surface of the earth, the temperature was 140°F. On a mountain near Cuzco, Peru, the temperature was 40°F. Determine the difference in temperature between these two filming sites. *Source:* History channel Web site

134. ***Temperature Change*** The most unusual temperature change according to the *Guinness Book of World*

Records occurred from 7:30 A.M. to 7:32 A.M. on January 22, 1943, in Spearfish, South Dakota. During these 2 minutes the temperature changed from −4°F to 45°F. Determine the increase in temperature in these 2 minutes.

135. ***Submarine Dive*** A submarine dives 358.9 feet. A short time later the submarine comes up 210.7 feet. Find the submarine's final depth from its starting point. (Consider distance in a downward direction as negative.)

136. ***Checking Account*** Sharon Koch had a balance of −$32.64 in her checking account when she deposited a check for $99.38. What is her new balance?

137. *Book Contract* Harvey Mutchner signed a contract with a publishing company that called for an advance payment of $60,000 on the sale of his book *Brooklyn, the Bronx, and Other Places I Love.* When the book is published and sales begin, the publishers will automatically deduct this advance from the author's royalties.

a) Six months after the release of the book, the author's royalties totaled $47,600 before the advance was deducted. Determine how much money he will receive or owe the publisher.

b) After 1 year, the author's royalties are $87,500. Determine how much money he will receive or owe to the publishing company.

138. *Extreme Temperatures* The lowest temperature ever recorded in the United States was $-79.8°F$ on January 23, 1971, in Prospect Creek, Alaska. The lowest temperature in the contiguous states (all states except Alaska and Hawaii) was $-69.7°F$ on January 20, 1954, in Rogers Pass, Montana. Find the difference in these temperatures.

139. *Estimated Taxes* In 2002, Joanne Beebe made four quarterly estimated income tax payments of $3000 each. When she completed her year 2002 income tax forms, she found her total tax was $10,125.

a) Will Joanne be entitled to a refund or will she owe more taxes? Explain.

b) How much of a refund will she receive or how much more in taxes will she owe?

140. *Stock Prices* On September 21, 2001, Ron Breitfelder purchased 100 shares of Home Depot stock at $30.30 per share. On January 8, 2002, Ron sold all 100 shares for a price of $51.10 per share. What was Ron's total gain or loss for this transaction?

141. Write your own realistic word problem that involves subtracting a positive number from a negative number. Indicate the answer to your word problem.

142. Write your own realistic word problem that involves subtracting a negative number from a negative number. Indicate the answer to your problem.

143. *Balance of Payments* The graph shows the balance of payments in 2000 between the federal government and the 50 states. In some states, there was a large surplus because the federal government spent significantly more in those states than it collected from them in taxes. On the other hand, some states had big deficits, paying much more in taxes than the federal government returned to them.

See Exercise 143
The Balance of Payments in 2000

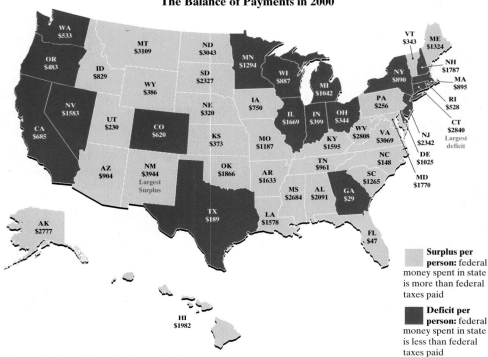

Source: Taubman Center for State & Local Government, Harvard University

Determine the difference in the per capita balance of payments between the following states.

a) NM and CT   **b)** OK and MI   **c)** VA and NJ   **d)** TX and CA

144. *Small Businesses* The chamber of commerce in West Union, Illinois, studied the success and failure of small start-up businesses (under five employees) in their town. They found the average first-year expenditures and the average first-year incomes, as shown on the bar graph to the right. Estimate the average first-year profit by subtracting the average first-year expenditures from the average first-year income.

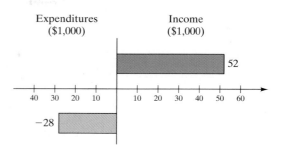

Challenge Problems

145. Evaluate $1 - 2 + 3 - 4 + \cdots + 99 - 100$. (*Hint:* Group in pairs of two numbers.)

146. Evaluate $1 + 2 - 3 + 4 + 5 - 6 + 7 + 8 - 9 + 10 + 11 - 12 + \cdots + 22 + 23 - 24$. (*Hint:* Examine in groups of three numbers.)

147. Evaluate $\dfrac{(1) \cdot |-2| \cdot (-3) \cdot |4| \cdot (-5)}{|-1| \cdot (-2) \cdot |-3| \cdot (4) \cdot |-5|}$.

148. Evaluate $\dfrac{(1)(-2)(3)(-4)(5) \cdots (97)(-98)}{(-1)(2)(-3)(4)(-5) \cdots (-97)(98)}$.

Cumulative Review Exercises

[1.2] **149.** Answer true or false: Every irrational number is a real number.

150. List the set of natural numbers.

151. Consider the set $\left\{3, 4, -2, \frac{5}{6}, \sqrt{3}, 0\right\}$. List the elements that are

a) integers,

b) rational numbers,

c) irrational numbers,

d) real numbers.

152. $A = \{4, 7, 9, 12\}$; $B = \{1, 4, 7, 15\}$. Find

a) $A \cup B$,

b) $A \cap B$.

153. Illustrate $\{x | -4 < x \le 6\}$ on a number line.

1.4 ORDER OF OPERATIONS

1   Evaluate exponential expressions.
2   Evaluate square and higher roots.
3   Evaluate expressions using the order of operations.
4   Evaluate expressions containing variables.
5   Evaluate expressions on a graphing calculator.

Before we discuss the order of operations we need to speak briefly about exponents and roots. We will discuss exponents in greater depth in Sections 1.5 and 7.2.

1   Evaluate Exponential Expressions

In a multiplication problem, the numbers or expressions that are multiplied are called **factors**. If $a \cdot b = c$, then a and b are factors of c. For example, since $2 \cdot 3 = 6$, both 2 and 3 are factors of 6. The number 1 is a factor of every number and expression. Can you explain why?

The quantity 3^2 is called an **exponential expression**. In the expression, the 3 is called the **base** and the 2 is called the **exponent**. The expression 3^2 is read "three squared" or "three to the second power." Note that

$$3^2 = \underbrace{3 \cdot 3}_{2 \text{ factors of } 3}$$

The expression 5^3 is read "five cubed" or "five to the third power." Note that

$$5^3 = \underbrace{5 \cdot 5 \cdot 5}_{3 \text{ factors of } 5}$$

In general, the base b to the nth power is written b^n. For any natural number n,

$$b^n = \underbrace{b \cdot b \cdot b \cdot b \cdots \cdot b}_{n \text{ factors of } b}$$

Note that 0^0 is *undefined*.

EXAMPLE 1   Evaluate.   **a)** $(0.5)^3$ **b)** $(-3)^5$ **c)** 1^{25} **d)** $\left(-\dfrac{4}{7}\right)^3$

Solution   **a)** $(0.5)^3 = (0.5)(0.5)(0.5) = 0.125$
b) $(-3)^5 = (-3)(-3)(-3)(-3)(-3) = -243$
c) $1^{25} = 1$; 1 raised to any power will equal 1. Why?
d) $\left(-\dfrac{4}{7}\right)^3 = \left(-\dfrac{4}{7}\right)\left(-\dfrac{4}{7}\right)\left(-\dfrac{4}{7}\right) = -\dfrac{64}{343}$

HELPFUL HINT

STUDY TIP

Be very careful when writing or copying exponents. Since exponents are small it is very easy to write or copy an exponent and then later not recognize what you have written. Some exponents that may be easily confused if not clearly written are 1 and 7, 2 and 3, 3 and 5, 4 and 9, 5 and 6, and 5 and 8.

It is not necessary to write exponents of 1. Whenever we encounter a numerical value or a variable without an exponent, we assume that it has an exponent of 1. Thus, 3 means 3^1, x means x^1, x^3y means x^3y^1, and $-xy$ means $-x^1y^1$.

Students often evaluate expressions containing $-x^2$ incorrectly. The expression $-x^2$ means $-(x^2)$, not $(-x)^2$. Note that -5^2 means $-(5^2) = -(5 \cdot 5) = -25$ while $(-5)^2$ means $(-5)(-5) = 25$. In general, $-x^m$ means $-(x^m)$, not $(-x)^m$. The expression $-x^2$ is read *negative x squared* or *the opposite of x^2*. The expression $(-x)^2$ is read *negative x quantity squared*.

EXAMPLE 2   Evaluate $-x^2$ for each value of x.   **a)** 3   **b)** -3

Solution   **a)** $-x^2 = -(3)^2 = -9$
b) $-x^2 = -(-3)^2 = -(9) = -9$

EXAMPLE 3   Evaluate $-5^2 + (-5)^2 - 4^3 + (-4)^3$.

Solution   First, we evaluate each exponential expression. Then we add or subtract, working from left to right.

$$-5^2 + (-5)^2 - 4^3 + (-4)^3 = -(5^2) + (-5)^2 - (4^3) + (-4)^3$$
$$= -25 + 25 - 64 + (-64)$$
$$= -25 + 25 - 64 - 64$$
$$= -128$$

NOW TRY EXERCISE 59

Using Your Calculator

Evaluating Exponential Expressions on a Scientific and a Graphing Calculator

On both scientific and graphing calculators the $\boxed{x^2}$ key can be used to square a number. Below we show the sequence of keys to press to evaluate 5^2.

Scientific calculator: 5 $\boxed{x^2}$ 25 ← answer displayed

Graphing calculator: 5 $\boxed{x^2}$ $\boxed{\text{ENTER}}$ 25 ← answer displayed

To evaluate exponential expressions with other exponents, you can use the $\boxed{y^x}$ or $\boxed{\wedge}$ key. Most scientific calculators have a $\boxed{y^x}$ key,* whereas graphing calculators use the $\boxed{\wedge}$ key. To evaluate exponential expressions using these keys, first enter the base, then press either the $\boxed{y^x}$ or $\boxed{\wedge}$ key, and then enter the exponent. For example, to evaluate 6^4 we do the following:

Scientific calculator: 6 $\boxed{y^x}$ 4 $\boxed{=}$ 1296 ← answer displayed

Graphing calculator: 6 $\boxed{\wedge}$ 4 $\boxed{\text{ENTER}}$ 1296 ← answer displayed

* Some calculators have $\boxed{x^y}$ or $\boxed{a^b}$ keys instead of an $\boxed{y^x}$ key.

2 Evaluate Square and Higher Roots

The symbol used to indicate a root, $\sqrt{}$, is called a **radical sign**. The number or expression inside the radical sign is called the **radicand**. In $\sqrt{25}$, the radicand is 25. The **principal or positive square root** of a positive number a, written $\sqrt{a}$, is the positive number that when multiplied by itself gives a. For example, the principal square root of 4 is 2, written $\sqrt{4} = 2$, because $2 \cdot 2 = 4$. In general, $\sqrt{a} = b$ if $b \cdot b = a$. Whenever we use the words *square root*, we are referring to the "principal square root."

EXAMPLE 4   Evaluate.   **a)** $\sqrt{25}$ **b)** $\sqrt{\dfrac{9}{4}}$ **c)** $\sqrt{0.64}$ **d)** $-\sqrt{49}$

Solution   **a)** $\sqrt{25} = 5$, since $5 \cdot 5 = 25$

b) $\sqrt{\dfrac{9}{4}} = \dfrac{3}{2}$, since $\dfrac{3}{2} \cdot \dfrac{3}{2} = \dfrac{9}{4}$

c) $\sqrt{0.64} = 0.8,$ since $(0.8)(0.8) = 0.64$

d) $-\sqrt{49}$ means $-(\sqrt{49})$. We determine that $\sqrt{49} = 7,$ since $7 \cdot 7 = 49$. Therefore, $-\sqrt{49} = -7.$ ✳

The square root of 4, $\sqrt{4}$, is a rational number since it is equal to 2. The square roots of other numbers, such as $\sqrt{2}$, $\sqrt{3}$, and $\sqrt{5}$, are irrational numbers. The decimal values of such numbers can never be given exactly since irrational numbers are nonterminating, nonrepeating decimal numbers. The approximate value of $\sqrt{2}$ and other irrational numbers can be found with a calculator.

$$\sqrt{2} \approx 1.414213562 \qquad \textit{From a calculator}$$

In this section we introduce square roots; cube roots, symbolized by $\sqrt[3]{}$; and higher roots. The number used to indicate the root is called the **index**.

The index of a square root is 2. However, we generally do not show the index 2. Therefore, $\sqrt{a} = \sqrt[2]{a}$.

The concept used to explain square roots can be expanded to explain cube roots and higher roots. The cube root of a number a is written $\sqrt[3]{a}$.

$$\sqrt[3]{a} = b \quad \text{if} \quad \underbrace{b \cdot b \cdot b}_{3 \text{ factors of } b} = a$$

For example, $\sqrt[3]{8} = 2$, because $2 \cdot 2 \cdot 2 = 8$. The expression $\sqrt[n]{a}$ is read "the nth root of a."

$$\sqrt[n]{a} = b \quad \text{if} \quad \underbrace{b \cdot b \cdot b \cdots b}_{n \text{ factors of } b} = a$$

EXAMPLE 5   Evaluate.   **a)** $\sqrt[3]{125}$ **b)** $\sqrt[4]{81}$ **c)** $\sqrt[5]{32}$

Solution   **a)** $\sqrt[3]{125} = 5$, since $5 \cdot 5 \cdot 5 = 125$

b) $\sqrt[4]{81} = 3$, since $3 \cdot 3 \cdot 3 \cdot 3 = 81$

c) $\sqrt[5]{32} = 2$, since $2 \cdot 2 \cdot 2 \cdot 2 \cdot 2 = 32$ ✳

EXAMPLE 6   Evaluate.   **a)** $\sqrt[4]{81}$ **b)** $\sqrt[3]{\dfrac{1}{27}}$ **c)** $\sqrt[3]{-8}$ **d)** $-\sqrt[3]{8}$

Solution   **a)** $\sqrt[4]{81} = 3$, since $3 \cdot 3 \cdot 3 \cdot 3 = 81$

b) $\sqrt[3]{\dfrac{1}{27}} = \dfrac{1}{3}$, since $\left(\dfrac{1}{3}\right)\left(\dfrac{1}{3}\right)\left(\dfrac{1}{3}\right) = \dfrac{1}{27}$

c) $\sqrt[3]{-8} = -2$, since $(-2)(-2)(-2) = -8$

d) $-\sqrt[3]{8}$ means $-(\sqrt[3]{8})$. We determine that $\sqrt[3]{8} = 2$, since $2 \cdot 2 \cdot 2 = 8$. Therefore,

NOW TRY EXERCISE 25 $-\sqrt[3]{8} = -2.$ ✳

Note that in Example 6 **c)** the cube root of a negative number is negative. Why is this so? We will discuss radicals in more detail in Chapter 7.

Using Your Calculator

Evaluating Roots on a Scientific Calculator

The square roots of numbers can be found on calculators with a square-root key, $\boxed{\sqrt{x}}$. To evaluate $\sqrt{25}$ on calculators that have this key, press

answer displayed

$$25 \quad \boxed{\sqrt{x}} \quad 5$$

Higher roots can be found on calculators that contain either the $\boxed{\sqrt[x]{y}}$ key or the $\boxed{y^x}$ key.* To evaluate $\sqrt[4]{625}$ on a calculator with a $\boxed{\sqrt[x]{y}}$ key, do the following:

answer displayed

$$625 \quad \boxed{\sqrt[x]{y}} \quad 4 \quad \boxed{=} \quad 5$$

Note that the number within the radical sign (the radicand), 625, is entered, then the $\boxed{\sqrt[x]{y}}$ key is pressed, and then the root (or index) 4 is entered. When the $\boxed{=}$ key is pressed, the answer 5 is displayed.

To evaluate $\sqrt[4]{625}$ on a calculator with a $\boxed{y^x}$ key, use the inverse key as follows:

answer displayed

$$625 \quad \boxed{\text{INV}} \quad \boxed{y^x} \quad 4 \quad \boxed{=} \quad 5$$

* Calculator keys vary. Some calculators have $\boxed{x^y}$ or $\boxed{a^b}$ keys instead of the $\boxed{y^x}$ key, and some calculators have a $\boxed{2^{nd}}$ or $\boxed{\text{shift}}$ key instead of the $\boxed{\text{INV}}$ key.

Using Your Graphing Calculator

Evaluating Roots on a Graphing Calculator

To find the square root on a graphing calculator, use $\sqrt{}$. The $\sqrt{}$ appears above the $\boxed{x^2}$ key, so you will need to press the $\boxed{2^{nd}}$ key to evaluate square roots. For example, to evaluate $\sqrt{25}$ press

$$\boxed{2^{nd}} \quad \boxed{x^2} \quad 25 \quad \boxed{\text{ENTER}} \quad 5 \longleftarrow \text{answer displayed}$$

When you press $\boxed{2^{nd}}$ $\boxed{x^2}$, the Texas Instruments TI-83 Plus generates $\sqrt{}($. Then you insert the radicand and press $\boxed{\text{ENTER}}$. To learn how to find cube and higher roots, refer to your graphing calculator manual.

With the TI-83 Plus, you can use the $\boxed{\text{MATH}}$ key. When you press this key you get a number of options including 4 and 5, which are shown below.

$$4:\sqrt[3]{}(\qquad 5:\sqrt[x]{}$$

Option 4 can be used to find cube roots and option 5 can be used to find higher roots, as shown in the following examples.

EXAMPLE Evaluate $\sqrt[3]{120}$.

Solution

answer displayed

$$\boxed{\text{MATH}} \quad 4 \quad \underset{\substack{\uparrow \\ \text{enter} \\ \text{radicand}}}{120} \quad \boxed{\text{ENTER}} \quad 4.932424149$$

$$\underset{\substack{\text{select} \\ \text{option 4}}}{}$$

(continued on the next page)

To find the root with an index greater than 3, first enter the index, then press the $\boxed{\text{MATH}}$ key, and then press option 5.

EXAMPLE   Evaluate $\sqrt[4]{625}$.

Solution

We will show another way to find roots on a graphing calculator in Section 7.2 when we discuss rational exponents.

NOW TRY EXERCISE 37

3   Evaluate Expressions Using the Order of Operations

You will often have to evaluate expressions containing multiple operations. To do so, follow the **order of operations** indicated below.

Order of Operations

To evaluate mathematical expressions, use the following order:

1. First, evaluate the expressions within grouping symbols, including parentheses, ( ), brackets, [ ], and braces, { }. If the expression contains nested grouping symbols (one pair of grouping symbols within another pair), evaluate the expression in the innermost grouping symbols first.

2. Next, evaluate all terms containing exponents and roots.

3. Next, evaluate all multiplications or divisions in the order in which they occur, working from left to right.

4. Finally, evaluate all additions or subtractions in the order in which they occur, working from left to right.

It should be noted that a *fraction bar* acts as a grouping symbol. Thus, when evaluating expressions containing a fraction bar, we work separately above and below the fraction bar.

Brackets are often used in place of parentheses to help avoid confusion. For example, the expression $7((5 \cdot 3) + 6)$ is easier to follow when written $7[(5 \cdot 3) + 6]$. Remember to evaluate the innermost group first.

EXAMPLE 7   Evaluate $8 + 3 \cdot 5^2 - 7$.

Solution   We will use shading to indicate the order in which the operations are to be evaluated. Since there are no parentheses, we first evaluate 5^2.

$$8 + 3 \cdot \boxed{5^2} - 7 = 8 + \boxed{3 \cdot 25} - 7$$

Next, we perform multiplications or divisions from left to right.

$$= \boxed{8 + 75} - 7$$

Finally, we perform additions or subtractions from left to right.

$$= 83 - 7$$
$$= 76$$

EXAMPLE 8   Evaluate $10 + \{6 - [4(5 - 2)]\}^2$.

Solution   First, evaluate the expression within the innermost parentheses. Then continue according to the order of operations.

$$10 + \{6 - [4(5 - 2)]\}^2 = 10 + \{6 - [\,4(3)\,]\}^2$$
$$= 10 + [6 - (12)]^2$$
$$= 10 + (-6)^2$$
$$= 10 + 36$$
$$= 46$$

NOW TRY EXERCISE 77

EXAMPLE 9   Evaluate $\dfrac{6 \div \frac{1}{2} + 5|7 - 3|}{1 + (3 - 5) \div 2}$.

Solution   Remember that the fraction bar acts as a grouping symbol. Work separately above the fraction bar and below the fraction bar.

$$\frac{6 \div \frac{1}{2} + 5|7 - 3|}{1 + (3 - 5) \div 2} = \frac{6 \div \frac{1}{2} + 5|4|}{1 + (-2) \div 2}$$
$$= \frac{12 + 20}{1 + (-1)}$$
$$= \frac{32}{0}$$

Since division by 0 is not possible, the original expression is **undefined**.

4   Evaluate Expressions Containing Variables

To evaluate mathematical expressions, we use the order of operations just given. Example 10 is an application problem in which we use the order of operations.

EXAMPLE 10   **Alternate Remedies**   Frustration with traditional medicine has led many Americans to try alternate remedies such as vitamins, herbs, and other supplements available without a doctor's prescription. The approximate sales of such supplements between 1995 and 2002, in billions of U.S. dollars, can be estimated by the equation

$$\text{sales} = -0.063x^2 + 1.62x + 9.5$$

where x represents years since 1995. In the expression on the right side of the equal sign, substitute 1 for x to estimate the sales of supplements in 1996, 2 for x to estimate the sales of supplements in 1997, and so on.

Estimate the sales of supplements during the years **a)** 1996 and **b)** 2000.

Solution   **a)** We will substitute 1 for x to estimate the sales of supplements in 1996.

$$\text{sales} = -0.063x^2 + 1.62x + 9.5$$
$$= -0.063(1)^2 + 1.62(1) + 9.5$$
$$= -0.063 + 1.62 + 9.5$$
$$= 11.057$$

Therefore, in 1996 approximately \$11.057 billion worth of supplements was sold in the United States.

b) The year 2000 corresponds to the number 5. We can obtain the 5 by subtracting 1995 from 2000. Therefore, to estimate the sales of supplements in 2000, we substitute 5 for x in the equation.

$$\text{sales} = -0.063x^2 + 1.62x + 9.5$$
$$= -0.063(5)^2 + 1.62(5) + 9.5$$
$$= -0.063(25) + 8.1 + 9.5$$
$$= 16.025$$

The answer is reasonable: From the information given we expected to see an increase. In 2000 approximately $16.025 billion worth of supplements was sold in the United States.

NOW TRY EXERCISE 115

EXAMPLE 11 Evaluate $-x^3 - xy - y^2$ when $x = -2$ and $y = 5$.

Solution Substitute -2 for each x and 5 for each y in the expression. Then evaluate.

$$-x^3 - xy - y^2 = -(-2)^3 - (-2)(5) - (5)^2$$
$$= -(-8) - (-10) - 25$$
$$= 8 + 10 - 25$$
$$= -7$$

NOW TRY EXERCISE 101

5 Evaluate Expressions on a Graphing Calculator

Throughout this book, the material presented on **graphing calculators** (or graphers) will often reinforce the concepts presented. Therefore, even if you do not have or use a graphing calculator, you should read the material related to graphing calculators whenever it appears. You may find that it truly helps your understanding of the concepts. Some graphing calculator information will be given as regular text, and other information about graphing calculators will be given in Using Your Graphing Calculator boxes such as the one on page 33.

The information presented in this book is not meant to replace the manual that comes with your graphing calculator. Because of space limitations in this book, your graphing calculator manual may provide more detailed information on some tasks we discuss. Your manual will also illustrate many other uses for your grapher beyond what we discuss in this course. Keystrokes to use vary from calculator to calculator. When we illustrate keystrokes and screens, they will be for a Texas Instruments TI-83 Plus calculator. *We suggest that you carefully read the manual that came with your graphing calculator to determine the sequence of keystrokes to use to accomplish specific tasks.*

Many graphing calculators can store an expression (or equation) and then evaluate the expression for various values of the variable or variables without your having to reenter the expression each time. This is very valuable in both science and mathematics courses. For example, in Chapter 3 when we graph, we will need to evaluate an expression for various values of the variable.

Figure 1.9 displays the screen of a TI-83 Plus graphing calculator showing the expression $\frac{2}{3}x^2 + 2x - 4$ being evaluated for $x = 6$ and $x = -2.3$.

On this calculator screen, $6 \rightarrow X$ shows that we assigned a value of 6 to X. The expression being evaluated, $(2/3)X^2 + 2X - 4$, is shown after the colon. The 32 shown on the right side of the screen (or window) is the value of the expression when $X = 6$. On the next line, on the left side of the screen, we see $-2.3 \rightarrow X$, which shows that a value of -2.3 has been assigned to X. We see that the value of

```
6→X: (2/3)X²+2X-4
                 32
-2.3→X: (2/3)X²+2
X-4
          -5.073333333
```

FIGURE 1.9

the expression is -5.073333333 when $X = -2.3$. After you have entered the expression to be evaluated it is not necessary to reenter the expression to evaluate it for a different value of the variable. Read your graphing calculator manual to learn how to evaluate an expression for various values of the variable without having to reenter the expression each time. On the TI-83 Plus, after evaluating an expression for one value of the variable, you can press $\boxed{2^{\text{nd}}}$ $\boxed{\text{ENTER}}$ to display the previously assigned value and the expression to be evaluated. Then you can replace the value that was previously assigned to X with the new value to be assigned to X. After doing this and pressing $\boxed{\text{ENTER}}$ the new answer will be displayed.

The calculator screen displayed in Figure 1.9 illustrates two important points about graphing calculators.

1. Notice the parentheses around the 2/3. Some graphing calculators interpret $2/3x^2$ as $2/(3x^2)$. To evaluate $\frac{2}{3}x^2$ on such calculators, you must use parentheses around the 2/3. You should learn how your calculator evaluates expressions like $2/3x^2$. *Whenever you are in doubt, use parentheses to prevent possible errors.*

2. In the display, you will notice that the negative sign preceding the 2.3 is slightly smaller and higher than the subtraction sign preceding the 4 in the expression. Graphing calculators generally have both a negative sign key, $\boxed{(-)}$, and a subtraction sign key, $\boxed{-}$. You must be sure to use the correct key or you will get an error. The negative sign key is used to enter a negative number. The subtraction key is used to subtract one quantity from another. To enter the expression $-x - 4$ on a graphing calculator, you might press

Remember that $-x - 4$ means $-1x - 4$. By beginning with $\boxed{(-)}$ you are entering the coefficient -1. Different calculators use different keys to enter the variable x. The key shown after the negative sign is the key used on the TI-83 Plus calculator.

EXAMPLE 12

Average Sale Price of Homes Low interest rates, easy credit, and strong demand from the middle class drove up the average sales price of homes around the country from 1992 to 2001. The average sales price of a home, in thousands of U.S. dollars, during this time can be estimated by

$$\text{average sales price} = 0.71x^2 + 2.16x + 145.39$$

where x represents years since 1992. In the expression on the right side of the equal sign, substitute 1 for x to estimate the average sales price of a home in 1993, 2 for x to estimate the average sales price of a home in 1994, and so on. Use a graphing calculator, if available, to estimate the average sales price of a home in **a)** 1995 and **b)** 2000. *Source:* National Association of Realtors

```
3→X:0.71X²+2.16X
+145.39
            158.26
8→X:0.71X²+2.16X
+145.39
            208.11
```

FIGURE 1.10

Solution

a) The year 1995 corresponds to $x = 3$, so we first assign x a value of 3, then enter the expression, and press $\boxed{\text{ENTER}}$. Figure 1.10 shows the screen for a TI-83 Plus calculator with the expression evaluated for $x = 3$. From the screen we see that the average sales price of a home in 1995 was approximately $158.26 thousand or $158,260.

b) Since $2000 - 1992 = 8$, the year 2000 corresponds to $x = 8$. We first assign x a value of 8, then reenter the expression, and press $\boxed{\text{ENTER}}$. From Figure 1.10 we see that the average sales price of a home in 2000 was approximately $208.11 thousand or $208,110.

HELPFUL HINT

Always review your calculator screen to make sure that no keys were pressed incorrectly and no keys were omitted. Note that it is not necessary to enter the 0 before the decimal point in terms such as $-0.71x^2$.

Exercise Set 1.4

Concept/Writing Exercises

1. Consider the expression a^n.
   a) What is the a called?
   b) What is the n called?
2. What is the meaning of a^n?
3. Consider the radical expression $\sqrt[n]{a}$.
   a) What is the n called?
   b) What is the a called?
4. What does it mean if $\sqrt[n]{a} = b$?
5. What is the principal square root of a positive number?
6. Explain why $\sqrt{-4}$ cannot be a real number.
7. Explain why an odd root of a negative number will be negative.
8. Explain why an odd root of a positive number will be positive.

9. In your own words, explain the order of operations to follow when evaluating a mathematical expression.

10. a) In your own words, explain step-by-step how you would evaluate
$$\frac{5 - 18 \div 3^2}{4 - 3 \cdot 2}$$
   b) Evaluate the expression.

11. a) In your own words, explain step-by-step how you would evaluate $16 \div 2^2 + 6 \cdot 4 - 24 \div 6$.
   b) Evaluate the expression.

12. a) In your own words, explain step-by-step how you would evaluate $\{5 - [4 - (3 - 8)]\}^2$.
   b) Evaluate the expression.

Practice the Skills

Evaluate each expression without using a calculator.

13. 3^2
14. $(-4)^3$
15. -3^2
16. -4^3
17. $(-3)^2$
18. $\left(\frac{1}{2}\right)^3$
19. $-\left(\frac{3}{5}\right)^4$
20. $(0.3)^2$
21. $-\left(-\frac{5}{6}\right)^2$
22. $\sqrt{144}$
23. $-\sqrt{36}$
24. $\sqrt{0.64}$
25. $\sqrt[3]{-27}$
26. $\sqrt[3]{\frac{-216}{343}}$
27. $\sqrt[3]{0.001}$
28. $\sqrt[4]{\frac{1}{16}}$

Use a calculator to evaluate each expression. Round answers to the nearest thousandth.

29. $(0.35)^4$
30. $-(1.7)^{3.9}$
31. $\left(\frac{5}{7}\right)^7$
32. $\left(-\frac{13}{12}\right)^8$
33. $(6.721)^{5.9}$
34. $\sqrt{92}$
35. $\sqrt[3]{5}$
36. $-\sqrt[4]{72.8}$
37. $\sqrt[5]{362.65}$
38. $-\sqrt{\frac{8}{9}}$
39. $-\sqrt[3]{\frac{20}{53}}$
40. $\sqrt[3]{-\frac{15}{19}}$

Evaluate **a)** x^2 *and* **b)** $-x^2$ *for each given value of* x.

41. 3

42. 4

43. 1

44. -2

45. -1

46. -5

47. $\dfrac{1}{3}$

48. $-\dfrac{2}{5}$

Evaluate **a)** x^3 *and* **b)** $-x^3$ *for each given value of* x.

49. 3

50. -5

51. -3

52. -1

53. -2

54. -4

55. $\dfrac{2}{3}$

56. $-\dfrac{3}{4}$

Evaluate each expression.

57. $3^2 + 2^3 - 2^2 - 3^3$

58. $(-1)^2 + (-1)^3 - 1^4 + 1^5$

59. $-2^2 - 2^3 + 1^{10} + (-2)^3$

60. $(-3)^3 - 2^2 - (-2)^2 + (4 - 4)^2$

61. $(1.5)^2 - (3.9)^2 + (-2.1)^3$

62. $(3.7)^2 - (0.8)^2 + (2.4)^3$

63. $\left(-\dfrac{1}{2}\right)^4 - \left(\dfrac{1}{2}\right)^2 + \left(-\dfrac{1}{2}\right)^3$

64. $\left(\dfrac{3}{4}\right)^2 - \dfrac{1}{4} - \left(-\dfrac{3}{8}\right)^2 + \left(\dfrac{1}{4}\right)^3$

Evaluate each expression.

65. $2 + 5 \cdot 8$

66. $(2 - 7) \div 5 + 1$

67. $18 - 6 \div 6 + 5$

68. $4 \cdot 3 \div 6 - 2^2$

69. $\dfrac{3}{4} \div \dfrac{1}{2} - 2 + 2 \div 4$

70. $3 \cdot 4 \div 12 + \dfrac{4}{5}$

71. $\dfrac{1}{2} \cdot \dfrac{2}{3} \div \dfrac{3}{4} - \dfrac{1}{6} \cdot \left(-\dfrac{1}{3}\right)$

72. $3[4 + (-2)(8)] + 3^3$

73. $10 \div [(3 + 2^2) - (2^4 - 8)]$

74. $[3 - (4 - 2^3)^2]^2$

75. $5(\sqrt[3]{27} + \sqrt[4]{16}) \div \dfrac{\sqrt{100}}{2}$

76. $\{3 + [4^2 - 3(2 - 7)] - 3\}^2$

77. $\{[(12 - 15) - 3] - 2\}^2$

78. $3\{6 - [(25 \div 5) - 2]\}^3$

79. $4[5(13 - 3) \div (25 \div 5)^2]^2$

80. $\dfrac{15 \div 3 + 2 \cdot 2}{\sqrt{25} \div 5 + 8 \div 2}$

81. $\dfrac{4 - (2 + 3)^2 - 6}{4(3 - 2) - 3^2}$

82. $\dfrac{2(-3) + 4 \cdot 5 - 3^2}{-6 + \sqrt{4}(2^2 - 1)}$

83. $\dfrac{8 + 4 \div 2 \cdot 3 + 4}{5^2 - 3^2 \cdot 2 - 7}$

84. $-2\left|-3 - \dfrac{2}{3}\right| + 4$

85. $\dfrac{8 - [4 - (3 - 1)^2]}{5 - (-3)^2 + 4 \div 2}$

86. $12 - 15 \div |5| - (|4| - 2)^2$

87. $-2|-3| - \sqrt{36} \div |2| + 3^2$

88. $\dfrac{4 - |-12| \div |3|}{2(4 - |5|) + 9}$

89. $\dfrac{6 - |-4| - 4|6 - 3|}{5 - 6 \cdot 2 \div |-6|}$

90. $-\dfrac{1}{4}[8 - |-6| \div 3 - 4]^2$

91. $\dfrac{2}{5}[\sqrt[3]{27} - |-9| + 4 - 3^2]^2$

92. $\dfrac{3(5 - 2)^2}{-3^2} - \dfrac{2(3^2 - 4^2)}{4 - (-2)}$

93. $\dfrac{24 - 5 - 4^2}{|-8| + 4 - 2(3)} + \dfrac{4 - (-3)^2 + |4|}{3^2 - 4 \cdot 3 + |-7|}$

94. $\dfrac{-2 - 8 \div 4^2 \cdot |8|}{|8| - \sqrt{64}} + \dfrac{[(8 - 3)^2 - 4]^2}{2^2 + 16}$

Evaluate each expression for the given value or values.

95. $5x^2 + 3x$ when $x = 2$

96. $5x^2 - 2x + 5$ when $x = 3$

97. $-9x^2 + 3x - 29$ when $x = -1$

98. $3(x - 2)^2$ when $x = \dfrac{1}{4}$

99. $16(x + 4)^3 - 25(x + 4)$ when $x = -3$

100. $-6x + 3y^2$ when $x = 2, y = 4$

101. $6x^2 + 3y^3 - 5$ when $x = 1$, $y = -3$

102. $4x^2 - 3y - 5$ when $x = 4$, $y = -2$

103. $3(a + b)^2 + 4(a + b) - 6$ when $a = 4$, $b = -1$

104. $-3 - \{2x - [5x - (2x + 1)]\}$ when $x = 3$

105. $-6 - \{x - [2x - (x - 3)]\}$ when $x = 4$

106. $\dfrac{(x - 3)^2}{9} + \dfrac{(y + 5)^2}{16}$ when $x = 4$, $y = 3$

107. $\dfrac{-b + \sqrt{b^2 - 4ac}}{2a}$ when $a = 6$, $b = -11$, $c = 3$

108. $\dfrac{-b - \sqrt{b^2 - 4ac}}{2a}$ when $a = 2$, $b = 1$, $c = -10$

Problem Solving

In Exercises 109–113 write an algebraic expression for each problem. Then evaluate the expression for the given value of the variable or variables.

109. Multiply the variable y by 7. From this product subtract 14. Now divide this difference by 2. Find the value of this expression when $y = 6$.

110. Subtract 4 from z. Multiply this difference by 5. Now square this product. Find the value of this expression when $z = 10$.

111. Six is added to the product of 3 and x. This expression is then multiplied by 6. Nine is then subtracted from this product. Find the value of the expression when $x = 3$.

112. The sum of x and y is multiplied by 2. Then 5 is subtracted from this product. This expression is then squared. Find the value of the expression when $x = 2$ and $y = -3$.

113. Three is added to x. This sum is divided by twice y. This quotient is then squared. Finally, 3 is subtracted from this expression. Find the value of the expression when $x = 5$ and $y = 2$.

Use a calculator to answer Exercises 114–121.

114. ***Centenarians*** People who live to be 100 years old or older are known as centenarians. According to the U.S. Census Bureau, centenarians are the world's fastest-growing age group. The approximate number of centenarians living in the United States between the years of 1995 and 2050, in thousands, can be estimated by

number of centenarians $= 0.30x^2 - 3.69x + 92.04$

where x represents years since 1995. Substitute 1 for x to find the number of centenarians in 1996, 2 for x to find the number of centenarians in 1997, and so on.

a) Estimate the number of centenarians that were living in the United States in 2000.

b) Estimate the number of centenarians that will be living in the United States in 2050.
Source: U.S. Census Bureau

115. ***Public Transportation*** Increased gas prices and growing congestion in America's major cities have caused a boom in public transportation usage. Between 1992 and 2001, the approximate number of public transportation trips per year in the United States, in billions, can be estimated using

number of trips $= 0.065x^2 - 0.39x + 8.47$

where x represents years since 1992. Substitute 1 for x to estimate the number of trips made in 1993, 2 for x to estimate the number of trips made in 1994, and so on.

a) Estimate the number of trips made using public transportation in 2000.

b) Assume this trend continues. Estimate the number of trips that will be made in 2010.
Source: American Public Transportation Association

116. ***Carbon Dioxide*** Since 1905 the amount of carbon dioxide (CO_2) has been increasing. The total production of CO_2 from all countries except the United States, Canada, and western Europe (measured in millions of metric tons) can be approximated by

$$CO_2 = 0.073x^2 - 0.39x + 0.55$$

where x represents each 10-year period since 1905. Substitute 1 for x to calculate the CO_2 production in 1915, 2 for x to calculate the CO_2 production in 1925, 3 for x in 1935, and so on.

a) Find the approximate amount of CO_2 produced by all countries except the United States, Canada, and western Europe in 1945.

b) Assume this trend continues, find the approximate amount of CO_2 produced by all countries except the United States, Canada, and western Europe in 2005.

See Exercise 115.

117. *Latchkey Kids* The number of *latchkey kids*, children who care for themselves while their parents are working, increases with age. The percent of children of different ages, from 5 to 14 years old, who are latchkey kids can be approximated by

percent of children = $0.23x^2 - 1.98x + 4.42$

The x value represents the age of the children. For example, substitute 5 for x to get the percent of all 5-year-olds who are latchkey kids, substitute 6 for x to get the percent of all 6-year-olds who are latchkey kids, and so on.

a) Find the percent of all 10-year-olds who are latchkey kids.

b) Find the percent of all 14-year-olds who are latchkey kids.

118. *Newspaper Readers* The number of Americans who read a daily newspaper is steadily going down. The percent reading a daily newspaper can be approximated by

percent = $-6.2x + 82.2$

where x represents each 10-year period since 1960. Substitute 1 for x to get the percent for 1970, 2 for x to get the percent for 1980, 3 for x to get the percent for 1990, and so forth.

a) Find the percent of U.S. adults who read a daily newspaper in 1970.

b) Assuming that this trend continues, find the percent of U.S. adults who will read a daily newspaper in 2010.

119. *Organically Grown Food* Increasing fears of pesticides and genetically altered crops have led many people to purchase organically grown food. From 1990 to

2001, sales, in billions of U.S. dollars, of organically grown food can be estimated by

sales = $0.062x^2 + 0.020x + 1.18$

where x represents years since 1990. Substitute 1 for x to estimate the sales of organically grown food in 1991, 2 for x to estimate sales in 1992, and so on.

a) Estimate the sales of organically grown food in 1991.

b) Estimate the sales of organically grown food in 2001.

120. *Union Membership* Membership in labor unions in the United States has been changing. From 1983 to 2001, labor union membership, as a percent of the overall workforce, can be estimated using

membership percent = $0.016x^2 - 0.62x + 19.69$

where x represents years since 1983. Substitute 1 for x to estimate the labor union membership as a percent of the overall workforce in 1984, 2 for x to estimate membership as a percent in 1985, and so on.

a) Estimate union membership as a percent of the overall workforce in 1991.

b) Estimate union membership as a percent of the overall workforce in 2001.

Source: Bureau of Labor Statistics

121. *Cellular Phones* Use of cellular phones is currently on the rise. The number of cellular subscribers, in millions, can be approximated by

number of subscribers = $0.42x^2 - 3.44x + 5.80$

where x represents years since 1982. Substitute 1 for x to get the number of subscribers in 1983, 2 for x to get the number of subscribers in 1984, and so on.

a) Find the number of people who used cell phones in 1989.

b) Find the number of people who used cell phones in 2002.

Cumulative Review Exercises

[1.2] **122.** $A = \{a, b, c, d, f\}$, $B = \{b, c, f, g, h\}$. Find
    a) $A \cap B$,
    b) $A \cup B$.

[1.3] *In Exercises 123–125, the letter a represents a real number. For what values of a will each statement be true?*

123. $|a| = |-a|$

124. $|a| = a$

125. $|a| = 4$

126. List from smallest to largest: $-|6|, -4, |-5|, -|-2|, 0$.

127. Name the following property: $(2 + 3) + 5 = 2 + (3 + 5)$.

1.5 EXPONENTS

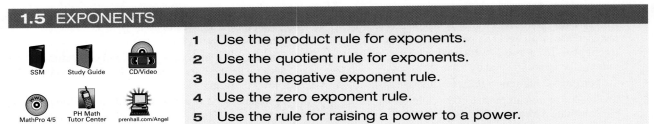

1   Use the product rule for exponents.
2   Use the quotient rule for exponents.
3   Use the negative exponent rule.
4   Use the zero exponent rule.
5   Use the rule for raising a power to a power.
6   Use the rule for raising a product to a power.
7   Use the rule for raising a quotient to a power.

In the previous section we introduced exponents. In this section we discuss the rules of exponents. We begin with the product rule for exponents.

1   Use the Product Rule for Exponents

Consider the multiplication $x^3 \cdot x^5$. We can simplify this expression as follows:

$$x^3 \cdot x^5 = (x \cdot x \cdot x) \cdot (x \cdot x \cdot x \cdot x \cdot x) = x^8$$

This problem could also be simplified using the **product rule for exponents**.*

> **Product Rule for Exponents**
>
> If m and n are natural numbers and a is any real number, then
>
> $$a^m \cdot a^n = a^{m+n}$$

To multiply exponential expressions, maintain the common base and add the exponents.

$$x^3 \cdot x^5 = x^{3+5} = x^8$$

EXAMPLE 1   Simplify.   **a)** $2^3 \cdot 2^4$ **b)** $d^2 \cdot d^5$ **c)** $h \cdot h^6$

Solution   **a)** $2^3 \cdot 2^4 = 2^{3+4} = 2^7 = 128$ **b)** $d^2 \cdot d^5 = d^{2+5} = d^7$
c) $h \cdot h^6 = h^1 \cdot h^6 = h^{1+6} = h^7$

2   Use the Quotient Rule for Exponents

Consider the division $x^7 \div x^4$. We can simplify this expression as follows:

$$\frac{x^7}{x^4} = \frac{\overset{1}{\cancel{x}} \cdot \overset{1}{\cancel{x}} \cdot \overset{1}{\cancel{x}} \cdot \overset{1}{\cancel{x}} \cdot x \cdot x \cdot x}{\underset{1}{\cancel{x}} \cdot \underset{1}{\cancel{x}} \cdot \underset{1}{\cancel{x}} \cdot \underset{1}{\cancel{x}}} = x \cdot x \cdot x = x^3$$

This problem could also be simplified using the **quotient rule for exponents**.

> **Quotient Rule for Exponents**
>
> If a is any nonzero real number and m and n are nonzero integers, then
>
> $$\frac{a^m}{a^n} = a^{m-n}$$

*The rules given in this section also apply for rational or fractional exponents. Rational exponents will be discussed in Section 7.2. We will review these rules again at that time.

To divide expressions in exponential form, maintain the common base and subtract the exponents.

$$\frac{x^7}{x^4} = x^{7-4} = x^3$$

EXAMPLE 2 Simplify. **a)** $\dfrac{5^4}{5^2}$ **b)** $\dfrac{x^5}{x^2}$ **c)** $\dfrac{y^2}{y^5}$

Solution **a)** $\dfrac{5^4}{5^2} = 5^{4-2} = 5^2 = 25$ **b)** $\dfrac{x^5}{x^2} = x^{5-2} = x^3$ **c)** $\dfrac{y^2}{y^5} = y^{2-5} = y^{-3}$ ✳

3 Use the Negative Exponent Rule

Notice in Example 2 **c)** that the answer contains a negative exponent. Let's do part **c)** again by dividing out common factors.

$$\frac{y^2}{y^5} = \frac{\overset{1}{\cancel{y}} \cdot \overset{1}{\cancel{y}}}{\underset{1}{\cancel{y}} \cdot \underset{1}{\cancel{y}} \cdot y \cdot y \cdot y} = \frac{1}{y^3}$$

By dividing out common factors and using the result from Example 2 **c)**, we can reason that $y^{-3} = 1/y^3$. This is an example of the negative exponent rule.

Negative Exponent Rule

For any nonzero real number a and any whole number m,

$$a^{-m} = \frac{1}{a^m}$$

An expression raised to a negative exponent is equal to 1 divided by the expression with the sign of the exponent changed.

EXAMPLE 3 Write each expression without negative exponents.

a) 5^{-2} **b)** $7a^{-4}$ **c)** $\dfrac{1}{c^{-4}}$

Solution **a)** $5^{-2} = \dfrac{1}{5^2} = \dfrac{1}{25}$ **b)** $7a^{-4} = 7 \cdot \dfrac{1}{a^4} = \dfrac{7}{a^4}$

c) $\dfrac{1}{c^{-4}} = 1 \div c^{-4} = 1 \div \dfrac{1}{c^4} = \dfrac{1}{1} \cdot \dfrac{c^4}{1} = c^4$

NOW TRY EXERCISE 37 ✳

HELPFUL HINT

In Example 3 **c)** we showed that $\dfrac{1}{c^{-4}} = c^4$. In general, for any nonzero real number a and any whole number m, $\dfrac{1}{a^{-m}} = a^m$. When a factor of the numerator or the denominator is raised to any power, the factor can be moved to the other side of the fraction bar provided the sign of the exponent is changed. Thus, for example

$$\frac{2a^{-3}}{b^2} = \frac{2}{a^3 b^2} \qquad \frac{a^{-2} b^4}{c^{-3}} = \frac{b^4 c^3}{a^2}$$

Generally, we do not leave exponential expressions with negative exponents. *When we indicate that an exponential expression is to be simplified, we mean that the answer should be written without negative exponents.*

EXAMPLE 4 Simplify. **a)** $\dfrac{3xz^2}{y^{-4}}$ **b)** $4^{-2}x^{-1}y^2$ **c)** $-3^3x^2y^{-3}$

Solution **a)** $\dfrac{3xz^2}{y^{-4}} = 3xy^4z^2$

b) $4^{-2}x^{-1}y^2 = \dfrac{1}{4^2} \cdot \dfrac{1}{x^1} \cdot y^2 = \dfrac{y^2}{16x}$

c) $-3^3x^2y^{-3} = -(3^3)x^2 \cdot \dfrac{1}{y^3} = -\dfrac{27x^2}{y^3}$

NOW TRY EXERCISE 61

Notice that the expressions in Example 4 do not involve addition or subtraction. The presence of a plus sign or a minus sign makes for a very different problem as we will see in our next example.

EXAMPLE 5 Simplify. **a)** $3^{-1} + 6^{-1}$ **b)** $2 \cdot 3^{-2} + 5 \cdot 6^{-2}$

Solution **a)** $3^{-1} + 6^{-1} = \dfrac{1}{3} + \dfrac{1}{6}$ *Negative exponent rule*

$= \dfrac{2}{6} + \dfrac{1}{6}$ *Rewrite with the LCD, 6.*

$= \dfrac{2+1}{6} = \dfrac{3}{6} = \dfrac{1}{2}$

b) $2 \cdot 3^{-2} + 5 \cdot 6^{-2} = 2 \cdot \dfrac{1}{3^2} + 5 \cdot \dfrac{1}{6^2}$ *Negative exponent rule*

$= \dfrac{2}{1} \cdot \dfrac{1}{9} + \dfrac{5}{1} \cdot \dfrac{1}{36}$

$= \dfrac{2}{9} + \dfrac{5}{36}$

$= \dfrac{8}{36} + \dfrac{5}{36}$ *Rewrite with the LCD, 36.*

$= \dfrac{8+5}{36} = \dfrac{13}{36}$

NOW TRY EXERCISE 75

4 Use the Zero Exponent Rule

The next rule we will study is the **zero exponent rule**. Any nonzero number divided by itself is 1. Therefore,

$$\frac{x^5}{x^5} = 1$$

By the quotient rule for exponents,

$$\frac{x^5}{x^5} = x^{5-5} = x^0$$

Since $x^0 = \dfrac{x^5}{x^5}$ and $\dfrac{x^5}{x^5} = 1$, by the transitive property of equality,

$$x^0 = 1$$

Zero Exponent Rule

If a is any nonzero real number, then

$$a^0 = 1$$

The zero exponent rule illustrates that *any nonzero real number with an exponent of 0 equals 1.* We must specify that $a \ne 0$ because 0^0 is not a real number.

EXAMPLE 6 Simplify (assume that the base is not 0).

    a) 200^0 **b)** $7y^0$ **c)** $-y^0$ **d)** $-(7x + 9y)^0$

Solution   **a)** $200^0 = 1$

    b) $7y^0 = 7 \cdot y^0 = 7 \cdot 1 = 7$

    c) $-y^0 = -1 \cdot y^0 = -1 \cdot 1 = -1$

    d) $-(7x + 9y)^0 = -1 \cdot (7x + 9y)^0 = -1 \cdot 1 = -1$

5 Use the Rule for Raising a Power to a Power

Consider the expression $\left(x^3\right)^2$. We can simplify this expression as follows:

$$\left(x^3\right)^2 = x^3 \cdot x^3 = x^{3+3} = x^6$$

This problem could also be simplified using the rule for **raising a power to a power** (also called the **power rule**).

Raising a Power to a Power (the Power Rule)

If a is a real number and m and n are integers, then

$$\left(a^m\right)^n = a^{m \cdot n}$$

To raise an exponential expression to a power, maintain the base and multiply the exponents.

$$\left(x^3\right)^2 = x^{3 \cdot 2} = x^6$$

EXAMPLE 7 Simplify (assume that the base is not 0).

    a) $\left(2^2\right)^4$ **b)** $\left(z^{-3}\right)^4$ **c)** $\left(2^{-3}\right)^2$

Solution   **a)** $\left(2^2\right)^4 = 2^{2 \cdot 4} = 2^8 = 256$

    b) $\left(z^{-3}\right)^4 = z^{-3 \cdot 4} = z^{-12} = \dfrac{1}{z^{12}}$

    c) $\left(2^{-3}\right)^2 = 2^{-3 \cdot 2} = 2^{-6} = \dfrac{1}{2^6} = \dfrac{1}{64}$

NOW TRY EXERCISE 81

Students often confuse the *product rule*

$$a^m \cdot a^n = a^{m+n}$$

with the *power rule*

$$(a^m)^n = a^{m \cdot n}$$

For example, $(x^3)^2 = x^6$, not x^5.

6 Use the Rule for Raising a Product to a Power

Consider the expression $(xy)^2$. We can simplify this expression as follows:

$$(xy)^2 = (xy)(xy) = x \cdot x \cdot y \cdot y = x^2 y^2$$

This expression could also be simplified using the rule for **raising a product to a power**.

Raising a Product to a Power

If a and b are real numbers and m is an integer, then

$$(ab)^m = a^m b^m$$

To raise a product to a power, raise all factors within the parentheses to the power outside the parentheses.

EXAMPLE 8 Simplify.   **a)** $(-4x^3)^2$ **b)** $(3x^{-2}y^3)^{-3}$

Solution **a)** $(-4x^3)^2 = (-4)^2(x^3)^2 = 16x^6$

b) $(3x^{-2}y^3)^{-3} = 3^{-3}(x^{-2})^{-3}(y^3)^{-3}$ *Raise a product to a power.*

$$= \frac{1}{3^3} \cdot x^6 \cdot y^{-9}$$ *Negative exponent rule, Power rule*

$$= \frac{1}{27} \cdot x^6 \cdot \frac{1}{y^9}$$ *Negative exponent rule*

$$= \frac{x^6}{27y^9}$$

NOW TRY EXERCISE 93

7 Use the Rule for Raising a Quotient to a Power

Consider the expression $\left(\dfrac{x}{y}\right)^2$. We can simplify this expression as follows:

$$\left(\frac{x}{y}\right)^2 = \frac{x}{y} \cdot \frac{x}{y} = \frac{x \cdot x}{y \cdot y} = \frac{x^2}{y^2}$$

This expression could also be simplified using the rule for **raising a quotient to a power**.

> ### Raising a Quotient to a Power
>
> If a and b are real numbers and m is an integer, then
>
> $$\left(\frac{a}{b}\right)^m = \frac{a^m}{b^m}, \qquad b \neq 0$$

To raise a quotient to a power, raise all factors in the parentheses to the exponent outside the parentheses.

EXAMPLE 9   Simplify.   **a)** $\left(\dfrac{2}{x^2}\right)^3$ **b)** $\left(\dfrac{4x^{-2}}{y^3}\right)^{-2}$

Solution   **a)** $\left(\dfrac{2}{x^2}\right)^3 = \dfrac{2^3}{(x^2)^3} = \dfrac{8}{x^6}$

     b) $\left(\dfrac{4x^{-2}}{y^3}\right)^{-2} = \dfrac{4^{-2}(x^{-2})^{-2}}{(y^3)^{-2}}$ *Raise a quotient to a power.*

        $= \dfrac{4^{-2}x^4}{y^{-6}}$ *Power rule*

        $= \dfrac{x^4 y^6}{4^2}$ *Negative exponent rule*

        $= \dfrac{x^4 y^6}{16}$

NOW TRY EXERCISE 99

Consider $\left(\dfrac{a}{b}\right)^{-n}$. Using the rule for raising a quotient to a power, we get

$$\left(\frac{a}{b}\right)^{-n} = \frac{a^{-n}}{b^{-n}} = \frac{b^n}{a^n} = \left(\frac{b}{a}\right)^n$$

Using this result, we see that when we have a rational number raised to a negative exponent, we can take the reciprocal of the base and change the sign of the exponent as follows.

$$\left(\frac{5}{9}\right)^{-3} = \left(\frac{9}{5}\right)^3 \qquad \left(\frac{x^2}{y^3}\right)^{-4} = \left(\frac{y^3}{x^2}\right)^4$$

Now we will work some examples that combine a number of properties. Whenever the same variable appears above and below the fraction bar, we generally move the variable with the *lesser exponent*. This will result in the exponent on the variable being positive when the product rule is applied. Examples 10 and 11 illustrate this procedure.

EXAMPLE 10   Simplify.   **a)** $\left(\dfrac{6x^2 y^4}{2x^2 y}\right)^2$ **b)** $\left(\dfrac{3x^4 y^{-2}}{6xy^3 z^{-1}}\right)^{-3}$

Solution   Exponential expressions can often be simplified in more than one way. In general, it will be easier to first simplify the expression within the parentheses.

     a) $\left(\dfrac{6x^2 y^4}{2x^2 y}\right)^2 = (3y^3)^2 = 9y^6$

b) $\left(\dfrac{3x^4 y^{-2}}{6xy^3 z^{-1}} \right)^{-3} = \left(\dfrac{x^4 \cdot x^{-1} z}{2y^3 \cdot y^2} \right)^{-3}$ *Move x, y^{-2}, and z^{-1} to the other side of the fraction bar and change the signs of their exponents.*

$= \left(\dfrac{x^3 z}{2y^5} \right)^{-3}$ *Product rule.*

$= \left(\dfrac{2y^5}{x^3 z} \right)^{3}$ *Take the reciprocal of the expression inside the parentheses and change the sign of the exponent.*

$= \dfrac{2^3 y^{5 \cdot 3}}{x^{3 \cdot 3} z^3}$ *Raise a quotient to a power.*

$= \dfrac{8y^{15}}{x^9 z^3}$

NOW TRY EXERCISE 109   ✳

EXAMPLE 11   Simplify $\dfrac{(2p^{-3}q^4)^{-2}}{(p^{-5}q^4)^{-3}}$.

Solution   First, use the power rule. Then simplify further.

$\dfrac{(2p^{-3}q^4)^{-2}}{(p^{-5}q^4)^{-3}} = \dfrac{2^{-2} p^6 q^{-8}}{p^{15} q^{-12}}$ *Power rule.*

$= \dfrac{q^{-8} \cdot q^{12}}{2^2 p^{15} \cdot p^{-6}}$ *Move 2^{-2}, p^6, and q^{-12} to the other side of the fraction bar and change the signs of their exponents.*

$= \dfrac{q^{-8+12}}{4p^{15-6}}$ *Product rule.*

$= \dfrac{q^4}{4p^9}$

NOW TRY EXERCISE 115   ✳

Summary of Rules of Exponents

For all real numbers a and b and all integers m and n:

Product rule	$a^m \cdot a^n = a^{m+n}$	
Quotient rule	$\dfrac{a^m}{a^n} = a^{m-n},$	$a \neq 0$
Negative exponent rule	$a^{-m} = \dfrac{1}{a^m},$	$a \neq 0$
Zero exponent rule	$a^0 = 1,$	$a \neq 0$
Raising a power to a power	$(a^m)^n = a^{m \cdot n}$	
Raising a product to a power	$(ab)^m = a^m b^m$	
Raising a quotient to a power	$\left(\dfrac{a}{b} \right)^m = \dfrac{a^m}{b^m},$	$b \neq 0$

Exercise Set 1.5

Concept/Writing Exercises

1. **a)** Give the product rule for exponents.

   b) Explain the product rule.

2. **a)** Give the quotient rule for exponents.

   b) Explain the quotient rule.

3. **a)** Give the zero exponent rule.

   b) Explain the zero exponent rule.

4. **a)** Give the negative exponent rule.

   b) Explain the negative exponent rule.

5. **a)** Give the rule for raising a product to a power.

   b) Explain the rule for raising a product to a power.

6. **a)** Give the rule for raising a power to a power.

   b) Explain the rule for raising a power to a power.

7. **a)** Give the rule for raising a quotient to a power.

   b) Explain the rule for raising a quotient to a power.

8. What is the exponent on a variable or coefficient if none is shown?

9. If $x^{-1} = 5$, what is the value of x? Explain.

10. If $x^{-1} = y^2$, what is x equal to? Explain.

11. **a)** Explain the difference between the opposite of x and the reciprocal of x.
    For parts **b)** and **c)** consider

    $$x^{-1}, \quad -x, \quad \frac{1}{x}, \quad \frac{1}{x^{-1}},$$

    b) Which represent (or are equal to) the *reciprocal* of x?

    c) Which represent the *opposite* (or *additive inverse*) of x?

12. Explain why $-2^{-2} \neq \dfrac{1}{(-2)^2}$.

Practice the Skills

Evaluate each expression.

13. $2^3 \cdot 2^2$

14. $3^2 \cdot 3^3$

15. $\dfrac{3^7}{3^5}$

16. $\dfrac{7^4}{7^3}$

17. 6^{-2}

18. 4^{-3}

19. $\dfrac{1}{2^{-3}}$

20. $\dfrac{1}{3^{-2}}$

21. 13^0

22. 17^0

23. $(2^3)^2$

24. $(3^2)^2$

25. $(2 \cdot 3)^2$

26. $(3 \cdot 5)^2$

27. $\left(\dfrac{2}{3}\right)^4$

28. $\left(\dfrac{3}{5}\right)^2$

Evaluate each expression.

29. **a)** 3^{-2} **b)** $(-3)^{-2}$ **c)** -3^{-2} **d)** $-(-3)^{-2}$

30. **a)** 5^{-3} **b)** $(-5)^{-3}$ **c)** -5^{-3} **d)** $-(-5)^{-3}$

31. **a)** $\left(\dfrac{1}{2}\right)^{-1}$ **b)** $\left(-\dfrac{1}{2}\right)^{-1}$ **c)** $-\left(\dfrac{1}{2}\right)^{-1}$ **d)** $-\left(-\dfrac{1}{2}\right)^{-1}$

32. **a)** $\left(\dfrac{3}{4}\right)^{-2}$ **b)** $\left(-\dfrac{3}{4}\right)^{-2}$ **c)** $-\left(\dfrac{3}{4}\right)^{-2}$ **d)** $-\left(-\dfrac{3}{4}\right)^{-2}$

Simplify each expression and write the answer without negative exponents. Assume that all bases represented by variables are nonzero.

33. a) $5x^0$ **b)** $-5x^0$ **c)** $(-5x)^0$ **d)** $-(-5x)^0$

34. a) $3y^0$ **b)** $(3y)^0$ **c)** $-3y^0$ **d)** $(-3y)^0$

35. a) $3xyz^0$ **b)** $(3xyz)^0$ **c)** $3x(yz)^0$ **d)** $3(xyz)^0$

36. a) $x^0 + y^0$ **b)** $(x + y)^0$ **c)** $x + y^0$ **d)** $x^0 + y$

Simplify each expression and write the answer without negative exponents.

37. $5y^{-3}$ **38.** $\dfrac{1}{x^{-1}}$ **39.** $\dfrac{1}{x^{-4}}$ **40.** $\dfrac{3}{5y^{-2}}$

41. $\dfrac{2a}{b^{-3}}$ **42.** $\dfrac{6x^4}{y^{-1}}$ **43.** $\dfrac{5m^{-2}n^{-3}}{2}$ **44.** $\dfrac{4x^{-3}}{z^4}$

45. $\dfrac{5x^{-2}y^{-3}}{z^{-4}}$ **46.** $\dfrac{10ab^5}{2c^{-3}}$ **47.** $\dfrac{6^{-1}x^{-1}}{y}$ **48.** $\dfrac{5^{-1}z}{x^{-1}y^{-1}}$

Simplify each expression and write the answer without negative exponents.

49. $2^5 \cdot 2^{-6}$ **50.** $a^3 \cdot a^4$ **51.** $x^6 \cdot x^{-2}$ **52.** $x^{-4} \cdot x^3$

53. $\dfrac{6^5}{6^3}$ **54.** $\dfrac{4^2}{4^{-2}}$ **55.** $\dfrac{7^{-5}}{7^{-3}}$ **56.** $\dfrac{x^{-9}}{x^2}$

57. $\dfrac{m^{-5}}{m^5}$ **58.** $\dfrac{p^0}{p^{-7}}$ **59.** $\dfrac{5w^{-2}}{w^{-7}}$ **60.** $\dfrac{x^{-3}}{x^{-5}}$

61. $3a^{-2} \cdot 4a^{-5}$ **62.** $(-7v^4)(-3v^{-5})$ **63.** $(-3p^{-2})(-p^3)$ **64.** $(2x^{-3}y^{-4})(6x^{-4}y^7)$

65. $(5r^2s^{-2})(-2r^5s^2)$ **66.** $(-3p^{-4}q^6)(2p^3q)$ **67.** $(2x^4y^7)(4x^3y^{-5})$ **68.** $\dfrac{24x^3y^2}{8xy}$

69. $\dfrac{27x^5y^{-4}}{9x^3y^2}$ **70.** $\dfrac{6x^{-2}y^3z^{-2}}{-2x^4y}$ **71.** $\dfrac{9xy^{-4}z^3}{-3x^{-2}yz}$ **72.** $\dfrac{(x^{-2})(4x^2)}{x^3}$

Evaluate each expression.

73. a) $3(a + b)^0$ **b)** $3a^0 + 3b^0$ **c)** $(3a + 3b)^0$ **d)** $-3a^0 + 3b^0$

74. a) $-2^0 + (-2)^0$ **b)** $-2^0 - (-2)^0$ **c)** $-2^0 + 2^0$ **d)** $-2^0 - 2^0$

75. a) $4^{-1} - 3^{-1}$ **b)** $4^{-1} + 3^{-1}$ **c)** $2 \cdot 4^{-1} + 3 \cdot 5^{-1}$ **d)** $(2 \cdot 4)^{-1} + (3 \cdot 5)^{-1}$

76. a) $5^{-2} + 4^{-1}$ **b)** $5^{-2} - 4^{-1}$ **c)** $3 \cdot 5^{-2} + 2 \cdot 4^{-1}$ **d)** $(3 \cdot 5)^{-2} - (2 \cdot 4)^{-1}$

Simplify each expression and write the answer without negative exponents.

77. $(3^2)^2$ **78.** $(5^2)^{-1}$ **79.** $(3^2)^{-2}$ **80.** $(x^2)^{-3}$

81. $(b^{-3})^{-2}$ **82.** $(-c)^4$ **83.** $(-c)^3$ **84.** $(-x)^{-2}$

85. $(-3x^{-3})^2$ **86.** $-3(x^{-3})^2$ **87.** $5^{-1} + 2^{-1}$ **88.** $4^{-2} + 8^{-1}$

89. $3 \cdot 4^{-2} + 5 \cdot 8^{-1}$

90. $5 \cdot 2^{-3} + 7 \cdot 4^{-2}$

91. $\left(\dfrac{4b}{3}\right)^{-2}$

92. $(-2m^3 n^2)^3$

93. $(4x^2 y^{-2})^2$

94. $(4x^2 y^3)^{-3}$

95. $(3p^2 q^{-4})^{-3}$

96. $(7s^{-3} t^{-4})^2$

97. $(-2g^{-4} h^3)^{-3}$

98. $5(x^2 y^{-1})^{-4}$

99. $\left(\dfrac{3j}{4k^2}\right)^2$

100. $\left(\dfrac{3x^2 y^4}{z}\right)^3$

101. $\left(\dfrac{2r^4 s^5}{r^2}\right)^3$

102. $\left(\dfrac{3m^5 n^6}{6m^4 n^7}\right)^3$

103. $\left(\dfrac{4xy}{y^3}\right)^{-3}$

104. $\left(\dfrac{3x^{-2}}{xy}\right)^{-2}$

105. $\left(\dfrac{4x^{-2} y}{x^{-5}}\right)^3$

106. $\left(\dfrac{4x^2 y}{x^{-5}}\right)^{-3}$

107. $\left(\dfrac{6x^2 y}{3xz}\right)^{-3}$

108. $\left(\dfrac{5xy}{z^{-2}}\right)^3$

109. $\left(\dfrac{x^6 y^{-2}}{x^{-2} y^3}\right)^2$

110. $\left(\dfrac{x^2 y^{-3} z^4}{x^{-1} y^2 z^3}\right)^{-1}$

111. $\left(\dfrac{4x^{-1} y^{-2} z^3}{2xy^2 z^{-3}}\right)^{-2}$

112. $\left(\dfrac{6x^4 y^{-6} z^4}{2xy^{-6} z^{-2}}\right)^{-2}$

113. $\left(\dfrac{-a^3 b^{-1} c^{-3}}{2ab^3 c^{-4}}\right)^{-3}$

114. $\dfrac{(2x^{-1} y^{-2})^{-3}}{(3x^{-1} y^3)^2}$

115. $\dfrac{(3x^{-4} y^2)^3}{(2x^3 y^5)^3}$

116. $\dfrac{(2xy^2 z^{-3})^2}{(3x^{-1} yz^2)^{-1}}$

Problem Solving

Simplify each expression. Assume that all variables represent nonzero integers.

117. $x^{2a} \cdot x^{5a+3}$

118. $y^{2m+3} \cdot y^{5m-7}$

119. $w^{2a-5} \cdot w^{3a-2}$

120. $d^{-4x+7} \cdot d^{5x-6}$

121. $\dfrac{x^{2w+3}}{x^{w-4}}$

122. $\dfrac{y^{5m-1}}{y^{7m-1}}$

123. $(x^{3p+5})(x^{2p-3})$

124. $(s^{2t-3})(s^{-t+5})$

125. $x^{-m}(x^{3m+2})$

126. $y^{3b+2} \cdot y^{2b+4}$

127. $\dfrac{25m^{a+b} n^{b-a}}{5m^{a-b} n^{a+b}}$

128. $\dfrac{20x^{c+3} y^{d+4}}{4x^{c-4} y^{d+6}}$

129. a) For what values of x is $x^4 > x^3$?
  b) For what values of x is $x^4 < x^3$?
  c) For what values of x is $x^4 = x^3$?
  d) Why can you not say that $x^4 > x^3$?

130. Is 3^{-8} greater than or less than 2^{-8}? Explain.

131. a) Explain why $(-1)^n = 1$ for any even number n.
  b) Explain why $(-1)^n = -1$ for any odd number n.

132. a) Explain why $(-12)^{-8}$ is positive.
  b) Explain why $(-12)^{-7}$ is negative.

133. a) Is $\left(-\dfrac{2}{3}\right)^{-2}$ equal to $\left(\dfrac{2}{3}\right)^{-2}$?
  b) Will $(x)^{-2}$ equal $(-x)^{-2}$ for all real numbers x except 0? Explain your answer.

134. a) Is $\left(-\dfrac{2}{3}\right)^{-3}$ equal to $\left(\dfrac{2}{3}\right)^{-3}$?
  b) Will $(x)^{-3}$ equal $(-x)^{-3}$ for any nonzero real number x? Explain.
  c) What is the relationship between $(-x)^{-3}$ and $(x)^{-3}$ for any nonzero real number x?

Determine what exponents must be placed in the shaded area to make each expression true. Each shaded area may represent a different exponent. Explain how you determined your answer.

135. $\left(\dfrac{x^2 y^{-2}}{x^{-3} y^{\blacksquare}}\right)^2 = x^{10} y^2$

136. $\left(\dfrac{x^{-2} y^3 z}{x^4 y^{\blacksquare} z^{-3}}\right)^3 = \dfrac{z^{12}}{x^{18} y^6}$

137. $\left(\dfrac{x^{\blacksquare} y^5 z^{-2}}{x^4 y^{\blacksquare} z}\right)^{-1} = \dfrac{x^5 z^3}{y^2}$

Challenge Problems

We will learn in Section 7.2 that the rules of exponents given in this section also apply when the exponents are rational numbers. Using this information and the rules of exponents, evaluate each expression.

138. $\left(\dfrac{x^{1/2}}{x^{-1}}\right)^{3/2}$

139. $\left(\dfrac{x^{5/8}}{x^{1/4}}\right)^{3}$

140. $\left(\dfrac{x^{4}}{x^{-1/2}}\right)^{-1}$

141. $\dfrac{x^{1/2}y^{-3/2}}{x^{5}y^{5/3}}$

142. $\left(\dfrac{x^{1/2}y^{4}}{x^{-3}y^{5/2}}\right)^{2}$

Group Activity

Discuss and answer Exercise 143 as a group.

143. **Doubling a Penny** On day 1 you are given a penny. On each following day, you are given double the amount you were given on the previous day.

  a) Write down the amounts you would be given on each of the first 6 days.

  b) Express each of these numbers as an exponential expression with a base of 2.

  c) By looking at the pattern, determine an exponential expression for the number of cents you will receive on day 10.

  d) Write a general exponential expression for the number of cents you will receive on day *n*.

  e) Write an exponential expression for the number of cents you will receive on day 30.

  f) Calculate the value of the expression in part **e)**. Use a calculator if one is available.

  g) Determine the amount found in part **f)** in dollars.

  h) Write a general exponential expression for the number of dollars you will receive on day *n*.

Cumulative Review Exercises

[1.2] **144.** If $A = \{3, 4, 6\}$ and $B = \{1, 2, 5, 8\}$, find
  a) $A \cup B$ and **b)** $A \cap B$.

145. Illustrate the following set on the number line: $\{x \mid -3 \le x < 2\}$.

[1.4] **146.** Evaluate $6 + |12| \div |-3| - 4 \cdot 2^2$.

147. Evaluate $\sqrt[3]{-125}$.

1.6 SCIENTIFIC NOTATION

SSM   Study Guide   CD/Video

MathPro 4/5   PH Math Tutor Center   prenhall.com/Angel

1 Write numbers in scientific notation.

2 Change numbers in scientific notation to decimal form.

3 Use scientific notation in problem solving.

1 Write Numbers in Scientific Notation

Scientists and engineers often deal with very large and very small numbers. For example, the frequency of an FM radio signal may be 14,200,000,000 hertz (or cycles per second) and the diameter of a hydrogen atom is about 0.0000000001 meter. Because it is difficult to work with many zeros, scientists often express such numbers with exponents. For example, the number 14,200,000,000 might be written as 1.42×10^{10}, and 0.0000000001 as 1×10^{-10}. Numbers such as 1.42×10^{10} and 1×10^{-10} are in a form called **scientific notation**. In scientific notation, numbers appear as a number greater than or equal to 1 and less than 10 multiplied by some power of 10. The exponent on the 10 must be an integer. When a power of 10 has

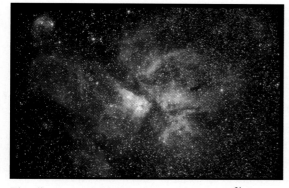

The diameter of this galaxy is about 1×10^{21} meters.

The diameter of these viruses (shown in green) is about 1×10^{-7} meters.

no numerical coefficient showing, as in 10^5, we assume that the numerical coefficient is 1. Thus, 10^5 means 1×10^5 and 10^{-4} means 1×10^{-4}.

Examples of Numbers in Scientific Notation

$$3.2 \times 10^6 \qquad 4.176 \times 10^3 \qquad 2.64 \times 10^{-2}$$

The following shows the number 32,400 changed to scientific notation.

$$32,400 = 3.24 \times 10,000$$
$$= 3.24 \times 10^4 \qquad (10,000 = 10^4)$$

There are four zeros in 10,000, the same number as the exponent in 10^4. The procedure for writing a number in scientific notation follows.

To Write a Number in Scientific Notation

1. Move the decimal point in the number to the right of the first nonzero digit. This gives a number greater than or equal to 1 and less than 10.
2. Count the number of places you moved the decimal point in step 1. If the original number is 10 or greater, the count is to be considered positive. If the original number is less than 1, the count is to be considered negative.
3. Multiply the number obtained in step 1 by 10 raised to the count (power) found in step 2.

EXAMPLE 1 Write the following numbers in scientific notation.

a) 68,900 **b)** 0.000572 **c)** 0.0074

Solution **a)** The decimal point in 68,900 is to the right of the last zero.

$$68,900. = 6.89 \times 10^4$$

The decimal point is moved four places. Since the original number is greater than 10, the exponent is positive.

b) $0.000572 = 5.72 \times 10^{-4}$

The decimal point is moved four places. Since the original number is less than 1, the exponent is negative.

c) $0.0074 = 7.4 \times 10^{-3}$

NOW TRY EXERCISE 11

2 Change Numbers in Scientific Notation to Decimal Form

Occasionally, you may need to convert a number written in scientific notation to its decimal form. The procedure to do so follows.

> ### To Convert a Number in Scientific Notation to Decimal Form
>
> 1. Observe the exponent on the base 10.
> 2. a) If the exponent is positive, move the decimal point in the number to the right the same number of places as the exponent. It may be necessary to add zeros to the number. This will result in a number greater than or equal to 10.
>   b) If the exponent is 0, the decimal point in the number does not move from its present position. Drop the factor 10^0. This will result in a number greater than or equal to 1 but less than 10.
>   c) If the exponent is negative, move the decimal point in the number to the left the same number of places as the exponent. It may be necessary to add zeros. This will result in a number less than 1.

EXAMPLE 2   Write the following numbers without exponents.
a) 2.1×10^4 **b)** 8.73×10^{-3} **c)** 1.45×10^8

Solution   **a)** Move the decimal point four places to the right.

$$2.1 \times 10^4 = 2.1 \times 10{,}000 = 21{,}000$$

b) Move the decimal point three places to the left.

$$8.73 \times 10^{-3} = 0.00873$$

c) Move the decimal point eight places to the right.

NOW TRY EXERCISE 25

$$1.45 \times 10^8 = 145{,}000{,}000$$

3 Use Scientific Notation in Problem Solving

We can use the rules of exponents when working with numbers written in scientific notation, as illustrated in the following applications.

EXAMPLE 3   **Public Debt per Person**   The public debt is the total amount owed by the U.S. federal government to lenders in the form of government bonds. On January 15, 2002, the U.S. public debt was approximately $5,894,000,000,000 (5 trillion, 894 billion dollars). The U.S. population on that date was approximately 286,000,000.
a) Find the average U.S. debt per person in the United States (the per capita debt).
b) On September 30, 1982, the U.S. debt was approximately $1,142,000,000,000. How much larger was the debt in 2002 than in 1982?
c) How many times greater was the debt in 2002 than in 1982?

Solution   **a)** To find the per capita debt, we divide the public debt by the population.

$$\frac{5{,}894{,}000{,}000{,}000}{286{,}000{,}000} = \frac{5.894 \times 10^{12}}{2.86 \times 10^8} \approx 2.06 \times 10^{12-8} \approx 2.06 \times 10^4 \approx 20{,}600$$

Thus, the per capita debt was about $20,600. This means that if the citizens of the United States wished to "chip in" and pay off the federal debt, it would take about $20,600 for every man, woman, and child in the United States.

b) We need to find the difference in the debt between 2002 and 1982.

$$5{,}894{,}000{,}000{,}000 - 1{,}142{,}000{,}000{,}000 = 5.894 \times 10^{12} - 1.142 \times 10^{12}$$
$$= (5.894 - 1.142) \times 10^{12}$$
$$= 4.752 \times 10^{12}$$
$$= 4{,}752{,}000{,}000{,}000$$

The U.S. public debt was $4,752,000,000,000 greater in 2002 than in 1982.

c) To find out how many times greater the 2002 public debt was, we divide the 2002 debt by the 1982 debt as follows:

$$\frac{5{,}894{,}000{,}000{,}000}{1{,}142{,}000{,}000{,}000} = \frac{5.984 \times 10^{12}}{1.142 \times 10^{12}} \approx 5.2$$

Thus, the 2002 public debt was about 5.2 times greater than the 1982 public debt. ✳

EXAMPLE 4   **Tax Collections**   The data for the graphs in Figure 1.11 were taken from the 2001 edition of the *Statistical Abstract of the United States*. The graphs show the cumulative state government tax collection in 1970 and 2000. We have given the amounts collected in scientific notation.

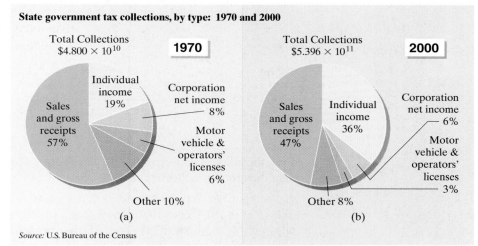

FIGURE 1.11   *Source:* U.S. Bureau of the Census

a) Determine, using scientific notation, how much money was collected from individual income (tax) in 2000.

b) Determine, using scientific notation, how many more dollars were collected, in total taxes, in 2000 than in 1970.

c) Determine, using scientific notation, how many times greater the amount collected in 2000 was than the amount collected in 1970.

d) Describe what you see as major differences in the taxes collected in 2000 versus 1970.

Solution   **a)** In 2000, 36% of the 5.396×10^{11} was collected from individual income taxes. In decimal form, 36% is 0.36, and in scientific notation 36% is 3.6×10^{-1}. To determine 36% of 5.396×10^{11}, we multiply using scientific notation as follows:

$$\text{income tax collected} = (3.6 \times 10^{-1})(5.396 \times 10^{11})$$
$$= (3.6 \times 5.396)(10^{-1} \times 10^{11})$$
$$= 19.4256 \times 10^{-1+11}$$
$$= 19.4256 \times 10^{10}$$
$$= 1.94256 \times 10^{11}$$

Thus, about $\$1.94256 \times 10^{11}$ or $\$194,256,000,000$ was collected from individual taxes in 2000.

b) In 1970, $\$4.800 \times 10^{10}$ in taxes was collected. In 2000, $\$5.396 \times 10^{11}$ in taxes was collected. To determine the difference in these amounts, we will subtract the amount collected in 1970 from the amount collected in 2000. We will write 4.800×10^{10} as 4.80×10^{10} and subtract as follows:

$$\begin{aligned} \text{difference in taxes collected} &= 5.396 \times 10^{11} - 4.80 \times 10^{10} \\ &= 53.96 \times 10^{10} - 4.80 \times 10^{10} \\ &= (53.96 - 4.80) \times 10^{10} \\ &= 49.16 \times 10^{10} \\ &= 4.916 \times 10^{11} \end{aligned}$$

Therefore, $\$4.916 \times 10^{11}$ or $\$491,600,000,000$ more was collected in 2000 than in 1970.

c) To determine the number of times greater the amount collected in 2000 was than the amount collected in 1970, we will divide as follows:

$$\begin{aligned} \text{number of times greater} &= \frac{\text{amount collected in 2000}}{\text{amount collected in 1970}} \\ &= \frac{5.396 \times 10^{11}}{4.80 \times 10^{10}} \\ &= \frac{5.396}{4.80} \times 10^{11-10} \\ &\approx 1.1242 \times 10^{1} \end{aligned}$$

Thus, the amount collected in 2000 was about 1.1242×10^{1} or 11.242 times greater than the amount collected in 1970.

d) We know from parts **b)** and **c)** that a great deal more taxes were collected in 2000 than in 1970. The graphs show that a smaller percentage in 2000 comes from sales tax (47% versus 57%) and a larger percentage comes from individual income tax (36% versus 19%). There are also smaller changes in the percentages of the other sectors. ✳

Using Your Calculator

On a scientific or graphing calculator the product of $(8,000,000)(400,000)$ might be displayed as $3.2^{\ 12}$ or 3.2E12. Both of these represent 3.2×10^{12}, which is $3,200,000,000,000$.

To enter numbers in scientific notation on either a scientific or graphing calculator, you generally use the $\boxed{\text{EE}}$ or $\boxed{\text{EXP}}$ keys. To enter 4.6×10^{8}, you would press either $4.6\ \boxed{\text{EE}}\ 8$ or $4.6\ \boxed{\text{EXP}}\ 8$. Your calculator screen might then show $4.6^{\ 08}$ or 4.6E8.

On the TI-83 Plus the EE appears above the $\boxed{,}$ key. So to enter $(8,000,000)(400,000)$ in scientific notation you would press

Exercise Set 1.6

Concept/Writing Exercises

1. What is the form of a number in scientific notation?

2. Can 1×10^n ever be a negative number for any positive integer n? Explain.

3. Which is greater, 1×10^{-2} or 1×10^{-3}? Explain.

4. Can 1×10^{-n} ever be a negative number for any positive integer n? Explain.

Practice the Skills

Express each number in scientific notation.

5. 4700

6. 560

7. 0.031

8. 0.000000718

9. 360,000

10. 5,260,000,000

11. 0.00000186

12. 0.00000914

13. 5,780,000

14. 0.0000773

15. 0.000101

16. 998,000,000

Express each number without exponents.

17. 3.1×10^4

18. 5×10^8

19. 2.13×10^{-5}

20. 5.78×10^{-5}

21. 9.17×10^{-1}

22. 5.3×10^1

23. 9×10^6

24. 7.3×10^4

25. 2.07×10^5

26. 9.35×10^{-6}

27. 1×10^6

28. 1×10^{-8}

Express each value without exponents.

29. $(4 \times 10^5)(6 \times 10^2)$

30. $(5.6 \times 10^{-3})(1.2 \times 10^{-1})$

31. $\dfrac{8.4 \times 10^{-6}}{4 \times 10^{-4}}$

32. $\dfrac{25 \times 10^3}{5 \times 10^{-2}}$

33. $\dfrac{6.75 \times 10^{-3}}{2.5 \times 10^2}$

34. $(6.7 \times 10^{-3})(4.1 \times 10^5)$

35. $(8.2 \times 10^5)(1.3 \times 10^{-2})$

36. $(6.3 \times 10^4)(3.7 \times 10^{-8})$

37. $\dfrac{1.68 \times 10^4}{5.6 \times 10^7}$

38. $\dfrac{4.8 \times 10^{-2}}{2.4 \times 10^{-6}}$

39. $(9.1 \times 10^{-4})(6.3 \times 10^{-4})$

40. $\dfrac{6.2 \times 10^{-8}}{3.1 \times 10^{-6}}$

Express each value in scientific notation.

41. $(0.03)(0.0005)$

42. $(2500)(7000)$

43. $\dfrac{35,000,000}{7000}$

44. $\dfrac{560,000}{0.0008}$

45. $\dfrac{0.00046}{23,000}$

46. $\dfrac{0.000012}{0.000006}$

47. $(47,000)(35,000,000)$

48. $\dfrac{0.0000282}{0.00141}$

49. $\dfrac{672}{0.0021}$

50. $\dfrac{0.018}{480}$

51. $\dfrac{0.00153}{0.00051}$

52. $(0.0015)(0.00036)$

Express each value in scientific notation. Round decimal numbers to the nearest thousandth.

53. $(4.78 \times 10^9)(1.96 \times 10^5)$

54. $\dfrac{3.33 \times 10^3}{1.11 \times 10^1}$

55. $(7.23 \times 10^{-3})(1.37 \times 10^5)$

56. $(5.71 \times 10^5)(4.7 \times 10^{-3})$

57. $\dfrac{4.36 \times 10^{-4}}{8.17 \times 10^{-7}}$

58. $\dfrac{8.45 \times 10^{25}}{4.225 \times 10^{15}}$

59. $(4.89 \times 10^{15})(7.91 \times 10^{-41})$

60. $(4.36 \times 10^{-6})(1.07 \times 10^{-6})$

61. $(7.71 \times 10^3)(9.14 \times 10^{-31})$

62. $\dfrac{3.71 \times 10^{11}}{4.72 \times 10^{-9}}$

63. $\dfrac{1.50 \times 10^{35}}{4.5 \times 10^{-26}}$

64. $(3.7 \times 10^5)(1.347 \times 10^{31})$

Problem Solving

65. Explain how you can quickly divide a number given in scientific notation by
   a) 10,
   b) 100,
   c) 1 million.
   d) Divide 6.58×10^{-4} by 1 million. Leave your answer in scientific notation.

66. Explain how you can quickly multiply a number given in scientific notation by
   a) 10,
   b) 100,
   c) 1 million.
   d) Multiply 7.59×10^7 by 1 million. Leave your answer in scientific notation.

67. *Science Experiment* During a science experiment you find that the correct answer is 5.25×10^4.
   a) If you mistakenly write the answer as 4.25×10^4, by how much is your answer off?
   b) If you mistakenly write the answer as 5.25×10^5, by how much is your answer off?
   c) Which of the two errors is the more serious? Explain.

68. *Earth Orbit*
   a) Earth completes its 5.85×10^8-mile orbit around the sun in 365 days. Find the distance traveled per day.
   b) Earth's speed is about eight times faster than a bullet. Estimate the speed of a bullet in miles per hour.

69. *Distance to the Sun* The distance to the sun is 93,000,000 miles. If a spacecraft travels at a speed of 3100 miles per hour, how long will it take for it to reach the sun?

70. *Milky Way* We have proof that there are at least 1 sextillion, 10^{21}, stars in the Milky Way.
   a) Write that number without exponents.
   b) How many million stars is this? Explain how you determined your answer to part **b)**.

71. *U.S. and World Populations* The population of the United States on census day, April 1, 2000, was about 2.81×10^8. On this day the population of the world was about 6.14×10^9. *Source:* U.S. Census Bureau
   a) How many people lived outside of the United States in 2000?
   b) What percentage of the world's population lived in the United States in 2000?

72. *Gross Domestic Product* Gross Domestic Product (GDP) is a measure of economic activity. GDP is the total amount of goods and services produced in a country in a year. In 2000, the GDP for the United States was about $9.8729 trillion and the population of the United States was about 281.42 million. *Source:* U.S. Treasury Web site
   a) Write each of these two numbers in scientific notation.
   b) Determine the GDP *per capita* by dividing the GDP by the population of the United States.

73. *New River George Bridge* The New River George bridge, shown below, is 3030.5 feet long. It was completed in 1977 near Fayetteville, West Virginia, and is the world's longest single arch steel span. Its total weight is 8.80×10^7 pounds and its heaviest single piece weighs 1.84×10^5 pounds.
   a) How many times greater is the total weight of the bridge than the weight of the heaviest single piece?
   b) What is the difference in weight between the total weight of the bridge and the weight of the heaviest single piece?

74. *Recycling Plastic* In the United States only about 5% of the 4.2×10^9 pounds of used plastic is recycled annually.

    a) How many pounds are recycled annually?

    b) How many pounds are not recycled annually?

75. *Airports* The top five U.S. airports in 2001, ranked by passengers boarding an airplane (or enplaning) are given in the following circle graph. About 139, 200,000 or 1.392×10^8 passengers enplaned in these five airports. The graph also provides the percent of the total number of passengers enplaned for these five airports.

Top five U.S. airports in enplaned passengers (percent)

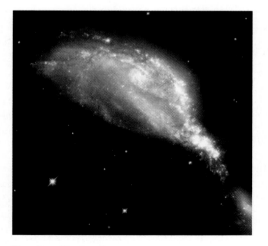

Dallas/Ft. Worth 18.2%
Chicago (O'Hare) 22.2%
Los Angeles 20.7%
Phoenix (Sky Harbor) 12.4%
Atlanta (Hartsfield) 26.5%

Source: Airports Council International

    a) How many passengers enplaned at Chicago's O'Hare airport?

    b) How many passengers enplaned at Atlanta's Hartsfield airport?

    c) How many times greater is the number who enplaned at Hartsfield than the number who enplaned at Sky Harbor airport?

76. *Distance to Proxima Centauri* The distance from earth to the sun is approximately 150 million kilometers. The next closest star to earth is Proxima Centauri.

Proxima Centauri

It is about 268,000 times farther away from earth than is the sun. Approximate the distance of Proxima Centauri from earth. Write your answer in scientific notation. *Source:* NASA Web site

77. *Most Populous Countries* In 2001, the six most populous countries accounted for 3,114,000,000 people out of the world's total population of 6,137,000,000. The six most populous countries in 2001 are shown in the following graph, along with each country's population.

Six most populous countries (population in millions)

Pakistan 145
Brazil 172
Indonesia 206
United States 285
China 1273
India 1033

Source: U.S. Census Bureau   *Note:* China includes mainland China and Taiwan.

    a) How many more people lived in China than in the United States?

    b) What percent of the world's population lived in China?

    c) If the area of China is 3.70×10^6 square miles, determine China's population density (people per square mile).

    d) If the area of the United States is 3.62×10^6 square miles, determine the population density in the United States.*

78. *World Population* The entire span of human history was required for the world's population to reach 6.14×10^9 people in the year 2000. At current rates, the world's population will double in about 54 years.

    a) Estimate the world's population in 2054.

    b) Assuming 365 days in a year, estimate the average number of additional people added to the world's population each day between 2000 and 2054.

---

*The area with the greatest population density is Macau, with a population density of 80,426 people per square mile. The country with the greatest population density is Monaco, with a population density of 41,076 people per square mile.

Challenge Problems

79. **Light-Year** A *light-year* is the distance that light travels in 1 year.

   a) Find the number of miles in a light-year if light travels at 1.86×10^5 miles *per second*.

   b) If earth is 93,000,000 miles from the sun, how long does it take for light from the sun to reach earth?

   c) Our galaxy, the Milky Way, is about 6.25×10^{16} miles across. If a spacecraft could travel at half the speed of light, how long would it take for the craft to travel from one end of the galaxy to the other?

CHAPTER SUMMARY

Key Words and Phrases

1.2
Constant
Counting numbers
Elements of a set
Endpoints
Expression
Inequalities
Integers
Intersection of sets
Irrational numbers
Natural numbers
Null (or empty) set
Operations
Rational numbers

Real numbers
Roster form
Set
Set builder notation
Subset
Union of sets
Variable
Whole numbers

1.3
Absolute value
Additive identity
   element
Additive inverse

Associative properties
Commutative properties
Distributive property
Multiplacitive identity
   element
Multiplicative inverse
Opposites
Properties of real
   numbers
Reciprocal

1.4
Base
Exponent

Exponential expression
Factor
Graphing calculator
   (or grapher)
Index
Order of operations
Principal square root
Radical sign
Radicand
Undefined expression

1.6
Scientific notation

IMPORTANT FACTS

Sets of Numbers

Real numbers	$\{x\,|\,x$ is a point on a number line$\}$
Natural or counting numbers	$\{1, 2, 3, 4, 5, \ldots\}$
Whole numbers	$\{0, 1, 2, 3, 4, \ldots\}$
Integers	$\{\ldots, -3, -2, -1, 0, 1, 2, 3, \ldots\}$
Rational numbers	$\left\{\dfrac{p}{q}\,\middle|\,p$ and q are integers, $q \neq 0\right\}$
Irrational numbers	$\{x\,|\,x$ is a real number that is not rational$\}$

Inequalities on the Real Number Line

$\{x\,|\,x > a\}$

$\{x\,|\,x \leq a\}$

$\{x\,|\,a \leq x < b\}$

(continued on the next page)

Properties of the Real Number System

Commutative properties	$a + b = b + a, \quad ab = ba$
Associative properties	$(a + b) + c = a + (b + c), \quad (ab)c = a(bc)$
Identity properties	$a + 0 = 0 + a = a, \quad a \cdot 1 = 1 \cdot a = a$
Inverse properties	$a + (-a) = (-a) + a = 0, \quad a \cdot \dfrac{1}{a} = \dfrac{1}{a} \cdot a = 1 \quad (a \neq 0)$
Distributive property	$a(b + c) = ab + ac$
Extended distributive property	$a(b + c + d + \cdots + n) = ab + ac + ad + \cdots + an$
Multiplicative property of 0	$a \cdot 0 = 0 \cdot a = 0$
Double negative property	$-(-a) = a$

Absolute Value: $|a| = \begin{cases} a & a \geq 0 \\ -a & a < 0 \end{cases}$

Exponents and Roots

$$b^n = \underbrace{b \cdot b \cdot b \cdot \cdots \cdot b}_{n \text{ factors of } b} \qquad\qquad \sqrt[n]{a} = b \text{ if } \underbrace{b \cdot b \cdot b \cdot \cdots \cdot b}_{n \text{ factors of } b} = a$$

Order of Operations

1. Parentheses or other grouping symbols
2. Exponents and roots
3. Multiplication or division from left to right
4. Addition or subtraction from left to right

Rules for Exponents

1. $a^m \cdot a^n = a^{m+n}$ *Product rule*

2. $\dfrac{a^m}{a^n} = a^{m-n}, a \neq 0$ *Quotient rule*

3. $a^{-m} = \dfrac{1}{a^m}, a \neq 0$ *Negative exponent rule*

4. $a^0 = 1, a \neq 0$ *Zero exponent rule*

5. $(a^m)^n = a^{mn}$ *Raising a power to a power (Power rule)*

    $(ab)^m = a^m b^m$ *Raising a product to a power*

    $\left(\dfrac{a}{b}\right)^m = \dfrac{a^m}{b^m}, b \neq 0$ *Raising a quotient to a power*

Scientific Notation

In scientific notation, numbers appear as a number greater than or equal to 1 and less than 10 multiplied by some power of 10. The exponent on the 10 must be an integer.

Chapter Review Exercises

[1.2] List each set in roster form.

1. $A = \{x|x$ is a natural number between 3 and 8$\}$

2. $B = \{x|x$ is a whole number multiple of 3$\}$

Let N = set of natural numbers, W = set of whole numbers, I = set of integers, Q = set of rational numbers, H = set of irrational numbers, and $\mathbb{R}$ = set of real numbers. Determine whether the first set is a subset of the second set for each pair of sets.

3. $\mathbb{N}, W$ 　　　　4. $Q, \mathbb{R}$ 　　　　5. $H, \mathbb{R}$ 　　　　6. Q, H

Consider the set of numbers $\left\{-2, 4, 6, \frac{1}{2}, \sqrt{7}, \sqrt{3}, 0, \frac{15}{27}, -\frac{1}{5}, 1.47\right\}$. List the elements of the set that are

7. natural numbers. 　　　8. whole numbers. 　　　9. integers.

10. rational numbers. 　　11. irrational numbers. 　　12. real numbers.

Indicate whether each statement is true or false.

13. $\frac{0}{1}$ is not a real number.

14. $0, \frac{3}{5}, -2$, and 4 are all rational numbers.

15. A real number cannot be divided by 0.

16. Every rational number and every irrational number is a real number.

Find $A \cup B$ and $A \cap B$ for each set A and B.

17. $A = \{1, 2, 3, 4, 5, 6\}, B = \{2, 4, 6, 8, 10\}$

18. $A = \{3, 5, 7, 9\}, B = \{2, 4, 6, 8\}$

19. $A = \{1, 3, 5, 7, \ldots\}, B = \{2, 4, 6, 8, \ldots\}$

20. $A = \{4, 6, 9, 10, 11\}, B = \{3, 5, 9, 10, 12\}$

Illustrate each set on the number line.

21. $\{x|x > 5\}$ 　　22. $\{x|x \leq -2\}$ 　　23. $\{x|-1.3 < x \leq 2.4\}$ 　　24. $\left\{x \left| \frac{2}{3} \leq x < 4 \text{ and } x \in N \right.\right\}$

[1.3] Insert either $<, >$, or $=$ in the shaded area between the two numbers to make each statement true.

25. $-8 \; \blacksquare \; 0$ 　　26. $-4 \; \blacksquare \; -3.9$ 　　27. $1.06 \; \blacksquare \; 1.6$ 　　28. $|-3| \; \blacksquare \; 3$

29. $|-4| \; \blacksquare \; |-6|$ 　　30. $13 \; \blacksquare \; |-5|$ 　　31. $\left|-\frac{2}{3}\right| \; \blacksquare \; \frac{3}{5}$ 　　32. $-|-2| \; \blacksquare \; -5$

Write the numbers in each list from smallest to largest.

33. $\pi, -\pi, -3, 3$

34. $0, \frac{3}{5}, 2.3, |-3|$

35. $|-7|, |-5|, 3, -2$

36. $|-3|, -7, |-7|, -3$

37. $-4, 6, -|-3|, 5$

38. $|1.6|, |-2.3|, -3, 0$

Name each property illustrated.

39. $-7(x + 4) = -7x - 28$

40. $mn = nm$

41. $(x + 3) + 2 = x + (3 + 2)$

42. $q + 0 = 0$

43. $5(rs) = (5r)s$

44. $-(-5) = 5$

45. $5(0) = 0$

46. $a + (-a) = 0$

47. $x \cdot \dfrac{1}{x} = 1$

48. $k + l = 1 \cdot (k + l)$

[1.3, 1.4] *Evaluate.*

49. $7 + 3^2 - \sqrt{36} \div 2$

50. $-4 \div (-2) + 16 - \sqrt{49}$

51. $(4 - 6) - (-3 + 5) + 12$

52. $2|-7| - 4|-6| + 3$

53. $(6 - 9) \div (9 - 6) + 1$

54. $|6 - 3| \div 3 + 4 \cdot 8 - 12$

55. $\sqrt{9} + \sqrt[3]{64} + \sqrt[4]{16}$

56. $3^2 - 6 \cdot 9 + 4 \div 2^2 - 3$

57. $4 - (2 - 9)^0 + 3^2 \div 1 + 3$

58. $5^2 + (-2 + 2^2)^3 + 1^4$

59. $-3^2 + 14 \div 2 \cdot 3 - 6$

60. $\{[(9 \div 3)^2 - 1]^2 \div 8\}^3$

61. $\dfrac{5 + 7 \div (3^2 - 2) + 4 \cdot 1}{\sqrt{81} + \sqrt{1} - 10}$

62. $\dfrac{-(4 - 6)^2 - 3(-2) + |-6|}{18 - 9 \div 3 \cdot 5}$

Evaluate.

63. Evaluate $2x^2 + 3x + 1$ when $x = 2$

64. Evaluate $5a^2 - 7b^2$ when $a = -3$ and $b = -4$.

65. ***Political Campaigning*** The cost of political campaigning has changed dramatically since 1952. The amount spent, in millions of dollars, on all U.S. elections—including local, state, and national offices; political parties; political action committees; and ballot issues—is approximated by

  dollars spent $= 50.86x^2 - 316.75x + 541.48$

where x represents each 4-year period since 1948. Substitute 1 for x to get the amount spent in 1952, 2 for x to get the amount spent in 1956, 3 for x to get the amount spent in 1960, and so on.

a) Find the amount spent for elections in 1976.

b) Find the amount spent for elections in 2004.

66. ***Railroad Traffic*** Railroad traffic has been increasing steadily, and has doubled since 1965. Most of this is due to the increase in trains used to transport goods by container. We can approximate the amount of freight carried in ton-miles (1 ton-mile equals 1 ton of freight hauled 1 mile) by

  freight hauled $= 14.04x^2 + 1.96x + 712.05$

where x represents each 5-year period since 1960. Substitute 1 for x to get the amount hauled in 1965, 2 for x to get the amount hauled in 1970, 3 for x for 1975, and so forth.

a) Find the amount of freight hauled by trains in 1980.

b) Find the amount of freight hauled by trains in 2000.

[1.5] *Simplify each expression and write the answer without negative exponents.*

67. $2^3 \cdot 2^2$

68. $x^2 \cdot x^3$

69. $\dfrac{a^{12}}{a^4}$

70. $\dfrac{y^{12}}{y^3}$

71. $\dfrac{b^7}{b^{-2}}$

72. $c^3 \cdot c^{-5}$

73. $5^{-2} \cdot 5^{-1}$

74. $3x^0$

75. $(-2m^3)^2$

76. $\left(\dfrac{5}{7}\right)^{-1}$

77. $\left(\dfrac{2}{3}\right)^{-3}$

78. $\left(\dfrac{x}{y^2}\right)^{-1}$

79. $(5xy^3)(-3x^2y)$

80. $(2v^3w^{-4})(3v^{-6}w)$

81. $\dfrac{6x^{-3}y^5}{2x^2y^{-2}}$

82. $\dfrac{12x^{-3}y^{-4}}{4x^{-2}y^5}$

83. $\dfrac{g^3h^{-6}j^{-9}}{g^{-2}h^{-1}j^5}$

84. $\dfrac{21m^{-3}n^{-2}}{7m^{-4}n^2}$

85. $\left(\dfrac{5a^2b}{a}\right)^3$

86. $\left(\dfrac{x^5y}{-3y^2}\right)^2$

87. $\left(\dfrac{p^3q^{-1}}{p^{-4}q^5}\right)^2$

88. $\left(\dfrac{-2ab^{-3}}{c^2}\right)^3$

89. $\left(\dfrac{6xy^3}{z^2}\right)^{-2}$

90. $\left(\dfrac{9m^{-2}n}{3mn}\right)^{-3}$

91. $(-2m^2n^{-3})^{-2}$

92. $\left(\dfrac{15x^5y^{-3}z^{-2}}{-3x^4y^{-4}z^3}\right)^4$

93. $\left(\dfrac{2x^{-1}y^5z^4}{3x^4y^{-2}z^{-2}}\right)^{-2}$

94. $\left(\dfrac{8x^{-2}y^{-2}z}{-x^4y^{-4}z^3}\right)^{-1}$

[1.6] Express each number in scientific notation.

95. 0.0000742   **96.** 260,000   **97.** 183,000   **98.** 0.000001

Simplify each expression and express the answer without exponents.

99. $(25 \times 10^{-3})(1.2 \times 10^6)$ **100.** $\dfrac{18 \times 10^3}{9 \times 10^5}$ **101.** $\dfrac{4{,}000{,}000}{0.02}$ **102.** $(0.004)(500{,}000)$

103. *Television Advertising* Car manufacturers spend millions of dollars each year on television advertising. In the second quarter of 2001, "the big three" U.S. automakers spent the following amounts on television advertising: General Motors spent 2.64×10^8, Ford spent 1.51×10^8, and Daimler Chrysler spent 9.2×10^7.

    a) How much more did General Motors spend than Ford?

    b) How much more did Ford spend than Daimler Chrysler?

    c) How many times larger is the amount General Motors spent than the amount Daimler Chrysler spent?

104. *Voyager* On February 17, 1998, the *Voyager 1* spacecraft became the most distant explorer in the solar system, breaking the record of the *Pioneer 10*. The 20-year-old *Voyager 1* has traveled more than 1.04×10^{10} kilometers from earth (about 70 times the distance from earth to the sun).

    a) Represent 1.04×10^{10} as a decimal number.

    b) How many billion kilometers has *Voyager 1* traveled?

    c) Assuming that *Voyager 1* traveled about the same number of kilometers in each of the 20 years, how many kilometers did it average in a year?

    d) If 1 kilometer ≈ 0.6 miles, how far, in miles, did *Voyager 1* travel?

Chapter Practice Test

1. List $A = \{x \mid x$ is a natural number greater than or equal to $6\}$ in roster form.

Indicate whether each statement is true or false.

2. Every real number is a rational number.

3. The union of the set of rational numbers and the set of irrational numbers is the set of real numbers.

Consider the set of numbers $\left\{-\frac{3}{5}, 2, -4, 0, \frac{19}{12}, 2.57, \sqrt{8}, \sqrt{2}, -1.92\right\}$. List the elements of the set that are

4. rational numbers.

5. real numbers.

Find $A \cup B$ and $A \cap B$ for sets A and B.

6. $A = \{8, 10, 11, 14\}, B = \{5, 7, 8, 9, 10\}$

7. $A = \{1, 3, 5, 7, \ldots\}, B = \{3, 5, 7, 9, 11\}$

In Exercises 8 and 9, illustrate each set on the number line.

8. $\{x | -2.3 \leq x < 5.2\}$

9. $\left\{ x \middle| -\dfrac{5}{2} < x < \dfrac{6}{5} \text{ and } x \in I \right\}$

10. List from smallest to largest: $|3|, -|4|, -2, 6$.

Name each property illustrated.

11. $(x + y) + 3 = x + (y + 3)$

12. $3x + 4y = 4y + 3x$

Evaluate each expression.

13. $\{4 - [7 - 3^2 \div (3^2 - 2 \cdot 3)]\}$

14. $2^4 + 4^2 \div 2^3 \cdot \sqrt{25} + 7$

15. $\dfrac{-3|4 - 8| \div 2 + 4}{-\sqrt{36} + 18 \div 3^2 + 4}$

16. $\dfrac{-6^2 + 3(4 - |6|) \div 6}{4 - (-3) + 12 \div 4 \cdot 5}$

17. Evaluate $-x^2 + 2xy + y^2$ when $x = 2$ and $y = 3$.

18. *Tax Refund* Since 1990 the average federal tax refund has been rising. The average federal tax refund from 1990 through 2002 can be estimated by

$$\text{average federal tax refund} = 6.42x^2 + 13.9x + 970$$

where x represents the number of years since 1990. Substitute 1 for x to estimate the average federal tax refund in 1991, 2 for x to estimate the average federal tax refund in 1992, and so on.

a) Estimate the average federal tax refund in 1995.

b) Estimate the average federal tax refund in 2002.

Simplify each expression and write the answer without negative exponents.

19. 3^{-2}

20. $\left(\dfrac{4m^{-3}}{n^2} \right)^2$

21. $\dfrac{24a^2 b^{-3} c^0}{30a^3 b^2 c^{-2}}$

22. $\left(\dfrac{-3x^3 y^{-2}}{x^{-1} y^5} \right)^{-3}$

23. Convert 242,000,000 to scientific notation.

24. Simplify $\dfrac{3.12 \times 10^6}{1.2 \times 10^{-2}}$ and write the number without exponents.

25. *Employment by Gender* The total number of people employed in the United States in 2000 was about 1.41×10^8. The graph shows the men/women breakdown.

a) How many men were employed in the United States?

b) How many women were employed in the United States?

c) How many more men than women were employed?

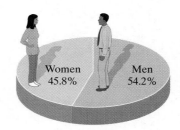

Women 45.8%   Men 54.2%

Chapter 2

Equations and Inequalities

2.1 Solving Linear Equations

2.2 Problem Solving and Using Formulas

2.3 Applications of Algebra

2.4 Additional Application Problems

2.5 Solving Linear Inequalities

2.6 Solving Equations and Inequalities Containing Absolute Values

Chapter Summary
Chapter Review Exercises
Chapter Practice Test
Cumulative Review Test

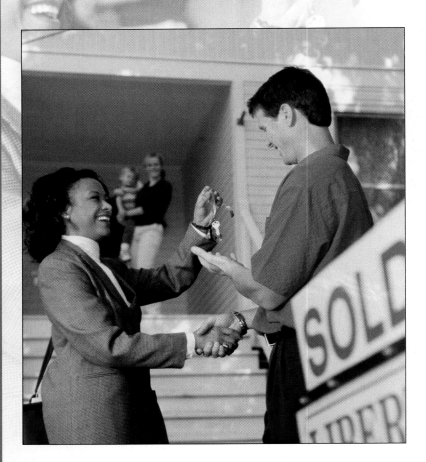

For most people, buying a home is the biggest purchase of their life. After negotiating a price for the house, you must also choose a loan. How do you choose a loan? Do you look for the loan that has no application fee, the fewest points, the lowest interest rate, or the one that offers free airline tickets? Is there anything else to consider? On page 97, we compare the costs of two loans by writing equations to describe the cost of each loan and then finding the month when the costs are the same. While doing this, we learn that the length of time you own the home is a key factor to consider when selecting a loan.

SSM   Study Guide   CD/Video   MathPro 4/5   PH Math   prenhall.com/Angel
                                                             Tutor Center

A Look Ahead

In this chapter, we focus on solving linear equations and inequalities, and on using linear equations, formulas, and inequalities to solve real-life problems. After reviewing the process of solving equations in Section 2.1, we introduce a powerful problem-solving technique in Section 2.2. We use this technique throughout Sections 2.2, 2.3, and 2.4 and the rest of this textbook. In Section 2.5, we introduce linear inequalities, and in Section 2.6, we introduce equations and inequalities using absolute value. Throughout the chapter, we will witness the power of algebra as a problem-solving tool in a multitude of areas including real estate, chemistry, business, banking, physics, and personal finance.

2.1 SOLVING LINEAR EQUATIONS

SSM   Study Guide   CD/Video

MathPro 4/5   PH Math Tutor Center   prenhall.com/Angel

1   Identify the reflexive, symmetric, and transitive properties.
2   Combine like terms.
3   Solve linear equations.
4   Solve equations containing fractions.
5   Identify conditional equations, contradictions, and identities.
6   Understand the concepts to solve equations.

1   Identify the Reflexive, Symmetric, and Transitive Properties

In elementary algebra you learned how to solve linear equations. We review these procedures briefly in this section. Before we do so, we need to introduce three useful properties of equality: the *reflexive property*, the *symmetric property*, and the *transitive property*.

Properties of Equality

For all real numbers a, b, and c:

1. $a = a$.   **reflexive property**
2. If $a = b$, then $b = a$.   **symmetric property**
3. If $a = b$ and $b = c$, then $a = c$.   **transitive property**

Examples of the Reflexive Property

$$7 = 7$$
$$x + 3 = x + 3$$

Examples of the Symmetric Property

If $x = 3$, then $3 = x$.

If $y = x + 4$, then $x + 4 = y$.

Examples of the Transitive Property

If $x = a$ and $a = 4y$, then $x = 4y$.

If $a + b = c$ and $c = 4r$, then $a + b = 4r$.

In this book, we will often use these properties without referring to them by name.

2 Combine Like Terms

When an algebraic expression consists of several parts, the parts that are added are called the **terms** of the expression. The expression $3x^2 - 6x - 2$, which can be written $3x^2 + (-6x) + (-2)$, has three terms; $3x^2, -6x$, and -2. The expression

$$6x^2 - 3(x + y) - 4 + \frac{x + 2}{5}$$

has four terms: $6x^2, -3(x + y), -4$, and $\frac{x + 2}{5}$.

Expression	Terms
$\frac{1}{2}x^2 - 3x - 7$	$\frac{1}{2}x^2, \quad -3x, \quad -7$
$-5x^3 + 3x^2y - 2$	$-5x^3, \quad 3x^2y, \quad -2$
$4(x + 3) + 2x + 5(x - 2) + 1$	$4(x + 3), \quad 2x, \quad 5(x - 2), \quad 1$

The numerical part of a term that precedes the variable is called its **numerical coefficient** or simply its **coefficient**. In the term $6x^2$, the 6 is the numerical coefficient. When the numerical coefficient is 1 or -1, we generally do not write the numeral 1. For example, x means $1x$, $-x^2y$ means $-1x^2y$, and $(x + y)$ means $1(x + y)$.

Term	Numerical Coefficient
$\frac{5k}{7}$	$\frac{5}{7}$
$-4(x + 2)$	-4
$\frac{x - 2}{3}$	$\frac{1}{3}$
$-(x + y)$	-1

Note that $\frac{x - 2}{3}$ *means* $\frac{1}{3}(x - 2)$ *and* $-(x + y)$ *means* $-1(x + y)$.

When a term consists of only a number, that number is called a **constant**. For example, in the expression $x^2 - 4$, the -4 is a constant.

The **degree of a term** with whole number exponents is the sum of the exponents on the variables in the term. For example, $3x^2$ is a second-degree term, and $-4x$ is a first-degree term ($-4x$ means $-4x^1$). The number 3 can be written as $3x^0$, so the number 3 (and every other nonzero constant) has degree 0. The term 0 is said to have no degree. The term $4xy^5$ is a sixth-degree term since the sum of the exponents is $1 + 5$ or 6. The term $6x^3y^5$ is an eighth-degree term since $3 + 5 = 8$.

Like terms are terms that have the same variables with the same exponents. For example, $3x$ and $5x$ are like terms, $2x^2$ and $-3x^2$ are like terms, and $3x^2y$ and $-2x^2y$ are like terms. Terms that are not like terms are said to be **unlike terms**. All constants are considered like terms.

To **simplify an expression** means to combine all like terms in the expression. To combine like terms, we can use the distributive property.

Examples of Combining Like Terms

$$5x - 2x = (5 - 2)x = 3x$$
$$3x^2 - 5x^2 = (3 - 5)x^2 = -2x^2$$
$$-7x^2y + 3x^2y = (-7 + 3)x^2y = -4x^2y$$
$$4(x - y) - (x - y) = 4(x - y) - 1(x - y) = (4 - 1)(x - y) = 3(x - y)$$

When simplifying expressions, we can rearrange the terms by using the commutative and associative properties discussed in Chapter 1.

EXAMPLE 1 Simplify. If an expression cannot be simplified, so state.

a) $-2x + 5 + 3x - 7$ b) $7x^2 - 2x^2 + 3x + 4$ c) $2x - 3y + 5x - 6y + 3$

Solution a) $-2x + 5 + 3x - 7 = \underbrace{-2x + 3x}_{x} \underbrace{+ 5 - 7}_{-2}$ *Place like terms together.*

This expression simplifies to $x - 2$.

b) $7x^2 - 2x^2 + 3x + 4 = 5x^2 + 3x + 4$

c) $2x - 3y + 5x - 6y + 3 = 2x + 5x - 3y - 6y + 3$ *Place like terms together.*
$$= 7x - 9y + 3$$

EXAMPLE 2 Simplify $-2(a + 7) - [-3(a - 1) + 5]$.

Solution
$$-2(a + 7) - [-3(a - 1) + 5] = -2(a + 7) - 1[-3(a - 1) + 5]$$
$$= -2a - 14 - 1[-3a + 3 + 5]$$

Distributive property

$$= -2a - 14 - 1[-3a + 8]$$ *Combine like terms.*
$$= -2a - 14 + 3a - 8$$ *Distributive property*
$$= a - 22$$ *Combine like terms.*

NOW TRY EXERCISE 55

3 Solve Linear Equations

An **equation** is a mathematical statement of equality. *An equation must contain an equal sign* and a mathematical expression on each side of the equal sign.

Examples of Equations

$$x + 4 = -7$$
$$2x^2 - 4 = -3x + 5$$

The numbers that make an equation a true statement are called the **solutions** of the equation. The **solution set** of an equation is the set of real numbers that make the equation true.

Equation	Solution	Solution Set
$2x + 3 = 9$	3	$\{3\}$

Two or more equations with the same solution set are called **equivalent equations**. Equations are generally solved by starting with the given equation and producing a series of simpler equivalent equations.

Example of Equivalent Equations

Equations	Solution Set
$2x + 3 = 9$	$\{3\}$
$2x = 6$	$\{3\}$
$x = 3$	$\{3\}$

In this section, we will discuss how to solve **linear equations in one variable**. A linear equation is an equation that can be written in the form $ax + b = c, a \neq 0$.

To solve equations, we use the addition and multiplication properties of equality to isolate the variable on one side of the equal sign.

Addition Property of Equality

If $a = b$, then $a + c = b + c$ for any $a, b,$ and c.

The addition property of equality states that the same number can be added to both sides of an equation without changing the solution to the original equation. Since subtraction is defined in terms of addition, *the addition property of equality also allows us to subtract the same number from both sides of an equation.*

Multiplication Property of Equality

If $a = b$, then $a \cdot c = b \cdot c$ for any $a, b,$ and c.

The multiplication property of equality states that both sides of an equation can be multiplied by the same number without changing the solution. Since division is defined in terms of multiplication, *the multiplication property of equality also allows us to divide both sides of an equation by the same nonzero number.*

To solve an equation, we will often have to use a combination of properties to isolate the variable. Our goal is to get the variable all by itself on one side of the equation (to isolate the variable). A general procedure to solve linear equations follows.

To Solve Linear Equations

1. **Clear fractions.** If the equation contains fractions, eliminate the fractions by multiplying both sides of the equation by the least common denominator.

2. **Simplify each side separately.** Simplify each side of the equation as much as possible. Use the distributive property to clear parentheses and combine like terms as needed.

3. **Isolate the variable term on one side.** Use the addition property to get all terms with the variable on one side of the equation and all constant terms on the other side. It may be necessary to use the addition property a number of times to accomplish this.

4. **Solve for the variable.** Use the multiplication property to get an equation having just the variable (with a coefficient of 1) on one side.

5. **Check.** Check by substituting the value obtained in step 4 back into the original equation.

EXAMPLE 3   Solve the equation $2x + 4 = 9$.

Solution

$$2x + 4 = 9$$

$$2x + 4 \; \boxed{- \; 4} = 9 \; \boxed{- \; 4} \qquad \textit{Subtract 4 from both sides.}$$

$$2x = 5$$

$$\frac{\overset{1}{\cancel{2}}x}{\underset{1}{\cancel{2}}} = \frac{5}{2} \qquad \textit{Divide both sides by 2.}$$

$$x = \frac{5}{2}$$

Check

$$2x + 4 = 9$$

$$2\left(\frac{5}{\cancel{2}}\right) + 4 \overset{?}{=} 9$$

$$5 + 4 \overset{?}{=} 9$$

$$9 = 9 \qquad \textit{True}$$

Since the value checks, the solution is $\frac{5}{2}$.   ✴

Whenever an equation contains like terms on the same side of the equal sign, combine the like terms before using the addition or multiplication properties.

EXAMPLE 4   Solve the equation $-2b + 5 = 3b - 10$.

Solution

$$-2b + 5 = 3b - 10$$

$$-2b \; \boxed{+ \; 2b} + 5 = 3b \; \boxed{+ \; 2b} - 10 \qquad \textit{Add 2b to both sides.}$$

$$5 = 5b - 10$$

$$5 \; \boxed{+ \; 10} = 5b - 10 \; \boxed{+ \; 10} \qquad \textit{Add 10 to both sides.}$$

$$15 = 5b$$

$$\frac{15}{5} = \frac{5b}{5} \qquad \textit{Divide both sides by 5.}$$

NOW TRY EXERCISE 63

$$3 = b \qquad\qquad ✴$$

Example 5 contains decimal numbers. We work this problem following the procedure given earlier.

EXAMPLE 5   Solve the equation $4(x - 3.1) = 2.1(x - 4) + 3.5x$.

Solution

$$4(x - 3.1) = 2.1(x - 4) + 3.5x$$

$$4(x) - 4(3.1) = 2.1(x) - 2.1(4) + 3.5x \qquad \textit{Distributive property}$$

$$4x - 12.4 = 2.1x - 8.4 + 3.5x$$

$$4x - 12.4 = 5.6x - 8.4 \qquad \textit{Combine like terms.}$$

$$4x - 12.4 \; \boxed{+ \; 8.4} = 5.6x - 8.4 \; \boxed{+ \; 8.4} \qquad \textit{Add 8.4 to both sides.}$$

$$4x - 4.0 = 5.6x$$

$$4x - 4x - 4.0 = 5.6x - 4x$$

Subtract 4x from both sides.

$$-4.0 = 1.6x$$

$$\frac{-4.0}{1.6} = \frac{1.6x}{1.6}$$

Divide both sides by 1.6.

$$-2.5 = x$$

The solution is -2.5.

To save space, we will not always show the check of our answers. You should, however, check all your answers. When the equation contains decimal numbers, using a calculator to solve and check the equation may save you some time.

Using Your Calculator

Checking Solutions by Substitution

Solutions to equations may be checked using a calculator. To check, substitute your solution into both sides of the equation to see whether you get the same value (there may sometimes be a slight difference in the last digits). The graphing calculator screen in Figure 2.1 shows that both sides of the equation given in Example 5 equal -22.4 when -2.5 is substituted for x. Thus the solution -2.5 checks.

$$4(x - 3.1) = 2.1(x - 4) + 3.5x$$
$$4(-2.5 - 3.1) = 2.1(-2.5 - 4) + 3.5(-2.5)$$

← Value of the left side of the equation

← Value of the right side of the equation

FIGURE 2.1

Exercises

Use your calculator to determine whether the given number is a solution to the equation.

1. $5.2(x - 3.1) = 2.3(x - 5.2)$; 1.4
2. $-2.3(4 - x) = 3.5(x - 6.1)$; 10.125

How we will work an example that contains nested parentheses.

EXAMPLE 6 Solve the equation $7c - 15 = -2[6(c - 3) - 4(2 - c)]$.

Solution

$$7c - 15 = -2[6(c - 3) - 4(2 - c)]$$
$$7c - 15 = -2[6c - 18 - 8 + 4c]$$ *Distributive property*
$$7c - 15 = -2[10c - 26]$$ *Combine like terms.*
$$7c - 15 = -20c + 52$$ *Distributive property*
$$7c + 20c - 15 = -20c + 20c + 52$$ *Add 20c to both sides.*
$$27c - 15 = 52$$
$$27c - 15 + 15 = 52 + 15$$ *Add 15 to both sides.*
$$27c = 67$$
$$\frac{27c}{27} = \frac{67}{27}$$ *Divide both sides by 27.*
$$c = \frac{67}{27}$$

NOW TRY EXERCISE 91

Notice that the solutions to Examples 5 and 6 are not integers. You should not expect the solutions to equations always to be integer values.

In solving equations, some of the intermediate steps can often be omitted. Now we will illustrate how this may be done.

<table>
<tr><td></td><td align="center">Solution</td><td></td><td align="center">Shortened Solution</td></tr>
<tr><td>**a)**</td><td align="center">$x + 4 = 6$</td><td></td><td>**a)** $x + 4 = 6$</td></tr>
<tr><td></td><td align="center">$x + 4 \boxed{-4} = 6 \boxed{-4}$ ⟵ Do this step mentally.</td><td></td><td align="center">$x = 2$</td></tr>
<tr><td></td><td align="center">$x = 2$</td><td></td><td></td></tr>
<tr><td>**b)**</td><td align="center">$3x = 6$</td><td></td><td>**b)** $3x = 6$</td></tr>
<tr><td></td><td align="center">$\dfrac{3x}{3} = \dfrac{6}{3}$ ⟵ Do this step mentally.</td><td></td><td align="center">$x = 2$</td></tr>
<tr><td></td><td align="center">$x = 2$</td><td></td><td></td></tr>
</table>

4 Solve Equations Containing Fractions

When an equation contains fractions, we begin by multiplying *both* sides of the equation by the least common denominator. The **least common denominator** (LCD) of a set of denominators (also called the **least common multiple**, LCM) is the smallest number that each of the denominators divides into without remainder. For example, if the denominators of two fractions are 5 and 6, then 30 is the least common denominator since 30 is the smallest number that both 5 and 6 divide into without remainder.

When you multiply both sides of the equation by the LCD, *each term* in the equation will be multiplied by the least common denominator. *After this step is performed, the equation should not contain any fractions.*

EXAMPLE 7 Solve the equation $5 - \dfrac{2x}{3} = -9$.

Solution The least common denominator is 3. Multiply both sides of the equation by 3 and then use the distributive property on the left side of the equation. *This process will eliminate all fractions from the equation.*

$$5 - \frac{2x}{3} = -9$$

$$\boxed{3}\left(5 - \frac{2x}{3}\right) = \boxed{3}(-9) \qquad \textit{Multiply both sides by 3.}$$

$$3(5) - \overset{1}{3}\left(\frac{2x}{\underset{1}{3}}\right) = -27 \qquad \textit{Distributive property}$$

$$15 - 2x = -27$$

$$15 \boxed{-15} - 2x = -27 \boxed{-15} \qquad \textit{Subtract 15 from both sides.}$$

$$-2x = -42$$

$$\frac{-2x}{-2} = \frac{-42}{-2} \qquad \textit{Divide both sides by } -2.$$

$$x = 21$$

EXAMPLE 8   Solve the equation $\frac{1}{2}(x + 4) = \frac{1}{3}x$.

Solution   Begin by multiplying both sides of the equation by 6, the LCD of 2 and 3.

$$6\left[\frac{1}{2}(x + 4)\right] = 6\left(\frac{1}{3}x\right) \qquad \text{\textit{Multiply both sides by 6.}}$$
$$3(x + 4) = 2x \qquad \text{\textit{Simplify.}}$$
$$3x + 12 = 2x \qquad \text{\textit{Distributive property}}$$
$$3x \;-\; 2x + 12 = 2x \;-\; 2x \qquad \text{\textit{Subtract 2x from both sides.}}$$
$$x + 12 = 0$$
$$x + 12 \;-\; 12 = 0 \;-\; 12 \qquad \text{\textit{Subtract 12 from both sides.}}$$
$$x = -12$$

We will be discussing equations containing fractions further in Section 6.4.

HELPFUL HINT

The equation in Example 8 may also be written as $\dfrac{x + 4}{2} = \dfrac{x}{3}$.
Can you explain why?

Using Your Graphing Calculator

Equations in one variable may be solved graphically using a graphing calculator. In Section 3.3 we discuss how this is done. You may wish to review that material now.

---

5   Identify Conditional Equations, Contradictions, and Identities

All equations discussed so far have been true for only one value of the variable. Such equations are called **conditional equations**. Some equations are never true and have no solution; these are called **contradictions**. Other equations, called **identities** have an infinite number of solutions. Table 2.1 summarizes these types of linear equations and their corresponding number of solutions.

TABLE 2.1	
Type of linear equation	**Solution**
Conditional equation	One
Contradiction	None (solution set: $\varnothing$)
Identity	Infinite number (solution set: $\mathbb{R}$)

The solution set of a conditional equation contains the solution given in set braces. For example, the solution set to Example 8 is $\{-12\}$. The solution set of a contradiction is the empty set or null set, $\{\ \}$ or $\varnothing$. The solution set of an identity is the set of real numbers, $\mathbb{R}$.

EXAMPLE 9   Determine whether the equation $5(d - 7) + 4d + 3 = 3(3d - 10) - 2$ is a conditional equation, a contradiction, or an identity. Give the solution set for the equation.

Solution

$$5(d - 7) + 4d + 3 = 3(3d - 10) - 2$$
$$5d - 35 + 4d + 3 = 9d - 30 - 2 \qquad \textit{Distributive property}$$
$$9d - 32 = 9d - 32 \qquad \textit{Combine like terms.}$$

NOW TRY EXERCISE 121

Since we obtain the same expression on both sides of the equation, it is an identity. This equation is true for all real numbers. Its solution set is $\mathbb{R}$. ❋

EXAMPLE 10 Determine whether $2(3x + 1) = 6x + 3$ is a conditional equation, a contradiction, or an identity. Give the solution set for the equation.

Solution

$$2(3x + 1) = 6x + 3$$
$$6x + 2 = 6x + 3 \qquad \textit{Distributive property}$$
$$6x - 6x + 2 = 6x - 6x + 3 \qquad \textit{Subtract 6x from both sides.}$$
$$2 = 3$$

NOW TRY EXERCISE 115

Since $2 = 3$ is never a true statement, this equation is a contradiction. Its solution set is $\varnothing$. ❋

6 Understand the Concepts to Solve Equations

The numbers or variables that appear in equations do not affect the procedures used to solve the equations. In the following example, which does not contain any numbers or letters, we will solve the equation using the concepts and procedures that have been presented.

EXAMPLE 11 Assume in the following equation that $\odot$ represents the variable that we are solving for and that all the other symbols represent nonzero real numbers. Solve the equation for $\odot$.

$$\square \odot + \triangle = \#$$

Solution To solve for $\odot$ we need to isolate the $\odot$. We use the addition and multiplication properties to solve for $\odot$.

$$\square \odot + \triangle = \#$$
$$\square \odot + \triangle - \triangle = \# - \triangle \qquad \textit{Subtract } \triangle \textit{ from both sides.}$$
$$\square \odot = \# - \triangle$$
$$\frac{\square \odot}{\square} = \frac{\# - \triangle}{\square} \qquad \textit{Divide both sides by } \square.$$
$$\odot = \frac{\# - \triangle}{\square}$$

Thus the solution is $\odot = \dfrac{\# - \triangle}{\square}$. ❋

Consider the equation $5x + 7 = 12$. If we let $5 = \square$, $x = \odot$, $7 = \triangle$, and $12 = \#$, the equation has the same form as the equation in Example 11. Therefore, the solution will be of the same form.

Equation	Solution

$$\Box \odot + \triangle = \#$$

$$\odot = \frac{\# - \triangle}{\Box}$$

$$5x + 7 = 12$$

$$x = \frac{12 - 7}{5} = \frac{5}{5} = 1$$

If you solve the equation $5x + 7 = 12$, you will see that its solution is 1. Thus the procedure used to solve an equation is not dependent on the numbers or variables given in the equation.

NOW TRY EXERCISE 139

Exercise Set 2.1

Concept/Writing Exercises

1. What are the terms of an expression?
2. Determine the coefficient of each term.
   a) $x^2 y^5$ b) $-a^3 b^7$ c) $-\frac{a - 7b}{5}$
3. Determine the coefficient of each term.
   a) $\frac{x + y}{4}$ b) $-(x + 3)$ c) $-\frac{3(x + 2)}{5}$
4. How do you find the degree of a term?
5. a) What are like terms?
   b) Are $3x$ and $3x^2$ like terms? Explain.
6. What is an equation?
7. Is 4 the solution to the equation $2x + 3 = x + 5$? Explain.
8. Is $\{8\}$ the solution set to the equation $x + 1 = 2x - 7$? Explain.

9. State the addition property of equality.
10. State the multiplication property of equality.
11. a) How many solutions does an identity have?
    b) If a linear equation is an identity, what is its solution set?
12. a) What is a contradiction?
    b) What is the solution set of a contradiction?
13. a) In your own words, explain step-by-step how you would solve the equation
    $$5x - 2(x - 4) = 2(x - 2)$$
    b) Solve this equation.
14. a) In your own words, explain step-by-step how you would solve the equation
    $$\frac{1}{6} = \frac{2}{3}x - \frac{1}{8}$$
    b) Solve this equation.

Practice the Skills

Name each indicated property.

15. If $x = 13$, then $13 = x$.
16. If $x + 2 = 3$, then $3 = x + 2$.
17. If $b = c$ and $c = 7$, then $b = 7$.
18. If $x + 1 = a$ and $a = 2y$, then $x + 1 = 2y$.
19. $a + c = a + c$
20. If $x = 4$, then $x + 3 = 4 + 3$.
21. If $x = 8$, then $x - 8 = 8 - 8$.
22. If $2x = 4$, then $3(2x) = 3(4)$.
23. If $5x = 4$, then $\frac{1}{5}(5x) = \frac{1}{5}(4)$.
24. If $x + 2 = 4$, then $x + 2 - 2 = 4 - 2$.
25. If $\frac{t}{4} + \frac{1}{3} = \frac{5}{6}$, then $12\left(\frac{t}{4} + \frac{1}{3}\right) = 12\left(\frac{5}{6}\right)$.
26. If $x - 3 = x + y$ and $x + y = z$, then $x - 3 = z$.

Give the degree of each term.

27. $5c^3$
28. $-6x^2$
29. $3xy$
30. $\frac{1}{2}x^4 y$

31. 6	**32.** -3	**33.** $-5x$	**34.** $18x^2y^3$
35. $5a^2b^4c$	**36.** x^4y^6	**37.** $3x^5y^6z$	**38.** $-2x^4y^7z^8$

Simplify each expression. If an expression cannot be simplified, so state.

39. $7r + 3b - 11x + 12y$

40. $3x^2 + 4x + 5$

41. $5x^2 - 3x + 2x - 5$

42. $11a - 12b - 4c + 5a$

43. $10.6c^2 - 2.3c + 5.9c - 1.9c^2$

44. $7y + 3x - 7 + 4x - 2y$

45. $w^3 + w^2 - w + 1$

46. $b + b^2 - 4b + b^2 + 3b$

47. $6pq - 7pq + p + q$

48. $7x^3y^2 + 11y^3x^2$

49. $12\left(\dfrac{1}{6} + \dfrac{d}{4}\right) - d$

50. $4.3 - 3.2x - 2(x - 2)$

51. $3\left(x + \dfrac{1}{2}\right) - \dfrac{1}{3}x + 5$

52. $6n + 0.6(n - 3) - 5(n + 0.7)$

53. $4 - [6(3x + 2) - x] + 4$

54. $3(x + y) - 4(x + y) - 3$

55. $4x - [3x - (5x - 4y)] + y$

56. $-2[3x - (2y - 1) - 5x] + y$

57. $5b - \{7[2(3b - 2) - (4b + 9)] - 2\}$

58. $2\{[3a - (2b - 5a)] - 3(2a - b)\}$

59. $-\{[2rs - 3(r + 2s)] - 2(2r^2 - s)\}$

60. $p^2q + 4pq - [-(pq + 4p^2q) + pq]$

Solve each equation.

61. $5a - 1 = 14$

62. $2x + 3 + x = 9$

63. $5x - 9 = 3(x - 2)$

64. $5s - 3 = 2s + 6$

65. $4x - 8 = -4(2x - 3) + 4$

66. $8w + 7 = -3w - 4$

67. $-6(z - 1) = -5(z + 2)$

68. $7(x - 1) = 4(x + 2)$

69. $-3(t - 5) = 2(t - 5)$

70. $4(2x - 4) = -2(x + 3)$

71. $3x + 4(x - 2) = 4x - 5$

72. $6(q - 3) = -4(q + 2)$

73. $2 - (x + 5) = 4x - 8$

74. $4x - 2(3x - 7) = 2x - 6$

75. $p - (p + 4) = 4(p - 1) + 2p$

76. $8x + 2(x - 4) = 8x + 10$

77. $-3(y - 1) + 2y = 4(y - 3)$

78. $5r - 13 - 6r = 3(r + 5) - 16$

79. $6 - (n + 3) = 3n + 5 - 2n$

80. $8 - 3(2x - 4) = 5 + 3x - 4x$

81. $4(2x - 2) - 3(x + 7) = -4$

82. $-2(3w + 6) - (4w - 3) = 21$

83. $-4(3 - 4x) - 2(x - 1) = 12x$

84. $-4(2z - 6) = -3(z - 4) + z$

85. $5(a + 3) - a = -(4a - 6) + 1$

86. $3(2x - 4) + 3(x + 1) = 9$

87. $5(x - 2) - 14x = x - 5$

88. $3[6 - (h + 2)] - 6 = 4(-h + 7)$

89. $2[3x - (4x - 6)] = 5(x - 6)$

90. $-z - 6z + 3 = 4 - [6 - z - (3 - 2z)]$

91. $4\{2 - [3(c + 1) - 2(c + 1)]\} = -2c$

92. $3\{[(x - 2) + 4x] - (x - 3)\} = 4 - (x - 12)$

93. $-\{4(d + 3) - 5[3d - 2(2d + 7)] - 8\} = -10d - 6$

94. $-3(6 - 4x) = 4 - \{5x - [6x - (4x - (3x + 2))]\}$

Solve each equation. Leave your answer as a fraction if it is not an integer value.

95. $\dfrac{s}{4} = -16$

96. $\dfrac{5x + 3}{4} = 2$

97. $\dfrac{4x - 2}{3} = -6$

98. $\dfrac{1}{2}(6x - 10) = 7$

99. $\dfrac{3}{4}t + \dfrac{7}{8}t = 39$

100. $\dfrac{1}{4}(x - 2) = \dfrac{1}{3}(2x + 6)$

101. $4 - \dfrac{3}{4}x = 7$

102. $x - 2 = \dfrac{3}{4}(x + 4)$

103. $\dfrac{1}{2} = \dfrac{4}{5}x - \dfrac{1}{4}$

104. $\dfrac{5}{6}m - \dfrac{5}{12} = \dfrac{7}{8}m + \dfrac{2}{3}$

Solve each equation. Round answers to the nearest hundredth.

105. $0.4n + 4.7 = 5.1n$

106. $0.2(x - 30) = 1.6x$

107. $4.7x - 3.6(x - 1) = 4.9$

108. $6.1p - 4.5(3 - 2p) = 15.7$

109. $5(z + 3.41) = -7.89(2z - 4) - 5.67$

110. $0.05(2000 + 2x) = 0.04(2500 - 6x)$

111. $0.6(500 - 2.4x) = 3.6(2x - 4000)$

112. $0.42x - x = 5.1(x + 3)$

113. $1000(7.34q + 14.78) = 100(3.91 - 4.21q)$

114. $0.6(14x - 8000) = -0.4(20x + 12{,}000) + 20.6x$

Find the solution set for each exercise. Then indicate whether the equation is conditional, an identity, or a contradiction.

115. $3(y + 3) - 4(2y - 7) = -5y + 2$

116. $4x + 12 - 8x = -6(x - 2) + 2x$

117. $4(2x - 3) + 5 = -6(x - 4) + 12x - 31$

118. $-5(c + 3) + 4(c - 2) = 2(c + 2)$

119. $-(-b + 7) - 6(b + 3) = -5(b + 5)$

120. $-[4 - (x - 2)] = 2x - 2 - x$

121. $6(x - 1) = -3(2 - x) + 3x$

122. $6(z + 5) - 5(z + 2) = -3(z + 1) + 4(z - 5)$

123. $-5(d - 4) + 3d - 5 = 3(d + 1 - 2d) + d$

124. $4(2 - 3x) = -[6x - (8 - 6x)]$

Problem Solving

125. *Sleeping Babies* Dr. Richard Ferber, a pediatric sleep expert, has developed a method* to help children, 6 months of age or older, sleep through the night. Often called "Ferberizing," it calls for parents to wait for increasing lengths of time before entering the child's room at night to comfort the crying child. The suggested waiting time depends on how many nights the parents have been using the method and may be found using the equation

$$W = 5n + 5$$

where W is the waiting time in minutes and n is the number of the night. For example, on the first night, $n = 1$, on the second night, $n = 2$, and so on.

a) How long should parents wait on the first night?

b) How long should parents wait on the fourth night?

c) On what night should parents wait 30 minutes?

d) On what night should parents wait 40 minutes?

126. *Population Density* The population density of the United States has been steadily increasing since 1990. The population density of the United States can be estimated using the equation

$$P = 0.81t + 70.4$$

where P is the population density, measured in people per square mile, and t is the number of years since 1990.

Use $t = 1$ for 1991, $t = 2$ for 1992, and so on. If the population density continues to increase at its current rate,

a) determine the population density of the United States in 2005.

b) during what year will the population density of the United States reach 100 people per square mile?

127. *American Automakers' Market Share* Since 1993 American automakers have been losing market share to Asian and European automakers. The percent of the total cars sold in the United States made by American automakers can be estimated using the equation

$$M = -1.26x + 75.34$$

where M is the percent of the total cars sold in the United States made by American automakers and x is the number of years since 1993. Use $x = 1$ for 1994, $x = 2$ for 1995, and so on.

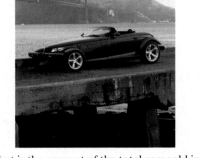

a) What is the percent of the total cars sold in the United States made by American automakers in 2003?

b) During what year will the percent of total cars sold in the United States made by American automakers be 58.96%?

*Before trying this method, parents should first consult with their pediatrician.

128. *Annuities* Annuities are life insurance contracts that guarantee future payments. One type of annuity, called a variable annuity, is a retirement account that lets someone invest in mutual funds and defer payment of taxes until withdrawals are taken at a later time. The number of variable annuities sold has been growing steadily over the past few years. Variable annuity sales can be approximated by the equation

$$S = 10x + 20$$

where S represents total sales of variable annuities (in billions of dollars) and x is the number of years since 1991. Use $x = 1$ for 1992, $x = 2$ for 1993, etc.

a) Find the total sales of variable annuities in 2001.

b) In what year will annuity sales reach the 140-billion-dollar mark?

129. *Population of Jamestown* The population of the Jamestown township in Wisconsin has been slowly growing since 1996. The population can be estimated using the equation

$$P = 7x + 2170$$

where P is the population of the Jamestown township and x is the number of years since 1996. Use $x = 1$ for 1997, $x = 2$ for 1998, and so on.

a) What was the population in 2003?

b) If the growth continues at the same rate, during what year will the population reach 2240?

130. Consider the equation $x = 4$. Give three equivalent equations. Explain why the equations are equivalent.

131. Consider the equation $2x = 5$. Give three equivalent equations. Explain why the equations are equivalent.

132. Make up an equation that is an identity. Explain how you created the equation.

133. Make up an equation that is a contradiction. Explain how you created the contradiction.

134. Create an equation with three terms to the left of the equal sign and two terms to the right of the equal sign that is equivalent to the equation $3x + 1 = x + 5$.

135. Create an equation with two terms to the left of the equal sign and three terms to the right of the equal sign that is equivalent to the equation $\frac{1}{2}x + 3 = 6$.

136. Consider the equation $-3(x + 2) + 5x + 12 = n$. What real number must n be for the solution of the equation to be 6? Explain how you determined your answer.

137. Consider the equation $2(x + 5) + n = 4x - 8$. What real number must n be for the solution of the equation to be -2? Explain how you determined your answer.

138. Consider the equation $\frac{n}{6} + \frac{x}{4} = 2$. What real number must n be for the solution of the equation to be $x = 2$? Explain how you determined your answer.

Solve each equation for the given symbol. Assume that the symbol you are solving for represents the variable and that all other symbols represent nonzero real numbers. See Example 11.

139. Solve $\ast\triangle - \square = \odot$ for $\triangle$.

140. Solve $\triangle(\odot + \square) = \otimes$ for $\triangle$.

141. Solve $\odot\square + \triangle = \otimes$ for $\odot$.

142. Solve $\triangle(\odot + \square) = \otimes$ for $\square$.

Cumulative Review Exercises

[1.3] 143. a) In your own words, explain how to find the absolute value of a number.

b) Write the definition of absolute value.

[1.4] *Evaluate.*

144. a) -3^2 **b)** $(-3)^2$

145. $\sqrt[3]{-64}$

146. $\left(-\frac{3}{4}\right)^3$

1 Use the problem-solving procedure.

2 Solve for a variable in an equation or formula.

SSM Study Guide CD/Video

MathPro 4/5 PH Math Tutor Center prenhall.com/Angel

1   Use the Problem-Solving Procedure

George Polya

One of the main reasons for studying mathematics is that we can use it to solve everyday problems. To solve most real-life application problems mathematically, we need to be able to express the problem in mathematical symbols using expressions or equations, and when we do this, we are creating a **mathematical model** of the situation.

In this section, we present a problem-solving procedure and discuss formulas. A **formula** is an equation that is a mathematical model of a real-life situation. Throughout the book we will be problem solving. When we do so, we will determine an equation or formula that represents or models the real-life situation.

We will now give the general five-step problem-solving procedure developed by George Polya and presented in his book *How to Solve It*. You can approach any problem by following this general procedure.

Guidelines for Problem Solving

1. **Understand the problem.**
   - Read the problem **carefully** at least twice. In the first reading, get a general overview of the problem. In the second reading, determine a) exactly what you are being asked to find and b) what information the problem provides.
   - If possible, make a sketch to illustrate the problem. Label the information given.
   - List the information in a table if it will help in solving the problem.

2. **Translate the problem into mathematical language.**
   - This will generally involve expressing the problem algebraically.
   - Sometimes this involves selecting a particular formula to use, whereas other times it is a matter of generating your own equation. It may be necessary to check other sources for the appropriate formula to use.

3. **Carry out the mathematical calculations necessary to solve the problem.**

4. **Check the answer obtained in step 3.**
   - Ask yourself: "Does the answer make sense?" "Is the answer reasonable?" If the answer is not reasonable, recheck your method for solving the problem and your calculations.
   - Check the solution in the original problem if possible.

5. **Answer the question.** Make sure you have answered the question asked. State the answer clearly.

The following examples show how to apply the guidelines for problem solving. We will sometimes provide the steps in the examples to illustrate the five-step process. However, in some problems it may not be possible or necessary to list every step.

As was stated in step 2 of the problem-solving process—*translate the problem into mathematical language*—we will sometimes need to find and use a *formula*. We will show how to do that in this section. In Section 2.3 we will explain how to develop *equations* to solve real-life application problems.

EXAMPLE 1   **A Personal Loan**   Sharon Avery makes a $5000, 6% simple interest personal loan to her brother, Sam Runde, for a period of 3 years.

a) At the end of 3 years, what interest will Sam pay Sharon?

b) When Sam settles his loan at the end of 3 years, how much money, in total, must he pay Sharon?

Solution   **a)** Understand   When a person borrows money using a simple interest loan, the person must repay the simple interest and principal (the original amount borrowed) at the maturity date of the loan. For example, if a simple interest loan is for 3 years, then 3 years after the loan is made, the principal plus the interest must be repaid. We are told in the problem that the simple interest rate is 6% and the loan is for 3 years.

Translate   Many business mathematics and investment books include the **simple interest formula**:

$$\text{interest} = \text{principal} \cdot \text{rate} \cdot \text{time} \quad \text{or} \quad i = prt$$

which can be used to find the simple interest, i. In the formula, p is the principal, r is the simple interest rate (always changed to decimal form when used in the formula), and t is time. The time and rate must be in the same units. For example, if the rate is 6% per *year*, then the time must be in *years*. In this problem, $p = \$5000$, $r = 0.06$, and $t = 3$. We obtain the simple interest, i, by substituting these values in the simple interest formula.

$$i = prt$$

Carry Out
$$= 5000(0.06)(3)$$
$$= 900$$

Check   The answer appears reasonable in that Sam will pay $900 for the use of $5000 of Sharon's money for 3 years.

Answer   **a)** The simple interest owed is $900.

b) At the end of 3 years Sam must pay the principal he borrowed, $5000, plus the interest determined in part **a)**, $900. (The principal plus interest owed is called the amount owed, A.) Thus, when Sam settles his loan, he must pay Sharon $5900.

NOW TRY EXERCISE 67 $5900.   ✳

EXAMPLE 2   **A Money Market Account**   Catherine Cant receives a tax refund of $1425 and invests this money to help pay for her son's first semester of college. She invests this money in a certificate of deposit at a 3% annual interest rate compounded monthly for 18 months.

a) How much money will the certificate of deposit be worth in 18 months?

b) How much interest will she earn during the 18 months?

Solution   **a)** Understand   Before you can understand the problem, you must understand what compound interest is. Compound interest means that you get interest on your investment for one period. Then in the next period you get interest paid on your investment, plus interest paid on the interest that was paid in the first period. This process then continues for each period. In many real-life situations, and in the workplace, you may need to do some research to answer questions asked.

     We are given that $1425 is invested for 18 months and the interest rate is 3% compounded monthly.

Translate   If you look in a business mathematics book or speak to a person involved with finance, you will learn of the compound interest formula:

$$A = p\left(1 + \frac{r}{n}\right)^{nt}$$

The compound interest formula is used by financial institutions to compute the accumulated amount (or the balance), A, in a savings account or other investment that earns compound interest. In the formula, p represents the principal (or the initial investment), r represents the interest rate written in decimal form, n represents the number of times per year the interest is compounded, and t represents time measured in years. In this problem, $p = \$1425$, $r = 0.03$, $t = 1.5$ (18 months is 1.5 years), and since the interest is compounded monthly, $n = 12$. Substitute these values into the formula and evaluate.

$$A = p\left(1 + \frac{r}{n}\right)^{nt}$$

$$= 1425\left(1 + \frac{0.03}{12}\right)^{12(1.5)}$$

Carry Out $= 1425(1 + 0.0025)^{18}$

$= 1425(1.0025)^{18}$

$= 1425(1.04596912)$ *From a calculator*

$= 1490.51$ *Rounded to the nearest cent*

Check   The answer $1490.51 is reasonable, since it is more than Catherine originally invested.

Answer   Catherine's certificate of deposit will be worth $1490.51 at the end of 18 months.

b) Understand   The interest will be the difference between the original amount invested and the value of the certificate of deposit at the end of 18 months.

Translate   interest $= \left(\begin{array}{c}\text{value of the certificate} \\ \text{of deposit after 18 months}\end{array}\right) - \left(\begin{array}{c}\text{amount originally} \\ \text{invested}\end{array}\right)$

Carry Out $= 1490.51 - 1425 = 65.51$

Check   The amount of interest is reasonable and the arithmetic is easily checked.

NOW TRY EXERCISE 77   Answer   The interest gained in the 18-month period will be $65.51.   ✳

Often a formula contains subscripts. **Subscripts** are numbers (or other variables) placed below and to the right of variables. They are used to help clarify a formula. For example, if a formula contains two velocities, the original velocity and the final velocity, these velocities might be symbolized as V_0 and V_f, respectively. Subscripts are read using the word "sub." For example, V_f is read "V sub f" and x_2 is read "x sub 2." The formula in Example 3 contains subscripts.

EXAMPLE 3   **Comparing Investments**   Sharon Griggs is in the 27% federal income tax bracket. She is trying to decide whether to invest in tax-free municipal bonds with a rate of 2.24% or in a taxable certificate of deposit with a rate of 3.70%.

a) Determine the taxable rate equivalent to a 2.24% tax-free rate for Sharon.

b) If both investments were for the same period of time, which investment would provide Sharon with the greater return on her investment?

Solution   **a)** Understand   Some interest we receive, such as from municipal bonds, is tax free. This means that we do not have to pay federal taxes on the interest received. Other interest we receive, such as from savings accounts or certificates of deposit, is taxable on our federal income tax returns. Paying taxes on the interest has the effect of reducing the amount of money we actually get to keep from the interest. We need to find the taxable interest rate that is equivalent to a 2.24% tax-free rate for Sharon (or for anyone in a 27% federal income tax bracket).

Translate   A formula found in many investment books and some government publications that may be used to compare taxable and tax-free interest rates is

$$T_f = T_a(1 - F)$$

where T_f is the tax-free rate, T_a is the taxable rate, and F is the federal income tax bracket. To determine the equivalent taxable rate, T_a, we substitute the appropriate values in the formula and solve for T_a.

$$T_f = T_a(1 - F)$$
$$0.0224 = T_a(1 - 0.27)$$

Carry Out
$$0.0224 = T_a(0.73)$$
$$\frac{0.0224}{0.73} = T_a$$
$$0.0307 \approx T_a \qquad \text{\textit{Rounded to four places}}$$

Check   The answer appears reasonable because it is larger than 2.24%, which is what we expected.

Answer   The equivalent taxable rate for Sharon is about 3.07%. This means that after taxes, a taxable investment yielding about 3.07% would give Sharon about the same interest as a 2.24% tax-free investment.

b) We are asked to determine which investment will provide Sharon with the greater return on her investment. We can compare the equivalent taxable rate of the municipal bonds to the taxable interest rate of the certificate of deposit. The rate that is higher will provide Sharon with the greater return on her investment.

As we saw in part **a)**, the equivalent taxable rate of the municipal bonds is 3.07%. The taxable rate of the certificate of deposit is 3.70%. Therefore the certificate of deposit paying 3.70% will give Sharon a greater return on her investment than the tax-free municipal bond paying 2.24%.

NOW TRY EXERCISE 83

✳

2 Solve for a Variable in an Equation or Formula

There are many occasions when you might be given an equation or formula that is solved for one variable but you want to solve it for a different variable. Suppose in Example 3 we wanted to find the equivalent taxable rate, T_a, for many different tax-free rates and many different income tax brackets. We could solve each individual problem as was done in Example 3. However, it would be much quicker to solve the formula $T_f = T_a(1 - F)$ for T_a and then substitute the appropriate values in the formula solved for T_a. We will do this in Example 8.

We will begin by solving equations for the variable y. We will need to do this in Chapter 3 when we discuss graphing. Since formulas are equations, the same procedure we use to solve for a variable in an equation will be used to solve for a variable in a formula.

When you are given an equation (or formula) that is solved for one variable and you want to solve it for a different variable, treat each variable in the equation, except the one you are solving for, as if it were a constant. Then *isolate the variable you are solving for* using the procedures similar to those used to solve equations.

EXAMPLE 4   Solve the equation $5x - 8y = 16$ for y.

Solution   We will solve for the variable y by isolating the term containing the y on the left side of the equation.

$$5x - 8y = 16$$

$$5x - 5x - 8y = -5x + 16 \qquad \text{\textit{Subtract 5x from both sides.}}$$

$$-8y = -5x + 16$$

$$\frac{-8y}{-8} = \frac{-5x + 16}{-8} \qquad \text{\textit{Divide both sides by } -8.}$$

$$y = \frac{-5x + 16}{-8}$$

$$y = \frac{-1(-5x + 16)}{-1(-8)} \qquad \text{\textit{Multiply the numerator and the}}$$
$$\text{\textit{denominator by } -1.}$$

NOW TRY EXERCISE 29

$$y = \frac{5x - 16}{8} \quad \text{or} \quad y = \frac{5}{8}x - 2 \qquad ✳$$

EXAMPLE 5   Solve the equation $2y - 3 = \frac{1}{2}(x + 3y)$ for y.

Solution   Since this equation contains a fraction, we begin by multiplying both sides of the equation by the least common denominator, 2. We then isolate the variable y by collecting all terms containing the variable y on one side of the equation and all other terms on the other side of the equation.

$$2y - 3 = \frac{1}{2}(x + 3y)$$

$$2(2y - 3) = 2\left[\frac{1}{2}(x + 3y)\right] \qquad \text{\textit{Multiply both sides by the LCD, 2.}}$$

$$4y - 6 = x + 3y \qquad \text{\textit{Distributive property}}$$

$$4y - 3y - 6 = x + 3y - 3y \qquad \text{\textit{Subtract 3y from both sides.}}$$

$$y - 6 = x$$

$$y - 6 + 6 = x + 6 \qquad \text{\textit{Add 6 to both sides.}}$$

$$y = x + 6 \qquad ✳$$

Now let's solve for a variable in a formula. Remember: Our goal is to isolate the variable for which we are solving. We use the same general procedure we used in Examples 4 and 5.

EXAMPLE 6   The formula for the perimeter of a rectangle is $P = 2l + 2w$, where l is the length and w is the width of the rectangle (see Fig. 2.2). Solve this formula for the width, w.

Solution   Since we are solving for w, we must isolate the w on one side of the equation.

Rectangle

FIGURE 2.2

$$P = 2l + 2w$$
$$P \;\boxed{-\; 2l} = 2l \;\boxed{-\; 2l} + 2w \quad \textit{Subtract 2l from both sides.}$$
$$P - 2l = 2w$$
$$\frac{P - 2l}{2} = \frac{2w}{2} \quad \textit{Divide both sides by 2.}$$
$$\frac{P - 2l}{2} = w$$

Thus, $w = \dfrac{P - 2l}{2}$ or $w = \dfrac{P}{2} - \dfrac{2l}{2} = \dfrac{P}{2} - l.$

EXAMPLE 7 A formula used to find the area of a trapezoid is $A = \frac{1}{2}h(b_1 + b_2)$, where h is the height and b_1 and b_2 are the lengths of the bases of the trapezoid (see Fig. 2.3). Solve this formula for b_2.

Solution We begin by multiplying both sides of the equation by the LCD, 2, to clear fractions.

Trapezoid

FIGURE 2.3

$$A = \frac{1}{2}h(b_1 + b_2)$$
$$\boxed{2} \cdot A = \boxed{2}\left[\frac{1}{2}h(b_1 + b_2)\right] \quad \textit{Multiply both sides by 2.}$$
$$2A = h(b_1 + b_2)$$
$$\frac{2A}{h} = \frac{h(b_1 + b_2)}{h} \quad \textit{Divide both sides by h.}$$
$$\frac{2A}{h} = b_1 + b_2$$
$$\frac{2A}{h} \;\boxed{-\; b_1} = b_1 \;\boxed{-\; b_1} + b_2 \quad \textit{Subtract } b_1 \textit{ from both sides.}$$
$$\frac{2A}{h} - b_1 = b_2$$

NOW TRY EXERCISE 57

EXAMPLE 8 In Example 3 we introduced the formula $T_f = T_a(1 - F)$.
a) Solve this formula for T_a.
b) John and Dorothy Cutter are in the 35% income tax bracket. What is the equivalent taxable yield of a 3% tax-free yield?

Solution **a)** We wish to solve this formula for T_a. Therefore we treat all other variables in the equation as if they were constants. Since T_a is multiplied by $(1 - F)$, to isolate T_a we divide both sides of the equation by $1 - F$.

$$T_f = T_a(1 - F)$$
$$\frac{T_f}{\boxed{1 - F}} = \frac{T_a(1 - F)}{\boxed{1 - F}} \quad \textit{Divide both sides by } 1 - F.$$
$$\frac{T_f}{1 - F} = T_a \quad \text{or} \quad T_a = \frac{T_f}{1 - F}$$

b) Substitute the appropriate values into the formula found in part **a)**.

$$T_a = \frac{T_f}{1 - F}$$
$$T_a = \frac{0.03}{1 - 0.35} = \frac{0.03}{0.65} \approx 0.046$$

NOW TRY EXERCISE 63 Thus, the equivalent taxable yield would be about 4.6%.

Exercise Set 2.2

Concept/Writing Exercises

1. What is a formula?

2. What is a mathematical model?

3. Outline the five-step problem-solving process that we will use to work out problems.

4. When we are solving a formula for a variable, we need to isolate the variable. Explain what this means.

5. Consider the equation $16 = 2l + 2(3)$ and the formula $P = 2l + 2w$.

   a) Solve the equation for l.

   b) Solve the formula for l.

   c) Was the procedure used to solve the formula for l any different from the procedure used to solve the equation for l?

   d) In the formula solved for l in part b) substitute 16 for P and 3 for w, and then find the value of l. How does it compare with your answer in part a)? Explain why this is so.

6. a) What are subscripts?

   b) How is x_0 read?

   c) How is v_f read?

Practice the Skills

Evaluate the following formulas for the values given. Use the π key on your calculator for π when needed. Round answers to the nearest hundredth.

7. $E = IR$ when $I = 1.2$, $R = 100$ (a formula known as *Ohm's Law* used when studying electricity)

8. $C = 2\pi r$ when $r = 12$ (a formula for finding the circumference of a circle)

9. $R = R_1 + R_2$ when $R_1 = 100$, $R_2 = 200$ (a formula used when studying electricity)

10. $A = \dfrac{1}{2}bh$ when $b = 7$, $h = 6$ (a formula for finding the area of a triangle)

11. $A = \pi r^2$ when $r = 8$ (a formula for finding the area of a circle)

12. $P_1 = \dfrac{T_1 P_2}{T_2}$ when $T_1 = 250$, $T_2 = 500$, $P_2 = 300$ (a chemistry formula that relates temperature and pressure of gases)

13. $\bar{x} = \dfrac{x_1 + x_2 + x_3}{3}$ when $x_1 = 40$, $x_2 = 120$, $x_3 = 80$ (a formula for finding the average of three numbers)

14. $A = \dfrac{1}{2}h(b_1 + b_2)$ when $h = 10$, $b_1 = 20$, $b_2 = 30$ (a formula for finding the area of a trapezoid)

15. $A = P + Prt$ when $P = 200$, $r = 0.05$, $t = 2$ (a banking formula that yields the total amount in an account after the interest is added)

16. $E = a_1 p_1 + a_2 p_2$ when $a_1 = 10$, $p_1 = 0.2$, $a_2 = 100$, $p_2 = 0.3$ (a statistics formula for finding the expected value of an event)

17. $m = \dfrac{y_2 - y_1}{x_2 - x_1}$ when $y_2 = 4$, $y_1 = -3$, $x_2 = -2$, $x_1 = -6$ (a formula for finding the slope of a straight line; we will discuss this formula in Chapter 3)

18. $F = G\dfrac{m_1 m_2}{r^2}$ when $G = 0.5$, $m_1 = 100$, $m_2 = 200$, $r = 4$ (a physics formula that gives the force of attraction between two masses that are separated by a distance, r)

19. $R_T = \dfrac{R_1 R_2}{R_1 + R_2}$ when $R_1 = 100$, $R_2 = 200$ (a formula in electronics for finding the total resistance in a parallel circuit containing two resistors)

20. $d = \sqrt{(x_2 - x_1)^2 + (y_2 - y_1)^2}$ when $x_2 = 5$, $x_1 = -3$, $y_2 = -6$, $y_1 = 3$ (a formula for finding the distance between two points on a straight line; we will discuss this formula in Chapter 10)

21. $x = \dfrac{-b + \sqrt{b^2 - 4ac}}{2a}$ when $a = 2$, $b = -5$, $c = -12$ (from the quadratic formula, we will discuss the quadratic formula in Chapter 8)

22. $x = \dfrac{-b - \sqrt{b^2 - 4ac}}{2a}$ when $a = 2$, $b = -5$, $c = -12$ (from the quadratic formula)

23. $A = p\left(1 + \dfrac{r}{n}\right)^{nt}$ when $p = 100$, $r = 0.06$, $n = 1$, $t = 3$ (the compound interest formula; see Example 2)

24. $z = \dfrac{\bar{x} - \mu}{\dfrac{\sigma}{\sqrt{n}}}$ when $\bar{x} = 80$, $\mu = 70$, $\sigma = 15$, $n = 25$ (a statistics formula for finding the standard, or z score, of a sample mean, $\bar{x}$)

Solve each equation for y (see Examples 4 and 5).

25. $3x + y = 5$

26. $7x + 3y = 9$

27. $x - 4y = 13$

28. $-3x + 5y = 25$

29. $6x - 2y = 16$

30. $6x = 7y + 23$

31. $\frac{3}{4}x - y = 1$

32. $\frac{x}{4} - \frac{y}{6} = 1$

33. $3(x - 2) + 3y = 6x$

34. $y - 4 = \frac{2}{3}(x + 6)$

35. $y + 1 = -\frac{4}{3}(x - 9)$

36. $\frac{1}{5}(x + 3y) = \frac{4}{7}(2x - 1)$

Solve each equation for the indicated variable (see Examples 6–8).

37. $d = rt$, for t

38. $C = \pi d$, for d

39. $i = prt$, for t

40. $A = lw$, for l

41. $P = 2l + 2w$, for l

42. $P = 2l + 2w$, for w

43. $V = lwh$, for h

44. $A = \frac{1}{2}bh$, for b

45. $V = \pi r^2 h$, for h

46. $Ax + By = C$, for y

47. $V = \frac{1}{3}lwh$, for l

48. $A = P + Prt$, for r

49. $y = mx + b$, for m

50. $IR + Ir = E$, for R

51. $y - y_1 = m(x - x_1)$, for m

52. $z = \frac{x - \mu}{\sigma}$, for σ

53. $z = \frac{x - \mu}{\sigma}$, for μ

54. $y = \frac{kx}{z}$, for z

55. $P_1 = \frac{T_1 P_2}{T_2}$, for T_2

56. $F = \frac{mv^2}{r}$, for m

57. $A = \frac{1}{2}h(b_1 + b_2)$, for h

58. $A = \frac{x_1 + x_2 + x_3}{n}$, for n

59. $S = \frac{n}{2}(f + l)$ for n

60. $S = \frac{n}{2}(f + l)$, for l

61. $C = \frac{5}{9}(F - 32)$, for F

62. $F = \frac{9}{5}C + 32$, for C

63. $F = \frac{km_1 m_2}{d^2}$, for m_1

64. $A = \frac{1}{2}h(b_1 + b_2)$, for b_1

Problem Solving

In Exercises 65–88, round your answer to two decimal places, when appropriate.

65. *Currency Exchange* a) According to the Universal Converter Internet Web site, on January 23, 2002, $1 of U.S. currency could be exchanged for 9.15 Mexican pesos. Write a formula using d for U.S. dollars and p for Mexican pesos that can be used to convert from dollars to pesos.

b) Write a formula that can be used to convert from Mexican pesos to U.S. dollars.

c) Explain how you determined your answers to parts **a)** and **b)**.

66. *Speed of the Titanic* Ships at sea measure their speed in knots. For example, when the *Titanic* struck the iceberg, its speed was about 20.5 knots. One knot is 1 nautical mile per hour. A nautical mile is about 6076 feet. When measuring speed in miles per hour, a mile is 5280 ft.

a) Determine a formula for converting a speed in knots (k) to a speed in miles per hour (m).

b) Explain how you determined this formula.

c) Determine the speed, in miles per hour, at which the *Titanic* struck the iceberg.

In Exercises 67–70, use the simple interest formula $i = prt$. See Example 1.

67. *A Personal Loan* David Jessee loaned his colleague, Ken Pothoven, $550 at a simple interest rate of 7% per year for 4 years. Determine the simple interest Ken must pay David when he repays the loan at the end of 4 years.

68. *Determine the Rate* Jeff Hilt borrowed $250 from his credit union for 2 years. The simple interest that he paid was $26.45. What simple interest rate was Jeff charged?

69. *Determine the Length of the Loan* Jacque Bryant loaned her daughter, Dawn $20,000 at a simple interest rate of 3.75% per year. At the end of the loan period, Dawn repaid Jacque the original $20,000 plus $4875 interest. Determine the length of the loan.

70. *A Certificate of Deposit* Fred Zerla received $2000 for speaking at a financial planning seminar. Fred invested the money in a certificate of deposit for 2 years. When he redeemed the certificate of deposit, he received $2166. What simple interest rate did Fred receive on this certificate of deposit?

In exercises 71–76, if you are not sure of the formula to use, refer to Appendix A.

71. *Dartboard Area* George Young, dart-throwing champion in the state of Michigan, practices on a dartboard with concentric circles as shown in the figure.

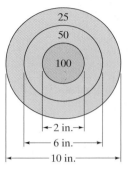

a) Find the area of the circle marked 100.

b) Find the area of the entire dartboard.

72. *Planning a Sandbox* Lisa Coyle is planning to build a rectangular sandbox for her daughter. She has 38 feet of wood to use for the sides. If the length is to be 11 feet, what is the width to be?

73. *Concrete Driveway Volume* Bill Leonard, is laying concrete for a driveway. The driveway is to be 15 feet long by 10 feet wide by 6 inches deep.

a) Find the volume of concrete needed in cubic feet.

b) If 1 cubic yard = 27 cubic feet, how many cubic yards of concrete are needed?

c) If the concrete costs $35 per cubic yard, what is the cost of the concrete? Concrete must be purchased in whole cubic yards.

74. *Helipad Area* A helipad in Raleigh, North Carolina, has two concentric circles as shown in the figure.

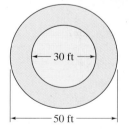

Find the area of the blue region in the figure.

75. *Ice Cream Containers* Gil and Lori's Delicious Ice Cream Company sells ice cream in two containers, a cylindrical tub and a rectangular solid box as shown in the diagram on the top of the next page. Which container holds more ice cream, and what is the difference in their volumes?

76. Bucket Capacity Belen Poltorak has a bucket in which she wishes to mix some detergent. The dimensions of the bucket are shown in the figure.

a) Find the capacity of the bucket in cubic inches.

b) If 231 cubic inches = 1 gallon, what is the capacity of the bucket in gallons?

c) If the instructions on the bottle of detergent say to add 1 ounce per gallon of water, how much detergent should Belen add to a full bucket of water?

For Exercises 77–80, refer to Example 2.

77. A Savings Account Beth Rechsteiner invested $10,000 in a savings account paying 6% interest compounded quarterly. How much money will she have in her account at the end of 2 years?

78. Monthly Compounding Elizabeth Wood invested $8500 in a savings account paying 6.5% interest compounded monthly. How much money will she have in her account at the end of 4 years?

79. A Certificate of Deposit Darrell Abney invests $4390 in a certificate of deposit paying 4.1% interest compounded semiannually. How much will the certificate of deposit be worth after 36 months?

80. Comparing Accounts Nabil Husni has $1500 to invest for 1 year. He has a choice of a credit union account that pays 4.5% annual simple interest and a bank account that pays 4% interest compounded quarterly. Determine which account would pay more interest and by how much.

For Exercises 81–84, refer to Example 3.

81. Equivalent Taxable Rate Donna Jovanovich is a student and is in the 15% federal income tax bracket. She is considering investing $1500 in a tax-free bond mutual fund paying 3.5% simple interest. Determine the taxable rate equivalent to a 3.5% tax-free rate.

82. Comparing Investments Lana Gael is in the 38.6% federal income tax bracket and is considering two investments: a tax-free municipal bond paying 3% simple interest or a taxable certificate of deposit paying 3.5% simple interest. Which investment yields the greatest return?

83. Father and Son Investing Carlos Menendez is in the 38.6% federal income tax bracket and his son, Antonio, is in the 27% federal income tax bracket. They are each considering a tax-free mutual fund currently yielding 4.6% simple interest.

a) Determine the taxable rate equivalent to a 4.6% tax-free rate for Carlos.

b) Determine the taxable rate equivalent to a 4.6% tax-free rate for Antonio.

84. Investment Comparison Marissa Felberty is considering investing $9200 in a 6.75% taxable account or in a $5\frac{1}{2}$% tax-free account. If she is in the 27% tax bracket, which investment will yield the greater return?

Exercises 85–88 contain a variety of situations. Solve each exercise.

85. Weight Loss A nutritionist explains to Jodi Thomas that a person loses weight by burning more calories than he or she eats. For example, Jodi, a 5′6″ woman who weighs 132 pounds, will stay about the same weight with normal exercise if she has a daily diet of 2400 calories. If she burns more than 2400 calories daily, her weight loss can be approximated by the mathematical model $w = 0.02c$, where w is the *weekly* weight loss and c is the number of calories per *day* burned *above* 2400 calories.

a) Find Jodi's weekly weight loss if she exercises and burns off 2600 calories per day.

b) How many calories would Jodi have to burn off in a day to lose 2 pounds in a week?

86. ***Stress Test*** A person given a stress test is generally told that should the heart rate reach a certain point, the test will be stopped. The maximum allowable heart rate depends on the person's age. The maximum allowable heart rate, m, in beats per minute, can be approximated by the equation $m = -0.875x + 190$, where x represents the patient's age from 1 through 99. Using this mathematical model, find

    a) the maximum heart rate for a 50-year-old.

    b) the age of a person whose maximum heart rate is 160 beats per minute.

87. ***Balancing a Portfolio*** Some financial planners recommend the following rule of thumb for investors. The percent of stocks in your total portfolio should be equal to 100 minus your age. The remainder is to be put into bonds or cash.

    a) Construct mathematical models for the percent to be kept in stocks (use S for percent in stock and a for a person's age).

    b) Using this rule of thumb, find the percent to be kept in stocks for a 60-year-old.

88. ***Body Mass Index*** The body mass index is a standard way of evaluating a person's body weight in relation to their height. To determine your body mass index (BMI) using metric measurements, divide your weight, in kilograms, by your height, in meters squared. An accurate shortcut for calculating BMI using pounds and inches, is to multiply your weight in pounds by 705, then divide by the square of your height, in inches.

    a) Create a formula for finding a person's BMI using kilograms and meters.

    b) Create a formula for finding a person's BMI when the weight is given in pounds and the height is given in inches.

    c) Determine your BMI.

Challenge Problem

89. Solve the formula $r = \dfrac{s/t}{t/u}$ for **a)** s, **b)** u.

Cumulative Review Exercises

[1.4] **90.** Evaluate $-\sqrt{3^2 + 4^2} + |3 - 4| - 5^2$.

91. Evaluate $\dfrac{7 + 9 \div (2^3 + 4 \div 4)}{|3 - 7| + \sqrt{5^2 - 3^2}}$.

92. Evaluate $a^3 - 3a^2b + 3ab^2 - b^3$ when $a = -2$, $b = 3$.

[2.1] **93.** Solve the equation $\dfrac{1}{4}t + \dfrac{1}{2} = 1 - \dfrac{1}{8}t$.

2.3 APPLICATIONS OF ALGEBRA

SSM Study Guide CD/Video
MathPro 4/5 PH Math Tutor Center prenhall.com/Angel

1 Translate a verbal statement into an algebraic expression or equation.
2 Use the problem-solving procedure.

1 Translate a Verbal Statement into an Algebraic Expression or Equation

The next few sections will present some of the many uses of algebra in real-life situations. Whenever possible, we include other relevant applications throughout the text.

Perhaps the most difficult part of solving a word problem is translating the problem into an equation. This is step 2 in the problem-solving procedure presented in Section 2.2. Before we represent problems as equations, we give some examples of phrases represented as algebraic expressions.

Phrase	Algebraic Expression
a number increased by 4	$x + 4$
twice a number	$2x$
5 less than a number	$x - 5$
one-eighth of a number	$\frac{1}{8}x$ or $\frac{x}{8}$
2 more than 3 times a number	$3x + 2$
4 less than 6 times a number	$6x - 4$
3 times the sum of a number and 5	$3(x + 5)$

The variable x was used in these algebraic expressions, but any variable could have been used to represent the unknown quantity.

EXAMPLE 1   Express each phrase as an algebraic expression.
a) the radius, r, decreased by 2 centimeters
b) 5 less than twice the distance, d
c) 7 times a number, n, increased by 4

Solution   **a)** $r - 2$ **b)** $2d - 5$ **c)** $7n + 4$

HELPFUL HINT

STUDY TIP

It is important that you prepare for the remainder of the chapter carefully. Make sure to read the book and examples carefully. *Attend class every day, and, most of all, work all the exercises assigned to you.*

As you read through the examples in the rest of the chapter, think about how they can be expanded to other, similar problems. For example, in Example 1 **a)** we stated that the radius, r, decreased by 2 centimeters, can be represented by $r - 2$. You can generalize this to other, similar problems. For example, a weight, w, decreased by 15 pounds, can be represented as $w - 15$.

EXAMPLE 2   Write each phrase as an algebraic expression.
a) the cost of purchasing x shirts at $4 each
b) the distance traveled in t hours at 55 miles per hour
c) the number of cents in n nickels
d) an 8% commission on sales of x dollars

Solution   **a)** We can reason like this: one shirt would cost $1(4)$ dollars; two shirts, $2(4)$ dollars; three shirts, $3(4)$ dollars; four shirts, $4(4)$ dollars, and so on. Continuing this reasoning process, we can see that x shirts would cost $x(4)$ or $4x$ dollars. We can use the same reasoning process to complete each of the other parts.
b) $55t$
c) $5n$
d) $0.08x$ (8% is written as 0.08 in decimal form.)

HELPFUL HINT

When we are asked to find a percent, we are always finding the percent of some quantity. Therefore, when a percent is listed, it is **always** multiplied by a number or a variable. In the following examples we use the variable c, but any letter could be used to represent the variable.

(continued on the next page)

Phrase	How Written
6% of a number	$0.06c$
the cost of an item increased by a 7% tax	$c + 0.07c$
the cost of an item reduced by 25%	$c - 0.25c$

Sometimes in a problem two numbers are related to each other. We often represent one of the numbers as a variable and the other as an expression containing that variable. We generally let the less complicated description be represented by the variable and write the second (more complex expression) in terms of the variable. In the following examples, we use x for the variable.

Phrase	One Number	Second Number
Dawn's age now and Dawn's age in 6 years	x	$x + 6$
one number is 4 times the other	x	$4x$
one number is 5 less than the other	x	$x - 5$
a number and the number increased by 7%	x	$x + 0.07x$
a number and the number decreased by 10%	x	$x - 0.10x$
the sum of two numbers is 10	x	$10 - x$
a 6-foot board cut into two lengths	x	$6 - x$
$10,000 shared by two people	x	$10,000 - x$

The last three examples may not be obvious. Consider "The sum of two numbers is 10." When we add x and $10 - x$, we get $x + (10 - x) = 10$. When a 6-foot board is cut into two lengths, the two lengths will be x and $6 - x$. For example, if one length is 2 feet, the other must be $6 - 2$ or 4 feet.

HELPFUL HINT

Suppose you read the following sentence in an application problem: "A 12-foot rope is cut into two pieces." You probably know you should let x (or some other variable) represent the length of the first piece. What you may not be sure of is whether you should let $x - 12$ or $12 - x$ represent the length of the second piece. To help you decide, it can be helpful to use specific numbers to establish a pattern. In this example, you might use a pattern similar to the one below to help you.

If the First Piece Is...	Then the Second Piece Is...
2 feet	10 feet = 12 feet − 2 feet
5 feet	7 feet = 12 feet − 5 feet

From this pattern you can see that if the first piece is x feet, then the second piece is $12 - x$ feet.

EXAMPLE 3 For each relation, select a variable to represent one quantity and express the second quantity in terms of the first.

a) The speed of the second train is 1.2 times the speed of the first.

b) $90 is shared by David and his brother.

c) It takes Tom 3 hours longer than Roberta to complete the task.

d) Hilda has $4 more than twice the amount of money Hector has.

e) The length of a rectangle is 2 units less than 3 times its width.

Solution   **a)** speed of first train, s; speed of second train, $1.2s$

       b) amount David has, x; amount brother has, $90 - x$

       c) Roberta, t; Tom, $t + 3$

       d) Hector, x; Hilda, $2x + 4$

       e) width, w; length, $3w - 2$

The word *is* in a word problem often means **is equal to** and is represented by an equal sign, $=$.

Verbal Statement	Algebraic Equation
4 less than 3 times a number *is* 5	$3x - 4 = 5$
a number decreased by 4 *is* 3 more than twice the number	$x - 4 = 2x + 3$
the product of two consecutive integers *is* 20	$x(x + 1) = 20$
a number increased by 15% *is* 90	$x + 0.15x = 90$
a number decreased by 12% *is* 38	$x - 0.12x = 38$
the sum of a number and the number increased by 4% *is* 204	$x + (x + 0.04x) = 204$
the cost of renting a VCR for x days at \$15 per day *is* \$120	$15x = 120$

2   Use the Problem-Solving Procedure

There are many types of word problems, and the general problem-solving procedure given in Section 2.2 can be used to solve all types. We now present the five-step problem-solving procedure again so you can easily refer to it. We have included some additional information under step 2, since in this section we are going to emphasize translating word problems into equations.

Problem-Solving Procedure for Solving Application Problems

1. **Understand the problem.** Identify the quantity or quantities you are being asked to find.

2. **Translate the problem into mathematical language** (express the problem as an equation).

     a) Choose a variable to represent one quantity, and **write down exactly what it represents**. Represent any other quantity to be found in terms of this variable.

     b) Using the information from step a), write an equation that represents the word problem.

3. **Carry out the mathematical calculations** (solve the equation).

4. **Check the answer** (using the original wording of the problem).

5. **Answer the question asked.**

Sometimes we will combine steps or not show some steps in the problem-solving procedure due to space limitations. Even if we do not show a check to a problem, you should always check the problem to make sure that your answer is reasonable and makes sense.

EXAMPLE 4

Long Distance Plans Verizon's Best Times telephone long-distance rate plan requires the customer to pay a $4.75 monthly fee and then 7¢ per minute for any long-distance call made. Verizon's e-Values rate plan does not have a monthly fee, but the customer pays 9¢ per minute for any long-distance call made.

Paul Jason is considering changing to one of these plans. Determine the number of minutes he would need to spend on long-distance calls for the two plans to cost the same.

Solution Understand We are given two plans in which one has a monthly fee and the other does not. We are asked to find the *number of minutes* of long-distance calls that would result in both plans having the same total cost. To solve the problem, we will set the cost of the two plans equal to one another and solve for the number of minutes.

Translate let n = number of minutes of long-distance calling

then $0.07 n$ = cost for n minutes at 7¢ per minute

and $0.09 n$ = cost for n minutes at 9¢ per minute

cost of the Best Times plan = cost of the e-Values plan

monthly fee + calling cost = calling cost

$$4.75 + 0.07n = 0.09n$$

Carry Out
$$4.75 + 0.07n = 0.09n$$
$$4.75 = 0.02n$$
$$\frac{4.75}{0.02} = \frac{0.02n}{0.02}$$
$$237.5 = n$$

Check The number of minutes is reasonable and the arithmetic is easily checked.

Answer If about 238 minutes were used per month, both plans would have about the same total cost.

NOW TRY EXERCISE 15

EXAMPLE 5

CDC Spending The Centers for Disease Control and Prevention (CDC) is a federal agency whose task is to protect the health and safety of the American public. In 2002, CDC had a budget of $4.093 billion. This was a 22.5% increase from the 2000 CDC budget, but a 2.6% decrease from the 2001 CDC budget.

a) Determine the 2000 CDC budget.

b) Determine the 2001 CDC budget.

Solution

a) Understand We need to determine the 2000 CDC budget. We will use the facts that the budget increased by 22.5% from 2000 to 2002 and that the 2002 budget was $4.093 billion to solve this problem.

Translate let x = the 2000 CDC budget

then $0.225x$ = the increase in the CDC budget from 2000 to 2002

$$\left(\text{2000 CDC budget}\right) + \left(\begin{array}{c}\text{increase in the budget}\\\text{from 2000 to 2002}\end{array}\right) = \left(\text{2002 CDC budget}\right)$$

$$x \qquad + \qquad 0.225x \qquad = \qquad 4.093$$

Carry Out
$$x + 0.225x = 4.093$$
$$1.225x = 4.093$$
$$x \approx 3.341$$

Check and Answer   The number obtained is less than the budget of 2002, which is what we expected. The 2000 CDC budget was about \$3.341 billion.

b) Understand   We must find the 2001 CDC budget. We are told that the CDC budget decreased by 2.6% from 2001 to 2002. We will translate this information into an equation using an approach similar to that in part **a)**.

Translate
let x = 2001 CDC budget
then $0.026x$ = the decrease in the CDC budget from 2001 to 2002

$$\left(\begin{array}{c} \text{2001 CDC budget} \end{array} \right) - \left(\begin{array}{c} \text{decrease in the budget} \\ \text{from 2001 to 2002} \end{array} \right) = \left(\begin{array}{c} \text{2002 CDC budget} \end{array} \right)$$

$$x \qquad\qquad - \qquad\qquad 0.026x \qquad\qquad = \qquad\qquad 4.093$$

Carry Out
$$x - 0.026x = 4.093$$
$$0.974x = 4.093$$
$$x \approx 4.202$$

Check and Answer   The number obtained is greater than the 2002 CDC budget, which is what we expected. Therefore the 2001 CDC budget was \$4.202 billion.   ✳

Mathematics in Action

Apollo 13

"Houston, we have a problem." These words, uttered by astronaut Jim Lovell (played by Tom Hanks) in the movie *Apollo 13*, mark the beginning of a long string of problems faced by the crew on the damaged moonbound spacecraft. The "problem" Lovell refers to is an explosion that changes the mission from landing on the moon into one of safely returning the astronauts to earth. Using some very unique problem-solving activities, the crew and team at Mission Control in Houston are able to overcome incredibly difficult obstacles to return *Apollo 13* safely to earth. Throughout this movie, you will witness a great deal of mathematical problem solving. One scene has an entire team of engineers calculating and recalculating the solution to an equation using slide rules. Another has the astronauts doing calculations by hand in their oxygen-deprived spacecraft. Although the mathematics being performed is beyond this textbook, the systematic approach to solving this dangerous situation could have been taken from the pages of George Polya's book *How to Solve It* (see page 80). We hope your studies in mathematics can be inspired by another famous line in the movie. When rallying the engineers as they begin the final phase of their revamped mission, NASA flight controller Gene Kranz (played by Ed Harris) bellows, "Failure is not an option!"

EXAMPLE 6   **Shopping in the Tampa Bay Area**   In January 2001, there were 36 major department stores in the Greater Tampa Bay Area. There was the same number of Sears stores as there was Burdines stores. There was one more Dillards than Sears, and there was one more JC Penney than Sears. The number of Montgomery Wards stores was nine less than twice the number of Sears Stores. There were six fewer Saks than Sears. How many of each of these major department stores were there?

Solution   **Understand**   We are asked to determine the number of Sears, Burdines, Dillards, JC Penney, Montgomery Wards, and Saks stores. Notice that the numbers of stores are all given in terms of the number of Sears stores. Therefore, we will let our unknown variable be the number of Sears stores. Then we can represent the number of the other stores in terms of our variable. Also notice that the sum of the number of stores is 36.

Translate   let n = the number of Sears stores

then n = the number of Burdines stores

and $n + 1$ = the number of Dillards stores

and $n + 1$ = the number of JC Penney stores

and $2n - 9$ = the number of Montgomery Wards stores

and $n - 6$ = the number of Saks stores

$$\left(\begin{array}{c}\text{number}\\\text{of Sears}\end{array}\right) + \left(\begin{array}{c}\text{number}\\\text{of Burdines}\end{array}\right) + \left(\begin{array}{c}\text{number}\\\text{of Dillards}\end{array}\right) + \left(\begin{array}{c}\text{number}\\\text{of JC Penney}\end{array}\right) + \left(\begin{array}{c}\text{number of}\\\text{Montgonery Wards}\end{array}\right) + \left(\begin{array}{c}\text{number}\\\text{of Saks}\end{array}\right) = 36$$

$$n \quad + \quad n \quad + \quad (n + 1) \quad + \quad (n + 1) \quad + \quad (2n - 9) \quad + \quad (n - 6) \quad = 36$$

Carry Out
$$n + n + n + 1 + n + 1 + 2n - 9 + n - 6 = 36$$
$$7n - 13 = 36$$
$$7n = 49$$
$$n = 7$$

Check and Answer   There were 7 Sears stores and 7 Burdines stores. There were $n + 1$ or $7 + 1$ or 8 Dillards stores. There were $n + 1$ or $7 + 1$ or 8 JC Penney stores. There were $2n - 9$ or $2(7) - 9$ or 5 Montgomery Wards stores. There were $n - 6$ or $7 - 6$ or 1 Saks store. If we add the numbers of each of the six stores, we get $7 + 7 + 8 + 8 + 5 + 1 = 36$ stores. Thus the answer checks. ✳

NOW TRY EXERCISE 27

EXAMPLE 7   **The Magic Kingdom**   Dave Visser took his family to visit the Magic Kingdom at Walt Disney World. They stayed for one night at the Holiday Inn in Kissimmee. When they made their hotel reservation, they were quoted a rate of $95 per night before tax. When they checked out, their total bill was $110.85, which included the room tax and a $3.50 charge for a candy bar (from the in-room bar). Determine the tax rate for the room.

Solution   **Understand**   Their total bill consists of their room rate, the room tax, and the $3.50 cost for the candy bar. The room tax is determined by multiplying the cost of the room by the tax rate for the room. We are asked to find the tax rate for the room.

Translate   let t = tax rate for the room

then $0.01t$ = room tax rate as a decimal

room cost + room tax + candy bar = total

$$95 \quad + \quad 95(0.01t) + \quad 3.50 \quad = 110.85$$

Carry Out
$$95 + 0.95t + 3.50 = 110.85$$
$$0.95t + 98.50 = 110.85$$
$$0.95t = 12.35$$
$$t = 13$$

NOW TRY EXERCISE 23

Check and Answer If you substitute 13 for t in the equatio... answer checks. The room tax rate is 13%.

EXAMPLE 8 **Home Mortage** Lori Paduchowski is buying her first home ... ing two banks for a $60,000 mortgage. Citicorp is charging a ... with no points for a 30-year loan. (A point is a one-time ch... amount of the mortgage.) The monthly mortgage payments for ... gage would be $379.24. Citicorp is also charging a $200 applica... America Corporation is charging a 6.00% interest rate with 2 poi... loan. The monthly mortgage payments for Bank of America would ... the cost of the points that Lori would need to pay at the tim... $0.02(\$60,000) = \$1200$. Bank of America has waived its applicatio...

a) How long would it take for the total payments of the Citicorp ... equal the total payments of the Bank of America mortgage?

b) If Lori plans to keep her house for 20 years, which mortgage would ... lower total cost?

Solution **a)** Understand Citicorp is charging a higher interest rate and a small applicati... fee but no points. Bank of America is charging a lower rate and no application fee but 2 points. We need to determine the number of months when the total payments of the two loans would be equal.

Translate
$$\text{let } x = \text{number of months}$$
$$\text{then } 379.24x = \text{cost of mortgage payments for } x \text{ months}$$
$$\text{with the Citicorp mortgage}$$
$$\text{and } 359.73x = \text{cost of mortgage payments for } x \text{ months}$$
$$\text{with the Bank of America Corporation}$$

total cost with Citicorp			=	total cost with Bank of America		
mortgage payments	+	application fee	=	mortgage payments	+	points
$379.24x$	+	200	=	$359.73x$	+	1200

Carry Out
$$379.24x + 200 = 359.73x + 1200$$
$$379.24x = 359.73x + 1000$$
$$19.51x = 1000$$
$$x \approx 51.26$$

Answer The cost would be the same in about 51.26 months or about 4.3 years.

b) The total cost would be the same at about 4.3 years. Prior to the 4.3 years, the cost of the loan with Bank of America would be more because of the initial $1200 charge for points. However, after the 4.3 years the cost with Bank of America would be less because of the lower monthly payment. If we evaluate the total cost with Citicorp over 20 years (240 monthly payments), we obtain $91,217.60. If we evaluate the total cost with Bank of America over 20 years, we obtain $87,535.20. Therefore, Lori will save $3682.40 over the 20-year period with Bank of America.

'CISE 31

Let's now look at two examples that involve angles. In Example 9 we use complementary angles. **Complementary angles** are two angles whose sum measures 90° (see Fig. 2.4).

In Figure 2.4, angle x (symbolized $\angle x$) and angle y ($\angle y$) are complementary angles since their sum measures 90°.

EXAMPLE 9   **Complementary Angles** If angle A and angle B are complementary angles and angle B is 42° greater than angle A, determine the measures of angle A and angle B.

Solution   Understand   The sum of the measures of the two angles must be 90° because they are complementary angles. We will use this fact to set up an equation. Since angle B is described in terms of angle A, we will let x represent the measure of angle A.

Translate

$$\text{let } x = \text{the measure of angle } A$$

$$\text{then } x + 42 = \text{the measure of angle } B$$

$$\text{the measure of angle } A + \text{the measure of angle } B = 90°$$

$$x \qquad + \qquad x + 42 \qquad = 90$$

Carry Out

$$2x + 42 = 90$$
$$2x = 48$$
$$x = 24$$

Check and Answer   Since $x = 24$, the measure of angle A is 24°. The measure of angle $B = x + 42 = 24 + 42 = 66$. Thus, angle B has a measure of 66°. Note that angle B is 42° greater than angle A, and the sum of the measures of both angles is 90° $(24° + 66° = 90°)$.

NOW TRY EXERCISE 1

FIGURE 2.5

In Example 10 we use supplementary angles. **Supplementary angles** are two angles whose sum measures 180° (see Fig. 2.5).

In Figure 2.5, angle x and angle y are supplementary angles since their sum measures 180°.

EXAMPLE 10   **Supplementary Angles** If angles C and D are supplementary angles and the measure of angle C is 6° greater than twice the measure of angle D, determine the measures of angle C and angle D.

Solution   Understand   The sum of the measures of the two angles must be 180° because they are supplementary angles. Since angle C is described in terms of angle D, we will let x represent the measure of angle D.

Translate

$$\text{let } x = \text{the measure of angle } D$$

$$\text{then } 2x + 6 = \text{the measure of angle } C$$

$$\text{the measure of angle } C + \text{the measure of angle } D = 180°$$

$$2x + 6 \qquad + \qquad x \qquad = 180$$

Carry Out

$$3x + 6 = 180$$
$$3x = 174$$
$$x = 58$$

Check and Answer   Since $x = 58$, the measure of angle D is 58°. The m of angle $C = 2x + 6 = 2(58) + 6 = 122$. Thus the measure of angle C Note that the measure of angle C is 6° more than twice the measure of a and the sum of the measures of the angles is 180° $(122° + 58° = 180°)$.

NOW TRY EXERCISE 3

NOW TRY EXERCISE 23

Check and Answer If you substitute 13 for *t* in the equation, you will see that the answer checks. The room tax rate is 13%.

EXAMPLE 8

Home Mortage Lori Paduchowski is buying her first home and she is considering two banks for a $60,000 mortgage. Citicorp is charging a 6.50% interest rate with no points for a 30-year loan. (A point is a one-time charge of 1% of the amount of the mortgage.) The monthly mortgage payments for the Citicorp mortgage would be $379.24. Citicorp is also charging a $200 application fee. Bank of America Corporation is charging a 6.00% interest rate with 2 points for a 30-year loan. The monthly mortgage payments for Bank of America would be $359.73 and the cost of the points that Lori would need to pay at the time of closing is $0.02(\$60,000) = \$1200$. Bank of America has waived its application fee.

a) How long would it take for the total payments of the Citicorp mortgage to equal the total payments of the Bank of America mortgage?

b) If Lori plans to keep her house for 20 years, which mortgage would result in the lower total cost?

Solution

a) Understand Citicorp is charging a higher interest rate and a small application fee but no points. Bank of America is charging a lower rate and no application fee but 2 points. We need to determine the number of months when the total payments of the two loans would be equal.

Translate

let x = number of months

then $379.24x$ = cost of mortgage payments for x months with the Citicorp mortgage

and $359.73x$ = cost of mortgage payments for x months with the Bank of America Corporation

total cost with Citicorp = total cost with Bank of America

| mortgage payments | + | application fee | = | mortgage payments | + | points |
| 379.24x | + | 200 | = | 359.73x | + | 1200 |

Carry Out

$$379.24x + 200 = 359.73x + 1200$$
$$379.24x = 359.73x + 1000$$
$$19.51x = 1000$$
$$x \approx 51.26$$

Answer The cost would be the same in about 51.26 months or about 4.3 years.

b) The total cost would be the same at about 4.3 years. Prior to the 4.3 years, the cost of the loan with Bank of America would be more because of the initial $1200 charge for points. However, after the 4.3 years the cost with Bank of America would be less because of the lower monthly payment. If we evaluate the total cost with Citicorp over 20 years (240 monthly payments), we obtain $91,217.60. If we evaluate the total cost with Bank of America over 20 years, we obtain $87,535.20. Therefore, Lori will save $3682.40 over the 20-year period with Bank of America.

NOW TRY EXERCISE 31

Let's now look at two examples that involve angles. In Example 9 we use complementary angles. **Complementary angles** are two angles whose sum measures 90° (see Fig. 2.4).

In Figure 2.4, angle *x* (symbolized ⦟*x*) and angle *y* (⦟*y*) are complementary angles since their sum measures 90°.

FIGURE 2.4

EXAMPLE 9   **Complementary Angles** If angle A and angle B are complementary angles and angle B is 42° greater than angle A, determine the measures of angle A and angle B.

Solution   Understand   The sum of the measures of the two angles must be 90° because they are complementary angles. We will use this fact to set up an equation. Since angle B is described in terms of angle A, we will let x represent the measure of angle A.

Translate   let x = the measure of angle A

then $x + 42$ = the measure of angle B

the measure of angle A + the measure of angle B = 90°

$$x \quad\quad + \quad\quad x + 42 \quad\quad = 90$$

Carry Out

$$2x + 42 = 90$$
$$2x = 48$$
$$x = 24$$

Check and Answer   Since $x = 24$, the measure of angle A is 24°. The measure of angle $B = x + 42 = 24 + 42 = 66$. Thus, angle B has a measure of 66°. Note that angle B is 42° greater than angle A, and the sum of the measures of both angles is 90° ($24° + 66° = 90°$).

NOW TRY EXERCISE 1

In Example 10 we use supplementary angles. **Supplementary angles** are two angles whose sum measures 180° (see Fig. 2.5).

In Figure 2.5, angle x and angle y are supplementary angles since their sum measures 180°.

FIGURE 2.5

EXAMPLE 10   **Supplementary Angles** If angles C and D are supplementary angles and the measure of angle C is 6° greater than twice the measure of angle D, determine the measures of angle C and angle D.

Solution   Understand   The sum of the measures of the two angles must be 180° because they are supplementary angles. Since angle C is described in terms of angle D, we will let x represent the measure of angle D.

Translate   let x = the measure of angle D

then $2x + 6$ = the measure of angle C

the measure of angle C + the measure of angle D = 180°

$$2x + 6 \quad\quad + \quad\quad x \quad\quad = 180$$

Carry Out

$$3x + 6 = 180$$
$$3x = 174$$
$$x = 58$$

Check and Answer   Since $x = 58$, the measure of angle D is 58°. The measure of angle $C = 2x + 6 = 2(58) + 6 = 122$. Thus the measure of angle $C = 122°$. Note that the measure of angle C is 6° more than twice the measure of angle D, and the sum of the measures of the angles is 180° ($122° + 58° = 180°$).

NOW TRY EXERCISE 3

HELPFUL HINT	Here are some suggestions if you are having some difficulty with application problems.
	1. Instructor—Make an appointment to see your instructor. Make sure you have read the material in the book and attempted all the homework problems. Go with specific questions for your instructor.
STUDY TIP	**2.** Videotapes—Find out if the videotapes that accompany this book are available at your college. If so, view the videotapes that go with this chapter. Using the pause control, you can watch the videotapes at your own pace.
	3. Student's Study Guide—If a copy of the Student's Study Guide is available, you may wish to read the material related to this chapter.
	4. Tutoring—If your college learning center offers free tutoring, as many colleges do, you may wish to take advantage of tutoring.
	5. Study Group—Form a study group with classmates. Exchange phone numbers and e-mail addresses. You may be able to help one another.
	6. Student's Solutions Manual—If you get stuck on an exercise, you may want to use the Student's Solutions Manual to help you understand a problem. Do not use the Solutions Manual in place of working the exercises. In general, the Solutions Manual should be used only to check your work.
	7. Web Site—If a computer is available, visit the Prentice Hall—Allen Angel Web site at prenhall.com/angel and study the material that relates to this chapter. You will find additional worked-out examples and exercises.
	8. Math Pro—If Math Pro is available, it can provide you with valuable help in understanding the material. Math Pro provides unlimited practice and gives you immediate feedback on your work, including an immediate review of your work via integrated video. Check with your instructor to determine if Math Pro is available.
	9. Prentice Hall Mathematics Tutor—Once the program has been initiated by your instructor, you can get individual tutoring by phone, fax, or email.
	It is important that you keep trying! Remember, the more you practice, the better you will become at solving application problems.

Exercise Set 2.3

Practice the Skills and Problem Solving

In Exercises 1–45, write an equation that can be used to solve the problem. Find the solution to the problem.

1. **Complementary Angles** Angles A and B are complementary angles. Determine the measures of angles A and B if angle A is four times the size of angle B. See Example 9.

2. **Complementary Angles** Angles C and D are complementary angles. Determine the measures of angles C and D if angle D is 15° less than twice angle C.

3. **Supplementary Angles** Angles A and B are supplementary angles. Find the measures of angles A and B if angle B is 4 times the size of angle A. See Example 10.

4. **Supplementary Angles** Angle A and angle B are supplementary angles. Find the measure of each angle if angle A is 30° greater than angle B.

5. **Angles in a Triangle** The sum of the measures of the angles of a triangle is 180°. Find the three angles of a triangle if one angle is 20° greater than the smallest angle and the third angle is twice the smallest angle.

6. **Angles in a Triangle** Find the measures of the three angles of a triangle if one angle is twice the smallest angle and the third angle is 60° greater than the smallest angle.

7. **History Honor Society** Thomas Penna is a member of Phi Alpha Theta, the national history honor society. One benefit of membership is a 25% discount on all history magazine subscriptions. Thomas used this discount to order an annual subscription to *American Heritage* magazine and paid $24. What was the cost of a regular subscription?

8. *A New Suit* Chuck Chapman is shopping for a new suit. At K & G Menswear, he finds that the sale price of a suit that was reduced by 25% is $187.50. Find the regular price of the suit.

9. *Bus Pass* Kate Spence buys a monthly bus pass, which entitles the owner to unlimited bus travel, for $45 per month. Without the pass each bus ride costs $1.80. How many rides per month would Kate have to take so that the cost of the rides without the bus pass is equal to the total cost of the rides with the bus pass.

10. *Laundry Costs* It costs Mike Sutton $12.50 a week to wash and dry his clothes at the corner laundry. If a washer and dryer cost a total of $940, how many weeks will it take for the laundry cost to equal the cost of a washer and dryer? (Disregard energy cost.)

11. *Truck Rental* The cost of renting a truck is $35 a day plus 20¢ per mile. How far can Tanya Richardson drive in 1 day if she has only $80?

12. *George Washington Bridge Tolls* Going into New York using the George Washington Bridge, customers must pay a toll (there is no toll to return to New Jersey). They can pay $6 cash or they can pay $4 (in nonpeak hours) using the E Z Pass system. The E Z Pass system is a prepaid plan that also requires a one-time $10 account activation fee. How many trips to New York would a person need to make (in nonpeak hours) so that the total amount spent with the E Z Pass is equal to the amount spent for tolls without the E Z Pass?

13. *Sales Tax* The sales tax rate in Missouri is 4.225%. What is the maximum price that Don and Betty Lichtenberg can spend on a new computer desk if the total cost of the desk, including sales tax, is $650?

14. *Toll Bridge* Mr. and Mrs. Ose live in a resort island community attached to the mainland by a toll bridge. The toll is $2.50 per car going to the island, but there is no toll coming from the island. Island residents can purchase a monthly pass for $20, which permits them to cross the toll bridge from the mainland for only 50¢ each time. How many times a month would the Oses have to go to the island from the mainland for the cost of the monthly pass to equal the regular toll cost?

15. *Playing Golf* Steve Korcheck wishes to join the Elton Country Club to play golf. He has two membership options. A social membership costs $1775 in annual dues. In addition, he would pay a $50 green fee and a $25 golf cart fee every time he played. A golf membership costs $2425 in annual dues. With this membership, Steve would only pay a $25 golf cart fee when he played. How many times per year would Steve need to golf for the two options to cost the same amount of money?

16. *Waitress Pay* Kathy Fields is a banquet waitress. She is paid $2.63 per hour plus 15% of the total cost of the food and beverages she serves during the banquet. If, during a 5-hour shift, Kathy earns $400, what was the total cost of the food and beverages she served?

17. *Mutual Funds* As of February 4, 2002, the two no-load stock mutual funds with the highest year-to-date returns were the Wasatch Micro Cap (WMC) and the Schroder Ultra Investor (SUI) funds. WMC assets were about 2.7 times the assets of SUI. If the sum of the assets of these two mutual funds was about $636.4 million, determine the assets (in dollars) of WMC and of SUI.

18. *Retirement Funds* Rich Rowe makes regular contributions of $5000 annually to the Teachers Insurance and Annuity Association/College Retirement Equity Fund (TIAA/CREF) 403-b retirement plan. He has some of his contribution going to the CREF stock fund and some going to the CREF global equities fund. His contribution to the stock fund is $250 less than twice what he contributes to the global equities fund. How much does he contribute to each fund?

19. *NASA Budget* In 2002, the budget for the National Aeronautics and Space Administration (NASA) was about $14.51 billion. Two categories made up 99.8% of this entire budget. Human space flight (HSF) involves all space shuttle missions and space station operations. Science, aeronautics, and technology (SAT) involves all research done for NASA across the United States. If NASA spent $0.10 billion more on HSF than on SAT, determine how much they spent on HSF and how much they spent on SAT. Round your answers to two decimal places.

20. *Pollen* There are 57 major sources of pollen in the United States. These pollen sources are categorized as grasses, weeds, and trees. If the number of weeds is 5 less than twice the number of grasses, and the number of trees is 2 more than twice the number of grasses, find the number of grasses, weeds, and trees that are major pollen sources.

21. *Bones and Steel* According to *Health* magazine, the stress a bone can withstand in pounds per square inch is 6000 pounds more than 3 times the amount that steel can withstand. If the difference between the amount of stress a bone and steel can withstand is 18,000 pounds per square inch, find the stress that both steel and a bone can withstand.

22. *Car Antitheft System* By purchasing and installing a LoJack antitheft system, Pola Sommers can save 15% of the price of her auto insurance. The LoJack system costs $743.65 to purchase and install. If Pola's annual insurance before installing the LoJack system is $849.44, in how many years would the LoJack system pay for itself?

23. *Ordering Lunch* After Carol Pharo is seated in a restaurant, she realizes that she only has $20.00. If she must pay 7% sales tax and wishes to leave a 15% tip on the total bill (meal plus tax), what is the maximum price of the lunch she can order?

24. *Payment Plan* The Midtown Tennis Club offers two payment plans for its members. Plan 1 is a monthly fee of $25 plus $10 per hour of court rental time. Plan 2 has no monthly fee, but court time costs $18.50 per hour. How many hours would Mrs. Levin have to play per month so that plan 1 becomes advantageous?

25. *European Gasoline Taxes* In September, 2000, Europeans protested the price of gasoline and called for governments to lower the high gasoline taxes. In the United Kingdom, customers paid $4.29 per gallon (in U.S. dollars). This price represented a 68% increase over the pretax price of gasoline.
  a) Determine the pretax price of gasoline.
  b) Determine the amount of taxes paid on each gallon of gasoline.

26. *Hotel Tax Rate* While on vacation in Milwaukee, the Browns were quoted a hotel room price of $85 per night plus tax. If after staying one night the Brown's bill came to $97.41, what was the tax rate?

27. *Olympic Medals* In the 2000 Summer Olympics, the United States, Russia, China, Australia, and Germany won a total of 359 medals. The United States won 19 fewer medals than twice the number of medals Australia won. Russia won 28 fewer medals than twice the number Australia won. China won 1 more medal than Australia won. Germany won 1 fewer medal than Australia won. Determine the number of medals won by the United States, Russia, China, Australia, and Germany in the 2000 Summer Olympics.

28. *Minimum Wage Increase* From 1980 to 2002, the minimum hourly wage has increased about 66.13% to $5.15 per hour. What was the minimum hourly wage in 1980?

29. *Monthly Rental Fees* The average monthly rent for a two-bedroom apartments in San Jose, California, increased by about 13.3% from 2001 to 2002. If the average monthly rent in 2002 was $1199, determine the average monthly rent in 2001?

30. *Comparing Mortgages* The Sanchezes are purchasing a new home and are considering a 30-year $70,000 mortgage with two different banks. Madison Savings is charging 9.0% with 0 points and First National is charging 8.5% with 2 points. First National is also charging a $200 application fee, whereas Madison is charging none. The monthly mortgage payments with Madison would be $563.50 and the monthly mortgage payments with First National would be $538.30.
  a) After how many months would the total payments for the two banks be the same?
  b) If the Sanchezes plan to keep their house for 30 years, which mortgage would have the lower total cost? (See Example 8.)

🔒 **31.** *Refinancing a Mortgage* Dung Nguyen is considering refinancing his house at a lower interest rate. He has an 11.875% mortgage, is presently making monthly principal and interest payments of $510, and has 20 years left on his mortgage. Because interest rates have dropped, Countrywide Mortgage Corporation is offering him a rate of 9.5%, which would result in principal and interest payments of $420.50 for 20 years. However, to get this mortgage, his closing cost would be $2500.

a) How many months after refinancing would he spend the same amount on his new mortgage plus closing cost as he would have spent on the original mortgage?

b) If he plans to spend the next 20 years in the house, would he save money by refinancing?

32. *Dinner Seminars* Amy June, a financial planner, is sponsoring dinner seminars. She must pay for the dinners of those attending out of her own pocket. She chooses a restaurant that seats 40 people and charges her $9.50 per person. If she earns 12% commission of sales made, determine how much in sales she must make from these 40 people

a) to break even.

b) to make a profit of $500.

33. *Perimeter of a Triangle* John is developing a game that contains a triangular game board. The perimeter of the triangular board is 36 inches. Find the length of the three sides if one side is 3 inches greater than the smallest side, and the third side is 3 inches less than twice the length of the smallest side.

34. *Triangular Garden* The perimeter of a triangular garden is 60 feet. Find the length of the three sides if one side is 4 feet greater than twice the length of the smallest side, and the third side is 4 feet less than 3 times the length of the smallest side.

35. *Angles of a Triangle* An $8\frac{1}{2}$ by 11-inch piece of paper is cut from opposite corners to form a triangle. One angle of the triangle measures 12° greater than the smallest angle. The third angle measures 27° less than three times the smallest angle. If the sum of the interior angles of a triangle measure 180°, determine the measures of the three angles.

36. *Stairway Railing* A stairway railing has a design that contains triangles. In one of the triangles one angle

measures 20° less than twice the smallest angle. The third angle measures 25° greater than twice the smallest angle. Determine the measures of the three angles.

37. *Sandbox Construction* Edie Hall is planning to build a rectangular sandbox for her children. She wants its length to be 3 feet more than its width. Find the length and width of the sandbox if only 22 feet of lumber are available to form the frame. Use $P = 2l + 2w$.

38. *Fence Dimensions* Craig Campanella, a landscape architect, wishes to fence in two equal areas as illustrated in the figure. If both areas are squares and the total length of fencing used is 91 meters, find the dimensions of each square.

39. *Bookcase Dimensions* Joe Murray wishes to build a bookcase with four shelves (including the top) as shown in the figure. The width of the bookcase is to be 3 feet more than the height. If only 30 feet of wood are available to build the bookcase, what will be the dimensions of the bookcase?

40. *Fence Dimensions* Donshay Latrell wishes to fence in three rectangular areas along a river bank as illustrated in the figure. Each rectangle is to have the same dimensions, and the length of each rectangle is to be 1 meter greater than its width (along the river). Find the length and width of each rectangle if the total amount of fencing used is 114 meters.

🔒 **41.** *Price Reductions* During the first week of a going-out-of-business sale, Sam's General Store reduces all prices by 10%. The second week of the sale, Sam's reduces all items by an additional $5. If Sivle Yelserp bought a calculator for $49 during the second week of the sale, find the original price of the calculator.

42. *Farm Divisions* Daniella Popoff's farm is divided into three regions. The area of one region is twice as large as the area of the smallest region, and the area of the third region is 4 acres less than three times the area of the smallest region. If the total acreage of the farm is 512 acres, find the area of each of the three regions.

43. *Comparing Toy Sales* Joy Kristin is shopping for a Little Tikes Victorian Kitchen for her niece and knows that Toys "R" Us and Wal-Mart sell the kitchen for the same price. On December 26, Toys "R" Us has the kitchen on sale for 37% off the original price and Wal-Mart has the kitchen on sale for $50 off the original price. After visiting both stores, Joy discovers that the sale prices for the kitchen at the two stores are also the same.

a) Determine the original price of the kitchen.

b) Determine the sale price of the kitchen.

44. *Selling Paintings* J. P. Richardson sells each of his paintings for $500. The gallery where he displays his work charges him $1350 a month plus a 10% commission on sales. How many paintings must J. P. sell in a month to break even?

45. *Incandescent Bulbs* The cost of purchasing incandescent bulbs for use over a 9750-hour period is $9.75. The energy cost of incandescent bulbs over this period is $73. The cost of one equivalent fluorescent bulb that lasts about 9750 hours is $20. By using a fluorescent bulb instead of incandescent bulbs for 9750 hours, the total savings of purchase price plus energy cost is $46.75. What is the energy cost of using the fluorescent bulb over this period?

46. *Dinner Bill* The five members of the Newton family are going out to dinner with the three members of the Lee family. Before dinner, they decide that the Newtons

will pay $\frac{5}{8}$ of the bill (before tip) and the Lees will pay $\frac{3}{8}$ plus the entire 15% tip. If the total bill including the 15% tip comes to $184.60, how much will be paid by each family?

47. *Plants and Animals* Approximately 1,500,000 species worldwide have been categorized as either plants, animals, or insects. Insects are often subdivided into beetles and insects that are not beetles. There are about 100,000 more plant than animal species. There are 290,000 more nonbeetle insects than animals. The number of beetles is 140,000 less than twice the number of animals. Find the number of animal, plant, nonbeetle insect, and beetle species.

48. *Earning an A* To find the average of a set of test scores, we divide the sum of the test scores by the number of test scores. On her first four algebra tests, Paula West's scores were 87, 93, 97, and 96.

a) Write an equation that can be used to determine the grade Paula needs to obtain on her fifth test to have a 90 average.

b) Explain how you determined your equation.

c) Solve the equation and determine the score.

49. *Physics Exam Average* Philip Kowalski's grades on five physics exams were 70, 83, 97, 84, and 74.

a) If the final exam will count twice as much as each exam, what grade does Philip need on the final exam to have an 80 average?

b) If the highest possible grade on the final exam is 100 points, is it possible for Philip to obtain a 90 average? Explain.

50. a) Make up your own realistic word problem involving percents. Represent this word problem as an equation.

b) Solve the equation and answer the word problem.

51. a) Make up your own realistic word problem involving money. Represent this word problem as an equation.

b) Solve the equation and answer the world problem.

Challenge Problems

52. *Money Market* On Monday Sophia Murkovic purchased shares in a money market fund. On Tuesday the value of the shares went up 5%, and on Wednesday the value fell 5%. How much did Sophia pay for the shares on Monday if she sold them on Thursday for $59.85?

53. *Truck Rental* The Elmers Truck Rental Agency charges $28 per day plus 15 ¢ a mile. If Denise Turcotte rented a small truck for 3 days and the total bill was $121.68, including a 4% sales tax, how many miles did she drive?

Group Activity

Discuss and answer Exercise 54 as a group.

54. a) Have each member of the group pick a number. Then multiply the number by 2, add 33, subtract 13, divide by 2, and subtract the number each started with. Record each answer.

b) Now compare answers. If you did not all get the same answer, check each other's work.

c) As a group, explain why this procedure will result in an answer of 10 for any real number n selected.

Cumulative Review Exercises

[1.3] Evaluate.

55. $2 + \left| -\dfrac{3}{5} \right|$

56. $-6.4 - (-3.7)$

57. $\left| -\dfrac{5}{8} \right| \div |-2|$

58. $5 - |-3| - |7|$

[1.5] **59.** Simplify $(2x^4 y^{-6})^{-3}$.

2.4 ADDITIONAL APPLICATION PROBLEMS

1 Solve motion problems.

2 Solve mixture problems.

In this section we discuss two additional types of application problems, motion and mixture problems. We have placed them in the same section because they are solved using similar procedures.

1 Solve Motion Problems

A formula with many useful applications is

$$\text{amount} = \text{rate} \cdot \text{time}$$

The "amount" in this formula can be a measure of many different quantities, depending on the rate. For example, if the rate is measuring *distance* per unit time, the amount will be distance. If the rate is measuring *volume* per unit time, the amount will be volume, and so on.

When applying this formula, we must make sure that the units are consistent. For example, when speaking about a copier, if the rate is given in copies per *minute*, the time must be given in *minutes*. Problems that can be solved using this formula are called **motion problems** because they involve movement, at a constant rate, for a certain period of time.

A nurse giving a patient an intravenous injection may use this formula to determine the drip rate of the fluid being injected. A company drilling for oil or water may use this formula to determine the amount of time needed to reach its goal.

When the motion formula is used to calculate distance, the word *amount* is replaced with the word *distance* and the formula is called the **distance formula**.

> **The distance formula is**
>
> $$\text{distance} = \text{rate} \cdot \text{time}$$
> $$\text{or} \quad d = rt$$

When a motion problem has two different rates, it is often helpful to put the information into a table to help analyze the problem.

EXAMPLE 1

34.5 mph

20.2 mph

100 mi

FIGURE 2.6

Ships at Sea The aircraft carrier USS *John F. Kennedy* and the nuclear-powered submarine USS *Memphis* leave from the Puget Sound Naval Yard at the same time heading for the same destination in the Indian Ocean. The aircraft carrier travels at its maximum speed of 34.5 miles per hour and the submarine travels submerged at its maximum speed of 20.2 miles per hour. The aircraft carrier and submarine are to travel at these speeds until they are 100 miles apart, at which time they will receive further communications and instructions from the naval base. How long will it take for the aircraft carrier and submarine to be 100 miles apart? (See Fig. 2.6.)

Solution Understand We wish to find out how long it will take for the difference between their distances to be 100 miles. To solve this problem, we will use the distance formula, $d = rt$. When we first introduced the problem-solving procedure, we indicated that to help understand a problem it might be useful to put the given information into a table, and that is what we will do now.

Let t = time.

	Rate	Time	Distance
Aircraft carrier	34.5	t	$34.5t$
Submarine	20.2	t	$20.2t$

Translate The difference between their distances is 100 miles. Thus,

$$\text{aircraft carrier's distance} - \text{submarine's distance} = 100$$
$$34.5t - 20.2t = 100$$

Carry Out
$$14.3t = 100$$
$$t \approx 6.99$$

Answer The aircraft carrier and submarine will be 100 miles apart in about 7 hours.

EXAMPLE 2 **Running Home** To get in shape for the upcoming track season, Juan and Pedro Santiago begin running home after school. Juan runs at a rate of 6 mph and Pedro runs at a rate of 4 mph. When they leave the same school at the same time, Juan arrives home $\frac{1}{2}$ hour before Pedro, see Figure 2.7.

School   Home

Pedro   Juan
4 mph   6 mph

Juan arrives at home
$\frac{1}{2}$ hour before Pedro

FIGURE 2.7

a) How long does it take for Pedro to reach home?

b) How far do Juan and Pedro live from school?

Solution   **a)** Understand   Both boys will run the same distance. However, since Juan runs at a faster rate than Pedro does, Juan's time will be less than Pedro's time by $\frac{1}{2}$ hour.

$$\text{let } t = \text{Pedro's time to run home}$$

$$\text{then } t - \frac{1}{2} = \text{Juan's time to run home}$$

Runner	Rate	Time	Distance
Pedro	4	t	$4t$
Juan	6	$t - \frac{1}{2}$	$6\left(t - \frac{1}{2}\right)$

Translate   When the boys are home they will both have run the same distance from school. So

$$\text{Pedro's distance} = \text{Juan's distance}$$

$$4t = 6\left(t - \frac{1}{2}\right)$$

Carry Out

$$4t = 6t - 3$$
$$-2t = -3$$
$$t = \frac{3}{2}$$

Answer   Pedro will take $1\frac{1}{2}$ hours to reach home.

b) The distance can be determined using either Juan's or Pedro's rate and time. We will multiply Pedro's rate by Pedro's time to find the distance.

$$d = rt = 4\left(\frac{3}{2}\right) = \frac{12}{2} = 6 \text{ miles}$$

NOW TRY EXERCISE 9   Therefore, Juan and Pedro live 6 miles from their school.   ❋

In Example 2, would the answer have changed if we had let t represent the time Juan is running rather than the time Pedro is running? Try it and see.

EXAMPLE 3   **Soda Pop Production**   A Coca-Cola bottling machine fills and caps bottles. The machine can be run at two different rates. At the faster rate the machine fills and caps 600 more bottles per hour than it does at the slower rate. The machine is turned on for 4.8 hours on the slower rate, then it is changed to the faster rate for another 3.2 hours. During these 8 hours a total of 25,920 bottles were filled and capped. Find the slower rate and the faster rate.

Solution   Understand   This problem uses a number of bottles, an amount, in place of a distance. However, the problem is worked in a similar manner. We will use the formula amount = rate · time. We are given that there are two different rates, and we are asked to find these rates. We will use the fact that the amount filled at the slower rate plus the amount filled at the faster rate equals the total amount filled.

$$\text{let } r = \text{the slower rate}$$

$$\text{then } r + 600 = \text{the faster rate}$$

	Rate	Time	Amount
Slower rate	r	4.8	$4.8r$
Faster rate	$r + 600$	3.2	$3.2(r + 600)$

Translate

$$\text{amount filled at slower rate} + \text{amount filled at faster rate} = 25{,}920$$
$$4.8r \qquad + \qquad 3.2(r + 600) \qquad = 25{,}920$$
$$4.8r + 3.2r + 1920 = 25{,}920$$

Carry Out

$$8r + 1920 = 25{,}920$$
$$8r = 24{,}000$$
$$r = 3000$$

Answer The slower rate is 3000 bottles per hour. The faster rate is $r + 600$ or $3000 + 600 = 3600$ bottles per hour.

NOW TRY EXERCISE 11

2 Solve Mixture Problems

Any problem where two or more quantities are combined to produce a different quantity or where a single quantity is separated into two or more different quantities may be considered a **mixture problem**. As we did when working with motion problems, we will use tables to help organize the information. Examples 4 and 5 are mixture problems that involve money.

EXAMPLE 4 **Two Investments** Bettie Truitt sold her boat for $15,000. Bettie loaned some of this money to her friend Kathy Testone. The loan was for 1 year with a simple interest rate of 4.5%. Bettie put the balance into a money market account at her credit union that yielded 3.75% simple interest. A year later, while working on her taxes, Bettie found that she had earned a total of $637.50 from the two investments, but could not remember how much money she had loaned to Kathy. Determine the amount Bettie loaned to Kathy.

Solution Understand and Translate To work this problem, we will use the simple interest formula, interest = principal · rate · time. We know that part of the investment is made at 4.5% and the rest is made at 3.75% simple interest. We are asked to determine the amount that Bettie loaned to Kathy.

$$\text{let } p = \text{amount loaned to Kathy at 4.5\%}$$
$$\text{then } 15{,}000 - p = \text{amount invested at 3.75\%}$$

Notice that the sum of the two amounts equals the total amount invested, $15,000. We will find the amount loaned to Kathy with the aid of a table.

Investment	Principal	Rate	Time	Interest
Loan to Kathy	p	0.045	1	$0.045p$
Money market	$15{,}000 - p$	0.0375	1	$0.0375(15{,}000 - p)$

Since the total interest collected is $637.50, we write:

$$\text{interest from 4.5\% loan} + \text{interest from 3.75\% account} = \text{total interest}$$
$$0.045p \qquad + \qquad 0.0375(15{,}000 - p) \qquad = \qquad 637.50$$

Carry Out

$$0.045p + 0.0375(15{,}000 - p) = 637.50$$
$$0.045p + 562.50 - 0.0375p = 637.50$$
$$0.0075p + 562.50 = 637.50$$
$$0.0075p = 75$$
$$p = 10{,}000$$

Answer   Therefore, the loan was for $10,000, and $15,000 − p or $15,000 − $10,000 = $5000 was invested in the money market account. ✳

EXAMPLE 5   **Hot Dog Stand Sales**   Matt's Hot Dog Stand in Chicago sells hot dogs for $2.00 each and beef tacos for $2.25 each. If the sales for the day total $585.50 and 278 items were sold, how many of each item were sold?

Solution   Understand and Translate   We are asked to find the number of hot dogs and beef tacos sold.

$$\text{let } x = \text{number of hot dogs sold}$$
$$\text{then } 278 - x = \text{number of beef tacos sold}$$

Item	Cost of Item	Number of Items	Total Sales
Hot dogs	2.00	x	$2.00x$
Beef tacos	2.25	$278 - x$	$2.25(278 - x)$

$$\text{total sales of hot dogs} + \text{total sales of beef tacos} = \text{total sales}$$
$$2.00x + 2.25(278 - x) = 585.50$$

Carry Out

$$2.00x + 625.50 - 2.25x = 585.50$$
$$-0.25x + 625.50 = 585.50$$
$$-0.25x = -40$$
$$x = \frac{-40}{-0.25} = 160$$

Answer   Therefore, 160 hot dogs and $278 - 160 = 118$ beef tacos were sold. ✳

In Example 5 we could have multiplied both sides of the equation by 100 to eliminate the decimal numbers, and then solved the equation.

Example 6 is a mixture problem that involves the mixing of two solutions.

EXAMPLE 6   **Medicine Mixture**   Jamal Reynolds, a chemist, has both 6% and 15% lithium citrate solutions. He wishes to make 0.5 liter of an 8% lithium citrate solution. How much of each solution must he mix?

Solution   Understand and Translate   We are asked to find the amount of each solution to be mixed.

$$\text{let } x = \text{number of liters of 6\% solution}$$
$$\text{then } 0.5 - x = \text{number of liters of 15\% solution}$$

The amount of lithium citrate in a solution is found by multiplying the percent strength of lithium citrate in the solution by the volume of the solution. We will draw a sketch of the problem (see Fig. 2.8) and then construct a table.

FIGURE 2.8

Solution	Strength of Solution	Number of Liters	Amount of Lithium Citrate
1	0.06	x	$0.06x$
2	0.15	$0.5 - x$	$0.15(0.5 - x)$
Mixture	0.08	0.5	$0.08(0.5)$

$$\begin{pmatrix} \text{amount of} \\ \text{lithium citrate} \\ \text{in 6\% solution} \end{pmatrix} + \begin{pmatrix} \text{amount of} \\ \text{lithium citrate} \\ \text{in 15\% solution} \end{pmatrix} = \begin{pmatrix} \text{amount of lithium citrate} \\ \text{in mixture} \end{pmatrix}$$

$$0.06x \quad + \quad 0.15(0.5 - x) \quad = \quad 0.08(0.5)$$

Carry Out

$$0.06x + 0.15(0.5 - x) = 0.08(0.5)$$
$$0.06x + 0.075 - 0.15x = 0.04$$
$$0.075 - 0.09x = 0.04$$
$$-0.09x = -0.035$$
$$x = \frac{-0.035}{-0.09} = 0.39 \quad \begin{pmatrix} \text{to nearest} \\ \text{hundredth} \end{pmatrix}$$

NOW TRY EXERCISE 21

Jamal must mix 0.39 liter of the 6% solution and $0.5 - x$ or $0.5 - 0.39 = 0.11$ liter of the 15% solution to make 0.5 liter of an 8% solution. ✳

Exercise Set 2.4

Practice the Skills and Problem Solving

In Exercises 1–14, write an equation that can be used to solve the motion problem. Solve the equation and answer the question asked.

1. *A Hike in the Rockys* Two friends, Don O'Neal and Judy McElroy, go hiking in the Rocky Mountains. While hiking they come across Bear Lake. They wonder what the distance around the lake is and decide to find out. Don knows he walks at 5 mph and Judy knows she walks at 4.5 mph. If they start walking at the same time in opposite directions around the lake,

and meet in 1.2 hours, what is the distance around the lake?

2. *Earthquake Shock Waves* An earthquake occurs in a desert of California. The shock waves travel outward in a circular path, similar to when a pebble is dropped into a pond. If the *P*-wave (one kind of shock wave) travels at 2.4 miles per second, how long would it take

for the wave to have a diameter of 60 miles? (See the figure.)

← 60 miles →

3. **Balloon Flights** Each year Albuquerque, New Mexico, has a hot air balloon festival during which people can obtain rides in hot air balloons. Suppose part of the Diaz family goes in one balloon and other members of the family go in another balloon. Because they fly at different altitudes and carry different weights, one balloon travels at 16 miles per hour while the other travels in the same direction at 14 miles per hour. In how many hours will they be 4 miles apart?

4. **Trains** A passenger train leaves Norfolk, Virginia, 1.2 hours after a freight train leaves. The passenger train is traveling 18 mph faster than the freight train on a parallel set of tracks. The trains are traveling toward each other. The two trains will be at the same location 3 hours after the passenger train leaves. Find the speed of each train.

5. **Cornfield** Rodney Joseph and Dennis Clarence are gleaning (or collecting) corn from a cornfield that is 1.5 miles long. Rodney starts on one end and is gleaning corn at a rate of 0.15 miles per hour. Dennis starts on the other end and is gleaning corn at a rate of 0.10 miles per hour. If they start at the same time and continue to work at the same rate, how long will it be before Rodney and Dennis meet?

6. **Photocopying** To make a large number of copies, Sandra Jessee uses two photocopiers. One copier can produce copies at a rate of 35 copies per minute. The other copier can produce copies at a rate of 40 copies per minute. If Sandra starts both machines at the same time, how long will it take the two machines to produce a total of 1050 copies?

7. **Charity Race** The Alpha Delta Pi Sorority raises money for the Ronald McDonald House by holding an annual "Roll for Ronald" race in College Station, Texas. Laura Mann rides a bicycle and travels at twice the speed of Wayne Siegert, who is on rollerblades. Laura and Wayne begin the race at the same time and after 3 hours, Laura is 18 miles ahead of Wayne.

a) What is Wayne's speed?

b) What is Laura's speed?

8. **Snooty the Manatee** The South Florida Museum in Bradenton is home to Snooty the manatee, who lives in a 60,000-gallon tank. Once a year, his keepers drain and refill Snooty's tank. The tank has two inlet valves that have the same flow rate. To fill the tank, the first valve is opened for a period of 17 hours. During this 17-hour period the second valve is opened for 7 hours. Determine the flow rate of the two inlet valves in gallons per hour.

9. **Canyon Hiking** Marcia Kleinz hikes down to the bottom of Bryce Canyon, camps overnight, and returns the next day. Her hiking speed down averages 3.4 miles per hour and her return trip averages 1.2 miles per hour. If she spent a total of 12 hours hiking, find

a) how long it took her to reach the bottom of the canyon?

b) the total distance traveled.

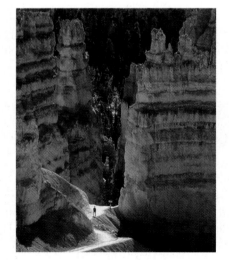

10. **Catching Up** Nicholas Thompson begins a long-distance walk at a rate of 4 mph. Forty-five minutes after he leaves, his wife, Kristin, realizes that Nicholas

has forgotten his wallet. Kristin gets on her bicycle and begins riding at a rate of 24 mph along the same path that Nicholas took.

a) How long will it take for Kristin to catch Nicholas?

b) How far from their house will Kristin catch Nicholas?

11. _Packing Spaghetti_ Two machines are packing spaghetti into boxes. The smaller machine can package 400 boxes per hour and the larger machine can package 600 boxes per hour. If the larger machine is on for 2 hours before the smaller machine is turned on, how long after the smaller machine is turned on will a total of 15,000 boxes of spaghetti be boxed?

12. _Snail Races_ As part of their preschool science project, Mrs. Joy Pribble's class is holding snail races. The first snail, Zippy, is known to move at a rate of 5 inches per hour. The second snail, Lightning, is known to move at 4.5 inches per hour. If the snails race along a straight path, and if Zippy finishes the race 0.25 hour before Lightning,

a) determine the time it takes Lightning to run the race.

b) determine the time it takes Zippy to run the race.

c) what is the distance covered by both snails?

13. _Airport Trip_ Lavon Burton began driving to an airport at a speed of 35 miles per _hour_. Fifteen _minutes_ after she left, her parents realized that she had forgotten to take her airline tickets. They then tried to catch up to her in their car. If the parents traveled at 50 miles per hour, how long did it take them to catch Lavon?

14. _Walkie-Talkie Range_ A set of Maxon RS446 walkie-talkies has a range of about two miles. Alice Burstein and Donna Gerken begin walking along a nature trail heading in opposite directions carrying their walkie-talkies. If Alice walks at a rate of 3.5 mph and Donna walks at a rate of 4.5 mph, how long will it take for them to be out of the range of the walkie-talkies?

In Exercises 15–28, write an equation that can be used to solve the mixture problem. Solve each equation and answer the question asked.

15. _Two Investments_ Bill Palow invested $30,000 in two separate accounts paying 3% and 4.1% annual simple interest. If Bill earned a total of $1091.73 from the two investments, how much was invested in each account.

16. _Two Investments_ Thea Prettyman invested $10,000 for 1 year, part at 7% and part at $6\frac{1}{4}$%. If she earned a total interest of $656.50, how much was invested at each rate?

17. _Mixing Coffee_ Joan Smith is the owner of a Starbucks Coffee Shop. She has many varieties of coffee in the shop including a Kona variety that sells for $6.20 per pound and an amaretto coffee that sells for $5.80 per pound. She finds that by mixing these two blends she creates a coffee that sells well. If she uses 18 pounds of amaretto coffee in the blend and wishes to sell the mixture at $6.10 per pound, how many pounds of the Kona coffee should she mix with the amaretto coffee?

18. _Mixing Nuts_ J.B. Davis owns a nut shop. He sells almonds for $6 per pound and walnuts for $5.20 per pound. He receives a special request from a customer who wants to purchase 30 pounds of a mixture of almonds and walnuts but does not want to pay more than $165. J.B. used algebra to determine the amount of each to mix so that he could mix the maximum amount of almonds without having the mixture's value exceed $165. Determine how many pounds of almonds and walnuts J.B. mixed.

19. _Investing an Inheritance_ Bart Nelson has inherited $250,000 and wishes to invest his inheritance in shares of Johnson & Johnson stock and AOL Time Warner stock. He wishes to buy twice as many shares of AOL as shares of Johnson & Johnson. On February 11, 2002, the price of Johnson & Johnson was $56.88 per share and the price of AOL was $27.36.

a) If Bart wishes to buy shares in blocks of 100, how many shares of each company can he buy?

b) How much money will he have left after making the purchases?

20. _Acetic Acid Solutions_ Fred Feldon, a chemistry teacher, needs a 10% acetic acid solution for an upcoming chemistry laboratory. When checking the storeroom, he realizes he has only 16 ounces of a 25% acetic acid solution. There is not enough time for him to order more, so he decides to make a 10% acetic acid solution by very carefully adding water to the 25% acetic acid solution. Knowing algebra, he figures out just how much water to add. Determine how much water Fred must add to the 25% solution to reduce it to a 10% solution.

21. _Vinegar Solutions_ Distilled white vinegar purchased in supermarkets generally has a 5% acidity level. To make her sauerbraten, Chef Judy Ackermary marinates veal overnight in a special 8% distilled vinegar that she creates herself. To create the 8% solution she mixes a regular 5% vinegar solution with a 12% vinegar solution that she purchases by mail. How many ounces of the 12% vinegar solution should she add to 40 ounces of the 5% vinegar solution to get the 8% vinegar solution?

22. _Hydrogen Peroxide Solution_ Tim Houtakker works as a chemical engineer for US Peroxide Corporation. Tim has 2500 gallons of a commercial-grade hydrogen peroxide solution that is 60% pure hydrogen peroxide. How much distilled water (which is 0% hydrogen peroxide) will Tim need to add to this solution to create a new solution that is 25% pure hydrogen peroxide?

23. **Horseradish Sauces** Jennifer Thornton has a recipe that calls for horseradish sauce that is 45% pure horseradish. At the grocery store she finds one horseradish sauce that is 30% pure horseradish and another that is 80% pure horseradish. How many teaspoons of each of these horseradish sauces should Jennifer mix together to get 4 teaspoons of horseradish sauce that is 45% pure horseradish?

24. **Grass Seed Mixture** The Pearlman Nursery sells two types of grass seeds in bulk. The lower quality seeds have a germination rate of 76%, but the germination rate of the higher quality seeds is not known. Twelve pounds of the higher quality seeds are mixed with 16 pounds of the lower quality seeds. If a later analysis of the mixture reveals that the mixture's germination rate was 82%, what is the germination rate of the higher quality seed?

25. **Acid Solutions** Two acid solutions are available to a chemist. One is a 20% sulfuric acid solution, but the label that indicates the strength of the other sulfuric acid solution is missing. Two hundred milliliters of the 20% acid solution and 100 milliliters of the solution with the unknown strength are mixed together. Upon analysis, the mixture was found to have a 25% sulfuric acid concentration. Find the strength of the solution with the missing label.

26. **A Tax Strategy** Some states allow a husband and wife to file individual state tax returns even though they file a joint federal return. It is usually to the taxpayers' advantage to do this when both the husband and wife work. The smallest amount of tax owed (or the largest refund) will occur when the husband's and wife's taxable incomes are the same.

Mr. Juenger's 2002 taxable income was $28,200 and Mrs. Juengers' taxable income for that year was $32,450. The Juengers' total tax deductions for the year were $6400. This deduction can be divided between Mr. and Mrs. Juenger in any way they wish. How should the $6400 be divided between them to result in each person's having the same taxable income?

27. **Sunflower Seed Mixture** The Agway Nursery is selling two types of sunflower seed. The striped sunflower seeds cost $1.20 per pound, while the black oil sunflower seeds cost $1.60 per pound. How many pounds of each should the nursery mix to get a 20-pound mixture that sells for $30?

28. **Octane Ratings** The octane rating of gasoline indicates the percent of pure octane in the gasoline. For example, most regular gasoline has an 87-octane rating, which means that this gasoline is 87% octane (and 13% of some other non-octane fuel such as pentane). Ollie Kieler owns a gas station and has 150 gallons of 87-octane gasoline. How many gallons of 93-octane gasoline must he mix with this 87-octane gasoline to obtain 89-octane gasoline?

See Exercise 27

In Exercises 29–46, write an equation that can be used to solve the motion or mixture problem. Solve each equation and answer the questions asked.

29. **Route 66** The famous highway U.S. Route 66 connects Chicago and Los Angeles and extends 2448 miles. Judy Kasabian begins in Chicago and drives at an average rate of 45 mph on Route 66 toward Los Angeles. At the same time Kamilia Nemri begins in Los Angeles and drives on Route 66 at an average rate of 50 mph toward Chicago. If Judy and Kamilia maintain these average speeds, how long after they start will they meet?

30. **Meeting at a Restaurant** Mike Mears and Scott Greenhalgh live 225 miles from each other. They frequently meet for lunch at a restaurant that is between Mike's house and Scott's house. Leaving at the same time from their respective houses, Mike takes 1 hour and 45 minutes and Scott takes 1 hour and 15 minutes to get to the restaurant. If they each drive at the same speed,

a) determine their speed,

b) how far from Scott's house is the restaurant?

31. **Sump Pump Rates** Gary Egan needs to empty his 15,000-gallon inground swimming pool, so he calls the fire department. The fire department agrees to lend him two pumps to drain the pool. One drains 10 gallons of water a minute while the other drains 20 gallons of water a minute. If the pumps are turned on at the same time and remain on until the pool is drained, how long will it take for the pool to be drained?

32. **Two Investments** Chuy Carreon invested $8000 for 1 year, part at 6% and part at 10% simple interest. How much was invested in each account if the same amount of interest was received from each account?

33. ***Antifreeze Solution*** How many quarts of pure antifreeze should Doreen Kelly add to 10 quarts of a 20% antifreeze solution to make a 50% antifreeze solution?

34. ***Trip to Hawaii*** A jetliner flew from Chicago to Los Angeles at an average speed of 500 miles per hour. Then it continued on over the Pacific Ocean to Hawaii at an average speed of 550 miles per hour. If the entire trip covered 5200 miles and the part over the ocean took twice as long as the part over land, how long did the entire trip take?

35. ***Refueling a Jet*** An Air Force jet is going on a long-distance flight and will need to be refueled in midair over the Pacific Ocean. A refueling plane that carries fuel can travel much farther than the jet but flies at a slower speed. The refueling plane and jet will leave from the same base, but the refueling plane will leave 2 hours before the jet. The jet will fly at 800 mph and the refueling plane will fly at 520 miles per hour.

a) How long after the jet takes off will the two planes meet?

b) How far from the base will the refueling take place?

36. ***Working Two Jobs*** Hal Turziz works two part-time jobs. One job pays $7.00 an hour and the other pays $7.75 an hour. Last week Hal earned a total of $190.25 and worked for a total of 26 hours. How many hours did Hal work at each job?

37. ***Painting Sales*** Peter Paul Rubin, an artist, sells both large paintings and small paintings at Jackson Square in New Orleans. He sells his small paintings for $60 and his large paintings for $180. At the end of the week he determined that the total amount he made by selling 12 paintings was $1200. Determine the number of small and the number of large paintings that he sold.

38. ***Trip to Work*** Irwin Metviner lives 28 miles from work. Due to construction he must drive the first 20 minutes at a speed 14 mph slower than he does on the rest of the trip. If the entire trip takes him 35 minutes, determine Irwin's speed on each part of the trip to work.

39. ***Alcohol Solution*** Herb Garrett has an 80% methyl alcohol solution. He wishes to make a gallon of windshield washer solution by mixing his methyl alcohol solution with water. If 128 ounces, or a gallon, of windshield washer fluid should contain 6% methyl alcohol, how much of the 80% methyl alcohol solution and how much water must be mixed?

40. ***Mowing the Lawn*** Tim Kent mows part of his lawn in second gear and part in third gear. It took him 2 hours to mow the entire lawn and the odometer on his tractor shows that he covered 13.8 miles while cutting the grass. If he averages 4.2 miles per hour in second gear and 7.8 miles per hour in third gear, how long did he cut in each gear?

41. ***Meatloaf*** Sybil Geraud is making a meat loaf by combining chopped sirloin with veal. The sirloin contains 1.2 grams of fat per ounce, and the veal contains 0.3 gram of fat per ounce. If she wants her 64-ounce mixture to have only 0.8 gram of fat per ounce, find how much sirloin and how much veal she must use.

42. ***Milk Mixture*** Sundance Dairy has 400 quarts of whole milk containing 5% butterfat. How many quarts of low-fat milk containing 1.5% butterfat should be added to produce milk containing 2% butterfat?

43. ***Comparing Transportation*** Tyrone Whitfield can ride his bike to work in $\frac{3}{4}$ hour. If he takes his car to work, the trip takes $\frac{1}{6}$ hour. If Tyrone drives his car an average of 14 miles per hour faster than he rides his bike, determine the distance he travels to work.

44. ***Milk Carton Machine*** An old machine that folds and seals milk cartons can produce 50 milk cartons per minute. A new machine can produce 70 milk cartons per minute. The old machine has made 200 milk cartons when the new machine is turned on. If both machines then continue working, how long after the new machine is turned on will the new machine produce the same total number of milk cartons as the old machine?

45. ***Ocean Salinity*** The salinity (salt content) of the Atlantic Ocean averages 37 parts per thousand. If 64 ounces of water is collected and placed in the sun, how many ounces of pure water would need to evaporate to raise the salinity to 45 parts per thousand? (Only the pure water is evaporated; the salt is left behind.)

46. ***Two Rockets*** Two rockets are launched from the Kennedy Space Center. The first rocket, launched at noon, will travel at 8000 miles per hour. The second rocket will be launched some time later and travel at 9500 miles per hour. When should the second rocket be launched if the rockets are to meet at a distance of 38,000 miles from earth?

a) Explain how to find the solution to this problem.

b) Find the solution to the problem.

47. a) Make up your own realistic motion problem that can be represented as an equation.

b) Write the equation that represents your problem.

c) Solve the equation, and then find the answer to your problem.

48. a) Make up your own realistic mixture problem that can be represented as an equation.

b) Write the equation that represents your problem.

c) Solve the equation, and then find the answer to your problem.

Challenge Problems

49. *Distance to Calais* The Chunnel (the underwater tunnel from Folkestone, England, to Calais, France) is 31 miles long. A person can board France's TGV bullet train in Paris and travel nonstop through the Chunnel and arrive in London 3 hours later. The TGV averages about 130 miles per hour from Paris to Calais. It then reduces its speed to an average of 90 miles per hour through the 31-mile Chunnel. When leaving the Chunnel in Folkestone it travels only at an average of about 45 miles per hour for the 68-mile trip from Folkestone to London because of outdated tracks. Using this information, determine the distance from Paris to Calais, France.

50. *Race Cars* Two cars labeled A and B are in a 500-lap race. Each lap is 1 mile. The lead car, A, is averaging 125 miles per hour when it reaches the halfway point. Car B is exactly 6.2 laps behind.

a) Find the average speed of car B.

b) When car A reaches the halfway point, how far behind, in seconds, is car B from car A?

51. *Antifreeze Solution* The radiator of an automobile has a capacity of 16 quarts. It is presently filled with a 20% antifreeze solution. How many quarts must be drained and replaced with pure antifreeze to make the radiator contain a 50% antifreeze solution?

Cumulative Review Exercises

[1.6] **52.** Express the quotient in scientific notation. $\dfrac{2.7 \times 10^{15}}{4.5 \times 10^4}$

Solve.

[2.1] **53.** $0.6x + 0.22 = 0.4(x - 2.3)$

54. $\dfrac{2}{9}x + 3 = x + \dfrac{1}{5}$

[2.2] **55.** Solve the equation $\dfrac{3}{5}(x - 2) = \dfrac{2}{7}(2x + 3y)$ for y.

[2.3] **56. *Truck Rental*** Hertz/Penske Truck Rental Agency charges $30 per day plus 14¢ a mile. Budget Truck Rental Agency charges $16 per day plus 24¢ a mile for the same truck. What distance would you have to drive in 1 day to make the cost of renting from Hertz/Penske equal to the cost of renting from Budget?

2.5 SOLVING LINEAR INEQUALITIES

SSM Study Guide CD/Video

MathPro 4/5 PH Math Tutor Center prenhall.com/Angel

1 Solve inequalities.
2 Graph solutions on a number line, interval notation and solution sets.
3 Solve compound inequalities involving "and."
4 Solve compound inequalities involving "or."

1 Solve Inequalities

Inequalities and set builder notation were introduced in Section 1.2. You may wish to review that section now. The inequality symbols follow.*

Inequality Symbols

$>$	is greater than
$\geq$	is greater than or equal to
$<$	is less than
$\leq$	is less than or equal to

A mathematical expression containing one or more of these symbols is called an **inequality**. The direction of the inequality symbol is sometimes called the **order** or **sense** of the inequality.

Examples of Inequalities in One Variable

$$2x + 3 \leq 5 \qquad 4x > 3x - 5 \qquad 1.5 \leq -2.3x + 4.5 \qquad \frac{1}{2}x + 3 \geq 0$$

To solve an inequality, we must isolate the variable on one side of the inequality symbol. To isolate the variable, we use the same basic techniques used in solving equations.

Properties Used to Solve Inequalities

1. If $a > b$, then $a + c > b + c$.
2. If $a > b$, then $a - c > b - c$.
3. If $a > b$, and $c > 0$, then $ac > bc$.
4. If $a > b$, and $c > 0$, then $\dfrac{a}{c} > \dfrac{b}{c}$.
5. If $a > b$, and $c < 0$, then $ac < bc$.
6. If $a > b$, and $c < 0$, then $\dfrac{a}{c} < \dfrac{b}{c}$.

The first two properties state that the same number can be added to or subtracted from both sides of an inequality. The third and fourth properties state that

*$\neq$, is not equal to, is also an inequality. $\neq$ means $<$ or $>$. Thus, $2 \neq 3$ means $2 < 3$ or $2 > 3$.

both sides of an inequality can be multiplied or divided by any positive real number. The last two properties indicate that **when both sides of an inequality are multiplied or divided by a negative number, the direction of the inequality symbol reverses**.

Example of Multiplication by a Negative Number	Example of Division by a Negative Number

Multiply both sides of the inequality by −1 and reverse the direction of the inequality symbol.

$$4 > -2$$
$$-1(4) < -1(-2)$$
$$-4 < 2$$

$$10 \geq -4$$
$$\frac{10}{-2} \leq \frac{-4}{-2}$$
$$-5 \leq 2$$

Divide both sides of the inequality by −2 and reverse the direction of the inequality symbol.

HELPFUL HINT

Do not forget to reverse the direction of the inequality symbol when multiplying or dividing both sides of the inequality by a negative number.

Inequality	Direction of Inequality Symbol
$-3x < 6$	$\dfrac{-3x}{-3} > \dfrac{6}{-3}$
$-\dfrac{x}{2} > 5$	$(-2)\left(-\dfrac{x}{2}\right) < (-2)(5)$

EXAMPLE 1 Solve the inequality $5t - 7 \geq -22$.

Solution

$$5t - 7 \geq -22$$
$$5t - 7 \boxed{+\ 7} \geq -22 \boxed{+\ 7} \qquad \text{Add 7 to both sides.}$$
$$5t \geq -15$$
$$\frac{5t}{5} \geq \frac{-15}{5} \qquad \text{Divide both sides by 5.}$$
$$t \geq -3$$

NOW TRY EXERCISE 17

The solution set is $\{t \mid t \geq -3\}$. Any real number greater than or equal to −3 will satisfy the inequality.   ✳

2 Graph Solutions on a Number Line, Interval Notation, and Solution Sets

The solution to an inequality can be indicated on a number line or written as a solution set, as was explained in Section 1.2. The solution can also be written in interval notation, as illustrated on the next page. Most instructors have a preferred way to indicate the solution to an inequality.

Recall that *a solid circle on the number line indicates that the endpoint is part of the solution, and an open circle indicates that the endpoint is not part of the solution. In interval notation, brackets,* [ ]*, are used to indicate that the endpoints are part of the solution and parentheses,* ( )*, indicate that the endpoints are not part of the solution.* The symbol ∞ is read "infinity"; it indicates that the solution set continues indefinitely. Whenever ∞ is used in interval notation, a *parenthesis* must be used on the corresponding side of the interval notation.

Solution of Inequality	Solution Set Indicated on Number Line	Solution Set Represented in Interval Notation
$x > a$		(a, ∞)
$x \geq a$		$[a, \infty)$
$x < a$		$(-\infty, a)$
$x \leq a$		$(-\infty, a]$
$a < x < b$		(a, b)
$a \leq x \leq b$		$[a, b]$
$a < x \leq b$		$(a, b]$
$a \leq x < b$		$[a, b)$
$x \geq 5$		$[5, \infty)$
$x < 3$		$(-\infty, 3)$
$2 < x \leq 6$		$(2, 6]$
$-6 \leq x \leq -1$		$[-6, -1]$

In the next example, we will solve an inequality which has fractions.

EXAMPLE 2 Solve the following inequality and give the solution both on a number line and in interval notation.

$$\frac{1}{4}z - \frac{1}{2} < \frac{2z}{3} + 2$$

Solution We can eliminate fractions from an inequality by multiplying both sides of the inequality by the least common denominator, LCD, of the fractions. In this case we multiply both sides of the inequality by 12. We then solve the resulting inequality as we did in the previous example.

$$\frac{1}{4}z - \frac{1}{2} < \frac{2z}{3} + 2$$

$$12\left(\frac{1}{4}z - \frac{1}{2}\right) < 12\left(\frac{2z}{3} + 2\right) \qquad \textit{Multipli both sides by the LCD, 12.}$$

$$3z - 6 < 8z + 24 \qquad \textit{Distributive property}$$

$$3z - 8z - 6 < 8z - 8z + 24 \qquad \textit{Subtract 8z from both sides.}$$

$$-5z - 6 < 24$$

$$-5z - 6 + 6 < 24 + 6 \qquad \textit{Add 6 to both sides.}$$

$$-5z < 30$$

$$\frac{-5z}{-5} > \frac{30}{-5} \qquad \textit{Divide both sides by} -5 \textit{ and change direction of inequality symbol.}$$

$$z > -6$$

Number Line

Interval Notation

$(-6, \infty)$

NOW TRY EXERCISE 25   The solution set is $\{z | z > -6\}$.

In Example 2 we illustrated the solution on a number line, in interval notation, and as a solution set. Your instructor may indicate which form he or she prefers.

EXAMPLE 3   Solve the inequality $2(3p - 4) + 9 \leq 8(p + 1) - 2(p - 3)$.

Solution

$$2(3p - 4) + 9 \leq 8(p + 1) - 2(p - 3)$$
$$6p - 8 + 9 \leq 8p + 8 - 2p + 6$$
$$6p + 1 \leq 6p + 14$$
$$6p - 6p + 1 \leq 6p - 6p + 14$$
$$1 \leq 14$$

Since 1 is always less than or equal to 14, the inequality is true for all real numbers. When an inequality is true for all real numbers, the solution set is *the set of all real numbers*, $\mathbb{R}$. The solution set to this example can also be indicated on a number line or given in interval notation.

NOW TRY EXERCISE 23

or $(-\infty, \infty)$

If Example 3 had resulted in the expression $1 \geq 14$, the inequality would never have been true, since 1 is never greater than or equal to 14. When an inequality is never true, it has no solution. The solution set of an inequality that has no solution is the *empty or null set*, $\{\ \}$ or $\varnothing$. We will represent the empty set on the number line as follows, .

HELPFUL HINT

Generally, when writing a solution to an inequality, we write the variable on the left. For example, when solving an inequality, if we obtain $5 \geq y$ we would write the solution as $y \leq 5$.

For example,

$-6 < x$ means $x > -6$ (inequality symbol points to -6 in both cases)

$-3 > x$ means $x < -3$ (inequality symbol points to x in both cases)

$a < x$ means $x > a$ (inequality symbol points to a in both cases)

$a > x$ means $x < a$ (inequality symbol points to x in both cases)

EXAMPLE 4   **Boxes on a Plane**   A small single-engine airplane can carry a maximum weight of 1500 pounds. Millie Johnson, the pilot, has to transport boxes weighing 80.4 pounds.

a) Write an inequality that can be used to determine the maximum number of boxes that Millie can safely place on her plane if she weighs 125 pounds.

b) Find the maximum number of boxes that Millie can transport.

Solution   **a)** Understand and Translate   Let n = number of boxes.

$$\text{Millie's weight} + \text{weight of } n \text{ boxes} \leq 1500$$
$$125 \quad + \quad 80.4n \quad \leq 1500$$

b) Carry Out

$$125 + 80.4n \leq 1500$$
$$80.4n \leq 1375$$
$$n \leq 17.1$$

Answer Therefore, Millie can transport up to 17 boxes per trip.

EXAMPLE 5

Bowling Alley Rates At the Corbin Bowl bowling alley in Tarzana, California, it costs $2.50 to rent bowling shoes and it costs $4.00 per game bowled.

a) Write an inequality that can be used to determine the maximum number of games that Ricky Olson can bowl if he has only $20.

b) Find the maximum number of games that Ricky can bowl.

Solution **a)** Understand and Translate

$$\text{let } g = \text{number of games bowled}$$
$$\text{then } 4.00g = \text{cost of bowling } g \text{ games}$$

cost of shoe rental + cost of bowling g games $\leq$ money Ricky has

$$2.50 \qquad + \qquad 4.00g \qquad \leq \qquad 20$$

b) Carry Out

$$2.50 + 4.00g \leq 20$$
$$4.00g \leq 17.50$$
$$\frac{4.00g}{4.00} \leq \frac{17.50}{4.00}$$
$$g \leq 4.375$$

Answer and Check Since he can't play a portion of a game, the maximum number of games that he can afford to bowl is 4. If Ricky were to bowl 5 games, he would owe $2.50 + 5($4.00) = $22.50, which is more money than the $20 that he has.

NOW TRY EXERCISE 65

EXAMPLE 6

Profit For a business to realize a profit, its revenue (or income), R, must be greater than its cost, C. That is, a profit will be obtained when $R > C$ (the company breaks even when $R = C$). A company that produces playing cards has a weekly cost equation of $C = 1525 + 1.7x$ and a weekly revenue equation of $R = 4.2x$, where x is the number of decks of playing cards produced and sold in a week. How many decks of cards must be produced and sold in a week for the company to make a profit?

Solution Understand and Translate The company will make a profit when $R > C$, or

$$4.2x > 1525 + 1.7x$$

Carry Out

$$2.5x > 1525$$
$$x > \frac{1525}{2.5}$$
$$x > 610$$

Answer The company will make a profit when more than 610 decks are produced and sold in a week.

NOW TRY EXERCISE 69

EXAMPLE 7

Tax Tables The Tax Relief Act of 2001 lowered the tax rates paid by most Americans. The 2001 tax rate schedule for married couples filing jointly is shown in the table on the next page.

a) Write, in interval notation, the amounts of taxable income (amount on Form 1040, line 39) that makes up each of the five listed tax brackets, that is, the 15%, 27.5%, 30.5%, 35.5%, and 39.1% tax brackets.

SCHEDULE Y-1 Use if your filing status is **Married filing jointly** or **Qualifying widow(er)**			
If the Amount on Form 1040, Line 39, Is: Over—	But Not Over—	Enter on Form 1040, Line 40	of the Amount Over—
$0	$45,200	15%	$0
45,200	109,250	$6,780.00 + 27.5%	45,200
109,250	166,500	24,393.75 + 30.5%	109,250
166,500	297,350	41,855.00 + 35.5%	166,500
297,350		88,306.75 + 39.1%	297,350

b) Determine the tax for a married couple filing jointly if their taxable income (line 39) is $36,000.

c) Determine the tax for a married couple filing jointly if their taxable income is $136,000.

Solution **a)** The words "But Not Over" mean "less than or equal to." The taxable incomes that make up the six tax brackets are:

$(0, 45,200]$ for the 15% tax bracket

$(45,200, 109,250]$ for the 27.5% tax bracket

$(109,250, 166,500]$ for the 30.5% tax bracket

$(166,500, 297,350]$ for the 35.5% tax bracket

$(297,350, \infty)$ for the 39.1% tax bracket

b) The tax for a married couple filing jointly with taxable income of $36,000 is 15% of $36,000. Therefore,

$$\text{tax} = 0.15(36,000) = \$5400$$

The tax is $5400.

c) A taxable income of $136,000 places a married couple filing jointly in the 30.5% tax bracket. The tax is $24,393.75 + 30.5% of the taxable income over $109,250. The taxable income over $109,250 is $136,000 − $109,250 = $26,750. Therefore,

$$\text{tax} = 24,393.75 + 0.305(26,750) = 24,393.75 + 8158.75 = 32,552.50$$

The tax is $32,552.50.

3 Solve Compound Inequalities Involving "And"

A **compound inequality** is formed by joining two inequalities with the word *and* or *or*. Sometimes the word *and* is implied without being written.

Examples of Compound Inequalities

$$3 < x \quad \text{and} \quad x < 5$$
$$x + 4 > 3 \quad \text{or} \quad 2x - 3 < 6$$
$$4x - 6 \geq -3 \quad \text{and} \quad x - 6 < 5$$

In this objective, we discuss compound inequalities that use or imply the word *and*. The solution of a compound inequality using the word *and* is all the numbers that make *both* parts of the inequality true. Consider

$$3 < x \quad \text{and} \quad x < 5$$

What are the numbers that satisfy both inequalities? The numbers that satisfy both inequalities may be easier to see if we graph the solution to each inequality on a number line (see Fig. 2.9). Now we can see that the numbers that satisfy both inequalities are the numbers between 3 and 5. The solution set is $\{x | 3 < x < 5\}$.

$3 < x$ (or $x > 3$) ⟵ ┼┼┼┼┼┼┼┼┼┼┼┼ ⟶
$\phantom{3 < x}$ $-3\ -2\ -1\ \ 0\ \ 1\ \ 2\ \ 3\ \ 4\ \ 5\ \ 6\ \ 7\ \ 8\ \ 9$

$x < 5$ ⟵ ┼┼┼┼┼┼┼┼┼┼┼┼ ⟶
$\phantom{x < 5}$ $-3\ -2\ -1\ \ 0\ \ 1\ \ 2\ \ 3\ \ 4\ \ 5\ \ 6\ \ 7\ \ 8\ \ 9$

FIGURE 2.9 $\qquad$ Solution: $3 < x < 5$ ⟵ ┼┼┼┼┼┼┼┼┼┼┼┼ ⟶
$\phantom{FIGURE 2.9 Solution: 3 < x < 5}$ $-3\ -2\ -1\ \ 0\ \ 1\ \ 2\ \ 3\ \ 4\ \ 5\ \ 6\ \ 7\ \ 8\ \ 9$

Recall from Chapter 1 that the intersection of two sets is the set of elements common to both sets. *To find the solution set of an inequality containing the word* **and**, *take the* **intersection** *of the solution sets of the two inequalities.*

EXAMPLE 8 $\quad$ Solve $x + 2 \le 5$ and $2x - 4 > -2$.

Solution $\quad$ Begin by solving each inequality separately.

$$x + 2 \le 5 \quad \text{and} \quad 2x - 4 > -2$$
$$x \le 3 \qquad\qquad\qquad 2x > 2$$
$$ x > 1$$

Now take the intersection of the sets $\{x | x \le 3\}$ and $\{x | x > 1\}$. When we find $\{x | x \le 3\} \cap \{x | x > 1\}$, we are finding the values of x common to both sets. Figure 2.10 illustrates that the solution set is $\{x | 1 < x \le 3\}$. In interval notation, the solution is $(1, 3]$.

$x \le 3$ ⟵ ┼┼┼┼┼┼┼┼┼┼┼┼ ⟶
 $-3\ -2\ -1\ \ 0\ \ 1\ \ 2\ \ 3\ \ 4\ \ 5\ \ 6\ \ 7\ \ 8\ \ 9$

$x > 1$ ⟵ ┼┼┼┼┼┼┼┼┼┼┼┼ ⟶
 $-3\ -2\ -1\ \ 0\ \ 1\ \ 2\ \ 3\ \ 4\ \ 5\ \ 6\ \ 7\ \ 8\ \ 9$

NOW TRY EXERCISE 57 $\quad$ FIGURE 2.10 $\qquad$ Solution: $1 < x \le 3$ ⟵ ┼┼┼┼┼┼┼┼┼┼┼┼ ⟶
$\phantom{NOW TRY EXERCISE 57 FIGURE 2.10 Solution: 1 < x \le 3}$ $-3\ -2\ -1\ \ 0\ \ 1\ \ 2\ \ 3\ \ 4\ \ 5\ \ 6\ \ 7\ \ 8\ \ 9$

Sometimes a compound inequality using the word *and* can be written in a shorter form. For example, $3 < x$ and $x < 5$ can be written as $3 < x < 5$. The word *and* does not appear when the inequality is written in this form, but it is implied. The compound inequality $1 < x + 5$ and $x + 5 \le 7$ can be written $1 < x + 5 \le 7$.

EXAMPLE 9 $\quad$ Solve $1 < x + 5 \le 7$.

Solution $\quad$ $1 < x + 5 \le 7$ means $1 < x + 5$ and $x + 5 \le 7$. Solve each inequality separately.

$$1 < x + 5 \quad \text{and} \quad x + 5 \le 7$$
$$-4 < x \qquad\qquad\qquad x \le 2$$

Remember that $-4 < x$ means $x > -4$. Figure 2.11 illustrates that the solution set is $\{x | -4 < x \leq 2\}$. In interval notation, the solution is $(-4, 2]$.

$$-4 < x \text{ (or } x > -4)$$
$$x \leq 2$$
Solution: $-4 < x \leq 2$

FIGURE 2.11

The inequality in Example 9, $1 < x + 5 \leq 7$, can be solved in another way. We can still use the properties discussed earlier to solve compound inequalities. However, when working with such inequalities, whatever we do to one part we must do to all three parts. In Example 9, we could have subtracted 5 from all three parts to isolate the variable in the middle and solve the inequality.

$$1 < x + 5 \leq 7$$
$$1 - 5 < x + 5 - 5 \leq 7 - 5$$
$$-4 < x \leq 2$$

Note that this is the same solution as obtained in Example 9.

EXAMPLE 10 Solve the inequality $-3 \leq 2x - 7 < 8$.

Solution We wish to isolate the variable x. We begin by adding 7 to all three parts of the inequality.

$$-3 \leq 2x - 7 < 8$$
$$-3 + 7 \leq 2x - 7 + 7 < 8 + 7$$
$$4 \leq 2x < 15$$

Now divide all three parts of the inequality by 2.

$$\frac{4}{2} \leq \frac{2x}{2} < \frac{15}{2}$$
$$2 \leq x < \frac{15}{2}$$

The solution may also be illustrated on a number line, written in interval notation, or written as a solution set. Below we show each form.

NOW TRY EXERCISE 35 The answer in interval notation is $\left[2, \frac{15}{2}\right)$. The solution set is $\left\{x \middle| 2 \leq x < \frac{15}{2}\right\}$.

EXAMPLE 11 Solve the inequality $-2 < \dfrac{4 - 3x}{5} < 8$.

Solution Multiply all three parts by 5 to eliminate the denominator.

$$-2 < \frac{4 - 3x}{5} < 8$$
$$-2(5) < 5\left(\frac{4 - 3x}{5}\right) < 8(5)$$

$$-10 < 4 - 3x < 40$$
$$-10 \boxed{- 4} < 4 \boxed{- 4} - 3x < 40 \boxed{- 4}$$
$$-14 < -3x < 36$$

Now divide all three parts of the inequality by -3. Remember that when we multiply or divide an inequality by a negative number, the direction of the inequality symbols reverse.

$$\frac{-14}{-3} > \frac{-3x}{-3} > \frac{36}{-3}$$
$$\frac{14}{3} > x > -12$$

Although $\frac{14}{3} > x > -12$ is correct, we generally write compound inequalities with the smaller value on the left. We will, therefore, rewrite the solution as

$$-12 < x < \frac{14}{3}$$

The solution may also be illustrated on a number line, written in interval notation, or written as a solution set.

NOW TRY EXERCISE 43   The solution in interval notation is $\left(-12, \frac{14}{3}\right)$. The solution set is $\left\{x \mid -12 < x < \frac{14}{3}\right\}$. ✳

HELPFUL HINT

You must be very careful when writing the solution to a compound inequality. In Example 11 we can change the solution from

$$\frac{14}{3} > x > -12 \quad \text{to} \quad -12 < x < \frac{14}{3}$$

This is correct since both say that x is greater than -12 and less than $\frac{14}{3}$. Notice that the inequality symbol in both cases is pointing to the smaller number.

In Example 11, had we written the answer $\frac{14}{3} < x < -12$, we would have given the incorrect solution. Remember that the inequality $\frac{14}{3} < x < -12$ means that $\frac{14}{3} < x$ and $x < -12$. There is no number that is both greater than $\frac{14}{3}$ and less than -12. Also, by examining the inequality $\frac{14}{3} < x < -12$, it appears as if we are saying that -12 is a greater number than $\frac{14}{3}$, which is obviously incorrect.

It would also be incorrect to write the answer as

$$\cancel{-12 < x > \frac{14}{3}} \quad \text{or} \quad \cancel{\frac{14}{3} < x > -12}$$

EXAMPLE 12 **Calculating Grades** In an anatomy and physiology course, an average score greater than or equal to 80 and less than 90 will result in a final grade of B. Steve Reinquist received scores of 85, 90, 68, and 70 on his first four exams. For Steve to receive a final grade of B in the course, between which two scores must his fifth (and last) exam fall?

Solution Let $x = $ Steve's last exam score.

$$80 \le \text{average of five exams} < 90$$
$$80 \le \frac{85 + 90 + 68 + 70 + x}{5} < 90$$

$$80 \leq \frac{313 + x}{5} < 90$$

$$400 \leq 313 + x < 450$$

$$400 - 313 \leq 313 - 313 + x < 450 - 313$$

$$87 \leq x < 137$$

Steve would need a minimum score of 87 on his last exam to obtain a final grade of B. If the highest score he could receive on the test is 100, is it possible for him to obtain a final grade of A (90 average or higher)? Explain. ✳

4 Solve Compound Inequalities Involving "Or"

The solution to a compound inequality using the word *or* is all the numbers that make *either* of the inequalities a true statement. Consider the compound inequality

$$x > 3 \quad \text{or} \quad x < 5$$

What are the numbers that satisfy the compound inequality? Let's graph the solution to each inequality on the number line (see Fig. 2.12). Note that every real number satisfies at least one of the two inequalities. Therefore, the solution set to the compound inequality is the set of all real numbers, $\mathbb{R}$.

FIGURE 2.12

Recall from Chapter 1 that the *union* of two sets is the set of elements that belong to *either* of the sets. *To find the solution set of an inequality containing the word **or**, take the **union** of the solution sets of the two inequalities that comprise the compound inequality.*

EXAMPLE 13 Solve $x + 3 \leq -1$ or $-4x + 3 < -5$.

Solution Solve each inequality separately.

$$x + 3 \leq -1 \quad \text{or} \quad -4x + 3 < -5$$
$$x \leq -4 \qquad\qquad -4x < -8$$
$$\qquad\qquad\qquad x > 2$$

Now graph each solution on number lines and then find the union (see Fig. 2.13). The union is $x \leq -4$ or $x > 2$.

FIGURE 2.13

The solution set is $\{x | x \leq -4\} \cup \{x | x > 2\}$, which can be written as $\{x | x \leq -4 \text{ or } x > 2\}$. In interval notation, the solution is $(-\infty, -4] \cup (2, \infty)$. ✳

NOW TRY EXERCISE 59

We often encounter inequalities in our daily lives. For example, on a highway the minimum speed may be 45 miles per hour and the maximum speed 65 miles per hour. A restaurant may have a sign stating that maximum capacity is 300 people, and the minimum takeoff speed of an airplane may be 125 miles per hour.

HELPFUL HINT

There are various ways to write the solution to an inequality problem. Be sure to indicate the solution to an inequality problem in the form requested by your professor. Examples of various forms follow.

Inequality	Number Line	Interval Notation	Solution Set
$x < \dfrac{5}{3}$		$\left(-\infty, \dfrac{5}{3}\right)$	$\left\{x \mid x < \dfrac{5}{3}\right\}$
$-4 < x \le \dfrac{5}{3}$		$\left(-4, \dfrac{5}{3}\right]$	$\left\{x \mid -4 < x \le \dfrac{5}{3}\right\}$

Exercise Set 2.5

Concept/Writing Exercises

1. When solving an inequality, when is it necessary to change the direction of the inequality symbol?

2. Explain the difference between $x < 7$ and $x \le 7$.

3. a) When indicating a solution on a number line, when do you use open circles?

b) When do you use closed circles?

c) Give an example of an inequality whose solution on a number line would contain an open circle.

d) Give an example of an inequality whose solution on a number line would contain a closed circle.

4. What is a compound inequality? Give one example.

5. What does the inequality $a < x < b$ mean?

6. Explain why $\{x \mid 5 < x < 3\}$ is not an acceptable solution set for an inequality.

Practice the Skills

*Express each inequality **a)** using a number line, **b)** in interval notation, and **c)** as a solution set (use set builder notation).*

7. $x > -2$

8. $x > \dfrac{5}{2}$

9. $w \le \pi$

10. $-2 < x < 3$

11. $-3 < q \le \dfrac{3}{4}$

12. $x \ge -\dfrac{6}{5}$

13. $-7 < x \le -4$

14. $-2\dfrac{7}{8} \le k < -1\dfrac{2}{3}$

Solve each inequality and graph the solution on the number line.

15. $x - 7 > -4$

16. $2x + 3 > 4$

17. $3 - x < -4$

18. $9b - 5 \le 5b + 7$

19. $4.7x - 5.48 \ge 11.44$

20. $1.4x + 2.2 < 2.6x - 0.2$

21. $4(x - 2) \le 4x - 8$

22. $15.3 > 3(a - 1.4)$

23. $5b - 6 \ge 3(b + 3) + 2b$

24. $-7(d + 2) < -9d + 2(d - 4)$

25. $\dfrac{y}{3} + \dfrac{2}{5} \le 4$

26. $2y - 6y + 10 \le 2(-2y + 3)$

Solve each inequality and give the solution in interval notation.

27. $4 + \dfrac{4x}{3} < 6$

28. $4 - 3x < 7 + 2x + 4$

29. $\dfrac{v - 5}{3} - v \ge -3(v - 1)$

30. $\dfrac{h}{2} - \dfrac{5}{6} < \dfrac{7}{8} + h$

31. $\dfrac{t}{3} - t + 2 \le -\dfrac{4t}{3} + 3$

32. $\dfrac{3(x - 2)}{5} > \dfrac{5(2 - x)}{3}$

33. $-3x + 1 < 3[(x + 2) - 2x] - 1$

34. $4[x - (3x - 2)] > 3(x + 2) - 6$

Solve each inequality and give the solution in interval notation.

35. $-2 \le q + 3 < 4$

36. $-7 < p - 5 \le -5$

37. $-15 \le -3z \le 12$

38. $-16 < 5 - 3n \le 10$

39. $4 \le 2x - 4 < 7$

40. $-12 < 3x - 5 \le -4$

41. $14 \le 2 - 3g < 20$

42. $\dfrac{1}{2} < 3x + 4 < 6$

Solve each inequality and give the solution set.

43. $5 \le \dfrac{3x + 1}{2} < 11$

44. $\dfrac{3}{5} < \dfrac{-x - 5}{3} < 6$

45. $6 \le -3(2x - 4) < 12$

46. $-6 < \dfrac{4 - 3x}{2} < \dfrac{2}{3}$

47. $0 \le \dfrac{3(u - 4)}{7} \le 1$

48. $-15 < \dfrac{3(x - 2)}{5} \le 0$

Solve each inequality and indicate the solution set.

49. $c \le 2$ and $c > -3$

50. $d > 0$ or $d \le 5$

51. $x < 2$ and $x > 4$

52. $w \le -1$ or $w > 5$

53. $x + 1 < 3$ and $x + 1 > -4$

54. $5x - 3 \le 7$ or $-x + 3 < -5$

Solve each inequality and give the solution in interval notation.

55. $2s + 3 < 7$ or $-3s + 4 \le -17$

56. $2a + 3 > 7$ and $-3a + 4 \le -17$

57. $4x + 5 \ge 5$ and $3x - 4 \le 2$

58. $5x - 3 > 10$ and $5 - 3x < -3$

59. $4 - x < -2$ or $3x - 1 < -1$

60. $-x + 3 < 0$ or $2x - 5 \ge 3$

61. $2k + 5 > -1$ and $7 - 3k \le 7$

62. $2q - 7 \le -3$ or $2 - 3q < 11$

Problem Solving

63. **UPS Packages** The length plus the girth of a package to be shipped by United Parcel Service (UPS) can be no larger than 130 inches.

   a) Write an inequality that expresses this information, using l for the length and g for the girth.

   b) UPS has defined girth as twice the width plus twice the depth. Write an inequality using length, l, width, w, and depth, d, to indicate the maximum allowable dimensions of a package that may be shipped by UPS.

c) If the length of a package is 40 inches and the width of a package is 20.5 inches, find the maximum allowable depth of the package.

64. **Carry-On Luggage** As of October 8, 2001, many airlines have limited the size of luggage passengers may carry onboard on domestic flights. The length, l, plus the width, w, plus the depth, d, of the carry-on luggage must not exceed 45 inches.

    a) Write an inequality that describes this restriction, using l, w, and d as described above.

    b) If Mike Schacter's luggage is 26 inches long and 12 inches wide, what is the maximum depth it can have and still be carried on the plane?

In Exercises 65–79, set up an inequality that can be used to solve the problem. Solve the problem and find the desired value.

65. **Weight Limit** Cal Worth, a janitor, must move a large shipment of books from the first floor to the fifth floor. The sign on the elevator reads "maximum weight 800 pounds." If each box of books weighs 70 pounds, find the maximum number of boxes that Cal should place in the elevator.

66. **Elevator Limit** If the janitor in Exercise 65, weighing 170 pounds, must ride up with the boxes of books, find the maximum number of boxes that can be placed into the elevator.

67. **Long Distance** The telephone long-distance carrier Telecom-USA, which markets itself as 10-10-220, charges customers $0.99 for the first 20 minutes and then $0.07 for each minute (or any part thereof) beyond 20 minutes. If Patricia Lanz uses this carrier, how long can she talk for $5.00?

68. **Parking Garage** A downtown parking garage in Austin, Texas, charges $1.25 for the first hour and $0.75 for each additional hour or part thereof. What is the maximum length of time you can park in the garage if you wish to pay no more than $3.75?

69. **Book Profit** Miriam Ladlow is considering writing and publishing her own book. She estimates her revenue equation as $R = 6.42x$, and her cost equation as $C = 10,025 + 1.09x$, where x is the number of books she sells. Find the minimum number of books she must sell to make a profit. See Example 6.

70. **Dry Cleaning Profit** Peter Collinge is opening a dry cleaning store. He estimates his cost equation as $C = 8000 + 0.08x$ and his revenue equation as $R = 1.85x$, where x is the number of garments dry cleaned in a year. Find the minimum number of garments that must be dry cleaned in a year for Peter to make a profit.

71. **First-Class Mail** On July 1, 2002, the cost for mailing a package first class was $0.37 for the first ounce and $0.23 for each additional ounce. What is the maximum weight of a package that Richard Van Lommel can mail first class for $10.00?

72. **Presorted First-Class Mail** Companies can send pieces of mail weighing up to 1 ounce by using *presorted first-class* mail. The company must first purchase a bulk permit for $150 per year, and then pay $0.275 per piece sent. Without the permit, each piece would cost $0.37. Determine the minimum number of pieces of mail that would have to be mailed for it to be financially worthwhile for a company to use presorted first-class mail.

73. **Comparing Payment Plans** Shatronda Ware recently accepted a sales position in Ohio. She can select between two payment plans. Plan 1 is a salary of $300 per week plus a 10% commission on sales. Plan 2 is a salary of $400 per week plus an 8% commission on sales. For what amount of weekly sales would Shatronda earn more by plan 1?

74. **College Employment** To be eligible to continue her financial assistance for college, Nikita Maxwell can earn no more than $2000 during her 8-week summer employment. She already earns $90 per week as a day-care assistant. She is considering adding an evening job at a fast-food restaurant, where she will earn $6.25 per hour. What is the maximum number of hours she can work at the restaurant without jeopardizing her financial assistance?

75. **A Passing Grade** To pass a course, Maria Matute needs an average score of 60 or more. If Maria's scores are 65, 72, 90, 47, and 62, find the minimum score that Maria can get on her sixth and last exam and pass the course.

76. **Minimum Grade** To receive an A in a course, Ray Renardo must obtain an average score of 90 or higher on five exams. If Ray's first four exam scores are 90, 87, 96, and 79, what is the minimum score that Ray can receive on the fifth exam to get an A in the course?

77. **Averaging Grades** Calisha Mahoney's grades on her first four exams are 87, 92, 70, and 75. An average greater than or equal to 80 and less than 90 will result in a final grade of B. What range of grades on Calisha's fifth and last exam will result in a final grade of B? Assume a maximum grade of 100.

78. **Clean Air** For air to be considered "clean," the average of three pollutants must be less than 3.2 parts per million. If the first two pollutants are 2.7 and 3.42 ppm, what values of the third pollutant will result in clean air?

79. *Water Acidity* Bob Angel is checking the water acidity in a swimming pool. The water acidity is considered normal when the average pH reading of three daily measurements is greater than 7.2 and less than 7.8. If the first two pH readings are 7.48 and 7.15, find the range of pH values for the third reading that will result in the acidity level being normal.

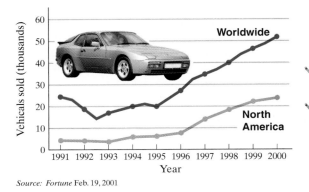

80. *Income Taxes* Refer to Example 7 on page 119. Su-hua and Ting-Fang Zheng file a joint tax return. Determine the 2001 income tax Su-hua and Ting-Fang will owe if their taxable income is

a) $78,221.

b) $301,233.

81. *Income Taxes* Refer to Example 7 on page 119. Jose and Mildred Battiste file a joint tax return. Determine the 2001 income tax Jose and Mildred will owe if their taxable income is

a) $128,479.

b) $175,248.

82. *Porsche Sales* Since the mid-1990s, sales of Porsche automobiles have rapidly increased. The following graph illustrates the number of Porsche vehicles sold (in thousands) from 1991 to 2000.

Porsche's Sales

Source: Fortune Feb. 19, 2001

a) In which years from 1991 to 2000 were fewer than 20,000 Porsche vehicles sold worldwide *and* fewer than 10,000 sold in North America? Explain how you determined your answer.

b) In which years from 1991 to 2000 were more than 30,000 Porsche vehicles sold worldwide *or* more than 20,000 Porsche vehicles sold in North America? Explain how you determined your answer.

83. *Comparing Debts* Fannie Mae and Freddie Mac are government-sponsored corporations designed to lend money to people wishing to purchase homes. Due to low interest rates and increased consumer purchasing power, since 1995 the debt of Fannie Mae and Freddie Mac has been sharply increasing. At the same time, the debt of the U.S. Treasury has been sharply decreasing. The following graph displays the projected debts of Fannie Mae and Freddie Mac as well as the projected debt of the U.S. Treasury from the years 2001 to 2005.

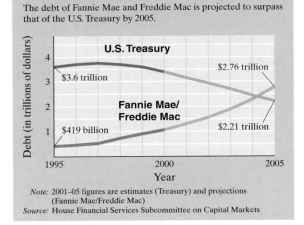

a) During which years from 1995 to 2005 was the Fannie Mae/Freddie Mac debt below $1 trillion *and* the U.S. Treasury debt above $3 trillion? Explain how you determined your answer.

b) During which years from 1995 to 2005 was the Fannie Mae/Freddie Mac debt above $1 trillion *or* the U.S. Treasury debt below $3 trillion? Explain how you determined your answer.

84. If $a > b$, will a^2 always be greater than b^2? Explain and give an example to support your answer.

85. *Insurance Policy* A Blue Cross/Blue Shield insurance policy has a $100 deductible, after which it pays 80% of the total medical cost, c. The customer pays 20% until the customer has paid a total of $500, after which the policy pays 100% of the medical cost. We can describe this policy as follows:

Blue Cross Pays

$$0, \quad \text{if } c \le \$100$$
$$0.80(c - 100), \quad \text{if } \$100 < c \le \$2100$$
$$c - 500, \quad \text{if } c > \$2100$$

Explain why this set of inequalities describes Blue Cross/Blue Shield's payment plan.

86. Explain why the inequality $a < bx + c < d$ cannot be solved for x unless additional information is given.

Challenge Problems

87. *Calculating Grades* Russell Meek's first five scores in European history were 82, 90, 74, 76, and 68. The final exam for the course is to count one-third in computing the final average. A final average greater than or equal to 80 and less than 90 will result in a final grade of B. What range of final exam scores will result in Russell's receiving a final grade of B in the course? Assume that a maximum score of 100 is possible.

*In Exercises 88–90, **a)** explain how to solve the inequality, and **b)** solve the inequality and give the solution in interval notation.*

88. $x < 3x - 10 < 2x$

89. $x < 2x + 3 < 2x + 5$

90. $x + 3 < x + 1 < 2x$

Cumulative Review Exercises

[1.2] **91.** For $A = \{1, 2, 6, 8, 9\}$ and $B = \{1, 3, 4, 5, 8\}$, find
  a) $A \cup B$.
  b) $A \cap B$.

92. For $A = \left\{ -3, 4, \dfrac{5}{2}, \sqrt{7}, 0, -\dfrac{29}{80} \right\}$, list the elements that are
  a) counting numbers.
  b) whole numbers.
  c) rational numbers.
  d) real numbers.

[1.3] *Name each illustrated property.*

93. $(3x + 6) + 4y = 3x + (6 + 4y)$

94. $3x + y = y + 3x$

[2.2] **95.** Solve the formula $R = L + (V - D)r$ for V.

2.6 SOLVING EQUATIONS AND INEQUALITIES CONTAINING ABSOLUTE VALUES

SSM Study Guide CD/Video

MathPro 4/5 PH Math Tutor Center prenhall.com/Angel

1 Understand the geometric interpretation of absolute value.
2 Solve equations of the form $|x| = a,\ a > 0$.
3 Solve inequalities of the form $|x| < a,\ a > 0$.
4 Solve inequalities of the form $|x| > a,\ a > 0$.
5 Solve inequalities of the form $|x| > a$ or $|x| < a,\ a < 0$.
6 Solve inequalities of the form $|x| > 0$ or $|x| < 0$.
7 Solve equations of the form $|x| = |y|$.

1 Understand the Geometric Interpretation of Absolute Value

In Section 1.3 we introduced the concept of absolute value. We stated that the absolute value of a number may be considered the distance (without sign) from the number 0 on the number line. The absolute value of 3, written $|3|$, is 3 since it is 3 units from 0 on the number line. Similarly, the absolute value of −3, written $|-3|$, is also 3 since it is 3 units from 0 on the number line.

Consider the equation $|x| = 3$; what values of x make this equation true? We know that $|3| = 3$ and $|-3| = 3$. The solutions to $|x| = 3$ are 3 and −3. When solving the equation $|x| = 3$, we are finding the values whose distances are exactly 3 units from 0 on the number line (see Fig. 2.14a).

FIGURE 2.14

Now consider the inequality $|x| < 3$. To solve this inequality, we need to find the set of values whose distances are less than 3 units from 0 on the number line. These are the values of x between −3 and 3 (see Fig. 2.14b).

To solve the inequality $|x| > 3$, we need to find the set of values whose distances are greater than 3 units from 0 on the number line. These are the values that are either less than −3 or greater than 3 (see Fig. 2.14c).

In this section we will solve equations and inequalities such as the following:

$$|2x - 1| = 5 \qquad |2x - 1| \le 5 \qquad |2x - 1| > 5$$

The geometric interpretation of $|2x - 1| = 5$ is similar to $|x| = 3$. When solving $|2x - 1| = 5$, we are determining the set of values that result in $2x - 1$ being exactly 5 units away from 0 on the number line.

The geometric interpretation of $|2x - 1| \le 5$ is similar to the geometric interpretation of $|x| \le 3$. When solving $|2x - 1| \le 5$, we are determining the set of values that result in $2x - 1$ being less than or equal to 5 units from 0 on the number line.

The geometric interpretation of $|2x - 1| > 5$ is similar to that of $|x| > 3$. When solving $|2x - 1| > 5$, we are determining the set of values that result in $2x - 1$ being greater than 5 units from 0 on the number line.

We will be solving absolute value equations and inequalities algebraically in the remainder of this section. We will first solve absolute value equations, then we will solve absolute value inequalities. We will end the section by solving absolute value equations where both sides of the equation contain an absolute value, for example, $|x + 3| = |2x - 5|$.

2 Solve Equations of the Form $|x| = a, \ a > 0$

When solving an equation of the form $|x| = a, a \ge 0$, we are finding the values that are exactly a units from 0 on the number line. The following procedure may be used to solve such problems.

To Solve Equations of the Form $\|x\| = a$
If $\|x\| = a$ and $a > 0$, then $x = a$ or $x = -a$.

EXAMPLE 1   Solve each equation   **a)** $|x| = 7$ **b)** $|x| = 0$ **c)** $|x| = -7$

Solution   **a)** Using the procedure, we get $x = 7$ or $x = -7$. The solution set is $\{-7, 7\}$.

b) The only real number whose absolute value equals 0 is 0. Thus, the solution set for $|x| = 0$ is $\{0\}$.

c) The absolute value of a number is never negative, so there are no solutions to this equation. The solution set is Ø.   ✳

EXAMPLE 2   Solve the equation $|2w - 1| = 5$.

Solution   At first this might not appear to be of the form $|x| = a$. However, if we let $2w - 1$ be x and 5 be a, you will see that the equation is of this form. We are looking for the values of w such that $2w - 1$ is exactly 5 units from 0 on a number line. Thus, the quantity $2w - 1$ must be equal to 5 or -5.

$$2w - 1 = 5 \quad \text{or} \quad 2w - 1 = -5$$
$$2w = 6 \qquad\qquad 2w = -4$$
$$w = 3 \qquad\qquad w = -2$$

Check

$w = 3$ $|2w - 1| = 5$ $w = -2$ $|2w - 1| = 5$
            $|2(3) - 1| \overset{?}{=} 5$ $|2(-2) - 1| \overset{?}{=} 5$
            $|6 - 1| \overset{?}{=} 5$ $|-4 - 1| \overset{?}{=} 5$
            $|5| \overset{?}{=} 5$ $|-5| \overset{?}{=} 5$
            $5 = 5$ *True* $5 = 5$ *True*

NOW TRY EXERCISE 21   The solutions 3 and -2 each result in $2w - 1$ being 5 units from 0 on the number line. The solution set is $\{-2, 3\}$.   ✳

Consider the equation $|2w - 1| - 3 = 2$. The first step in solving this equation is to isolate the absolute value term. We do this by adding 3 to both sides of the equation. This results in the equation we solved in Example 2.

3   Solve Inequalities of the Form $|x| < a, \ a > 0$

Now let's look at inequalities of the form $|x| < a$. Consider $|x| < 3$. This inequality represents the set of values that are less than 3 units from 0 on a number line (see Fig. 2.14b). The solution set is $\{x | -3 < x < 3\}$. The solution set to an inequality of the form $|x| < a$ is the set of values that are *less than a units from 0 on a number line*.

We can use the same reasoning process to solve more complicated problems, as shown in Example 3.

EXAMPLE 3   Solve the inequality $|2x - 3| < 5$.

Solution   The solution to this inequality will be the set of values such that the distance between $2x - 3$ and 0 on a number line will be less than 5 units (see Fig. 2.15). Using Figure 2.15, we can see that $-5 < 2x - 3 < 5$.

FIGURE 2.15

Solving, we get

$$-5 < 2x - 3 < 5$$
$$-2 < 2x < 8$$
$$-1 < x < 4$$

The solution set is $\{x|-1 < x < 4\}$. When x is any number between -1 and 4, the expression $2x - 3$ will represent a number that is less than 5 units from 0 on a number line (or a number between -5 and 5). ✳

To solve inequalities of the form $|x| < a$, we can use the following procedure.

To Solve Inequalities of the Form $\lvert x\rvert < a$
If $\lvert x\rvert < a$ and $a > 0$, then $-a < x < a$.

EXAMPLE 4 Solve the inequality $|3x - 4| \leq 5$ and graph the solution on a number line.

Solution Since this inequality is of the form $|x| \leq a$, we write

$$-5 \leq 3x - 4 \leq 5$$
$$-1 \leq 3x \leq 9$$
$$-\frac{1}{3} \leq x \leq 3$$

Any value of x greater than or equal to $-\frac{1}{3}$ and less than or equal to 3 would result in $3x - 4$ being less than or equal to 5 units from 0 on a number line. ✳

EXAMPLE 5 Solve the inequality $|5.3 - 2x| - 8.1 < 9.4$ and graph the solution on a number line.

Solution First isolate the absolute value by adding 8.1 to both sides of the inequality. Then solve as in the previous examples.

$$|5.3 - 2x| - 8.1 < 9.4$$
$$|5.3 - 2x| < 17.5$$
$$-17.5 < 5.3 - 2x < 17.5$$
$$-22.8 < -2x < 12.2$$
$$\frac{-22.8}{-2} > \frac{-2x}{-2} > \frac{12.2}{-2}$$
$$11.4 > x > -6.1 \quad \text{or} \quad -6.1 < x < 11.4$$

The solution set is $\{x|-6.1 < x < 11.4\}$. The solution set in interval notation is

NOW TRY EXERCISE 37 $(-6.1, 11.4)$. ✳

4 Solve Inequalities of the Form $|x| > a$, $a > 0$

Now we look at inequalities of the form $|x| > a$. Consider $|x| > 3$. This inequality represents the set of values that are greater than 3 units from 0 on a number line (see Fig. 2.14c). The solution set is $\{x | x < -3 \text{ or } x > 3\}$. The solution set to $|x| > a$ is the set of values that are *greater than a units from* 0 on a number line.

EXAMPLE 6 Solve the inequality $|2x - 3| > 5$ and graph the solution on a number line.

Solution The solution to $|2x - 3| > 5$ is the set of values such that the distance between $2x - 3$ and 0 on a number line will be greater than 5. The quantity $2x - 3$ must be either less than -5 or greater than 5 (see Fig. 2.16).

$$\leftarrow 2x - 3 \rightarrow| \qquad\qquad |\leftarrow 2x - 3 \rightarrow$$

FIGURE 2.16
$$-8\ -7\ -6\ -5\ -4\ -3\ -2\ -1\ \ 0\ \ 1\ \ 2\ \ 3\ \ 4\ \ 5\ \ 6\ \ 7\ \ 8$$

Since $2x - 3$ must be either less than -5 or greater than 5, we set up and solve the following compound inequality:

$$\begin{aligned}2x - 3 &< -5 \quad &\text{or} \quad 2x - 3 &> 5 \\ 2x &< -2 & 2x &> 8 \\ x &< -1 & x &> 4\end{aligned}$$

$$-5\ -4\ -3\ -2\ -1\ \ 0\ \ 1\ \ 2\ \ 3\ \ 4\ \ 5\ \ 6\ \ 7$$

The solution set to $|2x - 3| > 5$ is $\{x | x < -1 \text{ or } x > 4\}$. When x is any number less than -1 or greater than 4, the expression $2x - 3$ will represent a number that is greater than 5 units from 0 on a number line (or a number less than -5 or greater than 5). ✳

To solve inequalities of the form $|x| > a$, we can use the following procedure.

To Solve Inequalities of the Form $|x| > a$

If $|x| > a$ and $a > 0$, then $x < -a$ or $x > a$.

EXAMPLE 7 Solve the inequality $|2x - 1| \geq 7$ and graph the solution on a number line.

Solution Since this inequality is of the form $|x| \geq a$, we use the procedure given above.

$$\begin{aligned}2x - 1 &\leq -7 \quad &\text{or} \quad 2x - 1 &\geq 7 \\ 2x &\leq -6 & 2x &\geq 8 \\ x &\leq -3 & x &\geq 4\end{aligned}$$

$$-6\ -5\ -4\ -3\ -2\ -1\ \ 0\ \ 1\ \ 2\ \ 3\ \ 4\ \ 5\ \ 6$$

Any value of x less than or equal to -3, or greater than or equal to 4, would result in $2x - 1$ representing a number that is greater than or equal to 7 units from 0 on a number line. The solution set is $\{x | x \leq -3 \text{ or } x \geq 4\}$. In interval notation, the solution is $(-\infty, -3] \cup [4, \infty)$. ✳

EXAMPLE 8 Solve the inequality $\left|\dfrac{3x - 4}{2}\right| \geq 9$ and graph the solution on a number line.

Solution   Since this inequality is of the form $|x| \geq a$, we write

$$\frac{3x - 4}{2} \leq -9 \quad \text{or} \quad \frac{3x - 4}{2} \geq 9$$

Now multiply both sides of each inequality by the least common denominator, 2. Then solve each inequality.

$$2\left(\frac{3x - 4}{2}\right) \leq -9 \cdot 2 \quad \text{or} \quad 2\left(\frac{3x - 4}{2}\right) \geq 9 \cdot 2$$

$$3x - 4 \leq -18 \qquad\qquad 3x - 4 \geq 18$$
$$3x \leq -14 \qquad\qquad\quad 3x \geq 22$$
$$x \leq -\frac{14}{3} \qquad\qquad\quad x \geq \frac{22}{3}$$

NOW TRY EXERCISE 53

HELPFUL HINT

Some general information about equations and inequalities containing absolute value follows. For real numbers a, b, and c, where $a \neq 0$ and $c > 0$:

Form of Equation or Inequality	The Solution Will Be:	Solution on a Number Line:
$\lvert ax + b \rvert = c$	Two distinct numbers, p and q	
$\lvert ax + b \rvert < c$	The set of numbers between two numbers, $p < x < q$	
$\lvert ax + b \rvert > c$	The set of numbers less than one number or greater than a second number, $x < p$ or $x > q$	

---

5 Solve Inequalities of the Form $|x| > a$ or $|x| < a$, $a < 0$

We have solved inequalities of the form $|x| < a$ where $a > 0$. Now let us consider what happens in an absolute value inequality when $a < 0$. Consider the inequality $|x| < -3$. Since $|x|$ will always have a value greater than or equal to 0 for any real number x, this inequality can never be true, and the solution is the empty set, $\varnothing$. Whenever we have an absolute value inequality of this type, the solution will be the empty set.

EXAMPLE 9   Solve the inequality $|3x - 8| + 3 < 2$.

Solution   Begin by subtracting 3 from both sides of the inequality.

$$|3x - 8| + 3 < 2$$
$$|3x - 8| < -1$$

NOW TRY EXERCISE 41

Since $|3x - 8|$ will always be greater than or equal to 0 for any real number x, this inequality can never be true. Therefore, the solution is the empty set, $\varnothing$.

Now consider the inequality $|x| > -3$. Since $|x|$ will always have a value greater than or equal to 0 for any real number x, this inequality will always be true.

Since every value of x will make this inequality a true statement, the solution is the set of all real numbers, $\mathbb{R}$. Whenever we have an absolute value inequality of this type, the solution will be the set of all real numbers, $\mathbb{R}$.

EXAMPLE 10 Solve the inequality $|2x + 3| + 4 \geq -7$.

Solution Begin by subtracting 4 from both sides of the inequality.

$$|2x + 3| + 4 \geq -7$$
$$|2x + 3| \geq -11$$

Since $|2x + 3|$ will always be greater than or equal to 0 for any real number x, this inequality is true for all real numbers. Thus, the solution is the set of all real numbers, $\mathbb{R}$.

NOW TRY EXERCISE 59

6 Solve Inequalities of the Form $|x| > 0$ or $|x| < 0$

Now let us discuss inequalities where one side of the inequality is 0. The only value that satisfies the equation $|x - 5| = 0$ is 5, since 5 makes the expression inside the absolute value sign 0. Now consider $|x - 5| \leq 0$. Since the absolute value can never be negative, this inequality is true only when $x = 5$. The inequality $|x - 5| < 0$ has no solution. Can you explain why? What is the solution to $|x - 5| \geq 0$? Since any value of x will result in the absolute value being greater than or equal to 0, the solution is the set of all real numbers, $\mathbb{R}$. What is the solution to $|x - 5| > 0$? The solution is every real number except 5. Can you explain why 5 is excluded from the solution?

EXAMPLE 11 Solve each inequality.   **a)** $|x + 3| > 0$ **b)** $|3x - 4| \leq 0$

Solution **a)** The inequality will be true for every value of x except -3. The solution set is $\{x | x < -3 \text{ or } x > -3\}$.

b) Determine the number that makes the absolute value equal to 0 by setting the expression within the absolute value sign equal to 0 and solving for x.

$$3x - 4 = 0$$
$$3x = 4$$
$$x = \frac{4}{3}$$

The inequality will be true only when $x = \frac{4}{3}$. The solution set is $\left\{\frac{4}{3}\right\}$.

7 Solve Equations of the Form $|x| = |y|$

Now we will discuss absolute value equations where an absolute value appears on both sides of the equation. To solve equations of the form $|x| = |y|$, use the procedure that follows.

To Solve Equations of the Form $\|x\| = \|y\|$
If $\|x\| = \|y\|$, then $x = y$ or $x = -y$.

When solving an absolute value equation with an absolute value expression on each side of the equal sign, the two expressions must have the same absolute value. Therefore, *the expressions must be equal to each other or be opposites of each other.*

EXAMPLE 12 Solve the equation $|z + 3| = |2z - 7|$.

Solution If we let $z + 3$ be x and $2z - 7$ be y, this equation is of the form $|x| = |y|$. Using the procedure given on the previous page, we obtain the two equations

$$z + 3 = 2z - 7 \quad \text{or} \quad z + 3 = -(2z - 7)$$

Now solve each equation.

$$z + 3 = 2z - 7 \quad \text{or} \quad z + 3 = -(2z - 7)$$
$$3 = z - 7 \qquad\qquad z + 3 = -2z + 7$$
$$10 = z \qquad\qquad\quad 3z + 3 = 7$$
$$\qquad\qquad\qquad\qquad 3z = 4$$
$$\qquad\qquad\qquad\qquad z = \frac{4}{3}$$

Check

$$z = 10 \quad |z + 3| = |2z - 7| \qquad z = \frac{4}{3} \quad |z + 3| = |2z - 7|$$

$$|10 + 3| \overset{?}{=} |2(10) - 7| \qquad\qquad \left|\frac{4}{3} + 3\right| \overset{?}{=} \left|2\left(\frac{4}{3}\right) - 7\right|$$

$$|13| \overset{?}{=} |20 - 7| \qquad\qquad\qquad \left|\frac{13}{3}\right| \overset{?}{=} \left|\frac{8}{3} - \frac{21}{3}\right|$$

$$|13| \overset{?}{=} |13| \qquad\qquad\qquad\qquad \left|\frac{13}{3}\right| \overset{?}{=} \left|-\frac{13}{3}\right|$$

$$13 = 13 \quad \textit{True} \qquad\qquad\qquad \frac{13}{3} = \frac{13}{3} \quad \textit{True}$$

The solution set is $\left\{10, \frac{4}{3}\right\}$.   ✳

EXAMPLE 13 Solve the equation $|4x - 7| = |6 - 4x|$.

Solution
$$4x - 7 = 6 - 4x \quad \text{or} \quad 4x - 7 = -(6 - 4x)$$
$$8x - 7 = 6 \qquad\qquad\qquad 4x - 7 = -6 + 4x$$
$$8x = 13 \qquad\qquad\qquad\qquad -7 = -6 \qquad \textit{False}$$
$$x = \frac{13}{8}$$

Since the equation $4x - 7 = -(6 - 4x)$ results in a false statement, the absolute value equation has only one solution. A check will show that the solution set is $\left\{\frac{13}{8}\right\}$.

NOW TRY EXERCISE 63   ✳

Summary of Procedures for Solving Equations and Inequalities Containing Absolute Value

For $a > 0$,

$$\text{If } |x| = a, \quad \text{then} \quad x = a \quad \text{or} \quad x = -a.$$
$$\text{If } |x| < a, \quad \text{then} \quad -a < x < a.$$
$$\text{If } |x| > a, \quad \text{then} \quad x < -a \quad \text{or} \quad x > a.$$
$$\text{If } |x| = |y|, \quad \text{then} \quad x = y \quad \text{or} \quad x = -y.$$

Exercise Set 2.6

Concept/Writing Exercises

1. How do we solve equations of the form $|x| = a, a > 0$?
2. For each of the following equations, find the solution set and explain how you determined your answer.
   a) $|x| = -2$
   b) $|x| = 0$
   c) $|x| = 2$
3. How do we solve inequalities of the form $|x| < a$, $a > 0$?
4. How do you check to see whether -7 is a solution to $|2x + 3| = 11$? Is -7 a solution?
5. How do we solve inequalities of the form $|x| > a$, $a > 0$?
6. What is the solution to $|x| < 0$? Explain your answer.
7. What is the solution to $|x| > 0$? Explain your answer.
8. Suppose m and n $(m < n)$ are two distinct solutions to the equation $|ax + b| = c$. Indicate the solutions, using both inequality symbols and the number line, to each inequality. (See the Helpful Hint on page 134.)
   a) $|ax + b| < c$

   b) $|ax + b| > c$

9. Explain how to solve an equation of the form $|x| = |y|$.
10. How many solutions will $|ax + b| = k, a \neq 0$ have if
    a) $k < 0$,
    b) $k = 0$,
    c) $k > 0$?
11. How many solutions are there to the following equations or inequalities if $a \neq 0$ and $k > 0$?

    a) $|ax + b| = k$
    b) $|ax + b| < k$
    c) $|ax + b| > k$

12. Match each absolute value equation or inequality labeled **a)** through **e)** with the graph of its solution set, labeled A.–E.

    a) $|x| = 4$ A.
    b) $|x| < 4$ B.
    c) $|x| > 4$ C.
    d) $|x| \geq 4$ D.
    e) $|x| \leq 4$ E.

13. Match each absolute value equation or inequality, labeled **a)** through **e)**, with its solution set labeled A.–E..

    a) $|x| = 5$ A. $\{x | x \leq -5 \text{ or } x \geq 5\}$
    b) $|x| < 5$ B. $\{x | -5 < x < 5\}$
    c) $|x| > 5$ C. $\{x | -5 \leq x \leq 5\}$
    d) $|x| \leq 5$ D. $\{-5, 5\}$
    e) $|x| \geq 5$ E. $\{x | x < -5 \text{ or } x > 5\}$

14. Suppose $|x| < |y|$ and $x < 0$ and $y < 0$.

    a) Which of the following must be true: $x < y, x > y$, or $x = y$?

    b) Give an example to support your answer to part **a)**.

Practice the Skills

Find the solution set for each equation.

15. $|a| = 2$

16. $|b| = 13$

17. $|c| = \dfrac{1}{2}$

18. $|x| = 0$

19. $|d| = -\dfrac{3}{4}$

20. $|l + 4| = 6$

21. $|x + 5| = 7$

22. $|3 + y| = \dfrac{3}{5}$

23. $|4.5q + 22.5| = 0$

24. $|4.7 - 1.6z| = 14.3$

25. $|5 - 3x| = \dfrac{1}{2}$

26. $|3(y + 4)| = 12$

27. $\left|\dfrac{x - 3}{4}\right| = 5$

28. $\left|\dfrac{3z + 5}{6}\right| - 3 = 6$

29. $\left|\dfrac{x - 3}{4}\right| + 4 = 4$

30. $\left|\dfrac{5x - 3}{2}\right| + 2 = 6$

Find the solution set for each inequality.

31. $|w| < 11$

32. $|p| \leq 7$

33. $|q + 5| \leq 8$

34. $|7 - x| < 5$

35. $|5b - 15| < 10$

37. $|2x + 3| - 5 \leq 10$

39. $|3x - 7| + 5 < 11$

41. $|2x - 6| + 5 \leq 2$

43. $\left|\frac{1}{2}j + 3\right| < 6$

45. $\left|\frac{k}{4} - \frac{3}{8}\right| < \frac{7}{16}$

36. $|x - 3| - 2 < 3$

38. $|4 - 3x| - 4 < 11$

40. $|2x - 3| < -4$

42. $\left|\frac{2x - 1}{3}\right| \leq \frac{5}{3}$

44. $\left|\frac{x - 3}{2}\right| - 4 \leq -2$

46. $|2x + 3| < 0$

Find the solution set for each inequality.

47. $|y| > 7$

49. $|x + 4| > 5$

51. $|7 - 3b| > 5$

53. $\left|\frac{2h - 5}{3}\right| > 1$

55. $|0.1x - 0.4| + 0.4 > 0.6$

57. $\left|\frac{x}{2} + 4\right| \geq 5$

59. $|7w + 3| - 6 \geq -6$

61. $|4 - 2x| > 0$

48. $|a| \geq 9$

50. $|2b - 7| > 3$

52. $\left|\frac{6 + 2z}{3}\right| > 2$

54. $|2x - 1| - 4 \geq 8$

56. $|3.7d + 6.9| - 2.1 > -5.4$

58. $\left|4 - \frac{3x}{5}\right| \geq 9$

60. $|3 - 2x| \geq 0$

62. $|4c - 16| > 0$

Find the solution set for each equation.

63. $|3p - 5| = |2p + 10|$

65. $|6x| = |3x - 9|$

67. $\left|\frac{2r}{3} + \frac{5}{6}\right| = \left|\frac{r}{2} - 3\right|$

69. $\left|-\frac{3}{4}m + 8\right| = \left|7 - \frac{3m}{4}\right|$

64. $|6n + 3| = |4n - 13|$

66. $|5t - 10| = |10 - 5t|$

68. $|3x - 5| = |3x + 5|$

70. $\left|\frac{3}{2}r + 2\right| = \left|\frac{1}{2}r - 3\right|$

Find the solution set for each equation or inequality.

71. $|h| = 1$

73. $|q + 6| > 2$

75. $|2w - 7| \leq 9$

77. $|5a - 1| = 9$

79. $|5 + 2x| > 0$

81. $|4 + 3x| \leq 9$

83. $|3n + 8| - 4 = -10$

85. $\left|\frac{w + 4}{3}\right| - 1 < 3$

87. $\left|\frac{3x - 2}{4}\right| + 5 \geq 5$

89. $|2x - 8| = \left|\frac{1}{2}x + 3\right|$

91. $|2 - 3x| = \left|4 - \frac{5}{3}x\right|$

72. $|y| \leq 5$

74. $|9d + 7| \leq -1$

76. $|2z - 7| + 5 > 8$

78. $|2x - 4| + 2 = 10$

80. $|7 - 3b| = |5b + 15|$

82. $|2.4x + 4| + 4.9 > 1.9$

84. $|4 - 2x| - 5 = 5$

86. $\left|\frac{5t - 10}{6}\right| > \frac{5}{3}$

88. $\left|\frac{2x - 4}{5}\right| = 12$

90. $\left|\frac{1}{3}y + 3\right| = \left|\frac{2}{3}y - 1\right|$

92. $\left|\frac{-2u + 3}{7}\right| \leq 5$

Problem Solving

93. *Glass Thickness* Certain types of glass manufactured by PPG Industries ideally will have a thickness of 0.089 inches. However, due to limitations in the manufacturing process, the thickness is allowed to vary from the ideal thickness by up to 0.004 inch. If t represents the actual thickness of the glass, then the allowable range of thicknesses can be represented using the inequality $|t - 0.089| \leq 0.004$. *Source:* www.ppg.com

a) Solve this inequality for t (use interval notation).

b) What is the smallest thickness the glass is allowed to be?

c) What is the largest thickness the glass is allowed to be?

94. *Plywood Guarantee* Certain plywood manufactured by Lafor International is guaranteed to be $\frac{5}{8}$ inch thick with a tolerance of plus or minus $\frac{1}{56}$ of an inch. If t represents the actual thickness of the plywood, then the allowable range of thicknesses can be represented using the inequality $|t - \frac{5}{8}| \leq \frac{1}{56}$. *Source:* www.sticktrade.com

a) Solve this inequality for t (use interval notation)

b) What is the smallest thickness the plywood is allowed to be?

c) What is the largest thickness the plywood is allowed to be?

95. *Submarine Depth* A submarine is 160 feet below sea level. It has rock formations above and below it, and should not change its depth by more than 28 feet. Its distance below sea level, d, can be described by the inequality $|d - 160| \leq 28$.

a) Solve this inequality for d. Write your answer in interval notation.

b) Between what vertical distances, measured from sea level, may the submarine move?

96. *A Bouncing Spring* A spring hanging from a ceiling is bouncing up and down so that its distance, d, above the ground satisfies the inequality $|d - 4| \leq \frac{1}{2}$ foot (see the figure).

a) Solve this inequality for d. Write your answer in interval notation.

b) Between what distances, measured from the ground, will the spring oscillate?

In Exercises 97–100, determine an equation or inequality that has the given solution set.

97. $\{-5, 5\}$

98. $\{x | -5 < x < 5\}$

99. $\{x | x \leq -5 \text{ or } x \geq 5\}$

100. $\{x | -5 \leq x \leq 5\}$

101. For what value of x will the inequality $|ax + b| \leq 0$ be true? Explain.

102. For what value of x will the inequality $|ax + b| > 0$ *not* be true? Explain.

103. a) Explain how to find the solution to the equation $|ax + b| = c$. (Assume that $c > 0$ and $a \neq 0$.)

b) Solve this equation for x.

104. a) Explain how to find the solution to the inequality $|ax + b| < c$. (Assume that $a > 0$ and $c > 0$.)

b) Solve this inequality for x.

105. a) Explain how to find the solution to the inequality $|ax + b| > c$. (Assume that $a > 0$ and $c > 0$.)

b) Solve this inequality for x.

106. a) What is the first step in solving the inequality $-2|3x - 5| \leq -6$?

b) Solve this inequality and give the solution in interval notation.

Determine what values of x will make each equation true. Explain your answer.

107. $|x - 3| = |3 - x|$

108. $|x - 3| = -|x - 3|$

109. $|x| = x$

110. $|x + 2| = x + 2$

Solve. Explain how you determined your answer.

111. $|x + 1| = 2x - 1$ **112.** $|3x + 1| = x - 3$ **113.** $|x - 2| = -(x - 2)$

Challenge Problems

Solve by considering the possible signs for x.

114. $|x| + x = 6$ **115.** $x + |-x| = 6$ **116.** $|x| - x = 6$ **117.** $x - |x| = 6$

Group Activity

Discuss and answer Exercise 118 as a group.

118. Consider the equation $|x + y| = |y + x|$.

a) Have each group member select an x value and a y value and determine whether the equation holds. Repeat for two other pairs of x and y values.

b) As a group, determine for what values of x and y the equation is true. Explain your answer.

c) Now consider $|x - y| = -|y - x|$. Under what conditions will this equation be true?

Cumulative Review Exercises

Evaluate.

[1.4] **119.** $\dfrac{1}{3} + \dfrac{1}{4} \div \dfrac{2}{5}\left(\dfrac{1}{3}\right)^2$

120. $4(x + 3y) - 5xy$ when $x = 1, y = 3$

[2.4] **121.** *Swimming* Raul Sanchez swims across a lake averaging 2 miles an hour. Then he turns around and swims back across the lake, averaging 1.6 miles per hour. If his total swimming time is 1.5 hours, what is the width of the lake?

[2.5] **122.** Find the solution set to the inequality $3(x - 2) - 4(x - 3) > 2$.

CHAPTER SUMMARY

Key Words and Phrases

2.1
Coefficient (or numerical coefficient)
Conditional equation
Constant
Contradiction
Degree of a term
Equation
Equivalent equations
Identity
Least common denominator (LCD)

Least common multiple (LCM)
Like terms
Linear equations in one variable
Simplify an expression
Solution set
Solutions of an equation
Terms
Unlike terms

2.2
Formula
Mathematical model
Simple interest formula
Subscripts

2.3
Complementary angles
Supplementary angles

2.4
Distance formula
Mixture problem
Motion problem

2.5
And; intersection
Compound inequality
Inequality
Or; union
Order (or sense) of an inequality

IMPORTANT FACTS

Properties of Equality

Reflexive property: $a = a$

Symmetric property: If $a = b$, then $b = a$.

Transitive property: If $a = b$ and $b = c$, then $a = c$.

(continued on the next page)

Addition property of equality: If $a = b$, then $a + c = b + c$.
Multiplication property of equality: If $a = b$, then $ac = bc$.

To Solve Linear Equations

1. Clear fractions.
2. Simplify each side separately.
3. Isolate the variable term on one side.
4. Solve for the variable.
5. Check.

Problem-Solving Procedure

1. Understand the problem.
2. Translate the problem into mathematical language.
3. Carry out the mathematical calculations necessary to solve the problem.
4. Check the answer obtained in step 3.
5. Answer the question.

Distance Formula

distance = rate · time

or $d = rt$

Simple Interest Formula

interest = principal · rate · time

or $i = prt$

Properties Used to Solve Inequalities

1. If $a > b$, then $a + c > b + c$.
2. If $a > b$, then $a - c > b - c$.
3. If $a > b$ and $c > 0$, then $ac > bc$.
4. If $a > b$ and $c > 0$, then $\frac{a}{c} > \frac{b}{c}$.
5. If $a > b$ and $c < 0$, then $ac < bc$.
6. If $a > b$ and $c < 0$, then $\frac{a}{c} < \frac{b}{c}$.

Absolute Value, for $a > 0$

If $|x| = a$, then $x = a$ or $x = -a$.
If $|x| < a$, then $-a < x < a$.

If $|x| > a$, then $x < -a$ or $x > a$.
If $|x| = |y|$, then $x = y$ or $x = -y$.

Chapter Review Exercises

[2.1] State the degree of each term.

1. $23a^3b^5$
2. $6x$
3. $-4xyz^5$

Simplify each expression. If an expression cannot be simplified, so state.

4. $a(a + 3) - 4(a - 1)$
5. $x^2 + 2xy + 6x^2 - 4$
6. $b^2 + b - 7$
7. $2[-(x - y) + 3x] - 5y + 6$

Solve each equation. If an equation has no solution, so state.

8. $5(c + 4) - 2c = -(c - 4)$
9. $3(x + 2) - 6 = 4(x - 5)$
10. $3 + \frac{x}{2} = \frac{5}{6}$

11. $\frac{1}{2}(3t + 4) = \frac{1}{3}(4t + 1)$

12. $2\left(\frac{x}{2} - 4\right) = 3\left(x + \frac{1}{3}\right)$

13. $3x - 4 = 6x + 4 - 3x$

14. $2(x - 6) = 5 - \{2x - [4(x - 3) - 5]\}$

[2.2] Evaluate each formula for the given values.

15. $m = \dfrac{y_2 - y_1}{x_2 - x_1}$ when $y_2 = 5, y_1 = -2, x_2 = -8, x_1 = 6$

16. $x = \dfrac{-b + \sqrt{b^2 - 4ac}}{2a}$ when $a = 8, b = 10, c = -3$

17. $h = \dfrac{1}{2}at^2 + v_0t + h_0$ when $a = -32, v_0 = 0, h_0 = 80, t = 1$

18. $z = \dfrac{\overline{x} - \mu}{\dfrac{\sigma}{\sqrt{n}}}$ when $\overline{x} = 60, \mu = 64, \sigma = 5, n = 25$

Solve each equation for the indicated variable.

19. $E = IR$, for R

20. $A = \pi r^2 h$, for h

21. $P = 2l + 2w$, for w

22. $A = \dfrac{1}{2}bh$, for h

23. $y = mx + b$, for m

24. $2x - 3y = 5$, for y

25. $R_T = R_1 + R_2 + R_3$, for R_2

26. $S = \dfrac{3a + b}{2}$, for a

27. $K = 2(d + l)$, for l

[2.3] In Exercises 28–32, write an equation that can be used to solve each problem. Solve the problem and check your answer.

28. Calendar Sale On February 1, all Hallmark calendars go on sale for 75% off the original price. If Mary Beth Headlee purchases a calendar on sale for $5.50, what was the original price of the calendar?

29. Population Increase A small town's population is increasing by 350 people per year. If the present population is 4750, how long will it take for the population to reach 5800?

30. Commission Salary Dawn Clark's salary is $300 per week plus 6% commission of sales. How much in sales must Dawn make to earn $650 in a week?

31. Car Rental Comparison At the Kansas City airport, the cost to rent a Ford Focus from Hertz is $24.99 per day with unlimited mileage. The cost to rent the same car from Avis is $19.99 per day plus $0.10 per mile that the car is driven. If Cathy Panik needs to rent a car for 3 days, determine the number of miles she would need to drive in order for the cost of the car rental to be the same from both companies.

32. Sale At a going-out-of-business sale, furniture is selling at 40% off the regular price. In addition, green-tagged items are reduced by an additional $20. If Lalo Broyles purchased a green-tagged item and paid $120, find the item's regular price.

[2.4] In Exercises 33–37, solve the following motion and mixture problems.

33. Investing a Bonus After Ari Klein received a $5000 bonus at work, he invested some of the money in a money market account yielding 3.5% simple interest and the rest in a certificate of deposit yielding 4.0% simple interest. If the total amount of interest that Mr. Klein earned for the year was $187.15, determine the amount invested in each investment.

34. Fertilizer Solutions Dale Klitzke has liquid fertilizer solutions that are 20% and 60% nitrogen. How many gallons of each of these solutions should Dale mix to obtain 250 gallons of a solution that is 30% nitrogen?

35. Two Trains Two trains leave Portland, Oregon, at the same time traveling in opposite directions. One train travels at 60 miles per hour and the other at 90 miles per hour. In how many hours will they be 400 miles apart?

36. Space Shuttles Space Shuttle 2 takes off 0.5 hour after Shuttle 1 takes off. If Shuttle 2 travels 300 miles per hour faster than Shuttle 1 and overtakes Shuttle 1 exactly 5 hours after Shuttle 2 takes off, find

a) the speed of Shuttle 1 and

b) the distance from the launch pad when Shuttle 2 overtakes Shuttle 1.

37. *Mixing Coffee* Tom Tomlins, the owner of a gourmet coffee shop, has two coffees, one selling for $6.00 per pound and the other for $6.80 per pound. How many pounds of each type of coffee should he mix to make 40 pounds of coffee to sell for $6.50 per pound?

[2.3, 2.4] *Solve.*

38. *Electronics Sale* At Circuit City, the price of a cordless telephone has been reduced by 20%. If the sale price is $24, determine the original price.

39. *Jogging* Nicolle Ryba jogged for a distance and then turned around and walked back to her starting point. While jogging she averaged 7.2 miles per hour, and while walking she averaged 2.4 miles per hour. If the total time spent jogging and walking was 4 hours, find

a) how long she jogged, and

b) the total distance she traveled.

40. *Angle Measures* Find the measures of three angles of a triangle if one angle measures 25° greater than the smallest angle and the other angle measures 5° less than twice the smallest angle.

41. *Swimming Pool* Two hoses are filling a swimming pool. The hose with the larger diameter supplies 1.5 times as much water as the hose with the smaller diameter. The larger hose is on for 2 hours before the smaller hose is turned on. If 5 hours after the larger hose is turned on there are 3150 gallons of water in the pool, find the rate of flow from each hose.

42. *Complementary Angles* One complementary angle has a measure that is 15° less than twice the measure of the other angle. Determine the measures of the two angles.

43. *Blue Dye* A clothier has two blue dye solutions, both made from the same concentrate. One solution is 6% blue dye and the other is 20% blue dye. How many ounces of the 20% solution must be mixed with 10 ounces of the 6% solution to result in the mixture being a 12% blue dye solution?

44. *Two Investments* Ken Reysling invests $12,000 in two savings accounts. One account is paying 10% simple interest and the other account is paying 6% simple interest. If in one year, the same interest is earned on each account, how much was invested at each rate?

45. *Fitness Center* The West Ridge Fitness Center has two membership plans. The first plan is a flat $40-per-month fee plus $1.00 per visit. The second plan is $25 per month plus a $4.00 per visit charge. How many visits would Erick Ruston have to make per month to make it advantageous for him to select the first plan?

46. *Trains in Alaska* Two trains leave Anchorage at the same time along parallel tracks heading in opposite directions. The faster train travels 10 miles per hour faster than the slower train. Find the speed of each train if the trains are 270 miles apart after 3 hours.

[2.5] *Solve the inequality. Graph the solution on a real number line.*

47. $3z + 7 \leq 13$

48. $5 - 2w > -7$

49. $2x + 4 > 9$

50. $16 \leq 4x - 5$

51. $\dfrac{4x + 3}{5} > -3$

52. $2(x - 3) > 3x + 4$

53. $-4(x - 2) \geq 6x + 8 - 10x$

54. $\dfrac{x}{2} + \dfrac{3}{4} > x - \dfrac{x}{2} + 1$

Write an inequality that can be used to solve each problem. Solve the inequality and answer the question.

55. **Weight Limit** A canoe can safely carry a total of 500 pounds. If Joseph Streit weighs 180 pounds, what is the maximum number of 40-pound boxes of food he can safely carry in his canoe?

56. **Phone Booth Call** Jack Ruimpson, a telephone operator, informs a customer in a phone booth that the charge for calling Omaha, Nebraska, is $4.50 for the first 3 minutes and 95¢ for each additional minute or any part thereof. How long can the customer talk if he has $8.65?

57. **Fitness Center** A fitness center guarantees that customers will lose a minimum of 3 pounds the first week and $1\frac{1}{2}$ pounds each additional week. Find the maximum amount of time needed to lose 27 pounds.

58. **Exam Scores** Jekeila Ison's first four exam scores are 94, 73, 72, and 80. If a final average greater than or equal to 80 and less than 90 is needed to receive a final grade of B in the course, what range of scores on the fifth and last exam will result in Jekeila's receiving a B in the course? Assume a maximum score of 100.

Solve each inequality. Write the solution in interval notation.

59. $1 < x - 4 < 7$

60. $7 < p + 10 \leq 15$

61. $3 < 2x - 4 < 8$

62. $-12 < 6 - 3x < -2$

63. $-1 < \frac{5}{9}x + \frac{2}{3} \leq \frac{11}{9}$

64. $-8 < \frac{4 - 2x}{3} < 0$

Find the solution set to each compound inequality.

65. $h \leq 2$ and $7h - 4 > -25$

66. $2x - 1 > 5$ or $3x - 2 \leq 7$

67. $4x - 5 < 11$ and $-3x - 4 \geq 8$

68. $\frac{7 - 2g}{3} \leq -5$ or $\frac{3 - g}{8} > 1$

[2.5, 2.6] Find the solution set to each equation or inequality.

69. $|a| = 2$

70. $|x| < 3$

71. $|x| \geq 4$

72. $|l + 5| = 11$

73. $|x - 2| \geq 5$

74. $|4 - 2x| = 5$

75. $|-2q + 9| < 7$

76. $\left|\frac{2x - 3}{5}\right| = 1$

77. $\left|\frac{x - 4}{3}\right| < 6$

78. $|4d - 1| = |6d + 9|$

79. $|2x - 3| + 4 \geq -10$

Solve each inequality. Give the solution in interval notation.

80. $|3c + 8| - 5 \leq 2$

81. $3 < 2x - 5 \leq 9$

82. $-6 \leq \frac{3 - 2x}{4} < 5$

83. $2p - 5 < 7$ and $9 - 3p \leq 12$

84. $x - 3 \leq 4$ or $2x - 5 > 9$

85. $-10 < 3(x - 4) \leq 12$

Chapter Practice Test

1. State the degree of the term $-4a^2bc^4$.

Simplify.

2. $2p - 3q + 2pq - 6p(q - 3) - 4p$

3. $7q - \{2[3 - 4(q + 7)] + 5q\} - 6$

In Exercises 4–8, solve the equation.

4. $7(d + 2) = 3(2d - 4)$

5. $\frac{r}{6} + \frac{2}{3} = \frac{8}{9}$

6. $-2(x + 3) = 4\{3[x - (3x + 7)] + 2\}$

7. $7x - 6(2x - 4) = 3 - (5x - 6)$

8. $-\dfrac{1}{2}(4x - 6) = \dfrac{1}{3}(3 - 6x) + 2$

9. Find the value of S_n for the given values.

$$S_n = \frac{a_1(1 - r^n)}{1 - r}, \; a_1 = 3, r = \frac{1}{3}, n = 3$$

10. Solve $c = \dfrac{a - 3b}{2}$ for b.

11. Solve $A = \dfrac{1}{2}h(b_1 + b_2)$ for b_2.

In Exercises 12–16, write an equation that can be used to solve each problem. Solve the equation and answer the question asked.

12. **Golf Club Discount** Find the cost of a set of golf clubs, before tax, if the cost of the clubs plus 7% tax is $668.75.

13. **Health Club Costs** The cost of joining a health club is $240 per year, plus $2 per visit (for towel cleaning and toiletry expenses). If Jay Rampertaap wishes to spend a total of $400 per year for the health club, how many visits can he make?

14. **Bicycle Travel** Jeffrey Chang and Roberto Fernandez start at the same point and bicycle in opposite directions. Jeffrey's speed is 15 miles per hour and Roberto's speed is 20 miles per hour. In how many hours will the two men be 147 miles apart?

15. **Salt Solution** How many liters of a 12% salt solution must be added to 10 liters of a 25% salt solution to get a 20% salt solution?

16. **Two Investments** June White has $12,000 to invest. She places part of her money in a savings account paying 8% simple interest and the balance in a savings account paying 7% simple interest. If the total interest from the two accounts at the end of 1 year is $910, find the amount placed in each account.

Solve each inequality and graph the solution on a number line.

17. $3(2q + 4) < 5(q - 1) + 7$

18. $\dfrac{6 - 2x}{5} \geq -12$

Solve each inequality and write the solution in interval notation.

19. $x - 3 \leq 4$ and $2x - 4 > 5$

20. $1 \leq \dfrac{2u - 5}{3} < 7$

Find the solution set to the following equations.

21. $|2b + 5| = 9$

22. $|2x - 3| = \left|\dfrac{1}{2}x - 10\right|$

Find the solution set to the following inequalities.

23. $|4z + 12| = 0$

24. $|2x - 3| + 1 > 6$

25. $\left|\dfrac{2x - 3}{4}\right| \leq \dfrac{1}{2}$

Cumulative Review Test

Take the following test and check your answers with those that appear at the end of the test. Review any questions that you answered incorrectly. The section and objective where the material was covered are indicated after the answer.

1. If $A = \{1, 3, 5, 7, 9, 11, 13, 15\}$ and $B = \{2, 3, 5, 7, 11, 13\}$, find

   a) $A \cup B$.

   b) $A \cap B$.

2. Name each indicated property.

   a) $4x + y = y + 4x$

   b) $(2x)y = 2(xy)$

   c) $2(x + 3) = 2x + 6$

Evaluate.

3. $-4^2 + (-6)^2 \div (2^3 - 2)^2$

4. $a^2b^3 + ab^2 - 3b$ when $a = -1$ and $b = -2$

5. $\dfrac{8 - \sqrt[3]{27} \cdot 3 \div 9}{|-5| - [5 - (12 \div 4)]^2}$

In Exercises 6 and 7, simplify.

6. $(2x^4y^3)^{-2}$

7. $\left(\dfrac{3m^2n^{-4}}{m^{-3}n^2}\right)^2$

8. *Comparing State Sizes* Rhode Island has a land area of about 1.045×10^3 square miles. Alaska has a land area of about 5.704×10^5 square miles. How many times larger is the land area of Alaska than that of Rhode Island?

In Exercises 9–11, solve the equation.

9. $-3(y + 7) = 2(-2y - 8)$

10. $1.2(x - 3) = 2.4x - 4.98$

11. $\dfrac{2m}{3} - \dfrac{1}{6} = \dfrac{4}{9}m$

12. Explain the difference between a conditional linear equation, an identity, and a contradiction. Give an example of each.

13. Evaluate the formula $x = \dfrac{-b + \sqrt{b^2 - 4ac}}{2a}$ for $a = 3$, $b = -8$, and $c = -3$.

14. Solve the formula $y - y_1 = m(x - x_1)$ for x.

15. Solve the inequality $-4 < \dfrac{5x - 2}{3} < 2$ and give the answer:

    a) on a number line,

    b) as a solution set, and

    c) in interval notation.

In Exercises 16 and 17, find the solution set.

16. $|3h - 1| = 8$

17. $|2x - 4| - 6 \geq 18$

18. *Baseball Sale* One week after the World Series, Target marks the price of all baseball equipment down by 40%. If Maxwell Allen purchases a Louisville Slugger baseball bat for $21 on sale, what was the original price of the bat?

19. *Two Cars* Two cars leave Caldwell, New Jersey, at the same time traveling in opposite directions. The car traveling north is moving 10 miles per hour faster than the car traveling south. If the two cars are 270 miles apart after 3 hours, find the speed of each car.

20. *Mixed Nuts* Molly Fitzgerald, owner of Molly's Nut House, has cashews that cost $6.50 per pound and peanuts that cost $2.50 per pound. If she wishes to make 40 pounds of a mixture of cashews and peanuts that sells for $4.00 per pound, how many pounds of cashews and how many pounds of peanuts should Molly mix together?

Answers to Cumulative Review Test

1. a) $\{1, 2, 3, 5, 7, 9, 11, 13, 15\}$ **b)** $\{3, 5, 7, 11, 13\}$; [Sec. 1.2, Obj. 4]   **2. a)** commutative property of addition **b)** associative property of multiplication   **c)** distributive property; [Sec. 1.3, Obj. 6]   **3.** -15; [Sec. 1.4, Obj. 3]
4. -6; [Sec. 1.4, Obj. 4]   **5.** 7; [Sec. 1.4, Obj. 3]   **6.** $\dfrac{1}{4x^8y^6}$; [Sec. 1.5, Obj. 6]   **7.** $\dfrac{9m^{10}}{n^{12}}$; [Sec. 1.5, Obj. 7]   **8.** ≈ 545.8 times;
[Sec. 1.6, Obj. 3]   **9.** 5; [Sec. 2.1, Obj. 3]   **10.** 1.15 [Sec. 2.1, Obj. 3]   **11.** $\dfrac{3}{4}$; [Sec. 2.1, Obj. 4]   **12.** Conditional linear
equation is true for only one value, a linear equation that is an identity is always true, a linear equation that is a
contradiction is never true; [Sec. 2.1, Obj. 5]   **13.** 3; [Sec. 2.2, Obj. 1]   **14.** $x = \dfrac{y - y_1 + mx_1}{m}$; [Sec. 2.2, Obj. 2]

15. a)   **b)** $\left\{x \middle| -2 < x < \dfrac{8}{5}\right\}$ **c)** $\left(-2, \dfrac{8}{5}\right)$; [Sec. 2.5, Obj. 3]   **16.** $\left\{-\dfrac{7}{3}, 3\right\}$; [Sec. 2.6, Obj. 2]

17. $\{x | x \leq -10 \text{ or } x \geq 14\}$; [Sec. 2.6, Obj. 4]   **18.** $35; [Sec. 2.3, Obj. 2]   **19.** 40 miles per hour, 50 miles per hour;
[Sec. 2.4, Obj. 2]   **20.** Cashews: 15 pounds; peanuts: 25 pounds

Chapter 3

Graphs and Functions

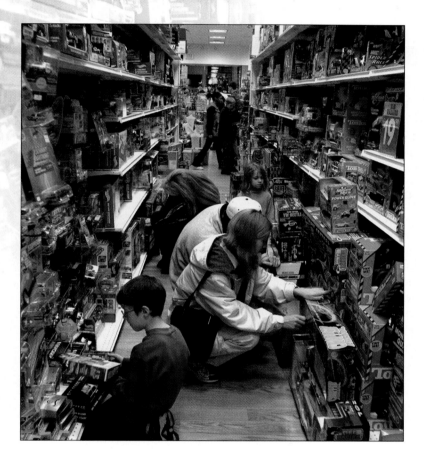

3.1 Graphs

3.2 Functions

3.3 Linear Functions: Graphs and Applications

3.4 The Slope–Intercept Form of a Linear Equation

3.5 The Point–Slope Form of a Linear Equation

3.6 The Algebra of Functions

3.7 Graphing Linear Inequalities

Chapter Summary
Chapter Review Exercises
Chapter Practice Test
Cumulative Review Test

Owning your own business is considered by many to be the ultimate American dream. If your business does well, you may be rewarded handsomely for your efforts. However, should your business do poorly then it may be only a matter of time before your business, like many new businesses, no longer exists. On page 185, we use a function to model the salary of a toy store owner. Following this example, in the exercises on pages 189 and 190, we use many other functions to model other real-life situations.

SSM   Study Guide   CD/Video   MathPro 4/5   PH Math Tutor Center   prenhall.com/Angel

A Look Ahead

Two primary goals of this book are to provide you with a good understanding of graphing and of functions. Graphing is a key element to this and many mathematics courses. We introduce graphing early in this book, and we use it to model real-life data that you see every day in newspapers and magazines. Functions are closely related to graphing, and graphing functions is a key element to many mathematics courses. We will use both graphing and functions throughout the rest of this book. Additional topics in this chapter will include graphing linear and non-linear equations and linear inequalities. We will revisit each of these topics in later chapters of the book.

3.1 GRAPHS

1. Plot points in the Cartesian coordinate system.
2. Draw graphs by plotting points.
3. Graph nonlinear equations.
4. Use a graphing calculator.
5. Interpret graphs.

1 Plot Points in the Cartesian Coordinate System

René Descartes

Many algebraic relationships are easier to understand if we can see a visual picture of them. A graph is a picture that shows the relationship between two or more variables in an equation. Before learning how to construct a graph, you must know the Cartesian coordinate system.

The **Cartesian** (or **rectangular**) **coordinate system**, named after the French mathematician and philosopher René Descartes (1596–1650), consists of two axes (or number lines) in a plane drawn perpendicular to each other (Fig. 3.1). Note how the two axes yield four **quadrants**, labeled with capital Roman numerals I, II, III, and IV.

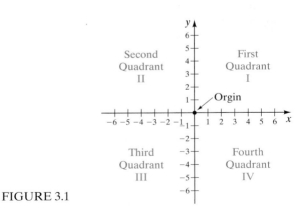

FIGURE 3.1

The horizontal axis is called the **x-axis**. The vertical axis is called the **y-axis**. The point of intersection of the two axes is called the **origin**. Starting from the origin and moving to the right, the numbers increase; moving to the left, the numbers decrease. Starting from the origin and moving up, the numbers increase; moving down, the numbers decrease. Note that the x-axis and y-axis are simply number lines, one horizontal and the other vertical.

An **ordered pair** (x, y) is used to give the two **coordinates** of a point. If, for example, the x-coordinate of a point is 2 and the y-coordinate is 3, the ordered pair representing that point is $(2, 3)$. The x-coordinate is always the first coordinate listed in the ordered pair. To plot a point, find the x-coordinate on the x-axis and the y-coordinate on the y-axis. Then suppose there was an imaginary vertical line from the x-coordinate and an imaginary horizontal line from the y-coordinate. The point is placed where the two imaginary lines intersect.

For example, the point corresponding to the ordered pair $(2, 3)$ is plotted in Figure 3.2. The phrase "the point corresponding to the ordered pair $(2, 3)$" is often abbreviated "the point $(2, 3)$." For example, if we write "the point $(-1, 5)$," it means the point corresponding to the ordered pair $(-1, 5)$. The ordered pairs A at $(-2, 3)$, B at $(0, 2)$, C at $(4, -1)$, and D at $(-4, 0)$ are plotted in Figure 3.3.

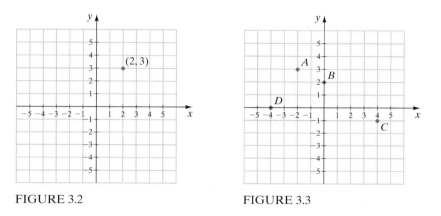

FIGURE 3.2   FIGURE 3.3

EXAMPLE 1   Plot the following points on the same set of axes.

a) $A(1, 4)$ **b)** $B(4, 1)$ **c)** $C(0, 2)$

d) $D(-3, 0)$ **e)** $E(-3, -1)$ **f)** $F(2, -4)$

Solution   See Figure 3.4. Notice that point $(1, 4)$ is a different point than $(4, 1)$. Also notice that when the x-coordinate is 0, as in part **c)**, the point is on the y-axis. When the y-coordinate is 0, as in part **d)**, the point is on the x-axis.

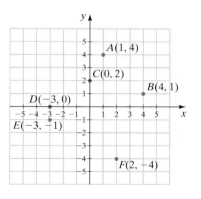

NOW TRY EXERCISE 7   FIGURE 3.4

2   Draw Graphs by Plotting Points

In Chapter 2, we solved equations that contained one variable. Now we will discuss equations that contain two variables. If an equation contains two variables, its solutions are pairs of numbers.

EXAMPLE 2 Determine whether the following ordered pairs are solutions of the equation $y = 2x - 3$.

a) $(1, -1)$ **b)** $\left(\dfrac{1}{2}, -2\right)$

c) $(3, 4)$ **d)** $(-1, -5)$

Solution We substitute the first number in the ordered pair for x and the second number for y. If the substitutions result in a true statement, the ordered pair is a solution to the equation. If the substitutions result in a false statement, the ordered pair is not a solution to the equation.

a) $y = 2x - 3$
$-1 \overset{?}{=} 2(1) - 3$
$-1 \overset{?}{=} 2 - 3$
$-1 = -1$ _True_

b) $y = 2x - 3$
$-2 \overset{?}{=} 2\left(\dfrac{1}{2}\right) - 3$
$-2 \overset{?}{=} 1 - 3$
$-2 = -2$ _True_

c) $y = 2x - 3$
$4 \overset{?}{=} 2(3) - 3$
$4 \overset{?}{=} 6 - 3$
$4 = 3$ _False_

d) $y = 2x - 3$
$-5 \overset{?}{=} 2(-1) - 3$
$-5 \overset{?}{=} -2 - 3$
$-5 \overset{?}{=} -5$ _True_

Thus the ordered pairs $(1, -1)$, $\left(\dfrac{1}{2}, -2\right)$, and $(-1, -5)$ are solutions to the equation $y = 2x - 3$. The ordered pair $(3, 4)$ is not a solution. ✳

There are many other solutions to the equation in Example 2. In fact, there are an infinite number of solutions. One method that may be used to find solutions to an equation like $y = 2x - 3$ is to substitute values for x and find the corresponding values of y. For example, to find the solution to the equation $y = 2x - 3$ when $x = 0$ we substitute 0 for x and solve for y.

$$y = 2x - 3$$
$$y = 2(0) - 3$$
$$y = 0 - 3$$
$$y = -3$$

Thus, another solution to the equation is $(0, -3)$.

A **graph** is an illustration of the set of points whose coordinates satisfy the equation. Sometimes when drawing a graph, we list some points that satisfy the equation in a table and then plot those points. We then draw a line through the points to obtain the graph. On the next page is a table of some points that satisfy the equation $y = 2x - 3$. The graph is drawn in Figure 3.5. Note that the equation $y = 2x - 3$ contains an infinite number of solutions and that the line continues indefinitely in both directions (as indicated by the arrows).

In Figure 3.5, the four points are in a straight line. Points that are in a straight line are said to be **collinear**. The graph is said to be **linear** because it is a straight line. Any equation whose graph is a straight line is called a **linear equation**. The equation $y = 2x - 3$ is an example of a linear equation. Linear equations are also called **first-degree equations** since the greatest exponent that appears on any variable is 1. In Examples 3 and 4, we graph linear equations.

x	y	(x, y)
-1	-5	$(-1, -5)$
0	-3	$(0, -3)$
$\frac{1}{2}$	-2	$\left(\frac{1}{2}, -2\right)$
1	-1	$(1, -1)$

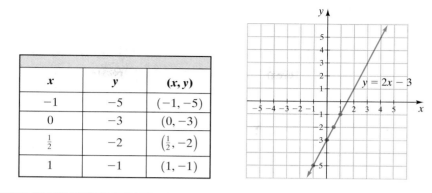

FIGURE 3.5

HELPFUL HINT

STUDY TIP

In this chapter, and in several upcoming chapters, you will be plotting points and drawing graphs using the Cartesian coordinate system. Sometimes students have trouble drawing accurate graphs. The following are some suggestions to improve the quality of the graphs you produce.

1. For your homework, use graph paper to draw your graphs. This will help you maintain a consistent scale throughout your graph. Ask your teacher if you may use graph paper on your tests and quizzes.

2. Use a ruler or straightedge to draw axes and lines. Your axes and lines will look much better and will be much more accurate if drawn with a ruler or straightedge.

3. If you are not using graph paper, use a ruler to make a consistent scale on your axes. It is impossible to get an accurate graph when axes are marked with an uneven scale.

4. Use a pencil instead of a pen. You may make a mistake as you draw your graph. A mistake can often be fixed quickly with a pencil and eraser, and you will not have to start over from the beginning.

5. You will need to practice to improve your graphing skills. Work all of the homework problems that you are assigned. To check your graphs on even-numbered exercises, you may want to use a graphing calculator.

EXAMPLE 3 Graph $y = x$.

Solution We first find some ordered pairs that are solutions by selecting values of x and finding the corresponding values of y. We will select 0, some positive values, and some negative values for x. We will also choose numbers close to 0, so that the ordered pairs will fit on the axes. The graph is illustrated in Figure 3.6.

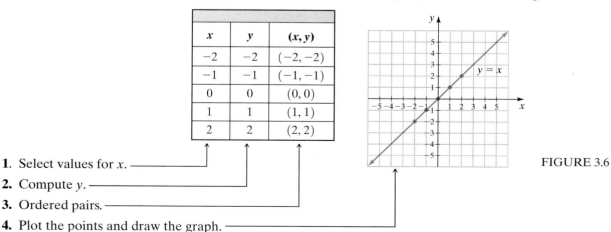

x	y	(x, y)
-2	-2	$(-2, -2)$
-1	-1	$(-1, -1)$
0	0	$(0, 0)$
1	1	$(1, 1)$
2	2	$(2, 2)$

FIGURE 3.6

1. Select values for x.
2. Compute y.
3. Ordered pairs.
4. Plot the points and draw the graph.

EXAMPLE 4   Graph $y = -\dfrac{1}{3}x + 1$.

Solution   We will select some values for x, find the corresponding values of y, and then draw the graph. When we select values for x, we will select some positive values, some negative values, and 0. The graph is illustrated in Figure 3.7. (To conserve space, we will not always list a column in the table for ordered pairs.)

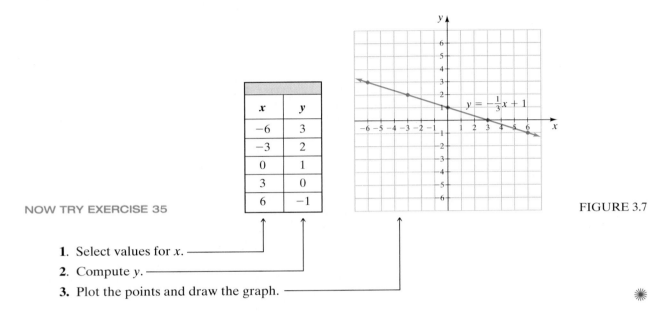

x	y
-6	3
-3	2
0	1
3	0
6	-1

NOW TRY EXERCISE 35

FIGURE 3.7

1. Select values for x.
2. Compute y.
3. Plot the points and draw the graph.

In Example 4, notice that we selected values of x that were multiples of 3 so we would not have to work with fractions.

If we are asked to graph an equation not solved for y, such as $x + 3y = 3$, our first step will be to solve the equation for y. For example, if we solve $x + 3y = 3$ for y using the procedure discussed in Section 2.2, we obtain

$$x + 3y = 3$$

$$3y = -x + 3 \qquad \textit{Subtract x from both sides.}$$

$$y = \frac{-x + 3}{3} \qquad \textit{Divide both sides by 3.}$$

$$y = \frac{-x}{3} + \frac{3}{3} = -\frac{1}{3}x + 1$$

The resulting equation, $y = -\frac{1}{3}x + 1$, is the same equation we graphed in Example 4. Therefore, the graph of $x + 3y = 3$ is also illustrated in Figure 3.7.

3   Graph Nonlinear Equations

There are many equations whose graphs are not straight lines. Such equations are called **nonlinear equations**. To graph nonlinear equations by plotting points, we follow the same procedure used to graph linear equations. However, since the graphs are not straight lines, we may need to plot more points to draw the graphs.

EXAMPLE 5   Graph $y = x^2 - 4$.

Solution   We select some values for x and find the corresponding values of y. Then we plot the points and connect them with a smooth curve. When we substitute values for x and evaluate the right side of the equation, we follow the order of operations discussed in Section 1.4. For example, if $x = -3$, then $y = (-3)^2 - 4 = 9 - 4 = 5$. The graph is shown in Figure 3.8.

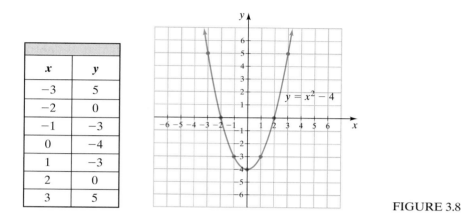

x	y
-3	5
-2	0
-1	-3
0	-4
1	-3
2	0
3	5

FIGURE 3.8

NOW TRY EXERCISE 41   If we substituted 4 for x, y would equal 12. When $x = 5$, $y = 21$. Notice that this graph rises steeply as x moves away from the origin.

EXAMPLE 6   Graph $y = \dfrac{1}{x}$.

Solution   We begin by selecting values for x and finding the corresponding values of y. We then plot the points and draw the graph. Notice that if we substitute 0 for x, we obtain $y = \frac{1}{0}$. Since $\frac{1}{0}$ is undefined, we cannot use 0 as a first coordinate. There will be no part of the graph at $x = 0$. We will plot points to the left of $x = 0$, and points to the right of $x = 0$ separately. Select points close to 0 to see what happens to the graph as x gets close to $x = 0$. Note, for example, that when $x = -\frac{1}{2}$, $y = \frac{1}{-1/2} = -2$. This graph has two branches, one to the left of the y-axis and one to the right of the y-axis, as shown in Figure 3.9.

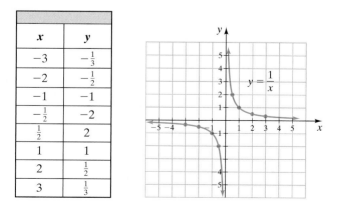

x	y
-3	$-\frac{1}{3}$
-2	$-\frac{1}{2}$
-1	-1
$-\frac{1}{2}$	-2
$\frac{1}{2}$	2
1	1
2	$\frac{1}{2}$
3	$\frac{1}{3}$

NOW TRY EXERCISE 51

FIGURE 3.9

In the graph for Example 6, notice that for values of x far to the right of 0, or far to the left of 0, the graph approaches the x-axis but does not touch it. For example when $x = 1000$, $y = 0.001$ and when $x = -1000$, $y = -0.001$. Can you explain why y can never have a value of 0?

EXAMPLE 7   Graph $y = |x|$.

Solution   Recall that $|x|$ is read "the absolute value of x." Absolute values were discussed in Section 1.3. To graph this absolute value equation, we select some values for x and find the corresponding values of y. For example, if $x = -4$, then $y = |-4| = 4$. Then we plot the points and draw the graph.
   Notice that this graph is V-shaped, as shown in Figure 3.10.

x	y
-4	4
-3	3
-2	2
-1	1
0	0
1	1
2	2
3	3
4	4

FIGURE 3.10

AVOIDING COMMON ERRORS

When graphing nonlinear equations, many students do not plot enough points to get a true picture of the graph. For example, when graphing $y = \dfrac{1}{x}$ many students consider only integer values of x. Following is a table of values for the equation and two graphs that contain the points indicated in the table.

x	-3	-2	-1	1	2	3
y	$-\frac{1}{3}$	$-\frac{1}{2}$	-1	1	$\frac{1}{2}$	$\frac{1}{3}$

Correct   Incorrect

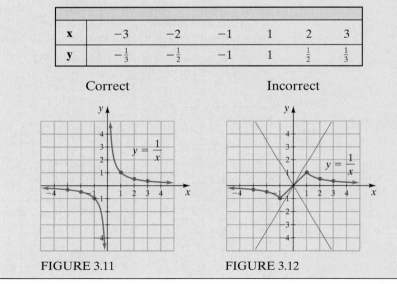

FIGURE 3.11   FIGURE 3.12

(continued on the next page)

If you select and plot fractional values of x near 0, as was done in Example 6, you get the graph in Figure 3.11. The graph in Figure 3.12 cannot be correct because the equation is not defined when x is 0 and therefore the graph cannot cross the y-axis. Whenever you plot a graph that contains a variable in the denominator, select values for the variable that are very close to the value that makes the denominator 0 and observe what happens. For example, when graphing $y = \dfrac{1}{x - 3}$ you should use values of x close to 3, such as 2.9 and 3.1 or 2.99 and 3.01, and see what values you obtain for y.

Also, when graphing nonlinear equations, it is a good idea to consider both positive and negative values. For example, if you used only positive values of x when graphing $y = |x|$, the graph would appear to be a straight line going through the origin, instead of the V-shaped graph shown in Figure 3.10.

4 Use a Graphing Calculator

If an equation is complex, finding ordered pairs can be time consuming. In this section we present a general procedure that can be used to graph equations using a **graphing calculator**.

A primary use of a graphing calculator is to graph equations. A **graphing calculator window** is the rectangular screen in which a graph is displayed. Figure 3.13 shows a TI-83 Plus calculator window with some information illustrated; Figure 3.14 shows the meaning of the information given in Figure 3.13.

FIGURE 3.13   FIGURE 3.14

The x-axis on the *standard calculator screen* goes from -10 (the minimum value of x, Xmin) to 10 (the maximum value of x, Xmax) with a scale of 1. Therefore each tick mark represents 1 unit (Xscl = 1). The y-axis goes from -10 (the minimum value of y, Ymin) to 10 (the maximum value of y, Ymax) with a scale of 1 (Yscl = 1).

Since the window is rectangular, the distances between tick marks on the standard window are greater on the horizontal axis than on the vertical axis.

When graphing you will often need to change these window values. Read your graphing calculator manual to learn how to change the window setting. On the TI-83 Plus, you press the WINDOW key and then change the settings.

Since the grapher does not display the x- and y-values in the window, we will occasionally list a set of values below the screen. Figure 3.15 shows a TI-83 Plus calculator window with the equation $y = -\frac{1}{2}x + 4$ graphed. Below the window we show six numbers, which represent in order: Xmin, Xmax, Xscl, Ymin, Ymax, and Yscl. The Xscl and Yscl represent the scale on the x- and y-axes, respectively. When we are showing the standard calculator window, we will generally not show these values below the window.

To graph the equation $y = -\frac{1}{2}x + 4$ on a TI-83 Plus, you would press

$\boxed{\text{Y=}}$ $\boxed{(-)}$ $\boxed{(}$ $\boxed{1}$ $\boxed{\div}$ $\boxed{2}$ $\boxed{)}$ $\boxed{\text{X, T, }\Theta, n}$ $\boxed{+}$ 4

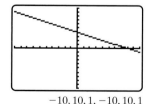

$-10, 10, 1, -10, 10, 1$

FIGURE 3.15

FIGURE 3.16

FIGURE 3.17

NOW TRY EXERCISE 93

Then when you press $\boxed{\text{GRAPH}}$, the equation is graphed. The $\boxed{\text{X, T, }\Theta, n}$ key can be used to enter any of the symbols shown on the key. In this book this key will always be used to enter the variable x.

Most graphing calculators offer a **TRACE feature** that allows you to investigate individual points after a graph is displayed. Often the $\boxed{\text{TRACE}}$ key is pressed to access this feature. After pressing the $\boxed{\text{TRACE}}$ key, you can move the flashing cursor along the line by pressing the arrow keys. As the flashing cursor moves along the line, the values of x and y change to correspond with the position of the cursor. Figure 3.16 shows the graph in Figure 3.15 after the $\boxed{\text{TRACE}}$ key has been pressed and the right arrow key has been pressed a few times.

Many graphing calculators also provide a **TABLE feature** that will illustrate a table of ordered pairs for any equation entered. On the TI-83 Plus, since TABLE appears above the $\boxed{\text{GRAPH}}$ key, to obtain a table you press $\boxed{2^{\text{nd}}}$ $\boxed{\text{GRAPH}}$. Figure 3.17 shows a table of values for the equation $y = -\frac{1}{2}x + 4$. You can scroll up and down the table by using the arrow keys.

Using TBLSET (for Table setup), you can control the x-values that appear in the table. For example, if you want the table to show values of x in tenths, you could do this using TBLSET.

This section is just a brief introduction to graphing equations, the TRACE feature, and the TABLE feature on a graphing calculator. You should read your graphing calculator manual to learn how to fully utilize these features.

5 Interpret Graphs

We see many different types of graphs daily in newspapers, in magazines, on television, and so on. Throughout this book, we present a variety of graphs. Since being able to draw and interpret graphs is very important, we will study this further in Section 3.2. In Example 8 you must understand and interpret graphs to answer the question.

EXAMPLE 8   When Jim Herring went to see his mother in Cincinnati, he boarded a Southwest Airlines plane. The plane sat at the gate for 20 minutes, taxied to the runway, and then took off. The plane flew at about 600 miles per hour for about 2 hours. It then reduced its speed to about 300 miles per hour and circled the Cincinnati Airport for about 15 minutes before it came in for a landing. After landing, the plane taxied to the gate and stopped. Which graph in Figure 3.18a–3.18d best illustrates this situation?

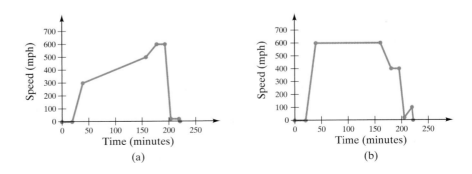

FIGURE 3.18 *(Figure continued on the next page)*

FIGURE 3.18   (c)   (d)

Solution   The graph that depicts the situation described is (c), reproduced with annotations in Figure 3.19. The graph shows speed versus time, with time on the horizontal axis. While the plane sat at the gate for 20 minutes its speed was 0 miles per hour (the horizontal line at 0 from 0 to 20 minutes). After 20 minutes the plane took off, and its speed increased to 600 miles per hour (the near-vertical line going from 0 to 600 mph). The plane then flew at about 600 miles per hour for 2 hours (the horizontal line at about 600 mph). It then slowed down to about 300 miles per hour (the near-vertical line from 600 mph to 300 mph). Next the plane circled at about 300 miles per hour for about 15 minutes (the horizontal line at about 300 mph). The plane then came in for a landing (the near-vertical line from about 300 mph to about 20 mph). It then taxied to the gate (the horizontal line at about 20 mph). Finally, it stopped at the gate (the near-vertical line when the speed dropped to 0 mph).

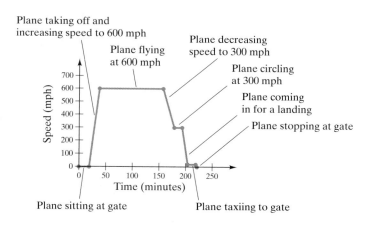

FIGURE 3.19

NOW TRY EXERCISE 81

Exercise Set 3.1

Concept/Writing Exercises

1. a) What does the graph of any linear equation look like?

    b) How many points are needed to graph a linear equation? Explain.

2. How many solutions does a linear equation in two variables have?

3. What does it mean when a set of points is collinear?

4. When graphing the equation $y = \dfrac{1}{x}$, what value cannot be substituted for x? Explain.

Practice the Skills

List the ordered pairs corresponding to the indicated points.

5.

6.

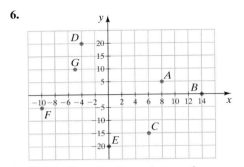

7. Graph the following points on the same axes.

$A(4, 2)$ $B(-6, 2)$ $C(0, -1)$ $D(-2, 0)$

8. Graph the following points on the same axes.

$A(-4, -2)$ $B(3, 2)$ $C(2, -3)$ $D(-3, 3)$

Determine the quadrant in which each point is located.

9. $(3, 5)$

10. $(-3, 1)$

11. $(4, -2)$

12. $(36, 41)$

13. $(-35, 18)$

14. $(-24, -8)$

15. $(-6, -19)$

16. $(8, -120)$

Determine whether the given ordered pair is a solution to the given equation.

17. $(2, 21)$; $y = 2x - 5$

18. $(1, 1)$; $2x + 3y = 6$

19. $(-4, -2)$; $y = |x| + 2$

20. $(1, 1)$; $y = x^2 + x - 1$

21. $(-2, 5)$; $s = 2r^2 - r - 5$

22. $\left(\dfrac{1}{2}, \dfrac{5}{2}\right)$; $y = |x - 3|$

23. $(2, 1)$; $-a^2 + 2b^2 = -2$

24. $(-10, -2)$; $|p| - 3|q| = 4$

25. $\left(\dfrac{1}{2}, \dfrac{3}{2}\right)$; $2x^2 + 4x - y = 0$

26. $\left(-3, \dfrac{11}{2}\right)$; $2n^2 + 3m = 2$

Graph each equation.

27. $y = x$

28. $y = 3x$

29. $y = \dfrac{1}{2}x$

30. $y = -\dfrac{1}{3}x$

31. $y = 2x + 4$

32. $y = x + 2$

33. $y = -3x - 5$

34. $y = -2x + 2$

35. $y = \dfrac{1}{2}x - 1$

36. $y = -\dfrac{1}{3}x + 2$

37. $y = -\dfrac{1}{2}x - 3$

38. $y = -\dfrac{1}{3}x + 4$

39. $y = x^2$

40. $y = x^2 - 4$

41. $y = -x^2$

42. $y = -x^2 + 4$

43. $y = |x| + 1$

44. $y = |x| + 2$

45. $y = -|x|$

46. $y = |x| - 2$

47. $y = x^3$

48. $y = -x^3$

49. $y = x^3 + 1$

50. $y = \dfrac{1}{x}$

51. $y = -\dfrac{1}{x}$

52. $x^2 = 1 + y$

53. $x = |y|$

54. $x = y^2$

In Exercises 55–62, use a calculator to obtain at least eight points that are solutions to the equation. Then graph the equation by plotting the points.

55. $y = x^3 - x^2 - x + 1$

56. $y = -x^3 + x^2 + x - 1$

57. $y = \dfrac{1}{x + 1}$

58. $y = \dfrac{1}{x} + 1$

59. $y = \sqrt{x}$

60. $y = \sqrt{x + 4}$

61. $y = \dfrac{1}{x^2}$

62. $y = \dfrac{|x^2|}{2}$

Problem Solving

63. Is the point represented by the ordered pair $\left(\dfrac{1}{3}, \dfrac{1}{12}\right)$ on the graph of the equation $y = \dfrac{x^2}{x + 1}$? Explain.

64. Is the point represented by the ordered pair $\left(-\dfrac{1}{2}, -\dfrac{3}{5}\right)$ on the graph of the equation $y = \dfrac{x^2 + 1}{x^2 - 1}$? Explain.

65. a) Plot the points $A(2, 7)$, $B(2, 3)$, and $C(6, 3)$, and then draw $\overline{AB}$, $\overline{AC}$, and $\overline{BC}$. ($\overline{AB}$ represents the line segment from A to B.)
   b) Find the area of the figure.

66. a) Plot the points $A(-4, 5)$, $B(2, 5)$, $C(2, -3)$, and $D(-4, -3)$, and then draw $\overline{AB}$, $\overline{BC}$, $\overline{CD}$, and $\overline{DA}$.
   b) Find the area of the figure.

67. *Personal Computer Sales* The following graph shows worldwide shipments of personal computers from 1999 to 2003. Shipments are measured in millions of computers.

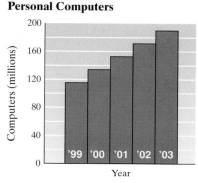

Worldwide Shipments of Personal Computers

Source: International Data Corporation

a) Estimate the worldwide shipments of computers in 1999.
b) Estimate the worldwide shipments of computers in 2003.
c) In which years were the worldwide shipments of personal computers greater than 140 million units?
d) Does the increase in worldwide shipments of personal computers from 1999 to 2003 appear to be approximately linear? Explain.

68. *Genetically Modified Crops* Worldwide production of genetically modified crops—in both developing nations and industrial nations—is rapidly increasing. The graph at the top of the next column shows the land area devoted to genetically modified crops for developing nations, industrial nations, and worldwide from 1995–2000. Land area is given in millions of hectares. A hectare is a metric system unit that is approximately equal to 2.471 acres.

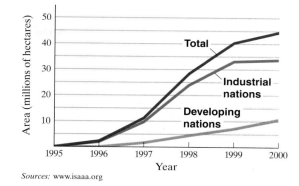

Global Area of Genetically Modified Crops

Sources: www.isaaa.org

a) Estimate the area in developing nations devoted to genetically modified crops in 1999.
b) Estimate the area in industrial nations devoted to genetically modified crops in 1999.
c) In what years from 1995 to 2000 was the total area devoted to genetically modified crops less than 20 million hectares?
d) In what years from 1995 to 2000 was the total area devoted to genetically modified crops greater than 35 million hectares?

We will discuss many of the concepts introduced in Exercises 69–76 in Section 3.4.

69. Graph $y = x + 1$, $y = x + 3$, and $y = x - 1$ on the same axes.
   a) What do you notice about the equations and the values where the graphs intersect the y-axis?
   b) Do all equations seem to have the same slant (or slope)?

70. Graph $y = \dfrac{1}{2}x$, $y = \dfrac{1}{2}x + 3$, and $y = \dfrac{1}{2}x - 4$ on the same axes.
   a) What do you notice about the equations and the values where the graphs intersect the y-axis?
   b) Do all of these graphs seem to have the same slant (or slope)?

71. Graph $y = 2x$. Determine the *rate of change* of y with respect to x. That is, by how many units does y change compared to each unit change in x?

72. Graph $y = 3x$. Determine the rate of change of y with respect to x.

73. Graph $y = 3x + 2$. Determine the rate of change of y with respect to x.

74. Graph $y = \frac{1}{2}x$. Determine the rate of change of y with respect to x.

75. The ordered pair $(3, -6)$ represents one point on the graph of a linear equation. If y increases 4 units for each unit increase in x on the graph, find two other solutions to the equation.

76. The ordered pair $(1, -4)$ represents one point on the graph of a linear equation. If y increases 3 units for each unit increase in x on the graph, find two other solutions to the equation.

Match Exercises 77–80 with the corresponding graph of elevation above sea level versus time, labeled a–d, below.

77. Mary Leeseberg walked for 5 minutes on level ground. Then for 5 minutes she climbed a slight hill. Then she walked on level ground for 5 minutes. Then for the next 5 minutes she climbed a steep hill. During the next 10 minutes she descended uniformly until she reached the height at which she had started.

78. Don Gordon walked on level ground for 5 minutes. Then he walked down a steep hill for 10 minutes. For the next 5 minutes he walked on level ground. For the next 5 minutes he walked back up to his starting height. For the next 5 minutes he walked on level ground.

79. Nancy Johnson started out by walking up a steep hill for 5 minutes. For the next 5 minutes she walked down a steep hill to an elevation lower than her starting point. For the next 10 minutes she walked on level ground. For the next 10 minutes she walked up a slight hill, at which time she reached her starting elevation.

80. James Condor started out by walking up a hill for 5 minutes. For the next 10 minutes he walked down a hill to an elevation equal to his starting elevation. For the next 10 minutes he walked on level ground. For the next 5 minutes he walked downhill.

(a)

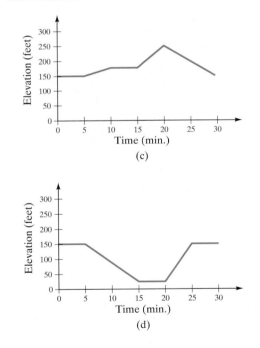

(c)

(b)

(d)

Match Exercises 81–84 with the corresponding graph of speed versus time, labeled a–d on page 161.

81. To go to work, Cletidus Hunt walked for 3 minutes, waited for the train for 5 minutes, rode the train for 15 minutes, then walked for 7 minutes.

82. To go to work, Tyrone Williams drove in stop-and-go traffic for 5 minutes, then drove on the expressway for 20 minutes, then drove in stop-and-go traffic for 5 minutes.

83. To go to work, Sheila Washington drove on a country road for 10 minutes, then drove on a highway for 12 minutes, then drove in stop-and-go traffic for 8 minutes.

84. To go to work, Brenda Pinkney rode her bike uphill for 10 minutes, then rode downhill for 15 minutes, then rode on a level street for 5 minutes.

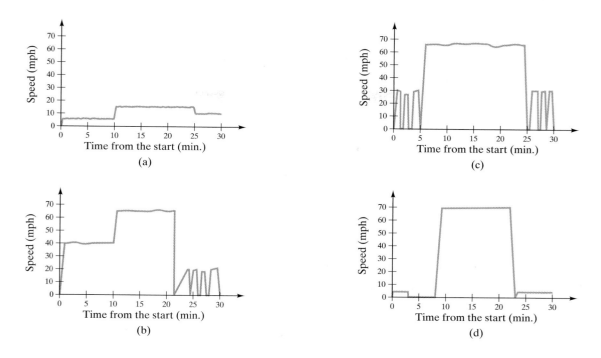

Match Exercises 85–88 with the corresponding graph of speed versus time, labeled a–d.

85. Christina Dwyer walked for 5 minutes to warm up, jogged for 20 minutes, and then walked for 5 minutes to cool down.

86. Annie Droullard went for a leisurely bike ride at a constant speed for 30 minutes.

87. Michael Odu took a 30-minute walk through his neighborhood. He stopped very brietly on 7 occasions to pick up trash.

88. Richard Dai walked through his neighborhood and stopped 3 times to chat with his neighbors. He was gone from his house a total of 30 minutes.

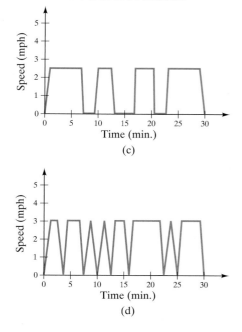

Match Exercises 89–92 with the corresponding graph of distance traveled versus time, labeled a–d. Recall from Chapter 2 that distance = rate × time. Selected distances are indicated on the graphs.

89. Train A traveled at a speed of 40 mph for 1 hour, then 80 mph for 2 hours, and then 60 mph for 3 hours.

90. Train C traveled at a speed of 80 mph for 2 hours, then stayed in a station for 1 hour, and then traveled 40 mph for 3 hours.

91. Train B traveled at a speed of 20 mph for 2 hours, then 60 mph for 3 hours, and then 80 mph for 1 hour.

92. Train D traveled at 30 mph for 1 hour, then 65 mph for 2 hours, and then 30 mph for 3 hours.

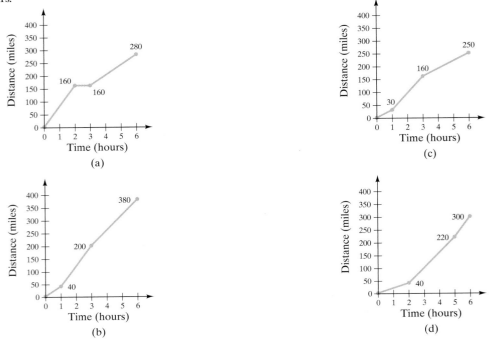

(a)

(b)

(c)

(d)

Use a graphing calculator to graph each function. Make sure you select values for the window that will show the curvature of the graph. Then, if your calculator can display tables, display a table of values in which the x-values extend by units, from 0 to 6.

93. $y = 2x - 3$

94. $y = \dfrac{1}{3}x + 2$

95. $y = x^2 - 2x - 8$

96. $y = -x^2 + 16$

97. $y = x^3 - 2x + 4$

98. $y = 2x^3 - 6x^2 - 1$

Challenge Problems

Graph each equation.

99. $y = |x - 2|$

100. $x = y^2 + 2$

Group Activity

Discuss and work Exercises 101–102 as a group.

101. a) Group member 1: Plot the points $(-2, 4)$ and $(6, 8)$. Determine the *midpoint* of the line segment connecting these points.
Group member 2: Follow the above instructions for the points $(-3, -2)$ and $(5, 6)$.
Group member 3: Follow the above instructions for the points $(4, 1)$ and $(-2, 4)$.

b) As a group, determine a formula for the midpoint of the line segment connecting the points (x_1, y_1) and (x_2, y_2). (*Note:* We will discuss the midpoint formula further in Chapter 10.)

102. Three points on a parallelogram are $A(3, 5)$, $B(8, 5)$, and $C(-1, -3)$.

a) Individually determine a fourth point D that completes the parallelogram.

b) Individually compute the area of your parallelogram.

c) Compare your answers. Did you all get the same answers? If not, why not?

d) Is there more than one point that can be used to complete the parallelogram? If so, give the points and find the corresponding areas of each parallelogram.

Cumulative Review Exercises

[2.2] **103.** Evaluate $\dfrac{-b + \sqrt{b^2 - 4ac}}{2a}$ for $a = 2$, $b = 7$, and $c = -15$.

[2.3] **104.** *Truck Rental* Hertz Truck Rental charges a daily fee of $60 plus 10¢ a mile. National Automobile Rental Agency charges a daily fee of $50 plus 24¢ a mile for the same truck. What distance would you have to drive in 1 day to make the cost of renting from Hertz equal to the cost of renting from National?

[2.5] **105.** Solve the inequality $-4 \le \dfrac{4 - 3x}{2} < 5$. Write the solution in set builder notation.

[2.6] **106.** Find the solution set for the inequality $|3x + 2| > 5$.

3.2 FUNCTIONS

SSM • Study Guide • CD/Video

MathPro 4/5 • PH Math Tutor Center • prenhall.com/Angel

1 Understand relations.
2 Recognize functions.
3 Use the vertical line test.
4 Understand function notation.
5 Applications of functions in daily life.

---

1 Understand Relations

In real life we often find that one quantity is related to a second quantity. For example, the amount you spend for oranges is related to the number of oranges you purchase. The speed of a sailboat is related to the speed of the wind. And the income tax you pay is related to the income you earn.

Suppose oranges cost 30 cents apiece. Then one orange costs 30 cents, two oranges cost 60 cents, three oranges cost 90 cents, and so on. We can list this information, or relationship, as a set of ordered pairs by listing the number of oranges first and the cost, in cents, second. The ordered pairs that represent this situation are $(1, 30)$, $(2, 60)$, $(3, 90)$, and so on. An equation that represents this situation is $c = 30n$, where c is the cost, in cents, and n is the number of oranges. Since the cost depends on the number of oranges, we say that the cost is the *dependent variable* and the number of oranges is the *independent variable*.

Now consider the equation $y = 2x + 3$. In this equation, the value obtained for y depends on the value selected for x. Therefore x is the *independent variable* and y is the *dependent variable*. Note that in this example, unlike with the oranges, there is no physical connection between x and y. The variable x is the independent variable and y is the dependent variable simply because of their placement in the equation.

For an equation in variables x and y, if the value of y depends on the value of x, then y is the **dependent variable** and x is the **independent variable**. Since related quantities can be represented as ordered pairs, the concept of a **relation** can be defined as follows.

DEFINITION	A **relation** is any set of ordered pairs.

2 Recognize Functions

We now develop the idea of a **function**—one of the most important concepts in mathematics. A function is a special type of relation in which each element in one set (called the domain) corresponds to *exactly one* element in a second set (called the range).

Consider the oranges that cost 30 cents apiece that we just discussed. We can illustrate the number of oranges and the cost of the oranges using Figure 3.20.

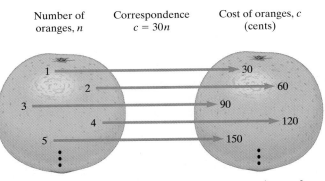

| Number of oranges, n | Correspondence $c = 30n$ | Cost of oranges, c (cents) |

FIGURE 3.20

Notice that each number in the set of numbers of oranges corresponds to (or is mapped to) exactly one number in the set of costs. Therefore, this correspondence is a function. The set consisting of the number of oranges, $\{1, 2, 3, 4, 5, \ldots\}$, is called the **domain**. The set consisting of the costs in cents, $\{30, 60, 90, 120, 150, \ldots\}$, is called the **range**. In general, the set of values for the independent variable is called the **domain**. The set of values for the dependent variable is called the **range**, see Figure 3.21.

FIGURE 3.21

EXAMPLE 1 Determine whether each correspondence is a function.

a) 1 ⟶ 1
2 ⟶ 4
3 ⟶ 9

b) ladybug
cricket ⟶ insect
wren ⟶ bird
hawk

c) JCPenney ⟶ Dallas
⟶ Milwaukee
Sears ⟶ Chicago

Solution **a)** For a correspondence to be a function, each element in the domain must correspond with exactly one element in the range. Here the domain is $\{1, 2, 3\}$ and the range is $\{1, 4, 9\}$. Since each element in the domain corresponds to exactly one element in the range, this correspondence is a function.

b) Here the domain is {ladybug, cricket, wren, hawk} and the range is {insect, bird}. Even though the domain has four elements and the range has two elements, each element in the domain corresponds with exactly one element in the range. Thus, this correspondence is a function.

c) Here the domain is {JCPenney, Sears} and the range is {Dallas, Milwaukee, Chicago}. Notice that JCPenney corresponds to both Dallas and Milwaukee. Therefore each element in the domain *does not* correspond to exactly one element in the range. Thus, this correspondence is a relation but *not* a function.

NOW TRY EXERCISE 17

Now we will formally define function.

DEFINITION A **function** is a correspondence between a first set of elements, the domain, and a second set of elements, the range, such that each element of the domain corresponds to *exactly one* element in the range.

EXAMPLE 2   Which of the following relations are functions?

a) $\{(1, 4), (2, 3), (3, 5), (-1, 3)(0, 6)\}$

b) $\{(-1, 3)(4, 2), (3, 1), (2, 6), (3, 5)\}$

Solution   **a)** The domain is the set of first coordinates in the set of ordered pairs, $\{1, 2, 3, -1, 0\}$, and the range is the set of second coordinates, $\{4, 3, 5, 6\}$. Notice that when listing the range, we only include the number 3 once, even though it appears in both $(2, 3)$ and $(-1, 3)$. Examining the set of ordered pairs, we see that each number in the domain corresponds with exactly one number in the range. For example, the 1 in the domain corresponds with only the 4 in the range, and so on. No x-value corresponds to more than one y-value. Therefore, this relation *is a function*.

b) The domain is $\{-1, 4, 3, 2\}$ and the range is $\{3, 2, 1, 6, 5\}$. Notice that 3 appears as the first coordinate in two ordered pairs even though it is listed only once in the set of elements that represent the domain. Since the ordered pairs $(3, 1)$ and $(3, 5)$ have *the same first coordinate* and a different second coordinate, each value in the domain does not correspond to exactly one value in the range. Therefore, this relation is *not a function*.

NOW TRY EXERCISE 23

Example 2 leads to an alternate definition of function.

DEFINITION | A **function** is a set of ordered pairs in which no *first* coordinate is repeated.

If the second coordinate in a set of ordered pairs repeats, the set of ordered pairs may still be a function, as in Example 2 **a)**. However, if two or more ordered pairs contain the same first coordinate, as in Example 2 **b)**, the set of ordered pairs is not a function.

3   Use the Vertical Line Test

The **graph of a function or relation** is the graph of its set of ordered pairs. The two sets of ordered pairs in Example 2 are graphed in Figures 3.22a and 3.22b. Notice that in the function in Figure 3.22a it is not possible to draw a vertical line that intersects two points. We should expect this because, in a function, each x-value must correspond to exactly one y-value. In Figure 3.22b we *can* draw a vertical line through the points $(3, 1)$ and $(3, 5)$. This shows that each x-value does not correspond to exactly one y-value, and the graph does not represent a function.

This method of determining whether a graph represents a function is called the **vertical line test.**

Function

Not a function

FIGURE 3.22   (a) First set of ordered pairs   (b) Second set of ordered pairs

Vertical Line Test

If a vertical line can be drawn through any part of the graph and the line intersects another part of the graph, the graph does not represent a function. If a vertical line cannot be drawn to intersect the graph at more than one point, the graph represents a function.

We use the vertical line test to show that Figure 3.23b represents a function and Figures 3.23a and 3.23c do not represent functions.

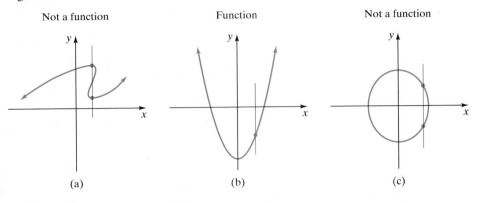

FIGURE 3.23   (a)   (b)   (c)

EXAMPLE 3   Use the vertical line test to determine whether the following graphs represent functions. Also determine the domain and range of each function or relation.

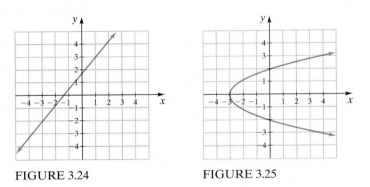

FIGURE 3.24   FIGURE 3.25

Solution   **a)** A vertical line cannot be drawn to intersect the graph in Figure 3.24 at more than one point. Thus this is the graph of a function. Since the line extends indefinitely in both directions, every value of x will be included in the domain. The domain is the set of real numbers.

$$\text{Domain:} \quad \mathbb{R} \quad \text{or} \quad (-\infty, \infty)$$

The range is also the set of real numbers since all values of y are included on the graph.

$$\text{Range:} \quad \mathbb{R} \quad \text{or} \quad (-\infty, \infty)$$

b) Since a vertical line can be drawn to intersect the graph in Figure 3.25 at more than one point, this is *not* the graph of a function. The domain of this relation is the set of values greater than or equal to -3.

$$\text{Domain:} \quad \{x | x \geq -3\} \quad \text{or} \quad [-3, \infty)$$

The range is the set of y-values, which can be any real number.

$$\text{Range:} \quad \mathbb{R} \quad \text{or} \quad (-\infty, \infty)$$

EXAMPLE 4   Consider the graph shown in Figure 3.26.

a) What member of the range is paired with 4 in the domain?

b) What members of the domain are paired with −2 in the range?

c) What is the domain of the function?

d) What is the range of the function?

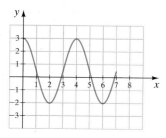

FIGURE 3.26

Solution   a) The range is the set of y-values. The y-value paired with the x-value of 4 is 3.

b) The domain is the set of x-values. The x-values paired with the y-value of −2 are 2 and 6.

c) The domain is the set of x-values, 0 through 7. Thus the domain is

$$\{x|0 \leq x \leq 7\} \quad \text{or} \quad [0, 7]$$

d) The range is the set of y-values, −2 through 3. Thus, the range is

NOW TRY EXERCISE 33

$$\{y|-2 \leq y \leq 3\} \quad \text{or} \quad [-2, 3]$$

✳

EXAMPLE 5   Figure 3.27 illustrates a graph of speed versus time of a man out for a walk and jog. Write a story about the man's outing that corresponds to this function.

FIGURE 3.27

Solution   Understand   The horizontal axis is time and the vertical axis is speed. When the graph is horizontal it means the person is traveling at the constant speed indicated on the vertical axis. The near-vertical lines that increase with time (or have a positive slope, as will be discussed later) indicate an increase in speed, whereas the near-vertical lines that decrease with time (or have a negative slope) indicate a decrease in speed.

Answer   Here is one possible interpretation of the graph. The man walks for about 5 minutes at a speed of about 2 miles per hour. Then the man speeds up to about 4 miles per hour and walks fast or jogs at about this speed for about 10 minutes. Then the man slows down and stops, and then rests for about 5 minutes. Finally, the man speeds up to about 5 miles per hour and jogs at this speed for about

NOW TRY EXERCISE 65   10 minutes.

✳

4   Understand Function Notation

In Section 3.1 we graphed a number of equations, as summarized in Table 3.1. If you examine each equation in the table, you will see that they are all functions, since their graphs pass the vertical line test.

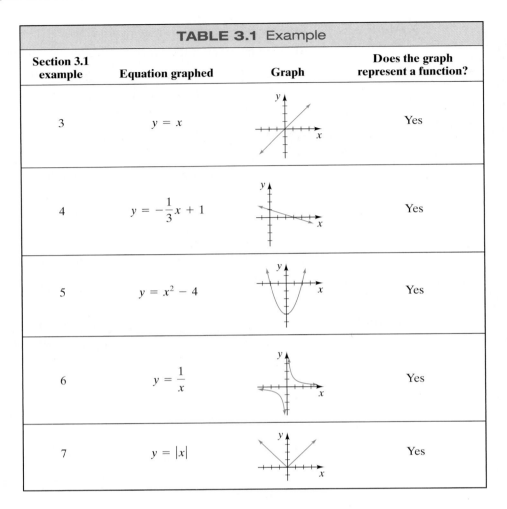

Section 3.1 example	Equation graphed	Graph	Does the graph represent a function?
3	$y = x$		Yes
4	$y = -\dfrac{1}{3}x + 1$		Yes
5	$y = x^2 - 4$		Yes
6	$y = \dfrac{1}{x}$		Yes
7	$y = |x|$		Yes

TABLE 3.1 Example

Since the graph of each equation shown represents a function, we may refer to each equation in the table as a function. When we refer to an equation in variables x and y as a function, it means that the graph of the equation satisfies the criteria for a function. That is, each x-value corresponds to exactly one y-value, and the graph of the equation passes the vertical line test.

Not all equations are functions, as you will see in Chapter 10, where we discuss equations of circles and ellipses. However, until we get to Chapter 10, all equations that we discuss will be functions.

Consider the equation $y = 3x + 2$. By applying the vertical line test to its graph (Fig. 3.28), we can see that the graph represents a function. When an equation in variables x and y is a function, we often write the equation using **function notation**, $f(x)$, read "f of x." Since the equation $y = 3x + 2$ is a function, and the value of y depends on the value of x, we say that **y is a function of x.** When we are given a linear equation in variables x and y, *that is solved for y*, we can write the equation in function notation by substituting $f(x)$ for y. In this case, we can write the equation in function notation as $f(x) = 3x + 2$. The notation $f(x)$ represents the dependent variable *and does not mean f times x.* Other letters may be used to indicate functions. For example, $g(x)$ and $h(x)$ also represent functions of x, and in Section 5.1 we will use $P(x)$ to represent polynomial functions.

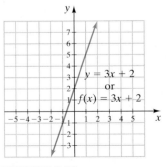

FIGURE 3.28

Functions written in function notation are also equations since they contain an equal sign. We may refer to $y = 3x + 2$ as either an equation or a function. Similarly, we may refer to $f(x) = 3x + 2$ as either a function or an equation.

If y is a function of x, the notation $f(5)$, read "f of 5," means the value of y when x is 5. To evaluate a function for a specific value of x, substitute that value for x in the function. For example, if $f(x) = 3x + 2$, then $f(5)$ is found as follows:

$$f(x) = 3x + 2$$
$$f(5) = 3(5) + 2 = 17$$

Therefore, when x is 5, y is 17. The ordered pair $(5, 17)$ would appear on the graph of $y = 3x + 2$.

HELPFUL HINT

Linear equations that are not solved for y can be written using function notation by solving the equation for y, then replacing y with $f(x)$. For example, the equation $-9x + 3y = 6$ becomes $y = 3x + 2$ when solved for y. We can therefore write $f(x) = 3x + 2$.

EXAMPLE 6   If $f(x) = -4x^2 + 3x - 2$, find

a) $f(2)$ **b)** $f(-1)$ **c)** $f(a)$

Solution   **a)** $f(x) = -4x^2 + 3x - 2$

$$f(2) = -4(2)^2 + 3(2) - 2 = -4(4) + 6 - 2 = -16 + 6 - 2 = -12$$

b) $f(-1) = -4(-1)^2 + 3(-1) - 2 = -4(1) - 3 - 2 = -4 - 3 - 2 = -9$

c) To evaluate the function at a, we replace each x in the function with an a.

$$f(x) = -4x^2 + 3x - 2$$
$$f(a) = -4a^2 + 3a - 2$$

✳

EXAMPLE 7   Determine each indicated function value.

a) $g(-2)$ for $g(t) = \dfrac{1}{t + 5}$

b) $h(5)$ for $h(s) = 2|s - 6|$

c) $j(-3)$ for $j(r) = \sqrt{6 - r}$

Solution   In each part, substitute the indicated value into the function and evaluate the function.

a) $g(-2) = \dfrac{1}{-2 + 5} = \dfrac{1}{3}$

b) $h(5) = 2|5 - 6| = 2|-1| = 2(1) = 2$

NOW TRY EXERCISE 45   **c)** $j(-3) = \sqrt{6 - (-3)} = \sqrt{6 + 3} = \sqrt{9} = 3$

✳

5   Applications of Functions in Daily Life

Many of the applications that we discussed in Chapter 2 were functions. However, we had not defined a function at that time. Now we examine additional applications of functions.

EXAMPLE 8   **The Masters Golf Tournament**   The Masters golf tournament is held every year in April at Augusta National Golf Club in Georgia. The graph in Figure 3.29 shows

Eyes on the Masters

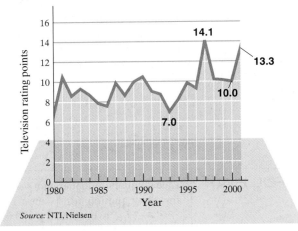

FIGURE 3.29

Source: NTI, Nielsen

Tiger Woods*

the television rating points for the Masters from 1980 through 2001. One rating point represents 1% of the U.S. households with televisions. For example, if a television program receives a rating of 14.1, it means that 14.1% of the U.S. households with televisions are tuned to that program.

a) Explain why the graph in Figure 3.29 represents a function.

b) Determine the television rating for the Masters in 1993.

c) Determine the percent increase in the television rating points from 1993 to 1997.

d) Determine the percent decrease in television rating points from 1997 to 2000.

Solution

a) This graph represents a function because each year corresponds to a specific number of television rating points. Notice that the graph passes the vertical line test.

b) In 1993 the Masters received about 7 television rating points. If we call the function f, then $f(1993) = 7$.

c) We will follow our problem-solving procedure to answer this question.

Understand and Translate   We are asked to find the percent increase in television rating points from 1993 to 1997. To do so, we use the formula

$$\text{percent change (increase or decrease)} = \frac{\left(\begin{array}{c}\text{amount in}\\\text{latest period}\end{array}\right) - \left(\begin{array}{c}\text{amount in}\\\text{previous period}\end{array}\right)}{\text{amount in previous period}}$$

The latest period is 1997 and the previous period is 1993. Substituting the corresponding values, we get

$$\text{percent change} = \frac{14.1 - 7.0}{7.0}$$

Carry Out
$$= \frac{7.1}{7.0} \approx 1.0143 \approx 101.4\%$$

Check and Answer   Our calculations appear correct. There was about a 101.4% increase in television rating points for the Masters from 1993 to 1997.

---

*It should be noted that in 1997 the extremely high rating was largely due to Tiger Woods winning the event.

d) To find the percent decrease from 1997 to 2000 we follow the same procedure as in part **c)**.

$$\text{percent change (increase or decrease)} = \frac{\left(\begin{array}{c}\text{amount in}\\ \text{latest period}\end{array}\right) - \left(\begin{array}{c}\text{amount in}\\ \text{previous period}\end{array}\right)}{\text{amount in previous period}}$$

$$= \frac{10.0 - 14.1}{14.1} = \frac{-4.1}{14.1} \approx -0.291 \approx -29.1\%$$

The negative sign preceding the 29.1% indicates a percent decrease. Thus, there was about a 29.1% decrease in television rating points for the Masters from 1997 to 2000.

EXAMPLE 9   **Immigration**   The size of the U.S. foreign-born population is at an all-time high. In 1890 the foreign-born population was 9 million people; in 1910 it was 14 million people; in 1930 it was 14 million people; in 1950 it was 10 million people; in 1970 it was 10 million people; in 1990 it was 20 million people; in 2000 it was 28 million people; and in 2005 it is projected to be 31 million people.

a) Represent this information on a graph.

b) Using your graph, explain why this set of points represents a function.

c) Using your graph, estimate the foreign-born population in 2003.

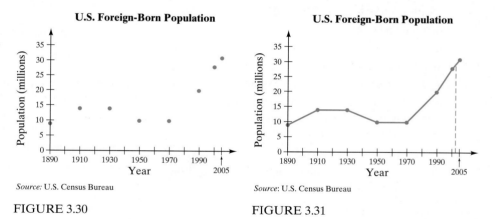

FIGURE 3.30   FIGURE 3.31

Solution   **a)** The set of points is plotted in Figure 3.30. We placed the year on the horizontal axis and the foreign-born population, in millions of people, on the vertical axis.

b) Since each year corresponds with exactly one population, this set of points represents a function. Notice that this graph passes the vertical line test.

c) We can connect the points with straight line segments as in Figure 3.31. Then we can estimate from the graph that there were about 30 million foreign-born Americans in 2003. If we call the function f, then $f(2003) = 30$.

NOW TRY EXERCISE 75

In Section 2.2 we learned to use formulas. Consider the formula for the area of a circle, $A = \pi r^2$. In the formula, π is a constant that is approximately 3.14. For each specific value of the radius, r, there corresponds exactly one area, A. Thus the area of a circle is a function of its radius. We may therefore write

$$A(r) = \pi r^2$$

Often formulas are written using function notation like this.

EXAMPLE 10 The Celsius temperature, C, is a function of the Fahrenheit temperature, F.

$$C(F) = \frac{5}{9}(F - 32)$$

Determine the Celsius temperature that corresponds to 50°F.

Solution We need to find $C(50)$. We do so by substitution.

$$C(F) = \frac{5}{9}(F - 32)$$

$$C(50) = \frac{5}{9}(50 - 32)$$

$$= \frac{5}{9}(18) = 10$$

Therefore, 50°F = 10°C.

In Example 10, F is the independent variable and C the dependent variable. If we solved the function for F, we would obtain $F(C) = \frac{9}{5}C + 32$. In this formula, C is the independent variable and F is the dependent variable.

NOW TRY EXERCISE 55

Exercise Set 3.2

Concept/Writing Exercises

1. What is a function?
2. What is a relation?
3. Are all functions also relations? Explain.
4. Are all relations also functions? Explain.
5. Explain how to use the vertical line test to determine if a relation is a function.
6. What is the domain of a function?
7. What is the range of a function?
8. What are the domain and range of the function $f(x) = 2x + 1$? Explain your answer.

9. Consider the function $y = \frac{1}{x}$. What is its domain and range? Explain.
10. What are the domain and range of a function of the form $f(x) = ax + b$, $a \neq 0$? Explain your answer.
11. Consider the absolute value function $y = |x|$. What is its domain and range? Explain.
12. What is a dependent variable?
13. What is an independent variable?
14. How is "$f(x)$" read?

Practice the Skills

In Exercises 15–20, a) determine if the relation illustrated is a function. b) Give the domain and range of each function or relation.

15. twice a number

$$3 \longrightarrow 6$$
$$5 \longrightarrow 10$$
$$10 \longrightarrow 20$$

16. nicknames

Robert $\longrightarrow$ Bobby
Robert $\longrightarrow$ Rob
Margaret $\longrightarrow$ Peggy
Margaret $\longrightarrow$ Maggie

17. number of siblings

Cameron $\longrightarrow$ 1
Tyrone $\longrightarrow$ 2
Vishnu $\longrightarrow$ 2

18. a number squared

$$4 \longrightarrow 16$$
$$5 \longrightarrow 25$$
$$6 \longrightarrow 36$$

19. cost of a stamp

$$1990 \longrightarrow 20$$
$$2001 \longrightarrow 34$$
$$2002 \longrightarrow 37$$

20. absolute value

$$|-3| \longrightarrow 3$$
$$|3| \longrightarrow 3$$
$$|0| \longrightarrow 0$$

In Exercises 21–28, **a)** *determine which of the following relations are also functions.* **b)** *Give the domain and range of each relation or function.*

21. $\{(1, 4), (2, 2), (3, 5), (4, 3), (5, 1)\}$

22. $\{(1, 1), (4, 2), (9, 3), (1, -1), (4, -2), (9, -3)\}$

23. $\{(3, -1), (5, 0), (1, 2), (4, 4), (2, 2), (7, 5)\}$

24. $\{(-1, 1), (0, -3), (3, 4), (4, 5), (-2, -2)\}$

25. $\{(1, 4), (2, 5), (3, 6), (2, 2), (1, 1)\}$

26. $\{(6, 3), (-3, 4), (0, 3), (5, 2), (3, 5), (2, 5)\}$

27. $\{(0, 3), (1, 3), (2, 2), (1, -1), (2, -7)\}$

28. $\{(3, 5), (2, 5), (1, 5), (0, 5), (-1, 5)\}$

In Exercises 29–40, **a)** *determine whether the graph illustrated represents a function.* **b)** *Give the domain and range of each function or relation.* **c)** *Approximate the value or values of x where y = 2.*

29.

30.

31.

32.

33.

34.

35.

36.

37.

38.

39.

40.

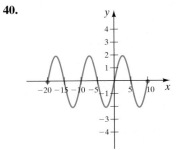

Evaluate each function at the indicated values.

41. $f(x) = -2x + 5$; find

    a) $f(2)$.

    b) $f(-3)$.

42. $f(a) = \dfrac{1}{3}a + 2$; find

    a) $f(0)$.

    b) $f(-6)$.

43. $h(x) = x^2 - x - 6$; find

    a) $h(0)$.

    b) $h(-1)$.

44. $g(x) = -2x^2 + x + 2$; find

    a) $g(2)$.

    b) $g\left(\dfrac{1}{2}\right)$.

45. $r(t) = -t^3 - 2t^2 + t + 4$; find

    a) $r(1)$.

    b) $r(-2)$.

46. $g(t) = 4 - 5t + 16t^2 - 2t^3$; find

    a) $g(0)$.

    b) $g(3)$.

47. $h(z) = |5 - 2z|$; find

    a) $h(6)$.

    b) $h\left(\dfrac{5}{2}\right)$.

48. $q(x) = -2|x + 3| - 3$; find

    a) $q(0)$.

    b) $q(-4)$.

49. $s(t) = \sqrt{t + 2}$; find

    a) $s(-2)$.

    b) $s(7)$.

50. $f(t) = \sqrt{5 - 2t}$; find

    a) $f(-2)$.

    b) $f(2)$.

51. $g(x) = \dfrac{x^3 - 2}{x - 2}$; find

    a) $g(0)$.

    b) $g(2)$.

52. $h(x) = \dfrac{x^2 + 4x}{x + 6}$; find

    a) $h(-3)$.

    b) $h\left(\dfrac{2}{5}\right)$.

Problem Solving

53. *Area of a Rectangle* The formula for the area of a rectangle is $A = lw$. If the length of a rectangle is 6 inches, then the area is a function of its width, $A(w) = 6w$. Find the area when the width is

a) 2 feet.

b) 4.5 feet.

54. *Simple Interest* The formula for the simple interest earned for a period of 1 year is $i = pr$ where p is the principal invested and r is the simple interest rate. If $1000 is invested, the simple interest earned in 1 year is a function of the simple interest rate, $i(r) = 1000r$. Determine the simple interest earned in 1 year if the interest rate is

a) 3%.   **b)** 4.25%.

55. *Area of a Circle* The formula for the area of a circle is $A = \pi r^2$. The area is a function of the radius.

a) Write this function using function notation.

b) Determine the area when the radius is 10 yards.

56. *Perimeter of a Square* The formula for the perimeter of a square is $P = 4s$ where s represents the length of any one of the sides of the square.

a) Write this function using function notation.

b) Determine the perimeter of a square with sides of length 3 meters.

57. *Temperature* The formula for changing Fahrenheit temperature into Celsius temperature is $C = \dfrac{5}{9}(F - 32)$.

The Celsius temperature is a function of Fahrenheit temperature.

a) Write this function using function notation.

b) Find the Celsius temperature that corresponds to $-40°F$.

58. *Volume of a Cylinder* The formula for the volume of a right circular cylinder is $V = \pi r^2 h$. If the height, h, is 3 feet, then the volume is a function of the radius, r.

a) Write this formula in function notation, where the height is 3 feet.

b) Find the volume if the radius is 2 feet.

59. *Sauna Temperature* The temperature, T, in degrees Celsius, in a sauna n minutes after being turned on is given by the function $T(n) = -0.03n^2 + 1.5n + 14$. Find the sauna's temperature after

a) 3 minutes.   **b)** 12 minutes.

60. *Stopping Distance* The stopping distance, d, in meters for a car traveling v kilometers per hour is given by the function $d(v) = 0.18v + 0.01v^2$. Find the stopping distance for the following speeds:

a) 50 km/hr   **b)** 25 km/hr

61. *Air Conditioning* When an air conditioner is turned on maximum in a bedroom at 80°, the temperature, T, in the room after A minutes can be approximated by the function $T(A) = -0.02A^2 - 0.34A + 80, 0 \le A \le 15$.

a) Estimate the room temperature 4 minutes after the air conditioner is turned on.

b) Estimate the room temperature 12 minutes after the air conditioner is turned on.

62. *Accidents* The number of accidents, n, in 1 month involving drivers x years of age can be approximated by the function $n(x) = 2x^2 - 150x + 4000$. Find the approximate number of accidents in 1 month that involved

a) 18-year-olds.   **b)** 25-year-olds.

63. *Oranges* The total number of oranges, T, in a square pyramid whose base is n by n oranges is given by the function

$$T(n) = \frac{1}{3}n^3 + \frac{1}{2}n^2 + \frac{1}{6}n$$

Find the number of oranges if the base is

a) 6 by 6 oranges.   **b)** 8 by 8 oranges.

64. *Rock Concert* If the cost of a ticket to a rock concert is increased by x dollars, the estimated increase in revenue, R, in thousands of dollars is given by the function $R(x) = 24 + 5x - x^2, x < 8$. Find the increase in revenue if the cost of the ticket is increased by

a) $1.   **b)** $4.

Review Example 5 before working Exercises 65–70

65. ***Heart Rate*** The following graph shows a person's heart rate while doing exercise. Write a story that this graph may represent.

66. ***Water Level*** The following graph shows the water level at a certain point during a flood. Write a story that this graph may represent.

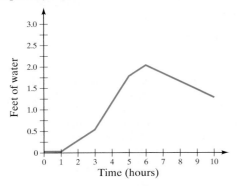

67. ***Height Above Sea Level*** The following graph shows height above sea level versus time when a man leaves his house and goes for a walk. Write a story that this graph may represent.

68. ***Water Level in a Bathtub*** The following graph shows the level of water in a bathtub versus time. Write a story that this graph may represent.

69. ***Speed of a Car*** The following graph shows the speed of a car versus time. Write a story that this graph may represent.

70. ***Distance Traveled*** The following graph shows the distance traveled by a person in a car versus time. Write a story that this graph may represent.

71. ***Computer Sales*** The following graph shows the amount American businesses have spent on computers and related equipment from 1995 through 2000.

Spending on Computers and Related Equipment

Source: Fortune, Dec. 18, 2000

a) Does this graph represent a function? Explain.

b) In this graph, what is the independent variable?

c) If f represents the function, determine $f(2000)$.

d) Determine the percent increase in the amount American businesses spent on computers and related equipment from 1997 through 2000.

72. ***Computer Exports*** The following graph shows the U.S. exports of computers and related equipment and parts from 1995 through 2000.

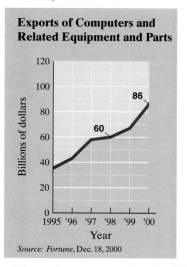

Exports of Computers and Related Equipment and Parts

Source: Fortune, Dec. 18, 2000

a) Does this graph represent a function? Explain.

b) In this graph, what is the independent variable?

c) If g represents the function, determine $g(2000)$.

d) Determine the percent increase in exports of computers and related equipment and parts from 1998 through 2000.

73. ***Trade Deficit with China*** The following graph shows that imports from China to the United States have been rapidly increasing while exports to China from the United States have been rising at a much slower rate.

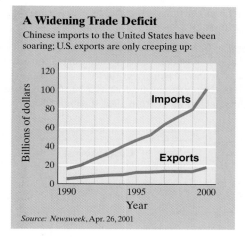

A Widening Trade Deficit
Chinese imports to the United States have been soaring; U.S. exports are only creeping up:

Source: Newsweek, Apr. 26, 2001

a) Does the graph of imports from China represent a function? Explain.

b) Does the graph of exports to China represent a function? Explain.

c) Does the graph of imports from China appear to be approximately linear? Explain.

d) Does the graph of exports to China appear to be approximately linear? Explain.

e) If f represents the function of imports from China and if t is the year, determine t if $f(t) = \$80$ billion.

f) If g represents the function of exports to China and if t is the year, determine t if $g(t) = \$18$ billion.

74. ***Online Travel*** The total amount of U.S. online bookings for travel (in billions of dollars) is shown in the following bar graph.

Online Travel is Growing Fast

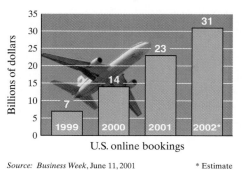

U.S. online bookings

Source: Business Week, June 11, 2001   * Estimate

a) Draw a line graph that displays this information.

b) Does the graph you drew in part **a)** appear to be approximately linear? Explain.

c) Assuming this trend continues, from the graph you drew, estimate the amount of online bookings for the year 2003.

d) Does the bar graph represent a function?

e) Does the line graph you drew in part **a)** represent a function?

75. *Super Bowl Commercials* The average price of the cost of a 30-second commercial during the Super Bowl has been increasing over the years. The following chart gives the approximate cost of a 30-second commercial for selected years from 1981 through 2001.

Year	Cost ($1000s)
1981	280
1985	500
1989	740
1993	970
1997	1300
2001	2300

a) Draw a line graph that displays this information.

b) Does the graph appear to be approximately linear? Explain.

c) From the graph, estimate the cost of a 30-second commercial in 2000.

76. *Household Expenditures* The average annual household expenditure is a function of the average annual household income. The average expenditure can be estimated by the function

$$f(i) = 0.6i + 5000 \quad \$3500 \le i \le \$50,000$$

where $f(i)$ is the average household expenditure and i is the average household income.

a) Draw a graph showing the relationship between average household income and the average household expenditure.

b) Estimate the average household expenditure for a family whose average household income is $30,000.

77. *Supply and Demand* The price of commodities, like soybeans, is determined by **supply and demand**. If too many soybeans are produced, the supply will be greater than the demand, and the price will drop. If not enough soybeans are produced, the demand will be greater than the supply, and the price of soybeans will rise. Thus the price of soybeans is a function of the number of bushels of soybeans produced. The price of a bushel of soybeans can be estimated by the function

$$f(Q) = -0.00004Q + 4.25, \quad 10,000 \le Q \le 60,000$$

where $f(Q)$ is the price of a bushel of soybeans and Q is the annual number of bushels of soybeans produced.

a) Construct a graph showing the relationship between the number of bushels of soybeans produced and the price of a bushel of soybeans.

b) Estimate the cost of a bushel of soybeans if 40,000 bushels of soybeans are produced in a given year.

Group Activity

*In many real-life situations, more than one function may be needed to represent a problem. This often occurs where two or more different rates are involved. For example, when discussing federal income taxes, there are different tax rates. When two or more functions are used to represent a problem, the function is called a **piecewise function**. Following are two examples of piecewise functions and their graphs.*

$$f(x) = \begin{cases} -x + 2, & 0 \le x < 4 \\ 2x - 10, & 4 \le x < 8 \end{cases}$$

$$f(x) = \begin{cases} 2x - 1, & -2 \le x < 2 \\ x - 2, & 2 \le x < 4 \end{cases}$$

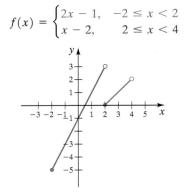

As a group, graph the following piecewise functions.

78. $f(x) = \begin{cases} x + 3, & -1 \le x < 2 \\ 7 - x, & 2 \le x < 4 \end{cases}$

79. $g(x) = \begin{cases} 2x + 3, & -3 < x < 0 \\ -3x + 1, & 0 \le x < 2 \end{cases}$

Cumulative Review Exercises

[2.1] **80.** Solve the equation $3x - 2 = \dfrac{1}{3}(3x - 3)$.

[2.2] **81.** Solve the following formula for p_2.

$$E = a_1p_1 + a_2p_2 + a_3p_3$$

[2.5] **82.** Solve the inequality $\dfrac{3}{5}(x - 3) > \dfrac{1}{4}(3 - x)$ and indicate the solution **a)** on the number line; **b)** in interval notation; and **c)** in set builder notation.

[2.6] **83.** Solve the equation $\left|\dfrac{x - 4}{3}\right| + 2 = 4$.

3.3 LINEAR FUNCTIONS: GRAPHS AND APPLICATIONS

SSM Study Guide CD/Video

MathPro 4/5 PH Math Tutor Center prenhall.com/Angel

1 Graph linear functions.
2 Graph linear functions using intercepts.
3 Graph equations of the form $x = a$ and $y = b$.
4 Study applications of functions.
5 Solve linear equations in one variable graphically.

---

1 Graph Linear Functions

In Section 3.1 we graphed linear equations. To graph the linear equation $y = 2x + 4$, we can make a table of values, plot the points, and draw the graph, as shown in Figure 3.32. Notice that this graph represents a function since it passes the vertical line test.

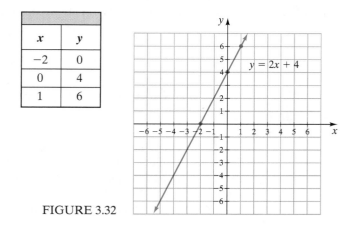

x	y
-2	0
0	4
1	6

FIGURE 3.32

We may write the equation graphed in Figure 3.32 using function notation as $f(x) = 2x + 4$. This is an example of a linear function. A **linear function** is a function of the form $f(x) = ax + b$. The graph of any linear function is a straight line. The domain of any function is the set of real numbers for which the function is a real number. The domain of any linear function is the set of all real numbers, $\mathbb{R}$: Any real number, x, substituted into a linear function will result in $f(x)$ being a real number. We will discuss domains of functions further in Section 3.6.

To graph a linear function, we treat $f(x)$ as y and follow the same procedure used to graph linear equations.

EXAMPLE 1   Graph $f(x) = \dfrac{1}{2}x - 1$.

Solution   We construct a table of values by substituting values for x and finding corresponding values of $f(x)$ or y. Then we plot the points and draw the graph, as illustrated in Figure 3.33.

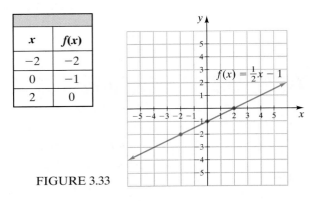

x	$f(x)$
-2	-2
0	-1
2	0

FIGURE 3.33

Note that the vertical axis in Figure 3.33 may also be labeled as $f(x)$ instead of y. In this book we will continue to label it y.

2   Graph Linear Functions Using Intercepts

Linear equations are not always given in the form $y = ax + b$. The equation $2x + 3y = 6$ is an example of a linear equation given in *standard form*.

DEFINITION

The **standard form of a linear equation** is

$$ax + by = c$$

where $a, b,$ and c are real numbers, and a and b are not both 0.

Examples of Linear Equations in Standard Form

NOW TRY EXERCISE 13

$$2x + 3y = 4 \qquad -x + 5y = -2$$

Sometimes when an equation is given in standard form it may be easier to draw the graph using the x- and y-intercepts. Let's examine two points on the graph shown in Figure 3.32. Note that the graph crosses the x-axis at the point $(-2, 0)$. Therefore, $(-2, 0)$ is called the **x-intercept**. Sometimes we say that the x-intercept is *at* -2 (on the x-axis), the x-coordinate of the ordered pair.

The graph crosses the y-axis at the point $(0, 4)$. Therefore, $(0, 4)$ is called the **y-intercept**. Sometimes we say that the y-intercept is *at* 4 (on the y-axis), the y-coordinate of the ordered pair.

Below we explain how the x- and y-intercepts may be determined algebraically.

To Find the *x*- and *y*-Intercepts

To find the y-intercept, set $x = 0$ and solve for y.
To find the x-intercept, set $y = 0$ and solve for x.

To graph a linear equation using the x- and y-intercepts, find the intercepts and plot the points. Then draw a straight line through the points. When graphing linear equations using the intercepts, you must be very careful. If either of your points is plotted wrong, your graph will be wrong.

EXAMPLE 2 Graph the equation $5x = 10y - 20$ by plotting the x- and y-intercepts.

Solution To find the y-intercept (the point where the graph crosses the y-axis), set $x = 0$ and solve for y.

$$5x = 10y - 20$$
$$5(0) = 10y - 20$$
$$0 = 10y - 20$$
$$20 = 10y$$
$$2 = y$$

The graph crosses the y-axis at $y = 2$. The ordered pair representing the y-intercept is $(0, 2)$.

To find the x-intercept (the point where the graph crosses the x-axis), set $y = 0$ and solve for x.

$$5x = 10y - 20$$
$$5x = 10(0) - 20$$
$$5x = -20$$
$$x = -4$$

The graph crosses the x-axis at $x = -4$. The ordered pair representing the x-intercept is $(-4, 0)$. Now plot the intercepts and draw the graph (Fig. 3.34).

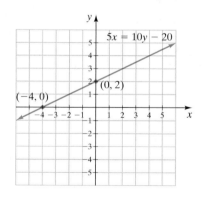

NOW TRY EXERCISE 23   FIGURE 3.34

EXAMPLE 3 Graph $f(x) = -\dfrac{1}{3}x - 1$ by plotting the x- and y-intercepts.

Solution Treat $f(x)$ the same as y. To find the y-intercept, set $x = 0$ and solve for $f(x)$.

$$f(x) = -\frac{1}{3}x - 1$$

$$f(x) = -\frac{1}{3}(0) - 1 = -1$$

The y-intercept is $(0, -1)$.

To find the *x*-intercept, set $f(x) = 0$ and solve for *x*.

$$f(x) = -\frac{1}{3}x - 1$$

$$0 = -\frac{1}{3}x - 1$$

$$3(0) = 3\left(-\frac{1}{3}x - 1\right) \qquad \textit{Multiply both sides by 3.}$$

$$0 = -x - 3 \qquad \textit{Distributive property}$$

$$x = -3 \qquad \textit{Add x to both sides.}$$

The *x*-intercept is $(-3, 0)$. The graph is shown in Figure 3.35.

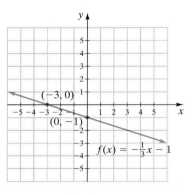

NOW TRY EXERCISE 17   FIGURE 3.35

Graphs of the form $ax + by = 0$ go through the origin and have the same *x*- and *y*-intercept, $(0, 0)$. To graph such equations we can use the intercept as one point and substitute values for *x* and find the corresponding values of *y* to get other points on the graph.

EXAMPLE 4   Graph $-6x + 4y = 0$.

Solution   If we substitute $x = 0$ we find that $y = 0$. Thus the graph goes through the origin. We will select $x = -2$ and $x = 2$ and substitute these values into the equation, one at a time, to find two other points on the graph.

Let $x = -2$.	Let $x = 2$.
$-6x + 4y = 0$	$-6x + 4y = 0$
$-6(-2) + 4y = 0$	$-6(2) + 4y = 0$
$12 + 4y = 0$	$-12 + 4y = 0$
$4y = -12$	$4y = 12$
$y = -3$	$y = 3$
ordered pairs: $(-2, -3)$	$(2, 3)$

Two other points on the graph are at $(-2, -3)$ and $(2, 3)$. The graph of $-6x + 4y = 0$ is shown in Figure 3.36.

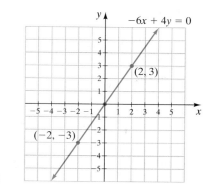

NOW TRY EXERCISE 35   FIGURE 3.36

Using Your Graphing Calculator

Sometimes it may be difficult to estimate the intercepts of a graph accurately. When this occurs, you might want to use a graphing calculator. We demonstrate how in the following example.

EXAMPLE   Determine the x- and y-intercepts of the graph of $y = 1.3(x - 3.2)$.

Solution   Press the $\boxed{Y=}$ key, and then assign $1.3(x - 3.2)$ to y. Then press the $\boxed{\text{GRAPH}}$ key to graph the function $y_1 = 1.3(x - 3.2)$, as shown in Figure 3.37a.

From the graph it may be difficult to determine the intercepts. One way to find the y-intercept is to use the TRACE feature, which was discussed in Section 3.1. Figure 3.37b shows a TI-83 Plus screen after the $\boxed{\text{TRACE}}$ key is pressed. Notice the y-intercept is at -4.16.

FIGURE 3.37a   FIGURE 3.37b

Some graphing calculators have the ability to find the x-intercepts of a function by pressing just a few keys. A **zero** (or **root**) of a function is a value of x such that $f(x) = 0$. A zero (or root) of a function is the x-coordinate of the x-intercept of the graph of the function. Read your calculator manual to learn how to find the zeros or roots of a function. On a TI-83 Plus you press the keys $\boxed{2^{\text{nd}}}$ $\boxed{\text{TRACE}}$ to get to the CALC menu (which stands for calculate). Then you choose option 2, *zero*. Once the zero feature has been selected, the calculator will display

Left bound?

At this time, move the cursor along the curve until it is to the *left* of the zero. Then press $\boxed{\text{ENTER}}$. The calculator now displays

Right bound?

Move the cursor along the curve until it is to the *right* of the zero. Then press $\boxed{\text{ENTER}}$. The calculator now displays

Guess?

Now press $\boxed{\text{ENTER}}$ for the third time and the zero is displayed at the bottom of the screen, as in Figure 3.38. Thus the x-intercept of the function is at 3.2. For practice at finding the intercepts on your calculator, work Exercises 69–72.

FIGURE 3.38

3 Graph Equations of the Form $x = a$ and $y = b$

Examples 5 and 6 illustrate how equations of the form $x = a$ and $y = b$, where a and b are constants, are graphed.

EXAMPLE 5 Graph the equation $y = 3$.

Solution This equation can be written as $y = 3 + 0x$. Thus, for any value of x selected, y is 3. The graph of $y = 3$ is illustrated in Figure 3.39. ✳

> The graph of any equation of the form $y = b$ will always be a horizontal line for any real number b.

Notice that the graph of $y = 3$ is a function since it passes the vertical line test. For each value of x selected, the value of y, or the value of the function, is 3. This is an example of a **constant function**. We may write

$$f(x) = 3$$

FIGURE 3.39

Any equation of the form $y = b$ or $f(x) = b$, where b represents a constant, is a constant function.

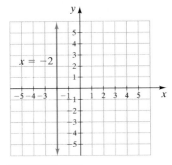

EXAMPLE 6 Graph the equation $x = -2$.

Solution This equation can be written as $x = -2 + 0y$. Thus, for every value of y selected, x will have a value of -2 (Fig. 3.40). ✳

> The graph of any equation of the form $x = a$ will always be a vertical line for any real number a.

FIGURE 3.40

NOW TRY EXERCISE 41

Notice that the graph of $x = -2$ does not represent a function since it does not pass the vertical line test. For $x = -2$ there is more than one value of y. In fact, when $x = -2$ there are an infinite number of values for y.

4 Study Applications of Functions

Graphs are often used to show the relationship between variables. The axes of a graph do not have to be labeled x and y. They can be any designated variables. Consider the following example.

EXAMPLE 7 **Tire Store Profit** The yearly profit, p, of a tire store can be estimated by the function $p(n) = 20n - 30{,}000$, where n is the number of tires sold per year.

a) Draw a graph of profit versus tires sold for up to and including 6000 tires.

b) Estimate the number of tires that must be sold for the company to break even.

c) Estimate the number of tires sold if the company has a \$40,000 profit.

Solution **a)** Understand The profit, p, is a function of the number of tires sold, n. The horizontal axis will therefore be labeled Number of tires sold (the independent variable) and the vertical axis will be labeled Profit (the dependent variable). Since the minimum number of tires that can be sold is 0, negative values do not have to be listed on the horizontal axis. The horizontal axis will therefore go from 0 to 6000 tires.

We will graph this equation by determining and plotting the intercepts.

Translate and Carry Out   To find the p-intercept, we set $n = 0$ and solve for $p(n)$.

$$p(n) = 20n - 30{,}000$$
$$p(n) = 20(0) - 30{,}000 = -30{,}000$$

Thus, the p-intercept is $(0, -30{,}000)$.

To find the n-intercept, we set $p(n) = 0$ and solve for n.

$$p(n) = 20n - 30{,}000$$
$$0 = 20n - 30{,}000$$
$$30{,}000 = 20n$$
$$1500 = n$$

Thus the n-intercept is $(1500, 0)$.

Answer   Now we use the p- and n-intercepts to draw the graph (see Fig. 3.41).

b) The break-even point is the number of tires that must be sold for the company to have neither a profit nor a loss. The break-even point is where the graph intersects the n-axis, for this is where the profit, p, is 0. To break even, approximately 1500 tires must be sold.

c) To make $40,000, approximately 3500 tires must be sold (shown by the dashed red line in Fig. 3.41).

Sometimes it is difficult to read an exact answer from a graph. To determine the exact number of tires needed to break even in Example 7, substitute 0 for $p(n)$ in the function $p(n) = 20n - 30{,}000$ and solve for n. To determine the exact number of tires needed to obtain a $40,000 profit, substitute 40,000 for $p(n)$ and solve the equation for n.

FIGURE 3.41

EXAMPLE 8

Toy Store Sales   Andrew Gestrich is the owner of a toy store. His monthly salary consists of $200 plus 10% of the store's sales for that month.

a) Write a function expressing his monthly salary, m, in terms of the store's sales, s.

b) Draw a graph of his monthly salary for sales up to and including $20,000.

c) If the store's sales for the month of April are $15,000, what will Andrew's salary be for April?

Solution

a) Andrew's monthly salary is a function of sales. His monthly salary, m, consists of $200 plus 10% of the sales, s. Ten percent of s is $0.10s$. Thus the function for finding his salary is

$$m(s) = 200 + 0.10s$$

b) Since monthly salary is a function of sales, Sales will be represented on the horizontal axis and Monthly salary will be represented on the vertical axis. Since sales can never be negative, the monthly salary can never be negative. Thus both axes will be drawn with only positive numbers. We will draw this graph by plotting points. We select values for s, find the corresponding values of m, and then draw the graph. We can select values of s that are between $0 and $20,000 (Fig. 3.42).

c) By reading our graph carefully, we can estimate that when the store's sales are $15,000, Andrew's monthly salary is about $1700.

s	m
0	200
10,000	1200
20,000	2200

FIGURE 3.42

NOW TRY EXERCISE 53

Mathematics in Action

Polluted Air

The relationship between pollutants and heightened levels of respiratory and cardiopulmonary disease has become clearer over the past few years. People living in urban areas with high levels of pollution have a higher risk of mortality than those living in less polluted cities. The pollutants most directly linked to the increased incidence of disease and death include ozone, particulates, carbon monoxide, sulfur dioxide, volatile organic compounds, and oxides of nitrogen.

Particulates—material suspended in the air—with the lowest range in diameter (10 microns or less) are called fine particles. They are able to pass through the natural filtering system of the nose and throat and can penetrate deep into the lungs and do serious damage. Sulfate aerosols constitute the largest percentage of fine particles in the eastern United States; they come from sulfur dioxide produced by coal and oil combustion. Nitrate aerosols, which make up about one-third of fine particles in Los Angeles, come from vehicle emissions.

A study by Harvard University researchers covered six cities over 16 years and tracked the health of more than 8000 people. The results, published in 1993, showed a nearly linear relationship between the concentrations of particles and increased mortality rates; even relatively low levels of fine particle pollution had a measurable effect on health.

Research that correlates disease and mortality with causative factors, such as air pollution in all forms, is critical information for the public and for lawmakers as policy is being formulated with regard to limiting automobile and smokestack emissions. A graph with a straight line essentially matching up thousands of deaths with tons of particulates in the air sends a message that is pretty hard to disregard.

5 Solve Linear Equations in One Variable Graphically

Earlier we discussed the graph of $f(x) = 2x + 4$. In Figure 3.43 on the next page we illustrate the graph of $f(x)$ along with the graph of $g(x) = 0$. Notice that the two graphs intersect at $(-2, 0)$. We can obtain the x-coordinate of the ordered pair by solving the equation $f(x) = g(x)$. Remember $f(x)$ and $g(x)$ both represent y, and by solving this equation for x we are obtaining the value of x where the y's are equal.

$$f(x) = g(x)$$
$$\overbrace{2x + 4} = \overbrace{0}$$
$$2x = -4$$
$$x = -2$$

Note that we obtain -2, the x-coordinate in the ordered pair at the point of intersection.

Now let's find the x-coordinate of the point at which the graphs of $f(x) = 2x + 4$ and $g(x) = 2$ intersect. We solve the equation $f(x) = g(x)$.

$$f(x) = g(x)$$
$$\overbrace{2x + 4} = \overbrace{2}$$
$$2x = -2$$
$$x = -1$$

FIGURE 3.43   FIGURE 3.44

FIGURE 3.45

NOW TRY EXERCISE 65

The x-coordinate of the point of intersection of the two graphs is -1, as shown in Figure 3.44. Notice that $f(-1) = 2(-1) + 4 = 2$.

In general, if we are given an equation in one variable, we can regard each side of the equation as a separate function. To obtain the solution to the equation, we can graph the two functions. The x-coordinate of the point of intersection will be the solution to the equation.

EXAMPLE 9   Find the solution to the equation $3x + 2 = 4x - 4$ graphically.

Solution   Let $f(x) = 3x + 2$ and $g(x) = 4x - 4$. The graph of these functions is illustrated in Figure 3.45. The x-coordinate of the point of intersection is 6. Thus, the solution to the equation is 6. Check the solution now.   ✳

Using Your Graphing Calculator

In Example 9, we solved an equation in one variable by graphing two functions. In the following example, we explain how to find the point of intersection of two functions on a graphing calculator.

EXAMPLE   Use a graphing calculator to find the solution to $2(x + 3) = \dfrac{1}{2}x + 4$.

Solution   Assign $2(x + 3)$ to Y_1 and assign $\dfrac{1}{2}x + 4$ to Y_2 to get

$$Y_1 = 2(x + 3)$$

$$Y_2 = \frac{1}{2}x + 4$$

FIGURE 3.46

Now press the $\boxed{\text{GRAPH}}$ key to graph the functions. The graph of the functions is shown in Figure 3.46.

By examining the graph can you determine the x-coordinate of the point of intersection? Is it -1, or -1.5, or some other value? We can determine the point of intersection in a number of different ways. One method involves using the TRACE and **ZOOM** features. Figure 3.47 shows the window of a TI-83 Plus after the TRACE feature has been used and the cursor has been moved close to the point of intersection. (Note that pressing the up and down arrows switches the cursor from one function to the other.)

FIGURE 3.47

(continued on the next page)

At the bottom of the screen in Figure 3.47, you see the x- and y-coordinates at the cursor. To get a closer view around the area of the cursor, you can *zoom in* using the ZOOM key. After you zoom in, you can move the cursor closer to the point of intersection and get a better reading (Fig. 3.48). You can do this over and over until you get as accurate an answer as you need. It appears from Figure 3.48 that the x-coordinate of the intersection is about -1.33.

Graphing calculators can also display the intersection of two graphs with the use of certain keys. The keys to press depend on your calculator. Read your calculator manual to determine how to do this. This procedure is generally quicker and easier to use to find the point of intersection of two graphs.

FIGURE 3.48   FIGURE 3.49

On the TI-83 Plus, select option 5: INTERSECT—from the CALC menu to find the intersection. Once the INTERSECT feature has been selected, the calculator will display

First curve?

At this time, move the cursor along the first curve until it is close to the point of intersection. Then press ENTER . The calculator will next display

Second curve?

The cursor will then appear on the second curve. If the cursor is not close to the point of intersection, move it along this curve until this happens. Then press ENTER . Next the calculator will display

Guess?

Now press ENTER again, and the point of intersection will be displayed.

Figure 3.49 shows the window after this procedure has been done. We see that the x-coordinate of the point of intersection is $-1.333\ldots$ or $-1\frac{1}{3}$ and the y-coordinate of the point of intersection is $3.333\ldots$ or $3\frac{1}{3}$.

For practice in using a graphing calculator to solve an equation in one variable, work Exercises 65–68.

Exercise Set 3.3

Concept/Writing Exercises

1. What is the standard form of a linear equation?

2. If you are given a linear equation in standard form, and wish to write the equation using function notation, how would you do it?

3. Explain how to find the x- and y-intercepts of the graph of an equation.

4. What terms do graphing calculators use to indicate the x-intercepts?

5. What will the graph of $x = a$ look like for any real number a?

6. What will the graph of $y = b$ look like for any real number b?

7. What will the graph of $f(x) = b$ look like for any real number b?

8. Is the graph of $x = a$ a function? Explain.

9. Explain how to solve an equation in one variable graphically.

10. Explain how to solve the equation $2(x - 1) = 3x - 5$ graphically.

Practice the Skills

Write each equation in standard form.

11. $y = -2x + 4$

12. $2x = 3y - 6$

13. $3(x - 2) = 4(y - 5)$

14. $\frac{1}{3}y = 2(x - 3) + 4$

Graph each equation using the x- and y-intercepts.

15. $y = -2x + 4$

16. $y = x - 5$

17. $f(x) = 2x + 3$

18. $f(x) = -6x + 5$

19. $2y = 4x + 6$

20. $x + 2y = 4$

21. $\frac{4}{3}x = y - 3$

22. $\frac{1}{2}x + 2y = 4$

23. $15x + 30y = 60$

24. $0.2x - 0.3y = 1.2$

25. $0.25x + 0.50y = 1.00$

26. $-1.6y = 0.4x + 9.6$

27. $120x - 360y = 720$

28. $125 = 25x - 25y$

29. $\frac{1}{3}x + \frac{1}{4}y = 12$

30. $\frac{1}{6}x + \frac{1}{2}y = -1$

Graph each equation.

31. $y = -2x$

32. $y = \frac{1}{2}x$

33. $f(x) = \frac{1}{3}x$

34. $g(x) = 4x$

35. $2x + 4y = 0$

36. $-6x + 3y = 0$

37. $4x - 6y = 0$

38. $15x + 5y = 0$

Graph each equation.

39. $y = 4$

40. $x = 4$

41. $x = -4$

42. $y = -4$

43. $y = -1.5$

44. $f(x) = -3$

45. $x = \frac{5}{2}$

46. $g(x) = 0$

47. $x = 0$

48. $x = -3.25$

Problem Solving

49. *Distance* Using the distance formula

$$\text{distance} = \text{rate} \cdot \text{time, or } d = rt$$

draw a graph of distance versus time for a constant rate of 30 miles per hour.

50. *Simple Interest* Using the simple interest formula

$$\text{interest} = \text{principal} \cdot \text{rate} \cdot \text{time, or } i = prt$$

draw a graph of interest versus time for a principal of $500 and a rate of 3%.

51. *Bicycle Profit* The profit of a bicycle manufacturer can be approximated by the function $p(x) = 60x - 80,000$, where x is the number of bicycles produced and sold.

   a) Draw a graph of profit versus the number of bicycles sold (for up to 5000 bicycles).

   b) Estimate the number of bicycles that must be sold for the company to break even.

   c) Estimate the number of bicycles that must be sold for the company to make $150,000 profit.

52. *Taxi Operating Costs* Raul Lopez's weekly cost of operating a taxi is $75 plus 15¢ per mile.

   a) Write a function expressing Raul's weekly cost, c, in terms of the number of miles, m.

   b) Draw a graph illustrating weekly cost versus the number of miles, for up to 200, driven per week.

   c) If during 1 week, Raul drove the taxi 150 miles, what would be the cost?

   d) How many miles would Raul have to drive for the weekly cost to be $147?

53. *Salary Plus Commission* Jayne Haydack's weekly salary at Charter Network is $500 plus 15% commission on her weekly sales.

   a) Write a function expressing Jayne's weekly salary, s, in terms of her weekly sales, x.

   b) Draw a graph of Jayne's weekly salary versus her weekly sales, for up to $5000 in sales.

   c) What is Jayne's weekly salary if her sales were $2500?

   d) If Jayne's weekly salary for the week was $1025, what were her weekly sales?

54. *Salary Plus Commission* Lynn Hicks, a real estate agent, makes $150 per week plus a 1% sales commission on each property she sells.

   a) Write a function expressing her weekly salary, s, in terms of sales, x.

b) Draw a graph of her salary versus her weekly sales, for sales up to $100,000.

c) If she sells one house per week for $80,000, what will her weekly salary be?

55. **Weight of Girls** The following graph shows weight, in kilograms, for girls (up to 36 months of age) versus length (or height), in centimeters. The red line is the average weight for all girls of the given length, and the green lines represent the upper and lower limits of the normal range.

Girls: Birth to 36 Months Physical Growth

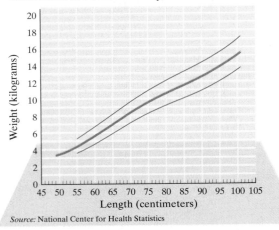

Source: National Center for Health Statistics

a) Explain why the red line represents a function.

b) What is the independent variable? What is the dependent variable?

c) Is the graph of weight versus length approximately linear?

d) What is the weight in kilograms of the average girl who is 85 centimeters long?

e) What is the average length in centimeters of the average girl with a weight of 7 kilograms?

f) What weights are considered normal for a girl 95 centimeters long?

g) What is happening to the normal range as the lengths increase? Is this what you would expect to happen? Explain.

56. **Compound Interest** The graph at the top of the right-hand column illustrates the effect of compound interest.

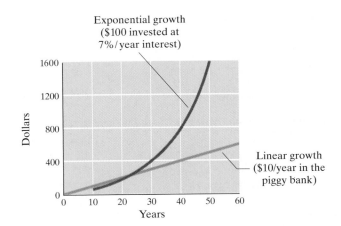

If a child puts $10 each year in a piggy bank, the savings will grow linearly, as shown by the lower curve. If, at age 10, the child invests $100 at 7% interest compounded annually, that $100 will grow exponentially.

a) Explain why both graphs represent functions.

b) What is the independent variable? What is the dependent variable?

c) Using the linear growth curve, determine how long it would take to save $600.

d) Using the exponential growth curve, which begins at year 10, determine how long after the account is opened would the amount reach $600?

e) Starting at year 20, how long would it take for the money growing at a linear rate to double?

f) Starting at year 20, how long would it take for the money growing exponentially to double? (Exponential growth will be discussed at length in Chapter 9.)

57. When, if ever, will the x- and y-intercepts of a graph be the same? Explain.

58. Write two linear functions whose x- and y-intercepts are both $(0, 0)$.

59. Write a function whose graph will have no x-intercept but will have a y-intercept at $(0, 4)$.

60. Write an equation whose graph will have no y-intercept but will have an x-intercept at -2.

61. If the x- and y-intercepts of a linear function are at 1 and -3, respectively, what will be the new x- and y-intercepts if the graph is moved (or translated) up 3 units?

62. If the x- and y-intercepts of a linear function are -1 and 3, respectively, what will be the new x- and y-intercepts if the graph is moved (or translated) down 4 units?

In Exercises 63 and 64, we give two ordered pairs, which are the x- and y-intercepts of a graph. ***a)*** *Plot the points and draw the line through the points.* ***b)*** *Find the change in y, or the vertical change, between the intercepts.* ***c)*** *Find the change in x, or the horizontal change, between the intercepts.* ***d)*** *Find the ratio of the vertical change to the horizontal change between these two points. Do you know what this ratio represents? (We will discuss this further in Section 3.4.)*

63. $(0, 2)$ and $(-4, 0)$

64. $(3, 5)$ and $(-1, -1)$

Solve each equation for x as done in Example 9. Use a graphing calculator if one is available. If not, draw the graphs yourself.

65. $3x + 2 = 2x + 3$

66. $-2(x - 2) = 3(x + 6) + 1$

67. $0.3(x + 5) = -0.6(x + 2)$

68. $2x + \dfrac{1}{4} = 5x - \dfrac{1}{2}$

Find the x- and y-intercepts of the graph of each equation using your graphing calculator.

69. $y = 2(x + 3.2)$

70. $5x - 2y = 7$

71. $-4x - 3.2y = 8$

72. $y = \dfrac{3}{5}x - \dfrac{1}{2}$

Cumulative Review Exercises

[1.4] **73.** Evaluate $4\{2 - 3[(1 - 4) - 5]\} - 2$.

[2.1] **74.** Solve $\dfrac{1}{3}y - 3y = 6(y + 2)$.

*[2.6] In Exercises 75–77 **a)** explain the procedure to solve the equation or inequality for x (assume that b > 0) and **b)** solve the equation or inequality.*

75. $|x - a| = b$

76. $|x - a| < b$

77. $|x - a| > b$

78. Solve the equation $|x - 4| = |2x - 2|$.

3.4 THE SLOPE–INTERCEPT FORM OF A LINEAR EQUATION

SSM Study Guide CD/Video

MathPro 4/5 PH Math Tutor Center prenhall.com/Angel

1   Understand translations of graphs.
2   Find the slope of a line.
3   Recognize slope as a rate of change.
4   Write linear equations in slope–intercept form.
5   Graph linear equations using the slope and the *y*-intercept.
6   Use the slope–intercept form to construct models from graphs.

1   Understand Translations of Graphs

In this section we discuss the translations of graphs, the concept of slope, and the slope–intercept form of a linear equation.

Consider the three equations

$$y = 2x + 3$$
$$y = 2x$$
$$y = 2x - 3$$

Each equation is graphed in Figure 3.50.

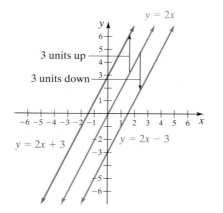

FIGURE 3.50

What are the y-intercepts of $y = 2x + 3$, $y = 2x$ (or $y = 2x + 0$), and $y = 2x - 3$? The y-intercepts are at $(0, 3)$, $(0, 0)$, and $(0, -3)$, respectively. Notice that the graph of $y = 2x + 3$ is the graph of $y = 2x$ shifted, or **translated**, 3 units up and $y = 2x - 3$ is the graph of $y = 2x$ translated 3 units down. All three lines are **parallel**; that is, they do not intersect no matter how far they are extended.

Using this information, can you guess what the y-intercept of $y = 2x + 4$ will be? How about the y-intercept of $y = 2x - \frac{5}{3}$? If you answered $(0, 4)$ and $\left(0, -\frac{5}{3}\right)$, respectively, you answered correctly. In fact, the graph of an equation of the form $y = 2x + b$ will have a y-intercept of $(0, b)$.

Now consider the graphs of the equations $y = -\frac{1}{3}x + 4$, $y = -\frac{1}{3}x$, and $y = -\frac{1}{3}x - 2$, shown in Figure 3.51. The y-intercepts of the three lines are $(0, 4)$, $(0, 0)$, and $(0, -2)$, respectively. The graph of $y = -\frac{1}{3}x + b$ will have a y-intercept of $(0, b)$.

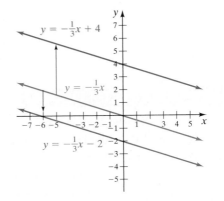

FIGURE 3.51

By looking at the preceding equations, their graphs, and y-intercepts, can you determine the y-intercept of the graph of $y = mx + b$ where m and b are real numbers? If you answered $(0, b)$, you answered correctly. In general, the graph of $y = mx + b$, where m and b are real numbers, has a y-intercept $(0, b)$.

If we look at the graphs in Figure 3.50, we see that the slopes (or slants) of the three lines appear to be the same. If we look at the graphs in Figure 3.51, we see that the slopes of those three lines appear to be the same, but their slope is different from the slope of the three lines in Figure 3.50.

If we consider the equation $y = mx + b$, where the b determines the y-intercept of the line, we can reason that the m is responsible for the slope (or the slant) of the line.

2 Find the Slope of a Line

Now let's speak about slope. The **slope of a line** is the ratio of the vertical change (or rise) to the horizontal change (or run) between any two points on a line. Consider the graph of $y = 2x$ (the blue line in Figure 3.50, repeated in Figure 3.52a). Two points on this line are $(1, 2)$ and $(3, 6)$. Let's find the slope of the line through these points. If we draw a line parallel to the x-axis through the point $(1, 2)$ and a line parallel to the y-axis through the point $(3, 6)$, the two lines intersect at $(3, 2)$. (See Fig. 3.52b.)

From Figure 3.52b we can determine the slope of the line. The vertical change (along the y-axis) is $6 - 2$, or 4 units. The horizontal change (along the x-axis) is $3 - 1$, or 2 units.

$$\text{slope} = \frac{\text{vertical change}}{\text{horizontal change}} = \frac{4}{2} = 2$$

FIGURE 3.52   (a)   (b)

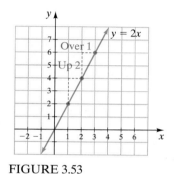

FIGURE 3.53

Thus, the slope of the line through the points $(3, 6)$ and $(1, 2)$ is 2. By examining the line connecting these two points, we can see that for each 2 units the graph moves up the y-axis, it moves 1 unit to the right on the x-axis (Fig. 3.53).

We have determined that the slope of the graph of $y = 2x$ is 2. If you were to compute the slope of the other two lines in Figure 3.50, you would find that the graphs of $y = 2x + 3$ and $y = 2x - 3$ also have a slope of 2.

Can you guess what the slope of the graphs of the equations $y = -3x + 2$, $y = -3x$, and $y = -3x - 2$ is? The slope of all three lines is -3. In general, the slope of an equation of the form $y = mx + b$ is m.*

Now let's determine the procedure to find the slope of a line passing through the two points (x_1, y_1) and (x_2, y_2). Consider Figure 3.54. The vertical change can be found by subtracting y_1 from y_2. The horizontal change can be found by subtracting x_1 from x_2.

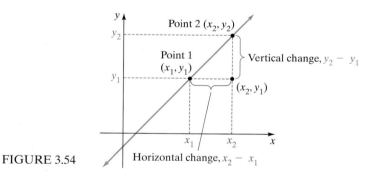

FIGURE 3.54

DEFINITION   The **slope** of the line through the distinct points (x_1, y_1) and (x_2, y_2) is

$$\text{slope} = \frac{\text{change in } y \text{ (vertical change)}}{\text{change in } x \text{ (horizontal change)}} = \frac{y_2 - y_1}{x_2 - x_1}$$

provided that $x_1 \neq x_2$.

It makes no difference which two points on the line are selected when finding the slope of a line. It also makes no difference which point you label (x_1, y_1) or (x_2, y_2). As mentioned before, the letter m is used to represent the slope of a line. The Greek capital letter delta, Δ, is used to represent the words "the change in."

---

*The letter m is traditionally used for slope. It is believed m comes from the French word *monter*, which means to climb.

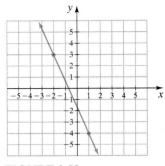

FIGURE 3.55

NOW TRY EXERCISE 35

Thus, the slope is sometimes indicated as

$$m = \frac{\Delta y}{\Delta x} = \frac{y_2 - y_1}{x_2 - x_1}$$

EXAMPLE 1   Find the slope of the line in Figure 3.55.

Solution   Two points on the line are $(-2, 3)$ and $(1, -4)$. Let $(x_2, y_2) = (-2, 3)$ and $(x_1, y_1) = (1, -4)$. Then

$$m = \frac{y_2 - y_1}{x_2 - x_1} = \frac{3 - (-4)}{-2 - 1} = \frac{3 + 4}{-3} = -\frac{7}{3}$$

The slope of the line is $-\frac{7}{3}$. Note that if we had let $(x_1, y_1) = (-2, 3)$ and $(x_2, y_2) = (1, -4)$, the slope would still be $-\frac{7}{3}$. Try it and see.   ✳

A line that rises going from left to right (Fig. 3.56a) has a **positive slope**. A line that neither rises nor falls going from left to right (Fig. 3.56b) has **zero slope**. A line that falls going from left to right (Fig. 3.56c) has a **negative slope**.

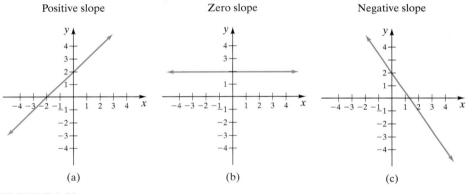

Positive slope	Zero slope	Negative slope
(a)	(b)	(c)

FIGURE 3.56

Slope is undefined.

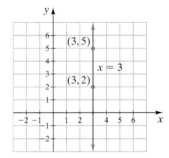

FIGURE 3.57

NOW TRY EXERCISE 19

Consider the graph of $x = 3$ (Fig. 3.57). What is its slope? The graph is a vertical line and goes through the points $(3, 2)$ and $(3, 5)$. Let the point $(3, 5)$ represent (x_2, y_2) and let $(3, 2)$ represent (x_1, y_1). Then the slope of the line is

$$m = \frac{y_2 - y_1}{x_2 - x_1} = \frac{5 - 2}{3 - 3} = \frac{3}{0}$$

Since it is meaningless to divide by 0, we say that the slope of this line is undefined. **The slope of any vertical line is undefined.**

HELPFUL HINT

When students are asked to give the slope of a horizontal or a vertical line, they often answer incorrectly. When asked for the slope of a horizontal line, your response should be "the slope is 0." If you give your answer as "no slope," your instructor may well mark it wrong since these words may have various interpretations. When asked for the slope of a vertical line, your answer should be "the slope is undefined." Again, if you use the words "no slope," this may be interpreted differently by your instructor and marked wrong.

3   Recognize Slope as a Rate of Change

Sometimes it is helpful to describe slope as a *rate of change*. Consider a slope of $\frac{5}{3}$. This means that the *y*-value increases 5 units for each 3-unit increase in *x*. Equivalently, we can say that the *y*-value increases $\frac{5}{3}$ units, or $1.\overline{6}$ units, for each 1-unit increase in *x*. When we give the change in *y* per unit change in *x* we are giving the slope as a **rate of change**. When discussing real-life situations or when creating mathematical models, it is often useful to discuss slope as a rate of change.

EXAMPLE 2   **Public Debt**   The following table of values and the corresponding graph (Fig. 3.58) illustrate the U.S. public debt in billions of dollars from 1910 through 2002.

Year	U.S. public debt (billions of dollars)
1910	1.1
1930	16.1
1950	256.1
1970	370.1
1990	3323.3
2002	5957.2

Source: U.S. Dept. of the Treasury, Bureau of Public Debt.

FIGURE 3.58

a) Determine the slope of the line segments between 1910 and 1930 and between 1990 and 2002.

b) Compare the two slopes found in part **a)** and explain what this means in terms of the U.S. public debt.

Solution   **a)** Understand   To find the slope between any 2 years, find the ratio of the change in debt to the change in years.

Slope from 1910 to 1930

$$m = \frac{16.1 - 1.1}{1930 - 1910} = \frac{15}{20} = 0.75$$

The U.S. public debt from 1910 to 1930 increased at a rate of $0.75 billion per year.

Slope from 1990 to 2002

$$m = \frac{5957.2 - 3323.3}{2002 - 1990} = \frac{2633.9}{12} \approx 219.49$$

The U.S. public debt from 1990 to 2002 increased at a rate of about $219.49 billion per year.

b) Slope measures a rate of change. Comparing the slopes for the two periods shows that there was a much greater increase in the average rate of change in the public debt from 1990 to 2002 than from 1910 to 1930. The slope of the line segment from 1990 to 2002 is greater than the slope of any other line segment on the graph. This indicates that the public debt from 1990 to 2002 grew at a faster rate than at any other time period illustrated.

NOW TRY EXERCISE 67

4 Write Linear Equations in Slope–Intercept Form

We have already shown that for an equation of the form $y = mx + b$, m represents the slope and b represents the y-intercept. For this reason a linear equation written in the form $y = mx + b$ is said to be in **slope–intercept form**.

DEFINITION	The **slope–intercept form of a linear equation** is
	$$y = mx + b$$
	where **m is the slope** of the line and **$(0, b)$ is the y-intercept** of the line.

Examples of Equations in Slope–Intercept Form

$$y = 3x - 6 \qquad\qquad y = \frac{1}{2}x + \frac{3}{2}$$

$$\underset{\text{Slope}}{} \; y = mx + b \;\underset{y\text{-intercept is } (0, b)}{}$$

Equation	Slope	y-Intercept
$y = 3x - 6$	3	$(0, -6)$
$y = \frac{1}{2}x + \frac{3}{2}$	$\frac{1}{2}$	$\left(0, \frac{3}{2}\right)$

To write an equation in slope–intercept form, solve the equation for y.

EXAMPLE 3 Determine the slope and y-intercept of the equation $-5x + 2y = 6$.

Solution Write the equation in slope–intercept form by solving the equation for y.

$$-5x + 2y = 6$$
$$2y = 5x + 6$$
$$y = \frac{5x + 6}{2}$$
$$y = \frac{5x}{2} + \frac{6}{2}$$
$$y = \frac{5}{2}x + 3$$

NOW TRY EXERCISE 45 The slope is $\frac{5}{2}$; the y-intercept is $(0, 3)$. ✳

5 Graph Linear Equations Using the Slope and the y-Intercept

One reason for studying the slope–intercept form of a line is that it can be useful in drawing the graph of a linear equation, as illustrated in Example 4.

EXAMPLE 4   Graph $2y + 4x = 6$ using the y-intercept and slope.

Solution   Begin by solving for y to get the equation in slope–intercept form.

$$2y + 4x = 6$$
$$2y = -4x + 6$$
$$y = -2x + 3$$

FIGURE 3.59

The slope is -2 and the y-intercept is $(0, 3)$. Place a point at 3 on the y-axis (Fig. 3.59). Then use the slope to obtain a second point. The slope is negative; therefore, the graph must fall as it goes from left to right. Since the slope is -2, the ratio of the vertical change to the horizontal change must be 2 to 1 (remember, 2 means $\frac{2}{1}$). Thus, if you start at $y = 3$ and move down 2 units and to the right 1 unit, you will obtain a second point on the graph.

　　　　Continue this process of moving 2 units down and 1 unit to the right to get a third point. Now draw a line through the three points to get the graph. ✳

　　　　In Example 4, we chose to move down and to the right to get the second and third points. We could have also chosen to move up and to the left to get the second and third points.

EXAMPLE 5   Graph $f(x) = \dfrac{4}{3}x - 3$ using the y-intercept and slope.

Solution   Since $f(x)$ is the same as y, this function is in slope–intercept form. The y-intercept is $(0, -3)$ and the slope is $\frac{4}{3}$. Place a point at -3 on the y-axis. Then, since the slope is positive, obtain the second and third points by moving up 4 units and to the right 3 units. The graph is shown in Figure 3.60.

NOW TRY EXERCISE 51 ✳

FIGURE 3.60

6   Use the Slope–Intercept Form to Construct Models from Graphs

Often we can use the slope–intercept form of a linear equation to determine a function that models a real-life situation. Example 6 shows how this may be done.

EXAMPLE 6   **Newspapers** Consider the graph in Figure 3.61, which shows the declining number of adults who read the daily newspaper. Notice that the graph is somewhat linear.

a) Write a linear function whose graph approximates the graph shown.

b) Assuming this trend continues, estimate the percent of adults who will read a daily newspaper in 2005 using the function determined in part **a)**.

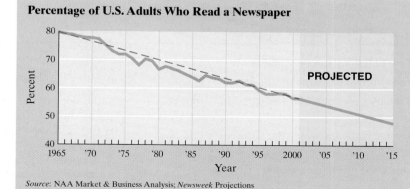

FIGURE 3.61   *Source*: NAA Market & Business Analysis; *Newsweek* Projections

Solution   **a)** To make the numbers easier to work with, we will select 1965 as a *reference year*. Then we can replace 1965 with 0, 1966 with 1, 1967 with 2, and so on. Then 2001 would be 36 (see Fig. 3.62).

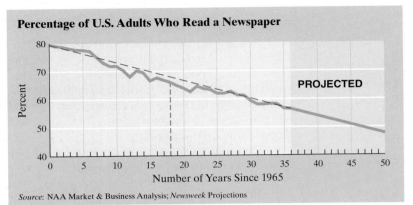

FIGURE 3.62

Source: NAA Market & Business Analysis; *Newsweek* Projections

If we call the vertical axis y and the horizontal axis x, then the y-intercept is 80. The ordered pair that represents the y-intercept is $(0, 80)$. In 2001 it appears that about 57% of the adult population read a daily newspaper. Let's select $(36, 57)$ as a second point on the graph of the straight line we drew in Figure 3.62. We designate $(36, 57)$ as (x_2, y_2) and $(0, 80)$ as (x_1, y_1).

$$\text{slope} = \frac{\text{change in percent}}{\text{change in year}} = \frac{y_2 - y_1}{x_2 - x_1} = \frac{57 - 80}{36 - 0} = -\frac{23}{36} \approx -0.64$$

Since the slope is approximately -0.64 and the y-intercept is $(0, 80)$, the equation of the straight line is $y = -0.64x + 80$. This equation in function notation is $f(x) = -0.64x + 80$. To use this function remember that $x = 0$ represents 1965, $x = 1$ represents 1966, and so on. Note that $f(x)$, the percent, is a function of x, the number of years since 1965.

b) To determine the approximate percent of readers in 2005, and since $2005 - 1965 = 40$, we substitute 40 for x in the function.

$$f(x) = -0.64x + 80$$
$$f(40) = -0.64(40) + 80$$
$$= -25.6 + 80$$
$$= 54.4$$

Thus, if the current trend continues, about 54% of adults will read a daily newspaper in 2005 (Fig. 3.62).

NOW TRY EXERCISE 71

Exercise Set 3.4

Concept/Writing Exercises

1. Explain how to find the slope of a line from its graph.

2. Explain what it means when the slope of a line is negative.

3. Explain what it means when the slope of a line is positive.

4. What is the slope of a horizontal line? Explain.

5. Why is the slope of a vertical line undefined?

6. **a)** Using the slope formula, $m = \dfrac{y_2 - y_1}{x_2 - x_1}$, determine the slope of the line that contains the points $(3, 4)$ and $(4, 6)$. Use $(3, 4)$ as (x_1, y_1).

   b) Calculate the slope again, but this time use $(4, 6)$ as (x_1, y_1).

   c) When finding the slope using the formula, will your answer be the same regardless of which of the two points you designate as (x_1, y_1)? Explain.

7. Explain how to write an equation given in standard form in slope–intercept form.

8. In the equation $y = mx + b$, what does the m represent? What does the b represent?

9. a) What does it mean when a graph is translated down 3 units?

   b) If the y-intercept of a graph is $(0, -3)$ and the graph is translated down 3 units, what will be its new y-intercept?

10. a) What does it mean when a graph is translated up 4 units?

   b) If the y-intercept of a graph is $(0, 2)$ and the graph is translated up 4 units, what will be its new y-intercept?

11. What does it mean when slope is given as a rate of change?

12. Explain how to graph a linear equation using its slope and y-intercept.

Practice the Skills

Find the slope of the line through the given points. If the slope of the line is undefined, so state.

13. $(2, 5)$ and $(0, 9)$

14. $(2, 3)$ and $(5, 4)$

15. $(5, 2)$ and $(1, 4)$

16. $(-3, 5)$ and $(5, -3)$

17. $(-3, 5)$ and $(2, 0)$

18. $(2, 3)$ and $(2, -3)$

19. $(4, 2)$ and $(4, -1)$

20. $(8, -4)$ and $(-1, -2)$

21. $(-3, 4)$ and $(-1, 4)$

22. $(2, 3)$ and $(-5, 3)$

23. $(-2, 3)$ and $(7, -3)$

24. $(2, -4)$ and $(-5, -3)$

Solve for the given variable if the line through the two given points is to have the given slope.

25. $(3, 2)$ and $(4, b), m = 1$

26. $(-4, 3)$ and $(-2, b), m = -3$

27. $(5, 3)$ and $(1, k), m = \frac{1}{2}$

28. $(5, d)$ and $(9, 2), m = -\frac{3}{4}$

29. $(x, 2)$ and $(3, -4), m = 2$

30. $(-2, -3)$ and $(x, 4), m = \frac{1}{2}$

31. $(2, -2)$ and $(r, -1), m = -\frac{1}{2}$

32. $(-4, -1)$ and $(x, 2), m = -\frac{3}{5}$

Find the slope of the line in each of the figures. If the slope of the line is undefined, so state. Then write an equation of the given line.

33.

34.

35.

36.

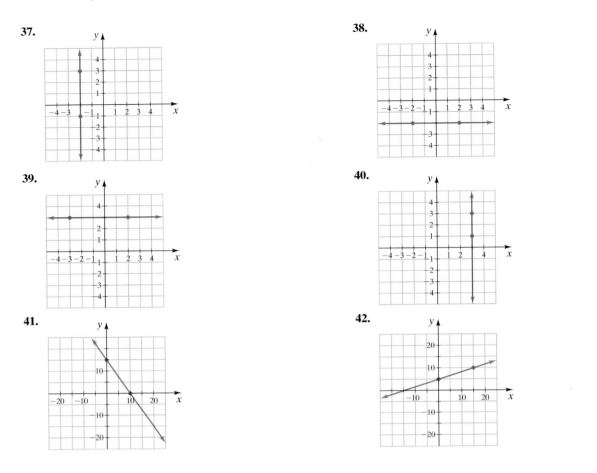

37.

38.

39.

40.

41.

42.

Write each equation in slope–intercept form (if not given in that form). Determine the slope and the y-intercept and use them to draw the graph of the linear equation.

43. $y = -x + 2$

44. $-3x + y = 6$

45. $5x + 15y = 30$

46. $-2x = 3y + 6$

47. $-50x + 20y = 40$

48. $60x = -30y + 60$

Use the slope and y-intercept to graph each function.

49. $f(x) = -2x + 1$

50. $g(x) = \dfrac{2}{3}x - 4$

51. $h(x) = -\dfrac{3}{4}x + 2$

52. $h(x) = -\dfrac{2}{5}x + 4$

Problem Solving

53. Given the equation $y = mx + b$, for the values of m and b given, match parts **a)**–**d)** with the appropriate graphs labeled 1–4.

a) $m > 0, b < 0$
b) $m < 0, b < 0$
c) $m < 0, b > 0$
d) $m > 0, b > 0$

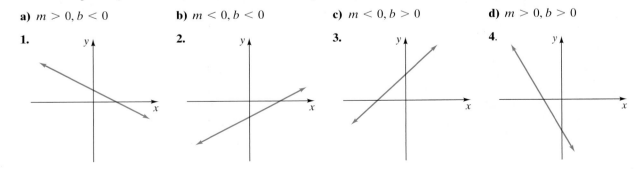

1.

2.

3.

4.

54. Given the equation $y = mx + b$, for the values of m and b given, match parts **a)–d)** with the appropriate graphs labeled 1–4.

a) $m = 0, b > 0$ **b)** $m = 0, b < 0$ **c)** m is undefined, x-intercept < 0 **d)** m is undefined, x-intercept > 0

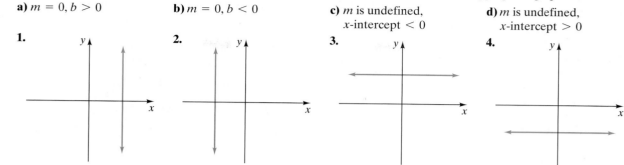

1. **2.** **3.** **4.**

55. We will be discussing parallel lines in the next section. Based on what you have read in this section, explain how you could determine (without graphing) that the graphs of two equations are parallel.

56. How can you determine whether two straight lines are parallel?

57. If one point on a graph is $(6, 3)$ and the slope of the line is $\frac{4}{3}$, find the y-intercept of the graph.

58. If one point on a graph is $(6, 1)$ and the slope of the line is $m = \frac{2}{3}$, find the y-intercept of the graph.

59. In the following graph, the green line is a translation of the blue line.
  a) Determine the equation of the blue line.
  b) Use the equation of the blue line to determine the equation of the green line.

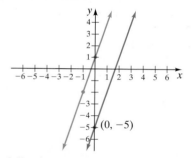

$(0, -5)$

60. In the following graph, the red line is a translation of the blue line.
  a) Determine the equation of the blue line.
  b) Use the equation of the blue line to determine the equation of the red line.

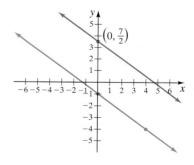

$\left(0, \frac{7}{2}\right)$

61. The graph of $y = x - 1$ is translated up 3 units. Determine
  a) the slope of the translated graph.
  b) the y-intercept of the translated graph.
  c) the equation of the translated graph.

62. The graph of $y = -\frac{3}{2}x + 3$ is translated down 4 units. Determine
  a) the slope of the translated graph.
  b) the y-intercept of the translated graph.
  c) the equation of the translated graph.

63. The graph of $3x - 2y = 6$ is translated down 4 units. Find the equation of the translated graph.

64. The graph of $-3x - 5y = 10$ is translated up 2 units. Find the equation of the translated graph.

65. If a line passes through the points $(6, 4)$ and $(-4, 2)$, find the change of y with respect to a 1-unit change in x.

66. If a line passes through the points $(-3, -4)$ and $(5, 2)$, find the change of y with respect to a 1-unit change in x.

67. *Amtrak Expenses* The National Railroad and Passenger Corporation, better known as Amtrak, continues to face economic struggles. Expenses since 1985 have grown at a much faster rate than revenues. The following table gives the expenses, in millions of dollars, of Amtrak for selected years.

Year	Amtrak expenses (in millions of dollars)
1985	$1600
1990	$2012
1995	$2257
2000	$2876
Source: Amtrak Fiscal Year 2000 Annual Report	

a) Plot these points on a graph.
b) Connect these points using line segments.

c) Determine the slopes of each of the three line segments.

d) During which period was there the greatest average rate of change? Explain.

68. Speed of Computers Each year computers become faster and more powerful. The following table shows the record speed, in billions of operations per second, for supercomputers for selected years.

Year	Operations per second (billions)
1994	143
1996	303
1997	1070
2001	7226

Source: U.S. Dept. of Energy

a) Plot these points on a graph.

b) Connect the points using line segments.

c) Determine the slope of each of the three line segments.

d) During which period was there the greatest average rate of change? Explain.

69. Heart Rate The following bar graph shows the maximum recommended heart rate, in beats per minute, under stress for men of different ages. The bars are connected by a straight line.

a) Use the straight line to determine a function that can be used to estimate the maximum recommended heart rate, h, for $0 \leq x \leq 50$, where x is the number of years after age 20.

b) Using the function from part **a)**, determine the maximum recommended heart rate for a 34-year-old man.

Heart Rate vs. Age

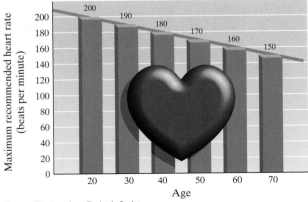

Source: The American Geriatric Society

70. Poverty Threshold The federal government defines the poverty threshold as an estimate of the annual family income necessary to have what society defines as a minimally acceptable standard of living. The following bar graph shows the poverty threshold for a family of four for the years 1995 through 2000.

U.S. Poverty Threshold for a Family of Four

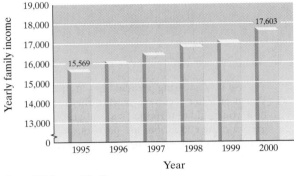

Source: U.S. Bureau of the Census

a) Determine a linear function that can be used to estimate the poverty threshold for a family of four, P, from 1995 through 2000. Let t represent the number of years since 1995. (In other words, 1995 corresponds to $t = 0$, 1996 corresponds to $t = 1$, and so on.)

b) Using the function from part **a)**, determine the poverty threshold in 1997. Compare your answer with the graph to see whether the graph supports your answer.

c) Assuming this trend continues, determine the poverty threshold for a family of four in the year 2005.

d) Assuming this trend continues, in which year will the poverty threshold for a family of four reach $20,000?

71. Number of People Below Poverty Threshold The poverty threshold was defined in Exercise 70. The following graph shows the number of people in the United States, in millions, who were below the poverty threshold in the years 1996 through 2000.

Persons Below the Poverty Threshold

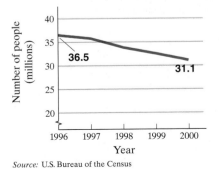

Source: U.S. Bureau of the Census

a) Using 1996 as a reference year, determine a linear function that can be used to estimate the number of people, N, below the poverty threshold for the years 1996 through 2000. In the function, let t represent the number of years since 1996.

b) Using the function from part **a)**, estimate the number of people below the poverty threshold for the year 1998. Compare your answer with the graph to see whether the graph supports your answer.

c) Assuming this trend continues, how many people will be below the poverty threshold in 2005?

d) Assuming this trend continues, during what year will the number of people below the poverty threshold reach 25 million people?

72. *Purchasing Power of the Dollar* The purchasing power of the dollar is measured by comparing the current price of items to the price of those same items in 1982. From the chart below you will see that the purchasing power of the dollar has steadily declined for the years 1990 through 2000. This means that $1 buys less each year.

Purchasing Power of the Dollar

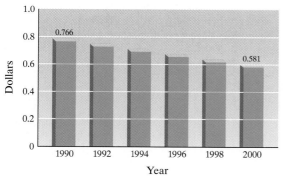

Source: U.S. Bureau of Economic Analysis

a) Using 1990 as the reference year, determine a linear function that can be used to estimate the purchasing power, P, for the years 1990 through 2000. In the function, let t represent the number of years since 1990.

b) Using the function from part **a)**, estimate the purchasing power of the dollar in 1994. Compare your answer with the graph to see whether the graph supports your answer.

c) Assuming this trend continues, what would be the purchasing power of the dollar in 2006?

d) Assuming this trend continues, when would the purchasing power of the dollar reach $0.45?

73. *U. S. Hospitals* The number of hospitals in the United States has been declining approximately linearly since 1975. In 1975 there were 7156 hospitals. In 2000 there were 5890 hospitals. Let n be the number of hospitals in the United States and t be the number of years since 1975. (Let $t = 0$ correspond to 1975 and $t = 25$ correspond to 2000.) *Source:* American Hospital Association

a) Find a linear function $n(t)$ that fits this data.

b) Use the function from part **a)** to estimate the number of hospitals in the United States in 1995.

c) If this trend continues, estimate the number of hospitals in the United States in 2005.

d) If this trend continues, in which year will the number of hospitals in the United States reach 5000?

74. *Tetanus Decline* Largely due to widespread vaccination campaigns, the disease tetanus, commonly called lockjaw, has rapidly declined in the United States. This decline has been approximately linear since 1990. There were 64 reported cases of tetanus in 1990. There were 26 reported cases in 2000. Let C be the number of cases of tetanus in the United States and t be the number of years since 1990. *Source:* Centers for Disease Control and Prevention

a) Find a linear function $C(t)$ that fits this data.

b) Use the function found in part **a)** to estimate the number of reported cases in 1998.

c) If this trend continues, estimate the number of cases of tetanus that will be reported in 2005.

d) If this trend continues, estimate the year in which no cases of tetanus will be reported.

75. *Median Home Sale Price* The median home sale price in the United States has been rising linearly since 1995. The median home sale price in 1995 was $110,500. The median home sale price in 2000 was $139,000. Let P be the median home sale price and let t be the number of years since 1995. *Source:* National Association of Realtors

a) Determine a function $P(t)$ that fits this data.

b) Use the function from part **a)** to estimate the median home sale price in 1997.

c) If this trend continues, estimate the median home sale price in 2010.

d) If this trend continues, in which year will the median home sale price reach $200,000?

76. *Social Security* The number of workers per social security beneficiary has been declining approximately linearly since 1970. In 1970 there were 3.7 workers per beneficiary. In 2050 it is projected there will be 2.0 workers per beneficiary. Let W be the workers per social security beneficiary and t be the number of years since 1970.

a) Find a function $W(t)$ that fits the data.

b) Estimate the number of workers per beneficiary in 2020.

Suppose you are attempting to graph the equations shown and you get the screens shown. Explain how you know that you have made a mistake in entering each equation. The standard window setting is used on each graph.

77. $y = 3x + 6$ **78.** $y = -2x - 4$ **79.** $y = \frac{1}{2}x + 4$ **80.** $y = -4x - 1$

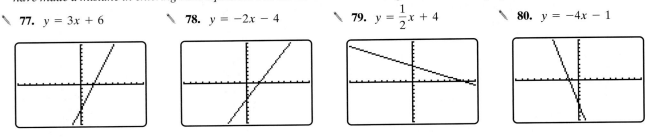

Challenge Problems

81. The photo below is the Castle at Chichén Itzá, Mexico. Each side of the castle has a stairway consisting of 91 steps. The steps of the castle are quite narrow and steep, which makes them hard to climb. The total vertical distance of the 91 steps is 1292.2 inches. If a straight line were to be drawn connecting the tips of the steps, the absolute value of the slope of this line would be 2.21875. Find the average height and width of a step.

82. A **tangent line** is a straight line that touches a curve at a single point (the tangent line may cross the curve at a different point if extended). Figure 3.63 shows three tangent lines to the curve at points *a*, *b*, and *c*. Note that the tangent line at point *a* has a positive slope, the tangent line at point *b* has a slope of 0, and the tangent line at point *c* has a negative slope. Now consider the curve in Figure 3.64. Imagine that tangent lines are drawn at all points on the curve except at endpoints *a* and *e*. Where on the curve in Figure 3.64 would the tangent lines have a positive slope, a slope of 0, and a negative slope?

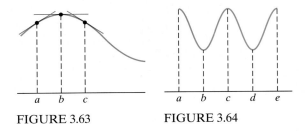

FIGURE 3.63   FIGURE 3.64

Group Activity

83. The following graph from *Consumer Reports* shows the depreciation on a typical car. The initial purchase price is represented as 100%.

a) Group member 1: Determine the 1-year period in which a car depreciates most. Estimate from the graph the percent a car depreciates during this period.

b) Group member 2: Determine between which years the depreciation appears linear or nearly linear.

c) Group member 3: Determine between which 2 years the depreciation is the lowest.

d) As a group, estimate the slope of the line segment from year 0 to year 1. Explain what this means in terms of rate of change.

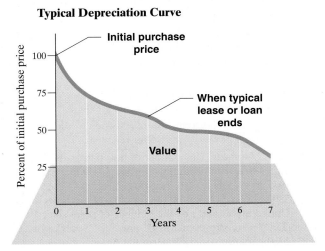

Cumulative Review Exercises

[1.4] **84.** Evaluate $\dfrac{-6^2 - 16 \div 2 \div |-4|}{5 - 3 \cdot 2 - 4 \div 2^2}$.

Solve each equation.

[2.1] **85.** $\dfrac{3}{4}x + \dfrac{1}{5} = \dfrac{2}{3}(x - 2)$

86. $2.6x - (-1.4x + 3.4) = 6.2$

[2.4] **87.** *Trains* Two trains leave Chicago, Illinois, traveling in the same direction along parallel tracks. The first train leaves 3 hours before the second, and its speed is 15 miles per hour faster than the second. Find the speed of each train if they are 270 miles apart 3 hours after the second train leaves Chicago.

[2.6] **88.** Solve
a) $|2x + 1| > 3$, **b)** $|2x + 1| < 3$.

3.5 THE POINT–SLOPE FORM OF A LINEAR EQUATION

1 Understand the point–slope form of a linear equation.
2 Use the point–slope form to construct models from graphs.
3 Recognize parallel and perpendicular lines.

SSM   Study Guide   CD/Video

MathPro 4/5   PH Math Tutor Center   prenhall.com/Angel

1 Understand the Point–Slope Form of a Linear Equation

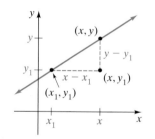

FIGURE 3.65

In the previous section we learned how to use the *slope–intercept form* of a line to determine the equation of a line when the slope and y-intercept of the line are known. In this section we learn how to use the **point–slope form** of a line to determine the equation of a line when the slope and a point on the line are known. The point–slope form can be developed from the expression for the slope between any two points (x, y) and (x_1, y_1) on a line, as shown in Figure 3.65.

$$m = \frac{y - y_1}{x - x_1}$$

Multiplying both sides of the equation by $x - x_1$, we obtain

$$y - y_1 = m(x - x_1)$$

DEFINITION

The **point–slope form of a linear equation** is

$$y - y_1 = m(x - x_1)$$

where m **is the slope** of the line and (x_1, y_1) is a point on the line.

EXAMPLE 1 Write, in slope–intercept form, the equation of the line that passes through the point $(1, 4)$ and has slope -2.

Solution Since we are given the slope of the line and a point on the line, we can write the equation in point–slope form. We can then solve the equation for y to write the equation in slope–intercept form. The slope, m, is -2. The point on the line, (x_1, y_1), is $(1, 4)$. Substitute -2 for m, 1 for x_1, and 4 for y_1 in the point–slope form.

$$y - y_1 = m(x - x_1)$$
$$y - 4 = -2(x - 1) \quad \textit{Point–slope form}$$
$$y - 4 = -2x + 2$$
$$y = -2x + 6 \quad \textit{Slope–intercept form}$$

The graph of $y = -2x + 6$ has a slope of -2 and passes through the point $(1, 4)$.

NOW TRY EXERCISE 5

In Example 1 we used the point–slope form to get the equation of a line when we were given a point on the line and the slope of the line. The point–slope form can also be used to find the equation of a line when we are given two points on the line. We show how to do this in Example 2.

EXAMPLE 2 Write, in slope–intercept form, the equation of the line that passes through the points $(2, 3)$ and $(1, 4)$.

Solution Although we are not given the slope of the line, we can use the two given points to determine the slope. We can then proceed as we did in Example 1. We will let $(2, 3)$ be (x_1, y_1) and $(1, 4)$ be (x_2, y_2).

$$m = \frac{y_2 - y_1}{x_2 - x_1} = \frac{4 - 3}{1 - 2} = \frac{1}{-1} = -1$$

The slope, m, is -1. Now we must choose one of the two given points to use as (x_1, y_1) in the point–slope form of the equation of a line. We will choose $(2, 3)$. Substitute -1 for m, 2 for x_1, and 3 for y_1 in the point–slope form.

$$y - y_1 = m(x - x_1)$$
$$y - 3 = -1(x - 2)$$
$$y - 3 = -x + 2$$
$$y = -x + 5$$

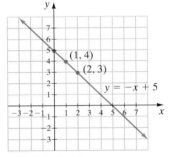

FIGURE 3.66

The graph of $y = -x + 5$ is shown in Figure 3.66. Notice that the y-intercept of this line is at 5, the slope is -1, and the line goes through the points $(2, 3)$ and $(1, 4)$.

Note that we could have also chosen the point $(1, 4)$ to substitute into the point–slope form. Had we done this we would still have obtained the equation $y = -x + 5$. You should verify this now.

NOW TRY EXERCISE 11

2 Use the Point–Slope Form to Construct Models from Graphs

Now let's look at an application where we use the point–slope form to determine a function that models a given situation.

EXAMPLE 3 **Burning Calories** For those trying to burn calories and lose weight, exercise specialists recommend participating in exercise that can be sustained consistently for long periods of time. The number of calories burned in 1 hour riding a bicycle is a linear function of the speed of the bicycle. The average person riding at 12 mph will burn about 564 calories in 1 hour and while riding at 18 mph will burn about 846 calories in 1 hour. This information is shown in the following graph (Fig. 3.67).

a) Determine a linear function that can be used to estimate the number of calories, C, burned in 1 hour when a bicycle is ridden at r mph, for $6 \leq r \leq 24$.

Calories Burned While Bike Riding

Source: American Heart Association

FIGURE 3.67

b) Use the function determined in part **a)** to estimate the number of calories burned in 1 hour when a bicycle is ridden at 20 mph.

c) Use the function determined in part **a)** to estimate the speed at which a bicycle should be ridden to burn 1000 calories in 1 hour.

Solution   **a)** Understand and Translate   In this example, instead of using the variables x and y as we used in Examples 1 and 2, we use the variables r (for rate or speed) and C (for calories). Regardless of the variables used, the procedure used to determine the equation of the line remains the same. To find the necessary function, we will use the points $(12, 564)$ and $(18, 846)$ and proceed as we did in Example 2. We will first calculate the slope and then use the point–slope form to determine the equation of the line.

Carry Out

$$m = \frac{C_2 - C_1}{r_2 - r_1}$$

$$= \frac{846 - 564}{18 - 12} = \frac{282}{6} = 47$$

Now we write the equation using the point–slope form. We will choose the point $(12, 564)$ for (r_1, C_1).

$$C - C_1 = m(r - r_1)$$
$$C - 564 = 47(r - 12) \quad \text{\textit{Point–slope form}}$$
$$C - 564 = 47r - 564$$
$$C = 47r \quad \text{\textit{Slope–intercept form}}$$

Answer   Since the number of calories burned, C, is a function of the rate, r, the function we are seeking is

$$C(r) = 47r$$

b) To estimate the number of calories burned in 1 hour while riding at 20 mph, we substitute 20 for r in the function.

$$C(r) = 47r$$
$$C(20) = 47(20) = 940$$

Therefore 940 calories are burned while riding at 20 mph for 1 hour.

c) To estimate the speed at which a bicycle should be ridden to burn 1000 calories in 1 hour, we substitute 1000 for $C(r)$ in the function.

$$C(r) = 47r$$
$$1000 = 47r$$
$$\frac{1000}{47} = r$$
$$r \approx 21.28$$

NOW TRY EXERCISE 53

Thus the bicycle would need to be ridden at about 21.28 mph to burn 1000 calories in 1 hour. ✳

In Example 3, the function we determined was $C(r) = 47r$. The graph of this function has a slope of 47 and a y-intercept at $(0, 0)$. If the graph in Figure 3.67 was extended to the left, it would intersect the origin. This makes sense since a rate of 0 miles per hour would result in 0 calories being burned by riding in 1 hour.

3 Recognize Parallel and Perpendicular Lines

Parallel lines

FIGURE 3.68

Figure 3.68 illustrates two *parallel* lines.

> Two lines are **parallel** when they have the same slope.

All vertical lines are parallel even though their slope is undefined.
Figure 3.69 illustrates perpendicular lines. Two lines are *perpendicular* when they cross at right (or 90°) angles.

> Two lines are **perpendicular** when their slopes are *negative reciprocals*.

For any nonzero number a, its **negative reciprocal** is $\frac{-1}{a}$ or $-\frac{1}{a}$. For example, the negative reciprocal of 2 is $\frac{-1}{2}$ or $-\frac{1}{2}$. The product of any nonzero number and its negative reciprocal is -1.

$$a\left(-\frac{1}{a}\right) = -1$$

Perpendicular lines

FIGURE 3.69

Note that any vertical line is perpendicular to any horizontal line even though the negative reciprocal cannot be applied. (Why not?)

EXAMPLE 4   Two points on l_1 are $(6, 3)$ and $(2, -3)$. Two points on l_2 are $(0, 2)$ and $(6, -2)$. Determine whether l_1 and l_2 are parallel lines, perpendicular lines, or neither.

Solution   Determine the slopes of l_1 and l_2.

$$m_1 = \frac{3 - (-3)}{6 - 2} = \frac{6}{4} = \frac{3}{2} \qquad m_2 = \frac{2 - (-2)}{0 - 6} = \frac{4}{-6} = -\frac{2}{3}$$

Since their slopes are different, l_1 and l_2 are not parallel. To see whether the lines are perpendicular, we need to determine whether the slopes are negative reciprocals. If $m_1m_2 = -1$, the slopes are negative reciprocals and the lines are perpendicular.

$$m_1m_2 = \frac{3}{2}\left(-\frac{2}{3}\right) = -1$$

NOW TRY EXERCISE 15   Since the product of the slopes equals -1, the lines are perpendicular.   ✳

EXAMPLE 5   Consider the equation $2x + 4y = 8$. Determine the equation of the line that has a y-intercept of 5 and is **a)** parallel to the given line and **b)** perpendicular to the given line.

Solution   **a)** If we know the slope of a line and its y-intercept, we can use the slope–intercept form, $y = mx + b$, to write the equation. We begin by solving the given equation for y.

$$2x + 4y = 8$$
$$4y = -2x + 8$$
$$y = \frac{-2x + 8}{4}$$
$$y = -\frac{1}{2}x + 2$$

Two lines are parallel when they have the same slope. Therefore, the slope of the line parallel to the given line must be $-\frac{1}{2}$. Since its slope is $-\frac{1}{2}$ and its y-intercept is 5, its equation must be

$$y = -\frac{1}{2}x + 5$$

The graphs of $2x + 4y = 8$ and $y = -\frac{1}{2}x + 5$ are shown in Figure 3.70.

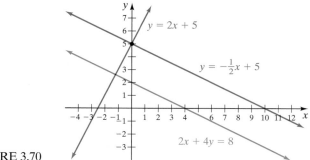

FIGURE 3.70

b) Two lines are perpendicular when their slopes are negative reciprocals. We know that the slope of the given line is $-\frac{1}{2}$. Therefore, the slope of the perpendicular line must be $-1/\left(-\frac{1}{2}\right)$ or 2. The line perpendicular to the given line has a y-intercept of 5. Thus the equation is

$$y = 2x + 5$$

Figure 3.70 also shows the graph of $y = 2x + 5$.   ✳

EXAMPLE 6   Consider the equation $5y = -10x + 7$.

a) Determine the equation of a line that passes through $\left(4, \frac{1}{3}\right)$ that is perpendicular to the graph of the given equation. Write the equation in standard form.

b) Write the equation determined in part **a)** using function notation.

Solution   **a)** Determine the slope of the given line by solving the equation for y.

$$5y = -10x + 7$$

$$y = \frac{-10x + 7}{5}$$

$$y = -2x + \frac{7}{5}$$

Since the slope of the given line is -2, the slope of a line perpendicular to it must be the negative reciprocal of -2, which is $\frac{1}{2}$. The line we are seeking must pass through the point $\left(4, \frac{1}{3}\right)$. Using the point–slope form, we obtain

$$y - y_1 = m(x - x_1)$$

$$y - \frac{1}{3} = \frac{1}{2}(x - 4) \qquad \textit{Point–slope form}$$

Now multiply both sides of the equation by the least common denominator, 6, to eliminate fractions.

$$6\left(y - \frac{1}{3}\right) = 6\left[\frac{1}{2}(x - 4)\right]$$

$$6y - 2 = 3(x - 4)$$

$$6y - 2 = 3x - 12$$

Now write the equation in standard form.

$$-3x + 6y - 2 = -12$$

$$-3x + 6y = -10 \qquad \textit{Standard form}$$

FIGURE 3.71

Note that $3x - 6y = 10$ is also an acceptable answer (see Fig. 3.71).

b) To write the equation using function notation, we solve the equation determined in part **a)** for y, and then replace y with $f(x)$.

We will leave it to you to show that the function is $f(x) = \frac{1}{2}x - \frac{5}{3}$.

NOW TRY EXERCISE 39

HELPFUL HINT

The following chart summarizes the three forms of a linear equation we have studied and mentions when each may be useful.

Standard form:	Useful when finding the intercepts of a graph
$ax + by = c$	Will be used in Chapter 4, Systems of Equations and Inequalities

(continued on the next page)

Slope–intercept form:	Used to find the slope and y-intercept of a line
$y = mx + b$	Used to find the equation of a line given its slope and y-intercept
	Used to determine if two lines are parallel or perpendicular
	Used to graph a linear equation
Point–slope form:	Used to find the equation of a line when given the slope of a line and a point on the line
$y - y_1 = m(x - x_1)$	Used to find the equation of a line when given two points on a line

Exercise Set 3.5

Concept/Writing Exercises

1. Give the point–slope form of a linear equation.

2. How can we determine whether two lines are parallel?

3. How can we determine whether two lines are perpendicular?

4. Why can't the negative reciprocal test be used to determine whether a vertical line is perpendicular to a horizontal line?

Practice the Skills

Use the point–slope form to find the equation of a line with the properties given. Then write the equation in slope–intercept form.

5. Slope = 2, through $(1, 1)$

6. Slope = -1, through $(-2, 3)$

7. Slope = $-\dfrac{1}{2}$, through $(4, -1)$

8. Slope = $-\dfrac{7}{8}$, through $(-8, -2)$

9. Slope = $\dfrac{1}{2}$, through $(-1, -5)$

10. Slope = $-\dfrac{2}{3}$, through $(-1, -2)$

11. Through $(-4, 6)$ and $(4, -6)$

12. Through $(4, -2)$ and $(1, 9)$

13. Through $(4, -3)$ and $(6, -2)$

14. Through $(4, 3)$ and $(-1, 2)$

Two points on l_1 and two points on l_2 are given. Determine whether l_1 is parallel to l_2, l_1 is perpendicular to l_2, or neither.

15. l_1: $(2, 0)$ and $(0, 2)$; l_2: $(5, 0)$ and $(0, 5)$

16. l_1: $(3, 2)$ and $(-1, 5)$; l_2: $(5, -1)$ and $(9, -4)$

17. l_1: $(1, 1)$ and $(5, 7)$; l_2: $(-1, -1)$ and $(1, 4)$

18. l_1: $(-3, 4)$ and $(4, -3)$; l_2: $(-5, -6)$ and $(6, -5)$

19. l_1: $(3, 2)$ and $(-1, -2)$; l_2: $(2, 0)$ and $(3, -1)$

20. l_1: $(0, 2)$ and $(6, -2)$; l_2: $(4, 0)$ and $(6, 3)$

Determine whether the two equations represent lines that are parallel, perpendicular, or neither.

21. $y = \dfrac{1}{5}x + 1$
$y = -5x + 2$

22. $2x + 3y = 6$
$y = -\dfrac{2}{3}x + 5$

23. $4x + 2y = 8$
$8x = 4 - 4y$

24. $2x - y = 4$
$2x + 4y = 8$

25. $4x + 2y = 6$
$-x + 4y = 4$

26. $6x + 2y = 8$
$4x - 9 = -y$

27. $y = \dfrac{1}{2}x - 6$
$-3y = 6x + 9$

28. $2y - 6 = -5x$
$y = -\dfrac{5}{2}x - 2$

29. $y = \dfrac{1}{2}x + 3$
$-2x + 4y = 8$

30. $-4x + 6y = 12$
$2x - 3y = 6$

31. $x - 3y = -9$
$y = 3x + 6$

32. $\dfrac{1}{2}x - \dfrac{3}{4}y = 1$
$\dfrac{3}{5}x + \dfrac{2}{5}y = -1$

Find the equation of a line with the properties given. Write the equation in the form indicated.

33. Through $(2, 5)$ and parallel to the graph of $y = 2x + 4$ (slope–intercept form)

34. Through $(-3, 2)$ and parallel to the graph of $4x - 2y = 6$ (slope–intercept form)

35. Through $(-3, -5)$ and parallel to the graph of $2x - 5y = 7$ (standard form)

36. Through $(-1, 3)$ and perpendicular to the graph of $y = -2x - 1$ (standard form)

37. With x-intercept $(3, 0)$ and y-intercept $(0, 5)$ (slope–intercept form)

38. Through $(-2, -1)$ and perpendicular to the graph of $f(x) = -\dfrac{1}{5}x + 1$ (function notation)

39. Through $(5, -1)$ and perpendicular to the graph of $y = \dfrac{1}{3}x + 1$ (function notation)

40. Through $(-3, 4)$ and perpendicular to the line with x-intercept $(2, 0)$ and y-intercept $(0, 2)$ (standard form)

41. Through $(6, 2)$ and perpendicular to the line with x-intercept $(2, 0)$ and y-intercept $(0, -3)$ (slope–intercept form)

42. Through the point $(2, 1)$ and parallel to the line through the points $(3, 5)$ and $(-2, 3)$ (function notation)

Problem Solving

43. *Treadmill* The number of calories burned in 1 hour on a treadmill is a function of the speed of the treadmill. The average person walking on a treadmill (at $0°$ incline) at a speed of 2.5 miles per hour will burn about 210 calories. At 6 miles per hour the average person will burn about 370 calories. Let C be the calories burned in 1 hour and s be the speed of the treadmill.

a) Determine a linear function $C(s)$ that fits the data.

b) Estimate the calories burned by the average person on a treadmill in 1 hour at a speed of 5 miles per hour.

44. *Inclined Treadmill* The number of calories burned for 1 hour on a treadmill going at a constant speed is a function of the incline of the treadmill. At 4 miles per hour an average person on a $5°$ incline will burn 525 calories. At 4 mph on a $15°$ incline the average person will burn 880 calories. Let C be the calories burned and d be the degrees of incline of the treadmill.

a) Determine a linear function $C(d)$ that fits the data.

b) Determine the number of calories burned by the average person in 1 hour on a treadmill going 4 miles per hour and at a $7°$ incline.

45. *Demand for DVD Players* The *demand* for a product is the number of items the public is willing to buy at a given price. Suppose the demand, d, for DVD players sold in 1 month is a linear function of the price, p, for $\$150 \le p \le \$400$. If the price is $\$200$, then 50 DVD players will be sold each month. If the price is $\$300$, only 30 DVD players will be sold.

a) Using ordered pairs of the form (p, d), write an equation for the demand, d, as a function of price, p.

b) Using the function from part **a)**, determine the demand when the price of the DVD players is $\$260$.

c) Using the function from part **a)**, determine the price charged if the demand for DVD players is 45.

46. *Demand for New Sandwiches* The marketing manager of Arby's restaurants determines that the demand, d, for a new chicken sandwich is a linear function of the price, p, for $\$0.80 \le p \le \$4.00$. If the price is $\$1.00$, then 530 chicken sandwiches will be sold each month. If the price is $\$2.00$, only 400 chicken sandwiches will be sold each month.

a) Using ordered pairs of the form (p, d), write an equation for the demand, d, as a function of price, p.

b) Using the function from part **a)**, determine the demand when the price of the chicken sandwich is $\$1.50$.

c) Using the function from part **a)**, determine the price charged if the demand for chicken sandwiches is 205 chicken sandwiches.

47. *Supply of Kites* The *supply* of a product is the number of items a seller is willing to sell at a given price. The maker of a new kite for children determines that the number of kites she is willing to supply, s, is a linear function of the selling price p for $2.00 \leq p \leq$ $4.00. If a kite sells for $2.00, then 130 per month will be supplied. If a kite sells for $4.00, then 320 per month will be supplied.

a) Using ordered pairs of the form (p, s), write an equation for the supply, s, as a function of price, p.

b) Using the function from part **a)**, determine the supply when the price of a kite is $2.80.

c) Using the function from part **a)**, determine the price paid if the supply is 225 kites.

48. *Supply of Baby Strollers* The manufacturer of baby strollers determines that the supply, s, is a linear function of the selling price, p, for $200 \leq p \leq$ $300. If a stroller sells for $210.00, then 20 strollers will be supplied per month. If a stroller sells for $230.00, then 30 strollers will be supplied per month.

a) Using ordered pairs of the form (p, s), write an equation for the supply, s, as a function of price, p.

b) Using the function from part **a)**, determine the supply when the price of a stroller is $206.00.

c) Using the function from part **a)**, determine the selling price if the supply is 35 strollers.

49. *Gas Mileage of a Buick Park Avenue* The gas mileage, m, of a Buick Park Avenue is a linear function of the speed, s, at which the car is driven, for $40 \leq s \leq 90$. If the car is driven at 45 mph, the car's gas mileage is 40 miles per gallon. If the car is driven at 90 mph, the car's gas mileage is 25 miles per gallon.
Source: http://physics.nadn.navy.mil/physics/faculty/schneider/buick.htm

a) Use this data to write the gas mileage, m, of the Buick Park Avenue as a function of speed, s.

b) Using the function from part **a)**, determine the gas mileage of the Buick Park Avenue if the car is driven at a speed of 60 mph.

c) Using the function from part **a)**, determine the speed at which the Buick Park Avenue must be driven to get gas mileage of 30 miles per gallon.

50. *Gas Mileage of a Honda Civic* The gas mileage, m, of a Honda Civic is a linear function of the speed, s, at which the car is driven, for $40 \leq s \leq 90$. If the car is driven at a rate of 45 mph, the car's gas mileage is 50 miles per gallon. If the car is driven at 90 mph, the car's gas mileage is 20 miles per gallon.

a) Use this data to write the gas mileage, m, of the Honda Civic as a function of speed, s.

b) Using the function from part **a)**, determine the gas mileage of the Honda Civic if the car is driven at a speed of 60 mph.

c) Using the function from part **a)**, determine the speed at which the Honda Civic must be driven to get gas mileage of 30 miles per gallon.

51. *Chief Warrant Officer Pay* The monthly pay for an Army chief warrant officer, level W-4, is a linear function of the years of service by the officer. An officer with 10 years of service is paid $3477 per month and an officer with 20 years of service is paid $4168 per month.

a) Use this data to write monthly pay, p, as a function of years of service, s.

b) Using the function from part **a)**, determine the monthly salary of a chief warrant officer with 18 years of service.

c) Using the function from part **a)**, determine the number of years of service necessary for a chief warrant officer to have a monthly salary of $4000.

52. *Lecturer Salary* The annual salary of a lecturer at the University of Portsmouth is a linear function of the number of years of teaching experience. A lecturer with 9 years of teaching experience is paid $26,350. A lecturer with 15 years of teaching experience is paid $31,687.

a) Use this data to write the annual salary of a lecturer, s, as a function of the number of years of teaching experience, n.

b) Using the function from part **a)**, determine the annual salary of a lecturer with 10 years of teaching experience.

c) Using the function from part **a)**, estimate the number of years of teaching experience a lecturer must have to obtain an annual salary of $30,000.

53. *Life Expectancy* As seen in the following graph, the expected number of remaining years of life of a person, y, *approximates* a linear function. The expected number of remaining years is a function of the person's current age, a, for $30 \leq a \leq 80$. For example, from the graph we see that a person who is currently 50 years old has a life expectancy of 36.0 *more* years.

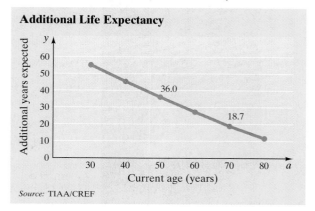

Additional Life Expectancy

Source: TIAA/CREF

a) Using the two points on the graph, determine the function $y(a)$ that can be used to approximate the graph.

b) Using the function from part **a)**, estimate the life expectancy of a person who is currently 37 years old.

c) Using the function from part **a)**, estimate the current age of a person who has a life expectancy of 25 years.

54. *Guarneri del Gesù Violin* Handcrafted around 1735, Guarneri del Gesù violins are extremely rare and extremely valuable. The graph below shows that the projected value, v, of a Guarneri del Gesù violin is a linear function of the age, a, in years, of the violin, for $261 \le a \le 290$.

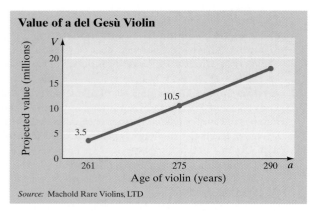

Value of a del Gesù Violin

Source: Machold Rare Violins, LTD

a) Determine the function $v(a)$ represented by this line.

b) Using the function from part **a)**, determine the projected value of a 280-year-old Guarneri del Gesù violin.

c) Using the function from part **a)**, determine the age of a Guarneri del Gesù violin with a projected value of $15 million.

Guarneri del Gesù, "Sainton," 1741

55. *Boys' Weights* Parents may recognize the following diagram from visits to the pediatrician's office. The diagram shows percentiles for boys' heights and weights from birth to age 36 months. Overall, the graphs shown are not linear functions. However, certain portions of the graphs can be approximated with a linear function. For example, the graph representing the 95th percentile of boys' weights (the top red line) from age 18 months to age 36 months is approximately linear.

Boys: Birth to 36 months
Length-for-Age and Weight-for-Age Percentiles

Source: National Center for Health Statistics

a) Use the points shown on the graph of the 95th percentile to write weight, w, as a linear function of age, a, for boys between 18 and 36 months old.

b) Using the function from part **a)**, estimate the weight of a 22-month-old boy who is in the 95th percentile for weight. Compare your answer with the graph to see whether the graph supports your answer.

56. *Boys' Lengths* The diagram in Exercise 55 shows that the graph representing the 95th percentile of boys' lengths (the top yellow line) from age 18 months to age 36 months is approximately linear.

a) Use the points shown on the graph of the 95th percentile to write length, l, as a linear function of age, a, for boys between age 18 and 36 months.

b) Using the function from part **a)**, estimate the length of a 21-month-old boy who is in the 95th percentile. Compare your answer with the graph to see whether the graph supports your answer.

Group Activity

57. The following graph shows the growth of the circumference of a girl's head. The red line is the average head circumference of all girls for the given age while the green lines represent the upper and lower limits of the normal range. Discuss and answer the following questions as a group.

a) Explain why the graph of the average head circumference represents a function.

b) What is the independent variable? What is the dependent variable?

c) What is the domain of the graph of the average head circumference? What is the range of the average head circumference graph?

d) What interval is considered normal for girls of age 18?

e) For this graph, is head circumference a function of age or is age a function of head circumference? Explain your answer.

f) Estimate the average girl's head circumference at age 10 and at age 14.

g) This graph appears to be nearly linear. Determine an equation or function that can be used to estimate the orange line between (2, 48) and (18, 55).

Head Circumference

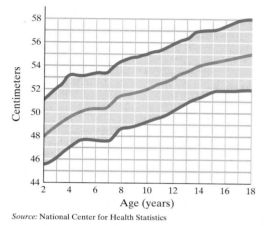

Source: National Center for Health Statistics

Cumulative Review Exercises

[2.5] **58.** Solve the inequality $4 - \frac{1}{2}x > 2x + 3$ and indicate the solution in interval notation.

59. What must you do when multiplying or dividing both sides of an inequality by a negative number?

[3.2] **60. a)** What is a relation?

b) What is a function?

c) Draw a graph that is a relation but not a function.

61. Find the domain and range of the function $\{(4, 3), (5, -2), (3, 2), (6, -1)\}$.

3.6 THE ALGEBRA OF FUNCTIONS

SSM   Study Guide   CD/Video

MathPro 4/5   PH Math Tutor Center   prenhall.com/Angel

1 Find the sum, difference, product, and quotient of functions.
2 Graph the sum of functions.

1 Find the Sum, Difference, Product, and Quotient of Functions

Let's discuss some ways that functions can be combined. If we let $f(x) = x + 2$ and $g(x) = x^2 + 2x$, we can find $f(5)$ and $g(5)$ as follows.

$$f(x) = x + 2 \qquad\qquad g(x) = x^2 + 2x$$
$$f(5) = 5 + 2 = 7 \qquad\qquad g(5) = 5^2 + 2(5) = 35$$

If we add $f(x) + g(x)$, we get

$$f(x) + g(x) = (x + 2) + (x^2 + 2x)$$
$$= x^2 + 3x + 2$$

This new function formed by the sum of $f(x)$ and $g(x)$ is designated as $(f + g)(x)$. Therefore we may write

$$(f + g)(x) = x^2 + 3x + 2$$

We find $(f + g)(5)$ as follows.

$$(f + g)(5) = 5^2 + 3(5) + 2$$
$$= 25 + 15 + 2 = 42$$

Notice that

$$f(5) + g(5) = (f + g)(5)$$
$$7 + 35 = 42 \qquad \textit{True}$$

In fact, for any real number substituted for x you will find that

$$f(x) + g(x) = (f + g)(x)$$

Similar notation exists for subtraction, multiplication, and division of functions.

Operations on Functions

If $f(x)$ represents one function, $g(x)$ represents a second function, and x is in the domain of both functions, then the following operations on functions may be performed:

Sum of functions: $(f + g)(x) = f(x) + g(x)$

Difference of functions: $(f - g)(x) = f(x) - g(x)$

Product of functions: $(f \cdot g)(x) = f(x) \cdot g(x)$

Quotient of functions: $(f/g)(x) = \dfrac{f(x)}{g(x)}$, provided that $g(x) \neq 0$

EXAMPLE 1 If $f(x) = x^2 + x - 6$ and $g(x) = x - 2$, find

a) $(f + g)(x)$ **b)** $(f - g)(x)$

c) $(g - f)(x)$ **d)** Does $(f - g)(x) = (g - f)(x)$?

Solution To answer parts **a)**–**c)**, we perform the indicated operation.

a) $(f + g)(x) = f(x) + g(x)$
$$= (x^2 + x - 6) + (x - 2)$$
$$= x^2 + x - 6 + x - 2$$
$$= x^2 + 2x - 8$$

b) $(f - g)(x) = f(x) - g(x)$
$$= (x^2 + x - 6) - (x - 2)$$
$$= x^2 + x - 6 - x + 2$$
$$= x^2 - 4$$

c) $(g - f)(x) = g(x) - f(x)$
$$= (x - 2) - (x^2 + x - 6)$$
$$= x - 2 - x^2 - x + 6$$
$$= -x^2 + 4$$

d) By comparing the answers to parts **b)** and **c)**, we see that

$$(f - g)(x) \neq (g - f)(x)$$

NOW TRY EXERCISE 11

EXAMPLE 2 If $f(x) = x^2 - 4$ and $g(x) = x - 2$, find

a) $(f - g)(6)$ **b)** $(f \cdot g)(4)$ **c)** $(f/g)(8)$

Solution

a) $(f - g)(x) = f(x) - g(x)$
$$= (x^2 - 4) - (x - 2)$$
$$= x^2 - x - 2$$
$(f - g)(6) = 6^2 - 6 - 2$
$$= 36 - 6 - 2$$
$$= 28$$

We could have also found the solution as follows:

$$f(x) = x^2 - 4 \qquad\qquad g(x) = x - 2$$
$$f(6) = 6^2 - 4 = 32 \qquad\qquad g(6) = 6 - 2 = 4$$
$$(f - g)(6) = f(6) - g(6)$$
$$= 32 - 4 = 28$$

b) We will find $(f \cdot g)(4)$ using the fact that

$$(f \cdot g)(4) = f(4) \cdot g(4)$$
$$f(x) = x^2 - 4 \qquad\qquad g(x) = x - 2$$
$$f(4) = 4^2 - 4 = 12 \qquad\qquad g(4) = 4 - 2 = 2$$

Thus $f(4) \cdot g(4) = 12 \cdot 2 = 24$. Therefore $(f \cdot g)(4) = 24$. We could have also found $(f \cdot g)(4)$ by multiplying $f(x) \cdot g(x)$ and then substituting 4 into the product. We will discuss how to do this in Section 5.2.

c) We will find $(f/g)(8)$ by using the fact that

$$(f/g)(8) = f(8)/g(8)$$
$$f(x) = x^2 - 4 \qquad\qquad g(x) = x - 2$$
$$f(8) = 8^2 - 4 = 60 \qquad\qquad g(8) = 8 - 2 = 6$$

Then $f(8)/g(8) = 60/6 = 10$. Therefore, $(f/g)(8) = 10$. We could have also found $(f/g)(8)$ by dividing $f(x)/g(x)$ and then substituting 8 into the quotient. We will discuss how to do this in Chapter 5.

NOW TRY EXERCISE 31

Notice that we included the phrase "and x is in the domain of both functions" in the Operations on Functions box on page 216. As we stated earlier, the domain of a function is the set of values that can be used for the independent variable. For example, the domain of the function $f(x) = 2x^2 - 6x + 5$ is all real numbers, because when x is any real number $f(x)$ will also be a real number. The domain of $g(x) = \dfrac{1}{x - 3}$ is all real numbers except 3, because when x is any real number except 3, the function $g(x)$ is a real number. When x is 3, the function is not a real number because $\frac{1}{0}$ is undefined. We will discuss the domain of functions further in Section 6.1.

2   Graph the Sum of Functions

Now we will explain how we can graph the sum, difference, product, or quotient of two functions. Figure 3.72 on page 218 shows two functions, $f(x)$ and $g(x)$.

To graph the sum of $f(x)$ and $g(x)$, or $(f + g)(x)$, we use $(f + g)(x) = f(x) + g(x)$. The table on the next page gives the integer values of x from -2 to 4, the values of $f(-2)$ through $f(4)$, and the values of $g(-2)$ through $g(4)$. These values are taken directly from Figure 3.72. The values of $(f + g)(-2)$ through $(f + g)(4)$ are determined by adding the values of $f(x)$ and $g(x)$. The graph of $(f + g)(x) = f(x) + g(x)$ is illustrated in green in Figure 3.73.

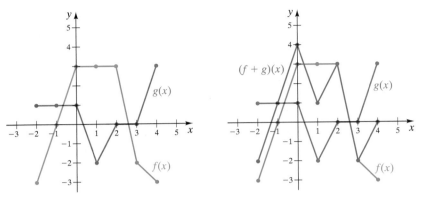

FIGURE 3.72   FIGURE 3.73

x	$f(x)$	$g(x)$	$(f + g)(x)$
−2	−3	1	−3 + 1 = −2
−1	0	1	0 + 1 = 1
0	3	1	3 + 1 = 4
1	3	−2	3 + (−2) = 1
2	3	0	3 + 0 = 3
3	−2	0	−2 + 0 = −2
4	−3	3	−3 + 3 = 0

We could graph the difference, product, or quotient of the two functions using a similar technique. For example, to graph the product function $(f \cdot g)(x)$, we would evaluate $(f \cdot g)(-2)$ as follows:

$$(f \cdot g)(-2) = f(-2) \cdot g(-2)$$
$$= (-3)(1) = -3$$

Thus, the graph of $(f \cdot g)(x)$ would have an ordered pair at $(-2, -3)$. Other ordered pairs would be determined by the same procedure.

NOW TRY EXERCISE 43

In newspapers and magazines we often find graphs that show the sum of two functions. Graphs that show the sum of two functions are generally illustrated in one of two ways. Example 3 shows one way and Example 4 shows the second way.

EXAMPLE 3   **Mutual Fund Accounts**   The number of mutual fund accounts has risen dramatically since 1980. The graph on the next page shows the number of mutual fund accounts separated into each of three categories, and the total of these three categories for selected years from 1980 through 2000.

a) How is the graph of the total number of accounts, T, determined from the graphs of the stock, S, bond/hybrid, B, and money market, M, accounts?

b) In which 5-year period did the number of stock accounts increase the most?

c) If y represents the year, describe what the function $(B + M)(y)$ represents.

Solution   **a)** In Figure 3.74, the graphs for stock, bond/hybrid, and money market accounts are shown separately on the same axes. The graph for the total of these accounts is obtained by adding the number of stock, bond/hybrid, and money market accounts together. For example, in 1995 there were about 70 million stock accounts,

Mutual Fund Accounts

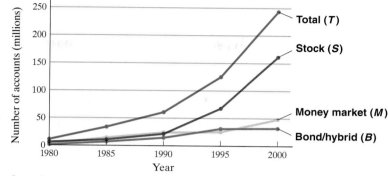

FIGURE 3.74   *Source:* Investment Company Institute

about 30 million bond/hybrid accounts, and about 25 million money market accounts. The sum of these numbers is 125 million, which is about the total number of accounts shown on the graph for 1995. Other points on the total account graph are determined in the same way.

b) From 1980 to 1985 and from 1985 to 1990, the increase in the number of stock accounts was much less than 50 million. From 1990 to 1995 the increase in the number of stock accounts was about 50 million. From 1995 to 2000 the increase in the number of stock accounts was much higher than 50 million. Therefore, the largest increase in the number of stock accounts took place from 1995 to 2000.

c) If y represents the year, then $B(y)$ represents the number of bond/hybrid accounts and $M(y)$ represents the number of money market accounts, in year y. The function $(B + M)(y)$ is equal to $B(y) + M(y)$. Therefore, $(B + M)(y)$ refers to the sum of the number of bond/hybrid accounts and the number of money market accounts in year y.

NOW TRY EXERCISE 57

    In Example 4, which also shows a sum of functions, the categories are "piggy-backed" on top of one another.

EXAMPLE 4   **Natural Gas Use**   The graph in Figure 3.75 shows U.S. natural gas use broken down into three categories: residential/commercial, industrial, and utility/transportation for select years from 1950 through 2000. The amount of natural gas used in each category is a function of the year. The total amount of natural gas used, indicated by the red line, is also a function of the year.

U.S. Natural Gas Use

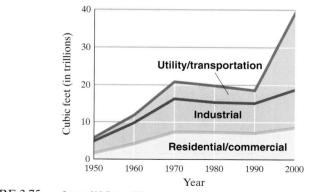

FIGURE 3.75   *Source:* U.S. Dept. of Energy

a) Estimate the amount of natural gas used for residential/commercial purposes in 2000.

b) Estimate the amount of natural gas used for industrial purposes in 2000.

c) Estimate the amount of natural gas used for utility/transportation purposes in 2000.

d) Estimate the total amount of natural gas used in 2000.

Solution **a)** By reading the graph, we see that the amount of natural gas used for residential purposes in 2000 (the gold color) was about 8 trillion cubic feet.

b) The blue color on the graph represents the amount of natural gas used for industrial purposes. In 2000, the blue shaded area starts at about 8 trillion and ends at about 19 trillion. The difference between these two values, 19 trillion − 8 trillion, is 11 trillion. Therefore, about 11 trillion cubic feet of natural gas was used for industrial purposes in 2000.

c) The peach color on the graph represents the amount of natural gas used for utility/transportation purposes. In 2000, the peach shaded area starts at about 19 trillion and ends at about 39 trillion. The difference between these two values, 39 trillion − 19 trillion, is 20 trillion. Therefore about 20 trillion cubic feet of natural gas was used for utility/transportation purposes in 2000.

d) In 2000, the total amount of natural gas used in the United States was about 39 trillion cubic feet. This can be read directly from the graph. Also notice that 39 trillion is the sum of the quantities determined in parts **a)**, **b)**, and **c)**.   ✳

NOW TRY EXERCISE 59

In Example 4, the total amount of natural gas used in any year is the sum of the natural gas used in the three categories. For example, if we add the answers obtained in parts a), b), and c) we get 8 + 11 + 20 = 39. Thus, in 2000 about 39 trillion cubic feet of natural gas was used. The red line at the top of the graph shows the total amount of natural gas used.

Using Your Graphing Calculator

Graphing calculators can graph the sums, differences, products, and quotients of functions. One way to do this is to enter the individual functions. Then, following the instructions that come with your calculator, you can add, subtract, multiply, or divide the functions. For example, the screen in Figure 3.76 shows a TI-83 Plus ready to graph $Y_1 = x - 3$, $Y_2 = 2x + 4$, and the sum of the functions, $Y_3 = Y_1 + Y_2$. On the TI-83 Plus, to get $Y_3 = Y_1 + Y_2$, you press the $\boxed{\text{VARS}}$ key. Then you move the cursor to Y-VARS, and then you select 1: Function. Next you press $\boxed{1}$ to enter Y_1. Next you press $\boxed{+}$. Then press $\boxed{\text{VARS}}$ and go to Y-VARS, and choose 1: Function. Finally, press $\boxed{2}$ to enter Y_2. Figure 3.77 shows the graphs of the two functions, and the graph of the sum of the functions.

FIGURE 3.76   FIGURE 3.77

Exercise Set 3.6

Concept/Writing Exercises

1. Does $f(x) + g(x) = (f + g)(x)$ for all values of x?

2. Does $f(x) - g(x) = (f - g)(x)$ for all values of x?

3. What restriction is placed on the property $f(x)/g(x) = (f/g)(x)$? Explain.

4. Does $(f + g)(x) = (g + f)(x)$ for all values of x? Explain and give an example to support your answer.

5. Does $(f - g)(x) = (g - f)(x)$ for all values of x? Explain and give an example to support your answer.

6. If $f(2) = 9$ and $g(2) = -3$, determine

a) $(f + g)(2)$ **b)** $(f - g)(2)$
c) $(f \cdot g)(2)$ **d)** $(f/g)(2)$

7. If $f(-2) = -3$ and $g(-2) = 5$, find

a) $(f + g)(-2)$ **b)** $(f - g)(-2)$
c) $(f \cdot g)(-2)$ **d)** $(f/g)(-2)$

8. If $f(7) = 6$ and $g(7) = 0$, determine

a) $(f + g)(7)$ **b)** $(f - g)(7)$
c) $(f \cdot g)(7)$ **d)** $(f/g)(7)$

Practice the Skills

*For each pair of functions, find **a)** $(f + g)(x)$, **b)** $(f + g)(a)$, and **c)** $(f + g)(2)$.*

9. $f(x) = x + 1, g(x) = x^2 + x$

10. $f(x) = x^2 - x - 2, g(x) = x^2 + 1$

11. $f(x) = -3x^2 + x - 4, g(x) = x^3 + 3x^2$

12. $f(x) = 4x^3 + 2x^2 - x - 1, g(x) = x^3 - x^2 + 2x + 3$

13. $f(x) = 4x^3 - 3x^2 - x, g(x) = 3x^2 + 4$

14. $f(x) = 3x^2 - x + 4, g(x) = 6 - 4x^2$

Let $f(x) = x^2 - 4$ and $g(x) = -5x + 3$. Find the following.

15. $f(3) + g(3)$ **16.** $f(7) + g(7)$ **17.** $f(-2) - g(-2)$

18. $f\left(\frac{1}{4}\right) - g\left(\frac{1}{4}\right)$ **19.** $f(3) \cdot g(3)$ **20.** $f(-4) \cdot g(-4)$

21. $\dfrac{f\left(\frac{3}{5}\right)}{g\left(\frac{3}{5}\right)}$ **22.** $f(-2)/g(-2)$ **23.** $g(-3) - f(-3)$

24. $g(6) \cdot f(6)$ **25.** $g(0)/f(0)$ **26.** $f(2)/g(2)$

Let $f(x) = 2x^2 - x$ and $g(x) = x - 6$. Find the following.

27. $(f + g)(x)$ **28.** $(f + g)(a)$ **29.** $(f + g)(0)$
30. $(f + g)(-1)$ **31.** $(f - g)(-3)$ **32.** $(f - g)(1)$
33. $(f \cdot g)(0)$ **34.** $(f \cdot g)(-5)$ **35.** $(f/g)(-1)$
36. $(f/g)(6)$ **37.** $(g/f)(5)$ **38.** $(g - f)(3)$
39. $(g - f)(x)$ **40.** $(g - f)(r)$

Problem Solving

Using the graph, find the value of the following.

41. $(f + g)(0)$ **42.** $(f - g)(0)$
43. $(f \cdot g)(2)$ **44.** $(f/g)(4)$
45. $(g - f)(-1)$ **46.** $(g + f)(-3)$
47. $(g/f)(4)$ **48.** $(g \cdot f)(-3)$

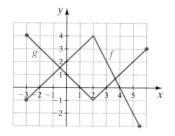

Using the graph below, find the value of the following.

49. $(f + g)(3)$ **50.** $(f - g)(3)$

51. $(f \cdot g)(1)$ **52.** $(g - f)(2)$

53. $(f/g)(4)$ **54.** $(g/f)(5)$

55. $(g/f)(2)$ **56.** $(g \cdot f)(0)$

57. *Health Care Expenditures* The following graph shows private, public, and total expenditures for health care in the United States for select years from 1970 through 2000.

National Expenditure for Health Care

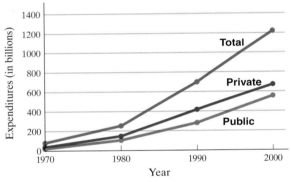

Source: U.S. Department of Health and Human Services

a) How is the graph for the total expenditures determined from the graphs of the private expenditures and the public expenditures?

b) During which 10-year period did the total amount of expenditures for health care increase the least?

c) During which 10-year period did the total amount of expenditures for health care increase the most?

58. *Sources of Electricity* The following graph shows the sources of electricity generated in the United States for select years from 1960 through 2000.

a) How is the graph of the total electricity generated determined from the graphs of electricity generated by fossil fuels, nuclear, and other?

b) During which 10-year period did the amount of electricity generated by nuclear power increase the most?

c) Which of the 4 categories indicated increased the least from 1960 through 2000?

Sources of Electricity

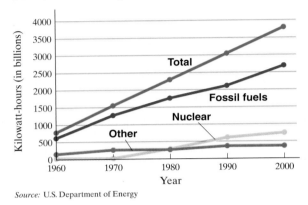

Source: U.S. Department of Energy

59. *Medicare Beneficiaries* The graph below shows the number of Medicare beneficiaries receiving hospital insurance who qualified because they were disabled and who qualified because of age for years 1980 through 2000.

Beneficiaries of Medicare
Hospital Insurance

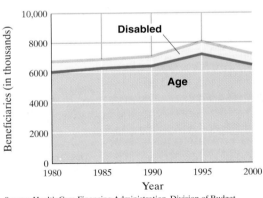

Source: Health Care Financing Administration, Division of Budget

a) Estimate the number of people receiving Medicare hospital insurance who qualified because of age in 2000.

b) Estimate the number of people receiving Medicare hospital insurance who qualified because they were disabled in 2000.

c) Estimate the total number of people receiving Medicare hospital insurance in 2000.

60. *Federal Food Assistance* The graph on the right shows the number of participants in the federal food assistance programs, categorized according to: food stamps, school programs (including lunch, breakfast, and day-care food programs), and women-infants-children (WIC).

a) Estimate the number of participants in the food stamps program in 2000.

b) Estimate the number of participants in the school programs in 2000.

c) Estimate the number of participants in the WIC programs in 2000.

d) Estimate the total number of all food assistance programs in 2000.

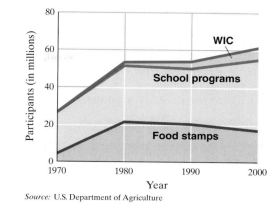

Food Assistance Programs

Source: U.S. Department of Agriculture

For Exercises 61–66, let f and g represent two functions that are graphed on the same axes.

61. If, at a, $(f + g)(a) = 0$, what must be true about $f(a)$ and $g(a)$?

62. If, at a, $(f \cdot g)(a) = 0$, what must be true about $f(a)$ and $g(a)$?

63. If, at a, $(f - g)(a) = 0$, what must be true about $f(a)$ and $g(a)$?

64. If, at a, $(f - g)(a) < 0$, what must be true about $f(a)$ and $g(a)$?

65. If, at a, $(f/g)(a) < 0$, what must be true about $f(a)$ and $g(a)$?

66. If, at a, $(f \cdot g)(a) < 0$, what must be true about $f(a)$ and $g(a)$?

Graph the following functions on your graphing calculator.

67. $y_1 = 2x + 3$
$y_2 = -x + 4$
$y_3 = y_1 + y_2$

68. $y_1 = x - 3$
$y_2 = 2x$
$y_3 = y_1 - y_2$

69. $y_1 = x$
$y_2 = x + 5$
$y_3 = y_1 \cdot y_2$

70. $y_1 = 2x^2 - 4$
$y_2 = x$
$y_3 = y_1/y_2$

Group Activity

71. *SAT Scores* The following graph shows the average math and verbal scores of entering college classes on the SAT college entrance exam for the years 1992 through 2002. Let f represent the math scores, and g represent the verbal scores, and let t represent the year. As a group, draw a graph that represents $(f + g)(t)$.

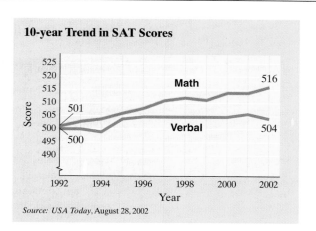

10-year Trend in SAT Scores

Source: USA Today, August 28, 2002

Cumulative Review Exercises

[1.5] **72.** Evaluate $(-3)^{-3}$.

[1.6] **73.** Express 1,630,000 in scientific notation.

[2.2] **74.** Solve the formula $A = \frac{1}{2}bh$ for h.

[2.3] **75.** *Washing Machine* The cost of a washing machine, including a 6% sales tax, is $477. Determine the pre-tax cost of the washing machine.

[3.1] **76.** Graph $y = |x| - 2$.

[3.3] **77.** Graph $3x - 4y = 12$.

3.7 GRAPHING LINEAR INEQUALITIES

SSM   Study Guide   CD/Video

MathPro 4/5   PH Math Tutor Center   prenhall.com/Angel

1 Graph linear inequalities in two variables.

1 Graph Linear Inequalities in Two Variables

A **linear inequality** results when the equal sign in a linear equation is replaced with an inequality sign.

Examples of Linear Inequalities in Two Variables

$$2x + 3y > 2 \qquad\qquad 3y < 4x - 6$$
$$-x - 2y \le 3 \qquad\qquad 5x \ge 2y - 3$$

A line divides a plane into three regions: the line itself and the two **half-planes** on either side of the line. The line is called the **boundary**. Consider the linear equation $2x + 3y = 6$. The graph of this line, the boundary line, divides the plane into the set of points that satisfy the inequality $2x + 3y < 6$ from the set of points that satisfy the inequality $2x + 3y > 6$. An inequality may or may not include the boundary line. Since the inequality $2x + 3y \le 6$ means $2x + 3y < 6$ or $2x + 3y = 6$, the inequality $2x + 3y \le 6$ contains the boundary line. Similarly, the inequality $2x + 3y \ge 6$ contains the boundary line. The graph of the inequalities $2x + 3y < 6$ and $2x + 3y > 6$ do not contain the boundary line. Now let's discuss how to graph linear inequalities.

To Graph a Linear Inequality in Two Variables

1. Replace the inequality symbol with an equal sign.

2. Draw the graph of the equation in step 1. If the original inequality contains a $\ge$ or $\le$ symbol, draw the graph using a solid line. If the original inequality contains a $>$ or $<$ symbol, draw the graph using a dashed line.

3. Select any point not on the line and determine if this point is a solution to the original inequality. If the point selected is a solution, shade the region on the side of the line containing this point. If the selected point does not satisfy the inequality, shade the region on the side of the line not containing this point.

In step 3 we are deciding which set of points satisfies the given inequality.

EXAMPLE 1 Graph the inequality $y < \frac{2}{3}x - 3$.

Solution First graph the equation $y = \frac{2}{3}x - 3$. Since the original inequality contains a less than sign, $<$, use a dashed line when drawing the graph (Fig. 3.78). The dashed line indicates that the points on this line are not solutions to the inequality $y < \frac{2}{3}x - 3$. Select a point not on the line and determine if this point satisfies the inequality. Often the easiest point to use is the origin, $(0, 0)$.

FIGURE 3.78

Check Point $(0, 0)$

$$y < \frac{2}{3}x - 3$$

$$0 \overset{?}{<} \frac{2}{3}(0) - 3$$

$$0 \overset{?}{<} 0 - 3$$

$$0 < -3 \qquad \textit{False}$$

Since 0 is not less than -3, the point $(0, 0)$ does not satisfy the inequality. The solution will be all points on the other side of the line from the point $(0, 0)$. Shade in this region (Fig. 3.79). Every point in the shaded area satisfies the given inequality. Let's check a few selected points A, B, and C.

FIGURE 3.79

NOW TRY EXERCISE 15

Point A	Point B	Point C
$(6, 0)$	$(3, -3)$	$(0, -4)$
$y < \frac{2}{3}x - 3$	$y < \frac{2}{3}x - 3$	$y < \frac{2}{3}x - 3$
$0 \overset{?}{<} \frac{2}{3}(6) - 3$	$-3 \overset{?}{<} \frac{2}{3}(3) - 3$	$-4 \overset{?}{<} \frac{2}{3}(0) - 3$
$0 \overset{?}{<} 4 - 3$	$-3 \overset{?}{<} 2 - 3$	$-4 \overset{?}{<} 0 - 3$
$0 < 1$ *True*	$-3 < -1$ *True*	$-4 < -3$ *True*

EXAMPLE 2 Graph the inequality $y \geq -\frac{1}{2}x$.

Solution First, we graph the equation $y = -\frac{1}{2}x$. Since the inequality is $\geq$, we use a solid line to indicate that the points on the line are solutions to the inequality (Fig. 3.80). Since the point $(0, 0)$ is on the line, we cannot select that point to find the solution. Let's arbitrarily select the point $(3, 1)$.

FIGURE 3.80

Checkpoint $(3, 1)$

$$y \geq -\frac{1}{2}x$$

$$1 \overset{?}{\geq} -\frac{1}{2}(3)$$

$$1 \geq -\frac{3}{2} \qquad \textit{True}$$

Since the point $(3, 1)$ satisfies the inequality, every point on the same side of the line as $(3, 1)$ will also satisfy the inequality $y \geq -\frac{1}{2}x$. Shade this region as indicated. Every point in the shaded region, as well as every point on the line, satisfies the inequality.

EXAMPLE 3 Graph the inequality $3x - 2y < -6$.

Solution First, we graph the equation $3x - 2y = -6$. Since the inequality is $<$, we use a dashed line when drawing the graph (Fig. 3.81). Substituting the check-point $(0, 0)$ into the inequality results in a false statement.

$$\text{Checkpoint } (0, 0)$$
$$3x - 2y < -6$$
$$3(0) - 2(0) \overset{?}{<} -6$$
$$0 < -6 \quad \textit{False}$$

FIGURE 3.81

NOW TRY EXERCISE 23

The solution is, therefore, that part of the plane that does not contain the origin.

Using Your Graphing Calculator

Graphers can also display graphs of inequalities. The procedure to display the graphs varies from calculator to calculator. In Figure 3.82, we show the graph of $y > 2x + 3$. Read your graphing calculator manual and learn how to display graphs of inequalities.

FIGURE 3.82

Exercise Set 3.7

Concept/Writing Exercises

1. When graphing an inequality containing $>$ or $<$, why are points on the line not solutions to the inequality?

2. When graphing an inequality containing $\geq$ or $\leq$, why are points on the line solutions to the inequality?

3. When graphing a linear inequality, when can $(0, 0)$ not be used as a check point?

4. When graphing a linear inequality of the form $y > ax + b$ where a and b are real numbers, will the solution always be above the line? Explain.

Practice the Skills

Graph each inequality.

5. $x > 1$

6. $x \geq \dfrac{3}{4}$

7. $y < -2$

8. $y < x$

9. $y \geq -\dfrac{1}{2}x$

10. $y < \dfrac{1}{2}x$

11. $y < 2x + 1$

12. $y \geq 3x - 1$

13. $y > 2x - 1$

14. $y \leq -x + 2$

15. $y \geq \dfrac{1}{2}x - 3$

16. $y < 3x + 5$

17. $2x - 3y \geq 12$ **18.** $2x + 3y > 6$ **19.** $y \leq -3x + 5$ **20.** $y \leq \frac{2}{3}x + 3$

21. $2x + y < 4$ **22.** $3x - 4y \leq 12$ **23.** $10 \geq 5x - 2y$ **24.** $-x - 2y > 4$

Problem Solving

25. *Life Insurance* The monthly rates for $100,000 of life insurance for women from the American General Financial Group increases approximately linearly from age 35 through age 50. The rate for a 35-year-old woman is $10.15 per month and the rate for a 50-year-old woman is $16.45 per month. *Source:* R. K. Reynolds Insurance Services

a) Draw a graph that fits this data.

b) On the graph, darken the part of the graph where the rate is less than or equal to $15 per month.

c) Estimate the age at which the rate first exceeds $15 per month.

26. *Consumer Price Index* The consumer price index (CPI) is a measure of inflation. Since 1990, the CPI has been increasing approximately linearly. The CPI in 1990 was 130.7 and in 2000 the CPI was 172.2. *Source:* U.S. Bureau of the Census

a) Draw a graph that fits this data.

b) On the graph, darken the part of the graph where the CPI is greater than or equal to 150.

c) Estimate the first year in which the CPI was greater than or equal to 150.

27. *Hourly Compensation* The hourly compensation is the total annual amount of expenses required to employ an individual divided by the annual number of hours the employee works. In the United States, the average hourly compensation for all employees has been increasing approximately linearly since 1975. In 1975, the average hourly compensation was $6.36 per hour. In 2000, the average hourly compensation was $19.86 per hour. *Source:* U.S. Bureau of Labor Statistics

a) Draw a graph that fits this data.

b) On the graph, darken the part of the graph where the average hourly compensation is greater than or equal to $10 per hour.

c) Estimate the first year that the average hourly compensation exceeded $10 per hour.

28. *California Farmland* The amount of land in California used as farmland has been declining approximately linearly since 1980. In 1980, California had about 34 million acres of farmland. In 2000, California had about 28 million acres of farmland. *Source:* U.S. Dept. of Agriculture

a) Draw a graph that fits this data.

b) On the graph, darken the part of the graph where the amount of farmland is less than or equal to 30 million acres.

c) Estimate the first year in which the amount of farmland was less than or equal to 30 million acres.

29. a) Graph $f(x) = 2x - 4$.

b) On the graph, shade the region bounded by $f(x)$, $x = 2$, $x = 4$, and the x-axis.

30. a) Graph $g(x) = -x + 4$.

b) On the graph, shade the region bounded by $g(x)$, $x = 1$, and the x- and y-axes.

Challenge Problems

Graph each inequality.

31. $y < |x|$ **32.** $y \geq x^2$ **33.** $y < x^2 - 4$

Cumulative Review Exercises

[2.1] **34.** Solve the equation $4 - \dfrac{5x}{3} = -6$.

[2.2] **35.** If $C = \bar{x} + Z\dfrac{\sigma}{\sqrt{n}}$, find C when $\bar{x} = 80$, $Z = 1.96$, $\sigma = 3$, and $n = 25$.

[2.3] **36.** *Store Sale* Olie's Records and Stuff is going out of business. The first week all items are being reduced by 10%. The second week all items are being reduced by an additional $2. If during the second week Bob Frieble purchases a CD for $12.15, find the original cost of the CD.

[3.2] **37.** $f(x) = -x^2 + 3$; find $f(-1)$

[3.3] **38.** Write an equation of the line that passes through the point $(6, -2)$ and is perpendicular to the line whose equation is $2x - y = 4$.

[3.4] **39.** Determine the slope of the line through $(-4, 7)$ and $(2, -1)$.

CHAPTER SUMMARY

Key Words and Phrases

3.1
Cartesian coordinate system
Collinear points
Coordinates
First-degree equation
Graph
Graphing calculator
Linear equation
Midpoint
Nonlinear equation
Ordered pair
Origin
Quadrant
Rectangular coordinate system
TABLE feature
TRACE feature
Window of a graphing calculator

x-axis
y-axis

3.2
Dependent variable
Domain
Function
Function notation
Graph of a function or relation
Independent variable
Piecewise function
Range
Relation
Supply and demand
Vertical line test
y is a function of x

3.3
Constant function
Linear function
Root
Standard form of a linear equation
x-intercept
y-intercept
Zero
ZOOM feature

3.4
Negative slope
Parallel lines
Positive slope
Rate of change
Slope of a line
Slope–intercept form of a linear equation

Tangent line
Translated graph
Zero slope

3.5
Negative reciprocal
Perpendicular line
Point–slope form of a linear equation

3.6
Difference of functions
Product of functions
Quotient of functions
Sum of functions

3.7
Linear inequality

IMPORTANT FACTS

Slope of a Line $\qquad m = \dfrac{\Delta y}{\Delta x} = \dfrac{y_2 - y_1}{x_2 - x_1}$

Forms of a Linear Equation

Standard form: $ax + by = c$
Slope–intercept form: $y = mx + b$
Point–slope form: $y - y_1 = m(x - x_1)$
To find the x-intercept, set $y = 0$ and solve the equation for x.
To find the y-intercept, set $x = 0$ and solve the equation for y.
To write an equation in slope–intercept form, solve the equation for y.

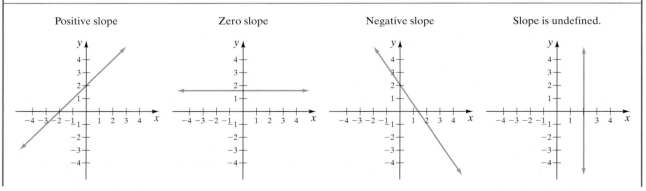

Positive slope   Zero slope   Negative slope   Slope is undefined.

(continued on the next page)

Operations on Functions

Sum of functions: $(f + g)(x) = f(x) + g(x)$
Difference of functions: $(f - g)(x) = f(x) - g(x)$
Product of functions: $(f \cdot g)(x) = f(x) \cdot g(x)$
Quotient of functions: $(f/g)(x) = \dfrac{f(x)}{g(x)}, g(x) \neq 0$

Chapter Review Exercises

[3.1]   **1.** Plot the ordered pairs on the same axes.

     a) $A\,(5, 3)$ **b)** $B\,(0, 4)$ **c)** $C\!\left(5, \dfrac{1}{2}\right)$ **d)** $D(-4, 3)$ **e)** $E(-6, -1)$ **f)** $F(-2, 0)$

Graph each equation.

2. $y = \dfrac{1}{2}x$ **3.** $y = -2x - 1$ **4.** $y = \dfrac{1}{2}x + 3$ **5.** $y = -\dfrac{3}{2}x + 1$ **6.** $y = x^2$

7. $y = x^2 - 1$ **8.** $y = |x|$ **9.** $y = |x| - 1$ **10.** $y = x^3$ **11.** $y = x^3 + 4$

[3.2] **12.** Define function.

     13. Is every relation a function? Is every function a relation? Explain.

Determine whether the following relations are functions. Explain your answers.

14.

$$a \searrow 1$$
$$b \nearrow 2$$
$$c \longrightarrow 3$$

15. $\{(2, 5), (3, -4), (5, 11), (6, -1), (2, -5)\}$

*For Exercises 16–19, **a)** determine whether the following graphs represent functions; **b)** determine the domain and range of each.*

16.

17.

18.

19.

20. If $f(x) = -x^2 + 3x - 5$, find
   a) $f(2)$ and **b)** $f(h)$.

21. If $g(t) = 2t^3 - 3t^2 + 1$, find
   a) $g(-1)$ and **b)** $g(a)$.

22. *Speed of Car* Jane Covillion goes for a ride in a car. The following graph shows the car's speed as a function of time. Make up a story that corresponds to this graph.

23. *Apple Orchard* The number of baskets of apples, N, that are produced by x trees in a small orchard ($x \leq 100$) is given by the function $N(x) = 40x - 0.2x^2$. How many baskets of apples are produced by

   a) 20 trees?

   b) 50 trees?

24. *Falling Ball* If a ball is dropped from the top of a 100-foot building, its height above the ground, h, at any time, t, can be found by the function $h(t) = -16t^2 + 100$, $0 \leq t \leq 2.5$. Find the height of the ball at

   a) 1 second.

   b) 2 seconds.

[3.3] Graph each equation using intercepts.

25. $3x - 4y = 6$

26. $\dfrac{2}{3}x = \dfrac{1}{4}y + 20$

Graph each equation or function.

27. $f(x) = 4$

28. $x = -2$

29. *Bagel Company* The yearly profit, p, of a bagel company can be estimated by the function $p(x) = 0.1x - 5000$, where x is the number of bagels sold per year.

   a) Draw a graph of profits versus bagels sold for up to 250,000 bagels.

   b) Estimate the number of bagels that must be sold for the company to break even.

   c) Estimate the number of bagels sold if the company has a $20,000 profit.

30. *Interest* Draw a graph illustrating the interest on a $12,000 loan for a 1-year period for various interest rates up to 20%. Use interest = principal · rate · time.

[3.4] Determine the slope and y-intercept of the graph represented by the given equation.

31. $y = \dfrac{1}{2}x - 3$

32. $f(x) = -2x + 1$

33. $3x + 5y = 12$

34. $3x + 4y = 10$

35. $x = -2$

36. $f(x) = 6$

Determine the slope of the line through the two given points.

37. $(2, 5), (-2, 7)$

38. $(-2, 3)(4, 1)$

Find the slope of each line. If the slope is undefined, so state. Then write the equation of the line.

39.
40.
41.

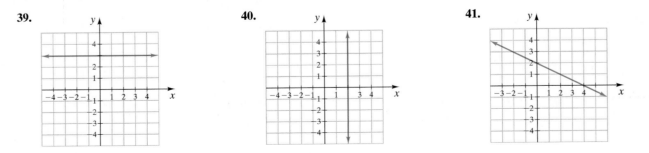

42. If the graph of $y = -2x + 3$ is translated down 4 units, determine
  a) the slope of the translated graph.
  b) the y-intercept of the translated graph.
  c) the equation of the translated graph.

43. If one point on a graph is $(-6, -8)$ and the slope is $\frac{4}{3}$, find the y-intercept of the graph.

44. *Typhoid Fever* The following chart shows the number of reported cases of typhoid fever in the United States for select years from 1970 through 2000.
  a) Plot each point and draw line segments from point to point.
  b) Compute the slope of the line segments.
  c) During which 10-year period did the number of reported cases of typhoid fever increase the most?

Year	Number of reported typhoid fever cases
1970	346
1980	510
1990	552
2000	317
Source: U.S. Dept. of Health and Human Services	

45. *Social Security* The following graph shows the number of social security beneficiaries from 1980 projected through 2070. Use the slope-intercept form to find the function $n(t)$ (represented by the straight line) that can be used to represent this data.

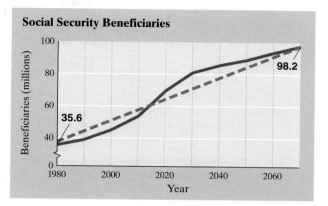

Social Security Beneficiaries

[3.5] *Determine whether the two given lines are parallel, perpendicular, or neither.*

46. $2x - 3y = 10$
    $y = \dfrac{2}{3}x - 4$

47. $2x - 3y = 9$
    $-3x - 2y = 6$

48. $4x - 2y = 10$
    $-2x + 4y = -8$

Find the equation of the line with the properties given. Write each answer in slope–intercept form.

49. Slope $= \dfrac{1}{2}$, through $(4, 5)$

50. Through $(-3, 1)$ and $(2, -4)$

51. Through $(0, 4)$ and parallel to the graph of $y = -\dfrac{2}{3}x + 1$

52. Through $(2, 3)$ and parallel to the graph whose equation is $5x - 2y = 7$

53. Through $(-3, 1)$ and perpendicular to the graph whose equation is $y = \dfrac{3}{5}x + 5$

54. Through $(4, 2)$ and perpendicular to the graph whose equation is $4x - 2y = 8$

Two points on l_1 and two points on l_2 are given. Determine whether l_1 is parallel to l_2, l_1 is perpendicular to l_2, or neither.

55. l_1: $(4, 3)$ and $(0, -3)$; l_2: $(1, -1)$ and $(2, -2)$

56. l_1: $(3, 2)$ and $(2, 3)$; l_2: $(4, 1)$ and $(1, 4)$

57. l_1: $(4, 0)$ and $(1, 3)$; l_2: $(5, 2)$ and $(6, 3)$

58. l_1: $(-3, 5)$ and $(2, 3)$; l_2: $(-4, -2)$ and $(-1, 2)$

59. *Insurance Rates* The monthly rates for $100,000 of life insurance from the General Financial Group for men increases approximately linearly from age 35 through age 50. The rate for a 35-year-old man is $10.76 per month and the rate for a 50-year-old man is $19.91 per month. Let r be the rate and let a be the age of a man between 35 and 50 years of age.
  a) Determine a linear function $r(a)$ that fits this data.
  b) Using the function in part **a)**, estimate the monthly rate for a 42-year-old man.

60. *Burning Calories* The number of calories burned in 1 hour of swimming, when swimming between 20 and 50 yards per minute, is a linear function of the speed of the swimmer. A person swimming at 30 yards per minute will burn about 489 calories in 1 hour. While swimming at 50 yards per minute a person will burn about 525 calories in 1 hour. This information is shown in the graph on page 232.

Calories Burned while Swimming

Source: *Health Magazine* Web Site, www.health.com

a) Determine a linear function that can be used to estimate the number of calories, *C*, burned in 1 hour when a person swims at *r* yards per minute.

b) Use the function determined in part **a)** to determine the number of calories burned in 1 hour when a person swims at 40 yards per minute.

c) Use the function determined in part **a)** to estimate the speed at which a person needs to swim to burn 600 calories in 1 hour.

[3.6] Given $f(x) = x^2 - 3x + 4$ and $g(x) = 2x - 5$, find the following.

61. $(f + g)(x)$

62. $(f + g)(3)$

63. $(g - f)(x)$

64. $(g - f)(-1)$

65. $(f \cdot g)(-1)$

66. $(f \cdot g)(5)$

67. $(f/g)(1)$

68. $(f/g)(2)$

69. *Daily Newspapers* The following graph shows the number of daily newspapers in the United States for select years from 1960 through 2000.

Daily U.S. Newspapers

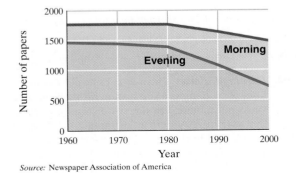

Source: Newspaper Association of America

a) Estimate the number of morning newspapers in 1960.

b) Estimate the number of morning newspapers in 2000.

c) Estimate the number of evening newspapers in 1960.

d) Estimate the number of evening newspapers in 2000.

e) Estimate the total number of daily newspapers in 1960.

f) Estimate the total number of daily newspapers in 2000.

70. *Motor Vehicle Registrations* The following graph shows the number of worldwide car registrations, the number of worldwide truck/bus registrations, and the worldwide total number of motor vehicle registrations for select years from 1970 through 2000. Let function *c* represent the number of car registrations and let function *t* represent the number of truck/bus registrations. Estimate

a) $c(2000)$

b) $t(2000)$

c) $(c + t)(2000)$

World Motor Vehicle Registrations

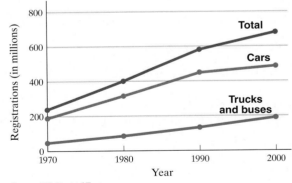

Source: U.S. Dept. of Energy

[3.7] Graph each inequality.

71. $y \geq -3$

72. $x < 4$

73. $y \leq 4x - 3$

74. $y < \dfrac{1}{3}x - 2$

Chapter Practice Test

1. Graph $y = -2x + 1$.

2. Graph $y = \sqrt{x}$.

3. Graph $y = x^2 - 4$.

4. Graph $y = |x|$.

5. Define function.

6. Is the following set of ordered pairs a function? Explain your answer.

$$\{(3, 1), (-2, 6), (4, 6), (5, 2), (6, 3)\}$$

In Exercises 7 and 8, determine whether the following graphs represent functions. Give the domain and range of the relation or function.

7.

8.

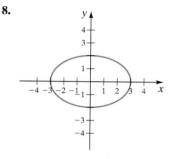

9. If $f(x) = 3x^2 - 6x + 2$, find $f(-2)$.

In Exercises 10 and 11, graph the equation using the x- and y-intercepts.

10. $-10x + 5y = 20$

11. $\dfrac{x}{5} - \dfrac{y}{4} = 1$

12. Graph $f(x) = -3$.

13. Graph $x = 4$.

14. **Profit Graph** The yearly profit, p, for Zico Publishing Company on the sales of a particular book can be estimated by the function $p(x) = 10.2x - 50{,}000$, where x is the number of books produced and sold.

    a) Draw a graph of profit versus books sold for up to 30,000 books.

    b) Use function $p(x)$ to estimate the number of books that must be sold for the company to break even.

    c) Use function $p(x)$ to estimate the number of books that the company must sell to make a $100,000 profit.

15. Determine the slope and y-intercept of the graph of the equation $4x - 3y = 9$.

16. Determine the equation, in slope–intercept form, of the line that goes through the points $(2, 1)$ and $(3, 4)$.

17. Determine the equation, in slope–intercept form, of the line that goes through the point $(5, -3)$ and is perpendicular to the graph of $y = \dfrac{1}{2}x + 1$.

18. **U.S. Population Projections** Determine the function represented by the red line on the graph that can be used to estimate the projected U.S. population, p, from 2000 through 2050. Let 2000 be the reference year so that 2000 is represented by $t = 0$.

U.S. Population Projections 2000–2050

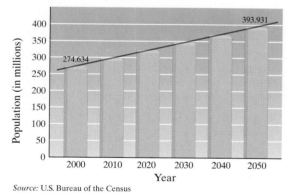

Source: U.S. Bureau of the Census

19. Determine whether the graphs of the two equations are parallel, perpendicular, or neither. Explain your answer.

$$2x - 3y = 6$$
$$4x + 8 = 6y$$

20. **Heart Disease** Although the rate of death due to heart disease is still much higher in the United States than in many other countries, the death rate due to heart disease in the United States has been declining approximately linearly since about 1970. The bar graph below indicates the number of deaths due to heart disease, per 100,000 deaths, in selected years since 1970.

a) Let r be the number of deaths due to heart disease per 100,000 deaths and let t represent the years since 1970. Write a linear function $r(t)$ that represents this data.

b) Using the function from part **a)**, determine the death rate due to heart disease in 1995.

c) Assuming that this trend continues, determine the death rate due to heart disease in 2010.

In Exercises 21–23, if $f(x) = 2x^2 - x$ and $g(x) = x - 5$, find

21. $(f + g)(3)$

22. $(f/g)(-1)$

23. $f(a)$

24. *Paper Use* The following graph shows paper use in 1995 and projected paper use from 1995 through 2015.

a) Estimate the total number of tons of paper to be used in 2010.

b) Estimate the number of tons of paper to be used by businesses in 2010.

c) Estimate the number of tons of paper to be used for reference, print media, and household use in 2010.

Heart Disease Death Rate

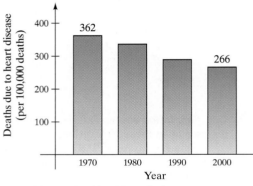

Source: U.S. Dept. of Health and Human Services

Paper Use

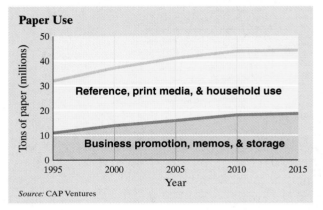

Source: CAP Ventures

25. Graph $y < 3x - 2$.

Cumulative Review Test

Take the following test and check your answers with those that appear at the end of the test. Review any questions that you answered incorrectly. The section and objective where the material was covered are indicated after the answer.

1. For $A = \{1, 3, 5, 7, 9\}$ and $B = \{2, 3, 5, 7, 11, 13\}$, determine

a) $A \cap B$.

b) $A \cup B$.

2. Consider the set $\left\{-6, -4, \frac{1}{3}, 0, \sqrt{3}, 4.67, \frac{37}{2}, -\sqrt{2}\right\}$ List the elements of the set that are

a) natural numbers.

b) real numbers.

3. Evaluate $2 - \{3[6 - 4(6^2 \div 4)]\}$.

Simplify.

4. $\left(\dfrac{4x^2}{y^{-3}}\right)^2$

5. $\left(\dfrac{2x^4 y^{-2}}{4xy^3}\right)^3$

6. *Baltimore Revenue* The total amount of revenue available to the city of Baltimore, Maryland, in 2001 was

1.576×10^9. The following graph shows a breakdown of the sources of this money.

a) How much money was obtained from property taxes?

b) How much money was obtained through federal grants?

c) How much more money was obtained from state shared taxes than from state grants?

Where the Money Comes From
($1.576 billion)

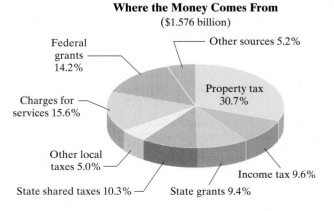

Source: Baltimore City Municipal Dept. of Finance

In Exercises 7 and 8, solve the equations.

7. $2(x + 4) - 5 = -3[x - (2x + 1)]$

8. $\dfrac{4}{5} - \dfrac{x}{3} = 10$

9. Simplify $5x - \{4 - [2(x - 4)] - 5\}$.

10. Solve $A = \dfrac{1}{2}h(b_1 + b_2)$ for b_1.

11. ***Hydrogen Peroxide Solutions*** How many gallons of 15% hydrogen peroxide solution must be mixed with 10 gallons of 4% hydrogen peroxide solution to get a 10% hydrogen peroxide solution?

12. Solve the inequality $3(x - 4) < 6(2x + 3)$.

13. Solve the inequality $-4 < 3x - 7 < 8$.

14. Determine the solution set of $|3x + 5| = |2x - 10|$.

15. Determine the solution set of $|2x - 1| \leq 3$.

16. Graph $y = -\dfrac{3}{2}x - 4$.

17. a) Determine whether the following graph represents a function.

b) Find the domain and range of the graph.

18. Determine the slope of the line through the points $(-5, 3)$ and $(4, -1)$.

19. Determine whether the graphs of the two given equations are parallel, perpendicular, or neither.
$$2x - 5y = 6$$
$$5x - 2y = 9$$

20. If $f(x) = x^2 + 3x - 2$ and $g(x) = 4x - 6$, find $(f + g)(x)$

Answers to Cumulative Review Test

1. a) $\{3, 5, 7\}$ **b)** $\{1, 2, 3, 5, 7, 9, 11, 13\}$; [Sec. 1.2, Obj. 4] **2. a)** None **b)** $-6, -4, \dfrac{1}{3}, 0, \sqrt{3}, 4.67, \dfrac{37}{2}, -\sqrt{2}$; [Sec. 1.2,

Obj. 5] **3.** 92; [Sec. 1.4, Obj. 3] **4.** $16x^4y^6$; [Sec. 1.5, Obj. 7] **5.** $\dfrac{x^9}{8y^{15}}$; [Sec. 1.5, Obj. 7] **6. a)** $\$4.83832 \times 10^8$ or $\$483,832,000$

b) $\$2.23792 \times 10^8$ or $\$223,792,000$ **c)** $\$1.4184 \times 10^7$ or $\$14,184,000$; [Sec. 1.5, Obj. 7] **7.** 0; [Sec. 2.1, Obj. 3]

8. $-\dfrac{138}{5}$; [Sec. 2.1, Obj. 4] **9.** $7x - 7$; [Sec. 2.1, Obj. 2] **10.** $b_1 = \dfrac{2A}{h} - b_2$; [Sec. 2.2, Obj. 2] **11.** 12 gal; [Sec. 2.4, Obj. 2]

12. $x > -\dfrac{10}{3}$; [Sec. 2.5, Obj. 1] **13.** $1 < x < 5$; [Sec. 2.5, Obj. 3] **14.** $\{-15, 1\}$; [Sec. 2.6, Obj. 7]

15. $\{x|-1 \leq x \leq 2\}$; [Sec. 2.6, Obj. 3] **16.** [Sec. 3.1, Obj. 2]

17. a) Not a function **b)** Domain: $\{x|x \leq 2\}$; Range: $\mathbb{R}$; [Sec. 3.2, Obj. 3] **18.** $-\dfrac{4}{9}$; [Sec. 3.4, Obj. 2]

19. Neither; [Sec. 3.5, Obj. 3] **20.** $x^2 + 7x - 8$; [Sec. 3.6, Obj. 1]

Chapter 4

Systems of Equations and Inequalities

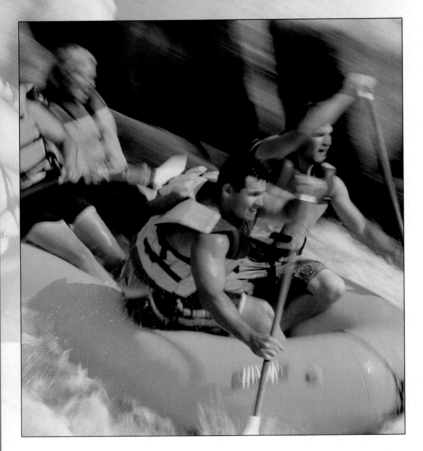

4.1 Solving Systems of Linear Equations in Two Variables

4.2 Solving Systems of Linear Equations in Three Variables

4.3 Systems of Linear Equations: Applications and Problem Solving

4.4 Solving Systems of Equations Using Matrices

4.5 Solving Systems of Equations Using Determinants and Cramer's Rule

4.6 Solving Systems of Linear Inequalities

Chapter Summary
Chapter Review Exercises
Chapter Practice Test
Cumulative Review Test

Business owners strive to run their companies at full capacity to maximize their resources. Mathematics can be used to find the most effective way to distribute resources. On pages 262–264, we solve a system of equations involving the construction of inflatable boats. Later in the exercises starting on page 265, we present similar problems involving the production of furniture, animal food, metal alloys, and other products. The field of mathematics that analyzes business operations is called *operations research*.

SSM

Study Guide

CD/Video

MathPro 4/5

PH Math Tutor Center

prenhall.com/Angel

A Look Ahead

In this chapter we solve systems of linear equations using the following methods: by graphing, by substitution, by the addition method, by using matrices, and by using determinants and Cramer's rule. We also solve systems of linear *inequalities*. Throughout the chapter, especially in Section 4.3, there are many real-life applications. The chapter covers essential topics used by businesses to consider the relationships among variables involved in the day-to-day operations of a business.

4.1 SOLVING SYSTEMS OF LINEAR EQUATIONS IN TWO VARIABLES

SSM   Study Guide   CD/Video

MathPro 4/5   PH Math Tutor Center   prenhall.com/Angel

1   Solve systems of linear equations graphically.
2   Solve systems of linear equations by substitution.
3   Solve systems of linear equations using the addition method.

It is often necessary to find a common solution to two or more linear equations. We refer to these equations as a **system of linear equations**. For example,

$$\left. \begin{array}{l} (1)\ y = x + 5 \\ (2)\ y = 2x + 4 \end{array} \right\} \quad \textit{System of linear equations}$$

A **solution to a system of equations** is an ordered pair or pairs that satisfy all equations in the system. The only solution to the system above is $(1, 6)$.

Check in Equation (1)	Check in Equation (2)
$(1, 6)$	$(1, 6)$
$y = x + 5$	$y = 2x + 4$
$6 \overset{?}{=}\ + 5$	$6 \overset{?}{=} 2(\ ) + 4$
$6 = 6$ *True*	$6 = 6$ *True*

The ordered pair $(1, 6)$ satisfies *both* equations and is the solution to the system of equations.

A system of equations may consist of more than two equations. If a system consists of three equations in three variables, such as x, y, and z, the solution will be an **ordered triple** of the form (x, y, z). If the ordered triple (x, y, z) is a solution to the system, it must satisfy all three equations in the system. Systems with three variables are discussed in Section 4.2. Systems of equations may have more than three variables, but we will not discuss them in this book.

1   Solve Systems of Linear Equations Graphically

To solve a system of linear equations in two variables graphically, graph both equations in the system on the same axes. The solution to the system will be the ordered pair (or pairs) common to both lines, or the point of intersection of both lines in the system.

When two lines are graphed, three situations are possible, as illustrated in Figure 4.1 on page 238. In Figure 4.1a, lines 1 and 2 intersect at exactly one point. This system of equations has *exactly one solution*. This is an example of a *consistent* system of equations. A **consistent system of equations** is a system of equations that has a solution.

Lines 1 and 2 of Figure 4.1b are different but parallel lines. The lines do not intersect, and this system of equations has *no solution*. This is an example of an *inconsistent* system of equations. An **inconsistent system of equations** is a system of equations that has no solution.

In Figure 4.1c, lines 1 and 2 are actually the same line. In this case, every point on the line satisfies both equations and is a solution to the system of equations. This system has *an infinite number of solutions*. This is an example of a *dependent* system of equations. In a dependent system of linear equations both equations represent the same line. A **dependent system of equations** is a system of equations that has an infinite number of solutions. *Note that a dependent system is also a consistent system since it has solutions.*

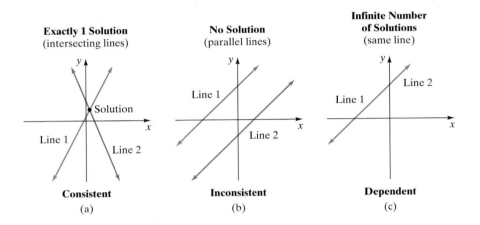

FIGURE 4.1

We can determine if a system of linear equations is consistent, inconsistent, or dependent by writing each equation in slope–intercept form and comparing the slopes and y-intercepts. If the slopes of the lines are different (Fig. 4.1a), the system is consistent. If the slopes are the same but the y-intercepts are different (Fig. 4.1b), the system is inconsistent, and if both the slopes and the y-intercepts are the same (Fig. 4.1c), the system is dependent.

EXAMPLE 1 Without graphing the equations, determine whether the following system of equations is consistent, inconsistent, or dependent.

$$3x - 4y = 8$$
$$-6x + 8y = -16$$

Solution Write each equation in slope–intercept form.

$$3x - 4y = 8 \qquad\qquad -6x + 8y = -16$$
$$-4y = -3x + 8 \qquad\qquad 8y = 6x - 16$$
$$y = \frac{3}{4}x - 2 \qquad\qquad y = \frac{3}{4}x - 2$$

Since the equations have the same slope, $\frac{3}{4}$, and the same y-intercept, $(0, -2)$, the equations represent the same line. Therefore, the system is dependent, and there are an infinite number of solutions.

NOW TRY EXERCISE 19

EXAMPLE 2 Solve the following system of equations graphically.

$$y = x + 2$$
$$y = -x + 4$$

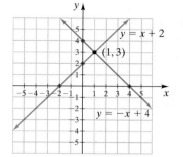

Solution Graph both equations on the same axes (Fig. 4.2). The solution is the point of intersection of the two lines, $(1, 3)$.

NOW TRY EXERCISE 25

The system of equations in Example 2 could be represented in function notation as

$$f(x) = x + 2$$
$$g(x) = -x + 4$$

FIGURE 4.2

Using Your Graphing Calculator

In the Using Your Graphing Calculator box on page 187, Section 3.3, we discussed using a graphing calculator to find the intersection of two graphs. Let's use the information provided in that Graphing Calculator box to solve a system of equations.

EXAMPLE Use your graphing calculator to solve the system of equations. Round the solution to the nearest hundredth.

$$-2.6x - 5.2y = -15.3 \qquad\qquad -8.6x + 3.7y = -12.5$$

Solution First solve each equation for y.

$$-2.6x - 5.2y = -15.3$$
$$-2.6x = 5.2y - 15.3$$
$$-2.6x + 15.3 = 5.2y$$
$$\frac{-2.6x + 15.3}{5.2} = y$$

$$-8.6x + 3.7y = -12.5$$
$$3.7y = 8.6x - 12.5$$
$$y = \frac{8.6x - 12.5}{3.7}$$

Now let $y_1 = \dfrac{-2.6x + 15.3}{5.2}$ and $y_2 = \dfrac{8.6x - 12.5}{3.7}$. The graphs y_1 and y_2 are illustrated in Figure 4.3.

FIGURE 4.3   FIGURE 4.4

Figure 4.4 shows that the intersection of the two graphs occurs at $(2.24, 1.82)$, rounded to the nearest hundredth.

Use your graphing calculator to determine the solution to each system. Round your answers to the nearest hundredth.

1. $2x + 3y = 8$
   $-3x + 4y = -5$

2. $5x - 6y = 9$
   $-3x + 5y = 8$

3. $3.4x - 5.6y = 10.2$
   $5.8x + 1.4y = -33.6$

4. $-2.3x + 7.9y = 88.3$
   $-5.3x - 2.7y = -16.5$

2 Solve Systems of Linear Equations by Substitution

Often, an exact solution to a system of equations may be difficult to find on a graph. A graphing calculator may not give an exact answer either. When an exact answer is necessary, the system should be solved algebraically, by either the substitution method or the addition (elimination) method. We discuss **substitution** first.

> **To Solve a Linear System of Equations by Substitution**
>
> 1. Solve for a variable in either equation. (If possible, solve for a variable with a numerical coefficient of 1 to avoid working with fractions.)
> 2. Substitute the expression found for the variable in step 1 into the other equation. This will result in an equation containing only one variable.
> 3. Solve the equation obtained in step 2 to find the value of this variable.
> 4. Substitute the value found in step 3 into the equation from step 1. Solve the equation to find the remaining variable.
> 5. Check your solution in *all* equations in the system.

EXAMPLE 3   Solve the following system of equations by substitution.

$$y = 3x - 13$$
$$y = -4x + 1$$

Solution   Since both equations are already solved for y, we can substitute $3x - 13$ for y in the second equation and then solve for the remaining variable, x.

$$3x - 13 = -4x + 1$$
$$7x - 13 = 1$$
$$7x = 14$$
$$x = 2$$

Now find y by substituting $x = 2$ into either of the original equations. We will use the first equation.

$$y = 3x - 13$$
$$y = 3(2) - 13$$
$$y = 6 - 13 = -7$$

A check will show that the solution to the system of equations is $(2, -7)$. ✳

EXAMPLE 4   Solve the following system of equations by substitution.

$$2x + y = 11$$
$$x + 3y = 18$$

Solution   Begin by solving for one of the variables in either of the equations. You may solve for either of the variables; however, if you solve for a variable with a numerical coefficient of 1, you may avoid working with fractions. In this system the y-term in $2x + y = 11$ and the x-term in $x + 3y = 18$ both have numerical coefficient 1. Let's solve for y in $2x + y = 11$.

$$2x + y = 11$$
$$y = -2x + 11$$

Next, substitute $-2x + 11$ for y in the *other equation*, $x + 3y = 18$, and solve for the remaining variable, x.

$$x + 3y = 18$$
$$x + 3(\overbrace{-2x + 11}) = 18 \qquad \text{\textit{Substitute} } -2x + 11 \text{ \textit{for} } y.$$
$$x - 6x + 33 = 18$$
$$-5x + 33 = 18$$
$$-5x = -15$$
$$x = 3$$

Finally, substitute $x = 3$ into the equation $y = -2x + 11$ and solve for y.

$$y = -2x + 11$$
$$y = -2(3) + 11 = 5$$

NOW TRY EXERCISE 41 The solution is the ordered pair $(3, 5)$. Check this solution. ✳

If, when solving a system of equations by either substitution or the addition method, you arrive at an equation that is false, such as $5 = 6$ or $0 = 3$, the system is inconsistent and has no solution. If you obtain an equation that is always true, such as $6 = 6$ or $0 = 0$, the system is dependent and has an infinite number of solutions.

HELPFUL HINT Students sometimes successfully solve for one of the variables and forget to solve for the other. Remember that a solution must contain a numerical value for each variable in the system.

---

3 Solve Systems of Linear Equations Using the Addition Method

A third and often the easiest method of solving a system of equations is the **addition (or elimination) method**. The object of this process is to obtain two equations whose sum will be an equation containing only one variable. Keep in mind that your immediate goal is to obtain one equation containing only one unknown.

EXAMPLE 5 Solve the following system of equations using the addition method.

$$2x + 5y = -1$$
$$3x - 5y = 11$$

Solution Note that one equation contains $+5y$ and the other contains $-5y$. By adding the equations, we can eliminate the variable y and obtain one equation containing only one variable, x.

$$2x + 5y = -1$$
$$\underline{3x - 5y = 11}$$
$$5x \qquad = 10$$

Now solve for the remaining variable, x.

$$\frac{5x}{5} = \frac{10}{5}$$
$$x = 2$$

Finally, solve for y by substituting 2 for x in either of the original equations.

$$2x + 5y = -1$$
$$2(2) + 5y = -1$$
$$4 + 5y = -1$$
$$5y = -5$$
$$y = -1$$

A check will show that the solution is $(2, -1)$. ✳

To Solve a Linear System of Equations Using the Addition (or Elimination) Method

1. If necessary, rewrite each equation in standard form, that is, with the terms containing variables on the left side of the equal sign and the constant on the right side of the equal sign.

2. If necessary, multiply one or both equations by a constant(s) so that when the equations are added, the sum will contain only one variable.

3. Add the respective sides of the equations. This will result in a single equation containing only one variable.

4. Solve for the variable in the equation obtained in step 3.

5. Substitute the value found in step 4 into either of the original equations. Solve that equation to find the value of the remaining variable.

6. Check your solution in all equations in the system.

In step 2 of the procedure, we indicate that it may be necessary to multiply both sides of an equation by a constant. To help avoid confusion, we will number our equations using parentheses, such as *(eq. 1)* or *(eq. 2)*.

In Example 6 we will solve the same system we solved in Example 4, but this time we will use the addition method.

EXAMPLE 6 Solve the following system of equations using the addition method.

$$2x + y = 11 \quad (eq.\ 1)$$
$$x + 3y = 18 \quad (eq.\ 2)$$

Solution The object of the addition process is to obtain two equations whose sum will be an equation containing only one variable. To eliminate the variable x, we multiply *(eq. 2)* by -2 and then add the two equations.

$$2x + y = 11 \quad (eq.\ 1)$$
$$-2x - 6y = -36 \quad (eq.\ 2) \quad \textit{Multiplied by } -2$$

Now add.

$$\begin{array}{r} 2x + y = 11 \\ -2x - 6y = -36 \\ \hline -5y = -25 \\ y = 5 \end{array}$$

Now solve for x by substituting 5 for y in either of the original equations.

$$2x + y = 11$$
$$2x + 5 = 11 \quad \text{\textit{Substitute 5 for y.}}$$
$$2x = 6$$
$$x = 3$$

The solution is $(3, 5)$. Note that we could have first eliminated the variable y by multiplying (*eq.* 1) by -3 and then adding. ✳

Sometimes both equations must be multiplied by different numbers in order for one of the variables to be eliminated. This procedure is illustrated in Example 7.

EXAMPLE 7 Solve the following system of equations using the addition method.

$$4x + 3y = 7 \quad (eq.\ 1)$$
$$3x - 7y = -3 \quad (eq.\ 2)$$

Solution We can eliminate the variable x by multiplying (*eq.* 1) by -3 and (*eq.* 2) by 4.

$$-12x - 9y = -21 \quad (eq.\ 1) \quad \textit{Multiplied by -3}$$
$$\underline{12x - 28y = -12} \quad (eq.\ 2) \quad \textit{Multiplied by 4}$$
$$-37y = -33 \quad \textit{Sum of equations}$$
$$y = \frac{33}{37}$$

We can now find x by substituting $\frac{33}{37}$ for y in one of the original equations and solving for x. If you try this, you will see that, although it can be done, it gets messy. An easier method to solve for x is to go back to the original equations and eliminate the variable y.

$$28x + 21y = 49 \quad (eq.\ 1) \quad \textit{Multiplied by 7}$$
$$\underline{9x - 21y = -9} \quad (eq.\ 2) \quad \textit{Multiplied by 3}$$
$$37x = 40 \quad \textit{Sum of equations}$$
$$x = \frac{40}{37}$$

NOW TRY EXERCISE 67 The solution is $\left(\frac{40}{37}, \frac{33}{37}\right)$. ✳

In Example 7, the same solution could be obtained by multiplying (*eq.* 1) by 3 and (*eq.* 2) by -4 and then adding. Try it now and see.

EXAMPLE 8 Solve the following system of equations using the addition method.

$$0.2x + 0.1y = 1.1 \quad (eq.\ 1)$$
$$\frac{x}{18} + \frac{y}{6} = 1 \quad (eq.\ 2)$$

Solution When a system of equations contains fractions or decimal numbers, it is generally best to *clear*, or remove, the fractions or decimals. In (*eq.* 1), if we multiply both sides of the equation by 10 we obtain

$$10\,(0.2x) + 10\,(0.1y) = 10\,(1.1)$$
$$2x + y = 11 \quad\quad (eq.\ 3)$$

In (*eq. 2*), if we multiply both sides of the equation by the least common denominator, 18, we obtain

$$18\left(\frac{x}{18}\right) + 18\left(\frac{y}{6}\right) = 18\,(1)$$

$$x + 3y = 18 \qquad (eq.\ 4)$$

The system of equations is now simplified to

$$2x + y = 11 \quad (eq.\ 3)$$
$$x + 3y = 18 \quad (eq.\ 4)$$

This is the same system of equations we solved in Example 6. Thus the solution to this system is $(3, 5)$, the same as obtained in Example 6. ✳

EXAMPLE 9 Solve the following system of equations using the addition method.

$$x - 3y = 4 \quad (eq.\ 1)$$
$$-2x + 6y = 1 \quad (eq.\ 2)$$

Solution
$$2x - 6y = 8 \quad (eq.\ 1) \qquad \textit{Multiplied by 2}$$
$$\underline{-2x + 6y = 1} \quad (eq.\ 2)$$
$$0 = 9 \qquad\qquad \textit{False}$$

Since $0 = 9$ is a false statement, this system has no solution. The system is inconsistent and the graphs of these equations are parallel lines. ✳

NOW TRY EXERCISE 59

EXAMPLE 10 Solve the following system of equations using the addition method.

$$x - \frac{1}{2}y = 2$$
$$y = 2x - 4$$

Solution First, align the *x*- and *y*-terms on the left side of the equation.

$$x - \frac{1}{2}y = 2 \quad (eq.\ 1)$$
$$2x - y = 4 \quad (eq.\ 2)$$

Now proceed as in previous examples.

$$-2x + y = -4 \quad (eq.\ 1) \qquad \textit{Multiplied by } -2$$
$$\underline{2x - y = 4} \quad (eq.\ 2)$$
$$0 = 0 \qquad\qquad \textit{True}$$

Since $0 = 0$ is a true statement, the system is dependent and has an infinite number of solutions. Both equations represent the same line. Notice that if you multiply both sides of (*eq. 1*) by 2 you obtain (*eq. 2*). ✳

We have illustrated three methods that can be used to solve a system of linear equations: graphing, substitution, and the addition method. When you are given a system of equations, which method should you use to solve the system? When you need an exact solution, graphing may not be the best to use. Of the two algebraic methods, the addition method may be the easiest to use if there are no numerical coefficients of 1 in the system. If one or more of the variables have a coefficient of 1, you may wish to use either method. We will present a fourth method, using matrices, in Section 4.4, and a fifth method, using determinants, in Section 4.5.

Exercise Set 4.1

Concept/Writing Exercises

1. What is a solution to a system of linear equations?

2. What is the solution to a system of linear equations in three variables called?

3. What is a dependent system of equations?

4. What is an inconsistent system of equations?

5. What is a consistent system of equations?

6. Explain how to determine the solution of a linear system graphically.

7. Explain how you can determine, without graphing or solving, whether a system of two linear equations is consistent, inconsistent or dependent.

8. When solving a system of linear equations using the addition (or elimination) method, what is the object of the process?

9. When solving a linear system by addition, how can you tell if the system is dependent?

10. When solving a linear system by addition, how can you tell if the system is inconsistent?

Practice the Skills

Determine which, if any, of the given ordered pairs or ordered triples satisfy the system of linear equations.

11. $y = 2x + 4$
    $y = 2x - 1$
    a) $(0, 4)$
    b) $(3, 10)$

12. $3x - 5y = 12$
    $y = \dfrac{3}{4}x - 3$
    a) $(4, 0)$ **b)** $(8, 3)$

13. $x + y = 25$
    $0.25x + 0.45y = 7.50$
    a) $(5, 20)$
    b) $(18.75, 6.25)$

14. $y = \dfrac{x}{3} - \dfrac{7}{3}$
    $5x - 35 = 15y$
    a) $(1, -2)$
    b) $(4, -1)$

15. $x + 2y - z = -5$
    $2x - y + 2z = 8$
    $3x + 3y + 4z = 5$
    a) $(1, 3, -2)$
    b) $(1, -2, 2)$

16. $4x + y - 3z = 1$
    $2x - 2y + 6z = 11$
    $-6x + 3y + 12z = -4$
    a) $(2, -1, -2)$
    b) $\left(\dfrac{1}{2}, 2, 1\right)$

Write each equation in slope–intercept form. Without graphing the equations, state whether the system of equations is consistent, inconsistent, or dependent. Also indicate whether the system has exactly one solution, no solution, or an infinite number of solutions.

17. $-6x + 3y = 1$
    $4y + 12 = -6x$

18. $x - \dfrac{1}{2}y = 5$
    $2x - y = 7$

19. $\dfrac{x}{3} + \dfrac{y}{4} = 1$
    $4x + 3y = 12$

20. $\dfrac{x}{3} + \dfrac{y}{4} = 1$
    $4x - y = 12$

21. $3x - 3y = 9$
    $2x - 2y = -4$

22. $2x = 3y + 4$
    $6x - 9y = 12$

23. $y = \dfrac{3}{2}x + \dfrac{1}{2}$
    $3x - 2y = -\dfrac{1}{2}$

24. $x - y = 3$
    $\dfrac{1}{2}x - 2y = -6$

Determine the solution to each system of equations graphically. If the system is inconsistent or dependent, so state.

25. $y = x + 5$
    $y = -x + 3$

26. $y = 2x + 4$
    $y = -3x - 6$

27. $y = 4x - 1$
    $2y = 8x + 6$

28. $x + y = 1$
    $3x - y = -5$

29. $2x + 3y = 6$
    $4x = -6y + 12$

30. $y = -2x - 1$
    $x + 2y = 4$

31. $x + 3y = 4$
    $x = 1$

32. $2x - 5y = 10$
    $y = \dfrac{2}{5}x - 2$

33. $y = -5x + 5$
    $y = 2x - 2$

34. $4x - y = 9$
    $x - 3y = 16$

35. $2x - y = -4$
    $2y = 4x - 6$

36. $y = -\dfrac{1}{3}x - 1$
    $3y = 4x - 18$

Find the solution to each system of equations by substitution.

37. $x + 3y = -1$
$y = x + 1$

38. $3x - 2y = -7$
$y = 2x - 3$

39. $x = 2y + 3$
$y = x$

40. $y = 3x - 14$
$x = y$

41. $a + 3b = 5$
$2a - b = 3$

42. $6s + 3t = 4$
$s = \frac{1}{2}t$

43. $y + \frac{1}{2}x = 0$
$x + 4y - 1 = 0$

44. $x = 0.5y + 1.7$
$10x - y = 1$

45. $a - \frac{1}{2}b = 2$
$b = 2a - 4$

46. $x + 3y = -2$
$y = -\frac{1}{3}x - \frac{2}{3}$

47. $5x - 2y = -7$
$y = \frac{5}{2}x + 1$

48. $y = \frac{2}{3}x - 1$
$2x - 3y = 5$

49. $5x - 4y = -7$
$x - \frac{3}{5}y = -2$

50. $m + 2n = 4$
$m + \frac{1}{2}n = 4$

51. $\frac{1}{2}x - \frac{1}{3}y = 2$
$\frac{1}{4}x + \frac{2}{3}y = 6$

52. $\frac{1}{2}x + \frac{1}{3}y = 13$
$\frac{1}{5}x + \frac{1}{8}y = 5$

Solve each system of equations using the addition method.

53. $x + y = 7$
$x - y = -3$

54. $-x + y = 4$
$x - 2y = 6$

55. $4x - 3y = 1$
$5x + 3y = -10$

56. $2x - 5y = 6$
$-4x + 10y = -1$

57. $10m - 2n = 6$
$-5m + n = -3$

58. $4r - 3s = 2$
$2r + s = 6$

59. $2c - 5d = 1$
$-4c + 10d = 6$

60. $2v - 3w = 8$
$3v - 6w = 1$

61. $3p - 2q = 1$
$2p + 5q = 7$

62. $5a - 10b = 15$
$a = 2b + 3$

63. $5s - 3t = 7$
$t = s + 1$

64. $2x - 7y = 3$
$-5x + 3y = 7$

65. $2x - y = 8$
$3x + y = 6$

66. $3x + 4y = 2$
$2x = -5y - 1$

67. $3x - 4y = 5$
$2x = 5y - 3$

68. $4x + 5y = 3$
$2x - 3y = 4$

69. $0.2x + 0.5y = 1.6$
$-0.3x + 0.4y = -0.1$

70. $0.15x - 0.40y = 0.65$
$0.60x + 0.25y = -1.1$

71. $2.1m - 0.6n = 8.4$
$-1.5m - 0.3n = -6.0$

72. $-0.25x + 0.10y = 1.05$
$-0.40x - 0.625y = -0.675$

73. $\frac{1}{2}x - \frac{1}{3}y = 1$
$\frac{1}{4}x - \frac{1}{9}y = \frac{2}{3}$

74. $\frac{1}{3}x = 4 - \frac{1}{4}y$
$3x = 4y$

75. $\frac{1}{5}x + \frac{1}{2}y = 4$
$\frac{2}{3}x - y = \frac{8}{3}$

76. $\frac{2}{3}x - 4 = \frac{1}{2}y$
$x - 3y = \frac{1}{3}$

Problem Solving

77. a) Write a system of equations that would be most easily solved by substitution.
b) Explain why substitution would be the easiest method to use.
c) Solve the system by substitution.

78. a) Write a system of equations that would be most easily solved using the addition method.
b) Explain why the addition method would be the easiest method to use.
c) Solve the system using the addition method.

79. *Aging Newlyweds* According to the 2000 U.S. census, both men and women are waiting longer to get married. The following graph shows the average age at first marriage for both men and women.

Average Age at First Marriage

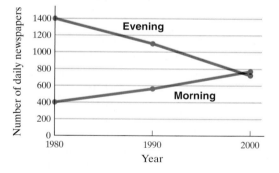

The average age at first marriage for women can be approximated by the function $W(t) = 0.12t + 20.3$ and the average age at first marriage for men can be approximated by the function $M(t) = 0.1t + 22.8$ where t is years since 1960. If the trend continues, determine the year that the average age at first marriage for men and women are the same.

80. *Newspapers* The following graph shows that since 1980 the number of evening U.S. newspapers has declined nearly linearly while the number of morning U.S. newspapers has increased nearly linearly.

Daily Newspapers

The number of evening newspapers, $E(t)$, can be approximated by the function $E(t) = -33.05t + 1388$ and the number of morning newspapers can be approximated by the function $M(t) = 18.95t + 387$ where t represents the number of years since 1980. Based on these functions, during what year did the number of evening newspapers equal the number of morning newspapers?

81. Explain how you can tell by observation that the following system is dependent.

$$2x + 3y = 1$$
$$4x + 6y = 2$$

82. Explain how you can tell by observation that the following system is inconsistent.

$$-x + 3y = 5$$
$$2x - 6y = -13$$

83. The solutions of a system of linear equations include $(-4, 3)$ and $(-6, 11)$.

    a) How many other solutions does the system have? Explain.

    b) Determine the slope of the line containing $(-4, 3)$ and $(-6, 11)$. Determine the equation of the line containing these points. Then determine the y-intercept.

    c) Does this line represent a function?

84. The solutions of a system of linear equations include $(-5, 1)$ and $(-5, -4)$.

    a) How many other solutions does the system have? Explain.

    b) Determine the slope of the line containing $(-5, 1)$ and $(-5, -4)$. Determine the equation of the line containing these points. Does this graph have a y-intercept? Explain.

    c) Does this line represent a function?

85. Construct a system of equations that is dependent. Explain how you created your system.

86. Construct a system of equations that is inconsistent. Explain how you created your system.

In Exercises 87 and 88, ***a)*** *create a system of linear equations that has the solution indicated and* ***b)*** *explain how you determined your solution.*

87. $(2, 5)$

88. $(-3, 4)$

89. The solution to the following system of equations is $(2, -3)$. Find A and B.

$$Ax + 4y = -8$$
$$3x - By = 21$$

90. The solution to the following system of equations is $(-5, 3)$. Find A and B.

$$3x + Ay = -3$$
$$Bx - 2y = -16$$

91. If $(2, 6)$ and $(-1, -6)$ are two solutions of $f(x) = mx + b$, find m and b.

92. If $(3, -5)$ and $(-2, 10)$ are two solutions of $f(x) = mx + b$, find m and b.

93. Suppose you graph a system of two linear equations on your graphing calculator, but only one line shows in the window. What are two possible explanations for this?

94. Suppose you graph a system of linear equations on your graphing calculator and get the following.

a) By observing the window, can you be sure that this system is inconsistent? Explain.

b) What can you do on your graphing calculator to determine whether the system is inconsistent?

Challenge Problems

Solve each system of equations.

95. $\dfrac{x+2}{2} - \dfrac{y+4}{3} = 4$

$\dfrac{x+y}{2} = \dfrac{1}{2} + \dfrac{x-y}{3}$

96. $\dfrac{5x}{2} + 3y = \dfrac{9}{2} + y$

$\dfrac{1}{4}x - \dfrac{1}{2}y = 6x + 12$

Solve each system of equations. (Hint: $\dfrac{3}{a} = 3 \cdot \dfrac{1}{a} = 3x$ *if* $x = \dfrac{1}{a}$.)

97. $\dfrac{3}{a} + \dfrac{4}{b} = -1$

$\dfrac{1}{a} + \dfrac{6}{b} = 2$

98. $\dfrac{6}{x} + \dfrac{1}{y} = -1$

$\dfrac{3}{x} - \dfrac{2}{y} = -3$

By solving for x and y, determine the solution to each system of equations. In all equations, a ≠ 0 and b ≠ 0. The solution will contain either a, b, or both letters.

99. $4ax + 3y = 19$

$-ax + y = 4$

100. $ax = 2 - by$

$-ax + 2by - 1 = 0$

Group Activity

Discuss and answer Exercise 101 as a group.

101. Trends The graph below appeared in both the *Journal of the American Medical Association* and *Scientific American*. The red line indicates the long-term trend of firearm deaths, and the purple line indicates the long-term trend in motor vehicle deaths. The black lines indicate the short-term trends in deaths from firearms and motor vehicles.

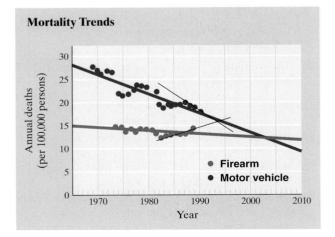

Mortality Trends

a) Discuss the long-term trend in motor vehicle deaths.

b) Discuss the long-term trend in firearm deaths.

c) Discuss the short-term trend in motor vehicle deaths compared with the long-term trend in motor vehicle deaths.

d) Discuss the short-term trend in firearm deaths compared with the long-term trend in firearm deaths.

e) Using the long-term trends, estimate when the number of deaths from firearms will equal the number of deaths from motor vehicles.

f) Repeat part **e)** using the short-term trends.

g) Determine a function, $M(t)$, that can be used to estimate the number of deaths per 100,000 people (long term) from motor vehicles from 1965 through 2010.

h) Determine a function, $F(t)$, that can be used to estimate the number of deaths per 100,000 people (long term) from firearms from 1965 through 2010.

i) Solve the system of equations formed in parts **g)** and **h)**. Does the solution agree with the solution in part **e)**? If not, explain why.

Cumulative Review Exercises

[1.2] **102.** Explain the difference between a rational number and an irrational number.

[1.2] **103.** **a)** Are all rational numbers real numbers?

b) Are all irrational numbers real numbers?

[2.1] **104.** Solve the equation $\frac{1}{2}(x - 7) = \frac{3}{4}(2x + 1)$.

[2.2] **105.** Find all numbers such that $|x - 4| = |4 - x|$.

[2.2] **106.** Evaluate $A = p\left(1 + \frac{r}{n}\right)^{t}$, when $p = 500$, $r = 0.08$, $n = 2$, and $t = 1$.

[3.5] **107.** Is the relation below a function? Explain your answer. $\{(-3, 4), (7, 2), (-4, 5), (5, 0), (-3, 2)\}$

[3.6] **108.** Let $f(x) = x + 3$ and $g(x) = x^2 - 9$. Find $(f/g)(3)$.

4.2 SOLVING SYSTEMS OF LINEAR EQUATIONS IN THREE VARIABLES

SSM   Study Guide   CD/Video

MathPro 4/5   PH Math Tutor Center   prenhall.com/Angel

1   Solve systems of linear equations in three variables.

2   Learn the geometric interpretation of a system of equations in three variables.

3   Recognize inconsistent and dependent systems.

1   Solve Systems of Linear Equations in Three Variables

The equation $2x - 3y + 4z = 8$ is an example of a linear equation in three variables. The solution to a linear equation in three variables is an *ordered triple* of the form (x, y, z). One solution to the equation given is $(1, 2, 3)$. Check now to verify that $(1, 2, 3)$ is a solution to the equation.

To solve systems of linear equations with three variables, we can use either substitution or the addition method, both of which were discussed in Section 4.1.

EXAMPLE 1   Solve the following system by substitution.

$$x = -3$$
$$3x + 4y = 7$$
$$-2x - 3y + 5z = 19$$

Solution   Since we know that $x = -3$, we substitute -3 for x in the equation $3x + 4y = 7$ and solve for y.

$$3x + 4y = 7$$
$$3(-3) + 4y = 7$$
$$-9 + 4y = 7$$
$$4y = 16$$
$$y = 4$$

Now we substitute $x = -3$ and $y = 4$ into the last equation and solve for z.

$$-2x - 3y + 5z = 19$$
$$-2(-3) - 3(4) + 5z = 19$$
$$6 - 12 + 5z = 19$$
$$-6 + 5z = 19$$
$$5z = 25$$
$$z = 5$$

Check $\quad x = -3, y = 4, z = 5$. The solution must be checked in *all three* original equations.

$$x = -3 \qquad\qquad 3x + 4y = 7 \qquad\qquad -2x - 3y + 5z = 19$$

$$-3 = -3 \;\; \textit{True} \qquad 3(-3) + 4(4) \overset{?}{=} 7 \qquad -2(-3) - 3(4) + 5(5) \overset{?}{=} 19$$

$$7 = 7 \;\; \textit{True} \qquad\qquad 19 = 19 \;\; \textit{True}$$

The solution is the ordered triple $(-3, 4, 5)$. Remember that the ordered triple lists the x-value first, the y-value second, and the z-value third.

NOW TRY EXERCISE 3

Not every system of linear equations in three variables can be solved by substitution. When such a system cannot be solved using substitution, we can find the solution by the addition method, as illustrated in Example 2.

EXAMPLE 2 Solve the following system of equations using the addition method.

$$3x + 2y + z = 4 \quad (eq.\ 1)$$
$$2x - 3y + 2z = -7 \quad (eq.\ 2)$$
$$x + 4y - z = 10 \quad (eq.\ 3)$$

Solution To solve this system of equations, we must first obtain two equations containing the same two variables. We do so by selecting two equations and using the addition method to eliminate one of the variables. For example, by adding $(eq.\ 1)$ and $(eq.\ 3)$, the variable z will be eliminated. Then we use a different pair of equations [either $(eq.\ 1)$ and $(eq.\ 2)$ or $(eq.\ 2)$ and $(eq.\ 3)$] and use the addition method to eliminate the *same* variable that was eliminated previously. If we multiply $(eq.\ 1)$ by -2 and add it to $(eq.\ 2)$, the variable z will again be eliminated. We will then have two equations containing only two unknowns. Let us begin by adding $(eq.\ 1)$ and $(eq.\ 3)$.

$$3x + 2y + z = 4 \quad (eq.\ 1)$$
$$\underline{x + 4y - z = 10} \quad (eq.\ 3)$$
$$4x + 6y \qquad = 14 \quad \textit{Sum of equations, } (eq.\ 4)$$

Now let's use a different pair of equations and again eliminate the variable z.

$$-6x - 4y - 2z = -8 \quad (eq.\ 1) \;\; \textit{Multiplied by } -2$$
$$\underline{2x - 3y + 2z = -7} \quad (eq.\ 2)$$
$$-4x - 7y \qquad = -15 \quad \textit{Sum of equations, } (eq.\ 5)$$

We now have a system consisting of two equations with two unknowns, $(eq.\ 4)$ and $(eq.\ 5)$. If we add these two equations, the variable x will be eliminated.

$$4x + 6y = 14 \quad (eq.\ 4)$$
$$\underline{-4x - 7y = -15} \quad (eq.\ 5)$$
$$-y = -1 \quad \textit{Sum of equations}$$
$$y = 1$$

Next we substitute $y = 1$ into either one of the two equations containing only two variables [$(eq.\ 4)$ or $(eq.\ 5)$] and solve for x.

$$4x + 6y = 14 \quad (eq.\ 4)$$
$$4x + 6(1) = 14 \quad \textit{Substitute 1 for y in } (eq.\ 4).$$
$$4x + 6 = 14$$
$$4x = 8$$
$$x = 2$$

Finally, we substitute $x = 2$ and $y = 1$ into any of the original equations and solve for z.

$$
\begin{aligned}
3x + 2y + z &= 4 \quad \text{(eq. 1)} \\
3(2) + 2(1) + z &= 4 \quad \text{\textit{Substitute 2 for x}} \\
& \text{\textit{and 1 for y in (eq. 1).}} \\
6 + 2 + z &= 4 \\
8 + z &= 4 \\
z &= -4
\end{aligned}
$$

The solution is the ordered triple $(2, 1, -4)$. Check this solution in *all three* original equations.   ✴

In Example 2 we chose first to eliminate the variable z by using (*eq. 1*) and (*eq. 3*) and then (*eq. 1*) and (*eq. 2*). We could have elected to eliminate either the variable x or the variable y first. For example, we could have eliminated variable x by multiplying (*eq. 3*) by -2 and then adding it to (*eq. 2*). We could also eliminate the variable x by multiplying (*eq. 3*) by -3 and then adding it to (*eq. 1*). Try solving the system in Example 2 by first eliminating the variable x.

EXAMPLE 3 Solve the following system of equations.

$$
\begin{aligned}
2x - 3y + 2z &= -1 \quad \text{(eq. 1)} \\
x + 2y \phantom{{}+ 2z} &= 14 \quad \text{(eq. 2)} \\
x \phantom{{}+ 2y} - 3z &= -5 \quad \text{(eq. 3)}
\end{aligned}
$$

Solution The third equation does not contain y. We will therefore work to obtain another equation that does not contain y. We will use (*eq. 1*) and (*eq. 2*) to do this.

$$
\begin{aligned}
4x - 6y + 4z &= -2 \quad \text{(eq. 1) \textit{Multiplied by 2}} \\
\underline{3x + 6y \phantom{{}+ 4z}} &= \underline{42} \quad \text{(eq. 2) \textit{Multiplied by 3}} \\
7x \phantom{{}+ 6y} + 4z &= 40 \quad \text{\textit{Sum of equations, (eq. 4)}}
\end{aligned}
$$

We now have two equations containing only the variables x and z.

$$
\begin{aligned}
7x + 4z &= 40 \quad \text{(eq. 4)} \\
x - 3z &= -5 \quad \text{(eq. 3)}
\end{aligned}
$$

Let's now eliminate the variable x.

$$
\begin{aligned}
7x + 4z &= 40 \quad \text{(eq. 4)} \\
\underline{-7x + 21z} &= \underline{35} \quad \text{(eq. 3) \textit{Multiplied by} -7} \\
25z &= 75 \quad \text{\textit{Sum of equations}} \\
z &= 3
\end{aligned}
$$

Now we solve for x by using one of the equations containing only the variables x and z. We substitute 3 for z in (*eq. 3*).

$$
\begin{aligned}
x - 3z &= -5 \quad \text{(eq. 3)} \\
x - 3(3) &= -5 \quad \text{\textit{Substitute 3 for z in (eq. 3).}} \\
x - 9 &= -5 \\
x &= 4
\end{aligned}
$$

Finally, we solve for y using any of the original equations that contains y.

$$x + 2y = 14 \quad (eq.\ 2)$$
$$4 + 2y = 14 \quad \text{\textit{Substitute 4 for x in (eq. 2).}}$$
$$2y = 10$$
$$y = 5$$

The solution is the ordered triple $(4, 5, 3)$.

Check   (eq. 1)   (eq. 2)   (eq. 3)

$$
\begin{array}{lll}
2x - 3y + 2z = -1 & x + 2y = 14 & x - 3z = -5 \\
2(4) - 3(5) + 2(3) \overset{?}{=} -1 & 4 + 2(5) \overset{?}{=} 14 & 4 - 3(3) \overset{?}{=} -5 \\
8 - 15 + 6 \overset{?}{=} -1 & 4 + 10 \overset{?}{=} 14 & 4 - 9 \overset{?}{=} -5 \\
-1 = -1 & 14 = 14 & -5 = -5
\end{array}
$$

NOW TRY EXERCISE 11   *True*   *True*   *True*   ✳

HELPFUL HINT | If an equation in a system contains fractions, eliminate the fractions by multiplying each term in the equation by the least common denominator. Then continue to solve the system. If, for example, one equation in the system is $\frac{3}{4}x - \frac{5}{8}y + z = \frac{1}{2}$, multiply both sides of the equation by 8 to obtain the equivalent equation $6x - 5y + 8z = 4$.

2 Learn the Geometric Interpretation of a System of Equations in Three Variables

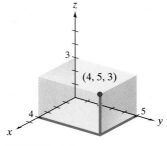

FIGURE 4.5

When we have a system of linear equations in two variables, we can find its solution graphically using the Cartesian coordinate system. A linear equation in three variables, x, y, and z, can be graphed on a coordinate system with three axes drawn perpendicular to each other (see Fig. 4.5).

A point plotted in this three-dimensional system would appear to be a point in space. If we were to graph an equation such as $x + 2y + 3z = 4$, we would find that its graph would be a plane, not a line. In Example 3 we indicated the solution to be the ordered triple $(4, 5, 3)$. This means that the three planes, one from each of the three given equations, all intersect at the point $(4, 5, 3)$. Figure 4.5 shows the location of this point of intersection of the three planes. The drawing in Exercise 39 illustrates three planes intersecting at a point.

3 Recognize Inconsistent and Dependent Systems

We discussed inconsistent and dependent systems of equations in Section 4.1. Systems of linear equations in three variables may also be inconsistent or dependent. When solving a system of linear equations in three variables, if you obtain a false statement like $3 = 0$, the system is inconsistent and has no solution. This means that at least two of the planes are parallel, so the three planes cannot intersect. (See Exercises 37 and 38.)

When solving a system of linear equations in three variables, if you obtain the true statement $0 = 0$, it indicates that the system is dependent and has an infinite number of solutions. This may happen when all three equations represent the same plane or when the intersection of the planes is a line, as in the drawing in Exercise 40. Examples 4 and 5 illustrate an inconsistent system and a dependent system, respectively.

EXAMPLE 4 Solve the following system of equations.

$$-3x + 5y + z = -3 \quad (eq.\ 1)$$
$$6x - 10y - 2z = 1 \quad (eq.\ 2)$$
$$7x - 4y + 11z = -6 \quad (eq.\ 3)$$

Solution We will begin by eliminating the variable x from $(eq.\ 1)$ and $(eq.\ 2)$.

$$-6x + 10y + 2z = -6 \quad (eq.\ 1) \quad \text{Multiplied by 2}$$
$$\underline{6x - 10y - 2z = 1 \quad (eq.\ 2)}$$
$$0 = -5 \qquad \text{False}$$

NOW TRY EXERCISE 31

Since we obtained the false statement $0 = -5$, this system is inconsistent and has no solution.

EXAMPLE 5 Solve the following system of equations.

$$x - y + z = 1 \quad (eq.\ 1)$$
$$x + 2y - z = 1 \quad (eq.\ 2)$$
$$x - 4y + 3z = 1 \quad (eq.\ 3)$$

Solution We will begin by eliminating the variable x from $(eq.\ 1)$ and $(eq.\ 2)$ and then from $(eq.\ 1)$ and $(eq.\ 3)$.

$$-x + y - z = -1 \quad (eq.\ 1) \quad \text{Multiplied by } -1$$
$$\underline{x + 2y - z = 1 \quad (eq.\ 2)}$$
$$3y - 2z = 0 \quad \text{Sum of equations, } (eq.\ 4)$$

$$x - y + z = 1 \quad (eq.\ 1)$$
$$\underline{-x + 4y - 3z = -1 \quad (eq.\ 3) \quad \text{Multiplied by } -1}$$
$$3y - 2z = 0 \quad \text{Sum of equations, } (eq.\ 5)$$

Now we eliminate the variable y using $(eq.\ 4)$ and $(eq.\ 5)$.

$$-3y + 2z = 0 \quad (eq.\ 4) \quad \text{Multiplied by } -1$$
$$\underline{3y - 2z = 0 \quad (eq.\ 5)}$$
$$0 = 0 \qquad \text{True}$$

Since we obtained the true statement $0 = 0$, this system is dependent and has an infinite number of solutions.

NOW TRY EXERCISE 33

Recall from Section 4.1 that systems of equations that are dependent are also consistent since they have a solution.

Exercise Set 4.2

Concept/Writing Exercises

1. What will be the graph of an equation such as $3x - 4y + 2z = 1$?

2. Assume that the solution to a system of linear equations in three variables is $(1, 3, 5)$. What does this mean geometrically?

Practice the Skills

Solve by substitution.

3. $x = 1$
$2x - y = 4$
$-3x + 2y - 2z = 1$

4. $-x + 3y - 5z = 7$
$2y - z = 4$
$z = 2$

5. $5x - 6z = -17$
$3x - 4y + 5z = -1$
$2z = -6$

6. $2x - 5y = 12$
$-3y = -9$
$2x - 3y + 4z = 8$

7. $x + 2y = 6$
$3y = 9$
$x + 2z = 12$

8. $x - y + 5z = -4$
$3x - 2z = 6$
$4z = 2$

Solve using the addition method.

9. $x - 2y = -3$
$3x + 2y = 7$
$2x - 4y + z = -6$

10. $2x + y - 8 = 0$
$3x - 4z = -3$
$2x - 3z = 1$

11. $2y + 4z = 6$
$x + y + 2z = 0$
$2x + y + z = 4$

12. $x - y + 2z = 1$
$y - 4z = 2$
$-2x + 2y - 5z = 2$

13. $3p + 2q = 11$
$4q - r = 6$
$2p + 2r = 2$

14. $-4s + 3t = 16$
$2t - 2u = 2$
$-s + 6u = -2$

15. $p + q + r = 4$
$p - 2q - r = 1$
$2p - q - 2r = -1$

16. $x - 2y + 3z = -7$
$2x - y - z = 7$
$-x + 3y + 2z = -8$

17. $2x - 2y + 3z = 5$
$2x + y - 2z = -1$
$4x - y - 3z = 0$

18. $2x - y - 2z = 3$
$x - 3y - 4z = 2$
$x + y + 2z = -1$

19. $r - 2s + t = 2$
$2r + 2s - t = -2$
$2r - s - 2t = 1$

20. $3a - 3b + 4c = -1$
$a - 2b + 2c = 2$
$2a - 2b - c = 3$

21. $2a + 2b - c = 2$
$3a + 4b + c = -4$
$5a - 2b - 3c = 5$

22. $x - 2y + 2z = 3$
$2x - 3y + 2z = 5$
$x + y + z = -1$

23. $-x + 3y + z = 0$
$-2x + 4y - z = 0$
$3x - y + 2z = 0$

24. $x + y + z = 0$
$-x - y + z = 0$
$-x + y + z = 0$

25. $-\dfrac{1}{4}x + \dfrac{1}{2}y - \dfrac{1}{2}z = -2$
$\dfrac{1}{2}x + \dfrac{1}{3}y - \dfrac{1}{4}z = 2$
$\dfrac{1}{2}x - \dfrac{1}{2}y + \dfrac{1}{4}z = 1$

26. $\dfrac{2}{3}x + y - \dfrac{1}{3}z = \dfrac{1}{3}$
$\dfrac{1}{2}x + y + z = \dfrac{5}{2}$
$\dfrac{1}{4}x - \dfrac{1}{4}y + \dfrac{1}{4}z = \dfrac{3}{2}$

27. $x - \dfrac{2}{3}y - \dfrac{2}{3}z = -2$
$\dfrac{2}{3}x + y - \dfrac{2}{3}z = \dfrac{1}{3}$
$-\dfrac{1}{4}x + y - \dfrac{1}{4}z = \dfrac{3}{4}$

28. $\dfrac{1}{4}x + \dfrac{1}{2}y + z = 3$
$\dfrac{1}{3}x + \dfrac{1}{4}y + z = \dfrac{17}{6}$
$-\dfrac{1}{4}x + \dfrac{1}{3}y - \dfrac{1}{2}z = -\dfrac{5}{6}$

29. $0.2x + 0.3y + 0.3z = 1.1$
$0.4x - 0.2y + 0.1z = 0.4$
$-0.1x - 0.1y + 0.3z = 0.4$

30. $0.3x - 0.4y + 0.2z = 1.6$
$-0.1x - 0.2y + 0.3z = 0.9$
$-0.2x - 0.1y - 0.3z = -1.2$

Determine whether the following systems are inconsistent, dependent, or neither.

31. $2x + y + 2z = 1$
$x - 2y - z = 0$
$3x - y + z = 2$

32. $5a - 4b + 2c = 5$
$-5a + 4b - 2c = -5$
$-7a - 4b + c = 7$

33. $x - 4y - 3z = -1$
$2x - 10y - 7z = 5$
$-3x + 12y + 9z = 3$

34. $2p - 4q + 6r = 8$
$-p + 2q - 3r = 6$
$3p + 4q + 5r = 8$

35. $x + 3y + 2z = 6$
$x - 2y - z = 8$
$-3x - 9y - 6z = -4$

36. $2x - 2y + 4z = 2$
$-3x + y = -9$
$2x - y + z = 5$

Problem Solving

An equation in three variables represents a plane. Consider a system of equations consisting of three equations in three variables. Answer the following questions.

37. If the three planes are parallel to one another as illustrated in the figure, how many points will be common to all three planes? Is the system consistent or inconsistent? Explain your answer.

38. If two of the planes are parallel to each other and the third plane intersects each of the other two planes, how many points will be common to all three planes? Is the system consistent or inconsistent? Explain your answer.

39. If the three planes are as illustrated in the figure, how many points will be common to all three planes? Is the system consistent or inconsistent? Explain your answer.

40. If the three planes are as illustrated in the figure, how many points will be common to all three planes? Is the system dependent? Explain your answer.

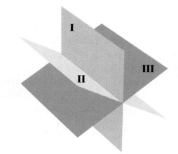

41. Is it possible for a system of linear equations in three variables to have exactly

  a) no solution,

  b) one solution,

  c) two solutions? Explain your answer.

42. In a system of linear equations in three variables, if the graphs of two equations are parallel planes, is it possible for the system to be

  a) consistent,

  b) dependent,

  c) inconsistent? Explain your answer.

43. Three solutions to the equation $Ax + By + Cz = 1$ are $(-1, 2, -1)$, $(-1, 1, 2)$, and $(1, -2, 2)$. Determine the values of A, B, and C and write the equation using the numerical values found.

44. Three solutions to the equation $Ax + By + Cz = 14$ are $(3, -1, 2)$, $(2, -2, 1)$, and $(-5, 3, -24)$. Find the values of A, B, and C and write the equation using the numerical values found.

Write a system of linear equations in three variables that has the given solution. Explain how you determined your answer.

45. $(3, 1, 6)$

46. $(-2, 5, 3)$

47. a) Find the values of a, b, and c such that the points $(1, -1)$, $(-1, -5)$, and $(3, 11)$ lie on the graph of $y = ax^2 + bx + c$.

  b) Find the quadratic equation whose graph passes through the three points indicated. Explain how you determined your answer.

48. a) Find the values of a, b, and c such that the points $(1, 7)$, $(-2, -5)$, and $(3, 5)$ lie on the graph of $y = ax^2 + bx + c$.

  b) Find the quadratic equation whose graph passes through the three points indicated. Explain how you determined your answer.

Challenge Problems

Find the solution to the following systems of equations.

49. $3p + 4q = 11$
$2p + r + s = 9$
$q - s = -2$
$p + 2q - r = 2$

50. $3a + 2b - c = 0$
$2a + 2c + d = 5$
$a + 2b - d = -2$
$2a - b + c + d = 2$

Cumulative Review Exercises

[2.2] **51.** *Cross-Country Skiing* Margie Steiner begins skiing along a trail at 3 miles per hour. Ten minutes $\left(\frac{1}{6} \text{ hour}\right)$ later, her husband, David, begins skiing along the same trail at 5 miles per hour.

a) How long after David leaves will he catch up to Margie?

b) How far from the starting point will they be when they meet?

[2.6] *Determine each solution set.*

52. $\left|4 - \dfrac{2x}{3}\right| > 5$

53. $\left|\dfrac{3x - 4}{2}\right| - 1 < 5$

54. $\left|2x - \dfrac{1}{2}\right| = -5$

4.3 SYSTEMS OF LINEAR EQUATIONS: APPLICATIONS AND PROBLEM SOLVING

1 Use systems of equations to solve applications.
2 Use linear systems in three variables to solve applications.

1 Use Systems of Equations to Solve Applications

Many of the applications solved in earlier chapters using only one variable can now be solved using two variables. Following are some examples showing how applications can be represented by systems of equations.

EXAMPLE 1 **A Changing Workforce** The graph in Figure 4.6 indicates that the percent of males in the workforce is constantly decreasing while the percent of females is constantly increasing. The function $m(t) = -0.25t + 85.4$, where $t = $ years since 1955, can be used to estimate the percent of males in the workforce, and the function $w(t) = 0.52t + 35.7$ can be used to estimate the percent of women in the workforce. If this trend continues, determine when the percent of women in the workforce will equal the percent of men.

Solution Understand and Translate Consider the two functions given above as the system of equations. To determine when the percent of women will equal the percent of men, we can set the two functions equal to one another and solve for time, t.

Women and Men in the Workforce
(Percent of population in the civilian labor force)

FIGURE 4.6

Carry Out

$$\text{percent of women} = \text{percent of men}$$
$$0.52t + 35.7 = -0.25t + 85.4$$
$$0.77t = 49.7$$
$$t \approx 64.5$$

Answer Thus the percent of women in the workforce will equal the percent of men about 64.5 years after 1955. Since $1955 + 64.5 = 2019.5$, the percents will be equal in the year 2019.

NOW TRY EXERCISE 37

EXAMPLE 2

Canoe Speed The Burnhams are canoeing on the Suwannee River. They travel at an average speed of 4.75 miles per hour when padding with the current and 2.25 miles per hour when padding against the current. Determine the speed of the canoe in still water and the speed of the current.

Solution Understand When they are traveling with the current, the canoe's speed is the canoe's speed in still water *plus* the current's speed. When traveling against the current, the canoe's speed is the canoe's speed in still water *minus* the current's speed.

Translate   let s = speed of the canoe in still water

c = speed of the current

The system of equations is:

speed of the canoe traveling with the current: $s + c = 4.75$
speed of the canoe traveling against the current: $s - c = 2.25$

Carry Out We will use the addition method, as we discussed in Section 4.1, to solve this system of equations.

$$s + c = 4.75$$
$$\underline{s - c = 2.25}$$
$$2s \quad\;\; = 7$$
$$s = 3.5$$

The speed of the canoe in still water is 3.5 miles per hour. We now determine the speed of the current.

$$s + c = 4.75$$
$$3.5 + c = 4.75$$
$$c = 1.25$$

Answer The speed of the current is 1.25 miles per hour, and the speed of the canoe in still water is 3.5 miles per hour.

EXAMPLE 3

Salary Yamil Bermudez, a salesman at Hancock Appliances, receives a weekly salary plus a commission, which is a percentage of his sales. One week, with sales of $3000, his total take-home pay was $850. The next week, with sales of $4000, his total take-home pay was $1000. Find his weekly salary and his commission rate.

Solution Understand Yamil's take-home pay consists of his weekly salary plus commission. We are given information about two specific weeks that we can use to find his weekly salary and his commission rate.

Translate   let s = his weekly salary

r = his commission rate

In week 1, his commission on \$3000 is $3000r$, and in week 2 his commission on \$4000 is $4000r$. Thus the system of equations is

salary + commission = take-home salary

$$\left.\begin{array}{ll}\text{1st week} & s + 3000r = 850\\ \text{2nd week} & s + 4000r = 1000\end{array}\right\} \text{ System of equations}$$

Carry Out

$$\begin{array}{ll}-s - 3000r = -850 & \textit{1st week multiplied by } -1\\ \underline{s + 4000r = 1000} & \textit{2nd week}\\ 1000r = 150 & \textit{Sum of equations}\end{array}$$

$$r = \frac{150}{1000}$$

$$r = 0.15$$

Yamil's commission rate is 15%. Now we find his weekly salary by substituting 0.15 for r in either equation.

$$s + 3000r = 850$$
$$s + 3000(0.15) = 850 \qquad \textit{Substitute 0.15 for r in the 1st-week equation.}$$
$$s + 450 = 850$$
$$s = 400$$

NOW TRY EXERCISE 11   **Answer** Yamil's weekly salary is \$400 and his commission rate is 15%. ✳

EXAMPLE 4 **Riding Horses** Ben Campbell leaves his ranch riding on his horse at 5 miles per hour. One-half hour later, Joe Campbell leaves the same ranch and heads along the same route on his horse at 8 miles per hour.

a) How long after Joe leaves the ranch will he catch up to Ben?

b) When Joe catches up to Ben, how far from the ranch will they be?

Solution a) Understand When Joe catches up to Ben, they both will have traveled the same distance. Joe will have traveled the distance in $\frac{1}{2}$ hour less time since he left $\frac{1}{2}$ hour after Ben. We will use the formula distance = rate · time to solve this problem.

Translate   let b = time traveled by Ben

j = time traveled by Joe

We will set up a table to organize the given information.

	Rate	Time	Distance
Ben	5	b	$5b$
Joe	8	j	$8j$

Since both Ben and Joe cover the same distance, we write

Ben's distance = Joe's distance

$$5b = 8j$$

Our second equation comes from the fact that Joe is traveling for $\frac{1}{2}$ hour less time than Ben. Therefore, $j = b - \frac{1}{2}$. Thus our system of equations is:

$$5b = 8j$$
$$j = b - \frac{1}{2}$$

Carry Out We will solve this system of equations using substitution. Since $j = b - \frac{1}{2}$, substitute $b - \frac{1}{2}$ for j in the first equation and solve for b.

$$5b = 8j$$
$$5b = 8\left(b - \frac{1}{2}\right)$$
$$5b = 8b - 4$$
$$-3b = -4$$
$$b = \frac{-4}{-3} = 1\frac{1}{3}$$

Therefore, the time Ben has been traveling is $1\frac{1}{3}$ hours. To get the time Joe has been traveling, we will subtract $\frac{1}{2}$ hour from Ben's time.

$$j = b - \frac{1}{2}$$
$$j = 1\frac{1}{3} - \frac{1}{2}$$
$$j = \frac{4}{3} - \frac{1}{2} = \frac{8}{6} - \frac{3}{6} = \frac{5}{6}$$

Answer Joe will catch up to Ben $\frac{5}{6}$ of an hour (or 50 minutes) after Joe leaves the ranch.

b) We can use either Ben's or Joe's distance to determine the distance traveled from the ranch. We will use Joe's distance.

$$d = 8j = 8\left(\frac{5}{6}\right) = \frac{8}{1} \cdot \frac{5}{6} = \frac{20}{3} = 6\frac{2}{3}$$

Thus, Joe will catch up to Ben when they are $6\frac{2}{3}$ miles from the ranch.

EXAMPLE 5 **Mixing Solutions** Chingshou Kong, a chemist for S.C. Johnson, wishes to create a new household cleaner containing 30% trisodium phosphate (TSP). Chingshou needs to mix a 16% TSP solution with a 72% TSP solution to get 6 liters of a 30% TSP solution. How many liters of the 16% solution and of the 72% solution will he need to mix?

Solution **Understand** To solve this problem we use the fact that the amount of TSP in a solution is found by multiplying the percent strength of the solution by the number of liters (the volume) of the solution. Chingshou needs to mix a 16% solution and a 72% solution to obtain 6 liters of a solution whose strength, 30%, is between the strengths of the two solutions being mixed.

Translate let x = number of liters of the 16% solution

y = number of liters of the 72% solution

We will draw a sketch (Fig. 4.7) and then use a table to help analyze the problem.

FIGURE 4.7

Solution	Strength of solution	Number of liters	Amount of TSP
16% solution	0.16	x	$0.16x$
72% solution	0.72	y	$0.72y$
Mixture	0.30	6	$0.30(6)$

Since the sum of the volumes of the 16% solution and the 72% solution is 6 liters, our first equation is

$$x + y = 6$$

The second equation comes from the fact that the solutions are mixed.

$$\left(\begin{array}{c}\text{amount of TSP}\\ \text{in 16\% solution}\end{array}\right) + \left(\begin{array}{c}\text{amount of TSP}\\ \text{in 72\% solution}\end{array}\right) = \left(\begin{array}{c}\text{amount of TSP}\\ \text{in mixture}\end{array}\right)$$
$$0.16x \qquad + \qquad 0.72y \qquad = \qquad 0.30(6)$$

Therefore, the system of equations is

$$x + y = 6$$
$$0.16x + 0.72y = 0.30(6)$$

Carry Out Solving $x + y = 6$ for y, we get $y = -x + 6$. Substituting $-x + 6$ for y in the second equation gives us

$$0.16x + 0.72y = 0.30(6)$$
$$0.16x + 0.72(-x + 6) = 0.30(6)$$
$$0.16x - 0.72x + 4.32 = 1.8$$
$$-0.56x + 4.32 = 1.8$$
$$-0.56x = -2.52$$
$$x = \frac{-2.52}{-0.56} = 4.5$$

Therefore, Chingshou must use 4.5 liters of the 16% solution. Since the two solutions must total 6 liters, he must use $6 - 4.5$ or 1.5 liters of the 72% solution. ✳

In Example 5, the equation $0.16x + 0.72y = 0.30(6)$ could have been simplified by multiplying both sides of the equation by 100. This would give the equation $16x + 72y = 30(6)$ or $16x + 72y = 180$. Then the system of equations

NOW TRY EXERCISE 13 would be $x + y = 6$ and $16x + 72y = 180$. If you solve this system, you should obtain the same solution. Try it and see.

2 Use Linear Systems in Three Variables to Solve Applications

Now let us look at some applications that involve three equations and three variables.

EXAMPLE 6 **Bank Loans** Tiny Tots Toys must borrow $25,000 to pay for an expansion. They are not able to obtain a loan for the total amount from a single bank, so they take out three loans from three different banks. They borrow some of the money at a bank that charges them 8% interest. At the second bank, they borrow $2000 more than one-half the amount borrowed from the first bank. The interest rate at the second bank is 10%. The balance of the $25,000 is borrowed from a third bank where they pay 9% interest. The total annual interest Tiny Tots Toys pays for the three loans is $2220. How much do they borrow at each rate?

Solution **Understand** We are asked to determine how much is borrowed at each of the three different rates. Therefore, this problem will contain three variables, one for each amount borrowed. Since the problem will contain three variables, we will need to determine three equations to use in our system of equations.

Translate

$\quad$ let x = amount borrowed at first bank

$\quad\quad y$ = amount borrowed at second bank

$\quad\quad z$ = amount borrowed at third bank

Since the total amount borrowed is $25,000 we know that

$$x + y + z = 25{,}000 \quad \text{\textit{Total amount borrowed is \$25,000.}}$$

At the second bank, Tiny Tots Toys borrows $2000 more than one-half the amount borrowed from the first bank. Therefore, our second equation is

$$y = \frac{1}{2}x + 2000 \quad \text{\textit{Second, y, is \$2000 more than } } \tfrac{1}{2} \text{\textit{ of first, x.}}$$

Our last equation comes from the fact that the total annual interest charged by the three banks is $2220. The interest at each bank is found by multiplying the interest rate by the amount borrowed.

$$0.08x + 0.10y + 0.09z = 2220 \quad \text{\textit{Total interest is \$2220.}}$$

Thus, our system of equations is

$$x + y + z = 25{,}000 \qquad (1)$$

$$y = \frac{1}{2}x + 2000 \qquad (2)$$

$$0.08x + 0.10y + 0.09z = 2220 \quad (3)$$

Both sides of equation (2) can be multiplied by 2 to remove fractions.

$$2\,(y) = 2\left(\frac{1}{2}x + 2000\right)$$

$$2y = x + 4000 \qquad\qquad \text{\textit{Distributive property}}$$

$$-x + 2y = 4000 \qquad\qquad \text{\textit{Subtract x from both sides.}}$$

The decimals in equation (3) can be removed by multiplying both sides of the equation by 100. This gives

$$8x + 10y + 9z = 222{,}000$$

Our simplified system of equations is therefore

$$x + y + z = 25{,}000 \quad (eq.\ 1)$$
$$-x + 2y = 4000 \quad (eq.\ 2)$$
$$8x + 10y + 9z = 222{,}000 \quad (eq.\ 3)$$

Carry Out There are various ways of solving this system. Let's use (*eq.* 1) and (*eq.* 3) to eliminate the variable z.

$$-9x - 9y - 9z = -225{,}000 \quad (eq.\ 1) \ \text{Multiplied by } -9$$
$$\underline{8x + 10y + 9z = 222{,}000} \quad (eq.\ 3)$$
$$-x + y = -3000 \quad \text{Sum of equations, } (eq.\ 4)$$

Now we use (*eq.* 2) and (*eq.* 4) to eliminate the variable x and solve for y.

$$x - 2y = -4000 \quad (eq.\ 2) \ \text{Multiplied by } -1$$
$$\underline{-x + y = -3000} \quad (eq.\ 4)$$
$$-y = -7000 \quad \text{Sum of equations}$$
$$y = 7000$$

Now that we know the value of y we can solve for x.

$$-x + 2y = 4000 \quad (eq.\ 2)$$
$$-x + 2(7000) = 4000 \quad \text{Substitute 7000 for } y \text{ in } (eq.\ 2).$$
$$-x + 14{,}000 = 4000$$
$$-x = -10{,}000$$
$$x = 10{,}000$$

Finally, we solve for z.

$$x + y + z = 25{,}000 \quad (eq.\ 1)$$
$$10{,}000 + 7000 + z = 25{,}000$$
$$17{,}000 + z = 25{,}000$$
$$z = 8000$$

Answer Tiny Tots Toys borrows $10,000 at 8%, $7000 at 10%, and $8000 at 9% interest. ✳

EXAMPLE 7 **Inflatable Boats** Hobson, Inc., has a small manufacturing plant that makes three types of inflatable boats: one-person, two-person, and four-person models. Each boat requires the service of three departments: cutting, assembly, and packaging. The cutting, assembly, and packaging departments are allowed to use a total of 380, 330, and 120 person-hours per week, respectively. The time requirements for each boat and department are specified in the following table. Determine how many of each type of boat Hobson must produce each week for its plant to operate at full capacity.

Department	Time (person-hr)		
	One-person boat	Two-person boat	Four-person boat
Cutting	0.6	1.0	1.5
Assembly	0.6	0.9	1.2
Packaging	0.2	0.3	0.5

Solution   Understand   We are told that three different types of boats are produced and we are asked to determine the number of each type produced. Since this problem involves three amounts to be found, the system will contain three equations in three variables.

Translate   We will use the information given in the table.

$$\text{let } x = \text{number of one-person boats}$$
$$y = \text{number of two-person boats}$$
$$z = \text{number of four-person boats}$$

The total number of cutting hours for the three types of boats must equal 380 person-hours.

$$0.6x + 1.0y + 1.5z = 380$$

The total number of assembly hours must equal 330 person-hours.

$$0.6x + 0.9y + 1.2z = 330$$

The total number of packaging hours must equal 120 person-hours.

$$0.2x + 0.3y + 0.5z = 120$$

Therefore, the system of equations is

$$0.6x + 1.0y + 1.5z = 380$$
$$0.6x + 0.9y + 1.2z = 330$$
$$0.2x + 0.3y + 0.5z = 120$$

Multiplying each equation in the system by 10 will eliminate the decimal numbers and give a simplified system of equations.

$$6x + 10y + 15z = 3800 \quad (eq.\ 1)$$
$$6x + 9y + 12z = 3300 \quad (eq.\ 2)$$
$$2x + 3y + 5z = 1200 \quad (eq.\ 3)$$

Carry Out   Let's first eliminate the variable x using (*eq.* 1) and (*eq.* 2), and then (*eq.* 1) and (*eq.* 3).

$$6x + 10y + 15z = 3800 \quad (eq.\ 1)$$
$$\underline{-6x - 9y - 12z = -3300} \quad (eq.\ 2) \quad \textit{Multiplied by } -1$$
$$y + 3z = 500 \quad \textit{Sum of equations, } (eq.\ 4)$$

$$6x + 10y + 15z = 3800 \quad (eq.\ 1)$$
$$\underline{-6x - 9y - 15z = -3600} \quad (eq.\ 3) \quad \text{Multiplied by } -3$$
$$y = 200 \quad \text{Sum of equations, } (eq.\ 5)$$

Note that when we added the last two equations, both variables x and z were eliminated at the same time. Now we know the value of y and can solve for z.

$$y + 3z = 500 \quad (eq.\ 4)$$
$$200 + 3z = 500 \quad \text{Substitute 200 for } y.$$
$$3z = 300$$
$$z = 100$$

Finally, we find x.

$$6x + 10y + 15z = 3800 \quad (eq.\ 1)$$
$$6x + 10(200) + 15(100) = 3800$$
$$6x + 2000 + 1500 = 3800$$
$$6x + 3500 = 3800$$
$$6x = 300$$
$$x = 50$$

Answer Hobson should produce 50 one-person boats, 200 two-person boats, and 100 four-person boats per week.

NOW TRY EXERCISE 51

Mathematics in Action

Zap! Ping! Bam! Math!

As serious as mathematics can seem, it figures heavily in a fun activity enjoyed by millions of people all over the world—video games. When you aim death ray pulses at the seven-legged Woogie as he (or she) attemps to destroy your interstellar neutronic energy station, your success or failure depends on large numbers of equations embedded in thousands of lines of program code controlling the video action.

One group of equations may determine the movement of the Woogie, consisting of basic moves that are influenced by variables like the clock time, the proximity of other objects, and the current score. Similarly, the trajectory of the death ray pulse is controlled by equations involving the position of the weapon, whether you have purchased additional laser energy, and whether the Woogie has erected a solargistic deflector shield.

Whatever.

Ultimately it all comes down to equations expressed as computer algorithms.

Some equations may be solved simultaneously, as a group, in systems of equations; others may be solved in rapid succession with calculated values passed from one equation to the next. Hundreds of thousands of calculations result in the positioning of visual objects according to the Cartesian coordinates of the video screen. When the numbers work out that the Woogie and your death ray pulse occupy the same coordinate location at the same instant, then it's good-bye Woogie.

Exercise Set 4.3

Practice the Skills/Problem Solving

1. ***Theme Parks*** The two most visited theme parks in the United States in 2001 were Walt Disney's Magic Kingdom in Florida and Disneyland in California. The total number of visitors to these parks was 27.1 million people. The number of people who visited the Magic Kingdom was 2.5 million more than the number of people who visited Disneyland. How many people visited each of these parks in 2001? *Source:* www.saferparks.org

2. ***Cable Viewers*** For the week ending March 23, 2002, two very popular cable television shows were *The Osbournes* and *SpongeBob*. About 150,000 more viewers watched *The Osbournes* than *SpongeBob*. The total number of viewers for both shows was 5,842,000. How many viewers watched each of these shows that week? *Source:* www.tv.zap2it.com

3. ***Fat Content*** A nutritionist finds that a large order of fries at McDonald's has more fat than their quarter-pound hamburger. The fries have 4 grams more than three times the amount of fat that the hamburger has. The difference in the fat content between the fries and the hamburger is 46 grams. Find the fat content of the hamburger and of the fries.

4. ***Baseball Game Cost*** The baseball fan cost index (FCI) includes the cost of four average tickets, two small beers, four small sodas, four hot dogs, parking, two programs, and two caps. In 2001, the major-league team with the lowest FCI was the Montreal Expos and the major-league team with the highest FCI was the Boston Red Sox. The FCI for the Red Sox was $25.92 less than three times the FCI for the Expos. The difference between the Red Sox's FCI and the Expos' FCI was $134.24. Determine the FCI for the Red Sox and the FCI for the Expos.

5. ***Complementary Angles*** Two angles are **complementary angles** if the sum of their measures is 90°. (See Section 2.3.) If the measure of the larger of two complementary angles is 15° more than two times the measure of the smaller angle, find the measures of the two angles.

6. ***Complementary Angles*** The difference between the measures of two complementary angles is 58°. Determine the measures of the two angles.

7. ***Supplementary Angles*** Two angles are **supplementary angles** if the sum of their measures is 180°. (See Section 2.3.) Find the measures of two supplementary angles if the measure of one angle is 28° less than three times the measure of the other.

8. ***Supplementary Angles*** Determine the measures of two supplementary angles if the measure of one angle is three and one half times larger than the measure of the other angle.

9. ***Rowing Speed*** The Heart O' Texas Rowing Team, while practicing in Austin, Texas rowed an average of 15.6 miles per hour with the current and 8.8 miles per hour against the current. Determine the team's rowing speed in still water and the speed of the current.

10. ***Flying Speed*** Daniel LaBonne, in his Piper Cub airplane, flew an average of 121 miles per hour with the wind and 87 miles per hour against the wind. Determine the speed of the airplane in still air and the speed of the wind.

11. ***Salary Plus Commission*** Rickardo Wallenhorst, an office equipment sales representative, earns a weekly salary plus a commission on his sales. One week his total compensation on sales of $4000 was $660. The next week his total compensation on sales of $6000 was $740. Find Rickardo's weekly salary and his commission rate.

12. ***Truck Rental*** A truck rental agency charges a daily fee plus a mileage fee. Julie was charged $85 for 2 days and 100 miles and Christina was charged $165 for 3 days and 400 miles. What is the agency's daily fee, and what is the mileage fee?

13. **Lavender Oil** Pola Sommers, a massage therapist, needs 3 ounces of a 20% lavender oil solution. She has only 5% and 30% lavender oil solutions available. How many ounces of each should Pola mix to obtain the desired solution?

14. **Fertilizer Solutions** Daryl Hinderman needs to apply a 10% liquid nitrogen solution to his rose garden, but he only has a 4% liquid nitrogen solution and a 20% liquid nitrogen solution available. How much of the 4% solution and how much of the 20% solution should Daryl mix together to get 10 gallons of the 10% solution?

15. **Weed Killer** Round-Up Concentrate Grass and Weed Killer consists of an 18% active ingredient glyphosate (and 82% inactive ingredients). The concentrate is to be mixed with water and the mixture applied to weeds. If the final mixture is to contain 0.9% active ingredient, how much concentrate and how much water should be mixed to make 200 gallons of the final mixture?

16. **Lawn Fertilizer** Scott's Winterizer Lawn Fertilizer is 22% nitrogen. Schultz's Lime with Lawn Fertilizer is 4% nitrogen. William Weaver, owner of Weaver's Nursery, wishes to mix these two fertilizers to make 400 pounds of a special 10% nitrogen mixture for midseason lawn feeding. How much of each fertilizer should he mix?

17. **Birdseed** Birdseed costs $0.59 a pound and sunflower seeds cost $0.89 a pound. Angela Leinenbachs' pet store wishes to make a 40-pound mixture of birdseed and sunflower seeds that sells for $0.76 per pound. How many pounds of each type of seed should she use?

18. **Coffee** John Bronson runs a grocery store. He wishes to mix 30 pounds of coffee to sell for a total cost of $170. To obtain the mixture, he will mix coffee that sells for $5.20 per pound with coffee that sells for $6.30 per pound. How many pounds of each coffee should he use?

19. **Little League** Admission tickets to the Braden River Little League all-star game cost $4.00 for adults and $1.50 for children. A total of 225 tickets was sold and a total of $500 was collected from ticket sales. How many adult tickets and how many children's tickets were sold?

20. **Buffalo Wings** The Wing House sells both regular orders and jumbo orders of buffalo wings. The regular order of wings costs $5.99 and the jumbo order of wings costs $8.99. One Saturday a total of 134 orders of wings was taken, and the receipts from the wings were $1024.66. How many regular orders and how many jumbo orders were filled?

21. **Savings Accounts** Mr. and Mrs. Abdullah invest a total of $10,000 in two savings accounts. One account pays 5% interest and the other 6%. Find the amount placed in each account if the accounts receive a total of $540 in interest after 1 year. Use interest = principal · rate · time.

22. **Investments** Louis Okonkwo invested $30,000, part at 9% and part at 5%. If he had invested the entire amount at 6.5%, his total annual interest would be the same as the sum of the annual interest received from the two other accounts. How much was invested at each interest rate?

23. **Milk** Becky Slaats is a plant engineer at Velda Farms Dairy Cooperative. She wishes to mix whole milk, which is 3.25% fat, and skim milk, which has no fat, to obtain 260 gallons of a mixture of milk that contains 2% fat. How many gallons of whole milk and how many gallons of skim milk should Becky mix to obtain the desired mixture?

24. **Quiche Lorraine** Lambert Marx's recipe for quiche lorraine calls for 2 cups (16 ounces) of light cream that is 20% butterfat. It is often difficult to find light cream with 20% butterfat at the supermarket. What is commonly found is heavy cream, which is 36% butterfat, and half-and-half, which is 10.5% butterfat. How much of the heavy cream and how much of the half-and-half should Lambert mix to obtain the mixture necessary for the recipe?

25. **Birdseed** The kindergarten class at Sandridge School wishes to buy 20 pounds of birdseed to attract birds to the school courtyard. To attract as many birds as possible they wish to buy two varieties of birdseed. By ordering directly through www.birdseed.com, they can purchase Season's Choice birdseed for $1.79 per pound and Garden Mix birdseed for $1.19 per pound. If the class wishes to spend $28 on birdseed, how many pounds of each type should they buy?

26. **Juice** The Healthy Favorites Juice Company sells apple juice for 8.3¢ an ounce and raspberry juice for 9.3¢ an ounce. The company wishes to market and sell 8-ounce

cans of apple-raspberry juice for 8.7¢ an ounce. How many ounces of each should be mixed?

27. ***Car Travel*** Two cars start at the same point in Alexandria, Virginia, and travel in opposite directions. One car travels 5 miles per hour faster than the other car. After 4 hours, the two cars are 420 miles apart. Find the speed of each car.

28. ***Road Construction*** Kip Ortiz drives from Atlanta to Louisville, a distance of 430 miles. Due to road construction and heavy traffic, during the first part of his trip, Kip drives at an average rate of 50 miles per hour. During the rest of his trip he drives at an average rate of 70 miles per hour. If his total trip takes 7 hours, how many hours does he drive at each speed?

29. ***Avon Conference*** Cabrina Wilson and Dabney Jefferson are Avon representatives who are attending a conference in Seattle. After the conference, Cabrina drives home to Boise at an average speed of 65 miles per hour and Dabney drives home to Portland at an average speed of 50 miles per hour. If the sum of their driving times is 11.4 hours and if the sum of the distances driven is 690 miles, determine the time each representative spent driving home.

30. ***Exercise*** For her exercise routine, Rita Sanchez rides a bicycle for half an hour and then jogs for half an hour. Rita rides the bicycle at a speed that is four times the speed at which she jogs. If the total distance covered by Rita is 12.5 miles, determine the speeds at which she bikes and jogs.

31. ***Animal Diet*** Animals in an experiment are on a strict diet. Each animal is to receive, among other nutrients, 20 grams of protein and 6 grams of carbohydrates. The scientist has only two food mixes available of the following compositions. How many grams of each mix should be used to obtain the right diet for a single animal?

Mix	Protein (%)	Carbohydrate (%)
Mix *A*	10	6
Mix *B*	20	2

32. ***Chair Manufacturing*** A company makes two models of chairs. Information about the construction of the chairs is given in the table above and to the right. On a particular day the company allocated 46.4 person-hours

for assembling and 8.8 person-hours for painting. How many of each chair can be made?

Model	Time to assemble	Time to paint
Model *A*	1 hr	0.5 hr
Model *B*	3.2 hr	0.4 hr

33. ***Brass Alloy*** By weight, one alloy of brass is 70% copper and 30% zinc. Another alloy of brass is 40% copper and 60% zinc. How many grams of each of these alloys need to be melted and combined to obtain 300 grams of a brass alloy that is 60% copper and 40% zinc?

34. ***Silver Alloy*** Sterling silver is 92.5% pure silver. How many grams of pure (100%) silver and how many grams of sterling silver must be mixed to obtain 250 g of a 94% silver alloy?

35. ***Texas Driving*** Tom Johnson and Melissa Acino started driving at the same time in different cars from Oklahoma City. They both traveled south on Route 35. When Melissa reached the Dallas/Ft. Worth area, a distance of 150 miles, Tom had only reached Denton, Texas, a distance of 120 miles. If Melissa averaged 15 miles per hour faster than Tom, find the average speed of each car.

36. ***Walking and Jogging*** Jerry Johnson tries to exercise every day. He walks at 3 miles per hour and then jogs at 5 miles per hour. If it takes him 0.9 hours to travel a total of 3.5 miles, how long does he jog?

37. ***Tax Returns*** The following graph shows the percent of U.S. Federal tax returns filed electronically and by paper for the years 1996 through 2001. If t represents the number of years since 1996, the percent of electronic tax returns can be estimated with the function $E(t) = 3.62t + 12.6$ and the percent of paper tax returns can be estimated with the function $P(t) = -3.62t + 87.4$. Assuming this trend continues, in what year will the percent of electronic returns equal the percent of paper returns?

Federal Tax Return Method

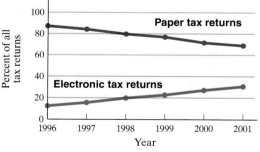

Source: Internal Revenue Service

38. *Doctorate Degrees* The number of men receiving doctorate degrees, M, in thousands, in the United States, can be approximated by the function $M(t) = -0.86t + 27.1$, where t is the number of years since 1999. The number of women receiving doctorate degrees, W, in thousands, can be approximated by the function $W(t) = 0.43t + 17.5$. Estimate the year when the number of women receiving doctorate degrees will equal the number of men receiving doctorate degrees.

39. *Long-Distance Plans* The Sprint Nickel Any Time long-distance-rate plan includes a monthly fee of $8.95 plus 5¢ per minute of phone use. The Sprint 7¢ Any Time plan includes a monthly fee of $5.95 plus 7¢ per minute of phone use.

a) Let C represent the total monthly cost, and let t represent the number of minutes of phone use. Write a system of equations in which the equations represent these long-distance-rate plans.

b) Graph this system of equations for values of t from 0 to 180 minutes.

c) From the graph, estimate the number of minutes necessary for the cost of the two plans to be equal.

d) Solve this system algebraically. If your answer does not agree with your answer in part **c)**, explain why.

40. *Photocopy Costs* At a local copy center two plans are available.
Plan 1: 10¢ per copy
Plan 2: an annual charge of $120 plus 4¢ per copy

a) Represent this information as a system of equations.

b) Graph the system of equations for up to 4000 copies made.

c) From the graph, estimate the number of copies a person would have to make in a year for the two plans to have the same total cost.

d) Solve the system algebraically. If your answer does not agree with your answer in part **c)**, explain why.

In Exercises 41–52, **a)** *express the problem as a linear system in three variables, and* **b)** *solve the problem.*

41. *Mail Volume* The average American household receives 24 pieces of mail each week. The number of bills and statements is two less than twice the number of pieces of personal mail. The number of advertisements is two more than five times the number of pieces of personal mail. How many pieces of personal mail, bills and statements, and advertisements does the average family get each week? *Source*: Arthur D. Little, Inc.

42. *Submarine Personnel* A 141-man crew is standard on a Los Angeles class submarine. The number of chief petty officers (enlisted) is four more than the number of commissioned officers. The number of other enlisted men is three less than eight times the number of commissioned officers. Determine the number of commis-

sioned officers, chief petty officers, and other enlisted men on the submarine.

43. *Land Mines* The countries that have the largest number of land mines are, in order, Iran, Angola, and Iraq. The total number in these three countries is estimated to be 41 million. The number of land mines in Iran is about 14 million fewer than three times the number in Iraq. The number in Angola is about 5 million fewer than twice the number in Iraq. Find the estimated number of land mines in Iraq, Angola, and Iran.

44. *Concert Tickets* Three kinds of tickets for a Soggy Bottom Boys concert are available: up-front, main-floor, and balcony. The most expensive tickets, the up-front tickets, are twice as expensive as balcony tickets. Balcony tickets are $10 less than main-floor tickets and $30 less than up-front tickets. Determine the price of each kind of ticket.

45. *Triangle* The sum of the measures of the angles of a triangle is 180°. The smallest angle of the triangle has a measure $\frac{2}{3}$ the measure of the second smallest angle. The largest angle has a measure that is 30° less than three times the measure of the second smallest angle. Determine the measure of each angle.

46. Triangle The largest angle of a triangle has a measure that is 10° less than three times the measure of the second smallest angle. The measure of the smallest angle is equal to the difference between the measure of the largest angle and twice the measure of the second smallest angle. Determine the measures of the three angles of the triangle.

47. Investments Marion Monrow received a check for $10,000. She decided to divide the money (not equally) into three different investments. She placed part of her money in a savings account paying 3% interest. The second amount, which was twice the first amount, was placed in a certificate of deposit paying 5% interest. She placed the balance in a money market fund paying 6% interest. If Marion's total interest over the period of 1 year was $525.00, how much was placed in each account?

48. Bonus Michael Kuzak, an attorney, divided his $15,000 holiday bonus check among three different investments. With some of the money, he purchased a municipal bond paying 5.5% simple interest. He invested twice the amount of money that he paid for the municipal bond in a certificate of deposit paying 4.5% simple interest. Michael placed the balance of the money in a money market account paying 3.75% simple interest. If Michael's total interest for 1 year was $692.50, how much was placed in each account?

49. Hydrogen Peroxide A 10% solution, a 12% solution, and a 20% solution of hydrogen peroxide are to be mixed to get 8 liters of a 13% solution. How many liters of each must be mixed if the volume of the 20% solution must be 2 liters less than the volume of the 10% solution?

50. Sulfuric Acid An 8% solution, a 10% solution, and a 20% solution of sulfuric acid are to be mixed to get 100 milliliters of a 12% solution. If the *volume of acid* from the 8% solution is to equal half the *volume of acid* from the other two solutions, how much of each solution is needed?

51. Furniture Manufacturing Donaldson Furniture Company produces three types of rocking chairs: the children's model, the standard model, and the executive model. Each chair is made in three stages: cutting, construction, and finishing. The time needed for each stage of each chair is given in the following chart. During a specific week the company has available a maximum of 154 hours for cutting, 94 hours for construction, and 76 hours for finishing. Determine how many of each type of chair the company should make to be operating at full capacity.

Stage	Children's	Standard	Executive
Cutting	5 hr	4 hr	7 hr
Construction	3 hr	2 hr	5 hr
Finishing	2 hr	2 hr	4 hr

52. Bicycle Manufacturing The Jamis Bicycle Company produces three models of bicycles: Dakar, Komodo, and Aragon. Each bicycle is made in three stages: welding, painting, and assembling. The time needed for each stage of each bicycle is given in the following chart. During a specific week, the company has available a maximum of 133 hours for welding, 78 hours for painting, and 96 hours for assembling. Determine how many of each type of bicycle the company should make to be operating at full capacity.

Stage	Dakar	Komodo	Aragon
Welding	2	3	4
Painting	1	2	2.5
Assembling	1.5	2	3

53. Current Flow In electronics it is necessary to analyze current flow through paths of a circuit. In three paths (A, B, and C) of a circuit, the relationships are the following:

$$I_A + I_B + I_C = 0$$
$$-8I_B + 10I_C = 0$$
$$4I_A - 8I_B = 6$$

where I_A, I_B, and I_C represent the current in paths A, B, and C, respectively. Determine the current in each path of the circuit.

54. Forces on a Beam In physics we often study the forces acting on an object. For three forces, F_1, F_2, and F_3, acting on a beam, the following equations were obtained.

$$3F_1 + F_2 - F_3 = 2$$
$$F_1 - 2F_2 + F_3 = 0$$
$$4F_1 - F_2 + F_3 = 3$$

Find the three forces.

Group Activity

Discuss and answer Exercise 55 as a group.

55. **Two Cars** A *nonlinear system of equations* is a system of equations containing at least one equation that is not linear. (Nonlinear systems of equations will be discussed in Chapter 10.) The graph shows a nonlinear system of equations. The curves represent speed versus time for two cars.

   a) Are the two curves functions? Explain.

   b) Discuss the meaning of this graph.

   c) At time $t = 0.5$ hr, which car is traveling at a greater speed? Explain your answer.

   d) Assume the two cars start at the same position and are traveling in the same direction. Which car, A or B, traveled farther in 1 hour? Explain your answer.

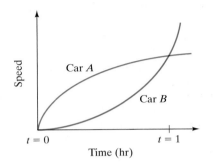

Cumulative Review Exercises

[1.4] 56. Evaluate $\frac{1}{2}x + \frac{2}{5}xy + \frac{1}{8}y$ when $x = -2$, $y = 5$.

57. Solve $4 - 2[(x - 5) + 2x] = -(x + 6)$.

58. Explain how to determine whether a graph represents a function.

59. Write an equation of the line that passes through points $(6, -4)$ and $(2, -8)$.

4.4 SOLVING SYSTEMS OF EQUATIONS USING MATRICES

1 Write an augmented matrix.

2 Solve systems of linear equations.

3 Solve systems of linear equations in three variables.

4 Recognize inconsistent and dependent systems.

SSM Study Guide CD/Video

MathPro 4/5 PH Math Tutor Center prenhall.com/Angel

1 Write an Augmented Matrix

A **matrix** is a rectangular array of numbers within brackets. The plural of *matrix* is **matrices**. Examples of matrices are

$$\begin{bmatrix} 4 & 6 \\ 9 & -2 \end{bmatrix} \qquad \begin{bmatrix} 5 & 7 & 2 \\ -1 & 3 & 4 \end{bmatrix}$$

The numbers inside the brackets are referred to as **elements** of the matrix.

The matrix on the left contains 2 rows and 2 columns and is called a 2 by 2 (2×2) matrix. The matrix on the right contains 2 rows and 3 columns and is a 2 by 3 (2×3) matrix. The number of rows is the first dimension given, and the number of columns is the second dimension given when describing the dimensions of a matrix. A **square matrix** has the same number of rows as columns. Thus, the matrix on the left is a square matrix.

In this section, we will use matrices to solve systems of linear equations. The first step in solving a system of two linear equations using matrices is to write each equation in the form $ax + by = c$. The next step is to write the **augmented matrix**, which is made up of two smaller matrices separated by a vertical line. The numbers on the left of the vertical line are the coefficients of the variables in the system of equations, and the numbers on the right are the constants. For the system of equations

$$a_1x + b_1y = c_1$$
$$a_2x + b_2y = c_2$$

the augmented matrix is written

$$\left[\begin{array}{cc|c} a_1 & b_1 & c_1 \\ a_2 & b_2 & c_2 \end{array}\right]$$

Following is a system of equations and its augmented matrix.

System of Equations

$$-x + \frac{1}{2}y = 4$$
$$-3x - 5y = -\frac{1}{2}$$

Augmented Matrix

$$\left[\begin{array}{cc|c} -1 & \frac{1}{2} & 4 \\ -3 & -5 & -\frac{1}{2} \end{array}\right]$$

Notice that the bar in the augmented matrix separates the numerical coefficients from the constants. Since the matrix is just a shortened way of writing the system of equations, we can solve a linear system using matrices in a manner very similar to solving a system of equations using the addition method.

2   Solve Systems of Linear Equations

To solve a system of two linear equations using matrices, we rewrite the augmented matrix in **triangular form**,

where the a, p, and q are constants. From this type of augmented matrix we can write an equivalent system of equations. This matrix represents the linear system

$$\begin{array}{cc} 1x + ay = p & \\ 0x + 1y = q & \end{array} \quad \text{or} \quad \begin{array}{c} x + ay = p \\ y = q \end{array}$$

For example,

$$\left[\begin{array}{cc|c} 1 & 3 & 4 \\ 0 & 1 & 2 \end{array}\right] \quad \text{represents} \quad \begin{array}{c} x + 3y = 4 \\ y = 2 \end{array}$$

Note that the system above on the right can be easily solved by substitution. Its solution is $(-2, 2)$.

We use **row transformations** to rewrite the augmented matrix in triangular form. We will use three row transformation procedures.

Procedures for Row Transformations

1. All the numbers in a row may be multiplied (or divided) by any nonzero real number. (This is the same as multiplying both sides of an equation by a nonzero real number.)
2. All the numbers in a row may be multiplied by any nonzero real number. These products may then be added to the corresponding numbers in any other row. (This is equivalent to eliminating a variable from a system of equations using the addition method.)
3. The order of the rows may be switched. (This is equivalent to switching the order of the equations in a system of equations.)

Generally, when changing an element in the augmented matrix to 1 we use row transformation procedure 1, and when changing an element to 0 we use row transformation procedure 2. *Work by columns starting from the left.* Start with the first column, first row.

EXAMPLE 1 Solve the following system of equations using matrices.

$$2x - 3y = 10$$
$$2x + 2y = 5$$

Solution First we write the augmented matrix.

$$\begin{bmatrix} 2 & -3 & | & 10 \\ 2 & 2 & | & 5 \end{bmatrix}$$

Our goal is to obtain a matrix of the form $\begin{bmatrix} 1 & a & | & p \\ 0 & 1 & | & q \end{bmatrix}$. We begin by using row transformation procedure 1 to change the 2 in the first column, first row, to 1. To do so, we multiply the first row of numbers by $\frac{1}{2}$. (We abbreviate this multiplication as $\frac{1}{2}R_1$ and place it to the right of the matrix in the same row where the operation *was* performed. This may help you follow the process more clearly.)

$$\begin{bmatrix} 2\left(\frac{1}{2}\right) & -3\left(\frac{1}{2}\right) & | & 10\left(\frac{1}{2}\right) \\ 2 & 2 & | & 5 \end{bmatrix} \quad \frac{1}{2}R_1$$

This gives

$$\begin{bmatrix} 1 & -\frac{3}{2} & | & 5 \\ 2 & 2 & | & 5 \end{bmatrix}$$

The next step is to obtain 0 in the first column, second row. At present, 2 is in this position. We do this by multiplying the numbers in row 1 by -2, and adding the products to the numbers in row 2. (This is abbreviated $-2R_1 + R_2$.)

The numbers in the first row multiplied by -2 are

$$1(-2) \quad -\frac{3}{2}(-2) \quad 5(-2)$$

Now we add these products to their respective numbers in the second row. This gives

$$\begin{bmatrix} 1 & -\frac{3}{2} & | & 5 \\ 2 + 1(-2) & 2 + \left(-\frac{3}{2}\right)(-2) & | & 5 + 5(-2) \end{bmatrix} \quad -2R_1 + R_2$$

Now we have

$$\left[\begin{array}{cc|c} 1 & -\frac{3}{2} & 5 \\ 0 & 5 & -5 \end{array}\right]$$

To obtain 1 in the second column, second row, we multiply the second row of numbers by $\frac{1}{5}$.

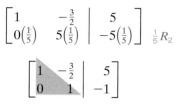

The matrix is now in the form we are seeking. The equivalent triangular system of equations is

$$x - \frac{3}{2}y = 5$$
$$y = -1$$

Now we can solve for x using substitution.

$$x - \frac{3}{2}y = 5$$
$$x - \frac{3}{2}(-1) = 5$$
$$x + \frac{3}{2} = 5$$
$$x = \frac{7}{2}$$

NOW TRY EXERCISE 19   A check will show that the solution to the system is $\left(\frac{7}{2}, -1\right)$.   ✳

3 Solve Systems of Linear Equations in Three Variables

Now we will use matrices to solve a system of three linear equations in three variables. We use the same row transformation procedures used when solving a system of two linear equations. Our goal is to obtain an augmented matrix in the triangular form

$$\left[\begin{array}{ccc|c} 1 & a & b & p \\ 0 & 1 & c & q \\ 0 & 0 & 1 & r \end{array}\right]$$

where a, b, c, and p, q and r are constants. This matrix represents the following system of equations.

$$
\begin{array}{lll}
1x + ay + bz = p & & x + ay + bz = p \\
0x + 1y + cz = q & \text{or} & y + cz = q \\
0x + 0y + 1z = r & & z = r
\end{array}
$$

When constructing the augmented matrix, *work by columns, from the left-hand column to the right-hand column. Always complete one column before moving to the next column. In each column, first obtain the 1 in the indicated position, and then obtain the zeros. Example 2 illustrates this procedure.*

When using matrices, be careful to keep all the numbers lined up neatly in rows and columns. One slight mistake in copying numbers from one matrix to another will lead to an incorrect, and often frustrating, attempt at solving a system of equations.

For example, the system of equations, $\begin{aligned} x - 3y + z &= 3 \\ 4x + 2y - 5z &= 20 \\ -5x - y - 4z &= 13 \end{aligned}$, when correctly

represented with the augmented matrix, $\begin{bmatrix} 1 & -3 & 1 & | & 3 \\ 4 & 2 & -5 & | & 20 \\ -5 & -1 & -4 & | & 13 \end{bmatrix}$, leads to the solution

$(1, -2, -4)$. However, a matrix that looks quite similar, $\begin{bmatrix} 1 & -3 & 1 & | & 3 \\ 4 & -1 & -5 & | & 20 \\ -5 & 2 & -4 & | & 13 \end{bmatrix}$, leads

to the incorrect ordered triple of $\left(-\dfrac{25}{53}, -\dfrac{130}{53}, -\dfrac{206}{53} \right)$.

EXAMPLE 2 Use matrices to solve the following system of equations.

$$\begin{aligned} x - 2y + 3z &= -7 \\ 2x - y - z &= 7 \\ -x + 3y + 2z &= -8 \end{aligned}$$

Solution First write the augmented matrix.

$$\begin{bmatrix} 1 & -2 & 3 & | & -7 \\ 2 & -1 & -1 & | & 7 \\ -1 & 3 & 2 & | & -8 \end{bmatrix}$$

Our next step is to use row transformations to change the first column to $\begin{smallmatrix} 1 \\ 0 \\ 0 \end{smallmatrix}$. Since the number in the first column, first row, is already a 1, we will work with the 2 in the first column, second row. Multiplying the numbers in the first row by -2 and adding those products to the respective numbers in the second row will result in the 2 changing to 0. The matrix is now

$$\begin{bmatrix} 1 & -2 & 3 & | & -7 \\ 0 & 3 & -7 & | & 21 \\ -1 & 3 & 2 & | & -8 \end{bmatrix} \quad -2R_1 + R_2$$

Continuing down the first column, we now change the -1 in the third row to 0. By multiplying the numbers in the first row by 1, and then adding the products to the third row, we get

$$\begin{bmatrix} 1 & -2 & 3 & | & -7 \\ 0 & 3 & -7 & | & 21 \\ 0 & 1 & 5 & | & -15 \end{bmatrix} \quad 1R_1 + R_3$$

Now we work with the second column. We wish to change the numbers in the second column to the form $\begin{smallmatrix} a \\ 1 \\ 0 \end{smallmatrix}$ where a represents a number. Since there is presently a 1 in the third row, second column, and we want a 1 in the second row, second column, we switch the second and third rows of the matrix. This gives

$$\begin{bmatrix} 1 & -2 & 3 & | & -7 \\ 0 & 1 & 5 & | & -15 \\ 0 & 3 & -7 & | & 21 \end{bmatrix} \quad \textit{Switch } R_2 \textit{ and } R_3.$$

Continuing down the second column, we now change the 3 in the third row to 0 by multiplying the numbers in the second row by -3 and adding those products to the third row. This gives

$$\begin{bmatrix} 1 & -2 & 3 & | & -7 \\ 0 & 1 & 5 & | & -15 \\ 0 & 0 & -22 & | & 66 \end{bmatrix} \quad -3R_2 + R_3$$

Now we work with the third column. We wish to change the numbers in the third column to the form $\begin{smallmatrix} b \\ c \\ 1 \end{smallmatrix}$ where b and c represent numbers. We must change the -22 in the third row to 1. We can do this by multiplying the numbers in the third row by $-\frac{1}{22}$. This results in the following.

$$\begin{bmatrix} 1 & -2 & 3 & | & -7 \\ 0 & 1 & 5 & | & -15 \\ 0 & 0 & 1 & | & -3 \end{bmatrix} \quad -\frac{1}{22} R_3$$

This matrix is now in the desired form. From this matrix we obtain the system of equations

$$x - 2y + 3z = -7$$
$$y + 5z = -15$$
$$z = -3$$

The third equation gives us the value of z in the solution. Now we can solve for y by substituting -3 for z in the second equation.

$$y + 5z = -15$$
$$y + 5(-3) = -15$$
$$y - 15 = -15$$
$$y = 0$$

Now we solve for x by substituting 0 for y and -3 for z in the first equation.

$$x - 2y + 3z = -7$$
$$x - 2(0) + 3(-3) = -7$$
$$x - 0 - 9 = -7$$
$$x - 9 = -7$$
$$x = 2$$

The solution is $(2, 0, -3)$. Check this now by substituting the appropriate values

NOW TRY EXERCISE 33 into each of the original equations.

4 Recognize Inconsistent and Dependent Systems

When solving a system of two equations, if you obtain an augmented matrix in which one row of numbers on the left side of the vertical line is all zeros but a zero does not appear in the same row on the right side of the vertical line, the system is inconsistent and has no solution. For example, a system of equations that yields the following augmented matrix is an inconsistent system.

$$\begin{bmatrix} 1 & 2 & | & 5 \\ 0 & 0 & | & 4 \end{bmatrix}$$ ⟵ Inconsistent system

The second row of the matrix represents the equation

$$0x + 0y = 4$$

which is never true.

If you obtain a matrix in which a 0 appears across an entire row, the system of equations is dependent. For example, a system of equations that yields the following augmented matrix is a dependent system.

$$\begin{bmatrix} 1 & -3 & | & -2 \\ 0 & 0 & | & 0 \end{bmatrix}$$ ⟵ Dependent system

The second row of the matrix represents the equation

$$0x + 0y = 0$$

which is always true.

Similar rules hold for systems with three equations.

$$\begin{bmatrix} 1 & 2 & 4 & | & 5 \\ 0 & 0 & 0 & | & -1 \\ 0 & 1 & -2 & | & 3 \end{bmatrix}$$ ⟵ Inconsistent system

$$\begin{bmatrix} 1 & 3 & -1 & | & 2 \\ 0 & 0 & 0 & | & 0 \\ 0 & 4 & 1 & | & -3 \end{bmatrix}$$ ⟵ Dependent system

NOW TRY EXERCISE 27

Using Your Graphing Calculator

Many graphing calculators have the ability to work with matrices. Such calculators have the ability to perform row operations on matrices. These graphing calculators can therefore be used to solve systems of equations using matrices.

Read the instruction manual that came with your graphing calculator to see if it can handle matrices. If so, learn how to use your graphing calculator to solve systems of equations using matrices.

Exercise Set 4.4

Concept/Writing Exercises

1. What is a square matrix?
2. Explain how to construct an augmented matrix.
3. If you obtain the following augmented matrix when solving a system of equations, what would be your next step in completing the process? Explain.

$$\begin{bmatrix} 1 & 3 & | & 6 \\ 0 & -1 & | & 4 \end{bmatrix}$$

4. If you obtained the following augmented matrix when solving a system of equations, what would be your

next step in completing the process? Explain your answer.

$$\begin{bmatrix} 1 & 3 & 7 & | & -1 \\ 0 & -1 & 5 & | & 3 \\ 2 & 4 & 6 & | & 8 \end{bmatrix}$$

5. If you obtained the following augmented matrix when solving a system of linear equations, what would be your next step in completing the process? Explain your answer.

$$\begin{bmatrix} 1 & 4 & -7 & | & 7 \\ 0 & 5 & 2 & | & -1 \\ 0 & 1 & 4 & | & 2 \end{bmatrix}$$

6. If you obtained the following augmented matrix when solving a system of linear equations, what would be your next step in completing the process? Explain your answer.

$$\begin{bmatrix} 1 & 3 & -2 & | & 1 \\ 0 & 1 & 2 & | & -3 \\ 0 & 0 & 4 & | & -12 \end{bmatrix}$$

7. When solving a system of linear equations by matrices, if two rows are identical, will the system be consistent, dependent, or inconsistent?

8. When solving a system of equations using matrices, how will you know if the system is
a) dependent,
b) inconsistent?

Practice the Skills

Perform each row transformation indicated and write the new matrix.

9. $\begin{bmatrix} 5 & -10 & | & -15 \\ 3 & 1 & | & -4 \end{bmatrix}$ Multiply numbers in the first row by $\frac{1}{5}$.

10. $\begin{bmatrix} 1 & 8 & | & 3 \\ 0 & 4 & | & -3 \end{bmatrix}$ Multiply numbers in the second row by $\frac{1}{4}$.

11. $\begin{bmatrix} 4 & 7 & 2 & | & -1 \\ 3 & 2 & 1 & | & -5 \\ 1 & 1 & 3 & | & -8 \end{bmatrix}$ Switch row 1 and row 3.

12. $\begin{bmatrix} 1 & 5 & 7 & | & 2 \\ 0 & 6 & -1 & | & 5 \\ 0 & 1 & 3 & | & -4 \end{bmatrix}$ Switch row 2 and row 3.

13. $\begin{bmatrix} 1 & 3 & | & 12 \\ -3 & 8 & | & -6 \end{bmatrix}$ Multiply numbers in the first row by 3 and add the products to the second row.

14. $\begin{bmatrix} 1 & 5 & | & 6 \\ \frac{1}{2} & 10 & | & -4 \end{bmatrix}$ Multiply numbers in the first row by $-\frac{1}{2}$ and add the products to the second row.

15. $\begin{bmatrix} 1 & 0 & 8 & | & \frac{1}{4} \\ 5 & 2 & 2 & | & -2 \\ 6 & -3 & 1 & | & 0 \end{bmatrix}$ Multiply numbers in the first row by -5 and add the products to the second row.

16. $\begin{bmatrix} 1 & 2 & -1 & | & 6 \\ 0 & 1 & 5 & | & 0 \\ 0 & 0 & 2 & | & 4 \end{bmatrix}$ Multiply numbers in the third row by $\frac{1}{2}$.

Solve each system using matrices.

17. $x + 3y = 3$
$-x + y = -3$

18. $x + 2y = 5$
$3x - y = 1$

19. $x + 3y = 4$
$-4x - y = 6$

20. $3x - 6y = 15$
$2x - y = 4$

21. $5a - 10b = -10$
$2a + b = 1$

22. $3s - 2t = 1$
$-2s + 4t = -6$

23. $2x - 5y = -6$
$-4x + 10y = 12$

24. $-2m - 4n = 7$
$3m + 6n = -11$

25. $12x + 10y = -14$
$4x - 3y = -11$

26. $4r + 2s = -10$
$-2r + s = -7$

27. $-3x + 6y = 5$
$2x - 4y = 8$

28. $6x - 3y = 9$
$-2x + y = -3$

29. $9x - 8y = 4$
$-3x + 4y = -1$

30. $2x - 3y = 3$
$-3x + 9y = -3$

31. $10m = 8n + 15$
$16n = -15m - 2$

32. $8x = 9y + 4$
$24x + 6y = 1$

Solve each system using matrices

33. $x - 3y + 2z = 5$
$2x + 5y - 4z = -3$
$-3x + y - 2z = -11$

34. $a - 3b + 4c = 13$
$4a + b + c = -4$
$-2a - 3b - 5c = -2$

35. $x + 2y = 5$
$y - z = -1$
$2x - 3z = 0$

36. $4a + 3c = -12$
$a + 2b = -1$
$7b - 4c = 7$

37. $x - 2y + 4z = 5$
$-3x + 4y - 2z = -8$
$4x + 5y - 4z = -3$

38. $3x - 5y + 2z = 8$
$-x - y - z = -3$
$3x - 2y + 4z = 10$

39. $2x - 5y + z = 1$
$3x - 5y + z = 2$
$-4x + 10y - 2z = -2$

40. $x + 2y + 3z = 1$
$4x + 5y + 6z = -3$
$7x + 8y + 9z = 0$

41. $4p - q + r = 4$
$-6p + 3q - 2r = -5$
$2p + 5q - r = 7$

42. $-4r + 3s - 6t = 14$
$4r + 2s - 2t = -3$
$2r - 5s - 8t = -23$

43. $2x - 4y + 3z = -12$
$3x - y + 2z = -3$
$-4x + 8y - 6z = 10$

44. $3x - 2y + z = -1$
$12x - 10y - 3z = 2$
$-9x + 8y - 4z = 5$

45. $5x - 3y + 4z = 22$
$-x - 15y + 10z = -15$
$-3x + 9y - 12z = -6$

46. $9x - 4y + 5z = -2$
$-9x + 5y - 10z = -1$
$9x + 3y + 10z = 1$

Problem Solving

47. When solving a system of linear equations using matrices, if two rows of matrices are switched, will the solution to the system change? Explain.

48. You can tell whether a system of two equations in two variables is consistent, dependent, or inconsistent by comparing the slopes and y-intercepts of the graphs of the equations. Can you tell, without solving, if a system of three equations in three variables is consistent, dependent, or inconsistent? Explain.

Solve using matrices.

49. *Angles of a Roof* In a triangular cross section of a roof, the largest angle is 55° greater than the smallest angle. The largest angle is 20° greater than the remaining angle. Find the measure of each angle.

51. *Bananas* Sixty-five percent of the world's bananas are controlled by Chiquita, Dole, or Del Monte (all American companies). Chiquita, the largest, controls 12% more bananas than Del Monte. Dole, the second largest, controls 3% less than twice the percent that Del Monte controls. Determine the percents to be placed in each sector of the circle graph shown.

World's Bananas

50. *Right Angle* A right angle is divided into three smaller angles. The largest of the three angles is twice the smallest. The remaining angle is 10° greater than the smallest angle. Find the measure of each angle.

52. ***Web Browsers*** The following graph shows that in 2002, Microsoft and Netscape controlled major shares of the web browser market. Microsoft controlled about 49% more of the market than Netscape, and Netscape controlled about eight times the amount of all others. Determine the percent to be placed in each sector of the circle shown.

Web Browser Market Share

Source: www.webreview.com

Cumulative Review Exercises

[1.2] **53.** $A = \{1, 2, 4, 6, 9\}$; $B = \{3, 4, 5, 6, 10\}$. Find
  a) $A \cup B$;
  b) $A \cap B$.

[2.5] **54.** Indicate the inequality $-2 < x \le 4$
  a) on a number line,

  b) as a solution set, and
  c) in interval notation.

[3.2] **55.** What does a graph represent?

[3.4] **56.** If $f(x) = -2x^2 + 4x - 6$, find $f(-5)$.

4.5 SOLVING SYSTEMS OF EQUATIONS USING DETERMINANTS AND CRAMER'S RULE

SSM   Study Guide   CD/Video

MathPro 4/5   PH Math Tutor Center   prenhall.com/Angel

1   Evaluate a determinant of a 2 × 2 matrix.
2   Use Cramer's rule.
3   Evaluate a determinant of a 3 × 3 matrix.
4   Use Cramer's rule with systems in three variables.

1   Evaluate a Determinant of a 2 × 2 Matrix

We have discussed various ways of solving a system of linear equations, including: graphing, substitution, the addition (or elimination) method, and matrices. A system of linear equations may also be solved using determinants.

Associated with every square matrix is a number called its **determinant**. For a 2 × 2 matrix, its determinant is defined as follows.

DEFINITION

The **determinant** of a 2 × 2 matrix $\begin{bmatrix} a_1 & b_1 \\ a_2 & b_2 \end{bmatrix}$ is denoted $\begin{vmatrix} a_1 & b_1 \\ a_2 & b_2 \end{vmatrix}$ and is evaluated as

$$\begin{vmatrix} a_1 & b_1 \\ a_2 & b_2 \end{vmatrix} = a_1 b_2 - a_2 b_1$$

EXAMPLE 1 Evaluate each determinant.

  a) $\begin{vmatrix} 2 & -1 \\ 3 & 5 \end{vmatrix}$ **b)** $\begin{vmatrix} -2 & 3 \\ -1 & 4 \end{vmatrix}$

Solution **a)** $a_1 = 2, a_2 = 3, b_1 = -1, b_2 = 5$

$$\begin{vmatrix} 2 & -1 \\ 3 & 5 \end{vmatrix} = 2(5) - (3)(-1) = 10 + 3 = 13$$

NOW TRY EXERCISE 7 **b)** $\begin{vmatrix} -2 & 3 \\ -1 & 4 \end{vmatrix} = (-2)(4) - (-1)(3) = -8 + 3 = -5$

2 Use Cramer's Rule

If we begin with the equations

$$a_1 x + b_1 y = c_1$$
$$a_2 x + b_2 y = c_2$$

we can use the addition method to show that

$$x = \frac{c_1 b_2 - c_2 b_1}{a_1 b_2 - a_2 b_1} \quad \text{and} \quad y = \frac{a_1 c_2 - a_2 c_1}{a_1 b_2 - a_2 b_1}$$

(see Challenge Problem 61). Notice that the *denominators* of x and y are both $a_1 b_2 - a_2 b_1$. Following is the determinant that yields this denominator. We have labeled this denominator D.

$$D = \begin{vmatrix} a_1 & b_1 \\ a_2 & b_2 \end{vmatrix} = a_1 b_2 - a_2 b_1$$

The *numerators* of x and y are different. Following are two determinants, labeled D_x and D_y, that yield the numerators of x and y.

$$D_x = \begin{vmatrix} c_1 & b_1 \\ c_2 & b_2 \end{vmatrix} = c_1 b_2 - c_2 b_1 \qquad D_y = \begin{vmatrix} a_1 & c_1 \\ a_2 & c_2 \end{vmatrix} = a_1 c_2 - a_2 c_1$$

We use determinants D, D_x, and D_y in Cramer's rule. **Cramer's rule** can be used to solve systems of linear equations.

Cramer's Rule for Systems of Linear Equations

For a system of linear equations of the form

$$a_1 x + b_1 y = c_1$$
$$a_2 x + b_2 y = c_2$$

$$x = \frac{\begin{vmatrix} c_1 & b_1 \\ c_2 & b_2 \end{vmatrix}}{\begin{vmatrix} a_1 & b_1 \\ a_2 & b_2 \end{vmatrix}} = \frac{D_x}{D} \quad \text{and} \quad y = \frac{\begin{vmatrix} a_1 & c_1 \\ a_2 & c_2 \end{vmatrix}}{\begin{vmatrix} a_1 & b_1 \\ a_2 & b_2 \end{vmatrix}} = \frac{D_y}{D}, \quad D \neq 0$$

HELPFUL HINT

The elements in determinant D are the numerical coefficients of the x and y terms in the two given equations, listed in the same order they are listed in the equations. To obtain the determinant D_x from determinant D, replace the coefficients of the x-terms (the values in the first column) with the constants of the two given equations. To obtain the determinant D_y from determinant D, replace the coefficients of the y-terms (the values in the second column) with the constants of the two given equations.

EXAMPLE 2   Use Cramer's rule to solve the following system.

$$3x + 5y = 7$$
$$4x - y = -6$$

Solution   Both equations are given in the desired form, $ax + by = c$. When labeling a, b, and c, we will refer to $3x + 5y = 7$ as equation 1 and $4x - y = -6$ as equation 2 (in the subscripts).

$$
\begin{matrix}
a_1 & b_1 & c_1 \\
\downarrow & \downarrow & \downarrow \\
3x + & 5y = & 7 \\
4x - & 1y = & -6 \\
\uparrow & \uparrow & \uparrow \\
a_2 & b_2 & c_2
\end{matrix}
$$

We now determine D, D_x, D_y.

$$D = \begin{vmatrix} a_1 & b_1 \\ a_2 & b_2 \end{vmatrix} = \begin{vmatrix} 3 & 5 \\ 4 & -1 \end{vmatrix} = 3(-1) - 4(5) = -3 - 20 = -23$$

$$D_x = \begin{vmatrix} c_1 & b_1 \\ c_2 & b_2 \end{vmatrix} = \begin{vmatrix} 7 & 5 \\ -6 & -1 \end{vmatrix} = 7(-1) - (-6)(5) = -7 + 30 = 23$$

$$D_y = \begin{vmatrix} a_1 & c_1 \\ a_2 & c_2 \end{vmatrix} = \begin{vmatrix} 3 & 7 \\ 4 & -6 \end{vmatrix} = 3(-6) - 4(7) = -18 - 28 = -46$$

Now we find the values of x and y.

$$x = \frac{D_x}{D} = \frac{23}{-23} = -1$$

$$y = \frac{D_y}{D} = \frac{-46}{-23} = 2$$

Thus the solution is $x = -1$, $y = 2$ or the ordered pair $(-1, 2)$. A check will show that this ordered pair satisfies both equations.

NOW TRY EXERCISE 15

When the determinant $D = 0$, Cramer's rule cannot be used since division by 0 is undefined. You must then use a different method to solve the system. Or you may evaluate D_x and D_y to determine whether the system is dependent or inconsistent.

> If $D = 0$, $D_x = 0$, $D_y = 0$, then the system is dependent.
> If $D = 0$ and either $D_x \neq 0$ or $D_y \neq 0$, then the system is inconsistent.

3 Evaluate a Determinant of a 3 × 3 Matrix

For the determinant

$$\begin{vmatrix} a_1 & b_1 & c_1 \\ a_2 & b_2 & c_2 \\ a_3 & b_3 & c_3 \end{vmatrix}$$

the **minor determinant** of a_1 is found by crossing out the elements in the same row and column in which the element a_1 appears. The remaining elements form the minor determinant of a_1. The minor determinants of other elements are found similarly.

$$\begin{vmatrix} a_1 & b_1 & c_1 \\ a_2 & b_2 & c_2 \\ a_3 & b_3 & c_3 \end{vmatrix} \qquad \begin{vmatrix} b_2 & c_2 \\ b_3 & c_3 \end{vmatrix} \qquad \textit{Minor determinant of } a_1$$

$$\begin{vmatrix} a_1 & b_1 & c_1 \\ a_2 & b_2 & c_2 \\ a_3 & b_3 & c_3 \end{vmatrix} \qquad \begin{vmatrix} b_1 & c_1 \\ b_3 & c_3 \end{vmatrix} \qquad \textit{Minor determinant of } a_2$$

$$\begin{vmatrix} a_1 & b_1 & c_1 \\ a_2 & b_2 & c_2 \\ a_3 & b_3 & c_3 \end{vmatrix} \qquad \begin{vmatrix} b_1 & c_1 \\ b_2 & c_2 \end{vmatrix} \qquad \textit{Minor determinant of } a_3$$

To evaluate determinants of a 3×3 matrix, we use minor determinants. The following box shows how such a determinant may be evaluated by **expansion by the minors of the first column**.

Expansion of the Determinant by the Minors of the First Column

$$\begin{vmatrix} a_1 & b_1 & c_1 \\ a_2 & b_2 & c_2 \\ a_3 & b_3 & c_3 \end{vmatrix} = a_1 \overset{\substack{\textit{Minor} \\ \textit{determinant} \\ \textit{of } a_1 \\ \downarrow}}{\begin{vmatrix} b_2 & c_2 \\ b_3 & c_3 \end{vmatrix}} - a_2 \overset{\substack{\textit{Minor} \\ \textit{determinant} \\ \textit{of } a_2 \\ \downarrow}}{\begin{vmatrix} b_1 & c_1 \\ b_3 & c_3 \end{vmatrix}} + a_3 \overset{\substack{\textit{Minor} \\ \textit{determinant} \\ \textit{of } a_3 \\ \downarrow}}{\begin{vmatrix} b_1 & c_1 \\ b_2 & c_2 \end{vmatrix}}$$

EXAMPLE 3 Evaluate $\begin{vmatrix} 4 & -2 & 6 \\ 3 & 5 & 0 \\ 1 & -3 & -1 \end{vmatrix}$ using expansion by the minors of the first column.

Solution We will follow the procedure given in the box.

$$\begin{vmatrix} 4 & -2 & 6 \\ 3 & 5 & 0 \\ 1 & -3 & -1 \end{vmatrix} = 4 \begin{vmatrix} 5 & 0 \\ -3 & -1 \end{vmatrix} - 3 \begin{vmatrix} -2 & 6 \\ -3 & -1 \end{vmatrix} + 1 \begin{vmatrix} -2 & 6 \\ 5 & 0 \end{vmatrix}$$

$$= 4[5(-1) - (-3)0] - 3[(-2)(-1) - (-3)6] + 1[(-2)0 - 5(6)]$$

$$= 4(-5 + 0) - 3(2 + 18) + 1(0 - 30)$$

$$= 4(-5) - 3(20) + 1(-30)$$

$$= -20 - 60 - 30$$

$$= -110$$

NOW TRY EXERCISE 13   The determinant has a value of -110.

4   Use Cramer's Rule with Systems in Three Variables

Cramer's rule can be extended to systems of equations in three variables as follows.

Cramer's Rule for a System of Equations in Three Variables

To evaluate the system

$$a_1x + b_1y + c_1z = d_1$$
$$a_2x + b_2y + c_2z = d_2$$
$$a_3x + b_3y + c_3z = d_3$$

with

$$D = \begin{vmatrix} a_1 & b_1 & c_1 \\ a_2 & b_2 & c_2 \\ a_3 & b_3 & c_3 \end{vmatrix} \qquad D_x = \begin{vmatrix} d_1 & b_1 & c_1 \\ d_2 & b_2 & c_2 \\ d_3 & b_3 & c_3 \end{vmatrix}$$

$$D_y = \begin{vmatrix} a_1 & d_1 & c_1 \\ a_2 & d_2 & c_2 \\ a_3 & d_3 & c_3 \end{vmatrix} \qquad D_z = \begin{vmatrix} a_1 & b_1 & d_1 \\ a_2 & b_2 & d_2 \\ a_3 & b_3 & d_3 \end{vmatrix}$$

then

$$x = \frac{D_x}{D} \qquad y = \frac{D_y}{D} \qquad z = \frac{D_z}{D}, \quad D \neq 0$$

Note that the denominators of the expressions for x, y, and z are all the same determinant, D. Note that the d's replace the a's, the numerical coefficients of the x-terms, in D_x. The d's replace the b's, the numerical coefficients of the y-terms, in D_y. And the d's replace the c's, the numerical coefficients of the z-terms, in D_z.

EXAMPLE 4   Solve the following system of equations using determinants.

$$3x - 2y - z = -6$$
$$2x + 3y - 2z = 1$$
$$x - 4y + z = -3$$

Solution

$$\begin{array}{llll} a_1 = 3 & b_1 = -2 & c_1 = -1 & d_1 = -6 \\ a_2 = 2 & b_2 = 3 & c_2 = -2 & d_2 = 1 \\ a_3 = 1 & b_3 = -4 & c_3 = 1 & d_3 = -3 \end{array}$$

We will use expansion by the minor determinants of the first column to evaluate D, D_x, D_y, and D_z.

$$D = \begin{vmatrix} 3 & -2 & -1 \\ 2 & 3 & -2 \\ 1 & -4 & 1 \end{vmatrix} = 3\begin{vmatrix} 3 & -2 \\ -4 & 1 \end{vmatrix} - 2\begin{vmatrix} -2 & -1 \\ -4 & 1 \end{vmatrix} + 1\begin{vmatrix} -2 & -1 \\ 3 & -2 \end{vmatrix}$$

$$= 3(-5) - 2(-6) + 1(7)$$
$$= -15 + 12 + 7 = 4$$

$$D_x = \begin{vmatrix} -6 & -2 & -1 \\ 1 & 3 & -2 \\ -3 & -4 & 1 \end{vmatrix} = -6\begin{vmatrix} 3 & -2 \\ -4 & 1 \end{vmatrix} - 1\begin{vmatrix} -2 & -1 \\ -4 & 1 \end{vmatrix} + (-3)\begin{vmatrix} -2 & -1 \\ 3 & -2 \end{vmatrix}$$

$$= -6(-5) - 1(-6) - 3(7)$$

$$= 30 + 6 - 21 = 15$$

$$D_y = \begin{vmatrix} 3 & -6 & -1 \\ 2 & 1 & -2 \\ 1 & -3 & 1 \end{vmatrix} = 3\begin{vmatrix} 1 & -2 \\ -3 & 1 \end{vmatrix} - 2\begin{vmatrix} -6 & -1 \\ -3 & 1 \end{vmatrix} + 1\begin{vmatrix} -6 & -1 \\ 1 & -2 \end{vmatrix}$$

$$= 3(-5) - 2(-9) + 1(13)$$

$$= -15 + 18 + 13 = 16$$

$$D_z = \begin{vmatrix} 3 & -2 & -6 \\ 2 & 3 & 1 \\ 1 & -4 & -3 \end{vmatrix} = 3\begin{vmatrix} 3 & 1 \\ -4 & -3 \end{vmatrix} - 2\begin{vmatrix} -2 & -6 \\ -4 & -3 \end{vmatrix} + 1\begin{vmatrix} -2 & -6 \\ 3 & 1 \end{vmatrix}$$

$$= 3(-5) - 2(-18) + 1(16)$$

$$= -15 + 36 + 16 = 37$$

We found that $D = 4$, $D_x = 15$, $D_y = 16$, and $D_z = 37$. Therefore,

$$x = \frac{D_x}{D} = \frac{15}{4} \qquad y = \frac{D_y}{D} = \frac{16}{4} = 4 \qquad z = \frac{D_z}{D} = \frac{37}{4}$$

The solution to the system is $\left(\frac{15}{4}, 4, \frac{37}{4}\right)$. Note the ordered triple lists x, y, and z in

NOW TRY EXERCISE 33   this order.   ✳

When we have a system of equations in three variables in which one or more equations are missing a variable, we insert the variable with a coefficient of 0. Thus,

$$\begin{aligned} 2x - 3y + 2z &= -1 \\ x + 2y \qquad &= 14 \\ x \qquad - 3z &= -5 \end{aligned} \quad \text{is written} \quad \begin{aligned} 2x - 3y + 2z &= -1 \\ x + 2y + 0z &= 14 \\ x + 0y - 3z &= -5 \end{aligned}$$

HELPFUL HINT

When evaluating determinants, if any two rows (or columns) are identical, or identical except for opposite signs, the determinant has a value of 0. For example,

$$\begin{vmatrix} 5 & -2 \\ 5 & -2 \end{vmatrix} = 0 \quad \text{and} \quad \begin{vmatrix} 5 & -2 \\ -5 & 2 \end{vmatrix} = 0$$

$$\begin{vmatrix} 5 & -3 & 4 \\ 2 & 6 & 5 \\ 5 & -3 & 4 \end{vmatrix} = 0 \quad \text{and} \quad \begin{vmatrix} 5 & -3 & 4 \\ -5 & 3 & -4 \\ 6 & 8 & 2 \end{vmatrix} = 0$$

As with determinants of a 2×2 matrix, when the determinant $D = 0$, Cramer's rule cannot be used since division by 0 is undefined. You must then use a different method to solve the system. Or you may evaluate D_x, D_y, and D_z to determine whether the system is dependent or inconsistent.

If $D = 0$, $D_x = 0$, $D_y = 0$, and $D_z = 0$, then the system is dependent.
If $D = 0$ and $D_x \neq 0$, $D_y \neq 0$, or $D_z \neq 0$, then the system is inconsistent.

Using Your Graphing Calculator

In Section 4.4 we mentioned that some graphing calculators can handle matrices. Graphing calculators with matrix capabilities can also find determinants of square matrices. Read your graphing calculator manual to learn if your calculator can find determinants. If so, learn how to do so on your calculator.

Exercise Set 4.5

Concept/Writing Exercises

1. Explain how to evaluate a 2×2 determinant.

2. Explain how to evaluate a 3×3 determinant by expansion by the minors of the first column.

3. Explain how you can determine whether a system of three linear equations is inconsistent using determinants.

4. Explain how you can determine whether a system of three linear equations is dependent using determinants.

5. While solving a system of two linear equations using Cramer's rule, you determine that $D = 4$, $D_x = -8$, and $D_y = -2$. What is the solution to this system?

6. While solving a system of three linear equations using Cramer's rule, you determine that $D = -2$, $D_x = 6$, $D_y = 10$, and $D_z = -2$. What is the solution to this system?

Practice the Skills

Evaluate each determinant.

7. $\begin{vmatrix} 2 & 3 \\ 1 & 5 \end{vmatrix}$

8. $\begin{vmatrix} 3 & 5 \\ -1 & -1 \end{vmatrix}$

9. $\begin{vmatrix} \frac{1}{2} & 3 \\ 2 & -4 \end{vmatrix}$

10. $\begin{vmatrix} 5 & -\frac{2}{3} \\ -1 & 0 \end{vmatrix}$

11. $\begin{vmatrix} 3 & 2 & 0 \\ 0 & 5 & 3 \\ -1 & 4 & 2 \end{vmatrix}$

12. $\begin{vmatrix} 4 & 1 & 1 \\ 0 & 0 & 3 \\ 2 & 2 & 7 \end{vmatrix}$

13. $\begin{vmatrix} 2 & 3 & 1 \\ 1 & -3 & -6 \\ -4 & 5 & 9 \end{vmatrix}$

14. $\begin{vmatrix} 5 & -8 & 6 \\ 3 & 0 & 4 \\ -5 & -2 & 1 \end{vmatrix}$

Solve each system of equations using determinants.

15. $x + 3y = 1$
$-2x - 3y = 4$

16. $2x + 4y = -2$
$-5x - 2y = 13$

17. $x - 2y = -1$
$x + 3y = 9$

18. $2r + 3s = -2$
$3r + 5s = -2$

19. $5p - 7q = -21$
$-4p + 3q = 22$

20. $6x + 3y = -4$
$9x + 5y = -6$

21. $4x = -5y - 2$
$-2x = y + 4$

22. $4x = 3y + 5$
$8x - 2 = -2y$

23. $x + 5y = 3$
$2x + 10y = 6$

24. $9x + 6y = -3$
$6x + 4y = -2$

25. $3r = -4s - 6$
$3s = -5r + 1$

26. $x = y - 1$
$3y = 2x + 8$

27. $5x - 5y = 3$
$x - y = -2$

28. $2x - 5y = -3$
$-4x + 10y = 1$

29. $6.3x - 4.5y = -9.9$
$-9.1x + 3.2y = -2.2$

30. $-1.1x + 8.3y = 36.5$
$3.5x + 1.6y = -4.1$

Solve each system using determinants.

31. $x + y + z = 2$
$-3y + 4z = 11$
$-3x + 4y - 2z = -11$

32. $2x + 3y = 4$
$3x + 7y - 4z = -3$
$x - y + 2z = 9$

33. $3x - 5y - 4z = -4$
$4x + 2y = 1$
$6y - 4z = -11$

34. $2x + 8y + 3z = 1$
$6x - 9y = 5$
$-3y + z = 2$

35. $x + 4y - 3z = -6$
$2x - 8y + 5z = 12$
$3x + 4y - 2z = -3$

36. $2x + y - 2z = 4$
$2x + 2y - 4z = 1$
$6x + 8y - 4z = 1$

🔒 **37.** $a - b + 2c = 3$
$a - b + c = 1$
$2a + b + 2c = 2$

38. $2x + y - 2 = 0$
$3x + 2y + z = 3$
$x - 3y - 5z = 5$

39. $a + 2b + c = 1$
$a - b + c = 1$
$2a + b + 2c = 2$

40. $4x - 2y + 6z = 2$
$-6x + 3y - 9z = -3$
$2x - 7y + 11z = -5$

41. $1.1x + 2.3y - 4.0z = -9.2$
$-2.3x + 4.6z = 6.9$
$-8.2y - 7.5z = -6.8$

42. $4.6y - 2.1z = 24.3$
$-5.6x + 1.8y = -5.8$
$2.8x - 4.7y - 3.1z = 7.0$

43. $-6x + 3y - 9z = -8$
$5x + 2y - 3z = 1$
$2x - y + 3z = -5$

44. $x - 2y + z = 2$
$4x - 6y + 2z = 1$
$2x - 3y + z = 0$

45. $2x + \frac{1}{2}y - 3z = 5$
$-3x + 2y + 2z = 1$
$4x - \frac{1}{4}y - 7z = 4$

46. $\frac{1}{4}x - \frac{1}{2}y + 3z = -3$
$2x - 3y + 2z = -1$
$\frac{1}{6}x + \frac{1}{3}y - \frac{1}{3}z = 1$

47. $0.2x - 0.1y - 0.3z = -0.1$
$0.2x - 0.1y + 0.1z = -0.9$
$0.1x + 0.2y - 0.4z = 1.7$

48. $0.6u - 0.4v + 0.5w = 3.1$
$0.5u + 0.2v + 0.2w = 1.3$
$0.1u + 0.1v + 0.1w = 0.2$

Problem Solving

✎ **49.** Given a determinant of the form $\begin{vmatrix} a_1 & b_1 \\ a_2 & b_2 \end{vmatrix}$, how will the value of the determinant change if the a's are switched with each other and the b's are switched with each other, $\begin{vmatrix} a_2 & b_2 \\ a_1 & b_1 \end{vmatrix}$? Explain your answer.

✎ **50.** Given a determinant of the form $\begin{vmatrix} a_1 & b_1 \\ a_2 & b_2 \end{vmatrix}$, how will the value of the determinant change if the a's are switched with the b's, $\begin{vmatrix} b_1 & a_1 \\ b_2 & a_2 \end{vmatrix}$? Explain your answer.

51. In a 2×2 determinant, if the rows are the same, what is the value of the determinant?

52. If all the elements in one row or one column of a 2×2 determinant are 0, what is the value of the determinant?

53. If all the elements in one row or one column of a 3×3 determinant are 0, what is the value of the determinant?

✎ **54.** Given a 3×3 determinant, if all the elements in one row are multiplied by -1, will the value of the determinant change? Explain.

✎ **55.** Given a 3×3 determinant, if the first and second rows are switched, will the value of the determinant change? Explain.

56. In a 3×3 determinant, if any two rows are the same, can you make a generalization about the value of the determinant?

Solve for the given letter.

🔒 **57.** $\begin{vmatrix} 4 & 6 \\ -2 & y \end{vmatrix} = 32$

58. $\begin{vmatrix} b - 2 & -4 \\ b + 3 & -6 \end{vmatrix} = 14$

59. $\begin{vmatrix} 4 & 7 & y \\ 3 & -1 & 2 \\ 4 & 1 & 5 \end{vmatrix} = -35$

60. $\begin{vmatrix} 3 & x & -2 \\ 0 & 5 & -6 \\ -1 & 4 & -7 \end{vmatrix} = -31$

Challenge Problems

61. Use the addition method to solve the following system for **a)** x and **b)** y

$$a_1 x + b_1 y = c_1$$
$$a_2 x + b_2 y = c_2$$

Cumulative Review Exercises

[2.5] 62. Solve the inequality $3(x - 2) < \frac{4}{5}(x - 4)$ and indicate the solution in interval notation.

Graph $3x + 4y = 8$ *using the indicated method.*

[3.2] 63. By plotting points

64. Using the x- and y-intercepts

[3.3] 65. Using the slope and y-intercept

4.6 SOLVING SYSTEMS OF LINEAR INEQUALITIES

1 Solve systems of linear inequalities.
2 Solve linear programming problems.
3 Solve systems of linear inequalities containing absolute value.

1 Solve Systems of Linear Inequalities

In Section 3.7 we showed how to graph linear inequalities in two variables. In Section 4.1 we learned how to solve systems of equations graphically. In this section we show how to solve **systems of linear inequalities** graphically.

To Solve a System of Linear Inequalities

Graph each inequality on the same axes. The solution is the set of points whose coordinates satisfy all the inequalities in the system.

EXAMPLE 1   Determine the solution to the following system of inequalities.

$$y < -\frac{1}{2}x + 2$$

$$x - y \le 4$$

Solution   First graph the inequality $y < -\frac{1}{2}x + 2$ (Fig. 4.8). Now on the same axes graph the inequality $x - y \le 4$ (Fig. 4.9). The solution is the set of points common to the graphs of both inequalities. It is the part of the graph that contains both shadings. The dashed line is not part of the solution, but the part of the solid line that satisfies both inequalities is part of the solution.

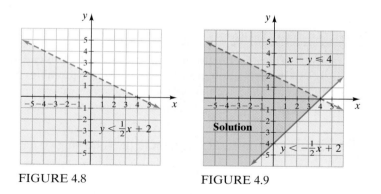

FIGURE 4.8   FIGURE 4.9

EXAMPLE 2   Determine the solution to the following system of inequalities.

$$3x - y < 6$$

$$2x + 2y \ge 5$$

Solution   Graph $3x - y < 6$ (see Fig. 4.10). Graph $2x + 2y \ge 5$ on the same axes (Fig. 4.11). The solution is the part of the graph that contains both shadings and the part of the solid line that satisfies both inequalities.

NOW TRY EXERCISE 7   FIGURE 4.10   FIGURE 4.11

EXAMPLE 3   Determine the solution to the following system of inequalities.

$$y > -1$$
$$x \le 4$$

Solution   The solution is illustrated in Figure 4.12.

NOW TRY EXERCISE 15   FIGURE 4.12

2 Solve Linear Programming Problems

There is a mathematical process called **linear programming** for which you often have to graph more than two linear inequalities on the same axes. These inequalities are called **constraints**. The following two examples illustrate how to determine the solution to a system of more than two inequalities.

EXAMPLE 4   Determine the solution to the following system of inequalities.

$$x \ge 0$$
$$y \ge 0$$
$$2x + 3y \le 12$$
$$2x + \ y \le 8$$

Solution   The first two inequalities, $x \ge 0$ and $y \ge 0$, indicate that the solution must be in the first quadrant because that is the only quadrant where both x and y are positive. Figure 4.13 illustrates the graphs of the four inequalities.

FIGURE 4.13

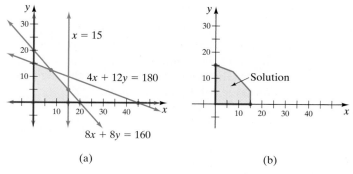

FIGURE 4.14

Figure 4.14 illustrates the graphs on the same axes and the solution to the system of inequalities. Note that every point in the shaded area and every point on the lines that form the polygonal region is part of the answer.

EXAMPLE 5   Determine the solution to the following system of inequalities.

$$x \geq 0$$
$$y \geq 0$$
$$x \leq 15$$
$$8x + 8y \leq 160$$
$$4x + 12y \leq 180$$

Solution   The first two inequalities indicate that the solution must be in the first quadrant. The third inequality indicates that x must be a value less than or equal to 15. Figure 4.15a indicates the graphs of corresponding equations and shows the region that satisfies all the inequalities in the system. Figure 4.15b indicates the solution to the system of inequalities.

NOW TRY EXERCISE 29

FIGURE 4.15   (a)   (b)

3   Solve Systems of Linear Inequalities Containing Absolute Value

Now we will graph *systems of linear inequalities containing absolute value* in the Cartesian coordinate system. Before we do some examples, let us recall the rules for absolute value inequalities that we learned in Section 2.6. Recall that

If $|x| < a$ and $a > 0$, then $-a < x < a$.
If $|x| > a$ and $a > 0$, then $x < -a$ or $x > a$.

EXAMPLE 6   Graph $|x| < 3$ in the Cartesian coordinate system.

Solution   From the rules given, we know that $|x| < 3$ means $-3 < x < 3$. We draw dashed vertical lines through -3 and 3 and shade the area between the two (Fig. 4.16).

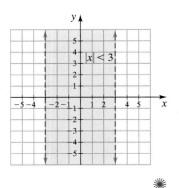

FIGURE 4.16

EXAMPLE 7   Graph $|y + 1| > 3$ in the Cartesian coordinate system.

Solution   From the rules given above, we know that $|y + 1| > 3$ means $y + 1 < -3$ or $y + 1 > 3$. First we solve each inequality.

$$y + 1 < -3 \quad \text{or} \quad y + 1 > 3$$
$$y < -4 \qquad\qquad y > 2$$

Now we graph both inequalities and take the *union* of the two graphs. The solution is the shaded area in Figure 4.17.

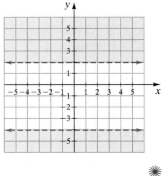

FIGURE 4.17

EXAMPLE 8   Graph the following system of inequalities.

$$|x| < 3$$
$$|y + 1| > 3$$

Solution   We draw both inequalities on the same axes. Therefore, we combine the graph drawn in Example 6 with the graph drawn in Example 7 (see Fig. 4.18). The points common to both inequalities form the solution to the system.

FIGURE 4.18

NOW TRY EXERCISE 41

Exercise Set 4.6

Concept/Writing Exercises

1. Explain how to find the solution to a system of linear inequalities graphically.

2. If in a system of two inequalities, one inequality contains $<$ and the other inequality contains $\geq$, is the point of intersection of the two boundary lines of the inequalities in the solution set? Explain.

3. If in a system of two inequalities, one inequality contains $\leq$ and the other inequality contains $\geq$, is the point of intersection of the two boundary lines of the inequalities in the solution set? Explain.

4. If in a system of two inequalities, one inequality contains $<$ and the other inequality contains $>$, is the point of intersection of the two boundary lines of the inequalities in the solution set? Explain.

Practice the Skills

Determine the solution to each system of inequalities.

5. $2x - y < 4$
$\quad y \geq -x + 2$

6. $y \leq -2x + 1$
$\quad y > -3x$

7. $y < 3x - 2$
$\quad y \leq -2x + 3$

8. $y \geq 2x - 5$
$\quad y > -3x + 5$

9. $y < x$
$\quad y \geq 3x + 2$

10. $-2x + 3y < -5$
$\quad 3x - 8y > 4$

11. $-3x + 2y \geq -5$
$\quad y \leq -4x + 7$

12. $-4x + 3y \geq -4$
$\quad y > -3x + 3$

13. $-4x + 5y < 20$
$\quad x \geq -3$

14. $y \geq -\dfrac{2}{3}x + 1$
$\quad y > -4$

15. $x \leq 4$
$\quad y \geq -2$

16. $x \geq 0$
$\quad x - 3y < 6$

17. $5x + 2y > 10$
$\quad 3x - y > 3$

18. $3x + 2y > 8$
$\quad x - 5y < 5$

19. $-2x > y + 4$
$\quad -x < \dfrac{1}{2}y - 1$

20. $y \leq 3x - 2$
$\quad \dfrac{1}{3}y < x + 1$

21. $y < 3x - 4$
$\quad 6x \geq 2y + 8$

22. $\dfrac{1}{2}x + \dfrac{1}{2}y \geq 2$
$\quad 2x - 3y \leq -6$

Determine the solution to each system of inequalities. Use the method discussed in Examples 4 and 5.

23. $x \geq 0$
$\quad y \geq 0$
$\quad 2x + 3y \leq 6$
$\quad 4x + y \leq 4$

24. $x \geq 0$
$\quad y \geq 0$
$\quad 2x + 3y \leq 8$
$\quad 4x + 2y \leq 8$

25. $x \geq 0$
$\quad y \geq 0$
$\quad x + y \leq 6$
$\quad 7x + 4y \leq 28$

26. $x \geq 0$
$\quad y \geq 0$
$\quad 3x + y \leq 9$
$\quad 2x + 5y \leq 10$

27. $x \geq 0$
$\quad y \geq 0$
$\quad 3x + 2y \leq 18$
$\quad 2x + 4y \leq 20$

28. $x \geq 0$
$\quad y \geq 0$
$\quad 5x + 4y \leq 16$
$\quad x + 6y \leq 18$

29. $x \geq 0$
$\quad y \geq 0$
$\quad x \leq 4$
$\quad x + y \leq 6$
$\quad x + 2y \leq 8$

30. $x \geq 0$
$\quad y \geq 0$
$\quad x \leq 4$
$\quad 2x + 3y \leq 18$
$\quad 4x + 2y \leq 20$

31. $x \geq 0$
$\quad y \geq 0$
$\quad x \leq 15$
$\quad 40x + 25y \leq 1000$
$\quad 5x + 30y \leq 900$

32. $x \geq 0$
$\quad y \geq 0$
$\quad x \leq 15$
$\quad 30x + 25y \leq 750$
$\quad 10x + 40y \leq 800$

Determine the solution to each system of inequalities.

33. $|x| > 1$
$\quad y < x$

34. $|x| > 2$
$\quad |y| \leq 4$

35. $|x| \geq 1$
$\quad |y| \geq 2$

36. $|x| < 2$
$\quad |y| \geq 3$

37. $|y| > 2$
$\quad y \leq x + 3$

38. $|x| > 1$
$\quad y \leq 3x + 2$

39. $|y| < 4$
$\quad y \geq -2x + 2$

40. $|x - 2| \leq 3$
$\quad x - y > 2$

41. $|x + 2| < 3$
$\quad |y| > 4$

42. $|x - 2| > 1$
$\quad y > -2$

43. $|x - 3| \leq 4$
$\quad |y + 2| \leq 1$

44. $|x + 1| \leq 2$
$\quad |y - 3| \leq 1$

Problem Solving

45. Is it possible for a system of linear inequalities to have no solution? Explain. Make up an example to support your answer.

46. Is it possible for a system of two linear inequalities to have exactly one solution? Explain. If you answer yes, make up an example to support your answer.

Without graphing, determine the number of solutions in each indicated system of inequalities. Explain your answers.

47. $2x + y < 6$
$2x + y > 6$

48. $3x - y \le 4$
$3x - y > 4$

49. $5x - 2y \le 3$
$5x - 2y \ge 3$

50. $2x - y < 7$
$3x - y < 4$

51. $5x - 3y > 5$
$5x - 3y > 6$

52. $x + y \le 0$
$x - y \ge 0$

Challenge Problems

Determine the solution to each system of inequalities.

53. $y \ge x^2$
$y \le 4$

54. $y < 4 - x^2$
$y > -5$

55. $y < |x|$
$y < 4$

56. $y \ge |x - 2|$
$y \le -|x - 2|$

Cumulative Review Exercises

[2.2] **57.** A formula for levers in physics is $f_1d_1 + f_2d_2 = f_3d_3$. Solve this formula for f_2

[3.2] *State the domain and range of each function.*

58. $\{(4, 3), (5, -2), (-1, 2), (0, -5)\}$

60.

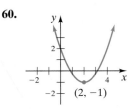

(2, −1)

59. $f(x) = \frac{2}{3}x - 4$

CHAPTER SUMMARY
Key Words and Phrases

4.1
Addition (or
 elimination) method
Consistent system of
 equations
Dependent system of
 equations
Inconsistent system of
 equations
Ordered triple
Solution to a system of
 equations
Substitution

System of linear
 equations

4.2
Geometric
 interpretation of a
 system of linear
 equations in three
 variables

4.3
Complementary angles
Supplementary angles

4.4
Augmented matrix
Elements
Matrix
Row transformations
Square matrix
Triangular form

4.5
Cramer's rule
Determinant
Expansion of the
 determinant by minors

Minor determinant

4.6
Constraints
Linear programming
Systems of linear
 inequalities
Systems of linear
 inequalities
 containing absolute
 value

(continued on the next page)

IMPORTANT FACTS

Augmented Matrices

The matrix $\begin{bmatrix} 1 & a & | & p \\ 0 & 1 & | & q \end{bmatrix}$ represents the system $\begin{aligned} x + a &= p \\ y &= q \end{aligned}$

The matrix $\begin{bmatrix} 1 & a & b & | & p \\ 0 & 1 & c & | & q \\ 0 & 0 & 1 & | & r \end{bmatrix}$ represents the system $\begin{aligned} x + ay + bz &= p \\ y + cz &= q \\ z &= r \end{aligned}$

Value of a Second-Order Determinant

$\begin{vmatrix} a_1 & b_1 \\ a_2 & b_2 \end{vmatrix} = a_1 b_2 - a_2 b_1$

Cramer's Rule:

For a system of equations of the form

$$a_1 x + b_1 y = c_1$$
$$a_2 x + b_2 y = c_2$$

$$x = \frac{\begin{vmatrix} c_1 & b_1 \\ c_2 & b_2 \end{vmatrix}}{\begin{vmatrix} a_1 & b_1 \\ a_2 & b_2 \end{vmatrix}} = \frac{D_x}{D} \quad \text{and} \quad y = \frac{\begin{vmatrix} a_1 & c_1 \\ a_2 & c_2 \end{vmatrix}}{\begin{vmatrix} a_1 & b_1 \\ a_2 & b_2 \end{vmatrix}} = \frac{D_y}{D}, \qquad D \neq 0$$

Value of a Third-Order Determinant

$$\begin{vmatrix} a_1 & b_1 & c_1 \\ a_2 & b_2 & c_2 \\ a_3 & b_3 & c_3 \end{vmatrix} = a_1 \underset{\substack{\uparrow \\ \text{Minor} \\ \text{determinant} \\ \text{of } a_1}}{\begin{vmatrix} b_2 & c_2 \\ b_3 & c_3 \end{vmatrix}} - a_2 \underset{\substack{\uparrow \\ \text{Minor} \\ \text{determinant} \\ \text{of } a_2}}{\begin{vmatrix} b_1 & c_1 \\ b_3 & c_3 \end{vmatrix}} + a_3 \underset{\substack{\uparrow \\ \text{Minor} \\ \text{determinant} \\ \text{of } a_3}}{\begin{vmatrix} b_1 & c_1 \\ b_2 & c_2 \end{vmatrix}}$$

Cramer's Rule:

For a system of equations of the form

$$a_1 x + b_1 y + c_1 z = d_1$$
$$a_2 x + b_2 y + c_2 z = d_2$$
$$a_3 x + b_3 y + c_3 z = d_3$$

$$x = \frac{\begin{vmatrix} d_1 & b_1 & c_1 \\ d_2 & b_2 & c_2 \\ d_3 & b_3 & c_3 \end{vmatrix}}{\begin{vmatrix} a_1 & b_1 & c_1 \\ a_2 & b_2 & c_2 \\ a_3 & b_3 & c_3 \end{vmatrix}} = \frac{D_x}{D}, \quad y = \frac{\begin{vmatrix} a_1 & d_1 & c_1 \\ a_2 & d_2 & c_2 \\ a_3 & d_3 & c_3 \end{vmatrix}}{\begin{vmatrix} a_1 & b_1 & c_1 \\ a_2 & b_2 & c_2 \\ a_3 & b_3 & c_3 \end{vmatrix}} = \frac{D_y}{D}, \quad z = \frac{\begin{vmatrix} a_1 & b_1 & d_1 \\ a_2 & b_2 & d_2 \\ a_3 & b_3 & d_3 \end{vmatrix}}{\begin{vmatrix} a_1 & b_1 & c_1 \\ a_2 & b_2 & c_2 \\ a_3 & b_3 & c_3 \end{vmatrix}} = \frac{D_z}{D}, \qquad D \neq 0$$

Chapter Review Exercises

[4.1] *Write each equation in slope-intercept form. Without graphing or solving the system of equations, state whether the system of linear equations is consistent, inconsistent, or dependent. Also indicate whether the system has exactly one solution, no solution, or an infinite number of solutions.*

1. $2x - 3y = -1$
  $-4x + 6y = 1$

2. $2x - 5y = 8$
  $3x + 4y = 9$

3. $y = \dfrac{1}{2}x + 4$
  $x + 2y = 8$

4. $6x = 4y - 8$
  $4x = 6y + 8$

Determine the solution to each system of equations graphically. If the system is inconsistent or dependent, so state.

5. $y = x + 3$
  $y = 2x + 5$

6. $x = -2$
  $y = 3$

7. $2x + 2y = 8$
  $2x - y = -4$

8. $2y = 2x - 6$
  $\dfrac{1}{2}x - \dfrac{1}{2}y = \dfrac{3}{2}$

Find the solution to each system of equations by subsitution.

9. $y = -4x + 2$
  $y = 3x - 12$

10. $4x - 3y = -1$
  $y = -3x - 4$

11. $a = 2b - 8$
  $2b - 5a = 0$

12. $3x + y = 17$
  $\dfrac{1}{2}x - \dfrac{3}{4}y = 1$

Find the solution to each system of equations using the addition method.

13. $x - 2y = 5$
  $2x + 2y = 4$

14. $-2x - y = 5$
  $2x + 2y = 6$

15. $2a + 3b = 7$
  $a - 2b = -7$

16. $0.4x - 0.3y = 1.8$
  $-0.7x + 0.5y = -3.1$

17. $4r - 3s = 8$
  $2r + 5s = 8$

18. $-2m + 3n = 15$
  $3m + 3n = 10$

19. $x + \dfrac{2}{5}y = \dfrac{9}{5}$
  $x - \dfrac{3}{2}y = -2$

20. $2x + 2y = 8$
  $y = 4x - 3$

21. $y = -\dfrac{3}{4}x + \dfrac{5}{2}$
  $x + \dfrac{5}{4}y = \dfrac{7}{2}$

22. $2x - 5y = 12$
  $x - \dfrac{4}{3}y = -2$

23. $2x + y = 4$
  $x + \dfrac{1}{2}y = 2$

24. $2x = 4y + 5$
  $2y = x - 6$

[4.2] *Determine the solution to each system of equations using substitution or the addition method.*

25. $x - 2y - 4z = 13$
  $3y + 2z = -2$
  $5z = -20$

26. $2a + b - 2c = 5$
  $3b + 4c = 1$
  $3c = -6$

27. $x + 2y + 3z = 3$
  $-2x - 3y - z = 5$
  $4x + 2y + 5z = -8$

28. $-x - 4y + 2z = 1$
  $2x + 2y + z = 0$
  $-3x - 2y - 5z = 5$

29. $3y - 2z = -4$
  $3x - 5z = -7$
  $2x + y = 6$

30. $3a + 2b - 5c = 19$
  $2a - 3b + 3c = -15$
  $5a - 4b - 2c = -2$

31. $x - y + 3z = 1$
  $-x + 2y - 2z = 1$
  $x - 3y + z = 2$

32. $-2x + 2y - 3z = 6$
  $4x - y + 2z = -2$
  $2x + y - z = 4$

[4.3] *Express each problem as a system of linear equations and use the method of your choice to find the solution to the problem.*

33. *Ages* Dennis McGinnes is 10 years older than his niece, Jennifer Miesen. If the sum of their ages is 66, find Dennis's age and Jennifer's age.

34. *Air Speed* An airplane can travel 560 miles per hour with the wind and 480 miles per hour against the wind. Determine the speed of the plane in still air and the speed of the wind.

35. *Mixing Solutions* James Curtis has two acid solutions, as illustrated. How much of each must he mix to get 6 liters of a 40% acid solution?

36. *Ice Hockey* The admission at an ice hockey game is $15 for adults and $11 for children. A total of 650 tickets were sold. Determine how many children's tickets and how many adult tickets were sold if a total of $8790 was collected.

37. *Return to Space* John Glenn was the first American astronaut to go into orbit around the Earth. Many years later he returned to space. The second time he returned to space he was 5 years younger than twice his age when he went into space for the first time. The sum of his ages for both times he was in space is 118. Find his age each time he was in space.

38. *Savings Accounts* Mary Weiner has a total of $40,000 invested in three different savings accounts. She has some money invested in one account that gives 7% interest. The second account has $5000 less than the first account and gives 5% interest. The third account gives 3% interest. If the total annual interest that Mary receives in a year is $2300, find the amount in each account.

[4.4] *Solve each system of equations using matrices.*

39. $x + 5y = 1$
$-2x - 8y = -6$

40. $2x - 3y = 3$
$2x + 4y = 10$

41. $y = 2x - 4$
$4x = 2y + 8$

42. $2x - y - z = 5$
$x + 2y + 3z = -2$
$3x - 2y + z = 2$

43. $3a - b + c = 2$
$2a - 3b + 4c = 4$
$a + 2b - 3c = -6$

44. $x + y + z = 3$
$3x + 2y = 1$
$y - 3z = -10$

[4.5] *Solve each system of equations using determinants.*

45. $7x - 8y = -10$
$-5x + 4y = 2$

46. $x + 4y = 5$
$-2x - 2y = 2$

47. $4m + 3n = 2$
$7m - 2n = -11$

48. $p + q + r = 5$
$2p + q - r = -5$
$-p + 2q - 3r = -4$

49. $-2a + 3b - 4c = -7$
$a + b + c = 4$
$-2a - 3b + 4c = 3$

50. $y + 3z = 4$
$-x - y + 2z = 0$
$x + 2y + z = 1$

[4.6] *Graph the solution to each system of inequalities.*

51. $-x + 3y > 6$
$2x - y \leq 2$

52. $5x - 2y \leq 10$
$3x + 2y > 6$

53. $y > 2x + 3$
$y < -x + 4$

54. $x > -2y + 4$
$y < -\dfrac{1}{2}x - \dfrac{3}{2}$

Determine the solution to the system of inequalities.

55. $x \geq 0$
$y \geq 0$
$x + y \leq 6$
$4x + y \leq 8$

56. $x \geq 0$
$y \geq 0$
$2x + y \leq 6$
$4x + 5y \leq 20$

57. $|x| \leq 3$
$|y| > 2$

58. $|x| > 4$
$|y - 2| \leq 3$

Chapter Practice Test

1. Define **a)** a consistent system of equations, **b)** a dependent system of equations, and **c)** an inconsistent system of equations.

Determine, without solving the system, whether the system of equations is consistent, inconsistent, or dependent. State whether the system has exactly one solution, no solution, or an infinite number of solutions.

2. $5x + 2y = 4$
   $5x = 3y - 7$

3. $5x + 3y = 9$
   $2y = -\dfrac{10}{3}x + 6$

4. $5x - 4y = 6$
   $-10x + 8y = -10$

Solve each system of equations by the method indicated.

5. $y = 3x - 2$
   $y = -2x + 8$
   graphically

6. $y = -x + 6$
   $y = 2x + 3$
   graphically

7. $y = -3x + 4$
   $y = 5x - 4$
   substitution

8. $2a + 4b = 2$
   $5a + b = -13$
   substitution

9. $4x + 3y = 10$
   $6x + y = 1$
   addition

10. $0.3x = 0.2y + 0.4$
    $-1.2x + 0.8y = -1.6$
    addition

11. $\dfrac{3}{2}a + b = 6$
    $a - \dfrac{5}{2}b = -4$
    addition

12. $x + y + z = 2$
    $-2x - y + z = 1$
    $x - 2y - z = 1$
    addition

13. Write the augmented matrix for the following system of equations.
    $-2x + 3y + 7z = 5$
    $3x - 2y + z = -2$
    $x - 6y + 5z = -13$

14. Consider the following augmented matrix.
    $$\begin{bmatrix} 6 & -2 & 4 & | & 4 \\ 4 & 3 & 5 & | & 6 \\ 2 & -1 & 4 & | & -3 \end{bmatrix}$$
    Show the results obtained by multiplying the elements in the third row by -2 and adding the products to their corresponding elements in the second row.

Solve each system of equations using matrices.

15. $x - 3y = 7$
    $3x + 5y = 7$

16. $x - 2y + z = 7$
    $-2x - y - z = -7$
    $3x - 2y + 2z = 15$

Evaluate each determinant.

17. $\begin{vmatrix} 3 & -1 \\ 4 & -2 \end{vmatrix}$

18. $\begin{vmatrix} 8 & 2 & -1 \\ 3 & 0 & 5 \\ 6 & -3 & 4 \end{vmatrix}$

Solve each system of equations using determinants and Cramer's rule.

19. $4x + 3y = -6$
    $-2x + 5y = 16$

20. $2r - 4s + 3t = -1$
    $-3r + 5s - 4t = 0$
    $-2r + s - 3t = -2$

Use the method of your choice to find the solution to each problem.

21. **Peanuts and Cashews** Dick Fields has cashews that sell for $7 a pound and peanuts that sell for $5.50 a pound. How much of each must he mix to get 20 pounds of a mixture that sells for $6.00 per pound?

22. *Mixing Solutions* Tyesha Blackwell, a chemist, has 6% and 15% solutions of sulfuric acid. How much of each solution should she mix to get 10 liters of a 9% solution?

23. *Sum of Numbers* The sum of three numbers is 25. The greatest number is three times the smallest number. The remaining number is 1 more than twice the smallest number. Find the three numbers.

Determine the solution to each system of inequalities.

24. $3x + 2y < 9$
$-2x + 5y \leq 10$

25. $|x| > 3$
$|y| \leq 1$

Cumulative Review Test

Take the following test and check your answers with those that appear at the end of the test. Review any questions that you answered incorrectly. The section and objective where the material was covered is indicated after the answer.

1. Evaluate $16 \div \left\{ 4\left[3 + \left(\dfrac{5 + 10}{5} \right)^2 \right] - 32 \right\}$.

2. Consider the following set of numbers.

$$\left\{ \frac{1}{2}, -4, 9, 0, \sqrt{3}, -4.63, 1 \right\}$$

List the elements of the set that are
a) natural numbers;
b) rational numbers;
c) real numbers.

3. Write the following numbers in order from smallest to largest.

$$-1, |-4|, \frac{3}{4}, \frac{5}{8}, -|-8|, |-10|$$

Solve.

4. $-[3 - 2(x - 4)] = 3(x - 6)$

5. $\dfrac{2}{3}x - \dfrac{5}{6} = 2$

6. $|2x - 3| - 4 = 5$

7. Solve the formula $M = \dfrac{1}{2}(a + x)$ for x.

8. Find the solution set of the inequality.

$$0 < \frac{3x - 2}{4} \leq 8$$

9. Simplify $\left(\dfrac{3x^2 y^{-2}}{y^3} \right)^{-2}$.

10. Graph $2y = 3x - 8$.

11. Write in slope-intercept form the equation of the line that is parallel to the graph of $2x - 3y = 8$ and passes through the point $(2, 3)$

12. Graph the inequality $6x - 3y < 12$.

13. Determine which of the following graphs represent functions. Explain.

a)   b)   c)

14. If $f(x) = \dfrac{x + 3}{x^2 - 9}$, find **a)** $f(-4)$ **b)** $f(h)$ **c)** $f(3)$

Solve each system of equations.

15. $3x + y = 6$
$y = 2x + 1$

16. $2p + 3q = 11$
$-3p - 5q = -16$

17. $x - 2y = 0$
$2x + z = 7$
$y - 2z = -5$

18. *Angles of a Triangle* If the largest angle of a triangle is nine times the measure of the smallest angle, and the middle-sized angle is 70° greater than the measure of the smallest angle, find the measure of the three angles.

19. *Walking and Jogging* Dawn Davis walks at 4 miles per hour and Judy Bolin jogs at 6 miles per hour. Dawn begins walking $\frac{1}{2}$ hour before Judy starts jogging. If Judy jogs on the same path that Dawn walks, how long after Judy begins jogging will she catch up to Dawn?

20. *Rock Concert* There are two different prices of seats at a rock concert. The higher priced seats sell for $20 and the less expensive seats sell for $16. If a total of 1000 tickets are sold and the total ticket sales are $18,400, how many of each type of ticket are sold?

Answers to Cumulative Review Test

1. 1; [Sec. 1.4, Obj. 3]   **2. a)** 9, 1   **b)** $\frac{1}{2}$, -4, 9, 0, -4.63, 1   **c)** $\frac{1}{2}$, -4, 9, 0, $\sqrt{3}$, -4.63, 1; [Sec. 1.2, Obj. 5]

3. $-|-8|$, -1, $\frac{5}{8}$, $\frac{3}{4}$, $|-4|$, $|-10|$; [Sec. 1.3, Obj. 1]   **4.** 7; [Sec. 2.1, Obj. 3]   **5.** $\frac{17}{4}$; [Sec. 2.1, Obj. 4]   **6.** 6, -3; [Sec. 2.6, Obj. 2]

7. $x = 2M - a$; [Sec. 2.2, Obj. 2]   **8.** $\left\{ x \left| \frac{2}{3} < x \leq \frac{34}{3} \right. \right\}$; [Sec. 2.5, Obj. 3]   **9.** $\frac{y^{10}}{9x^4}$; [Sec. 1.5, Obj. 7]

10.   ; [Sec. 3.3, Obj. 2]   **11.** $y = \frac{2}{3}x + \frac{5}{3}$; [Sec. 3.5, Obj. 3]   **12.** ;

[Sec. 3.7, Obj. 1]   **13. a)** function   **b)** function   **c)** not a function; [Sec. 3.2, Obj. 3]   **14. a)** $-\frac{1}{7}$ **b)** $\frac{h + 3}{h^2 - 9}$

c) undefined; [Sec. 3.2, Obj. 4]   **15.** (1, 3); [Sec. 4.1, Obj. 2]   **16.** (7, -1); [Sec. 4.1, Ob. 3]   **17.** (2, 1, 3); [Sec. 4.2, Obj. 1]

18. 10°, 80°, 90°; [Sec. 2.3, Obj. 2]   **19.** 1 hour; [Sec. 2.4, Obj. 1]   **20.** 600 at $20, 400 at $16; [Sec. 4.3, Obj. 1]

Chapter 5

Polynomials and Polynomial Functions

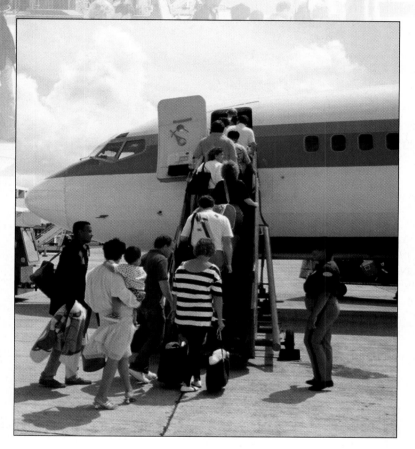

5.1 Addition and Subtraction of Polynomials

5.2 Multiplication of Polynomials

5.3 Division of Polynomials and Synthetic Division

5.4 Factoring a Monomial from a Polynomial and Factoring by Grouping

5.5 Factoring Trinomials

5.6 Special Factoring Formulas

5.7 A General Review of Factoring

5.8 Polynomial Equations

Chapter Summary
Chapter Review Exercises
Chapter Practice Test
Cumulative Review Test

The Internet is changing the way we make airline reservations and purchases. To meet the needs of their customers in a rapidly expanding environment, travel companies use mathematical models to predict the number of Internet clients they can expect. On page 302 we evaluate a polynomial function that predicts the annual online travel revenue. As the online travel revenue increases, the kinds of employees needed will also change. What are some qualities that you think an online travel company will look for when hiring new employees?

SSM | Study Guide | CD/Video | MathPro 4/5 | PH Math Tutor Center | prenhall.com/Angel

A Look Ahead

In this chapter we discuss polynomials, polynomial functions, and factoring. In the first few sections we add, subtract, multiply, and divide polynomials and polynomial functions. Since graphing is such an important part of this course, make sure you understand the graphs of polynomial functions.

After discussing polynomials, we turn our attention to factoring. *You must have a thorough understanding of factoring to work the problems in many of the remaining chapters.* In Section 5.8 we explain how to solve quadratic equations by factoring and show how applications are solved using factoring. Pay particular attention to how to use factoring to find the *x*-intercepts of a quadratic function. We will refer back to this topic later in the course.

5.1 ADDITION AND SUBTRACTION OF POLYNOMIALS

SSM   Study Guide   CD/Video

MathPro 4/5   PH Math Tutor Center   prenhall.com/Angel

1   Find the degree of a polynomial.
2   Evaluate polynomial functions.
3   Understand graphs of polynomial functions.
4   Add and subtract polynomials.

1   Find the Degree of a Polynomial

Recall from Chapter 2 that the parts that are added or subtracted in a mathematical expression are called **terms**. The **degree of a term** with whole number exponents is the sum of the exponents of the variables, if there are variables. Nonzero constants have degree 0, and the term 0 has no degree.

A **polynomial** is a finite sum of terms in which all variables have whole number exponents and no variable appears in a denominator. The expression $3x^2 + 2x + 6$ is an example of a *polynomial in one variable, x*. The expression $x^2y - 2x + 3$ is an example of a *polynomial in two variables, x and y*. The expressions $x^{1/2}$ and $\frac{1}{x}$ (or x^{-1}) are *not* polynomials because the variables do not have whole number exponents. The expression $\frac{1}{x-1}$ is *not* a polynomial because a variable appears in the denominator.

The **leading term** of a polynomial is the term of highest degree. The **leading coefficient** is the coefficient of the leading term.

EXAMPLE 1   For each polynomial give the number of terms, the degree of the polynomial, the leading term, and the leading coefficient.

a) $2x^5 - 3x^2 + 6x - 4$ b) $2x^2y^4 - 6xy^3 + 3xy^2z^4$

Solution   We will organize the answers in tabular form.

Polynomial	Number of terms	Degree of polynomial	Leading term	Leading coefficient
a) $2x^5 - 3x^2 + 6x - 4$	4	5 (from $2x^5$)	$2x^5$	2
b) $2x^2y^4 - 6xy^3 + 3xy^2z^4$	3	7 (from $3xy^2z^4$)	$3xy^2z^4$	3

NOW TRY EXERCISE 29

Polynomials are classified according to the number of terms they have, as indicated in the following chart.

Polynomial type	Description	Examples
Monomial	A polynomial of one term	$4x^2$, $6x^2y$, 3, $-2xyz^5$, 7
Binomial	A polynomial of two terms	$x^2 + 1$, $2x^2 - y$, $6x^3 - 5y^2$
Trinomial	A polynomial of three terms	$x^3 + 6x - 4$, $x^2y - 6x + y^2$

Polynomials containing more than three terms are not given specific names. *Poly* is a prefix meaning *many*. A polynomial is called **linear** if it is of degree 0 or 1. A polynomial in one variable is called **quadratic** if it is of degree 2, and **cubic** if it is of degree 3.

Type of Polynomial	Examples
Linear	$2x - 4$, 5
Quadratic	$3x^2 + x - 6$, $4x^2 - 6$
Cubic	$-4x^3 + 3x^2 + 5$, $2x^3 + 6x$

The polynomials $2x^3 + 4x^2 - 6x + 3$ and $4x^2 - 3xy + 5y^2$ are examples of polynomials in **descending order** of the variable x because the exponents on the variable x descend (or get lower) as the terms go from left to right. Polynomials are often written in descending order of a given variable.

EXAMPLE 2   Write each polynomial in descending order of the variable x.
   a) $5x + 4x^2 - 6$ **b)** $xy - 6x^2 + 3y^2$

Solution   **a)** $5x + 4x^2 - 6 = 4x^2 + 5x - 6$
   b) $xy - 6x^2 + 3y^2 = -6x^2 + xy + 3y^2$ ✳

2   Evaluate Polynomial Functions

The expression $2x^3 + 6x^2 + 3$ is a polynomial. If we write $P(x) = 2x^3 + 6x^2 + 3$, then we have a polynomial function. In a **polynomial function**, the expression used to describe the function is a polynomial. To evaluate a polynomial function, we use substitution, just as we did to evaluate other functions in Chapter 3.

EXAMPLE 3   For the polynomial function $P(x) = 4x^3 - 6x^2 - 2x + 8$, find
   a) $P(0)$ **b)** $P(3)$ **c)** $P(-2)$

Solution   **a)** $P(x) = 4x^3 - 6x^2 - 2x + 8$
   $P(0) = 4(0)^3 - 6(0)^2 - 2(0) + 8$
   $= 0 - 0 - 0 + 8 = 8$

   b) $P(3) = 4(3)^3 - 6(3)^2 - 2(3) + 8$
   $= 4(27) - 6(9) - 6 + 8 = 56$

   c) $P(-2) = 4(-2)^3 - 6(-2)^2 - 2(-2) + 8$
   $= 4(-8) - 6(4) + 4 + 8 = -44$ ✳

NOW TRY EXERCISE 35

Businesses, governments, and other organizations often need to track and make projections about things such as sales, profits, changes in the population, effectiveness of new drugs, and so on. To do so, they often use graphs and functions. Example 4 gives one such example.

EXAMPLE 4

Online Travel Revenues

Source: Travel Industry Association of America

FIGURE 5.1

Online Travel Revenue The bar graph in Figure 5.1 shows the projected online travel revenue (travel plans made on the Internet) in billions of dollars from 1996 through 2002. A polynomial function that can be used to approximate the revenues is $R(t) = 0.18t^2 + 0.37t + 0.28$, where t is years since 1996, $0 \le t \le 6$, and R is revenue in billions of dollars.

a) Using the function, estimate the online travel revenue in 1996.

b) Using the function, estimate the revenue in 2002.

Solution **a)** Understand We first need to determine what value of t to substitute into the function. Since t is years since 1996, the year 1996 corresponds to $t = 0$. Thus to estimate the revenue in 1996, we evaluate $R(0)$.

Translate and Carry Out

$$R(t) = 0.18t^2 + 0.37t + 0.28$$
$$R(0) = 0.18(0)^2 + 0.37(0) + 0.28$$
$$= 0 + 0 + 0.28$$
$$= 0.28$$

Check and Answer Thus the revenue in 1996 is about $0.28 billion. The graph supports this answer.

b) Understand The year 2002 is 6 years since 1996 ($2002 - 1996 = 6$). Therefore to estimate the revenue in 2002, we evaluate $R(6)$.

Translate and Carry Out

$$R(t) = 0.18t^2 + 0.37t + 0.28$$
$$R(6) = 0.18(6)^2 + 0.37(6) + 0.28$$
$$= 6.48 + 2.22 + 0.28$$
$$= 8.98$$

NOW TRY EXERCISE 97

Check and Answer The estimated revenue in 2002 is about $8.98 billion. This is consistent with the amount shown on the graph for the year 2002. ✳

3 Understand Graphs of Polynomial Functions

The graphs of all polynomial functions are smooth, continuous curves. Figure 5.2 shows a graph of a quadratic polynomial function. The graphs of all quadratic polynomial functions with a *positive leading coefficient* will have the shape of the graph in Figure 5.2.

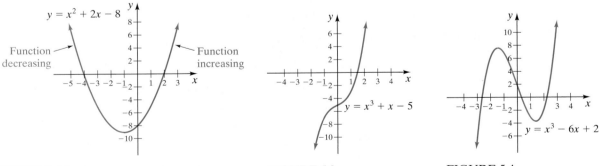

FIGURE 5.2 **FIGURE 5.3** **FIGURE 5.4**

The graph of a cubic polynomial function with a *positive leading coefficient* may have the shape of the graph in either Figure 5.3 or Figure 5.4. Notice that *whenever the leading coefficient in a polynomial function is positive, the polynomial function will increase (or move upward as x increases—the green part of the curve) to the right of some value of x.* For example, in Figure 5.2, the graph continues increasing to the right of $x = -1$. In Figure 5.3, the graph is continuously increasing, and in Figure 5.4, the graph is increasing to the right of about $x = 1.4$.

Polynomial functions with a negative leading coefficient will decrease (or move downward as x increases—the red part of the curve) to the right of some value of x. A quadratic polynomial function with a negative leading coefficient is shown in Figure 5.5, and cubic polynomial functions with negative leading coefficients are shown in Figure 5.6 and Figure 5.7. In Figure 5.5, the quadratic function is decreasing to the right of $x = 2$. In Figure 5.6, the cubic function is continuously decreasing, and in Figure 5.7 the cubic function is decreasing to the right of about $x = 1.2$.

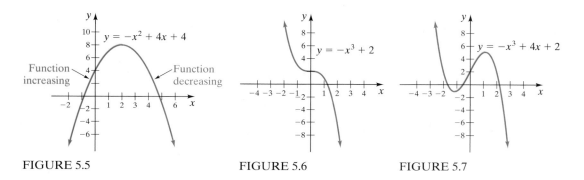

FIGURE 5.5   FIGURE 5.6   FIGURE 5.7

Why does the leading coefficient determine whether a function will increase or decrease to the right of some value of x? The leading coefficient is the coefficient of the term with the greatest exponent on the variable. As x increases, this term will eventually dominate all the other terms in the function. So if the coefficient of this term is positive, the function will *eventually* increase as x increases. If the leading coefficient is negative, the function will *eventually* decrease as x increases. This information, along with checking the y-intercept of the graph, can be useful in determining whether a graph is correct or complete. Read the Using Your Graphing Calculator box that follows, even if you are not using a graphing calculator. Also work Exercises 93 through 96.

Using Your Graphing Calculator

Whenever you graph a polynomial function on your grapher, make sure your screen shows every change in direction of your graph. For example, suppose you graph $y = 0.1x^3 - 2x^2 + 5x - 8$ on your grapher. Using the standard window, you get the graph shown in Figure 5.8 on page 304.

However, from our preceding discussion you should realize that since the leading coefficient, 0.1, is positive, the graph must increase to the right of some value of x. The graph in Figure 5.8 does not show this. If you change your window as shown in Figure 5.9, you will get the graph shown. Now you can see how the graph increases to the right of about $x = 12$. When graphing, the y-intercept is often helpful in determining the values to use for the range. Recall that to find the y-intercept, we set $x = 0$ and solve for y. For example, if graphing $y = 4x^3 + 6x^2 + x - 180$ the y-intercept will be at -180.

(continued on the next page)

FIGURE 5.8   FIGURE 5.9 $[-10, 30, 2, -100, 60, 10]$

Exercises

Use your grapher to graph each polynomial. Make sure your window shows all changes in direction of the graph.

1. $y = 0.2x^3 + 5.1x^2 - 6.2x + 9.3$
2. $y = 4.1x^3 - 19.6x^2 + 5.4x - 60.2$

4   Add and Subtract Polynomials

When we found sums and differences of functions in Section 3.6, we added and subtracted polynomials, although they were not called polynomials at that time. To *add or subtract polynomials*, remove parentheses if any are present, then combine like terms.

EXAMPLE 5   Simplify $(4x^2 - 6x + 3) + (2x^2 + 5x - 1)$.

Solution
$$(4x^2 - 6x + 3) + (2x^2 + 5x - 1)$$
$$= 4x^2 - 6x + 3 + 2x^2 + 5x - 1 \qquad \textit{Remove parentheses.}$$
$$= \underline{4x^2 + 2x^2} \; \underline{- 6x + 5x} \; \underline{+ 3 - 1} \qquad \textit{Rearrange terms.}$$
$$= \quad 6x^2 \qquad -x \qquad +2 \qquad \textit{Combine like terms.} \; ✳$$

EXAMPLE 6   Simplify $(3x^2y - 4xy + y) + (x^2y + 2xy + 3y - 2)$.

Solution
$$(3x^2y - 4xy + y) + (x^2y + 2xy + 3y - 2)$$
$$= 3x^2y - 4xy + y + x^2y + 2xy + 3y - 2 \qquad \textit{Remove parentheses.}$$
$$\underline{3x^2y + x^2y} \; \underline{- 4xy + 2xy} \; \underline{+ y + 3y} - 2 \qquad \textit{Rearrange terms.}$$
$$= \quad 4x^2y \qquad -2xy \qquad +4y \quad - 2 \qquad \textit{Combine like terms.} \; ✳$$

HELPFUL HINT

Recall that $-x$ means $-1 \cdot x$. Thus $-(2x^2 - 4x + 6)$ means $-1(2x^2 - 4x + 6)$, and the distributive property applies. When you subtract one polynomial from another, the *signs of every term* of the polynomial being subtracted must change. For example,
$$x^2 - 6x + 3 - (2x^2 - 4x + 6) = x^2 - 6x + 3 - 1(2x^2 - 4x + 6)$$
$$= x^2 - 6x + 3 - 2x^2 + 4x - 6$$
$$= -x^2 - 2x - 3$$

EXAMPLE 7   Subtract $(-x^2 - 2x + 3)$ from $(x^3 + 4x + 6)$.

Solution
$$(x^3 + 4x + 6) - (-x^2 - 2x + 3)$$
$$= (x^3 + 4x + 6) - 1(-x^2 - 2x + 3) \qquad \textit{Insert 1.}$$
$$= x^3 + 4x + 6 + x^2 + 2x - 3 \qquad \textit{Distributive property}$$
$$= x^3 + x^2 + 4x + 2x + 6 - 3 \qquad \textit{Rearrange terms.}$$
$$= x^3 + x^2 + 6x + 3 \qquad \textit{Combine like terms.} \; ✳$$

EXAMPLE 8

Solution

Simplify $x^2y - 4xy^2 + 5 - (2x^2y - 3y^2 + 4)$.

$$x^2y - 4xy^2 + 5 - 1(2x^2y - 3y^2 + 4) \quad \textit{Insert 1.}$$
$$= x^2y - 4xy^2 + 5 - 2x^2y + 3y^2 - 4 \quad \textit{Distributive property}$$
$$= x^2y - 2x^2y - 4xy^2 + 3y^2 + 5 - 4 \quad \textit{Rearrange terms.}$$
$$= -x^2y - 4xy^2 + 3y^2 + 1 \quad \textit{Combine like terms.}$$

Note that $-x^2y$ and $-4xy^2$ are not like terms since the variables have different exponents. Also, $-4xy^2$ and $3y^2$ are not like terms since $3y^2$ does not contain the variable x.

NOW TRY EXERCISE 45

EXAMPLE 9

Solution

Perimeter Find an expression for the perimeter of the quadrilateral in Figure 5.10.

The perimeter is the sum of the lengths of the sides of the figure. For a quadrilateral, the perimeter is the sum of the lengths of the four sides.

$$\text{Perimeter} = (x^2 + 2x + 3) + (x^2 + 1) + (5x + 3) + (3x + 2) \quad \textit{Sum of the sides}$$
$$= x^2 + 2x + 3 + x^2 + 1 + 5x + 3 + 3x + 2 \quad \textit{Remove parentheses.}$$
$$= x^2 + x^2 + 2x + 5x + 3x + 3 + 1 + 3 + 2 \quad \textit{Rearrange terms.}$$
$$= 2x^2 + 10x + 9 \quad \textit{Combine like terms.}$$

FIGURE 5.10

NOW TRY EXERCISE 79

The perimeter of the quadrilateral is $2x^2 + 10x + 9$.

Mathematics in Action

Tracking Trade with China

The United States has many trade agreements with countries all over the world. Goods sold to these countries are called exports while goods bought from them are called imports. The chart shown here represents our trade with China for the period 1990–2000.

The U.S.-China Trade Gap

Source: U.S. Bureau of the Census, Foreign Trade Division

The blue bars show imports from China to the United States while the yellow bars show the exports

from the United States to China. The red bars show the difference between the exports and the imports. The negative numbers on the left side of the chart indicate that imports exceed exports, creating what is called a negative balance of trade, also called a trade deficit.

The increasing length of the red bars confirm what your eyes can clearly see in the upper portion of the graph: imports from China have skyrocketed in the period while exports to China have risen only slightly.

The imports, $I(t)$, in billions of dollars, can be estimated by the function

$$I(t) = 0.37t^2 + 3.97t + 15.35$$

where t is the number of years since 1990.

The balance of trade, $B(t)$, in billions of dollars, can be estimated by the function

$$B(t) = -(0.386t^2 + 3.07t + 8.38)$$

where t is the number of years since 1990.

The minus sign in front of the $B(t)$ quadratic function indicates a *negative* balance of trade. We can distribute the minus sign to express $B(t)$ as

$$B(t) = -0.386t^2 - 3.07t - 8.38$$

(continued on the next page)

Trade information can be represented on spreadsheets, with pie charts, and any number of other graphic forms. Capturing import–export information mathematically, however, opens the way to the sophisticated analysis needed by policymakers to negotiate trade agreements. Done correctly, the result can be a fairer balance of exports and imports between the United States and China, or any other country with whom we trade.

Exercise Set 5.1

Concept/Writing Exercises

1. What are terms of a mathematical expression?
2. What is the degree of a nonzero constant?
3. What is a polynomial?
4. What is the leading term of a polynomial?
5. What is the leading coefficient of a polynomial?
6. **a)** How do you determine the degree of a term?
   b) What is the degree of $6x^4y^3z$?
7. **a)** How do you determine the degree of a polynomial?
   b) What is the degree of $-4x^4 + 6x^3y^4 + z^5$?
8. What does it mean when a polynomial is in descending order of the variable x?
9. **a)** When is a polynomial linear?

   b) Give an example of a linear polynomial.
10. **a)** When is a polynomial quadratic?
    b) Give an example of a quadratic polynomial.
11. **a)** When is a polynomial cubic?
    b) Give an example of a cubic polynomial.
12. When one polynomial is being subtracted from another, what happens to the signs of all the terms of the polynomial being subtracted?
13. Write a fifth-degree trinomial in x in descending order that lacks fourth-, third-, and second-degree terms.
14. Write a seventh-degree polynomial in y in descending order that lacks fifth-, third-, and second-degree terms.

Practice the Skills

Determine whether each expression is a polynomial. If the polynomial has a specific name, for example, "monomial" or "binomial," give the name. If the expression is not a polynomial, explain why it is not.

15. -6
16. $2x^{-1}$
17. $5y$
18. $5x^2 - 6x + 9$
19. $5x^{-3}$
20. $8x^2 - 2x + 8y^2$
21. $3x^{1/2} + 2xy$
22. $2xy + 5y^2$

Write each polynomial in descending order of the variable x. If the polynomial is already in descending order, so state. Give the degree of each polynomial.

23. $-5 + 4x - x^2$
24. $3x - 4 - x^2$
25. $9y^2 + 3xy + 10x^2$
26. $-2 + x - 8x^2 + 4x^3$
27. $-2x^4 + 5x^2 - 4$
28. $5xy^2 + 3x^2y - 6 - 2x^3$

Give **a)** *the degree of each polynomial and* **b)** *its leading coefficient.*

29. $x^4 + 3x^6 - 2x - 10$

30. $-2x^4 + 6x^5 - x^7 + 5x^3$

31. $4x^2y^3 + 6xy^4 + 9xy^5$

32. $-a^4b^3c^2 + 7a^8b^9c^4 - 5a^7c^{20}$

33. $-\dfrac{1}{3}m^4n^5p^8 + \dfrac{3}{5}m^3p^6 - \dfrac{5}{9}n^4p^6q$

34. $-0.6x^2y^3z^2 - 2.9xyz^9 - 1.3x^8y^4$

Evaluate each polynomial function at the given value.

35. Find $P(2)$ if $P(x) = x^2 - 6x + 1$.

36. Find $P(-1)$ if $P(x) = 4x^2 + 6x + 12$.

37. Find $P\left(\dfrac{1}{2}\right)$ if $P(x) = 2x^2 - 3x - 6$.

38. Find $P\left(\dfrac{1}{3}\right)$ if $P(x) = \dfrac{1}{2}x^3 - x^2 + 6$.

39. Find $P(0.4)$ if $P(x) = 0.2x^3 + 1.6x^2 - 2.3$.

40. Find $P(-1.2)$ if $P(x) = -1.6x^3 - 4.6x^2 - 0.1x$.

Simplify.

41. $(x^2 + 3x - 7) + (6x - 5)$

42. $(5b^2 - 4b + 7) - (2b^2 - 3b - 5)$

43. $(x^2 - 8x + 2) - (5x + 9)$

44. $(2x - 5) - (3x^2 - 4x + 16)$

45. $(4y^2 + 9y - 1) - (2y^2 + 10)$

46. $(5n^2 - 7) + (2n^2 + 3n + 12)$

47. $\left(-\dfrac{5}{9}a + 8\right) + \left(-\dfrac{2}{3}a^2 - \dfrac{1}{4}a - 1\right)$

48. $(6y^2 - 9y + 4) - (-2y^2 - y - 8)$

49. $(1.4x^2 + 1.6x - 8.3) - (4.9x^2 + 3.7x + 11.3)$

50. $(-12.4x^2y - 6.2xy + 9.3y^2) - (-5.3x^2y + 1.6xy - 10.4y^2)$

51. $\left(-\dfrac{1}{3}x^3 + \dfrac{1}{4}x^2y + 8xy^2\right) + \left(-x^3 - \dfrac{1}{2}x^2y + xy^2\right)$

52. $\left(-\dfrac{3}{5}xy^2 + \dfrac{5}{8}\right) - \left(-\dfrac{1}{2}xy^2 + \dfrac{3}{5}\right)$

53. $(3a - 6b + 5c) - (-2a + 4b - 8c)$

54. $(6r + 7s - t) + (-2r - 2s - 5t)$

55. $(3a^2b - 6ab + 5b^2) - (4ab - 6b^2 - 5a^2b)$

56. $(3x^2 - 5y^2 - 2xy) - (4x^2 + 8y^2 - 9xy)$

57. $(8r^2 - 5t^2 + 2rt) + (-6rt + 2t^2 - r^2)$

58. $(a^2 - b^2 + 5ab) + (-3b^2 - 2ab + a^2)$

59. $6x^2 - 2x - [3x - (4x^2 - 9)]$

60. $3xy^2 - 2x - [-(4xy^2 + 3x) - 5xy]$

61. $5w - 6w^2 - [(3w - 2w^2) - (4w + w^2)]$

62. $-[-(5r^2 - 3r) - (2r - 3r^2) - 2r^2]$

63. Subtract $(4x - 11)$ from $(7x + 3)$.

64. Subtract $(-x^2 + 3x + 5)$ from $(4x^2 - 6x + 2)$.

65. Add $-2x^2 + 4x - 12$ and $-x^2 - 2x$.

66. Subtract $(5x^2 - 6)$ from $(2x^2 - 4x + 8)$.

67. Subtract $0.2a^2 - 3.9a + 26.4$ from $-4.2a^2 - 9.6a$.

68. Add $6x^2 + 3xy$ and $-2x^2 + 4xy + 3y$.

69. Subtract $\left(5x^2y + \dfrac{5}{9}\right)$ from $\left(-\dfrac{1}{2}x^2y + xy^2 + \dfrac{3}{5}\right)$.

70. Subtract $(6x^2y + 3xy)$ from $(2x^2y + 12xy)$.

Simplify. Assume that all exponents represent natural numbers.

71. $(3x^{2r} - 7x^r + 1) + (2x^{2r} - 3x^r + 2)$

72. $(6x^{2r} - 5x^r + 4) + (2x^{2r} + x^r + 3)$

73. $(x^{2s} - 8x^s + 6) - (2x^{2s} - 4x^s - 9)$

74. $(5a^{2m} - 6a^m + 4) - (2a^{2m} + 7)$

75. $(7b^{4n} - 5b^{2n} + 1) - (3b^{3n} - b^{2n})$

76. $(-3r^{3a} + r^a - 6) - (-2r^{3a} - 5r^{2a} + 6)$

Problem Solving

In Exercises 77–82, find an expression for the perimeter of each figure. See Example 9.

77.

Square

$x^2 + 2x + 5$

78.

Rectangle $3x + 11$

$x^2 - x + 7$

79.

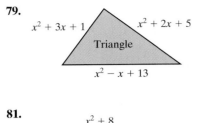

$x^2 + 3x + 1$ $x^2 + 2x + 5$
Triangle
$x^2 - x + 13$

80.

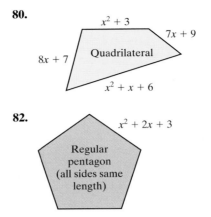

$x^2 + 3$
$7x + 9$
$8x + 7$ Quadrilateral
$x^2 + x + 6$

81.

$x^2 + 8$
$5x - 1$ $4x + 1$
Trapezoid
$x^2 + 3x + 1$

82.

$x^2 + 2x + 3$
Regular
pentagon
(all sides same
length)

83. Is the sum of two trinomials always a trinomial? Explain, and give an example to support your answer.

84. Is the sum of two binomials always a binomial? Explain, and give an example to support your answer.

85. Is the sum of two quadratic polynomials always a quadratic polynomial? Explain, and give an example to support your answer.

86. Is the difference of two cubic polynomials always a cubic polynomial? Explain, and give an example to support your answer.

87. *Area* The area of a circle is a function of its radius, where $A(r) = \pi r^2$. Find the area of a circle if its radius is 6 inches. Use the $\boxed{\pi}$ key on your calculator.

88. *Balloon* The volume of a sphere is a function of its radius where $V(r) = \frac{4}{3}\pi r^3$. A circular balloon is being blown up. Find its volume when its radius is 4 inches.

89. *Height* When an object is dropped from the Empire State Building (height 1250 feet), the object's height, h, in feet, from the ground at time, t, in seconds, after being dropped can be determined by

$$h = P(t) = -16t^2 + 1250$$

Find the distance an object is from the ground 6 seconds after being dropped.

90. *Spelling Bee* The number of ways that the first-, second-, and third-place winners in a spelling bee can be selected from n participants is given by $P(n) = n^3 - 3n^2 + 2n$. If there are six participants, how many ways can the winner and first and second runners-up be selected?

Profit *The profit of a company is found by subtracting its cost from its revenue. In Exercises 91 and 92, $R(x)$ represents the company's revenue when selling x items and $C(x)$ represents the company's cost when producing x items.* **a)** *Determine a profit function $P(x)$.* **b)** *Evaluate $P(x)$ when $x = 100$.*

91. $R(x) = 2x^2 - 60x$,
$C(x) = 8050 - 420x$

92. $R(x) = 5.5x^2 - 80.3x$
$C(x) = 1.2x^2 + 16.3x + 12{,}040.6$

*In Exercises 93–96, determine which of the graphs—**a)**, **b)**, or **c)**—is the graph of the given equation. Explain how you determined your answer.*

93. $y = x^2 + 3x - 4$

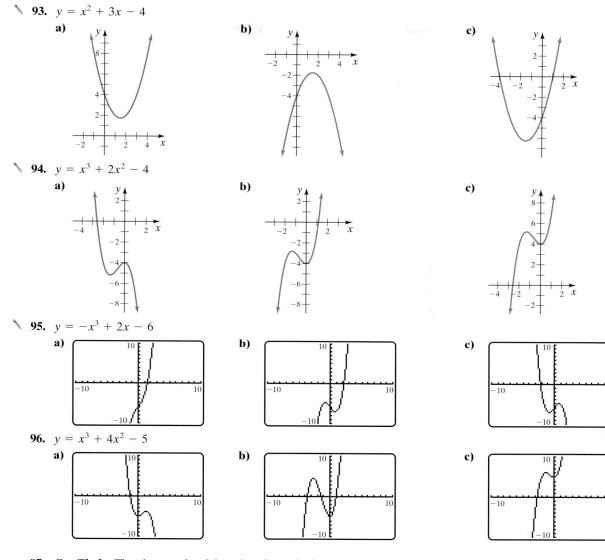

94. $y = x^3 + 2x^2 - 4$

95. $y = -x^3 + 2x - 6$

96. $y = x^3 + 4x^2 - 5$

97. **Car Thefts** The chart to the right, taken from the January 1, 2002, issue of the *New York Times*, shows that the number of car thefts in New York City has been declining since 1993.

    The function $f(t) = 1.55t^2 - 22.03t + 113.65$, where t is the number of years since 1993, $0 \le t \le 10$, can be used to estimate the number of car thefts, in thousands.

    a) Use this function to estimate the number of car thefts in New York City in 2001.

    b) Compare your answer from part **a)** to the bar graph. Does the graph support your answer?

New York City Car Thefts

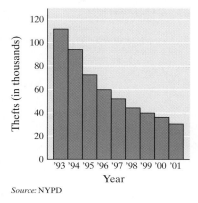

Source: NYPD

98. *Inclined Plane* A ball rolls down an inclined plane. The distance, $d(t)$, in feet, the ball has traveled is given by the function

$$d(t) = 2.36t^2$$

where t is time in seconds, $0 \le t \le 5$.

Find the distance the ball has traveled down the inclined plane in

a) 1 second.

b) 3 seconds.

c) 5 seconds.

99. *Inflation* Inflation erodes our purchasing power. Because of inflation, we will pay more for the same goods in the future than we pay for them today. The function $C(t) = 0.31t^2 + 0.59t + 9.61$, where t is years since 1997, approximates the cost, in thousands of dollars, for purchasing in the future what $10,000 would purchase in 1997. This function is based on a 6% annual inflation rate and $0 \le t \le 25$. Estimate the cost in 2012 for goods that cost $10,000 in 1997.

100. *Drug-free Schools* The function $f(a) = -2.32a^2 + 76.85a - 559.87$ can be used to estimate the percent of students who say their school is not drug free. In this function, a represents the student's age, where $12 \le a \le 17$. Use this function to estimate the percent of 13-year-olds who say their school is not drug free.

*Answer Exercises 101 and 102 using a graphing calculator if you have one. If you do not have a graphing calculator, draw the graphs in part **a)** by plotting points. Then answer parts **b)** through **e)**.*

101. a) Graph

$$y_1 = x^3$$
$$y_2 = x^3 - 3x^2 - 3$$

b) In both graphs, for values of $x > 3$, do the functions increase or decrease as x increases?

c) When the leading term of a polynomial function is x^3, the polynomial must increase for $x > a$, where a is some real number greater than 0. Explain why this must be so.

d) In both graphs, for values of $x < -3$, do the functions increase or decrease as x decreases?

e) When the leading term of a polynomial function is x^3, the polynomial must decrease for $x < a$, where a is some real number less than 0. Explain why this must be so.

102. a) Graph

$$y_1 = x^4$$
$$y_2 = x^4 - 6x^2$$

b) In each graph, for values of $x > 3$, are the functions increasing or decreasing as x increases?

c) When the leading term of a polynomial function is x^4, the polynomial must increase for $x > a$, where a is some real number greater than 0. Explain why this must be so.

d) In each graph, for values of $x < -3$, are the functions increasing or decreasing as x decreases?

e) When the leading term of a polynomial function is x^4, the polynomial must increase for $x < a$, where a is some real number less than 0. Explain why this must be so.

Challenge Problems

*Determine which of the graphs—**a)**, **b)**, or **c)**—is the graph of the given equation. Explain how you determined your answer.*

103. $y = -x^4 + 3x^3 - 5$

a) **b)** **c)**

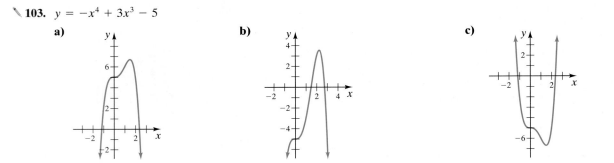

104. $y = 2x^4 + 9x^2 - 5$

a)

b)

c)

Group Activity

Discuss and answer Exercises 105 and 106 as a group.

105. If the leading term of a polynomial function is $3x^3$, which of the following could possibly be the graph of the polynomial? Explain. Consider what happens for large positive values of x and for large negative values of x.

a)

b)

c)

106. If the leading term of a polynomial is $-2x^4$, which of the following could possibly be the graph of the polynomial? Explain.

a)

b)

c)

Cumulative Review Exercises

[1.4] **107.** Evaluate $\sqrt[4]{81}$.

[2.1] **108.** Solve $\dfrac{1}{2} = \dfrac{4}{5}x - \dfrac{1}{4}$.

[2.4] **109.** *Molding Machines* An older molding machine can produce 40 plastic buckets in 1 hour. A newer machine can produce 50 buckets in 1 hour. How long will it take the two machines to produce a total of 540 buckets?

[3.4] **110.** Find the slope of the line through the points $(8, -4)$ and $(-1, -2)$.

[4.2] **111.** Solve the system of equations.

$$-4s + 3t = 16$$
$$2t - 2u = 2$$
$$-s + 6u = -2$$

5.2 MULTIPLICATION OF POLYNOMIALS

SSM Study Guide CD/Video
MathPro 4/5 PH Math Tutor Center prenhall.com/Angel

1 Multiply a monomial by a polynomial.
2 Multiply a binomial by a binomial.
3 Multiply a polynomial by a polynomial.
4 Find the square of a binomial.
5 Find the product of the sum and difference of the same two terms.
6 Find the product of polynomial functions.

1 Multiply a Monomial by a Polynomial

In Section 3.6 we added and subtracted functions, but we did not multiply polynomial functions. After this section you will be able to find the product of functions, that is, $(f \cdot g)(x)$.

To multiply polynomials, you must remember that *each term of one polynomial must multiply each term of the other polynomial*. This results in monomials multiplying monomials. To multiply monomials, we use the product rule of exponents, which was presented in Section 1.5.

HELPFUL HINT	In this chapter you will be working with exponents. The rules for exponents were covered in Section 1.5. The rules for exponents that you will need to work problems in this chapter are presented again for your convenience, along with examples, prior to where you need to use them. Below we present the Product Rule for Exponents, and in Section 5.3 we present the Quotient Rule for Exponents and the Zero Exponent Rule. If, after reading the examples, you would like additional examples of the use of these rules, review Section 1.5.
STUDY TIP	

Product Rule for Exponents: $a^m \cdot a^n = a^{m+n}$

In Example 1, we review how to multiply monomials using the Product Rule for Exponents. Also, in Example 1, we use the word *factors*. Recall that any expressions that are *multiplied* are called *factors*.

EXAMPLE 1 Multiply. **a)** $(4x^2)(5x^6)$ **b)** $(3x^2 y)(4x^5 y^3)$ **c)** $(-2a^4 b^7)(-3a^8 b^3 c)$

Solution We use the Product Rule for Exponents to multiply the factors.

a) $(4x^2)(5x^6) = 4 \cdot 5 \cdot x^2 \cdot x^6$ *Remove parentheses and rearrange terms.*

$\qquad\qquad = 20x^{2+6}$ *Product Rule,* $x^2 \cdot x^6 = x^{2+6}$

$\qquad\qquad = 20x^8$

b) $(3x^2 y)(4x^5 y^3) = 3 \cdot 4 \cdot x^2 \cdot x^5 \cdot y \cdot y^3$ *Remove parentheses, rearrange terms.*

$\qquad\qquad\qquad = 12x^{2+5} y^{1+3}$ *Product Rule*

$\qquad\qquad\qquad = 12x^7 y^4$

c) $(-2a^4 b^7)(-3a^8 b^3 c) = (-2)(-3)a^4 \cdot a^8 \cdot b^7 \cdot b^3 \cdot c$ *Remove parentheses, rearrange terms.*

$\qquad\qquad\qquad\qquad = 6a^{4+8} b^{7+3} c$ *Product rule*

$\qquad\qquad\qquad\qquad = 6a^{12} b^{10} c$ ❋

In Example 1**a)**, both $4x^2$ and $5x^6$ are *factors* of the product $20x^8$. In Example 1**b)**, both $3x^2 y$ and $4x^5 y^3$ are *factors* of the product $12x^7 y^4$.

When multiplying a monomial by a binomial, we can use the distributive property. When multiplying a monomial by a polynomial that contains more than two terms we can use the **expanded form of the distributive property**.

Distributive Property, Expanded Form

$$a(b + c + d + \cdots + n) = ab + ac + ad + \cdots + an$$

EXAMPLE 2 Multiply.

a) $3x^2\left(\dfrac{1}{6}x^3 - 5x^2\right)$ b) $2xy(3x^2y + 6xy^2 + 4)$ c) $0.4x(0.3x^3 + 0.7xy^2 - 0.2y^4)$

Solution   a) $3x^2\left(\dfrac{1}{6}x^3 - 5x^2\right) = 3x^2\left(\dfrac{1}{6}x^3\right) - 3x^2(5x^2) = \dfrac{1}{2}x^5 - 15x^4$

b) $2xy(3x^2y + 6xy^2 + 4) = (2xy)(3x^2y) + (2xy)(6xy^2) + (2xy)(4)$
$$= 6x^3y^2 + 12x^2y^3 + 8xy$$

c) $0.4x(0.3x^3 + 0.7xy^2 - 0.2y^4)$
$$= (0.4x)(0.3x^3) + (0.4x)(0.7xy^2) - (0.4x)(0.2y^4)$$

NOW TRY EXERCISE 13 $= 0.12x^4 + 0.28x^2y^2 - 0.08xy^4$ ✳

2   Multiply a Binomial by a Binomial

Consider multiplying $(a + b)(c + d)$. Treating $(a + b)$ as a single term and using the distributive property, we get

$$(a + b)(c + d) = (a + b)c + (a + b)d$$
$$= ac + bc + ad + bd$$

When multiplying a binomial by a binomial, each term of the first binomial must be multiplied by each term of the second binomial, and then all like terms are combined.

Binomials can be multiplied vertically as well as horizontally.

EXAMPLE 3 Multiply $(3x + 2)(x - 5)$.

Solution   We will multiply vertically. List the binomials in descending order of the variable one beneath the other. It makes no difference which one is placed on top. Then multiply each term of the top binomial by each term of the bottom binomial, as shown. Remember to align like terms so that they can be added.

$$
\begin{array}{r}
3x + 2 \\
x - 5 \\
\hline
\end{array}
$$

$-5(3x + 2) \longrightarrow -15x - 10$ *Multiply the top binomial by -5.*
$x(3x + 2) \longrightarrow 3x^2 + 2x$ *Multiply the top binomial by x.*
$$3x^2 - 13x - 10$$ *Add like terms in columns.*   ✳

In Example 3, the binomials $3x + 2$ and $x - 5$ are *factors* of the trinomial $3x^2 - 13x - 10$.

The FOIL Method

A convenient way to multiply two binomials is called the **FOIL method**. To multiply two binomials using the FOIL method, list the binomials side by side. The word FOIL indicates that you multiply the **F**irst terms, **O**uter terms, **I**nner terms, and **L**ast terms of the two binomials. This procedure is illustrated in Example 4, where we multiply the same two binomials we multiplied in Example 3.

EXAMPLE 4   Multiply $(3x + 2)(x - 5)$ using the FOIL method.

Solution

$$
\begin{array}{ccccccc}
\text{F} & & \text{O} & & \text{I} & & \text{L} \\
(3x)(x) & + & (3x)(-5) & + & (2)(x) & + & (2)(-5)
\end{array}
$$

NOW TRY EXERCISE 21 $= \quad 3x^2 \quad - \quad 15x \quad + \quad 2x \quad - \quad 10 \quad = 3x^2 - 13x - 10$ ✳

We performed the multiplications following the FOIL order. However, any order could be followed as long as each term of one binomial is multiplied by each term of the other binomial. We use FOIL rather than OILF or any other combination of letters because it is easier to remember.

3   Multiply a Polynomial by a Polynomial

When multiplying a trinomial by a binomial or a trinomial by a trinomial, every term of the first polynomial must be multiplied by every term of the second polynomial. It is helpful for aligning terms to place each polynomial in descending order, if not given that way.

EXAMPLE 5   Multiply $x^2 + 1 - 4x$ by $2x^2 - 3$.

Solution   Since the trinomial is not in descending order, rewrite it as $x^2 - 4x + 1$.
      Place the longer polynomial on top, then multiply. Make sure you align like terms as you multiply so that the terms can be added more easily.

$$
\begin{array}{lr}
 & x^2 - 4x + 1 \qquad \text{\textit{Trinomial written in descending order}}\\
 & 2x^2 - 3 \\
\hline
-3\,(x^2 - 4x + 1) \longrightarrow \quad -3x^2 + 12x - 3 & \text{\textit{Multiply top expression by} } -3.\\
2x^2\,(x^2 - 4x + 1) \longrightarrow 2x^4 - 8x^3 + 2x^2 & \text{\textit{Multiply top expression by} } 2x^2.\\
\hline
2x^4 - 8x^3 - x^2 + 12x - 3 & \text{\textit{Add like terms in columns.}}
\end{array}
$$

✳

EXAMPLE 6   Multiply $3x^2 + 6xy - 5y^2$ by $x + 3y$.

Solution

$$
\begin{array}{lr}
 & 3x^2 + 6xy - 5y^2 \\
 & x + 3y \\
\hline
3y\,(3x^2 + 6xy - 5y^2) \longrightarrow \quad 9x^2y + 18xy^2 - 15y^3 & \text{\textit{Multiply top expression by} } 3y.\\
x\,(3x^2 + 6xy - 5y^2) \longrightarrow 3x^3 + 6x^2y - 5xy^2 & \text{\textit{Multiply top expression by} } x.\\
\hline
3x^3 + 15x^2y + 13xy^2 - 15y^3 & \text{\textit{Add like terms in columns.}}
\end{array}
$$

NOW TRY EXERCISE 31

✳

4 Find the Square of a Binomial

Now we will study some special formulas. We must often *square a binomial*, so we have special formulas for doing so.

Square of a Binomial

$$(a + b)^2 = a^2 + 2ab + b^2$$
$$(a - b)^2 = a^2 - 2ab + b^2$$

If you forget the formulas, you can easily derive them by multiplying $(a + b)(a + b)$ and $(a - b)(a - b)$.

Examples 7 and 8 illustrate the use of the square of a binomial formula.

EXAMPLE 7 Expand. **a)** $(3x + 5)^2$ **b)** $(4x^2 - 3y)^2$

Solution **a)** $(3x + 5)^2 = (3x)^2 + 2(3x)(5) + (5)^2$
$$= 9x^2 + 30x + 25$$

b) $(4x^2 - 3y)^2 = (4x^2)^2 - 2(4x^2)(3y) + (3y)^2$
$$= 16x^4 - 24x^2y + 9y^2$$

Squaring binomials, as in Example 7, can also be done using the FOIL method.

AVOIDING COMMON ERRORS

Remember the middle term when squaring a binomial.

CORRECT	INCORRECT
$(x + 2)^2 = (x + 2)(x + 2)$	$\cancel{(x + 2)^2 = x^2 + 4}$
$\qquad = x^2 + 4x + 4$	
$(x - 3)^2 = (x - 3)(x - 3)$	$\cancel{(x - 3)^2 = x^2 + 9}$
$\qquad = x^2 - 6x + 9$	

EXAMPLE 8 Expand $[x + (y - 1)]^2$.

Solution This problem looks more complicated than the previous example, but it is worked the same way as any other square of a binomial. Treat x as the first term and $(y - 1)$ as the second term. Use the formula twice.

$$[x + (y - 1)]^2 = (x)^2 + 2(x)(y - 1) + (y - 1)^2$$
$$= x^2 + (2x)(y - 1) + y^2 - 2y + 1$$
$$= x^2 + 2xy - 2x + y^2 - 2y + 1$$

None of the six terms are like terms, so no terms can be combined. Note that $(y - 1)^2$ is also the square of a binomial and was expanded as such.

NOW TRY EXERCISE 51

5 Find the Product of the Sum and Difference of the Same Two Terms

Below we multiply $(x + 6)(x - 6)$ using the FOIL method.

$$(x + 6)(x - 6) = x^2 - 6x + 6x - (6)(6) = x^2 - 6^2$$

Note that the outer and inner products add to 0. By examining this example, we see that the product of the sum and difference of the same two terms is the difference of the squares of the two terms.

> **Product of the Sum and Difference of the Same Two Terms**
>
> $$(a + b)(a - b) = a^2 - b^2$$

In other words, to multiply two binomials that differ only in the sign between their two terms, subtract the square of the second term from the square of the first term. Note that $a^2 - b^2$ represents a **difference of two squares**.

EXAMPLE 9 Multiply. **a)** $\left(3x + \dfrac{2}{5}\right)\left(3x - \dfrac{2}{5}\right)$ **b)** $(0.2x + 0.3z^2)(0.2x - 0.3z^2)$

Solution Each is a product of the sum and difference of the same two terms. Therefore,

a) $\left(3x + \dfrac{2}{5}\right)\left(3x - \dfrac{2}{5}\right) = (3x)^2 - \left(\dfrac{2}{5}\right)^2 = 9x^2 - \dfrac{4}{25}$

b) $(0.2x + 0.3z^2)(0.2x - 0.3z^2) = (0.2x)^2 - (0.3z^2)^2$
$$= 0.04x^2 - 0.09z^4$$ ✳

EXAMPLE 10 Multiply $(5x + y^3)(5x - y^3)$.

Solution $$(5x + y^3)(5x - y^3) = (5x)^2 - (y^3)^2 = 25x^2 - y^6$$ ✳

EXAMPLE 11 Multiply $[4x + (3y + 2)][4x - (3y + 2)]$.

Solution We treat $4x$ as the first term and $3y + 2$ as the second term. Then we have the sum and difference of the same two terms.

$$[4x + (3y + 2)][4x - (3y + 2)] = (4x)^2 - (3y + 2)^2$$
$$= 16x^2 - (9y^2 + 12y + 4)$$
$$= 16x^2 - 9y^2 - 12y - 4$$ ✳

NOW TRY EXERCISE 55

EXAMPLE 12 **Area** Figure 5.11 consists of a square and two rectangles. Find a polynomial expression for the total area of the figure.

FIGURE 5.11

Solution To find the total area, we find the areas of the three regions and then add.

$$\text{Area of the orange square} = x \cdot x = x^2$$
$$\text{Area of the blue rectangle} = x \cdot 4 = 4x$$
$$\text{Area of the green rectangle} = x \cdot 5 = 5x$$

The total area is the sum of these three quantities.

NOW TRY EXERCISE 85 Total area $= x^2 + 4x + 5x = x^2 + 9x$. ✳

6 Find the Product of Polynomial Functions

Earlier we learned that for functions $f(x)$ and $g(x)$, $(f \cdot g)(x) = f(x) \cdot g(x)$. Let's work one example involving multiplication of polynomial functions now.

EXAMPLE 13 Let $f(x) = x + 4$ and $g(x) = x - 2$. Find
a) $f(3) \cdot g(3)$ **b)** $(f \cdot g)(x)$ **c)** $(f \cdot g)(3)$

Solution **a)** Both $f(x)$ and $g(x)$ are polynomial functions since the expressions to the right of the equal signs are polynomials.

$$f(x) = x + 4 \qquad\qquad g(x) = x - 2$$
$$f(3) = 3 + 4 = 7 \qquad\qquad g(3) = 3 - 2 = 1$$
$$f(3) \cdot g(3) = 7 \cdot 1 = 7$$

b) From Section 3.6, we know that

$$(f \cdot g)(x) = f(x) \cdot g(x)$$
$$= (x + 4)(x - 2)$$
$$= x^2 - 2x + 4x - 8$$
$$= x^2 + 2x - 8$$

c) To evaluate $(f \cdot g)(3)$, substitute 3 for each x in $(f \cdot g)(x)$.

$$(f \cdot g)(x) = x^2 + 2x - 8$$
$$(f \cdot g)(3) = 3^2 + 2(3) - 8$$
$$= 9 + 6 - 8 = 7$$

NOW TRY EXERCISE 79

Note that in part **c)** we found $(f \cdot g)(3) = 7$ and in part **a)** we found $f(3) \cdot g(3) = 7$. Thus $(f \cdot g)(3) = f(3) \cdot g(3)$, which is what we expected from what we learned in Section 3.6.

$f(x) = x + 4$

$(f \cdot g) = x^2 + 2x - 8$

$g(x) = x - 2$

FIGURE 5.12

In Example 13, we found that if $f(x) = x + 4$ and $g(x) = x - 2$, then $(f \cdot g)(x) = x^2 + 2x - 8$. The graphs of $y = f(x) = x + 4$, $y = g(x) = x - 2$, and $y = (f \cdot g)(x) = x^2 + 2x - 8$ are shown in Figure 5.12. We see from the graphs that $f(3) = 7, g(3) = 1$, and $(f \cdot g)(3) = 7$, which is what we expected from Example 13. Every point on $y = x^2 + 2x - 8$ can be determined the same way. For example, $f(-4) = 0$ and $g(-4) = -6$. Since $0(-6) = 0$, $(f \cdot g)(-4) = 0$. Also $f(2) = 6$ and $g(2) = 0$; thus $(f \cdot g)(2) = 6 \cdot 0 = 0$. Notice in Figure 5.12 that the product of two linear functions gives a quadratic function.

Exercise Set 5.2

Concept/Writing Exercises

1. a) Explain how to multiply two binomials using the FOIL method.

b) Make up two binomials and multiply them using the FOIL method.

c) Multiply the same two binomials using LIOF (last, inner, outer, first).

d) Compare the results of parts **b)** and **c)**. If they are not the same, explain why.

2. a) Explain how to multiply a monomial by a polynomial.

b) Multiply $3x(4x^2 - 6x - 5)$ using your procedure from part **a)**.

3. a) Explain how to multiply a polynomial by a polynomial.

b) Multiply $4 + x$ by $x^2 - 6x + 3$ using your procedure from part **a)**.

4. a) Explain how to expand $(2x - 3)^2$ using the formula for the square of a binomial.

b) Expand $(2x - 3)^2$ using your procedure from part **a)**.

5. **a)** What is meant by the product of the sum and difference of the same two terms?

   b) Give an example of a problem that is the product of the sum and difference of the same two terms.

   c) How do you multiply the product of the sum and difference of the same two terms?

   d) Multiply the example you gave in part **b)** using your procedure from part **c)**.

6. Will the product of two binomials always be a
   a) binomial?
   b) trinomial? Explain.

7. Will the product of two first-degree polynomials always be a second-degree polynomial? Explain.

8. **a)** Given $f(x)$ and $g(x)$, explain how you would find $(f \cdot g)(x)$.
   b) If $f(x) = x - 2$ and $g(x) = x + 2$, find $(f \cdot g)(x)$.

Practice the Skills

Multiply.

9. $(4xy)(6xy^4)$

10. $(-2xy^4)(3x^4y^6)$

11. $\left(\frac{5}{9}x^2y^5\right)\left(\frac{1}{5}x^5y^3z^2\right)$

12. $2y^3(3y^2 + 2y - 6)$

13. $-3x^2y(-2x^4y^2 + 3xy^3 + 4)$

14. $3x^4(2xy^2 + 5x^7 - 6y)$

15. $\frac{2}{3}yz(3x + 4y - 9y^2)$

16. $\frac{1}{2}x^2y(4x^5y^2 + 3x - 6y^2)$

17. $0.3(2x^2 - 5x + 7y)$

18. $0.8(0.2a + 0.9b - 1.3c)$

19. $0.3a^5b^4(9.5a^6b - 4.6a^4b^3 + 1.2ab^5)$

20. $4.6m^2n(1.3m^4n^2 - 2.6m^3n^3 + 5.9n^4)$

Multiply the following binomials.

21. $(4x - 6)(3x - 5)$

22. $(2x - 1)(5x + 7)$

23. $(4 - x)(3 + 2x^2)$

24. $(5x + y)(6x - y)$

25. $\left(\frac{1}{2}x + 2y\right)\left(2x - \frac{1}{3}y\right)$

26. $\left(\frac{1}{3}a + \frac{1}{4}b\right)\left(\frac{1}{2}a - b\right)$

27. $(0.3a + 5b)(2a - 0.7b)$

28. $(4.6r - 5.8s)(0.2r - 2.3s)$

Multiply the following polynomials.

29. $(x^2 + 3x + 1)(x - 2)$

30. $(x + 3)(2x^2 - x - 9)$

31. $(a - 3b)(2a^2 - ab + 2b^2)$

32. $(7p - 3)(-2p^2 - 4p + 1)$

33. $(x^3 - x^2 + 3x + 7)(x + 1)$

34. $(2x - 1)(x^3 + 3x^2 - 5x + 6)$

35. $(5x^3 + 4x^2 - 6x + 2)(x + 5)$

36. $(a^3 - 2a^2 + 5a - 6)(2a^2 - 5a - 2)$

37. $(3m^2 - 2m + 4)(m^2 - 3m - 5)$

38. $(2a^2 - 6a + 3)(3a^2 - 5a - 2)$

39. $(2x - 1)^3$

40. $(3x + y)^3$

41. $(5r^2 - rs + 2s^2)(2r^2 - s^2)$

42. $(4x^2 - 5xy + y^2)(x^2 - 2y^2)$

Multiply using either the formula for the square of a binomial or for the product of the sum and difference of the same two terms.

43. $(x + 2)(x + 2)$

44. $(y - 4)(y - 4)$

45. $(2x - 3)(2x - 3)$

46. $(3z + 5)(3z + 5)$

47. $(4x - 3y)^2$

48. $(2a + 5b)^2$

49. $(5m^2 + 2n)(5m^2 - 2n)$

50. $(5p^2 + 6q^2)(5p^2 + 6q^2)$

51. $[y + (4 - 2x)]^2$

52. $[(a + b) + 9]^2$

53. $[5x + (2y + 3)]^2$

54. $[4 - (p - 3q)]^2$

55. $[a + (b + 2)][a - (b + 2)]$

56. $[2x + (y + 5)][2x - (y + 5)]$

Multiply.

57. $2xy(x^2 + xy + 3y^2)$

58. $3a^2b^2\left(\dfrac{1}{3}ab - \dfrac{1}{9}b^4\right)$

59. $\dfrac{1}{2}xy^2(4x^2 + 3xy - 7y^2)$

60. $-\dfrac{3}{5}x^2y\left(-\dfrac{2}{3}xy^4 + \dfrac{1}{9}xy + 3\right)$

61. $-\dfrac{3}{5}xy^3z^2\left(-xy^2z^5 - 5xy + \dfrac{1}{6}xz^7\right)$

62. $\dfrac{2}{3}x^2y^4\left(\dfrac{3}{5}xy^3 - \dfrac{1}{4}x^4y + 2xy^3z^5\right)$

63. $(3a + 4)(7a - 6)$

64. $(5p - 9q)(4p - 11q)$

65. $\left(8x + \dfrac{1}{4}\right)\left(8x - \dfrac{1}{4}\right)$

66. $\left(7a - \dfrac{1}{6}\right)\left(7a + \dfrac{1}{6}\right)$

67. $(2x - 9y)^2$

68. $\left(2x + \dfrac{1}{3}\right)^3$

69. $(x + 3)(2x^2 + 4x - 3)$

70. $(5a + 4)(a^2 - a + 4)$

71. $(2p - 3q)(3p^2 + 4pq - 2q^2)$

72. $(2m + n)(3m^2 - mn + 2n^2)$

73. $[(3x + 2) + y][(3x + 2) - y]$

74. $[a + (3b + 4)][a - (3b + 4)]$

75. $(a + b)(a - b)(a^2 - b^2)$

76. $(2a + 3)(2a - 3)(4a^2 + 9)$

77. $(x - 4)(6 + x)(2x - 8)$

78. $(3x - 5)(5 - 2x)(3x + 8)$

For the functions given, find **a)** $(f \cdot g)(x)$ *and* **b)** $(f \cdot g)(4)$.

79. $f(x) = x - 5, g(x) = x + 4$

80. $f(x) = 2x - 3, g(x) = x - 6$

81. $f(x) = 2x^2 + 6x - 4, g(x) = 5x + 3$

82. $f(x) = 4x^2 + 7, g(x) = 2 - x$

83. $f(x) = -x^2 + 3x, g(x) = x^2 + 2$

84. $f(x) = -x^2 + 2x + 7, g(x) = x^2 - 1$

Problem Solving

In Exercises 85–88, find a polynomial expression for the total area of each figure.

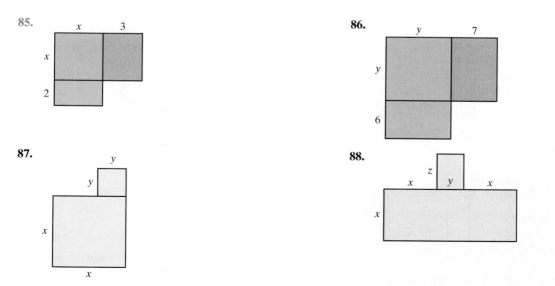

85.

86.

87.

88.

In Exercises 89 and 90, **a)** *find the area of the rectangle by finding the area of the four sections and adding them, and* **b)** *multiply the two sides and compare the product with your answer to part* **a)**.

89.

90.

Write a polynomial expression for the area of each figure. All angles in the figures are right angles.

91.

$6 - x$

$6 + x$

92.

$5 - x$

$10 + 2x$

In Exercises 93 and 94, **a)** *write a polynomial expression for the area of the shaded portion of the figure.* **b)** *The area of the shaded portion is indicated above each figure. Find the area of the larger and smaller rectangles.*

93.

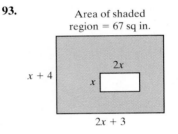

Area of shaded
region = 67 sq in.

$x + 4$

$2x$

x

$2x + 3$

94.

Area of shaded
region = 139 sq in.

$2x + 4$

$2x - 1$

$3x - 1$

$3x + 6$

95. Write two binomials whose product is $x^2 - 25$. Explain how you determined your answer.

96. Write two binomials whose product is $4x^2 - 9$. Explain how you determined your answer.

97. Write two binomials whose product is $x^2 + 12x + 36$. Explain how you determined your answer.

98. Write two binomials whose product is $4y^2 - 12y + 9$. Explain how you determined your answer.

99. Consider the expression $a(x - n)^3$. Write this expression as a product of factors.

100. Consider the expression $P(1 - r)^4$. Write this expression as a product of factors.

101. *Area* The expression $(a + b)^2$ can be represented by the following figure.

a b

a

b

a) Explain why this figure represents $(a + b)^2$.

b) Find $(a + b)^2$ using the figure by finding the area of each of the four parts of the figure, then adding the areas together.

c) Simplify $(a + b)^2$ by multiplying $(a + b)(a + b)$.

d) How do the answers in parts **b)** and **c)** compare? If they are not the same, explain why.

102. *Volume* The expression $(a + b)^3$ can be represented by the following figure.

a

b

a b b a

a) Explain why this figure represents $(a + b)^3$.

b) Find $(a + b)^3$ by adding the volume of the eight parts of the figure.

c) Simplify $(a + b)^3$ by multiplying.

d) How do the answers in parts **b)** and **c)** compare? If they are not the same, explain why.

103. ***Compound Interest*** The compound interest formula is

$$A = P\left(1 + \frac{r}{n}\right)^{nt}$$

where A is the amount, P is the principal invested, r is the annual rate of interest, n is the number of times the interest is compounded annually, and t is the time in years.

a) Simplify this formula for $n = 1$.

b) Find the value of A if $P = \$1000, n = 1, r = 6\%$, and $t = 2$ years.

104. ***Compound Interest*** Use the formula given in Exercise 103 to find A if $P = \$4000, n = 2, r = 8\%$, and $t = 2$ years.

105. If $f(x) = x^2 - 3x + 5$, find $f(a+b)$ by substituting $(a+b)$ for each x in the function.

106. If $f(x) = 2x^2 - x + 3$, find $f(a+b)$.

In Exercises 107–112, simplify. Assume that all variables represent natural numbers.

107. $3x^t(5x^{2t-1} + 6x^{3t})$

108. $5k^{r+2}(4k^{r+2} - 3k^r - k)$

109. $(6x^m - 5)(2x^{2m} - 3)$

110. $(x^{3n} - y^{2n})(x^{2n} + 2y^{4n})$

111. $(y^{a-b})^{a+b}$

112. $(a^{m+n})^{m+n}$

In Exercises 113 and 114, perform the polynomial multiplication.

113. $(x - 3y)^4$

114. $(2a - 4b)^4$

115. a) Explain how a multiplication in one variable, such as $(x^2 + 2x + 3)(x + 2) = x^3 + 4x^2 + 7x + 6$, may be checked using a graphing calculator.

b) Check the multiplication given in part **a)** using your grapher.

116. a) Show that the multiplication $(x^2 - 4x - 5)(x - 1) \ne x^3 + 6x^2 - 5x + 5$ using your grapher.

b) Multiply $(x^2 - 4x - 5)(x - 1)$.

c) Check your answer in part **b)** on your grapher.

Challenge Problems

Multiply.

117. $[(y + 1) - (x + 2)]^2$

118. $[(a - 2) - (a + 1)]^2$

Cumulative Review Exercises

[1.3] **119.** Evaluate $\dfrac{4}{5} - \left(\dfrac{3}{4} - \dfrac{2}{3}\right)$.

[1.5] **120.** Simplify $\left(\dfrac{2r^4s^5}{r^2}\right)^3$.

[2.5] **121.** Solve the inequality $-12 < 3x - 5 \le -4$ and give the solution in interval notation.

[3.2] **122.** If $g(x) = -x^2 - 2x + 3$, find $g\left(\dfrac{1}{2}\right)$.

5.3 DIVISION OF POLYNOMIALS AND SYNTHETIC DIVISION

SSM   Study Guide   CD/Video

MathPro 4/5   PH Math Tutor Center   prenhall.com/Angel

1 Divide a polynomial by a monomial.

2 Divide a polynomial by a binomial.

3 Divide polynomials using synthetic division.

4 Use the Remainder Theorem.

1 Divide a Polynomial by a Monomial

In division of polynomials, division by 0 is not permitted. When we are given a division problem containing a variable in the denominator, *we always assume that the denominator is nonzero.*

To divide a polynomial by a monomial, we use the fact that

$$\frac{A + B}{C} = \frac{A}{C} + \frac{B}{C}$$

If the polynomial has more than two terms, we expand this procedure.

To Divide a Polynomial by a Monomial

Divide each term of the polynomial by the monomial.

To divide a polynomial by a monomial, we will need to use two rules of exponents that were presented in Section 1.5, the Quotient Rule for Exponents and the Zero Exponent Rule. We state the two rules below and then give examples reviewing both rules.

$$\text{Quotient Rule for Exponents:} \quad \frac{a^m}{a^n} = a^{m-n}, \quad a \neq 0$$

$$\text{Zero Exponent Rule:} \quad a^0 = 1, \quad a \neq 0$$

EXAMPLE 1 Divide **a)** $\dfrac{x^7}{x^4}$ **b)** $\dfrac{5x^3y^5}{2xy^2}$.

Solution We will use the Quotient Rule to divide.

a) $\dfrac{x^7}{x^4} = x^{7-4}$ *Quotient Rule*

$\qquad = x^3$

b) $\dfrac{5x^3y^5}{2xy^2} = \dfrac{5}{2} \cdot \dfrac{x^3}{x} \cdot \dfrac{y^5}{y^2}$

$\qquad = \dfrac{5}{2}x^{3-1}y^{5-2}$ *Quotient Rule*

$\qquad = \dfrac{5x^2y^3}{2}$

EXAMPLE 2 Divide **a)** $\dfrac{p^4}{p^4}$ **b)** $\dfrac{8r^5s^7}{3rs^7}$.

Solution We will use both the Quotient Rule and the Zero Exponent Rule to divide.

a) $\dfrac{p^4}{p^4} = p^{4-4}$ *Quotient Rule*

$\qquad = p^0$

$\qquad = 1$ *Zero Exponent Rule*

b) $\dfrac{8r^5s^7}{3rs^7} = \dfrac{8}{3} \cdot \dfrac{r^5}{r} \cdot \dfrac{s^7}{s^7}$

$\qquad = \dfrac{8}{3}r^{5-1}s^{7-7}$ *Quotient Rule*

$\qquad = \dfrac{8}{3}r^4s^0$

$\qquad = \dfrac{8}{3}r^4(1)$ *Zero Exponent Rule*

$\qquad = \dfrac{8}{3}r^4$ or $\dfrac{8r^4}{3}$

In Example 2, both $\dfrac{8}{3}r^4$ and $\dfrac{8r^4}{3}$ are acceptable answers. Now we are ready to divide a polynomial by a monomial.

EXAMPLE 3 Divide $\dfrac{4x^2 - 8x - 3}{2x}$.

Solution

$$\frac{4x^2 - 8x - 3}{2x} = \frac{4x^2}{2x} - \frac{8x}{2x} - \frac{3}{2x}$$

$$= 2x - 4 - \frac{3}{2x}$$

EXAMPLE 4 Divide $\dfrac{4y - 6x^4y^3 - 3x^5y^2 + 5x}{2xy^2}$.

Solution

$$\frac{4y - 6x^4y^3 - 3x^5y^2 + 5x}{2xy^2} = \frac{4y}{2xy^2} - \frac{6x^4y^3}{2xy^2} - \frac{3x^5y^2}{2xy^2} + \frac{5x}{2xy^2}$$

$$= \frac{2}{xy} - 3x^3y - \frac{3x^4}{2} + \frac{5}{2y^2}$$

NOW TRY EXERCISE 17

2 Divide a Polynomial by a Binomial

We divide a polynomial by a binomial in much the same way as we perform long division. In a division problem, the expression we are dividing is called the *dividend* and the expression we are dividing by is called the *divisor*.

EXAMPLE 5 Divide $\dfrac{x^2 + 7x + 10}{x + 2}$.

Solution Rewrite the division problem as

$$x + 2 \overline{)x^2 + 7x + 10}$$

Divide x^2 (the first term in the dividend $x^2 + 7x + 10$) by x (the first term in the divisor $x + 2$).

$$\frac{x^2}{x} = x$$

Place the quotient, x, above the term containing x in the dividend.

$$\begin{array}{r} x \phantom{{}+7x+10} \\ x + 2 \overline{)x^2 + 7x + 10} \end{array}$$

Next, multiply the x by $x + 2$ as you would do in long division and place the product under the dividend, aligning like terms.

$$\begin{array}{r} \textbf{Times} \qquad\qquad x \\ x + 2 \overline{)x^2 + 7x + 10} \\ \textbf{Equals} \quad x^2 + 2x \quad \longleftarrow x(x + 2) \end{array}$$

Now subtract $x^2 + 2x$ from $x^2 + 7x$.

$$\begin{array}{r} x \\ x + 2\overline{)\,x^2 + 7x + 10} \\ \underline{-(x^2 + 2x)} \\ 5x \end{array}$$

Now bring down the next term, $+10$.

$$\begin{array}{r} x \\ x + 2\overline{)\,x^2 + 7x + 10} \\ \underline{x^2 + 2x} \\ 5x + 10 \end{array}$$

Divide $5x$ by x.

$$\frac{5x}{x} = +5$$

Place $+5$ above the constant in the dividend and multiply 5 by $x + 2$. Finish the problem by subtracting.

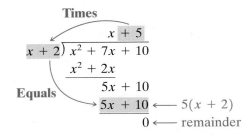

Thus, $\dfrac{x^2 + 7x + 10}{x + 2} = x + 5$. There is no remainder.

In Example 5, there was no remainder. Therefore, $x^2 + 7x + 10 = (x + 2)(x + 5)$. Note that $x + 2$ and $x + 5$ are both *factors* of $x^2 + 7x + 10$. In a division problem, if there is no remainder, then both the divisor and quotient are factors of the dividend.

When writing an answer in a division problem when there is a remainder, write the remainder over the divisor and add this expression to the quotient. For example, suppose that the remainder in Example 5 was 4. Then the answer would be written $x + 5 + \dfrac{4}{x + 2}$. If the remainder in Example 5 was -7, the answer would be written $x + 5 + \dfrac{-7}{x + 2}$, which we would rewrite as $x + 5 - \dfrac{7}{x + 2}$.

EXAMPLE 6 Divide $\dfrac{6x^2 - 7x + 3}{2x + 1}$.

Solution In this example we will subtract mentally and not show the change of sign in the subtractions.

$$
\begin{array}{r}
3x - 5 \\
2x + 1\overline{)6x^2 - 7x + 3} \\
\end{array}
$$

$$
\begin{array}{r}
6x^2 + 3x \quad \longleftarrow \quad 3x(2x + 1) \\
-10x + 3 \\
-10x - 5 \quad \longleftarrow \quad -5(2x + 1) \\
\hline
8 \quad \longleftarrow \quad \text{Remainder}
\end{array}
$$

NOW TRY EXERCISE 31

Thus, $\dfrac{6x^2 - 7x + 3}{2x + 1} = 3x - 5 + \dfrac{8}{2x + 1}$.

✸

When dividing a polynomial by a binomial, the answer may be *checked* by multiplying the divisor by the quotient and then adding the remainder. You should obtain the polynomial you began with. To check Example 6, we do the following:

$$(2x + 1)(3x - 5) + 8 = 6x^2 - 10x + 3x - 5 + 8 = 6x^2 - 7x + 3$$

Since we got the polynomial we began with, our division is correct.

When you are dividing a polynomial by a binomial, you should list both the polynomial and binomial in descending order. If a term of any degree is missing, it is often helpful to include that term with a numerical coefficient of 0. For example, when dividing $(6x^2 + x^3 - 4) \div (x - 2)$, we rewrite the problem as $(x^3 + 6x^2 + 0x - 4) \div (x - 2)$ before beginning the division.

EXAMPLE 7 Divide $(4x^2 - 12x + 3x^5 - 17)$ by $(-2 + x^2)$.

Solution Write both the dividend and divisor in descending powers of the variable x. This gives $(3x^5 + 4x^2 - 12x - 17) \div (x^2 - 2)$. Where a power of x is missing, add that power of x with a coefficient of 0, then divide.

$$
\begin{array}{r}
3x^3 \qquad\qquad + 6x + 4 \\
x^2 + 0x - 2\overline{)3x^5 + 0x^4 + 0x^3 + 4x^2 - 12x - 17} \\
\end{array}
$$

$$
\begin{array}{r}
3x^5 + 0x^4 - 6x^3 \quad \longleftarrow \qquad 3x^3(x^2 + 0x - 2) \\
\hline
6x^3 + 4x^2 - 12x \\
6x^3 + 0x^2 - 12x \quad \longleftarrow \quad 6x(x^2 + 0x - 2) \\
\hline
4x^2 + 0x - 17 \\
4x^2 + 0x - 8 \quad \longleftarrow \quad 4(x^2 + 0x - 2) \\
\hline
-9 \quad \longleftarrow \quad \text{Remainder}
\end{array}
$$

In obtaining the answer, we performed the divisions

$$\frac{3x^5}{x^2} = 3x^3 \qquad \frac{6x^3}{x^2} = 6x \qquad \frac{4x^2}{x^2} = 4$$

The quotients $3x^3$, $6x$, and 4 were placed above their like terms in the dividend. The answer is $3x^3 + 6x + 4 - \dfrac{9}{x^2 - 2}$. You should check this answer for yourself by

NOW TRY EXERCISE 45
multiplying the divisor by the quotient and adding the remainder.

✸

3 Divide Polynomials Using Synthetic Division

When a polynomial is divided by a binomial of the form $x - a$, the division process can be greatly shortened by a process called **synthetic division**. Consider the following examples. In the example on the right, we use only the numerical coefficients.

$$
\begin{array}{r}
2x^2 + 5x - 4 \\
x - 3\overline{)2x^3 - x^2 - 19x + 15} \\
\underline{2x^3 - 6x^2} \\
5x^2 - 19x \\
\underline{5x^2 - 15x} \\
-4x + 15 \\
\underline{-4x + 12} \\
3
\end{array}
\qquad
\begin{array}{r}
2 + 5 - 4 \\
1 - 3\overline{)2 - 1 - 19 + 15} \\
\underline{2 - 6} \\
5 - 19 \\
\underline{5 - 15} \\
-4 + 15 \\
\underline{-4 + 12} \\
3
\end{array}
$$

Note that the variables do not play a role in determining the numerical coefficients of the quotient. This division problem can be done more quickly and easily using synthetic division.

Following is an explanation of how we use synthetic division. Consider again the division

$$\frac{2x^3 - x^2 - 19x + 15}{x - 3}$$

1. Write the dividend in descending powers of x. Then list the numerical coefficients of each term in the dividend. If a term of any degree is missing, place 0 in the appropriate position to serve as a placeholder. In the problem above, the numerical coefficients of the dividend are

$$2 \quad -1 \quad -19 \quad 15$$

2. When dividing by a binomial of the form $x - a$, place a to the left of the line of numbers from step 1. In this problem, we are dividing by $x - 3$; thus, $a = 3$. We write

$$3\rfloor \quad 2 \quad -1 \quad -19 \quad 15$$

3. Leave some space under the row of coefficients, then draw a horizontal line. Bring down the first coefficient on the left as follows:

$$
\begin{array}{c|cccc}
3 & 2 & -1 & -19 & 15 \\
\hline
& 2 & & &
\end{array}
$$

4. Multiply the 3 by the number brought down, the 2, to get 6. Place the 6 under the next coefficient, the -1. Then add $-1 + 6$ to get 5.

$$
\begin{array}{c|cccc}
3 & 2 & -1 & -19 & 15 \\
& & 6 & & \\
\hline
& 2 & 5 & &
\end{array}
$$

5. Multiply the 3 by the sum 5 to get 15. Place 15 under -19. Then add to get -4. Repeat this procedure as illustrated.

In the last row, the first three numbers are the numerical coefficients of the quotient, as shown in the long division. The last number, 3, is the remainder obtained by long division. The quotient must be one less than the degree of the dividend since we are dividing by $x - 3$. The original dividend was a third-degree polynomial. Therefore, the quotient must be a second-degree polynomial. Use the first three numbers from the last row as the coefficients of a second-degree polynomial in x. This gives $2x^2 + 5x - 4$, which is the quotient. The last number, 3, is the remainder. Therefore,

$$\frac{2x^3 - x^2 - 19x + 15}{x - 3} = 2x^2 + 5x - 4 + \frac{3}{x - 3}$$

EXAMPLE 8 Use synthetic division to divide.

$$(6 - x^2 + x^3) \div (x + 2)$$

Solution First, list the terms of the dividend in descending order of x.

$$(x^3 - x^2 + 6) \div (x + 2)$$

Since there is no first-degree term, insert 0 as a placeholder when listing the numerical coefficients. Since $x + 2 = x - (-2)$, $a = -2$.

$$
\begin{array}{r|rrrr}
-2 & 1 & -1 & 0 & 6 \\
   &   & -2 & 6 & -12 \\
\hline
   & 1 & -3 & 6 & -6 \quad \leftarrow \textit{Remainder}
\end{array}
$$

Since the dividend is a third-degree polynomial, the quotient must be a second-degree polynomial. The answer is $x^2 - 3x + 6 - \dfrac{6}{x + 2}$. ✳

EXAMPLE 9 Use synthetic division to divide.

$$(3x^4 + 11x^3 - 20x^2 + 7x + 35) \div (x + 5)$$

Solution

$$
\begin{array}{r|rrrrr}
-5 & 3 & 11 & -20 & 7 & 35 \\
   &   & -15 & 20 & 0 & -35 \\
\hline
   & 3 & -4 & 0 & 7 & 0 \quad \leftarrow \textit{Remainder}
\end{array}
$$

Since the dividend is of the fourth degree, the quotient must be of the third degree. The quotient is $3x^3 - 4x^2 + 0x + 7$ with no remainder. This can be simplified to $3x^3 - 4x^2 + 7$. ✳

NOW TRY EXERCISE 61

In Example 9, since there was no remainder, both $x + 5$ and $3x^3 - 4x + 7$ are *factors* of $3x^4 + 11x^3 - 20x^2 + 7x + 35$. Furthermore, since both are factors,

$$(x + 5)(3x^3 - 4x^2 + 7) = 3x^4 + 11x^3 - 20x^2 + 7x + 35$$

4 Use the Remainder Theorem

In Example 8, when we divided $x^3 - x^2 + 6$ by $x + 2$, we found that the remainder was -6. If we write $x + 2$ as $x - (-2)$ and evaluate the polynomial function $P(x) = x^3 - x^2 + 6$ at -2, we obtain -6.

$$P(x) = x^3 - x^2 + 6$$
$$P(-2) = (-2)^3 - (-2)^2 + 6 = -8 - 4 + 6 = -6$$

Is it just a coincidence that $P(-2)$, the value of the function at -2, is the same as the remainder when the function $P(x)$ is divided by $x - (-2)$? The answer is no.

It can be shown that for any polynomial function $P(x)$, the value of the function at a, $P(a)$, has the same value as the remainder when $P(x)$ is divided by $x - a$.

To obtain the remainder when a polynomial $P(x)$ is divided by a binomial of the form $x - a$, we can use the **Remainder Theorem**.

Remainder Theorem

If the polynomial $P(x)$ is divided by $x - a$, the remainder is equal to $P(a)$.

EXAMPLE 10 Use the Remainder Theorem to find the remainder when $3x^4 + 6x^3 - 2x + 4$ is divided by $x + 4$.

Solution First we write the divisor $x + 4$ in the form $x - a$. Since $x + 4 = x - (-4)$, we evaluate $P(-4)$.

$$P(x) = 3x^4 + 6x^3 - 2x + 4$$
$$P(-4) = 3(-4)^4 + 6(-4)^3 - 2(-4) + 4$$
$$= 3(256) + 6(-64) + 8 + 4$$
$$= 768 - 384 + 8 + 4 = 396$$

Thus, when $3x^4 + 6x^3 - 2x + 4$ is divided by $x + 4$, the remainder is 396. ✳

Using synthetic division, we will show that the remainder in Example 10 is indeed 396.

$$
\begin{array}{r|rrrrr}
-4 & 3 & 6 & 0 & -2 & 4 \\
   &   & -12 & 24 & -96 & 392 \\
\hline
   & 3 & -6 & 24 & -98 & 396 \quad \leftarrow \textit{Remainder}
\end{array}
$$

If we were to graph the polynomial $P(x) = 3x^4 + 6x^3 - 2x + 4$, the value of $P(x)$, or y, at $x = -4$ would be 396.

EXAMPLE 11 Use the Remainder Theorem to determine whether $x - 5$ is a factor of $6x^2 - 25x - 25$.

Solution Let $P(x) = 6x^2 - 25x - 25$. If $P(5) = 0$, then the remainder of $(6x^2 - 25x - 25)/(x - 5)$ is 0 and $x - 5$ is a factor of the polynomial. If $P(5) \neq 0$, then there is a remainder and $x - 5$ is not a factor.

$$P(x) = 6x^2 - 25x - 25$$
$$P(5) = 6(5)^2 - 25(5) - 25$$
$$= 6(25) - 25(5) - 25$$
$$= 150 - 125 - 25 = 0$$

Since $P(5) = 0$, $x - 5$ is a factor of $6x^2 - 25x - 25$. Note that

NOW TRY EXERCISE 79 $6x^2 - 25x - 25 = (x - 5)(6x + 5)$. ✳

Exercise Set 5.3

Concept/Writing Exercises

1. a) Explain how to divide a polynomial by a monomial.

b) Divide $\dfrac{5x^4 - 6x^3 - 4x^2 - 12x + 7}{3x}$ using the procedure you gave in part **a)**.

2. a) Explain how to divide a trinomial in x by a binomial in x.

b) Divide $2x^2 - 12 + 5x$ by $x + 4$ using the procedure you gave in part **a)**.

3. A trinomial divided by a binomial has a remainder of 0. Is the quotient a factor of the trinomial? Explain.

4. a) Explain how the answer may be checked when dividing a polynomial by a binomial.

    b) Use your explanation in part **a)** to check whether the following division is correct.

$$\frac{8x^2 + 2x - 15}{4x - 5} = 2x + 3$$

    c) Check to see whether the following division is correct.

$$\frac{6x^2 - 23x + 14}{3x - 4} = 2x - 5 - \frac{8}{3x - 4}$$

5. When dividing a polynomial by a polynomial, before you begin the division, what should you do to the polynomials?

6. Explain why $\frac{x - 1}{x}$ is not a polynomial.

7. a) Describe in your own words how to divide a polynomial by $(x - a)$ using synthetic division.

    b) Divide $x^2 + 3x - 4$ by $x - 5$ using the procedure you gave in part **a)**.

8. a) State the Remainder Theorem in your own words.

    b) Find the remainder when $x^2 - 6x - 4$ is divided by $x - 1$, using the procedure you stated in part **a)**.

9. In the division problem $\frac{x^2 + 11x + 21}{x + 2} = x + 9 + \frac{3}{x + 2}$, is $x + 9$ a factor of $x^2 + 11x + 21$? Explain.

10. In the division problem $\frac{x^2 - 3x - 28}{x + 4} = x - 7$, is $x - 7$ a factor of $x^2 - 3x - 28$? Explain.

Practice the Skills

Divide.

11. $\dfrac{4x + 18}{2}$

12. $\dfrac{9x + 17}{3}$

13. $\dfrac{4x^2 + 2x}{2x}$

14. $\dfrac{12x^2 - 8x - 20}{4}$

15. $\dfrac{5y^3 + 6y^2 - 9y}{3y}$

16. $\dfrac{18y^5 + 12y^2}{6y^4}$

17. $\dfrac{4x^5 - 6x^4 + 12x^3 - 8x^2}{4x^2}$

18. $\dfrac{15x^3y - 25xy^3}{5xy}$

19. $\dfrac{8x^2y^2 - 10xy^3 - 5y}{2y^2}$

20. $\dfrac{4x^{13} + 12x^9 - 7x^7}{4x^6}$

21. $\dfrac{9x^2y - 12x^3y^2 + 7y^3}{2xy^2}$

22. $\dfrac{a^2b^2c - 6abc^2 + 5a^3b^5}{2abc^2}$

23. $\dfrac{3xyz + 6xyz^2 - 9x^3y^5z^7}{6xy}$

24. $\dfrac{6abc^3 - 5a^2b^3c^4 + 8ab^5c}{3ab^2c^3}$

Divide using long division.

25. $\dfrac{x^2 + 3x + 2}{x + 1}$

26. $\dfrac{x^2 + 2x - 15}{x + 5}$

27. $\dfrac{2x^2 + 11x + 12}{x + 4}$

28. $\dfrac{6x^2 + 16x + 8}{3x + 2}$

29. $\dfrac{6x^2 + x - 2}{2x - 1}$

30. $\dfrac{12x^2 + 25x + 7}{3x + 1}$

31. $\dfrac{x^2 + 6x + 7}{x + 1}$

32. $\dfrac{a^2 - a - 13}{a + 3}$

33. $\dfrac{2b^2 + b - 9}{b - 2}$

34. $\dfrac{2c^2 + c - 1}{2c + 5}$

35. $\dfrac{8x^2 + 6x - 25}{2x - 3}$

36. $\dfrac{8z^2 - 18z - 9}{4z + 1}$

37. $\dfrac{4x^2 - 25}{2x - 5}$

38. $\dfrac{16p^2 - 9}{4p + 3}$

39. $\dfrac{x^3 + 3x^2 + 5x + 4}{x + 1}$

40. $\dfrac{-a^3 - 6a^2 + 2a - 3}{a - 1}$

41. $\dfrac{9b^3 - 3b^2 - 3b + 4}{3b + 2}$

42. $\dfrac{4y^3 + 12y^2 + 7y - 8}{2y + 3}$

43. $(2x^3 + 6x - 4) \div (x + 4)$

44. $(4a^3 - 5a) \div (2a - 1)$

45. $\dfrac{3x^5 + 4x^2 - 12x - 8}{x^2 - 2}$

46. $\dfrac{4b^5 - 18b^3 + 8b^2 + 18b - 12}{2b^2 - 3}$

47. $\dfrac{3x^4 + 4x^3 - 32x^2 - 5x - 20}{3x^3 - 8x^2 - 5}$

48. $\dfrac{3a^4 - 9a^3 + 13a^2 - 11a + 4}{a^2 - 2a + 1}$

49. $\dfrac{2c^4 - 8c^3 + 19c^2 - 33c + 15}{c^2 - c + 5}$

50. $\dfrac{2y^5 + 2y^4 - 3y^3 - 15y^2 + 18}{2y^2 - 3}$

Use synthetic division to divide.

51. $(x^2 + 7x + 6) \div (x + 1)$

52. $(x^2 - 7x + 6) \div (x - 1)$

53. $(x^2 + 5x + 6) \div (x + 2)$

54. $(x^2 - 5x + 6) \div (x - 2)$

55. $(x^2 - 11x + 28) \div (x - 4)$

56. $(x^2 + x - 72) \div (x + 9)$

57. $(x^2 + 5x - 12) \div (x - 3)$

58. $(x^2 - 2x - 37) \div (x + 5)$

59. $(3x^2 - 7x - 10) \div (x - 4)$

60. $(2b^2 - 9b + 15) \div (b - 6)$

61. $(4x^3 - 3x^2 + 2x) \div (x - 1)$

62. $(z^3 - 7z^2 - 13z + 15) \div (z - 2)$

63. $(3c^3 + 7c^2 - 4c + 18) \div (c + 3)$

64. $(3y^4 - 25y^2 - 29) \div (y - 3)$

65. $(y^4 - 1) \div (y - 1)$

66. $(a^4 - 16) \div (a - 2)$

67. $\dfrac{x^4 + 16}{x + 4}$

68. $\dfrac{z^4 + 81}{z + 3}$

69. $\dfrac{x^5 + x^4 - 10}{x + 1}$

70. $\dfrac{a^7 - 2a^6 + 19}{a - 2}$

71. $\dfrac{b^5 + 4b^4 - 12}{b + 1}$

72. $\dfrac{z^5 - 3z^3 - 7z}{z - 2}$

73. $(3x^3 + 2x^2 - 4x + 1) \div \left(x - \dfrac{1}{3}\right)$

74. $(8x^3 - 6x^2 - 5x + 3) \div \left(x + \dfrac{3}{4}\right)$

75. $(2x^4 - x^3 + 2x^2 - 3x + 1) \div \left(x - \dfrac{1}{2}\right)$

76. $(9y^3 + 9y^2 - y + 2) \div \left(y + \dfrac{2}{3}\right)$

Determine the remainder for the following divisions using the Remainder Theorem. If the divisor is a factor of the dividend, so state.

77. $(4x^2 - 5x + 4) \div (x - 2)$

78. $(-2x^2 + 3x - 2) \div (x + 3)$

79. $(x^3 - 2x^2 + 4x - 8) \div (x - 2)$

80. $(-3x^3 + 4x - 12) \div (x + 4)$

81. $(-2x^3 - 6x^2 + 2x - 4) \div \left(x - \dfrac{1}{2}\right)$

82. $(-5x^3 - 6) \div \left(x - \dfrac{1}{5}\right)$

Problem Solving

83. *Area* The area of a rectangle is $6x^2 - 8x - 8$. If the length is $2x - 4$, find the width.

84. *Area* The area of a rectangle is $15x^2 - 29x - 14$. If the width is $5x + 2$, find the length.

In Exercises 85 and 86, how many times greater is the area or volume of the figure on the right than the figure on the left? Explain how you determined your answer.

85.

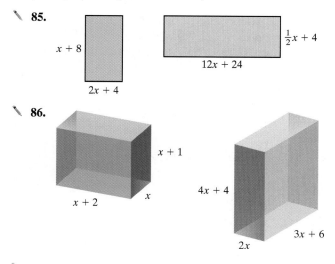

86.

87. Is it possible to divide a binomial by a monomial and obtain a monomial as a quotient? Explain.

88. a) Is the sum, difference, and product of two polynomials always a polynomial?

   b) Is the quotient of two polynomials always a polynomial? Explain.

89. Explain how you can determine using synthetic division if an expression of the form $x - a$ is a factor of a polynomial in x.

90. Given $P(x) = ax^2 + bx + c$ and a value d such that $P(d) = 0$, explain why d is a solution to the equation $ax^2 + bx + c = 0$.

91. If $\dfrac{P(x)}{x - 4} = x + 2$, find $P(x)$.

92. If $\dfrac{P(x)}{2x + 4} = x - 3$, find $P(x)$.

93. If $\dfrac{P(x)}{x + 4} = x + 5 + \dfrac{4}{x + 4}$, find $P(x)$.

94. If $\dfrac{P(x)}{2x - 3} = 2x - 1 - \dfrac{7}{2x - 3}$, find $P(x)$.

In Exercises 95 and 96, divide.

95. $\dfrac{2x^3 - x^2y - 7xy^2 + 2y^3}{x - 2y}$

96. $\dfrac{x^3 + y^3}{x + y}$

In Exercises 97 and 98, divide. The answers contain fractions.

97. $\dfrac{2x^2 + 2x - 2}{2x - 3}$

98. $\dfrac{3x^3 - 5}{3x - 2}$

99. *Volume* The volume of the box that follows is $2r^3 + 4r^2 + 2r$. Find w in terms of r.

100. *Volume* The volume of the box that follows is $6a^3 + a^2 - 2a$. Find b in terms of a.

101. When a polynomial is divided by $x - 3$, the quotient is $x^2 - 3x + 4 + \dfrac{2}{x - 3}$. What is the polynomial? Explain how you determined your answer.

102. When a polynomial is divided by $2x - 3$, the quotient is $2x^2 + 6x - 5 + \dfrac{5}{2x - 3}$. What is the polynomial? Explain how you determined your answer.

In Exercises 103 and 104, divide. Assume that all variables in the exponents are natural numbers.

103. $\dfrac{4x^{n+1} + 2x^n - 3x^{n-1} - x^{n-2}}{2x^n}$

104. $\dfrac{3x^n + 6x^{n-1} - 2x^{n-2}}{2x^{n-1}}$

105. Is $x - 1$ a factor of $x^{100} + x^{99} + \cdots + x^1 + 1$? Explain.

106. Is $x + 1$ a factor of $x^{100} + x^{99} + \cdots + x^1 + 1$? Explain.

107. Is $x + 1$ a factor of $x^{99} + x^{98} + \cdots + x^1 + 1$? Explain.

108. Divide $0.2x^3 - 4x^2 + 0.32x - 0.64$ by $x - 0.4$.

109. Synthetic division can be used to divide polynomials by binomials of the form $ax - b, a \neq 1$. To perform this division, divide $ax - b$ by a to obtain $x - \dfrac{b}{a}$. Then place $\dfrac{b}{a}$ to the left of the numerical coefficients of the polynomial. Work the problem as explained previously.

After summing the numerical values below the line, divide all of them, except the remainder, by a. Then write the quotient of the problem using these numbers.

a) Use this procedure to divide $9x^3 + 9x^2 + 5x + 12$ by $3x + 5$.

b) Explain why we do not divide the remainder by a.

Cumulative Review Exercises

[1.6] **110.** Divide $\dfrac{8.45 \times 10^{25}}{4.225 \times 10^{15}}$ and express the answer in scientific notation.

[2.3] **111.** *Triangle* Find the three angles of a triangle if one angle is twice the smallest angle and the third angle is 60° greater than the smallest angle.

[2.6] **112.** Find the solution set for $\left|\dfrac{5x - 3}{2}\right| + 3 = 7$.

[3.6] **113.** Let $f(x) = x^2 - 4$ and $g(x) = -5x + 3$. Find $f(6) \cdot g(6)$.

[5.1] **114.** Add $(6r + 5s - t) + (-3r - 2s - 5t)$.

5.4 FACTORING A MONOMIAL FROM A POLYNOMIAL AND FACTORING BY GROUPING

1 Find the greatest common factor.

2 Factor a monomial from a polynomial.

3 Factor a common binomial factor.

4 Factor by grouping.

Factoring is the opposite of multiplying. To factor an expression means to write it as a product of other expressions. For example, in Section 5.2 we learned to perform the following multiplications:

$$3x^2(6x + 3y + 5x^3) = 18x^3 + 9x^2y + 15x^5$$

and

$$(6x + 3y)(2x - 5y) = 12x^2 - 24xy - 15y^2$$

In this section, we learn how to determine the *factors* of a given expression. For example, we will learn how to perform each factoring illustrated below.

$$18x^3 + 9x^2y + 15x^5 = 3x^2(6x + 3y + 5x^3)$$

and

$$12x^2 - 24xy - 15y^2 = (6x + 3y)(2x - 5y)$$

1 Find the Greatest Common Factor

To factor a monomial from a polynomial, we factor the **greatest common factor** (GCF) from each term in the polynomial. The GCF is the product of the factors common to all terms in the polynomial. For example, the GCF for $6x + 15$ is 3 since 3 is the largest number that is a factor of both $6x$ and 15. To factor, we use the distributive property.

$$6x + 15 = 3(2x + 5)$$

The 3 and the $2x + 5$ are *factors* of the polynomial $6x + 15$.

Consider the terms $x^3, x^4, x^5,$ and x^6. The GCF of these terms is x^3, since x^3 is the highest power of x that divides all four terms.

EXAMPLE 1 Find the GCF of the following terms.

a) y^{12}, y^4, y^9, y^7 **b)** x^3y^2, xy^4, x^4y^5 **c)** $6x^2y^3z, 9x^3y^4, 24x^2z^5$

Solution **a)** Note that y^4 is the highest power of y common to all four terms. The GCF is, therefore, y^4.

b) The highest power of x that is common to all three terms is x (or x^1). The highest power of y that is common to all three terms is y^2. Thus, the GCF of the three terms is xy^2.

c) The GCF is $3x^2$. Since y does not appear in $24x^2z^5$, it is not part of the GCF. Since z does not appear in $9x^3y^4$, it is not part of the GCF.   ✳

EXAMPLE 2 Find the GCF of the following terms.

$$6(x - 3)^2, 5(x - 3), 18(x - 3)^4$$

Solution The three numbers 6, 5, and 18 have no common factor other than 1. The highest power of $(x - 3)$ common to all three terms is $(x - 3)$. Thus, the GCF of the three terms is $(x - 3)$.   ✳

2 Factor a Monomial from a Polynomial

When we factor a monomial from a polynomial, we are factoring out the greatest common factor. *The first step in any factoring problem is to determine and then factor out the GCF.*

> **To Factor a Monomial from a Polynomial**
>
> 1. Determine the greatest common factor of all terms in the polynomial.
> 2. Write each term as the product of the GCF and another factor.
> 3. Use the distributive property to *factor out* the GCF.

EXAMPLE 3 Factor $15x^4 - 5x^3 + 20x^2$.

Solution The GCF is $5x^2$. Write each term as the product of the GCF and another product. Then factor out the GCF.

$$15x^4 - 5x^3 + 20x^2 = 5x^2 \cdot 3x^2 - 5x^2 \cdot x + 5x^2 \cdot 4$$
$$= 5x^2(3x^2 - x + 4)$$ ✳

To check the factoring process, multiply the factors using the distributive property. The product should be the expression with which you began. For instance, in Example 3,

Check $5x^2(3x^2 - x + 4) = 5x^2(3x^2) + 5x^2(-x) + 5x^2(4)$
$$= 15x^4 - 5x^3 + 20x^2$$

EXAMPLE 4   Factor $20x^3y^3 + 6x^2y^4 - 12xy^5$.

Solution   The GCF is $2xy^3$. Write each term as the product of the GCF and another product. Then factor out the GCF.

$$20x^3y^3 + 6x^2y^4 - 12xy^5 = \mathbf{2xy^3} \cdot 10x^2 + \mathbf{2xy^3} \cdot 3xy - \mathbf{2xy^3} \cdot 6y^2$$
$$= 2xy^3(10x^2 + 3xy - 6y^2)$$

NOW TRY EXERCISE 19   Check $2xy^3(10x^2 + 3xy - 6y^2) = 20x^3y^3 + 6x^2y^4 - 12xy^5$ ✳

When the leading coefficient of a polynomial is negative, we generally factor out a common factor with a negative coefficient. This results in the leading coefficient of the remaining polynomial being positive.

EXAMPLE 5   Factor.   **a)** $-12a - 18$ **b)** $-2b^3 + 6b^2 - 18b$

Solution   Since the leading coefficients in parts **a)** and **b)** are negative, we factor out common factors with a negative coefficient.

a) $-12a - 18 = -6(2a + 3)$ *Factor out −6.*

b) $-2b^3 + 6b^2 - 18b = -2b(b^2 - 3b + 9)$ *Factor out −2b.*   ✳

EXAMPLE 6   ***Throwing a Ball*** When a ball is thrown upward with a velocity of 32 feet per second from the top of a 160-foot-tall building, its distance, d, from the ground at any time, t, can be determined by the function $d(t) = -16t^2 + 32t + 160$.

a) Determine the ball's distance from the ground after 3 seconds—that is, find $d(3)$.

b) Factor out the GCF from the right side of the function.

c) Evaluate $d(3)$ in factored form.

d) Compare your answers to parts **a)** and **c)**.

Solution   **a)** $d(t) = -16t^2 + 32t + 160$

$d(3) = -16(3)^2 + 32(3) + 160$ *Substitute 3 for t.*

$ = -16(9) + 96 + 160$

$ = 112$

The distance is 112 feet.

b) Factor -16 from the three terms on the right side of the equal sign.

$$d(t) = -16(t^2 - 2t - 10)$$

c) $d(t) = -16(t^2 - 2t - 10)$

$d(3) = -16[3^2 - 2(3) - 10]$ *Substitute 3 for t.*

$ = -16(9 - 6 - 10)$

$ = -16(-7)$

$ = 112$

d) The answers are the same. You may find the calculations in part **c)** easier than

NOW TRY EXERCISE 65   the calculations in part **a)**.   ✳

3 Factor a Common Binomial Factor

Sometimes factoring involves factoring a binomial as the greatest common factor, as illustrated in Examples 7 through 9.

EXAMPLE 7   Factor $3x(5x - 2) + 4(5x - 2)$.

Solution   The GCF is $(5x - 2)$. Factoring out the GCF gives

$$3x(5x - 2) + 4(5x - 2) = (5x - 2)(3x + 4)$$ ✳

In Example 7, we could have also placed the common factor on the right to obtain

$$3x(5x - 2) + 4(5x - 2) = (3x + 4)(5x - 2)$$

The factored forms $(5x - 2)(3x + 4)$ and $(3x + 4)(5x - 2)$ are equivalent because of the commutative property of multiplication and both are correct. Generally, when we list the answer to an example or exercise, we will place the common term that has been factored out on the left.

EXAMPLE 8   Factor $9(2x - 5) + 6(2x - 5)^2$.

Solution   The GCF is $3(2x - 5)$. Rewrite each term as the product of the GCF and another factor.

$$\begin{aligned} 9(2x - 5) + 6(2x - 5)^2 &= 3(2x - 5) \cdot 3 + 3(2x - 5) \cdot 2(2x - 5) \\ &= 3(2x - 5)[3 + 2(2x - 5)] \quad \text{\textit{Factor out }} 3(2x - 5). \\ &= 3(2x - 5)[3 + 4x - 10] \quad \text{\textit{Distributive property}} \\ &= 3(2x - 5)(4x - 7) \quad \text{\textit{Simplify.}} \end{aligned}$$ ✳

EXAMPLE 9   Factor $(2x - 5)(a + b) - (x - 1)(a + b)$.

Solution   The binomial $a + b$ is the GCF of the two terms. We therefore factor it out.

$$\begin{aligned} (2x - 5)(a + b) - (x - 1)(a + b) &= (a + b)[(2x - 5) - (x - 1)] \quad \text{\textit{Factor out }} (a+b). \\ &= (a + b)(2x - 5 - x + 1) \quad \text{\textit{Simplify.}} \\ &= (a + b)(x - 4) \quad \text{\textit{Factors}} \end{aligned}$$ ✳

NOW TRY EXERCISE 43

EXAMPLE 10   **Area**   In Figure 5.13, the area of the large rectangle is $7x(2x + 9)$ and the area of the small rectangle is $3(2x + 9)$. Find an expression, in factored form, for the difference of the areas for these two rectangles.

$A = 7x(2x + 9)$

$A = 3(2x + 9)$

FIGURE 5.13

Solution   To find the difference of the areas, subtract the area of the small rectangle from the area of the large rectangle.

$$7x(2x + 9) - 3(2x + 9) \quad \text{\textit{Subtract areas.}}$$
$$= (2x + 9)(7x - 3) \quad \text{\textit{Factor out }} (2x + 9).$$

NOW TRY EXERCISE 59   The difference of the areas for the two rectangles is $(2x + 9)(7x - 3)$. ✳

4   Factor by Grouping

When a polynomial contains *four terms*, it may be possible to factor the polynomial by grouping. To **factor by grouping**, remove common factors from groups of terms. This procedure is illustrated in the following example.

EXAMPLE 11 Factor $ax + ay + bx + by$.

Solution There is no factor (other than 1) common to all four terms. However, a is common to the first two terms and b is common to the last two terms. Factor a from the first two terms and b from the last two terms.

$$\boxed{a}x + \boxed{a}y + \boxed{b}x + \boxed{b}y = \boxed{a}(x + y) + \boxed{b}(x + y)$$

Now $(x + y)$ is common to both terms. Factor out $(x + y)$.

$$\boxed{a}(x + y) + \boxed{b}(x + y) = (x + y)(\boxed{a} + \boxed{b})$$

Thus, $ax + ay + bx + by = (x + y)(a + b)$ or $(a + b)(x + y)$. ✳

> ### To Factor Four Terms by Grouping
> 1. Arrange the four terms into two groups of two terms each. Each group of two terms must have a GCF.
> 2. Factor the GCF from each group of two terms.
> 3. If the two terms formed in step 2 have a GCF, factor it out.

EXAMPLE 12 Factor $x^3 - 5x^2 + 2x - 10$ by grouping.

Solution There are no factors common to all four terms. However, x^2 is common to the first two terms and 2 is common to the last two terms.

$$x^3 - 5x^2 + 2x - 10 = x^2(x - 5) + 2(x - 5)$$
$$= (x - 5)(x^2 + 2)$$ ✳

In Example 12, $(x^2 + 2)(x - 5)$ is also an acceptable answer. Would the answer to Example 12 change if we switch the order of $2x$ and $-5x^2$? Let's try it in Example 13.

EXAMPLE 13 Factor $x^3 + 2x - 5x^2 - 10$.

Solution Factor x from the first two terms and -5 from the last two terms.

$$x^3 + 2x - 5x^2 - 10 = x(x^2 + 2) - 5(x^2 + 2)$$
$$= (x^2 + 2)(x - 5)$$ ✳

NOW TRY EXERCISE 51

Notice that we got equivalent results in Examples 12 and 13.

HELPFUL HINT
> When factoring four terms by grouping, if the *first* and *third* terms are positive, you must factor a positive expression from both the first two terms and the last two terms to obtain a factor common to the remaining two terms (see Example 12). If the *first* term is positive and the *third* term is negative, you must factor a positive expression from the first two terms and a negative expression from the last two terms to obtain a factor common to the remaining two terms (see Example 13).

The first step in any factoring problem is to determine whether all the terms have a common factor. If so, begin by factoring out the common factor. For instance, to factor $x^4 - 5x^3 + 2x^2 - 10x$, we first factor out x from each term. Then we factor the remaining four terms by grouping, as was done in Example 12.

$$x^4 - 5x^3 + 2x^2 - 10x = x(x^3 - 5x^2 + 2x - 10)$$
Factor out the GCF, x, from all four terms.
$$= x(x - 5)(x^2 + 2)$$
Factors from Example 12

Exercise Set 5.4

Concept/Writing Exercises

1. What is the first step in *any* factoring problem?

2. What is the greatest common factor of the terms of an expression?

3. a) Explain how to find the greatest common factor of the terms of a polynomial.

   b) Using your procedure from part a), find the greatest common factor of the polynomial
   $$6x^2y^5 - 2x^3y + 12x^9y^3$$

   c) Factor the polynomial in part b).

4. When a term of a polynomial is itself the GCF, what is written in place of that term when the GCF is factored out? Explain.

5. a) Explain how to factor a polynomial of four terms by grouping.

   b) Factor $6x^3 - 2xy^3 + 3x^2y^2 - y^5$ by your procedure from part a).

6. What is the first step when factoring $-x^2 + 8x - 15$? Explain your answer.

7. Determine the GCF of the following terms:
   $$x^4y^6, x^3y^5, xy^6, x^2y^4$$
   Explain how you determined your answer.

8. Determine the GCF of the following terms:
   $$12(x - 4)^3, 6(x - 4)^4, 3(x - 4)^5$$
   Explain how you determined your answer.

Practice the Skills

Factor out the greatest common factor.

9. $7n + 7$

10. $15p + 25$

11. $2x^2 - 4x + 8$

12. $6x^2 - 12x + 21$

13. $12y^2 - 16y + 24$

14. $12x^3 - 8x^2 - 6x$

15. $9x^4 - 3x^3 + 11x^2$

16. $45y^{12} + 30y^{10}$

17. $-24a^7 + 9a^6 - 3a^2$

18. $-16c^5 - 12c^4 + 6c^3$

19. $3x^2y + 6x^2y^2 + 3xy$

20. $24a^2b^2 + 16ab^4 + 64ab^3$

21. $80a^5b^4c - 16a^4b^2c^2 + 8a^2c$

22. $36xy^2z^3 + 36x^3y^2z + 9x^2yz$

23. $9p^4q^5r - 3p^2q^2r^2 + 6pq^5r^3$

24. $24m^6 + 8m^4 - 4m^3n$

25. $-52p^2q^2 - 16pq^3 + 26r$

26. $-14y^3z^5 - 28y^3z^6 + 9xy^2z^2$

Factor out a factor with a negative coefficient.

27. $-8x + 4$

28. $-20a - 10$

29. $-x^2 + 4x - 12$

30. $-y^5 - 6y^2 - 4$

31. $-3r^2 - 6r + 9$

32. $-12t^2 + 48t - 36$

33. $-6r^4s^3 + 4r^2s^4 + 2rs^5$

34. $-5p^6q^3 - 10p^4q^4 + 25pq^7$

35. $-a^4b^2c + 5a^3bc^2 + a^2b$

36. $-20x^5y^3z - 4x^4yz^2 - 8x^2y^5$

Factor.

37. $x(a + 3) + 1(a + 3)$

38. $y(b - 2) - 5(b - 2)$

39. $3c(x - 4) + 2(x - 4)$

40. $4d(y + 1) - 7(y + 1)$

41. $(x - 2)(3x + 5) - (x - 2)(5x - 4)$

42. $(z + 4)(z + 3) + (z - 1)(z + 3)$

43. $(2a + 4)(a - 3) - (2a + 4)(2a - 1)$

44. $(6b - 1)(b + 4) + (6b - 1)(2b + 5)$

45. $x^2 + 3x - 5x - 15$

46. $a^2 + 3a - 2a - 6$

47. $8y^2 - 4y - 20y + 10$

48. $18m^2 + 30m + 9m + 15$

49. $ax + ay + bx + by$

50. $cx - cy - dx + dy$

51. $x^3 - 3x^2 + 4x - 12$

52. $2z^3 + 4z^2 - 5z - 10$

53. $10m^2 - 12mn - 25mn + 30n^2$

54. $12x^2 + 9xy - 4xy - 3y^2$

55. $5a^3 + 15a^2 - 10a - 30$

56. $2r^4 - 2r^3 - 7r^2 + 7r$

57. $c^5 - c^4 + c^3 - c^2$

58. $b^4 - b^3 - b + b^2$

Problem Solving

In Exercises 59–62, A represents an expression for the area of the figure. Find an expression, in factored form, for the difference of the areas of the geometric figures. See Example 10.

59.

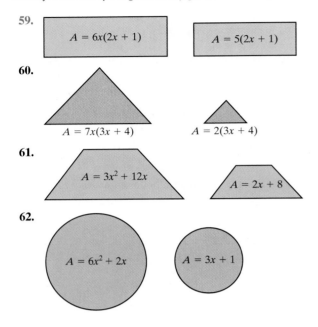

$A = 6x(2x + 1)$

$A = 5(2x + 1)$

60.

$A = 7x(3x + 4)$

$A = 2(3x + 4)$

61.

$A = 3x^2 + 12x$

$A = 2x + 8$

62.

$A = 6x^2 + 2x$

$A = 3x + 1$

In Exercises 63 and 64, V represents an expression for the volume of the figure. Find an expression, in factored form, for the difference of the volumes of the geometric solids.

63.

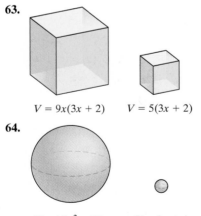

$V = 9x(3x + 2)$ $V = 5(3x + 2)$

64.

$V = 18x^2 + 24x$ $V = 3x + 4$

65. *Flare* When a flare is shot upward with a velocity of 128 feet per second, its height, h, in feet, above the ground at t seconds can be found by the function $h(t) = -16t^2 + 128t$.

  a) Find the height of the flare 3 seconds after it was shot.

  b) Express the function with the right side in factored form.

  c) Evaluate $h(3)$ using the factored form from part **b)**.

66. *Jump Shot* When a basketball player shoots a jump shot, the height, h, in feet, of the ball above the ground at any time t, under certain circumstances, may be found by the function $h(t) = -16t^2 + 20t + 8$.

  a) Find the height of the ball at 1 second.

  b) Express the function with the right side in factored form.

  c) Evaluate $h(1)$ using the factored form in part **b)**.

67. *Skating Rink* The area of the skating rink with semicircular ends shown is $A = \pi r^2 + 2rl$.

  a) Find A when $r = 20$ feet and $l = 40$ feet

  b) Write the area A in factored form.

  c) Find A when $r = 20$ feet and $l = 40$ feet using the factored form in part **b)**.

68. *Area* The formula for finding the area of a trapezoid may be given as $A = \frac{1}{2}hb_1 + \frac{1}{2}hb_2$. Express this formula in factored form.

69. *Price of Cars* When the 2003 cars came out, the list price of one model increased by 6% over the list price of the 2002 model. Then in a special sale, the prices of all 2003 cars were reduced by 6%. The sale price can be represented by $(x + 0.06x) - 0.06(x + 0.06x)$, where x is the list price of the 2002 model.

  a) Factor out $(x + 0.06x)$ from each term.

  b) Is the sale price more or less than the price of the 2002 model?

Read Exercise 69 before working Exercises 70–72.

70. ***Price of Dress*** A dress is reduced by 10%, and then the sale price is reduced by another 10%.

  a) Write an expression for the final price of the dress.

  b) How does the final price compare with the regular price of the dress? Use factoring in obtaining your answer.

71. ***Price of Mower*** The price of a Toro lawn mower is increased by 15%. Then at a Fourth of July sale the price is reduced by 20%.

  a) Write an expression for the final price of the lawn mower.

  b) How does the sale price compare with the regular price? Use factoring in obtaining your answer.

72. ***Find Price*** In which of the following, **a)** or **b)**, will the final price be lower, and by how much?

  a) Decreasing the price of an item by 6%, then increasing that price by 8%

  b) Increasing the price of an item by 6%, then decreasing that price by 8%

Factor.

73. $5a(3x-2)^5 + 4(3x-2)^4$

75. $4x^2(x-3)^3 - 6x(x-3)^2 + 4(x-3)$

77. $ax^2 + 2ax - 3a + bx^2 + 2bx - 3b$

74. $4p(2r-3)^7 - 3(2r-3)^6$

76. $12(p+2q)^4 - 40(p+2q)^3 + 12(p+2q)^2$

78. $6a^2 - a^2c + 18a - 3ac + 6ab - abc$

Factor. Assume that all variables in the exponents represent natural numbers.

79. $x^{6m} - 2x^{4m}$

81. $3x^{4m} - 2x^{3m} + x^{2m}$

83. $a^r b^r + c^r b^r - a^r d^r - c^r d^r$

80. $x^{2mn} + x^{4mn}$

82. $r^{y+4} + r^{y+3} + r^{y+2}$

84. $6a^k b^k - 2a^k c^k - 9b^k + 3c^k$

85. a) Does $6x^3 - 3x^2 + 9x = 3x(2x^2 - x + 3)$?

  b) If the above factoring is correct, what should be the value of $6x^3 - 3x^2 + 9x - [3x(2x^2 - x + 3)]$ for any value of x? Explain.

  c) Select a value for x and evaluate the expression in part **b)**. Did you get what you expected? If not, explain why.

86. a) Determine whether the following factoring is correct.

$$3(x-2)^2 - 6(x-2) = 3(x-2)[(x-2)-2]$$
$$= 3(x-2)(x-4)$$

  b) If the above factoring is correct, what should be the value of $3(x-2)^2 - 6(x-2) - [3(x-2)(x-4)]$ for any value of x? Explain.

  c) Select a value for x and evaluate the expression in part **b)**. Did you get what you expected? If not, explain why.

87. Consider the factoring $8x^3 - 16x^2 - 4x = 4x(2x^2 - 4x - 1)$.

  a) If we let

$$y_1 = 8x^3 - 16x^2 - 4x$$
$$y_2 = 4x(2x^2 - 4x - 1)$$

and graph each function, what should happen? Explain.

  b) On your graphing calculator, graph y_1 and y_2 as given in part **a)**.

  c) Did you get the results you expected?

  d) When checking a factoring process by this technique, what does it mean if the graphs do not overlap? Explain.

88. Consider the factoring $2x^4 - 6x^3 - 8x^2 = 2x^2(x^2 - 3x - 4)$.

  a) Enter

$$y_1 = 2x^4 - 6x^3 - 8x^2$$
$$y_2 = 2x^2(x^2 - 3x - 4)$$

into your calculator.

  b) If you use the TABLE feature of your calculator, how would you expect the table of values for y_1 to compare with the table of values for y_2? Explain.

  c) Use the TABLE feature to show the values for y_1 and y_2 for values of x from 0 to 6.

  d) Did you get the results you expected?

  e) When checking a factoring process using the TABLE feature, what does it mean if the values of y_1 and y_2 are different?

Cumulative Review Exercises

[1.4] 89. Evaluate $\dfrac{\left(\left|\frac{1}{2}\right| - \left|-\frac{1}{3}\right|\right)^2}{-\left|\frac{1}{3}\right| \cdot \left|-\frac{2}{5}\right|}$.

[2.1] 90. Solve $3(2x-4) + 3(x+1) = 9$.

[3.1] 91. Graph $y = x^2 - 1$.

[4.3] 92. ***Exercise*** Jason Richter tries to exercise every day. He walks at 3 mph and then jogs at 5 mph. If it takes him 0.9 hr to travel a total of 3.5 mi, how long does he jog?

[5.2] 93. Multiply $(7a-3)(-2a^2 - 4a + 1)$.

5.5 FACTORING TRINOMIALS

1 Factor trinomials of the form $x^2 + bx + c$.

2 Factor out a common factor.

3 Factor trinomials of the form $ax^2 + bx + c, a \neq 1$, using trial and error.

4 Factor trinomials of the form $ax^2 + bx + c, a \neq 1$, using grouping.

5 Factor trinomials using substitution.

SSM   Study Guide   CD/Video

MathPro 4/5   PH Math Tutor Center   prenhall.com/Angel

1 Factor Trinomials of the Form $x^2 + bx + c$

In this section we learn how to **factor trinomials** of the form $ax^2 + bx + c, a \neq 0$.

Trinomials	Coefficients
$3x^2 + 2x - 5$	$a = 3, \quad b = 2, \quad c = -5$
$-\dfrac{1}{2}x^2 - 4x + 3$	$a = -\dfrac{1}{2}, \quad b = -4, \quad c = 3$

To Factor Trinomials of the Form $x^2 + bx + c$ (note: $a = 1$)

1. Find two numbers (or factors) whose product is c and whose sum is b.

2. The factors of the trinomial will be of the form

$$(x + \blacksquare)(x + \blacksquare)$$

    ↑   ↑

   One factor *Other factor*

   determined *determined*

   in step 1 *in step 1*

If the numbers determined in step 1 are, for example, 3 and -5, the factors would be written $(x + 3)(x - 5)$. This procedure is illustrated in the following examples.

EXAMPLE 1   Factor $x^2 - x - 12$.

Solution $a = 1, b = -1, c = -12$. We must find two numbers whose product is c, which is -12, and whose sum is b, which is -1. We begin by listing the factors of -12, trying to find a pair whose sum is -1.

Factors of −12	Sum of Factors
$(1)(-12)$	$1 + (-12) = -11$
$(2)(-6)$	$2 + (-6) = -4$
$(3)(-4)$	$3 + (-4) = -1$
$(4)(-3)$	$4 + (-3) = 1$
$(6)(-2)$	$6 + (-2) = 4$
$(12)(-1)$	$12 + (-1) = 11$

The numbers we are seeking are 3 and -4 because their product is -12 and their sum is -1. Now we factor the trinomial using the 3 and -4.

$$x^2 - x - 12 = (x \underbrace{+ 3}_{})(x \underbrace{- 4}_{})$$

One factor   Other factor
of -12 of -12

Notice in Example 1 that we listed all the factors of -12. However, after the two factors whose product is c and whose sum is b are found, there is no need to go further in listing the factors. The factors were listed here to show, for example, that $(2)(-6)$ is a different set of factors than $(-2)(6)$. Note that as the positive factor increases the sum of the factors increases.

HELPFUL HINT

Consider the factors $(2)(-6)$ and $(-2)(6)$ and the sums of these factors.

Factors	Sum of Factors
$2(-6)$	$2 + (-6) = -4$
$-2(6)$	$-2 + 6 = 4$

Notice that if the signs of each number in the product are changed, the sign of the sum of factors is changed. We can use this fact to more quickly find the factors we are seeking. If, when seeking a specific sum, you get the opposite of that sum, change the sign of each factor to get the sum you are seeking.

EXAMPLE 2 Factor $p^2 - 7p + 6$.

Solution We must find two numbers whose product is 6 and whose sum is -7. Since the sum of two negative numbers is a negative number, and the product of two negative numbers is a positive number, both numbers must be negative numbers. The negative factors of 6 are $(-1)(-6)$ and $(-2)(-3)$. As shown below, the numbers we are looking for are -1 and -6.

Factors of 6	Sum of Factors
$(-1)(-6)$	$-1 + (-6) = -7$
$(-2)(-3)$	$-2 + (-3) = -5$

Therefore,

$$p^2 - 7p + 6 = (p - 1)(p - 6)$$

Since the factors may be placed in any order, $(p - 6)(p - 1)$ is also an acceptable answer.

NOW TRY EXERCISE 23

HELPFUL HINT

Checking Factoring

Factoring problems can be checked by multiplying the factors obtained. If the factoring is correct, you will obtain the polynomial you started with. To check Example 2, we will multiply the factors using the FOIL method.

$$(p - 1)(p - 6) = p^2 - 6p - p + 6 = p^2 - 7p + 6$$

Since the product of the factors is the trinomial we began with, our factoring is correct. You should always check your factoring.

The procedure used to factor trinomials of the form $x^2 + bx + c$ can be used on other trinomials, as in the following example.

EXAMPLE 3   Factor $x^2 + 2xy - 15y^2$.

Solution   We must find two numbers whose product is -15 and whose sum is 2. The two numbers are 5 and -3.

Factors of -15	Sum of Factors
$5(-3)$	$5 + (-3) = 2$

Since the last term of the trinomial contains y^2, the second term of each factor must contain y.

$$x^2 + 2xy - 15y^2 = (x + 5y)(x - 3y)$$

Check $(x + 5y)(x - 3y) = x^2 - 3xy + 5xy - 15y^2$
$$= x^2 + 2xy - 15y^2 \qquad ✻$$

2   Factor Out a Common Factor

The first step when factoring any trinomial is to determine whether all three terms have a common factor. If so, factor out that common factor. Then factor the remaining polynomial.

EXAMPLE 4   Factor $3x^4 - 6x^3 - 72x^2$.

Solution   The factor $3x^2$ is common to all three terms of the trinomial. Factor it out first.

$$3x^4 - 6x^3 - 72x^2 = 3x^2(x^2 - 2x - 24) \quad \textit{Factor out } 3x^2.$$

The $3x^2$ that was factored out is a part of the answer but plays no further part in the factoring process. Now continue to factor $x^2 - 2x - 24$. Find two numbers whose product is -24 and whose sum is -2. The numbers are -6 and 4.

$$3x^2(x^2 - 2x - 24) = 3x^2(x - 6)(x + 4)$$

NOW TRY EXERCISE 33   Therefore, $3x^4 - 6x^3 - 72x^2 = 3x^2(x - 6)(x + 4)$. ✻

3   Factor Trinomials of the Form $ax^2 + bx + c$, $a \neq 1$, Using Trial and Error

Now we will look at some examples of factoring trinomials of the form

$$ax^2 + bx + c, \quad a \neq 1$$

Two methods of factoring this type of trinomial will be illustrated. The first method, trial and error, involves trying various combinations until the correct combination is found. The second method makes use of factoring by grouping, a procedure that was presented in Section 5.4.

Let's first discuss the trial-and-error method of factoring trinomials. This procedure is sometimes called the FOIL (or Reverse FOIL) method. As an aid in our explanation, we will multiply $(2x + 3)(x + 1)$ using the FOIL method.

Therefore, if you are factoring the trinomial $2x^2 + 5x + 3$, you should realize that the product of the first terms of the factors must be $2x^2$, the product of the last terms must be 3, and the sum of the products of the outer and inner terms must be $5x$.

To factor $2x^2 + 5x + 3$, we begin as shown here.

$$2x^2 + 5x + 3 = (2x \quad)(x \quad) \qquad \textit{The product of the first terms is } 2x^2.$$

Now we fill in the second terms using positive integers whose product is 3. Only positive integers will be considered since the product of the last terms is positive, and the sum of the products of the outer and inner terms is also positive. The two possibilities are

$$\left. \begin{array}{l} (2x + 1)(x + 3) \\ (2x + 3)(x + 1) \end{array} \right\} \begin{array}{l} \textit{The product of} \\ \textit{the last terms is 3.} \end{array}$$

To determine which factoring is correct, we find the sum of the products of the outer and inner terms. If either has a sum of $5x$, the middle term of the trinomial, that factoring is correct.

$$(2x + 1)(x + 3) = 2x^2 + 6x + x + 3 = 2x^2 + 7x + 3 \qquad \textit{Wrong middle term}$$
$$(2x + 3)(x + 1) = 2x^2 + 2x + 3x + 3 = 2x^2 + 5x + 3 \qquad \textit{Correct middle term}$$

Therefore, the factors of $2x^2 + 5x + 3$ are $2x + 3$ and $x + 1$. Thus,

$$2x^2 + 5x + 3 = (2x + 3)(x + 1)$$

Note that if we had begun factoring by writing

$$2x^2 + 5x + 3 = (x \quad)(2x \quad)$$

we could have also obtained the correct factors.

Following are guidelines for the **trial and error** method of factoring a trinomial where $a \neq 1$ and the three terms have no common factors.

To Factor Trinomials of the Form $ax^2 + bx + c$, $a \neq 1$, Using Trial and Error

1. Write all pairs of factors of the coefficient of the squared term, a.
2. Write all pairs of factors of the constant, c.
3. Try various combinations of these factors until the correct middle term, bx, is found.

EXAMPLE 5 Factor $3t^2 - 13t + 10$.

Solution First we determine that the three terms have no common factor. Next we determine that a is 3 and the only factors of 3 are 1 and 3. Therefore, we write

$$3t^2 - 13t + 10 = (3t \quad)(t \quad)$$

The number 10 has both positive and negative factors. However, since the product of the last terms must be positive $(+10)$, and the sum of the products of the outer and inner terms must be negative (-13), the two factors of 10 must both be negative. (Why?) The negative factors of 10 are $(-1)(-10)$ and $(-2)(-5)$. Below is a list of the possible factors. We look for the factors that give us the correct middle term, $-13t$.

Possible Factors	Sum of Products of Outer and Inner Terms
$(3t - 1)(t - 10)$	$-31t$
$(3t - 10)(t - 1)$	$-13t$ ⟵ *Correct middle term*
$(3t - 2)(t - 5)$	$-17t$
$(3t - 5)(t - 2)$	$-11t$

Thus, $3t^2 - 13t + 10 = (3t - 10)(t - 1)$. ✳

The following Helpful Hint is very important. Study it carefully.

HELPFUL HINT

Factoring by Trial and Error

When factoring a trinomial of the form $ax^2 + bx + c$, the sign of the constant term, c, is very helpful in finding the solution. If $a > 0$, then:

1. When the constant term, c, is positive, and the numerical coefficient of the x term, b, is positive, both numerical factors will be positive.

   Example $\quad x^2 + 7x + 12 = (x + 3)(x + 4)$

   Positive Positive   Positive Positive

2. When c is positive and b is negative, both numerical factors will be negative.

   Example $\quad x^2 - 5x + 6 = (x - 2)(x - 3)$

   Negative Positive   Negative Negative

Whenever the constant, c, is positive (as in the two examples above) the sign in both factors will be the same as the sign in the x-term of the trinomial.

3. When c is negative, one of the numerical factors will be positive and the other will be negative.

   Example $\quad x^2 + x - 6 = (x + 3)(x - 2)$

   Negative   Positive Negative

EXAMPLE 6 Factor $8x^2 + 8x - 30$.

Solution First we check to see whether the three terms have a common factor. We notice that 2 can be factored out.

$$8x^2 + 8x - 30 = 2(4x^2 + 4x - 15)$$

The factors of 4, the leading coefficient, are $4 \cdot 1$ and $2 \cdot 2$. Therefore the factoring will be of the form $(4x \quad)(x \quad)$ or $(2x \quad)(2x \quad)$. It makes no difference whether you start with the first set of factors or the last set of factors. We will generally start with the medium-sized factors first, so we will start with $(2x \quad)(2x \quad)$. If using these factors does not give our answer, we will work with the other set of factors. The factors of -15 are $(1)(-15)$, $(3)(-5)$, $(5)(-3)$, and $(15)(-1)$. We want our middle term to be $4x$.

Possible Factors	Sum of Products of Outer and Inner Terms
$(2x + 1)(2x - 15)$	$-28x$
$(2x + 3)(2x - 5)$	$-4x$
$(2x + 5)(2x - 3)$	$4x$

Since we found the set of factors that gives the correct x-term, we can stop. Thus,

$$8x^2 + 8x - 30 = 2(2x + 5)(2x - 3)$$ ✳

In Example 6, if we compare the second and third set of factors we see they are the same except for the signs of the second terms. Notice that when the signs of the second term in each factor are switched the sum of the products of the outer and inner terms also changes sign.

Using Your Graphing Calculator

The graphing calculator can be used to check factoring problems. To check the factoring in Example 6,

$$8x^2 + 8x - 30 = 2(2x + 5)(2x - 3)$$

we let $y_1 = 8x^2 + 8x - 30$ and $y_2 = 2(2x + 5)(2x - 3)$. Then we use the TABLE feature to compare results, as in Figure 5.14.

FIGURE 5.14

Since y_1 and y_2 have the same values for each value of x, a mistake has not been made. This procedure can only tell you if a mistake has been made; it cannot tell you if you have factored completely. For example, $8x^2 + 8x - 30$ and $(4x + 10)(2x - 3)$ will give the same set of values.

Exercises

Use your grapher to determine whether each trinomial is factored correctly.

1. $30x^2 + 37x - 84 \stackrel{?}{=} (6x - 7)(5x + 12)$

2. $72x^2 + 20x - 35 \stackrel{?}{=} (9x - 5)(8x + 7)$

EXAMPLE 7 Factor $6x^2 - 11xy - 10y^2$.

Solution The factors of 6 are either $6 \cdot 1$ or $2 \cdot 3$. Therefore, the factors of the trinomial may be of the form $(6x \quad)(x \quad)$ or $(2x \quad)(3x \quad)$. We will begin with the middle-sized factors. We write

$$6x^2 - 11xy - 10y^2 = (2x \quad)(3x \quad)$$

The factors of -10 are $(-1)(10)$, $(1)(-10)$, $(-2)(5)$, and $(2)(-5)$. Since there are eight factors of -10, there will be eight pairs of possible factors to try. Can you list them? The correct factorization is

$$6x^2 - 11xy - 10y^2 = (2x - 5y)(3x + 2y)$$ ✳

In Example 7, we were fortunate to find the correct factors by using the form $(2x \quad)(3x \quad)$. If we had not found the correct factors using these, we would have tried $(6x \quad)(x \quad)$.

When factoring a trinomial whose leading coefficient is negative, we start by factoring out a negative number. For example,

$$-24x^3 - 60x^2 + 36x = -12x(2x^2 + 5x - 3) \quad \text{Factor out } -12x.$$
$$= -12x(2x - 1)(x + 3)$$

and

$$-3x^2 + 8x + 16 = -1(3x^2 - 8x - 16) \quad \text{Factor out } -1.$$
$$= -(3x + 4)(x - 4)$$

NOW TRY EXERCISE 49

EXAMPLE 8 **Area of shaded region** In Figure 5.15, find an expression, in factored form, for the area of the shaded region.

Solution To find the area of the shaded region, we need to subtract the area of the small rectangle from the area of the large rectangle. Recall that the area of a rectangle is length · width.

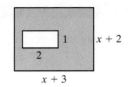

FIGURE 5.15

$$\text{Area of large rectangle} = (x + 3)(x + 2)$$
$$= x^2 + 2x + 3x + 6$$
$$= x^2 + 5x + 6$$

$$\text{Area of small rectangle} = (2)(1) = 2$$

$$\text{Area of shaded region} = \text{large area} - \text{small area}$$
$$= x^2 + 5x + 6 - 2$$
$$= x^2 + 5x + 4 \quad \text{Simplify.}$$
$$= (x + 4)(x + 1) \quad \text{Factor.}$$

NOW TRY EXERCISE 89 The area of the shaded region is $(x + 4)(x + 1)$.

4 Factor Trinomials of the Form $ax^2 + bx + c$, $a \neq 1$, Using Grouping

Now we will discuss the **grouping** method of factoring trinomials of the form $ax^2 + bx + c$, $a \neq 1$.

> **To Factor Trinomials of the Form $ax^2 + bx + c$, $a \neq 1$, Using Grouping**
>
> 1. Find two numbers whose product is $a \cdot c$ and whose sum is b.
> 2. Rewrite the middle term, bx, using the numbers found in step 1.
> 3. Factor by grouping.

EXAMPLE 9 Factor $2x^2 - 5x - 12$.

Solution We see that $a = 2, b = -5$, and $c = -12$. We must find two numbers whose product is $a \cdot c$ or $2(-12) = -24$, and whose sum is b, -5. The two numbers are -8 and 3 because $(-8)(3) = -24$ and $-8 + 3 = -5$. Now rewrite the middle term, $-5x$, using $-8x$ and $3x$.

$$2x^2 - 5x - 12 = 2x^2 \overset{-5x}{\overbrace{- 8x + 3x}} - 12$$

Now factor by grouping as explained in Section 5.4. Factor out $2x$ from the first two terms and 3 from the last two terms.

$$2x^2 - 5x - 12 = 2x(x - 4) + 3(x - 4)$$
$$= (x - 4)(2x + 3) \qquad \text{Factor out } (x - 4). \;\; *$$

Note in Example 9 that we wrote $-5x$ as $-8x + 3x$. As we show below, the same factors would be obtained if we wrote $-5x$ as $3x - 8x$. Therefore, it makes no difference which factor is listed first when factoring by grouping. Below we factor x from the first two terms and -4 from the last two terms.

$$2x^2 - 5x - 12 = 2x^2 \overbrace{+ 3x - 8x}^{-5x} - 12$$
$$= x(2x + 3) - 4(2x + 3)$$
$$= (2x + 3)(x - 4) \qquad \text{Factor out } (2x + 3).$$

EXAMPLE 10 Factor $12a^2 - 19ab + 5b^2$.

Solution We must find two numbers whose product is $(12)(5) = 60$ and whose sum is -19. Since the product of the numbers is positive and their sum is negative, the two numbers must both be negative. (Why?)

The two numbers are -15 and -4 because $(-15)(-4) = 60$ and $-15 + (-4) = -19$. Now rewrite the middle term, $-19ab$, using $-15ab$ and $-4ab$. Then factor by grouping.

$$12a^2 - 19ab + 5b^2 = 12a^2 \overbrace{- 15ab - 4ab}^{-19ab} + 5b^2$$
$$= 3a(4a - 5b) - b(4a - 5b)$$
NOW TRY EXERCISE 37 $$= (4a - 5b)(3a - b) \qquad\qquad *$$

Try Example 10 again, this time writing $-19ab$ as $-4ab - 15ab$. If you do it correctly, you should get the same factors.

It is important for you to realize that not every trinomial can be factored by the methods presented in this section. In Sections 8.1 and 8.2, we will give some procedures that can be used to factor polynomials that cannot be factored using only the integers (or over the set of integers). A polynomial that cannot be factored (over a specific set of numbers) is called a **prime polynomial**.

EXAMPLE 11 Factor $2x^2 + 6x + 5$.

Solution When you try to factor this polynomial, you will see that it cannot be factored using either trial and error or grouping. This polynomial is prime over the set of integers. $\qquad *$

5 Factor Trinomials Using Substitution

Sometimes a more complicated trinomial can be factored by substituting one variable for another. The next three examples illustrate **factoring using substitution**.

EXAMPLE 12 Factor $y^4 - y^2 - 6$.

Solution If we can rewrite this expression in the form $ax^2 + bx + c$, it will be easier to factor. Since $(y^2)^2 = y^4$, if we substitute x for y^2, the trinomial becomes

$$y^4 - y^2 - 6 = (y^2)^2 - y^2 - 6$$
$$= x^2 - x - 6 \qquad \text{Substitute } x \text{ for } y^2.$$

Now factor $x^2 - x - 6$.

$$= (x + 2)(x - 3)$$

Finally, substitute y^2 in place of x to obtain

$$= (y^2 + 2)(y^2 - 3) \quad \textit{Substitute } y^2 \textit{ for } x.$$

Thus, $y^4 - y^2 - 6 = (y^2 + 2)(y^2 - 3)$. Note that x was substituted for y^2, and then y^2 was substituted back for x. ✳

EXAMPLE 13 Factor $3z^4 - 17z^2 - 28$.

Solution Let $x = z^2$. Then the trinomial can be written

$$3z^4 - 17z^2 - 28 = 3(z^2)^2 - 17z^2 - 28$$
$$= 3x^2 - 17x - 28 \quad \textit{Substitute } x \textit{ for } z^2.$$
$$= (3x + 4)(x - 7) \quad \textit{Factor.}$$

Now substitute z^2 for x.

$$= (3z^2 + 4)(z^2 - 7) \quad \textit{Substitute } z^2 \textit{ for } x.$$

NOW TRY EXERCISE 69   Thus, $3z^4 - 17z^2 - 28 = (3z^2 + 4)(z^2 - 7)$. ✳

EXAMPLE 14 Factor $2(x + 5)^2 - 5(x + 5) - 12$.

Solution We will again use a substitution, as in Examples 12 and 13. By substituting $a = x + 5$ in the equation, we obtain

$$2(x + 5)^2 - 5(x + 5) - 12$$
$$= 2a^2 - 5a - 12 \quad \textit{Substitute } a \textit{ for } (x + 5).$$

Now factor $2a^2 - 5a - 12$.

$$= (2a + 3)(a - 4)$$

Finally, replace a with $x + 5$ to obtain

$$= [2(x + 5) + 3][(x + 5) - 4] \quad \textit{Substitute } (x + 5) \textit{ for } a.$$
$$= [2x + 10 + 3][x + 1]$$
$$= (2x + 13)(x + 1)$$

NOW TRY EXERCISE 73   Thus, $2(x + 5)^2 - 5(x + 5) - 12 = (2x + 13)(x + 1)$. Note that a was substituted for $x + 5$, and then $x + 5$ was substituted back for a. ✳

In Examples 12 and 13 we used x in our substitution, whereas in Example 14 we used a. The letter selected does not affect the final answer.

Exercise Set 5.5

Concept/Writing Exercises

1. When factoring any trinomial, what should the first step always be?

2. On a test, Tom Phu wrote the following factoring and did not receive full credit. Explain why Tom's factoring is not complete.

   $$15x^2 - 21x - 18 = (5x + 3)(3x - 6)$$

3. a) Explain in your own words the step-by-step procedure to factor $6x^2 - x - 12$.

   b) Factor $6x^2 - x - 12$ using the procedure you explained in part a).

4. a) Explain in your own words the step-by-step procedure to factor $8x^2 - 26x + 6$.

b) Factor $8x^2 - 26x + 6$ using the procedure you explained in part **a)**.

5. Has $2x^2 + 8x + 6 = (x + 3)(2x + 2)$ been factored completely? If not, give the complete factorization. Explain.

6. Has $x^3 - 3x^2 - 10x = (x^2 + 2x)(x - 5)$ been factored completely? If not, give the complete factorization. Explain.

7. Has $3x^3 + 6x^2 - 24x = x(x + 4)(3x - 6)$ been factored completely? If not, give the complete factorization. Explain.

8. Has $x^4 + 11x^3 + 30x^2 = x^2(x + 5)(x + 6)$ been factored completely? If not, give the complete factorization. Explain.

When factoring a trinomial of the form $ax^2 + bx + c$, what will be the signs between the terms in the binomial factors if:

9. $a > 0, b > 0,$ and $c > 0$

10. $a > 0, b > 0,$ and $c < 0$

11. $a > 0, b < 0,$ and $c < 0$

12. $a > 0, b < 0,$ and $c > 0$

Practice the Skills

Factor each trinomial completely. If the polynomial is prime, so state.

13. $x^2 + 7x + 12$

14. $a^2 - 2a - 15$

15. $b^2 - 10b + 9$

16. $y^2 - 9y + 20$

17. $c^2 - 12c + 36$

18. $z^2 + 4z + 4$

19. $y^2 - 18y + 81$

20. $r^2 + 22r + 121$

21. $x^2 - 34x + 64$

22. $x^2 + 11x - 210$

23. $x^2 - 13x - 30$

24. $p^2 - 6p - 19$

25. $-a^2 + 18a - 45$

26. $-x^2 - 15x - 56$

27. $x^2 + xy + 6y^2$

28. $a^2 + 6ab + 8b^2$

29. $m^2 - 7mn + 10n^2$

30. $-3x^2 - 12x - 9$

31. $4r^2 + 12r - 16$

32. $b^2 - 12bc - 45c^2$

33. $x^3 - 3x^2 - 18x$

34. $x^3 + 14x^2 + 33x$

35. $5a^2 - 8a + 3$

36. $4w^2 + 13w + 3$

37. $3x^2 - 3x - 6$

38. $-3b^2 - 14b + 5$

39. $6c^2 - 13c - 63$

40. $30z^2 - 71z + 35$

41. $8b^2 - 2b - 3$

42. $4a^2 + 43a + 30$

43. $6c^2 + 11c - 10$

44. $5z^2 - 11z + 6$

45. $16p^2 - 16pq - 12q^2$

46. $6r^4 + 5r^3 - 4r^2$

47. $4x^2 + 4xy + 9y^2$

48. $32x^2 - 22xy + 3y^2$

49. $18a^2 + 18ab - 8b^2$

50. $6r^2 + 7rs + 8s^2$

51. $9y^2 - 104y - 48$

52. $8x^2 + 30xy - 27y^2$

53. $100b^2 - 90b + 20$

54. $x^5y - 3x^4y - 18x^3y$

55. $a^3b^5 - a^2b^5 - 12ab^5$

56. $a^3b + 2a^2b - 35ab$

57. $3b^4c - 18b^3c^2 + 27b^2c^3$

58. $6p^3q^2 - 24p^2q^3 - 30pq^4$

59. $8m^8n^3 + 4m^7n^4 - 24m^6n^5$

60. $18x^2 + 9x - 20$

61. $30x^2 - x - 20$

62. $36x^2 - 23x - 8$

63. $8x^4y^4 + 24x^3y^4 - 32x^2y^4$

64. $8b^3c^2 + 28b^2c^3 + 12bc^4$

Factor each trinomial completely.

65. $x^4 + x^2 - 6$

66. $y^4 + 7y^2 + 12$

67. $b^4 + 9b^2 + 20$

68. $c^4 - 8c^2 + 12$

69. $6a^4 + 5a^2 - 25$

70. $(2x + 1)^2 + 2(2x + 1) - 15$

71. $4(x + 1)^2 + 8(x + 1) + 3$

72. $(2y + 3)^2 - (2y + 3) - 6$

73. $6(a + 2)^2 - 7(a + 2) - 5$

74. $6(p - 5)^2 + 11(p - 5) + 3$

75. $x^2y^2 + 9xy + 14$

76. $a^2b^2 + 6ab - 27$

77. $2x^2y^2 - 9xy - 11$

78. $3b^2c^2 - bc - 2$

79. $2y^2(5 - y) - 7y(5 - y) + 5(5 - y)$

80. $2y^2(y + 2) + 13y(y + 2) + 15(y + 2)$

81. $2p^2(p - 3) + 7p(p - 3) + 6(p - 3)$

82. $3x^2(x - 2) + 5x(x - 2) - 2(x - 2)$

83. $a^6 - 7a^3 - 30$

84. $2y^6 - 9y^3 - 5$

85. $x^2(x + 3) + 3x(x + 3) + 2(x + 3)$

86. $x^2(x + 6) - x(x + 6) - 30(x + 6)$

87. $5a^5b^2 - 8a^4b^3 + 3a^3b^4$

88. $2x^2y^6 + 3xy^5 - 9y^4$

Problem Solving

In Exercises 89–92, find an expression, in factored form, for the area of the shaded region. See Example 8.

89.

90.

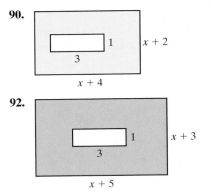

91.

92.

93. If the factors of a polynomial are $(2x + 3y)$ and $(x - 4y)$, find the polynomial. Explain how you determined your answer.

94. If the factors of a polynomial are $3, (4x - 5)$, and $(2x - 3)$, find the polynomial. Explain how you determined your answer.

95. If we know that one factor of the polynomial $x^2 + 3x - 18$ is $x - 3$, how can we find the other factor? Find the other factor.

96. If we know that one factor of the polynomial $x^2 - xy - 6y^2$ is $x - 3y$, how can we find the other factor? Find the other factor.

97. a) Which of the following do you think would be more difficult to factor by trial and error? Explain your answer.

$$30x^2 + 23x - 40 \quad \text{or} \quad 49x^2 - 98x + 13$$

b) Factor both trinomials.

98. a) Which of the following do you think would be more difficult to factor by trial and error? Explain your answer.

$$48x^2 + 26x - 35 \quad \text{or} \quad 35x^2 - 808x + 69$$

b) Factor both trinomials.

99. Find all integer values of b for which $2x + bx - 5$ is factorable.

100. Find all integer values of b for which $3x^2 + bx - 7$ is factorable.

101. If $x^2 + bx + 5$ is factorable, what are the only two possible values of b? Explain.

102. If $x^2 + bx + c$ is factorable and c is a prime number, what are the only two possible factors of b? Explain.

*Consider the trinomial $ax^2 + bx + c$. We will learn later in the course that if the expression $b^2 - 4ac$, called the **discriminant**, is not a perfect square, the trinomial cannot be factored over the set of integers. **Perfect squares** are 1, 4, 9, 16, 25, 49, and so on. The square root of a perfect square is a whole number. For Exercises 103–106, a) find the value of $b^2 - 4ac$. b) If $b^2 - 4ac$ is a perfect square, factor the polynomial; if $b^2 - 4ac$ is not a perfect square, indicate that the polynomial cannot be factored.*

103. $x^2 - 8x + 15$

104. $6y^2 - 5y - 6$

105. $x^2 - 4x + 6$

106. $3t^2 - 6t + 2$

107. Construct a trinomial of the form $x^2 + (c + 1)x + c$, where c is a real number, that is factorable.

108. Construct a trinomial of the form $x^2 - (c + 1)x + c$, where c is a real number, that is factorable.

Factor completely. Assume that the variables in the exponents represent positive integers.

109. $4a^{2n} - 4a^n - 15$

110. $a^2(a + b) - 2ab(a + b) - 3b^2(a + b)$

111. $x^2(x + y)^2 - 7xy(x + y)^2 + 12y^2(x + y)^2$

112. $3m^2(m - 2n) - 4mn(m - 2n) - 4n^2(m - 2n)$

113. $x^{2n} + 3x^n - 10$

114. $9r^{4y} + 3r^{2y} - 2$

115. Consider $x^2 + 2x - 8 = (x + 4)(x - 2)$.
 a) Explain how you can check this factoring using graphs on your graphing calculator.
 b) Check the factoring as you explained in part a) to see whether it is correct.

116. Consider $6x^3 - 11x^2 - 10x = x(2x - 5)(3x + 2)$.
 a) Explain how you can check this factoring using the TABLE feature of a graphing calculator.
 b) Check the factoring as you explained in part a) to see whether it is correct.

Cumulative Review Exercises

[2.2] **117.** Solve $F = \dfrac{9}{5}C + 32$ for C.

[3.3] **118.** Graph $y = -3x + 4$.

[4.5] **119.** Evaluate the determinant $\begin{vmatrix} 3 & -2 & -1 \\ 2 & 3 & -2 \\ 1 & -4 & 1 \end{vmatrix}$.

[5.2] **120.** Multiply $[(x + y) + 5]^2$.

[5.3] **121.** Factor $2x^3 + 4x^2 - 5x - 10$.

5.6 SPECIAL FACTORING FORMULAS

 1 Factor the difference of two squares.
 2 Factor perfect square trinomials.
 3 Factor the sum and difference of two cubes.

1 Factor the Difference of Two Squares

In this section we present some special formulas for factoring the difference of two squares, perfect square trinomials, and the sum and difference of two cubes. It will be to your advantage to memorize these formulas.

The expression $x^2 - 9$ is an example of the difference of two squares.

$$x^2 - 9 = (x)^2 - (3)^2$$

To factor the difference of two squares, it is convenient to use the **difference of two squares formula**. This formula was first presented in Section 5.2.

> **Difference of Two Squares**
> $$a^2 - b^2 = (a + b)(a - b)$$

EXAMPLE 1 Factor the following using the difference of two squares formula.
 a) $x^2 - 16$ **b)** $25x^2 - 9y^2$

Solution Rewrite each expression as a difference of two squares. Then use the formula.
 a) $x^2 - 16 = (x)^2 - (4)^2$
 $= (x + 4)(x - 4)$
 b) $25x^2 - 9y^2 = (5x)^2 - (3y)^2$
 $= (5x + 3y)(5x - 3y)$

EXAMPLE 2 Factor the following differences of squares.
 a) $x^6 - y^4$ **b)** $2z^4 - 162x^6$

Solution Rewrite each expression as a difference of two squares. Then use the formula.
 a) $x^6 - y^4 = (x^3)^2 - (y^2)^2$
 $= (x^3 + y^2)(x^3 - y^2)$

b) $2z^4 - 162x^6 = 2(z^4 - 81x^6)$
$$= 2[(z^2)^2 - (9x^3)^2]$$
$$= 2(z^2 + 9x^3)(z^2 - 9x^3)$$

NOW TRY EXERCISE 17

EXAMPLE 3 Factor $x^4 - 16y^4$.

Solution
$$x^4 - 16y^4 = (x^2)^2 - (4y^2)^2$$
$$= (x^2 + 4y^2)(x^2 - 4y^2)$$

Note that $(x^2 - 4y^2)$ is also a difference of two squares. We use the difference of two squares formula a second time to obtain
$$= (x^2 + 4y^2)[(x)^2 - (2y)^2]$$
$$= (x^2 + 4y^2)(x + 2y)(x - 2y)$$

EXAMPLE 4 Factor $(x - 5)^2 - 49$ using the formula for the difference of two squares.

Solution First we express $(x - 5)^2 - 49$ as a difference of two squares.
$$(x - 5)^2 - 49 = (x - 5)^2 - 7^2$$
$$= [(x - 5) + 7][(x - 5) - 7]$$
$$= (x + 2)(x - 12)$$

NOW TRY EXERCISE 25

Note: It is not possible to factor the sum of two squares of the form $a^2 + b^2$ over the set of real numbers.

For example, it is not possible to factor $x^2 + 4$ since $x^2 + 4 = x^2 + 2^2$, which is a sum of two squares.

2 Factor Perfect Square Trinomials

In Section 5.2 we saw that
$$(a + b)^2 = a^2 + 2ab + b^2$$
$$(a - b)^2 = a^2 - 2ab + b^2$$

If we reverse the left and right sides of these two formulas, we obtain two **special factoring formulas**.

Perfect Square Trinomials
$$a^2 + 2ab + b^2 = (a + b)^2$$
$$a^2 - 2ab + b^2 = (a - b)^2$$

These two trinomials are called **perfect square trinomials** since each is the square of a binomial. *To be a perfect square trinomial, the first and last terms must be the squares of some expression and the middle term must be twice the product of the first and last terms.* When you are given a trinomial to factor, determine whether it is a perfect square trinomial before you attempt to factor it by the procedures explained in Section 5.5. If it is a perfect square trinomial, you can factor it using the formulas given above

Examples of Perfect Square Trinomials
$$y^2 + 6y + 9 \quad \text{or} \quad y^2 + 2(y)(3) + 3^2$$
$$9a^2b^2 - 24ab + 16 \quad \text{or} \quad (3ab)^2 - 2(3ab)(4) + 4^2$$
$$(r + s)^2 + 10(r + s) + 25 \quad \text{or} \quad (r + s)^2 + 2(r + s)(5) + 5^2$$

Now let's factor some perfect square trinomials.

EXAMPLE 5 Factor $x^2 - 8x + 16$.

Solution Since the first and last terms, x^2 and 4^2, are squares, this trinomial might be a perfect square trinomial. To determine whether it is, take twice the product of x and 4 to see if you obtain $8x$.

$$2(x)(4) = 8x$$

Since $8x$ is the middle term and since the sign of the middle term is negative, factor as follows:

NOW TRY EXERCISE 29
$$x^2 - 8x + 16 = (x - 4)^2 \qquad ✳$$

EXAMPLE 6 Factor $9x^4 - 12x^2 + 4$.

Solution The first term is a square, $(3x^2)^2$, as is the last term, 2^2. Since $2(3x^2)(2) = 12x^2$, we factor as follows:

$$9x^4 - 12x^2 + 4 = (3x^2 - 2)^2 \qquad ✳$$

EXAMPLE 7 Factor $(a + b)^2 + 6(a + b) + 9$.

Solution The first term, $(a + b)^2$, is a square. The last term, 9 or 3^2, is a square. The middle term is $2(a + b)(3) = 6(a + b)$. Therefore, this is a perfect square trinomial. Thus,

NOW TRY EXERCISE 39
$$(a + b)^2 + 6(a + b) + 9 = [(a + b) + 3]^2 = (a + b + 3)^2 \qquad ✳$$

EXAMPLE 8 Factor $x^2 - 6x + 9 - y^2$.

Solution Since $x^2 - 6x + 9$ is a perfect square trinomial, which can be expressed as $(x - 3)^2$, we write

$$(x - 3)^2 - y^2$$

Now $(x - 3)^2 - y^2$ is a difference of two squares; therefore

$$(x - 3)^2 - y^2 = [(x - 3) + y][(x - 3) - y]$$
$$= (x - 3 + y)(x - 3 - y)$$

NOW TRY EXERCISE 45 Thus, $x^2 - 6x + 9 - y^2 = (x - 3 + y)(x - 3 - y)$. ✳

The polynomial in Example 8 has four terms. In Section 5.4 we learned to factor polynomials with four terms by grouping. If you study Example 8, you will see that no matter how you arrange the four terms they cannot be arranged so that the first two terms have a common factor and the last two terms have a common factor. Whenever a polynomial of four terms cannot be factored by grouping, try to rewrite three of the terms as the square of a binomial and then factor using the difference of two squares formula.

EXAMPLE 9 Factor $4a^2 + 12ab + 9b^2 - 25$.

Solution We first notice that this polynomial of four terms cannot be factored by grouping. We next look to see if three terms of the polynomial can be expressed as the square of a binomial. Since this can be done, we write the three terms as the square of a binomial. We complete our factoring using the difference of two squares formula.

$$4a^2 + 12ab + 9b^2 - 25 = (2a + 3b)^2 - 5^2$$
$$= [(2a + 3b) + 5][(2a + 3b) - 5]$$
$$= (2a + 3b + 5)(2a + 3b - 5) \qquad ✳$$

3 Factor the Sum and Difference of Two Cubes

Earlier in this section we factored the difference of two squares. Now we will factor the sum and difference of two cubes. Consider the product of $(a + b)(a^2 - ab + b^2)$.

$$
\begin{array}{r}
a^2 - ab + b^2 \\
a + b \\
\hline
a^2b - ab^2 + b^3 \\
a^3 - a^2b + ab^2 \\
\hline
a^3 \qquad\qquad + b^3
\end{array}
$$

Thus, $a^3 + b^3 = (a + b)(a^2 - ab + b^2)$. Using multiplication, we can also show that $a^3 - b^3 = (a - b)(a^2 + ab + b^2)$. Formulas for factoring **the sum and the difference of two cubes** appear in the following boxes.

Sum of Two Cubes

$$a^3 + b^3 = (a + b)(a^2 - ab + b^2)$$

Difference of Two Cubes

$$a^3 - b^3 = (a - b)(a^2 + ab + b^2)$$

EXAMPLE 10 Factor $x^3 + 64$.

Solution Rewrite $x^3 + 64$ as a sum of two cubes, $x^3 + 4^3$. Let x correspond to a and 4 to b. Then factor using the sum of two cubes formula.

$$
\begin{aligned}
a^3 + b^3 &= (a + b)(a^2 - a\,b + b^2) \\
x^3 + 4^3 &= (x + 4)[x^2 - x(4) + 4^2] \\
&= (x + 4)(x^2 - 4x + 16)
\end{aligned}
$$

Thus, $x^3 + 64 = (x + 4)(x^2 - 4x + 16)$. ✳

EXAMPLE 11 Factor $27x^3 - 8y^6$.

Solution We first observe that $27x^3$ and $8y^6$ have no common factors other than 1. Since we can express both $27x^3$ and $8y^6$ as cubes, we can factor using the difference of two cubes formula.

$$
\begin{aligned}
27x^3 - 8y^6 &= (3x)^3 - (2y^2)^3 \\
&= (3x - 2y^2)[(3x)^2 + (3x)(2y^2) + (2y^2)^2] \\
&= (3x - 2y^2)(9x^2 + 6xy^2 + 4y^4)
\end{aligned}
$$

NOW TRY EXERCISE 57 Thus, $27x^3 - 8y^6 = (3x - 2y^2)(9x^2 + 6xy^2 + 4y^4)$. ✳

EXAMPLE 12 Factor $8y^3 - 64x^6$.

Solution First factor out 8, which is common to both terms.

$$8y^3 - 64x^6 = 8(y^3 - 8x^6)$$

Next factor $y^3 - 8x^6$ by writing it as a difference of two cubes.

$$8(y^3 - 8x^6) = 8[(y)^3 - (2x^2)^3]$$
$$= 8(y - 2x^2)[y^2 + y(2x^2) + (2x^2)^2]$$
$$= 8(y - 2x^2)(y^2 + 2x^2y + 4x^4)$$

Thus, $8y^3 - 64x^6 = 8(y - 2x^2)(y^2 + 2x^2y + 4x^4)$. ✳

EXAMPLE 13   Factor $(x - 2)^3 + 125$.

Solution   Write $(x - 2)^3 + 125$ as a sum of two cubes, then use the sum of two cubes formula to factor.

$$(x - 2)^3 + (5)^3 = [(x - 2) + 5][(x - 2)^2 - (x - 2)(5) + (5)^2]$$
$$= (x - 2 + 5)(x^2 - 4x + 4 - 5x + 10 + 25)$$
$$= (x + 3)(x^2 - 9x + 39)$$

NOW TRY EXERCISE 65

✳

HELPFUL HINT

The square of a binomial has a 2 as part of the middle term of the trinomial.

$$(a + b)^2 = a^2 + 2ab + b^2$$
$$(a - b)^2 = a^2 - 2ab + b^2$$

The sum or the difference of two cubes has a factor similar to the trinomial in the square of the binomial. However, the middle term does not contain a 2.

$$a^3 + b^3 = (a + b)(a^2 - ab + b^2)$$
$$a^3 - b^3 = (a - b)(a^2 + ab + b^2)$$

$\underbrace{\qquad}_{\text{not } 2ab}$

EXAMPLE 14   **Volume**   Using the cubes in Figure 5.16, find an expression, in factored form, for the difference of the volumes.

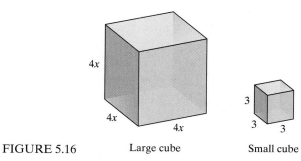

FIGURE 5.16   Large cube   Small cube

Solution   To find the difference of the volumes, subtract the volume of the small cube from the volume of the large cube.

$$\text{Volume of large cube} = (4x)^3$$
$$\text{Volume of small cube} = 3^3$$

$$\text{Difference of the volumes} = (4x)^3 - 3^3 \qquad \textit{Subtract volumes.}$$
$$= (4x - 3)[(4x)^2 + (4x)3 + 3^2] \qquad \textit{Factor.}$$
$$= (4x - 3)(16x^2 + 12x + 9) \qquad \textit{Simplify.}$$

NOW TRY EXERCISE 87   The difference of the volumes for the two cubes is $(4x - 3)(16x^2 + 12x + 9)$. ✳

Exercise Set 5.6

Concept/Writing Exercises

1. **a)** Explain in your own words how to factor the difference of two squares.

   b) Factor $x^2 - 16$ using the procedure you explained in part **a)**.

2. Explain why a sum of two squares, $a^2 + b^2$, cannot be factored over the set of real numbers.

3. Explain in your own words how to determine whether a trinomial is a perfect square trinomial.

4. **a)** Explain in your own words how to factor a perfect square trinomial.

   b) Factor $x^2 + 12x + 36$ using the procedure you explained in part **a)**.

5. Give the formula for factoring the sum of two cubes.

6. Give the formula for factoring the difference of two cubes.

7. Is $x^2 + 14x - 49 = (x + 7)(x - 7)$ factored correctly? Explain.

8. Is $x^2 + 14x + 49 = (x + 7)^2$ factored correctly? Explain.

9. Is $x^2 - 81 = (x - 9)^2$ factored correctly? Explain.

10. Is $x^2 - 64 = (x + 8)(x - 8)$ factored correctly? Explain.

Practice the Skills

Use the difference of two squares formula or the perfect square trinomial formula to factor each polynomial.

11. $x^2 - 81$

12. $x^2 - 25$

13. $a^2 - 100$

14. $1 - 4x^2$

15. $1 - 36b^2$

16. $x^2 - 81y^2$

17. $25 - 16y^4$

18. $a^6 - 144b^4$

19. $\dfrac{1}{25} - y^2$

20. $\dfrac{1}{49} - z^2$

21. $x^2y^2 - 121c^2$

22. $4a^2c^2 - 16x^2y^2$

23. $0.04x^2 - 0.09$

24. $0.16p^2 - 0.81q^2$

25. $36 - (x - 6)^2$

26. $144 - (a + b)^2$

27. $a^2 - (3b + 2)$

28. $(2c + 3)^2 - 9$

29. $x^2 + 10x + 25$

30. $49 - 14t + t^2$

31. $4 + 4a + a^2$

32. $b^2 - 18b + 81$

33. $4x^2 - 20xy + 25y^2$

34. $36p^2q^2 + 12pq + 1$

35. $0.81x^2 - 0.36x + 0.04$

36. $0.25x^2 - 0.40x + 0.16$

37. $y^4 + 4y^2 + 4$

38. $b^4 - 12b^2 + 36$

39. $(x + y)^2 + 2(x + y) + 1$

40. $(a + 1)^2 + 6(a + 1) + 9$

41. $(y - 3)^2 + 8(y - 3) + 16$

42. $a^4 - 2a^2b^2 + b^4$

43. $x^2 + 6x + 9 - y^2$

44. $p^2 + 2pq + q^2 - 16r^2$

45. $25 - (x^2 + 4x + 4)$

46. $9a^2 - 12ab + 4b^2 - 9$

47. $9 - (c^2 - 8c + 16)$

48. $(4a - 3b)^2 - (2a + 5b)^2$

49. $y^4 - 6y^2 + 9$

50. $z^6 - 14z^3 + 49$

Factor using the sum or difference of two cubes formula.

51. $x^3 - 27$

52. $a^3 + 125$

53. $64 - a^3$

54. $8 - b^3$

55. $p^3 - 27a^3$

56. $w^3 - 216$

57. $27y^3 - 8x^3$

58. $5x^3 + 40y^3$

59. $32a^3 - 108b^3$

60. $x^6 + y^9$

61. $2b^3 - 250c^3$

62. $16x^6 - 250y^3$

63. $(x + 1)^3 + 1$

64. $(a - 3)^3 + 8$

65. $(a - b)^3 - 27$

66. $(2x + y)^3 - 64$

67. $b^3 - (b + 3)^3$

68. $(m - n)^3 - (m + n)^3$

Factor using a special factoring formula.

69. $121y^4 - 49x^2$

70. $a^4 - 4b^4$

71. $16y^2 - 81x^2$

72. $49 - 64x^2y^2$

73. $25x^4 - 81y^6$

74. $(x + y)^2 - 16$

75. $x^3 - 64$

76. $2a^2 - 24a + 72$

77. $9x^2y^2 + 24xy + 16$

78. $a^4 + 12a^2 + 36$ 🔒 **79.** $a^4 + 2a^2b^2 + b^4$ **80.** $8y^3 - 125x^6$
81. $x^2 - 2x + 1 - y^2$ **82.** $4r^2 + 4rs + s^2 - 9$ **83.** $(x + y)^3 + 1$
84. $9x^2 - 6xy + y^2 - 4$ **85.** $(m + n)^2 - (2m - n)^2$ **86.** $(r + p)^3 + (r - p)^3$

Problem Solving

In Exercises 87–90, find an expression, in factored form, for the difference of the volumes of the two cubes. See Example 14.

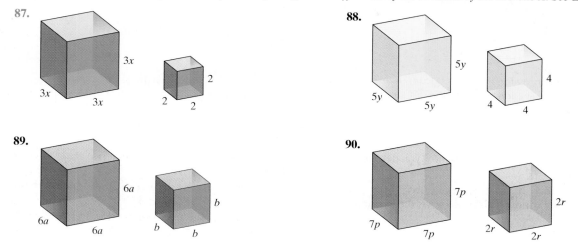

87.   **88.**

89.   **90.**

In Exercises 91 and 92, find an expression, in factored form, for the sum of the volumes of the two cubes.

91.   **92.**

In Exercises 93–97, **a)** *find the area or volume of the shaded figure by subtracting the smaller area or volume from the larger. The formula to find the area or volume is given under the figure.* **b)** *Write the expression obtained in part* **a)** *in factored form. Part of the GCF in Exercises 94, 96 and 97 is π.*

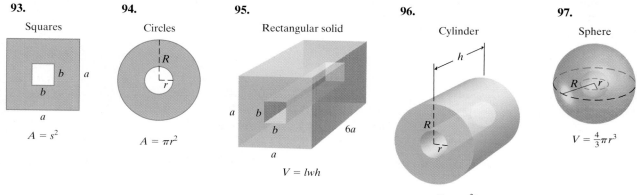

93.   **94.**   **95.**   **96.**   **97.**
Squares   Circles   Rectangular solid   Cylinder   Sphere

$A = s^2$ $A = \pi r^2$ $V = lwh$ $V = \pi r^2 h$ $V = \frac{4}{3}\pi r^3$

98. *Area and Volume* A circular hole is cut from a cube of wood, as shown in the figure.

a) Write an expression in factored form in terms of x for the cross-sectional area of the remaining wood.

b) Write an expression in factored form in terms of x for the volume of the remaining wood.

99. Find two values of b that will make $4x^2 + bx + 9$ a perfect square trinomial. Explain how you determined your answer.

100. Find two values of c that will make $16x^2 + cx + 4$ a perfect square trinomial. Explain how you determined your answer.

101. Find the value of c that will make $25x^2 + 20x + c$ a perfect square trinomial. Explain how you determined your answer.

102. Find the value of d that will make $49x^2 - 42x + d$ a perfect square trinomial. Explain how you determined your answer.

103. *Area* A formula for the area of a square is $A = s^2$, where s is a side. Suppose the area of a square is as given below,

$$A(x) = 25x^2 - 30x + 9$$

a) explain how to find the length of side x, $s(x)$,

b) find $s(x)$,

c) find $s(2)$.

Factor completely.

111. $64x^{4a} - 9y^{6a}$

113. $a^{2n} - 16a^n + 64$

115. $x^{3n} - 8$

104. *Area* The formula for the area of a circle is $A = \pi r^2$, where r is the radius. Suppose the area of a circle is as given below,

$$A(x) = 9\pi x^2 + 12\pi x + 4\pi$$

a) explain how to find the radius, $r(x)$,

b) find $r(x)$,

c) find $r(4)$.

105. Factor $x^4 + 64$ by writing the expression as $(x^4 + 16x^2 + 64) - 16x^2$, which is a difference of two squares.

106. Factor $x^4 + 4$ by adding and subtracting $4x^2$. (See Exercise 105.)

107. If $P(x) = x^2$, use the difference of two squares to simplify $P(a + h) - P(a)$.

108. If $P(x) = x^2$, use the difference of two squares to simplify $P(a + 1) - P(a)$.

109. *Sum of Areas* The figure shows how we *complete the square*. The sum of the areas of the three parts of the square that are shaded in blue and purple is

$$x^2 + 4x + 4x \quad \text{or} \quad x^2 + 8x$$

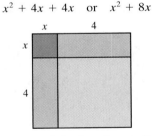

a) Find the area of the fourth part (in pink) to complete the square.

b) Find the sum of the areas of the four parts of the square.

c) This process has resulted in a perfect square trinomial in part **b)**. Write this perfect square trinomial as the square of a binomial.

110. Factor $(m - n)^3 - (9 - n)^3$.

112. $16p^{8w} - 49p^{6w}$

114. $144r^{8k} + 48r^{4k} + 4$

116. $27x^{3m} + 64x^{6m}$

🖩 *In Exercises 117 and 118, use your graphing calculator to check the factoring. Indicate if the factoring is correct or not correct. Explain your answers.*

117. $2x^2 - 18 \overset{?}{=} 2(x + 3)(x - 3)$

118. $8x^3 + 27 \overset{?}{=} 2x(4x^2 + 5x + 9)$

Challenge Problem

119. The expression $x^6 - 1$ can be factored using either the difference of two squares or the difference of two cubes. At first the factors do not appear the same. But with a little algebraic manipulation they can be shown to be equal. Factor $x^6 - 1$ using **a)** the difference of two squares and **b)** the difference of two cubes. **c)** Show that these two answers are equal by factoring the answer obtained in part **a)** completely. Then multiply the two binomials by the two trinomials.

Group Activity

Discuss and answer Exercise 120 as a group.

120. Later in the book we will need to construct perfect square trinomials. Examine some perfect square trinomials with a leading coefficient of 1.

   a) Explain how b and c are related if the trinomial $x^2 + bx + c$ is a perfect square trinomial.

   b) Construct a perfect square trinomial if the first two terms are $x^2 + 6x$.

   c) Construct a perfect square trinomial if the first two terms are $x^2 - 10x$.

   d) Construct a perfect square trinomial if the first two terms are $x^2 - 14x$.

Cumulative Review Exercises

[2.1] **121.** Simplify $-2[3x - (2y - 1) - 5x] + y$.

[3.6] **122.** If $f(x) = x^2 - 3x + 4$ and $g(x) = 2x - 5$, find $(g - f)(-1)$.

[4.4] **123.** *Angles* A right angle is divided into three smaller angles. The largest of the three angles is twice the smallest. The remaining angle is 10° greater than the smallest angle. Find the measure of each angle.

[5.4] **124.** Factor out the greatest common factor of $45y^{12} + 30y^{10}$.

   125. Factor $12x^2 - 9xy + 4xy - 3y^2$.

5.7 A GENERAL REVIEW OF FACTORING

SSM   Study Guide   CD/Video

MathPro 4/5   PH Math Tutor Center   prenhall.com/Angel

1 Factor polynomials using a combination of techniques.

1 Factor Polynomials Using a Combination of Techniques

We have presented a number of factoring methods. Now we will combine problems and techniques from the previous sections.

A general procedure to factor any polynomial follows.

To Factor a Polynomial

1. Determine whether all the terms in the polynomial have a greatest common factor other than 1. If so, factor out the GCF.

2. If the polynomial has two terms, determine whether it is a difference of two squares or a sum or difference of two cubes. If so, factor using the appropriate formula.

(continued on the next page)

3. If the polynomial has three terms, determine whether it is a perfect square trinomial. If so, factor accordingly. If it is not, factor the trinomial using trial and error, grouping, or substitution as explained in Section 5.5.

4. If the polynomial has more than three terms, try factoring by grouping. If that does not work, see if three of the terms are the square of a binomial.

5. As a final step, examine your factored polynomial to see if any factors listed have a common factor and can be factored further. If you find a common factor, factor it out at this point.

The following examples illustrate how to use the procedure.

EXAMPLE 1 Factor $3x^4 - 48x^2$.

Solution First, check for a greatest common factor other than 1. Since $3x^2$ is common to both terms, factor it out.

$$3x^4 - 48x^2 = 3x^2(x^2 - 16) = 3x^2(x + 4)(x - 4)$$

Note that $x^2 - 16$ is factored as a difference of two squares.

EXAMPLE 2 Factor $3x^2y^2 - 24xy^2 + 48y^2$.

Solution Begin by factoring the GCF, $3y^2$, from each term.

$$3x^2y^2 - 24xy^2 + 48y^2 = 3y^2(x^2 - 8x + 16) = 3y^2(x - 4)^2$$

Note that $x^2 - 8x + 16$ is a perfect square trinomial. If you did not recognize this, you would still obtain the correct answer by factoring the trinomial into
NOW TRY EXERCISE 27 $(x - 4)(x - 4)$.

EXAMPLE 3 Factor $24x^2 - 6xy + 16xy - 4y^2$.

Solution As always, begin by determining if all the terms in the polynomial have a common factor. In this example, 2 is common to all terms. Factor out the 2; then factor the remaining four-term polynomial by grouping.

$$24x^2 - 6xy + 16xy - 4y^2 = 2(12x^2 - 3xy + 8xy - 2y^2)$$
$$= 2[3x(4x - y) + 2y(4x - y)]$$
$$= 2(4x - y)(3x + 2y)$$

EXAMPLE 4 Factor $10a^2b - 15ab + 20b$.

Solution $$10a^2b - 15ab + 20b = 5b(2a^2 - 3a + 4)$$

Since $2a^2 - 3a + 4$ cannot be factored, we stop here.

EXAMPLE 5 Factor $2x^4y + 54xy$.

Solution $$2x^4y + 54xy = 2xy(x^3 + 27)$$
$$= 2xy(x + 3)(x^2 - 3x + 9)$$

NOW TRY EXERCISE 19 Note that $x^3 + 27$ was factored as a sum of two cubes.

EXAMPLE 6 Factor $6x^2 - 3x + 6y^2 - 9$.

Solution First, we factor 3 from all four terms.

$$6x^2 - 3x + 6y^2 - 9 = 3(2x^2 - x + 2y^2 - 3)$$

Now we determine whether the four terms within parentheses can be factored by grouping. Since these four terms cannot be factored by grouping, we check to see whether any three of the terms can be written as the square of a binomial. No matter how we rearrange the terms, this cannot be done. We conclude that this expression cannot be factored further. Thus,

$$6x^2 - 3x + 6y^2 - 9 = 3(2x^2 - x + 2y^2 - 3)$$

EXAMPLE 7   Factor $3x^2 - 18x + 27 - 3y^2$.

Solution   Factor out 3 from all four terms.

$$3x^2 - 18x + 27 - 3y^2 = 3(x^2 - 6x + 9 - y^2)$$

Now try factoring by grouping. Since the four terms within parentheses cannot be factored by grouping, check to see whether any three of the terms can be written as the square of a binomial. Since this can be done, express $x^2 - 6x + 9$ as $(x - 3)^2$ and then use the difference of two squares formula. Thus,

$$3x^2 - 18x + 27 - 3y^2 = 3[(x - 3)^2 - y^2]$$
$$= 3[(x - 3 + y)(x - 3 - y)]$$
$$= 3(x - 3 + y)(x - 3 - y)$$

NOW TRY EXERCISE 43

HELPFUL HINT

STUDY TIP

In this section, we have reviewed all the techniques for factoring expressions. If you are still having difficulty with factoring, you should study the material in sections 5.4–5.6 again.

Exercise Set 5.7

Concept/Writing Exercises

1. Explain the possible procedures that may be used to factor a polynomial of **a)** two terms **b)** three terms **c)** four terms.

2. What is the first step in the factoring process?

Practice the Skills

Factor each polynomial completely.

3. $3x^2 - 75$

4. $3x^2 - 24x + 48$

5. $10s^2 + 19s - 15$

6. $6x^3y^2 + 10x^2y^3 + 8x^2y^2$

7. $-8r^2 + 26r - 15$

8. $4x^3 - 16x^2 - 48x$

9. $0.4x^2 - 0.036$

10. $0.5x^2 - 0.08$

11. $6x^5 - 54x$

12. $7x^2y^2z^2 - 28x^2y^2$

13. $3x^6 - 3x^5 + 12x^5 - 12x^4$

14. $2x^2y^2 + 6xy^2 - 10xy^2 - 30y^2$

15. $5x^4y^2 + 20x^3y^2 + 15x^3y^2 + 60x^2y^2$

16. $6x^2 - 15x - 9$

17. $x^4 - x^2y^2$

18. $6x^3 + 162$

19. $x^7y^2 - x^4y^2$

20. $x^4 - 81$

21. $x^5 - 16x$

22. $12x^2y^2 + 33xy^2 - 9y^2$

23. $2x^6 + 16y^3$

24. $8x^4 - 4x^3 - 4x^3 + 2x^2$

25. $2(a + b)^2 - 50$

26. $12x^3y^2 + 4x^2y^2 - 40xy^2$

27. $6x^2 + 36xy + 54y^2$

28. $3x^2 - 30x + 75$

29. $(x + 2)^2 - 4$

30. $4y^4 - 36x^6$

31. $(2a + b)(2a - 3b) - (2a + b)(a - b)$

32. $pq - 6q + pr - 6r$

33. $(y + 5)^2 + 4(y + 5) + 4$

34. $b^4 + 2b^2 + 1$

35. $45a^4 - 30a^3 + 5a^2$

36. $(x + 1)^2 - (x + 1) - 6$

37. $x^3 + \dfrac{1}{64}$

38. $8y^3 - \dfrac{1}{27}$

39. $3x^3 + 2x^2 - 27x - 18$

40. $6y^3 + 14y^2 + 4y$

41. $a^3b - 64ab^3$

42. $x^6 + y^6$

43. $81 - (x^2 + 2xy + y^2)$

44. $x^2 - 2xy + y^2 - 49$

45. $24x^2 - 34x + 12$

46. $40x^2 + 52x - 12$

47. $16x^2 + 34x - 15$

48. $7(a - b)^2 + 4(a - b) - 3$

49. $x^4 - 16$

50. $(x + 2)^2 - 12(x + 2) + 36$

51. $5bc - 10cx - 7by + 14xy$

52. $16y^4 - 9y^2$

🔒 **53.** $3x^4 - x^2 - 4$

54. $x^2 + 16x + 64 - 100y^2$

55. $y^2 - (x^2 - 12x + 36)$

56. $4a^3 + 32$

57. $2(y + 4)^2 + 5(y + 4) - 12$

58. $x^6 + 11x^3 + 30$

59. $a^2 + 12ab + 36b^2 - 16c^2$

60. $y - y^3$

61. $6x^4y + 15x^3y - 9x^2y$

62. $4x^2y^2 + 12xy + 9$

🔒 **63.** $x^4 - 2x^2y^2 + y^4$

64. $6r^2s^2 + rs - 1$

Problem Solving

Match Exercises 65–72 with the items labeled **a)** *through* **h)** *on the right.*

65. $a^2 + b^2$

66. $a^2 + 2ab + b^2$

67. $a^3 + b^3$

68. $a^2 - b^2$

69. $a^3 - b^3$

70. $a^2 - 2ab + b^2$

71. a factor of $a^3 + b^3$

72. a factor of $a^3 - b^3$

a) $(a + b)(a^2 - ab + b^2)$

b) $(a - b)^2$

c) $a^2 - ab + b^2$

d) $(a + b)^2$

e) not factorable

f) $(a - b)(a^2 + ab + b^2)$

g) $a^2 + ab + b^2$

h) $(a + b)(a - b)$

In Exercises 73 and 74, find an expression, in factored form, for the perimeter of each figure.

73. $5x + 3$, $x^2 + 3$

74. $7x + 13$, $5x + 12$, $x^2 + 11$

In Exercises 75–78, find an expression, in factored form, for the area of the shaded region for each figure.

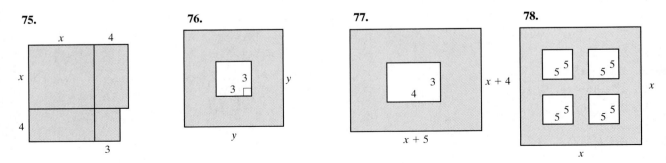

75. x, 4, x, 4, 3

76. 3, 3, y, y

77. 3, 4, $x + 4$, $x + 5$

78. 5, 5, 5, 5, 5, 5, 5, 5, $x + 4$, x, x

In Exercises 79 and 80, find an expression, in factored form, for the difference in the volumes of the cubes.

79.   **80.**

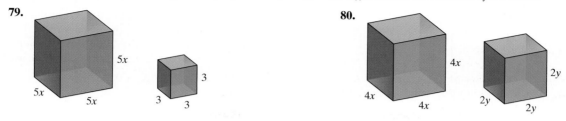

In Exercises 81–84, **a)** *write an expression for the shaded area of the figure, and* **b)** *write the expression in factored form.*

81.   **82.**

83.   **84.**

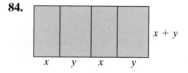

85. *Surface Area*
   a) Write an expression for the surface area of the four
      sides of the box shown (omit top and bottom).
   b) Write the expression in factored form.

86. Explain how the formula for factoring the *difference* of
   two cubes can be used to factor $x^3 + 27$.

87. a) Explain how to construct a perfect square trinomial.
   b) Construct a perfect square trinomial and then show
      its factors.

Challenge Problems

We have worked only with integer exponents in this chapter. However, fractional exponents may also be factored out of an expression. The expressions below are not polynomials. **a)** *Factor out the variable with the lowest (or most negative) exponent from each expression. (Fractional exponents are discussed in Section 7.2.)* **b)** *Factor completely.*

88. $x^{-2} - 5x^{-3} + 6x^{-4}$, factor out x^{-4}

89. $x^{-3} - 2x^{-4} - 3x^{-5}$, factor out x^{-5}

90. $x^{5/2} + 3x^{3/2} - 4x^{1/2}$, factor out $x^{1/2}$

91. $5x^{1/2} + 2x^{-1/2} - 3x^{-3/2}$, factor out $x^{-3/2}$

Cumulative Review Exercises

[2.1] 92. Solve $6(x + 4) - 4(3x + 3) = 6$.

[2.6] 93. Find the solution set for $\left|\dfrac{6 + 2z}{3}\right| > 2$.

[4.3] 94. *Mixing Coffees* John Murphy runs a grocery store. He wishes to mix 30 pounds of coffee to sell for a total cost of $170. To obtain the mixture, he will mix coffee that sells for $5.20 per pound with coffee that sells for $6.30 per pound. How many pounds of each coffee should he use?

[5.2] 95. Multiply $(5x + 4)(x^2 - x + 4)$.

[5.4] 96. Factor $2x^3 + 4x^2 - 5x - 10$.

5.8 POLYNOMIAL EQUATIONS

1 Use the zero-factor property to solve equations.
2 Use factoring to solve equations.
3 Use factoring to solve applications.
4 Use factoring to find the *x*-intercepts of a quadratic function.

SSM   Study Guide   CD/Video

MathPro 4/5   PH Math Tutor Center   prenhall.com/Angel

Whenever two polynomials are set equal to each other, we have a **polynomial equation**.

Examples of Polynomial Equations

$$x^2 + 2x = x - 5$$
$$y^3 + 3y - 2 = 0$$
$$4x^4 + 2x^2 = -3x + 2$$

The **degree of a polynomial equation** is the same as the degree of its highest term. For example, the three equations above have degree 2, 3, and 4, respectively. A second-degree equation in one variable is often called a **quadratic equation**.

Examples of Quadratic Equations

$$3x^2 + 6x - 4 = 0$$
$$5x = 2x^2 - 4$$
$$(x + 4)(x - 3) = 0$$

Any quadratic equation can be written in **standard form**.

> ### Standard Form of a Quadratic Equation
>
> $$ax^2 + bx + c = 0, \quad a \neq 0$$
>
> where a, b, and c are real numbers.

Before going any further, make sure that you can convert each of the three quadratic equations given above to standard form, with $a > 0$.

1 Use the Zero-Factor Property to Solve Equations

To solve equations using factoring, we use the **zero-factor property**.

> ### Zero-Factor Property
>
> For all real numbers a and b, if $a \cdot b = 0$, then either $a = 0$ or $b = 0$, or both a and $b = 0$.

The zero-factor property states that *if the product of two factors equals 0, one (or both) of the factors must be 0.*

EXAMPLE 1 Solve the equation $(x + 5)(x - 2) = 0$.

Solution Since the product of the factors equals 0, according to the zero-factor property, one or both factors must equal 0. Set each factor equal to 0 and solve each equation separately.

$$x + 5 = 0 \quad \text{or} \quad x - 2 = 0$$
$$x = -5 \qquad\qquad x = 2$$

Thus, if x is either -5 or 2, the product of the factors is 0.

Check $x = -5$ $x = 2$
$$(x + 5)(x - 3) = 0 \qquad\qquad (x + 5)(x - 2) = 0$$
$$(-5 + 5)(-5 - 3) \overset{?}{=} 0 \qquad\qquad (2 + 5)(2 - 2) \overset{?}{=} 0$$
$$0(-8) \overset{?}{=} 0 \qquad\qquad 7(0) \overset{?}{=} 0$$
$$0 = 0 \quad \textit{True} \qquad\qquad 0 = 0 \quad \textit{True}$$

2 Use Factoring to Solve Equations

Following is a procedure that can be used to obtain the solution to an equation by factoring.

To Solve an Equation by Factoring

1. Use the addition property to remove all terms from one side of the equation. This will result in one side of the equation being equal to 0.
2. Combine like terms in the equation and then factor.
3. Set each factor *containing a variable* equal to 0, solve the equations, and find the solutions.
4. Check the solutions in the *original* equation.

EXAMPLE 2 Solve the equation $2x^2 = 12x$.

Solution First, make the right side of the equation equal to 0 by subtracting $12x$ from both sides of the equation. Then factor the left side of the equation.

$$2x^2 - 12x = 0$$
$$2x(x - 6) = 0$$

Now set each factor equal to 0 and solve for x.

$$2x = 0 \quad \text{or} \quad x - 6 = 0$$
$$x = 0 \qquad\qquad x = 6$$

A check will show that the numbers 0 and 6 both satisfy the equation $2x^2 = 12x$.

AVOIDING COMMON ERRORS

The zero-factor property can be used only when one side of the equation is equal to 0.

CORRECT   INCORRECT
$$(x - 4)(x + 3) = 0 \qquad (x - 4)(x + 3) = 2$$
$$x - 4 = 0 \quad \text{or} \quad x + 3 = 0 \qquad x - 4 = 2 \quad \text{or} \quad x + 3 = 2$$
$$x = 4 \qquad\qquad x = -3 \qquad x = 6 \qquad\qquad x = -1$$

(continued on the next page)

In the incorrect process illustrated on the right at the bottom of the previous page, the zero-factor property cannot be used since the right side of the equation is not equal to 0. Example 3 shows how to solve such problems correctly.

EXAMPLE 3   Solve the equation $(x - 1)(3x + 2) = 4x$.

Solution   Since the right side of the equation is not 0, we cannot use the zero-factor property yet. Instead, we begin by multiplying the factors on the left side of the equation. Then we subtract $4x$ from both sides of the equation to obtain 0 on the right side. Then we can factor and solve the equation.

$$(x - 1)(3x + 2) = 4x$$
$$3x^2 - x - 2 = 4x \qquad \textit{Multiply the factors.}$$
$$3x^2 - 5x - 2 = 0 \qquad \textit{Make one side 0.}$$
$$(3x + 1)(x - 2) = 0 \qquad \textit{Factor the trinomial.}$$
$$3x + 1 = 0 \quad \text{or} \quad x - 2 = 0 \quad \textit{Zero-factor property}$$
$$3x = -1 \qquad\qquad x = 2 \quad \textit{Solve the equations.}$$
$$x = -\frac{1}{3}$$

The solutions are $-\frac{1}{3}$ and 2. Check these values in the original equation.   ✳

EXAMPLE 4   Solve the equation $3x^2 + 2x - 12 = -7x$.

Solution

$$3x^2 + 2x - 12 = -7x$$
$$3x^2 + 9x - 12 = 0 \qquad \textit{Make one side 0.}$$
$$3(x^2 + 3x - 4) = 0 \qquad \textit{Factor out 3.}$$
$$3(x + 4)(x - 1) = 0 \qquad \textit{Factor the trinomial.}$$
$$x + 4 = 0 \quad \text{or} \quad x - 1 = 0 \quad \textit{Zero-factor property}$$
$$x = -4 \qquad\qquad x = 1 \quad \textit{Solve for x.}$$

Since the factor 3 does not contain a variable, we do not have to set it equal to 0. Only the numbers -4 and 1 satisfy the equation $3x^2 + 2x - 12 = -7x$.   ✳

NOW TRY EXERCISE 45

HELPFUL HINT

When solving an equation whose leading term has a negative coefficient, we generally make the coefficient positive by multiplying both sides of the equation by -1. This makes the factoring process easier, as in the following example.

$$-x^2 + 5x + 6 = 0$$
$$-1(-x^2 + 5x + 6) = -1 \cdot 0$$
$$x^2 - 5x - 6 = 0$$

Now we can solve the equation $x^2 - 5x - 6 = 0$ by factoring.

$$(x - 6)(x + 1) = 0$$
$$x - 6 = 0 \quad \text{or} \quad x + 1 = 0$$
$$x = 6 \qquad\qquad x = -1$$

The numbers 6 and -1 both satisfy the original equation, $-x^2 + 5x + 6 = 0$.

The equations in Examples 1 through 4 were all quadratic equations that were rewritten in the form $ax^2 + bx + c = 0$ and solved by factoring. Other methods that can be used to solve quadratic equations include completing the square and the quadratic formula; we discuss these methods in Chapter 8.

The zero-factor property can be extended to three or more factors, as illustrated in Example 5.

EXAMPLE 5 Solve the equation $2p^3 + 7p^2 + 3p = 0$.

Solution First factor, then set each factor containing a p equal to 0.

$$2p^3 + 7p^2 + 3p = 0$$
$$p(2p^2 + 7p + 3) = 0 \qquad \text{\textit{Factor out p.}}$$
$$p(2p + 1)(p + 3) = 0 \qquad \text{\textit{Factor the trinomial.}}$$

$p = 0 \quad$ or $\quad 2p + 1 = 0 \quad$ or $\quad p + 3 = 0 \qquad$ *Zero-factor property*

$\qquad\qquad\qquad 2p = -1 \qquad\qquad p = -3 \qquad$ *Solve for p.*

$$p = -\frac{1}{2}$$

The numbers 0, $-\frac{1}{2}$, and -3 are all solutions to the equation. ✳

Note that the equation in Example 5 is not a quadratic equation because the exponent in the leading term is 3, not 2. This is a **cubic**, or **third-degree, equation**.

EXAMPLE 6 For the function $f(x) = 2x^2 - 13x - 16$, find all values of a for which $f(a) = 8$.

Solution First we rewrite the function as $f(a) = 2a^2 - 13a - 16$. Since $f(a) = 8$, we write

$$2a^2 - 13a - 16 = 8 \qquad \text{\textit{Set f(a) equal to 8.}}$$
$$2a^2 - 13a - 24 = 0 \qquad \text{\textit{Make one side 0.}}$$
$$(2a + 3)(a - 8) = 0 \qquad \text{\textit{Factor the trinomial.}}$$

$2a + 3 = 0 \quad$ or $\quad a - 8 = 0 \qquad$ *Zero-factor property*

$\quad 2a = -3 \qquad\qquad a = 8 \qquad$ *Solve for a.*

$$a = -\frac{3}{2}$$

NOW TRY EXERCISE 69 If you check these answers, you will find that $f\left(-\frac{3}{2}\right) = 8$ and $f(8) = 8$. ✳

3 Use Factoring to Solve Applications

Now let us look at some applications that use factoring in their solution.

EXAMPLE 7 **Triangle** At an exhibition, a large canvas tent is to have an entrance in the shape of a triangle (see Fig. 5.17).

Find the base and height of the entrance if the height is to be 3 feet less than twice the base and the total area of the entrance is 27 square feet.

Solution Understand Let's draw a picture of the entrance and label it with the given information (Fig. 5.18).

FIGURE 5.17

Area = 27ft^2

$2x - 3$

x

FIGURE 5.18

Translate   We use the formula for the area of a triangle to solve the problem.

$$A = \frac{1}{2}(\text{base})(\text{height})$$

$$27 = \frac{1}{2}(x)(2x - 3)$$ *Substitute expressions for base, height, and area.*

Carry Out

$$2(27) = 2\left[\frac{1}{2}(x)(2x - 3)\right]$$ *Multiply both sides by 2 to remove fractions.*

$$54 = x(2x - 3)$$

$$54 = 2x^2 - 3x$$

$$\text{or} \quad 2x^2 - 3x - 54 = 0$$ *Make one side 0.*

$$(2x + 9)(x - 6) = 0$$ *Factor the trinomial.*

$$2x + 9 = 0 \quad \text{or} \quad x - 6 = 0$$ *Zero-factor property*

$$2x = -9 \qquad\qquad x = 6$$ *Solve for x.*

$$x = -\frac{9}{2}$$

Answer   Since the dimensions of a geometric figure cannot be negative, we can eliminate $x = -\frac{9}{2}$ as an answer to our problem. Therefore,

$$\text{base} = x = 6 \text{ feet}$$

$$\text{height} = 2x - 3 = 2(6) - 3 = 9 \text{ feet}$$ ✳

EXAMPLE 8   **Height of a Cannonball**   A cannon sits on top of a 288-foot cliff overlooking a lake. A cannonball is fired upward at a speed of 112 feet per second. The height, h, in feet, of the cannonball above the lake at any time, t, is determined by the function

$$h(t) = -16t^2 + 112t + 288$$

Find the time it takes for the cannonball to hit the water after the cannon is fired.

Solution   Understand   We need to draw a picture to help analyze the problem (see Fig. 5.19). When the cannonball hits the water, its height above the water is 0 feet.

FIGURE 5.19

Translate   To solve the problem we need to find the time, t, when $h(t) = 0$. To do so we set the given function equal to 0 and solve for t.

$$-16t^2 + 112t + 288 = 0$$ *Set h(t) = 0.*

$$-16(t^2 - 7t - 18) = 0$$ *Factor out −16.*

$$-16(t + 2)(t - 9) = 0$$ *Factor the trinomial.*

$$t + 2 = 0 \quad \text{or} \quad t - 9 = 0$$ *Zero-factor property*

$$t = -2 \qquad\qquad t = 9$$ *Solve for t.*

NOW TRY EXERCISE 105

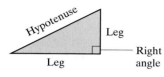

FIGURE 5.20

Answer Since t is the number of seconds, -2 is not a possible answer. The cannonball will hit the water in 9 seconds. ✴

Pythagorean Theorem

Our next application uses the Pythagorean Theorem. Consider a right triangle (see Fig. 5.20). The two shorter sides of a right triangle are called the **legs** and the side opposite the right angle is called the **hypotenuse**. The **Pythagorean Theorem** expresses the relationship between the legs of the triangle and its hypotenuse.

Pythagorean Theorem

The square of the length of the hypotenuse of a right triangle is equal to the sum of the squares of the lengths of the two legs; that is,

$$\text{leg}^2 + \text{leg}^2 = \text{hyp}^2$$

If a and b represent the lengths of the legs and c represents the length of the hypotenuse, then

$$a^2 + b^2 = c^2$$

EXAMPLE 9

Tree Wires Jack Keating places a guy wire on a tree to help it grow straight. The location of the stake and where the wire attaches to the tree are given in Figure 5.21. Find the length of the wire. Notice that the length of the wire is the hypotenuse.

Solution Understand To solve this problem we use the Pythagorean Theorem. From the figure, we see that the legs are x and $x + 1$ and the hypotenuse is $x + 2$.

FIGURE 5.21

Translate
$$\text{leg}^2 + \text{leg}^2 = \text{hyp}^2 \quad \textit{Pythagorean Theorem}$$
$$x^2 + (x + 1)^2 = (x + 2)^2 \quad \textit{Substitute expressions for legs and hypotenuse.}$$

Carry Out
$$x^2 + x^2 + 2x + 1 = x^2 + 4x + 4 \quad \textit{Square terms.}$$
$$2x^2 + 2x + 1 = x^2 + 4x + 4 \quad \textit{Simplify.}$$
$$x^2 - 2x - 3 = 0 \quad \textit{Make one side 0.}$$
$$(x - 3)(x + 1) = 0 \quad \textit{Factor.}$$
$$x - 3 = 0 \quad \text{or} \quad x + 1 = 0 \quad \textit{Solve.}$$
$$x = 3 \qquad\qquad x = -1$$

Answer From the diagram we know that x cannot be a negative value. Therefore the only possible answer is 3. The stake is placed 3 feet from the tree. The wire attaches to the tree $x + 1$ or 4 feet from the ground. The length of the wire is $x + 2$ or 5 feet. ✴

NOW TRY EXERCISE 99

 Using Your Graphing Calculator

The applications given in this section and the exercise set have been written so that the quadratic equations are factorable. In real life, quadratic equations generally are not factorable (over the set of integers) and need to be solved in other ways. We will discuss methods to solve quadratic equations that are not factorable in Sections 8.1 and 8.2.

(continued on the next page)

You can find approximate solutions to quadratic equations that are not factorable using your graphing calculator. Consider the following real-life example.

EXAMPLE *Cellular Antennas* The number of relay antennas for cellular phones in the United States has been increasing. From 1996 through 2002, the number of cellular relay antennas, N, in thousands, in the United States can be closely approximated by the function

$$N(t) = -1.45t^2 + 21.88t + 25.44$$

where t is the number of years since 1996. Determine the year in which the number of cellular antennas reached 80,000.

Solution Understand and Translate To answer this question we need to set the function $N(t)$ equal to 80 and solve for t.

$$-1.45t^2 + 21.88t + 25.44 = 80 \quad \text{Set } N(t) = 80.$$

We cannot solve this equation by factoring, but we can solve it using a graphing calculator. To do so, let's call one side of the equation y_1 and the other side y_2.

$$y_1 = -1.45x^2 + 21.88x + 25.44$$

$$y_2 = 80$$

Carry Out Now graph the two functions on your graphing calculator and use the $\boxed{\text{TRACE}}$ and $\boxed{\text{ZOOM}}$ keys, or other keys (for example the $\boxed{\text{CALC}}$ key with option 5, *intersect*, on the TI-83 Plus) to obtain your answer. Figure 5.22 illustrates the screen a TI-83 Plus showing that the x-coordinate of the intersection of the equations is about $x = 3.1520$.

Answer Therefore, there will be about 80,000 relay antennas about 3 years after 1996, or in 1999.

$[0, 6, 1, 0, 120, 30]$

FIGURE 5.22

4 Use Factoring to Find the *x*-Intercepts of a Quadratic Function

Consider the graph in Figure 5.23.

At the x-intercepts, the value of the function, or y, is 0. Thus if we wish to find the **x-intercepts of a graph**, we can set the function equal to 0 and solve for x.

EXAMPLE 10 Find the x-intercepts of the graph of $y = x^2 - 2x - 8$.

Solution At the x-intercepts y has a value of 0. Thus to find the x-intercepts we write

$$x^2 - 2x - 8 = 0$$

$$(x - 4)(x + 2) = 0$$

$$x - 4 = 0 \quad \text{or} \quad x + 2 = 0$$

$$x = 4 \qquad\qquad x = -2$$

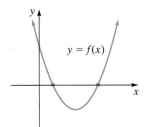

FIGURE 5.23

The solutions of $x^2 - 2x - 8 = 0$ are 4 and -2. The x-intercepts of the graph of $y = x^2 - 2x - 8$ are $(-2, 0)$ and $(4, 0)$, as illustrated in Figure 5.24.

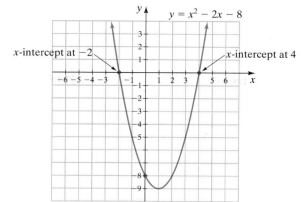

FIGURE 5.24

NOW TRY EXERCISE 75

If we know the x-intercepts of a graph, we can work backward to find the equation of the graph. Read the Using Your Graphing Calculator box that follows to learn how this is done.

Using Your Graphing Calculator

In Example 10, we saw that the graph of $y = x^2 - 2x - 8$ had x-intercepts at -2 and 4. If we know the x-intercepts of a graph, we can find equations whose graphs have those intercepts. For example,

x-Intercepts at	Factors	Equation
-2 and 4	$(x + 2)(x - 4)$	$y = (x + 2)(x - 4)$ or $y = x^2 - 2x - 8$

Note that other equations may have graphs with the same x-intercepts. For example, the graph of $y = 2(x^2 - 2x - 8)$ or $y = 2x^2 - 4x - 16$ also has x-intercepts at 4 and -2. In fact, the graph of $y = a(x^2 - 2x - 8)$, for any nonzero real number a, will have x-intercepts at 4 and -2.

Consider the graph in Figure 5.25.

$$[-10, 10, 1, -10, 20, 2]$$

FIGURE 5.25   FIGURE 5.26

If we assume the intercepts are integer values, then the x-intercepts are at 2 and 8. Therefore,

x-Intercepts at	Factors	Possible equation of graph
2 and 8	$(x - 2)(x - 8)$	$y = (x - 2)(x - 8)$ or $y = x^2 - 10x + 16$

If we change the window, we see that the y-intercepts of the graph is at 16 (see Fig. 5.26). Therefore, $y = x^2 - 10x + 16$ is the equation of the graph. If, for example, the y-intercept of the graph was at 32, then the equation of the graph would be $y = 2(x^2 - 10x + 16)$ or $y = 2x^2 - 20x + 32$.

(continued on the next page)

Exercises

Write the equation of each graph illustrated. Assume that all x-intercepts are integer values and that the standard window is shown.

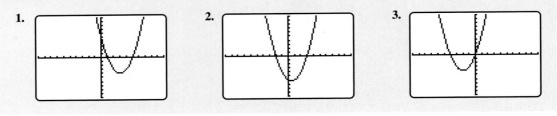

1.

2.

3.

Exercise Set 5.8

Concept/Writing Exercises

1. How do you determine the degree of a polynomial function?

2. What is a quadratic equation?

3. What is the standard form of a quadratic equation?

4. a) Explain the zero-factor property in your own words.
   b) Solve the equation $(3x - 7)(2x + 3) = 0$ using the zero-factor property.

5. a) Explain why the equation $(x + 3)(x + 4) = 2$ *cannot* be solved by writing $x + 3 = 2$ or $x + 4 = 2$.
   b) Solve the equation $(x + 3)(x + 4) = 2$.

6. When a constant is factored out of an equation, why is it not necessary to set that constant equal to 0 when solving the equation?

7. a) Explain in your own words how to solve a polynomial equation using factoring.
   b) Solve the equation $-x - 20 = -12x^2$ using the procedure in part **a)**.

8. a) What is the first step in solving the equation $-x^2 + 2x + 35 = 0$?
   b) Solve the equation in part **a)**.

9. a) What are the two shorter sides of a right triangle called?
   b) What is the longest side of a right triangle called?

10. Give the Pythagorean Theorem and explain its meaning.

11. If the graph of $y = x^2 + 10x + 16$ has x-intercepts at -8 and -2, what is the solution to the equation $x^2 + 10x + 16 = 0$? Explain.

12. If the solutions to the equation $2x^2 - 15x + 18 = 0$ are $\frac{3}{2}$ and 6, what are the x-intercepts of the graph of $y = 2x^2 - 15x + 18$? Explain.

13. Is it possible for a quadratic function to have no x-intercepts? Explain.

14. Is it possible for a quadratic function to have only one x-intercept? Explain.

15. Is it possible for a quadratic function to have two x-intercepts? Explain.

16. Is it possible for a quadratic function to have three x-intercepts? Explain.

Practice the Skills

Solve.

17. $x(x - 4) = 0$

18. $x(x + 2) = 0$

19. $3x(x - 1) = 0$

20. $5x(x + 6) = 0$

21. $2(x + 1)(x - 7) = 0$

22. $3(a - 5)(a + 2) = 0$

23. $x(x - 9)(x + 4) = 0$

24. $2a(a + 3)(a + 8) = 0$

25. $(2x + 3)(4x + 5) = 0$

26. $(3x - 2)(7x - 1) = 0$

27. $4x^2 = 12x$

28. $3y^2 = -24y$

29. $x^2 + 5x = 0$

30. $2a^2 - 8a = 0$

31. $-x^2 + 6x = 0$

32. $-3x^2 - 9x = 0$

33. $3x^2 = 15x$

34. $9a^2 = -18a$

35. $x^2 - 6x + 5 = 0$

36. $a^2 + 6a + 5 = 0$

37. $x^2 + x - 12 = 0$

38. $b^2 + b - 72 = 0$

39. $x^2 + 8x + 16 = 0$

40. $c^2 - 10c = -25$

41. $x(x - 12) = -20$

42. $b(b - 2) = 48$

43. $2x^2 = -14x - 12$

44. $3a^2 = -a + 2$

45. $3x^2 - 6x - 72 = 0$

46. $4a^2 + 36a + 80 = 0$

47. $x^3 - 3x^2 = 18x$

48. $x^3 = -19x^2 + 42x$

49. $12a^2 = 16a + 3$

50. $3b^3 - 8b^2 - 3b = 0$

51. $4c^3 + 4c^2 - 48c = 0$

52. $18z^3 = 15z^2 + 12z$

53. $x^2 - 36 = 0$

54. $6y^2 = 16y$

55. $4x^2 = 9$

56. $49c^2 = 25$

57. $25x^3 - 16x = 0$

58. $3x^4 - 48x^2 = 0$

59. $-x^2 = 2x - 99$

60. $-x^2 + 16x = 63$

61. $(x + 4)^2 - 16 = 0$

62. $(x - 4)^2 - 4 = 0$

63. $(2x + 5)^2 - 9 = 0$

64. $(x + 1)^2 - 3x = 7$

65. $6a^2 - 12 - 4a = 19a - 32$

66. $4(a^2 - 3) = 6a + 4(a + 3)$

67. $2b^3 + 16b^2 = -30b$

68. $(a - 1)(3a + 2) = 4a$

69. For $f(x) = 3x^2 + 7x + 9$, find all values of a for which $f(a) = 7$.

70. For $f(x) = 4x^2 - 11x$, find all values of a for which $f(a) = -6$.

71. For $g(x) = 10x^2 - 31x + 19$, find all values of a for which $g(a) = 4$.

72. For $g(x) = 6x^2 + x - 3$, find all values of a for which $g(a) = -2$.

73. For $r(x) = x^2 - x$, find all values of a for which $r(a) = 30$.

74. For $r(x) = 10x^2 - 19x + 2$, find all values of a for which $r(a) = -4$.

Use factoring to find the x-intercepts of the graphs of each equation (see Example 10).

75. $y = x^2 + 2x - 24$

76. $y = x^2 - 13x + 42$

77. $y = x^2 + 16x + 64$

78. $y = 15x^2 - 14x - 8$

79. $y = 6x^3 - 23x^2 + 20x$

80. $y = 12x^3 - 39x^2 + 30x$

In Exercises 81–86, use the Pythagorean Theorem to find x.

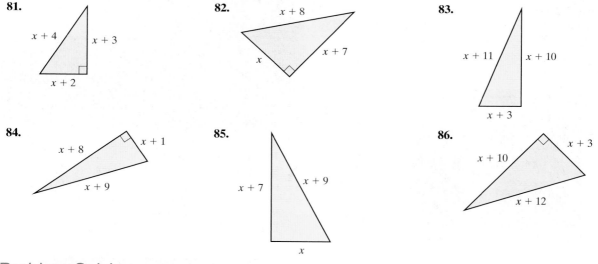

81. $x + 4$, $x + 3$, $x + 2$

82. $x + 8$, x, $x + 7$

83. $x + 11$, $x + 10$, $x + 3$

84. $x + 8$, $x + 1$, $x + 9$

85. $x + 7$, $x + 9$, x

86. $x + 3$, $x + 10$, $x + 12$

Problem Solving

In Exercises 87–90, determine the x-intercepts of each graph; then match the equation with the appropriate graph labeled **a)–d)**.

87. $y = x^2 - 5x + 6$

88. $y = x^2 - x - 6$

89. $y = x^2 + 5x + 6$

90. $y = x^2 + x - 6$

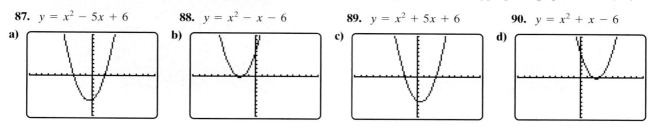

a)   b)   c)   d)

Write an equation whose graph will have x-intercepts at the given values.

91. 2 and 5

92. 3 and -7

93. 4 and -1

94. $\frac{3}{2}$ and 6

95. $-\frac{5}{6}$ and 2

96. -0.4 and 2.6

97. *Rectangular Coffee Table* A coffee table is rectangular. If the length of its surface area is 1 foot greater than twice its width and the surface area of the tabletop is 10 square feet, find its length and width.

98. *Rectangular Shed* The floor of a shed has an area of 54 square feet. Find the length and width if the length is 3 feet less than twice its width.

99. *Triangular Sail* A sailboat sail is triangular with a height 6 feet greater than its base. If the sail's area is 80 square feet, find its base and height.

100. *Triangular Tent* A triangular tent has a height that is 4 feet less than its base. If the area of a side is 70 square feet, find the base and height of the tent.

101. *Picture Frame* The outside dimensions of a picture frame are 28 cm and 23 cm. The area of the picture is 414 square centimeters. Find the width of the frame.

102. *Rectangle* Frank Bullock's garden is surrounded by a uniform-width walkway. The garden and the walkway together cover an area of 320 square feet. If the dimensions of the garden are 12 feet by 16 feet, find the width of the walkway.

103. *Vegetable Garden* Sally Yang's rectangular vegetable garden is 20 feet by 30 feet. In addition to mulching her garden, she wants to put mulch around the outside of her garden in a uniform width. If she has enough mulch to cover an area of 936 square feet, how wide should the mulch border be?

104. *Square Garden* Danny Tucker has a square garden. He adds a 2-foot-wide walkway around his garden. If the total area of the walkway and garden is 196 square feet, find the dimensions of the garden.

105. *Water Sculpture* In a building at Navy Pier in Chicago, a water fountain jet shoots short spurts of water over a walkway. The water spurts reach a maximum height, then come down into a pond of water on the other side of the walkway. The height above the jet, h, of a spurt of water t seconds after leaving the jet can be found by the function $h(t) = -16t^2 + 32t$. Find the time it takes for the spurt of water to return to the jet's height; that is, when $h(t) = 0$.

106. *Projectile* A model rocket will be launched from a hill 80 feet above sea level. The launch site is next to the ocean (sea level), and the rocket will fall into the ocean. The rocket's distance, s, above sea level at any time, t, is found by the equation $s(t) = -16t^2 + 64t + 80$. Find the time it takes for the rocket to strike the ocean.

107. Bicycle Riding Two cyclists, Bob and Tim, start at the same point. Bob rides west and Tim rides north. At some time, they are 13 miles apart. If Bob traveled 7 miles farther than Tim, determine how far each person traveled.

108. Picture Frame April is making a rectangular picture frame for her mother. The diagonal of the frame is 10 inches. Find the dimensions of the frame if its length is 2 inches greater than its width.

109. Tent Wires A tent has wires attached to it to help stabilize it. A wire is attached to the ground 12 feet from the tent. The length of wire used is 8 feet greater than the height from the ground where the wire is attached. How long is the wire?

110. Room Speakers Speakers are being wired where the walls meet the ceiling in the corners of a rectangular room. The wires are to go through the attic and be connected at point A, as shown in the figure. The width of the room is 12 feet. The distance from the corner to point A is 3 feet less than twice the distance from point A to the wall. Find the length of a wire from point A to one opposite corner of the room (a blue line in the figure).

111. Bicycle Shop The Energy Conservatory Bicycle Shop has a monthly revenue equation $R(x) = 60x - x^2$ and a monthly cost equation $C(x) = 7x + 150$, where x is the number of bicycles sold and $x \geq 10$. Find the number of bicycles that must be sold for the company to break even; that is, where revenue equals costs.

112. Silk Plant Edith Hall makes silk plants and sells them to various outlets. Her company has a revenue equation $R(x) = 30x - x^2$ and a cost equation $C(x) = 4x + 25$, where x is the number of plants sold and $x \geq 5$. Find the number of plants that must be sold for the company to break even.

113. Making a Box Carol Jones is making a box by cutting out 2-in.-by-2-in. squares from a square piece of cardboard and folding up the edges to make a 2-inch-high box.

What size piece of cardboard does Carol need to make a 2-inch-high box with a volume of 162 cubic inches?

114. Making a Box A rectangular box is to be formed by cutting squares from each corner of a rectangular piece of tin and folding up the sides. The box is to be 3 inches high, the length is to be twice the width, and the volume of the box is to be 96 cubic inches. Find the length and width of the box.

115. Cube A solid cube with dimensions a^3 has a rectangular solid with dimensions ab^2 removed.

a) Write a formula for the remaining volume, V.

b) Factor the right side of the formula in part **a)**.

c) If the volume is 1620 cubic inches and a is 12 in., find b.

116. Circular Steel Blade A circular steel blade has a hole cut out of its center as shown in the figure.

a) Write a formula for the remaining area of the blade.

b) Factor the right side of the formula in part **a)**.

c) Find A if $R = 10$ cm and $r = 3$ cm.

117. Consider the following graph of a quadratic function.

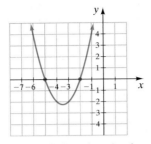

a) Write a quadratic function that has the x-intercepts indicated.

b) Write a quadratic equation in one variable that has solutions of −2 and −5.

c) How many different quadratic functions can have x-intercepts of −2 and −5? Explain.

d) How many different quadratic equations in one variable can have solutions of −2 and −5? Explain.

118. The graph of the equation $y = x^2 + 4$ is illustrated below.

a) How many x-intercepts does the graph have?

b) How many real solutions does the equation $x^2 + 4 = 0$ have? Explain your answer.

119. Consider the quadratic function

$$P(x) = ax^2 + bx + c, a > 0.$$

a) The graph of this type of function may have no x-intercepts, one x-intercept, or two x-intercepts. Sketch each of these possibilities.

b) How many possible real solutions may the equation $ax^2 + bx + c = 0, a > 0$ have? Explain your answer to part **b)** by using the sketches in part **a)**.

120. *Stopping Distance* A typical car's stopping distance on dry pavement, d, in feet, can be approximated by the function $d(s) = 0.034s^2 + 0.56s − 17.11$, where s is the speed of the car before braking and $60 \le s \le 80$ miles per hour. How fast is the car going if it requires 190 ft to stop after the brakes are applied?

121. *Stopping Distance* A typical car's stopping distance on wet pavement, d, in feet, can be approximated by the function $d(s) = −0.31s^2 + 59.82s − 2180.22$, where s is the speed of the car before braking and $60 \le s \le 80$ miles per hour. How fast is the car going if it requires 545 feet for the car to stop after the brakes are applied?

Challenge Problems

Solve.

122. $x^4 − 5x^2 + 4 = 0$

123. $x^4 − 13x^2 = −36$

124. $x^6 − 9x^3 + 8 = 0$

Group Activity

In more advanced mathematics courses you may need to solve an equation for y' (read "y prime"). When doing so, treat the y' as a different variable from y. Individually solve each equation for y'. Compare your answers and as a group obtain the correct answers.

125. $xy' + yy' = 1$

126. $xy − xy' = 3y' + 2$

127. $2xyy' − xy = x − 3y'$

Cumulative Review Exercises

[1.5] **128.** Simplify $(4x^{-2}y^3)^{-2}$.

[2.5] **129.** Solve the inequality and graph the solution on the number line.

$$-1 < \frac{4(3x − 2)}{3} \le 5$$

[4.1] **130.** Solve the system of equations.

$$3x + 4y = 2$$
$$2x = −5y − 1$$

[5.2] **131.** If $f(x) = −x^2 + 3x$ and $g(x) = x^2 + 2$, find $(f \cdot g)(4)$.

[5.7] **132.** Factor $(x + 1)^2 − (x + 1) − 6$.

CHAPTER SUMMARY

Key Words and Phrases

5.1
Addition of polynomials
Cubic polynomial
Degree of a term
Descending order
Leading coefficient
Leading term
Linear polynomial
Polynomial
Polynomial function
Quadratic polynomial
Subtraction of
  polynomials
Terms

5.2
Difference of two
  squares

Expanded form of the
  distributive property
Factors of a trinomial
FOIL method
Multiplication of
  polynomials
Square of a binomial

5.3
Division of polynomials
Remainder Theorem
Synthetic division

5.4
Factor a monomial from
  a polynomial
Factor by grouping
Greatest common factor

5.5
Factor by grouping
Factor by trial and error
Factor trinomials
Factor using substitution
Prime polynomial

5.6
Difference of two cubes
Difference of two
  squares
Perfect square
  trinomials
Special factoring
  formulas
Sum of two cubes

5.8
Cubic equation
Degree of a polynomial
  equation
Equations quadratic in
  form
Hypotenuse of a right
  triangle
Leg of a right triangle
Polynomial equation
Pythagorean Theorem
Quadratic equation
Solution to a polynomial
  equation
Standard form of a
  quadratic equation
x-intercepts of a graph
Zero-factor property

IMPORTANT FACTS

FOIL Method to Multiply Binomials

F— Multiply First terms.
O— Multiply Outer terms.
I— Multiply Inner terms.
L— Multiply Last terms.

$$ac + ad + bc + bd$$

Special Product Formulas

$(a + b)^2 = a^2 + 2ab + b^2$
$(a - b)^2 = a^2 - 2ab + b^2$ } *Square of a binomial*

$(a + b)(a - b) = a^2 - b^2$ *Product of sum and difference of same two terms (or difference of two squares)*

Remainder Theorem

If the polynomial $P(x)$ is divided by $x - a$, the remainder is equal to $P(a)$.

Special Factoring Formulas

$a^2 - b^2 = (a + b)(a - b)$ *Difference of two squares*

$a^2 + 2ab + b^2 = (a + b)^2$
$a^2 - 2ab + b^2 = (a - b)^2$ } *Perfect square trinomials*

$a^3 + b^3 = (a + b)(a^2 - ab + b^2)$ *Sum of two cubes*

$a^3 - b^3 = (a - b)(a^2 + ab + b^2)$ *Difference of two cubes*

Note: The sum of two squares, $a^2 + b^2$, cannot be factored over the set of real numbers.

(continued on the next page)

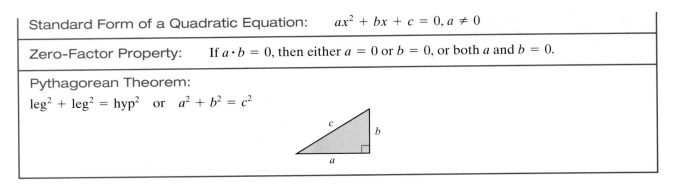

| Standard Form of a Quadratic Equation: | $ax^2 + bx + c = 0, a \neq 0$ |

Zero-Factor Property: If $a \cdot b = 0$, then either $a = 0$ or $b = 0$, or both a and $b = 0$.

Pythagorean Theorem:
$\text{leg}^2 + \text{leg}^2 = \text{hyp}^2$ or $a^2 + b^2 = c^2$

Chapter Review Exercises

*[5.1] Determine whether each expression is a polynomial. If the expression is a polynomial, **a)** give the special name of the polynomial if it has one, **b)** write the polynomial in descending order of the variable x, and **c)** give the degree of the polynomial.*

1. $3x^2 + 2$

2. $5x + 4x^3 - 9$

3. $8x - x^{-1} + 6$

4. $-3 - 10x^2y + 6xy^3 + 2x^4$

Perform each indicated operation.

5. $(x^2 - 5x + 6) + (2x + 3)$

6. $(7x^2 + 2x - 5) - (2x^2 - 9x - 1)$

7. $(2a - 3b - 2) - (-a + 5b - 8)$

8. $(4x^3 - 4x^2 - 2x) + (2x^3 + 4x^2 - 7x + 10)$

9. $(3x^2y + 6xy - 5y^2) - (4y^2 + 3xy)$

10. $(-8ab + 2b^2 - 3a) + (-b^2 + 5ab + a)$

11. Add $x^2 - 3x + 5$ and $4x^2 + 10x - 9$.

12. Subtract $3a^2b - 5ab$ from $-7a^2b - ab$.

13. Find $P(2)$ if $P(x) = 2x^2 - 3x + 13$.

14. Find $P(-3)$ if $P(x) = x^3 - 3x^2 + 4x - 9$.

In Exercises 15 and 16, find a polynomial expression for the perimeter of each figure.

15.

$x^2 - x + 7$

$x^2 + 1$

$x^2 + x + 19$

16.

$x^2 + 7$

$13x + 8$

$9x + 5$

$x^2 + 2x + 3$

We use the following graph in Exercises 17 and 18 on page 379. The graph shows social security receipts and outlays from 1997 through 2025.

Social Security Receipts and Outlays

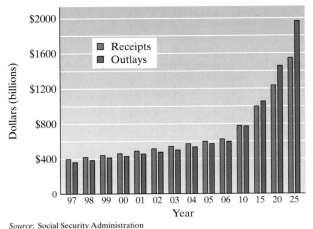

■ Receipts
■ Outlays

Dollars (billions)

Year

Source: Social Security Administration

17. *Social Security Receipts* The function $R(t) = 0.78t^2 + 20.28t + 385.0$, where t is years since 1997 and $0 \le t \le 28$, gives an approximation of social security receipts, $R(t)$, in billions of dollars.

   a) Using the function provided, estimate the receipts in 2010.

   b) Compare your answer in part **a)** with the graph on page 378. Does the graph support your answer?

18. *Social Security Outlays* The function $G(t) = 1.74t^2 + 7.32t + 383.91$, where t is years since 1997 and $0 \le t \le 28$, gives an approximation of social security outlays, $G(t)$, in billions of dollars.

   a) Using the function provided, estimate the outlays in 2010.

   b) Compare your answer in part **a)** with the graph on page 378. Does the graph support your answer?

[5.2] *Multiply.*

19. $2x(3x^2 - 7x + 5)$

20. $-3xy^2(x^3 + xy^4 - 6y^5)$

21. $(3x - 5)(2x + 1)$

22. $(5a + 9)(10a - 3)$

23. $(x + 8y)^2$

24. $(a - 11b)^2$

25. $(2xy - 1)(5x + 4y)$

26. $(2pq - r)(3pq + 7r)$

27. $(2a + 9b)^2$

28. $(3x - 2y)^2$

29. $(7x + 5y)(7x - 5y)$

30. $(2a - 5b^2)(2a + 5b^2)$

31. $(5xy + 6)(5xy - 6)$

32. $(9a^2 - 2b^2)(9a^2 + 2b^2)$

33. $[(x + 3y) + 2]^2$

34. $[(2p - q) - 5]^2$

35. $(3x^2 + 4x - 6)(2x - 3)$

36. $(4x^3 + 6x - 5)(x + 3)$

In Exercises 37 and 38, find an expression for the total area of each figure.

37.

38.

For each pair of functions, find **a)** $(f \cdot g)(x)$ *and* **b)** $(f \cdot g)(3)$.

39. $f(x) = x + 2, g(x) = x - 3$

40. $f(x) = 2x - 4, g(x) = x^2 - 3$

41. $f(x) = x^2 + x - 3, g(x) = x - 2$

42. $f(x) = x^2 - 2, g(x) = x^2 + 2$

[5.3] *Divide.*

43. $\dfrac{6x^2 + 9x + 12}{3}$

44. $\dfrac{7a^2 - 16a + 20}{4}$

45. $\dfrac{21y^3 + 6y + 2}{3y}$

46. $\dfrac{45pq - 25q^2 - 10q}{5q}$

47. $\dfrac{4x^3y^2 + 8x^2y^3 + 12xy^4}{8xy^3}$

48. $(8x^2 + 14x - 15) \div (2x + 5)$

49. $(2x^4 - 3x^3 + 4x^2 + 17x + 7) \div (2x + 1)$

50. $(4a^4 - 7a^2 - 5a + 4) \div (2a - 1)$

51. $(x^2 + x - 18) \div (x - 3)$

52. $(4x^3 + 12x^2 + x - 10) \div (2x + 3)$

Use synthetic division to obtain each quotient.

53. $(3x^3 - 2x^2 + 10) \div (x - 3)$

54. $(2y^5 - 10y^3 + y - 1) \div (y + 1)$

55. $(x^5 - 20) \div (x - 2)$

56. $(2x^3 + x^2 + 5x - 3) \div \left(x - \dfrac{1}{2}\right)$

Determine the remainder of each division problem using the Remainder Theorem. If the divisor is a factor of the dividend, so state.

57. $(x^2 - 4x + 11) \div (x - 3)$

58. $(2x^3 - 6x^2 + 3x) \div (x + 4)$

59. $(3x^3 - 6) \div \left(x - \dfrac{1}{3}\right)$

60. $(2x^4 - 6x^2 - 8) \div (x + 2)$

[5.4] *Factor out the greatest common factor in each expression.*

61. $4x^2 + 8x + 24$

62. $15x^4 + 6x^3 - 12x^4y^3$

63. $10a^3b^3 - 12a^2b^6$

64. $12xy^4z^3 + 6x^2y^3z^2 - 15x^3y^2z^3$

Factor by grouping.

65. $5x^2 - xy + 20xy - 4y^2$

66. $12a^2 - 8ab + 15ab - 10b^2$

67. $(2x - 3)(2x + 1) - (2x - 3)(x - 8)$

68. $7x(3x - 5) + 3(3x - 5)^2$

In Exercises 69 and 70, A represents the area of the figure. Find an expression, in factored form, for the difference of the areas of the geometric figures.

69.

$A = 13x(5x + 2)$

$A = 7(5x + 2)$

70.

$A = 14x^2 + 18x$

$A = 7x + 9$

In Exercises 71 and 72, V represents the volume of the figure. Find an expression, in factored form, for the difference of the volumes of the geometric figures.

71.

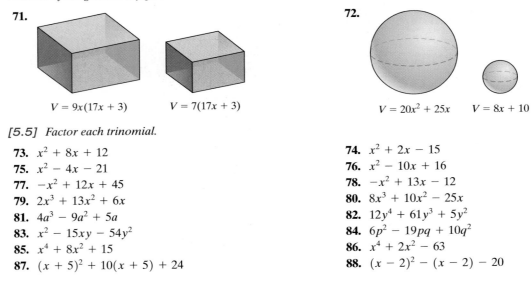

$V = 9x(17x + 3)$ $V = 7(17x + 3)$

72.

$V = 20x^2 + 25x$ $V = 8x + 10$

[5.5] *Factor each trinomial.*

73. $x^2 + 8x + 12$

74. $x^2 + 2x - 15$

75. $x^2 - 4x - 21$

76. $x^2 - 10x + 16$

77. $-x^2 + 12x + 45$

78. $-x^2 + 13x - 12$

79. $2x^3 + 13x^2 + 6x$

80. $8x^3 + 10x^2 - 25x$

81. $4a^3 - 9a^2 + 5a$

82. $12y^4 + 61y^3 + 5y^2$

83. $x^2 - 15xy - 54y^2$

84. $6p^2 - 19pq + 10q^2$

85. $x^4 + 8x^2 + 15$

86. $x^4 + 2x^2 - 63$

87. $(x + 5)^2 + 10(x + 5) + 24$

88. $(x - 2)^2 - (x - 2) - 20$

In Exercises 89 and 90, find an expression, in factored form, for the area of the shaded region in each figure.

89.

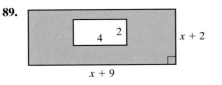

4   2

$x + 2$

$x + 9$

90.

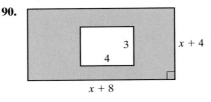

3

4

$x + 4$

$x + 8$

[5.6] *Use a special factoring formula to factor the following.*

91. $x^2 - 49$

92. $x^2 - 100$

93. $x^4 - 16$

94. $x^4 - 81$

95. $4a^2 + 4a + 1$

96. $4y^2 - 12y + 9$

97. $(x + 2)^2 - 9$

98. $(3y - 1)^2 - 25$

99. $p^4 + 16p^2 + 64$

100. $b^4 - 14b^2 + 49$

101. $x^2 + 8x + 16 - y^2$

102. $a^2 + 6ab + 9b^2 - 4c^2$

103. $9x^2 + 6xy + y^2$

104. $36b^2 - 60bc + 25c^2$

105. $x^3 - 27$

106. $y^3 + 64$

107. $125x^3 - 1$

108. $8a^3 - 27b^3$

109. $(x + 1)^3 - 8$

110. $(x - 2)^3 - 27$

111. $y^3 - 64z^3$

112. $(a + 3)^3 + 1$

In Exercises 113 and 114, find an expression, in factored form, for the area of the shaded region in each figure.

113.

114.

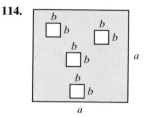

115. *Volume* Find an expression, in factored form, for the difference in volumes of the two cubes below.

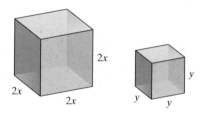

116. *Volume* Find an expression, in factored form, for the volume of the shaded region in the figure below.

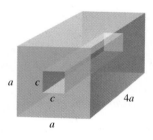

[5.4–5.7] Factor completely.

117. $x^2 y^2 - 2xy^2 - 15y^2$

118. $3x^3 - 18x^2 + 24x$

119. $3x^3 y^4 + 18x^2 y^4 - 6x^2 y^4 - 36xy^4$

120. $3y^5 - 27y$

121. $2x^3 y + 16y$

122. $5x^4 y + 20x^3 y + 20x^2 y$

123. $6x^3 - 21x^2 - 12x$

124. $x^2 + 10x + 25 - y^2$

125. $3x^3 + 24y^3$

126. $x^2(x + 4) + 3x(x + 4) - 4(x + 4)$

127. $4(2x + 3)^2 - 12(2x + 3) + 5$

128. $4x^4 + 4x^2 - 3$

129. $(x - 1)x^2 - (x - 1)x - 2(x - 1)$

130. $9ax - 3bx + 12ay - 4by$

131. $6p^2 q^2 - 5pq - 6$

132. $9x^4 - 12x^2 + 4$

133. $4y^2 - (x^2 + 4x + 4)$

134. $6(2a + 3)^2 - 7(2a + 3) - 3$

135. $6x^4 y^4 + 9x^3 y^4 - 27x^2 y^4$

136. $x^3 - \dfrac{8}{27} y^6$

Area In Exercises 137–142, find an expression, in factored form, for the area of the shaded region in each figure.

137.

138.

139.

140.

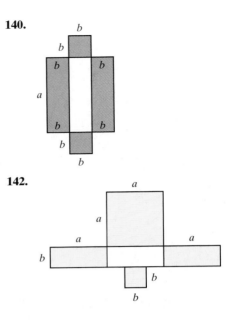

141.

$a + 3b$

a b a b

142.

a

a

a a

b

b

b

[5.8] Solve.

143. $(x - 3)(4x + 1) = 0$

144. $(2x + 5)(3x + 7) = 0$

145. $2x^2 = 4x$

146. $15x^2 + 20x = 0$

147. $x^2 + 7x + 12 = 0$

148. $a^2 + a - 30 = 0$

149. $x^2 = 8x - 7$

150. $c^3 - 6c^2 + 8c = 0$

151. $12d^2 = 13d + 4$

152. $20p^2 - 6 = 7p$

153. $5x^2 = 20$

154. $x(x + 3) = 2(x + 4) - 2$

Use factoring to find the x-intercepts of the graph of each equation.

155. $y = 2x^2 - 2x - 60$

156. $y = 20x^2 - 49x + 30$

Write an equation whose graph will have x-intercepts at the given values.

157. -3 and 6

158. $-\dfrac{3}{2}$ and $-\dfrac{5}{6}$

In Exercises 159–163, answer the question.

159. *Carpeting* The area of Fred Bank's rectangular carpet is 99 square feet. Find the length and width of the carpet if the length is 2 feet greater than the width.

160. *Triangular Sign* The base of a large triangular sign is 3 feet more than twice the height. Find the base and height if the area of the triangle is 22 square feet.

161. *Square* One square has a side 4 inches longer than the side of a second square. If the area of the larger square is 49 square inches, find the length of a side of each square.

162. *Velocity* A rocket is projected upward from the top of a 144-foot-tall building with a velocity of 128 feet per second. The rocket's distance from the ground, s, at any time, t, in seconds, is given by the formula $s(t) = -16t^2 + 128t + 144$. Find the time it takes for the rocket to strike the ground.

163. *Telephone Pole* Two guy wires are attached to a telephone pole to help stabilize it. The wire is attached to the ground x feet from the base of the pole. The height of the pole is $x + 31$ and the length of the wire is $x + 32$. Find x.

$x + 31$

$x + 32$

x

Chapter Practice Test

1. a) Give the specific name of the following polynomial.
$$-4x^2 + 2x - 6x^4$$
   b) Write the polynomial in descending powers of the variable x.

   c) State the degree of the polynomial.

   d) What is the leading coefficient of the polynomial?

Perform each operation.

2. $(7x^2y - 5y^2 + 2x) - (3x^2y + 9y^2 - 6y)$

3. $2x^3y^2(-4x^5y + 10x^3y^2 - 6x)$

4. $(2a - 3b)(5a + b)$

5. $(2x^2 + 3xy - 6y^2)(2x + y)$

6. $(12x^6 - 15x^2y + 21) \div 3x^2$

7. $(2x^2 - 7x + 10) \div (2x + 3)$

8. Use synthetic division to obtain the quotient.
$$(3x^4 - 12x^3 - 60x + 4) \div (x - 5)$$

9. Use the Remainder Theorem to find the remainder when $2x^3 - 6x^2 - 5x + 4$ is divided by $x + 3$.

Factor completely.

10. $12x^3y + 10x^2y^4 - 8xy^3$

11. $x^3 - 2x^2 - 3x$

12. $2a^2 + 4ab + 3ab + 6b^2$

13. $2b^4 + 5b^2 - 18$

14. $4(x - 3)^2 + 20(x - 3)$

15. $(x + 5)^2 + 2(x + 5) - 3$

16. $27p^3q^6 - 8q^6$

17. If $f(x) = 3x - 4$ and $g(x) = x - 5$, find **a)** $(f \cdot g)(x)$ and **b)** $(f \cdot g)(2)$

In Exercises 18 and 19, find an expression, in factored form, for the area of the shaded region.

18.

19.

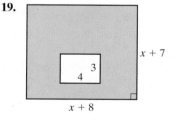

Solve.

20. $7x^2 + 25x - 12 = 0$

21. $x^3 + 3x^2 - 4x = 0$

22. Use factoring to find the x-intercepts of the graph of the equation $y = 8x^2 + 10x - 3$.

23. Find an equation whose graph has x-intercepts at 2 and 6.

24. *Area* The area of a triangle is 28 square meters. If the base of the triangle is 2 meters greater than 3 times the height, find the base and height of the triangle.

25. *Baseball* A baseball is projected upward from the top of a 448-foot-tall building with an initial velocity of 48 feet per second. The distance, s, of the baseball from the ground at any time, t, in seconds, is given by the equation $s(t) = -16t^2 + 48t + 448$. Find the time it takes for the baseball to strike the ground.

Cumulative Review Test

Take the following test and check your answers with those that appear at the end of the test. Review any questions that you answered incorrectly. The section and objective where the material was covered is indicated after the answer.

1. Find $A \cup B$ for $A = \{2, 4, 6, 8\}$ and $B = \{3, 5, 6, 8\}$.

2. Illustrate $\{x | x \leq -5\}$ on a real number line.

3. Divide $\left| \dfrac{3}{8} \right| \div (-2)$.

4. Evaluate $(-3)^3 - 2^2 - (-2)^2 + (7 - 7)^2$.

5. Simplify $\left(\dfrac{2r^4 s^5}{r^2} \right)^3$.

6. Solve $4(2x - 2) - 3(x + 7) = -4$.

7. Solve $k = 2(d + e)$ for e.

8. *Landscaping* Craig Campanella, a landscape architect, wishes to fence in two equal areas as illustrated in the figure. If both areas are squares and the total length of fencing used is 91 meters, find the dimensions of each square.

9. *Making Copies* Cecil Winthrop has a manuscript. He needs to make 6 copies before sending it to his editor in Boston. The first copy costs 15 cents per page and each additional copy costs 5 cents per page. If the total bill before tax is $248, how many pages are in the manuscript?

10. *Test Average* Todd Garner's first four test grades are 68, 72, 90, and 86. What range of grades on his fifth test will result in an average greater than or equal to 70 and less than 80?

11. Is $(4, 1)$ a solution to the equation $3x + 2y = 9$?

12. Write the equation $2 = 6x - 3y$ in standard form.

13. Find the slope of the line passing through the points $(8, -4)$ and $(-1, -2)$.

14. If $f(x) = 2x^3 - 4x^2 + x - 16$, find $f(-4)$.

15. Graph the inequality $2x - y \leq 6$.

16. Solve the system of equations.
$$\frac{1}{5}x + \frac{1}{2}y = 4$$
$$\frac{2}{3}x - y = \frac{8}{3}$$

17. Solve the system of equations.
$$x - 2y = 2$$
$$2x + 3y = 11$$
$$-y + 4z = 7$$

18. Evaluate the determinant.
$$\begin{vmatrix} 7 & 3 \\ -2 & 1 \end{vmatrix}$$

19. Divide $(2x^3 - 9x + 15) \div (x - 6)$.

20. Factor $64x^3 - 27y^3$.

Answers to Cumulative Review Test

1. $A \cup B = \{2, 3, 4, 5, 6, 8\}$; [Sec. 1.2, Obj. 4]   2. ⟵————●————⟶ [Sec. 1.2, Obj. 3]   3. $-\dfrac{3}{16}$; [Sec. 1.3, Obj. 5]
                                                                         -5

4. -35; [Sec. 1.4, Obj. 3]   5. $8r^6 s^{15}$; [Sec. 1.5, Obj. 7]   6. 5; [Sec. 2.1, Obj. 3]   7. $e = \dfrac{k - 2d}{2}$; [Sec. 2.2, Obj. 2]   8. 13 meters

by 13 meters; [Sec. 2.3, Obj. 2]   9. 620 pages; [Sec. 2.3, Obj. 2]   10. $34 \leq x < 84$; [Sec. 2.5, Obj. 3]   11. No; [Sec. 3.1, Obj. 2]

12. $6x - 3y = 2$; [Sec. 3.3, Obj. 2]   13. $-\dfrac{2}{9}$; [Sec. 3.4, Obj. 2]   14. -212; [Sec. 3.6, Obj. 1]   15.

[Sec. 3.7, Obj. 1]   16. $(10, 4)$; [Sec. 4.1, Obj. 3]   17. $(4, 1, 2)$; [Sec. 4.2, Obj. 1]

18. 13; [Sec. 4.5, Obj. 1]   19. $2x^2 + 12x + 63 + \dfrac{393}{x - 6}$; [Sec. 5.3, Obj. 2]

20. $(4x - 3y)(16x^2 + 12xy + 9y^2)$; [Sec. 5.6, Obj. 3]

Chapter 6

Rational Expressions and Equations

6.1 The Domains of Rational Functions and Multiplication and Division of Rational Expressions

6.2 Addition and Subtraction of Rational Expressions

6.3 Complex Fractions

6.4 Solving Rational Equations

6.5 Rational Equations: Applications and Problem Solving

6.6 Variation

Chapter Summary
Chapter Review Exercises
Chapter Practice Test
Cumulative Review Test

S ometimes the context of a problem makes it seem more difficult than it really is. If this happens to you, try to write a similar problem using a context with which you are more familiar. For instance, on page 437 you are asked to determine how far a space station is from NASA headquarters by writing an equation based on the relative time it takes two shuttles traveling at different speeds to reach the station. Although the specific numbers will be different, the problem is the same as determining how far your college is from your home based on the relative time it takes you and your parents to get to the school if you are driving at different speeds.

SSM   Study Guide   CD/Video   MathPro 4/5   PH Math Tutor Center   prenhall.com/Angel

A Look Ahead

In this chapter we explain how to add, subtract, multiply, and divide rational expressions. Rational expressions are expressions that contain fractions. We also explain how to solve equations containing rational expressions. Equations containing rational expressions are sometimes referred to as rational equations.

In Section 6.1, we introduce rational expressions and discuss the domains of rational expressions and rational functions. If you take a higher mathematics course, you will probably discuss rational functions in much more detail and graph them.

To succeed in Sections 6.1 and 6.2, you must use the factoring techniques presented in Chapter 5.

In Chapter 2, we solved equations containing fractions. In Section 6.4, we present and solve many types of equations containing rational expressions. Applications of equations containing rational expressions are included in Sections 6.4 and 6.5.

A knowledge of variation is important in many science courses. In Section 6.6, we discuss various types of variations, including direct, inverse, joint, and combined variations.

6.1 THE DOMAINS OF RATIONAL FUNCTIONS AND MULTIPLICATION AND DIVISION OF RATIONAL EXPRESSIONS

SSM   Study Guide   CD/Video

MathPro 4/5   PH Math Tutor Center   prenhall.com/Angel

1 Find the domains of rational functions.
2 Simplify rational expressions.
3 Multiply rational expressions.
4 Divide rational expressions.

1 Find the Domains of Rational Functions

To understand rational expressions, you must have a thorough understanding of the factoring techniques discussed in Chapter 5. **A rational expression** is an expression of the form p/q, where p and q are polynomials and $q \neq 0$.

Examples of Rational Expressions

$$\frac{2}{x}, \quad \frac{x+3}{x}, \quad \frac{x^2+4x}{x-3}, \quad \frac{a}{a^2-4}, \quad \frac{t^2-5t+7}{t^3+t^2-3t}$$

Note that the denominator of a rational expression cannot equal 0 because division by 0 is undefined. In the expression $\frac{x+3}{x}$, x cannot equal 0, since the denominator would then equal 0. In $\frac{x^2+4x}{x-3}$ x cannot equal 3 because that would result in the denominator having a value of 0. What values of a cannot be used in the expression $\frac{a}{a^2-4}$ If you answered 2 and -2, you answered correctly.

Whenever we write a rational expression containing a variable in the denominator, we always assume that the value or values of the variable that make the denominator 0 are excluded. For example, if we write $\frac{5}{x-3}$, we assume $x \neq 3$, even though we do not specifically write it.

In Section 5.1, we discussed polynomial functions. Now we introduce rational functions. **A rational function** is a function of the form $f(x) = p/q$ or $y = p/q$, where p and q are polynomials and $q \neq 0$.

Examples of Rational Functions

$$f(x) = \frac{5}{x} \qquad y = \frac{x^2 + 9}{x + 3} \qquad T(a) = \frac{a + 1}{a^2 - 4} \qquad h(x) = \frac{7x - 3}{2x + 1}$$

The **domain** of a rational function will be the set of values that can be used to replace the variable. For example, in the rational function $f(x) = \frac{x + 2}{x - 3}$, the domain will be all real numbers except 3, written $\{x | x \neq 3\}$. If x were 3, the denominator would be 0, and division by 0 is undefined.

EXAMPLE 1   For the given functions $f(x)$ and $g(x)$, find the domain of $\left(\frac{f}{g}\right)(x)$.

a) $f(x) = x^2, g(x) = x^2 - 4$

b) $f(x) = x - 2, g(x) = x^2 + 3x - 10$

c) $f(x) = x, g(x) = x^2 + 6$

Solution   **a)** Since $f(x)$ and $g(x)$ are polynomial functions, each of their domains is the set of all real numbers. The domain of the quotient of functions $(f/g)(x)$ will therefore be all real numbers for which the denominator of the quotient is not equal to 0. From Section 3.6 we know that

$$\left(\frac{f}{g}\right)(x) = \frac{f(x)}{g(x)}.$$

Therefore, $\left(\frac{f}{g}\right)(x) = \dfrac{x^2}{x^2 - 4}$ *Substitute expressions for $f(x)$ and $g(x)$.*

$$= \frac{x^2}{(x + 2)(x - 2)} \qquad \textit{Factor the denominator.}$$

From this factored form we see that x cannot be 2 or -2. Therefore the domain is all real numbers except 2 or -2. The domain may be expressed as $\{x | x \neq 2 \text{ and } x \neq -2\}$.

b) $\left(\frac{f}{g}\right)(x) = \dfrac{f(x)}{g(x)}$

$$= \frac{x - 2}{x^2 + 3x - 10} \qquad \textit{Substitute expressions for } f(x) \textit{ and } g(x).$$

$$= \frac{x - 2}{(x - 2)(x + 5)} \qquad \textit{Factor the denominator.}$$

Notice that the $x - 2$ in the numerator will divide out with the $x - 2$ in the denominator. However, when determining the domain of the quotient of functions, we do so *before* we simplify the expression. Since the denominator cannot be 0, x cannot have values of 2 or -5. The domain is $\{x | x \neq 2 \text{ and } x \neq -5\}$.

c) $\left(\frac{f}{g}\right)(x) = \dfrac{f(x)}{g(x)}$

$$= \frac{x}{x^2 + 6}$$

NOW TRY EXERCISE 21

Since there is no value of x that makes the denominator 0, the domain is all real numbers. The domain can be written as $\{x|x$ is a real number$\}$. ✳

Using Your Graphing Calculator

If you have a graphing calculator, you may wish to experiment by graphing some rational functions. This will give you some idea of the wide variety of graphs of rational functions.

If you graph $y = \dfrac{x^2}{x^2 - 4}$ from Example 1**a)** on a graphing calculator, the display might look like that in Figure 6.1.

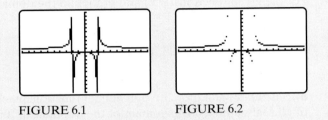

FIGURE 6.1   FIGURE 6.2

The domain of this function is all real numbers except 2 and -2.

Notice what appear to be vertical lines at $x = -2$ and $x = 2$, the values of x where the function is un-defined. This calculator is in a mode called *connected mode*. When a calculator is in connected mode, it con-nects all points it plots, going from the point with the smallest x-coordinate to the next larger one. Just to the left of -2, the value of y is a large positive number, and just to the right of -2, the value of y is a large neg-ative number. The vertical line is the calculator's attempt to connect the point with this large positive y-value to the point with this large negative y-value. A similar situation occurs at $x = 2$.

You may sometimes wish to have your calculator in *dot mode*. When the calculator is in dot mode it dis-plays unconnected points that have been calculated. Read the manual that comes with your calculator to learn how to change from connected mode to dot mode, and vice versa. The graph in Figure 6.2 shows the same graph as in Figure 6.1 except that this time the calculator is in dot mode.

2 Simplify Rational Expressions

When we work problems containing rational expressions, we must make sure that we write the answer in lowest terms. A rational expression is **simplified** when the numerator and denominator have no common factors other than 1. The fraction $\frac{6}{9}$ is not simplified because the 6 and 9 both contain the common factor of 3. When the 3 is factored out, the simplified fraction is $\frac{2}{3}$.

$$\frac{6}{9} = \frac{\overset{1}{\cancel{3}} \cdot 2}{\underset{1}{\cancel{3}} \cdot 3} = \frac{2}{3}$$

The rational expression $\dfrac{ab - b^2}{2b}$ is not simplified because both the numerator and denominator have a common factor, b. To simplify this expression, factor b from each term in the numerator; then divide it out.

$$\frac{ab - b^2}{2b} = \frac{\cancel{b}(a - b)}{2\cancel{b}} = \frac{a - b}{2}$$

Thus $\dfrac{ab - b^2}{2b}$ becomes $\dfrac{a - b}{2}$ when simplified.

> ### To Simplify Rational Expressions
>
> 1. Factor both numerator and denominator as completely as possible.
> 2. Divide both the numerator and the denominator by any common factors.

EXAMPLE 2 Simplify. **a)** $\dfrac{x^2 + 4x + 3}{x + 3}$ **b)** $\dfrac{3x^3 - 3x^2}{x^3 - x}$

Solution **a)** Factor the numerator; then divide out the common factor.

$$\frac{x^2 + 4x + 3}{x + 3} = \frac{\cancel{(x + 3)}(x + 1)}{\cancel{x + 3}} = x + 1$$

The rational expression simplifies to $x + 1$.

b) Factor the numerator and denominator. Then divide out common factors.

$$\frac{3x^3 - 3x^2}{x^3 - x} = \frac{3x^2(x - 1)}{x(x^2 - 1)}$$

$$= \frac{3\overset{x}{\cancel{x^2}}\cancel{(x - 1)}}{\cancel{x}(x + 1)\cancel{(x - 1)}} \qquad \textit{Factor } x^2 - 1.$$

$$= \frac{3x}{x + 1}$$

The rational expression simplifies to $\dfrac{3x}{x + 1}$.

When the terms in a numerator differ only in sign from the terms in a denominator, we can factor out -1 from either the numerator or denominator. *When -1 is factored from a polynomial, the sign of each term in the polynomial changes.* For example,

$$-2x + 3 = -1(2x - 3) = -(2x - 3)$$
$$6 - 5x = -1(-6 + 5x) = -(5x - 6)$$
$$-3x^2 + 5x - 6 = -1(3x^2 - 5x + 6) = -(3x^2 - 5x + 6)$$

EXAMPLE 3 Simplify $\dfrac{27x^3 - 8}{2 - 3x}$.

Solution

$$\frac{27x^3 - 8}{2 - 3x} = \frac{(3x)^3 - (2)^3}{2 - 3x} \qquad \textit{Write the numerator as a difference of two cubes.}$$

$$= \frac{(3x - 2)(9x^2 + 6x + 4)}{2 - 3x} \qquad \textit{Factor; recall that } a^3 - b^3 = (a - b)(a^2 + ab + b^2).$$

$$= \frac{\cancel{(3x - 2)}(9x^2 + 6x + 4)}{-1\cancel{(3x - 2)}} \qquad \textit{Factor } -1 \textit{ from the denominator and divide out common factors.}$$

$$= \frac{9x^2 + 6x + 4}{-1}$$

NOW TRY EXERCISE 41
$$= -(9x^2 + 6x + 4) \quad \text{or} \quad -9x^2 - 6x - 4$$

AVOIDING COMMON ERRORS

Remember that you can divide out only common **factors**. Only when expressions are *multiplied* can they be factors. Neither of the expressions above can be simplified from their original form.

3 Multiply Rational Expressions

Now that we know how to simplify a rational expression, we can discuss multiplication of rational expressions.

> ### To Multiply Rational Expressions
>
> To multiply rational expressions, use the following rule
>
> $$\frac{a}{b} \cdot \frac{c}{d} = \frac{a \cdot c}{b \cdot d}, \quad b \neq 0, d \neq 0$$
>
> To multiply rational expressions, follow these steps.
>
> 1. Factor all numerators and denominators as far as possible.
> 2. Divide out any common factors.
> 3. Multiply using the above rule.
> 4. Simplify the answer when possible. (This step is not necessary if step 2 is done completely!)

If all common factors were factored out in step 2, your answer in step 4 should be in simplified form. However, if you missed a common factor in step 2, you can factor it out in step 4 to obtain an answer that is simplified.

EXAMPLE 4 Multiply.   **a)** $\dfrac{x-5}{6x} \cdot \dfrac{x^2-2x}{x^2-7x+10}$ **b)** $\dfrac{2x-5}{x-4} \cdot \dfrac{x^2-8x+16}{5-2x}$

Solution   **a)** $\dfrac{x-5}{6x} \cdot \dfrac{x^2-2x}{x^2-7x+10} = \dfrac{x-5}{6x} \cdot \dfrac{x(x-2)}{(x-2)(x-5)}$ *Factor; divide out common factors.*

$$= \frac{1}{6}$$

b) $\dfrac{2x-5}{x-4} \cdot \dfrac{x^2-8x+16}{5-2x} = \dfrac{2x-5}{x-4} \cdot \dfrac{(x-4)(x-4)}{5-2x}$ *Factor.*

$$= \frac{2x-5}{x-4} \cdot \frac{(x-4)(x-4)}{-1(2x-5)}$$ *Factor -1 from denominator; divide out common factors.*

$$= \frac{x-4}{-1}$$

$$= -(x-4) \quad \text{or} \quad -x+4 \quad \text{or} \quad 4-x$$

EXAMPLE 5 Multiply $\dfrac{x^2 - y^2}{x + y} \cdot \dfrac{x + 2y}{2x^2 - xy - y^2}$.

Solution $\dfrac{x^2 - y^2}{x + y} \cdot \dfrac{x + 2y}{2x^2 - xy - y^2} = \dfrac{\cancel{(x + y)}(x - y)}{\cancel{x + y}} \cdot \dfrac{x + 2y}{(2x + y)\cancel{(x - y)}}$ *Factor; divide out common factors.*

$$= \dfrac{x + 2y}{2x + y}$$

NOW TRY EXERCISE 55

EXAMPLE 6 Multiply $\dfrac{ab - ac + bd - cd}{ab + ac + bd + cd} \cdot \dfrac{b^2 + bc + bd + cd}{b^2 + bd - bc - cd}$.

Solution Factor both numerators and denominators by grouping. Then divide out common factors.

$$\dfrac{ab - ac + bd - cd}{ab + ac + bd + cd} \cdot \dfrac{b^2 + bc + bd + cd}{b^2 + bd - bc - cd}$$

$$= \dfrac{a(b - c) + d(b - c)}{a(b + c) + d(b + c)} \cdot \dfrac{b(b + c) + d(b + c)}{b(b + d) - c(b + d)}$$ *Distributive property*

$$= \dfrac{\cancel{(b - c)}\cancel{(a + d)}}{\cancel{(b + c)}\cancel{(a + d)}} \cdot \dfrac{\cancel{(b + c)}\cancel{(b + d)}}{\cancel{(b + d)}\cancel{(b - c)}} = 1$$ *Factor completely; divide out common factors.*

4 Divide Rational Expressions

Now let's discuss division of rational expressions.

> ### To Divide Rational Expressions
>
> To divide rational expressions, use the following rule:
>
> $$\dfrac{a}{b} \div \dfrac{c}{d} = \dfrac{a}{b} \cdot \dfrac{d}{c} = \dfrac{a \cdot d}{b \cdot c}, \quad b \neq 0, c \neq 0, d \neq 0$$
>
> To divide rational expressions, invert the divisor (the second or bottom fraction) and proceed as when multiplying rational expressions.

EXAMPLE 7 Divide $\dfrac{12x^4}{5y^3} \div \dfrac{3x^5}{25y}$.

Solution $\dfrac{12x^4}{5y^3} \div \dfrac{3x^5}{25y} = \dfrac{\cancel{12}x^4}{\cancel{5}y^3} \cdot \dfrac{\cancel{25}y}{\cancel{3}x^5}$ *Invert divisor; divide out common factors.*

$$= \dfrac{4 \cdot 5}{y^2 x} = \dfrac{20}{xy^2}$$

In Example 7, all the numerators and denominators were monomials. When the numerators or denominators are binomials or trinomials, we factor them, if possible, in order to divide out common factors. This procedure is illustrated in Example 8.

EXAMPLE 8 Divide.   **a)** $\dfrac{x^2 - 25}{x + 4} \div \dfrac{x - 5}{x + 4}$ **b)** $\dfrac{12a^2 - 22a + 8}{3a} \div \dfrac{3a^2 + 2a - 8}{2a^2 + 4a}$

Solution **a)** $\dfrac{x^2 - 25}{x + 4} \div \dfrac{x - 5}{x + 4} = \dfrac{x^2 - 25}{x + 4} \cdot \dfrac{x + 4}{x - 5}$ *Invert divisor.*

$\qquad\qquad\qquad = \dfrac{(x + 5)\cancel{(x - 5)}}{\cancel{x + 4}} \cdot \dfrac{\cancel{x + 4}}{\cancel{x - 5}}$ *Factor; divide out common factors.*

$\qquad\qquad\qquad = x + 5$

b) $\dfrac{12a^2 - 22a + 8}{3a} \div \dfrac{3a^2 + 2a - 8}{2a^2 + 4a}$

$\qquad = \dfrac{12a^2 - 22a + 8}{3a} \cdot \dfrac{2a^2 + 4a}{3a^2 + 2a - 8}$ *Invert divisor.*

$\qquad = \dfrac{2(6a^2 - 11a + 4)}{3a} \cdot \dfrac{2a(a + 2)}{(3a - 4)(a + 2)}$ *Factor.*

$\qquad = \dfrac{2\cancel{(3a - 4)}(2a - 1)}{3\cancel{a}} \cdot \dfrac{2\cancel{a}\cancel{(a + 2)}}{\cancel{(3a - 4)}\cancel{(a + 2)}}$ *Factor further; divide out common factors.*

$\qquad = \dfrac{4(2a - 1)}{3}$

EXAMPLE 9   Divide $\dfrac{x^4 - y^4}{x - y} \div \dfrac{x^2 + xy}{x^2 - 2xy + y^2}$.

Solution $\dfrac{x^4 - y^4}{x - y} \div \dfrac{x^2 + xy}{x^2 - 2xy + y^2}$

$\qquad = \dfrac{x^4 - y^4}{x - y} \cdot \dfrac{x^2 - 2xy + y^2}{x^2 + xy}$ *Invert divisor.*

$\qquad = \dfrac{(x^2 + y^2)(x^2 - y^2)}{x - y} \cdot \dfrac{(x - y)(x - y)}{x(x + y)}$ *Factor.*

$\qquad = \dfrac{(x^2 + y^2)\cancel{(x + y)}\cancel{(x - y)}}{\cancel{x - y}} \cdot \dfrac{(x - y)(x - y)}{x\cancel{(x + y)}}$ *Factor further; divide out common factors.*

$\qquad = \dfrac{(x^2 + y^2)(x - y)^2}{x}$

NOW TRY EXERCISE 69

HELPFUL HINT

STUDY TIP

Throughout this chapter, you will need to factor polynomials. It is important that you understand the factoring techniques covered in Chapter 5. If you are having difficulty factoring, review factoring now.

Exercise Set 6.1

Concept/Writing Exercises

1. a) What is a rational expression?
   b) Give your own example of a rational expression.

2. Explain why $\dfrac{\sqrt{x}}{x + 1}$ is not a rational expression.

3. a) What is a rational function?
   b) Give your own example of a rational function.

4. Explain why $f(x) = \dfrac{2}{\sqrt{x} + 1}$ is not a rational function.

5. a) What is the domain of a rational function?
   b) What is the domain of $f(x) = \dfrac{3}{x^2 - 16}$?

6. a) Explain how to simplify a rational expression.

b) Using the procedure stated in part **a)**, simplify

$$\frac{6x^2 + 23x + 20}{4x^2 - 25}$$

7. a) Explain how to simplify a rational expression where the numerator and denominator differ only in sign.

b) Using the procedure explained in part **a)**, simplify

$$\frac{3x^2 - 2x - 8}{-3x^2 + 2x + 8}$$

8. a) Explain how to multiply rational expressions.

b) Using the procedure given in part **a)**, multiply

$$\frac{6a^2 + a - 1}{3a^2 + 2a - 1} \cdot \frac{3a^2 + 4a + 1}{6a^2 + 5a + 1}$$

9. a) Explain how to divide rational expressions.

b) Using the procedure given in part **a)**, divide

$$\frac{r + 2}{r^2 + 9r + 18} \div \frac{(r + 2)^2}{r^2 + 5r + 6}$$

10. Consider $f(x) = \dfrac{x}{x}$. Will $f(x) = 1$ for all values of x? Explain.

Practice the Skills

Determine the values that are excluded in the following expressions.

11. $\dfrac{4x}{3x - 12}$

12. $\dfrac{x + 2}{x^2 - 64}$

13. $\dfrac{4}{2x^2 - 15x + 25}$

14. $\dfrac{2}{(x - 5)^2}$

15. $\dfrac{x - 3}{x^2 + 4}$

16. $\dfrac{-2}{16 - r^2}$

17. $\dfrac{x^2 + 36}{x^2 - 36}$

18. $\dfrac{x^2 - 36}{x^2 + 36}$

Determine the domain of each function.

19. $f(p) = \dfrac{p + 1}{p - 2}$

20. $f(z) = \dfrac{3}{-8z + 4}$

21. $y = \dfrac{5}{x^2 + x - 6}$

22. $y = \dfrac{x - 3}{x^2 + 4x - 21}$

23. $f(a) = \dfrac{3a^2 - 6a + 4}{2a^2 + 3a - 2}$

24. $f(x) = \dfrac{4 - 2x}{x^3 + 8x}$

25. $g(x) = \dfrac{x^2 - x + 2}{x^2 + 1}$

26. $h(x) = \dfrac{x^3 - 64x}{x^2 + 100}$

27. $m(a) = \dfrac{a^2 + 49}{a^2 - 49}$

28. $k(b) = \dfrac{b^2 - 49}{b^2 + 49}$

Simplify each rational expression.

29. $\dfrac{x - xy}{x}$

30. $\dfrac{x^2 - 2x}{x}$

31. $\dfrac{5x^2 - 10xy}{15x}$

32. $\dfrac{x^2 + 3x}{x^2 - 2x}$

33. $\dfrac{x^3 - x}{x^2 - 1}$

34. $\dfrac{4x^2y + 12xy + 18x^3y^3}{8xy^2}$

35. $\dfrac{5r - 4}{4 - 5r}$

36. $\dfrac{4x^2 - 9}{2x^2 - x - 3}$

37. $\dfrac{p^2 - 2p - 24}{6 - p}$

38. $\dfrac{4x^2 - 16x^4 + 6x^5y}{8x^3y^2}$

39. $\dfrac{a^2 - 3a - 10}{a^2 + 5a + 6}$

40. $\dfrac{y^2 - 10yz + 24z^2}{y^2 - 5yz + 4z^2}$

41. $\dfrac{8x^3 - 125y^3}{2x - 5y}$

42. $\dfrac{x(x - 3) + x(x - 4)}{2x - 7}$

43. $\dfrac{(x + 5)(x - 3) + (x + 5)(x - 2)}{2(x + 5)}$

44. $\dfrac{(2x - 5)(x + 4) - (2x - 5)(x + 1)}{3(2x - 5)}$

45. $\dfrac{a^2 + 3a - ab - 3b}{a^2 - ab + 5a - 5b}$

46. $\dfrac{xy - yw + xz - zw}{xy + yw + xz + zw}$

47. $\dfrac{a^3 - b^3}{a^2 - b^2}$

48. $\dfrac{x^2 + x - 6}{x^3 + 27}$

Multiply or divide as indicated. Simplify all answers.

49. $\dfrac{2x}{3y} \cdot \dfrac{y^3}{6}$

50. $\dfrac{16x^2}{y^4} \cdot \dfrac{5x^3}{4y^2}$

51. $\dfrac{9x^3}{4} \div \dfrac{3}{16y^2}$

52. $\dfrac{80m^4}{49x^5y^7} \cdot \dfrac{14x^{12}y^5}{25m^5}$

53. $\dfrac{3 - r}{r - 3} \cdot \dfrac{r - 6}{6 - r}$

54. $\dfrac{7a + 7b}{3} \div \dfrac{a^2 - b^2}{a - b}$

55. $\dfrac{p^2 + 7p + 10}{p + 5} \cdot \dfrac{1}{p + 2}$

56. $\dfrac{x^2 + 3x - 10}{4x} \cdot \dfrac{x^2 - 3x}{x^2 - 5x + 6}$

57. $\dfrac{r^2 + 10r + 21}{r + 7} \div (r^2 - 5r - 24)$

58. $(x - 3) \div \dfrac{x^2 + 3x - 18}{x^2}$

59. $\dfrac{x^2 + 12x + 35}{x^2 + 4x - 5} \div \dfrac{x^2 + 3x - 28}{x - 1}$

60. $\dfrac{x + 1}{x^2 - 17x + 30} \div \dfrac{3x + 3}{x^2 + 7x - 18}$

61. $\dfrac{a - b}{9a + 9b} \div \dfrac{a^2 - b^2}{a^2 + 2a + 1}$

62. $\dfrac{2x^2 + 8xy + 8y^2}{x^2 + 4xy + 4y^2} \cdot \dfrac{2x^2 + 7xy + 6y^2}{4x^2 + 14xy + 12y^2}$

63. $\dfrac{3x^2 - x - 4}{4x^2 + 5x + 1} \cdot \dfrac{2x^2 - 5x - 12}{6x^2 + x - 12}$

64. $\dfrac{6x^3 - x^2 - x}{2x^2 + x - 1} \cdot \dfrac{x^2 - 1}{x^3 - 2x^2 + x}$

65. $\dfrac{x + 2}{x^3 - 8} \cdot \dfrac{(x - 2)^2}{x^2 + 4}$

66. $\dfrac{x^2 - y^2}{x^2 - 2xy + y^2} \div \dfrac{(x + y)^2}{(x - y)^2}$

67. $\dfrac{x^4 - y^8}{x^2 + y^4} \div \dfrac{x^2 - y^4}{x^4}$

68. $\dfrac{(x^2 - y^2)^2}{(x^2 - y^2)^3} \div \dfrac{x^2 + y^2}{x^4 - y^4}$

69. $\dfrac{2x^4 + 4x^2}{6x^2 + 14x + 4} \div \dfrac{x^2 + 2}{3x^2 + x}$

70. $\dfrac{8a^3 - 1}{4a^2 + 2a + 1} \div \dfrac{a^2 - 2a + 1}{(a - 1)^2}$

71. $\dfrac{(a - b)^3}{a^3 - b^3} \cdot \dfrac{a^2 - b^2}{(a - b)^2}$

72. $\dfrac{r^2 - 9}{r^3 - 27} \div \dfrac{r^2 + 6r + 9}{r^2 + 3r + 9}$

73. $\dfrac{4x + y}{5x + 2y} \cdot \dfrac{10x^2 - xy - 2y^2}{8x^2 - 2xy - y^2}$

74. $\dfrac{2x^3 - 7x^2 + 3x}{x^2 + 2x - 3} \cdot \dfrac{x^2 + 3x}{(x - 3)^2}$

75. $\dfrac{ac - ad + bc - bd}{ac + ad + bc + bd} \cdot \dfrac{pc + pd - qc - qd}{pc - pd + qc - qd}$

76. $\dfrac{3r^2 + 17rs + 10s^2}{6r^2 + 13rs - 5s^2} \div \dfrac{6r^2 + rs - 2s^2}{6r^2 - 5rs + s^2}$

77. $\dfrac{x^3 - 4x^2 + x - 4}{x^4 - x^3 + x^2 - x} \cdot \dfrac{2x^3 + 2x^2 + x + 1}{2x^3 - 8x^2 + x - 4}$

78. $\dfrac{2p^2 + 2pq - pq^2 - q^3}{p^3 + p^2 + pq^2 + q^2} \div \dfrac{p^3 + p + p^2q + q}{p^3 + p + p^2 + 1}$

Problem Solving

79. Make up a rational expression that is undefined at $x = 2$ and $x = -3$. Explain how you determined your answer.

80. Make up a rational expression that is undefined at $x = 4$ and $x = -2$. Explain how you determined your answer.

81. Consider the rational function $f(x) = \dfrac{1}{x}$. Explain why this function can never equal 0.

82. Consider the rational function $g(x) = \dfrac{2}{x + 3}$. Explain why this function can never equal 0.

83. Consider the rational function $f(x) = \dfrac{x - 4}{x^2 - 4}$. For what value of x, if any, will this function **a)** equal 0? **b)** be undefined? Explain.

84. Consider the function $f(x) = \dfrac{x - 2}{x^2 - 81}$. For what value of x, if any, will this function **a)** equal 0? **b)** be undefined? Explain.

85. Give a function that is undefined at $x = 3$ and $x = -1$ and has a value of 0 at $x = 2$. Explain how you determined your answer.

86. Give a function that is undefined at $x = -4$ and $x = -2$ and has a value of 0 at $x = 5$. Explain how you determined your answer.

Determine the polynomial to be placed in the shaded area to give a true statement. Explain how you determined your answer.

87. $\dfrac{\rule[0.5ex]{2cm}{0.4ex}}{x^2 + 2x - 15} = \dfrac{1}{x - 3}$

88. $\dfrac{\rule[0.5ex]{2cm}{0.4ex}}{3x + 4} = x - 3$

89. $\dfrac{y^2 - y - 20}{\rule[0.5ex]{2cm}{0.4ex}} = \dfrac{y + 4}{y + 1}$

90. $\dfrac{\rule[0.5ex]{2cm}{0.4ex}}{6p^2 + p - 15} = \dfrac{2p - 1}{2p - 3}$

Determine the polynomial to be placed in the shaded area to give a true statement. Explain how you determined your answer.

91. $\dfrac{x^2 - x - 12}{x^2 + 2x - 3} \cdot \dfrac{\rule[0.5ex]{2cm}{0.4ex}}{x^2 - 2x - 8} = 1$

92. $\dfrac{x^2 - 4}{(x + 2)^2} \cdot \dfrac{2x^2 + x - 6}{\rule[0.5ex]{1.5cm}{0.4ex}} = \dfrac{x - 2}{2x + 5}$

93. $\dfrac{x^2 - 9}{2x^2 + 3x - 2} \div \dfrac{2x^2 - 9x + 9}{\rule[0.5ex]{1.5cm}{0.4ex}} = \dfrac{x + 3}{2x - 1}$

94. $\dfrac{4r^2 - r - 18}{\rule[0.5ex]{1.5cm}{0.4ex}} \div \dfrac{4r^3 - 9r^2}{6r^2 - 9r + 3} = \dfrac{3(r - 1)}{r^2}$

95. Consider the rectangle below. Its area is $3a^2 + 7ab + 2b^2$ and its length is $2a + 4b$. Find its width, w, in terms of a and b, by dividing its area by its length.

$2a + 4b$

96. Consider the rectangle below. Its area is $a^2 + 2ab + b^2$ and its length is $3a + 3b$. Find its width, w, in terms of a and b, by dividing its area by its length.

$3a + 3b$

97. Consider the triangle below. If its area is $a^2 + 4ab + 3b^2$ and its base is $a + 3b$, find its height h. Use area $= \frac{1}{2}$(base)(height).

$a + 3b$

98. Consider the trapezoid below. If its area is $a^2 + 2ab + b^2$, find its height, h. Use area $= \frac{1}{2}h(a+b)$.

a

Perform each indicated operation.

99. $\left(\dfrac{2x^2 - 3x - 14}{2x^2 - 9x + 7} \div \dfrac{6x^2 + x - 15}{3x^2 + 2x - 5} \right) \cdot \dfrac{6x^2 - 7x - 3}{2x^2 - x - 3}$

100. $\left(\dfrac{a^2 - b^2}{2a^2 - 3ab + b^2} \cdot \dfrac{2a^2 - 7ab + 3b^2}{a^2 + ab} \right) \div \dfrac{ab - 3b^2}{a^2 + 2ab + b^2}$

101. $\dfrac{5x^2(x - 1) - 3x(x - 1) - 2(x - 1)}{10x^2(x - 1) + 9x(x - 1) + 2(x - 1)} \cdot \dfrac{2x + 1}{x + 3}$

102. $\dfrac{x^2(3x - y) - 5x(3x - y) - 24(3x - y)}{x^2(3x - y) - 9x(3x - y) + 8(3x - y)} \cdot \dfrac{x - 1}{x + 3}$

103. $\dfrac{(x - p)^n}{x^{-2}} \div \dfrac{(x - p)^{2n}}{x^{-4}}$

104. $\dfrac{x^{-3}}{(a - b)^r} \div \dfrac{x^{-5}}{(a - b)^{r+2}}$

Simplify.

105. $\dfrac{x^{5y} + 3x^{4y}}{3x^{3y} + x^{4y}}$

106. $\dfrac{m^{2x} - m^x - 2}{m^{2x} - 4}$

For Exercises 107–110,

a) *Determine the domain of the function.*

b) *Graph the function in connected mode.*

c) *Is the function increasing, decreasing, or remaining the same as x gets closer and closer to 2, approaching 2 from the* left *side?*

d) *Is the function increasing, decreasing, or remaining the same as x gets closer and closer to 2, approaching 2 from the* right *side?*

107. $f(x) = \dfrac{1}{x - 2}$

108. $f(x) = \dfrac{x}{x - 2}$

109. $f(x) = \dfrac{x^2}{x - 2}$

110. $f(x) = \dfrac{x - 2}{x - 2}$

111. Consider the rational function $f(x) = \dfrac{1}{x}$.

    a) Determine the domain of the function.

    b) Complete the following table for the function.

x	-10	-1	-0.5	-0.1	-0.01	0.01	0.1	0.5	1	10
y										

    c) Draw the graph of $f(x) = \dfrac{1}{x}$. Consider what happens to the function as x gets closer and closer to 0, approaching 0 from both the left and right sides.

    d) Can this graph ever have a value of 0? Explain your answer.

 ## Group Activity

112. Consider the rational function $f(x) = \dfrac{x^2 - 4}{x - 2}$.

    a) As a group, determine the domain of this function.

    b) Have each member of the group individually complete the following table for the function.

x	-2	-1	0	1	1.9	1.99	2.01	2.1	3	4	5	6
y												

    c) Compare your answers to part **b)** and agree on the correct table values.

    d) As a group draw the graph of $f(x) = \dfrac{x^2 - 4}{x - 2}$. Is the function defined when $x = 2$?

    e) Can this graph ever have a value of 0? If so, for what value(s) of a is $f(a) = 0$?

Cumulative Review Exercises

[2.2] **113.** Solve $3(x - 2) + 3y = 6x$ for y.

[2.5] **114.** Solve $4 + \dfrac{4x}{3} < 6$ and give the answer in interval notation.

[2.6] **115.** Solve $\left| \dfrac{2x - 4}{5} \right| = 12$.

[3.2] **116.** Let $f(x) = |6 - 3x| - 2$. Find $f(1.3)$.

[4.1] **117.** Solve the system of equations.

$$3x + 4y = 2$$
$$2x + 5y = -1$$

[5.6] **118.** Factor $9x^2 - 6xy + y^2 - 4$.

6.2 ADDITION AND SUBTRACTION OF RATIONAL EXPRESSIONS

SSM **Study Guide** **CD/Video** **MathPro 4/5** **PH Math Tutor Center** **prenhall.com/Angel**

  1 Add and subtract expressions with a common denominator.

  2 Find the least common denominator (LCD).

  3 Add and subtract expressions with unlike denominators.

  4 Study on application of rational expressions.

1 Add and Subtract Expressions with a Common Denominator

When adding (or subtracting) two rational expressions with a common denominator, we add (or subtract) the numerators while keeping the common denominator.

To Add or Subtract Rational Expressions

To add or subtract rational expressions, use the following rules.

Addition	Subtraction
$\dfrac{a}{c} + \dfrac{b}{c} = \dfrac{a + b}{c}, \quad c \neq 0$	$\dfrac{a}{c} - \dfrac{b}{c} = \dfrac{a - b}{c}, \quad c \neq 0$

(continued on the next page)

To add or subtract rational expressions with a common denominator,

1. Add or subtract the expressions using the rules given on the bottom of the previous page.
2. Simplify the expression if possible.

EXAMPLE 1   Add.   **a)** $\dfrac{3}{x + 2} + \dfrac{x - 4}{x + 2}$ **b)** $\dfrac{x^2 + 3x - 2}{(x + 5)(x - 3)} + \dfrac{4x + 12}{(x + 5)(x - 3)}$

Solution   **a)** Since the denominators are the same, we add the numerators and keep the common denominator.

$$\frac{3}{x + 2} + \frac{x - 4}{x + 2} = \frac{3 + (x - 4)}{x + 2} \qquad \textit{Add numerators.}$$

$$= \frac{x - 1}{x + 2}$$

b) $\dfrac{x^2 + 3x - 2}{(x + 5)(x - 3)} + \dfrac{4x + 12}{(x + 5)(x - 3)} = \dfrac{x^2 + 3x - 2 + (4x + 12)}{(x + 5)(x - 3)}$ $\qquad$ *Add numerators.*

$$= \frac{x^2 + 7x + 10}{(x + 5)(x - 3)} \qquad \textit{Combine like terms.}$$

$$= \frac{\cancel{(x + 5)}(x + 2)}{\cancel{(x + 5)}(x - 3)} \qquad \textit{Factor; divide out common factors.}$$

$$= \frac{x + 2}{x - 3}$$

When subtracting rational expressions, be sure to subtract the entire numerator of the fraction being subtracted. Read the Avoiding Common Errors box that follows very carefully.

AVOIDING COMMON ERRORS

The error presented here is sometimes made by students. Study the information presented so that you will not make this error.

How do you simplify this problem?

$$\frac{4x}{x - 2} - \frac{2x + 1}{x - 2}$$

The procedure on the right side is incorrect because the *entire numerator*, $2x + 1$, must be subtracted from $4x$. Instead, only $2x$ was subtracted. Note that **the sign of each term (not just the first term) in the numerator of the fraction being subtracted must change**. Note that $-(2x + 1) = -2x - 1$, by the distributive property.

EXAMPLE 2   Subtract $\dfrac{a}{a - 6} - \dfrac{a^2 - 4a - 6}{a - 6}$.

Solution

$$\dfrac{a}{a - 6} - \dfrac{a^2 - 4a - 6}{a - 6} = \dfrac{a - (a^2 - 4a - 6)}{a - 6} \qquad \textit{Subtract numerators.}$$

$$= \dfrac{a - a^2 + 4a + 6}{a - 6}$$

$$= \dfrac{-a^2 + 5a + 6}{a - 6} \qquad \textit{Combine like terms.}$$

$$= \dfrac{-(a^2 - 5a - 6)}{a - 6} \qquad \textit{Factor out } -1.$$

$$= \dfrac{-(a - 6)(a + 1)}{a - 6} \qquad \begin{array}{l}\textit{Factor; divide out}\\\textit{common factors.}\end{array}$$

NOW TRY EXERCISE 13

$$= -(a + 1) \text{ or } -a - 1 \qquad \text{✳}$$

2 Find the Least Common Denominator

To add or subtract two numerical fractions with *unlike denominators*, we must first obtain a common denominator. Obtaining a common denominator may involve writing numerical values as products of prime numbers. A **prime number** is a number greater than 1 that has only two divisors, itself and 1. Some prime numbers are 2, 3, 5, 7, 11, 13, and 17. Following we show how the numbers 36 and 48 are written as a product of prime numbers,

$$36 = 2 \cdot 2 \cdot 3 \cdot 3 = 2^2 \cdot 3^2$$
$$48 = 2 \cdot 2 \cdot 2 \cdot 2 \cdot 3 = 2^4 \cdot 3$$

We may need to write numerical coefficients as products of prime numbers to find the **least common denominator** of a rational expression.

> **To Find the Least Common Denominator (LCD) of Rational Expressions**
>
> 1. Write each nonprime coefficient (other than 1) of monomials in denominators as a product of prime numbers.
> 2. Factor each denominator completely. Any factors that occur more than once should be expressed as powers. For example, $(x + 5)(x + 5)$ should be expressed as $(x + 5)^2$.
> 3. List all different factors (other than 1) that appear in any of the denominators. When the same factor appears in more than one denominator, write the factor with the highest power that appears.
> 4. The least common denominator is the product of all the factors found in step 2.

EXAMPLE 3   Find the LCD of each expression.

a) $\dfrac{3}{5x} - \dfrac{2}{x^2}$ **b)** $\dfrac{1}{18x^3 y} + \dfrac{5}{27x^2 y^3}$ **c)** $\dfrac{3}{x} - \dfrac{2y}{x + 5}$ **d)** $\dfrac{7}{x^2(x + 1)} + \dfrac{3z}{x(x + 1)^3}$

Solution   **a)** The factors that appear in the denominators are 5 and x. List each factor with its highest power. The LCD is the product of these factors.

Highest power of x

$$\text{LCD} = 5 \cdot x^2 = 5x^2$$

b) The numerical coefficients written as products of prime numbers are $18 = 2 \cdot 3^2$ and $27 = 3^3$. The variable factors are x and y. Using the highest powers of the factors, we obtain the LCD.

$$\text{LCD} = 2 \cdot 3^3 \cdot x^3 y^3 = 54 x^3 y^3$$

c) The factors are x and $x + 5$. Note that the x in the second denominator, $x + 5$, is not a factor of that denominator since the operation is addition rather than multiplication.

$$\text{LCD} = x(x + 5)$$

d) The factors are x and $x + 1$. The highest power of x is 2 and the highest power of $x + 1$ is 3.

$$\text{LCD} = x^2(x + 1)^3 \qquad \text{✳}$$

Sometimes it is necessary to factor all denominators to obtain the LCD. This is illustrated in the next example.

EXAMPLE 4 Find the LCD of each expression.

a) $\dfrac{3}{2x^2 - 4x} + \dfrac{7x}{x^2 - 4x + 4}$ **b)** $\dfrac{5x}{x^2 - x - 12} - \dfrac{6x^2}{x^2 - 7x + 12}$

Solution **a)** Factor both denominators.

$$\frac{3}{2x^2 - 4x} + \frac{7x}{x^2 - 4x + 4} = \frac{3}{2x(x - 2)} + \frac{7x}{(x - 2)^2}$$

The factors are $2, x$, and $x - 2$. Multiply the factors raised to the highest power that appears for each factor.

$$\text{LCD} = 2 \cdot x \cdot (x - 2)^2 = 2x(x - 2)^2$$

b) Factor both denominators.

$$\frac{5x}{x^2 - x - 12} - \frac{6x^2}{x^2 - 7x + 12} = \frac{5x}{(x + 3)(x - 4)} - \frac{6x^2}{(x - 3)(x - 4)}$$

$$\text{LCD} = (x + 3)(x - 4)(x - 3)$$

Note that although $x - 4$ is a common factor of each denominator, the highest power of the factor that appears in either denominator is 1. ✳

3 Add and Subtract Expressions with Unlike Denominators

The procedure used to add or subtract rational expressions with unlike denominators is given below.

> ### To Add or Subtract Rational Expressions with Unlike Denominators
>
> 1. Determine the LCD.
> 2. Rewrite each fraction as an equivalent fraction with the LCD. This is done by multiplying both the numerator and denominator of each fraction by any factors needed to obtain the LCD.
> 3. Leave the denominator in factored form, but multiply out the numerator.
> 4. Add or subtract the numerators while maintaining the LCD.
> 5. When it is possible to reduce the fraction by factoring the numerator, do so.

EXAMPLE 5 Add. **a)** $\dfrac{2}{x} + \dfrac{7}{y}$ **b)** $\dfrac{5}{4a^2} + \dfrac{3}{14ab^3}$

Solution **a)** First we determine the LCD.

$$\text{LCD} = xy$$

Now we write each fraction with the LCD. We do this by multiplying *both* numerator and denominator of each fraction by any factors needed to obtain the LCD.

In this problem, the first fraction must be multiplied by $\dfrac{y}{y}$ and the second fraction must be multiplied by $\dfrac{x}{x}$.

$$\frac{2}{x} + \frac{7}{y} = \frac{y}{y} \cdot \frac{2}{x} + \frac{7}{y} \cdot \frac{x}{x} = \frac{2y}{xy} + \frac{7x}{xy}$$

By multiplying both the numerator and denominator by the same factor, we are in effect multiplying by 1, which does not change the value of the fraction, only its appearance. Thus, the new fraction is equivalent to the original fraction.

Now we add the numerators while leaving the LCD alone.

$$\frac{2y}{xy} + \frac{7x}{xy} = \frac{2y + 7x}{xy} \quad \text{or} \quad \frac{7x + 2y}{xy}$$

Therefore, $\dfrac{2}{x} + \dfrac{7}{y} = \dfrac{7x + 2y}{xy}$.

b) The LCD of 4 and 14 is 28. The LCD of the two fractions is $28a^2b^3$. We must write each fraction with the denominator $28a^2b^3$. To do this, we multiply the first fraction by $\dfrac{7b^3}{7b^3}$ and the second fraction by $\dfrac{2a}{2a}$.

$$\frac{5}{4a^2} + \frac{3}{14ab^3} = \frac{7b^3}{7b^3} \cdot \frac{5}{4a^2} + \frac{3}{14ab^3} \cdot \frac{2a}{2a} \qquad \textit{Multiply to obtain LCD.}$$

$$= \frac{35b^3}{28a^2b^3} + \frac{6a}{28a^2b^3}$$

$$= \frac{35b^3 + 6a}{28a^2b^3} \qquad \textit{Add numerators.} \quad \text{✳}$$

NOW TRY EXERCISE 39

EXAMPLE 6 Subtract $\dfrac{x + 2}{x - 4} - \dfrac{x + 5}{x + 4}$.

Solution The LCD is $(x - 4)(x + 4)$. Write each fraction with the denominator $(x - 4)(x + 4)$.

$$\frac{x + 2}{x - 4} - \frac{x + 5}{x + 4} = \frac{x + 4}{x + 4} \cdot \frac{x + 2}{x - 4} - \frac{x + 5}{x + 4} \cdot \frac{x - 4}{x - 4} \qquad \textit{Multiply to obtain LCD.}$$

$$= \frac{(x + 4)(x + 2)}{(x + 4)(x - 4)} - \frac{(x + 5)(x - 4)}{(x + 4)(x - 4)}$$

$$= \frac{x^2 + 6x + 8}{(x + 4)(x - 4)} - \frac{x^2 + x - 20}{(x + 4)(x - 4)} \qquad \textit{Multiply binomials in numerators.}$$

$$= \frac{x^2 + 6x + 8 - (x^2 + x - 20)}{(x + 4)(x - 4)} \qquad \textit{Subtract numerators.}$$

$$= \frac{x^2 + 6x + 8 - x^2 - x + 20}{(x + 4)(x - 4)}$$

$$= \frac{5x + 28}{(x + 4)(x - 4)} \qquad \textit{Combine like terms.} \quad ✳$$

EXAMPLE 7 Add $\dfrac{2}{x - 3} + \dfrac{x + 5}{3 - x}$.

Solution Note that each denominator is the opposite, or additive inverse, of the other. (The terms of one denominator differ only in sign from the terms of the other denominator.) When this special situation arises, we can multiply the numerator and denominator of either one of the fractions by -1 to obtain the LCD.

$$\frac{2}{x - 3} + \frac{x + 5}{3 - x} = \frac{2}{x - 3} + \frac{-1}{-1} \cdot \frac{(x + 5)}{(3 - x)} \qquad \textit{Multiply to obtain LCD.}$$

$$= \frac{2}{x - 3} + \frac{-x - 5}{x - 3}$$

$$= \frac{2 - x - 5}{x - 3} \qquad \textit{Add numerators.}$$

$$= \frac{-x - 3}{x - 3} \qquad \textit{Combine like terms.}$$

Because there are no common factors in both the numerator and denominator, $\dfrac{-x - 3}{x - 3}$ cannot be simplified further.

NOW TRY EXERCISE 43   ✳

EXAMPLE 8 Subtract $\dfrac{3x + 4}{2x^2 - 5x - 12} - \dfrac{2x - 3}{5x^2 - 18x - 8}$.

Solution Factor the denominator of each expression.

$$\frac{3x + 4}{2x^2 - 5x - 12} - \frac{2x - 3}{5x^2 - 18x - 8} = \frac{3x + 4}{(2x + 3)(x - 4)} - \frac{2x - 3}{(5x + 2)(x - 4)}$$

The LCD is $(2x + 3)(x - 4)(5x + 2)$.

$$\frac{3x + 4}{(2x + 3)(x - 4)} - \frac{2x - 3}{(5x + 2)(x - 4)}$$

$$= \frac{5x + 2}{5x + 2} \cdot \frac{3x + 4}{(2x + 3)(x - 4)} - \frac{2x - 3}{(5x + 2)(x - 4)} \cdot \frac{2x + 3}{2x + 3} \qquad \textit{Multiply to obtain LCD.}$$

$$= \frac{15x^2 + 26x + 8}{(5x + 2)(2x + 3)(x - 4)} - \frac{4x^2 - 9}{(5x + 2)(2x + 3)(x - 4)} \qquad \textit{Multiply numerators.}$$

$$= \frac{15x^2 + 26x + 8 - (4x^2 - 9)}{(5x + 2)(2x + 3)(x - 4)} \qquad \textit{Subtract numerators.}$$

$$= \frac{15x^2 + 26x + 8 - 4x^2 + 9}{(5x + 2)(2x + 3)(x - 4)}$$

$$= \frac{11x^2 + 26x + 17}{(5x + 2)(2x + 3)(x - 4)} \qquad \textit{Combine like terms.} \quad ✳$$

EXAMPLE 9   Perform the indicated operations.

$$\frac{x-1}{x-2} - \frac{x+1}{x+2} + \frac{x-6}{x^2-4}$$

Solution   First, factor $x^2 - 4$. The LCD of the three fractions is $(x+2)(x-2)$.

$$\frac{x-1}{x-2} - \frac{x+1}{x+2} + \frac{x-6}{x^2-4}$$

$$= \frac{x-1}{x-2} - \frac{x+1}{x+2} + \frac{x-6}{(x+2)(x-2)}$$

$$= \frac{x+2}{x+2} \cdot \frac{x-1}{x-2} - \frac{x+1}{x+2} \cdot \frac{x-2}{x-2} + \frac{x-6}{(x+2)(x-2)} \qquad \textit{Multiply to obtain LCD.}$$

$$= \frac{x^2+x-2}{(x+2)(x-2)} - \frac{x^2-x-2}{(x+2)(x-2)} + \frac{x-6}{(x+2)(x-2)} \qquad \textit{Multiply numerators.}$$

$$= \frac{x^2+x-2-(x^2-x-2)+(x-6)}{(x+2)(x-2)} \qquad \textit{Subtract and add numerators.}$$

$$= \frac{x^2+x-2-x^2+x+2+x-6}{(x+2)(x-2)}$$

$$= \frac{3x-6}{(x+2)(x-2)} \qquad \textit{Combine like terms.}$$

$$= \frac{3\cancel{(x-2)}}{(x+2)\cancel{(x-2)}} \qquad \textit{Factor; divide out common factors.}$$

$$= \frac{3}{x+2}$$

NOW TRY EXERCISE 67

Using Your Graphing Calculator

In Example 9, we found that

$$\frac{x-1}{x-2} - \frac{x+1}{x+2} + \frac{x-6}{x^2-4} = \frac{3}{x+2}$$

Suppose we let

$$y_1 = \frac{x-1}{x-2} - \frac{x+1}{x+2} + \frac{x-6}{x^2-4}$$

$$y_2 = \frac{3}{x+2}$$

If we use the TABLE feature on a graphing calculator, how will the values of y_1 and y_2 compare? The function y_1 is not defined at $x = -2$ and $x = 2$. The function y_2 is not defined at $x = -2$. For all values of x other than -2 and 2, the values of y_1 and y_2 should be the same if we have not made a mistake. Following is a table of values for y_1 and y_2, for values of x from -3 to 3.

(continued on the next page)

The graphs of y_1 and y_2 are shown in Figures 6.3 and 6.4, respectively. We illustrated the graphs in this format (rather than on a graphing calculator screen) to show more detail. The open circle on the graph in Figure 6.3 is not shown on a grapher. Notice that the graph of y_1 has an open circle at 2 because y_1 is not defined at $x = 2$. Since y_2 is defined at $x = 2$, Figure 6.4 does not include this open circle. Neither function is defined at $x = -2$.

$$y_1 = \frac{x-1}{x-2} - \frac{x+1}{x+2} + \frac{x-6}{x^2-4}$$

$$y_2 = \frac{3}{x+2}$$

FIGURE 6.3   FIGURE 6.4

HELPFUL HINT

STUDY TIP

Now that we have discussed the operations of addition, subtraction, multiplication, and division of rational expressions, let's quickly summarize the procedures.

To add or subtract rational expressions, obtain the LCD. Express each fraction with the LCD. Then add or subtract the numerators and write this result over the LCD.

To multiply rational expressions, factor each expression completely, divide out common factors, multiply numerators, and multiply denominators.

To divide rational expressions, multiply the first (or top) fraction by the reciprocal of the second (or bottom) fraction. Then, factor each expression completely, divide out common factors, multiply the numerators, and multiply the denominators.

4 Study an Application of Rational Expressions

Section 6.5 deals with applications of rational expressions. At this time, we will introduce an application that involves adding and subtracting rational expressions or functions.

In economics, we study revenue, cost, and profit. If $R(x)$ is a revenue function and $C(x)$ is a cost function, then the profit function, $P(x)$, is

$$P(x) = R(x) - C(x)$$

where x is the number of items manufactured and sold by a company. We will use this information in Example 10.

EXAMPLE 10   **Sailboats**   The Don Perrione Sailboat Company builds and sells at least six sailboats each week.

Suppose $R(x) = \dfrac{6x - 7}{x + 2}$ and $C(x) = \dfrac{4x - 13}{x + 3}$

where x is the number of sailboats sold. Determine the profit function.

Solution   Understand and Translate   To determine the profit function, we subtract the cost function from the revenue function.

$$P(x) = R(x) - C(x)$$

$$P(x) = \frac{6x - 7}{x + 2} - \frac{4x - 13}{x + 3}$$

The LCD is $(x + 2)(x + 3)$.

Carry Out

$$= \frac{x + 3}{x + 3} \cdot \frac{6x - 7}{x + 2} - \frac{4x - 13}{x + 3} \cdot \frac{x + 2}{x + 2} \qquad \textit{Multiply to obtain LCD.}$$

$$= \frac{6x^2 + 11x - 21}{(x + 3)(x + 2)} - \frac{4x^2 - 5x - 26}{(x + 3)(x + 2)} \qquad \textit{Multiply numerators.}$$

$$= \frac{(6x^2 + 11x - 21) - (4x^2 - 5x - 26)}{(x + 3)(x + 2)} \qquad \textit{Subtract numerators.}$$

$$= \frac{6x^2 + 11x - 21 - 4x^2 + 5x + 26}{(x + 3)(x + 2)}$$

$$= \frac{2x^2 + 16x + 5}{(x + 3)(x + 2)} \qquad \textit{Combine like terms.}$$

NOW TRY EXERCISE 81

Answer   The profit function is $P(x) = \dfrac{2x^2 + 16x + 5}{(x + 3)(x + 2)}$.

Exercise Set 6.2

Concept/Writing Exercises

1. a) What is the least common denominator of two or more rational expressions?
b) Explain how to find the LCD.
c) Using the procedure you gave in part **b)**, find the LCD of

$$\frac{5}{64x^2 - 121} \quad \text{and} \quad \frac{1}{8x^2 - 27x + 22}$$

2. a) Explain how to add or subtract two rational expressions.
b) Add $\dfrac{4}{x + 2} + \dfrac{x}{3x^2 - 4x - 20}$ following the procedure you gave in part **a)**.

*In Exercises 3 and 4, **a)** explain why the subtraction is not correct, and **b)** show the correct subtraction.*

3. $\dfrac{x^2 - 4x}{(x + 3)(x - 2)} - \dfrac{x^2 + x - 2}{(x + 3)(x - 2)} \neq \dfrac{x^2 - 4x - x^2 + x - 2}{(x + 3)(x - 2)}$

4. $\dfrac{x - 5}{(x + 4)(x - 3)} - \dfrac{x^2 - 6x + 5}{(x + 4)(x - 3)} \neq \dfrac{x - 5 - x^2 - 6x + 5}{(x + 4)(x - 3)}$

Practice the Skills

Add or subtract.

5. $\dfrac{3x}{x + 2} + \dfrac{5}{x + 2}$

6. $\dfrac{3x}{x + 2} + \dfrac{6}{x + 2}$

7. $\dfrac{7x}{x - 5} - \dfrac{2}{x - 5}$

8. $\dfrac{10x}{x - 5} - \dfrac{50}{x - 5}$

9. $\dfrac{x}{x + 3} + \dfrac{9}{x + 3} - \dfrac{2}{x + 3}$

10. $\dfrac{2x}{x + 3} + \dfrac{9}{x + 3} - \dfrac{3}{x + 3}$

11. $\dfrac{5x - 6}{x - 2} + \dfrac{2x - 5}{x - 2}$

12. $\dfrac{-4x + 6}{x^2 + x - 6} + \dfrac{5x - 3}{x^2 + x - 6}$

🔒 **13.** $\dfrac{x^2 - 2}{x^2 + 6x - 7} - \dfrac{-4x + 19}{x^2 + 6x - 7}$

14. $\dfrac{-x^2}{x^2 + 5xy - 14y^2} + \dfrac{x^2 + xy + 7y^2}{x^2 + 5xy - 14y^2}$

15. $\dfrac{3r^2 + 15r}{r^3 + 2r^2 - 8r} + \dfrac{2r^2 + 5r}{r^3 + 2r^2 - 8r}$

16. $\dfrac{x^3 - 12x^2 + 45x}{x(x - 8)} - \dfrac{x^2 + 5x}{x(x - 8)}$

17. $\dfrac{3x^2 - x}{2x^2 - x - 21} + \dfrac{3x - 8}{2x^2 - x - 21} - \dfrac{x^2 - x + 27}{2x^2 - x - 21}$

18. $\dfrac{2x^2 + 9x - 15}{2x^2 - 13x + 20} - \dfrac{3x + 10}{2x^2 - 13x + 20} - \dfrac{3x - 5}{2x^2 - 13x + 20}$

Find the least common denominator.

19. $\dfrac{7}{2a^2} + \dfrac{9}{3a^3}$

20. $\dfrac{1}{9x^2} - \dfrac{5}{6x^3}$

21. $\dfrac{-4}{8x^2y^2} + \dfrac{7}{5x^4y^5}$

22. $\dfrac{x + 3}{16x^2y} - \dfrac{x^2}{3x^3}$

23. $\dfrac{2}{3a^4b^2} + \dfrac{7}{2a^3b^5}$

24. $\dfrac{1}{x + 1} - \dfrac{x}{x - 3}$

25. $\dfrac{4x}{x + 3} + \dfrac{6}{x + 2}$

26. $\dfrac{4}{(r - 7)(r + 3)} - \dfrac{r + 8}{r - 7}$

27. $5z^2 + \dfrac{9z}{z - 4}$

28. $\dfrac{b^2 + 3}{18b} - \dfrac{b - 7}{12(b + 5)}$

29. $\dfrac{x}{x^4(x - 2)} - \dfrac{x + 9}{x^2(x - 2)^3}$

30. $\dfrac{x + 2}{(x - 3)^3(x + 4)^2} + \dfrac{x - 7}{(x + 4)^4(x - 5)}$

🔒 **31.** $\dfrac{a - 2}{a^2 - 5a - 24} + \dfrac{3}{a^2 + 11a + 24}$

32. $\dfrac{3x - 5}{6x^2 + 13xy + 6y^2} + \dfrac{3}{3x^2 + 5xy + 2y^2}$

33. $\dfrac{3}{x^2 + 3x - 4} - \dfrac{4}{4x^2 + 5x - 9} + \dfrac{x + 2}{4x^2 + 25x + 36}$

34. $\dfrac{x}{2x^2 - 7x + 3} + \dfrac{x - 3}{4x^2 + 4x - 3} - \dfrac{x^2 + 1}{2x^2 - 3x - 9}$

Add or subtract.

35. $\dfrac{2}{3r} + \dfrac{5}{r}$

36. $\dfrac{6}{x^2} + \dfrac{3}{2x}$

37. $\dfrac{7}{12x} - \dfrac{1}{4x^2}$

38. $\dfrac{3x}{4y} + \dfrac{7}{6xy}$

39. $\dfrac{5}{8x^4y} + \dfrac{1}{5x^2y^3}$

40. $\dfrac{3}{4xy^3} + \dfrac{1}{6x^2y}$

41. $\dfrac{8}{b - 2} + \dfrac{3b}{2 - b}$

42. $\dfrac{4x}{3xy} + 2$

43. $\dfrac{a}{a - b} - \dfrac{a}{b - a}$

44. $\dfrac{b}{a - b} - \dfrac{a + b}{b}$

45. $\dfrac{4x}{x - 4} + \dfrac{x + 3}{x + 1}$

46. $\dfrac{x}{x^2 - 9} - \dfrac{4(x - 3)}{x + 3}$

47. $\dfrac{3}{a + 2} + \dfrac{a + 1}{a^2 + 4a + 4}$

48. $\dfrac{2m + 1}{m - 5} - \dfrac{4}{m^2 - 3m - 10}$

49. $\dfrac{x}{x^2 + 2x - 8} + \dfrac{x + 3}{x^2 - 3x + 2}$

50. $\dfrac{-x^2 + 5x}{(x - 5)^2} + \dfrac{x + 1}{x - 5}$

51. $\dfrac{5x}{x^2 - 9x + 8} - \dfrac{3(x + 2)}{x^2 - 6x - 16}$

52. $\dfrac{4}{(2p - 3)(p + 4)} - \dfrac{3}{(p + 4)(p - 4)}$

53. $5 - \dfrac{x - 1}{x^2 + 3x - 10}$

54. $\dfrac{3x}{2x - 3} + \dfrac{3x + 6}{2x^2 + x - 6}$

55. $\dfrac{3a - 4}{4a + 1} + \dfrac{3a + 6}{4a^2 + 9a + 2}$

56. $\dfrac{7}{3q^2 + q - 4} + \dfrac{9q + 2}{3q^2 - 2q - 8}$

57. $\dfrac{x - y}{x^2 - 4xy + 4y^2} + \dfrac{x - 3y}{x^2 - 4y^2}$

58. $\dfrac{x + 2y}{x^2 - xy - 2y^2} - \dfrac{y}{x^2 - 3xy + 2y^2}$

🔒 **59.** $\dfrac{2r}{r - 4} - \dfrac{2r}{r + 4} + \dfrac{64}{r^2 - 16}$

60. $\dfrac{4}{p + 1} + \dfrac{3}{p - 1} + \dfrac{p + 4}{p^2 - 1}$

61. $\dfrac{-4}{x^2 + 2x - 3} - \dfrac{1}{x + 3} + \dfrac{1}{x - 1}$

62. $\dfrac{2}{x^2 - 16} + \dfrac{x + 1}{x^2 + 8x + 16} + \dfrac{3}{x - 4}$

63. $\dfrac{3}{3x - 2} - \dfrac{1}{x - 4} + 5$

64. $\dfrac{x}{3x + 4} + \dfrac{3x + 2}{x - 5} - \dfrac{7x^2 + 24x + 28}{3x^2 - 11x - 20}$

65. $3 - \dfrac{1}{8r^2 + 2r - 15} + \dfrac{r + 2}{4r - 5}$

66. $\dfrac{x}{x^2 - 10x + 24} - \dfrac{3}{x - 6} + 1$

67. $\dfrac{3}{5x + 6} + \dfrac{x^2 - x}{5x^2 - 4x - 12} - \dfrac{4}{x - 2}$

68. $\dfrac{3}{x^2 - 13x + 36} + \dfrac{4}{2x^2 - 7x - 4} + \dfrac{1}{2x^2 - 17x - 9}$

69. $\dfrac{m}{6m^2 + 13mn + 6n^2} + \dfrac{2m}{4m^2 + 8mn + 3n^2}$

70. $\dfrac{(x - y)^2}{x^3 - y^3} + \dfrac{2}{x^2 + xy + y^2}$

71. $\dfrac{5r - 2s}{25r^2 - 4s^2} - \dfrac{2r - s}{10r^2 - rs - 2s^2}$

72. $\dfrac{6}{(2r - 1)^2} + \dfrac{2}{2r - 1} - 5$

73. $\dfrac{2}{2x + 3y} - \dfrac{4x^2 - 6xy + 9y^2}{8x^3 + 27y^3}$

74. $\dfrac{4}{4x - 5y} - \dfrac{3x^2 + 2y^2}{64x^3 - 125y^3}$

Problem Solving

75. When two rational expressions are being added or subtracted, should the numerators of the expressions being added or subtracted be factored? Explain.

76. Are the fractions $\dfrac{x - 3}{5 - x}$ and $-\dfrac{x - 3}{x - 5}$ equivalent? Explain.

77. Are the fractions $\dfrac{8 - x}{3 - x}$ and $\dfrac{x - 8}{x - 3}$ equivalent? Explain.

78. If $f(x)$ and $g(x)$ are both rational functions, will $(f + g)(x)$ always be a rational function?

79. If $f(x) = \dfrac{x + 2}{x - 3}$ and $g(x) = \dfrac{x}{x + 4}$, find

a) the domain of $f(x)$.
b) the domain of $g(x)$.
c) $(f + g)(x)$.
d) the domain of $(f + g)(x)$.

80. If $f(x) = \dfrac{x + 1}{x^2 - 9}$ and $g(x) = \dfrac{x}{x - 3}$, find

a) the domain of $f(x)$.
b) the domain of $g(x)$.
c) $(f + g)(x)$.
d) the domain of $(f + g)(x)$.

In Exercises 81–84, find the profit function, P(x). (See Example 10.)

81. $R(x) = \dfrac{4x - 5}{x + 1}$ and $C(x) = \dfrac{2x - 7}{x + 2}$

82. $R(x) = \dfrac{5x - 2}{x + 2}$ and $C(x) = \dfrac{3x - 4}{x + 1}$

83. $R(x) = \dfrac{8x - 3}{x + 2}$ and $C(x) = \dfrac{5x - 8}{x + 3}$

84. $R(x) = \dfrac{7x - 10}{x + 3}$ and $C(x) = \dfrac{5x - 8}{x + 4}$

*The dashed red lines in the figures below are called **asymptotes**. The asymptotes are not a part of the graph but are used to show values that the graph approaches, but does not touch. In Exercises 85 and 86, determine the domain and range of the rational function shown.*

85.

86.

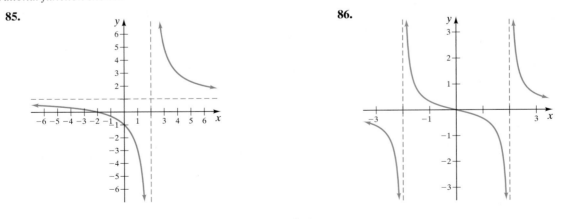

In Exercises 87–90, use $f(x) = \dfrac{x}{x^2 - 4}$ and $g(x) = \dfrac{2}{x^2 + x - 6}$. Find the following.

87. $(f + g)(x)$

88. $(f - g)(x)$

89. $(f \cdot g)(x)$

90. $(f/g)(x)$

91. Show that $\dfrac{a}{b} + \dfrac{c}{d} = \dfrac{ad + bc}{bd}$.

92. Show that $x^{-1} + y^{-1} = \dfrac{x + y}{xy}$.

*Consider the rectangles given below. Find **a)** the perimeter, and **b)** the area.*

93.

$\dfrac{a + b}{a}$

$\dfrac{a - b}{a}$

94.

$\dfrac{a + 2b}{b}$

$\dfrac{-a + 2b}{b}$

Determine the polynomial to be placed in the shaded area to give a true statement. Explain how you determined your answer.

95. $\dfrac{5x^2 - 6}{x^2 - x - 1} - \dfrac{\rule{1.5cm}{0.3cm}}{x^2 - x - 1} = \dfrac{-2x^2 + 6x - 12}{x^2 - x - 1}$

96. $\dfrac{r^2 - 6}{r^2 - 5r + 6} - \dfrac{\rule{1.5cm}{0.3cm}}{r^2 - 5r + 6} = \dfrac{1}{r - 2}$

Perform the indicated operations.

97. $\left(3 + \dfrac{1}{x + 3}\right)\left(\dfrac{x + 3}{x - 2}\right)$

98. $\left(\dfrac{3}{r + 1} - \dfrac{4}{r - 2}\right)\left(\dfrac{r - 2}{r + 10}\right)$

99. $\left(\dfrac{5}{a - 5} - \dfrac{2}{a + 3}\right) \div (3a + 25)$

100. $\left(\dfrac{x^2 + 4x - 5}{2x^2 + x - 3} \cdot \dfrac{2x + 3}{x + 1}\right) - \dfrac{2}{x + 2}$

101. $\left(\dfrac{x + 5}{x - 3} - x\right) \div \dfrac{1}{x - 3}$

102. $\left(\dfrac{x + 5}{x^2 - 25} + \dfrac{1}{x + 5}\right)\left(\dfrac{2x^2 - 13x + 15}{4x^2 - 6x}\right)$

103. The weighted average of two values a and b is given by $a\left(\dfrac{x}{n}\right) + b\left(\dfrac{n - x}{n}\right)$, where $\dfrac{x}{n}$ is the weight given to a and $\dfrac{n - x}{n}$ is the weight given to b.

a) Express this sum as a single fraction.

b) On exam a you received a grade of 60 and on exam b you received a grade of 92. If exam a counts $\frac{2}{5}$ of your final grade and exam b counts $\frac{3}{5}$, determine your weighted average.

104. Show that $\left(\dfrac{x}{y}\right)^{-1} + \left(\dfrac{y}{x}\right)^{-1} + (xy)^{-1} = \dfrac{x^2 + y^2 + 1}{xy}$.

In Exercises 105 and 106, perform the indicated operation.

105. $(a - b)^{-1} + (a - b)^{-2}$

106. $\left(\dfrac{a - b}{a}\right)^{-1} - \left(\dfrac{a + b}{a}\right)^{-1}$

Use your graphing calculator to determine whether the following additions are correct.

107. $\dfrac{x - 3}{x + 4} + \dfrac{x}{x^2 - 2x - 24} \stackrel{?}{=} \dfrac{x^2 - 10x + 18}{(x + 4)(x - 6)}$

108. $\dfrac{x - 2}{x^2 - 25} + \dfrac{x - 2}{2x^2 + 17x + 35} \stackrel{?}{=} \dfrac{3x^2 - 4x - 4}{(x + 5)(x - 5)(2x + 7)}$

Challenge Problems

109. Express each sum as a single fraction.

a) $1 + \dfrac{1}{x}$

b) $1 + \dfrac{1}{x} + \dfrac{1}{x^2}$

c) $1 + \dfrac{1}{x} + \dfrac{1}{x^2} + \dfrac{1}{x^3} + \dfrac{1}{x^4}$

d) $1 + \dfrac{1}{x} + \dfrac{1}{x^2} + \cdots + \dfrac{1}{x^n}$

110. Let $f(x) = \dfrac{1}{x}$. Find $f(a + h) - f(a)$.

111. Let $g(x) = \dfrac{1}{x + 1}$. Find $g(a + h) - g(a)$.

Cumulative Review Exercises

[2.4] **112.** ***Filling Bottles*** A bottling machine fills and caps bottles at a rate of 80 per minute. Then the machine is slowed down and fills and caps bottles at a rate of 60 per minute. If the sum of the two time periods is 14 minutes and the number of bottles filled and capped at the higher rate is the same as the number filled and capped at the lower rate, determine **a)** how long the machine is used at the faster rate, and **b)** the total number of bottles filled and capped over the 14-minute period.

[2.6] **113.** Solve for x and give the solution in set notation.
$$|x - 3| - 2 < 3$$

[3.4] **114.** Find the slope of the line passing through the points $(-2, 3)$ and $(7, -3)$.

[4.5] **115.** Evaluate the determinant $\begin{vmatrix} -1 & 3 \\ 5 & 6 \end{vmatrix}$.

[5.3] **116.** Divide $\dfrac{6x^2 - 5x + 6}{2x + 3}$.

[5.8] **117.** Solve $3p^2 = 22p - 7$.

6.3 COMPLEX FRACTIONS

SSM   Study Guide   CD/Video

MathPro 4/5   PH Math Tutor Center   prenhall.com/Angel

1 Recognize complex fractions.

2 Simplify complex fractions by multiplying by a common denominator.

3 Simplify complex fractions by simplifying the numerator and denominator.

1 Recognize Complex Fractions

A **complex fraction** is one that has a fractional expression in its numerator or its denominator or both its numerator and denominator.

Examples of Complex Fractions

$$\frac{\frac{2}{3}}{5}, \quad \frac{\frac{x+1}{x}}{3x}, \quad \frac{\frac{x}{y}}{x+1}, \quad \frac{\frac{a+b}{a}}{\frac{a-b}{b}}, \quad \frac{3+\frac{1}{x}}{\frac{1}{x^2}+\frac{3}{x}}$$

The expression above the **main fraction line** is the numerator, and the expression below the main fraction line is the denominator of the complex fraction.

$\dfrac{a+b}{a}$ ⟵ numerator of complex fraction

——— ⟵ main fraction line

$\dfrac{a-b}{b}$ ⟵ denominator of complex fraction

We will explain two methods that can be used to simplify complex fractions. To simplify a complex fraction means to write the expression without a fraction in its numerator and its denominator.

2 Simplify Complex Fractions by Multiplying by a Common Denominator

The first method involves multiplying both the numerator and denominator of the complex fraction by a common denominator.

> ### To Simplify a Complex Fraction by Multiplying by a Common Denominator
>
> 1. Find the least common denominator of all fractions appearing within the complex fraction. This is the LCD of the complex fraction.
> 2. Multiply both the numerator and denominator of the complex fraction by the LCD of the complex fraction found in step 1.
> 3. Simplify when possible.

In step 2, you are actually multiplying the complex fraction by $\frac{\text{LCD}}{\text{LCD}}$, which is equivalent to multiplying the fraction by 1.

EXAMPLE 1 Simplify $\dfrac{\dfrac{2}{x^2} - \dfrac{3}{x}}{\dfrac{x^2}{5}}$.

Solution The denominators in the complex fraction are x^2, x, and 5. Therefore, the LCD of the complex fraction is $5x^2$. Multiply the numerator and denominator by $5x^2$.

$$\frac{\dfrac{2}{x^2} - \dfrac{3}{x}}{\dfrac{x^2}{5}} = \frac{5x^2\left(\dfrac{2}{x^2} - \dfrac{3}{x}\right)}{5x^2\left(\dfrac{x^2}{5}\right)}$$ *Multiply the numerator and denominator by $5x^2$.*

$$= \frac{5x^2\left(\dfrac{2}{x^2}\right) - 5x^2\left(\dfrac{3}{x}\right)}{5x^2\left(\dfrac{x^2}{5}\right)}$$ *Distributive property*

$$= \frac{10 - 15x}{x^4}$$ *Simplify.*

EXAMPLE 2 Simplify $\dfrac{a + \dfrac{3}{b}}{b + \dfrac{3}{a}}$.

Solution Multiply the numerator and denominator of the complex fraction by its LCD, ab.

$$\frac{a + \dfrac{3}{b}}{b + \dfrac{3}{a}} = \frac{ab\left(a + \dfrac{3}{b}\right)}{ab\left(b + \dfrac{3}{a}\right)}$$

Multiply the numerator and denominator by ab.

$$= \frac{a^2b + 3a}{ab^2 + 3b}$$

Distributive property

$$= \frac{a\cancel{(ab + 3)}}{b\cancel{(ab + 3)}} = \frac{a}{b}$$

Factor and simplify. ✳

NOW TRY EXERCISE 13

EXAMPLE 3 Simplify $\dfrac{a^{-1} + ab^{-2}}{ab^{-2} - a^{-2}b^{-1}}$.

Solution First rewrite each expression without negative exponents.

$$\frac{a^{-1} + ab^{-2}}{ab^{-2} - a^{-2}b^{-1}} = \frac{\dfrac{1}{a} + \dfrac{a}{b^2}}{\dfrac{a}{b^2} - \dfrac{1}{a^2b}}$$

$$= \frac{a^2b^2\left(\dfrac{1}{a} + \dfrac{a}{b^2}\right)}{a^2b^2\left(\dfrac{a}{b^2} - \dfrac{1}{a^2b}\right)}$$

Multiply the numerator and denominator by a^2b^2, the LCD of the complex fraction.

$$= \frac{\cancel{a^2}b^2\left(\dfrac{1}{\cancel{a}}\right) + a^2\cancel{b^2}\left(\dfrac{a}{\cancel{b^2}}\right)}{a^2\cancel{b^2}\left(\dfrac{a}{\cancel{b^2}}\right) - \cancel{a^2}\cancel{b^2}\left(\dfrac{1}{\cancel{a^2}\cancel{b}}\right)}$$

Distributive property

$$= \frac{ab^2 + a^3}{a^3 - b}$$ ✳

NOW TRY EXERCISE 37

In Example 3, although we could factor an a from both terms in the numerator of the answer, we could not simplify the answer further by dividing out common factors. So, we leave the answer in the form given.

3 Simplify Complex Fractions by Simplifying the Numerator and Denominator

Complex fractions can also be simplified as follows:

> **To Simplify a Complex Fraction by Simplifying the Numerator and the Denominator**
> 1. Add or subtract as necessary to get one rational expression in the numerator.
> 2. Add or subtract as necessary to get one rational expression in the denominator.
> 3. Invert the denominator of the complex fraction and multiply by the numerator of the complex fraction.
> 4. Simplify when possible.

Example 4 will show how Example 1 can be simplified by this second method.

EXAMPLE 4 Simplify $\dfrac{\dfrac{2}{x^2} - \dfrac{3}{x}}{\dfrac{x^2}{5}}$.

Solution Subtract the fractions in the numerator to get one rational expression in the numerator. The common denominator of the fractions in the numerator is x^2.

$$\frac{\dfrac{2}{x^2} - \dfrac{3}{x}}{\dfrac{x^2}{5}} = \frac{\dfrac{2}{x^2} - \dfrac{3}{x} \cdot \dfrac{x}{x}}{\dfrac{x^2}{5}} \qquad \textit{Obtain common denominator in numerator.}$$

$$= \frac{\dfrac{2}{x^2} - \dfrac{3x}{x^2}}{\dfrac{x^2}{5}}$$

$$= \frac{\dfrac{2 - 3x}{x^2}}{\dfrac{x^2}{5}}$$

$$= \frac{2 - 3x}{x^2} \cdot \frac{5}{x^2} \qquad \textit{Invert denominator and multiply.}$$

$$= \frac{5(2 - 3x)}{x^4}$$

$$\text{or} \quad \frac{10 - 15x}{x^4}$$

This is the same answer obtained in Example 1.

HELPFUL HINT

Some students prefer to use the second method when the complex fraction consists of a single fraction over a single fraction, such as

$$\frac{\dfrac{x + 3}{18}}{\dfrac{x - 8}{6}}$$

For more complex fractions, many students prefer the first method because you do not have to add fractions.

NOW TRY EXERCISE 7

Concept/Writing Exercises

1. What is a complex fraction?

2. We have indicated two procedures for evaluating complex fractions. Which procedure do you prefer? Why?

Practice the Skills

Simplify.

🔒 **3.** $\dfrac{\dfrac{15a}{b^2}}{\dfrac{b^3}{5}}$

4. $\dfrac{\dfrac{10x^2y^4}{3z^3}}{\dfrac{5xy}{9z^5}}$

5. $\dfrac{1 - \dfrac{x}{y}}{3x}$

6. $\dfrac{\dfrac{x^2y}{4}}{\dfrac{2}{x}}$

7. $\dfrac{\dfrac{36x^4}{5y^4z^5}}{\dfrac{9xy^2}{15z^5}}$

8. $\dfrac{x + \dfrac{7}{y}}{1 + \dfrac{x}{y}}$

9. $\dfrac{x - \dfrac{x}{y}}{\dfrac{3 + x}{y}}$

10. $\dfrac{\dfrac{4}{x} + \dfrac{2}{x^2}}{2 + \dfrac{1}{x}}$

🔒 **11.** $\dfrac{\dfrac{2}{a} + \dfrac{1}{2a}}{a + \dfrac{a}{2}}$

12. $\dfrac{4 - \dfrac{1}{y}}{2 - \dfrac{1}{y}}$

13. $\dfrac{\dfrac{x}{y} - \dfrac{y}{x}}{\dfrac{x + y}{x}}$

14. $\dfrac{\dfrac{a^2}{b} - b}{\dfrac{b^2}{a} - a}$

15. $\dfrac{\dfrac{1}{m} + \dfrac{2}{m^2}}{2 + \dfrac{1}{m^2}}$

16. $\dfrac{\dfrac{a}{b} - 5}{\dfrac{-a}{b} + 5}$

17. $\dfrac{\dfrac{4x + 8}{3x^2}}{\dfrac{4x^3}{6}}$

18. $\dfrac{\dfrac{x^2 - y^2}{x}}{\dfrac{x + y}{x^3}}$

19. $\dfrac{\dfrac{a}{a + 1} - 1}{\dfrac{2a + 1}{a - 1}}$

20. $\dfrac{\dfrac{x}{4} - \dfrac{1}{x}}{1 + \dfrac{x + 4}{x}}$

🔒 **21.** $\dfrac{1 + \dfrac{x}{x + 1}}{\dfrac{2x + 1}{x - 1}}$

22. $\dfrac{\dfrac{2}{x - 1} + 2}{\dfrac{2}{x + 1} - 2}$

23. $\dfrac{\dfrac{a + 1}{a - 1} + \dfrac{a - 1}{a + 1}}{\dfrac{a + 1}{a - 1} - \dfrac{a - 1}{a + 1}}$

24. $\dfrac{\dfrac{a - 2}{a + 2} - \dfrac{a + 2}{a - 2}}{\dfrac{a - 2}{a + 2} + \dfrac{a + 2}{a - 2}}$

🔒 **25.** $\dfrac{\dfrac{5}{5 - x} + \dfrac{6}{x - 5}}{\dfrac{3}{x} + \dfrac{2}{x - 5}}$

26. $\dfrac{\dfrac{2}{m} + \dfrac{1}{m^2} + \dfrac{3}{m - 1}}{\dfrac{2}{m - 1}}$

27. $\dfrac{\dfrac{3}{x^2} - \dfrac{1}{x} + \dfrac{2}{x - 2}}{\dfrac{1}{x}}$

28. $\dfrac{\dfrac{2}{a^2 - 3a + 2} + \dfrac{2}{a^2 - a - 2}}{\dfrac{2}{a^2 - 1} + \dfrac{2}{a^2 + 4a + 3}}$

29. $\dfrac{\dfrac{2}{x^2 + x - 20} + \dfrac{3}{x^2 - 6x + 8}}{\dfrac{2}{x^2 + 3x - 10} + \dfrac{3}{x^2 + 2x - 24}}$

30. $\dfrac{\dfrac{1}{x^2 + 5x + 4} + \dfrac{2}{x^2 + 2x - 8}}{\dfrac{2}{x^2 - x - 2} + \dfrac{1}{x^2 - 5x + 6}}$

Simplify.

31. $3a^{-2} + b$

32. $2a^{-2} + b^{-1}$

33. $(a^{-1} + b^{-1})^{-1}$

34. $\dfrac{a^{-1} + b^{-1}}{\dfrac{4}{ab}}$

🔒 **35.** $\dfrac{a^{-1} + 1}{b^{-1} - 1}$

36. $\dfrac{\dfrac{2a}{b} + a^{-1}}{\dfrac{b}{a} + a^{-1}}$

37. $\dfrac{x^{-1} - y^{-1}}{x^{-1} + y^{-1}}$

38. $\dfrac{x^{-2} + \dfrac{3}{x}}{3x^{-1} + x^{-2}}$

39. $\dfrac{a^{-1} + b^{-1}}{(a + b)^{-1}}$

40. $\dfrac{3a^{-1} - b^{-1}}{(a - b)^{-1}}$

41. $2x^{-1} - (3y)^{-1}$

42. $\dfrac{\dfrac{4}{x} + \dfrac{1}{y}}{(x - y)^{-1}}$

43. $\dfrac{\dfrac{2}{xy} - \dfrac{3}{y} + \dfrac{5}{x}}{3x^{-1} - 4y^{-2}}$

44. $\dfrac{4m^{-1} + 3n^{-1} + (2mn)^{-1}}{\dfrac{5}{m} + \dfrac{9}{n}}$

Problem Solving

For Exercises 45–48, the area and width of each rectangle are given. Find the length, l, by dividing the area, A, by the width, w.

45.

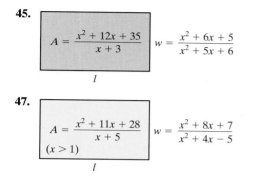

$A = \dfrac{x^2 + 12x + 35}{x + 3}$ $w = \dfrac{x^2 + 6x + 5}{x^2 + 5x + 6}$

l

46.

$A = \dfrac{x^2 + 10x + 16}{x + 4}$ $w = \dfrac{x^2 + 11x + 24}{x^2 + 3x - 4}$
$(x > 1)$

l

47.

$A = \dfrac{x^2 + 11x + 28}{x + 5}$ $w = \dfrac{x^2 + 8x + 7}{x^2 + 4x - 5}$
$(x > 1)$

l

48.

$A = \dfrac{x^2 + 17x + 72}{x + 3}$ $w = \dfrac{x^2 + 11x + 18}{x^2 + x - 6}$
$(x > 2)$

l

49. *Automobile Jack* The efficiency of a jack, E, is given by the formula

$$E = \dfrac{\dfrac{1}{2}h}{h + \dfrac{1}{2}}$$

where h is determined by the pitch of the jack's thread.

Pitch

Determine the efficiency of a jack whose values of h are:

a) $\dfrac{2}{3}$ **b)** $\dfrac{4}{5}$

50. *Resistors* If two resistors with resistances R_1 and R_2 are connected in parallel, their combined resistance, R_T, can be found from the formula

$$R_T = \dfrac{1}{\dfrac{1}{R_1} + \dfrac{1}{R_2}}$$

Simplify the right side of the formula.

51. *Resistors* If three resistors with resistances R_1, R_2, and R_3 are connected in parallel, their combined resistance, R_T, can be found from the formula

$$R_T = \dfrac{1}{\dfrac{1}{R_1} + \dfrac{1}{R_2} + \dfrac{1}{R_3}}$$

Simplify the right side of this formula.

52. *Optics* A formula used in the study of optics is

$$f = (p^{-1} + q^{-1})^{-1}$$

where p is the object's distance from a lens, q is the image distance from the lens, and f is the focal length of the lens. Express the right side of the formula without any negative exponents.

53. If $f(x) = \dfrac{1}{x}$, find $f(f(a))$.

54. If $f(x) = \dfrac{2}{x + 2}$, find $f(f(a))$.

Challenge Problems

For each function, find $\dfrac{f(a + h) - f(a)}{h}$.

55. $f(x) = \dfrac{1}{x}$

56. $f(x) = \dfrac{7}{x}$

57. $f(x) = \dfrac{1}{x + 1}$

58. $f(x) = \dfrac{4}{x - 1}$

59. $f(x) = \dfrac{1}{x^2}$

60. $f(x) = \dfrac{5}{x^2}$

Simplify.

61. $\dfrac{1}{2a + \dfrac{1}{2a + \dfrac{1}{2a}}}$

62. $\dfrac{1}{x + \dfrac{1}{x + \dfrac{1}{x + 1}}}$

63. $\dfrac{1}{2 + \dfrac{1}{2 + \dfrac{1}{2}}}$

Cumulative Review Exercises

[1.4] **64.** Evaluate $\dfrac{\left|-\dfrac{3}{9}\right| - \left(-\dfrac{5}{9}\right) \cdot \left|-\dfrac{3}{8}\right|}{|-5 - (-3)|}$.

[2.5] **65.** Solve $\dfrac{3}{5} < \dfrac{-x - 5}{3} < 6$ and give the solution in interval notation.

[2.6] **66.** Solve $|x - 1| = |2x - 4|$.

[3.5] **67.** Determine if the two lines represented by the following equations are parallel, perpendicular, or neither.

$$6x + 2y = 8$$
$$4x - 9 = -y$$

6.4 SOLVING RATIONAL EQUATIONS

SSM Study Guide CD/Video

MathPro 4/5 PH Math Tutor Center prenhall.com/Angel

1 Solve rational equations.
2 Check solutions.
3 Solve proportions.
4 Solve problems involving rational functions.
5 Solve applications using rational expressions.
6 Solve for a variable in a formula containing rational expressions.

1 Solve Rational Equations

In Sections 6.1 through 6.3 we presented techniques to add, subtract, multiply, and divide rational expressions. In this section, we present a method for solving rational equations. A **rational equation** is an equation that contains at least one rational expression.

To Solve Rational Equations

1. Determine the LCD of all rational expressions in the equation.
2. Multiply *both* sides of the equation by the LCD. This will result in every term in the equation being multiplied by the LCD.
3. Remove any parentheses and combine like terms on each side of the equation.
4. Solve the equation using the properties discussed in earlier sections.
5. Check the solution in the *original* equation.

In step 2, we multiply both sides of the equation by the LCD to eliminate fractions. In some examples we will not show the check to save space.

EXAMPLE 1   Solve $\dfrac{3x}{4} + \dfrac{1}{2} = \dfrac{x-1}{2}$.

Solution   Multiply both sides of the equation by the LCD, 4. Then use the distributive property, which results in each term in the equation being multiplied by the LCD.

$$4\left(\frac{3x}{4} + \frac{1}{2}\right) = \frac{x-1}{2} \cdot 4 \qquad \textit{Multiply both sides by 4.}$$

$$4\left(\frac{3x}{4}\right) + 4\left(\frac{1}{2}\right) = 2(x-1) \qquad \textit{Distributive property}$$

$$3x + 2 = 2x - 2$$

$$x + 2 = -2$$

$$x = -4 \qquad \text{\LARGE ✳}$$

2   Check Solutions

Whenever a variable appears in any denominator, you must check your apparent solution in the original equation. When checking, if an apparent solution makes any denominator equal to 0, that value is not a solution to the equation. Such values are called **extraneous roots** or **extraneous solutions**. An extraneous root is a number obtained when solving an equation that is not a solution to the original equation.

EXAMPLE 2   Solve $2 - \dfrac{4}{x} = \dfrac{1}{3}$.

Solution   Multiply both sides of the equation by the LCD, $3x$.

$$3x\left(2 - \frac{4}{x}\right) = \left(\frac{1}{3}\right) \cdot 3x \qquad \textit{Multiply both sides by 3x.}$$

$$3x(2) - 3x\left(\frac{4}{\cancel{x}}\right) = \left(\frac{1}{\cancel{3}}\right)\cancel{3}x \qquad \textit{Distributive property}$$

$$6x - 12 = x$$

$$5x - 12 = 0$$

$$5x = 12$$

$$x = \frac{12}{5}$$

Check

$$2 - \frac{4}{x} = \frac{1}{3}$$

$$2 - \frac{4}{(12/5)} \stackrel{?}{=} \frac{1}{3} \qquad \textit{Substitute } \frac{12}{5} \textit{ for x.}$$

$$2 - \frac{20}{12} \stackrel{?}{=} \frac{1}{3}$$

$$\frac{1}{3} = \frac{1}{3} \qquad \textit{True} \qquad \text{\LARGE ✳}$$

EXAMPLE 3 Solve $x + \dfrac{6}{x} = -7$.

Solution

$$\boxed{x} \cdot \left(x + \dfrac{6}{x}\right) = -7 \cdot \boxed{x} \qquad \textit{Multiply both sides by the LCD, x.}$$

$$x(x) + x\left(\dfrac{6}{\cancel{x}}\right) = -7x \qquad \textit{Distributive property}$$

$$x^2 + 6 = -7x$$

$$x^2 + 7x + 6 = 0$$

$$(x + 1)(x + 6) = 0$$

$$x + 1 = 0 \qquad \text{or} \qquad x + 6 = 0$$

$$x = -1 \qquad\qquad\qquad x = -6$$

NOW TRY EXERCISE 35 Checks of -1 and -6 will show that they are both solutions to the equation. ✳

EXAMPLE 4 Solve $\dfrac{3x}{x^2 - 4} + \dfrac{1}{x - 2} = \dfrac{2}{x + 2}$.

Solution First factor the denominator $x^2 - 4$, then find the LCD.

$$\dfrac{3x}{(x + 2)(x - 2)} + \dfrac{1}{x - 2} = \dfrac{2}{x + 2}$$

The LCD is $(x + 2)(x - 2)$. Multiply both sides of the equation by the LCD, and then use the distributive property. This process will eliminate the fractions from the equation.

$$(x + 2)(x - 2) \cdot \left[\dfrac{3x}{(x + 2)(x - 2)} + \dfrac{1}{x - 2}\right] = \dfrac{2}{x + 2} \cdot (x + 2)(x - 2)$$

$$\cancel{(x + 2)}\cancel{(x - 2)} \cdot \dfrac{3x}{\cancel{(x + 2)}\cancel{(x - 2)}} + (x + 2)\cancel{(x - 2)} \cdot \dfrac{1}{\cancel{x - 2}} = \dfrac{2}{\cancel{x + 2}} \cdot \cancel{(x + 2)}(x - 2)$$

$$3x + (x + 2) = 2(x - 2)$$

$$4x + 2 = 2x - 4$$

$$2x + 2 = -4$$

$$2x = -6$$

$$x = -3$$

NOW TRY EXERCISE 39 A check will show that -3 is the solution. ✳

EXAMPLE 5 Solve the equation $\dfrac{22}{2p^2 - 9p - 5} - \dfrac{3}{2p + 1} = \dfrac{2}{p - 5}$.

Solution Factor the denominator, then determine the LCD.

$$\dfrac{22}{(2p + 1)(p - 5)} - \dfrac{3}{2p + 1} = \dfrac{2}{p - 5}$$

Multiply both sides of the equation by the LCD, $(2p + 1)(p - 5)$.

$$\cancel{(2p + 1)}\cancel{(p - 5)} \cdot \dfrac{22}{\cancel{(2p + 1)}\cancel{(p - 5)}} - \cancel{(2p + 1)}(p - 5) \cdot \dfrac{3}{\cancel{2p + 1}} = (2p + 1)\cancel{(p - 5)} \cdot \dfrac{2}{\cancel{p - 5}}$$

$$22 - 3(p - 5) = 2(2p + 1)$$

$$22 - 3p + 15 = 4p + 2$$
$$37 - 3p = 4p + 2$$
$$35 = 7p$$
$$5 = p$$

The solution appears to be 5. However, since a variable appears in a denominator, this solution must be checked.

Check

$$\frac{22}{2p^2 - 9p - 5} - \frac{3}{2p + 1} = \frac{2}{p - 5}$$

$$\frac{22}{2(5)^2 - 9(5) - 5} - \frac{3}{2(5) + 1} \stackrel{?}{=} \frac{2}{5 - 5} \qquad \textit{Substitute 5 for p.}$$

$$\textit{Undefined} \longrightarrow \frac{22}{0} - \frac{3}{11} = \frac{2}{0} \longleftarrow \textit{Undefined}$$

NOW TRY EXERCISE 43   Since 5 makes a denominator 0 and division by 0 is undefined, 5 is an extraneous solution. Therefore, you should write "**no solution**" as your answer.   ✳

In Example 5, the only possible solution is 5. However, when $p = 5$, the denominator of $2/(p - 5)$ is 0. Therefore, 5 cannot be a solution. We did not actually have to show the complete check, but we did so in this example for clarity.

HELPFUL HINT

> Remember, whenever you solve an equation where a variable appears in any denominator, you must check the apparent solution to make sure it is not an extraneous solution. If the apparent solution makes any denominator 0, then it is an extraneous solution and not a true solution to the equation.

3   Solve Proportions

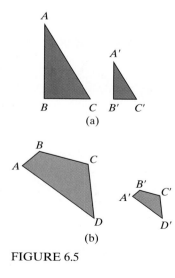

FIGURE 6.5

Proportions are equations of the form $\frac{a}{b} = \frac{c}{d}$. Proportions are one type of rational equation. Proportions may be solved by *cross multiplication* as follows. If $\frac{a}{b} = \frac{c}{d}$, then $ad = bc$, $b \neq 0$, $d \neq 0$. Proportions may also be solved by multiplying both sides of the proportion by the least common denominator. In Examples 6 and 7, we solve proportions by multiplying both sides by the LCD. We then ask you to determine the solutions, if possible, using cross multiplication. *When solving a proportion where the denominator of one or more of the ratios contains a variable, you must check to make sure that your solution is not extraneous.*

Proportions are often used when working with similar figures. **Similar figures** are figures whose corresponding angles are equal and whose corresponding sides are in proportion. Figure 6.5 illustrates two sets of similar figures.

In Figure 6.5a, the ratio of the length of side AB to the length of side BC is the same as the ratio of the length of side $A'B'$ to the length of side $B'C'$. That is,

$$\frac{AB}{BC} = \frac{A'B'}{B'C'}$$

In a pair of similar figures, if the length of a side is unknown, it can often be found by using proportions, as illustrated in Example 6.

EXAMPLE 6   **Similar Triangles**   Triangles ABC and $A'B'C'$ in Figure 6.6 are similar figures. Find the length of sides AB and $B'C'$.

Solution   We can set up a proportion and then solve for x. Then we can find the lengths.

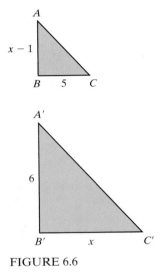

FIGURE 6.6

$$\frac{AB}{BC} = \frac{A'B'}{B'C'}$$

$$\frac{x-1}{5} = \frac{6}{x}$$

$$5x \cdot \frac{x-1}{5} = \frac{6}{x} \cdot 5x \qquad \text{Multiply both sides by the LCD, } 5x.$$

$$x(x-1) = 6 \cdot 5$$

$$x^2 - x = 30$$

$$x^2 - x - 30 = 0$$

$$(x-6)(x+5) = 0 \qquad \text{Factor the trinomial.}$$

$$x - 6 = 0 \qquad \text{or} \qquad x + 5 = 0$$

$$x = 6 \qquad\qquad x = -5$$

Since the length of the side of a triangle cannot be a negative number, -5 is not a possible answer. Substituting 6 for x, we see that the length of side $B'C'$ is 6 and the length of side AB is $6 - 1$ or 5.

Check

$$\frac{AB}{BC} = \frac{A'B'}{B'C'}$$

$$\frac{5}{5} \overset{?}{=} \frac{6}{6}$$

NOW TRY EXERCISE 49

$$1 = 1 \qquad \text{True}$$

∗

The answer to Example 6 could also be obtained using cross multiplication. Try solving Example 6 using cross multiplication now.

EXAMPLE 7   Solve $\dfrac{x^2}{x-4} = \dfrac{16}{x-4}$.

Solution   This equation is a proportion. We will solve this equation by multiplying both sides of the equation by the LCD, $x - 4$.

$$(x-4) \cdot \frac{x^2}{x-4} = \frac{16}{x-4} \cdot (x-4)$$

$$x^2 = 16$$

$$x^2 - 16 = 0$$

$$(x+4)(x-4) = 0 \qquad\qquad \text{Factor the difference of two squares.}$$

$$x + 4 = 0 \qquad \text{or} \qquad x - 4 = 0$$

$$x = -4 \qquad\qquad x = 4$$

Check $x = -4$ $\qquad\qquad\qquad$ $x = 4$

$$\frac{x^2}{x-4} = \frac{16}{x-4} \qquad\qquad \frac{x^2}{x-4} = \frac{16}{x-4}$$

$$\frac{(-4)^2}{-4-4} \overset{?}{=} \frac{16}{-4-4} \qquad\qquad \frac{4^2}{4-4} \overset{?}{=} \frac{16}{4-4}$$

$$\frac{16}{-8} \overset{?}{=} \frac{16}{-8} \qquad\qquad \frac{16}{0} \overset{?}{=} \frac{16}{0} \quad\longleftarrow \textit{Undefined}$$

$$-2 = -2 \qquad\qquad\qquad \textit{True}$$

Since $x = 4$ results in a denominator of 0, 4 is *not* a solution to the equation. It is an extraneous root. The only solution to the equation is -4. ✳

In Example 7, what would you obtain if you began by cross multiplying? Try it and see.

4 Solve Problems Involving Rational Functions

Now we will work a problem that involves a rational function.

EXAMPLE 8 Consider the function $f(x) = x - \dfrac{2}{x}$. Find all a for which $f(a) = 1$.

Solution Since $f(a) = a - \dfrac{2}{a}$, we need to find all values for which $a - \dfrac{2}{a} = 1, a \neq 0$. We begin by multiplying both sides of the equation by a, the LCD.

$$a \cdot \left(a - \frac{2}{a} \right) = a \cdot 1$$

$$a^2 - 2 = a$$

$$a^2 - a - 2 = 0$$

$$(a - 2)(a + 1) = 0$$

$$a - 2 = 0 \quad\text{or}\quad a + 1 = 0$$

$$a = 2 \qquad\qquad a = -1$$

Check $f(x) = x - \dfrac{2}{x}$

$$f(2) = 2 - \frac{2}{2} = 2 - 1 = 1$$

$$f(-1) = -1 - \frac{2}{(-1)} = -1 + 2 = 1$$

For $a = 2$ or $a = -1, f(a) = 1$. ✳

We used $f(x) = x - \dfrac{2}{x}$ in Example 8. Figure 6.7 shows the graph of $f(x) = x - \dfrac{2}{x}$. In this course you will not have to graph functions like this. We illustrate this graph to reinforce the answer obtained in Example 8.

Notice the function is undefined at $x = 0$. Also notice that when $x = -1$ or $x = 2$, it appears that $f(x) = 1$. This was what we should have expected from the results obtained in Example 8.

Example 8 also could have been solved using your graphing calculator by setting $y_1 = x - \dfrac{2}{x}$ and $y_2 = 1$ and finding the x-coordinate of the intersections of the two graphs.

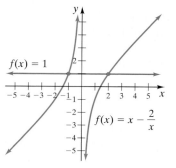

FIGURE 6.7

5 Solve Applications Using Rational Expressions

Now let's look at an application of rational equations.

EXAMPLE 9   **Total Resistance**   In electronics, the total resistance, R_T, of resistors connected in a parallel circuit is determined by the formula

$$\frac{1}{R_T} = \frac{1}{R_1} + \frac{1}{R_2} + \frac{1}{R_3} + \cdots + \frac{1}{R_n}$$

where $R_1, R_2, R_3, \ldots, R_n$ are the resistances of the individual resistors (measured in ohms, symbolized Ω) in the circuit. Find the total resistance if two resistors, one of 100 ohms and the other of 300 ohms, are connected in a parallel circuit.

Solution   Since there are only two resistances, use the formula

$$\frac{1}{R_T} = \frac{1}{R_1} + \frac{1}{R_2}$$

Let $R_1 = 100$ ohms and $R_2 = 300$ ohms; then

$$\frac{1}{R_T} = \frac{1}{100} + \frac{1}{300}$$

Multiply both sides of the equation by the LCD, $300R_T$.

$$300R_T \cdot \frac{1}{R_T} = 300R_T\left(\frac{1}{100} + \frac{1}{300}\right)$$

$$300R_T \cdot \frac{1}{R_T} = \overset{3}{300}R_T\left(\frac{1}{100}\right) + 300R_T\left(\frac{1}{300}\right)$$

$$300 = 3R_T + R_T$$

$$300 = 4R_T$$

$$R_T = \frac{300}{4} = 75$$

Thus, the total resistance of the parallel circuit is 75 ohms. Notice that the resistors actually have less resistance when connected in a parallel circuit than separately. ✳

6 Solve for a Variable in a Formula Containing Rational Expressions

Sometimes you may need to solve for a variable in a formula where the variable you are solving for occurs in more than one term. When this happens, it may be possible to solve for the variable by using factoring. To do so, collect all the terms containing the variable you are solving for on one side of the equation and all other terms on the other side of the equation. Then factor out the variable you are solving for. This process is illustrated in Examples 10 through 12.

EXAMPLE 10   **Optics**   A formula used in the study of optics is $\frac{1}{p} + \frac{1}{q} = \frac{1}{f}$. In the formula, p is the distance of an object from a lens or a mirror, q is the distance of the image from the lens or mirror, and f is the focal length of the lens or mirror. For people wearing glasses, the image distance is the distance from the lens to their retina. Solve this formula for f.

Solution

Our goal is to isolate the variable f. We begin by multiplying both sides of the equation by the least common denominator, pqf, to eliminate fractions.

$$\frac{1}{p} + \frac{1}{q} = \frac{1}{f}$$

$$pqf\left(\frac{1}{p} + \frac{1}{q}\right) = pqf\left(\frac{1}{f}\right) \quad \text{Multiply both sides by the LCD, } pqf.$$

$$pqf\left(\frac{1}{p}\right) + pqf\left(\frac{1}{q}\right) = pqf\left(\frac{1}{f}\right) \quad \text{Distributive property}$$

$$qf + pf = pq \quad \text{Simplify.}$$

$$f(q + p) = pq \quad \text{Factor out } f.$$

$$\frac{f(q+p)}{q+p} = \frac{pq}{q+p} \quad \text{Divide both sides by } q + p.$$

$$f = \frac{pq}{q+p} \quad \text{or} \quad f = \frac{pq}{p+q}$$

NOW TRY EXERCISE 69

EXAMPLE 11

Banking A formula used in banking is $A = P + Prt$, where A represents the amount that must be repaid to the bank when P dollars are borrowed at simple interest rate, r, for time, t, in years. Solve this equation for P.

Solution

Since both terms containing P are by themselves on the right side of the equation, we factor P from both terms.

$$A = P + Prt \quad \text{P is in both terms.}$$

$$A = P(1 + rt) \quad \text{Factor out } P.$$

$$\frac{A}{1+rt} = \frac{P(1+rt)}{1+rt} \quad \text{Divide both sides by } 1 + rt \text{ to isolate } P.$$

$$\frac{A}{1+rt} = P$$

Thus, $P = \dfrac{A}{1 + rt}$.

EXAMPLE 12

Physics A formula used for levers in physics is $d = \dfrac{fl}{f + w}$. Solve this formula for f.

Solution

We begin by multiplying both sides of the formula by $f + w$ to clear fractions. Then we will rewrite the expression with all terms containing f on one side of the equal sign and all terms not containing f on the other side of the equal sign.

$$d = \frac{fl}{f + w}$$

$$d(f + w) = \frac{fl}{(f+w)}(f+w) \quad \text{Multiply by } f + w \text{ to clear fractions.}$$

$$d(f + w) = fl$$

$$df + dw = fl \quad \text{Distributive property}$$

$$df - df + dw = fl - df \quad \text{Isolate terms containing } f \text{ on the right side of the equation.}$$

$$dw = fl - df$$
$$dw = f(l - d)$$
Factor out f.
$$\frac{dw}{l - d} = \frac{f(l - d)}{l - d}$$
Isolate f by dividing both sides by l − d.

NOW TRY EXERCISE 79

Thus, $f = \dfrac{dw}{l - d}$.

AVOIDING COMMON ERRORS

After solving equations, as was done in this section, some students forget to keep their common denominator when adding and subtracting rational expressions. Remember, we multiply both sides of an *equation* by the LCD to remove the common denominator. If we are *adding or subtracting rational expressions*, we write the fractions with the LCD then add or subtract the numerators while *keeping the common denominator*.

For example, consider the addition problem

$$\frac{x}{x + 7} + \frac{3}{x + 7}$$

CORRECT

$$\frac{x}{x + 7} + \frac{3}{x + 7} = \frac{x + 3}{x + 7}$$

INCORRECT

$$\frac{x}{x + 7} + \frac{3}{x + 7} = (x + 7)\left(\frac{x}{x + 7} + \frac{3}{x + 7}\right)$$
$$= x + 3$$

Mathematics in Action

The Doppler Effect

We have all heard it. The pitch of an ambulance siren goes higher as the vehicle approaches, then the pitch goes lower as the vehicle races away. This change is the result of a shift in the frequency of sound waves called the Doppler Effect. This effect can be represented by the rational equation

$$f_d = f_o \frac{v \pm v_o}{v \pm v_s}$$

where f_d is the Doppler effect (or perceived) frequency, f_o is the original (or actual) frequency, v is the speed of sound, v_o is the speed of the observer, and v_s is the speed of the source. The choice of using the plus (+) or minus (−) sign is made according to the convention that if the source and observer are moving toward each other, the perceived frequency

is higher than the actual frequency. If the source and observer are moving away from each other, the perceived frequency is lower than the actual frequency.

This formula (or a different form of it) lies at the heart of many technologies and devices, some of which affect our daily lives. Here is a short list:

Traffic radar—A device we often refer to as a "radar gun" sends out a radio wave of a known wavelength. Your car is moving and a shift in wavelength occurs as the radio wave bounces off it. Using Doppler calculations, a computer in the device is able to calculate the speed of your car.

Weather radar—Precipitation particles such as rain and snow show a different phase shift when they are moving away from the radar source than when they are approaching the radar source. The result of taking Doppler measurements is more accurate forecasting and more vivid visualizations, such as you see in TV weather reports.

Astronomical measurements—Astronomers use Doppler shifts to calculate how fast stars and other astronomical objects move toward or away from Earth. They do this by measuring the wavelengths of the light these objects emit. An increased wavelength indicates that the astronomical object is moving away from the earth; a decreased wavelength means it is moving toward the earth.

Exercise Set 6.4

Concept/Writing Exercises

1. What is an extraneous root?

2. Under what circumstances is it necessary to check your answers for extraneous roots?

3. Consider the equation $\dfrac{x}{4} - \dfrac{x}{3} = 2$ and the expression $\dfrac{x}{4} - \dfrac{x}{3} + 2$.

   a) What is the first step in solving the equation? Explain what effect the first step will have on the equation.

   b) Solve the equation.

   c) What is the first step in simplifying the expression? Explain what effect the first step has when simplifying the expression.

   d) Simplify the expression.

4. Consider the equation $\dfrac{x}{2} - \dfrac{x}{3} = 3$ and the expression $\dfrac{x}{2} - \dfrac{x}{3} + 3$.

   a) What is the first step in solving the equation?

   b) Solve the equation.

   c) What is the first step in simplifying the expression? Explain what effect the first step has when simplifying the expression.

   d) Simplify the expression.

5. What are similar figures?

6. a) Explain in your own words how to solve a rational equation.

   b) Solve $\dfrac{3}{x-4} + \dfrac{1}{x+4} = \dfrac{4}{x^2-16}$ following your procedure in part **a)**.

7. Tom Kelly solved an equation containing the term $\dfrac{7}{x-3}$ and obtained the answer $x = 3$. Can this be correct? Explain.

8. Geurfino Muldo solved an equation containing the term $\dfrac{21x}{x^2-16}$ and obtained the answer $x = 4$. Can this be correct? Explain.

Practice the Skills

Solve each equation and check your solution.

9. $\dfrac{7}{x} = 1$

10. $\dfrac{2}{x} = 3$

11. $\dfrac{10}{b} = 2$

12. $\dfrac{4x+5}{6} = \dfrac{7}{2}$

13. $\dfrac{1}{4} = \dfrac{z+2}{12}$

14. $\dfrac{a+2}{7} = \dfrac{a-3}{2}$

15. $\dfrac{6x+7}{5} = \dfrac{2x+9}{3}$

16. $\dfrac{y}{10} = 6 - \dfrac{y}{5}$

17. $\dfrac{z}{3} - \dfrac{3z}{4} = -\dfrac{5z}{12}$

18. $\dfrac{1}{4} - \dfrac{w}{5} = \dfrac{5}{4}$

19. $\dfrac{3}{4} - x = 2x$

20. $\dfrac{2}{y} + \dfrac{1}{2} = \dfrac{5}{2y}$

21. $\dfrac{3}{r} + \dfrac{5}{3r} = 1$

22. $\dfrac{x}{2} - \dfrac{x}{6} = \dfrac{1}{6}$

23. $\dfrac{x-1}{x-5} = \dfrac{4}{x-5}$

24. $\dfrac{c+3}{c+1} = \dfrac{3}{2}$

25. $\dfrac{5y-2}{7} = \dfrac{15y-2}{28}$

26. $\dfrac{3}{x+1} = \dfrac{2}{x-2}$

27. $\dfrac{5.6}{-p-6.2} = \dfrac{2}{p}$

28. $\dfrac{4.5}{y-3} = \dfrac{6.9}{y+3}$

29. $\dfrac{m+1}{m+10} = \dfrac{m-2}{m+4}$

30. $\dfrac{x-3}{x+1} = \dfrac{x-6}{x+5}$

31. $x - \dfrac{4}{3x} = -\dfrac{1}{3}$

32. $b - \dfrac{8}{b} = -7$

33. $\dfrac{2x-1}{3} - \dfrac{x}{4} = \dfrac{7.4}{6}$

34. $x + \dfrac{3}{x} = \dfrac{19}{x}$

35. $x + \dfrac{6}{x} = -5$

36. $\dfrac{15}{x} + \dfrac{9x-7}{x+2} = 9$

37. $2 - \dfrac{5}{2b} = \dfrac{2b}{b+1}$

38. $\dfrac{3z-2}{z+1} = 4 - \dfrac{z+2}{z-1}$

39. $\dfrac{1}{w-3} + \dfrac{1}{w+3} = \dfrac{-5}{w^2-9}$

40. $a - \dfrac{a}{4} + \dfrac{a}{5} = 19$

41. $\dfrac{8}{x^2-9} = \dfrac{2}{x-3} - \dfrac{4}{x+3}$

42. $\dfrac{2}{w-5} = \dfrac{22}{2w^2-9w-5} - \dfrac{3}{2w+1}$

43. $\dfrac{y}{2y+2} + \dfrac{2y-16}{4y+4} = \dfrac{2y-3}{y+1}$

44. $\dfrac{3}{x+3} + \dfrac{5}{x+4} = \dfrac{12x+15}{x^2+7x+12}$

45. $\dfrac{1}{x+2} + \dfrac{1}{x-2} = \dfrac{4}{x^2-4}$

46. $\dfrac{4r-1}{r^2+5r-14} = \dfrac{1}{r-2} - \dfrac{2}{r+7}$

47. $\dfrac{5}{x^2+4x+3} + \dfrac{2}{x^2+x-6} = \dfrac{3}{x^2-x-2}$

48. $\dfrac{2}{x^2+2x-8} - \dfrac{1}{x^2+9x+20} = \dfrac{4}{x^2+3x-10}$

For each pair of similar figures, find the length of the two unknown sides (that is, those two sides whose lengths involve the variable x).

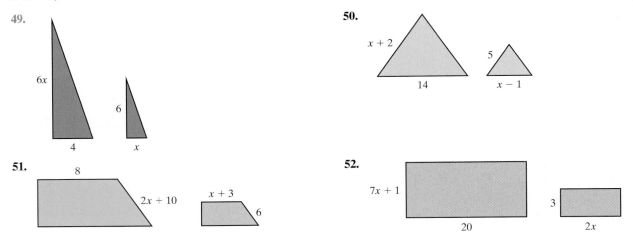

49.

6x, 6, 4, x

50.

x + 2, 14, 5, x − 1

51.

8, 2x + 10, x + 3, 6

52.

7x + 1, 20, 3, 2x

For each rational function given, find all values a for which f(a) has the indicated value.

53. $f(x) = 2x - \dfrac{4}{x}$, $f(a) = -2$

54. $f(x) = \dfrac{x-2}{x+5}$, $f(a) = \dfrac{3}{5}$

55. $f(x) = 3x - \dfrac{5}{x}$, $f(a) = -14$

56. $f(x) = \dfrac{6}{x} + \dfrac{6}{2x}$, $f(a) = 6$

57. $f(x) = \dfrac{x+3}{x+5}$, $f(a) = \dfrac{5}{7}$

58. $f(x) = \dfrac{4}{x} - \dfrac{3}{2x}$, $f(a) = 4$

Solve each formula for the indicated variable.

59. $\dfrac{V_1}{V_2} = \dfrac{P_2}{P_1}$, for P_2 (chemistry)

60. $T_a = \dfrac{T_f}{1-f}$, for f (investment formula)

61. $\dfrac{V_1}{V_2} = \dfrac{P_2}{P_1}$, for V_2 (chemistry)

62. $S = \dfrac{a}{1-r}$, for r (mathematics)

63. $m = \dfrac{y-y_1}{x-x_1}$, for y (slope)

64. $m = \dfrac{y-y_1}{x-x_1}$, for x_1 (slope)

65. $z = \dfrac{x-\bar{x}}{s}$, for x (statistics)

66. $z = \dfrac{x-\bar{x}}{s}$, for s (statistics)

67. $d = \dfrac{fl}{f+w}$, for w (physics)

68. $\dfrac{1}{p} + \dfrac{1}{q} = \dfrac{1}{f}$, for p (optics)

69. $\dfrac{1}{p} + \dfrac{1}{q} = \dfrac{1}{f}$, for q (optics)

70. $\dfrac{1}{R_T} = \dfrac{1}{R_1} + \dfrac{1}{R_2}$, for R_T (electronics)

71. $at_2 - at_1 + v_1 = v_2$, for a (physics)

72. $2P_1 - 2P_2 - P_1 P_c = P_2 P_c$, for P_c (economics)

73. $a_n = a_1 + nd - d$, for d (mathematics)

74. $S_n - S_n r = a_1 - a_1 r^n$, for S_n (mathematics)

75. $F = \dfrac{Gm_1 m_2}{d^2}$, for G (physics)

76. $\dfrac{P_1 V_1}{T_1} = \dfrac{P_2 V_2}{T_2}$, for T_2 (physics)

77. $\dfrac{P_1V_1}{T_1} = \dfrac{P_2V_2}{T_2}$, for T_1 (physics)

78. $A = \dfrac{1}{2}h(a + b)$, for h (mathematics)

79. $\dfrac{S - S_0}{V_0 + gt} = t$, for V_0 (physics)

80. $\dfrac{E}{e} = \dfrac{R + r}{r}$, for e (engineering)

*Simplify each expression in **a**) and solve each equation in **b**).*

81. a) $\dfrac{2}{x - 2} + \dfrac{3}{x^2 - 4}$

    b) $\dfrac{2}{x - 2} + \dfrac{3}{x^2 - 4} = 0$

82. a) $\dfrac{4}{x - 3} + \dfrac{5}{2x - 6} + \dfrac{1}{2}$

    b) $\dfrac{4}{x - 3} + \dfrac{5}{2x - 6} = \dfrac{1}{2}$

83. a) $\dfrac{b + 3}{b} - \dfrac{b + 4}{b + 5} - \dfrac{15}{b^2 + 5b}$

    b) $\dfrac{b + 3}{b} - \dfrac{b + 4}{b + 5} = \dfrac{15}{b^2 + 5b}$

84. a) $\dfrac{4x + 3}{x^2 + 11x + 30} - \dfrac{3}{x + 6} + \dfrac{2}{x + 5}$

    b) $\dfrac{4x + 3}{x^2 + 11x + 30} - \dfrac{3}{x + 6} = \dfrac{2}{x + 5}$

Problem Solving

85. What restriction must be added to the statement "If $ac = bc$, then $a = b$"? Explain.

86. Consider $\dfrac{x - 2}{x - 5} = \dfrac{3}{x - 5}$.

    a) Solve the equation.

    b) If you subtract $\dfrac{3}{x - 5}$ from both sides of the equation, you get $\dfrac{x - 2}{x - 5} - \dfrac{3}{x - 5} = 0$. Simplify the difference on the left side of the equation and solve the equation.

    c) Use the information obtained in parts **a)** and **b)** to construct another equation that has no solution.

87. Below are two graphs. One is the graph of $f(x) = \dfrac{x^2 - 9}{x - 3}$ and the other is the graph of $g(x) = x + 3$. Determine which graph is $f(x)$ and which graph is $g(x)$. Explain how you determined your answer.

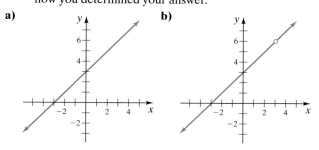

a)

b)

88. *Tax-free Investment* The formula $T_a = \dfrac{T_f}{1 - f}$ can be used to find the equivalent taxable yield, T_a, of a tax-free investment, T_f. In this formula, f is the individual's federal income tax bracket. Tran Du is in the 26% income tax bracket.

    a) Determine the equivalent taxable yield of an 8% tax-free-investment for Tran.

    b) Solve this equation for T_f.

    c) Determine the equivalent tax-free yield for a 5% taxable investment for Tran.

89. *Insurance* When a homeowner purchases a homeowner's insurance policy in which the dwelling is insured for less than 80% of the replacement value, the insurance company will not reimburse the homeowner in total for their loss. The following formula is used to determine the insurance company payout, I, when the dwelling is insured for less than 80% of the replacement value.

$$I = \dfrac{AC}{0.80R}$$

In the formula, A is the amount of insurance carried, C is the cost of repairing the damaged area, and R is the replacement value of the home. (There are some exceptions to when this formula is used.)

    a) Suppose Jan Burdett had a fire in her kitchen that caused $10,000 worth of damage. If she carried $50,000 of insurance on a home with $100,000 replacement value, what would the insurance company pay for the repairs?

    b) Solve this formula for R, the replacement value.

90. *Average Velocity* Average velocity is defined as a change in distance divided by a change in time, or

$$v = \dfrac{d_2 - d_1}{t_2 - t_1}$$

This formula can be used when an object at distance d_1 at time t_1 travels to a distance d_2 at time t_2.

    a) Assume $t_1 = 2$ hours, $d_1 = 118$ miles, $t_2 = 9$ hours, and $d_2 = 412$ miles. Find the average velocity.

    b) Solve the formula for t_2.

91. *Average Acceleration* Average acceleration is defined as a change in velocity divided by a change in time, or

$$a = \frac{v_2 - v_1}{t_2 - t_1}$$

This formula can be used when an object at velocity v_1 at time t_1 accelerates (or decelerates) to velocity v_2 at time t_2

a) Assume $v_1 = 20$ feet per minute, $t_1 = 20$ minutes, $v_2 = 60$ feet per minute, and $t_2 = 22$ minutes. Find the average acceleration. The units will be ft/min².

b) Solve the formula for t_1.

92. *Rate of Discount* The *rate of discount*, P, expressed as a fraction or decimal, can be found by the formula

$$P = 1 - \frac{R - D}{R}$$

where R is the regular price of an item and D is the discount (the amount saved off the regular price).

a) Determine the rate of discount on a purse with a regular price of $39.99 that is on sale for $30.99

b) Solve the formula above for D.

c) Solve the formula above for R.

93. *Economics* A formula for break-even analysis in economics is

$$Q = \frac{F + D}{R - V}$$

This formula is used to determine the number of units, Q, in an apartment building that must be rented for an investor to break even. In the formula, F is the monthly fixed expenses for the entire building, D is the monthly debt payment on the building, R is the rent per unit, and V is the variable expense per unit.

Assume an investor is considering investing in ABC Properties, a 50-unit building. Each two-bedroom apartment can be rented for $500 per month. Variable expenses are estimated to be $200 per month per unit, fixed expenses are estimated to be $2500 per month, and the monthly debt payment is $8000. How many apartments must be rented for ABC to break even?

Refer to Example 9 for Exercises 94–96.

94. *Total Resistance* What is the total resistance in the circuit if resistors of 200 ohms and 600 ohms are connected in parallel?

95. *Total Resistance* What is the total resistance in the circuit if resistors of 300 ohms, 500 ohms, and 3000 ohms are connected in parallel?

96. *Total Resistance* Three resistors of identical resistance are to be connected in parallel. What should be the resistance of each resistor if the circuit is to have a total resistance of 700 ohms?

Refer to Example 10 for Exercises 97 and 98.

97. *Focal Length* In a slide or movie projector, the film acts as the object whose image is projected on a screen. If a 100-mm-focal length (or 0.10 meter) lens is to project an image on a screen 7.5 meters away, how far from the lens should the film be?

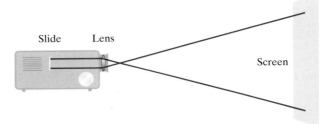

98. *Curved Mirror* A diamond ring is placed 20.0 cm from a concave (curved in) mirror whose focal length is 15.0 cm. Find the position of the image (or the *image distance*).

99. *Investments* Some investments, such as certain municipal bonds and municipal bond funds, are not only federally tax free but are also state and county or city tax free. When you wish to compare a taxable investment, T_a, with an investment that is federal, state, and county tax free, T_f, you can use the formula

$$T_a = \frac{T_f}{1 - [f + (s + c)(1 - f)]}$$

In the formula, s is your state tax bracket, c is your county or local tax bracket, and f is your federal income tax bracket. Howard Levy, who lives in Detroit, Michigan, is in a 4.6% state tax bracket, a 3% city tax bracket, and a 33% federal tax bracket. He is choosing between the Fidelity Michigan Triple Tax-Free Money Market Portfolio yielding 6.01% and the Fidelity Taxable Cash Reserve Money Market Fund yielding 7.68%.

a) Using his tax brackets, determine the taxable equivalent of the 6.01% tax-free yield.

b) Which investment should Howard make? Explain your answer.

100. *Periods of Planets* The synodic period of Mercury is the time required for swiftly moving Mercury to gain one lap on Earth in their orbits around the Sun. If the orbital periods (in Earth days) of the two planets are designated P_m and P_e, Mercury will be seen on the average to move $1/P_m$ of a revolution per day, while Earth moves $1/P_e$ of a revolution per day in pursuit. Mercury's daily gain on Earth is $(1/P_m) - (1/P_e)$ of a revolution, so that the time for Mercury to gain one complete revolution on Earth, the synodic period, s, may be found by the formula

$$\frac{1}{s} = \frac{1}{P_m} - \frac{1}{P_e}$$

If P_e is 365 days and P_m is 88 days, find the synodic period in Earth days.

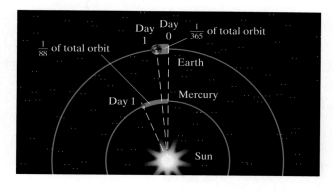

Challenge Problems

101. Make up an equation that cannot have 4 or -2 as a solution. Explain how you determined your answer.

102. Make up an equation containing the sum of two rational expressions in the variable x whose solution is the set of *real numbers*. Explain how you determined your answer.

103. Make up an equation in the variable x containing the sum of two rational expressions whose solution is the set of all real numbers except 0. Explain how you determined your answer.

Group Activity

104. *Focal Length* An 80-mm focal length lens is used to focus an image on the film of a camera. The maximum distance allowed between the lens and the film plane is 120 mm.

a) Group member 1: Determine how far ahead of the film the lens should be if the object to be photographed is 10.0 meters away.

b) Group member 2: Repeat part **a)** for a distance of 3 meters away.

c) Group member 3: Repeat part **a)** for a distance of 1 meter away.

d) Individually, determine the closest distance that an object could be photographed sharply.

e) Compare your answers to see whether they appear reasonable and consistent.

Cumulative Review Exercises

[2.5] **105.** Solve the inequality $-2 \le 4 - 2x < 6$.

[3.4] **106.** Find the slope and y-intercept of the graph of the equation $3(y - 4) = -(x - 2)$.

[5.1] **107.** Simplify
$3x^2y - 4xy + 2y^2 - (3xy + 6y^2 + 2x)$.

[5.8] **108.** *Landscaping* Jessyca Nino Aquino's garden is surrounded by a uniform-width walkway. The garden and the walkway together cover an area of 320 square feet. If the dimensions of the garden are 12 feet by 16 feet, find the width of the walkway.

6.5 RATIONAL EQUATIONS: APPLICATIONS AND PROBLEM SOLVING

1 Solve work problems.
2 Solve number problems.
3 Solve motion problems.

Some applications of equations with rational expressions were illustrated in Section 6.4 and Exercise Set 6.4. In this section, we examine some additional applications. We study work problems first.

1 Solve Work Problems

Problems where two or more machines or people work together to complete a certain task are sometimes referred to as **work problems**. To solve work problems we use the fact that the part of the work done by person 1 (or machine 1) plus the part of the work done by person 2 (or machine 2) is equal to the total amount of work done by both people (or both machines), or 1 (for 1 whole task completed).

$$\begin{pmatrix} \text{part of task done} \\ \text{by first person} \\ \text{or machine} \end{pmatrix} + \begin{pmatrix} \text{part of task done} \\ \text{by second person} \\ \text{or machine} \end{pmatrix} = \begin{pmatrix} 1 \\ \text{(one whole task} \\ \text{completed)} \end{pmatrix}$$

To determine the part of the task done by each person or machine, we use the formula

$$\text{part of task completed } = \text{ rate} \cdot \text{time}$$

This formula is very similar to the formula *amount = rate · time* that was discussed in Section 2.4.

Let's now discuss how to determine the rate. If, for example, John can do a particular task in 5 hours, he could complete $\frac{1}{5}$ of the task in 1 hour. Thus, his rate is $\frac{1}{5}$ of the task per hour. If Kishi can do a job in 6 hours, her rate is $\frac{1}{6}$ of the job per hour. Similarly, if Maria can do a job in x minutes, her rate is $\frac{1}{x}$ of the job per minute. *In general, if a person or machine can complete a task in x units of time, the rate is $\frac{1}{x}$.*

EXAMPLE 1 **Planting Flowers** Sana and Jerry Jenkins both work for a botanical garden. The botanical garden is adding a number of floral designs around its grounds. Sana, who has more experience, can plant the flowers and make the design by herself in 3 hours. It takes Jerry 5 hours working by himself to make the same design. If Sana and Jerry are assigned to work together, how long will it take them to make the design?

Solution Understand We need to find the time for both Sana and Jerry working together to make the floral design. Let x = time, in hours, for Sana and Jerry to make the floral design together. We will construct a table to assist us in finding the part of the task completed by each person.

Worker	Rate of work	Time worked	Part of task completed
Sana	$\frac{1}{3}$	x	$\frac{x}{3}$
Jerry	$\frac{1}{5}$	x	$\frac{x}{5}$

Translate

$$\begin{pmatrix} \text{part of floral design made} \\ \text{by Sana in } x \text{ hours} \end{pmatrix} + \begin{pmatrix} \text{part of floral design made} \\ \text{by Jerry in } x \text{ hours} \end{pmatrix} = 1(\text{entire floral design})$$

$$\frac{x}{3} \qquad + \qquad \frac{x}{5} \qquad = \quad 1$$

Carry Out We multiply both sides of the equation by the LCD, 15. Then we solve for x, the number of hours.

$$15\left(\frac{x}{3} + \frac{x}{5}\right) = 15 \cdot 1 \qquad \textit{Multiply by the LCD, 15.}$$

$$15\left(\frac{x}{3}\right) + 15\left(\frac{x}{5}\right) = 15 \qquad \textit{Distributive property}$$

$$5x + 3x = 15$$

$$8x = 15$$

$$x = \frac{15}{8} \qquad \textit{1.88 hours (to the nearest hundredth)}$$

Answer   Sana and Jerry together can make the floral design in about 1.88 hours. This answer is reasonable because this time is less than it takes either person to make the design by herself on himself.

NOW TRY EXERCISE 11

Sometimes a problem may involve decimals as shown in the next example.

EXAMPLE 2   **Filling a Tub**   Jim and Joy Love just moved into a house that has a Jacuzzi bathtub. When they turn on the tap to fill the tub, the water is cloudy. They wish to run as much water through the tub as possible until the water clears. To accomplish this they turn on the cold water tap and open the drain of the tub. The cold water tap can fill the tub in 7.6 minutes and the drain can empty the tub in 10.3 minutes. If the drain is open and the cold water faucet is turned on, how long will it take before the water fills the tub?

Solution   Understand   As water from the faucet is filling the tub, water going down the drain is emptying the tub. Thus, the faucet and drain are working against each other. Let x = amount of time needed to fill the tub.

	Rate of work	Time worked	Part of tub filled or emptied
Faucet filling tub	$\frac{1}{7.6}$	x	$\frac{x}{7.6}$
Drain emptying tub	$\frac{1}{10.3}$	x	$\frac{x}{10.3}$

Translate   Since the faucet and drain are working against each other, we will *subtract* the part of the water being emptied from the part of water being added to the tub.

$$\left(\begin{array}{c}\text{part of tub filled} \\ \text{in } x \text{ minutes}\end{array}\right) - \left(\begin{array}{c}\text{part of tub emptied} \\ \text{in } x \text{ minutes}\end{array}\right) = 1(\text{whole tub filled})$$

$$\frac{x}{7.6} - \frac{x}{10.3} = 1$$

Carry Out   Using a calculator, we can determine that the LCD is $(7.6)(10.3) = 78.28$. Now we multiply both sides of the equation by 78.28 to remove fractions.

$$78.28\left(\frac{x}{7.6} - \frac{x}{10.3}\right) = 78.28\,(1)$$

$$\overset{10.3}{\cancel{78.28}}\left(\frac{x}{\cancel{7.6}}\right) - \overset{7.6}{\cancel{78.28}}\left(\frac{x}{\cancel{10.3}}\right) = 78.28(1)$$

$$10.3x - 7.6x = 78.28$$
$$2.7x = 78.28$$
$$x \approx 28.99$$

NOW TRY EXERCISE 21   Answer   The tub will fill in about 29 minutes.   ✳

EXAMPLE 3   **Working at a Vineyard**   Chris Burdett and Mark Greenlaugh work at a vineyard in California. When Chris and Mark work together, they can check all the plants in a given field in 24 minutes. When Chris checks the plants by himself, it takes him 36 minutes. How long does it take Mark to check the plants by himself?

Solution   Understand   Let x = amount of time for Mark to check the plants by himself. We know that when working together they can check the plants in 24 minutes. We organize this information in the table as follows.

Worker	Rate of work	Time worked	Part of plants checked
Chris	$\dfrac{1}{36}$	24	$\dfrac{24}{36} = \dfrac{2}{3}$
Mark	$\dfrac{1}{x}$	24	$\dfrac{24}{x}$

Translate

$$\left(\begin{array}{c}\text{part of plants}\\\text{checked by Chris}\end{array}\right) + \left(\begin{array}{c}\text{part of plants}\\\text{checked by Mark}\end{array}\right) = 1\,(\text{whole field checked})$$

$$\frac{2}{3} \qquad\qquad + \qquad\qquad \frac{24}{x} \qquad\qquad = \quad 1$$

Carry Out $3x\left(\dfrac{2}{3} + \dfrac{24}{x}\right) = 3x \cdot 1$ *Multiply both sides by the LCD, 3x.*

$$2x + 72 = 3x$$
$$72 = x$$

NOW TRY EXERCISE 19   Answer   Mark can check the plants by himself in 72 minutes.   ✳

Note that in Example 3, we used $\frac{2}{3}$ rather than $\frac{24}{36}$ for the part of the plants checked by Chris. Always use simplified fractions when setting up and solving equations.

2   Solve Number Problems

Now let us look at a **number problem**, where we must find a number related to one or more other numbers.

EXAMPLE 4   **Number Problem**   When the reciprocal of 3 times a number is subtracted from 5, the result is the reciprocal of twice the number. Find the number.

Solution   Understand   Let x = unknown number. Then $3x$ is 3 times the number, and $\dfrac{1}{3x}$ is the reciprocal of 3 times the number. Twice the number is $2x$, and $\dfrac{1}{2x}$ is the reciprocal of twice the number.

Translate $5 - \dfrac{1}{3x} = \dfrac{1}{2x}$

Carry Out

$$6x \left(5 - \frac{1}{3x} \right) = 6x \cdot \frac{1}{2x}$$ *Multiply by the LCD, 6x.*

$$6x(5) - 6x \left(\frac{1}{3x} \right) = 6x \left(\frac{1}{2x} \right)$$

$$30x - 2 = 3$$

$$30x = 5$$

$$x = \frac{5}{30} = \frac{1}{6}$$

NOW TRY EXERCISE 29   Answer   A check will verify that the number is $\frac{1}{6}$.   ✳

3 Solve Motion Problems

The last type of problem we will look at is **motion problems**. Recall that we discussed motion problems in Section 2.4. In that section, we learned that distance = rate · time. Sometimes it is convenient to solve for the time when solving motion problems.

$$\text{time} = \frac{\text{distance}}{\text{rate}}$$

EXAMPLE 5   **Flying an Airplane**   Sally Sestani owns a single-engine Cessna airplane. When making her preflight plan, she finds that there is a 20-mile-per-hour wind moving from east to west at the altitude she plans to fly. If she travels west (with the wind), she will be able to travel 400 miles in the same amount of time that she would be able to travel 300 miles flying east (against the wind). Assuming that if it were not for the wind, the plane would fly at the same speed going east or west, find the speed of the plane in still air.

Solution   Understand   Let x = speed of the plane in still air. We will set up a table to help answer the question.

Wind (20mph)

West East

Flying with the wind, 400 miles

Flying against the wind, 300 miles

Plane	Distance	Rate	Time
Against wind	300	$x - 20$	$\dfrac{300}{x - 20}$
With wind	400	$x + 20$	$\dfrac{400}{x + 20}$

Translate   Since the times are the same, we set up and solve the following equation:

$$\frac{300}{x - 20} = \frac{400}{x + 20}$$

Carry Out

$$300(x + 20) = 400(x - 20)$$ *Cross multiply.*

$$300x + 6000 = 400x - 8000$$

$$6000 = 100x - 8000$$

$$14{,}000 = 100x$$

$$140 = x$$

NOW TRY EXERCISE 37   Answer   The speed of the plane in still air is 140 miles per hour.   ✳

EXAMPLE 6   **Paddling a Water Bike**   Becky and Al Ryckman go out on a water bike. When paddling against the current (going out from shore), they average 2 miles per hour. Coming back (going toward shore), paddling with the current, they average 3 miles per hour. If it takes $\frac{1}{4}$ hour longer to paddle out from shore than to paddle back, how far out did they paddle?

Solution

Understand   In this problem, the times going out and coming back are not the same. It takes $\frac{1}{4}$ hour longer to paddle out than to paddle back. Therefore, to make the times equal, we can add $\frac{1}{4}$ hour to the time to paddle back (or subtract $\frac{1}{4}$ hour from the time to paddle out). Let $x =$ the distance they paddle out from shore.

Bike	Distance	Rate	Time
Going out	x	2	$\dfrac{x}{2}$
Coming back	x	3	$\dfrac{x}{3}$

Translate

$$\text{time coming back} + \frac{1}{4} \text{ hour} = \text{time going out}$$

$$\frac{x}{3} + \frac{1}{4} = \frac{x}{2}$$

Carry Out

$$12\left(\frac{x}{3} + \frac{1}{4}\right) = 12 \cdot x \qquad \textit{Multiply by the LCD, 12.}$$

$$12\left(\frac{x}{3}\right) + 12\left(\frac{1}{4}\right) = 12\left(\frac{x}{2}\right) \qquad \textit{Distributive property}$$

$$4x + 3 = 6x$$

$$3 = 2x$$

$$1.5 = x$$

NOW TRY EXERCISE 49   Answer   Therefore, they paddle out 1.5 miles from shore.   ✳

EXAMPLE 7   **Taking a Trip**   Dawn Puppel lives in Buffalo, New York, and travels to college in South Bend, Indiana. The speed limit on some of the roads she travels is 55 miles per hour, while on others it is 65 miles per hour. The total distance traveled by Dawn is 490 miles. If Dawn follows the speed limits and the total trip takes 8 hours, how long does she drive at 55 miles per hour and how long does she drive at 65 miles per hour?

Solution   Understand and Translate   Let $x =$ number of miles driven at 55 mph

then $490 - x =$ number of miles driven at 65 mph

Speed limit	Distance	Rate	Time
55 mph	x	55	$\dfrac{x}{55}$
65 mph	$490 - x$	65	$\dfrac{490 - x}{65}$

Since the total time is 8 hours, we write

$$\frac{x}{55} + \frac{490 - x}{65} = 8$$

Carry Out   The LCD of 55 and 65 is 715.

$$715\left(\frac{x}{55} + \frac{490 - x}{65}\right) = 715 \cdot 8$$

$$715\left(\frac{x}{55}\right) + 715\left(\frac{490 - x}{65}\right) = 5720$$

$$13x + 11(490 - x) = 5720$$

$$13x + 5390 - 11x = 5720$$

$$2x + 5390 = 5720$$

$$2x = 330$$

$$x = 165$$

Answer   The number of miles driven at 55 mph is 165 miles. Then the time driven at 55 mph is $\frac{165}{55} = 3$ hours, and the time driven at 65 mph is $\frac{490 - 165}{65} = \frac{325}{65} = 5$ hours.

Notice that in Example 7 the answer to the question was not the value obtained for x. The value obtained was a distance, and the question asked us to find the time. *When working word problems, you must read and work the problems very carefully and make sure you answer the questions that were asked.*

Exercise Set 6.5

Practice the Skills and Problem Solving

1. *Painting a Wall* Two brothers take exactly the same time to paint a wall. If they paint the wall together, will the total time needed to paint the wall be less than $\frac{1}{2}$ the time, equal to $\frac{1}{2}$ the time, or greater than $\frac{1}{2}$ the time it takes each brother separately to paint the wall? Explain.

2. *Tractors in a Field* Two tractors, a larger one and a smaller one, work together to level a field. In the same amount of time, the larger tractor levels more land than the smaller one. Will the smaller tractor take more or less time working alone than the two take working together? Explain.

3. a) *Working Together* Bill and Bob are planning to do a task together. Bill can do the task in 7 hours and Bob can do it in 8 hours. Let $x = $ the time for Bill and Bob to do the task together. Complete the table below.

Worker	Rate of work	Time worked	Part of task completed
Bill		x	
Bob		x	

b) Study the examples in this section. Then, write the equation that can be used to solve for x. (Do not solve.)

c) Working together, will it take them more or less than 7 hours to complete the task? Explain.

4. a) *Working Together* Sarah and Sally are planning to do a task together. Sally can do the task in 3.6 hours and Sarah can do the task in 5.2 hours. Let $x = $ the time for Sally and Sarah to do the task together. Complete the table below.

Worker	Rate of work	Time worked	Part of task completed
Sally		x	
Sarah		x	

b) Study the examples in this section. Then, write the equation that can be used to solve for x. (Do not solve.)

c) Working together, will it take them more or less than 3.6 hours to complete the task? Explain.

Exercises 5–26 involve work problems. Answer the question asked. When necessary, round answers to the nearest hundredth.

5. ***Totem Pole*** It takes Margerie Miller 2 months to carve a totem pole. It takes Matthew Jones 6 months to carve the same totem pole. How long will it take them working together to carve the totem pole?

6. ***Window Washers*** Fran Thompson can wash windows in the lobby at the Days Inn in 3 hours. Jill Franks can wash the same windows in 4 hours. How long will it take them working together to wash the windows?

7. ***Shampooing a Carpet*** Jason La Rue can shampoo the carpet on the main floor of the Sheraton Hotel in 3 hours. Tom Lockheart can shampoo the same carpet in 6 hours. If they work together, how long will it take them to shampoo the carpet?

8. ***Printing Checks*** At the Merck Corporation it takes one computer 3 hours to print checks for its employees and a second computer 5 hours to complete the same job. How long will it take the two computers working together to complete the job?

9. ***Dairy Farm*** On a small dairy farm. Richard Gill can milk 10 cows in 30 minutes. His son, Duane, can milk the same cows in 50 minutes. How long will it take them working together to milk the 10 cows?

10. ***Mowing Lawns*** Kerry and Kim Johnson mow lawns during the summer months. Using a self-propelled hand mower, Kerry can mow a large area in 5 hours. Using a riding mower, Kim can mow the same lawn in 4 hours. How long will it take them working together to mow the lawn?

11. ***Plowing a Field*** Wanda Garner and Shawn Robinson are friends and neighbors, and each owns a farm in California. Wanda can plow a field used for corn in 4 hours using her John Deere tractor. Shawn can plow the same acreage in 6 hours using his Caterpillar tractor. One Saturday Wanda has to catch a plane but wishes to first plow her cornfield. She asks Shawn to help her. How long will it take them working together to plow the field?

12. ***Painting*** Sharon Kaur and Hephner Bennett are painters. Sharon can paint a living room in a Habitat for Humanity house in 6 hours. Hephner can paint the same room in 4.5 hours. How long will it take them working together to paint the living room?

13. ***Filling a Pool*** A $\frac{1}{2}$-inch-diameter hose can fill a swimming pool in 8 hours. A $\frac{4}{5}$-inch-diameter hose can fill the same pool in 5 hours. How long will it take to fill the pool when both hoses are used?

14. ***Milk Tank*** At a dairy plant, a milk tank can be filled in 6 hours (using the in-valve). Using the out-valve, the tank can be emptied in 8 hours. If both valves are open and milk is being pumped into the tank, how long will it take to fill the tank?

15. ***Oil Refinery*** An oil refinery has large tanks to hold oil. Each tank has an inlet valve and an outlet valve. The tank can be filled with oil in 20 hours when the inlet valve is wide open and the outlet valve is closed. The tank can be emptied in 25 hours when the outlet valve is wide open and the inlet valve is closed. If a new tank is placed in operation and both the inlet valve and outlet valve are wide open, how long will it take to fill the tank?

16. ***Cabinet Makers*** Laura Burton and Marcia Kleinz are cabinet makers. Laura can make a set of kitchen cabinets by herself in 10 hours. If Laura and Marcia work together, they can make the same set of cabinets in 6 hours. How long will it take Marcia working by herself to make the cabinets?

17. ***Archeology*** Dr. Indiana Jones and his father, Dr. Henry Jones, are working on a dig near the Forum in Rome. Indiana and his father working together can unearth a specific plot of land in 2.6 months. Indiana can unearth the entire area by himself in 3.9 months. How long would it take Henry to unearth the entire area by himself?

18. ***Digging a Trench*** Arthur Altshiller and Sally Choi work for General Telephone. Together it takes them 2.7 hours to dig a trench where a wire is to be laid. If Arthur can dig the trench by himself in 3.2 hours, how long would it take Sally by herself to dig the trench?

19. **Jellyfish Tanks** John Jerome and Shane Wheeler work together at the Monterey Aquarium. It takes John 50 minutes to clean the Jellyfish tanks. Since Shane is new to the job, it takes him longer to perform the same task. When working together, they can perform the task in 30 minutes. How long does it take Shane to do the task by himself?

20. **Irrigating Crops** A large tank is being used on Jed Saifer's farm to irrigate the crops. The tank has two inlet pipes and one outlet pipe. The two inlet pipes can fill the tank in 8 and 12 hours, respectively. The outlet pipe can empty the tank in 15 hours. If the tank is empty, how long would it take to fill the tank when all three valves are open?

21. **Filling a Washtub** When only the cold water valve is opened, a washtub will fill in 8 minutes. When only the hot water valve is opened, the washtub will fill in 12 minutes. When the drain of the washtub is open, it will drain completely in 7 minutes. If both the hot and cold water valves are open and the drain is open, how long will it take for the washtub to fill?

22. **Installing Windows** Adam, Frank, and Willy are experts at installing windows in houses. Adam can install five living room windows in a house in 10 hours. Frank can do the same job in 8 hours, and Willy can do it in 6 hours. If all three men work together, how long will it take them to install the windows?

🔒 **23.** **Pumping Water** The Rushville fire department uses three pumps to remove water from flooded basements. The three pumps can remove all the water from a flooded basement in 6 hours, 5 hours, and 4 hours, respectively. If all three pumps work together how long will it take to empty the basement?

24. **Planting Flowers** Justin Marshall and Ed Freeman plant petunias in their flower garden. It takes Justin twice as long as Ed to plant the flowers. Working together, they can plant the flowers in 10 hours. How long does it take Ed to plant the flowers by himself?

25. **Roofing a House** Gary Glaze requires 15 hours to put a new roof on a house. His apprentice, Anna Gandy, can reroof the house by herself in 20 hours. After working alone on a roof for 6 hours, Gary leaves for another job. Anna takes over and completes the job. How long does it take Anna to complete the job?

26. **Filling a Tank** Two pipes are used to fill an oil tanker. When the larger pipe is used alone, it takes 60 hours to fill the tanker. When the smaller pipe is used alone, it takes 80 hours to fill the tanker. The large pipe begins filling the tanker. After 20 hours, the large pipe is closed and the smaller pipe is opened. How much longer will it take to finish filling the tanker using only the smaller pipe?

Exercises 27-36 involve number problems. Answer the question asked.

27. What number multiplied by the numerator and added to the denominator of the fraction $\frac{4}{3}$ makes the resulting fraction $\frac{5}{2}$?

28. What number added to the numerator and multiplied by the denominator of the fraction $\frac{4}{5}$ makes the resulting fraction $\frac{1}{15}$?

🔒 **29.** One number is twice an other. The sum of their reciprocals is $\frac{3}{4}$. Find the numbers.

30. The sum of the reciprocals of two consecutive integers is $\frac{11}{30}$. Find the two integers.

31. The sum of the reciprocals of two consecutive even integers is $\frac{5}{12}$. Find the two integers.

32. When a number is added to both the numerator and denominator of the fraction $\frac{7}{9}$, the resulting fraction is $\frac{5}{6}$. Find the number added.

33. When 3 is added to twice the reciprocal of a number, the sum is $\frac{31}{10}$. Find the number.

34. The reciprocal of 3 less than a certain number is twice the reciprocal of 6 less than twice the number. Find the number(s).

35. If three times a number is added to twice the reciprocal of the number, the answer is 5. Find the number(s).

36. If three times the reciprocal of a number is subtracted from twice the reciprocal of the square of the number, the difference is −1. Find the number(s).

Exercises 37-57 involve motion problems. Answer the question asked. When necessary, round answers to the nearest hundredth.

37. **Canoeing** When Brett Burton and his classmate Truong Du go canoeing in still water (with no current), they travel at 4 miles per hour. When they paddle at the same rate in the Suannee River it takes them the same amount of time to travel 4.2 miles downstream (with the current) as it does to travel 3.5 miles upstream (against the current). Find the current of the river.

38. ***Auto Train*** The Amtrak Auto Train travels nonstop from Lorton, Virginia to Sanford, Florida. Wanda Garner wishes to bring two cars down to Florida for the winter, so she decides to send one by auto train and drive the other down. The train travels 600 kilometers in the time it takes for her to drive 400 kilometers. If the average speed of the train is 40 kilometers greater than the speed of Wanda's car, find the speeds of the train and car.

39. ***Moving Sidewalk*** The moving sidewalk at Chicago's O'Hare International Airport moves at a speed of 2.0 feet per second. Walking on the moving sidewalk, Nancy Killian walks 120 feet in the same time that it takes her to walk 52 feet without the moving sidewalk. How fast does Nancy walk?

40. ***Moving Sidewalk*** The moving sidewalk at the Denver International Airport moves at a speed of 1.8 feet per second. Nathan Nearpass walks 100 feet on the moving sidewalk, then turns around on the moving sidewalk and walks at the same speed 40 feet in the opposite direction. If the time walking in each direction was the same, find Nathan's walking speed.

41. ***Skiing*** Bonnie Hellier and Clide Vincent go cross-country skiing in the Adirondack Mountains. Clide is an expert skier who averages 10 miles per hour. Bonnie averages 6 miles per hour. If it takes Bonnie 1/2 hour longer to ski the trail, how long is the trail?

42. ***Outing in a Park*** Ruth and Jerry Mackin go for an outing in Memorial Park in Houston, Texas. Ruth jogs while Jerry rollerblades. Jerry rollerblades 2.9 miles per hour faster than Ruth jogs. When Jerry has rollerbladed 5.7 miles, Ruth has jogged 3.4 miles. Find Ruth's jogging speed.

43. ***Visiting a Resort*** Allen Angel drove 60 miles to Yosemite National Park. He spent twice as much time visiting the park as it took him to drive to the park. The total time for driving and visiting the park was 5 hours. Find the average speed driving to Yosemite National Park.

See Exercise 43. Photo shows Allen Angel, the author of this book.

44. ***Boating Trip*** Ray Packerd starts out on a boating trip at 8 A.M. Ray's boat can go 20 miles per hour in still water. How far downstream can Ray go if the current is 5 miles per hour and he wishes to go down and back in 4 hours?

45. ***Football Game*** At a football game the Green Bay Packers have the ball on their own 20-yard line. Brett Favre passed the ball to Terry Glenn, who catches the ball and runs into the end zone for a touchdown. Assume that the ball, when passed, traveled at 14.7 yards per second and that after Terry caught the ball, he ran at 5.8 yards per second into the end zone. If the play, from the time Brett released the ball to the time Terry reached the end zone, took 10.6 seconds, how far from where Brett threw the ball did Terry catch it? Assume the entire play went down the center of the field.

46. ***Skyline Drive*** In one day, Pauline Shannon drove from Front Royal, Virginia, to Asheville, North Carolina, a distance of 492 miles. For part of the trip she drove at a steady rate of 50 miles per hour, but in some areas she drove at a steady rate of 35 miles per hour. If the total time of the trip was 11.13 hours, how far did she travel at each speed?

47. ***Subway Trains*** The number 4 train in the New York City subway system goes from Woodlawn/Jerome Avenue in the Bronx to Flatbush Avenue in Brooklyn. The one-way distance between these two stops is 24.2 miles. On this route, two tracks run parallel to each other, one for the local train and the other for the express train. The local and express trains leave Woodlawn/Jerome Avenue at the same time. When the express reaches the end of the line at Flatbush Avenue, the local is at Wall Street, 7.8 miles from Flatbush. If the express averages 5.2 miles per hour faster than the local, find the speeds of the two trains.

48. *Riding a Horse* Each morning, Ron Lucky takes his horse, Beauty, for a ride on Pfeiffer Beach in Big Sur, California. He typically rides Beauty for 5.4 miles, then walks Beauty 2.3 miles. His speed when riding is 4.2 times his speed when walking. If his total outing takes 1.5 hours, find the rate at which he walks Beauty.

49. *Traveling* A train and car leave from the Pasadena, California, railroad station at the same time headed for the state fair in Sacramento. The car averages 50 miles per hour and the train averages 70 miles per hour. If the train arrives at the fair 2 hours ahead of the car, find the distance from the railroad station to the state fair.

50. *Traveling* A car and a train leave Union Station in Washington, D.C., at the same time for Rochester, New York, 390 miles away. If the speed of the car averages twice the speed of the train and the car arrives 6.5 hours before the train, find the speed of the car and the speed of the train.

51. *Traveling* Two friends drive from Dallas going to El Paso, a distance of 600 miles. Mary Ann Zilke travels by highway and arrives at the same destination 2 hours ahead of Carla Canola, who took a different route. If the average speed of Mary Ann's car was 10 miles per hour faster than Carla's car, find the average speed of Mary Ann's car.

52. *Sailboat Race* In a 30-mile sailboat race, the winning boat, the Buccaneer, finishes 10 minutes ahead of the second-place boat, the Raven. If the Buccaneer's average speed was 2 miles per hour faster than the Raven's, find the average speed of the Buccaneer.

53. *Helicopter Ride* Kathy Angel took a helicopter ride to the top of the Mt. Cook Glacier in New Zealand. The flight covered a distance of 60 kilometers. Kathy stayed on top of the glacier for $\frac{1}{2}$ hour. She then flew to the town of Te Anu, 140 kilometers away. The helicopter averaged 20 kilometers per hour faster going to Te Anu

than on the flight up to the top of the glacier. The total time involved in the outing was 2 hours. Find the average speed of the helicopter going to the glacier.

54. *Sailboats* Two sailboats, the *Serendipity* and the *Zerwilliker*, start at the same point and at the same time on Lake Michigan, heading to the same restaurant on the lake. The *Serendipity* sails an average of 5.2 miles per hour and the *Zerwilliker* sails an average of 4.6 miles per hour. If the *Serendipity* arrives at the restaurant 0.4 hour ahead of the *Zerwilliker*, find the distance from their starting point to the restaurant.

55. *Suspension Bridge* The Capilano suspension bridge in Vancouver, Canada, is 450 feet long. Phil and Heim started walking across the bridge at the same time. Heim's speed was 2 feet per minute faster than Phil's. If Heim finished walking across the bridge $2\frac{1}{2}$ minutes before Phil, find Phil's average speed in feet per minutes.

56. *Inclined Railroad* A trip up Mount Pilatus, near Lucerne, Switzerland, involves riding an inclined railroad up to the top of the mountain, spending time at the top, then coming down the opposite side of the mountain in an aerial tram.

The distance traveled up the mountain is 7.5 kilometers and the distance traveled down the mountain is 8.7 kilometers. The speed coming down the mountain is 1.2 times the speed going up. If the Lieblichs stayed at the top of the mountain for 3 hours and the total time of their outing was 9 hours, find the speed of the inclined railroad.

57. *Launching Rockets* Two rockets are to be launched at the same time from NASA headquarters in Houston, Texas, and are to meet at a space station many miles from Earth. The first rocket is to travel at 20,000 miles per hour and the second rocket will travel at 18,000 miles per hour. If the first rocket is scheduled to reach the space station 0.6 hour before the second rocket, how far is the space station from NASA headquarters?

58. Make up your own word problem and find the solution.

59. Make up your own motion problem and find the solution.

60. Make up your own number problem and find the solution.

Challenge Problem

61. An officer flying a California Highway Patrol aircraft determines that a car 10 miles ahead of her is speeding at 90 miles per hour.

    a) If the aircraft is traveling 150 miles per hour, how long, in minutes, will it take for the aircraft to reach the car?

    b) How far will the car have traveled in the time it takes the aircraft to reach it?

    c) If the pilot wishes to reach the car in exactly 8 minutes, how fast must the airplane fly?

Cumulative Review Exercises

[1.5] **62.** Simplify $\dfrac{(2x^{-1}y^{-2})^{-3}}{(3x^{-1}y^3)^2}$.

[1.6] **63.** Express 5,260,000,000 in scientific notation.

[2.3] **64.** *Weekly Salary* Sandy Ivey receives a flat weekly salary of $240 plus 12% commission on the total dollar volume of all sales she makes. What

must her dollar sales volume be in a week for her to earn $540?

[3.1] **65.** Graph $y = |x| - 2$.

[5.4] **66.** Factor $2a^4 - 2a^3 - 5a^2 + 5a$.

6.6 VARIATION

SSM   Study Guide   CD/Video

MathPro 4/5   PH Math Tutor Center   prenhall.com/Angel

1 Solve direct variation problems.

2 Solve inverse variation problems.

3 Solve joint variation problems.

4 Solve combined variation problems.

In Sections 6.4 and 6.5, we saw many applications of equations containing rational expressions. In this section, we see still more.

1 Solve Direct Variation Problems

Many scientific formulas are expressed as variations. A **variation** is an equation that relates one variable to one or more other variables using the operations of multiplication or division (or both operations). There are essentially three types of variation problems: direct, inverse, and joint variation.

In **direct variation**, the two related variables will both increase together or both decrease together; that is, as one increases so does the other, and as one decreases so does the other.

Consider a car traveling at 30 miles an hour. The car travels 30 miles in 1 hour, 60 miles in 2 hours, and 90 miles in 3 hours. Notice that as the time increases, the distance traveled increases.

The formula used to calculate distance traveled is

$$\text{distance} = \text{rate} \cdot \text{time}$$

Since the rate is a constant, 30 miles per hour, the formula can be written

$$d = 30t$$

We say that distance varies directly as time or that distance is directly proportional to time. This is an example of a direct variation.

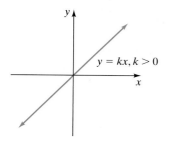

FIGURE 6.8

> ### Direct Variation
>
> If a variable y varies directly as a variable x, then
> $$y = kx$$
> where k is the **constant of proportionality** (or the variation constant).

The graph of $y = kx$, $k > 0$, is always a straight line that goes through the origin (see Fig. 6.8). The slope of the line depends on the value of k. The greater the value of k, the greater the slope.

EXAMPLE 1 **Circle** The circumference of a circle, C, is directly proportional to (or varies directly as) its radius, r. Write the equation for the circumference of a circle if the constant of proportionality, k, is 2π.

Solution
$$C = kr \ (C \text{ varies directly as } r)$$
$$C = 2\pi r \ (\text{constant of proportionality is } 2\pi)$$

EXAMPLE 2 **Administering a Drug** The amount, a, of the drug theophylline given to patients is directly proportional to the patient's mass, m, in kilograms.

a) Write this variation as an equation.

b) If 150 mg is given to a boy whose mass is 30 kg, find the constant of proportionality.

c) How much of the drug should be given to a patient whose mass is 62 kg?

Solution **a)** We are told this is a direct variation. That is, the greater a person's mass the more of the drug that will need to be given. We therefore set up a direct variation.
$$a = km$$

b) Understand and Translate   To determine the value of the constant of proportionality, we substitute the given values for the amount and mass. We then solve for k.

$$a = km$$
$$150 = k(30) \qquad \textit{Substitute the given values.}$$
Carry Out $\qquad\qquad 5 = k$

Answer   Thus, $k = 5$ mg. Five milligrams of the drug should be given for each kilogram of a person's mass.

c) Understand and Translate   Now that we know the constant of proportionality we can use it to determine the amount of the drug to use for a person's mass. We set up the variation and substitute the values of k and m.

$$a = km$$
$$a = 5(62) \qquad \textit{Substitute the given values.}$$
Carry Out $\qquad\qquad a = 310$

Answer   Thus, 310 mg of theophylline should be given to a person whose mass is 62 kg. ✳

EXAMPLE 3 y varies directly as the square of z. If y is 80 when z is 20, find y when z is 45.

Solution   Since y varies directly as the *square of z*, we begin with the formula $y = kz^2$. Since the constant of proportionality is not given, we must first find k using the given information.

$$y = kz^2$$
$$80 = k(20)^2 \quad \text{Substitute the given values.}$$
$$80 = 400k \quad \text{Solve for k.}$$
$$\frac{80}{400} = \frac{400k}{400}$$
$$0.2 = k$$

We now use $k = 0.2$ to find y when z is 45.

$$y = kz^2$$
$$y = 0.2(45)^2 \quad \text{Substitute the given values.}$$
$$y = 405$$

NOW TRY EXERCISE 35   Thus, when z equals 45, y equals 405. ✳

2 Solve Inverse Variation Problems

A second type of variation is **inverse variation**. When two quantities vary inversely, it means that as one quantity increases, the other quantity decreases, and vice versa.

To explain inverse variation, we again use the formula, distance = rate · time. If we solve for time, we get time = distance/rate. Assume that the distance is fixed at 120 miles; then

$$\text{time} = \frac{120}{\text{rate}}$$

At 120 miles per hour, it would take 1 hour to cover this distance. At 60 miles an hour, it would take 2 hours. At 30 miles an hour, it would take 4 hours. Note that as the rate (or speed) decreases the time increases, and vice versa.

The equation above can be written

$$t = \frac{120}{r}$$

This equation is an example of an inverse variation. The time and rate are inversely proportional. The constant of proportionality is 120.

> **Inverse Variation**
>
> If a variable y varies inversely as a variable x, then
>
> $$y = \frac{k}{x} \quad (\text{or } xy = k)$$
>
> where k is the constant of proportionality.

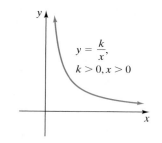

$y = \dfrac{k}{x}$,
$k > 0, x > 0$

FIGURE 6.9

Two quantities vary inversely, or are inversely proportional, when as one quantity increases the other quantity decreases. The graph of $y = k/x$, for $k > 0$ and $x > 0$, will have the shape illustrated in Figure 6.9. The graph of an inverse variation is not defined at $x = 0$ because 0 is not in the domain of the function $y = k/x$.

EXAMPLE 4 Melting Ice

The amount of time, t, it takes a block of ice to melt in water is inversely proportional to the water's temperature, T.

a) Write this variation as an equation.

b) If a block of ice takes 15 minutes to melt in 75°F water, determine the constant of proportionality.

c) Determine how long it will take a block of ice of the same size to melt in 90°F water.

Solution   **a)** The hotter the water temperature, the shorter the time for the block of ice to melt. The inverse variation is

$$t = \frac{k}{T}$$

b) Understand and Translate   To determine the constant of proportionality, we substitute the values for the temperature and time and solve for k.

$$t = \frac{k}{T}$$

$$15 = \frac{k}{75} \quad \textit{Substitute the given values.}$$

Carry Out $1125 = k$

Answer   The constant of proportionality is 1125.

c) Understand and Translate   Now that we know the constant of proportionality, we can use it to determine how long it will take for the same size block of ice to melt in 90°F water. We set up the proportion, substitute the values for k and T, and solve for t.

$$t = \frac{k}{T}$$

$$t = \frac{1125}{90} \quad \textit{Substitute the given values.}$$

Carry Out $t = 12.5$

Answer   It will take 12.5 minutes for the block of ice to melt in the 90°F water. ✳

EXAMPLE 5

Lighting   The illuminance, I, of a light source varies inversely as the square of the distance, d, from the source. Assuming that the illuminance is 75 units at a distance of 6 meters, find a formula that expresses the relationship between the illuminance and the distance.

Solution   Understand and Translate   Since the illuminance varies inversely as the *square* of the distance, the general form of the equation is

$$I = \frac{k}{d^2} \qquad (\text{or } Id^2 = k)$$

To find k, we substitute the given values for I and d.

$$75 = \frac{k}{6^2} \qquad \textit{Substitute the given values.}$$

Carry Out

$$75 = \frac{k}{36} \qquad \textit{Solve for } k.$$

$$(75)(36) = k$$

$$2700 = k$$

NOW TRY EXERCISE 63   Answer   A formula is $I = \dfrac{2700}{d^2}$.

3   Solve Joint Variation Problems

One quantity may vary as a product of two or more other quantities. This type of variation is called **joint variation**.

Joint Variation
If y varies jointly as x and z, then $$y = kxz$$ where k is the constant of proportionality.

EXAMPLE 6   **Area of a Triangle**   The area, A, of a triangle varies jointly as its base, b, and height, h. If the area of a triangle is 48 square inches when its base is 12 inches and its height is 8 inches, find the area of a triangle whose base is 15 inches and height is 40 inches.

Solution   Understand and Translate   First write the joint variation; then substitute the known values and solve for k.

$$A = kbh$$

$$48 = k(12)(8) \qquad \textit{Substitute the given values.}$$

Carry Out

$$48 = k(96) \qquad \textit{Solve for } k.$$

$$\frac{48}{96} = k$$

$$k = \frac{1}{2}$$

Now solve for the area of the given triangle.

$$A = kbh$$

$$= \frac{1}{2}(15)(40) \qquad \textit{Substitute the given values.}$$

$$= 300$$

NOW TRY EXERCISE 67   Answer   The area of the triangle is 300 square inches.

Summary of Variations		
Direct	Inverse	Joint
$y = kx$	$y = \dfrac{k}{x}$	$y = kxz$

4 Solve Combined Variation Problems

Often in real-life situations one variable varies as a combination of variables. The following examples illustrate the use of **combined variations**.

EXAMPLE 7   **Pretzel Shop**   The owners of the Freeport Pretzel Shop find that their weekly sales of pretzels, S, vary directly as their advertising budget, A, and inversely as their pretzel price, P. When their advertising budget is $400 and the price is $1, they sell 6200 pretzels.

a) Write an equation of variation expressing S in terms of A and P. Include the value of the constant.

b) Find the expected sales if the advertising budget is $600 and the price is $1.20.

Solution   **a) Understand and Translate**   We begin with the equation

$$S = \frac{kA}{P}$$

$$6200 = \frac{k(400)}{1} \qquad \textit{Substitute the given values.}$$

Carry Out
$$6200 = 400k \qquad \textit{Solve for } k.$$
$$15.5 = k$$

Answer   Therefore, the equation for the sales of pretzels is $S = \dfrac{15.5A}{P}$.

b) Understand and Translate   Now that we know the combined variation equation, we can use it to determine the expected sales for the given values.

$$S = \frac{15.5A}{P}$$

$$= \frac{15.5(600)}{1.20} \qquad \textit{Substitute the given values.}$$

Carry Out
$$= 7750$$

Answer   They can expect to sell 7750 pretzels.   ✳

EXAMPLE 8   **Electrostatic Force**   The electrostatic force, F, of repulsion between two positive electrical charges is jointly proportional to the two charges, q_1 and q_2, and inversely proportional to the square of the distance, d, between the two charges. Express F in terms of q_1, q_2, and d.

Solution
$$F = \frac{kq_1q_2}{d^2}$$ ✳

NOW TRY EXERCISE 69

Exercise Set 6.6

Concept/Writing Exercises

1. a) Explain what it means when two items vary directly.

b) Give your own example of two quantities that vary directly.

c) Write the direct variation for your example in part **b)**.

2. a) Explain what it means when two items vary inversely.

b) Give your own example of two quantities that vary inversely.

c) Write the inverse variation for your example in part **b)**.

3. What is meant by joint variation?

4. What is meant by combined variation?

5. **a)** In the equation $y = \dfrac{17}{x}$, as x increases, does the value for y increase or decrease?

b) Is this an example of direct or inverse variation? Explain.

6. **a)** In the equation $z = 0.8x^3$, as x increases, does the value for z increase or decrease?

b) Is this an example of direct or inverse variation? Explain.

Use your intuition to determine whether the variation between the indicated quantities is direct or inverse.

7. The speed and the distance covered by a person riding a bike on the Mount Vernon bike path in Alexandria, Virginia

8. The number of pages Tom can read in a 2-hour period and his reading speed

9. The speed of an athlete and the time it takes him to run a 10-kilometer race

10. Barbara's weekly salary and the amount of money withheld for state income taxes

11. The radius of a circle and its area

12. The side of a cube and its volume

13. The radius of a balloon and its volume

14. The diameter of a circle and its circumference

15. The diameter of a hose and the volume of water coming out of the hose

16. The weight of a rocket and its distance from Earth (due to Earth's gravity)

17. The time it takes an ice cube to melt in water and the temperature of the water

18. The distance between two cities on a map and the actual distance between the two cities

19. The shutter opening of a camera and the amount of sunlight that reaches the film

20. The cubic-inch displacement in liters and the horsepower of an engine

21. The length of a board and the force needed to break the board at the center

22. The number of calories eaten and the amount of exercise required to burn off those calories

23. The light illuminating an object and the distance the light is from the object

24. The number of calories in a cheeseburger and the size of the cheeseburger

Practice the Skills

*For Exercises 25–32, **a)** write the variation and **b)** find the quantity indicated.*

25. x varies directly as y. Find x when $y = 12$ and $k = 6$.

26. C varies directly as the square of Z. Find C when $Z = 9$ and $k = \frac{3}{4}$.

27. y varies directly as R. Find y when $R = 180$ and $k = 1.7$.

28. x varies inversely as y. Find x when $y = 25$ and $k = 5$.

29. R varies inversely as W. Find R when $W = 160$ and $k = 8$.

30. L varies inversely as the square of P. Find L when $P = 4$ and $k = 100$.

31. A varies directly as B and inversely as C. Find A when $B = 12$, $C = 4$, and $k = 3$.

32. A varies jointly as R_1 and R_2 and inversely as the square of L. Find A when $R_1 = 120$, $R_2 = 8$, $L = 5$, and $k = \frac{3}{2}$.

*For Exercises 33–40, **a)** write the variation and **b)** find the quantity indicated.*

33. x varies directly as y. If x is 12 when y is 3, find x when y is 5.

34. Z varies directly as W. If Z is 7 when W is 28, find Z when W is 140.

35. y varies directly as the square of R. If y is 5 when $R = 5$, find y when R is 10.

36. S varies inversely as G. If S is 12 when G is 0.4, find S when G is 5.

37. C varies inversely as J. If C is 7 when J is 0.7, find C when J is 12.

38. x varies inversely as the square of P. If $x = 4$ when P is 5, find x when $P = 2$.

39. F varies jointly as M_1 and M_2 and inversely as d. If F is 20 when $M_1 = 5$, $M_2 = 10$, and $d = 0.2$, find F when $M_1 = 10$, $M_2 = 20$, and $d = 0.4$.

40. F varies jointly as q_1 and q_2 and inversely as the square of d. If F is 8 when $q_1 = 2$, $q_2 = 8$, and $d = 4$, find F when $q_1 = 28$, $q_2 = 12$, and $d = 2$.

Problem Solving

41. Assume a varies directly as b. If b is doubled, how will it affect a? Explain.

42. Assume a varies directly as b^2. If b is doubled, how will it affect a? Explain.

43. Assume y varies inversely as x. If x is doubled, how will it affect y? Explain.

44. Assume y varies inversely as a^2. If a is doubled, how will it affect y? Explain.

In Exercises 45–50, use the formula $F = \dfrac{km_1m_2}{d^2}$.

45. If m_1 is doubled, how will it affect F?

46. If m_1 is quadrupled and d is doubled, how will it affect F?

47. If m_1 is doubled and m_2 is halved, how will it affect F?

48. If d is halved, how will it affect F?

49. If m_1 is halved and m_2 is quadrupled, how will it affect F?

50. If m_1 is doubled, m_2 is quadrupled, and d is quadrupled, how will it affect F?

In Exercises 51 and 52, determine if the variation is of the form $y = kx$ *or* $y = \dfrac{k}{x}$, *and find k.*

51.

x	y
2	$\frac{5}{2}$
5	1
10	$\frac{1}{2}$
20	$\frac{1}{4}$

52.

x	y
6	2
9	3
15	5
27	9

53. *Profit* The profit from selling lamps is directly proportional to the number of lamps sold. When 150 lamps are sold, the profit is $2542.50. Find the profit when 520 lamps are sold.

54. *Profit* The profit from selling stereos is directly proportional to the number of stereos sold. When 65 stereos are sold, the profit is $4056. Find the profit when 80 stereos are sold.

55. *Antibiotic* The recommended dosage, d, of the antibiotic drug vancomycin is directly proportional to a person's weight. If Phuong Kim, who is 132 pounds, is given 2376 milligrams, find the recommended dosage for Nathan Brown, who weighs 172 pounds.

56. *Dollars and Pesos* Converting American dollars to Mexican pesos is a direct variation. The more dollars you convert the more pesos you receive. Last week, Carlos Manuel converted $275 into 2433.75 pesos. Today, he received $400 from his aunt. If the conversion rate is unchanged when he converts the $400 into pesos, how many pesos will be receive?

57. *Building a Brick Wall* The time, t, required to build a brick wall varies inversely as the number of people, n, working on it. If it takes 8 hours for five bricklayers to build a wall, how long will it take four bricklayers to build a wall?

58. *Distance* When a car travels at a constant speed, the distance traveled, d, is directly proportional to the time, t. If a car travels 150 miles in 2.5 hours, how far will the same car travel in 4 hours?

59. *Pressure and Volume* The volume of a gas, V, varies inversely as its pressure, P. If the volume, V, is 800 cubic centimeters when the pressure is 200 millimeters (mm) of mercury, find the volume when the pressure is 25 mm of mercury.

60. *Hooke's Law* Hooke's law states that the length a spring will stretch, S, varies directly with the force (or weight), F, attached to the spring. If a spring stretches 1.4 inches when 20 pounds is attached, how far will it stretch when 15 pounds is attached?

61. *Running a Race* The time, t, it takes a runner to cover a specified distance is inversely proportional to the runner's speed. If Aimee Calhoun runs at an average of 6 miles per hour, she will finish a race in 2.6 hours. How long will it take Kim Martino, who runs at 5 miles per hour, to finish the same race?

62. *Pitching a Ball* When a ball is pitched in a professional baseball game, the time, t, it takes for the ball to reach home plate varies inversely with the speed, s, of the pitch.* A ball pitched at 90 miles per hour takes 0.459 second to reach the plate. How long will it take a ball pitched at 75 miles per hour to reach the plate?

63. *Tennis Ball* When a tennis player serves the ball, the time it takes for the ball to hit the ground in the service box is inversely proportional to the speed the ball is traveling. If Andy Roddick serves at 122 miles per hour, it takes 0.21 second for the ball to hit the ground after striking his racquet. How long will it take the ball to hit the ground if he serves at 80 miles per hour?

64. *Intensity of Light* The intensity, I, of light received at a source varies inversely as the square of the distance, d, from the source. If the light intensity is 20 foot-candles at 15 feet, find the light intensity at 10 feet.

65. *Stopping Distance* Assume that the stopping distance of a van varies directly with the square of the speed. A van traveling 40 miles per hour can stop in 60 feet. If the van is traveling 56 miles per hour, what is its stopping distance?

66. *Falling Rock* A rock is dropped from the top of a cliff. The distance it falls in feet is directly proportional to the square of the time in seconds. If the rock falls 4 feet in $\frac{1}{2}$ second, how far will it fall in 3 seconds?

67. *Mortgage Payment* The monthly mortgage payment, P, you pay on a mortgage varies jointly as the interest rate, r, and the amount of the mortgage, m. If the monthly mortgage payment on a \$50,000 mortgage at a 7% interest rate is \$332.50, find the monthly payment on a \$66,000 mortgage at 7%.

68. *Weight of an Object* The weight, w, of an object in Earth's atmosphere varies inversely with the square of the distance, d, between the object and the center of Earth. A 140-pound person standing on Earth is approximately 4000 miles from Earth's center. Find the weight (or gravitational force of attraction) of this person at a distance 100 miles from Earth's surface.

69. *Videotape Rental* The weekly videotape rentals, R, at Busterblock Video vary directly with their advertising budget, A, and inversely with the daily rental price, P. When their advertising budget is \$400 and the rental price is \$2 per day, they rent 4600 tapes per week. How many tapes would they rent per week if they increased their advertising budget to \$500 and raised their rental price to \$2.50?

70. *Wattage Rating* The wattage rating of an appliance, W, varies jointly as the square of the current, I, and the resistance, R. If the wattage is 3 watts when the current is 0.1 ampere and the resistance is 100 ohms, find the wattage when the current is 0.4 ampere and the resistance is 250 ohms.

71. *Electrical Resistance* The electrical resistance of a wire, R, varies directly as its length, L, and inversely as its cross-sectional area, A. If the resistance of a wire is 0.2 ohm when the length is 200 feet and its cross-sectional area is 0.05 square inch, find the resistance of a wire whose length is 5000 feet with a cross-sectional area of 0.01 square inch.

72. *Phone Calls* The number of phone calls between two cities during a given time period, N, varies directly as the populations p_1 and p_2 of the two cities and inversely as the distance, d, between them. If 100,000 calls are made between two cities 300 miles apart and the populations of the cities are 60,000 and 200,000, how many calls are made between two cities with populations of 125,000 and 175,000 that are 450 miles apart?

73. *Water Bill* In a specific region of the country, the amount of a customer's water bill, W, is directly proportional to the average daily temperature for the month, T, the lawn area, A, and the square root of F, where F is the family size, and inversely proportional to the number of inches of rain, R.

In one month, the average daily temperature is 78°F and the number of inches of rain is 5.6. If the

---

*A ball slows down on its way to the plate due to wind resistance. For a 95-mph pitch, the ball is about 8 mph faster when it leaves the pitcher's hand than when it crosses the plate.

average family of four, who has 1000 square feet of lawn, pays $68 for water, estimate the water bill in the same month for the average family of six, who has 1500 square feet of lawn.

74. *Intensity of Illumination* An article in the magazine *Outdoor and Travel Photography* states, "If a surface is illuminated by a point-source of light (a flash), the intensity of illumination produced is inversely proportional to the square of the distance separating them."

If the subject you are photographing is 4 feet from the flash, and the illumination on this subject is $\frac{1}{16}$ of the light of the flash, what is the intensity of illumination on an object that is 7 feet from the flash?

75. *Force of Attraction* One of Newton's laws states that the force of attraction, F, between two masses is directly proportional to the masses of the two objects, m_1 and m_2, and inversely proportional to the square of the distance, d, between the two masses.

a) Write the formula that represents Newton's law.

b) What happens to the force of attraction if one mass is doubled, the other mass is tripled, and the distance between the objects is halved?

76. *Pressure on an Object* The pressure, P, in pounds per square inch (psi) on an object x feet below the sea is 14.70 psi plus the product of a constant of proportionality, k, and the number of feet, x, the object is below sea level (see the figure). The 14.70 represents the weight, in pounds, of the column of air (from sea level to the top of the atmosphere) standing over a 1-inch-by-1-inch square of ocean. The kx represents the weight, in pounds, of a column of water 1 inch by 1 inch by x feet.

a) Write a formula for the pressure on an object x feet below sea level.

b) If the pressure gauge in a submarine 60 feet deep registers 40.5 psi, find the constant k.

c) A submarine is built to withstand a pressure of 160 psi. How deep can the submarine go?

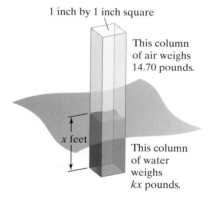

Cumulative Review Exercises

[2.2] **77.** Solve the formula $V = \frac{4}{3}\pi r^2 h$ for h.

[3.6] **78.** Let $f(x) = x^2 - 4$ and $g(x) = -5x + 3$. Find $f(-4) \cdot g(-4)$.

[5.2] **79.** Multiply $(7x - 3)(-2x^2 - 4x + 1)$.

[5.7] **80.** Factor $(x + 1)^2 - (x + 1) - 6$.

CHAPTER SUMMARY

Key Words and Phrases

6.1
Domain
Rational expression
Rational function
Simplified (or reduced to lowest terms)

6.2
Least common denominator (LCD)

Prime number

6.3
Complex fraction
Main fraction line

6.4
Extraneous roots or solutions
Proportion

Rational equation
Similar figures

6.5
Motion problem
Number problem
Work problem

6.6
Combined variations

Constant of proportionality
Direct variation
Inverse variation
Joint variation
Variation

(continued on the next page)

IMPORTANT FACTS

Types of Variation

Direct	Inverse	Joint
$y = kx$	$y = \dfrac{k}{x}$	$y = kxz$

Chapter Review Exercises

[6.1] *Determine the value or values of the variable that must be excluded in each rational expression.*

1. $\dfrac{3}{x - 5}$

2. $\dfrac{x}{x + 2}$

3. $\dfrac{-2x}{x^2 + 9}$

Determine the domain of each rational function.

4. $y = \dfrac{5}{(x + 1)^2}$

5. $f(x) = \dfrac{x + 6}{x^2}$

6. $f(x) = \dfrac{x^2 - 2}{x^2 + 3x - 10}$

Simplify each expression.

7. $\dfrac{x^2 + xy}{x + y}$

8. $\dfrac{x^2 - 49}{x + 7}$

9. $\dfrac{6 - 5x}{5x - 6}$

10. $\dfrac{x^2 + 3x - 4}{x^2 + 2x - 8}$

11. $\dfrac{2x^2 - 6x + 5x - 15}{2x^2 + 7x + 5}$

12. $\dfrac{a^3 - 8b^3}{a^2 - 4b^2}$

13. $\dfrac{27x^3 - y^3}{9x^2 - y^2}$

14. $\dfrac{2x^2 + x - 6}{x^3 + 8}$

[6.2] *Find the least common denominator of each expression.*

15. $\dfrac{6x}{x + 1} - \dfrac{3}{x}$

16. $\dfrac{3x + 1}{x + 2y} + \dfrac{7x - 2y}{x^2 - 4y^2}$

17. $\dfrac{19x - 5}{x^2 + 2x - 35} + \dfrac{3x - 2}{x^2 + 9x + 14}$

18. $\dfrac{3}{(x + 2)^2} - \dfrac{6(x + 3)}{x^2 - 4} - \dfrac{4x}{x + 1}$

[6.1, 6.2] *Perform each indicated operation.*

19. $\dfrac{15x^2y^3}{3z} \cdot \dfrac{6z^3}{5xy^3}$

20. $\dfrac{x}{x - 7} \cdot \dfrac{7 - x}{4}$

21. $\dfrac{16x^2y^4}{xz^5} \div \dfrac{2x^2y^4}{x^4z^{10}}$

22. $\dfrac{7}{3x} + \dfrac{2}{x^2}$

23. $\dfrac{4x - 4y}{x^2y} \cdot \dfrac{y^3}{8x}$

24. $\dfrac{4x^2 - 11x + 4}{x - 3} - \dfrac{x^2 - 4x + 10}{x - 3}$

25. $\dfrac{2}{xy} + \dfrac{3y}{5x^2}$

26. $\dfrac{x + 2}{x - 1} \cdot \dfrac{x^2 + 3x - 4}{x^2 + 6x + 8}$

27. $\dfrac{3x^2 - 7x + 4}{3x^2 - 14x - 5} + \dfrac{x^2 + 2x + 9}{3x^2 - 14x - 5}$

28. $3 + \dfrac{a + 2}{a + 1}$

29. $7 - \dfrac{b + 1}{b - 1}$

30. $\dfrac{a^2 - b^2}{a + b} \cdot \dfrac{a^2 + 2ab + b^2}{a^3 + a^2b}$

31. $\dfrac{1}{a^2 + 8a + 15} \div \dfrac{3}{a + 5}$

32. $\dfrac{a + c}{c} - \dfrac{a - c}{a}$

33. $\dfrac{4x^2 + 8x - 5}{2x + 5} \cdot \dfrac{x + 1}{4x^2 - 4x + 1}$

34. $(a + b) \div \dfrac{a^2 - 2ab - 3b^2}{a - 3b}$

35. $\dfrac{x^2 - 3xy - 10y^2}{6x} \div \dfrac{x + 2y}{12x^2}$

36. $\dfrac{a + 1}{2a} + \dfrac{3}{4a + 8}$

37. $\dfrac{x - 4}{x - 5} - \dfrac{3}{x + 5}$

38. $\dfrac{x + 4}{x^2 - 4} - \dfrac{3}{x - 2}$

39. $\dfrac{x + 1}{x - 3} \cdot \dfrac{x^2 + 2x - 15}{x^2 + 7x + 6}$

40. $\dfrac{1}{x^2 - x - 6} - \dfrac{1}{x^2 - 4}$

41. $\dfrac{4x^2 - 16y^2}{9} \div \dfrac{(x + 2y)^2}{12}$

42. $\dfrac{a^2 + 5a + 6}{a^2 + 4a + 4} \cdot \dfrac{3a + 6}{a^4 + 3a^3}$

43. $\dfrac{x+5}{x^2-15x+50} - \dfrac{x-2}{x^2-25}$

44. $\dfrac{x+2}{x^2-x-6} + \dfrac{x-3}{x^2-8x+15}$

45. $\dfrac{1}{x+3} - \dfrac{2}{x-3} + \dfrac{6}{x^2-9}$

46. $\dfrac{a-4}{a-5} - \dfrac{3}{a+5} - \dfrac{10}{a^2-25}$

47. $\dfrac{x^3+64}{2x^2-32} \div \dfrac{x^2-4x+16}{2x+12}$

48. $\dfrac{a^2-b^4}{a^2+2ab^2+b^4} \div \dfrac{3a-3b^2}{a^2+3ab^2+2b^4}$

49. $\left(\dfrac{x^2-x-56}{x^2+14x+49} \cdot \dfrac{x^2+4x-21}{x^2-9x+8}\right) + \dfrac{3}{x^2+8x-9}$

50. $\left(\dfrac{x^2-8x+16}{2x^2-x-6} \cdot \dfrac{2x^2-7x-15}{x^2-2x-24}\right) \div \dfrac{x^2-9x+20}{x^2+2x-8}$

51. If $f(x) = \dfrac{x+1}{x+2}$ and $g(x) = \dfrac{x}{x+4}$, find

   a) the domain of $f(x)$.
   b) the domain of $g(x)$.
   c) $(f+g)(x)$.
   d) the domain of $(f+g)(x)$.

52. If $f(x) = \dfrac{x}{x^2-9}$ and $g(x) = \dfrac{x+4}{x-3}$, find

   a) the domain of $f(x)$.
   b) the domain of $g(x)$.
   c) $(f+g)(x)$.
   d) the domain of $(f+g)(x)$.

[6.3] *Simplify each complex fraction.*

53. $\dfrac{\dfrac{9a^2b}{2c}}{\dfrac{6ab^4}{4c^3}}$

54. $\dfrac{\dfrac{1}{x}+\dfrac{1}{y}}{\dfrac{x}{y}+y^2}$

55. $\dfrac{\dfrac{3}{y}-\dfrac{1}{y^2}}{5+\dfrac{1}{y^2}}$

56. $\dfrac{a^{-1}+2}{a^{-1}+\dfrac{1}{a}}$

57. $\dfrac{x^{-2}+\dfrac{1}{x}}{\dfrac{1}{x^2}-\dfrac{1}{x}}$

58. $\dfrac{\dfrac{1}{x^2-3x-18}+\dfrac{2}{x^2-2x-15}}{\dfrac{3}{x^2-11x+30}+\dfrac{1}{x^2-9x+20}}$

In Exercises 59 and 60, the area and width of each rectangle are given. Find the length, l, by dividing the area, A, by the width, w.

59.

$A = \dfrac{x^2+5x+6}{x+4}$ $w = \dfrac{x^2+8x+15}{x^2+5x+4}$

l

60.

$A = \dfrac{x^2+10x+24}{x+5}$ $w = \dfrac{x^2+9x+18}{x^2+7x+10}$

l

[6.4] *In Exercises 61–70, solve each equation.*

61. $\dfrac{2}{x} = \dfrac{5}{9}$

62. $\dfrac{x}{1.5} = \dfrac{x-4}{4.5}$

63. $\dfrac{3x+4}{5} = \dfrac{2x-8}{3}$

64. $\dfrac{x}{4.8} + \dfrac{x}{2} = 1.7$

65. $\dfrac{2}{y} + \dfrac{1}{5} = \dfrac{3}{y}$

66. $\dfrac{2}{x+4} - \dfrac{3}{x-4} = \dfrac{-11}{x^2-16}$

67. $\dfrac{x}{x^2-9} + \dfrac{2}{x+3} = \dfrac{4}{x-3}$

68. $\dfrac{7}{x^2-5} + \dfrac{3}{x+5} = \dfrac{4}{x-5}$

69. $\dfrac{x-3}{x-2} + \dfrac{x+1}{x+3} = \dfrac{2x^2+x+1}{x^2+x-6}$

70. $\dfrac{x+1}{x+3} + \dfrac{x+2}{x-4} = \dfrac{2x^2-18}{x^2-x-12}$

71. Solve $\dfrac{1}{a} + \dfrac{1}{b} = \dfrac{1}{c}$ for b.

72. Solve $z = \dfrac{x-\bar{x}}{s}$ for $\bar{x}$.

73. ***Resistors*** Three resistors of 100, 200, and 600 ohms are connected in parallel. Find the total resistance of the circuit. Use the formula $\dfrac{1}{R_T} = \dfrac{1}{R_1} + \dfrac{1}{R_2} + \dfrac{1}{R_3}$.

74. ***Focal Length*** What is the focal length, f, of a curved mirror if the object distance, p, is 6 centimeters and the image distance, q, is 2 centimeters? Use the formula $\dfrac{1}{p} + \dfrac{1}{q} = \dfrac{1}{f}$.

In Exercises 75 and 76, each pair of triangles is similar. Find the lengths of the unknown sides.

75. *Triangles*

76.

[6.5] *In Exercises 77–82, answer the question asked. When necessary, round answers to the nearest hundredth.*

77. *Picking String Beans* Sanford and Jerome work on a farm near Oklahoma City, Oklahoma. Sanford can pick a basket of string beans in 40 minutes while Jerome can do the same task in 30 minutes. How long will it take them to pick a basket of string beans if they work together?

78. *Planting a Garden* Sam and Fran plan to plant a flower garden. Together, they can plant a garden in 4.2 hours. If Sam by himself can plant the same garden in 6 hours, how long will it take Fran to plant the garden by herself?

79. *Fractions* What number added to the numerator and subtracted from the denominator of the fraction $\frac{1}{11}$ makes the result equal to $\frac{1}{2}$?

80. *Fractions* When the reciprocal of twice a number is subtracted from 1, the result is the reciprocal of three times the number. Find the number.

81. *Traveling by Boat* Paul Webster's motorboat can travel 15 miles per hour in still water. Traveling with the current of a river, the boat can travel 20 miles in the same time it takes to go 10 miles against the current. Find the current.

82. *Flying a Plane* A small plane and a car start from the same location, at the same time, heading toward the same town 450 miles away. The speed of the plane is three times the speed of the car. The plane arrives at the town 6 hours ahead of the car. Find the speeds of the car and the plane.

[6.6] *Find each indicated quantity.*

83. x is directly proportional to the square of y. If $x = 45$ when $y = 3$, find x when $y = 2$.

84. W is directly proportional to the the square of L and inversely proportional to A. If $W = 4$ when $L = 2$ and $A = 10$, find W when $L = 5$ and $A = 20$.

85. z is jointly proportional to x and y and inversely proportional to the square of r. If $z = 12$ when $x = 20$, $y = 8$, and $r = 8$, find z when $x = 10$, $y = 80$, and $r = 3$.

86. *Electric Surcharge* The Potomac Electric Power Company places a space occupancy surcharge, s, on electric bills that is directly proportional to the amount of energy used, e. If the surcharge is $7.20 when 3600 kilowatt-hours are used, what is the surcharge when 4200 kilowatt-hours are used?

87. *Free Fall* The distance, d, an object falls in free fall is directly proportional to the square of the time, t. If a person falls 16 feet in 1 second, how far will the person fall in 10 seconds? Disregard wind resistance.

See Exercise 87.

88. *Area* The area, A, of a circle varies directly with the square of its radius, r. If the area is 78.5 when the radius is 5, find the area when the radius is 8.

89. *Melting Ice Cube* The time, t, for an ice cube to melt is inversely proportional to the temperature of the water it is in. If it takes an ice cube 1.7 minutes to melt in 70°F water, how long will it take an ice cube of the same size to melt in 50°F water?

Chapter Practice Test

1. Determine the values that are excluded in the expression $\dfrac{x + 4}{x^2 + 3x - 28}$

2. Determine the domain of the function $f(x) = \dfrac{x^2 + 7}{2x^2 + 7x - 4}$

Simplify each expression.

3. $\dfrac{8x^7y^2 + 16x^2y + 18x^3y^3}{2x^2y}$

4. $\dfrac{x^2 - 3xy - 10y^2}{x^2 + 3xy + 2y^2}$

In Exercise 5–14, perform the indicated operation.

5. $\dfrac{3xy^4}{6x^2y^3} \cdot \dfrac{2x^2y^4}{x^5y^7}$

6. $\dfrac{x + 1}{x^2 - 7x - 8} \cdot \dfrac{x^2 - x - 56}{x^2 + 9x + 14}$

7. $\dfrac{5a + 10b}{a^2 - 4b^2} \div \dfrac{a^3 + a^2b}{a^2 - 2ab}$

8. $\dfrac{x^3 + y^3}{x + y} \div \dfrac{x^2 - xy + y^2}{x^2 + y^2}$

9. $\dfrac{3}{x + 1} + \dfrac{2}{x^2}$

10. $\dfrac{x - 1}{x^2 - 9} - \dfrac{x}{x^2 - 2x - 3}$

11. $\dfrac{m}{12m^2 + 4mn - 5n^2} + \dfrac{2m}{12m^2 + 28mn + 15n^2}$

12. $\dfrac{x + 1}{4x^2 - 4x + 1} + \dfrac{3}{2x^2 + 5x - 3}$

13. $\dfrac{x^3 - 8}{x^2 + 5x - 14} \div \dfrac{x^2 + 2x + 4}{x^2 + 10x + 21}$

14. If $f(x) = \dfrac{x - 3}{x + 5}$ and $g(x) = \dfrac{x}{2x + 3}$, find **a)** $(f + g)(x)$. **b)** the domain of $(f + g)(x)$.

15. If the area of a rectangle is $\dfrac{x^2 + 11x + 30}{x + 2}$ and the length is $\dfrac{x^2 + 9x + 18}{x + 3}$, find the width.

Simplify.

16. $\dfrac{\dfrac{1}{x} + \dfrac{1}{y}}{\dfrac{1}{x} - \dfrac{1}{y}}$

17. $\dfrac{\dfrac{a^2 - b^2}{ab}}{\dfrac{a + b}{b^2}}$

18. $\dfrac{\dfrac{3}{x} - \dfrac{4}{x^2}}{5 - \dfrac{1}{x}}$

Solve each equation.

19. $\dfrac{x}{6} - \dfrac{x}{5} = -1$

20. $\dfrac{x}{x - 8} + \dfrac{6}{x - 2} = \dfrac{x^2}{x^2 - 10x + 16}$

21. Solve $A = \dfrac{2b}{C - d}$ for C.

22. ***Wattage Rating*** The wattage rating of an appliance, W, varies jointly as the square of the current, I, and the resistance, R. If the wattage is 10 when the current is 1 ampere and the resistance is 1000 ohms, find the wattage when the current is 0.5 ampere and the resistance is 300 ohms.

23. R varies directly as P and inversely as the square of T. If $R = 30$ when $P = 40$ and $T = 2$, find R when $P = 50$ and $T = 5$.

24. *Washing Windows* Paul Weston can wash the windows of a house in 10 hours. His friend, Nancy Delaney, can wash the same windows in 7 hours. How long will it take them together to wash the windows of this house?

25. *Jogging* Cameron Barnette and Ashley Elliot start rollerblading at the same time at the beginning of a trail. Cameron averages 8 miles per hour, while Ashley averages 5 miles per hour. If it takes Ashley $\frac{1}{2}$ hour

longer than Cameron to reach the end of the trail, how long is the trail?

Cumulative Review Test

Take the following test and check your answers with those that appear at the end of the test. Review any questions that you answered incorrectly. The section and objective where the material was covered are indicated after the answer.

1. Illustrate the set $\left\{ x \mid -\dfrac{5}{3} < x \le \dfrac{19}{4} \right\}$ on the number line.

2. Evaluate $-3x^3 - 2x^2y + \dfrac{1}{2}xy^2$ when $x = 2$ and $y = \dfrac{1}{2}$.

3. Solve the equation $2(x + 3) = \dfrac{1}{2}(x - 5)$

4. *Distance Learning* The Internet has made it possible for schools to offer degrees through online distance learning. The following diagram shows the degrees most often granted through online programs in 2003.

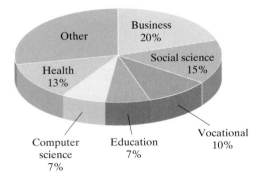

Source: The CEO Forum and Market Data Research

a) What percent makes up the "Other" category?

b) If approximately 220,000 degrees were granted through online programs, approximately how many were awarded in business?

5. Evaluate $4x^2 - 3y - 5$ when $x = 4$ and $y = -2$.

6. Simplify $\left(\dfrac{3x^5y^6}{6x^4y^7} \right)^3$.

7. Solve $F = \dfrac{mv^2}{r}$ for m.

8. *Simple Interest* Carmalla Banjanie invested $5000 in a certificate of deposit for 1 year. When she redeemed the certificate, she received $5300, what was the simple interest rate?

9. *Meeting for a Picnic* Dawn and Paula leave their homes at 8 A.M. planning to meet for a picnic at a point between them. If Dawn travels at 60 miles per hour and Paula travels at 50 miles per hour, and they live 330 miles apart, at what time will they meet?

10. Solve $\left| \dfrac{3x + 5}{3} \right| - 3 = 6$.

11. Graph $y = x^2 - 2$.

12. Let $f(x) = \sqrt{2x + 7}$. Evaluate $f(9)$.

13. Find the slope of the line passing through $(2, -4)$ and $(-5, -3)$.

14. Find an equation of a line through $\left(\dfrac{1}{2}, 3 \right)$ that is parallel to the graph of $2x + 3y - 9 = 0$. Write the equation in standard form.

15. Solve the system of equations:
$$2x - y = -2$$
$$4x + 3y = 11$$

16. Multiply $(3x^2 - 4y)(3x^2 + 4y)$

17. Factor $3x^2 - 30x + 75$.

18. Graph $y = |x| + 2$.

19. Add $\dfrac{7}{3x^2 + x - 4} + \dfrac{9x + 2}{3x^2 - 2x - 8}$.

20. Solve $\dfrac{3y - 2}{y + 1} = 4 - \dfrac{y + 2}{y - 1}$.

Answers to Cumulative Review Test

1. ; [Sec. 1.2, Obj. 3]   **2.** $-27\frac{3}{4}$; [Sec. 1.4, Obj. 4]   **3.** $\frac{7}{3}$; [Sec. 2.1, Obj. 4]

4. a) 28%; [Sec. 1.3, Obj. 3]   **b)** $\approx 44,000$; [Sec. 1.3, Obj. 4]   **5.** 65; [Sec. 1.4, Obj. 4]   **6.** $\dfrac{x^3}{8y^3}$; [Sec. 1.5, Obj. 7]

7. $m = \dfrac{rF}{v^2}$; [Sec. 2.2, Obj. 2]   **8.** 6%; [Sec. 2.2, Obj. 1]   **9.** 11 A.M. [Sec. 2.4, Obj. 1]   **10.** $\left\{-\frac{32}{3}, \frac{22}{3}\right\}$; [Sec. 2.6, Obj. 2]

11. ; [Sec. 3.1, Obj. 3]   **12.** 5; [Sec. 3.2, Obj. 4]   **13.** $-\frac{1}{7}$; [Sec. 3.4, Obj. 2]

14. $2x + 3y = 10$; [Sec. 3.5, Obj. 3]   **15.** $\left(\frac{1}{2}, 3\right)$; [Sec. 4.1, Obj. 3]

16. $9x^2 - 16y^2$; [Sec. 5.2, Obj. 5]   **17.** $3(x - 5)^2$; [Sec. 5.6, Obj. 2]

18. ; [Sec. 3.1, Obj. 3]   **19.** $\dfrac{3x - 4}{(x - 1)(x - 2)}$; [Sec. 6.2, Obj. 3]   **20.** 4; [Sec. 6.4, Obj. 2]

Chapter 7

Roots, Radicals, and Complex Numbers

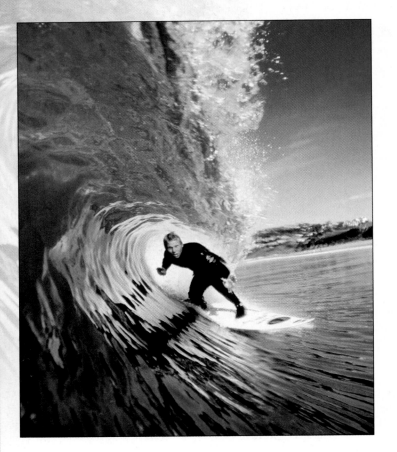

7.1 Roots and Radicals

7.2 Rational Exponents

7.3 Simplifying Radicals

7.4 Adding, Subtracting, and Multiplying Radicals

7.5 Dividing Radicals

7.6 Solving Radical Equations

7.7 Complex Numbers

Chapter Summary
Chapter Review Exercises
Chapter Practice Test
Cumulative Review Test

Many scientific formulas, including many that pertain to real-life situations, contain radical expressions. For example, when police investigate an accident, the speed of a car before braking can be estimated using a formula containing a radical expression. The velocity of an object free falling and the height of ocean waves can be determined using formulas containing radicals. On page 463, we will determine the velocity of the hammer of a pile driver and the height of ocean waves.

SSM

Study Guide

CD/Video

MathPro 4/5

PH Math
Tutor Center

prenhall.com/Angel

A Look Ahead

In this chapter, we explain how to add, subtract, multiply, and divide radical expressions. We also introduce imaginary numbers and complex numbers.

In Section 7.1, we graph square root and cube root functions. In Section 7.2 we change expressions from radical form to exponential form, and vice versa. The rules of exponents discussed in Section 1.5 still apply to rational exponents and we use those rules again here. In Sections 7.3 and 7.4 we discuss simplifying radicals.

In Section 7.5, we discuss rationalizing the denominator, which removes radicals from a denominator. Make sure that you understand the three requirements for a radical expression to be simplified, as discussed in Section 7.5.

In Section 7.6, we discuss how to solve equations that contain radical expressions. We will use these procedures again in Chapters 8 and 11. Section 7.6 also illustrates applications of radical equations.

Imaginary numbers and complex numbers are introduced in Section 7.7. These numbers play a very important role in higher mathematics courses. We will be using imaginary and complex numbers throughout Chapter 8.

7.1 ROOTS AND RADICALS

1 Find square roots.
2 Find cube roots.
3 Understand odd and even roots.
4 Evaluate radicals using absolute value.

In this chapter we expand on the concept of radicals introduced in Chapter 1. In the expression $\sqrt{x}$, the $\sqrt{}$ is called the **radical sign**. The expression within the radical sign is called the **radicand**.

$$\text{Radical sign} \searrow$$
$$\sqrt{x}$$
$$\uparrow \text{Radicand}$$

The entire expression, including the radical sign and radicand, is called the **radical expression**. Another part of the radical expression is its index. The **index** (plural *indices* or *indexes*) gives the "root" of the expression. Square roots have an index of 2. The index of a square root is generally not written. Thus,

$$\sqrt{x} \quad \text{means} \quad \sqrt[2]{x}$$

1 Find Square Roots

Every positive number has two square roots, a principal or positive square root and a negative square root. For any positive number x, the positive square root is written $\sqrt{x}$, and the negative square root is written $-\sqrt{x}$.

Number	Principal or Positive Square Root	Negative Square Root
25	$\sqrt{25}$	$-\sqrt{25}$
10	$\sqrt{10}$	$-\sqrt{10}$

DEFINITION

The **principal square root** of a positive number a, written $\sqrt{a}$, is the *positive* number b such that $b^2 = a$.

Examples

$$\sqrt{25} = 5 \qquad \text{since } 5^2 = 5 \cdot 5 = 25$$
$$\sqrt{0.49} = 0.7 \qquad \text{since } (0.7)^2 = (0.7)(0.7) = 0.49$$
$$\sqrt{\frac{4}{9}} = \frac{2}{3} \qquad \text{since } \left(\frac{2}{3}\right)^2 = \left(\frac{2}{3}\right)\left(\frac{2}{3}\right) = \frac{4}{9}$$

Remember that $-\sqrt{25}$ means the opposite of $\sqrt{25}$. Because $\sqrt{25} = 5$, $-\sqrt{25} = -5$.

In this book whenever we use the words *square root* we will be referring to the principal or positive square root. If you are asked to find the value of $\sqrt{25}$, your answer will be 5.

In Chapter 1, we indicated that a rational number is one that can be represented as either a terminating or repeating decimal number. If you use the square root key on your calculator, $\boxed{\sqrt{}}$, to evaluate the three examples above, you will find they are all terminating or repeating decimal numbers. Thus, they are all *rational numbers*. Many radicals, such as $\sqrt{2}$ and $\sqrt{10}$, are not rational numbers. When evaluating $\sqrt{2}$ and $\sqrt{10}$ on a calculator, the results are nonterminating, nonrepeating decimal numbers. Thus, $\sqrt{2}$ and $\sqrt{10}$ are *irrational numbers*.

Radical	Calculator Results	
$\sqrt{2}$	1.414213562	*Nonterminating, nonrepeating decimal*
$\sqrt{10}$	3.16227766	*Nonterminating, nonrepeating decimal*

Now let's consider $\sqrt{-25}$. Since the square of any real number will always be greater than or equal to 0, there is no real number that when squared equals -25. For this reason, $\sqrt{-25}$ is *not a real number*. Since the square of any real number cannot be negative, *the square root of a negative number is not a real number*. If you evaluate $\sqrt{-25}$ on a calculator, you will get an error message. We will discuss numbers like $\sqrt{-25}$ later in this chapter.

Radical	Calculator Results	
$\sqrt{-25}$	Error	*$\sqrt{-25}$ is not a real number.*
$\sqrt{-2}$	Error	*$\sqrt{-2}$ is not a real number.*

HELPFUL HINT

Do not confuse $-\sqrt{36}$ with $\sqrt{-36}$. Because $\sqrt{36} = 6, -\sqrt{36} = -6$. However, $\sqrt{-36}$ is not a real number. The square root of a negative number is not a real number.

$$\sqrt{36} = 6$$
$$-\sqrt{36} = -6$$
$$\sqrt{-36} \text{ is not a real number.}$$

The Square Root Function

When graphing square root functions, functions of the form $f(x) = \sqrt{x}$, we must always remember that the radicand, x, cannot be negative. Thus, the domain of $f(x) = \sqrt{x}$ is $\{x | x \geq 0\}$, or $[0, \infty)$ in interval notation. To graph $f(x) = \sqrt{x}$, we can select some convenient values of x and find the corresponding values of $f(x)$ or y and then plot the ordered pairs, as shown in Figure 7.1.

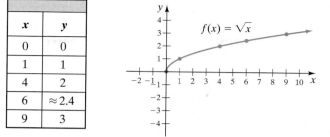

x	y
0	0
1	1
4	2
6	≈2.4
9	3

FIGURE 7.1

Since the value of $f(x)$ can never be negative, the range of $f(x) = \sqrt{x}$ is $\{y | y \geq 0\}$, or $[0, \infty)$ in interval notation.

By studying Figure 7.1, do you think you can graph $g(x) = -\sqrt{x}$? The graph of $g(x) = -\sqrt{x}$ would be similar to the graph in Figure 7.1, but the graph would be drawn below the x-axis. Can you explain why? What about the graph of $h(x) = \sqrt{x - 4}$? For the graph of $h(x) = \sqrt{x - 4}$, you would only select values of $x \geq 4$ since the radicand cannot be negative. The domain of $h(x) = \sqrt{x - 4}$ is $\{x | x \geq 4\}$ or $[4, \infty)$.

To evaluate radical functions, it may be necessary to use a calculator.

EXAMPLE 1   For each function, find the indicated value(s).

    a) $f(x) = \sqrt{12x + 4}, f(5)$ **b)** $g(r) = -\sqrt{-3r + 1}, g(-5)$ and $g(7)$

Solution   **a)** $f(5) = \sqrt{12(5) + 4}$ *Substitute 5 for x.*

           $= \sqrt{64}$

           $= 8$

    b) $g(-5) = -\sqrt{-3(-5) + 1}$ *Substitute −5 for r.*

           $= -\sqrt{16}$

           $= -4$

        $g(7) = -\sqrt{-3(7) + 1}$ *Substitute 7 for r.*

           $= -\sqrt{-20}$ *Not a real number.*

Thus, $g(7)$ is not a real number.   ✳

2 Find Cube Roots

The index of a cube root is 3. We introduced cube roots in Section 1.4, where we explained how to find cube roots on a calculator. You may wish to review that material now.

DEFINITION | The **cube root** of a number a, written $\sqrt[3]{a}$, is the number b such that $b^3 = a$.

Examples

$$\sqrt[3]{8} = 2 \qquad \text{since } 2^3 = 8$$
$$\sqrt[3]{-27} = -3 \qquad \text{since } (-3)^3 = -27$$

For each real number, there is only one cube root. The cube root of a positive number is positive and the cube root of a negative number is negative. The cube root function, $f(x) = \sqrt[3]{x}$, has all real numbers as its domain.

EXAMPLE 2   For each function, find the indicated value(s).

a) $f(x) = \sqrt[3]{10x + 34}, f(3)$ b) $g(r) = \sqrt[3]{12r - 20}, g(-4)$ and $g(1)$

Solution   a) $f(3) = \sqrt[3]{10(3) + 34}$ *Substitute 3 for x.*

$\qquad = 4$

b) $g(-4) = \sqrt[3]{12(-4) - 20}$ *Substitute −4 for r.*

$\qquad = \sqrt[3]{-68}$

$\qquad \approx -4.081655102$ *From a calculator*

$g(1) = \sqrt[3]{12(1) - 20}$ *Substitute 1 for r.*

$\qquad = \sqrt[3]{-8}$

NOW TRY EXERCISE 71 $\qquad = -2$

The Cube Root Function

Figure 7.2 shows the graph of $y = \sqrt[3]{x}$. To obtain this graph, we substituted values for x and found the corresponding values of $f(x)$ or y.

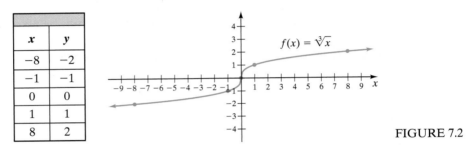

x	y
−8	−2
−1	−1
0	0
1	1
8	2

FIGURE 7.2

Notice that both the domain and range are all real numbers, $\mathbb{R}$. We will ask you to graph cube root functions on your graphing calculator in the exercise set.

3 Understand Odd and Even Roots

Up to this point we have discussed square and cube roots. Other radical expressions have different indices. For example, in the expression $\sqrt[5]{xy}$, (read "the fifth root of xy") the index is 5 and the radicand is xy.

Radical expressions that have indices of $2, 4, 6, \ldots$, or any even integer are **even roots**. Square roots are even roots since their index is 2. Radical expressions that have indices of $3, 5, 7, \ldots$, or any odd integer are **odd roots**.

Even Indices

The nth root of a, $\sqrt[n]{a}$, where n is an *even index* and a is a nonnegative real number, is the nonnegative real number b such that $b^n = a$.

Examples of Even Roots

$\sqrt{9} = 3$ since $3^2 = 3 \cdot 3 = 9$

$\sqrt[4]{16} = 2$ since $2^4 = 2 \cdot 2 \cdot 2 \cdot 2 = 16$

$\sqrt[6]{729} = 3$ since $3^6 = 3 \cdot 3 \cdot 3 \cdot 3 \cdot 3 \cdot 3 = 729$

Any real number when raised to an even power results in a positive real number. Thus, *when the index of a radical is even, the radicand must be nonnegative for the radical to be a real number.*

HELPFUL HINT

There is an important difference between $-\sqrt[4]{16}$ and $\sqrt[4]{-16}$. The number $-\sqrt[4]{16}$ is the opposite of $\sqrt[4]{16}$. Because $\sqrt[4]{16} = 2$, $-\sqrt[4]{16} = -2$. However, $\sqrt[4]{-16}$ is not a real number since no real number when raised to the fourth power equals -16.

$$-\sqrt[4]{16} = -(\sqrt[4]{16}) = -2$$

$\sqrt[4]{-16}$ is not a real number.

Odd Indices

The nth root of a, $\sqrt[n]{a}$, where n is an *odd index* and a is *any real number*, is the real number b such that $b^n = a$.

Examples of Odd Roots

$\sqrt[3]{8} = 2$ since $2^3 = 2 \cdot 2 \cdot 2 = 8$

$\sqrt[3]{-8} = -2$ since $(-2)^3 = (-2)(-2)(-2) = -8$

$\sqrt[5]{243} = 3$ since $3^5 = 3 \cdot 3 \cdot 3 \cdot 3 \cdot 3 = 243$

$\sqrt[5]{-243} = -3$ since $(-3)^5 = (-3)(-3)(-3)(-3)(-3) = -243$

$\sqrt[7]{128} = 2$ since $2^7 = 2 \cdot 2 \cdot 2 \cdot 2 \cdot 2 \cdot 2 \cdot 2 = 128$

An odd root of a positive number is a positive number, and an odd root of a negative number is a negative number.

It is important to realize that a radical with an even index must have a nonnegative radicand if it is to be a real number. A radical with an odd index will be a real number with any real number as its radicand. Note that $\sqrt[n]{0} = 0$, regardless of whether n is an odd or even index.

EXAMPLE 3   Indicate whether or not each radical expression is a real number. If the expression is a real number, find its value.

a) $\sqrt[4]{-81}$ **b)** $-\sqrt[4]{81}$ **c)** $\sqrt[5]{-32}$ **d)** $-\sqrt[5]{-32}$

Solution   **a)** Not a real number. Even roots of negative numbers are not real numbers.

b) Real number, $-\sqrt[4]{81} = -(\sqrt[4]{81}) = -(3) = -3$

c) Real number, $\sqrt[5]{-32} = -2$ since $(-2)^5 = -32$

d) Real number, $-\sqrt[5]{-32} = -(-2) = 2$

NOW TRY EXERCISE 21

Table 7.1 summarizes the information about even and odd roots.

TABLE 7.1		
	n is even	**n is odd**
$a > 0$	$\sqrt[n]{a}$ is a positive real number.	$\sqrt[n]{a}$ is a positive real number.
$a < 0$	$\sqrt[n]{a}$ is not a real number.	$\sqrt[n]{a}$ is a negative real number.
$a = 0$	$\sqrt[n]{0} = 0$	$\sqrt[n]{0} = 0$

4 Evaluate Radicals Using Absolute Value

You may think that $\sqrt{a^2} = a$, but this is not necessarily true. Below we evaluate $\sqrt{a^2}$ for $a = 2$ and $a = -2$. You will see that when $a = -2$, $\sqrt{a^2} \neq a$.

$a = 2$: $\qquad \sqrt{a^2} = \sqrt{2^2} = \sqrt{4} = 2 \qquad$ *Note that* $\sqrt{2^2} = 2$.

$a = -2$: $\qquad \sqrt{a^2} = \sqrt{(-2)^2} = \sqrt{4} = 2 \qquad$ *Note that* $\sqrt{(-2)^2} \neq -2$.

By examining these examples and other examples we could make up, we can reason that $\sqrt{a^2}$ will *always be a positive real number* for any nonzero real number a. Recall from Section 1.3 that the *absolute value* of any real number a, or $|a|$, is also a positive number for any nonzero number. We use these facts to reason that

> **For any real number *a*,**
> $$\sqrt{a^2} = |a|$$

This indicates that the principal square root of a^2 is the absolute value of a.

EXAMPLE 4 Use absolute value to evaluate. **a)** $\sqrt{9^2}$ **b)** $\sqrt{0^2}$ **c)** $\sqrt{(12.8)^2}$

Solution **a)** $\sqrt{9^2} = |9| = 9$ **b)** $\sqrt{0^2} = |0| = 0$ **c)** $\sqrt{(12.8)^2} = |12.8| = 12.8$ ✳

When simplifying a square root, if the radicand contains a variable and we are not sure that the radicand is positive, we need to use absolute value signs when simplifying.

EXAMPLE 5 Simplify. **a)** $\sqrt{(x + 7)^2}$ **b)** $\sqrt{9x^2}$ **c)** $\sqrt{25y^6}$ **d)** $\sqrt{a^2 - 6a + 9}$

Solution Each square root has a radicand that contains a variable. Since we do not know the value of the variable, we do not know whether it is positive or negative. Therefore, we must use absolute value signs when simplifying.

a) $\sqrt{(x + 7)^2} = |x + 7|$

b) Write $9x^2$ as $(3x)^2$, then simplify.
$$\sqrt{9x^2} = \sqrt{(3x)^2} = |3x|$$

c) Write $25y^6$ as $(5y^3)^2$, then simplify.
$$\sqrt{25y^6} = \sqrt{(5y^3)^2} = |5y^3|$$

d) Notice that $a^2 - 6a + 9$ is a perfect square trinomial. Write the trinomial as the square of a binomial, then simplify.

NOW TRY EXERCISE 63

$$\sqrt{a^2 - 6a + 9} = \sqrt{(a - 3)^2} = |a - 3|$$ ✳

If you have a square root whose radicand contains a variable and are given instructions like "Assume all variables represent positive values and the radicand is nonnegative," then it is not necessary to use the absolute value sign when simplifying.

EXAMPLE 6 Simplify. Assume all variables represent positive values and the radicand is nonnegative.

a) $\sqrt{25x^2}$ **b)** $\sqrt{81p^4}$ **c)** $\sqrt{36x^6}$ **d)** $\sqrt{4x^2 - 12xy + 9y^2}$

Solution **a)** $\sqrt{25x^2} = \sqrt{(5x)^2} = 5x$ $\qquad$ *Write $25x^2$ as $(5x)^2$.*

b) $\sqrt{81p^4} = \sqrt{(9p^2)^2} = 9p^2$ $\qquad$ *Write $81p^4$ as $(9p^2)^2$.*

c) $\sqrt{36x^6} = \sqrt{(6x^3)^2} = 6x^3$ $\qquad$ *Write $36x^6$ as $(6x^3)^2$.*

d) $\sqrt{4x^2 - 12xy + 9y^2} = \sqrt{(2x - 3y)^2}$ $\qquad$ *Write $4x^2 - 12xy + 9y^2$ as $(2x - 3y)^2$.*
$$= 2x - 3y$$ ✳

We only need to be concerned about adding absolute value signs when discussing square (and other even) roots. We do not need to use absolute value signs when the index is odd.

Exercise Set 7.1

Concept/Writing Exercises

1. a) How many square roots does every positive real number have? Name them.

   b) Find all square roots of the number 49.

   c) In this text, when we refer to "the square root," which square root are we referring to?

   d) Find the square root of 49.

2. a) What are even roots? Give an example of an even root.

   b) What are odd roots? Give an example of an odd root.

3. Explain why $\sqrt{-81}$ is not a real number.

4. Will a radical expression with an odd index and a real number as the radicand always be a real number? Explain your answer.

5. Will a radical expression with an even index and a real number as the radicand always be a real number? Explain your answer.

6. a) To what is $\sqrt{a^2}$ equal?

   b) To what is $\sqrt{a^2}$ equal if we know $a \geq 0$?

7. a) Evaluate $\sqrt{a^2}$ for $a = 1.3$.

   b) Evaluate $\sqrt{a^2}$ for $a = -1.3$.

8. a) Evaluate $\sqrt{a^2}$ for $a = 5.72$.

   b) Evaluate $\sqrt{a^2}$ for $a = -5.72$.

9. a) Evaluate $\sqrt[3]{27}$.

   b) Evaluate $-\sqrt[3]{27}$.

   c) Evaluate $\sqrt[3]{-27}$.

10. a) Evaluate $\sqrt[4]{81}$.

    b) Evaluate $-\sqrt[4]{81}$.

    c) Evaluate $\sqrt[4]{-81}$.

Practice the Skills

Evaluate each radical expression if it is a real number. Use a calculator to approximate irrational numbers to the nearest hundredth. If the expression is not a real number, so state.

11. $\sqrt{64}$

12. $-\sqrt{64}$

13. $\sqrt[3]{-64}$

14. $\sqrt[3]{125}$

15. $\sqrt[3]{-125}$

16. $-\sqrt[3]{-125}$

17. $\sqrt[5]{1}$

18. $\sqrt[5]{-1}$

19. $-\sqrt[5]{-1}$

20. $\sqrt[6]{64}$

21. $\sqrt[6]{-64}$

22. $\sqrt[3]{343}$

23. $\sqrt[3]{-343}$

24. $\sqrt{100}$

25. $\sqrt{-36}$

26. $\sqrt{45.3}$

27. $\sqrt{-45.3}$

28. $\sqrt{53.9}$

29. $\sqrt{\dfrac{1}{25}}$

30. $\sqrt{-\dfrac{1}{25}}$

31. $\sqrt[3]{\dfrac{1}{8}}$

32. $\sqrt[3]{-\dfrac{1}{8}}$

33. $\sqrt{\dfrac{4}{9}}$

34. $\sqrt[3]{\dfrac{1}{27}}$

35. $\sqrt[3]{-\dfrac{8}{27}}$

36. $\sqrt[4]{-6.2}$

37. $-\sqrt[4]{18.2}$

38. $\sqrt[5]{93}$

Use absolute value to evaluate.

39. $\sqrt{9^2}$

40. $\sqrt{(-9)^2}$

41. $\sqrt{(-19)^2}$

42. $\sqrt{19^2}$

43. $\sqrt{152^2}$

44. $\sqrt{(-152)^2}$

45. $\sqrt{(235.23)^2}$

46. $\sqrt{(-201.5)^2}$

47. $\sqrt{(0.06)^2}$

48. $\sqrt{(-0.19)^2}$

49. $\sqrt{\left(\dfrac{11}{13}\right)^2}$

50. $\sqrt{\left(-\dfrac{101}{319}\right)^2}$

Write as an absolute value.

51. $\sqrt{(x-9)^2}$

52. $\sqrt{(a+7)^2}$

53. $\sqrt{(x-3)^2}$

54. $\sqrt{(9a-11b)^2}$

55. $\sqrt{(3x^2-2)^2}$

56. $\sqrt{(7y^2-3y)^2}$

57. $\sqrt{(6a^3-5b^4)^2}$

58. $\sqrt{(7y^4-2z^3)^2}$

Use absolute value to simplify. You may need to factor first.

59. $\sqrt{a^{10}}$

60. $\sqrt{y^{18}}$

61. $\sqrt{z^{30}}$

62. $\sqrt{x^{102}}$

63. $\sqrt{a^2-8a+16}$

64. $\sqrt{x^2+12x+36}$

65. $\sqrt{9a^2+12ab+4b^2}$

66. $\sqrt{4x^2-20xy+25y^2}$

Find the indicated value of each function. Use your calculator to approximate irrational numbers. Round irrational numbers to the nearest thousandth.

67. $f(x)=\sqrt{5x-6},f(2)$

68. $f(a)=\sqrt{14a-36},f(4)$

69. $g(x)=\sqrt{64-8x},g(-3)$

70. $p(x)=\sqrt[3]{8x+9},p(2)$

71. $h(x)=\sqrt[3]{9x^2+4},h(4)$

72. $k(c)=\sqrt[4]{16c-5},k(4)$

73. $f(a)=\sqrt[3]{-2x^2+x-6},f(-3)$

74. $t(x)=\sqrt[4]{2x^3-3x^2+6x},t(2)$

Problem Solving

75. Find $f(81)$ if $f(x)=x+\sqrt{x}+5$.

76. Find $g(25)$ if $g(x)=x^2+\sqrt{x}-3$.

77. Find $t(18)$ if $t(x)=\dfrac{x}{2}+\sqrt{2x}-1$.

78. Find $m(36)$ if $m(x)=\dfrac{x}{3}+\sqrt{4x}+2$.

79. Find $k(8)$ if $k(x)=x^2+\sqrt{\dfrac{x}{2}}-11$.

80. Find $r(45)$ if $r(x)=\dfrac{x}{9}+\sqrt{\dfrac{x}{5}}+21$.

81. Select a value for x for which $\sqrt{(2x+1)^2}\ne 2x+1$.

82. Select a value for x for which $\sqrt{(5x-3)^2}\ne 5x-3$.

83. For what values of x will $\sqrt{(x-1)^2}=x-1$? Explain how you determined your answer.

84. For what values of x will $\sqrt{(x+3)^2}=x+3$? Explain how you determined your answer.

85. For what values of x will $\sqrt{(2x-6)^2}=2x-6$? Explain how you determined your answer.

86. For what values of x will $\sqrt{(4x-8)^2}=4x-8$? Explain how you determined your answer.

87. **a)** For what values of a is $\sqrt{a^2}=|a|$?

b) For what values of a is $\sqrt{a^2}=a$?

c) For what values of a is $\sqrt[3]{a^3}=a$?

88. Under what circumstances is the expression $\sqrt[n]{x}$ not a real number?

89. Explain why the expression $\sqrt[n]{x^n}$ is a real number for any real number x.

90. Under what circumstances is the expression $\sqrt[n]{x^m}$ not a real number?

91. Find the domain of $\dfrac{\sqrt{x+5}}{\sqrt[3]{x+5}}$. Explain how you determined your answer.

92. Find the domain of $\dfrac{\sqrt[3]{x-2}}{\sqrt[6]{x+1}}$. Explain how you determined your answer.

By considering the domains of the functions in Exercises 93 through 96, match each function with its graph, labeled a) through d).

93. $f(x)=\sqrt{x}$

94. $f(x)=\sqrt{x^2}$

95. $f(x)=\sqrt{x-5}$

96. $f(x)=\sqrt{x+5}$

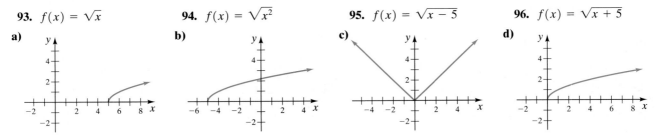

97. Give a radical function whose domain is $\{x|x \geq 6\}$.

98. Give a radical function whose domain is $\{x|x \leq 5\}$.

99. If $f(x) = -\sqrt{x}$, can $f(x)$ ever be **a)** greater than 0, **b)** equal to 0, **c)** less than 0? Explain.

100. If $f(x) = \sqrt{x + 5}$, can $f(x)$ ever be **a)** less than 0, **b)** equal to 0, **c)** greater than 0? Explain.

101. *Velocity of an Object* The velocity, V, of an object, in feet per second, after it has fallen a distance, h, in feet, can be found by the formula $V = \sqrt{64.4h}$. A pile driver is a large mass that is used as a hammer to drive pilings into soft earth to support a building or other structure.

With what velocity will the hammer hit the piling if it falls from **a)** 20 feet above the top of the piling? **b)** 40 feet above the top of the piling?

102. *Wave Action* Scripps Institute of Oceanography in La Jolla, California, developed the formula for relating wind speed, u, in knots, with the height, H, in feet, of the waves the wind produces in certain areas of the ocean. This formula is

$$u = \sqrt{\frac{H}{0.026}}$$

If waves produced by a storm have a height of 15 feet, what is the wind speed producing the waves?

103. Graph $f(x) = \sqrt{x + 1}$.

104. Graph $g(x) = -\sqrt{x}$.

105. Graph $g(x) = \sqrt{x} + 1$.

106. Graph $f(x) = \sqrt{x - 2}$.

For Exercises 107–112 use your graphing calculator.

107. Check the graph drawn in Exercise 103.

108. Check the graph drawn in Exercise 105.

109. Determine whether the domain you gave in Exercise 91 is correct.

110. Determine whether the domain you gave in Exercise 92 is correct.

111. Graph $y = \sqrt[3]{x + 4}$.

112. Graph $f(x) = \sqrt[3]{2x - 3}$.

Group Activity

In this activity, you will determine the conditions under which certain properties of radicals are true. We will discuss these properties later in this chapter. Discuss and answer these exercises as a group.

113. The property $\sqrt[n]{a} \cdot \sqrt[n]{b} = \sqrt[n]{ab}$, called the *multiplication property for radicals*, is true for certain real numbers a and b. By substituting values for a and b, determine under what conditions this property is true.

114. The property $\dfrac{\sqrt[n]{a}}{\sqrt[n]{b}} = \sqrt[n]{\dfrac{a}{b}}$, called the *division property for radicals*, is true for certain real numbers a and b. By substituting values for a and b, determine under what conditions this property is true.

Cumulative Review Exercises

Factor.

[5.4] 115. $9ax - 3bx + 12ay - 4by$

[5.5] 116. $3x^3 - 18x^2 + 24x$

117. $4x^4 + 4x^2 - 3$

[5.6] 118. $x^3 - \dfrac{8}{27}y^3$

7.2 RATIONAL EXPONENTS

1   Change a radical expression to an exponential expression.
2   Simplify radical expressions.
3   Apply the rules of exponents to rational and negative exponents.
4   Factor expressions with rational exponents.

1 Change a Radical Expression to an Exponential Expression

In this section we discuss changing radical expressions to exponential expressions and vice versa. When you see a rational exponent, you should realize that the expression can be written as a radical expression by using the following procedure.

$$\sqrt[n]{a} = a^{1/n}$$

When a is nonnegative, n can be any index.
When a is negative, n must be odd.

For the remainder of this chapter, unless you are instructed otherwise, assume that all variables in radicands represent nonnegative real numbers and that the radicand is nonnegative. With this assumption, we will not need to state that the variable is nonnegative whenever we have a radical with an even index. This will allow us to write many answers without absolute value signs.

EXAMPLE 1   Write each expression with rational exponents.

a) $\sqrt{7}$ **b)** $\sqrt[3]{13ab}$ **c)** $\sqrt[7]{-4x^2y^5}$ **d)** $\sqrt[8]{\dfrac{5x^7}{2z^{11}}}$

Solution   **a)** $\sqrt{7} = 7^{1/2}$ *Recall that the index of a square root is 2.*

b) $\sqrt[3]{13ab} = (13ab)^{1/3}$ **c)** $\sqrt[7]{-4x^2y^5} = (-4x^2y^5)^{1/7}$ **d)** $\sqrt[8]{\dfrac{5x^7}{2z^{11}}} = \left(\dfrac{5x^7}{2z^{11}}\right)^{1/8}$ ✳

Exponential expressions can be converted to radical expressions by reversing the procedure.

EXAMPLE 2   Write each expression without rational exponents.

a) $9^{1/2}$ **b)** $(-8)^{1/3}$ **c)** $y^{1/4}$ **d)** $(6x^2y)^{1/7}$ **e)** $5rs^{1/2}$

Solution   **a)** $9^{1/2} = \sqrt{9} = 3$ **b)** $(-8)^{1/3} = \sqrt[3]{-8} = -2$

NOW TRY EXERCISE 33   **c)** $y^{1/4} = \sqrt[4]{y}$ **d)** $(6x^2y)^{1/7} = \sqrt[7]{6x^2y}$ **e)** $5rs^{1/2} = 5r\sqrt{s}$ ✳

2 Simplify Radical Expressions

We can expand the preceding rule so that radicals of the form $\sqrt[n]{a^m}$ can be written as exponential expressions. Consider $a^{2/3}$. We can write $a^{2/3}$ as $(a^{1/3})^2$ or $(a^2)^{1/3}$. This suggests $a^{2/3} = (\sqrt[3]{a})^2 = \sqrt[3]{a^2}$.

For any nonnegative number a, and integers m and n,

$$\sqrt[n]{a^m} = \left(\sqrt[n]{a}\right)^m = a^{m/n} \leftarrow \text{Index}$$

Power

This rule can be used to change an expression from radical form to exponential form and vice versa. When changing a radical expression to exponential form, the *power* is placed in the *numerator*, and the *index or root* is placed in the *denominator* of the rational exponent. Thus, for example, $\sqrt[3]{x^4}$ can be written $x^{4/3}$. Also $(\sqrt[5]{y})^2$ can be written $y^{2/5}$. Additional examples follow.

Examples

$$\sqrt{y^3} = y^{3/2} \qquad \sqrt[3]{z^2} = z^{2/3} \qquad \sqrt[5]{2^8} = 2^{8/5}$$

$$(\sqrt{p})^3 = p^{3/2} \qquad (\sqrt[4]{x})^3 = x^{3/4} \qquad (\sqrt[4]{7})^3 = 7^{3/4}$$

By this rule, for nonnegative values of the variable we can write

$$\sqrt{x^5} = (\sqrt{x})^5 \qquad (\sqrt[4]{p})^3 = \sqrt[4]{p^3}$$

EXAMPLE 3   Write each expression with rational exponents, then simplify.

     a) $\sqrt[4]{x^{16}}$ **b)** $(\sqrt[3]{y})^{15}$ **c)** $(\sqrt[6]{x})^{12}$

Solution   **a)** $\sqrt[4]{x^{16}} = x^{16/4} = x^4$ **b)** $(\sqrt[3]{y})^{15} = y^{15/3} = y^5$ **c)** $(\sqrt[6]{x})^{12} = x^{12/6} = x^2$ ✳

Exponential expressions with rational exponents can be converted to radical expressions by reversing the procedure. The *numerator* of the rational exponent is the *power*, and the *denominator* of the rational exponent is the *index or root* of the radical expression. Here are some examples.

Examples

$$x^{1/2} = \sqrt{x} \qquad\qquad 5^{1/3} = \sqrt[3]{5}$$

$$6^{2/3} = \sqrt[3]{6^2} \text{ or } \left(\sqrt[3]{6}\right)^2 \qquad y^{3/10} = \sqrt[10]{y^3} \text{ or } \left(\sqrt[10]{y}\right)^3$$

$$x^{9/5} = \sqrt[5]{x^9} \text{ or } \left(\sqrt[5]{x}\right)^9 \qquad z^{10/3} = \sqrt[3]{z^{10}} \text{ or } \left(\sqrt[3]{z}\right)^{10}$$

Notice that you may choose, for example, to write $6^{2/3}$ as either $\sqrt[3]{6^2}$ or $\left(\sqrt[3]{6}\right)^2$.

EXAMPLE 4   Write each expression without rational exponents.

     a) $x^{2/5}$ **b)** $(6ab)^{5/4}$

Solution   **a)** $x^{2/5} = \sqrt[5]{x^2} \text{ or } \left(\sqrt[5]{x}\right)^2$ **b)** $(6ab)^{5/4} = \sqrt[4]{(6ab)^5} \text{ or } \left(\sqrt[4]{6ab}\right)^5$ ✳

EXAMPLE 5   Simplify.   **a)** $25^{3/2}$ **b)** $\sqrt[6]{(49)^3}$ **c)** $\sqrt[4]{(xy)^{20}}$ **d)** $\left(\sqrt[15]{z}\right)^5$

Solution   **a)** Sometimes an expression with a rational exponent can be simplified more easily by writing the expression as a radical, as illustrated.

$$25^{3/2} = \left(\sqrt{25}\right)^3 \qquad \textit{Write as a radical.}$$

$$= (5)^3$$

$$= 125$$

b) Sometimes a radical expression can be simplified more easily by writing the expression with rational exponents, as illustrated in parts **b)** through **d)**.

$$\sqrt[6]{(49)^3} = 49^{3/6} \qquad \textit{Write with a rational exponent.}$$

$$= 49^{1/2} \qquad \textit{Reduce exponent.}$$

$$= \sqrt{49} \qquad \textit{Write as a radical.}$$

$$= 7 \qquad \textit{Simplify.}$$

c) $\sqrt[4]{(xy)^{20}} = (xy)^{20/4} = (xy)^5$

NOW TRY EXERCISE 51   **d)** $\left(\sqrt[15]{z}\right)^5 = z^{5/15} = z^{1/3}$ or $\sqrt[3]{z}$ ✳

Now let's consider $\sqrt[5]{x^5}$. When written in exponential form, this is $x^{5/5} = x^1 = x$. This leads to the following rule.

> For any nonnegative real number a,
> $$\sqrt[n]{a^n} = \left(\sqrt[n]{a}\right)^n = a^{n/n} = a$$

In the preceding box, we specified that a was nonnegative. If n is an even index and a is a negative real number, $\sqrt[n]{a^n} = |a|$ and not a. For example, $\sqrt[6]{(-5)^6} = |-5| = 5$. *Since we are assuming, except where noted otherwise, that variables in radicands represent nonnegative real numbers*, we may write $\sqrt[6]{x^6} = x$ and not $|x|$. This assumption also lets us write $\sqrt{x^2} = x$ and $(\sqrt[4]{z})^4 = z$.

Examples

$$\sqrt{3^2} = 3 \qquad \sqrt[4]{y^4} = y$$
$$\sqrt[6]{(xy)^6} = xy \qquad (\sqrt[5]{z})^5 = z$$

3   Apply the Rules of Exponents to Rational and Negative Exponents

In Section 1.5, we introduced and discussed the rules of exponents. In that section we only used exponents that were whole numbers. The rules still apply when the exponents are rational numbers. Let's review those rules now.

Rules of Exponents	
For all real numbers a and b and all rational numbers m and n,	
Product rule	$a^m \cdot a^n = a^{m+n}$
Quotient rule	$\dfrac{a^m}{a^n} = a^{m-n}, \quad a \neq 0$
Negative exponent rule	$a^{-m} = \dfrac{1}{a^m}, \quad a \neq 0$
Zero exponent rule	$a^0 = 1, \quad a \neq 0$
Raising a power to a power	$(a^m)^n = a^{m \cdot n}$
Raising a product to a power	$(ab)^m = a^m b^m$
Raising a quotient to a power	$\left(\dfrac{a}{b}\right)^m = \dfrac{a^m}{b^m}, \quad b \neq 0$

Using these rules, we will now work some problems in which the exponents are rational numbers.

EXAMPLE 6   Evaluate.   **a)** $8^{-2/3}$ **b)** $(-27)^{-5/3}$ **c)** $(-32)^{-6/5}$

Solution   **a)** Begin by using the negative exponent rule.

$$8^{-2/3} = \frac{1}{8^{2/3}} \qquad \textit{Negative exponent rule}$$

$$= \frac{1}{(\sqrt[3]{8})^2} \quad \textit{Write the denominator as a radical.}$$

$$= \frac{1}{2^2} \quad \textit{Simplify the denominator.}$$

$$= \frac{1}{4}$$

b) $(-27)^{-5/3} = \frac{1}{(-27)^{5/3}} = \frac{1}{(\sqrt[3]{-27})^5} = \frac{1}{(-3)^5} = -\frac{1}{243}$

NOW TRY EXERCISE 81   **c)** $(-32)^{-6/5} = \frac{1}{(-32)^{6/5}} = \frac{1}{(\sqrt[5]{-32})^6} = \frac{1}{(-2)^6} = \frac{1}{64}$ ✳

Note that Example 6 **a)** could have been evaluated as follows:

$$8^{-2/3} = \frac{1}{8^{2/3}} = \frac{1}{\sqrt[3]{8^2}} = \frac{1}{\sqrt[3]{64}} = \frac{1}{4}$$

However, it is generally easier to evaluate the root before applying the power.
Consider the expression $(-16)^{3/4}$. This can be rewritten as $(\sqrt[4]{-16})^3$. Since $(\sqrt[4]{-16})^3$ is not a real number, the expression $(-16)^{3/4}$ is not a real number.
In Chapter 1, we indicated that

$$\left(\frac{a}{b}\right)^{-n} = \left(\frac{b}{a}\right)^n$$

We use this fact in the following example.

EXAMPLE 7   Evaluate.   **a)** $\left(\frac{9}{25}\right)^{-1/2}$ **b)** $\left(\frac{27}{8}\right)^{-1/3}$

Solution   **a)** $\left(\frac{9}{25}\right)^{-1/2} = \left(\frac{25}{9}\right)^{1/2} = \sqrt{\frac{25}{9}} = \frac{5}{3}$ **b)** $\left(\frac{27}{8}\right)^{-1/3} = \left(\frac{8}{27}\right)^{1/3} = \sqrt[3]{\frac{8}{27}} = \frac{2}{3}$ ✳

NOW TRY EXERCISE 83

HELPFUL HINT

How do the expressions $-25^{1/2}$ and $(-25)^{1/2}$ differ?
Recall that $-x^2$ means $-(x^2)$. The same principle applies here.

$$-25^{1/2} = -(25)^{1/2} = -\sqrt{25} = -5$$

$$(-25)^{1/2} = \sqrt{-25}, \text{ which is not a real number.}$$

EXAMPLE 8   Simplify each expression and write the answer without negative exponents.

a) $a^{1/2} \cdot a^{-2/3}$ **b)** $(7x^2 y^{-4})^{-1/2}$ **c)** $3.2x^{1/3}(2.4x^{1/2} + x^{-1/4})$ **d)** $\left(\frac{5x^{-4} z^{2/5}}{z^{-3/5}}\right)^{1/8}$

Solution   **a)** $a^{1/2} \cdot a^{-2/3} = a^{(1/2)-(2/3)}$ *Product rule*

$$= a^{-1/6} \quad \textit{Find the LCD and subtract the exponents.}$$

$$= \frac{1}{a^{1/6}} \quad \textit{Negative exponent rule}$$

b) $(7x^2y^{-4})^{-1/2} = 7^{-1/2}x^{2(-1/2)}y^{-4(-1/2)}$ *Raise the product to a power.*

$= 7^{-1/2}x^{-1}y^2$ *Multiply the exponents.*

$= \dfrac{y^2}{7^{1/2}x}\left(\text{or } \dfrac{y^2}{x\sqrt{7}}\right)$ *Negative exponent rule*

c) Begin by using the distributive property.

$3.2x^{1/3}(2.4x^{1/2}+x^{-1/4}) = (3.2x^{1/3})(2.4x^{1/2}) + (3.2x^{1/3})(x^{-1/4})$ *Distributive property*

$= (3.2)(2.4)(x^{(1/3)+(1/2)}) + 3.2x^{(1/3)-(1/4)}$ *Product rule*

$= 7.68x^{5/6} + 3.2x^{1/12}$

d) $\left(\dfrac{5x^{-4}z^{2/5}}{z^{-3/5}}\right)^{1/8} = (5x^{-4}z^{(2/5)-(-3/5)})^{1/8}$ *Quotient rule*

$= (5x^{-4}z)^{1/8}$ *Subtract the exponents.*

$= 5^{1/8}x^{-4(1/8)}z^{1/8}$ *Raise the product to a power.*

$= 5^{1/8}x^{-4/8}z^{1/8}$ *Multiply the exponents.*

$= 5^{1/8}x^{-1/2}z^{1/8}$

$= \dfrac{5^{1/8}z^{1/8}}{x^{1/2}}$ *Negative exponent rule*

NOW TRY EXERCISE 105

EXAMPLE 9   Simplify.   **a)** $\sqrt[15]{(6y)^5}$ **b)** $\left(\sqrt[4]{a^2b^3c}\right)^{20}$ **c)** $\sqrt[4]{\sqrt[3]{x}}$

Solution   **a)** $\sqrt[15]{(6y)^5} = (6y)^{5/15}$ *Write with a rational exponent.*

$= (6y)^{1/3}$ *Simplify the exponent.*

$= \sqrt[3]{6y}$ *Write as a radical.*

b) $\left(\sqrt[4]{a^2b^3c}\right)^{20} = (a^2b^3c)^{20/4}$ *Write with a rational exponent.*

$= (a^2b^3c)^5$

$= a^{10}b^{15}c^5$ *Raise the product to a power.*

c) $\sqrt[4]{\sqrt[3]{x}} = \sqrt[4]{x^{1/3}}$ *Write $\sqrt[3]{x}$ as $x^{1/3}$.*

$= (x^{1/3})^{1/4}$ *Write with a rational exponent.*

$= x^{1/12}$ *Raise the power to a power.*

$= \sqrt[12]{x}$ *Write as a radical.*

NOW TRY EXERCISE 53

0.00

Using Your Calculator

Finding Roots or Expressions with Rational Exponents on a Scientific or Graphing Calculator

There are often many ways to evaluate an expression like $(\sqrt[5]{845})^3$ or $845^{3/5}$ on a calculator. The procedure to use varies from calculator to calculator. One general method is to write the expression with a rational exponent and use the $\boxed{y^x}$ or $\boxed{a^x}$ or $\boxed{\wedge}$ key with parentheses keys, as shown below.*

Scientific Calculator

To evaluate $845^{3/5}$, press

Answer displayed

845 $\boxed{y^x}$ $\boxed{(}$ 3 $\boxed{\div}$ 5 $\boxed{)}$ $\boxed{=}$ 57.03139903

(continued on the next page)

To evaluate $845^{-3/5}$, press

845 $\boxed{y^x}$ $\boxed{(}$ 3 $\boxed{+/_-}$ $\boxed{\div}$ 5 $\boxed{)}$ $\boxed{=}$ 0.017534201 — Answer displayed

Graphing Calculator

To evaluate $845^{3/5}$, press the following keys.

845 $\boxed{\wedge}$ $\boxed{(}$ 3 $\boxed{\div}$ 5 $\boxed{)}$ $\boxed{\text{ENTER}}$ 57.03139903 — Answer displayed

To evaluate $845^{-3/5}$, press the following keys.

845 $\boxed{\wedge}$ $\boxed{(}$ $\boxed{(-)}$ 3 $\boxed{\div}$ 5 $\boxed{)}$ $\boxed{\text{ENTER}}$ .0175342008 — Answer displayed

* Keystrokes to use vary from calculator to calculator. Read your calculator manual to learn how to evaluate exponential expressions.

4 Factor Expressions with Rational Exponents

In higher-level math courses, you may have to factor out a variable with a rational exponent. To factor a rational expression, factor out the term with the smallest (or most negative) exponent.

EXAMPLE 10 Factor $x^{2/5} + x^{-3/5}$.

Solution The smallest of the two exponents is $-3/5$. Therefore, we will factor $x^{-3/5}$ from both terms. To find the new exponent on the variable that originally had the greater exponent, we subtract the exponent that was factored out from the original exponent.

$$x^{\frac{2}{5}} + x^{-\frac{3}{5}} = x^{-\frac{3}{5}}\left(x^{\frac{2}{5}-\left(-\frac{3}{5}\right)} + 1\right)$$

$$= x^{-\frac{3}{5}}(x^1 + 1)$$

$$= x^{-\frac{3}{5}}(x + 1)$$

$$= \frac{x + 1}{x^{3/5}}$$

We can check our factoring by multiplying.

$$x^{-3/5}(x + 1) = x^{-3/5} \cdot x + x^{-3/5} \cdot 1$$
$$= x^{(-3/5)+1} + x^{-3/5}$$
$$= x^{2/5} + x^{-3/5}$$

NOW TRY EXERCISE 139 Since we obtained the original expression, the factoring is correct.

Mathematics in Action

Brrrrrr...It's Cold Outside

Wind chill is an important issue, especially for people in regions where low temperatures and strong, gusty winds blow in the winter. Exposure to these conditions without proper protective clothing can lead to frostbite, loss of limbs, or loss of life.

Before November 1, 2001, the formula used for calculating wind chill was

$$0(v) = 0.0817(3.71v^{0.5} + 5.81)(t - 91.4) + 91.4$$

where t is the outdoor temperature in degrees Farenheit and v is the wind speed (velocity) in miles per hour. However, based on medical research, scientific study, and meteorological technology, it was determined that this formula was not a true reflection of the wind chill. On November 1, 2001, the National Weather Service adopted a new formula for determining wind chill:

$$N(v) = 35.74 + 0.6215t - 35.75v^{0.16} + 0.4275tv^{0.16}$$

Observe that both formulas have decimal exponents. The old formula, $0(v)$, has $v^{0.5}$ and the new formula, $N(v)$, has $v^{0.16}$.

If you live in a cold climate, it is possible for the outdoor temperature to dip to 35°F with the wind blowing at 20 mph. Using the old formula, the wind chill would have been 11°F. Using the new formula, it is 24°F.

Below is a comparison of the wind chill using both formulas where the outdoor temperature is fixed at 5°F and the wind speed varies from 1 mph to 105 mph.

Source: National Weather Service; NOAA

The National Weather Service has established the Web site *www.nws.noaa.gov/om/windchill*, which will give the wind chill using the new formula. To obtain this value, enter the outdoor temperature in °F and the wind speed in mph.

Exercise Set 7.2

Concept/Writing Exercises

1. **a)** Under what conditions is $\sqrt[n]{a}$ a real number?
   b) When $\sqrt[n]{a}$ is a real number, how can it be expressed with rational exponents?

2. **a)** Under what conditions is $\sqrt[n]{a^m}$ a real number?
   b) Under what conditions is $(\sqrt[n]{a})^m$ a real number?
   c) When $\sqrt[n]{a^m}$ is a real number, how can it be expressed with rational exponents?

3. **a)** Under what conditions is $\sqrt[n]{a^n}$ a real number?
   b) When n is even and $a \geq 0$, what is $\sqrt[n]{a^n}$ equal to?
   c) When n is odd, what is $\sqrt[n]{a^n}$ equal to?

4. **d)** When n is even and a may be any real number, what is $\sqrt[n]{a^n}$ equal to?

4. **a)** Explain the difference between $-16^{1/2}$ and $(-16)^{1/2}$.
   b) Evaluate each expression in part **a)** if possible.

5. **a)** Is $(xy)^{1/2} = xy^{1/2}$? Explain.
   b) Is $(xy)^{-1/2} = \dfrac{x^{1/2}}{y^{-1/2}}$? Explain.

6. **a)** Is $\sqrt[6]{(3y)^3} = (3y)^{6/3}$? Explain.
   b) Is $\sqrt{(ab)^4} = (ab)^2$? Explain.

Practice the Skills

In this exercise set, assume that all variables represent positive real numbers. Write each expression in exponential form.

7. $\sqrt{a^3}$ **8.** $\sqrt{y^7}$ **9.** $\sqrt{9^5}$ **10.** $\sqrt[3]{y}$

11. $\sqrt[3]{z^5}$ **12.** $\sqrt[3]{x^{11}}$ **13.** $\sqrt[3]{7^{10}}$ **14.** $\sqrt[5]{7^{11}}$

15. $\sqrt[4]{9^7}$ **16.** $(\sqrt{x})^7$ **17.** $(\sqrt[3]{y})^{13}$ **18.** $\sqrt{ab^5}$

19. $\sqrt[4]{a^3b}$ **20.** $\sqrt[3]{x^4y}$ **21.** $\sqrt[4]{x^9z^5}$ **22.** $\sqrt[6]{y^{11}z}$

23. $\sqrt[3]{5x^6y^7}$ **24.** $\sqrt[4]{7a^5b^3}$ **25.** $\sqrt[6]{3a+5b}$ **26.** $\sqrt[9]{3x+7z^4}$

Write each expression in radical form.

27. $a^{1/2}$ **28.** $b^{2/3}$ **29.** $c^{5/2}$ **30.** $15^{1/2}$

31. $17^{5/3}$ **32.** $y^{17/6}$ **33.** $(24x^3)^{1/2}$ **34.** $(85a^3)^{5/2}$

35. $(7b^2c)^{3/5}$ **36.** $(9x^3y^2)^{7/4}$ **37.** $(6a+5b)^{1/5}$ **38.** $(8x^2-9y)^{7/3}$

39. $(b^3-c)^{-1/3}$ **40.** $(7x^2+2y^3)^{-1/6}$

Simplify each radical expression by changing the expression to exponential form. Write the answer in radical form when appropriate.

41. $\sqrt{a^4}$ **42.** $\sqrt[4]{a^8}$ 🔒 **43.** $\sqrt[3]{x^9}$ **44.** $\sqrt[4]{x^{12}}$

45. $\sqrt[6]{y^2}$ **46.** $\sqrt{y^6}$ **47.** $\sqrt[8]{b^4}$ **48.** $\sqrt[12]{z^4}$

49. $(\sqrt{18.1})^2$ **50.** $\sqrt[4]{(6.83)^4}$ **51.** $\left(\sqrt[5]{xy^2}\right)^{15}$ **52.** $\left(\sqrt[4]{a^4bc^3}\right)^{40}$

🔒 **53.** $(\sqrt[8]{xyz})^4$ **54.** $\left(\sqrt[9]{a^2bc^4}\right)^3$ **55.** $\sqrt{\sqrt{x}}$ **56.** $\sqrt{\sqrt[3]{a}}$

57. $\sqrt{\sqrt[4]{y}}$ **58.** $\sqrt[3]{\sqrt[4]{b}}$ **59.** $\sqrt[3]{\sqrt[3]{x^2y}}$ **60.** $\sqrt[4]{\sqrt[3]{5y}}$

61. $\sqrt{\sqrt[5]{a^7}}$ **62.** $\sqrt[5]{\sqrt[4]{ab}}$

Evaluate if possible. If the expression is not a real number, so state.

63. $9^{1/2}$ **64.** $100^{1/2}$ **65.** $64^{1/3}$ **66.** $81^{1/4}$

67. $16^{3/2}$ **68.** $27^{2/3}$ **69.** $(-25)^{1/2}$ **70.** $(-64)^{1/4}$

71. $\left(\dfrac{16}{9}\right)^{1/2}$ **72.** $\left(\dfrac{36}{49}\right)^{1/2}$ **73.** $\left(\dfrac{1}{8}\right)^{1/3}$ **74.** $\left(\dfrac{1}{32}\right)^{1/5}$

75. $-36^{1/2}$ **76.** $(-36)^{1/2}$ 🔒 **77.** $-64^{1/3}$ **78.** $(-64)^{1/3}$

🔒 **79.** $64^{-1/3}$ **80.** $100^{-1/2}$ **81.** $64^{-2/3}$ **82.** $16^{-3/2}$

83. $\left(\dfrac{64}{27}\right)^{-1/3}$ **84.** $(-81)^{3/4}$ **85.** $(-100)^{3/2}$ **86.** $-\left(\dfrac{4}{49}\right)^{-1/2}$

87. $81^{1/2}+169^{1/2}$ **88.** $49^{-1/2}+36^{-1/2}$ **89.** $343^{-1/3}+4^{-1/2}$ **90.** $16^{-1/2}-256^{-3/4}$

Simplify. Write the answer in exponential form without negative exponents.

91. $x^4 \cdot x^{1/2}$ **92.** $x^{2/3} \cdot x^{3/8}$ **93.** $\dfrac{x^{1/2}}{x^{1/3}}$ **94.** $x^{-4/5}$

95. $(x^{1/2})^{-2}$ **96.** $(a^{-1/3})^{-1/2}$ **97.** $(7^{-1/3})^0$ **98.** $\dfrac{x^5}{x^{-1/2}}$

🔒 **99.** $\dfrac{5y^{-1/3}}{60y^{-2}}$ **100.** $x^{-1/2}x^{-2/5}$ **101.** $4x^{5/3}\,2x^{-7/2}$ **102.** $(x^{-4/5})^{1/3}$

103. $\left(\dfrac{8}{64x}\right)^{1/3}$ **104.** $\left(\dfrac{108}{4x^4}\right)^{1/3}$ 🔒 **105.** $\left(\dfrac{22x^{3/7}}{2x^{1/2}}\right)^2$ **106.** $\left(\dfrac{x^{-1/3}}{x^{-2}}\right)^2$

107. $\left(\dfrac{a^4}{5a^{-2/5}}\right)^{-3}$ **108.** $\left(\dfrac{81z^{1/4}y^3}{9z^{1/4}}\right)^{1/2}$ **109.** $\left(\dfrac{x^{3/4}y^{-2}}{x^{1/2}y^2}\right)^4$ **110.** $\left(\dfrac{250a^{-3/4}b^5}{2a^{-2}b^2}\right)^{2/3}$

Multiply.

111. $3z^{-1/2}(2z^4 - z^{1/2})$ **112.** $-3a^{-4/9}(2a^{1/9} - a^2)$ **113.** $5x^{-1}(x^{-4} + 4x^{-1/2})$

114. $-9z^{3/2}(z^{3/2} - z^{-3/2})$ **115.** $-4x^{5/3}(-2x^{1/2} + 7x^{1/3})$ **116.** $\dfrac{1}{2}x^{-2}(10x^{4/3} - 18x^{-1/2})$

Use a calculator to evaluate each expression. Give the answer to the nearest hundredth.

117. $\sqrt{140}$ **118.** $\sqrt[3]{168}$ **119.** $\sqrt[5]{402.83}$ **120.** $\sqrt[4]{1096}$

121. $45^{2/3}$ **122.** $69.1^{3/2}$ **123.** $1000^{-1/2}$ **124.** $8060^{-3/2}$

Problem Solving

125. Under what conditions will $\sqrt[n]{a^n} = (\sqrt[n]{a})^n = a$?

126. By selecting values for a and b, show that $(a^2 + b^2)^{1/2}$ *is not equal to* $a + b$.

127. By selecting values for a and b, show that $(a^{1/2} + b^{1/2})^2$ *is not equal to* $a + b$.

128. By selecting values for a and b, show that $(a^3 + b^3)^{1/3}$ is not equal to $a + b$.

129. By selecting values for a and b, show that $(a^{1/3} + b^{1/3})^3$ is not equal to $a + b$.

130. Determine whether $\sqrt[3]{\sqrt{x}} = \sqrt{\sqrt[3]{x}}, x \geq 0$.

Factor. Write the answer without negative exponents.

131. $x^{5/2} + x^{1/2}$ **132.** $x^{1/4} - x^{5/4}$ **133.** $y^{1/3} - y^{7/3}$

134. $x^{-1/2} + x^{1/2}$ **135.** $y^{-2/5} + y^{8/5}$ **136.** $a^{6/5} + a^{-4/5}$

In Exercises 137 through 142 use a calculator where appropriate.

137. *Growing Bacteria* The function, $B(t) = 2^{10} \cdot 2^t$, approximates the number of bacteria in a certain culture after t hours.

   a) The initial number of bacteria is determined when $t = 0$. What is the initial number of bacteria?

   b) How many bacteria are there after $\frac{1}{2}$ hour?

138. *Carbon Dating* Carbon dating is used by scientists to find the age of fossils, bones, and other items. The formula used in carbon dating is $P = P_0 2^{-t/5600}$, where P_0 represents the original amount of carbon 14 (C_{14}) present and P represents the amount of C_{14} present after t years. If 10 milligrams (mg) of C_{14} is present in an animal bone recently excavated, how many milligrams will be present in 5000 years?

139. *Retirement Plans* Each year more and more people contribute to their companies' 401(k) retirement plans. The total assets, $A(t)$, in U.S. 401(k) plans, in billions of dollars, can be approximated by the function $A(t) = 2.69t^{3/2}$, where t is years since 1993 and $1 \leq t \leq 14$. (Therefore, this function holds from 1994 through 2007.) Estimate the total assets in U.S. 401(k) plans in **a)** 2000 and **b)** 2007.

140. *Internet Sales* Retail Internet sales are increasing annually. The total amount, $I(t)$, in billions of dollars, of Internet sales can be approximated by the function $I(t) = 0.25t^{5/3}$, where t is years since 1999 and $1 \leq t \leq 8$. Find the total Internet sales in **a)** 2000 and **b)** 2007.

141. Evaluate $\left(3^{\sqrt{2}}\right)^{\sqrt{2}}$. Explain how you determined your answer.

142. **a)** On your calculator, evaluate 3^π.

   b) Explain why your value from part **a)** does or does not make sense.

143. Find the domain of $f(x) = (x - 5)^{1/2}(x + 3)^{-1/2}$.

144. Find the domain of $f(x) = (x + 4)^{1/2}(x - 2)^{-1/2}$.

145. Assume that x can be any real number. Simplify $\sqrt[n]{(x - 4)^{2n}}$ if

   a) n is even.

   b) n is odd.

Determine the index to be placed in the shaded area to make the statement true. Explain how you determined your answer.

146. $\sqrt[4]{\sqrt[\blacksquare]{\sqrt{x}}} = x^{1/24}$

147. $\sqrt[4]{\sqrt[5]{\sqrt[\blacksquare]{\sqrt[3]{z}}}} = z^{1/120}$

148. a) Write $f(x) = \sqrt{2x + 3}$ in exponential form.

b) On your grapher, check that the answer you gave in part **a)** is correct by graphing both $f(x)$ as given and the function you gave in exponential form.

Cumulative Review Exercises

[3.2] **149.** Determine which of the following relations are also functions.

a) **b)** **c)**

[6.3] **150.** Simplify $\dfrac{a^{-2} + ab^{-1}}{ab^{-2} - a^{-2}b^{-1}}$.

[6.4] **151.** Solve the equation $\dfrac{3x - 2}{x + 4} = \dfrac{2x + 1}{3x - 2}$.

[6.5] **152.** *Flying a Plane* Amy Mayfield can fly her plane 500 miles against the wind in the same time it takes her to fly 560 miles with the wind. If the wind blows at 25 miles per hour, find the speed of the plane in still air.

7.3 SIMPLIFYING RADICALS

SSM Study Guide CD/Video

MathPro 4/5 PH Math Tutor Center prenhall.com/Angel

1 Understand perfect powers.

2 Simplify radicals using the product rule for radicals.

3 Simplify radicals using the quotient rule for radicals.

1 Understand Perfect Powers

In this section, we will simplify radicals using the **product rule for radicals** and the **quotient rule for radicals**. However, before we introduce these rules let's look at **perfect powers**, which will help us with our discussion.

A number or expression is a **perfect square** if it is the square of an expression. Examples of perfect squares are illustrated below.

Perfect squares	1,	4,	9,	16,	25,	36...
	↓	↓	↓	↓	↓	↓
Square of a number	1^2,	2^2,	3^2,	4^2,	5^2,	6^2,...

Variables with exponents may also be perfect squares, as illustrated below.

Perfect squares	x^2,	x^4,	x^6,	x^8,	x^{10},...
	↓	↓	↓	↓	↓
Square of an expression	$(x)^2$,	$(x^2)^2$,	$(x^3)^2$,	$(x^4)^2$,	$(x^5)^2$,...

Notice that the exponents on the variables in the perfect squares are all multiples of 2.

Just as there are perfect squares, there are perfect cubes. A number or expression is a **perfect cube** if it can be written as the cube of an expression. Examples of perfect cubes are illustrated on the top of the next page.

Perfect cubes	1,	8,	27,	64,	125,	216,...
	↓	↓	↓	↓	↓	↓
Cube of a number	1^3,	2^3,	3^3,	4^3,	5^3,	6^3,...

Perfect cubes	x^3,	x^6,	x^9,	x^{12},	x^{15},...
	↓	↓	↓	↓	↓
Cube of an expression	$(x)^3$,	$(x^2)^3$,	$(x^3)^3$,	$(x^4)^3$,	$(x^5)^3$,...

Notice that the exponents on the variables in the perfect cubes are all multiples of 3.

We can expand our discussion to perfect powers of a variable for any radicand. In general, the radicand x^n is a perfect power *when n is a multiple of the index* of the radicand (or where *n* is divisible by the index).

Example

Perfect powers of x^n for index *n* x^n, x^{2n}, x^{3n}, x^{4n}, x^{5n},...

For example, if the index of a radical expression is 5, then $x^5, x^{10}, x^{15}, x^{20}$, etc., are perfect powers of the index.

HELPFUL HINT

A quick way to determine if a radicand x^n is a perfect power for an index is to determine if the exponent *n* is divisible by the index of the radical. For example, consider $\sqrt[5]{x^{20}}$. Since the exponent, 20, is divisible by the index, 5, x^{20} is a perfect fifth power. Now consider $\sqrt[6]{x^{20}}$. Since the exponent, 20, is not divisible by the index, 6, x^{20} is not a perfect sixth power. However, x^{18} and x^{24} are both perfect sixth powers since 6 divides both 18 and 24.

Notice that the square root of a perfect square simplifies to an expression without a radical sign, the cube root of a perfect cube simplifies to an expression without a radical sign, and so on.

Examples

$$\sqrt{16} = \sqrt{4^2} = 4^{2/2} = 4$$
$$\sqrt[3]{27} = \sqrt[3]{3^3} = 3^{3/3} = 3$$
$$\sqrt{x^6} = x^{6/2} = x^3$$
$$\sqrt[3]{z^{12}} = z^{12/3} = z^4$$
$$\sqrt[5]{n^{30}} = n^{30/5} = n^6$$

Now we are ready to discuss the product rule for radicals.

2 Simplify Radicals Using the Product Rule for Radicals

To introduce the **product rule for radicals**, observe that $\sqrt{4} \cdot \sqrt{9} = 2 \cdot 3 = 6$. Also, $\sqrt{4 \cdot 9} = \sqrt{36} = 6$. We see that $\sqrt{4} \cdot \sqrt{9} = \sqrt{4 \cdot 9}$. This is one example of the product rule for radicals.

Product Rule for Radicals

For nonnegative real numbers *a* and *b*,

$$\sqrt[n]{a} \cdot \sqrt[n]{b} = \sqrt[n]{ab}$$

Examples of the Product Rule for Radicals

$$\sqrt{20} = \begin{cases} \sqrt{1} \cdot \sqrt{20} \\ \sqrt{2} \cdot \sqrt{10} \\ \sqrt{4} \cdot \sqrt{5} \end{cases} \qquad \sqrt[3]{20} = \begin{cases} \sqrt[3]{1} \cdot \sqrt[3]{20} \\ \sqrt[3]{2} \cdot \sqrt[3]{10} \\ \sqrt[3]{4} \cdot \sqrt[3]{5} \end{cases}$$

$\sqrt{20}$ can be factored into any of these forms. $\sqrt[3]{20}$ can be factored into any of these forms.

$$\sqrt{x^7} = \begin{cases} \sqrt{x} \cdot \sqrt{x^6} \\ \sqrt{x^2} \cdot \sqrt{x^5} \\ \sqrt{x^3} \cdot \sqrt{x^4} \end{cases} \qquad \sqrt[3]{x^7} = \begin{cases} \sqrt[3]{x} \cdot \sqrt[3]{x^6} \\ \sqrt[3]{x^2} \cdot \sqrt[3]{x^5} \\ \sqrt[3]{x^3} \cdot \sqrt[3]{x^4} \end{cases}$$

$\sqrt{x^7}$ can be factored into any of these forms. $\sqrt[3]{x^7}$ can be factored into any of these forms.

Now that we have introduced the product rule for radicals, we will use this rule to simplify radicals. Here is a general procedure that can be used to simplify radicals using the product rule.

To Simplify Radicals Using the Product Rule

1. If the radicand contains a coefficient other than 1, write it as a product of two numbers, one of which is the largest perfect power for the index.

2. Write each variable factor as a product of two factors, one of which is the largest perfect power of the variable for the index.

3. Use the product rule to write the radical expression as a product of radicals. Place all the perfect powers (numbers and variables) under the same radical.

4. Simplify the radical containing the perfect powers.

If we are simplifying a *square* root, we will write the radicand as the product of the largest *perfect square* and another number. If we are simplifying a *cube* root, we will write the radicand as the product of the largest *perfect cube* and another number, and so on.

EXAMPLE 1   Simplify.   **a)** $\sqrt{32}$ **b)** $\sqrt{60}$ **c)** $\sqrt[3]{54}$ **d)** $\sqrt[4]{80}$

Solution   The radicands in this example contain no variables. We will follow step 1 of the procedure.

a) Since we are evaluating a square root, we look for the largest perfect square that divides 32. The largest perfect square that divides, or is a factor of, 32 is 16.

$$\sqrt{32} = \sqrt{16 \cdot 2} = \sqrt{16}\,\sqrt{2} = 4\sqrt{2}$$

b) The largest perfect square that is a factor of 60 is 4.

$$\sqrt{60} = \sqrt{4 \cdot 15} = \sqrt{4}\,\sqrt{15} = 2\sqrt{15}$$

c) The largest perfect cube that is a factor of 54 is 27.

$$\sqrt[3]{54} = \sqrt[3]{27 \cdot 2} = \sqrt[3]{27}\,\sqrt[3]{2} = 3\sqrt[3]{2}$$

d) The largest perfect fourth power that is a factor of 80 is 16.

$$\sqrt[4]{80} = \sqrt[4]{16 \cdot 5} = \sqrt[4]{16}\,\sqrt[4]{5} = 2\sqrt[4]{5}$$

NOW TRY EXERCISE 19

HELPFUL HINT

In Example 1 **a)**, if you first thought that 4 was the largest perfect square that divided 32, you could proceed as follows:

$$\sqrt{32} = \sqrt{4 \cdot 8} = \sqrt{4}\sqrt{8} = 2\sqrt{8}$$
$$= 2\sqrt{4 \cdot 2} = 2\sqrt{4}\sqrt{2} = 2 \cdot 2\sqrt{2} = 4\sqrt{2}$$

Note that the final result is the same, but you must perform more steps. The chart on pages 473 and 474 can help you determine the largest perfect square or perfect cube that is a factor of a radicand.

In Example 1 **b)**, $\sqrt{15}$ can be factored as $\sqrt{5 \cdot 3}$; however, since neither 5 nor 3 is a perfect square, $\sqrt{15}$ cannot be simplified.

When the radicand is a perfect power for the index, the radical can be simplified by writing it in exponential form, as in Example 2.

EXAMPLE 2 Simplify. **a)** $\sqrt{x^4}$ **b)** $\sqrt[3]{x^{12}}$ **c)** $\sqrt[5]{z^{35}}$

Solution **a)** $\sqrt{x^4} = x^{4/2} = x^2$ **b)** $\sqrt[3]{x^{12}} = x^{12/3} = x^4$ **c)** $\sqrt[5]{z^{35}} = z^{35/5} = z^7$

EXAMPLE 3 Simplify. **a)** $\sqrt{x^9}$ **b)** $\sqrt[5]{x^{23}}$ **c)** $\sqrt[4]{y^{33}}$

Solution Because the radicands have coefficients of 1, we start with step 2 of the procedure.

a) The largest perfect square less than or equal to x^9 is x^8.

$$\sqrt{x^9} = \sqrt{x^8 \cdot x} = \sqrt{x^8} \cdot \sqrt{x} = x^{8/2}\sqrt{x} = x^4\sqrt{x}$$

b) The largest perfect fifth power less than or equal to x^{23} is x^{20}.

$$\sqrt[5]{x^{23}} = \sqrt[5]{x^{20} \cdot x^3} = \sqrt[5]{x^{20}}\sqrt[5]{x^3} = x^{20/5}\sqrt[5]{x^3} = x^4\sqrt[5]{x^3}$$

c) The largest perfect fourth power less than or equal to y^{33} is y^{32}.

NOW TRY EXERCISE 35

$$\sqrt[4]{y^{33}} = \sqrt[4]{y^{32} \cdot y} = \sqrt[4]{y^{32}}\sqrt[4]{y} = y^{32/4}\sqrt[4]{y} = y^8\sqrt[4]{y}$$

If you observe the answers to Example 3, you will see that the exponent on the variable in the radicand is always less than the index. **When a radical is simplified, the radicand does not have a variable with an exponent greater than or equal to the index.**

In Example 3 **b)**, we simplified $\sqrt[5]{x^{23}}$. If we divide 23, the exponent in the radicand, by 5, the index, we obtain

$$5)\overline{23} \quad \begin{array}{l} 4 \leftarrow Quotient \\ \underline{20} \\ 3 \leftarrow Remainder \end{array}$$

Notice that $\sqrt[5]{x^{23}}$ simplifies to $x^4\sqrt[5]{x^3}$ and

$$Quotient \longrightarrow x^4\sqrt[5]{x^3} \leftarrow Remainder$$

When simplifying a radical, if you divide the exponent within the radical by the index, the quotient will be the exponent on the variable outside the radical sign and the remainder will be the exponent on the variable within the radical sign. Simplify Example 3 **c)** using this technique now.

EXAMPLE 4 Simplify. **a)** $\sqrt{x^{12}y^{17}}$ **b)** $\sqrt[4]{x^6 y^{23}}$

Solution **a)** x^{12} is a perfect square. The largest perfect square that is a factor of y^{17} is y^{16}. Write y^{17} as $y^{16} \cdot y$.

$$\sqrt{x^{12}y^{17}} = \sqrt{x^{12} \cdot y^{16} \cdot y} = \sqrt{x^{12}y^{16}}\sqrt{y}$$
$$= \sqrt{x^{12}}\sqrt{y^{16}}\sqrt{y}$$
$$= x^{12/2}y^{16/2}\sqrt{y}$$
$$= x^6 y^8 \sqrt{y}$$

b) We begin by finding the largest perfect fourth power factors of x^6 and y^{23}. For an index of 4, the largest perfect power that is a factor of x^6 is x^4. The largest perfect power that is a factor of y^{23} is y^{20}.

$$\sqrt[4]{x^6 y^{23}} = \sqrt[4]{x^4 \cdot x^2 \cdot y^{20} \cdot y^3}$$
$$= \sqrt[4]{x^4 y^{20} \cdot x^2 y^3}$$
$$= \sqrt[4]{x^4 y^{20}} \sqrt[4]{x^2 y^3}$$
$$= xy^5 \sqrt[4]{x^2 y^3}$$

Often the steps where we change the radical expression to exponential form are done mentally, and those steps are not illustrated. For instance, in Example 4 **b)**, we changed $\sqrt[4]{x^4 y^{20}}$ to xy^5 mentally and did not show the intermediate steps.

EXAMPLE 5 Simplify. **a)** $\sqrt{80x^5 y^{12} z^3}$ **b)** $\sqrt[3]{54x^{17}y^{25}}$

Solution **a)** The largest perfect square that is a factor of 80 is 16. The largest perfect square that is a factor of x^5 is x^4. The expression y^{12} is a perfect square. The largest perfect square that is a factor of z^3 is z^2. Place all the perfect squares under the same radical, and then simplify.

$$\sqrt{80x^5 y^{12} z^3} = \sqrt{16 \cdot 5 \cdot x^4 \cdot x \cdot y^{12} \cdot z^2 \cdot z}$$
$$= \sqrt{16x^4 y^{12} z^2 \cdot 5xz}$$
$$= \sqrt{16x^4 y^{12} z^2} \cdot \sqrt{5xz}$$
$$= 4x^2 y^6 z \sqrt{5xz}$$

b) The largest perfect cube that is a factor of 54 is 27. The largest perfect cube that is a factor of x^{17} is x^{15}. The largest perfect cube that is a factor of y^{25} is y^{24}.

$$\sqrt[3]{54x^{17}y^{25}} = \sqrt[3]{27 \cdot 2 \cdot x^{15} \cdot x^2 \cdot y^{24} \cdot y}$$
$$= \sqrt[3]{27x^{15}y^{24} \cdot 2x^2 y}$$
$$= \sqrt[3]{27x^{15}y^{24}} \cdot \sqrt[3]{2x^2 y}$$
$$= 3x^5 y^8 \sqrt[3]{2x^2 y}$$

NOW TRY EXERCISE 57

HELPFUL HINT

In Example 4 **b)**, we showed that

$$\sqrt[4]{x^6 y^{23}} = xy^5 \sqrt[4]{x^2 y^3}$$

As mentioned on page 476, this radical can also be simplified by dividing the exponents on the variables in the radicand, 6 and 23, by the index, 4, and observing the quotients and remainders.

(continued on the next page)

Can you explain why this procedure works? You may wish to use this procedure to work or check certain problems.

Now let's introduce the quotient rule for radicals.

3 Simplify Radicals Using the Quotient Rule for Radicals

In mathematics we sometimes need to simplify a quotient of two radicals. To do so we use the **quotient rule for radicals**.

> ### Quotient Rule for Radicals
>
> For nonnegative real numbers a and b,
>
> $$\frac{\sqrt[n]{a}}{\sqrt[n]{b}} = \sqrt[n]{\frac{a}{b}}, \qquad b \neq 0$$

Examples of the Quotient Rule for Radicals

$$\frac{\sqrt{18}}{\sqrt{2}} = \sqrt{\frac{18}{2}} \qquad\qquad \sqrt{\frac{9}{16}} = \frac{\sqrt{9}}{\sqrt{16}}$$

$$\frac{\sqrt{x^3}}{\sqrt{x}} = \sqrt{\frac{x^3}{x}} \qquad\qquad \sqrt{\frac{x^4}{y^2}} = \frac{\sqrt{x^4}}{\sqrt{y^2}}$$

$$\frac{\sqrt[3]{y^5}}{\sqrt[3]{y^2}} = \sqrt[3]{\frac{y^5}{y^2}} \qquad\qquad \sqrt[3]{\frac{z^9}{27}} = \frac{\sqrt[3]{z^9}}{\sqrt[3]{27}}$$

Examples 6 and 7 illustrate how to use the quotient rule to simplify radical expressions.

EXAMPLE 6 Simplify. **a)** $\dfrac{\sqrt{75}}{\sqrt{3}}$ **b)** $\dfrac{\sqrt[3]{24x}}{\sqrt[3]{3x}}$ **c)** $\dfrac{\sqrt[3]{x^4 y^7}}{\sqrt[3]{xy^{-2}}}$

Solution In each part we use the quotient rule to write the quotient of radicals as a single radical. Then we simplify.

a) $\dfrac{\sqrt{75}}{\sqrt{3}} = \sqrt{\dfrac{75}{3}} = \sqrt{25} = 5$

b) $\dfrac{\sqrt[3]{24x}}{\sqrt[3]{3x}} = \sqrt[3]{\dfrac{24x}{3x}} = \sqrt[3]{8} = 2$

c) $\dfrac{\sqrt[3]{x^4 y^7}}{\sqrt[3]{xy^{-2}}} = \sqrt[3]{\dfrac{x^4 y^7}{xy^{-2}}}$ *Quotient rule for radicals*

$= \sqrt[3]{x^3 y^9}$ *Simplify the radicand.*

NOW TRY EXERCISE 93 $= xy^3$

In Section 7.1, when we introduced radicals we indicated that $\sqrt{\dfrac{4}{9}} = \dfrac{2}{3}$ since $\dfrac{2}{3} \cdot \dfrac{2}{3} = \dfrac{4}{9}$. The quotient rule may be helpful in evaluating square roots containing fractions as illustrated in Example 7 **a)**.

EXAMPLE 7   Simplify.   **a)** $\sqrt{\dfrac{144}{25}}$ **b)** $\sqrt[3]{\dfrac{8x^4y}{27xy^{10}}}$ **c)** $\sqrt[4]{\dfrac{15xy^5}{3x^9y}}$

Solution   In each part we first simplify the radicand, if possible. Then we use the quotient rule to write the given radical as a quotient of radicals.

a) $\sqrt{\dfrac{144}{25}} = \dfrac{\sqrt{144}}{\sqrt{25}} = \dfrac{12}{5}$

b) $\sqrt[3]{\dfrac{8x^4y}{27xy^{10}}} = \sqrt[3]{\dfrac{8x^3}{27y^9}} = \dfrac{\sqrt[3]{8x^3}}{\sqrt[3]{27y^9}} = \dfrac{2x}{3y^3}$

c) $\sqrt[4]{\dfrac{15xy^5}{3x^9y}} = \sqrt[4]{\dfrac{5y^4}{x^8}} = \dfrac{\sqrt[4]{5y^4}}{\sqrt[4]{x^8}} = \dfrac{\sqrt[4]{y^4}\,\sqrt[4]{5}}{x^2} = \dfrac{y\sqrt[4]{5}}{x^2}$

NOW TRY EXERCISE 97

AVOIDING COMMON ERRORS

The following simplifications are correct because the numbers and variables divided out are not within square roots.

An expression within a square root cannot be divided by an expression not within the square root.

Exercise Set 7.3

Concept/Writing Exercises

1. a) How do you obtain the numbers that are perfect squares?

   b) List the first six perfect squares.

2. a) How do you obtain the numbers that are perfect cubes?

   b) List the first six perfect cube numbers.

3. a) How do you obtain numbers that are perfect fifth powers?

   b) List the first five perfect fifth-power numbers.

4. In your own words, state the product rule for radicals.

5. When we gave the product rule, we stated that for nonnegative real numbers a and b, $\sqrt[n]{a} \cdot \sqrt[n]{b} = \sqrt[n]{ab}$. Why is it necessary to specify that both a and b are nonnegative real numbers?

6. In your own words, state the quotient rule for radicals.

7. When we gave the quotient rule, we stated that for nonnegative real numbers a and b, $\dfrac{\sqrt[n]{a}}{\sqrt[n]{b}} = \sqrt[n]{\dfrac{a}{b}}$, $b \neq 0$.

Why is it necessary to state that both a and b are nonnegative real numbers?

8. In the quotient rule discussed in Exercise 7, why can the denominator never equal 0?

Practice the Skills

In this exercise set, assume that all variables represent positive real numbers.

Simplify.

9. $\sqrt{8}$ **10.** $\sqrt{24}$ **11.** $\sqrt{28}$ **12.** $\sqrt{18}$

13. $\sqrt{12}$ **14.** $\sqrt{48}$ **15.** $\sqrt{50}$ **16.** $\sqrt{300}$

17. $\sqrt{75}$ **18.** $\sqrt{72}$ **19.** $\sqrt{40}$ **20.** $\sqrt{500}$

21. $\sqrt[3]{16}$ **22.** $\sqrt[3]{24}$ **23.** $\sqrt[3]{54}$ **24.** $\sqrt[3]{81}$

25. $\sqrt[3]{32}$ **26.** $\sqrt[3]{40}$ **27.** $\sqrt[3]{108}$ **28.** $\sqrt[4]{80}$

29. $\sqrt[4]{162}$ **30.** $\sqrt[4]{48}$ **31.** $-\sqrt[5]{64}$ **32.** $-\sqrt[5]{243}$

33. $\sqrt{x^3}$ **34.** $-\sqrt{x^5}$ **35.** $\sqrt{a^{11}}$ **36.** $4\sqrt{y^{12}}$

37. $-\sqrt{z^{21}}$ **38.** $\sqrt[3]{x^6}$ **39.** $\sqrt[3]{a^7}$ **40.** $\sqrt[3]{b^9}$

41. $\sqrt[3]{b^{13}}$ **42.** $7\sqrt[3]{z^{32}}$ **43.** $\sqrt[4]{x^5}$ **44.** $\sqrt[4]{b^{23}}$

45. $\sqrt[5]{z^7}$ **46.** $\sqrt[6]{x^9}$ **47.** $3\sqrt[5]{y^{23}}$ **48.** $\sqrt[7]{y^{15}}$

49. $\sqrt{24x^3}$ **50.** $2\sqrt{50y^9}$ **51.** $\sqrt{8x^4y^7}$ **52.** $\sqrt{75a^7b^{11}}$

53. $-\sqrt{20x^6y^7z^{12}}$ **54.** $\sqrt{24x^{15}y^{20}z^{27}}$ **55.** $\sqrt[3]{x^3y^7}$ **56.** $\sqrt[3]{16x^3y^6}$

57. $\sqrt[3]{81a^6b^8}$ **58.** $\sqrt[3]{128a^{10}b^{11}c^{12}}$ **59.** $\sqrt[4]{32x^8y^9z^{19}}$ **60.** $\sqrt[4]{48x^{11}y^{21}}$

61. $\sqrt[4]{81a^8b^9}$ **62.** $-\sqrt[4]{32x^{18}y^{31}}$ **63.** $\sqrt[5]{32a^{10}b^{12}}$ **64.** $\sqrt[6]{64x^{12}y^{23}z^{50}}$

Simplify.

65. $\sqrt{\dfrac{36}{4}}$ **66.** $\sqrt{\dfrac{75}{3}}$ **67.** $\sqrt{\dfrac{4}{25}}$ **68.** $\sqrt{\dfrac{81}{100}}$

69. $\dfrac{\sqrt{27}}{\sqrt{3}}$ **70.** $\dfrac{\sqrt{98}}{\sqrt{2}}$ **71.** $\dfrac{\sqrt{3}}{\sqrt{48}}$ **72.** $\dfrac{\sqrt{15}}{\sqrt{60}}$

73. $\sqrt[3]{\dfrac{2}{16}}$ **74.** $\sqrt[3]{\dfrac{2}{54}}$ **75.** $\dfrac{\sqrt[3]{3}}{\sqrt[3]{81}}$ **76.** $\dfrac{\sqrt[3]{64}}{\sqrt[3]{8}}$

77. $\sqrt[4]{\dfrac{3}{48}}$ **78.** $\dfrac{\sqrt[4]{243}}{\sqrt[4]{3}}$ **79.** $\sqrt[5]{\dfrac{96}{3}}$ **80.** $\dfrac{\sqrt[5]{4}}{\sqrt[5]{128}}$

81. $\sqrt{\dfrac{r^4}{25}}$ **82.** $\sqrt{\dfrac{81a^8}{49b^6}}$ **83.** $\sqrt{\dfrac{36x^4}{25y^{10}}}$ **84.** $\sqrt{\dfrac{81a^8b^{10}}{121c^{14}}}$

85. $\sqrt[3]{\dfrac{c^6}{27}}$ **86.** $\sqrt[3]{\dfrac{64x^6}{y^{12}}}$ **87.** $\sqrt[4]{\dfrac{a^8b^{12}}{b^{-4}}}$ **88.** $\sqrt[4]{\dfrac{16x^{16}y^{32}}{81x^{-4}}}$

89. $\dfrac{\sqrt{24}}{\sqrt{3}}$ **90.** $\dfrac{\sqrt{64x^5}}{\sqrt{2x^3}}$ **91.** $\dfrac{\sqrt{27x^6}}{\sqrt{3x^2}}$ **92.** $\dfrac{\sqrt{72x^2y^5}}{\sqrt{8x^2y^7}}$

93. $\dfrac{\sqrt{40x^6y^9}}{\sqrt{5x^2y^4}}$ **94.** $\dfrac{\sqrt{150a^{10}b^{11}}}{\sqrt{2ab^2}}$ **95.** $\sqrt[3]{\dfrac{7xy}{8x^{13}}}$ **96.** $\sqrt[3]{\dfrac{64a^5b^{12}}{27a^{14}b^5}}$

97. $\sqrt[3]{\dfrac{25x^2y^9}{5x^8y^2}}$ **98.** $\sqrt[3]{\dfrac{54xy^4z^{17}}{18x^{13}z}}$ **99.** $\sqrt[4]{\dfrac{20x^4y}{81x^{-8}}}$ **100.** $\sqrt[4]{\dfrac{3a^6b^5}{16a^{-2}b^{13}}}$

Problem Solving

101. Prove $\sqrt{a \cdot b} = \sqrt{a}\sqrt{b}$ by converting $\sqrt{a \cdot b}$ to exponential form.

102. Will the product of two radicals always be a radical? Give an example to support your answer.

103. Will the quotient of two radicals always be a radical? Give an example to support your answer.

104. Prove $\sqrt[n]{\dfrac{a}{b}} = \dfrac{\sqrt[n]{a}}{\sqrt[n]{b}}$ by converting $\sqrt[n]{\dfrac{a}{b}}$ to exponential form

105. a) Will $\dfrac{\sqrt[n]{x}}{\sqrt[n]{x}}$ always equal 1?

b) If your answer to part **a)** was no, under what conditions does $\dfrac{\sqrt[n]{x}}{\sqrt[n]{x}}$ equal 1?

Cumulative Review Exercises

[2.2] **106.** Solve the formula $F = \dfrac{9}{5}C + 32$ for C.

[2.6] **107.** Solve for x: $\left| \dfrac{2x - 4}{5} \right| = 12$

[5.3] **108.** Divide $\dfrac{15x^{12} - 5x^9 + 30x^6}{5x^6}$.

[5.6] **109.** Factor $(x - 3)^3 + 8$.

7.4 ADDING, SUBTRACTING, AND MULTIPLYING RADICALS

1 Add and subtract radicals.
2 Multiply radicals.

SSM	Study Guide	CD/Video
MathPro 4/5	PH Math Tutor Center	prenhall.com/Angel

1 Add and Subtract Radicals

Like radicals are radicals having the same radicand and index. **Unlike radicals** are radicals differing in either the radicand or the index.

Examples of Like Radicals	Examples of Unlike Radicals	
$\sqrt{5}, 3\sqrt{5}$	$\sqrt{5}, \sqrt[3]{5}$	*Indices differ.*
$5\sqrt{7}, -2\sqrt{7}$	$\sqrt{5}, \sqrt{7}$	*Radicands differ.*
$\sqrt{x}, 5\sqrt{x}$	$\sqrt{x}, \sqrt{2x}$	*Radicands differ.*
$\sqrt[3]{2x}, -4\sqrt[3]{2x}$	$\sqrt{x}, \sqrt[3]{x}$	*Indices differ.*
$\sqrt[4]{x^2 y^5}, -\sqrt[4]{x^2 y^5}$	$\sqrt[3]{xy}, \sqrt[3]{x^2 y}$	*Radicands differ.*

Like radicals are added and subtracted in exactly the same way that like terms are added or subtracted. To add or subtract like radicals, add or subtract their numerical coefficients and multiply this sum or difference by the like radical.

Examples of Adding and Subtracting Like Radicals

$$3\sqrt{5} + 2\sqrt{5} = (3 + 2)\sqrt{5} = 5\sqrt{5}$$
$$5\sqrt{x} - 7\sqrt{x} = (5 - 7)\sqrt{x} = -2\sqrt{x}$$
$$\sqrt[3]{4x^2} + 5\sqrt[3]{4x^2} = (1 + 5)\sqrt[3]{4x^2} = 6\sqrt[3]{4x^2}$$
$$4\sqrt{5x} - y\sqrt{5x} = (4 - y)\sqrt{5x}$$

EXAMPLE 1 Simplify. **a)** $6 + 4\sqrt{2} - \sqrt{2} + 3$ **b)** $2\sqrt[3]{x} + 5x + 4\sqrt[3]{x} - 3$

Solution **a)** $6 + 4\sqrt{2} - \sqrt{2} + 3 = 6 + 3 + 4\sqrt{2} - \sqrt{2}$ *Place like terms together.*

$$= 9 + (4 - 1)\sqrt{2}$$
$$= 9 + 3\sqrt{2} \quad (\text{or } 3\sqrt{2} + 9)$$

NOW TRY EXERCISE 15 **b)** $2\sqrt[3]{x} + 5x + 4\sqrt[3]{x} - 3 = 6\sqrt[3]{x} + 5x - 3$

It is sometimes possible to convert unlike radicals into like radicals by simplifying one or more of the radicals, as was discussed in Section 7.3.

EXAMPLE 2 Simplify $\sqrt{3} + \sqrt{27}$.

Solution Since $\sqrt{3}$ and $\sqrt{27}$ are unlike radicals, they cannot be added in their present form. We can simplify $\sqrt{27}$ to obtain like radicals.

$$\sqrt{3} + \sqrt{27} = \sqrt{3} + \sqrt{9}\sqrt{3}$$
$$= \sqrt{3} + 3\sqrt{3} = 4\sqrt{3}$$

✳

To Add or Subtract Radicals

1. Simplify each radical expression.

2. Combine like radicals (if there are any).

EXAMPLE 3 Simplify. **a)** $4\sqrt{24} + \sqrt{54}$ **b)** $2\sqrt{45} - \sqrt{80} + \sqrt{20}$ **c)** $\sqrt[3]{27} + \sqrt[3]{81} - 4\sqrt[3]{3}$

Solution **a)** $4\sqrt{24} + \sqrt{54} = 4 \cdot \sqrt{4} \cdot \sqrt{6} + \sqrt{9} \cdot \sqrt{6}$
$$= 4 \cdot 2\sqrt{6} + 3\sqrt{6}$$
$$= 8\sqrt{6} + 3\sqrt{6} = 11\sqrt{6}$$

b) $2\sqrt{45} - \sqrt{80} + \sqrt{20} = 2 \cdot \sqrt{9} \cdot \sqrt{5} - \sqrt{16} \cdot \sqrt{5} + \sqrt{4} \cdot \sqrt{5}$
$$= 2 \cdot 3\sqrt{5} - 4\sqrt{5} + 2\sqrt{5}$$
$$= 6\sqrt{5} - 4\sqrt{5} + 2\sqrt{5} = 4\sqrt{5}$$

c) $\sqrt[3]{27} + \sqrt[3]{81} - 4\sqrt[3]{3} = 3 + \sqrt[3]{27} \cdot \sqrt[3]{3} - 4\sqrt[3]{3}$

NOW TRY EXERCISE 23
$$= 3 + 3\sqrt[3]{3} - 4\sqrt[3]{3} = 3 - \sqrt[3]{3}$$

✳

EXAMPLE 4 Simplify. **a)** $\sqrt{x^2} - \sqrt{x^2 y} + x\sqrt{y}$ **b)** $\sqrt[3]{x^{10} y^2} - \sqrt[3]{x^4 y^8}$

Solution **a)** $\sqrt{x^2} - \sqrt{x^2 y} + x\sqrt{y} = x - \sqrt{x^2} \cdot \sqrt{y} + x\sqrt{y}$
$$= x - x\sqrt{y} + x\sqrt{y}$$
$$= x$$

b) $\sqrt[3]{x^{10} y^2} - \sqrt[3]{x^4 y^8} = \sqrt[3]{x^9} \cdot \sqrt[3]{xy^2} - \sqrt[3]{x^3 y^6} \cdot \sqrt[3]{xy^2}$
$$= x^3 \sqrt[3]{xy^2} - xy^2 \sqrt[3]{xy^2}$$

Now factor out the common factor, $\sqrt[3]{xy^2}$.

NOW TRY EXERCISE 35
$$= (x^3 - xy^2)\sqrt[3]{xy^2}$$

✳

HELPFUL HINT

The product rule and quotient rule for radicals presented in Section 7.3 are

$$\sqrt[n]{a} \cdot \sqrt[n]{b} = \sqrt[n]{ab} \qquad \frac{\sqrt[n]{a}}{\sqrt[n]{b}} = \sqrt[n]{\frac{a}{b}}$$

Students often incorrectly assume similar properties exist for addition and subtraction. They do not. To illustrate this, let n be a square root (index 2), $a = 9$, and $b = 16$.

$$\sqrt[n]{a} + \sqrt[n]{b} \neq \sqrt[n]{a + b}$$
$$\sqrt{9} + \sqrt{16} \neq \sqrt{9 + 16}$$
$$3 + 4 \neq \sqrt{25}$$
$$7 \neq 5$$

We will now discuss multiplication of radicals.

2 Multiply Radicals

To multiply radicals, we use the product rule given earlier. After multiplying, we can often simplify the new radical (see Examples 5 and 6).

EXAMPLE 5 Multiply and simplify. **a)** $\sqrt{6x^3}\sqrt{8x^6}$ **b)** $\sqrt[3]{2x}\sqrt[3]{4x^2}$ **c)** $\sqrt[4]{8x^{11}y}\sqrt[4]{8x^6y^{22}}$

Solution **a)** $\sqrt{6x^3}\sqrt{8x^6} = \sqrt{6x^3 \cdot 8x^6}$ *Product rule for radicals*

$$= \sqrt{48x^9}$$
$$= \sqrt{16x^8}\sqrt{3x} \qquad 16x^8 \text{ is a perfect square.}$$
$$= 4x^4\sqrt{3x}$$

b) $\sqrt[3]{2x}\sqrt[3]{4x^2} = \sqrt[3]{2x \cdot 4x^2}$ *Product rule for radicals*

$$= \sqrt[3]{8x^3} \qquad 8x^3 \text{ is a perfect cube.}$$
$$= 2x$$

c) $\sqrt[4]{8x^{11}y}\sqrt[4]{8x^6y^{22}} = \sqrt[4]{8x^{11}y \cdot 8x^6y^{22}}$ *Product rule for radicals*

$$= \sqrt[4]{64x^{17}y^{23}}$$
$$= \sqrt[4]{16x^{16}y^{20}}\sqrt[4]{4xy^3} \qquad \textit{The largest perfect fourth root factors are } 16, x^{16}, \textit{ and } y^{20}.$$

NOW TRY EXERCISE 47
$$= 2x^4y^5\sqrt[4]{4xy^3} \qquad \text{✳}$$

Remember, as stated earlier, when a radical is simplified, the radicand does not have any variable with an exponent greater than or equal to the index.

EXAMPLE 6 Multiply and simplify $\sqrt{2x}(\sqrt{8x} - \sqrt{32})$.

Solution Begin by using the distributive property.

$$\sqrt{2x}(\sqrt{8x} - \sqrt{32}) = (\sqrt{2x})(\sqrt{8x}) + (\sqrt{2x})(-\sqrt{32})$$
$$= \sqrt{16x^2} - \sqrt{64x}$$
$$= 4x - \sqrt{64}\sqrt{x}$$
$$= 4x - 8\sqrt{x} \qquad \text{✳}$$

Note in Example 6 that the same result could be obtained by first simplifying $\sqrt{8x}$ and $\sqrt{32}$ and then multiplying. You may wish to try this now.

Now we will multiply two binomial factors. To multiply binomial factors, each term in one factor must be multiplied by each term in the other factor. This can be accomplished using the FOIL method that was discussed earlier.

EXAMPLE 7 Multiply $(\sqrt{x} - \sqrt{y})(\sqrt{x} - y)$.

Solution We will multiply using the FOIL method.

$$\begin{array}{cccc} \text{F} & \text{O} & \text{I} & \text{L} \\ \downarrow & \downarrow & \downarrow & \downarrow \end{array}$$

$$(\sqrt{x})(\sqrt{x}) \;+\; (\sqrt{x})(-y) \;+\; (-\sqrt{y})(\sqrt{x}) \;+\; (-\sqrt{y})(-y)$$
$$= \quad\; \sqrt{x^2} \quad\;-\quad\; y\sqrt{x} \quad\;-\quad\; \sqrt{xy} \quad\;+\quad\; y\sqrt{y}$$
$$= \quad x - y\sqrt{x} - \sqrt{xy} + y\sqrt{y} \qquad \text{✳}$$

EXAMPLE 8 Simplify. **a)** $(3\sqrt{6} - \sqrt{3})^2$ **b)** $\left(\sqrt[3]{x} - \sqrt[3]{2y^2}\right)\left(\sqrt[3]{x^2} - \sqrt[3]{8y}\right)$

Solution **a)** $(3\sqrt{6} - \sqrt{3})^2 = (3\sqrt{6} - \sqrt{3})(3\sqrt{6} - \sqrt{3})$

Now multiply the factors using the FOIL method.

$$
\overset{\text{F}}{(3\sqrt{6})(3\sqrt{6})} + \overset{\text{O}}{(3\sqrt{6})(-\sqrt{3})} + \overset{\text{I}}{(-\sqrt{3})(3\sqrt{6})} + \overset{\text{L}}{(-\sqrt{3})(-\sqrt{3})}
$$
$$
= 9(6) - 3\sqrt{18} - 3\sqrt{18} + 3
$$
$$
= 54 - 3\sqrt{18} - 3\sqrt{18} + 3
$$
$$
= 57 - 6\sqrt{18}
$$
$$
= 57 - 6\sqrt{9}\,\sqrt{2}
$$
$$
= 57 - 18\sqrt{2}
$$

b) Multiply the factors using the FOIL method.

$$
\left(\sqrt[3]{x} - \sqrt[3]{2y^2}\right)\left(\sqrt[3]{x^2} - \sqrt[3]{8y}\right) = \overset{\text{F}}{(\sqrt[3]{x})(\sqrt[3]{x^2})} + \overset{\text{O}}{(\sqrt[3]{x})(-\sqrt[3]{8y})} + \overset{\text{I}}{\left(-\sqrt[3]{2y^2}\right)(\sqrt[3]{x^2})} + \overset{\text{L}}{\left(-\sqrt[3]{2y^2}\right)(-\sqrt[3]{8y})}
$$
$$
= \sqrt[3]{x^3} - \sqrt[3]{8xy} - \sqrt[3]{2x^2y^2} + \sqrt[3]{16y^3}
$$
$$
= \sqrt[3]{x^3} - \sqrt[3]{8}\,\sqrt[3]{xy} - \sqrt[3]{2x^2y^2} + \sqrt[3]{8y^3}\,\sqrt[3]{2}
$$

NOW TRY EXERCISE 99
$$
= x - 2\sqrt[3]{xy} - \sqrt[3]{2x^2y^2} + 2y\sqrt[3]{2} \qquad ✳
$$

EXAMPLE 9 Multiply $(3 + \sqrt{8})(3 - \sqrt{8})$.

Solution We can multiply using the FOIL method.

$$
(3 + \sqrt{8})(3 - \sqrt{8}) = \overset{\text{F}}{3(3)} + \overset{\text{O}}{3(-\sqrt{8})} + \overset{\text{I}}{(\sqrt{8})(3)} + \overset{\text{L}}{(\sqrt{8})(-\sqrt{8})}
$$
$$
= 9 \quad - \quad 3\sqrt{8} \quad + \quad 3\sqrt{8} \quad - \quad \sqrt{64}
$$
$$
= 9 - \sqrt{64}
$$
$$
= 9 - 8 = 1 \qquad ✳
$$

Note that in Example 9, we multiplied *the sum and difference of the same two terms*. Recall from Section 5.6 that $(a + b)(a - b) = a^2 - b^2$. If we let $a = 3$ and $b = \sqrt{8}$, then we can multiply as follows.

$$
(a + b)(a - b) = a^2 - b^2
$$
$$
(3 + \sqrt{8})(3 - \sqrt{8}) = 3^2 - (\sqrt{8})^2
$$
$$
= 9 - 8
$$
$$
= 1
$$

When multiplying the sum and difference of the same two terms, you may obtain the answer using the difference of the squares of the two terms. We will see additional multiplications of this type in Section 7.5.

EXAMPLE 10 If $f(x) = \sqrt[3]{x^2}$ and $g(x) = \sqrt[3]{x^4} + \sqrt[3]{x^2}$, find

a) $(f \cdot g)(x)$ and   **b)** $(f \cdot g)(5)$.

Solution **a)** From Section 3.6, we know that $(f \cdot g)(x) = f(x) \cdot g(x)$.

$$
(f \cdot g)(x) = f(x) \cdot g(x)
$$
$$
= \sqrt[3]{x^2}\left(\sqrt[3]{x^4} + \sqrt[3]{x^2}\right) \qquad \textit{Substitute given values.}
$$
$$
= \sqrt[3]{x^2}\,\sqrt[3]{x^4} + \sqrt[3]{x^2}\,\sqrt[3]{x^2} \qquad \textit{Distributive property}
$$
$$
= \sqrt[3]{x^6} + \sqrt[3]{x^4} \qquad \textit{Product rule for radicals}
$$
$$
= x^2 + x\sqrt[3]{x} \qquad \textit{Simplify radicals.}
$$

b) To compute $(f \cdot g)(5)$, substitute 5 for x in the answer obtained in part **a)**.

$$(f \cdot g)(x) = x^2 + x\sqrt[3]{x}$$
$$(f \cdot g)(5) = 5^2 + 5\sqrt[3]{5} \quad \textit{Substitute 5 for x.}$$
$$= 25 + 5\sqrt[3]{5}$$

NOW TRY EXERCISE 77

EXAMPLE 11 Simplify $f(x)$ if **a)** $f(x) = \sqrt{x+2}\,\sqrt{x+2}$, $x \geq -2$ and

b) $f(x) = \sqrt{3x^2 - 30x + 75}$, assume that the variable may be any real number.

Solution **a)** $f(x) = \sqrt{x+2}\,\sqrt{x+2}$
$$= \sqrt{(x+2)(x+2)} \quad \textit{Product rule for radicals}$$
$$= \sqrt{(x+2)^2}$$
$$= x + 2$$

Since we are told that $x \geq -2$, we can use the product rule. Note that the radicand will be a nonnegative number for any $x \geq -2$ and we can write the answer as $x + 2$ rather than $|x+2|$.

b) $f(x) = \sqrt{3x^2 - 30x + 75}$
$$= \sqrt{3(x^2 - 10x + 25)} \quad \textit{Factor out 3.}$$
$$= \sqrt{3(x-5)^2} \quad \textit{Write as the square of a binomial.}$$
$$= \sqrt{3}\,\sqrt{(x-5)^2} \quad \textit{Product rule for radicals}$$
$$= \sqrt{3}\,|x-5|$$

Since the variable could be any real number, we write our answer with absolute value signs. If we had been told that $x - 5$ was nonnegative, then we could have written our answer as $\sqrt{3}(x-5)$.

NOW TRY EXERCISE 105

Exercise Set 7.4

Concept/Writing Exercises

1. What are like radicals?

2. a) Explain how to add like radicals.

  b) Using the procedure in part **a)**, add $\frac{3}{5}\sqrt{5} + \frac{5}{4}\sqrt{5}$.

3. Use a calculator to estimate $\sqrt{3} + 3\sqrt{2}$.

4. Use a calculator to estimate $2\sqrt{3} + \sqrt{5}$.

5. Does $\sqrt{a} + \sqrt{b} = \sqrt{a+b}$? Explain your answer and give an example supporting your answer.

6. Since $64 + 36 = 100$, does $\sqrt{64} + \sqrt{36} = \sqrt{100}$? Explain your answer.

Practice the Skills

In this exercise set, assume that all variables represent positive real numbers.
Simplify.

7. $\sqrt{5} - \sqrt{5}$

8. $2\sqrt{6} - \sqrt{6}$

9. $6\sqrt{3} - 2\sqrt{3}$

10. $5\sqrt{2} + 7\sqrt{2} - 11$

11. $2\sqrt{3} - 2\sqrt{3} - 4\sqrt{3} + 5$

12. $6\sqrt[3]{5} - 8\sqrt[3]{5}$

13. $3\sqrt[4]{y} - 9\sqrt[4]{y}$

14. $3\sqrt[5]{a} + 7 + 5\sqrt[5]{a} - 6$

15. $3\sqrt{5} - \sqrt[3]{x} + 4\sqrt{5} + 3\sqrt[3]{x}$

16. $6 + 4\sqrt[4]{a} - 7\sqrt[4]{a} + 5$

17. $5\sqrt{x} - 4\sqrt{y} + 3\sqrt{x} + 2\sqrt{y} - \sqrt{x}$

18. $3\sqrt{a} + 4\sqrt[3]{b} + 2\sqrt{a} - 12\sqrt[3]{b}$

Simplify.

19. $\sqrt{8} - \sqrt{12}$

20. $\sqrt{75} + \sqrt{108}$

21. $-6\sqrt{75} + 4\sqrt{125}$

22. $3\sqrt{250} + 5\sqrt{160}$

23. $-4\sqrt{90} + 3\sqrt{40} + 2\sqrt{10}$

24. $3\sqrt{40x^2y} + 2x\sqrt{490y}$

25. $\sqrt{500xy^2} + y\sqrt{320x}$

26. $5\sqrt{8} + 2\sqrt{50} - 3\sqrt{72}$

27. $2\sqrt{5x} - 3\sqrt{20x} - 4\sqrt{45x}$

28. $3\sqrt{27c^2} - 2\sqrt{108c^2} - \sqrt{48c^2}$

29. $3\sqrt{50a^2} - 3\sqrt{72a^2} - 8a\sqrt{18}$

30. $4\sqrt[3]{5} - 5\sqrt[3]{40}$

31. $\sqrt[3]{108} + 2\sqrt[3]{32}$

32. $2\sqrt[3]{16} + \sqrt[3]{54}$

33. $\sqrt[3]{27} - 5\sqrt[3]{8}$

34. $3\sqrt{45x^3} + \sqrt{5x}$

35. $2\sqrt[3]{a^4b^2} + 3a\sqrt[3]{ab^2}$

36. $3y\sqrt[4]{48x^5} - x\sqrt[4]{3x^5y^4}$

37. $\sqrt{4r^7s^5} + 3r^2\sqrt{r^3s^5} - 2rs\sqrt{r^5s^3}$

38. $x\sqrt[3]{27x^5y^2} - x^2\sqrt[3]{x^2y^2} + 2\sqrt[3]{x^8y^2}$

39. $\sqrt[3]{128x^9y^{10}} - 2x^2y\sqrt[3]{16x^3y^7}$

40. $5\sqrt[3]{320x^5y^8} + 2x\sqrt[3]{135x^2y^8}$

Simplify.

41. $\sqrt{50}\,\sqrt{2}$

42. $\sqrt[3]{2}\,\sqrt[3]{4}$

43. $\sqrt[3]{2}\,\sqrt[3]{28}$

44. $\sqrt[3]{3}\,\sqrt[3]{54}$

45. $\sqrt{9m^3n^7}\,\sqrt{3mn^4}$

46. $\sqrt[3]{5ab^2}\,\sqrt[3]{25a^4b^{12}}$

47. $\sqrt[3]{9x^7y^{10}}\,\sqrt[3]{6x^4y^3}$

48. $\left(\sqrt[3]{2x^3y^4}\right)^2$

49. $\sqrt[4]{3x^9y^{12}}\,\sqrt[4]{54x^4y^7}$

50. $\sqrt[5]{x^{24}y^{30}z^9}\,\sqrt[5]{x^{13}y^8z^7}$

51. $\sqrt[4]{8x^4yz^3}\,\sqrt[4]{2x^2y^3z^7}$

52. $\sqrt{2}(\sqrt{6} + \sqrt{2})$

53. $\sqrt{5}(\sqrt{5} - \sqrt{3})$

54. $\sqrt{3}(\sqrt{12} + \sqrt{6})$

55. $\sqrt[3]{y}\left(2\sqrt[3]{y} - \sqrt[3]{y^8}\right)$

56. $\sqrt{3y}\left(\sqrt{27y^2} - \sqrt{y}\right)$

57. $2\sqrt[3]{x^4y^5}\left(\sqrt[3]{8x^{12}y^4} + \sqrt[3]{16xy^9}\right)$

58. $\sqrt[5]{16x^7y^6}\left(\sqrt[5]{2x^6y^9} - \sqrt[5]{10x^3y^7}\right)$

59. $(5 + \sqrt{5})(5 - \sqrt{5})$

60. $(7 - \sqrt{5})(7 + \sqrt{5})$

61. $(\sqrt{6} + x)(\sqrt{6} - x)$

62. $(\sqrt{x} + y)(\sqrt{x} - y)$

63. $(\sqrt{5} - \sqrt{z})(\sqrt{5} + \sqrt{z})$

64. $(3\sqrt{a} - 7\sqrt{b})(3\sqrt{a} + 7\sqrt{b})$

65. $(\sqrt{3} + 4)(\sqrt{3} + 5)$

66. $(1 + \sqrt{5})(6 + \sqrt{5})$

67. $(3 - \sqrt{2})(4 - \sqrt{8})$

68. $(5\sqrt{6} + 3)(4\sqrt{6} - 2)$

69. $(4\sqrt{3} + \sqrt{2})(\sqrt{3} - \sqrt{2})$

70. $(\sqrt{3} + 4)^2$

71. $(2\sqrt{5} - 3)^2$

72. $(\sqrt{y} + \sqrt{6z})(\sqrt{2z} - \sqrt{8y})$

73. $(2\sqrt{3x} - \sqrt{y})(3\sqrt{3x} + \sqrt{y})$

74. $(\sqrt[3]{9} + \sqrt[3]{2})(\sqrt[3]{3} + \sqrt[3]{4})$

75. $(\sqrt[3]{4} - \sqrt[3]{6})(\sqrt[3]{2} - \sqrt[3]{36})$

76. $(\sqrt[3]{4x} - \sqrt[3]{2y})(\sqrt[3]{4x} + \sqrt[3]{10})$

In Exercises 77–82, $f(x)$ and $g(x)$ are given. Find $(f \cdot g)(x)$.

77. $f(x) = \sqrt{2x}, g(x) = \sqrt{8x} - \sqrt{32}$

78. $f(x) = \sqrt{5x}, g(x) = \sqrt{5x} - \sqrt{10x}$

79. $f(x) = \sqrt[3]{x}, g(x) = \sqrt[3]{x^5} + \sqrt[3]{x^4}$

80. $f(x) = \sqrt[3]{2x^2}, g(x) = \sqrt[3]{4x} + \sqrt[3]{32x^2}$

81. $f(x) = \sqrt[4]{3x^2}, g(x) = \sqrt[4]{9x^4} - \sqrt[4]{x^7}$

82. $f(x) = \sqrt[4]{2x^3}, g(x) = \sqrt[4]{8x^5} - \sqrt[4]{3x^6}$

Simplify. These exercises are a combination of the types of exercises presented earlier in this exercise set.

83. $\sqrt{24}$

84. $\sqrt{200}$

85. $\sqrt{125} + \sqrt{20}$

86. $3\sqrt{7} + 2\sqrt{63} - 2\sqrt{28}$

87. $(3\sqrt{2} - 4)(\sqrt{2} + 5)$

88. $(\sqrt{5} + \sqrt{2})(\sqrt{2} + \sqrt{20})$

89. $\sqrt{6}(4 - \sqrt{2})$

90. $2\sqrt[3]{81} + 4\sqrt[3]{24}$

91. $\sqrt{75}\,\sqrt{6}$

92. $\sqrt[4]{8}\,\sqrt[4]{10}$

93. $\sqrt[3]{80x^{11}}$

94. $\sqrt[3]{x^9y^{11}z}$

95. $\sqrt[6]{128ab^{17}c^9}$

96. $\sqrt[5]{14x^4y^2}\,\sqrt[5]{3x^4y^3}$

97. $2b\sqrt[4]{a^4b} + ab\sqrt[4]{16b}$

98. $2\sqrt[3]{24a^3y^4} + 4a\sqrt[3]{81y^4}$

99. $\left(\sqrt[3]{x^2} - \sqrt[3]{y}\right)\left(\sqrt[3]{x} - 2\sqrt[3]{y^2}\right)$

100. $\left(\sqrt[3]{a} + 5\right)\left(\sqrt[3]{a^2} - 3\right)$

101. $\sqrt[3]{3ab^2}\left(\sqrt[3]{4a^4b^3} - \sqrt[3]{8a^5b^4}\right)$

102. $\sqrt[4]{4st^2}\left(\sqrt[4]{2s^5t^6} + \sqrt[4]{5s^9t^2}\right)$

Simplify the following. In Exercises 105 and 106, assume the variable can be any real number. See Example 11.

103. $f(x) = \sqrt{2x + 5}\,\sqrt{2x + 5}, x \geq -\dfrac{5}{2}$

104. $g(a) = \sqrt{3a + 4}\,\sqrt{3a + 4}, a \geq -\dfrac{4}{3}$

105. $h(r) = \sqrt{4r^2 - 32r + 64}$

106. $f(b) = \sqrt{20b^2 + 60b + 45}$

Problem Solving

Find the perimeter and area of the following figures. Write the perimeter and area in radical form with the radicals simplified.

107.

$\sqrt{45}$

$\sqrt{80}$

108.

$\sqrt{54}$ $\sqrt{24}$ $\sqrt{96}$

$\sqrt{150}$

109.

$\sqrt{245}$

$\sqrt{80}$ $\sqrt{45}$ $\sqrt{180}$

110.

$\sqrt{40}$

$\sqrt{18}$ $\sqrt{8}$ $\sqrt{18}$

$\sqrt{160}$

111. Will the sum of two radicals always be a radical? Give an example to support your answer.

112. Will the difference of two radicals always be a radical? Give an example to support your answer.

113. ***Skid Marks*** Law enforcement officials sometimes use the formula $s = \sqrt{30FB}$ to determine a car's speed, s, in miles per hour, from a car's skid marks. The F in the formula represents the "road factor," which is determined by the road's surface, and the B represents the braking distance, in feet. Officer Jenkins is investigating an accident. Find the car's speed if the skid marks are 80 feet long and **a)** the road was dry asphalt, whose road factor is 0.85, **b)** the road was wet gravel, whose road factor is 0.52.

114. ***Water Through a Fire Hose*** The rate at which water flows through a particular fire hose, R, in gallons per minute, can be approximated by the formula $R = 28d^2\sqrt{P}$, where d is the diameter of the nozzle, in inches, and P is the nozzle pressure, in pounds per square inch. If a nozzle has a diameter of 2.5 inches and the nozzle pressure is 80 pounds per square inch, find the flow rate.

115. ***Height of Girls*** The function $f(t) = 3\sqrt{t} + 19$ can be used to approximate the median height, $f(t)$, in inches, for U.S. girls of age t, in months, where $1 \le t \le 60$. Estimate the median height of girls at age **a)** 36 months and **b)** 40 months.

116. ***Standard Deviation*** In statistics, the standard deviation of the population, σ, read "sigma," is a measure of the spread of a set of data about the mean of the data. The greater the spread, the greater the standard deviation. One formula used to determine sigma is $\sigma = \sqrt{npq}$, where n represents the sample size, p represents the percent chance (or probability) that something specific happens, and q represents the percent chance (or probability) that the specific thing does not happen. In a sample of 600 people who purchase airline tickets, the percent that showed up for their flight, p, was 0.93, and the percent that did not show up for their flight, q, was 0.07. Use this information to find σ.

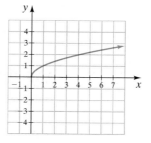

117. The graph of $f(x) = \sqrt{x}$ is shown.

a) If $g(x) = 2$, sketch the graph of $(f + g)(x)$.

b) What effect does adding 2 have to the graph of $f(t)$?

118. The graph of $f(x) = -\sqrt{x}$ is shown.

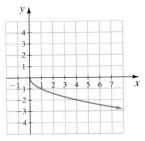

a) If $g(x) = 3$, sketch the graph of $(f + g)(x)$.

b) What effect does adding 3 have to the graph of $f(x)$?

119. You are given that $f(x) = \sqrt{x}$ and $g(x) = \sqrt{x} - 2$.

a) Sketch the graph of $(f - g)(x)$. Explain how you determined your answer.

b) What is the domain of $(f - g)(x)$?

120. You are given that $f(x) = \sqrt{x}$ and $g(x) = -\sqrt{x} - 3$.

a) Sketch the graph of $(f + g)(x)$. Explain how you determined your answer.

b) What is the domain of $(f + g)(x)$?

121. Graph the function $f(x) = \sqrt{x^2}$.

122. Graph the function $f(x) = \sqrt{x^2} - 4$.

Cumulative Review Exercises

[1.2] **123.** What is a rational number?

[1.3] **124.** What is a real number?

125. What is an irrational number?

126. What is the definition of $|a|$?

[2.2] **127.** Solve the formula $E = \frac{1}{2}mv^2$ for m.

[2.5] **128.** Solve the inequality $-4 < 2x - 3 \le 5$ and indicate the solution **a)** on the number line; **b)** in internal notation; **c)** in set builder notation.

7.5 DIVIDING RADICALS

1. Rationalize denominators.
2. Rationalize a denominator using the conjugate.
3. Understand when a radical is simplified.
4. Use rationalizing the denominator in an addition problem.
5. Divide radical expressions with different indices.

1 Rationalize Denominators

We introduced the quotient rule for radicals in Section 7.3. Now we will use the quotient rule to work other division problems and to rationalize denominators.

When the denominator of a fraction contains a radical, we generally simplify the expression by **rationalizing the denominator**. To rationalize a denominator is to remove all radicals from the denominator. When adding radicals, it may be necessary to rationalize denominators, as will be illustrated in Example 6.

To Rationalize a Denominator

Multiply both the numerator and the denominator of the fraction by a radical that will result in the radicand in the denominator becoming a perfect power.

When both the numerator and denominator are multiplied by the same radical expression, you are in effect multiplying the fraction by 1, which does not change its value.

EXAMPLE 1   Simplify.   **a)** $\dfrac{1}{\sqrt{5}}$ **b)** $\dfrac{x}{3\sqrt{2}}$ **c)** $\dfrac{3}{\sqrt{2x}}$ **d)** $\dfrac{\sqrt[3]{16a^4}}{\sqrt[3]{b}}$

Solution   To simplify each expression, we must rationalize the denominators. We do so by multiplying both the numerator and denominator by a radical that will result in the denominator becoming a perfect power for the given index.

a) $\dfrac{1}{\sqrt{5}} = \dfrac{1}{\sqrt{5}} \cdot \dfrac{\sqrt{5}}{\sqrt{5}} = \dfrac{\sqrt{5}}{\sqrt{25}} = \dfrac{\sqrt{5}}{5}$

b) $\dfrac{x}{3\sqrt{2}} = \dfrac{x}{3\sqrt{2}} \cdot \dfrac{\sqrt{2}}{\sqrt{2}} = \dfrac{x\sqrt{2}}{3 \cdot 2} = \dfrac{x\sqrt{2}}{6}$

c) There are two factors in the radicand, 2 and x. We must make each factor a perfect square. Since 2^2 or 4 is a perfect square and x^2 is a perfect square, we multiply both numerator and denominator by $\sqrt{2x}$.

$$\dfrac{3}{\sqrt{2x}} = \dfrac{3}{\sqrt{2x}} \cdot \dfrac{\sqrt{2x}}{\sqrt{2x}}$$
$$= \dfrac{3\sqrt{2x}}{\sqrt{4x^2}}$$
$$= \dfrac{3\sqrt{2x}}{2x}$$

d) There are no common factors in the numerator and denominator. Before we rationalize the denominator, let's simplify the numerator.

$$\dfrac{\sqrt[3]{16a^4}}{\sqrt[3]{b}} = \dfrac{\sqrt[3]{8a^3}\,\sqrt[3]{2a}}{\sqrt[3]{b}} \qquad \text{\textit{Product rule for radicals}}$$
$$= \dfrac{2a\sqrt[3]{2a}}{\sqrt[3]{b}} \qquad \text{\textit{Simplify the numerator.}}$$

Now we rationalize the denominator. Since the denominator is a cube root, we need to make the radicand a perfect cube. Since the denominator contains b and we want b^3, we need two more factors of b. We therefore multiply both numerator and denominator by $\sqrt[3]{b^2}$.

$$= \dfrac{2a\sqrt[3]{2a}}{\sqrt[3]{b}} \cdot \dfrac{\sqrt[3]{b^2}}{\sqrt[3]{b^2}}$$
$$= \dfrac{2a\sqrt[3]{2ab^2}}{\sqrt[3]{b^3}}$$
$$= \dfrac{2a\sqrt[3]{2ab^2}}{b}$$

NOW TRY EXERCISE 25

EXAMPLE 2   Simplify.   **a)** $\sqrt{\dfrac{3}{5}}$ **b)** $\sqrt[3]{\dfrac{x}{2y^2}}$ **c)** $\sqrt[4]{\dfrac{32x^9y^6}{3z^2}}$

Solution   In each part, we will use the quotient rule to write the radical as a quotient of two radicals.

a) $\sqrt{\dfrac{3}{5}} = \dfrac{\sqrt{3}}{\sqrt{5}} \cdot \dfrac{\sqrt{5}}{\sqrt{5}} = \dfrac{\sqrt{15}}{\sqrt{25}} = \dfrac{\sqrt{15}}{5}$

b) $\sqrt[3]{\dfrac{x}{2y^2}} = \dfrac{\sqrt[3]{x}}{\sqrt[3]{2y^2}}$

The denominator is $\sqrt[3]{2y^2}$ and we want to change it to $\sqrt[3]{2^3y^3}$. We now multiply both the numerator and denominator by the cube root of an expression that will make the radicand in the denominator $\sqrt[3]{2^3 y^3}$. Since $2 \cdot 2^2 = 2^3$ and $y^2 \cdot y = y^3$, we multiply both numerator and denominator by $\sqrt[3]{2^2 y}$.

$$\frac{\sqrt[3]{x}}{\sqrt[3]{2y^2}} = \frac{\sqrt[3]{x}}{\sqrt[3]{2y^2}} \cdot \frac{\sqrt[3]{2^2 y}}{\sqrt[3]{2^2 y}}$$

$$= \frac{\sqrt[3]{x} \, \sqrt[3]{4y}}{\sqrt[3]{2^3 y^3}}$$

$$= \frac{\sqrt[3]{4xy}}{2y}$$

c) After using the quotient rule, we simplify the numerator.

$$\sqrt[4]{\frac{32x^9 y^6}{3z^2}} = \frac{\sqrt[4]{32x^9 y^6}}{\sqrt[4]{3z^2}} \qquad \textit{Quotient rule for radicals}$$

$$= \frac{\sqrt[4]{16x^8 y^4} \, \sqrt[4]{2xy^2}}{\sqrt[4]{3z^2}} \qquad \textit{Product rule for radicals}$$

$$= \frac{2x^2 y \sqrt[4]{2xy^2}}{\sqrt[4]{3z^2}} \qquad \textit{Simplify the numerator.}$$

Now we rationalize the denominator. To make the radicand in the denominator a perfect fourth power, we need to get each factor to a power of 4. Since the denominator contains one factor of 3, we need 3 more factors of 3, or 3^3. Since there are two factors of z, we need 2 more factors of z, or z^2. Thus we will multiply both numerator and denominator by $\sqrt[4]{3^3 z^2}$.

$$= \frac{2x^2 y \sqrt[4]{2xy^2}}{\sqrt[4]{3z^2}} \cdot \frac{\sqrt[4]{3^3 z^2}}{\sqrt[4]{3^3 z^2}}$$

$$= \frac{2x^2 y \sqrt[4]{2xy^2} \, \sqrt[4]{27z^2}}{\sqrt[4]{3z^2} \, \sqrt[4]{3^3 z^2}}$$

$$= \frac{2x^2 y \sqrt[4]{54xy^2 z^2}}{\sqrt[4]{3^4 z^4}} \qquad \textit{Product rule for radicals}$$

$$= \frac{2x^2 y \sqrt[4]{54xy^2 z^2}}{3z}$$

Note: There are no perfect fourth power factors of 54, and each exponent in the radicand is less than the index.

NOW TRY EXERCISE 53

2 Rationalize a Denominator Using the Conjugate

When the denominator of a rational expression is a binomial that contains a radical, we rationalize the denominator. We do this by multiplying both the numerator and the denominator of the fraction by the **conjugate** of the denominator. The conjugate of a binomial is a binomial having the same two terms with the sign of the second term changed.

Expression	Conjugate
$3 + \sqrt{2}$	$3 - \sqrt{2}$
$2\sqrt{3} - \sqrt{5}$	$2\sqrt{3} + \sqrt{5}$
$\sqrt{x} + \sqrt{y}$	$\sqrt{x} - \sqrt{y}$
$a + \sqrt{b}$	$a - \sqrt{b}$

When a binomial is multiplied by its conjugate, the outer and inner products will sum to 0. We multiplied radicals containing binomial factors in Section 7.4. We will work one more example of multiplication of radical expressions in Example 3.

EXAMPLE 3 Multiply $(2 + \sqrt{3})(2 - \sqrt{3})$.

Solution Multiply using the FOIL method.

$$\overset{\text{F}}{} \quad \overset{\text{O}}{} \quad \overset{\text{I}}{} \quad \overset{\text{L}}{}$$
$$(2 + \sqrt{3})(2 - \sqrt{3}) = 2(2) + 2(-\sqrt{3}) + 2(\sqrt{3}) + \sqrt{3}(-\sqrt{3})$$
$$= 4 - 2\sqrt{3} + 2\sqrt{3} - \sqrt{9}$$
$$= 4 - \sqrt{9}$$
$$= 4 - 3$$
$$= 1$$

In Example 3, we would get the same result using the formula for the product of the sum and difference of the same two terms. The product results in the difference of two squares, $(a + b)(a - b) = a^2 - b^2$. In Example 3, if we let $a = 2$ and $b = \sqrt{3}$, then using the formula we get the following.

$$(a + b)(a - b) = a^2 - b^2$$
$$\downarrow \quad \downarrow \ \downarrow \quad \downarrow \quad \downarrow \quad \downarrow$$
$$(2 + \sqrt{3})(2 - \sqrt{3}) = 2^2 - (\sqrt{3})^2$$
$$= 4 - 3$$
$$= 1$$

Now let's work an example where we rationalize a denominator containing two terms.

EXAMPLE 4 Simplify. **a)** $\dfrac{5}{2 + \sqrt{3}}$ **b)** $\dfrac{6}{\sqrt{5} - \sqrt{2}}$ **c)** $\dfrac{x - \sqrt{y}}{x + \sqrt{y}}$

Solution In each part, we rationalize the denominator by multiplying the numerator and the denominator by the conjugate of the denominator.

a)
$$\frac{5}{2 + \sqrt{3}} = \frac{5}{2 + \sqrt{3}} \cdot \frac{2 - \sqrt{3}}{2 - \sqrt{3}}$$
$$= \frac{5(2 - \sqrt{3})}{(2 + \sqrt{3})(2 - \sqrt{3})}$$
$$= \frac{5(2 - \sqrt{3})}{4 - 3}$$
$$= \frac{5(2 - \sqrt{3})}{1} \quad \text{or} \quad 10 - 5\sqrt{3}$$

b)
$$\frac{6}{\sqrt{5} - \sqrt{2}} = \frac{6}{\sqrt{5} - \sqrt{2}} \cdot \frac{\sqrt{5} + \sqrt{2}}{\sqrt{5} + \sqrt{2}}$$
$$= \frac{6(\sqrt{5} + \sqrt{2})}{5 - 2}$$
$$= \frac{\overset{2}{\cancel{6}}(\sqrt{5} + \sqrt{2})}{\underset{1}{\cancel{3}}}$$
$$= 2(\sqrt{5} + \sqrt{2}) \quad \text{or} \quad 2\sqrt{5} + 2\sqrt{2}$$

c)

$$\frac{x - \sqrt{y}}{x + \sqrt{y}} = \frac{x - \sqrt{y}}{x + \sqrt{y}} \cdot \frac{x - \sqrt{y}}{x - \sqrt{y}}$$

$$= \frac{x^2 - x\sqrt{y} - x\sqrt{y} + \sqrt{y}^2}{x^2 - y}$$

$$= \frac{x^2 - 2x\sqrt{y} + y}{x^2 - y}$$

Remember that you cannot divide out x^2 or y because they are terms, not factors.

❋

NOW TRY EXERCISE 65

Now that we have illustrated how to rationalize denominators, let's discuss the criteria a radical must meet to be considered simplified.

3 Understand When a Radical Is Simplified

After you have simplified a radical expression, you should check it to make sure that it is simplified as far as possible.

> ### A Radical Expression Is Simplified When the Following Are All True
>
> 1. No perfect powers are factors of the radicand and all exponents in the radicand are less than the index.
> 2. No radicand contains a fraction.
> 3. No denominator contains a radical.

EXAMPLE 5 Determine whether the following expressions are simplified. If not, explain why. Simplify the expressions if not simplified.

a) $\sqrt{27x^5}$ **b)** $\sqrt{\dfrac{1}{2}}$ **c)** $\dfrac{1}{\sqrt{3}}$

Solution **a)** This expression is not simplified because 9 is a perfect square factor of 27 and because x^4 is a perfect square factor of x^5. Notice that the exponent on the variable in the radicand, 5, is greater than the index, 2. Whenever the exponent on the variable in the radicand is greater than the index, the radicand has a perfect power factor of the variable, and the radical needs to be simplified further. Let's simplify the radical.

$$\sqrt{27x^5} = \sqrt{9x^4 \cdot 3x} = \sqrt{9x^4} \cdot \sqrt{3x} = 3x^2\sqrt{3x}$$

b) This expression is not simplified since the radicand contains the fraction $\dfrac{1}{2}$. This violates item 2. We simplify by first using the quotient rule, and then we rationalize the denominator, as follows.

$$\sqrt{\frac{1}{2}} = \frac{\sqrt{1}}{\sqrt{2}} \cdot \frac{\sqrt{2}}{\sqrt{2}} = \frac{\sqrt{2}}{2}$$

c) This expression is not simplified since the denominator, $\sqrt{3}$, contains a radical. This violates item 3. We simplify by rationalizing the denominator, as follows.

$$\frac{1}{\sqrt{3}} = \frac{1}{\sqrt{3}} \cdot \frac{\sqrt{3}}{\sqrt{3}} = \frac{\sqrt{3}}{3}$$

❋

4 Use Rationalizing the Denominator in an Addition Problem

Now let's work an addition problem that requires rationalizing the denominator. This example makes use of the methods discussed in Sections 7.3 and 7.4 to add and subtract radicals.

EXAMPLE 6 Simplify $4\sqrt{2} - \dfrac{1}{\sqrt{8}} + \sqrt{32}$.

Solution Begin by rationalizing the denominator and by simplifying $\sqrt{32}$.

$$4\sqrt{2} - \frac{1}{\sqrt{8}} + \sqrt{32} = 4\sqrt{2} - \frac{1}{\sqrt{8}} \cdot \frac{\sqrt{2}}{\sqrt{2}} + \sqrt{16}\,\sqrt{2} \qquad \textit{Rationalize denominator.}$$

$$= 4\sqrt{2} - \frac{\sqrt{2}}{\sqrt{16}} + 4\sqrt{2} \qquad \textit{Product rule}$$

$$= 4\sqrt{2} - \frac{1}{4}\sqrt{2} + 4\sqrt{2} \qquad \textit{Write } \frac{\sqrt{2}}{\sqrt{16}} \textit{ as } \frac{1}{4}\sqrt{2}.$$

$$= \left(4 - \frac{1}{4} + 4\right)\sqrt{2} \qquad \textit{Simplify.}$$

$$= \frac{31\sqrt{2}}{4}$$

NOW TRY EXERCISE 105

5 Divide Radical Expressions with Different Indices

Now we will divide radical expressions where the radicals have different indices. To divide such problems, write each radical in exponential form. Then use the rules of exponents, with the rational exponents, as was discussed in Section 7.2, to simplify the expression. Example 7 illustrates this procedure.

EXAMPLE 7 Simplify. **a)** $\dfrac{\sqrt[5]{(m+n)^7}}{\sqrt[3]{(m+n)^4}}$ **b)** $\dfrac{\sqrt[3]{a^5b^4}}{\sqrt{a^2b}}$

Solution Begin by writing the numerator and denominator with rational exponents.

a) $\dfrac{\sqrt[5]{(m+n)^7}}{\sqrt[3]{(m+n)^4}} = \dfrac{(m+n)^{7/5}}{(m+n)^{4/3}}$ *Write with rational exponents.*

$$= (m+n)^{(7/5)-(4/3)} \qquad \textit{Quotient rule for exponents}$$

$$= (m+n)^{1/15}$$

$$= \sqrt[15]{m+n} \qquad \textit{Write as a radical.}$$

b) $\dfrac{\sqrt[3]{a^5b^4}}{\sqrt{a^2b}} = \dfrac{(a^5b^4)^{1/3}}{(a^2b)^{1/2}}$ *Write with rational exponents.*

$$= \frac{a^{5/3}b^{4/3}}{ab^{1/2}} \qquad \textit{Raise the product to a power.}$$

$$= a^{(5/3)-1}b^{(4/3)-(1/2)} \qquad \textit{Quotient rule for exponents}$$

$$= a^{2/3}b^{5/6}$$

$$= a^{4/6}b^{5/6} \qquad \textit{Write the fractions with denominator 6.}$$

$$= (a^4b^5)^{1/6} \qquad \textit{Rewrite using the laws of exponents.}$$

$$= \sqrt[6]{a^4b^5} \qquad \textit{Write as a radical.}$$

NOW TRY EXERCISE 123

Exercise Set 7.5

Concept/Writing Exercises

1. **a)** What is the conjugate of a binomial?
   b) What is the conjugate of $x - \sqrt{3}$?

2. What does it mean to rationalize a denominator?

3. **a)** Explain how to rationalize a denominator that contains a radical expression of one term.
   b) Rationalize $\dfrac{4}{\sqrt{3y}}$ using the procedure you specified in part **a)**.

4. **a)** Explain how to rationalize a denominator that contains a binomial in which one or both terms is a radical expression.

   b) Rationalize $\dfrac{\sqrt{2} + \sqrt{5}}{\sqrt{2} - \sqrt{5}}$ using the procedure you specified in part **a)**.

5. What are the three conditions that must be met for a radical expression to be simplified?

6. Explain why each of the following is not simplified.

   a) $\sqrt{x^3}$ **b)** $\sqrt{\dfrac{1}{3}}$ **c)** $\dfrac{1}{\sqrt{2}}$

Simplify. In this exercise set, assume all variables represent positive real numbers.

7. $\dfrac{1}{\sqrt{3}}$

8. $\dfrac{1}{\sqrt{6}}$

9. $\dfrac{3}{\sqrt{7}}$

10. $\dfrac{4}{\sqrt{5}}$

11. $\dfrac{1}{\sqrt{17}}$

12. $\dfrac{1}{\sqrt{z}}$

13. $\dfrac{7}{\sqrt{7}}$

14. $\dfrac{y}{\sqrt{y}}$

15. $\dfrac{p}{\sqrt{2}}$

16. $\dfrac{x}{\sqrt{13}}$

17. $\dfrac{\sqrt{y}}{\sqrt{5}}$

18. $\dfrac{\sqrt{17}}{\sqrt{q}}$

19. $\dfrac{5\sqrt{3}}{\sqrt{5}}$

20. $\dfrac{14x}{\sqrt{x}}$

21. $\dfrac{\sqrt{x}}{\sqrt{y}}$

22. $\dfrac{2\sqrt{3}}{\sqrt{w}}$

23. $\sqrt{\dfrac{5m}{8}}$

24. $\dfrac{2\sqrt{3}}{\sqrt{y^3}}$

25. $\dfrac{2n}{\sqrt{18n}}$

26. $\sqrt{\dfrac{120x}{4y^3}}$

27. $\sqrt{\dfrac{18x^4y^3}{2z^3}}$

28. $\sqrt{\dfrac{5pq^4}{2r}}$

29. $\sqrt{\dfrac{20y^4z^3}{3xy^{-2}}}$

30. $\sqrt{\dfrac{5xy^6}{6z}}$

31. $\sqrt{\dfrac{75x^6y^5}{3z^3}}$

32. $\sqrt{\dfrac{45y^{12}z^{10}}{2x}}$

Simplify.

33. $\dfrac{1}{\sqrt[3]{2}}$

34. $\dfrac{1}{\sqrt[3]{4}}$

35. $\dfrac{3}{\sqrt[3]{x}}$

36. $\dfrac{2}{\sqrt[3]{x^2}}$

37. $\dfrac{1}{\sqrt[4]{2}}$

38. $\dfrac{x}{\sqrt[4]{4}}$

39. $\dfrac{z}{\sqrt[4]{8}}$

40. $\dfrac{7}{\sqrt[4]{z}}$

41. $\dfrac{x}{\sqrt[4]{z^2}}$

42. $\dfrac{17}{\sqrt[4]{z^3}}$

43. $\dfrac{13}{\sqrt[5]{y^2}}$

44. $\dfrac{x}{\sqrt[5]{y^4}}$

45. $\dfrac{3}{\sqrt[7]{a^3}}$

46. $\sqrt[3]{\dfrac{5x}{y}}$

47. $\sqrt[3]{\dfrac{1}{4x}}$

48. $\sqrt[3]{\dfrac{5c}{9y^2}}$

49. $\dfrac{5m}{\sqrt[4]{2}}$

50. $\dfrac{3}{\sqrt[4]{a}}$

51. $\sqrt[4]{\dfrac{5}{3x^3}}$

52. $\sqrt[4]{\dfrac{2x^3}{4y^2}}$

53. $\sqrt[3]{\dfrac{3x^2}{2y^2}}$

54. $\sqrt[3]{\dfrac{15x^6y^7}{2z^2}}$

55. $\sqrt[3]{\dfrac{8xy^2}{2z^2}}$

56. $\sqrt[6]{\dfrac{r^4s^9}{2r^5}}$

Simplify by rationalizing the denominator.

57. $\dfrac{1}{\sqrt{3}+1}$

58. $\dfrac{5}{\sqrt{2}+1}$

59. $\dfrac{1}{2+\sqrt{3}}$

60. $\dfrac{2}{5-\sqrt{7}}$

61. $\dfrac{4}{\sqrt{2}-7}$

62. $\dfrac{2}{\sqrt{2}+\sqrt{3}}$

63. $\dfrac{\sqrt{5}}{2\sqrt{5}-\sqrt{6}}$

64. $\dfrac{1}{\sqrt{17}-\sqrt{8}}$

65. $\dfrac{2}{6+\sqrt{x}}$

66. $\dfrac{3\sqrt{5}}{\sqrt{a}-3}$

67. $\dfrac{4\sqrt{x}}{\sqrt{x}-y}$

🔒 **68.** $\dfrac{\sqrt{8x}}{x+\sqrt{y}}$

69. $\dfrac{\sqrt{2}-2\sqrt{3}}{\sqrt{2}+4\sqrt{3}}$

70. $\dfrac{\sqrt{c}-\sqrt{2d}}{\sqrt{c}-\sqrt{d}}$

71. $\dfrac{\sqrt{a^3}+\sqrt{a^7}}{\sqrt{a}}$

72. $\dfrac{2\sqrt{xy}-\sqrt{xy}}{\sqrt{x}+\sqrt{y}}$

73. $\dfrac{2}{\sqrt{x+2}-3}$

74. $\dfrac{5}{\sqrt{y-3}+6}$

Simplify. These exercises are a combination of the types of exercises presented earlier in this exercise set.

75. $\sqrt{\dfrac{x}{9}}$

76. $\sqrt[4]{\dfrac{x^4}{16}}$

77. $\sqrt{\dfrac{2}{5}}$

78. $\sqrt{\dfrac{a}{b}}$

79. $(\sqrt{5}+\sqrt{6})(\sqrt{5}-\sqrt{6})$

80. $\sqrt[3]{\dfrac{1}{3}}$

🔒 **81.** $\sqrt{\dfrac{24x^3y^6}{5z}}$

82. $\dfrac{6}{4-\sqrt{y}}$

83. $\sqrt{\dfrac{12xy^4}{2x^3y^4}}$

84. $\dfrac{4x}{\sqrt[3]{5y}}$

85. $\dfrac{1}{\sqrt{a}+3}$

86. $\dfrac{\sqrt{x}}{\sqrt{x}+5\sqrt{y}}$

87. $-\dfrac{7\sqrt{x}}{\sqrt{98}}$

88. $\sqrt{\dfrac{2xy^4}{18xy^2}}$

89. $\sqrt[4]{\dfrac{3y^2}{2x}}$

90. $\sqrt{\dfrac{25x^2y^5}{3z}}$

91. $\sqrt[3]{\dfrac{32y^{12}z^{10}}{2x}}$

92. $\dfrac{\sqrt{3}+2}{\sqrt{2}+\sqrt{3}}$

93. $\dfrac{\sqrt{ar}}{\sqrt{a}-2\sqrt{r}}$

94. $\sqrt[4]{\dfrac{2}{9x}}$

95. $\dfrac{\sqrt[3]{6x}}{\sqrt[3]{5xy}}$

96. $\dfrac{\sqrt[3]{16m^2n}}{\sqrt[3]{2mn^2}}$

97. $\sqrt[4]{\dfrac{2x^7y^{12}z^4}{3x^9}}$

98. $\dfrac{3}{\sqrt{y+3}-\sqrt{y}}$

Simplify.

99. $\dfrac{1}{\sqrt{2}}+\dfrac{\sqrt{2}}{2}$

100. $\dfrac{1}{\sqrt{3}}+\dfrac{\sqrt{3}}{3}$

101. $\sqrt{5}-\dfrac{1}{\sqrt{5}}$

102. $\dfrac{\sqrt{6}}{2}-\dfrac{1}{\sqrt{6}}$

103. $\sqrt{\dfrac{1}{6}}+\sqrt{24}$

104. $4\sqrt{3}-\dfrac{3}{\sqrt{3}}+2\sqrt{18}$

105. $3\sqrt{2}-\dfrac{2}{\sqrt{8}}+\sqrt{50}$

106. $\dfrac{1}{3}+\dfrac{1}{\sqrt{3}}+\sqrt{75}$

107. $\sqrt{\dfrac{1}{2}}+3\sqrt{2}+\sqrt{18}$

108. $\dfrac{1}{2}\sqrt{18}-\dfrac{3}{\sqrt{2}}-3\sqrt{50}$

109. $\dfrac{2}{\sqrt{50}}-3\sqrt{50}-\dfrac{1}{\sqrt{8}}$

110. $\dfrac{\sqrt{3}}{3}+\dfrac{2}{\sqrt{3}}+\sqrt{12}$

111. $\sqrt{\dfrac{3}{8}}+\sqrt{\dfrac{3}{2}}$

112. $2\sqrt{\dfrac{8}{3}}-4\sqrt{\dfrac{100}{6}}$

🔒 **113.** $-2\sqrt{\dfrac{x}{y}}+3\sqrt{\dfrac{y}{x}}$

114. $-5x\sqrt{\dfrac{y}{y^2}}+2x\sqrt{\dfrac{1}{y}}$

115. $\dfrac{3}{\sqrt{a}}-\sqrt{\dfrac{9}{a}}+\sqrt{a}$

116. $4\sqrt{x}+\dfrac{1}{\sqrt{x}}+\sqrt{\dfrac{1}{x}}$

Simplify.

🔒 **117.** $\dfrac{\sqrt{(a+b)^4}}{\sqrt[3]{a+b}}$

118. $\dfrac{\sqrt[3]{c+2}}{\sqrt[4]{(c+2)^3}}$

119. $\dfrac{\sqrt[5]{(a+2b)^4}}{\sqrt[3]{(a+2b)^2}}$

120. $\dfrac{\sqrt[6]{(r+3)^5}}{\sqrt[3]{(r+3)^5}}$

121. $\dfrac{\sqrt[3]{r^2 s^4}}{\sqrt{rs}}$

122. $\dfrac{\sqrt{a^2 b^4}}{\sqrt[3]{ab^2}}$

123. $\dfrac{\sqrt[5]{x^4 y^6}}{\sqrt[3]{(xy)^2}}$

124. $\dfrac{\sqrt[6]{4m^8 n^4}}{\sqrt[4]{m^4 n^2}}$

Problem Solving

125. *Illumination of a Light* Under certain conditions the formula

$$d = \sqrt{\dfrac{72}{I}}$$

is used to show the relationship between the illumination on an object, I, in lumens per meter, and the distance, d, in meters, the object is from the light source. If the illumination on a person standing near a light source is 5.3 lumens per meter, how far is the person from the light source?

126. *Strength of a Board* When sufficient pressure is applied to a particular particle board, the particle board will break (or rupture). The thicker the particle board the greater will be the pressure that will need to be applied before the board breaks. The formula

$$T = \sqrt{\dfrac{0.05\,LB}{M}}$$

relates the thickness of a specific particle board, T, in inches, the board's length, L, in inches, the board's load that will cause the board to rupture, B, in pounds, and the modulus of rupture, M, in pounds per square inch. The modulus of rupture is a constant determined by sample tests on the specific type of particle board.

Find the thickness of a 36-inch-long particle board if the modulus of rupture is 2560 pounds per square inch and the board ruptures when 800 pounds are applied.

127. *Volume of a Fish Tank* A new restaurant wants to have a spherical fish tank in its lobby. The radius, r, in inches, of a spherical tank can be found by the formula

$$r = \sqrt[3]{\dfrac{3V}{4\pi}}$$

where V is the volume of the tank in cubic inches. Find the radius of a spherical tank whose volume is 7238.23 cubic inches.

128. *Consecutive Numbers* If we consider the set of consecutive natural numbers $1, 2, 3, 4, \ldots, n$ to be the population, the standard deviation, σ, which is a measure of the spread of the data from the mean, can be calculated by the formula

$$\sigma = \sqrt{\dfrac{n^2 - 1}{12}}$$

where n represents the number of natural numbers in the population. Find the standard deviation for the first 100 consecutive natural numbers.

129. *U.S. Farms* The number of farms in the United States is declining annually (however, the size of the remaining farms is increasing). A function that can be used to estimate the number of farms, $N(t)$, in millions, is

$$N(t) = \dfrac{6.21}{\sqrt[4]{t}}$$

where t is years since 1959 and $1 \le t \le 48$. Estimate the number of farms in the United States in **a)** 1960 and **b)** 2007.

130. *Infant Mortality Rate* The U.S. infant mortality rate has been declining steadily. The infant mortality rate, $N(t)$, defined as deaths per 1000 live births, can be estimated by the function

$$N(t) = \dfrac{28.46}{\sqrt[3]{t^2}}$$

where t is years since 1969 and $1 \le t \le 37$. Estimate the infant mortality rate in **a)** 1970 and **b)** 2006.

131. Which is greater, $\dfrac{2}{\sqrt{2}}$ or $\dfrac{3}{\sqrt{3}}$? Explain.

132. Which is greater, $\dfrac{\sqrt{3}}{2}$ or $\dfrac{2}{\sqrt{3}}$? Explain.

133. Which is greater, $\dfrac{1}{\sqrt{3} + 2}$ or $2 + \sqrt{3}$? (Do not use a calculator.) Explain how you determined your answer.

134. Which is greater, $\dfrac{1}{\sqrt{3}} + \sqrt{75}$ or $\dfrac{2}{\sqrt{12}} + \sqrt{48} + 2\sqrt{3}$? (Do not use a calculator.) Explain how you determined your answer.

135. Consider the functions $f(x) = x^{a/2}$ and $g(x) = x^{b/3}$.
  a) List three values for a that will result in $x^{a/2}$ being a perfect square.
  b) List three values for b that will result in $x^{b/3}$ being a perfect cube.
  c) If $x \geq 0$, find $(f \cdot g)(x)$.
  d) If $x \geq 0$, find $(f/g)(x)$.

Rationalize each denominator.

136. $\dfrac{1}{\sqrt{a + b}}$

137. $\dfrac{3}{\sqrt{2a - 3b}}$

In higher math courses, it may be necessary to rationalize the numerators of radical expressions. Rationalize the numerators of the following expressions. (Your answers will contain radicals in the denominators.)

138. $\dfrac{\sqrt{6}}{3}$

139. $\dfrac{5 - \sqrt{5}}{6}$

140. $\dfrac{4\sqrt{x} - \sqrt{3}}{x}$

141. $\dfrac{\sqrt{x + h} - \sqrt{x}}{h}$

Group Activity

The following two exercises will reinforce many of the concepts presented in this chapter. Work each problem as a group. Make sure each member of the group understands each step in obtaining the solution. The figures in each exercise are similar. For each exercise, use a proportion to find the length of side x. Write the answer in radical form with a rationalized denominator.

142.

143.

Cumulative Review Exercises

[2.2] **144.** Solve the equation $A = \frac{1}{2}h(b_1 + b_2)$ for b_2.

[2.4] **145.** *Moving Vehicles* Two cars leave from West Point at the same time traveling in opposite directions. One travels 10 mile per hour faster than the other.

If the two cars are 270 miles apart after 3 hours, find the speed of each car.

[5.2] **146.** Multiply $(x - 2)(4x^2 + 9x - 2)$.

[6.4] **147.** Solve $\dfrac{x}{2} - \dfrac{4}{x} = -\dfrac{7}{2}$.

7.6 SOLVING RADICAL EQUATIONS

SSM

Study Guide

CD/Video

MathPro 4/5

PH Math Tutor Center

prenhall.com/Angel

1 Solve equations containing one radical.

2 Solve equations containing two radicals.

3 Solve equations containing two radical terms and a nonradical term.

4 Solve applications using radical equations.

5 Solve for a variable in a radicand.

1 Solve Equations Containing One Radical

A **radical equation** is an equation that contains a variable in a radicand.

Examples of Radical Equations

$$\sqrt{x} = 5, \qquad \sqrt[3]{y+4} = 9, \qquad \sqrt{x-2} = 7 + \sqrt{x+8}$$

To Solve Radical Equations

1. Rewrite the equation so that one radical containing a variable is by itself (isolated) on one side of the equation.
2. Raise each side of the equation to a power equal to the index of the radical.
3. Combine like terms.
4. If the equation still contains a term with a variable in a radicand, repeat steps 1 through 3.
5. Solve the resulting equation for the variable.
6. Check all solutions in the original equation for extraneous solutions.

Recall from Section 6.4 that an extraneous solution is a number obtained when solving an equation that is not a solution to the original equation.

The following examples illustrate the procedure for solving radical equations.

EXAMPLE 1 Solve the equation $\sqrt{x} = 7$.

Solution The square root containing the variable is already by itself on one side of the equation. Square both sides of the equation.

$$\sqrt{x} = 7$$
$$(\sqrt{x})^2 = (7)^2$$
$$x = 49$$

Check
$$\sqrt{x} = 7$$
$$\sqrt{49} \overset{?}{=} 7$$
$$7 = 7 \quad \textit{True}$$

✳

EXAMPLE 2 Solve. **a)** $\sqrt{x-4} - 6 = 0$ **b)** $\sqrt[3]{x} + 9 = 7$ **c)** $\sqrt{x} + 3 = 0$

Solution The first step in each part will be to isolate the term containing the radical.

a)
$$\sqrt{x-4} - 6 = 0$$
$$\sqrt{x-4} = 6 \qquad \textit{Isolate the radical containing the variable.}$$
$$(\sqrt{x-4})^2 = 6^2 \qquad \textit{Square both sides.}$$
$$x - 4 = 36 \qquad \textit{Solve for the variable.}$$
$$x = 40$$

A check will show that 40 is the solution.

b)
$$\sqrt[3]{x} + 9 = 7$$
$$\sqrt[3]{x} = -2 \qquad \textit{Isolate the radical containing the variable.}$$
$$(\sqrt[3]{x})^3 = (-2)^3 \qquad \textit{Cube both sides.}$$
$$x = -8$$

A check will show that -8 is the solution.

c)
$$\sqrt{x} + 3 = 0$$
$$\sqrt{x} = -3 \qquad \text{\textit{Isolate the radical containing the variable.}}$$
$$(\sqrt{x})^2 = (-3)^2 \quad \text{\textit{Square both sides.}}$$
$$x = 9$$

Check
$$\sqrt{x} + 3 = 0$$
$$\sqrt{9} + 3 \overset{?}{=} 0$$
$$3 + 3 \overset{?}{=} 0$$
$$6 = 0 \quad \text{\textit{False}}$$

A check shows that 9 is not a solution. The answer to part **c)** is "no real solution." You may have realized there was no real solution to the problem when you obtained the equation $\sqrt{x} = -3$, because $\sqrt{x}$ cannot equal a negative real number. ✳

NOW TRY EXERCISE 17

HELPFUL HINT

Don't forget to check your solutions in the original equation. When you raise both sides of an equation to a power you may introduce extraneous solutions.

Consider the equation $x = 2$. Note what happens when you square both sides of the equation.

$$x = 2$$
$$x^2 = 2^2$$
$$x^2 = 4$$

Note that the equation $x^2 = 4$ has two solutions, $+2$ and -2. Since the original equation $x = 2$ has only one solution, 2, we introduced the extraneous solution, -2.

EXAMPLE 3 Solve $\sqrt{2x - 3} = x - 3$.

Solution Since the radical is already isolated, we square both sides of the equation. Then we solve the resulting quadratic equation.

$$(\sqrt{2x - 3})^2 = (x - 3)^2$$
$$2x - 3 = (x - 3)(x - 3)$$
$$2x - 3 = x^2 - 6x + 9$$
$$0 = x^2 - 8x + 12$$

Now we factor and use the zero-factor property.

$$x^2 - 8x + 12 = 0$$
$$(x - 6)(x - 2) = 0$$
$$x - 6 = 0 \qquad \text{or} \qquad x - 2 = 0$$
$$x = 6 \qquad\qquad\qquad x = 2$$

Check
$$x = 6 \qquad\qquad\qquad\qquad x = 2$$
$$\sqrt{2x - 3} = x - 3 \qquad\qquad \sqrt{2x - 3} = x - 3$$
$$\sqrt{2(6) - 3} \overset{?}{=} 6 - 3 \qquad\qquad \sqrt{2(2) - 3} \overset{?}{=} 2 - 3$$
$$\sqrt{9} \overset{?}{=} 3 \qquad\qquad\qquad\qquad \sqrt{1} \overset{?}{=} -1$$
$$3 = 3 \quad \text{\textit{True}} \qquad\qquad\qquad 1 = -1 \quad \text{\textit{False}}$$

Thus, 6 is a solution, but 2 is not a solution to the equation. The 2 is an extraneous solution because 2 satisfies the equation $(\sqrt{2x - 3})^2 = (x - 3)^2$, but not the original equation, $\sqrt{2x - 3} = x - 3$. ✳

NOW TRY EXERCISE 43

Using Your Graphing Calculator

In Example 3, we found the solution to $\sqrt{2x-3}=x-3$ to be 6. If we let $y_1=\sqrt{2x-3}$ and $y_2=x-3$ and graph y_1 and y_2 on a graphing calculator, we get Figure 7.3. Notice the graphs appear to intersect at $x=6$, which is what we expect.

The table of values in Figure 7.4 shows that the y-coordinate at the point of intersection is 3. In the table, ERROR appears under y_1 for the values of x of 0 and 1. For any values less than $\frac{3}{2}$, the value of $2x-3$ is negative and therefore $\sqrt{2x-3}$ is not a real number. The domain of function y_1 is $\{x|x\geq\frac{3}{2}\}$, which may be found by solving the inequality $2x-3\geq 0$.

You can use your grapher to either solve or check radical equations.

FIGURE 7.3   FIGURE 7.4

Exercises

Use your graphing calculator to determine whether the indicated value is the solution to the radical equation. If it is not the solution, use your grapher to determine the solution.

1. $\sqrt{2x+9}=5(x-7),8$ **2.** $\sqrt{3x+4}=\sqrt{x+12},6$

EXAMPLE 4 Solve $x-2\sqrt{x}-8=0$.

Solution First, isolate the radical term by writing the radical term by itself on one side of the equation.

$$x-2\sqrt{x}-8=0$$
$$-2\sqrt{x}=-x+8$$
$$2\sqrt{x}=x-8$$

Now square both sides of the equation.

$$(2\sqrt{x})^2=(x-8)^2$$
$$4x=x^2-16x+64$$
$$0=x^2-20x+64$$
$$0=(x-16)(x-4)$$
$$x-16=0 \quad\text{or}\quad x-4=0$$
$$x=16 \qquad\qquad x=4$$

Check $x=4$ $x=16$

$x-2\sqrt{x}-8=0$ $x-2\sqrt{x}-8=0$

$4-2\sqrt{4}-8\overset{?}{=}0$ $16-2\sqrt{16}-8\overset{?}{=}0$

$4-2(2)-8\overset{?}{=}0$ $16-2(4)-8\overset{?}{=}0$

$4-4-8\overset{?}{=}0$ $16-8-8\overset{?}{=}0$

$-8=0$ *False* $8-8\overset{?}{=}0$

$0=0$ *True*

NOW TRY EXERCISE 41   The solution is 16. The value 4 is an extraneous solution.

2 Solve Equations Containing Two Radicals

Now we will look at some equations that contain two radicals.

EXAMPLE 5 Solve $\sqrt{9x^2 + 6} = 3\sqrt{x^2 + x - 2}$.

Solution Since the two radicals appear on different sides of the equation, we square both sides of the equation.

$$\left(\sqrt{9x^2 + 6}\right)^2 = \left(3\sqrt{x^2 + x - 2}\right)^2 \quad \textit{Square both sides.}$$
$$9x^2 + 6 = 9(x^2 + x - 2)$$
$$9x^2 + 6 = 9x^2 + 9x - 18 \quad \textit{Distributive property.}$$
$$6 = 9x - 18 \quad \textit{9x}^2 \textit{ was subtracted from both sides.}$$
$$24 = 9x$$
$$\frac{8}{3} = x$$

A check will show that $\dfrac{8}{3}$ is the solution.

In higher mathematics courses, equations are sometimes given using exponents rather than radicals. Example 6 illustrates such an equation.

EXAMPLE 6 For $f(x) = 3(x - 2)^{1/3}$ and $g(x) = (17x - 14)^{1/3}$, find all values of x for which $f(x) = g(x)$.

Solution You should realize that alternate ways of writing $f(x)$ and $g(x)$ are $f(x) = 3\sqrt[3]{x - 2}$ and $g(x) = \sqrt[3]{17x - 14}$. We could therefore work this example using radicals, but we will work instead with rational exponents. We set the two functions equal to each other and solve for x.

$$f(x) = g(x)$$
$$3(x - 2)^{1/3} = (17x - 14)^{1/3}$$
$$[3(x - 2)^{1/3}]^3 = [(17x - 14)^{1/3}]^3 \quad \textit{Cube both sides.}$$
$$3^3(x - 2) = 17x - 14$$
$$27(x - 2) = 17x - 14$$
$$27x - 54 = 17x - 14$$
$$10x - 54 = -14$$
$$10x = 40$$
$$x = 4$$

A check will show that the solution is 4. If you substitute 4 into both $f(x)$ and $g(x)$, you will find they both simplify to $3\sqrt[3]{2}$. Check this now.

NOW TRY EXERCISE 69 In Example 6, if you solve the equation $3\sqrt[3]{x - 2} = \sqrt[3]{17x - 14}$ you will obtain the solution 4. For additional practice, do this now.

3 Solve Equations Containing Two Radical Terms and a Nonradical Term

When a radical equation contains two radical terms and a third nonradical term, you will sometimes need to raise both sides of the equation to a given power twice to obtain the solution. First, isolate one radical term. Then raise both sides of the equation to the given power. This will eliminate one of the radicals. Next, isolate the remaining radical on one side of the equation. Then raise both sides of the equation to the given power a second time. This procedure is illustrated in Example 7.

EXAMPLE 7   Solve $\sqrt{5x - 1} - \sqrt{3x - 2} = 1$.

Solution   We must isolate one radical term on one side of the equation. We will begin by adding $\sqrt{3x - 2}$ to both sides of the equation to isolate $\sqrt{5x - 1}$. Then we will square both sides of the equation and combine like terms.

$$\sqrt{5x - 1} = 1 + \sqrt{3x - 2} \qquad \text{Isolate } \sqrt{5x - 1}.$$
$$(\sqrt{5x - 1})^2 = (1 + \sqrt{3x - 2})^2 \qquad \text{Square both sides.}$$
$$5x - 1 = (1 + \sqrt{3x - 2})(1 + \sqrt{3x - 2}) \qquad \text{Write as a product.}$$
$$5x - 1 = 1 + \sqrt{3x - 2} + \sqrt{3x - 2} + (\sqrt{3x - 2})^2 \quad \text{Multiply.}$$
$$5x - 1 = 1 + 2\sqrt{3x - 2} + 3x - 2 \qquad \text{Combine like terms; simplify.}$$
$$5x - 1 = 3x - 1 + 2\sqrt{3x - 2} \qquad \text{Combine like terms.}$$
$$2x = 2\sqrt{3x - 2} \qquad \text{Isolate the radical term.}$$
$$x = \sqrt{3x - 2} \qquad \text{Both sides were divided by 2.}$$

We have isolated the remaining radical term. We now square both sides of the equation again and solve for x.

$$x = \sqrt{3x - 2}$$
$$x^2 = (\sqrt{3x - 2})^2 \qquad \text{Square both sides.}$$
$$x^2 = 3x - 2$$
$$x^2 - 3x + 2 = 0$$
$$(x - 2)(x - 1) = 0$$
$$x - 2 = 0 \qquad \text{or} \qquad x - 1 = 0$$
$$x = 2 \qquad\qquad\qquad x = 1$$

NOW TRY EXERCISE 61   A check will show that both 2 and 1 are solutions of the equation. ✳

EXAMPLE 8   For $f(x) = \sqrt{5x - 1} - \sqrt{3x - 2}$, find all values of x for which $f(x) = 1$.

Solution   Substitute 1 for $f(x)$. This gives

$$1 = \sqrt{5x - 1} - \sqrt{3x - 2}$$

Since this is the same equation we solved in Example 7, the answers are $x = 2$ and $x = 1$. Verify for yourself that $f(2) = 1$ and $f(1) = 1$. ✳

AVOIDING COMMON ERRORS

In Chapter 5, we stated that $(a + b)^2 \neq a^2 + b^2$. Be careful when squaring a binomial like $1 + \sqrt{x}$. Look at the following computations carefully so that you do not make the mistake shown on the right.

CORRECT

$$(1 + \sqrt{x})^2 = (1 + \sqrt{x})(1 + \sqrt{x})$$
$$\overset{F \qquad O \qquad I \qquad L}{= 1 + \sqrt{x} + \sqrt{x} + \sqrt{x}\,\sqrt{x}}$$
$$= 1 + 2\sqrt{x} + x$$

INCORRECT

$$(1 + \sqrt{x})^2 = 1^2 + (\sqrt{x})^2$$
$$= 1 + x$$

4 Solve Applications Using Radical Equations

Now we will look at a few of the many applications of radicals.

EXAMPLE 9   **Baseball Diamond**   A regulation baseball diamond is a square with 90 feet between bases. How far is second base from home plate?

Solution   Understand   Figure 7.5 illustrates the problem. We need to find the distance from second base to home plate, which is the hypotenuse of a right triangle.

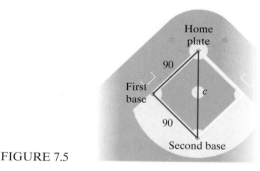

FIGURE 7.5

Translate   To solve the problem, we use the Pythagorean Theorem, which was discussed earlier, leg^2+leg^2=hyp^2, or $a^2 + b^2 = c^2$.

$$90^2 + 90^2 = c^2 \qquad \textit{Substitute known values.}$$

Carry Out
$$8100 + 8100 = c^2$$
$$16{,}200 = c^2$$

$$\sqrt{16{,}200} = \sqrt{c^2} \qquad \textit{Take the square root of both sides.}$$
$$\sqrt{16{,}200} = c \qquad \textit{*See footnote.}$$
$$127.28 \approx c$$

Answer   The distance from second base to home plate is about 127.28 feet. ✳

EXAMPLE 10   **Period of a Pendulum**   The length of time it takes for a pendulum to make one complete swing back and forth is called the *period* of the pendulum. See Figure 7.6. The period of a pendulum, T, in seconds, can be found by the formula $T = 2\pi\sqrt{\dfrac{L}{32}}$, where L is the length of the pendulum in feet. Find the period of a pendulum if its length is 6 feet.

Solution   Substitute 6 for L and 3.14 for π in the formula. If you have a calculator that has a [π] key, use it to enter π.

$$T = 2\pi\sqrt{\frac{L}{32}}$$
$$\approx 2(3.14)\sqrt{\frac{6}{32}}$$
$$\approx 2(3.14)\sqrt{0.1875} \approx 2.72$$

FIGURE 7.6

NOW TRY EXERCISE 103

Thus, the period is about 2.72 seconds. If you have a grandfather clock with a 6-foot pendulum, it will take about 2.72 seconds for it to swing back and forth. ✳

---

*$c^2 = 16{,}200$ has two solutions: $c = \sqrt{16{,}200}$ and $c = -\sqrt{16{,}200}$. Since we are solving for a length, which must be a positive quantity, we use the positive root.

5 Solve for a Variable in a Radicand

You may be given a formula and be asked to solve for a variable in a radicand. To do so, follow the same general procedure used to solve a radical equation. Begin by isolating the radical expression. Then raise both sides of the equation to the same power as the index of the radical. This procedure is illustrated in Example 11 **b)**.

EXAMPLE 11   **Error of Estimation**   A formula in statistics for finding the maximum error of estimation is $E = Z\dfrac{\sigma}{\sqrt{n}}$.

a) Find E if $Z = 1.28$, $\sigma = 10$, and $n = 36$.

b) Solve this equation for n.

Solution   **a)** $E = Z\dfrac{\sigma}{\sqrt{n}} = 1.28\left(\dfrac{10}{\sqrt{36}}\right) = 1.28\left(\dfrac{10}{6}\right) \approx 2.13$

b) First multiply both sides of the equation by $\sqrt{n}$ to eliminate fractions. Then isolate $\sqrt{n}$. Finally, solve for n by squaring both sides of the equation.

$$E = Z\frac{\sigma}{\sqrt{n}}$$

$$\sqrt{n}\,(E) = \left(Z\frac{\sigma}{\sqrt{n}}\right)\sqrt{n} \qquad \textit{Eliminate fractions.}$$

$$\sqrt{n}\,(E) = Z\sigma$$

$$\sqrt{n} = \frac{Z\sigma}{E} \qquad \textit{Isolate the radical term.}$$

$$(\sqrt{n})^2 = \left(\frac{Z\sigma}{E}\right)^2 \qquad \textit{Square both sides.}$$

$$n = \left(\frac{Z\sigma}{E}\right)^2 \quad \text{or} \quad n = \frac{Z^2\sigma^2}{E^2}$$

Exercise Set 7.6

Concept/Writing Exercises

1. a) Explain in your own words how to solve a radical equation.
  b) Solve $\sqrt{2x + 26} - 2 = 4$ using the procedure you gave in part **a)**.

2. Consider the equation $\sqrt{x + 3} = -\sqrt{2x - 1}$. Explain why this equation can have no real solution.

3. Consider the equation $-\sqrt{x^2} = \sqrt{(-x)^2}$. By studying the equation, can you determine its solution? Explain.

4. Consider the equation $\sqrt[3]{x^2} = -\sqrt[3]{x^2}$. By studying the equation, can you determine its solution? Explain.

5. Explain without solving the equation how you can tell that $\sqrt{x - 3} + 3 = 0$ has no solution.

6. Why is it necessary to check solutions to radical equations?

7. Does the equation $\sqrt{x} = 3$ have one or two solutions? Explain.

8. Does the equation $x^2 = 9$ have one or two solutions? Explain.

Practice the Skills

Solve and check your solution(s). If the equation has no real solution, so state.

9. $\sqrt{x} = 4$

10. $\sqrt{x} = 12$

11. $\sqrt{x} = -9$

12. $\sqrt[3]{x} = 4$

13. $\sqrt[3]{x} = -4$

14. $\sqrt{a - 3} + 6 = 7$

15. $-\sqrt{2x+4} = -6$

16. $\sqrt[3]{5x+14} = 4$

17. $\sqrt[3]{3x} + 4 = 7$

18. $2\sqrt{4x+5} = 14$

19. $\sqrt[3]{2x+29} = 3$

20. $\sqrt{8a-3} = \sqrt{7a+3}$

21. $\sqrt[4]{x} = 2$

22. $\sqrt[4]{x} = -2$

23. $\sqrt[4]{x+5} = 3$

24. $\sqrt[4]{3x+1} = 2$

25. $\sqrt[4]{2x+1} + 5 = 1$

26. $\sqrt{2x} + 3 = 9$

27. $\sqrt{x+8} = \sqrt{x-8}$

28. $\sqrt{r+5} + 6 = 9$

29. $\sqrt[3]{6t+1} = \sqrt[3]{2t+5}$

30. $2\sqrt[3]{x-1} = \sqrt[3]{x^2+2x}$

31. $\sqrt[4]{x+8} = \sqrt[4]{2x}$

32. $\sqrt[4]{3x-1} + 5 = 0$

33. $\sqrt{3x+1} - 4 = 0$

34. $\sqrt{x^2+9x+3} = -x$

35. $\sqrt{m^2+6m-4} = m$

36. $\sqrt{x^2+3x+9} = x$

37. $\sqrt{6c+1} - 11 = 0$

38. $\sqrt{b^2-2} = b+4$

39. $\sqrt{z^2+3} = z+1$

40. $\sqrt{x} + 2x = 1$

41. $\sqrt{2y+5} = y-5$

42. $\sqrt{4x+1} = \frac{1}{2}x + 2$

43. $\sqrt{5x+6} = 2x-6$

44. $\sqrt{8b-15} + b = 10$

45. $(2a+9)^{1/2} - a + 3 = 0$

46. $(3x+4)^{1/2} - x = -2$

47. $(2x^2+4x+6)^{1/2} = (2x^2+6)^{1/2}$

48. $(2x+1)^{1/2} + 7 = x$

49. $(r+2)^{1/3} = (3r+8)^{1/3}$

50. $(3x+6)^{1/3} + 3 = 0$

51. $(5x+8)^{1/4} = (9x+2)^{1/4}$

52. $(5b+2)^{1/4} = (2b+16)^{1/4}$

53. $\sqrt[4]{x+5} = -3$

54. $\sqrt{x^2+x-1} = -\sqrt{x+3}$

Solve. You will have to square both sides of the equation twice to eliminate all radicals.

55. $\sqrt{3x-5} = \sqrt{2x} - 1$

56. $\sqrt{3a+1} = \sqrt{a-4} + 3$

57. $2\sqrt{b} - 1 = \sqrt{b+16}$

58. $\sqrt{x+1} = 2 - \sqrt{x}$

59. $\sqrt{x+3} = \sqrt{x} - 3$

60. $\sqrt{y+2} = 2 + \sqrt{y-6}$

61. $\sqrt{x+7} = 6 - \sqrt{x-5}$

62. $\sqrt{b-3} = 4 - \sqrt{2b+1}$

63. $\sqrt{4x-3} = 2 + \sqrt{2x-5}$

64. $\sqrt{r+10} + 3 + \sqrt{r-5} = 0$

65. $\sqrt{y+1} = \sqrt{y+5} - 2$

66. $3 + \sqrt{x+1} = \sqrt{3x+12}$

For each pair of functions, find all real values of x where $f(x) = g(x)$.

67. $f(x) = \sqrt{x+5}, g(x) = \sqrt{2x-2}$

68. $f(x) = \sqrt{x^2-6x+11}, g(x) = \sqrt{x-1}$

69. $f(x) = \sqrt[3]{5x-17}, g(x) = \sqrt[3]{6x-21}$

70. $f(x) = (14x-8)^{1/2}, g(x) = 2(3x+2)^{1/2}$

71. $f(x) = 2(8x+24)^{1/3}, g(x) = 4(2x-2)^{1/3}$

72. $f(x) = 2\sqrt{x+7}, g(x) = 10 - \sqrt{x+14}$

Solve each formula for the indicated variable.

73. $p = \sqrt{2v}$, for v

74. $l = \sqrt{4r}$, for r

75. $v = \sqrt{2gh}$, for g

76. $v = \sqrt{\dfrac{2E}{m}}$, for E

77. $v = \sqrt{\dfrac{FR}{M}}$, for F

78. $\omega = \sqrt{\dfrac{a_0}{b_0}}$, for b_0

79. $x = \sqrt{\dfrac{m}{k}} V_0$, for m

80. $T = 2\pi\sqrt{\dfrac{L}{32}}$, for L

81. $r = \sqrt{\dfrac{A}{\pi}}$, for A

82. $r = \sqrt[3]{\dfrac{3V}{4\pi}}$, for V

Problem Solving

Use the Pythagorean Theorem to find the length of the unknown side of each triangle. Write the answer as a radical in simplified form.

83. **84.** **85.** **86.**

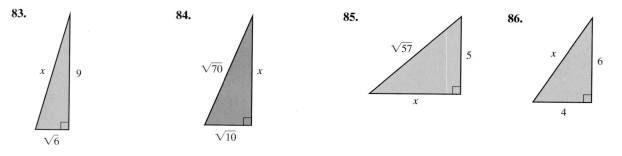

Solve. You will need to square both sides of the equation twice.

87. $\sqrt{x+5} - \sqrt{x} = \sqrt{x-3}$

88. $\sqrt{2x} - \sqrt{x-4} = \sqrt{12-x}$

89. $\sqrt{4y+6} + \sqrt{y+5} = \sqrt{y+1}$

90. $\sqrt{2b-2} + \sqrt{b-5} = \sqrt{4b}$

91. $\sqrt{a+2} - \sqrt{a-3} = \sqrt{a-6}$

92. $\sqrt{c+1} + \sqrt{c-2} = \sqrt{3c}$

93. $\sqrt{2t-1} + \sqrt{t-4} = \sqrt{3t+1}$

94. $\sqrt{r-1} - \sqrt{r+6} = \sqrt{r-9}$

Solve. You will need to square both sides of the equation twice.

95. $\sqrt{2 - \sqrt{x}} = \sqrt{x}$

96. $\sqrt{6 + \sqrt{x+4}} = \sqrt{2x-1}$

97. $\sqrt{2 + \sqrt{x+1}} = \sqrt{7-x}$

98. $\sqrt{1 + \sqrt{x-1}} = \sqrt{x-6}$

99. *Wire from Telephone Pole* A telephone pole is at a right, or 90°, angle with the ground as shown in the figure. Find the length of the wire that connects to the pole 40 feet above the ground and is anchored to the ground 20 feet from the base of the pole.

100. *Ladder against a House* Ms. Song Tran places an extension ladder against her house. The base of the ladder is 2 meters from the house and the ladder rests against the house 6 meters above the ground. How far is her ladder extended?

101. *Side of a Garden* When you are given the area of a square, the length of a side can be found by the formula $s = \sqrt{A}$. Find the side of Tom Kim's square garden if it has an area of 144 square feet.

102. *Radius of Basketball Hoop* When you are given the area of a circle, its radius can be found by the formula $r = \sqrt{A/\pi}$.

a) Find the radius of a basketball hoop if the area enclosed by the hoop is 254.47 square inches.

b) If the diameter of a basketball is 9 inches, what is the minimum distance possible between the hoop and the ball when the center of the ball is in the center of the hoop?

103. *Period of a Pendulum* The formula for the period of a pendulum is

$$T = 2\pi\sqrt{\frac{l}{g}}$$

where T is the period in seconds, l is its length in feet, and g is the acceleration of gravity. On Earth, gravity is 32 ft/sec². The formula when used on Earth becomes

$$T = 2\pi\sqrt{\frac{l}{32}}$$

a) Find the period of a pendulum whose length is 10 feet.

b) If the length of a pendulum is doubled, what effect will this have on the period? Explain.

c) The gravity on the Moon is 1/6 that on Earth. If a pendulum has a period of 2 seconds on Earth, what will be the period of the same pendulum on the Moon?

104. *Diagonal of a Suitcase* A formula for the length of a diagonal from the upper corner of a box to the opposite lower corner is $d = \sqrt{L^2 + W^2 + H^2}$, where L, W, and H are the length, width, and height, respectively.

a) Find the length of the diagonal of a suitcase of length 22 inches, width 15 inches, and height 9 inches.

b) If the length, width, and height are all doubled, how will the diagonal change?

c) Solve the formula for W.

105. *Blood Flowing in an Artery* The formula

$$r = \sqrt[4]{\frac{8\mu l}{\pi R}}$$

is used in determining movement of blood through arteries. In the formula, R represents the resistance to blood flow, μ is the viscosity of blood, l is the length of the artery, and r is the radius of the artery. Solve this equation for R.

106. *Falling Object* The formula

$$t = \frac{\sqrt{19.6s}}{9.8}$$

can be used to tell the time, t, in seconds, that an object has been falling if it has fallen s meters. Suppose an object has been dropped from a helicopter and has fallen 100 meters. How long has it been in free fall?

107. *Earth Days* For any planet in our solar system, its "year" is the time it takes for the planet to revolve once around the Sun. The number of Earth days in a given planet's year, N, is approximated by the formula $N = 0.2(\sqrt{R})^3$, where R is the mean distance of the planet to the Sun in millions of kilometers. Find the number of Earth days in the year of the planet Earth, whose mean distance to the Sun is 149.4 million kilometers.

108. *Earth Days* Find the number of Earth days in the year of the planet Mercury, whose mean distance to the Sun is 58 million kilometers. See Exercise 107.

109. *Forces on a Car* When two forces, F_1 and F_2, pull at right angles to each other as illustrated below, the resultant, or the effective force, R, can be found by the formula $R = \sqrt{F_1^2 + F_2^2}$. Two cars are trying to pull a third out of the mud, as illustrated. If car A is exerting a force of 60 pounds and car B is exerting a force of 80 pounds, find the resulting force on the car stuck in the mud.

110. *Escape Velocity* The escape velocity, or the velocity needed for a spacecraft to escape a planet's gravitational field, is found by the formula $v_e = \sqrt{2gR}$, where g is the

force of gravity of the planet and R is the radius of the planet. Find the escape velocity for Earth, in meters per second, where $g = 9.75$ meters per second squared and $R = 6{,}370{,}000$ meters.

111. *Motion of a Wave* A formula used in the study of shallow water wave motion is $c = \sqrt{gH}$, in which c is wave velocity, H is water depth, and g is the acceleration due to gravity. Find the wave velocity if the water's depth is 10 feet. (Use $g = 32$ ft/sec^2.)

112. *Diagonal of a Box* The top of a rectangular box measures 20 inches by 32 inches. Find the length of the diagonal for the top of the box.

113. *Flower Garden* A rectangular flower garden measures 25 meters by 32 meters. Find the length of the diagonal for the garden.

114. *Speed of Sound* When sound travels through air (or any gas), the velocity of the sound wave is dependent on the air (or gas) temperature. The velocity, v, in meters per second, at air temperature, t, in degrees Celsius, can be found by the formula

$$v = 331.3\sqrt{1 + \frac{t}{273}}$$

Find the speed of sound in air whose temperature is 20°C (equivalent to 68°F).

A formula that we have already mentioned and that we will be discussing in more detail shortly is the quadratic formula

$$x = \frac{-b \pm \sqrt{b^2 - 4ac}}{2a}$$

115. Find x when $a = 1, b = 0, c = -4$.

116. Find x when $a = 1, b = 1, c = -12$.

117. Find x when $a = 2, b = 5, c = -12$.

118. Find x when $a = -1, b = 4, c = 5$.

Given $f(x)$, find all values of x for which $f(x)$ is the indicated value.

119. $f(x) = \sqrt{x - 5}, f(x) = 4$

120. $f(x) = \sqrt[3]{2x + 3}, f(x) = 5$

121. $f(x) = \sqrt{3x^2 - 11} + 4, f(x) = 12$

122. $f(x) = 4 + \sqrt[3]{x^2 + 152}, f(x) = 10$

123. **a)** Consider the equation $\sqrt{4x - 12} = x - 3$. Setting each side of the equation equal to y yields the following system of equations.

$$y = \sqrt{4x - 12}$$
$$y = x - 3$$

The graphs of the equations in the system are illustrated in the figure.

From the graphs, determine the values that appear to be solutions to the equation $\sqrt{4x - 12} = x - 3$. Explain how you determined your answer.

b) Substitute the values found in part **a)** into the original equation and determine whether they are the solutions to the equation.

c) Solve the equation $\sqrt{4x - 12} = x - 3$ algebraically and see if your solution agrees with the values obtained in part **a)**.

124. If the graph of a radical function, $f(x)$, does not intersect the x-axis, then the equation $f(x) = 0$ has no real solutions. Explain why.

125. Suppose we are given a rational function $g(x)$. If $g(4) = 0$, then the graph of $g(x)$ must intersect the x-axis at 4. Explain why.

126. The graph of the equation $y = \sqrt{x - 3} + 2$ is illustrated in the figure.
   a) What is the domain of the function?
   b) How many real solutions does the equation $\sqrt{x - 3} + 2 = 0$ have? List all the real solutions. Explain how you determined your answer.

127. *Confidence Interval* In statistics, a "confidence interval" is a range of values that is likely to contain the true value of the population. For a "95% confidence interval," the lower limit of the range, L_1, and the upper limit of the range, L_2, can be found by the formulas

$$L_1 = p - 1.96\sqrt{\frac{p(1 - p)}{n}}$$

$$L_2 = p + 1.96\sqrt{\frac{p(1 - p)}{n}}$$

where p represents the percent obtained from a sample and n is the size of the sample. Francesco, a statistician, takes a sample of 36 families and finds that 60% of those surveyed use an answering machine in their home. He can be 95% certain that the true percent of families that use an answering machine in their home is between L_1 and L_2. Find the values of L_1 and L_2. Use $p = 0.60$ and $n = 36$ in the formulas.

128. *Quadratic Mean* The *quadratic mean* (or *root mean square, RMS*) is often used in physical applications. In power distribution systems, for example, voltages and currents are usually referred to in terms of their RMS values. The quadratic mean of a set of scores is obtained by squaring each score and adding the results (signified by Σx^2), then dividing the value obtained by the number of scores, n, and then taking the square root of this value. We may express the formula as

$$\text{quadratic mean} = \sqrt{\frac{\Sigma x^2}{n}}$$

Find the quadratic mean of the numbers 2, 4, and 10.

In Exercises 129 and 130 solve the equation.

129. $\sqrt{x^2 + 9} = (x^2 + 9)^{1/2}$
130. $\sqrt{x^2 - 4} = (x^2 - 4)^{1/2}$

In Exercises 131–134 use your graphing calculator to solve the equations. Round your solutions to the nearest tenth.

131. $\sqrt{x + 8} = \sqrt{3x + 5}$
132. $\sqrt{10x - 16} - 15 = 0$
133. $\sqrt[3]{5x^2 - 6} - 4 = 0$
134. $\sqrt[3]{5x^2 - 10} = \sqrt[3]{4x + 95}$

Challenge Problems

Solve.

135. $\sqrt{\sqrt{x + 25} - \sqrt{x}} = 5$

136. $\sqrt{\sqrt{x + 9} + \sqrt{x}} = 3$

Solve each equation for n.

137. $z = \dfrac{\overline{x} - \mu}{\dfrac{\sigma}{\sqrt{n}}}$

138. $z = \dfrac{p' - p}{\sqrt{\dfrac{pq}{n}}}$

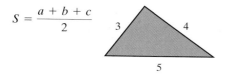 **Group Activity**

Discuss and answer Exercise 139 as a group.

139. *Heron's Formula* The area of a triangle is $A = \frac{1}{2}bh$. If the height is not known but we know the lengths of the three sides, we can use Heron's formula to find the area, A. Heron's formula is

$$A = \sqrt{S(S - a)(S - b)(S - c)}$$

where a, b, and c are the lengths of the three sides and

$$S = \frac{a + b + c}{2}$$

a) Have each group member use Heron's formula to find the area of a triangle whose sides are 3 inches, 4 inches, and 5 inches.

b) Compare your answers for part **a)**. If any member of the group did not get the correct answer, make sure they understand their error.

c) Have each member of the group do the following.

1. Draw a triangle on the following grid. Place each vertex of the triangle at the intersection of two grid lines.

2. Measure with a ruler the length of each side of your triangle.

3. Use Heron's formula to find the area of your triangle.

4. Compare and discuss your work from part **c)**.

Cumulative Review Exercises

[2.2] **140.** Solve the formula $P_1 P_2 - P_1 P_3 = P_2 P_3$ for P_2.

[6.1] **141.** Simplify $\dfrac{x(x - 3) + x(x - 4)}{2x - 7}$.

Perform each indicated operation.

[6.1] **142.** $\dfrac{4a^2 - 9b^2}{4a^2 + 12ab + 9b^2} \cdot \dfrac{6a^2 b}{8a^2 b^2 - 12ab^3}$

143. $(t^2 - t - 12) \div \dfrac{t^2 - 9}{t^2 - 3t}$

[6.2] **144.** $\dfrac{2}{x + 3} - \dfrac{1}{x - 3} + \dfrac{2x}{x^2 - 9}$

[6.4] **145.** Solve $2 + \dfrac{3x}{x - 1} = \dfrac{8}{x - 1}$.

7.7 COMPLEX NUMBERS

SSM   Study Guide   CD/Video

MathPro 4/5   PH Math Tutor Center   prenhall.com/Angel

1 Recognize a complex number.
2 Add and subtract complex numbers.
3 Multiply complex numbers.
4 Divide complex numbers.
5 Find powers of *i*.

1   Recognize a Complex Number

In Section 7.1 we mentioned that the square roots of negative numbers, such as $\sqrt{-4}$, are not real numbers. Numbers like $\sqrt{-4}$ are called **imaginary numbers**. Such numbers are called imaginary because when they were introduced many mathematicians refused to believe that they existed. Although they do not belong to the set of real numbers, the imaginary numbers, by definition, do exist and are very useful in mathematics and science.

Every imaginary number has $\sqrt{-1}$ as a factor. The $\sqrt{-1}$, called the **imaginary unit**, is often denoted by the letter i.

$$i = \sqrt{-1}$$

To write the square root of a negative number in terms of i, use the following property.

For any positive real number n,
$$\sqrt{-n} = \sqrt{-1}\,\sqrt{n} = i\sqrt{n}$$

Therefore, we can write

$$\sqrt{-4} = \sqrt{-1}\,\sqrt{4} = i2 \quad \text{or} \quad 2i$$
$$\sqrt{-9} = \sqrt{-1}\,\sqrt{9} = i3 \quad \text{or} \quad 3i$$
$$\sqrt{-7} = \sqrt{-1}\,\sqrt{7} = i\sqrt{7}$$

In this book we will generally write $i\sqrt{7}$ rather than $\sqrt{7}i$ to avoid confusion with $\sqrt{7i}$. Also, $3\sqrt{5}i$ is written as $3i\sqrt{5}$.

Examples

$$\sqrt{-81} = 9i \qquad \sqrt{-5} = i\sqrt{5}$$
$$\sqrt{-49} = 7i \qquad \sqrt{-10} = i\sqrt{10}$$

The real number system is a part of a larger number system, called the *complex number system*. Now we will discuss **complex numbers**.

DEFINITION

Every number of the form

$$a + bi$$

where a and b are real numbers, is a **complex number**.

Every real number and every imaginary number are also complex numbers. A complex number has two parts: a real part, a, and an imaginary part, b.

If $b = 0$, the complex number is a real number. If $a = 0$, the complex number is a *pure imaginary number*.

Examples of Complex Numbers

$$3 + 7i \qquad a = 3, b = 7$$
$$5 - i\sqrt{3} \qquad a = 5, b = -\sqrt{3}$$
$$6 \qquad a = 6, b = 0 \qquad \text{(real number, } b = 0)$$
$$8i \qquad a = 0, b = 8 \qquad \text{(imaginary number, } a = 0)$$
$$-i\sqrt{7} \qquad a = 0, b = -\sqrt{7} \qquad \text{(imaginary number, } a = 0)$$

We stated that all real numbers and imaginary numbers are also complex numbers. The relationship between the various sets of numbers is illustrated in Figure 7.7.

Complex Numbers		
Real Numbers		**Nonreal Numbers**
Rational numbers $\frac{1}{2}, -\frac{3}{5}, \frac{9}{4}$	Irrational numbers $\sqrt{2}, \sqrt{3}$	$2 + 3i$
Integers $-4, -9$	$-\sqrt{7}, \pi$	$6 - 4i$
Whole numbers $0, 4, 12$		$\sqrt{2} + i\sqrt{3}$
		$i\sqrt{5}$
		$6i$

FIGURE 7.7

EXAMPLE 1 Write each of the following complex numbers in the form $a + bi$.

a) $3 + \sqrt{-36}$ **b)** $5 - \sqrt{-12}$ **c)** 19 **d)** $\sqrt{-50}$ **e)** $6 + \sqrt{7}$

Solution

a) $3 + \sqrt{-36} = 3 + \sqrt{-1}\sqrt{36}$
$= 3 + i6$ or $3 + 6i$

b) $5 - \sqrt{-12} = 5 - \sqrt{-1}\sqrt{12}$
$= 5 - \sqrt{-1}\sqrt{4}\sqrt{3}$
$= 5 - i(2)\sqrt{3}$ or $5 - 2i\sqrt{3}$

c) $19 = 19 + 0i$

d) $\sqrt{-50} = 0 + \sqrt{-50}$
$= 0 + \sqrt{-1}\sqrt{25}\sqrt{2}$
$= 0 + i(5)\sqrt{2}$ or $0 + 5i\sqrt{2}$

e) Both 6 and $\sqrt{7}$ are real numbers. Written as a complex number, the answer is $(6 + \sqrt{7}) + 0i$.

NOW TRY EXERCISE 23

Complex numbers can be added, subtracted, multiplied, and divided. To perform these operations, we use the definitions that $i = \sqrt{-1}$ and

$$i^2 = -1$$

2 Add and Subtract Complex Numbers

We now explain how to add or subtract complex numbers.

> **To Add or Subtract Complex Numbers**
>
> 1. Change all imaginary numbers to bi form.
> 2. Add (or subtract) the real parts of the complex numbers.
> 3. Add (or subtract) the imaginary parts of the complex numbers.
> 4. Write the answer in the form $a + bi$.

EXAMPLE 2 Add $(7 + 15i) + (-6 - 2i) + 20$.

Solution $(7 + 15i) + (-6 - 2i) + 20 = 7 + 15i - 6 - 2i + 20$

$$= 7 - 6 + 20 + 15i - 2i \quad \textit{Rearrange terms.}$$
$$= 21 + 13i \quad \textit{Combine like terms.} \quad \text{✳}$$

EXAMPLE 3 Subtract $(5 - \sqrt{-27}) - (-3 + \sqrt{-48})$.

Solution $(5 - \sqrt{-27}) - (-3 + \sqrt{-48}) = (5 - \sqrt{-1}\sqrt{27}) - (-3 + \sqrt{-1}\sqrt{48})$

$$= (5 - \sqrt{-1}\sqrt{9}\sqrt{3}) - (-3 + \sqrt{-1}\sqrt{16}\sqrt{3})$$
$$= (5 - 3i\sqrt{3}) - (-3 + 4i\sqrt{3})$$
$$= 5 - 3i\sqrt{3} + 3 - 4i\sqrt{3}$$
$$= 5 + 3 - 3i\sqrt{3} - 4i\sqrt{3}$$
$$= 8 - 7i\sqrt{3}$$

NOW TRY EXERCISE 35

3 Multiply Complex Numbers

Now let's discuss how to multiply complex numbers.

> **To Multiply Complex Numbers**
>
> 1. Change all imaginary numbers to bi form.
> 2. Multiply the complex numbers as you would multiply polynomials.
> 3. Substitute -1 for each i^2.
> 4. Combine the real parts and the imaginary parts. Write the answer in $a + bi$ form.

EXAMPLE 4 Multiply. **a)** $5i(3 - 2i)$ **b)** $\sqrt{-9}(\sqrt{-3} + 7)$ **c)** $(2 - \sqrt{-18})(\sqrt{-2} + 5)$

Solution **a)** $5i(3 - 2i) = 5i(3) + 5i(-2i)$ *Distributive property*

$$= 15i - 10i^2$$
$$= 15i - 10(-1) \quad \textit{Replace } i^2 \textit{ with } -1.$$
$$= 15i + 10 \quad \text{or} \quad 10 + 15i$$

b) $\sqrt{-9}(\sqrt{-3} + 7) = 3i(i\sqrt{3} + 7)$ *Change imaginary numbers to bi form.*

$$= 3i(i\sqrt{3}) + 3i(7) \quad \textit{Distributive property}$$
$$= 3i^2\sqrt{3} + 21i$$
$$= 3(-1)\sqrt{3} + 21i \quad \textit{Replace } i^2 \textit{ with } -1.$$
$$= -3\sqrt{3} + 21i$$

c) $(2 - \sqrt{-18})(\sqrt{-2} + 5) = (2 - \sqrt{-1}\,\sqrt{18})(\sqrt{-1}\,\sqrt{2} + 5)$

$$= (2 - \sqrt{-1}\,\sqrt{9}\,\sqrt{2})(\sqrt{-1}\,\sqrt{2} + 5)$$

$$= (2 - 3i\sqrt{2})(i\sqrt{2} + 5)$$

Now use the FOIL method to multiply.

$(2 - 3i\sqrt{2})(i\sqrt{2} + 5) = (2)(i\sqrt{2}) + (2)(5) + (-3i\sqrt{2})(i\sqrt{2}) + (-3i\sqrt{2})(5)$

$$= 2i\sqrt{2} + 10 - 3i^2(2) - 15i\sqrt{2}$$

$$= 2i\sqrt{2} + 10 - 3(-1)(2) - 15i\sqrt{2}$$

$$= 2i\sqrt{2} + 10 + 6 - 15i\sqrt{2}$$

NOW TRY EXERCISE 45

$$= 16 - 13i\sqrt{2}$$

AVOIDING COMMON ERRORS

What is $\sqrt{-4} \cdot \sqrt{-2}$?

CORRECT

$\sqrt{-4} \cdot \sqrt{-2} = 2i \cdot i\sqrt{2}$

$= 2i^2\sqrt{2}$

$= 2(-1)\sqrt{2}$

$= -2\sqrt{2}$

INCORRECT

$\sqrt{-4} \cdot \sqrt{-2} = \sqrt{8}$

$= \sqrt{4} \cdot \sqrt{2}$

$= 2\sqrt{2}$

Recall that $\sqrt{a} \cdot \sqrt{b} = \sqrt{ab}$ only for *nonnegative* real numbers a and b.

4   Divide Complex Numbers

The **conjugate of a complex number** $a + bi$ is $a - bi$. For example,

Complex Number	Conjugate
$3 + 4i$	$3 - 4i$
$1 - i\sqrt{3}$	$1 + i\sqrt{3}$
$2i$ (or $0 + 2i$)	$-2i$ (or $0 - 2i$)
15 (or $15 + 0i$)	15 (or $15 - 0i$)

When a complex number is multiplied by its conjugate using the FOIL method, the inner and outer products will sum to 0, and the result is a real number. For example,

$$(5 + 3i)(5 - 3i) = 25 - 15i + 15i - 9i^2$$

$$= 25 - 9i^2$$

$$= 25 - 9(-1)$$

$$= 25 + 9 = 34$$

Now we explain how to divide complex numbers.

To Divide Complex Numbers

1. Change all imaginary numbers to bi form.
2. Rationalize the denominator by multiplying both the numerator and denominator by the conjugate of the denominator.
3. Write the answer in $a + bi$ form.

EXAMPLE 5 Divide $\dfrac{6+i}{i}$.

Solution Begin by multiplying both numerator and denominator by $-i$, the conjugate of i.

$$\frac{6+i}{i} \cdot \frac{-i}{-i} = \frac{(6+i)(-i)}{-i^2}$$

$$= \frac{-6i - i^2}{-i^2} \qquad \text{Distributive property}$$

$$= \frac{-6i - (-1)}{-(-1)} \qquad \text{Replace } i^2 \text{ with } -1.$$

$$= \frac{-6i + 1}{1}$$

$$= 1 - 6i \qquad ✳$$

EXAMPLE 6 Divide $\dfrac{3-2i}{4-i}$.

Solution Multiply both numerator and denominator by $4+i$, the conjugate of $4-i$.

$$\frac{3-2i}{4-i} \cdot \frac{4+i}{4+i} = \frac{12 + 3i - 8i - 2i^2}{16 - i^2}$$

$$= \frac{12 - 5i - 2(-1)}{16 - (-1)}$$

$$= \frac{14 - 5i}{17} \quad \text{or} \quad \frac{14}{17} - \frac{5}{17}i \qquad ✳$$

NOW TRY EXERCISE 65

EXAMPLE 7 **Impedance** A concept needed for the study of electronics is *impedance*. Impedance affects the current in a circuit. The impedance, Z, in a circuit is found by the formula $Z = \frac{V}{I}$, where V is voltage and I is current. Find Z when $V = 1.6 - 0.3i$ and $I = -0.2i$, where $i = \sqrt{-1}$.

Solution $Z = \dfrac{V}{I} = \dfrac{1.6 - 0.3i}{-0.2i}$. Now multiply both numerator and denominator by the conjugate of the denominator, $0.2i$.

$$Z = \frac{1.6 - 0.3i}{-0.2i} \cdot \frac{0.2i}{0.2i} = \frac{0.32i - 0.06i^2}{-0.04i^2}$$

$$= \frac{0.32i + 0.06}{0.04}$$

$$= \frac{0.32i}{0.04} + \frac{0.06}{0.04}$$

$$= 8i + 1.5 \quad \text{or} \quad 1.5 + 8i \qquad ✳$$

Most algebra books use i as the imaginary unit. However, most electronics books use j as the imaginary unit because i is often used to represent current.

5 Find Powers of i

Using $i = \sqrt{-1}$ and $i^2 = -1$, we can find other **powers of i**. For example,

$$i^3 = i^2 \cdot i = -1 \cdot i = -i \qquad\qquad i^6 = i^4 \cdot i^2 = 1(-1) = -1$$

$$i^4 = i^2 \cdot i^2 = (-1)(-1) = 1 \qquad i^7 = i^4 \cdot i^3 = 1(-i) = -i$$

$$i^5 = i^4 \cdot i^1 = 1 \cdot i = i \qquad\qquad i^8 = i^4 \cdot i^4 = (1)(1) = 1$$

Note that successive powers of i rotate through the four values i, -1, $-i$, and 1 (see Fig. 7.8).

FIGURE 7.8

EXAMPLE 8 Evaluate. **a)** i^{35} **b)** i^{101}

Solution Write each expression as a product of factors such that the exponent of one factor is the largest multiple of 4 less than or equal to the given exponent. Then write this factor as i^4 raised to some power. Since i^4 has a value of 1, the expression i^4 raised to a power will also have a value of 1.

a) $i^{35} = i^{32} \cdot i^3 = (i^4)^8 \cdot i^3 = 1 \cdot i^3 = 1(-i) = -i$

NOW TRY EXERCISE 101 **b)** $i^{101} = i^{100} \cdot i^1 = (i^4)^{25} \cdot i = 1 \cdot i = i$

HELPFUL HINT

A quick way of evaluating i^n is to divide the exponent by 4 and observe the remainder.

If the remainder is 0, the value is 1.   If the remainder is 2, the value is -1.

If the remainder is 1, the value is i.   If the remainder is 3, the value is $-i$.

For Example 8 **a)**
$$\begin{array}{r} 8 \\ 4\overline{)35} \\ \underline{32} \\ 3 \end{array}$$ ← Answer is $-i$.

For 8 **b)**
$$\begin{array}{r} 25 \\ 4\overline{)101} \\ \underline{8} \\ 21 \\ \underline{20} \\ 1 \end{array}$$ ← Answer is i.

EXAMPLE 9 Let $f(x) = x^2$. Find: **a)** $f(5i)$ **b)** $f(3 + 7i)$

Solution **a)** $f(x) = x^2$
$$f(5i) = (5i)^2 = 25i^2 = 25(-1) = -25$$

b) $f(x) = x^2$
$$f(3 + 7i) = (3 + 7i)^2 = (3)^2 + 2(3)(7i) + (7i)^2$$
$$= 9 + 42i + 49i^2$$
$$= 9 + 42i + 49(-1)$$
$$= 9 + 42i - 49$$

NOW TRY EXERCISE 115
$$= -40 + 42i$$

Exercise Set 7.7

1. a) What does i equal?

   b) What does i^2 equal?

2. Write $\sqrt{-n}$ using i.

3. Are all of the following complex numbers? If any are not complex numbers, explain why.

   a) 4 **b)** $-\frac{1}{6}$ **c)** $3 - \sqrt{-2}$

   d) $5 - 3i$ **e)** $4.2i$ **f)** $3 + \sqrt{7}$

4. What does i^4 equal?

5. Is every real and every imaginary number a complex number?

6. Is every complex number a real number?

7. What is the conjugate of $a + bi$?

8. a) Is $i \cdot i$ a real number? Explain.

    b) Is $i \cdot i \cdot i$ a real number? Explain.

9. List, if possible, a number that is *not*

a) a rational number.

b) an irrational number.

c) a real number.

d) an imaginary number.

e) a complex number.

10. Write a paragraph or two explaining the relationship between the real numbers, imaginary numbers, and complex numbers. Include in your discussion how the various sets of numbers relate to each other.

Practice the Skills

Write each expression as a complex number in the form a + bi.

11. 5

12. $2i$

13. $\sqrt{49}$

14. $\sqrt{-81}$

15. $21 - \sqrt{-36}$

16. $\sqrt{3} + \sqrt{-3}$

17. $\sqrt{-24}$

18. $\sqrt{49} - \sqrt{-49}$

19. $8 - \sqrt{-12}$

20. $\sqrt{-9} + \sqrt{-81}$

21. $1 + \sqrt{-98}$

22. $\sqrt{-4} + 7i$

23. $9 - \sqrt{-25}$

24. $\sqrt{75} + \sqrt{-32}$

25. $2i - \sqrt{-45}$

26. $\sqrt{144} + \sqrt{-96}$

Add or Subtract.

27. $(19 - i) + (2 + 6i)$

28. $(22 + i) - 5(11 - 3i)$

29. $(8 - 3i) + (-8 + 3i)$

30. $(6 - \sqrt{-4}) - (-1 - \sqrt{-16})$

31. $(1 + \sqrt{-1}) + (-13 - \sqrt{-169})$

32. $(15 - i\sqrt{3}) + (17 - \sqrt{-3})$

33. $(\sqrt{3} + \sqrt{2}) + (3\sqrt{2} - \sqrt{-8})$

34. $(7 - \sqrt{2}) - (5 + \sqrt{-15})$

35. $(3 - \sqrt{-72}) + (6 + \sqrt{-8})$

36. $(27 + \sqrt{-75}) + (\sqrt{-147})$

37. $(\sqrt{4} - \sqrt{-45}) + (-\sqrt{81} + \sqrt{-5})$

38. $(\sqrt{20} - \sqrt{-12}) + (2\sqrt{5} + \sqrt{-75})$

Multiply.

39. $2(5 - i)$

40. $-7(2 + 3i\sqrt{5})$

41. $i(2 + 9i)$

42. $3i(4 - i)$

43. $\sqrt{-9}(7 + 11i)$

44. $\dfrac{1}{2}i\left(\dfrac{1}{3} - 16i\right)$

45. $\sqrt{-16}(\sqrt{3} - 5i)$

46. $\sqrt{-27}(\sqrt{3} - \sqrt{-3})$

47. $\sqrt{-32}(\sqrt{2} + \sqrt{-8})$

48. $-\sqrt{-24}(\sqrt{6} - \sqrt{-3})$

49. $(3 + 2i)(1 + i)$

50. $(6 - 2i)(3 + i)$

51. $(20 - 3i)(20 + 3i)$

52. $(-4 + 3i)(2 - 5i)$

53. $(7 + \sqrt{-2})(5 - \sqrt{-8})$

54. $(\sqrt{4} - 3i)(4 + \sqrt{-4})$

55. $\left(\dfrac{1}{2} - \dfrac{1}{3}i\right)\left(\dfrac{1}{4} + \dfrac{2}{3}i\right)$

56. $\left(\dfrac{3}{5} - \dfrac{1}{4}i\right)\left(\dfrac{2}{3} + \dfrac{2}{5}i\right)$

Divide.

57. $\dfrac{2}{3i}$

58. $\dfrac{-7}{4i}$

59. $\dfrac{2 + 3i}{2i}$

60. $\dfrac{4 - 3i}{2i}$

61. $\dfrac{4}{2 - i}$

62. $\dfrac{3}{5 + i}$

63. $\dfrac{9}{1 - 2i}$

64. $\dfrac{11}{-3 - 4i}$

65. $\dfrac{6 - 3i}{4 + 2i}$

66. $\dfrac{4 - 3i}{4 + 3i}$

67. $\dfrac{4}{6 - \sqrt{-4}}$

68. $\dfrac{5}{3 + \sqrt{-5}}$

69. $\dfrac{\sqrt{2}}{5 + \sqrt{-12}}$

70. $\dfrac{\sqrt{6}}{\sqrt{3} - \sqrt{-9}}$

71. $\dfrac{\sqrt{10} + \sqrt{-3}}{5 - \sqrt{-20}}$

72. $\dfrac{12 - \sqrt{-12}}{\sqrt{3} + \sqrt{-5}}$

73. $\dfrac{\sqrt{-75}}{\sqrt{-3}}$

74. $\dfrac{\sqrt{-60}}{\sqrt{-2}}$

75. $\dfrac{\sqrt{-32}}{\sqrt{-18}\,\sqrt{2}}$

76. $\dfrac{\sqrt{-80}\,\sqrt{-10}}{\sqrt{-4}}$

Perform each indicated operation. These exercises are a combination of the types of exercises presented earlier in this exercise set.

77. $(4 - 2i) + (3 - 5i)$

78. $\left(\frac{1}{2} + i\right) - \left(\frac{3}{5} - \frac{2}{3}i\right)$

79. $(\sqrt{8} - \sqrt{2}) - (\sqrt{-12} - \sqrt{-48})$

80. $(8 - \sqrt{-6}) - (2 - \sqrt{-24})$

81. $5.2(4 - 3.2i)$

82. $\sqrt{-6}(\sqrt{3} - \sqrt{-10})$

83. $(9 + 2i)(3 - 5i)$

84. $(\sqrt{3} + 2i)(\sqrt{6} - \sqrt{-8})$

85. $\dfrac{5 + 4i}{2i}$

86. $\dfrac{1}{2 + 3i}$

87. $\dfrac{5}{\sqrt{3} - \sqrt{-4}}$

88. $\dfrac{5 - 2i}{3 + 2i}$

89. $\left(5 - \frac{5}{9}i\right) - \left(4 - \frac{3}{5}i\right)$

90. $\frac{4}{7}\left(4 - \frac{2}{5}i\right)$

🔒 **91.** $\left(\frac{2}{3} - \frac{1}{5}i\right)\left(\frac{3}{5} - \frac{3}{4}i\right)$

92. $\sqrt{\frac{4}{9}}\left(\sqrt{\frac{25}{36}} - \sqrt{-\frac{4}{25}}\right)$

93. $\dfrac{\sqrt{-96}}{\sqrt{-24}}$

94. $\dfrac{-1 - 2i}{2 + \sqrt{-5}}$

95. $(5.23 - 6.41i) - (9.56 + 4.5i)$

96. $(\sqrt{-6} + 3)(\sqrt{-15} + 5)$

For each imaginary number, indicate whether its value is i, -1, $-i$, or 1.

97. i^{10}

98. i^{63}

99. i^{200}

100. i^{231}

101. i^{93}

102. i^{103}

103. i^{807}

104. i^{1213}

Problem Solving

105. Consider the complex number $2 + 3i$.

  a) Find the additive inverse.

  b) Find the multiplicative inverse. Write the answer in simplified form.

106. Consider the complex number $4 - 5i$.

  a) Find the additive inverse.

  b) Find the multiplicative inverse. Write the answer in simplified form.

Answer true or false. Support your answer with an example.

107. The product of two pure imaginary numbers is always a real number.

108. The sum of two pure imaginary numbers is always an imaginary number.

109. The product of two complex numbers is always a real number.

110. The sum of two complex numbers is always a complex number.

111. What values of n will result in i^n being a real number? Explain.

112. What values of n will result in i^{2n} being a real number? Explain.

113. If $f(x) = x^2$, find $f(i)$.

114. If $f(x) = x^2$, find $f(4i)$.

🔒 **115.** If $f(x) = x^4 - 2x$, find $f(2i)$.

116. If $f(x) = x^3 - 3x^2$, find $f(5i)$.

117. If $f(x) = x^2 + x$, find $f(3 + i)$.

118. If $f(x) = \dfrac{x^2}{x - 2}$, find $f(4 - i)$.

Evaluate each expression for the given value of x.

119. $x^2 - 2x + 5$, $x = 1 - 2i$

120. $x^2 - 2x + 5$, $x = 1 + 2i$

121. $x^2 + 2x + 7$, $x = -1 + i\sqrt{5}$

122. $x^2 + 2x + 9$, $x = -1 - i\sqrt{5}$

In Exercises 123–126, determine whether the given value of x is a solution to the equation.

123. $x^2 - 4x + 5 = 0, x = 2 - i$

124. $x^2 - 4x + 5 = 0, x = 2 + i$

125. $x^2 - 6x + 11 = 0, x = -3 + i\sqrt{3}$

126. $x^2 - 6x + 15 = 0, x = 3 - i\sqrt{3}$

127. *Impedance* Find the impedance, Z, using the formula $Z = V/I$ when $V = 1.8 + 0.5i$ and $I = 0.6i$. See Example 7.

128. *Impedance* Refer to Exercise 127. Find the impedance when $V = 2.4 - 0.6i$ and $I = -0.4i$.

129. *Impedance* Under certain conditions, the total impedance, Z_T, of a circuit is given by the formula

$$Z_T = \frac{Z_1 Z_2}{Z_1 + Z_2}$$

Find Z_T when $Z_1 = 2 - i$ and $Z_2 = 4 + i$.

130. *Impedance* Refer to Exercise 129. Find Z_T when $Z_1 = 3 - i$ and $Z_2 = 5 + i$.

131. Determine whether i^{-1} is equal to $i, -1, -i$, or 1. Show your work.

132. Determine whether i^{-3} is equal to $i, -1, -i$, or 1. Show your work.

In Chapter 8, we will use the quadratic formula $x = \dfrac{-b \pm \sqrt{b^2 - 4ac}}{2a}$ to solve equations of the form $ax^2 + bx + c = 0$.

(a) Use the quadratic formula to solve the following quadratic equations. (b) Check each of the two solutions by substituting the values found for x (one at a time) back into the original equation. In these exercises, the $\pm$ (read "plus or minus") results in two distinct complex answers.

133. $x^2 - 4x + 6 = 0$

134. $x^2 - 2x + 6 = 0$

Given the complex numbers $a = 5 + 2i\sqrt{3}, b = 1 + i\sqrt{3}$, evaluate each expression.

135. $a + b$

136. $a - b$

137. ab

138. $\dfrac{a}{b}$

Cumulative Review Exercises

[4.3] 139. *Mixture* Berreda Coughlin, a grocer in Dallas, has two coffees, one selling for $5.50 per pound and the other for $6.30 per pound. How many pounds of each type of coffee should he mix to make 40 pounds of coffee to sell for $6.00 per pound?

[5.3] 140. Divide $\dfrac{8c^2 + 6c - 25}{4c + 9}$.

[6.2] 141. Add $\dfrac{b}{a - b} + \dfrac{a + b}{b}$.

[6.4] 142. Solve $\dfrac{x}{4} + \dfrac{1}{2} = \dfrac{x - 1}{2}$.

See Exercise 139.

CHAPTER SUMMARY

Key Terms and Phrases

7.1
Cube root
Even roots
Index
Odd roots

Principal square root
Radical expression
Radical sign
Radicand
Square root

7.2
Rational exponent

7.3
Perfect cubes

Perfect powers
Perfect squares
Product rule for radicals
Quotient rule for radicals

(continued on the next page)

7.4	7.5	7.6	Conjugate of a complex

7.4
Like radicals
Unlike radicals

7.5
Conjugate
Rationalizing the
denominator
Simplified radical

7.6
Radical equation

7.7
Complex number

Conjugate of a complex
number
Imaginary number
Imaginary unit
Powers of i

IMPORTANT FACTS

If n Is Even and $a \geq 0$:
$$\sqrt[n]{a} = b \text{ if } b \geq 0 \text{ and } b^n = a$$

If n Is Odd:
$$\sqrt[n]{a} = b \text{ if } b^n = a$$

Rules of Radicals

$$\sqrt{a^2} = |a|$$
$$\sqrt{a^2} = a, \quad a \geq 0$$
$$\sqrt[n]{a^n} = a, \quad a \geq 0$$
$$\sqrt[n]{a} = a^{1/n}, \quad a \geq 0$$

$$\sqrt[n]{a^m} = (\sqrt[n]{a})^m = a^{m/n}, \quad a \geq 0$$
$$\sqrt[n]{a} \cdot \sqrt[n]{b} = \sqrt[n]{ab}, \quad a \geq 0, b \geq 0$$
$$\frac{\sqrt[n]{a}}{\sqrt[n]{b}} = \sqrt[n]{\frac{a}{b}}, \quad a \geq 0, b > 0$$

A Radical Is Simplified When the Following Are All True:

1. No perfect powers are factors of any radicand, and all exponents in the radicand, are less than the index.

2. No radicand contains a fraction.
3. No denominator contains a radical.

Powers of i
$$i = \sqrt{-1}, i^2 = -1, i^3 = -i, i^4 = 1$$

Chapter Review Exercises

[7.1] Evaluate.

1. $\sqrt{49}$ **2.** $\sqrt[3]{-27}$ **3.** $\sqrt[4]{256}$ **4.** $\sqrt[3]{-125}$

Use absolute value to evaluate.

5. $\sqrt{(-7)^2}$ **6.** $\sqrt{(-93.4)^2}$

Write as an absolute value.

7. $\sqrt{x^2}$ **8.** $\sqrt{(x-2)^2}$ **9.** $\sqrt{(x-y)^2}$ **10.** $\sqrt{(x^2-4x+12)^2}$

11. Let $f(x) = \sqrt{10x+9}$. Find $f(4)$.
12. Let $k(x) = 2x + \sqrt{\dfrac{x}{3}}$. Find $k(27)$.

13. Let $g(x) = \sqrt[3]{2x+3}$. Find $g(4)$ and round the answer to the nearest tenth.
14. The area of a square is 121 square meters. Find the length of its side.

For the remainder of these review exercises, assume that all variables represent positive real numbers.

[7.2] Write in exponential form.

15. $\sqrt{x^5}$ **16.** $\sqrt[3]{x^5}$ **17.** $(\sqrt[4]{y})^{15}$ **18.** $\sqrt{5^2}$

Write in radical form.

19. $x^{1/2}$ **20.** $a^{4/5}$ **21.** $(3m^2n)^{7/4}$ **22.** $(x+3y)^{-5/3}$

Simplify each radical expression by changing the expression to exponential form. Write the answer in radical form when appropriate.

23. $\sqrt[3]{3^6}$ **24.** $\sqrt{x^{10}}$ **25.** $(\sqrt[4]{7})^8$ **26.** $\sqrt[20]{a^5}$

Evaluate if possible. If the expression is not a real number, so state.

27. $-25^{1/2}$ **28.** $(-25)^{1/2}$ **29.** $\left(\dfrac{64}{27}\right)^{-1/3}$ **30.** $64^{-1/2} + 8^{-2/3}$

Simplify. Write the answer without negative exponents.

31. $x^{3/5} \cdot x^{-1/3}$ **32.** $\left(\dfrac{64}{y^6}\right)^{1/3}$ **33.** $\left(\dfrac{a^{-6/5}}{a^{1/5}}\right)^{2/3}$ **34.** $\left(\dfrac{28x^5y^{-3}}{4y^{1/2}}\right)^2$

Multiply.

35. $a^{1/2}(5a^{3/2} - 2a^2)$ **36.** $4x^{-2/3}\left(x^{-1/2} + \dfrac{9}{4}x^{2/3}\right)$

Factor each expression. Write the answer without negative exponents.

37. $x^{2/5} + x^{7/5}$ **38.** $a^{-1/2} + a^{5/2}$

For each function, find the indicated value of the function. Use your calculator to evaluate irrational numbers. Round irrational numbers to the nearest thousandth.

39. If $f(x) = \sqrt{7x - 17}$, find $f(6)$. **40.** If $g(x) = \sqrt[3]{9r - 17}$, find $g(4)$.

Graph the following functions.

41. $f(x) = \sqrt{x}$ **42.** $f(x) = \sqrt{x} - 4$

[7.2–7.5] Simplify.

43. $\sqrt{75}$ **44.** $\sqrt[3]{128}$ **45.** $\sqrt{\dfrac{49}{4}}$ **46.** $\sqrt[3]{\dfrac{27}{125}}$

47. $-\sqrt{\dfrac{81}{64}}$ **48.** $\sqrt[3]{-\dfrac{8}{125}}$ **49.** $\sqrt{20}\sqrt{5}$ **50.** $\sqrt[3]{16}\sqrt[3]{4}$

51. $\sqrt{8x^2y^3z^4}$ **52.** $\sqrt{50x^3y^7}$ **53.** $\sqrt[3]{54a^7b^{10}}$ **54.** $\sqrt[3]{125x^8y^9z^{16}}$

55. $\left(\sqrt[6]{x^2y^3z^5}\right)^{42}$ **56.** $\left(\sqrt[5]{2ab^4c^6}\right)^{15}$ **57.** $\sqrt{5x}\sqrt{8x^5}$ **58.** $\sqrt[3]{2x^2y}\sqrt[3]{4x^9y^4}$

59. $\sqrt[3]{2x^4y^5}\sqrt[3]{16x^4y^4}$ **60.** $\sqrt[4]{8x^4y^7}\sqrt[4]{2x^5y^9}$ **61.** $\sqrt{3x}(\sqrt{12x} - \sqrt{20})$ **62.** $\sqrt[3]{2x^2y}\left(\sqrt[3]{4x^4y^7} + \sqrt[3]{9x}\right)$

63. $\sqrt{\sqrt{\sqrt{a^3b^2}}}$ **64.** $\sqrt{\sqrt[3]{x^5y^2}}$ **65.** $\left(\dfrac{3r^2p^{1/3}}{r^{1/2}p^{4/3}}\right)^3$ **66.** $\left(\dfrac{4y^{2/5}z^{1/3}}{x^{-1}y^{3/5}}\right)^{-1}$

67. $\sqrt{\dfrac{3}{5}}$ **68.** $\sqrt[3]{\dfrac{7}{4}}$ **69.** $\sqrt[4]{\dfrac{2}{3}}$ **70.** $\dfrac{x}{\sqrt{11}}$

71. $\dfrac{6}{\sqrt{x}}$ **72.** $\dfrac{a}{\sqrt[3]{25}}$ **73.** $\dfrac{9}{\sqrt[3]{y^2}}$ **74.** $\dfrac{7}{\sqrt[4]{z}}$

75. $\sqrt[3]{\dfrac{x^3}{8}}$ **76.** $\dfrac{\sqrt[3]{2x^9}}{\sqrt[3]{16x^6}}$ **77.** $\sqrt{\dfrac{32x^2y^5}{2x^4y}}$ **78.** $\sqrt[4]{\dfrac{32x^9y^{15}}{2xy^3}}$

79. $\sqrt{\dfrac{5x^4}{y}}$ **80.** $\sqrt{\dfrac{12a}{7b}}$ **81.** $\sqrt{\dfrac{18x^4y^5}{3z}}$ **82.** $\sqrt{\dfrac{125x^2y^5}{3z}}$

83. $\sqrt[3]{\dfrac{108x^3y^7}{2y^3}}$ **84.** $\sqrt[3]{\dfrac{3x}{5y}}$ **85.** $\sqrt[3]{\dfrac{4x^5y^3}{x^6}}$ **86.** $\sqrt[3]{\dfrac{y^6}{2x^2}}$

87. $\sqrt[4]{\dfrac{3a^2b^{11}}{a^5b}}$ **88.** $\sqrt[4]{\dfrac{2x^2y^6}{8x^3}}$ **89.** $(3 - \sqrt{2})(3 + \sqrt{2})$ **90.** $(\sqrt{x} + y)(\sqrt{x} - y)$

91. $(x - \sqrt{y})(x + \sqrt{y})$ **92.** $(\sqrt{3} + 5)^2$ **93.** $(\sqrt{x} - \sqrt{3y})(\sqrt{x} + \sqrt{5y})$

94. $(\sqrt[3]{2x} - \sqrt[3]{3y})(\sqrt[3]{3x} - \sqrt[3]{2y})$ **95.** $\dfrac{5}{2 + \sqrt{5}}$ **96.** $\dfrac{x}{3 + \sqrt{x}}$

97. $\dfrac{a}{7 - \sqrt{b}}$ **98.** $\dfrac{x}{\sqrt{y} - 8}$ **99.** $\dfrac{\sqrt{x}}{\sqrt{x} + \sqrt{y}}$

100. $\dfrac{\sqrt{x} - 2\sqrt{y}}{\sqrt{x} - \sqrt{y}}$ **101.** $\dfrac{3}{\sqrt{a - 1} - 2}$ **102.** $\dfrac{4}{\sqrt{y + 2} - 3}$

103. $\sqrt[3]{x} + 3\sqrt[3]{x} - 2\sqrt[3]{x}$ **104.** $\sqrt{3} + \sqrt{27} - \sqrt{192}$ **105.** $\sqrt[3]{16} - 5\sqrt[3]{54} + 2\sqrt[3]{64}$

106. $4\sqrt{2} - \dfrac{3}{\sqrt{32}} + \sqrt{50}$ **107.** $3\sqrt{x^5y^6} - \sqrt{16x^7y^8}$ **108.** $2\sqrt[3]{x^7y^8} - \sqrt[3]{x^4y^2} + 3\sqrt[3]{x^{10}y^2}$

In Exercises 109 and 110, $f(x)$ and $g(x)$ are given. Find $(f \cdot g)(x)$.

109. $f(x) = \sqrt{3x}, g(x) = \sqrt{6x} - \sqrt{10}$ **110.** $f(x) = \sqrt[3]{2x^2}, g(x) = \sqrt[3]{4x^4} + \sqrt[3]{8x^5}$

Simplify. In Exercise 112, assume the variable can be any real number.

111. $f(x) = \sqrt{2x + 4}\,\sqrt{2x + 4}, x \geq -2$ **112.** $g(a) = \sqrt{20a^2 + 100a + 125}$

Simplify.

113. $\dfrac{\sqrt[3]{(x + 5)^5}}{\sqrt{(x + 5)^3}}$ **114.** $\dfrac{\sqrt[3]{a^3b^2}}{\sqrt[4]{a^4b}}$

*For each figure, find **a)** the perimeter and **b)** the area. Write the perimeter and area in radical form with the radicals simplified.*

115.

Rectangle

116.

117. The graph of $f(x) = \sqrt{x} + 2$ is given.

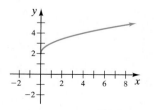

a) For $g(x) = -3$, sketch the graph of $(f + g)(x)$.
b) What is the domain of $(f + g)(x)$?

118. The graph of $f(x) = -\sqrt{x}$ is given.

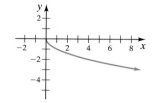

a) For $g(x) = \sqrt{x} + 2$, sketch the graph of $(f + g)(x)$.
b) What is the domain of $(f + g)(x)$?

522 • Chapter 7 • Roots, Radicals, and Complex Numbers

[7.6] Solve each equation and check your solutions.

119. $\sqrt{x} = 8$

120. $\sqrt{x} = -8$

121. $\sqrt[3]{x} = 5$

122. $\sqrt[3]{x} = -5$

123. $3 + \sqrt{x} = 10$

124. $8 + \sqrt[3]{x} = 12$

125. $\sqrt{3x + 4} = \sqrt{5x + 12}$

126. $\sqrt{x^2 + 2x - 4} = x$

127. $\sqrt[3]{x - 9} = \sqrt[3]{5x + 3}$

128. $(x^2 + 5)^{1/2} = x + 1$

129. $\sqrt{x + 3} = \sqrt{3x + 9}$

130. $\sqrt{6x - 5} - \sqrt{2x + 6} - 1 = 0$

For each pair of functions, find all values of x for which $f(x) = g(x)$.

131. $f(x) = \sqrt{3x + 4}, g(x) = 2\sqrt{2x - 4}$

132. $f(x) = (4x + 3)^{1/3}, g(x) = (6x - 9)^{1/3}$

Solve the following for the indicated variable.

133. $V = \sqrt{\dfrac{2L}{w}}$, for L

134. $r = \sqrt{\dfrac{A}{\pi}}$, for A

Find the length of the unknown side of each right triangle. Write the answer as a radical in simplified form.

135.

136.

Solve.

137. *Telephone Pole* How long a wire does a phone company need to reach the top of a 5-meter telephone pole from a point on the ground 2 meters from the base of the pole?

138. *Velocity* Use the formula $v = \sqrt{2gh}$ to find the velocity of an object after it has fallen 20 feet ($g = 32$ ft/s^2).

139. *Pendulum* Use the formula

$$T = 2\pi\sqrt{\dfrac{L}{32}}$$

to find the period of a pendulum, T, if its length, L, is 64 feet.

140. *Kinetic and Potential Energy* There are two types of energy: kinetic and potential. Potential energy is due to position and kinetic energy is due to motion. For example, if you hold a billiard ball above the ground it has potential energy. If you let go of the ball the potential energy is changed to kinetic energy as the ball drops. The formula

$$V = \sqrt{\dfrac{2K}{m}}$$

can be used to determine the velocity, V, in meters per second, when a mass, m, in kilograms, has a kinetic energy, K, in joules. A 0.145-kg baseball is thrown. If the kinetic energy of the moving ball is 45 joules, at what speed is the ball moving?

141. *Speed of Light* Albert Einstein found that if an object at rest, with mass m_0, is made to travel close to the speed of light, its mass increases to m, where

$$m = \dfrac{m_0}{\sqrt{1 - \dfrac{v^2}{c^2}}}$$

In the formula, v is the velocity of the moving object and c is the speed of light.* In an accelerator used for cancer therapy, particles travel at speeds of $0.98c$, that is, at 98% the speed of light. At a speed of $0.98c$, determine a particle's mass, m, in terms of its rest mass, m_0. Use $v = 0.98c$ in the above formula.

*The speed of light is 3.00×10^8 meters per second. However, you do not need this information to solve this problem.

[7.7] Write each expression as a complex number in the form a + bi.

142. 5   **143.** -6 **144.** $2 - \sqrt{-256}$ **145.** $3 + \sqrt{-16}$

Perform each indicated operation.

146. $(3 + 2i) + (4 - i)$ **147.** $(4 - 6i) - (3 - 4i)$ **148.** $(\sqrt{3} + \sqrt{-5}) + (2\sqrt{3} - \sqrt{-7})$

149. $\sqrt{-6}(\sqrt{6} + \sqrt{-6})$ **150.** $(4 + 3i)(2 - 3i)$ **151.** $(6 + \sqrt{-3})(4 - \sqrt{-15})$

152. $\dfrac{2}{3i}$ **153.** $\dfrac{2 + \sqrt{3}}{2i}$ **154.** $\dfrac{5}{3 + 2i}$ **155.** $\dfrac{\sqrt{3}}{5 - \sqrt{-6}}$

Evaluate each expression for the given value of x.

156. $x^2 - 2x + 9, \ x = 1 + 2i\sqrt{2}$ **157.** $x^2 - 2x + 12, \ x = 1 - 2i$

For each imaginary number, indicate whether its value is i, −1, −i, or 1.

158. i^{53} **159.** i^{19} **160.** i^{404} **161.** i^{5326}

Chapter Practice Test

1. Write $\sqrt{(5x - 1)^2}$ as an absolute value.

2. Simplify $\left(\dfrac{x^{2/5} \cdot x^{-1}}{x^{3/5}}\right)^2$.

3. Factor $x^{-2/3} + x^{4/3}$.

4. Graph $g(x) = \sqrt{x} + 1$.

In Exercises 5–14, simplify. Assume that all variables represent positive real numbers.

5. $\sqrt{48x^7 y^{10}}$ **6.** $\sqrt[3]{4x^5 y^2} \ \sqrt[3]{10x^6 y^8}$ **7.** $\sqrt{\dfrac{3x^6 y^3}{8z}}$ **8.** $\dfrac{4}{\sqrt[3]{x}}$

9. $\dfrac{\sqrt{3}}{3 + \sqrt{27}}$ **10.** $2\sqrt{24} - 5\sqrt{6} + 3\sqrt{54}$ **11.** $\sqrt[3]{8x^3 y^5} + 2\sqrt[3]{x^6 y^8}$ **12.** $(\sqrt{3} - 5)(6 - \sqrt{8})$

13. $\sqrt[4]{\sqrt{x^5 y^3}}$ **14.** $\dfrac{\sqrt[4]{(7x + 2)^5}}{\sqrt[3]{(7x + 2)^2}}$

In Exercises 15–17, solve the equation.

15. $\sqrt{3x + 4} = 5$

16. $\sqrt{x^2 - x - 12} = x + 3$

17. $\sqrt{a - 8} = \sqrt{a} - 2$

18. For $f(x) = (9x + 37)^{1/3}$ and $g(x) = 2(2x + 2)^{1/3}$, find all values of x such that $f(x) = g(x)$.

19. Solve the formula $w = \dfrac{\sqrt{2gh}}{4}$ for g.

20. *Falling Object* The velocity, V, in feet per second, after it has fallen a distance, h, in feet, can be found by the formula $V = \sqrt{64.4\,h}$. Find the velocity of a pen after it has fallen 200 feet.

21. *Ladder* A ladder is placed against a house. If the base of the ladder is 5 feet from the house and the ladder rests on the house 12 feet above the ground, find the length of the ladder.

See Exercise 21.

22. *Springs* A formula used in the study of springs is

$$T = 2\pi\sqrt{\dfrac{m}{k}}$$

where T is the period of a spring (the time for the spring to stretch and return to its rest point), m is the mass on the spring, in kilograms, and k is the spring's constant, in newtons/meter. A mass of 1400 kilograms rests on a spring. Find the period of the spring if the spring's constant is 65,000 newtons/meter.

23. Multiply $(6 - \sqrt{-4})(2 + \sqrt{-9})$.

24. Divide $\dfrac{5 - i}{7 + 2i}$.

25. Evaluate $x^2 + 6x + 12$ for $x = -3 + i$.

Cumulative Review Test

1. Solve $\frac{1}{5}(x - 3) = \frac{3}{4}(x + 3) - x$.

2. Solve $3(x - 4) = 6x - (4 - 5x)$.

3. *Sweater* When the price of a sweater is decreased by 60%, it costs \$20. Find the original price of the sweater.

4. Find the solution set of $|3 - 2x| < 7$.

5. Graph $y = \dfrac{3}{2}x - 3$.

6. Determine whether the graphs of the given equations are parallel lines, perpendicular lines, or neither.

$$y = 3x - 6$$
$$6y = 18x + 6$$

7. Given $f(x) = x^2 - 3x + 4$ and $g(x) = 2x - 5$, find $(g - f)(x)$.

8. Find the equation of the line through $(3, -4)$ that is perpendicular to the graph of $3x - 2y = 6$.

9. Solve the system of equations.

$$x + 2y = 12$$
$$4x = 8$$
$$3x - 4y + 5z = 20$$

10. Evaluate the determinant.

$$\begin{vmatrix} 3 & -6 & -1 \\ 2 & 1 & -2 \\ 1 & 3 & 1 \end{vmatrix}$$

11. *Volume* The volume of the box that follows is $6r^3 + 5r^2 + r$. Find w in terms of r.

12. Multiply $(5xy - 6)(5xy + 6)$.

13. Solve $\sqrt{2x^2 + 7} + 3 = 8$.

14. Factor $4x^3 - 9x^2 + 5x$.

15. Factor $(x + 1)^3 - 8$.

16. Solve $8x^2 - 3 = -10x$.

17. Multiply $\dfrac{4x + 4y}{x^2 y} \cdot \dfrac{y^3}{8x}$.

18. Add $\dfrac{x - 4}{x - 5} - \dfrac{3}{x + 5} - \dfrac{10}{x^2 - 25}$.

19. Solve $\dfrac{4}{x} - \dfrac{1}{6} = \dfrac{1}{x}$.

20. *Falling Object* The distance, d, an object drops in free fall is directly proportional to the square of the time, t. If an object falls 16 feet in 1 second, how far will the object fall in 5 seconds?

Answers to Cumulative Review Test

1. $\dfrac{57}{9}$; [Sec. 2.1, Obj. 4] **2.** -1; [Sec. 2.1, Obj. 3] **3.** \$50; [Sec. 2.3, Obj. 2] **4.** $\{x | -2 < x < 5\}$; [Sec. 2.6, Obj. 3]

5. ; [Sec. 3.4, Obj. 5] **6.** Parallel; [Sec. 3.5, Obj. 3] **7.** $-x^2 + 5x - 9$; [Sec. 3.6, Obj. 1]

8. $y = -\dfrac{2}{3}x - 2$; [Sec. 3.5, Obj. 3] **9.** $\left(2, 5, \dfrac{34}{5}\right)$; [Sec. 4.2, Obj. 1] **10.** 40; [Sec. 4.5, Obj. 3]

11. $w = 2r + 1$; [Sec. 5.3, Obj. 2] **12.** $25x^2y^2 - 36$; [Sec. 5.2, Obj. 5]

13. 3, -3; [Sec. 7.6, Obj. 1] **14.** $x(4x - 5)(x - 1)$; [Sec. 5.5, Obj. 3]

15. $(x - 1)(x^2 + 4x + 7)$; [Sec. 5.6, Obj. 3] **16.** $\dfrac{1}{4}, -\dfrac{3}{2}$; [Sec. 5.8, Obj. 2]

17. $\dfrac{(x + y)y^2}{2x^3}$; [Sec. 6.1, Obj. 3] **18.** $\dfrac{x + 3}{x + 5}$; [Sec. 6.2, Obj. 3] **19.** 18; [Sec. 6.4, Obj. 2]

20. 400 feet; [Sec. 6.6, Obj. 1]

Chapter 8

Quadratic Functions

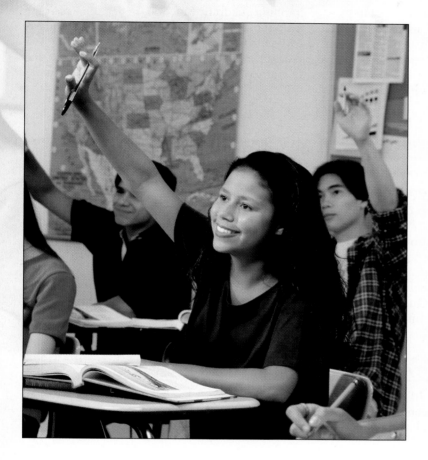

8.1 Solving Quadratic Equations by Completing the Square

8.2 Solving Quadratic Equations by the Quadratic Formula

8.3 Quadratic Equations: Applications and Problem Solving

8.4 Writing Equations in Quadratic Form

8.5 Graphing Quadratic Functions

8.6 Quadratic and Other Inequalities in One Variable

Chapter Summary
Chapter Review Exercises
Chapter Practice Test
Cumulative Review Test

There are many real-life situations that can be represented or approximated with quadratic equations. As you read through this chapter, you will see several real-life applications of quadratic functions. For example, in Exercise 34 on page 556 we use a quadratic equation to estimate a typical student's grade point average based on the number of hours the student studies per week.

SSM   Study Guide   CD/Video   MathPro 4/5   PH Math Tutor Center   prenhall.com/Angel

A Look Ahead

We introduced quadratic functions in Section 5.1. Now we expand on the concepts. In Section 8.1, we introduce completing the square and in Section 8.2, we discuss the quadratic formula. After studying these sections, you will know three techniques for solving quadratic equations: factoring (when possible), completing the square, and the quadratic formula.

In Section 8.3, we study applications of quadratic equations. In Section 8.4, we introduce equations that can be expressed and solved as if they were quadratic equations. Such equations are called equations that are quadratic in form.

We discuss techniques for graphing quadratic functions in Section 8.5. Some of what we discuss, such as the translation of parabolas and completing the square, will be used again in Section 10.1 when we discuss parabolas further.

In Section 8.6, we solve quadratic and other nonlinear inequalities in one variable. We will draw on what we learned in Section 2.5, where we solved linear inequalities in one variable.

8.1 SOLVING QUADRATIC EQUATIONS BY COMPLETING THE SQUARE

SSM   Study Guide   CD/Video

MathPro 4/5   PH Math Tutor Center   prenhall.com/Angel

1   Use the square root property to solve equations.

2   Understand perfect square trinomials.

3   Solve quadratic equations by completing the square.

In this section, we introduce two concepts, the square root property and completing the square. The square root property will be used in several sections of this book.

In Section 5.8, we solved quadratic, or second-degree, equations by factoring. Quadratic equations that cannot be solved by factoring can be solved by completing the square, or by the quadratic formula, which is presented in Section 8.2.

---

1   Use the Square Root Property to Solve Equations

In Section 7.1, we stated that every positive number has two square roots. Thus far, we have been using only the positive square root. In this section, we use both the positive and negative square roots of a number.

Positive Square Root of 25	Negative Square Root of 25
$\sqrt{25} = 5$	$-\sqrt{25} = -5$

A convenient way to indicate the two square roots of a number is to use the plus or minus symbol, $\pm$. For example, the square roots of 25 can be indicated by ± 5, read "plus or minus 5." The equation $x^2 = 25$ has two solutions, the two square roots of 25, which are ± 5. If you check each root, you will see that each value satisfies the equation. The **square root property** can be used to find the solutions to equations of the form $x^2 = a$.

Square Root Property

If $x^2 = a$, where a is a real number, then $x = \pm\sqrt{a}$.

EXAMPLE 1 Solve the following equations.   **a)** $x^2 - 9 = 0$ **b)** $x^2 + 5 = 80$

Solution **a)** Add 9 to both sides of the equation to isolate the variable.

$$x^2 - 9 = 0$$
$$x^2 = 9 \qquad \text{Isolate the variable.}$$
$$x = \pm\sqrt{9} \qquad \text{Square root property}$$
$$= \pm 3$$

Check the solutions in the original equation.

$$x = 3 \qquad\qquad x = -3$$
$$x^2 - 9 = 0 \qquad\qquad x^2 - 9 = 0$$
$$3^2 - 9 \stackrel{?}{=} 0 \qquad\qquad (-3)^2 - 9 \stackrel{?}{=} 0$$
$$0 = 0 \quad \text{True} \qquad\qquad 0 = 0 \quad \text{True}$$

In both cases the check is true, which means that both 3 and −3 are solutions to the equation.

b)
$$x^2 + 5 = 80$$
$$x^2 = 75 \qquad \text{Isolate the variable.}$$
$$x = \pm\sqrt{75} \qquad \text{Square root property}$$
$$= \pm\sqrt{25}\sqrt{3} \qquad \text{Simplify.}$$
$$= \pm 5\sqrt{3}$$

The solutions are $5\sqrt{3}$ and $-5\sqrt{3}$.

Not all quadratic equations have real solutions, as is illustrated in Example 2.

EXAMPLE 2 Solve the equation $x^2 + 7 = 0$.

Solution
$$x^2 + 7 = 0$$
$$x^2 = -7 \qquad \text{Isolate the variable.}$$
$$x = \pm\sqrt{-7} \qquad \text{Square root property}$$
$$= \pm i\sqrt{7}$$

The solutions are $i\sqrt{7}$ and $-i\sqrt{7}$, both of which are imaginary numbers.

EXAMPLE 3 Solve   **a)** $(a - 4)^2 = 32$ **b)** $(z + 3)^2 + 28 = 0$.

Solution **a)** Since the term containing the variable is already isolated, begin by using the square root property.

$$(a - 4)^2 = 32$$
$$a - 4 = \pm\sqrt{32} \qquad \text{Square root property}$$
$$a = 4 \pm \sqrt{32} \qquad \text{Add 4 to both sides.}$$
$$= 4 \pm \sqrt{16}\sqrt{2} \qquad \text{Simplify.}$$
$$= 4 \pm 4\sqrt{2}$$

NOW TRY EXERCISE 33   The solutions are $4 + 4\sqrt{2}$ and $4 - 4\sqrt{2}$.

b) Begin by subtracting 28 from both sides of the equation to isolate the term containing the variable.

$$(z + 3)^2 + 28 = 0$$
$$(z + 3)^2 = -28$$

Now use the square root property.

$$z + 3 = \pm\sqrt{-28}$$
$$z = -3 \pm \sqrt{-28} \qquad \textit{Subtract 3 from both sides.}$$
$$= -3 \pm \sqrt{28}\,\sqrt{-1}$$
$$= -3 \pm i\sqrt{4}\,\sqrt{7} \qquad \textit{Simplify } \sqrt{28} \textit{ and replace}$$
$$= -3 \pm 2i\sqrt{7} \qquad\qquad \sqrt{-1} \textit{ with } i.$$

The solutions are $-3 + 2i\sqrt{7}$ and $-3 - 2i\sqrt{7}$. Note that the solutions to the equation $(z + 3)^2 + 28 = 0$ are not real numbers. The solutions are complex numbers. ✳

2 Understand Perfect Square Trinomials

Now that we know the square root property we can focus our attention on completing the square. To understand this procedure, you need to know how to form perfect square trinomials. Perfect square trinomials were introduced in Section 5.6. Recall that a **perfect square trinomial** is a trinomial that can be expressed as the square of a binomial. Some examples follow.

Perfect Square Trinomials		Factors		Square of a Binomial
$x^2 + 8x + 16$	$=$	$(x + 4)(x + 4)$	$=$	$(x + 4)^2$
$x^2 - 8x + 16$	$=$	$(x - 4)(x - 4)$	$=$	$(x - 4)^2$
$x^2 + 10x + 25$	$=$	$(x + 5)(x + 5)$	$=$	$(x + 5)^2$
$x^2 - 10x + 25$	$=$	$(x - 5)(x - 5)$	$=$	$(x - 5)^2$

In a perfect square trinomial with a leading coefficient of 1, there is a relationship between the coefficient of the first-degree term and the constant term. In such trinomials the constant term is the square of one-half the coefficient of the first degree term.

Let's examine some perfect square trinomials for which the leading coefficient is 1.

$$x^2 + 8x + 16 = (x + 4)^2$$
$$\left[\tfrac{1}{2}(8)\right]^2 = (4)^2$$

$$x^2 - 10x + 25 = (x - 5)^2$$
$$\left[\tfrac{1}{2}(-10)\right]^2 = (-5)^2$$

When a perfect square trinomial with a leading coefficient of 1 is written as the square of a binomial, the constant in the binomial is one-half the coefficient of the first-degree term in the trinomial. For example,

$$x^2 + 8x + 16 = (x + 4)^2$$
$$\tfrac{1}{2}(8)$$

$$x^2 - 10x + 25 = (x - 5)^2$$
$$\tfrac{1}{2}(-10)$$

3   Solve Quadratic Equations by Completing the Square

Now we introduce completing the square. To solve a quadratic equation by **completing the square**, we add a constant to both sides of the equation so that the remaining trinomial is a perfect square trinomial. Then we use the square root property to solve the resulting equation. We will now summarize the procedure.

To Solve a Quadratic Equation by Completing the Square

1. Use the multiplication (or division) property of equality if necessary to make the leading coefficient 1.
2. Rewrite the equation with the constant by itself on the right side of the equation.
3. Take one-half the numerical coefficient of the first-degree term, square it, and add this quantity to both sides of the equation.
4. Replace the trinomial with the square of a binomial.
5. Use the square root property to take the square root of both sides of the equation.
6. Solve for the variable.
7. Check your solutions in the *original* equation.

EXAMPLE 4   Solve the equation $x^2 + 6x + 5 = 0$ by completing the square.

Solution   Since the leading coefficient is 1, step 1 is not necessary.

Step 2:   Move the constant, 5, to the right side of the equation by subtracting 5 from both sides of the equation.

$$x^2 + 6x + 5 = 0$$
$$x^2 + 6x = -5$$

Step 3:   Determine the square of one-half the numerical coefficient of the first-degree term, 6.

$$\frac{1}{2}(6) = 3, \qquad 3^2 = \boxed{9}$$

Add this value to both sides of the equation.

$$x^2 + 6x \boxed{+ 9} = -5 \boxed{+ 9}$$
$$x^2 + 6x + 9 = 4$$

Step 4:   By following this procedure, we produce a perfect square trinomial on the left side of the equation. The expression $x^2 + 6x + 9$ is a perfect square trinomial that can be expressed as $(x + 3)^2$.

$$\overbrace{(x + 3)}^{\frac{1}{2}\text{ the numerical coefficient of the first-degree term is }\frac{1}{2}(6) = +3.}{}^2 = 4$$

Step 5:   Use the square root property.

$$x + 3 = \pm\sqrt{4}$$
$$x + 3 = \pm 2$$

Step 6: Finally, solve for x by subtracting 3 from both sides of the equation.

$$x + 3 \boxed{- 3} = \boxed{-3} \pm 2$$
$$x = -3 \pm 2$$
$$x = -3 + 2 \quad \text{or} \quad x = -3 - 2$$
$$x = -1 \qquad\qquad x = -5$$

Step 7: Check both solutions in the original equation.

$x = -1$	$x = -5$
$x^2 + 6x + 5 = 0$	$x^2 + 6x + 5 = 0$
$(-1)^2 + 6(-1) + 5 \stackrel{?}{=} 0$	$(-5)^2 + 6(-5) + 5 \stackrel{?}{=} 0$
$1 - 6 + 5 \stackrel{?}{=} 0$	$25 - 30 + 5 \stackrel{?}{=} 0$
$0 = 0$ *True*	$0 = 0$ *True*

NOW TRY EXERCISE 49   Since each number checks, both -1 and -5 are solutions to the original equation.

HELPFUL HINT

When solving the equation $x^2 + bx + c = 0$ by completing the square, we obtain $x^2 + bx + \left(\dfrac{b}{2}\right)^2$ on the left side of the equation and a constant on the right side of the equation. We then replace $x^2 + bx + \left(\dfrac{b}{2}\right)^2$ with $\left(x + \dfrac{b}{2}\right)^2$. In the figure that follows we show why

$$x^2 + bx + \left(\frac{b}{2}\right)^2 = \left(x + \frac{b}{2}\right)^2$$

The figure is a square with sides of length $x + \dfrac{b}{2}$. The area is therefore $\left(x + \dfrac{b}{2}\right)^2$. The area of the square can also be determined by adding the areas of the four sections as follows:

$$x^2 + \frac{b}{2}x + \frac{b}{2}x + \left(\frac{b}{2}\right)^2 = x^2 + bx + \left(\frac{b}{2}\right)^2$$

Comparing the areas, we see that $x^2 + bx + \left(\dfrac{b}{2}\right)^2 = \left(x + \dfrac{b}{2}\right)^2$.

The area of this piece represents the term we add to each side of the equation when we complete the square.

EXAMPLE 5   Solve the equation $-x^2 = -3x - 18$ by completing the square.

Solution   The numerical coefficient of the squared term must be 1, not -1. Therefore, begin by multiplying both sides of the equation by -1 to make the coefficient of the squared term equal to 1.

$$-x^2 = -3x - 18$$
$$-1(-x^2) = -1(-3x - 18)$$
$$x^2 = 3x + 18$$

Now move all terms except the constant to the left side of the equation.

$$x^2 - 3x = 18$$

Take half the numerical coefficient of the x-term, square it, and add this product to both sides of the equation. Then rewrite the left side of the equation as the square of a binomial.

$$\frac{1}{2}(-3) = -\frac{3}{2} \qquad \left(-\frac{3}{2}\right)^2 = \frac{9}{4}$$

$$x^2 - 3x + \frac{9}{4} = 18 + \frac{9}{4} \qquad \textit{Complete the square.}$$

$$\left(x - \frac{3}{2}\right)^2 = 18 + \frac{9}{4} \qquad \textit{Rewrite the trinomial as the square of a binomial.}$$

$$\left(x - \frac{3}{2}\right)^2 = \frac{72}{4} + \frac{9}{4}$$

$$\left(x - \frac{3}{2}\right)^2 = \frac{81}{4}$$

$$x - \frac{3}{2} = \pm\sqrt{\frac{81}{4}} \qquad \textit{Square root property}$$

$$x - \frac{3}{2} = \pm\frac{9}{2} \qquad \textit{Simplify.}$$

$$x = \frac{3}{2} \pm \frac{9}{2} \qquad \textit{Add } \tfrac{3}{2} \textit{ to both sides.}$$

$$x = \frac{3}{2} + \frac{9}{2} \quad \text{or} \quad x = \frac{3}{2} - \frac{9}{2}$$

$$x = \frac{12}{2} = 6 \qquad\qquad x = -\frac{6}{2} = -3$$

NOW TRY EXERCISE 53 The solutions are 6 and -3. ✳

In the following examples we will not show some of the intermediate steps.

EXAMPLE 6 Solve the equation $x^2 - 6x + 17 = 0$.

Solution

$$x^2 - 6x + 17 = 0$$

$$x^2 - 6x = -17 \qquad \textit{Move the constant term to the right side.}$$

$$x^2 - 6x + 9 = -17 + 9 \qquad \textit{Complete the square.}$$

$$(x - 3)^2 = -8 \qquad \textit{Write the trinomial as the square of a binomial.}$$

$$x - 3 = \pm\sqrt{-8} \qquad \textit{Square root property}$$

$$x - 3 = \pm 2i\sqrt{2} \qquad \textit{Simplify.}$$

$$x = 3 \pm 2i\sqrt{2} \qquad \textit{Solve for } x.$$

NOW TRY EXERCISE 65 The solutions are $3 + 2i\sqrt{2}$ and $3 - 2i\sqrt{2}$. ✳

EXAMPLE 7 Solve the equation $-3m^2 + 6m + 24 = 0$ by completing the square.

Solution

$$-3m^2 + 6m + 24 = 0$$

$$-\frac{1}{3}(-3m^2 + 6m + 24) = -\frac{1}{3}(0) \quad \text{\textit{Multiply by} } -\frac{1}{3} \text{ \textit{to obtain} } \text{\textit{a leading coefficient of 1.}}$$

$$m^2 - 2m - 8 = 0$$

Now proceed as before.

$$m^2 - 2m = 8 \qquad \text{\textit{Move the constant term to the right side.}}$$

$$m^2 - 2m \boxed{+ 1} = 8 \boxed{+ 1} \qquad \text{\textit{Complete the square.}}$$

$$(m - 1)^2 = 9 \qquad \text{\textit{Write the trinomial as the square of a binomial.}}$$

$$m - 1 = \pm 3 \qquad \text{\textit{Square root property}}$$

$$m = 1 \pm 3 \qquad \text{\textit{Solve for m.}}$$

$$m = 1 + 3 \quad \text{or} \quad m = 1 - 3$$

$$m = 4 \qquad\qquad m = -2 \qquad \text{\Large ✳}$$

If you were asked to solve the equation $-\frac{1}{4}x^2 + 2x - 8 = 0$ by completing the square, what would you do first? If you answered, "Multiply both sides of the equation by -4 to make the leading coefficient 1," you answered correctly. To solve the equation $\frac{2}{3}x^2 + 3x - 5 = 0$, you would multiply both sides of the equation by $\frac{3}{2}$ to obtain a leading coefficient of 1.

Generally, quadratic equations that cannot be easily solved by factoring will be solved by the *quadratic formula*, which will be presented in the next section. We introduced completing the square because we use it to derive the quadratic formula in Section 8.2. We will use completing the square later in this chapter and in Chapter 11.

EXAMPLE 8 **Compound Interest** The compound interest formula $A = p\left(1 + \dfrac{r}{n}\right)^{nt}$ can be

used to find the amount, A, when an initial principal, p, is invested at an annual interest rate, r, compounded n times a year for t years.

a) Josh Adams initially invested $1000 in a savings account whose interest is compounded annually (once a year). If after 2 years the amount, or balance, in the account is $1102.50, find the annual interest rate, r.

b) Trisha McDowell initially invested $1000 in a savings account whose interest is compounded quarterly. If after 3 years the amount in the account is $1195.62, find the annual interest rate, r.

Solution **a)** Understand We are given the following information:

$$p = \$1000, \qquad A = \$1102.50, \qquad n = 1, \qquad t = 2$$

We are asked to find the annual rate, r. To do so, we substitute the appropriate values into the formula and solve for r.

Translate

$$A = p\left(1 + \frac{r}{n}\right)^{nt}$$

$$1102.50 = 1000\left(1 + \frac{r}{1}\right)^{1(2)}$$

Carry Out

$$1102.50 = 1000(1 + r)^2$$
$$1.10250 = (1 + r)^2 \qquad \textit{Divide both sides by 1000.}$$
$$\sqrt{1.10250} = 1 + r \qquad \textit{Square root property; use principal root since r must be positive.}$$
$$1.05 = 1 + r$$
$$0.05 = r \qquad \textit{Subtract 1 from both sides.}$$

Answer   The annual interest rate is 0.05 or 5%.

b) Understand   We are given

$$p = 1000, \qquad A = \$1195.62, \qquad n = 4, \qquad t = 3$$

To find r, we substitute the appropriate values into the formula and solve for r.

Translate

$$A = p\left(1 + \frac{r}{n}\right)^{nt}$$

$$1195.62 = 1000\left(1 + \frac{r}{4}\right)^{4(3)}$$

Carry Out

$$1.19562 = \left(1 + \frac{r}{4}\right)^{12} \qquad \textit{Divide both sides by 1000.}$$

$$\sqrt[12]{1.19562} = 1 + \frac{r}{4} \qquad \textit{Take the 12th root of both sides (or raise both sides to the } \frac{1}{12} \textit{ power).}$$

$$1.015 \approx 1 + \frac{r}{4} \qquad \textit{Approximate } \sqrt[12]{1.19562} \textit{ on a calculator.}$$

$$0.015 \approx \frac{r}{4} \qquad \textit{Subtract 1 from both sides.}$$

$$0.06 \approx r \qquad \textit{Multiply both sides by 4.}$$

NOW TRY EXERCISE 103   Answer   The annual interest rate is approximately 0.06 or 6%.

HELPFUL HINT

STUDY TIP

In this chapter, you will be working with roots and radicals. This material was discussed in Chapter 7. If you do not remember how to evaluate or simplify radicals, review Chapter 7 now.

Exercise Set 8.1

Concept/Writing Exercises

1. Write the two square roots of 36.
2. Write the two square roots of 17.
3. Write the square root property.
4. What is the first step in completing the square?
5. Explain how to determine whether a trinomial is a perfect square trinomial.
6. Write a paragraph explaining in your own words how to construct a perfect square trinomial.

7. **a)** Is $x = 9$ the solution to $x - 9 = 0$? If not, what is the correct solution? Explain.
   b) Is $x = 3$ the solution to $x^2 - 9 = 0$? If not, what is the correct solution? Explain.

8. **a)** Is $x = -7$ the solution to $x + 7 = 0$? If not, what is the correct solution? Explain.
   b) Is $x = \pm\sqrt{7}$ the solution to $x^2 + 7 = 0$? If not, what is the correct solution? Explain.

9. What is the first step in solving the equation $2x^2 + 3x = 9$ by completing the square? Explain.

10. What is the first step in solving the equation $\frac{1}{5}x^2 + 12x = -4$ by completing the square? Explain.

11. When solving the equation $x^2 - 6x = 17$ by completing the square, what number do we add to both sides of the equation? Explain.

12. When solving the equation $x^2 + 10x = 39$ by completing the square, what number do we add to both sides of the equation? Explain.

Practice the Skills

Use the square root property to solve each equation.

13. $x^2 = 49$

14. $x^2 - 49 = 0$

15. $x^2 + 49 = 0$

16. $x^2 - 48 = 0$

17. $x^2 + 48 = 0$

18. $y^2 - 11 = 50$

19. $y^2 + 11 = -50$

20. $(x - 3)^2 = 49$

21. $(p - 4)^2 = 16$

22. $(x + 3)^2 = 49$

23. $(x + 3)^2 + 49 = 0$

24. $(a - 4)^2 = 45$

25. $(a - 4)^2 + 45 = 0$

26. $(a + 4)^2 + 45 = 0$

27. $\left(b + \frac{1}{3}\right)^2 = \frac{4}{9}$

28. $\left(b - \frac{1}{3}\right)^2 = \frac{4}{9}$

29. $\left(b - \frac{1}{3}\right)^2 + \frac{4}{9} = 0$

30. $(x - 0.2)^2 = 0.64$

31. $(x + 1.8)^2 = 0.81$

32. $\left(x + \frac{1}{2}\right)^2 = \frac{4}{9}$

33. $(2a - 5)^2 = 12$

34. $(4y + 1)^2 = 8$

35. $\left(2y + \frac{1}{2}\right)^2 = \frac{4}{25}$

36. $\left(3x - \frac{1}{4}\right)^2 = \frac{9}{25}$

Solve each equation by completing the square.

37. $x^2 + 3x - 4 = 0$

38. $x^2 - 3x - 4 = 0$

39. $x^2 + 2x - 15 = 0$

40. $x^2 - 2x - 15 = 0$

41. $x^2 - 6x + 8 = 0$

42. $x^2 + 6x + 8 = 0$

43. $x^2 - 6x + 5 = 0$

44. $x^2 - 9x + 18 = 0$

45. $2x^2 + x - 1 = 0$

46. $3c^2 - 5c - 2 = 0$

47. $2z^2 - 7z - 4 = 0$

48. $4a^2 + 11a = 3$

49. $x^2 - 11x + 28 = 0$

50. $x^2 - x - 12 = 0$

51. $-x^2 + 3x + 4 = 0$

52. $-a^2 - 5a + 14 = 0$

53. $-z^2 + 9z - 20 = 0$

54. $-z^2 - 4z + 12 = 0$

55. $b^2 = 3b + 28$

56. $-x^2 = 4x - 32$

57. $x^2 + 9x = 10$

58. $-x^2 + 40 = -3x$

59. $x^2 - 4x - 10 = 0$

60. $x^2 - 6x + 2 = 0$

61. $r^2 + 8r + 5 = 0$

62. $a^2 + 4a - 1 = 0$

63. $c^2 - c - 3 = 0$

64. $p^2 - 5p = 8$

65. $x^2 + 3x + 6 = 0$

66. $z^2 - 5z + 7 = 0$

67. $2x^2 - 2x = 0$

68. $3y^2 + 12y = 0$

69. $-\frac{1}{4}b^2 - \frac{1}{2}b = 0$

70. $\frac{1}{3}a^2 - \frac{4}{3}a = 0$

71. $18z^2 - 6z = 0$

72. $x^2 = \frac{3}{2}x$

73. $-\frac{1}{2}p^2 - p + \frac{3}{2} = 0$

74. $2x^2 + 2x = 24$

75. $2x^2 = 8x + 90$

76. $2x^2 + 18x + 4 = 0$

77. $3x^2 + 33x + 72 = 0$

78. $\frac{2}{3}x^2 + \frac{4}{3}x + 1 = 0$

79. $3w^2 + 2w - 1 = 0$

80. $3c^2 - 8c + 4 = 0$

81. $2x^2 - x = -5$

82. $-3x^2 + 6x = 6$

83. $\frac{5}{2}x^2 + \frac{3}{2}x - \frac{5}{4} = 0$

84. $\frac{x^2}{3} = -\frac{2}{3}x - \frac{5}{3}$

Problem Solving

In Exercises 85–88, the area, A, of each rectangle is given. **a)** *Write an equation for the area.* **b)** *Solve the equation for x.*

85. A = 21, x − 2, x + 2

86. A = 35, x + 3, x + 5

87. A = 18, x + 2, x + 4

88. A = 23, x − 3, x − 1

89. *Stopping Distance on Snow* The formula for approximating the stopping distance, d, in feet, for a specific car on snow is $d = \frac{1}{6}x^2$, where x is the speed of the car, in miles per hour, before the brakes are applied. If the car's stopping distance was 24 feet, what was the car's speed before the brakes were applied?

90. *Stopping Distance on Dry Pavement* The formula for approximating the stopping distance, d, in feet, for a specific car on dry pavement is $d = \frac{1}{10}x^2$, where x is the speed of the car, in miles per hour, before the brakes are applied. If the car's stopping distance was 90 feet, what was the car's speed before the brakes were applied?

91. *Integers* The product of two consecutive positive odd integers is 63. Find the two odd integers.

92. *Integers* The larger of two integers is 2 more than twice the smaller. Find the two numbers if their product is 12.

93. *Rectangular Garden* Donna Simm has marked off an area in her yard where she will plant tomato plants. Find the dimensions of the rectangular area if the length is 2 feet more than twice the width and the area is 60 square feet.

94. *Driveway* Manuel Cortez is planning to blacktop his driveway. Find the dimensions of the rectangular driveway if its area is 381.25 square feet and its length is 18 feet greater than its width.

95. *Bedroom* Robert Thompson is designing a house. One of the bedrooms is a square whose diagonal is 6 feet longer than the length of a side. Find the dimensions of the bedroom.

96. *Wading Pool* The Lakeside Hotel is planning to build a shallow wading pool for children. If the pool is to be square and the diagonal of the square is 10 feet longer than a side, find the dimensions of the pool.

97. *Inscribed Triangle* When a triangle is inscribed in a semicircle where a diameter of the circle is a side of the triangle, the triangle formed is always a right triangle. If an isosceles triangle (two equal sides) is inscribed in a semicircle of radius 10 inches, find the length of the other two sides of the triangle.

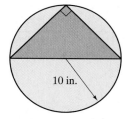

10 in.

98. *Inscribed Triangle* Refer to Exercise 97. Suppose a triangle is inscribed in a semicircle whose diameter is 12 meters. If one side of the inscribed triangle is 2 meters, find the third side.

99. *Area of Circle* The area of a circle is 24π square feet. Use the formula $A = \pi r^2$ to find the radius of the circle.

100. *Area of Circle* The area of a circle is 16.4π square meters. Find the radius of the circle.

Use the formula $A = p\left(1 + \dfrac{r}{n}\right)^{nt}$ *to answer Exercises 101–104.*

101. *Savings Account* Frank Dipalo initially invested $500 in a savings account whose interest is compounded annually. If after 2 years the amount in the account is $540.80, find the annual interest rate.

102. *Savings Account* Margret Chang initially invested $800 in a savings account whose interest is compounded annually. If after 2 years the amount in the account is $898.88, find the annual interest rate.

103. *Savings Account* Steve Rodi initially invested $1200 in a savings account whose interest is compounded semiannually. If after 3 years the amount in the account is $1432.86, find the annual interest rate.

104. *Savings Account* Angela Reyes initially invested $1500 in a savings account whose interest is compounded semiannually. If after 4 years the amount in the account is $2052.85, find the annual interest rate.

105. *Surface Area and Volume* The surface area, S, and volume, V, of a right circular cylinder of radius, r, and height, h, are given by the formulas

$$S = 2\pi r^2 + 2\pi rh, \quad V = \pi r^2 h$$

a) Find the surface area of the cylinder if its height is 10 inches and its volume is 160 cubic inches.

b) Find the radius if the height is 10 inches and the volume is 160 cubic inches.

c) Find the radius if the height is 10 inches and the surface area is 160 square inches.

Group Activity

Discuss and answer Exercise 106 as a group.

106. On the following grid, the points (x_1, y_1), (x_2, y_2), and (x_1, y_2) are plotted.

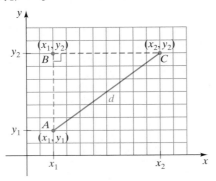

a) Explain why (x_1, y_2) is placed where it is and not somewhere else on the graph.

b) Express the length of the orange dashed line in terms of y_2 and y_1. Explain how you determined your answer.

c) Express the length of the green dashed line in terms of x_2 and x_1.

d) Using the Pythagorean Theorem and the right triangle ABC, derive a formula for the distance, d, between points (x_1, y_1) and (x_2, y_2).* Explain how you determined the formula.

e) Use the formula you determined in part **d)** to find the distance of the line segment between the points $(1, 4)$ and $(3, 7)$.

Cumulative Review Exercises

[2.1] **107.** Solve $-4(2z - 6) = -3(z - 4) + z$.

[2.4] **108.** *Investment* Thea Prettyman invested $10,000 for 1 year, part at 7% and part at $6\frac{1}{4}$%. If she earned a total interest of $656.50, how much was invested at each rate?

[2.6] **109.** Solve $|x + 3| = |2x - 7|$.

[3.4] **110.** Find the slope of the line through $(-3, 4)$ and $(-1, 4)$.

[5.2] **111.** Multiply $(x - 2)(4x^2 + 9x - 2)$.

8.2 SOLVING QUADRATIC EQUATIONS BY THE QUADRATIC FORMULA

SSM Study Guide CD/Video

MathPro 4/5 PH Math Tutor Center prenhall.com/Angel

1 Derive the quadratic formula.

2 Use the quadratic formula to solve equations.

3 Write a quadratic equation given its solutions.

4 Use the discriminant to determine the number of real solutions to a quadratic equation.

5 Study applications that use quadratic equations.

---

*The distance formula will be discussed in Section 10.1.

1   Derive the Quadratic Formula

The quadratic formula can be used to solve any quadratic equation. *It is the most useful and most versatile method of solving quadratic equations.* It is generally used in place of completing the square because of its efficiency.

The standard form of a quadratic equation is $ax^2 + bx + c = 0$, where a is the coefficient of the squared term, b is the coefficient of the first-degree term, and c is the constant.

Quadratic Equation in Standard Form	Values of Coefficients
$x^2 - 3x + 4 = 0$	$a = 1, \quad b = -3, \quad c = 4$
$1.3x^2 - 4.2 = 0$	$a = 1.3, \quad b = 0, \quad c = -4.2$
$-\dfrac{5}{6}x^2 + \dfrac{3}{5}x = 0$	$a = -\dfrac{5}{6}, \quad b = \dfrac{3}{5}, \quad c = 0$

We can derive the quadratic formula by starting with a quadratic equation in standard form and completing the square, as discussed in the preceding section.

$$ax^2 + bx + c = 0$$

$$\frac{ax^2}{a} + \frac{b}{a}x + \frac{c}{a} = 0 \qquad \textit{Divide both sides by } a.$$

$$x^2 + \frac{b}{a}x = -\frac{c}{a} \qquad \textit{Subtract } c/a \textit{ from both sides.}$$

$$x^2 + \frac{b}{a}x + \boxed{\frac{b^2}{4a^2}} = -\frac{c}{a} + \boxed{\frac{b^2}{4a^2}} \qquad \begin{array}{l}\textit{Take 1/2 of } b/a \textit{ (that is, } b/2a), \textit{ and}\\ \textit{square it to get } b^2/4a^2. \textit{ Then add}\\ \textit{this expression to both sides.}\end{array}$$

$$\left(x + \frac{b}{2a}\right)^2 = \frac{b^2}{4a^2} - \frac{c}{a} \qquad \begin{array}{l}\textit{Rewrite the left side of the equa-}\\ \textit{tion as the square of a binomial.}\end{array}$$

$$\left(x + \frac{b}{2a}\right)^2 = \frac{b^2 - 4ac}{4a^2} \qquad \begin{array}{l}\textit{Write the right side with a common}\\ \textit{denominator.}\end{array}$$

$$x + \frac{b}{2a} = \pm\sqrt{\frac{b^2 - 4ac}{4a^2}} \qquad \textit{Square root property}$$

$$x + \frac{b}{2a} = \pm\frac{\sqrt{b^2 - 4ac}}{2a} \qquad \textit{Quotient rule for radicals}$$

$$x = -\frac{b}{2a} \pm \frac{\sqrt{b^2 - 4ac}}{2a} \qquad \textit{Subtract } b/2a \textit{ from both sides.}$$

$$x = \frac{-b \pm \sqrt{b^2 - 4ac}}{2a} \qquad \begin{array}{l}\textit{Write with a common denominator}\\ \textit{to get the quadratic formula.}\end{array}$$

2   Use the Quadratic Formula to Solve Equations

Now that we have derived the quadratic formula, we will use it to solve quadratic equations.

> ## To Solve a Quadratic Equation by the Quadratic Formula
>
> **1.** Write the quadratic equation in standard form, $ax^2 + bx + c = 0$, and determine the numerical values for a, b, and c.
>
> **2.** Substitute the values for a, b, and c into the quadratic formula and then evaluate the formula to obtain the solution.
>
>
>
> The Quadratic Formula
>
> $$x = \frac{-b \pm \sqrt{b^2 - 4ac}}{2a}$$
>
>

EXAMPLE 1   Solve the equation $x^2 + 2x - 8 = 0$ by the quadratic formula.

Solution   In this equation, $a = 1$, $b = 2$, and $c = -8$.

$$x = \frac{-b \pm \sqrt{b^2 - 4ac}}{2a}$$

$$x = \frac{-(2) \pm \sqrt{(2)^2 - 4(1)(-8)}}{2(1)}$$

$$= \frac{-2 \pm \sqrt{4 + 32}}{2}$$

$$= \frac{-2 \pm \sqrt{36}}{2}$$

$$= \frac{-2 \pm 6}{2}$$

$$x = \frac{-2 + 6}{2} \quad \text{or} \quad x = \frac{-2 - 6}{2}$$

$$x = \frac{4}{2} = 2 \qquad\qquad x = \frac{-8}{2} = -4$$

A check will show that both 2 and −4 are solutions to the equation. Note that the solutions to the equation $x^2 + 2x - 8 = 0$ are two real numbers. ✳

The solution to Example 1 could also be obtained by factoring, as illustrated below.

$$x^2 + 2x - 8 = 0$$
$$(x + 4)(x - 2) = 0$$
$$x + 4 = 0 \quad \text{or} \quad x - 2 = 0$$
$$x = -4 \qquad\qquad x = 2$$

When you are given a quadratic equation to solve and the method to solve it has not been specified, you may try solving by factoring first (as we discussed in Section 5.8). If the equation cannot be easily factored, use the quadratic formula.

When solving a quadratic equation using the quadratic formula, the calculations may be easier if the leading coefficient, a, is positive. Thus, if solving the quadratic equation $-x^2 + 3x = 2$, you may wish to rewrite the equation as $x^2 - 3x + 2 = 0$.

EXAMPLE 2 Solve $-9x^2 = -6x + 1$ by the quadratic formula.

Solution Begin by adding $9x^2$ to both sides of the equation to obtain

$$0 = 9x^2 - 6x + 1$$

$$\text{or} \quad 9x^2 - 6x + 1 = 0$$

$$a = 9, \qquad b = -6, \qquad c = 1$$

$$x = \frac{-b \pm \sqrt{b^2 - 4ac}}{2a}$$

$$= \frac{-(-6) \pm \sqrt{(-6)^2 - 4(9)(1)}}{2(9)}$$

$$= \frac{6 \pm \sqrt{36 - 36}}{18} = \frac{6 \pm \sqrt{0}}{18} = \frac{6}{18} = \frac{1}{3}$$

Note that the solution to the equation $-9x^2 = -6x + 1$ is $\frac{1}{3}$, a single value. Some quadratic equations have just one value as the solution. This occurs when $b^2 - 4ac = 0$. ✳

AVOIDING COMMON ERRORS

The entire numerator of the quadratic formula must be divided by $2a$.

CORRECT

$$x = \frac{-b \pm \sqrt{b^2 - 4ac}}{2a}$$

INCORRECT

$$x = -b \pm \frac{\sqrt{b^2 - 4ac}}{2a}$$

$$x = \frac{-b}{2a} \pm \sqrt{b^2 - 4ac}$$

EXAMPLE 3 Solve the equation $p^2 + \dfrac{2}{5}p + \dfrac{1}{3} = 0$ by the quadratic formula.

Solution Do not let the change in variable worry you. The quadratic formula is used exactly the same way as when x is the variable.

We could solve this equation using the quadratic formula with $a = 1$, $b = \frac{2}{5}$, and $c = \frac{1}{3}$. However, when a quadratic equation contains fractions, it is generally easier to begin by multiplying both sides of the equation by the least common denominator. In this example, the least common denominator is 15.

$$15\left(p^2 + \frac{2}{5}p + \frac{1}{3} \right) = 15(0)$$

$$15p^2 + 6p + 5 = 0$$

Now we can use the quadratic formula with $a = 15$, $b = 6$, and $c = 5$.

$$p = \frac{-b \pm \sqrt{b^2 - 4ac}}{2a}$$

$$= \frac{-6 \pm \sqrt{(6)^2 - 4(15)(5)}}{2(15)}$$

$$= \frac{-6 \pm \sqrt{-264}}{30}$$

$$= \frac{-6 \pm \sqrt{-4}\sqrt{66}}{30}$$

$$= \frac{-6 \pm 2i\sqrt{66}}{30}$$

$$= \frac{\overset{1}{2}(-3 \pm i\sqrt{66})}{\underset{15}{\cancel{30}}}$$

$$= \frac{-3 \pm i\sqrt{66}}{15}$$

The solutions are $\dfrac{-3 + i\sqrt{66}}{15}$ and $\dfrac{-3 - i\sqrt{66}}{15}$. Note that neither solution is a real

NOW TRY EXERCISE 53 number. Both solutions are complex numbers. ✳

AVOIDING COMMON ERRORS

Some students use the quadratic formula correctly until the last step, where they make an error. Below are illustrated both the correct and incorrect procedures for simplifying an answer.

When *both* terms in the numerator *and* the denominator have a common factor, that common factor may be divided out, as follows:

CORRECT

$$\frac{2 + 4\sqrt{3}}{2} = \frac{\overset{1}{2}(1 + 2\sqrt{3})}{\underset{1}{\cancel{2}}} = 1 + 2\sqrt{3}$$

$$\frac{6 + 3\sqrt{3}}{6} = \frac{\overset{1}{3}(2 + \sqrt{3})}{\underset{2}{\cancel{6}}} = \frac{2 + \sqrt{3}}{2}$$

Below are some common errors. Study them carefully so you will not make them. Can you explain why each of the following procedures is incorrect?

INCORRECT

Note that $(2 + 3)/2$ simplifies to $5/2$. However, $(3 + 2\sqrt{5})/2$, $(3 + \sqrt{6})/2$, and $(4 + 3\sqrt{5})/2$ cannot be simplified any further.

EXAMPLE 4   Given $f(x) = 2x^2 + 4x$, find all real values of x for which $f(x) = 5$.

Solution   We wish to determine all real values of x for which

$$2x^2 + 4x = 5$$

We can solve this equation with the quadratic formula. First, write the equation in standard form.

$$2x^2 + 4x - 5 = 0$$

Now, use the quadratic formula with $a = 2$, $b = 4$, and $c = -5$.

$$x = \frac{-b \pm \sqrt{b^2 - 4ac}}{2a}$$

$$= \frac{-4 \pm \sqrt{(4)^2 - 4(2)(-5)}}{2(2)} = \frac{-4 \pm \sqrt{56}}{4} = \frac{-4 \pm 2\sqrt{14}}{4}$$

Next, factor out 2 from both terms in the numerator, and then divide out the common factor.

$$x = \frac{\overset{1}{\cancel{2}}(-2 \pm \sqrt{14})}{\underset{2}{\cancel{4}}} = \frac{-2 \pm \sqrt{14}}{2}\text{*}$$

Thus, the solutions are $\dfrac{-2 + \sqrt{14}}{2}$ and $\dfrac{-2 - \sqrt{14}}{2}$.

NOW TRY EXERCISE 69

Note that the expression in Example 4, $2x^2 + 4x - 5$, is not factorable. Therefore, Example 4 could not be solved by factoring. ✳

If all the numerical coefficients in a quadratic equation have a common factor, you should factor it out before using the quadratic formula. For example, consider the equation $3x^2 + 12x + 3 = 0$. Here $a = 3$, $b = 12$, and $c = 3$. If we use the quadratic formula, we would eventually obtain $x = -2 \pm \sqrt{3}$ as solutions. By factoring the equation before using the formula, we get

$$3x^2 + 12x + 3 = 0$$
$$3(x^2 + 4x + 1) = 0$$

If we consider $x^2 + 4x + 1 = 0$, then $a = 1$, $b = 4$, and $c = 1$. If we use these new values of a, b, and c in the quadratic formula, we will obtain the identical solution, $x = -2 \pm \sqrt{3}$. However, the calculations with these smaller values of a, b, and c are simplified. Solve both equations now using the quadratic formula to convince yourself.

3 Write a Quadratic Equation Given Its Solutions

If we are given the solutions of an equation, we can find the equation by working backward. This procedure is illustrated in Example 5.

EXAMPLE 5   Write an equation that has the following solutions:

a) -4 and 1   **b)** $3 + 2i$ and $3 - 2i$

Solution   **a)** If the solutions are -4 and 1 we write

$$x = -4 \quad \text{or} \quad x = 1$$
$$x + 4 = 0 \qquad x - 1 = 0 \qquad \textit{Set equations equal to 0.}$$
$$(x + 4)(x - 1) = 0 \qquad \textit{Zero-factor property}$$
$$x^2 - x + 4x - 4 = 0 \qquad \textit{Multiply factors.}$$
$$x^2 + 3x - 4 = 0 \qquad \textit{Combine like terms.}$$

Thus, the equation is $x^2 + 3x - 4 = 0$. Many other equations have solutions -4 and 1. In fact, any equation of the form $k(x^2 + 3x - 4) = 0$, where k is a constant, has those solutions. Can you explain why?

---

*Solutions will be given in this form in the Answer Section.

b) $x = 3 + 2i$ or $x = 3 - 2i$

$x - (3 + 2i) = 0$ $x - (3 - 2i) = 0$ *Set equations equal to 0.*

$[x - (3 + 2i)][x - (3 - 2i)] = 0$ *Zero-factor property*

$x \cdot x - x(3 - 2i) - x(3 + 2i) + (3 + 2i)(3 - 2i) = 0$ *Multiply.*

$x^2 - 3x + 2xi - 3x - 2xi + (9 - 4i^2) = 0$ *Distributive property; multiply*

$x^2 - 6x + 9 - 4i^2 = 0$ *Combine like terms.*

$x^2 - 6x + 9 - 4(-1) = 0$ *Substitute $i^2 = -1$.*

$x^2 - 6x + 13 = 0$ *Simplify.*

NOW TRY EXERCISE 75   The equation $x^2 - 6x + 13 = 0$ has the complex solutions $3 + 2i$ and $3 - 2i$.
✳

In Example 5 **a)**, we determined that the equation $x^2 + 3x - 4 = 0$ has solutions -4 and 1. Consider the graph of $f(x) = x^2 + 3x - 4$. The x-intercepts of the graph of $f(x)$ occur when $f(x) = 0$, or when $x^2 + 3x - 4 = 0$. Therefore, the x-intercepts of the graph of $f(x) = x^2 + 3x - 4$ are $(-4, 0)$ and $(1, 0)$, as shown in Figure 8.1. In Example 5 **b)**, we determined that the equation $x^2 - 6x + 13 = 0$ has no real solutions. Thus, the graph of $f(x) = x^2 - 6x + 13$ has no x-intercepts. The graph of $f(x) = x^2 - 6x + 13$ is shown in Figure 8.2.

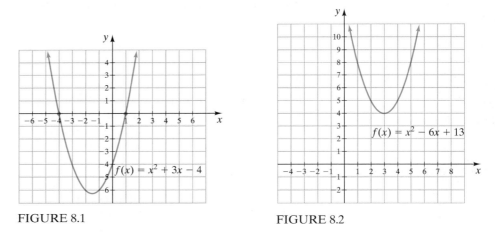

FIGURE 8.1   FIGURE 8.2

4 Use the Discriminant to Determine the Number of Real Solutions to a Quadratic Equation

The expression under the radical sign in the quadratic formula is called the **discriminant**.

$$\underbrace{b^2 - 4ac}_{\text{discriminant}}$$

The discriminant gives the number and kinds of solutions of a quadratic equation.

> ### Solutions of a Quadratic Equation
>
> For a quadratic equation of the form $ax^2 + bx + c = 0$, $a \neq 0$:
> If $b^2 - 4ac > 0$, the quadratic equation has two distinct real number solutions.
> If $b^2 - 4ac = 0$, the quadratic equation has a single real number solution.
> If $b^2 - 4ac < 0$, the quadratic equation has no real number solution.

EXAMPLE 6 **a)** Find the discriminant of the equation $x^2 - 8x + 16 = 0$.
b) How many real number solutions does the given equation have?
c) Use the quadratic formula to find the solution(s).

Solution **a)** $a = 1$, $b = -8$, $c = 16$

$$b^2 - 4ac = (-8)^2 - 4(1)(16)$$
$$= 64 - 64 = 0$$

b) Since the discriminant equals 0, there is a single real number solution.

c)
$$x = \frac{-b \pm \sqrt{b^2 - 4ac}}{2a}$$
$$= \frac{-(-8) \pm \sqrt{0}}{2(1)} = \frac{8 \pm 0}{2} = \frac{8}{2} = 4$$

The only solution is 4.

EXAMPLE 7 Without actually finding the solutions, determine whether the following equations have two distinct real number solutions, a single real number solution, or no real number solution.

a) $2x^2 - 4x + 6 = 0$ **b)** $x^2 - 5x - 8 = 0$ **c)** $4x^2 - 12x = -9$

Solution We use the discriminant of the quadratic formula to answer these questions.
a) $b^2 - 4ac = (-4)^2 - 4(2)(6) = 16 - 48 = -32$

Since the discriminant is negative, this equation has no real number solution.
b) $b^2 - 4ac = (-5)^2 - 4(1)(-8) = 25 + 32 = 57$

Since the discriminant is positive, this equation has two distinct real number solutions.
c) First, rewrite $4x^2 - 12x = -9$ as $4x^2 - 12x + 9 = 0$.

$$b^2 - 4ac = (-12)^2 - 4(4)(9) = 144 - 144 = 0$$

NOW TRY EXERCISE 15 Since the discriminant is 0, this equation has a single real number solution.

The discriminant can be used to find the number of real solutions to an equation of the form $ax^2 + bx + c = 0$. Since the x-intercepts of a quadratic function, $f(x) = ax^2 + bx + c$, occur where $f(x) = 0$, the discriminant can also be used to find the number of x-intercepts of a quadratic function. Figure 8.3 shows the relationship between the discriminant and the number of x-intercepts for a function of the form $f(x) = ax^2 + bx + c$.

Graphs of $f(x) = ax^2 + bx + c$

If $b^2 - 4ac > 0$, $f(x)$ has two distinct x-intercepts.   If $b^2 - 4ac = 0$, $f(x)$ has a single x-intercept.   If $b^2 - 4ac < 0$, $f(x)$ has no x-intercepts.

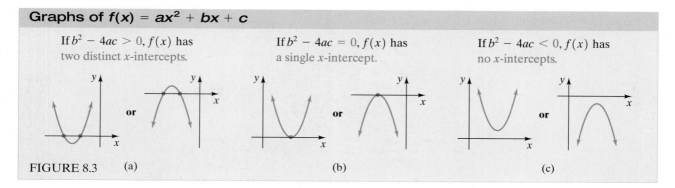

FIGURE 8.3   (a)   (b)   (c)

We will discuss graphing of quadratic functions in detail in Section 8.5.

5 Study Applications That Use Quadratic Equations

We will now look at some applications of quadratic equations.

EXAMPLE 8   **Cell Phones**   Mary Beth Orrange owns a business that manufactures and sells cell phones. The revenue, $R(n)$, from selling the cell phones is determined by multiplying the number of cell phones by the cost per phone. Suppose the revenue from selling n cell phones, $n \le 50$, is

$$R(n) = n(50 - 0.2n)$$

where $(50 - 0.2n)$ is the price per cell phone, in dollars.

a) Find the revenue when 40 cell phones are sold.

b) How many cell phones must be sold to have a revenue of $480?

Solution   **a)** To find the revenue when 40 cell phones are sold, we evaluate the revenue function for $n = 40$.

$$R(n) = n(50 - 0.2n)$$
$$R(40) = 40[50 - 0.2(40)]$$
$$= 40(50 - 8)$$
$$= 40(42)$$
$$= 1680$$

The revenue from selling 40 cell phones is $1680.

b) Understand   We want to find the number of cell phones that need to be sold to have $480 in revenue. Thus, we need to let $R(n) = 480$ and solve for n.

$$R(n) = n(50 - 0.2n)$$
$$480 = n(50 - 0.2n)$$
$$480 = 50n - 0.2n^2$$
$$0.2n^2 - 50n + 480 = 0$$

Now we can use the quadratic formula to solve the equation.

Translate $a = 0.2,\qquad b = -50,\qquad c = 480$

$$n = \frac{-b \pm \sqrt{b^2 - 4ac}}{2a}$$

$$= \frac{-(-50) \pm \sqrt{(-50)^2 - 4(0.2)(480)}}{2(0.2)}$$

Carry Out
$$= \frac{50 \pm \sqrt{2500 - 384}}{0.4}$$

$$= \frac{50 \pm \sqrt{2116}}{0.4}$$

$$= \frac{50 \pm 46}{0.4}$$

$$n = \frac{50 + 46}{0.4} = 240 \qquad \text{or} \qquad n = \frac{50 - 46}{0.4} = 10$$

Answer   Since the problem specified that $n \le 50$, the only acceptable solution is
NOW TRY EXERCISE 91 $n = 10$. Thus, to obtain $480 in revenue, Mary Beth must sell 10 cell phones.   ✳

An important formula in physics is $h = \frac{1}{2}gt^2 + v_0t + h_0$. When an object is projected upward from an initial height, h_0, with initial velocity of v_0, this formula can be used to find the height, h, of the object above the ground at any time, t. The g in the formula is the acceleration of gravity. Since the acceleration of Earth's gravity is -32 ft/sec^2, we use -32 for g in the formula when discussing Earth. This formula can also be used in describing projectiles on the Moon and other planets, but the value of g in the formula will need to change for each planetary body. We will use this formula in Example 9.

EXAMPLE 9 **Throwing a Ball** Betsy Farber is standing on top of a building and throws a ball upward from a height of 60 feet with an initial velocity of 30 feet per second. Use the formula $h = \frac{1}{2}gt^2 + v_0t + h_0$ to answer the following questions.

a) How long after the ball is thrown, to the nearest tenth of a second, will the ball be 25 feet above the ground?

b) How long after the ball is thrown will the ball strike the ground?

Solution **a)** Understand We will illustrate this problem with a diagram (see Fig. 8.4). Here $g = -32$, $v_0 = 30$, and $h_0 = 60$. We are asked to find the time, t, it takes for the ball to reach a height, h, of 25 feet above the ground. We substitute these values into the formula and then solve for t.

Translate
$$h = \frac{1}{2}gt^2 + v_0t + h_0$$

$$25 = \frac{1}{2}(-32)t^2 + 30t + 60$$

Carry Out Now we write the quadratic equation in standard form and solve for t by using the quadratic formula.

$$0 = -16t^2 + 30t + 35$$
$$\text{or} \quad -16t^2 + 30t + 35 = 0$$
$$a = -16, \qquad b = 30, \qquad c = 35$$
$$t = \frac{-b \pm \sqrt{b^2 - 4ac}}{2a}$$
$$= \frac{-30 \pm \sqrt{(30)^2 - 4(-16)(35)}}{2(-16)}$$
$$= \frac{-30 \pm \sqrt{3140}}{-32}$$
$$t = \frac{-30 + \sqrt{3140}}{-32} \quad \text{or} \quad t = \frac{-30 - \sqrt{3140}}{-32}$$
$$\approx -0.8 \qquad\qquad\qquad \approx 2.7$$

Answer Since time cannot be negative, the only reasonable solution is 2.7 seconds. Thus, about 2.7 seconds after the ball is thrown, it will be 25 feet above the ground.

b) Understand We wish to find the time at which the ball strikes the ground. When the ball strikes the ground, its distance above the ground is 0. We substitute $h = 0$ into the formula and solve for t.

Translate
$$h = \frac{1}{2}gt^2 + v_0t + h_0$$

$$0 = \frac{1}{2}(-32)t^2 + 30t + 60$$

60 ft

25 ft

FIGURE 8.4

Carry Out

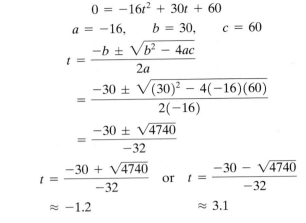

$$0 = -16t^2 + 30t + 60$$

$$a = -16, \qquad b = 30, \qquad c = 60$$

$$t = \frac{-b \pm \sqrt{b^2 - 4ac}}{2a}$$

$$= \frac{-30 \pm \sqrt{(30)^2 - 4(-16)(60)}}{2(-16)}$$

$$= \frac{-30 \pm \sqrt{4740}}{-32}$$

$$t = \frac{-30 + \sqrt{4740}}{-32} \quad \text{or} \quad t = \frac{-30 - \sqrt{4740}}{-32}$$

$$\approx -1.2 \qquad\qquad\qquad \approx 3.1$$

Answer   Since time cannot be negative, the only reasonable solution is 3.1 seconds. Thus, the ball will strike the ground approximately 3.1 seconds after it is thrown.

Exercise Set 8.2

Concept/Writing Exercises

1. Give the quadratic formula. (You should memorize this formula.)

2. To solve the equation $3x + 2x^2 - 7 = 0$ using the quadratic formula, what are the values for $a, b,$ and c?

3. To solve the equation $6x - 3x^2 + 5 = 0$ using the quadratic formula, what are the values for $a, b,$ and c?

4. To solve the equation $4x^2 - 5x = 7$ using the quadratic formula, what are the values for $a, b,$ and c?

5. Consider the two equations $-6x^2 + \frac{1}{2}x - 5 = 0$ and $6x^2 - \frac{1}{2}x + 5 = 0$. Must the solutions to these two equations be the same? Explain your answer.

6. Consider $12x^2 - 15x - 6 = 0$ and $3(4x^2 - 5x - 2) = 0$.

   a) Will the solution to the two equations be the same? Explain.

   b) Solve $12x^2 - 15x - 6 = 0$.

   c) Solve $3(4x^2 - 5x - 2) = 0$.

7. a) Explain how to find the discriminant.

   b) What is the discriminant for the equation $3x^2 - 6x + 20 = 0$?

   c) Write a paragraph or two explaining the relationship between the value of the discriminant and the number of real solutions to a quadratic equation. In your paragraph, explain *why* the value of the discriminant determines the number of real solutions.

8. Write a paragraph or two explaining the relationship between the value of the discriminant and the number of x-intercepts of $f(x) = ax^2 + bx + c$. In your paragraph, explain when the function will have no, one, and two x-intercepts.

Practice the Skills

Use the discriminant to determine whether each equation has two distinct real solutions, a single real solution, or no real solution.

9. $x^2 + 3x + 2 = 0$

10. $2x^2 + x + 1 = 0$

11. $3z^2 + 4z + 5 = 0$

12. $-a^2 + 3a - 6 = 0$

13. $5p^2 + 3p - 7 = 0$

14. $2x^2 = 16x - 32$

15. $-5x^2 + 5x - 6 = 0$

16. $4.1x^2 - 3.1x - 5.2 = 0$

17. $x^2 + 10.2x + 26.01 = 0$

18. $\frac{1}{2}x^2 + \frac{1}{3}x + 12 = 0$

19. $b^2 = -3b - \frac{9}{4}$

20. $\frac{x^2}{3} = \frac{2x}{5}$

Solve each equation by the quadratic formula.

21. $x^2 - 9x + 18 = 0$

22. $x^2 + 9x + 18 = 0$

23. $a^2 - 6a + 8 = 0$

24. $a^2 + 6a + 8 = 0$

25. $x^2 = -2x + 3$

26. $-a^2 + 3a + 10 = 0$

27. $-b^2 = 4b - 20$

28. $a^2 - 36 = 0$

29. $b^2 - 49 = 0$

30. $2x^2 = 4x + 1$

31. $3w^2 - 4w + 5 = 0$

32. $x^2 - 4x = 0$

33. $c^2 - 3c = 0$

34. $-t^2 - t - 1 = 0$

35. $4s^2 - 8s + 6 = 0$

36. $-3r^2 = 9r + 6$

37. $a^2 + 2a + 1 = 0$

38. $y^2 + 6y + 9 = 0$

39. $x^2 - 10x + 25 = 0$

40. $b^2 - 20b + 100 = 0$

41. $x^2 - 2x - 1 = 0$

42. $-n^2 = 3n + 6$

43. $2 - 3r^2 = -4r$

44. $-9d - 3d^2 = 5$

45. $2x^2 + 5x - 3 = 0$

46. $(2r - 6)(3r + 4) = -20$

🔒 **47.** $(2a + 3)(3a - 1) = 2$

48. $6x^2 = 21x + 27$

49. $\dfrac{1}{2}t^2 + t - 12 = 0$

50. $\dfrac{1}{3}x^2 = 4x - 9$

51. $9r^2 - 9r + 2 = 0$

52. $3x^2 - 6x - 3 = 0$

🔒 **53.** $\dfrac{1}{2}x^2 + 2x + \dfrac{2}{3} = 0$

54. $x^2 - \dfrac{11}{3}x = \dfrac{10}{3}$

55. $a^2 - \dfrac{a}{5} - \dfrac{1}{3} = 0$

56. $b^2 = -\dfrac{b}{2} + \dfrac{2}{3}$

57. $c = \dfrac{6 - c}{c - 4}$

58. $3y = \dfrac{5y + 6}{2y + 3}$

59. $2x^2 - 4x + 3 = 0$

60. $3a^2 - 4a = -5$

61. $2y^2 + y = -3$

62. $b^2 - \dfrac{7}{6}b + \dfrac{2}{3} = 0$

63. $0.1x^2 + 0.6x - 1.2 = 0$

64. $2.3x^2 - 5.6x - 0.4 = 0$

For each function, determine all real values of the variable for which the function has the value indicated.

65. $f(x) = x^2 - 2x + 4, f(x) = 4$

66. $g(x) = x^2 + 3x + 5, g(x) = 5$

67. $k(x) = x^2 - x - 10, k(x) = 20$

68. $p(r) = r^2 + 17r + 84, p(r) = 12$

69. $h(t) = 2t^2 - 7t + 1, h(t) = -3$

70. $t(x) = x^2 + 5x - 3, t(x) = 4$

🔒 **71.** $g(a) = 2a^2 - 3a + 16, g(a) = 14$

72. $h(x) = 6x^2 + 3x + 1, h(x) = -7$

Write a function that has the given solutions.

73. $2, 5$

74. $-3, 4$

75. $4, -6$

76. $-3, -5$

🔒 **77.** $-\dfrac{3}{5}, \dfrac{2}{3}$

78. $-\dfrac{2}{3}, -\dfrac{3}{4}$

79. $\sqrt{3}, -\sqrt{3}$

80. $\sqrt{5}, -\sqrt{5}$

81. $2i, -2i$

82. $6i, -6i$

83. $3 + \sqrt{2}, 3 - \sqrt{2}$

84. $5 - \sqrt{3}, 5 + \sqrt{3}$

85. $2 + 3i, 2 - 3i$

86. $5 - 4i, 5 + 4i$

*In Exercises 87–90, **a)** set up a revenue function, $R(n)$, that can be used to solve the problem, and then, **b)** solve the problem. See Example 8.*

87. **Selling Lamps** A business sells n lamps, $n \le 65$, at a price of $(10 - 0.02n)$ dollars per lamp. How many lamps must be sold to have a revenue of $450?

88. **Selling Batteries** A business sells n batteries, $n \le 26$, at a price of $(25 - 0.1n)$ dollars per battery. How many batteries must be sold to have a revenue of $460?

89. **Selling Chairs** A business sells n chairs, $n \le 50$, at a price of $(50 - 0.2n)$ dollars per chair. How many chairs must be sold to have a revenue of $1680?

90. **Selling Watches** A business sells n watches, $n \le 75$, at a price of $(30 - 0.15n)$ dollars per watch. How many watches must be sold to have a revenue of $1365?

91. Give your own example of a quadratic equation that can be solved by the quadratic formula but not by factoring over the set of integers.

92. Are there any quadratic equations that **a)** can be solved by the quadratic formula that cannot be solved by completing the square? **b)** can be solved by completing the square that cannot be solved by factoring over the set of integers?

93. When solving a quadratic equation by the quadratic formula, if the discriminant is a perfect square, must the equation be factorable over the set of integers?

94. When solving a quadratic equation by the quadratic formula, if the discriminant is a natural number, must the equation be factorable over the set of integers?

In Exercises 95–100, use a calculator as needed to give the solution in decimal form. Round irrational numbers to the nearest hundredth.

95. Numbers Twice the square of a positive number increased by three times the number is 14. Find the number.

96. Numbers Three times the square of a positive number decreased by twice the number is 21. Find the number.

97. Rectangular Garden The length of a rectangular garden is 2 feet less than 3 times its width. Find the length and width if the area of the garden is 21 square feet.

98. Rectangular Region Lora Wallman wishes to fence in a rectangular region along a riverbank by constructing fencing as illustrated in the diagram. If she has only 400 feet of fencing and wishes to enclose an area of 15,000 square feet, find the dimensions of the rectangular region.

River

Fence

99. Photo John Williams, a professional photographer, has a 6-inch-by-8-inch photo. He wishes to reduce the photo by the same amount on each side so that the resulting photo will have half the area of the original photo. By how much will he have to reduce the length of each side?

100. Rectangular Garden Bart Simmons has a 12-meter-by-9-meter flower garden. He wants to build a gravel path of uniform width along the inside of the garden on each side so that the resulting garden will have half the area of the original garden. What will be the width of the gravel path?

In Exercises 101 and 102, use the equation $h = \frac{1}{2}gt^2 + v_0t + h_0$ (refer to Example 9).

101. Throwing a Horseshoe A horseshoe is thrown upward from an initial height of 80 feet with an initial velocity of 60 feet per second. How long after the horseshoe is projected upward

a) will it be 20 feet from the ground?

b) will it strike the ground?

102. Gravity on the Moon Gravity on the Moon is about one-sixth of that on Earth. Suppose Neil Armstrong is standing on a hill on the Moon 60 feet high. If he jumps upward with a velocity of 40 feet per second, how long will it take for him to land on the ground below the hill?

Solve by the quadratic formula.

103. $x^2 - \sqrt{5}x - 10 = 0$

104. $x^2 + 5\sqrt{6}x + 36 = 0$

Challenge Problems

105. Heating a Metal Cube A metal cube expands when heated. If each edge increases 0.20 millimeter after being heated and the total volume increases by 6 cubic millimeters, find the original length of a side of the cube.

106. Six Solutions The equation $x^n = 1$ has n solutions (including the complex solutions). Find the six solutions to

$x^6 = 1$. (*Hint:* Rewrite the equation as $x^6 - 1 = 0$, then factor using the formula for the difference of two squares.)

107. Throwing a Rock Travis Hawley is on the fourth floor of an eight-story building and Courtney Prenzlow is on the roof. Travis is 60 feet above the ground while Courtney is 120 feet above the ground.

a) If Travis drops his rock out of a window, determine the time it takes for the rock to strike the ground.

b) If Courtney drops her rock off the roof, determine the time it takes for the rock to strike the ground.

c) If Travis throws a rock upward with an initial velocity of 100 feet per second at the same time that Courtney throws a rock upward at 60 feet per second, whose rock will strike the ground first? Explain.

d) Will the rocks ever be at the same distance above the ground? If so, at what times?

Cumulative Review Exercises

[1.6] **108.** Evaluate $\dfrac{3.33 \times 10^3}{1.11 \times 10^1}$.

[3.2] **109.** If $f(x) = x^2 + 2x - 5$, find $f(3)$.

[4.1] **110.** Solve the system of equations.

$$3x + 4y = 2$$
$$2x = -5y - 1$$

[6.3] **111.** Simplify $2x^{-1} - (3y)^{-1}$.

[7.6] **112.** Solve $\sqrt{x^2 - 6x - 4} = x$.

8.3 QUADRATIC EQUATIONS: APPLICATIONS AND PROBLEM SOLVING

SSM   Study Guide   CD/Video

MathPro 4/5   PH Math Tutor Center   prenhall.com/Angel

1 Solve additional applications.

2 Solve for a variable in a formula.

1 Solve Additional Applications

We have already discussed a few applications of quadratic equations. In this section, we will explore several more applications. We will also discuss solving for a variable in a formula. We start by investigating a profit for a new company.

EXAMPLE 1 **Company Profits**   Laserox, a start-up company, projects that its annual profits, $p(n)$, in thousands of dollars, over the first 8 years of operation can be approximated by the function $p(n) = 1.2n^2 + 4n - 8$, where n is the number of years completed.

a) Estimate the profit (or loss) of the company after the first year.

b) Estimate the profit (or loss) of the company after 8 years.

c) Estimate the time needed for the company to break even.

Solution **a)** To estimate the profit after 1 year, we evaluate the function at 1.

$$p(n) = 1.2n^2 + 4n - 8$$
$$p(1) = 1.2(1)^2 + 4(1) - 8 = -2.8$$

Thus, at the end of the first year the company projects a loss of $2.8 thousand or a loss of $2800.

b) $p(8) = 1.2(8)^2 + 4(8) - 8 = 100.8$

Thus, at the end of the eighth year the company's projected profit is $100.8 thousand, or a profit of $100,800.

c) Understand   The company will break even when the profit is 0. Thus, to find the break-even point (no profit or loss) we solve the equation

$$1.2n^2 + 4n - 8 = 0$$

We can use the quadratic formula to solve this equation.

Translate
$$a = 1.2, \quad b = 4, \quad c = -8$$

$$n = \frac{-b \pm \sqrt{b^2 - 4ac}}{2a}$$

$$= \frac{-4 \pm \sqrt{(4)^2 - 4(1.2)(-8)}}{2(1.2)}$$

Carry Out
$$= \frac{-4 \pm \sqrt{16 + 38.4}}{2.4}$$

$$= \frac{-4 \pm \sqrt{54.4}}{2.4}$$

$$\approx \frac{-4 \pm 7.376}{2.4}$$

$$n \approx \frac{-4 + 7.376}{2.4} \approx 1.4 \quad \text{or} \quad n \approx \frac{-4 - 7.376}{2.4} \approx -4.74$$

Answer Since time cannot be negative, the break-even time is about 1.4 years.

Now, let's consider another example that uses the quadratic formula to solve a quadratic equation.

EXAMPLE 2 **Life Expectancy** The function $N(t) = 0.0054t^2 - 1.46t + 95.11$ can be used to estimate the average number of years of life expectancy remaining for a person of age t years where $30 \le t \le 100$.

a) Estimate the remaining life expectancy of a person of age 50.

b) If a person has a remaining life expectancy of 14.3 years, estimate the age of the person.

Solution **a) Understand** We would expect that the older a person gets the shorter the remaining life expectancy. To determine the remaining life expectancy for a 50-year-old, we substitute 50 for t in the function and evaluate.

Translate
$$N(t) = 0.0054t^2 - 1.46t + 95.11$$
$$N(50) = 0.0054(50)^2 - 1.46(50) + 95.11$$

Carry Out
$$= 0.0054(2500) - 73.00 + 95.11$$
$$= 13.5 - 73.00 + 95.11$$
$$= 35.61$$

Answer and Check The answer appears reasonable. Thus, on the average, a 50-year-old can expect to live another 35.61 years to an age of 85.61 years.

b) Understand Here we are given the remaining life expectancy, $N(t)$, and asked to find the age of the person, t. To solve this problem, we substitute 14.3 for $N(t)$ and solve for t. To solve for t, we will use the quadratic formula.

Translate
$$N(t) = 0.0054t^2 - 1.46t + 95.11$$
$$14.3 = 0.0054t^2 - 1.46t + 95.11$$

Carry Out
$$0 = 0.0054t^2 - 1.46t + 80.81$$
$$a = 0.0054, \quad b = -1.46, \quad c = 80.81$$

$$t = \frac{-b \pm \sqrt{b^2 - 4ac}}{2a}$$

$$= \frac{-(-1.46) \pm \sqrt{(-1.46)^2 - 4(0.0054)(80.81)}}{2(0.0054)}$$

$$= \frac{1.46 \pm \sqrt{2.1316 - 1.745496}}{0.0108}$$

$$= \frac{1.46 \pm \sqrt{0.386104}}{0.0108}$$

$$\approx \frac{1.46 \pm 0.6214}{0.0108}$$

$$t \approx \frac{1.46 + 0.6214}{0.0108} \quad \text{or} \quad t \approx \frac{1.46 - 0.6214}{0.0108}$$

$$\approx 192.72 \qquad\qquad\qquad \approx 77.65$$

Answer Since 192.72 is not a reasonable age, we can exclude that as a possibility. Thus, the average person who has a life expectancy of 14.3 years is about 77.65 years old.

NOW TRY EXERCISE 31

Motion Problems

We first discussed motion problems in Section 2.4. The motion problem we give here is solved using the quadratic formula.

EXAMPLE 3

Motorboat Ride It is a beautiful day, so Charles Curtis decides to go for a relaxing ride in his motorboat on the Potomac River. His trip starts near the Clara Barton Natural Historic Site in Bethesda, Maryland. He travels 12 miles until he enters the Arlington, Virginia/Washington, D.C., area, where he turns around and heads back to his starting point. The total time of his trip is 5 hours. Along the way, he passes a digital sign that indicates the river current is 2 miles per hour. If during the entire trip he did not touch the throttle to change the speed, find the speed the boat would have been traveling in still water.

Solution Understand We are asked to find the rate of the boat in still water. Let $r =$ the rate of the boat in still water. We know that the total time of the trip is 5 hours. Thus, the time downriver plus the time upriver must sum to 5 hours. Since distance = rate · time, we can find the time by dividing the distance by the rate.

Direction	Distance	Rate	Time
Downriver (with current)	12	$r + 2$	$\dfrac{12}{r + 2}$
Upriver (against current)	12	$r - 2$	$\dfrac{12}{r - 2}$

Translate

time downriver + time upriver = total time

$$\frac{12}{r + 2} + \frac{12}{r - 2} = 5$$

Carry Out $(r + 2)(r - 2)\left(\dfrac{12}{r + 2} + \dfrac{12}{r - 2}\right) = (r + 2)(r - 2)(5)$ *Multiply by the LCD.*

$(r + 2)(r - 2)\left(\dfrac{12}{r + 2}\right) + (r + 2)(r - 2)\left(\dfrac{12}{r - 2}\right) = (r + 2)(r - 2)(5)$ *Distributive property*

$$12(r - 2) + 12(r + 2) = 5(r^2 - 4)$$
$$12r - 24 + 12r + 24 = 5r^2 - 20 \quad \text{\textit{Distributive property}}$$
$$24r = 5r^2 - 20 \quad \text{\textit{Simplify.}}$$
$$\text{or} \quad 5r^2 - 24r - 20 = 0$$

Using the quadratic formula with $a = 5$, $b = -24$, and $c = -20$, we obtain

$$r = \frac{24 \pm \sqrt{976}}{10}$$

$$r \approx 5.5 \quad \text{or} \quad r \approx -0.7$$

Answer Since the rate cannot be negative, the rate or speed of the boat in still water is about 5.5 miles per hour.

NOW TRY EXERCISE 43

Notice that in real-life situations most answers are not integral values.

Work Problems

Let's do an example involving a work problem. Work problems were discussed in Section 6.5. You may wish to review that section before studying the next example.

EXAMPLE 4   **Water Removal**   After a hurricane, the Duvals needed to remove water from their flooded basement. They had one sump pump (used to pump out water) and borrowed a second from their local fire department. The fire chief informed the Duvals he expected that with both pumps working together, their basement should empty in about 6 hours. He also informed them that the fire department's pump had a higher horsepower, and it would empty their basement by itself in 2 hours less time than the Duvals' pump could if it were working alone. How long would it take each pump to empty the basement if each were working alone?

Solution   *Understand*   Recall from Section 6.5 that the rate of work multiplied by the time worked gives the part of the task completed.

let t = number of hours for the slower pump to complete the job by itself

$t - 2$ = number of hours for the faster pump to complete the job by itself

Pump	Rate of work	Time worked	Part of task completed
Slower pump	$\dfrac{1}{t}$	6	$\dfrac{6}{t}$
Faster pump	$\dfrac{1}{t - 2}$	6	$\dfrac{6}{t - 2}$

Translate $\left(\begin{array}{c}\text{part of task} \\ \text{by slower pump}\end{array}\right) + \left(\begin{array}{c}\text{part of task} \\ \text{by faster pump}\end{array}\right) = 1$

$$\frac{6}{t} + \frac{6}{t-2} = 1$$

Carry Out

$$t(t-2)\left(\frac{6}{t} + \frac{6}{t-2}\right) = t(t-2)(1)$$

Multiply both sides by the LCD, $t(t-2)$.

$$t(t-2)\left(\frac{6}{t}\right) + t(t-2)\left(\frac{6}{t-2}\right) = t^2 - 2t$$

Distributive property

$$6(t-2) + 6t = t^2 - 2t$$

$$6t - 12 + 6t = t^2 - 2t$$

$$t^2 - 14t + 12 = 0$$

Using the quadratic formula, we obtain

$$t = \frac{14 \pm \sqrt{148}}{2}$$

$$t \approx 13.1 \quad \text{or} \quad t \approx 0.9$$

Answer   Both 13.1 and 0.9 satisfy the equation $\frac{6}{t} + \frac{6}{t-2} = 1$ (with some round-off involved). However, if we accept 0.9 as a solution, then the faster pump could complete the task in a negative time ($t - 2 = 0.9 - 2 = -1.1$ hours), which is not possible. Therefore, 0.9 hour is not an acceptable solution. The only solution is 13.1 hours. The slower pump takes approximately 13.1 hours by itself, and the faster pump takes approximately $13.1 - 2$ or 11.1 hours by itself to empty the basement.

NOW TRY EXERCISE 45

2   Solve for a Variable in a Formula

When the square of a variable appears in a *formula*, you may need to use the square root property to solve for the variable. However, *when you use the square root property in most formulas, you will use only the principal or positive root*, because you are generally solving for a quantity that cannot be negative.

EXAMPLE 5   **a)** The formula for the area of a circle is $A = \pi r^2$. Solve this equation for the radius, r.

b) *Newton's law of universal gravity* states that every particle in the universe attracts every other particle with a force proportional to the product of their masses and inversely proportional to the square of the distance between them. We may represent Newton's law as

$$F = G\frac{m_1 m_2}{r^2}$$

Solve the equation for r.

Solution   **a)**

$$A = \pi r^2$$

$$\frac{A}{\pi} = r^2$$

Isolate r^2 by dividing both sides by π.

$$\sqrt{\frac{A}{\pi}} = r$$

Square root property

b)

$$F = G\frac{m_1 m_2}{r^2}$$

$$r^2 F = G m_1 m_2 \qquad \textit{Multiply both sides of formula by } r^2.$$

$$r^2 = \frac{G m_1 m_2}{F} \qquad \textit{Isolate } r^2 \textit{ by dividing both sides by } F.$$

$$r = \sqrt{\frac{G m_1 m_2}{F}} \qquad \textit{Square root property} \qquad \ast$$

NOW TRY EXERCISE 23

In Example 5, since r must be greater than 0, when we used the square root property, we listed only the principal or positive square root.

EXAMPLE 6 **Diagonal of a Suitcase** The diagonal of a box can be calculated by the formula

$$d = \sqrt{L^2 + W^2 + H^2}$$

where L is the length, W is the width, and H is the height of the box. See Figure 8.5.

a) Find the diagonal of a suitcase of length 30 inches, width 15 inches, and height 10 inches.

b) Solve the equation for the width, W.

FIGURE 8.5

Solution **a)** Understand   To find the diagonal, we need to substitute the appropriate values into the formula and solve for the diagonal, d.

Translate
$$d = \sqrt{L^2 + W^2 + H^2}$$
$$d = \sqrt{(30)^2 + (15)^2 + (10)^2}$$

Carry Out
$$= \sqrt{900 + 225 + 100}$$
$$= \sqrt{1225}$$
$$= 35$$

Answer   Thus, the diagonal of the suitcase is 35 inches.

b) Our first step in solving for W is to square both sides of the formula.

$$d = \sqrt{L^2 + W^2 + H^2}$$
$$d^2 = \left(\sqrt{L^2 + W^2 + H^2}\right)^2 \quad \textit{Square both sides.}$$
$$d^2 = L^2 + W^2 + H^2 \qquad \textit{Use } (\sqrt{a})^2 = a, a \geq 0.$$
$$d^2 - L^2 - H^2 = W^2 \qquad \textit{Isolate } W^2.$$
$$\sqrt{d^2 - L^2 - H^2} = W \qquad \textit{Square root property} \ast$$

EXAMPLE 7 **Traffic Cones** The surface area of a right circular cone is

$$s = \pi r \sqrt{r^2 + h^2}$$

a) An orange traffic cone used on roads is 18 inches high with a radius of 12 inches. Find the surface area of the cone.

b) Solve the formula for h.

Solution　**a)** Understand and Translate　To find the surface area, we substitute the appropriate values into the formula.

$$s = \pi r \sqrt{r^2 + h^2}$$
$$= \pi(12)\sqrt{(12)^2 + (18)^2}$$

Carry Out

$$= 12\pi\sqrt{144 + 324}$$
$$= 12\pi\sqrt{468}$$
$$\approx 815.56$$

Answer　The surface area is about 815.56 square inches.

b) To solve for h we need to isolate h on one side of the equation. There are various ways to solve the equation for h.

$$s = \pi r \sqrt{r^2 + h^2}$$

$$\frac{s}{\pi r} = \sqrt{r^2 + h^2} \qquad \text{\textit{Divide both sides by } } \pi r.$$

$$\left(\frac{s}{\pi r}\right)^2 = \left(\sqrt{r^2 + h^2}\right)^2 \qquad \text{\textit{Square both sides.}}$$

$$\frac{s^2}{\pi^2 r^2} = r^2 + h^2 \qquad \text{\textit{Use }} (\sqrt{a})^2 = a, a \geq 0.$$

$$\frac{s^2}{\pi^2 r^2} - r^2 = h^2 \qquad \text{\textit{Subtract } } r^2 \text{ \textit{from both sides.}}$$

$$\sqrt{\frac{s^2}{\pi^2 r^2} - r^2} = h \qquad \text{\textit{Square root property}}$$

Other acceptable answers are $h = \sqrt{\dfrac{s^2 - \pi^2 r^4}{\pi^2 r^2}}$ and $h = \dfrac{\sqrt{s^2 - \pi^2 r^4}}{\pi r}$. Can you

NOW TRY EXERCISE 27　explain why?

Exercise Set 8.3

Concept/Writing Exercises

1. In general, when solving for a variable in a formula, whether you use the square root property or the quadratic formula, you use only the positive square root. Explain why.

2. Suppose $P = ☺^2 + □^2$ is a real formula. Solving for ☺ gives ☺ $= \sqrt{P - □^2}$. If ☺ is to be a real number, what relationship must exist between P and □?

Practice the Skills

Solve for the indicated variable. Assume the variable you are solving for must be greater than 0.

3. $A = s^2$, for s (area of a square)

4. $A = (s + 1)^2$, for s (area of a square)

5. $A = S^2 - s^2$, for S (area between two squares)

6. $d = 4.9t^2$, for t (distance an object has fallen)

7. $E = i^2 r$, for i (current in electronics)

8. $A = 4\pi r^2$, for r (surface area of a sphere)

9. $d = 16t^2$, for t (distance of a falling object)

10. $d = \dfrac{1}{7}x^2$, for x (stopping distance on pavement)

11. $E = mc^2$, for c (Einstein's famous energy formula)

12. $V = \pi r^2 h$, for r (volume of a right circular cylinder)

13. $V = \dfrac{1}{3}\pi r^2 h$, for r (volume of a right circular cone)

14. $d = \sqrt{L^2 + W^2}$, for L (diagonal of a rectangle)

15. $d = \sqrt{L^2 + W^2}$, for W (diagonal of a rectangle)

16. $a^2 + b^2 = c^2$, for a (Pythagorean Theorem)

🔒 17. $a^2 + b^2 = c^2$, for b (Pythagorean Theorem)

18. $d = \sqrt{L^2 + W^2 + H^2}$, for L (diagonal of a box)

19. $d = \sqrt{L^2 + W^2 + H^2}$, for H (diagonal of a box)

20. $A = P(1 + r)^2$, for r (compound interest formula)

21. $h = -16t^2 + s_0$, for t (height of an object)

22. $P = \dfrac{1}{2}kx^2$, for x (potential energy of a stretched spring)

🔒 23. $E = \dfrac{1}{2}mv^2$, for v (kinetic energy)

24. $f_x^2 + f_y^2 = f^2$, for f_x (forces acting on an object)

25. $a = \dfrac{v_2^2 - v_1^2}{2d}$, for v_1 (acceleration of a vehicle)

26. $A = 4\pi(R^2 - r^2)$, for R (surface area of two spheres)

27. $v' = \sqrt{c^2 - v^2}$, for c (relativity; v' is read "v prime")

28. $L = L_0\sqrt{1 - \dfrac{v^2}{c^2}}$, for v (art, a painting's contraction)

Problem Solving

29. ***Profit*** The profit for the Big Wheels Company, which sells tractors, is $P(n) = 2.4n^2 + 9n - 3$ where $P(n)$ is in hundreds of dollars.

   a) Find the profit when 6 tractors are sold.

   b) How many tractors should be sold to have a profit of $20,000?

30. ***Profit*** The profit for the S.W. Anderson Company, which sells refrigerators, is $P(n) = 6.2n^2 + 6n - 3$ where $P(n)$ is in dollars.

   a) Find the profit when 5 refrigerators are sold.

   b) How many refrigerators should be sold to have a profit of $675?

31. ***Temperature*** The temperature, T, in degrees Fahrenheit, in a car's radiator during the first 4 minutes of driving is a function of time, t. The temperature can be found by the formula $T = 6.2t^2 + 12t + 32$, $0 \le t \le 4$.

   a) What is the car radiator's temperature at the instant the car is turned on?

   b) What is the car radiator's temperature after the car has been driven for 1 minute?

   c) How long after the car has begun operating will the car radiator's temperature reach 120°F?

32. ***School Enrollment*** The function $N(t) = -0.043t^2 + 1.22t + 46.0$ can be used to estimate total U.S. elementary and secondary school enrollment, in millions, between the years 1990 and 2008. In the equation t is years since 1989 and $1 \le t \le 19$.

   a) Estimate total enrollment in 1995.

   b) In what years is the total enrollment 54 million students?

33. ***Cost of a House*** The cost, C, of a ranch house in Norfolk, Virginia, is a function of the number of square feet of the house, s. The cost of a house can be approximated by the formula

   $$C = -0.01s^2 + 80s + 20{,}000, \qquad 1200 \le s \le 4000$$

   a) Estimate the cost of a 1500-square-foot house.

   b) How large a house can Tom Boyle purchase if he has $150,000 to spend on a house?

34. ***Grade Point Average*** At a college, records show that the average person's grade point average, G, is a function of the number of hours he or she studies and does homework per week, h. The grade point average can

be estimated by the equation $G = 0.01h^2 + 0.2h + 1.2$, $0 \leq h \leq 8$.

a) What is the GPA of the average student who studies for 0 hours a week?

b) What is the GPA of the average student who studies 4 hours per week?

c) To obtain a 3.2 GPA, how many hours per week would the average student need to study?

35. *Roller Coaster* The roller coaster at Busch Gardens in Williamsburg, Virginia, has a number of steep drops. One drop has a vertical distance of 62 feet. The speed, s, of the last car, in feet per second, t seconds after it has begun its drop, can be calculated by the formula $s = 6.74t + 2.3$, $0 \leq t \leq 4$. The height of the last car from the bottom of the drop t seconds after it has begun this drop can be found by the formula $h = -3.3t^2 - 2.3t + 62$, $0 \leq t \leq 4$.

a) Find the time it takes for the last car to travel from the top of the drop to the bottom of the drop.

b) Find the speed of the last car when it reaches the bottom of the drop.

36. *Drug-Free Schools* The following graph summarizes data on the percent of students at various ages who say their school is not drug-free.

Students Who Say Their School Is Not Drug Free

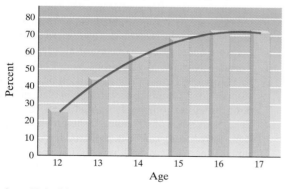

Source: National Center on Addiction and Substance Abuse

The function $f(a) = -2.32a^2 + 76.58a - 559.87$ can be used to estimate the percent of students who say their school is not drug free. In the function, a represents the student's age, where $12 \leq a \leq 17$. Use the function to answer the following questions.

a) Estimate the percent of 13-year-olds who say their school is not drug free.

b) At what age do 70% of the students say their school is not drug free?

37. *Veterinary Bills* The following graph illustrates veterinary expenses for dogs and cats.

Veterinary Expenditures

Source: American Veterinary Medical Association

The annual amount spent in the United States on veterinary bills for dogs, $m(t)$, in billions of dollars, can be estimated by the function $m(t) = 0.05t^2 - 0.32t + 3.15$, where t represents the number of years since 1982 and $1 \leq t \leq 28$.

a) Use the function to estimate the amount spent on veterinary bills for dogs in 2003.

b) In what year will $25 billion be spent on veterinary bills for dogs?

38. *Profit* A video store's weekly profit, P, in thousands of dollars, is a function of the rental price of the tapes, t. The profit equation is $P = 0.2t^2 + 1.5t - 1.2$, $0 \leq t \leq 5$.

a) What is the store's weekly profit or loss if they charge $1 per tape?

b) What is the weekly profit if they charge $5 per tape?

c) At what tape rental price will their weekly profit be 1.6 thousand?

39. *Playground* The area of a children's rectangular playground is 500 square meters. The length is 5 meters longer than the width. Find The length and width of the playground.

40. *Travel* Hana Juarez drove for 70 miles in heavy traffic. She then reached the highway where she drove 260 miles at an average speed that was 30 miles per hour greater than the average speed she drove in heavy traffic. If the total trip took 6 hours, determine her average speed in heavy traffic and on the highway.

41. *Drilling a Well* Paul and Rima Jones, who live in Cedar Rapids, Iowa, want a well on their property. They hired the Ruth Cardiff Drilling Company to drill the well. The company had to drill 64 feet to hit water. The company informed the Jones that they had just ordered new drilling equipment that drills at an average of 1 foot per hour faster, and that with their new equipment, they would have hit water in 3.2 hours less time. Find the rate at which their present equipment drills.

42. ***Car Carrier*** Frank Sims, a truck driver, was transporting a heavy load of new cars on a car carrier from Detroit, Michigan, to Indianapolis, Indiana. On his return trip to Detroit, since his truck was lighter, he averaged 10 miles per hour faster than on his trip to Indianapolis. If the total distance traveled each way was 300 miles and the total time he spent driving was 11 hours, find his average speed going and returning.

43. ***Runner*** Latoya Williams, a long-distance runner, starts jogging at her house. She jogs 6 miles and then turns around and jogs back to her house. The first part of her jog is mostly uphill, so her speed averages 2 miles per hour less than her returning speed. If the total time she spends jogging is $1\frac{3}{4}$ hours, find her speed going and her speed returning.

44. ***Garden of the Gods*** Kathy Nickell traveled from near Garden of the Gods in Colorado Springs, Colorado, to just outside Grand Junction, Colorado. The total distance she traveled was 300 miles. After she got to Grand Junction, she figured out that had she averaged 10 miles per hour faster, she would have arrived 1 hour earlier. Find the average speed that Kathy drove.

45. ***Build an Engine*** Two mechanics, Bonita Rich and Pamela Pearson, take 6 hours to rebuild an engine when they work together. If each worked alone, Bonita, the more experienced mechanic, could complete the job 1 hour faster than Pamela. How long would it take each of them to rebuild the engine working alone?

46. ***Riding a Bike*** Richard Semmler enjoys riding his bike from Washington, D.C., to Bethesda, Maryland, and back, a total of 30 miles on the Capital Crescent path. The trip to Bethesda is uphill most of the distance. The bike's average speed going to Bethesda is 5 miles per hour slower than the average speed returning to D.C. If the round-trip takes 4.5 hours, find the average speed in each direction.

47. ***Flying a Plane*** Dole Rohm flew his single-engine Cessna airplane 80 miles with the wind from Jackson Hole, Wyoming, to above Blackfoot, Idaho. He then turned around and flew back to Jackson Hole against the wind. If the wind was a constant 30 miles per hour, and the total time going and returning was 1.3 hours, find the speed of the plane in still air.

48. ***Ships*** After a small oil spill, two cleanup ships are sent to siphon off the oil floating in Baffin Bay. The newer ship can clean up the entire spill by itself in 3 hours less time than the older ship takes by itself. Working together the two ships can clean up the oil spill in 8 hours. How long will it take the newer ship by itself to clean up the spill?

49. ***Janitorial Service*** The O'Connors own a small janitorial service. John requires $\frac{1}{2}$ hour more time to clean the Moose Club by himself than Chris does working by herself. If together they can clean the club in 6 hours, find the time required by each to clean the club.

50. ***Electric Heater*** A small electric heater requires 6 minutes longer to raise the temperature in an unheated garage to a comfortable level than does a larger electric heater. Together the two heaters can raise the garage temperature to a comfortable level in 42 minutes. How long would it take each heater by itself to raise the temperature in the garage to a comfortable level?

51. ***Travel*** Shywanda Moore drove from Lubbock, Texas, to Plainview, Texas, a distance of 60 miles. She then stopped for 2.5 hours to see a friend in Plainview before continuing her journey from Plainview to Amarillo, Texas, a distance of 100 miles. If she drove 10 miles per hour faster from Lubbock to Plainview and the total time of the trip was 5.5 hours, find her average speed from Lubbock to Plainview.

52. ***Travel*** Lewis and his friend George are traveling from Memphis, Tennessee, to Richmond, Virginia. Lewis travels

Graceland in Memphis, TN

by car while George travels by train. The train and car leave Memphis at the same time from the same point. During the trip Lewis and George speak by cellular phone, and Lewis informs George that he has just stopped for the evening after traveling for 500 miles. One and two-thirds hours later, George calls Lewis and informs him that the train had just reached Richmond, a distance of 800 miles from Memphis. Assuming the train averaged 20 miles per hour faster than the car, find the average speed of the car and the train.

53. Write your own motion problem and solve it.

54. Write your own work problem and solve it.

Challenge Problems

55. *Area* The area of a rectangle is 18 square meters. When the length is increased by 2 meters and the width by 3 meters, the area becomes 48 square meters. Find the dimensions of the smaller rectangle.

56. *Area* The area of a rectangle is 24 square inches. When the length is decreased by 1 inch and the width is increased by 1 inch, the area of the new rectangle is 25 square inches. Find the dimensions of the original rectangle.

Cumulative Review Exercises

[1.4] **57.** Evaluate $-[4(5-3)^3] + 2^3$.

[2.2] **58.** Solve $IR + Ir = E$ for R.

[6.2] **59.** Add $\dfrac{2r}{r-4} - \dfrac{2r}{r+4} + \dfrac{64}{r^2-16}$.

[7.2] **60.** Simplify $\left(\dfrac{x^{3/4}y^{-2}}{x^{1/2}y^2}\right)^4$.

[7.6] **61.** Solve $\sqrt{x^2 + 3x + 9} = x$.

8.4 WRITING EQUATIONS IN QUADRATIC FORM

1 Solve equations that are quadratic in form.
2 Solve equations with rational exponents.

1 Solve Equations That Are Quadratic in Form

Sometimes we need to solve an equation that is not a quadratic equation but can be rewritten in the form of a quadratic equation. We can then solve the equation in quadratic form by factoring, completing the square, or the quadratic formula.

DEFINITION | An equation that can be written in the form $au^2 + bu + c = 0$ for $a \neq 0$, where u is an algebraic expression, is called **quadratic in form**.

When you are given an equation quadratic in form, make a substitution to get the equation in the form $au^2 + bu + c = 0$. In general, if the exponents are positive, the substitution to make is to let u be the middle term, without the numerical coefficient, when the expression is listed in descending order of the variable. For example,

Equation quadratic in form	Substitution	Equation with substitution
$y^4 - y^2 - 6 = 0$	$u = y^2$	$u^2 - u - 6 = 0$
$2(x+5)^2 - 5(x+5) - 12 = 0$	$u = x+5$	$2u^2 - 5u - 12 = 0$

To solve equations quadratic in form, we use the following procedure. We will illustrate this procedure in Example 1.

To Solve Equations Quadratic in Form

1. Make a substitution that will result in an equation of the form $au^2 + bu + c = 0$, $a \neq 0$, where u is a function of the original variable.

2. Solve the equation $au^2 + bu + c = 0$ for u.

3. Replace u with the function of the original variable from step 1 and solve the resulting equation for the original variable.

4. Check for extraneous solutions by substituting the apparent solutions into the original equation.

EXAMPLE 1 **a)** Solve the equation $x^4 - 5x^2 + 4 = 0$.

b) Find the x-intercepts of the function $f(x) = x^4 - 5x^2 + 4$.

Solution **a)** To obtain an equation that is quadratic in form, write x^4 as $(x^2)^2$.

$$x^4 - 5x^2 + 4 = 0$$

$$(x^2)^2 - 5x^2 + 4 = 0 \qquad \text{\textit{Replace } } x^4 \text{ \textit{with} } (x^2)^2 \text{ \textit{to obtain} an equation in desired form.}$$

Now let $u = x^2$. This gives an equation that is quadratic in form.

$$u^2 - 5u + 4 = 0 \qquad \text{\textit{Substitute } u \textit{ for } x^2.}$$

$$(u - 4)(u - 1) = 0 \qquad \text{\textit{Solve for } u.}$$

$$u - 4 = 0 \qquad \text{or} \qquad u - 1 = 0$$

$$u = 4 \qquad\qquad u = 1$$

$$x^2 = 4 \qquad\qquad x^2 = 1 \qquad \text{\textit{Replace } u \textit{ with } x^2.}$$

$$x = \pm\sqrt{4} \qquad\qquad x = \pm\sqrt{1} \qquad \text{\textit{Solve for } x.}$$

$$x = \pm 2 \qquad\qquad x = \pm 1$$

Check the four possible solutions in the original equation.

$x = 2$	$x = -2$	$x = 1$	$x = -1$
$x^4 - 5x^2 + 4 = 0$	$x^4 - 5x^2 + 4 = 0$	$x^4 - 5x^2 + 4 = 0$	$x^4 - 5x^2 + 4 = 0$
$2^4 - 5(2)^2 + 4 \overset{?}{=} 0$	$(-2)^4 - 5(-2)^2 + 4 \overset{?}{=} 0$	$1^4 - 5(1)^2 + 4 \overset{?}{=} 0$	$(-1)^4 - 5(-1)^2 + 4 \overset{?}{=} 0$
$16 - 20 + 4 \overset{?}{=} 0$	$16 - 20 + 4 \overset{?}{=} 0$	$1 - 5 + 4 \overset{?}{=} 0$	$1 - 5 + 4 \overset{?}{=} 0$
$0 = 0$	$0 = 0$	$0 = 0$	$0 = 0$
True	*True*	*True*	*True*

$-3, 3, 1, -3, 6, 1$

FIGURE 8.6

Thus, the solutions are 2, -2, 1, and -1.

b) The x-intercepts occur where $f(x) = 0$. Therefore, the graph will cross the x-axis at the solutions to the equation $x^4 - 5x^2 + 4 = 0$.

From part **a)**, we know the solutions are 2, -2, 1, and -1. Thus, the x-intercepts are $(2, 0)$, $(-2, 0)$, $(1, 0)$, and $(-1, 0)$. Figure 8.6 is the graph of $f(x) = x^4 - 5x^2 + 4 = 0$ as illustrated on a graphing calculator. Notice that the graph crosses the x-axis at $x = 2$, $x = -2$, $x = 1$, and $x = -1$. ✳

EXAMPLE 2 Solve the equation $p^4 + 2p^2 = 8$.

Solution

$$p^4 + 2p^2 - 8 = 0 \qquad \text{\textit{Set equation equal to 0.}}$$

$$(p^2)^2 + 2p^2 - 8 = 0 \qquad \text{\textit{Write } } p^4 \text{ \textit{as} } (p^2)^2 \text{ \textit{to obtain} equation in desired form.}}$$

Now let $u = p^2$. This gives an equation that is quadratic in form.

$$u^2 + 2u - 8 = 0 \qquad \text{\textit{Substitute u for } } p^2.$$
$$(u + 4)(u - 2) = 0 \qquad \text{\textit{Solve the equation for u.}}$$

$$u + 4 = 0 \qquad \text{or} \qquad u - 2 = 0$$
$$u = -4 \qquad\qquad\qquad u = 2$$
$$p^2 = -4 \qquad\qquad\qquad p^2 = 2 \qquad \text{\textit{Replace u with } } p^2.$$
$$p = \pm\sqrt{-4} \qquad\qquad\qquad p = \pm\sqrt{2} \qquad \text{\textit{Solve for p.}}$$
$$p = \pm 2i$$

Check the four possible solutions in the *original* equation.

$p = 2i$	$p = -2i$	$p = \sqrt{2}$	$p = -\sqrt{2}$
$p^4 + 2p^2 = 8$	$p^4 + 2p^2 = 8$	$p^4 + 2p^2 = 8$	$p^4 + 2p^2 = 8$
$(2i)^4 + 2(2i)^2 \stackrel{?}{=} 8$	$(-2i)^4 + 2(-2i)^2 \stackrel{?}{=} 8$	$(\sqrt{2})^4 + 2(\sqrt{2})^2 \stackrel{?}{=} 8$	$(-\sqrt{2})^4 + 2(-\sqrt{2})^2 \stackrel{?}{=} 8$
$2^4 i^4 + 2(2^2)(i^2) \stackrel{?}{=} 8$	$(-2)^4 i^4 + 2(-2)^2 i^2 \stackrel{?}{=} 8$	$4 + 2(2) \stackrel{?}{=} 8$	$4 + 2(2) \stackrel{?}{=} 8$
$16(1) + 8(-1) \stackrel{?}{=} 8$	$16(1) + 8(-1) \stackrel{?}{=} 8$	$8 = 8$	$8 = 8$
$16 - 8 = 8$	$16 - 8 = 8$	*True*	*True*
True	*True*		

Thus, the solutions are $2i$, $-2i$, $\sqrt{2}$, and $-\sqrt{2}$. ✳

The solutions to equations like $p^4 + 2p^2 = 8$ will always check unless a mistake has been made. In equations like this, extraneous solutions will not be introduced. However, extraneous solutions *may* be introduced when working with rational exponents, as will be shown in Example 6.

NOW TRY EXERCISE 17

HELPFUL HINT

> Students sometimes solve the equation for u but then forget to complete the problem by solving for the original variable. Remember that if the original equation is in x you must obtain values for x. If the original equation is in p (as in Example 2) you must obtain values for p, and so on.

EXAMPLE 3 Solve the equation $4(2w + 1)^2 - 16(2w + 1) + 15 = 0$.

Solution If we let $u = 2w + 1$, the equation becomes

$$4(2w + 1)^2 - 16(2w + 1) + 15 = 0$$
$$4u^2 - 16u + 15 = 0 \qquad \text{\textit{Substitute u for } } 2w + 1.$$

Now we can factor and solve.

$$(2u - 3)(2u - 5) = 0$$
$$2u - 3 = 0 \qquad \text{or} \qquad 2u - 5 = 0$$
$$2u = 3 \qquad\qquad\qquad 2u = 5$$
$$u = \frac{3}{2} \qquad\qquad\qquad u = \frac{5}{2}$$

We are not finished. Since the variable in the original equation is w, we must solve for w, not u. Therefore, we substitute back $2w + 1$ for u and solve for w.

$$u = \frac{3}{2} \qquad\qquad u = \frac{5}{2}$$

$$2w + 1 = \frac{3}{2} \qquad 2w + 1 = \frac{5}{2} \qquad \text{Substitute } 2w + 1 \text{ for } u.$$

$$2w = \frac{1}{2} \qquad\qquad 2w = \frac{3}{2}$$

$$w = \frac{1}{4} \qquad\qquad w = \frac{3}{4}$$

NOW TRY EXERCISE 29 A check will show that both $\frac{1}{4}$ and $\frac{3}{4}$ are solutions to the original equation. ✳

EXAMPLE 4 Find the x-intercepts of the function $f(x) = 2x^{-2} + x^{-1} - 1$.

Solution The x-intercepts occur where $f(x) = 0$. Therefore, to find the x-intercepts we must solve the equation

$$2x^{-2} + x^{-1} - 1 = 0$$

This equation can be expressed as

$$2(x^{-1})^2 + x^{-1} - 1 = 0$$

When we let $u = x^{-1}$, the equation becomes

$$2u^2 + u - 1 = 0$$
$$(2u - 1)(u + 1) = 0$$
$$2u - 1 = 0 \quad \text{or} \quad u + 1 = 0$$
$$u = \frac{1}{2} \qquad\qquad u = -1$$

Now we substitute x^{-1} for u.

$$x^{-1} = \frac{1}{2} \quad \text{or} \quad x^{-1} = -1$$
$$\frac{1}{x} = \frac{1}{2} \qquad\qquad \frac{1}{x} = -1$$
$$x = 2 \qquad\qquad x = -1$$

A check will show that both 2 and -1 are solutions to the original equation. Thus, the x-intercepts are $(2, 0)$ and $(-1, 0)$. ✳

The equation in Example 4 could also be expressed as

$$\frac{2}{x^2} + \frac{1}{x} - 1 = 0$$

A second method to solve this equation is to multiply both sides of the equation by the least common denominator, x^2, then simplify.

$$x^2\left(\frac{2}{x^2} + \frac{1}{x} - 1\right) = x^2 \cdot 0$$
$$2 + x - x^2 = 0$$
$$x^2 - x - 2 = 0$$
$$(x - 2)(x + 1) = 0$$
$$x - 2 = 0 \quad \text{or} \quad x + 1 = 0$$
$$x = 2 \qquad\qquad x = -1$$

NOW TRY EXERCISE 61 Many of the equations solved in this section may be solved by more than one method.

2 Solve Equations with Rational Exponents

When solving equations that are quadratic in form with rational exponents, we raise both sides of the equation to some power to eliminate the rational exponents. Recall that we did this in Section 7.6 when we solved radical equations. Whenever you raise both sides of an equation to a power, you may introduce extraneous solutions. **Therefore, whenever you raise both sides of an equation to a power, you must check all apparent solutions in the original equation to make sure that none are extraneous.** We will now work two examples showing how to solve equations that contain rational exponents. We use the same procedure as used earlier.

EXAMPLE 5   Solve the equation $x^{2/5} + x^{1/5} - 6 = 0$.

Solution   This equation can be rewritten as

$$(x^{1/5})^2 + x^{1/5} - 6 = 0$$

Let $u = x^{1/5}$. Then the equation becomes

$$u^2 + u - 6 = 0$$
$$(u + 3)(u - 2) = 0$$
$$u + 3 = 0 \quad \text{or} \quad u - 2 = 0$$
$$u = -3 \qquad\qquad u = 2$$

Now substitute $x^{1/5}$ for u and raise both sides of the equation to the fifth power to remove the rational exponents.

$$x^{1/5} = -3 \quad \text{or} \quad x^{1/5} = 2$$
$$(x^{1/5})^5 = (-3)^5 \qquad (x^{1/5})^5 = 2^5$$
$$x = -243 \qquad\qquad x = 32$$

The two *possible* solutions are -243 and 32. Remember that whenever you raise both sides of an equation to a power, as you did here, you need to check for extraneous solutions.

Check $x = -243$ $x = 32$

$$x^{2/5} + x^{1/5} - 6 = 0 \qquad\qquad\qquad x^{2/5} + x^{1/5} - 6 = 0$$
$$(-243)^{2/5} + (-243)^{1/5} - 6 \overset{?}{=} 0 \qquad (32)^{2/5} + (32)^{1/5} - 6 \overset{?}{=} 0$$
$$(\sqrt[5]{-243})^2 + \sqrt[5]{-243} - 6 \overset{?}{=} 0 \qquad (\sqrt[5]{32})^2 + \sqrt[5]{32} - 6 \overset{?}{=} 0$$
$$(-3)^2 - 3 - 6 \overset{?}{=} 0 \qquad\qquad\qquad 2^2 + 2 - 6 \overset{?}{=} 0$$
$$9 - 3 - 6 \overset{?}{=} 0 \qquad\qquad\qquad 4 + 2 - 6 \overset{?}{=} 0$$
$$0 = 0 \;\; \textit{True} \qquad\qquad\qquad 0 = 0 \;\; \textit{True}$$

NOW TRY EXERCISE 63   Since both values check, the solutions are -243 and 32.   ✳

EXAMPLE 6   Solve the equation $2p - \sqrt{p} - 10 = 0$.

Solution   We can express this equation as

$$2p - p^{1/2} - 10 = 0$$
$$2(p^{1/2})^2 - p^{1/2} - 10 = 0$$

If we let $u = p^{1/2}$, this equation is quadratic in form.

$$2u^2 - u - 10 = 0$$
$$(2u - 5)(u + 2) = 0$$

$$2u - 5 = 0 \quad \text{or} \quad u + 2 = 0$$
$$2u = 5 \qquad\qquad u = -2$$
$$u = \frac{5}{2}$$

However, since our original equation is in the variable p, we must solve for p. We substitute $p^{1/2}$ for u.

$$p^{1/2} = \frac{5}{2} \qquad\qquad p^{1/2} = -2$$

Now we square both sides of the equation.

$$(p^{1/2})^2 = \left(\frac{5}{2}\right)^2 \qquad (p^{1/2})^2 = (-2)^2$$

$$p = \frac{25}{4} \qquad\qquad p = 4$$

We must now check both apparent solutions in the original equation.

Check
$$p = \frac{25}{4} \qquad\qquad\qquad p = 4$$

$$2p - \sqrt{p} - 10 = 0 \qquad\qquad 2p - \sqrt{p} - 10 = 0$$

$$2\left(\frac{25}{4}\right) - \sqrt{\frac{25}{4}} - 10 \overset{?}{=} 0 \qquad\qquad 2(4) - \sqrt{4} - 10 \overset{?}{=} 0$$

$$\frac{25}{2} - \frac{5}{2} - 10 \overset{?}{=} 0 \qquad\qquad 8 - 2 - 10 \overset{?}{=} 0$$

$$0 = 0 \quad \text{\textit{True}} \qquad\qquad -4 = 0 \quad \text{\textit{False}}$$

NOW TRY EXERCISE 25   Since 4 does not check, it is an extraneous solution. The only solution is $\frac{25}{4}$.   ✳

Example 6 could also be solved by writing the equation as $\sqrt{p} = 2p - 10$ and squaring both sides of the equation. Try this now. If you have forgotten how to do this, review Section 7.6.

Exercise Set 8.4

Concept/Writing Exercises

1. Explain how you can determine whether a given equation can be expressed as an equation that is quadratic in form.

2. When solving an equation that is quadratic in form, when is it essential to check your answer for extraneous solutions? Explain why.

3. To solve the equation $3x^4 - 5x^2 + 1 = 0$, what is the correct choice for u to obtain an equation that is quadratic in form? Explain.

4. To solve the equation $2y^{2/3} + 9y^{1/3} - 7 = 0$, what is the correct choice for u to obtain an equation that is quadratic in form? Explain.

5. To solve the equation $z^{-2} - z^{-1} = 56$, what is the correct choice for u to obtain an equation that is quadratic in form? Explain.

6. To solve the equation $3\left(\frac{x+1}{x+3}\right)^2 + \left(\frac{x+1}{x+3}\right) - 9 = 0$, what is the correct choice for u to obtain an equation that is quadratic in form? Explain.

Practice the Skills

Solve each equation.

7. $x^4 - 10x^2 + 9 = 0$ **8.** $x^4 + 10x^2 + 9 = 0$ **9.** $x^4 - 26x^2 + 25 = 0$

10. $x^4 + 26x^2 + 25 = 0$ **11.** $x^4 - 13x^2 + 36 = 0$ **12.** $x^4 + 13x^2 + 36 = 0$

13. $a^4 - 7a^2 + 12 = 0$ **14.** $b^4 + 7b^2 + 12 = 0$ **15.** $4x^4 - 5x^2 + 1 = 0$

16. $9d^4 - 10d^2 + 1 = 0$ **17.** $r^4 - 8r^2 = -15$ **18.** $p^4 - 5p^2 = -6$

19. $z^4 - 7z^2 = 18$ **20.** $a^4 - a^2 = 30$ **21.** $-c^4 = 4c^2 - 5$

22. $9b^4 = 57b^2 - 18$ **23.** $\sqrt{x} = 2x - 6$ **24.** $2b + 7\sqrt{b} = 22$

25. $x + \sqrt{x} = 6$ **26.** $2x - 8 = -6\sqrt{x}$ **27.** $9x + 3\sqrt{x} = 2$

28. $8x + 2\sqrt{x} = 3$ **29.** $(x + 3)^2 + 2(x + 3) = 24$ **30.** $(x + 2)^2 + 4(x + 2) + 3 = 0$

31. $6(a - 2)^2 = -19(a - 2) - 10$ **32.** $10(z - 1)^2 = 3(z - 1) + 1$ **33.** $(x^2 - 1)^2 - (x^2 - 1) - 6 = 0$

34. $(a^2 - 1)^2 - 5(a^2 - 1) - 14 = 0$ **35.** $2(b + 2)^2 + 5(b + 2) - 3 = 0$ **36.** $(z^2 - 5)^2 + 3(z^2 - 5) - 10 = 0$

37. $18(x^2 - 5)^2 + 27(x^2 - 5) + 10 = 0$ **38.** $28(x^2 - 8)^2 - 23(x^2 - 8) - 15 = 0$ **39.** $x^{-2} + 10x^{-1} + 25 = 0$

40. $a^{-2} + 6a^{-1} + 9 = 0$ **41.** $6b^{-2} - 5b^{-1} + 1 = 0$ **42.** $5x^{-2} + 4x^{-1} - 1 = 0$

43. $2b^{-2} = 7b^{-1} - 3$ **44.** $12z^{-2} - 4z^{-1} - 1 = 0$ **45.** $x^{-2} + 9x^{-1} = 10$

46. $6a^{-2} = a^{-1} + 12$ **47.** $x^{-2} = 4x^{-1} + 12$ **48.** $x^{2/3} - 5x^{1/3} + 6 = 0$

49. $x^{2/3} - 3x^{1/3} = -2$ **50.** $x^{2/3} = 2x^{1/3} + 3$ **51.** $b^{2/3} + 11b^{1/3} + 28 = 0$

52. $c^{2/3} - 16 = 0$ **53.** $-2a - 5a^{1/2} + 3 = 0$ **54.** $r^{2/3} - 7r^{1/3} + 6 = 0$

55. $c^{2/5} - 3c^{1/5} + 2 = 0$ **56.** $x^{2/5} - 5x^{1/5} + 6 = 0$

Find all x-intercepts of each function.

57. $f(x) = x - 5\sqrt{x} + 4$ **58.** $g(x) = x - 13\sqrt{x} + 36$

59. $h(x) = x + 13\sqrt{x} + 36$ **60.** $k(x) = x + 8\sqrt{x} + 15$

61. $p(x) = 4x^{-2} - 19x^{-1} - 5$ **62.** $g(x) = 4x^{-2} + 12x^{-1} + 9$

63. $f(x) = x^{2/3} + x^{1/3} - 6$ **64.** $f(x) = x^{1/2} + 4x^{1/4} - 5$

65. $g(x) = (x^2 - 3x)^2 + 2(x^2 - 3x) - 24$ **66.** $g(x) = (x^2 - 6x)^2 - 5(x^2 - 6x) - 24$

67. $f(x) = x^4 - 20x^2 + 64$ **68.** $h(x) = x^4 - 6x^2 + 5$

Problem Solving

69. Give a general procedure for solving an equation of the form $ax^4 + bx^2 + c = 0$.

70. Give a general procedure for solving an equation of the form $ax^{2n} + bx^n + c = 0$.

71. Give a general procedure for solving an equation of the form $ax^{-2} + bx^{-1} + c = 0$.

72. Give a general procedure for solving an equation of the form $a(x - r)^2 + b(x - r) - c = 0$.

73. Write an equation of the form $ax^4 + bx^2 + c = 0$ that has solutions ± 2 and ± 4. Explain how you obtained your answer.

74. Write an equation of the form $ax^4 + bx^2 + c = 0$ that has solutions ± 3 and $\pm 2i$. Explain how you obtained your answer.

75. Write an equation of the form $ax^4 + bx^2 + c = 0$ that has solutions $\pm\sqrt{2}$ and $\pm\sqrt{3}$. Explain how you obtained your answer.

76. Write an equation of the form $ax^4 + bx^2 + c = 0$ that has solutions $\pm 2i$ and $\pm 5i$. Explain how you obtained your answer.

77. Is it possible for an equation of the form $ax^4 + bx^2 + c = 0$ to have exactly one imaginary solution? Explain.

78. Is it possible for an equation of the form $ax^4 + bx^2 + c = 0$ to have exactly one real solution? Explain.

79. Solve the equation $\dfrac{3}{x^2} - \dfrac{3}{x} = 60$ by

     a) multiplying both sides of the equation by the LCD.

     b) writing the equation with negative exponents.

80. Solve the equation $1 = \dfrac{2}{x} - \dfrac{2}{x^2}$ by

     a) multiplying both sides of the equation by the LCD.

     b) writing the equation with negative exponents.

Find all real solutions to each equation.

81. $15(r + 2) + 22 = -\dfrac{8}{r + 2}$

82. $2(p + 3) - 4 = \dfrac{3}{p + 3}$

83. $4 - (x - 2)^{-1} = 3(x - 2)^{-2}$

84. $3(x - 4)^{-2} = 16(x - 4)^{-1} + 12$

85. $x^6 - 9x^3 + 8 = 0$

86. $x^6 - 3x^3 - 40 = 0$

87. $(x^2 + 2x - 2)^2 - 7(x^2 + 2x - 2) + 6 = 0$

88. $(x^2 + 3x - 2)^2 - 10(x^2 + 3x - 2) + 16 = 0$

Find all solutions to each equation.

89. $2n^4 - 6n^2 - 3 = 0$

90. $3x^4 + 8x^2 - 1 = 0$

Cumulative Review Exercises

[1.3] **91.** Evaluate $\dfrac{4}{5} - \left(\dfrac{3}{4} - \dfrac{2}{3}\right)$.

[2.1] **92.** Solve $6(x + 4) - 4(3x + 3) = 6$.

[3.2] **93.** State the domain and range for $y = (x - 2)^2$.

[7.3] **94.** Simplify $\sqrt[3]{16x^3y^6}$.

[7.5] **95.** Add $\sqrt{75} + \sqrt{108}$.

8.5 GRAPHING QUADRATIC FUNCTIONS

SSM   Study Guide   CD/Video

MathPro 4/5   PH Math Tutor Center   prenhall.com/Angel

1 Determine when a parabola opens upward or downward.

2 Find the axis of symmetry, vertex, and *x*-intercepts of a parabola.

3 Graph quadratic functions using the axis of symmetry, vertex, and intercepts.

4 Solve maximum and minimum problems.

5 Understand translations of parabolas.

6 Write functions in the form $f(x) = a(x - h)^2 + k$.

We graphed quadratic equations by plotting points in Section 3.2, and we had a brief discussion of the *x*-intercepts of the graphs of quadratic functions in Section 5.8. In this section we study the graphs of quadratic functions, called **parabolas**, in more depth. In objective 3, we will explain how to graph quadratic functions using the axis of symmetry, vertex, and intercepts. In objective 5, we will study patterns in the graphs of parabolas and use these patterns to determine translations, or shifts, that can be used to graph parabolas.

1 Determine When a Parabola Opens Upward or Downward

Parabolas have a shape that resembles, but is not the same as, the letter U. Parabolas may open upward or downward. For a quadratic function of the form $f(x) = ax^2 + bx + c$, the *sign* of the leading coefficient, *a*, determines whether a parabola opens upward or downward. *When a > 0, the parabola opens upward* (see Fig. 8.7a). *When a < 0, the parabola opens downward* (see Fig. 8.7b).

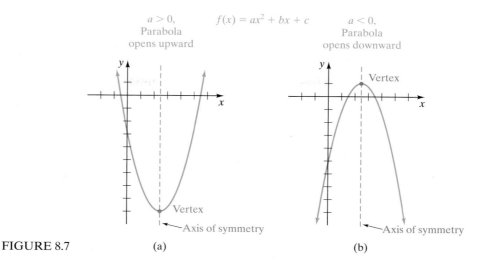

FIGURE 8.7   (a)   (b)

For a parabola that opens upward, the **vertex** is the lowest point on the curve. The minimum value of the function is the y-coordinate of the vertex. The minimum value is obtained when the x-coordinate of the vertex is substituted into the function. For a parabola that opens downward, the vertex is the highest point on the curve. The maximum value of the function is the y-coordinate of the vertex. The maximum value is obtained when the x-coordinate of the vertex is substituted into the function.

2 Find the Axis of Symmetry, Vertex, and x-Intercepts of a Parabola

Graphs of quadratic functions of the form $f(x) = ax^2 + bx + c$ will have **symmetry** about a vertical line through the vertex. This means that if we fold the paper along this imaginary line, called the **axis of symmetry**, the right and left sides of the graph will coincide (see Fig. 8.7). We will now give the equation for finding the axis of symmetry.

To Find the Axis of Symmetry

For a function of the form $f(x) = ax^2 + bx + c$, the equation of the **axis of symmetry**, of the parabola is

$$x = -\frac{b}{2a}$$

Now we will derive the formula for the axis of symmetry, and find the coordinates of the vertex of a parabola, by beginning with a quadratic function of the form $f(x) = ax^2 + bx + c$ and completing the square on the first two terms.

$$f(x) = ax^2 + bx + c$$
$$= a\left(x^2 + \frac{b}{a}x\right) + c \qquad \textit{Factor out a.}$$

Half the coefficient of x is $\dfrac{b}{2a}$. Its square is $\dfrac{b^2}{4a^2}$. Add and subtract this term inside the parentheses. The sum of these two terms is zero.

$$f(x) = a\left[x^2 + \frac{b}{2a}x + \frac{b^2}{4a^2} - \frac{b^2}{4a^2}\right] + c$$

Now rewrite the function as follows.

$$f(x) = \boxed{a}\left[x^2 + \frac{b}{a}x + \left(\frac{b^2}{4a^2}\right)\right] - \boxed{a}\left(\frac{b^2}{4a^2}\right) + c$$

$$= a\left(x + \frac{b}{2a}\right)^2 - \frac{b^2}{4a} + c \qquad \text{Replace the trinomial with the square of a binomial.}$$

$$= a\left(x + \frac{b}{2a}\right)^2 - \frac{b^2}{4a} + \frac{4ac}{4a} \qquad \text{Write fractions with a common denominator.}$$

$$= a\left(x + \frac{b}{2a}\right)^2 + \frac{4ac - b^2}{4a} \qquad \text{Combine the last two terms; write with the variable } a \text{ first.}$$

$$= a\left[x - \left(-\frac{b}{2a}\right)\right]^2 + \frac{4ac - b^2}{4a}$$

The expression $\left[x - \left(-\frac{b}{2a}\right)\right]^2$ will always be greater than or equal to 0. (Why?) If $a > 0$, the parabola will open upward and have a minimum value. Since $\left[x - \left(-\frac{b}{2a}\right)\right]^2$ will have a minimum value when $x = -\frac{b}{2a}$, the minimum value of the graph will occur when $x = -\frac{b}{2a}$. If $a < 0$, the parabola will open downward and have a maximum value. The maximum value will occur when $x = -\frac{b}{2a}$. To determine the lowest, or highest, point on a parabola, substitute $-\frac{b}{2a}$ for x in the function to find y. The resulting ordered pair will be the vertex of the parabola. Since the axis of symmetry is the vertical line through the vertex, its equation is found using the x-coordinate of the ordered pair. Thus, the equation of the axis of symmetry is $x = -\frac{b}{2a}$. Note that when $x = -\frac{b}{2a}$, the value of $f(x)$ is $\frac{4ac - b^2}{4a}$. Do you know why?

To Find the Vertex of a Parabola

The parabola represented by the function $f(x) = ax^2 + bx + c$ will have axis of symmetry $x = -\frac{b}{2a}$ and vertex

$$\left(-\frac{b}{2a}, \frac{4ac - b^2}{4a}\right)$$

Since we often find the y-coordinate of the vertex by substituting the x-coordinate of the vertex into $f(x)$, the vertex may also be designated as

$$\left(-\frac{b}{2a}, f\left(-\frac{b}{2a}\right)\right)$$

The parabola given by the function $f(x) = ax^2 + bx + c$ will open upward when a is greater than 0 and open downward when a is less than 0.

Recall that to find the x-intercept of the graph of $f(x) = ax^2 + bx + c$, we set $f(x) = 0$ and solve the equation

$$ax^2 + bx + c = 0$$

This equation may be solved by factoring, the quadratic formula, or completing the square.

As we mentioned in Section 8.2, the discriminant, $b^2 - 4ac$, may be used to determine the *number of x-intercepts*. The following table summarizes information about the discriminant.

Discriminant, $b^2 - 4ac$	Number of x-intercepts	Possible graphs of $f(x) = ax^2 + bx + c$
> 0	Two	
$= 0$	One	
< 0	None	

3 Graph Quadratic Functions Using the Axis of Symmetry, Vertex, and Intercepts

Now we will draw graphs of quadratic functions.

EXAMPLE 1 Consider the equation $y = -x^2 + 8x - 12$.

a) Determine whether the parabola opens upward or downward.

b) Find the y-intercept.

c) Find the vertex.

d) Find the x-intercepts, if any.

e) Draw the graph.

Solution **a)** Since a is -1, which is less than 0, the parabola opens downward.

b) To find the y-intercept, set $x = 0$ and solve for y.

$$y = -(0)^2 + 8(0) - 12 = -12$$

The y-intercept is $(0, -12)$.

c) First, find the x-coordinate, then find the y-coordinate of the vertex.

$$x = -\frac{b}{2a} = -\frac{8}{2(-1)} = 4$$

$$y = \frac{4ac - b^2}{4a} = \frac{4(-1)(-12) - 8^2}{4(-1)} = \frac{48 - 64}{-4} = 4$$

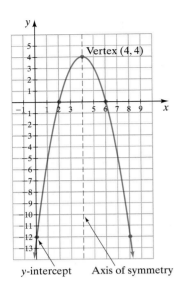

y-intercept   Axis of symmetry

FIGURE 8.8

The vertex is at $(4, 4)$. The y-coordinate of the vertex could also be found by substituting 4 for x in the function and finding the corresponding value of y, which is 4.

d) To find the x-intercepts, set $y = 0$.

$$0 = -x^2 + 8x - 12$$
$$\text{or} \quad x^2 - 8x + 12 = 0$$
$$(x - 6)(x - 2) = 0$$
$$x - 6 = 0 \quad \text{or} \quad x - 2 = 0$$
$$x = 6 \quad \text{or} \quad x = 2$$

Thus, the x-intercepts are $(2, 0)$ and $(6, 0)$. These values could also be found by the quadratic formula (or by completing the square).

e) Use all this information to draw the graph (Fig. 8.8). ✳

NOW TRY EXERCISE 15

If you obtain irrational values when finding x-intercepts by the quadratic formula, use your calculator to estimate these values, and then plot these decimal values. For example, if you obtain $x = (2 \pm \sqrt{10})/2$, you would evaluate $(2 + \sqrt{10})/2$ and $(2 - \sqrt{10})/2$ on your calculator and obtain 2.58 and -0.58, respectively, to the nearest hundredth. The x-intercepts would therefore be $(2.58, 0)$ and $(-0.58, 0)$.

EXAMPLE 2 Consider the function $f(x) = 2x^2 + 6x + 5$.

a) Determine whether the parabola opens upward or downward.

b) Find the y-intercept.

c) Find the vertex.

d) Find the x-intercepts, if any.

e) Draw the graph.

Solution

a) Since a is 2, which is greater than 0, the parabola opens upward.

b) Since $f(x)$ is the same as y, to find the y-intercept, set $x = 0$ and solve for $f(x)$, or y.

$$f(0) = 2(0)^2 + 6(0) + 5 = 5$$

The y-intercept is $(0, 5)$.

c)
$$x = -\frac{b}{2a} = -\frac{6}{2(2)} = -\frac{6}{4} = -\frac{3}{2}$$

$$y = \frac{4ac - b^2}{4a} = \frac{4(2)(5) - 6^2}{4(2)} = \frac{40 - 36}{8} = \frac{4}{8} = \frac{1}{2}$$

The vertex is $\left(-\frac{3}{2}, \frac{1}{2}\right)$. The y-coordinate of the vertex can also be found by evaluating $f\left(-\frac{3}{2}\right)$.

d) To find the x-intercepts, set $f(x) = 0$.

$$0 = 2x^2 + 6x + 5$$

This trinomial cannot be factored. To determine whether this equation has any real solutions, evaluate the discriminant.

$$b^2 - 4ac = 6^2 - 4(2)(5) = 36 - 40 = -4$$

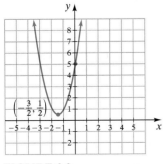

FIGURE 8.9

Since the discriminant is less than 0, this equation has no real solutions. You should have expected this answer because the y-coordinate of the vertex is a positive number and therefore above the x-axis. Since the parabola opens upward, it cannot intersect the x-axis.

NOW TRY EXERCISE 39   **e)** The graph is given in Figure 8.9 on the previous page.

4 Solve Maximum and Minimum Problems

A parabola that opens upward has a **minimum value** at its vertex, as illustrated in Figure 8.10a. A parabola that opens downward has a **maximum value** at its vertex, as shown in Figure 8.10b. If you are given a function of the form $f(x) = ax^2 + bx + c$, the maximum or minimum value will occur at $-\dfrac{b}{2a}$, and the value will be $\dfrac{4ac - b^2}{4a}$. There are many real-life problems that require finding maximum and minimum values.

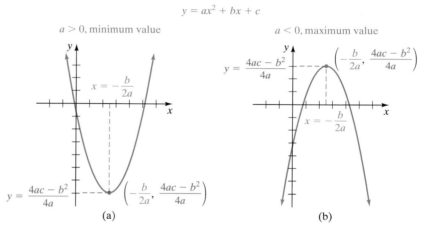

FIGURE 8.10   (a)   (b)

EXAMPLE 3   **Baseball**   Tommy Magee plays baseball with the Yorktown Cardinals. In the seventh inning of a game against the Arlington Blue Jays, he hit the ball to the outfield (see Fig. 8.11). He hit the ball at a height of 3 feet above the ground. For this particular hit, the height of the ball above the ground, $f(t)$, in feet, at time, t, in seconds, can be estimated by the formula

$$f(t) = -16t^2 + 52t + 3$$

a) Find the maximum height attained by the baseball.

b) Find the time it takes for the baseball to reach its maximum height.

c) Find the time it takes for the baseball to strike the ground.

FIGURE 8.11

Solution   **a)** Understand   The baseball will follow the path of a parabola that opens downward ($a < 0$). The baseball will rise to a maximum height, then begin its fall back to the ground due to gravity. To find the maximum height, we use the formula $y = \dfrac{4ac - b^2}{4a}$.

Translate $a = -16, \qquad b = 52, \qquad c = 3$

$$y = \frac{4ac - b^2}{4a}$$

Carry Out
$$= \frac{4(-16)(3) - (52)^2}{4(-16)}$$

$$= \frac{-192 - 2704}{-64}$$

$$= \frac{-2896}{-64}$$

$$= 45.25$$

Answer   The maximum height attained by the baseball is 45.25 feet.

b) The baseball reaches its maximum height at

$$t = -\frac{b}{2a} = -\frac{52}{2(-16)} = -\frac{52}{-32} = \frac{13}{8} \quad \text{or} \quad 1\frac{5}{8} \quad \text{or} \quad 1.625 \text{ seconds}$$

c) Understand and Translate   When the baseball strikes the ground, its height, y, above the ground is 0. Thus, to determine when the baseball strikes the ground, we solve the equation

$$-16t^2 + 52t + 3 = 0$$

We will use the quadratic formula to solve the equation.

$$t = \frac{-b \pm \sqrt{b^2 - 4ac}}{2a}$$

Carry Out
$$= \frac{-52 \pm \sqrt{(52)^2 - 4(-16)(3)}}{2(-16)}$$

$$= \frac{-52 \pm \sqrt{2704 + 192}}{-32}$$

$$= \frac{-52 \pm \sqrt{2896}}{-32}$$

$$\approx \frac{-52 \pm 53.81}{-32}$$

$$t \approx \frac{-52 + 53.81}{-32} \quad \text{or} \quad t \approx \frac{-52 - 53.81}{-32}$$

$$\approx -0.06 \text{ second} \qquad\qquad \approx 3.31 \text{ seconds}$$

Answer   The only acceptable value is 3.31 seconds. The baseball strikes the ground in about 3.31 seconds. Notice in part **b)** that the time it takes the baseball to reach its maximum height, 1.625 seconds, is not quite half the total time the baseball was in flight, 3.31 seconds. The reason for this is that the baseball was hit from a height of 3 feet and not at ground level.   ✳

EXAMPLE 4   **Area of a Rectangle**   Consider the rectangle below where the length is $x + 3$ and the width is $10 - x$.

$10 - x$

$x + 3$

a) Find an equation for the area, $A(x)$.

b) Find the value for x that gives the largest (maximum) area.

c) Find the maximum area.

Solution   **a)** Area is length times width. The area function is

$$A(x) = (x + 3)(10 - x)$$
$$= -x^2 + 7x + 30$$

b) Understand and Translate   The graph of the function is a parabola that opens downward. Thus, the maximum value occurs at the vertex. Therefore, the maximum area occurs at $x = -\dfrac{b}{2a}$.

Carry Out $$x = -\frac{b}{2a} = -\frac{7}{2(-1)} = \frac{7}{2} = 3.5$$

Answer   The maximum area occurs when x is 3.5 units.

c) To find the maximum area, substitute 3.5 for each x in the equation determined in part **a)**.

$$A(x) = -x^2 + 7x + 30$$
$$A(3.5) = -(3.5)^2 + 7(3.5) + 30$$
$$= -12.25 + 24.5 + 30$$
$$= 42.25$$

Observe that for this rectangle, the length is $x + 3 = 3.5 + 3 = 6.5$ units and the width is $10 - x = 10 - 3.5 = 6.5$ units. The rectangle is actually a square, and its area is $(6.5)(6.5) = 42.25$ square units. Therefore, the maximum area is 42.25 square units.

NOW TRY EXERCISE 73

✸

In Example 4**c)**, the maximum area could have been determined by using the formula $y = \dfrac{4ac - b^2}{4a}$. Determine the maximum area now using this formula. You should obtain the same answer, 42.25 square units.

EXAMPLE 5   **Rectangular Corral**   John W. Brown is building a corral in the shape of a rectangle for newborn calves (see Fig. 8.12). If he plans to use 100 meters of fencing, find the dimensions of the corral that will give the greatest area.

Solution   Understand   We are given the perimeter of the corral, 100 meters. The formula for the perimeter of a rectangle is $P = 2l + 2w$. For this problem, $100 = 2l + 2w$. We are asked to maximize the area, A, where

$$A = l \cdot w$$

We need to express the area in terms of one variable, not two. To express the area in terms of l, we solve the perimeter formula, $100 = 2l + 2w$, for w, then make a substitution.

FIGURE 8.12

Translate $$100 = 2l + 2w$$
$$100 - 2l = 2w$$
$$50 - l = w$$

Carry Out   Now we substitute $50 - l$ for w into $A = l \cdot w$. This gives

$$A = lw$$
$$A = l(50 - l)$$
$$A = -l^2 + 50l$$

In this quadratic equation, $a = -1$, $b = 50$, and $c = 0$. The maximum area will occur at

$$l = -\frac{b}{2a} = -\frac{50}{2(-1)} = 25$$

Answer   The length that will give the largest area is 25 meters. The width, $w = 50 - l$, will also be 25 meters. Thus, a square with dimensions 25 by 25 meters will give the largest area.

The largest area can be found by substituting $l = 25$ into the formula $A = l(50 - l)$ or by using $A = \dfrac{4ac - b^2}{4a}$. In either case, we obtain an area of 625 square meters.

NOW TRY EXERCISE 93

In Example 5, when we obtained the equation $A = -l^2 + 50l$, we could have completed the square as follows:

$$A = -(l^2 - 50l)$$
$$= -(l^2 - 50l + 625 - 625)$$
$$= -(l^2 - 50l + 625) + 625$$
$$= -(l - 25)^2 + 625$$

From this equation we can determine that the maximum area, 625 square meters, occurs when the length is 25 meters.

5   Understand Translations of Parabolas

Now we will look at another method used to graph parabolas. With this method, you start with a graph of the form $f(x) = ax^2$ and **translate**, or shift, the position of the graph to obtain the graph of the function you are seeking. As a reference, Figure 8.13a shows the graphs of $f(x) = x^2$, $g(x) = 2x^2$, and $h(x) = \frac{1}{2}x^2$. Figure 8.13b shows the graphs of $f(x) = -x^2$, $g(x) = -2x^2$, and $h(x) = -\frac{1}{2}x^2$.

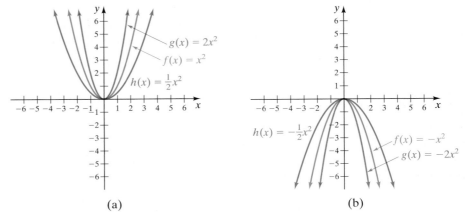

FIGURE 8.13

(a)   (b)

You can verify that each of the graphs is correct by plotting points. Notice that in Figures 8.13a and b the *value of a* in $f(x) = ax^2$ determines the width of the parabola. As $|a|$ gets larger, the parabola gets narrower and as $|a|$ gets smaller, the parabola gets wider.

Now let's consider the three functions $f(x) = x^2$, $g(x) = (x - 2)^2$, and $h(x) = (x + 2)^2$. These functions are graphed in Figure 8.14. (You can verify that these are graphs of the three functions by plotting points.)

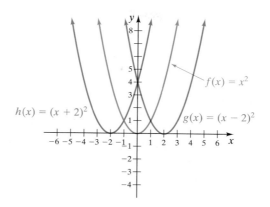

FIGURE 8.14

Notice that the graphs of $g(x)$ and $h(x)$ are identical to the graph of $f(x)$ except that $g(x)$ has been translated, or shifted, 2 units to the right and $h(x)$ has been translated 2 units to the left. In general, the graph of $g(x) = a(x - h)^2$ will have the same shape as the graph of $f(x) = ax^2$. The graph of an equation of the form $g(x) = a(x - h)^2$ will be shifted horizontally from the graph of $f(x) = ax^2$. *If h is a positive real number, the graph of $g(x) = a(x - h)^2$ will be shifted h units to the right of the graph of $f(x) = ax^2$. If h is a negative real number, the graph of $g(x) = a(x - h)^2$ will be shifted $|h|$ units to the left of the graph of $f(x) = ax^2$.*

Now consider the graphs of $f(x) = x^2$, $g(x) = x^2 + 3$ and $h(x) = x^2 - 3$ that are illustrated in Figure 8.15. You can verify that these are graphs of the three functions by plotting points.

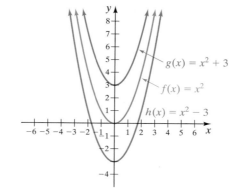

FIGURE 8.15

Notice that the graphs of $g(x)$ and $h(x)$ are identical to the graph of $f(x)$ except that $g(x)$ is translated 3 units up, and $h(x)$ is translated 3 units down. In general, *the graph of $g(x) = ax^2 + k$ is the graph of $f(x) = ax^2$ shifted k units up if k is a positive real number, and $|k|$ units down if k is a negative real number.*

Now consider the graphs of $f(x) = x^2$ and $g(x) = (x - 2)^2 + 3$, shown in Figure 8.16 at the top of the next page.

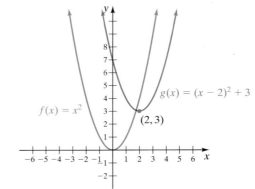

FIGURE 8.16

Notice that the graph of $g(x)$ has the same general shape as that of $f(x)$. The graph of $g(x)$ is the graph of $f(x)$ translated 2 units to the right and 3 units up. This graph and the discussion preceding it lead to the following important facts.

> For any function $f(x) = ax^2$, the graph of $g(x) = a(x - h)^2 + k$ will have the same shape as the graph of $f(x)$. The graph of $g(x)$ will be the graph of $f(x)$ shifted as follows:
>
> - If h is a positive real number, the graph will be shifted h units to the right.
> - If h is a negative real number, the graph will be shifted $|h|$ units to the left.
> - If k is a positive real number, the graph will be shifted k units up.
> - If k is a negative real number, the graph will be shifted $|k|$ units down.

Examine the graph of $g(x) = (x - 2)^2 + 3$ in Figure 8.16. Notice that its axis of symmetry is $x = 2$ and its vertex is $(2, 3)$.

> The graph of any function of the form
>
> $$f(x) = a(x - h)^2 + k$$
>
> will be a parabola with axis of symmetry $x = h$ and vertex at (h, k).

Example	Axis of Symmetry	Vertex	Parabola Opens
$f(x) = 2(x - 4)^2 + 5$	$x = 4$	$(4, 5)$	upward, $a > 0$
$f(x) = -\frac{1}{2}(x - 6)^2 - 4$	$x = 6$	$(6, -4)$	downward, $a < 0$

Now consider $f(x) = 2(x + 5)^2 + 3$. We can rewrite this as $f(x) = 2[x - (-5)]^2 + 3$. Therefore, h has a value of -5 and k has a value of 3. The graph of this function has axis of symmetry $x = -5$ and vertex at $(-5, 3)$.

Example	Axis of Symmetry	Vertex	Parabola Opens
$f(x) = 3(x + 4)^2 - 2$	$x = -4$	$(-4, -2)$	upward, $a > 0$
$f(x) = -\frac{1}{2}\left(x + \frac{1}{3}\right)^2 + \frac{1}{4}$	$x = -\frac{1}{3}$	$\left(-\frac{1}{3}, \frac{1}{4}\right)$	downward, $a < 0$

Now we are ready to graph parabolas using translations.

EXAMPLE 6   The graph of $f(x) = -2x^2$ is illustrated in Figure 8.17. Using this graph as a guide, graph $g(x) = -2(x + 3)^2 - 4$.

Solution   The function $g(x)$ may be written $g(x) = -2[x - (-3)]^2 - 4$. Therefore, in the function, h has a value of -3 and k has a value of -4. The graph of $g(x)$ will therefore be the graph of $f(x)$ translated 3 units to the left (because $h = -3$) and 4 units down (because $k = -4$). The graphs of $f(x)$ and $g(x)$ are illustrated in Figure 8.18.

FIGURE 8.17

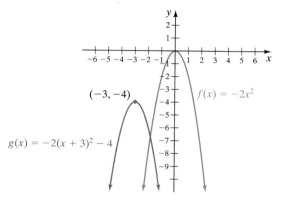

NOW TRY EXERCISE 49   FIGURE 8.18

In objective 2, we started with a function of the form $f(x) = ax^2 + bx + c$ and completed the square to obtain

$$f(x) = a\left[x - \left(-\frac{b}{2a}\right)\right]^2 + \frac{4ac - b^2}{4a}$$

We stated that the vertex of the parabola of this function is $\left(-\dfrac{b}{2a}, \dfrac{4ac - b^2}{4a}\right)$.

Suppose we substitute h for $-\dfrac{b}{2a}$ and k for $\dfrac{4ac - b^2}{4a}$ in the function. We then get

$$f(x) = a(x - h)^2 + k$$

which we know is a parabola with vertex at (h, k). Therefore, both functions $f(x) = ax^2 + bx + c$ and $f(x) = a(x - h)^2 + k$ yield the same vertex and axis of symmetry for any given function.

6   Write Functions in the Form $f(x) = a(x - h)^2 + k$

If we wish to graph parabolas using translations, we need to change the form of a function from $f(x) = ax^2 + bx + c$ to $f(x) = a(x - h)^2 + k$. To do this, we *complete the square* as was discussed in Section 8.1. By completing the square we obtain a perfect square trinomial, which we can represent as the square of a binomial. Examples 7 and 8 explain the procedure. We will use this procedure again in Chapter 11, when we discuss conic sections.

EXAMPLE 7   Given $f(x) = x^2 - 6x + 10$,
a) write $f(x)$ in the form $f(x) = a(x - h)^2 + k$.
b) graph $f(x)$.

Solution   **a)** We use the x^2 and $-6x$ terms to obtain a perfect square trinomial.

$$f(x) = (x^2 - 6x) + 10$$

578 • Chapter 8 • Quadratic Functions

Now we take half the coefficient of the x-term and square it.

$$\left[\frac{1}{2}(-6)\right]^2 = \boxed{9}$$

We then add this value, 9, within the parentheses. Since we are adding 9 within parentheses, we add -9 outside parentheses. Adding 9 and -9 to an expression is the same as adding 0, which does not change the value of the expression.

$$f(x) = (x^2 - 6x \boxed{+ 9}) \boxed{- 9} + 10$$

By doing this we have created a perfect square trinomial within the parentheses, plus a constant outside the parentheses. We express the perfect square trinomial as the square of a binomial.

$$f(x) = (x - 3)^2 + 1$$

The function is now in the form we are seeking.

b) Since $a = 1$, which is greater than 0, the parabola opens upward. The axis of symmetry of the parabola is $x = 3$ and the vertex is at $(3, 1)$. The y-intercept can be easily obtained by substituting $x = 0$ and finding $f(x)$. When $x = 0$, $f(x) = (-3)^2 + 1 = 10$. Thus, the y-intercept is at 10. By plotting the vertex, y-intercept, and a few other points, we obtain the graph in Figure 8.19. The figure also shows the graph of $y = x^2$ for comparison.

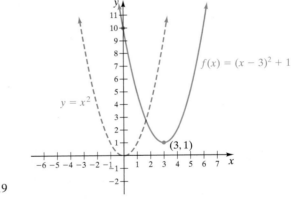

FIGURE 8.19

EXAMPLE 8   Given $f(x) = -2x^2 - 10x - 13$,

a) write $f(x)$ in the form $f(x) = a(x - h)^2 + k$.

b) graph $f(x)$.

Solution   **a)** When the leading coefficient is not 1, we factor out the leading coefficient from the terms containing the variable.

$$f(x) = -2(x^2 + 5x) - 13$$

Now we complete the square.

Half of coefficient of
x-term squared
↓

$$\left[\frac{1}{2}(5)\right]^2 = \frac{25}{4}$$

If we add $\frac{25}{4}$ within the parentheses, we are actually adding $-2\left(\frac{25}{4}\right)$ or $-\frac{25}{2}$, since each term in parentheses is multiplied by -2. Therefore, to compensate, we must add $\frac{25}{2}$ outside the parentheses.

$$f(x) = -2\left(x^2 + 5x + \boxed{\frac{25}{4}}\right) + \boxed{\frac{25}{2}} - 13$$

$$= -2\left(x + \frac{5}{2}\right)^2 - \frac{1}{2}$$

b) Since $a = -2$, the parabola opens downward. The axis of symmetry is $x = -\frac{5}{2}$ and the vertex is $\left(-\frac{5}{2}, -\frac{1}{2}\right)$. The y-intercept is at $f(0) = -13$. We plot a few points and draw the graph in Figure 8.20. In the figure, we also show the graph of $y = -2x^2$ for comparison.

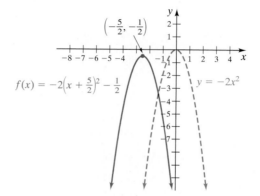

FIGURE 8.20

NOW TRY EXERCISE 63 Notice that $f(x) = -2\left(x + \frac{5}{2}\right)^2 - \frac{1}{2}$ has no x-intercepts. Therefore, there are no real values of x for which $f(x) = 0$. ✳

A second way to change the equation from $f(x) = ax^2 + bx + c$ to $f(x) = a(x - h)^2 + k$ form is to let $h = -\dfrac{b}{2a}$ and $k = \dfrac{4ac - b^2}{4a}$. Find the values for h and k and then substitute the values obtained into $f(x) = a(x - h)^2 + k$. For example, for the function $f(x) = -2x^2 - 10x - 13$, in Example 8, $a = -2$, $b = -10$, and $c = -13$. Then

$$h = -\frac{b}{2a} = -\frac{-10}{2(-2)} = -\frac{5}{2}$$

$$k = \frac{4ac - b^2}{4a} = \frac{4(-2)(-13) - (-10)^2}{4(-2)} = -\frac{1}{2}$$

Therefore,

$$f(x) = a(x - h)^2 + k$$

$$= -2\left[x - \left(-\frac{5}{2}\right)\right]^2 - \frac{1}{2}$$

$$= -2\left(x + \frac{5}{2}\right)^2 - \frac{1}{2}$$

This answer checks with that obtained in Example 8.

Mathematics in Action

Parabolas and Energy

An extremely diverse set of real-world applications of the parabola involves the idea of a three-dimensional parabolic reflector formed when a parabola is revolved about its axis. Mirrors in reflecting telescopes, automobile headlights, TV satellite dishes, and giant parabolic listening devices directed toward space all share a similar property. Parallel waves directed at the concave parabolic surface will intersect one another, or will be focused, at a point close to the surface of the parabolic "dish." If the parallel lines are light waves, the light becomes focused; if sound waves, the sound becomes focused.

A most interesting variant of the parabolic dish is the parabolic trough. Sighting down the surface at the bottom of a parabolic trough, the walls rise on both sides in the shape of parabolic curves. Parabolic troughs are used to gather solar energy for the generation of electricity.

Between the mid-1980s and early 1990s, private industry built commercial solar trough power plants in California capable of generating 354 megawatts of electricity per year, enough to satisfy the energy needs of about 500,000 people. The first of these was built in the Mojave Desert. The nine systems built in this period are still operating successfully, producing more than 90 percent of the world's thermal solar energy, and saving the equivalent of 2.3 million barrels of oil every year.

The way the trough system works is that curved solar collectors focus sunlight onto a receiver pipe through which a synthetic oil circulates. The heated oil travels through a heat exchanger and produces steam that drives a conventional electricity-generating turbine. As with other renewable technologies, no pollutants are emitted in generating this electricity.

As non-renewable resources such as oil and natural gas edge toward depletion, and the need to produce cleaner energy gets treated more seriously, solar generation using parabolic troughs will surely be one of the great areas of research and development in the coming decades.

Exercise Set 8.5

Concept/Writing Exercises

1. What is the graph of a quadratic equation called?

2. What is the vertex of a parabola?

3. What is the axis of symmetry of a parabola?

4. What is the equation of the axis of symmetry of the graph of $f(x) = ax^2 + bx + c$?

5. What is the vertex of the graph of $f(x) = ax^2 + bx + c$?

6. How many x-intercepts does a quadratic function have if the discriminant is **a)** < 0, **b)** $= 0$, **c)** > 0?

7. For $f(x) = ax^2 + bx + c$, will $f(x)$ have a maximum or a minimum if **a)** $a > 0$, **b)** $a < 0$? Explain.

8. Explain how to find the x-intercepts of the graph of a quadratic function.

9. Explain how to find the y-intercept of the graph of a quadratic function.

10. Consider the graph of $f(x) = ax^2$. Explain how the shape of $f(x)$ changes as $|a|$ increases and as $|a|$ decreases.

11. Consider the graph of $f(x) = ax^2$. What is the general shape of $f(x)$ if **a)** $a > 0$, **b)** $a < 0$?

12. Will the graphs of $f(x) = ax^2$ and $g(x) = -ax^2$ have the same vertex for any nonzero real number a? Explain.

13. Does the function $f(x) = 3x^2 - 5x + 2$ have a maximum or a minimum value? Explain.

14. Does the function $g(x) = -\frac{1}{5}x^2 + 2x - 7$ have a maximum or a minimum value? Explain.

Practice the Skills

a) *Determine whether the parabola opens upward or downward.* *b)* *Find the y-intercept.* *c)* *Find the vertex.* *d)* *Find the x-intercepts (if any).* *e)* *Draw the graph.*

15. $f(x) = x^2 + 8x + 15$

16. $f(x) = x^2 - 4x + 3$

17. $g(x) = x^2 + 2x - 3$

18. $h(x) = x^2 - 2x - 8$

19. $f(x) = -x^2 - 2x + 8$

20. $g(x) = -x^2 + 4x + 5$

21. $p(x) = -x^2 + 8x - 15$

22. $n(x) = -x^2 - 2x + 24$

23. $t(x) = -x^2 + 4x - 5$

24. $g(x) = x^2 + 6x + 13$

25. $f(x) = x^2 - 4x + 4$

26. $r(x) = -x^2 + 10x - 25$

27. $r(x) = x^2 + 2$

28. $l(x) = -x^2 + 5$

29. $g(x) = -x^2 + 6x$

30. $f(x) = x^2 + 4x$

31. $f(x) = -2x^2 + 4x - 8$

32. $m(x) = 3x^2 + 4x + 3$

33. $g(x) = -2x^2 - 6x + 4$

34. $p(x) = -2x^2 + 5x + 4$

35. $y = 3x^2 + 4x - 6$

36. $y = x^2 - 6x + 4$

37. $y = 2x^2 - x - 6$

38. $h(x) = -2x^2 + 4x - 5$

39. $f(x) = -x^2 + 3x - 5$

40. $g(x) = -4x^2 + 6x - 9$

Using the graphs in Figures 8.13 through 8.16 as a guide, graph each function and label the vertex.

41. $f(x) = (x - 3)^2$

42. $f(x) = (x - 4)^2$

43. $f(x) = (x + 1)^2$

44. $f(x) = (x + 3)^2$

45. $f(x) = x^2 + 3$

46. $f(x) = x^2 + 6$

47. $f(x) = x^2 - 1$

48. $f(x) = x^2 - 4$

49. $f(x) = (x - 2)^2 + 3$

50. $f(x) = (x - 3)^2 - 4$

51. $f(x) = (x + 4)^2 + 4$

52. $h(x) = (x + 4)^2 - 1$

53. $g(x) = -(x + 3)^2 - 2$

54. $g(x) = (x - 1)^2 + 4$

55. $y = -2(x - 2)^2 + 2$

56. $f(x) = -(x - 5)^2 + 2$

57. $h(x) = -2(x + 1)^2 - 3$

58. $y = -2(x - 3)^2 + 1$

In Exercises 59–68, *a)* *express each function in the form* $f(x) = a(x - h)^2 + k$ *and* *b)* *draw the graph of each function and label the vertex.*

59. $f(x) = x^2 - 6x + 8$

60. $g(x) = x^2 + 6x + 5$

61. $f(x) = x^2 - x + 1$

62. $g(x) = x^2 - x - 3$

63. $f(x) = -x^2 - 4x - 6$

64. $h(x) = -x^2 + 6x + 2$

65. $g(x) = x^2 - 4x - 5$

66. $p(x) = x^2 - 2x - 15$

67. $f(x) = 2x^2 + 5x - 3$

68. $k(x) = 2x^2 + 7x - 4$

Problem Solving

*Match the functions in Exercises 69–72 with the appropriate graphs labeled **a)** through **d)**.*

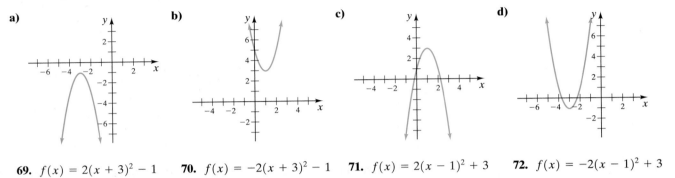

a)   **b)**   **c)**   **d)**

69. $f(x) = 2(x + 3)^2 - 1$ **70.** $f(x) = -2(x + 3)^2 - 1$ **71.** $f(x) = 2(x - 1)^2 + 3$ **72.** $f(x) = -2(x - 1)^2 + 3$

*For each rectangle, **a)** find the value for x that gives the maximum area and **b)** find the maximum area.*

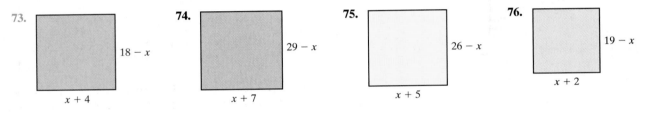

73. $18 - x$, $x + 4$

74. $29 - x$, $x + 7$

75. $26 - x$, $x + 5$

76. $19 - x$, $x + 2$

77. Selling Batteries The revenue function for selling n batteries is $R(n) = n(10 - 0.02n) = -0.02n^2 + 10n$. Find **a)** the number of batteries that must be sold to obtain the maximum revenue and **b)** the maximum revenue.

78. Selling Watches The revenue function for selling n watches is $R(n) = n(25 - 0.1n) = -0.1n^2 + 25n$. Find **a)** the number of watches that must be sold to obtain the maximum revenue and **b)** the maximum revenue.

79. Enrollment The enrollment in a high school in the Naplewood School District can be approximated by the function

$$N(t) = -0.043t^2 + 1.22t + 46.0$$

where t is the number of years since 1989 and $1 \le t \le 15$. In what year will the maximum enrollment be obtained?

80. Drug-Free Schools The percent of students in schools in the United States who say their school is not drug free can be approximated by the function

$$f(a) = -2.32a^2 + 76.58a - 559.87$$

where a is the student's age and $12 < a < 20$. What age has the highest percent of students who say their school is not drug free?

See Exercise 79

81. What is the distance between the vertices of the graphs of $f(x) = (x - 2)^2 + \frac{5}{2}$ and $g(x) = (x - 2)^2 - \frac{3}{2}$?

82. What is the distance between the vertices of the graphs of $f(x) = 2(x - 4)^2 - 3$ and $g(x) = -3(x - 4)^2 + 2$?

83. What is the distance between the vertices of the graphs of $f(x) = 2(x + 4)^2 - 3$ and $g(x) = -(x + 1)^2 - 3$?

84. What is the distance between the vertices of the graphs of $f(x) = -\frac{1}{3}(x - 3)^2 - 2$ and $g(x) = 2(x + 5)^2 - 2$?

85. Write the function whose graph has the shape of the graph of $f(x) = 2x^2$ and has a vertex at $(3, -2)$.

86. Write the function whose graph has the shape of the graph of $f(x) = -\frac{1}{2}x^2$ and has a vertex at $\left(\frac{2}{3}, -5\right)$.

87. Write the function whose graph has the shape of the graph of $f(x) = -4x^2$ and has a vertex at $\left(-\frac{3}{5}, -\sqrt{2}\right)$.

88. Write the function whose graph has the shape of the graph of $f(x) = \frac{3}{5}x^2$ and has a vertex at $(-\sqrt{3}, \sqrt{5})$.

89. Consider $f(x) = x^2 - 8x + 12$ and $g(x) = -x^2 + 8x - 12$.

　　a) Without graphing, can you explain how the graphs of the two functions compare?

　　b) Will the graphs have the same x-intercepts? Explain.

　　c) Will the graphs have the same vertex? Explain.

　　d) Graph both functions on the same axes.

90. By observing the leading coefficient in a quadratic function and by determining the coordinates of the vertex of its graph, explain how you can determine the number of x-intercepts the parabola has.

91. *Selling Tickets* The Riverside High School Theater Club is trying to set the price of tickets for a play. If the price is too low, they will not make enough money to cover expenses, and if the price is too high, not enough people will pay the price of a ticket. They estimate that their total income per concert, I, in hundreds of dollars, can be approximated by the formula

$$I = -x^2 + 24x - 44, 0 \le x \le 24$$

where x is the cost of a ticket.

　　a) Draw a graph of income versus the cost of a ticket.

　　b) Determine the minimum cost of a ticket for the theater club to break even.

　　c) Determine the maximum cost of a ticket that the theater club can charge and break even.

　　d) How much should they charge to receive the maximum income?

　　e) Find the maximum income.

92. *Throwing an Object* An object is projected upward with an initial velocity of 192 feet per second. The object's distance above the ground, d, after t seconds may be found by the formula $d = -16t^2 + 192t$.

　　a) Find the object's distance above the ground after 3 seconds.

　　b) Draw a graph of distance versus time.

　　c) What is the maximum height the object will reach?

　　d) At what time will it reach its maximum height?

　　e) At what time will the object strike the ground?

93. *Profit* The Holley-Berry Company earns a weekly profit according to the function $f(x) = -0.4x^2 + 80x - 200$, where x is the number of bird feeders built and sold.

　　a) Find the number of bird feeders that the company must sell in a week to obtain the maximum profit.

　　b) Find the maximum profit.

94. *Profit* The Trish Lanz Company earns a weekly profit according to the function $f(x) = -1.2x^2 + 180x - 280$, where x is the number of rocking chairs built and sold.

　　a) Find the number of rocking chairs that the company must sell in a week to obtain the maximum profit.

　　b) Find the maximum profit.

95. *Firing a Cannon* If a certain cannon is fired from a height of 9.8 meters above the ground, at a certain angle, the height of the cannonball above the ground, h, in meters, at time, t, in seconds, is found by the function

$$h(t) = -4.9t^2 + 24.5t + 9.8$$

　　a) Find the maximum height attained by the cannonball.

　　b) Find the time it takes for the cannonball to reach its maximum height.

　　c) Find the time it takes for the cannonball to strike the ground.

96. *Throwing a Ball* Ramon Loomis throws a ball into the air with an initial velocity of 32 feet per second. The height of the ball at any time t is given by the formula $h = 32t - 16t^2$. At what time does the ball reach its maximum height? What is the maximum height?

97. *Renting Office Space* The following graph, taken from the February 3, 2002, issue of *The Wall Street Journal*, indicates that the price of renting office space in 50 major U.S. markets has been on the rise in recent years. The function $C(x) = 0.19x^2 - 0.657x + 16.6$ can be used to estimate the annual rent per square foot, where x is the years since 1991.

Rent per Square Foot

Source: REIS Inc.

　　a) If we assume the current trend continues, estimate the annual rent per square foot in 2004.

　　b) In what year was the annual rent per square foot a minimum?

　　c) Estimate the minimum annual rent per square foot.

98. *Sports Apparel* The number of pieces of sports team apparel sold in the United States (excluding caps), in millions, can be approximated by the function

$$A(t) = -0.044t^2 + 0.163t + 4.865$$

where t is years since 1992.

a) In what year was the retail sales of sports apparel a maximum?

b) If this trend continues, estimate the retail sales of sports apparel in 2003.

99. *Room in a House* Jake Kishner is designing plans for his house. What is the maximum possible area of a room if its perimeter is to be 60 feet?

100. *Greatest Area* What are the dimensions of a garden that will have the greatest area if its perimeter is to be 70 feet?

101. *Minimum Product* What is the minimum product of two numbers that differ by 8? What are the numbers?

102. *Minimum Product* What is the minimum product of two numbers that differ by 10? What are the numbers?

103. *Maximum Product* What is the maximum product of two numbers that add to 40? What are the numbers?

104. *Maximum Product* What is the maximum product of two numbers that add to 9? What are the numbers?

*The profit of a company, in dollars, is the difference between the company's revenue and cost. Exercises 105 and 106 give cost, C(x), and revenue, R(x), functions for a particular company. The x represents the number of items produced and sold to distributors. Determine **a)** the maximum profit of the company and **b)** the number of items that must be produced and sold to obtain the maximum profit.*

105. $C(x) = 2000 + 40x$
$R(x) = 800x - x^2$

106. $C(x) = 5000 + 12x$
$R(x) = 2000x - x^2$

Challenge Problems

107. *Baseball* In Example 3 of this section, we used the function $f(t) = -16t^2 + 52t + 3$ to find that the maximum height, f, attained by a baseball hit by Tommy Magee was 45.25 feet. The ball reached this height at 1.625 seconds after the baseball was hit.
Review Example 3 now.

a) Write $f(t)$ in the form $f(t) = a(t - h)^2 + k$ by completing the square.

b) Using the function you obtained in part **a)**, determine the maximum height attained by the baseball, and the time after it was hit that the baseball attained its maximum value.

c) Are the answers you obtained in part **b)** the same answers obtained in Example 3? If not, explain why not.

Group Activity

Discuss and answer Exercise 108 as a group.

108. a) Group member 1: Write two quadratic functions $f(x)$ and $g(x)$ so that the functions will not intersect.

b) Group member 2: Write two quadratic functions $f(x)$ and $g(x)$ so that neither function will have x-intercepts, and the vertices of the functions are on opposite sides of the x-axis.

c) Group member 3: Write two quadratic functions $f(x)$ and $g(x)$ so that both functions have the same vertex but one function opens upward and the other opens downward.

d) As a group, review each answer in parts **a)–c)** and decide whether each answer is correct. Correct any answer that is incorrect.

Cumulative Review Exercises

[2.2] **109.** Find the area shaded blue in the figure.

[3.7] **110.** Graph $y \le \frac{2}{3}x + 3$.

[4.2] **111.** Solve the system of equations.
$$x - y = -5$$
$$2x + 2y - z = 0$$
$$x + y + z = 3$$

[4.5] **112.** Evaluate the determinant.
$$\begin{vmatrix} \frac{1}{2} & 3 \\ 2 & -4 \end{vmatrix}.$$

[6.1] **113.** Divide $(x - 3) \div \dfrac{x^2 + 3x - 18}{x}$.

8.6 QUADRATIC AND OTHER INEQUALITIES IN ONE VARIABLE

1 Solve quadratic inequalities.
2 Solve other polynomial inequalities.
3 Solve rational inequalities.

In Section 2.5, we discussed linear inequalities in one variable. Now we discuss quadratic inequalities in one variable.

When the equal sign in a quadratic equation of the form $ax^2 + bx + c = 0$ is replaced by an inequality sign, we get a **quadratic inequality**.

Examples of Quadratic Inequalities
$$x^2 + x - 12 > 0, \qquad 2x^2 - 9x - 5 \leq 0$$

The **solution to a quadratic inequality** is the set of all values that make the inequality a true statement. For example, if we substitute 5 for x in $x^2 + x - 12 > 0$, we obtain

$$x^2 + x - 12 > 0$$
$$5^2 + 5 - 12 \overset{?}{>} 0$$
$$18 > 0 \quad \textit{True}$$

The inequality is true when x is 5, so 5 satisfies the inequality. However, 5 is not the only solution. There are other values that satisfy (or are solutions to) the inequality. Does 4 satisfy the inequality? Does 2 satisfy the inequality?

1 Solve Quadratic Inequalities

A number of methods can be used to find the solutions to quadratic inequalities. We will begin by introducing a **sign graph**. Consider the function $f(x) = x^2 + x - 12$. Its graph is shown in Figure 8.21a. Figure 8.21b shows, in red, that when $x < -4$ or $x > 3$, $f(x) > 0$ or $x^2 + x - 12 > 0$. It also shows, in green, that when $-4 < x < 3$, $f(x) < 0$ or $x^2 + x - 12 < 0$.

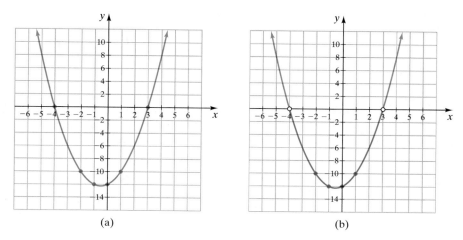

FIGURE 8.21   (a)   (b)

One way of finding the solutions to inequalities is to draw the graph and determine from the graph which values of the variable satisfy the inequality, as we just did. In many cases, it may be inconvenient or take too much time to draw the graph of a function, so we provide an alternate method to solve quadratic and other inequalities.

In Example 1, we will show how $x^2 + x - 12 > 0$ is solved using a number line. Then we will outline the procedure we used.

EXAMPLE 1 Solve the inequality $x^2 + x - 12 > 0$. Give the solution **a)** on a number line, **b)** in interval notation, and **c)** in set builder notation.

Solution Set the inequality equal to 0 and solve the equation.

$$x^2 + x - 12 = 0$$
$$(x + 4)(x - 3) = 0$$
$$x + 4 = 0 \quad \text{or} \quad x - 3 = 0$$
$$x = -4 \qquad\qquad x = 3$$

The numbers obtained are called **boundary values**. The boundary values are used to break a number line up into intervals. If the original inequality is $<$ or $>$, the boundary values are not part of the intervals. If the original inequality is $\leq$ or $\geq$, the boundary values are part of the intervals.

In Figure 8.22, we have labeled the intervals A, B, and C. Next, we select one test value in *each* interval. Then we substitute each of those numbers, one at a time, into either $x^2 + x - 12 > 0$ or $(x + 4)(x - 3) > 0$ and determine whether they result in a true statement. If the test value results in a true statement, all values in that interval will also satisfy the inequality. If the test value results in a false statement, no numbers in that interval will satisfy the inequality.

In this example, we will use the test values of -5 in interval A, 0 in interval B, and 4 in interval C (see Fig. 8.23).

A $(-\infty, 4)$ B $(-4, 3)$ C $(3, \infty)$

-4 3

FIGURE 8.22

A -5 B 0 C 4

-4 3

FIGURE 8.23

Interval A	Interval B	Interval C
Test value, -5	Test value, 0	Test value, 4
Is $x^2 + x - 12 > 0$?	Is $x^2 + x - 12 > 0$?	Is $x^2 + x - 12 > 0$?
$(-5)^2 - 5 - 12 \overset{?}{>} 0$	$0^2 + 0 - 12 \overset{?}{>} 0$	$4^2 + 4 - 12 \overset{?}{>} 0$
$8 > 0$	$-12 > 0$	$8 > 0$
True	*False*	*True*

Since the test values in both intervals A and C satisfy the inequality, the solution is all real numbers in intervals A or C. The inequality symbol is $>$. The values -4 and 3 are not included in the solution because they make the inequality equal to 0.

The answer to parts **a)**, **b)**, and **c)** follow.

a) The solution is illustrated on a number line in Figure 8.24.

b) The solution in interval notation is $(-\infty, -4) \cup (3, \infty)$.

c) The solution in set builder notation is $\{x | x < -4 \text{ or } x > 3\}$.

Note that the solution, in any form, is consistent with the graph in Figure 8.21b.

Solution

-4 3

FIGURE 8.24

NOW TRY EXERCISE 15

> ### To Solve Quadratic and Other Inequalities
> 1. Write the inequality as an equation and solve the equation.
> 2. If solving a rational inequality, determine the values that make any denominator 0.
> 3. Construct a number line. Mark each solution from step 1 and numbers obtained in step 2 on the number line. Mark the lowest value on the left, with values increasing from left to right.
> 4. Select a test value in each interval and determine whether it satisfies the inequality. Also test each boundary value.
> 5. Write the solution in the form requested by your instructor.

EXAMPLE 2 Solve the inequality $x^2 - 4x \geq -4$. Give the solution **a)** on a number line, **b)** in interval notation, and **c)** in set builder notation.

Solution Write the inequality as an equation, then solve the equation.

$$x^2 - 4x = -4$$
$$x^2 - 4x + 4 = 0$$
$$(x - 2)(x - 2) = 0$$
$$x - 2 = 0 \quad \text{or} \quad x - 2 = 0$$
$$x = 2 \qquad\qquad x = 2$$

FIGURE 8.25

Since both factors are the same there is only one boundary value, 2, (see Fig. 8.25). Both test values, 1 and 3, result in true statements.

Interval A	Interval B
Test value, 1	Test value, 3
$x^2 - 4x \geq -4$	$x^2 - 4x \geq -4$
$1^2 - 4(1) \overset{?}{\geq} -4$	$3^2 - 4(3) \overset{?}{\geq} -4$
$1 - 4 \overset{?}{\geq} -4$	$9 - 12 \overset{?}{\geq} -4$
$-3 \geq -4$ *True*	$-3 \geq -4$ *True*

The solution set includes both intervals and the boundary value, 2. The solution set is the set of real numbers, $\mathbb{R}$. The answer to parts **a)**, **b)**, and **c)** follow.

a) ◄————————|————————► **b)** $(-\infty, \infty)$ **c)** $\{x | -\infty < x < \infty\}$ ✳
             0

We can check the solution to Example 2 using graphing. Let $f(x) = x^2 - 4x$ and $g(x) = -4$. For $x^2 - 4x \geq -4$ to be true, we want $f(x) \geq g(x)$. The graphs of $f(x)$ and $g(x)$ are given in Figure 8.26.

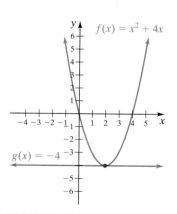

    Observe that $f(x) = g(x)$ at $x = 2$ and $f(x) > g(x)$ for all other values of x. Thus, $f(x) \geq g(x)$ for all values of x, and the solution set is the set of real numbers.

    In Example 2, if we rewrite the inequality $x^2 - 4x \geq -4$ as $x^2 - 4x + 4 \geq 0$ and then as $(x - 2)^2 \geq 0$ we can see that the solution must be the set of real numbers, since $(x - 2)^2$ must be greater than or equal to 0 for any real number x. The solution to $x^2 - 4x < -4$ is the empty set, $\varnothing$. Can, you explain why?

FIGURE 8.26

EXAMPLE 3 Solve the inequality $x^2 - 2x - 4 \leq 0$. Express the solution in interval notation.

Solution First we need to solve the equation $x^2 - 2x - 4 = 0$. Since this equation is not factorable, we use the quadratic formula to solve.

$$x = \frac{-b \pm \sqrt{b^2 - 4ac}}{2a}$$

$$= \frac{2 \pm \sqrt{4 - 4(1)(-4)}}{2(1)} = \frac{2 \pm \sqrt{20}}{2} = \frac{2 \pm 2\sqrt{5}}{2} = 1 \pm \sqrt{5}$$

The boundary values are $1 - \sqrt{5}$ and $1 + \sqrt{5}$. The value of $1 - \sqrt{5}$ is about -1.24 and the value of $1 + \sqrt{5}$ is about 3.24. We will select test values of -2, 0, and 4 (see Fig. 8.27)

FIGURE 8.27

Interval A	Interval B	Interval C
Test value, -2	Test value, 0	Test value, 4
$x^2 - 2x - 4 \leq 0$	$x^2 - 2x - 4 \leq 0$	$x^2 - 2x - 4 \leq 0$
$(-2)^2 - 2(-2) - 4 \overset{?}{\leq} 0$	$0^2 - 2(0) - 4 \overset{?}{\leq} 0$	$4^2 - 2(4) - 4 \overset{?}{\leq} 0$
$4 + 4 - 4 \overset{?}{\leq} 0$	$0 - 0 - 4 \overset{?}{\leq} 0$	$16 - 8 - 4 \overset{?}{\leq} 0$
$4 \leq 0$	$-4 \leq 0$	$4 \leq 0$
False	*True*	*False*

FIGURE 8.28

NOW TRY EXERCISE 19

The boundary values are part of the solution because the inequality symbol is $\leq$ and the boundary values make the inequality equal to 0. Thus, the solution in interval notation is $[1 - \sqrt{5}, 1 + \sqrt{5}]$. The solution is illustrated on the number line in Figure 8.28.

HELPFUL HINT

If $ax^2 + bx + c = 0$, with $a > 0$, has two distinct real solutions, then:

Inequality of form	Solution is	Solution on number line
$ax^2 + bx + c \geq 0$	End intervals	
$ax^2 + bx + c \leq 0$	Center interval	

Example 1 is an inequality of the form $ax^2 + bx + c > 0$, and Example 3 is an inequality of the form $ax^2 + bx + c \leq 0$. Example 2 does not have two distinct real solutions, so this Helpful Hint does not apply.

2 Solve Other Polynomial Inequalities

A procedure similar to the one used earlier can be used to solve other **polynomial inequalities**, as illustrated in the following examples.

EXAMPLE 4 Solve the inequality $(3x - 2)(x + 3)(x + 5) < 0$. Illustrate the solution on a number line and write the solution in both interval notation and set builder notation.

Solution We use the zero-factor property to solve the equation $(3x - 2)(x + 3)(x + 5) = 0$.

$$3x - 2 = 0 \quad \text{or} \quad x + 3 = 0 \quad \text{or} \quad x + 5 = 0$$
$$x = \frac{2}{3} \qquad\qquad x = -3 \qquad\qquad x = -5$$

FIGURE 8.29

The solutions -5, -3, and $\frac{2}{3}$ break the number line into four intervals (see Fig. 8.29). The test values we will use are -6, -4, 0, and 1. We show the results in the following table.

Interval	Test value	$(3x - 2)(x + 3)(x + 5)$	< 0
A	-6	-60	*True*
B	-4	14	*False*
C	0	-30	*True*
D	1	24	*False*

FIGURE 8.30

NOW TRY EXERCISE 27

Since the original inequality symbol is $<$, the boundary values are not part of the solution. The solution, intervals A and C, is illustrated on the number line in Figure 8.30. The solution in set builder notation is $\left\{x | x < -5 \text{ or } -3 < x < \frac{2}{3}\right\}$. The solution in interval notation is $(-\infty, -5) \cup \left(-3, \frac{2}{3}\right)$.

EXAMPLE 5 Given $f(x) = 3x^3 - 3x^2 - 6x$, find all values of x for which $f(x) \geq 0$.

Solution We need to solve the inequality

$$3x^3 - 3x^2 - 6x \geq 0$$

We start by solving the equation $3x^3 - 3x^2 - 6x = 0$.

$$3x(x^2 - x - 2) = 0$$
$$3x(x - 2)(x + 1) = 0$$
$$3x = 0 \quad \text{or} \quad x - 2 = 0 \quad \text{or} \quad x + 1 = 0$$
$$x = 0 \qquad\qquad x = 2 \qquad\qquad x = -1$$

FIGURE 8.31

The solutions -1, 0, and 2 break the number line into four intervals (see Fig. 8.31). The test values that we will use are -2, $-\frac{1}{2}$, 1, and 3.

Interval	Test value	$3x^3 - 3x^2 - 6x$	≥ 0
A	-2	-24	*False*
B	$-\dfrac{1}{2}$	$\dfrac{15}{8}$	*True*
C	1	-6	*False*
D	3	36	*True*

Since the original inequality is $\geq$, the boundary values are part of the solution. The solution, intervals B and D, is illustrated on the number line in Figure 8.32a on page 590. The solution in interval notation is $[-1, 0] \cup [2, \infty)$. Figure 8.32b on page 590 shows the graph of $f(x) = 3x^3 - 3x^2 - 6x$. Notice $f(x) \geq 0$ for $-1 \leq x \leq 0$ and for $x \geq 2$, which agrees with our solution.

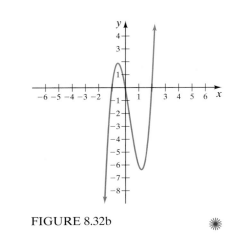

FIGURE 8.32a

FIGURE 8.32b

NOW TRY EXERCISE 41

In all the examples we have solved, the coefficient of the leading term has been a positive number.

Consider the inequality $-3x^3 + 3x^2 + 6x \leq 0$. Notice the coefficient of the leading term, $-3x^3$, is a negative number, -3. It is generally easier to solve an inequality where the coefficient of the leading term is a positive number. We can make the leading coefficient positive by multiplying both sides of the inequality by -1. When doing this, remember to reverse the inequality symbol.

$$-3x^3 + 3x^2 + 6x \leq 0$$
$$-1(-3x^3 + 3x^2 + 6x) \geq -1(0) \quad \textit{Reverse inequality symbol.}$$
$$3x^3 - 3x^2 - 6x \geq 0$$

This inequality was solved in Example 5.

3 Solve Rational Inequalities

In Examples 6 and 7, we solve **rational inequalities**, which are inequalities that contain a rational expression.

EXAMPLE 6 Solve the inequality $\dfrac{x - 1}{x + 3} \geq 2$ and graph the solution on a number line.

Solution Change the $\geq$ to $=$ and solve the resulting equation.

$$\frac{x - 1}{x + 3} = 2$$

$$\cancel{x + 3} \cdot \frac{x - 1}{\cancel{x + 3}} = 2\,(x + 3) \quad \textit{Multiply both sides by } x + 3.$$

$$x - 1 = 2x + 6$$
$$-1 = x + 6$$
$$-7 = x$$

When solving rational inequalities, we also need to determine the value or values that make the denominator 0. We set the denominator equal to 0 and solve.

$$x + 3 = 0$$
$$x = -3$$

FIGURE 8.33

We use the solution to the equation, -7, and the value that makes the denominator 0, -3, to determine the intervals, shown in Figure 8.33. We will use -8, -5, and 0 as our test values.

Interval A	Interval B	Interval C
Test value, -8	Test value, -5	Test value, 0

$$\frac{x-1}{x+3} \geq 2 \qquad \frac{x-1}{x+3} \geq 2 \qquad \frac{x-1}{x+3} \geq 2$$

$$\frac{-8-1}{-8+3} \overset{?}{\geq} 2 \qquad \frac{-5-1}{-5+3} \overset{?}{\geq} 2 \qquad \frac{0-1}{0+3} \overset{?}{\geq} 2$$

$$\frac{9}{5} \geq 2 \quad \textit{False} \qquad 3 \geq 2 \quad \textit{True} \qquad -\frac{1}{3} \geq 2 \quad \textit{False}$$

Only interval B satisfies the inequality. Whenever we have a rational inequality we must be very careful to determine which boundary values are included in the solution. Remember we can never include in our solution any value that makes the denominator 0. Now check the boundary values -7 and -3. Since -7 results in the inequality $-2 \geq -2$, which is true, -7 is a solution. Since division by 0 is not permitted, -3 is not a solution. Thus, the solution is $[-7, -3)$. The solution is illustrated on the number line in Figure 8.34. ✳

FIGURE 8.34

In Example 6, we solved $\dfrac{x-1}{x+3} \geq 2$. Suppose we graphed $f(x) = \dfrac{x-1}{x+3}$. For what values of x would $f(x) \geq 2$? If you answered $-7 \leq x < -3$ you answered correctly. Figure 8.35 shows the graph of $f(x) = \dfrac{x-1}{x+3}$ and the graph of $y = 2$. Notice $f(x) \geq 2$ when $-7 \leq x < -3$.

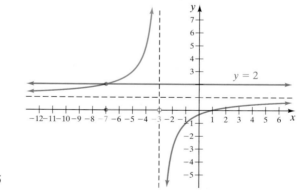

FIGURE 8.35

EXAMPLE 7 Solve the inequality $\dfrac{(x-3)(x+4)}{x+1} \geq 0$. Illustrate the solution on a number line and give the solution in interval notation.

Solution The solutions to the equation $\dfrac{(x-3)(x+4)}{x+1} = 0$ are 3 and -4 since these are the values that make the numerator equal to 0. The equation is not defined at -1. We therefore use the values 3, -4, and -1 to determine the intervals on the number line (see Fig. 8.36). Checking test values at -5, -2, 0, and 4, we find that the values in intervals B and D, $-4 < x < -1$ and $x > 3$, satisfy the inequality. Check the test values yourself to verify this. The values 3 and -4 make the inequality equal to 0 and are part of the solution. The inequality is not defined at -1, so -1 is not part of the solution. The solution is $[-4, -1) \cup [3, \infty)$. The solution is illustrated on the number line in Figure 8.37.

NOW TRY EXERCISE 71   FIGURE 8.36   FIGURE 8.37

Exercise Set 8.6

Concept/Writing Exercises

1. The graph of $f(x) = x^2 - 7x + 10$ is given. Find the solution to **a)** $f(x) > 0$ and **b)** $f(x) < 0$.

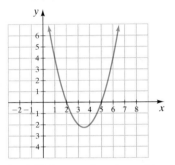

2. The graph of $f(x) = -x^2 - 4x + 5$ is given. Find the solution to **a)** $f(x) \geq 0$ and **b)** $f(x) \leq 0$.

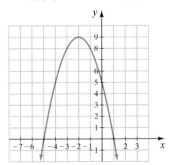

3. In solving the inequality $(x - 5)(x + 3) \geq 0$, are the boundary values 5 and -3 included in the solution set? Explain.

4. In solving the inequality $(x - 2)(x + 4) < 0$, are the boundary values 2 and -4 included in the solution set? Explain.

5. In solving the inequality $\dfrac{(x + 2)(x - 1)}{x + 1} \leq 0$, are the boundary values -2 and 1 included in the solution set? Is the boundary value -1 included in the solution set? Explain.

6. In solving the inequality $\dfrac{(x + 3)}{(x + 4)(x - 2)} \geq 0$, is the boundary value -3 included in the solution set? Are the boundary values -4 and 2 included in the solution set? Explain.

Practice the Skills

Solve each inequality and graph the solution on a number line.

7. $x^2 - 2x - 8 \geq 0$

8. $x^2 - 2x - 8 < 0$

9. $x^2 + 8x + 7 > 0$

10. $x^2 + 8x + 7 < 0$

11. $p^2 + 4p > 0$

12. $x^2 - 6x \geq 0$

13. $x^2 - 16 < 0$

14. $r^2 - 3r < 0$

15. $2x^2 + 5x - 3 \geq 0$

16. $3n^2 - 7n \leq 6$

17. $5x^2 + 19x \leq 4$

18. $3x^2 + 5x - 3 \leq 0$

19. $2x^2 - 12x + 9 \leq 0$

20. $5x^2 \leq -20x - 4$

Solve each inequality and give the solution in interval notation.

21. $(x - 2)(x + 1)(x + 3) \geq 0$

22. $(x - 2)(x + 2)(x + 3) \leq 0$

23. $(a - 3)(a + 2)(a + 4) < 0$

24. $(r - 1)(r + 2)(r + 5) < 0$

25. $(2c + 5)(3c - 6)(c + 6) > 0$

26. $(a - 4)(a - 2)(a + 6) > 0$

27. $(3x + 5)(x - 3)(x + 4) > 0$

28. $(3c - 1)(c + 4)(3c + 6) \leq 0$

29. $(x + 2)(x + 2)(3x - 8) \geq 0$

30. $(x + 3)^2 (4x - 5) \leq 0$

31. $x^3 - 4x^2 + 4x < 0$

32. $x^3 + 3x^2 - 18x > 0$

For each function provided, find all values of x for which $f(x)$ satisfies the indicated conditions. Graph the solution on a number line.

33. $f(x) = x^2 - 3x, f(x) \geq 0$

34. $f(x) = x^2 - 7x, f(x) > 0$

35. $f(x) = x^2 + 4x, f(x) > 0$

36. $f(x) = x^2 + 6x, f(x) \leq 0$

37. $f(x) = x^2 - 14x + 48, f(x) < 0$

38. $f(x) = x^2 + 2x - 15, f(x) < 0$

39. $f(x) = 2x^2 + 9x - 4, f(x) \leq 2$

40. $f(x) = x^2 + 5x - 4, f(x) \leq 3$

41. $f(x) = 2x^3 + 9x^2 - 35x, f(x) \geq 0$

42. $f(x) = x^3 - 5x, f(x) \leq 0$

Solve each inequality and give the solution in set builder notation.

43. $\dfrac{x + 2}{x - 4} > 0$

44. $\dfrac{x + 2}{x - 4} \geq 0$

45. $\dfrac{x - 3}{x + 5} < 0$

46. $\dfrac{x - 3}{x + 5} \leq 0$

47. $\dfrac{x + 1}{x - 1} \geq 0$

48. $\dfrac{x - 4}{x + 6} > 0$

49. $\dfrac{a - 4}{a + 5} < 0$

50. $\dfrac{b + 7}{b + 1} \leq 0$

51. $\dfrac{c - 8}{c - 4} > 0$

52. $\dfrac{2d - 6}{d - 1} < 0$

53. $\dfrac{3y + 6}{y + 6} \leq 0$

54. $\dfrac{4z - 8}{z - 7} \geq 0$

55. $\dfrac{3a + 6}{2a - 1} \geq 0$

56. $\dfrac{x + 4}{x - 4} \leq 0$

57. $\dfrac{3x + 4}{2x - 1} < 0$

58. $\dfrac{k + 3}{k} \geq 0$

59. $\dfrac{3x + 5}{x - 2} \leq 0$

60. $\dfrac{4x - 2}{2x - 4} > 0$

Solve each inequality and give the solution in interval notation.

61. $\dfrac{(x + 1)(x - 5)}{x + 3} < 0$

62. $\dfrac{(x + 1)(x - 5)}{x + 3} \leq 0$

63. $\dfrac{(x - 2)(x + 3)}{x - 4} > 0$

64. $\dfrac{(x - 2)(x + 3)}{x - 4} \geq 0$

65. $\dfrac{(a - 1)(a - 6)}{a + 2} \geq 0$

66. $\dfrac{(b - 2)(b + 4)}{b} < 0$

67. $\dfrac{c}{(c - 3)(c + 7)} \leq 0$

68. $\dfrac{z - 5}{(z + 6)(z - 7)} \geq 0$

🔒 **69.** $\dfrac{x - 6}{(x + 4)(x - 1)} \leq 0$

70. $\dfrac{x + 6}{(x - 2)(x + 4)} > 0$

71. $\dfrac{(x - 3)(2x + 5)}{x - 6} \geq 0$

72. $\dfrac{r(r - 3)}{2r + 6} < 0$

Solve each inequality and graph the solution on a number line.

73. $\dfrac{2}{x - 4} \geq 1$

74. $\dfrac{2}{x - 4} > 1$

75. $\dfrac{3}{x - 1} > -1$

76. $\dfrac{3}{x - 1} \geq -1$

77. $\dfrac{4}{x + 2} \leq 1$

78. $\dfrac{4}{x + 2} < 1$

79. $\dfrac{2p - 5}{p - 4} \leq 1$

80. $\dfrac{2}{2a - 1} > 2$

81. $\dfrac{4}{x - 2} \geq 2$

82. $\dfrac{x + 8}{x + 2} > 1$

🔒 **83.** $\dfrac{w}{3w - 2} > -2$

84. $\dfrac{x - 1}{2x + 6} \leq -3$

✏️ **85.** The graph of $y = \dfrac{x^2 - 4x + 4}{x - 4}$ is illustrated. Determine the solution to the following inequalities.

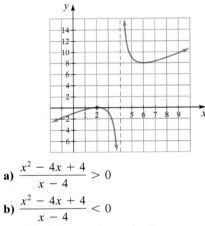

a) $\dfrac{x^2 - 4x + 4}{x - 4} > 0$

b) $\dfrac{x^2 - 4x + 4}{x - 4} < 0$

Explain how you determined your answer.

✏️ **86.** The graph of $y = \dfrac{x^2 + x - 6}{x - 4}$ is illustrated. Determine the solution to the following inequalities.

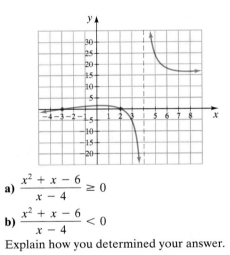

a) $\dfrac{x^2 + x - 6}{x - 4} \geq 0$

b) $\dfrac{x^2 + x - 6}{x - 4} < 0$

Explain how you determined your answer.

87. Write a quadratic inequality whose solution is

$\xleftarrow{\quad\overset{\circ}{}\quad\quad\overset{\circ}{}\quad}\rightarrow$
   −4   2

🔒 **89.** Write a rational inequality whose solution is

$\xleftarrow{\quad\overset{\circ}{}\quad\quad\overset{\circ}{}\quad}\rightarrow$
   −3   4

✏️ **91.** What is the solution to the inequality $(x + 3)^2(x - 1)^2 \geq 0$? Explain your answer.

✏️ **93.** What is the solution to the inequality $\dfrac{x^2}{(x + 1)^2} \geq 0$?

Explain your answer.

✏️ **95.** If $f(x) = ax^2 + bx + c$ where $a > 0$ and the discriminant is negative, what is the solution to $f(x) < 0$? Explain.

88. Write a quadratic inequality whose solution is

$\xleftarrow{\quad\overset{\circ}{}\quad\quad\overset{\circ}{}\quad}\rightarrow$
   −3   5

90. Write a rational inequality whose solution is

$\xleftarrow{\quad\overset{\circ}{}\quad\quad\overset{\circ}{}\quad}\rightarrow$
   −7   −1

✏️ **92.** What is the solution to the inequality $x^2(x - 3)^2(x + 4)^2 < 0$? Explain your answer.

✏️ **94.** What is the solution to the inequality $\dfrac{x^2}{(x - 3)^2} > 0$? Explain your answer.

✏️ **96.** If $f(x) = ax^2 + bx + c$ where $a < 0$ and the discriminant is negative, what is the solution to $f(x) > 2$? Explain.

Challenge Problems

Solve each inequality and graph the solution on the number line.

97. $(x + 1)(x - 3)(x + 5)(x + 9) \geq 0$

98. $\dfrac{(x - 4)(x + 2)}{x(x + 6)} \geq 0$

Write a quadratic inequality with the following solutions. Many answers are possible. Explain how you determined your answers.

99. $(-\infty, 0) \cup (3, \infty)$ **100.** $\{4\}$ **101.** $\varnothing$ **102.** $\mathbb{R}$

In Exercises 103 and 104, solve each inequality and give the solution in interval notation. Use techniques from Section 8.5 to help you find the solution.

103. $x^4 - 10x^2 + 9 > 0$

104. $x^4 - 17x^2 + 16 \leq 0$

In Exercises 105 and 106, solve each inequality using factoring by grouping. Give the solution in interval notation.

105. $x^3 + x^2 - 4x - 4 \geq 0$

106. $2x^3 + x^2 - 18y - 9 < 0$

Group Activity

Discuss and answer Exercises 107 and 108 as a group.

107. Consider the number line below, where $a, b,$ and c are distinct real numbers.

a) In which intervals will the real numbers satisfy the inequality $(x - a)(x - b)(x - c) > 0$? Explain.

b) In which intervals will the real numbers satisfy the inequality $(x - a)(x - b)(x - c) < 0$? Explain.

108. Consider the number line below where $a, b, c,$ and d are distinct real numbers.

Interval Interval Interval Interval Interval
   1   2   3   4   5

       a b c d

a) In which intervals do the real numbers satisfy the inequality $(x - a)(x - b)(x - c)(x - d) > 0$? Explain.

b) In which interval do the real numbers satisfy the inequality $(x - a)(x - b)(x - c)(x - d) < 0$? Explain.

Cumulative Review Exercises

[2.4] **109.** *Antifreeze* How many quarts of a 100% antifreeze solution should Paul Simmons add to 10 quarts of a 20% antifreeze solution to make a 50% antifreeze solution?

[3.2] **110.** If $h(x) = \dfrac{x^2 + 4x}{x + 6}$, find $h(-3)$.

[5.1] **111.** Add $(6r + 5s - t) + (-3r - 2s - 5t)$.

[6.3] **112.** Simplify $\dfrac{1 + \dfrac{x}{x + 1}}{\dfrac{2x + 1}{x - 1}}$.

[7.7] **113.** Multiply $(3 - 4i)(6 + 5i)$.

CHAPTER SUMMARY

Key Words and Phrases

8.1
Completing the square
Perfect square trinomial
Square root property

8.2
Discriminant

8.3
Solving for a variable in
   a formula

8.4
Equation quadratic in
   form

8.5
Axis of symmetry
Maximum value
Minimum value
Parabola
Symmetry
Translate a parabola
Vertex

8.6
Boundary values
Polynomial inequality
Quadratic inequality
Rational inequality
Sign graph
Solution to a quadratic
   inequality

IMPORTANT FACTS

Square Root Property: If $x^2 = a$, then $x = \pm\sqrt{a}$.

Quadratic Formula: $x = \dfrac{-b \pm \sqrt{b^2 - 4ac}}{2a}$

Discriminant: $b^2 - 4ac$

For an equation of the form $ax^2 + bx + c = 0$:

If $b^2 - 4ac > 0$, the quadratic equation has two distinct real solutions.

If $b^2 - 4ac = 0$, the quadratic equation has one real solution.

If $b^2 - 4ac < 0$, the quadratic equation has no real solution.

For $f(x) = ax^2 + bx + c$, the vertex of the parabola is $\left(-\dfrac{b}{2a}, \dfrac{4ac - b^2}{4a}\right)$ or $\left(-\dfrac{b}{2a}, f\left(-\dfrac{b}{2a}\right)\right)$.

For $f(x) = a(x - h)^2 + k$, the vertex of the parabola is (h, k).

If $f(x) = ax^2 + bx + c, a > 0$, the function will have a minimum value of $\dfrac{4ac - b^2}{4a}$ at $x = -\dfrac{b}{2a}$.

If $f(x) = ax^2 + bx + c, a < 0$, the function will have a maximum value of $\dfrac{4ac - b^2}{4a}$ at $x = -\dfrac{b}{2a}$.

Chapter Review Exercises

[8.1] Use the square root property to solve each equation.

1. $(x - 3)^2 = 24$

2. $(2x + 5)^2 = 60$

3. $\left(x - \dfrac{1}{3}\right)^2 = \dfrac{4}{9}$

4. $\left(2x - \dfrac{1}{2}\right)^2 = 4$

Solve each equation by completing the square.

5. $x^2 - 7x + 12 = 0$

6. $x^2 - 4x - 21 = 0$

7. $a^2 + 2a - 5 = 0$

8. $z^2 + 6z = 10$

9. $x^2 - 2x + 10 = 0$

10. $2r^2 - 8r = -64$

Area In Exercises 11 and 12, the area, A, of each rectangle is given. *a)* Write an equation for the area. *b)* Solve the equation for x.

11.

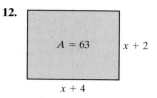

A = 32

$x + 1$

$x + 5$

12.

A = 63

$x + 2$

$x + 4$

13. *Consecutive Integers* The product of two consecutive positive integers is 72. Find the two integers.

14. *Living Room* Ronnie Sampson just moved into a new house where the living room is a square whose diagonal is 7 feet longer than the length of a side. Find the dimensions of the living room.

[8.2] Determine whether each equation has two distinct real solutions, a single real solution, or no real solution.

15. $2x^2 - 5x - 7 = 0$

16. $3x^2 + 2x = -4$

17. $r^2 + 12r = -36$

18. $5x^2 - x + 1 = 0$

19. $a^2 - 14a = -49$

20. $\frac{1}{2}x^2 - 3x = 10$

Solve each equation by the quadratic formula.

21. $3x^2 + 7x = 0$

22. $x^2 - 11x = -30$

23. $r^2 = 7r + 8$

24. $4x^2 = 9x$

25. $6a^2 + a - 15 = 0$

26. $4x^2 + 11x = 3$

27. $x^2 - 6x + 7 = 0$

28. $b^2 + 4b = 8$

29. $2x^2 + 4x - 3 = 0$

30. $3y^2 - 6y = 8$

31. $x^2 - x + 30 = 0$

32. $1.2r^2 + 5.7r = 2.3$

33. $x^2 - \frac{5}{6}x = \frac{25}{6}$

34. $2x^2 + \frac{5}{2}x - \frac{3}{4} = 0$

For the given function, determine all real values of the variable for which the function has the value indicated.

35. $f(x) = x^2 - 4x - 45, f(x) = 15$

36. $g(x) = 6x^2 + 5x, g(x) = 6$

37. $h(r) = 5r^2 - 7r - 6, h(r) = -4$

38. $f(x) = -2x^2 + 6x + 5, f(x) = -4$

Write a function that has the given solutions.

39. $4, -1$

40. $\frac{2}{3}, -2$

41. $-\sqrt{7}, \sqrt{7}$

42. $3 - 2i, 3 + 2i$

[8.1–8.3]

43. *Rectangular Garden* Sophia Yang is designing a rectangular flower garden. If the area is to be 88 square feet and the length is to be 3 feet greater than the width, find the dimensions of the garden.

44. *Triangle and Circle* Find the length of side x in the figure on the right.

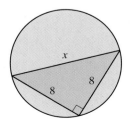

45. *Savings Account* Samuel Rivera invested $1200 in a savings account where the interest is compounded annually. If after 2 years the amount in the account is $1323, find the annual interest rate.

46. *Numbers* The larger of two positive numbers is 2 greater than the smaller. Find the two numbers if their product is 63.

47. *Rectangle* The length of a rectangle is 4 inches less than twice its width. Find the dimensions if its area is 96 square inches.

48. *Wheat Crop* The value, V, in dollars per acre of a wheat crop d days after planting is given by the formula $V = 12d - 0.05d^2$, $20 < d < 80$. Find the value of an acre of wheat 50 days after it has been planted.

49. *Falling Object* The distance, d, in feet, that an object is from the ground t seconds after being dropped from an airplane is given by the formula $d = -16t^2 + 784$.

　a) Find the distance the object is from the ground 3 seconds after it has been dropped.

　b) When will the object hit the ground?

50. *Throwing an Object* If an object is thrown upward from the top of a 100-foot-tall building, its height above the ground, $h(t)$, at any time, t, can be found by the function $h(t) = -16t^2 + 16t + 100$.

　a) Find the height of the object at 2 seconds.

　b) When will the object hit the ground?

51. *Oil Leak* A tractor has an oil leak. The amount of oil, $L(t)$, in milliliters per hour that leaks out is a function of the tractor's operating temperature, t, in degrees Celsius. The function is

$$L(t) = 0.0004t^2 + 0.16t + 20, \ 100°C \le t \le 160°C$$

　a) How many milliliters of oil will leak out in 1 hour if the operating temperature of the tractor is 100°C?

　b) If oil is leaking out at 53 milliliters per hour, what is the operating temperature of the tractor?

52. *Molding Machines* Two molding machines can complete an order in 12 hours. The larger machine can complete the order by itself in 1 hour less time than the smaller machine can by itself. How long will it take each machine to complete the order working by itself?

53. *Travel Time* Steve Forrester drove 25 miles at a constant speed, and then he increased his speed by 15 miles per hour for the next 65 miles. If the time required to travel 90 miles was 1.5 hours, find the speed he drove during the first 25 miles.

54. *Canoe Trip* Joan Banker canoed downstream going with the current for 3 miles, then turned around and canoed upstream against the current to her starting point. If the total time she spent canoeing was 4 hours and the current was 0.4 mile per hour, what is the speed she canoes in still water?

55. *Area* The area of a rectangle is 73 square units. If the length is x units and the width is $x - 2$ units, find the length and the width. Round your answer to the nearest tenth of a unit.

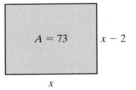

56. *Selling Tables* A business sells n tables, $n \le 40$, at a price of $(60 - 0.3n)$ dollars per table. How many tables must be sold to have a revenue of $1530?

In Exercises 57–60, solve each equation for the variable indicated.

57. $a^2 + b^2 = c^2$, for a (Pythagorean Theorem)

58. $h = -4.9t^2 + c$, for t (height of an object)

59. $v_x^2 + v_y^2 = v^2$, for v_y (vectors)

60. $a = \dfrac{v_2^2 - v_1^2}{2d}$, for v_2

[8.4] Solve each equation.

61. $x^4 - 13x^2 + 36 = 0$

62. $x^4 - 19x^2 + 48 = 0$

63. $a^4 = 5a^2 + 24$

64. $6y^{-2} + 11y^{-1} = 10$

65. $4r + 23\sqrt{r} - 6 = 0$

66. $2p^{2/3} - 7p^{1/3} + 6 = 0$

67. $6(x - 2)^{-2} = -13(x - 2)^{-1} + 8$

68. $10(r + 2) = \dfrac{12}{r + 2} - 7$

Find all x-intercepts of the given function.

69. $f(x) = x^4 - 29x^2 + 100$

70. $f(x) = 30x + 13\sqrt{x} - 10$

71. $f(x) = x - 6\sqrt{x} + 10$

72. $f(x) = (x^2 - 6x)^2 - 5(x^2 - 6x) - 24$

[8.5] **a)** *Determine whether the parabola opens upward or downward.* **b)** *Find the y-intercept.* **c)** *Find the vertex.* **d)** *Find the x-intercepts if they exist.* **e)** *Draw the graph.*

73. $y = x^2 + 5x$

74. $y = x^2 - 2x - 8$

75. $y = -x^2 - 9$

76. $y = -2x^2 - x + 15$

77. *Selling Tickets* The Hamilton High School Theater estimates that their total income, I, in hundreds of dollars, for their production of the play, *The King and I*, can be approximated by the formula $I = -x^2 + 22x - 30$, $2 \le x \le 20$, where x is the cost of a ticket.

a) How much should they charge to maximize their profit?

b) What is the maximum profit?

78. *Tossing a Ball* Josh Vincent tosses a ball upward from the top of a 60-foot building. The height, $s(t)$, of the ball at any time t can be determined by the function $s(t) = -16t^2 + 80t + 60$.

a) At what time will the ball attain its maximum height?

b) What is the maximum height?

Graph each function.

79. $f(x) = (x - 3)^2$

80. $f(x) = -(x + 2)^2 - 3$

81. $g(x) = -2(x + 4)^2 - 1$

82. $h(x) = \frac{1}{2}(x - 1)^2 + 3$

[8.6] *Graph the solution to each inequality on a number line.*

83. $x^2 + 7x + 6 \ge 0$

84. $x^2 + 3x - 10 \le 0$

85. $x^2 \le 11x - 20$

86. $3x^2 + 8x > 16$

87. $4x^2 - 9 \le 0$

88. $5x^2 - 25 > 0$

Solve each inequality and give the solution in set builder notation.

89. $\dfrac{x + 1}{x - 3} > 0$

90. $\dfrac{x - 5}{x + 2} \le 0$

91. $\dfrac{2x - 4}{x + 3} \ge 0$

92. $\dfrac{3x + 5}{x - 6} < 0$

93. $(x + 3)(x + 1)(x - 2) > 0$

94. $x(x - 3)(x - 5) \le 0$

Solve each inequality and give the solution in interval notation.

95. $(3x + 4)(x - 1)(x - 3) \ge 0$

96. $2x(x + 2)(x + 5) < 0$

97. $\dfrac{x(x - 4)}{x + 2} > 0$

98. $\dfrac{(x - 2)(x - 5)}{x + 3} < 0$

99. $\dfrac{x - 3}{(x + 2)(x - 5)} \ge 0$

100. $\dfrac{x(x - 5)}{x + 3} \le 0$

Solve each inequality and graph the solution on a number line.

101. $\dfrac{3}{x + 4} \ge -1$

102. $\dfrac{2x}{x - 2} \le 1$

103. $\dfrac{2x + 3}{3x - 5} < 4$

Chapter Practice Test

Solve by completing the square.

1. $x^2 + 2x - 15 = 0$

2. $x^2 - 4x = -17$

Solve by the quadratic formula.

3. $x^2 - 5x - 14 = 0$

4. $a^2 + 8a = -5$

Solve by the method of your choice.

5. $3r^2 + r = 2$

6. $p^2 + 5 = -7p$

7. Write a function that has x-intercepts $4, -\dfrac{2}{5}$.

8. Solve the formula $K = \frac{1}{2}mv^2$ for v.

9. *Cost* The cost, c, of a house in Duquoin, Illinois, is a function of the number of square feet, s, of the house. The cost of the house can be approximated by
$$c(s) = -0.01s^2 + 78s + 22{,}000, \quad 1300 \le s \le 3900$$
  a) Estimate the cost of a 2000-square-foot house.
  b) How large a house can Clarissa Skocy purchase if she wishes to spend $160,000 on a house?

10. *Trip in a Park* Tom Ficks drove his 4-wheel-drive Jeep from Anchorage, Alaska, to the Chena River State Recreation Park, a distance of 520 miles. Had he averaged 15 miles per hour faster, the trip would have taken 2.4 hours less. Find the average speed that Tom drove.

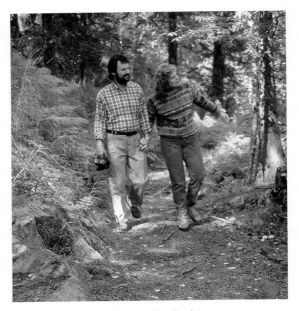

Chena River State Recreation Park

Solve.

11. $2x^4 + 15x^2 - 50 = 0$

12. $3r^{2/3} + 11r^{1/3} - 42 = 0$

13. Find all x-intercepts of $f(x) = 16x - 40\sqrt{x} + 25$.

Graph each function.

14. $f(x) = (x - 3)^2 + 2$

15. $h(x) = -\frac{1}{2}(x - 2)^2 - 2$

16. Determine whether $5x^2 = 4x + 3$ has two distinct real solutions, a single real solution, or no real solution. Explain your answer.

17. Consider the quadratic equation $y = x^2 + 2x - 8$.
  a) Determine whether the parabola opens upward or downward.
  b) Find the y-intercept.
  c) Find the vertex.
  d) Find the x-intercepts (if they exist).
  e) Draw the graph.

18. Write a quadratic function whose x-intercepts are $(-6, 0), \left(\frac{1}{2}, 0\right)$.

Solve each inequality and graph the solution on a number line.

19. $x^2 - x \ge 42$

20. $\dfrac{(x + 3)(x - 4)}{x + 1} \ge 0$

Solve the following inequality. Write the answer in a) interval notation and b) set builder notation.

21. $\dfrac{x+3}{x+2} \le -1$

22. *Carpet* The length of a rectangular Persian carpet is 4 feet greater than twice its width. Find the length and width of the carpet if its area is 48 square feet.

23. *Throwing a Ball* Jose Ramirez throws a ball upward from the top of a building. The distance, d, of the ball from the ground at any time, t, is $d = -16t^2 + 64t + 80$. How long will it take for the ball to strike the ground?

24. *Profit* The Leigh Ann Sims Company earns a weekly profit according to the function $f(x) = -1.4x^2 + 56x - 60$, where x is the number of wood carvings made and sold each week.

a) Find the number of carvings the company must sell in a week to maximize its profit.

b) What is its maximum weekly profit?

25. *Selling Brooms* A business sells n brooms, $n \le 32$, at a price of $(10 - 0.1n)$ dollars per broom. How many brooms must be sold to have a revenue of $160?

Cumulative Review Test

Take the following test and check your answers with those that appear at the end of the test. Review any questions that you answered incorrectly. The section and objective where the material was covered are indicated after the answer.

1. Evaluate $-4 \div (-2) + 16 - \sqrt{49}$.

2. Evaluate $2x^2 + 3x + 1$ when $x = 2$.

3. Express 183,000 in scientific notation.

4. Find the solution set for the equation $|4 - 2x| = 5$.

5. Simplify $4x - \{3 - [2(x-2) - 5x]\}$.

6. Solve the equation $-\dfrac{1}{2}(4x - 6) = \dfrac{1}{3}(3 - 6x) + 2$.

7. Solve the inequality $-4 < \dfrac{x+4}{2} < 8$. Write the solution in interval notation.

8. Find the slope and y-intercept of the graph of $9x + 7y = 15$.

9. *Small Orchard* The number of baskets of apples, N, that are produced by x trees in a small orchard is given by the function $N(x) = -0.2x^2 + 40x$. How many baskets of apples are produced by 50 trees?

10. Write the equation in slope–intercept form of a line passing through the points $(4, 3)$ and $(2, 1)$.

11. a) Determine whether the following graph represents a function. Explain your answer.

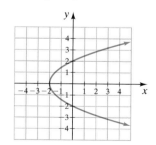

b) Determine the domain and range of the function or relation.

12. Graph each equation.

a) $x = -4$

b) $y = 2$

13. Evaluate the following determinant.

$$\begin{vmatrix} 4 & 0 & -2 \\ 3 & 5 & 1 \\ 1 & -1 & 7 \end{vmatrix}$$

14. Solve the system of equations.

$$4x - 3y = 10$$
$$2x + y = 5$$

15. Factor $(x+5)^2 + 10(x+5) + 24$.

16. a) Write an expression for the shaded area in the figure and **b)** write the expression in factored form.

17. Add $\dfrac{x+2}{x^2 - x - 6} + \dfrac{x-3}{x^2 - 8x + 15}$.

18. Solve the equation

$$\dfrac{1}{a-2} = \dfrac{4a-1}{a^2 + 5a - 14} + \dfrac{2}{a+7}$$

19. *Wattage Rating* The wattage rating of an appliance, w, varies jointly as the square of the current, I, and the resistance, R. If the wattage is 12 when the current is 2 amperes and the resistance is 100 ohms, find the wattage when the current is 0.8 ampere and the resistance is 600 ohms.

20. Simplify $\dfrac{3 - 4i}{2 + 3i}$.

Answers to Cumulative Review Test

1. 11; [Sec. 1.4, Obj. 3]   **2.** 15; [Sec. 1.4, Obj. 4]   **3.** 1.83×10^5; [Sec. 1.6, Obj. 1]   **4.** $\left\{ -\dfrac{1}{2}, \dfrac{9}{2} \right\}$; [Sec. 2.6, Obj. 2]

5. $x - 7$; [Sec. 2.1, Obj. 2]   **6.** All real numbers, $\mathbb{R}$; [Sec. 2.1, Obj. 5]   **7.** $(-12, 12)$; [Sec. 2.5, Obj. 3]

8. $m = -\dfrac{9}{7}, \left(0, \dfrac{15}{7} \right)$; [Sec. 3.4, Obj. 4]   **9.** 1500; [Sec. 3.2, Obj. 5]   **10.** $y = x - 1$; [Sec. 3.5, Obj. 1]   **11. a)** No, the graph

does not pass the vertical line test   **b)** Domain: $\{x | x \geq -2\}$, Range: $\mathbb{R}$; [Sec. 3.2, Obj. 3]

12. a)   **b)**   ; [Sec. 3.3, Obj. 3]   **13.** 160; [Sec. 4.5, Obj. 3]

14. $\left(\dfrac{5}{2}, 0 \right)$; [Sec. 4.1, Obj. 3]   **15.** $(x + 9)(x + 11)$; [Sec. 5.5, Obj. 5]   **16. a)** $a^2 + 2ab + b^2$ **b)** $(a + b)^2$; [Sec. 5.7, Obj. 1]

17. $\dfrac{2(x - 4)}{(x - 3)(x - 5)}$; [Sec. 6.2, Obj. 3]   **18.** $\dfrac{12}{5}$; [Sec. 6.4, Obj. 2]   **19.** 11.52 watts; [Sec. 6.6, Obj. 3]

20. $\dfrac{-6 - 17i}{13}$; [Sec. 7.7, Obj. 4]

Chapter 9

Exponential and Logarithmic Functions

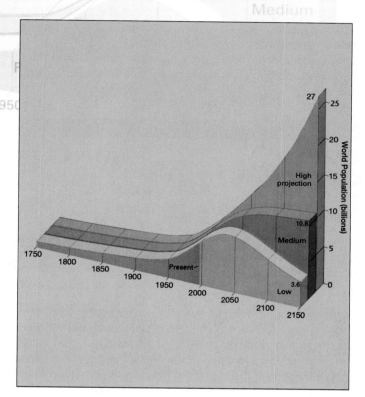

9.1 Composite and Inverse Functions

9.2 Exponential Functions

9.3 Logarithmic Functions

9.4 Properties of Logarithms

9.5 Common Logarithms

9.6 Exponential and Logarithmic Equations

9.7 Natural Exponential and Natural Logarithmic Functions

Chapter Summary
Chapter Review Exercises
Chapter Practice Test
Cumulative Review Test

The Population Division of the Department of Economic and Social Affairs uses mathematical models to make estimates and projections of the world population. The world population is growing at about 1.3% per year, down from 1.7% per year 10 years ago and 2.0% per year from 40 years ago. It is believed that the rate could be as low as 1.0% in 13 years. Since the population is growing by a percent rather than by a fixed amount, it is modeled by an exponential function rather than a linear one. In Exercise 79 on page 662, we will investigate the effect different rates have on the growth of the world population. We will also look at the doubling time for world population.

SSM

Study Guide

CD/Video

MathPro 4/5

PH Math
Tutor Center

prenhall.com/Angel

A Look Ahead

Exponential and logarithmic functions have a wide variety of uses, some of which you will see as you read through this chapter. You often read in newspaper and magazine articles that health care spending, use of the Internet, and the world population, to list just a few, are growing exponentially. By the time you finish this chapter you should have a clear understanding of just what this means.

In Section 9.1, we introduce composite and inverse functions. The remainder of the chapter is spent discussing logarithmic and exponential functions. Common logarithms, that is, logarithms to the base 10, are discussed in Section 9.5.

In Section 9.6, we introduce special functions, the natural exponential function and the natural logarithmic function. In both of these functions the base is *e*, an irrational number whose value is approximately 2.7183. Many natural phenomena, such as carbon dating, radioactive decay, and the growth of savings invested in an account compounding interest continuously, can be described by natural exponential functions.

9.1 COMPOSITE AND INVERSE FUNCTIONS

SSM · Study Guide · CD/Video · MathPro 4/5 · PH Math Tutor Center · prenhall.com/Angel

1 Find composite functions.
2 Understand one-to-one functions.
3 Find inverse functions.
4 Find the composition of a function and its inverse.

The focus of this chapter is logarithms. However, before we can study logarithms we discuss composite functions, one-to-one functions, and inverse functions. Let's start with composite functions.

1 Find Composite Functions

Often we come across situations in which one quantity is a function of one variable. That variable, in turn, is a function of some other variable. For example, the cost of advertising on a television show may be a function of the Nielsen rating of the show. The Nielsen rating, in turn, is a function of the number of people who watch the show. In the final outcome, the cost of advertising may be affected by the number of people who watch the show. Functions like this are called *composite functions*.

Let's consider another example. Suppose that 1 U.S. dollar can be converted into 1.30 Canadian dollars and that 1 Canadian dollar can be converted into 6.2 Mexican pesos. Using this information, we can convert 20 U.S. dollars into Mexican pesos. We have the following functions.

$$g(x) = 1.30x \text{ (U.S. dollars to Canadian dollars)}$$
$$f(x) = 6.2x \text{ (Canadian dollars to Mexican pesos)}$$

If we let $x = 20$, for \$20 U.S., then it can be converted into \$26 Canadian using function *g*:

$$g(x) = 1.30x$$
$$g(20) = 1.30(20) = \$26 \text{ Canadian}$$

The \$26 Canadian can, in turn, be converted into 161.20 Mexican pesos using function *f*:

$$f(x) = 6.2x$$
$$f(26) = 6.2(26) = 161.20 \text{ Mexican pesos}$$

Is there a way of finding this conversion without performing this string of calculations? The answer is yes. One U.S. dollar can be converted into Mexican pesos by substituting the $1.30x$ found in function $g(x)$ for the x in $f(x)$. This gives a new function, h, which converts U.S. dollars directly into Mexican pesos.

$$g(x) = 1.30x, \qquad f(x) = 6.2x$$
$$h(x) = f[g(x)]$$
$$= 6.2(1.30x) \quad \textit{Substitute } g(x) \textit{ for } x \textit{ in } f(x).$$
$$= 8.06x$$

Thus, for each U.S. dollar, x, we get 8.06 Mexican pesos. If we substitute \$20 for x, we get 161.20 pesos, which is what we expected.

$$h(x) = 8.06x$$
$$h(20) = 8.06(20) = 161.20$$

Function h, called a **composition of f with g**, is denoted $(f \circ g)$ and is read "f composed with g" or "f circle g." Figure 9.1 shows how the composite function h relates to functions f and g.

FIGURE 9.1

We now define **composite function**.

DEFINITION

Composite Function
The **composite function** $f \circ g$ is defined as

$$(f \circ g)(x) = f[g(x)]$$

When we are given $f(x)$ and $g(x)$, to find $(f \circ g)(x)$ we substitute $g(x)$ for x in $f(x)$ to get $f[g(x)]$.

EXAMPLE 1   Given $f(x) = x^2 - 2x + 3$ and $g(x) = x - 5$, find

a) $f(4)$ **b)** $f(a)$ **c)** $(f \circ g)(x)$ **d)** $(f \circ g)(3)$

Solution   **a)** To find $f(4)$, we substitute 4 for each x in $f(x)$.

$$f(x) = x^2 - 2x + 3$$
$$f(4) = 4^2 - 2 \cdot 4 + 3 = 16 - 8 + 3 = 11$$

b) To find $f(a)$, we substitute a for each x in $f(x)$.

$$f(x) = x^2 - 2x + 3$$
$$f(a) = a^2 - 2a + 3$$

c) $(f \circ g)(x) = f[g(x)]$. To find $(f \circ g)(x)$, we substitute $g(x)$, which is $x - 5$, for each x in $f(x)$.

$$f(\boxed{x}) = \boxed{x}^2 - 2\boxed{x} + 3$$
$$f[g(x)] = [g(x)]^2 - 2[g(x)] + 3$$

Since $g(x) = x - 5$, we substitute as follows

$$f[g(x)] = (x - 5)^2 - 2(x - 5) + 3$$
$$= (x - 5)(x - 5) - 2x + 10 + 3$$
$$= x^2 - 10x + 25 - 2x + 13$$
$$= x^2 - 12x + 38$$

Therefore, the composite function of f with g is $x^2 - 12x + 38$.

$$(f \circ g)(x) = f[g(x)] = x^2 - 12x + 38$$

d) To find $(f \circ g)(3)$, we substitute 3 for x in $(f \circ g)(x)$.

$$(f \circ g)(x) = x^2 - 12x + 38$$
$$(f \circ g)(3) = 3^2 - 12(3) + 38 = 11$$

How do you think we would determine $(g \circ f)(x)$ or $g[f(x)]$? If you answered, "Substitute $f(x)$ for each x in $g(x)$," you answered correctly. Using $f(x)$ and $g(x)$ as given in Example 1, we find $(g \circ f)(x)$ as follows.

$$g(x) = x - 5, \quad f(x) = x^2 - 2x + 3$$
$$g[f(x)] = (x^2 - 2x + 3) - 5$$
$$= x^2 - 2x + 3 - 5$$
$$= x^2 - 2x - 2$$

Therefore, the composite function of g with f is $x^2 - 2x - 2$.

$$(g \circ f)(x) = g[f(x)] = x^2 - 2x - 2$$

By comparing the illustrations above, we see that in this example, $f[g(x)] \neq g[f(x)]$.

EXAMPLE 2 Given $f(x) = x^2 + 4$ and $g(x) = \sqrt{x - 3}$, find

a) $(f \circ g)(x)$ **b)** $(g \circ f)(x)$

Solution **a)** To find $(f \circ g)(x)$, we substitute $g(x)$, which is $\sqrt{x - 3}$, for each x in $f(x)$. You should realize that $\sqrt{x - 3}$ is a real number only when $x \geq 3$.

$$f(x) = x^2 + 4$$
$$(f \circ g)(x) = f[g(x)] = (\sqrt{x - 3})^2 + 4 = x - 3 + 4 = x + 1, \; x \geq 3$$

Since values of $x < 3$ are not in the domain of $g(x)$, values of $x < 3$ are not in the domain of $(f \circ g)(x)$.

b) To find $(g \circ f)(x)$, we substitute $f(x)$, which is $x^2 + 4$, for each x in $g(x)$.

$$g(x) = \sqrt{x - 3}$$
$$(g \circ f)(x) = g[f(x)] = \sqrt{(x^2 + 4) - 3} = \sqrt{x^2 + 1}$$

EXAMPLE 3 Given $f(x) = x - 1$ and $g(x) = x + 7$, find

a) $(f \circ g)(x)$ **b)** $(f \circ g)(2)$ **c)** $(g \circ f)(x)$ **d)** $(g \circ f)(2)$

Solution **a)**
$$f(x) = x - 1$$
$$(f \circ g)(x) = f[g(x)] = (x + 7) - 1 = x + 6$$

b) We find $(f \circ g)(2)$ by substituting 2 for each x in $(f \circ g)(x)$.

$$(f \circ g)(x) = x + 6$$
$$(f \circ g)(2) = 2 + 6 = 8$$

c)
$$g(x) = \boxed{x} + 7$$
$$(g \circ f)(x) = g[f(x)] = (\boxed{x-1}) + 7 = x + 6$$

NOW TRY EXERCISE 11 **d)** Since $(g \circ f)(x) = x + 6$, $(g \circ f)(2) = 2 + 6 = 8$.

In general, $(f \circ g)(x) \neq (g \circ f)(x)$ as we saw at the end of Example 1. In Example 3, $(f \circ g)(x) = (g \circ f)(x)$, but this is only due to the specific functions used.

HELPFUL HINT

Caution Do not confuse finding the product of two functions with finding a composite function.

Product of functions f and g: $(fg)(x)$ or $(f \cdot g)(x)$

Composite function of f with g: $(f \circ g)(x)$

When multiplying functions f and g, we can use a dot between the f and g. When finding the composite function of f with g, we use a small *open* circle.

2 Understand One-to-One Functions

Consider the following two sets of ordered pairs.

$$A = \{(1, 2), (3, 5), (4, 6), (-2, 1)\}$$
$$B = \{(1, 2), (3, 5), (4, 6), (-2, 5)\}$$

Both sets of ordered pairs, A and B, are functions since each value of x has a unique value of y. In set A, each value of y also has a unique value of x, as shown in Figure 9.2. In set B, each value of y does not have a unique value of x. In the ordered pairs $(3, 5)$ and $(-2, 5)$, the y-value 5 corresponds with two values of x, as shown in Figure 9.3.

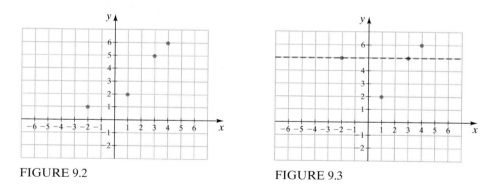

FIGURE 9.2   FIGURE 9.3

The set of ordered pairs in A is an example of a *one-to-one function*. The set of ordered pairs in B is not a one-to-one function. In a **one-to-one function**, each value in the range has a unique value in the domain. Thus, if y is a one-to-one function of x, in addition to each x-value having a unique y-value (the definition of a function), each y-value must also have a unique x-value.

DEFINITION

One-to-One Function
A function is a **one-to-one function** if each value in the range corresponds with exactly one value in the domain.

For a function to be a one-to-one function, its graph must pass not only a **vertical line test** (the test to ensure that it is a function) but also a **horizontal line test** (to test the one-to-one criteria).

Consider the function $f(x) = x^2$ (see Fig. 9.4). Note that it is a function since its graph passes the vertical line test. For each value of x, there is a unique value of y. Does each value of y also have a unique value of x? The answer is no, as illustrated in Figure 9.5. Note that for the indicated value of y there are two values of x, namely x_1 and x_2. If we limit the domain of $f(x) = x^2$ to values of x greater than or equal to 0, then each x-value has a unique y-value and each y-value also has a unique x-value (see Fig. 9.6). The function $f(x) = x^2$, $x \geq 0$, Figure 9.6, is an example of a one-to-one function.

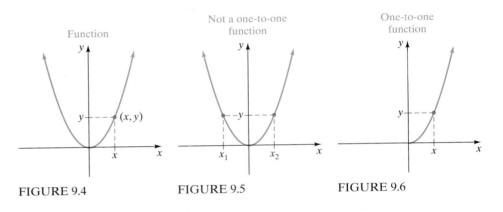

FIGURE 9.4   FIGURE 9.5   FIGURE 9.6

In Figure 9.7, the graphs in parts (a) through (e) are functions since they all pass the vertical line test. However, only the graphs in parts (a), (d), and (e) are one-to-one functions since they also pass the horizontal line test. The graph in part (f) is not a function; therefore, it is not a one-to-one function, even though it passes the horizontal line test.

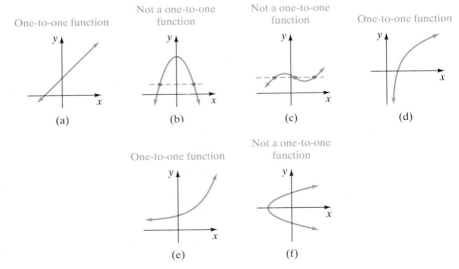

FIGURE 9.7

3 Find Inverse Functions

Now that we have discussed one-to-one functions, we can introduce inverse functions. **Only one-to-one functions have inverse functions.** If a function is one-to-one, its **inverse function** may be obtained by interchanging the first and second coordinates in each ordered pair of the function. Thus, for each ordered pair (x, y) in the function, the ordered pair (y, x) will be in the inverse function. For example,

Function: $\{(1, 4), (2, 0), (3, 7), (-2, 1), (-1, -5)\}$

Inverse function: $\{(4, 1), (0, 2), (7, 3), (1, -2), (-5, -1)\}$

Note that the domain of the function becomes the range of the inverse function, and the range of the function becomes the domain of the inverse function.

If we graph the points in the function and the points in the inverse function (Fig. 9.8), we see that the points are symmetric with respect to the line $y = x$.

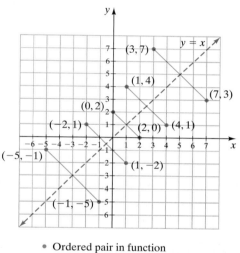

FIGURE 9.8

- Ordered pair in function
- Ordered pair in inverse function

For a function $f(x)$, the notation $f^{-1}(x)$ represents its inverse function. Note that the -1 in the notation is *not* an exponent. Therefore, $f^{-1}(x) \neq \dfrac{1}{f(x)}$.

DEFINITION

Inverse Function

If $f(x)$ is a one-to-one function with ordered pairs of the form (x, y), its **inverse function**, $f^{-1}(x)$, is a one-to-one function with ordered pairs of the form (y, x).

When a function $f(x)$ and its inverse function $f^{-1}(x)$ are graphed on the same axes, $f(x)$ *and* $f^{-1}(x)$ are *symmetric about the line* $y = x$ as seen in Figure 9.8.

When a one-to-one function is given as an equation, its inverse function can be found by the following procedure.

To Find the Inverse Function of a One-to-One Function

1. Replace $f(x)$ with y.
2. Interchange the two variables x and y.
3. Solve the equation for y.
4. Replace y with $f^{-1}(x)$ (this gives the inverse function using inverse function notation).

The following example will illustrate the procedure.

EXAMPLE 4   **a)** Find the inverse function of $f(x) = 4x + 2$.
b) On the same axes, graph both $f(x)$ and $f^{-1}(x)$.

Solution   **a)** This function is one-to-one, therefore we will follow the four-step procedure.

$$f(x) = 4x + 2 \qquad \textit{Original function}$$

Step 1 $\qquad\qquad y = 4x + 2 \qquad \textit{Replace } f(x) \textit{ with } y.$

Step 2 $\qquad\qquad x = 4y + 2 \qquad \textit{Interchange } x \textit{ and } y.$

Step 3 $\qquad\qquad x - 2 = 4y \qquad \textit{Solve for } y.$

$$\frac{x - 2}{4} = y$$

$$y = \frac{x - 2}{4}$$

Step 4 $\qquad\qquad f^{-1}(x) = \dfrac{x - 2}{4} \qquad \textit{Replace } y \textit{ with } f^{-1}(x).$

b) Below we show tables of values for $f(x)$ and $f^{-1}(x)$. The graphs of $f(x)$ and $f^{-1}(x)$ are shown in Figure 9.9.

x	$y = f(x)$
0	2
1	6

x	$y = f^{-1}(x)$
2	0
6	1

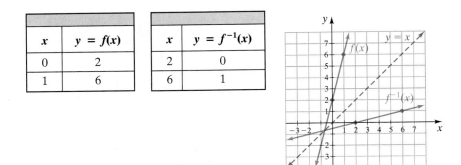

FIGURE 9.9

Note the symmetry of $f(x)$ and $f^{-1}(x)$ about the line $y = x$. Also note that both the domain and range of both $f(x)$ and $f^{-1}(x)$ are the set of real numbers, $\mathbb{R}$.

NOW TRY EXERCISE 67

In Chapter 7 when we solved equations containing cube roots, we cubed each side of the equation. To solve cubic equations, we raise each side of the equation to the one-third power, which is equivalent to taking the cube root of each side of the equation. Recall from Chapter 7 that $\sqrt[3]{a^3} = a$ for any real number a.

EXAMPLE 5   **a)** Find the inverse function of $f(x) = x^3 + 2$.

b) On the same axes, graph both $f(x)$ and $f^{-1}(x)$.

Solution   **a)** This function is one-to-one; therefore we will follow the four-step procedure to find its inverse.

$$f(x) = x^3 + 2 \qquad \text{\textit{Original function}}$$

Step 1 $$y = x^3 + 2 \qquad \text{\textit{Replace } } f(x) \text{ \textit{with } } y.$$

Step 2 $$x = y^3 + 2 \qquad \text{\textit{Interchange } } x \text{ \textit{and } } y.$$

Step 3 $$x - 2 = y^3 \qquad \text{\textit{Solve for } } y.$$
$$\sqrt[3]{x - 2} = \sqrt[3]{y^3} \qquad \text{\textit{Take the cube root of both sides.}}$$
$$\sqrt[3]{x - 2} = y$$
$$y = \sqrt[3]{x - 2}$$

Step 4 $$f^{-1}(x) = \sqrt[3]{x - 2} \qquad \text{\textit{Replace } } y \text{ \textit{with } } f^{-1}(x).$$

b) Below we show tables of values for $f(x)$ and $f^{-1}(x)$.

x	$y = f(x)$
-2	-6
-1	1
0	2
1	3
2	10

x	$y = f^{-1}(x)$
-6	-2
1	-1
2	0
3	1
10	2

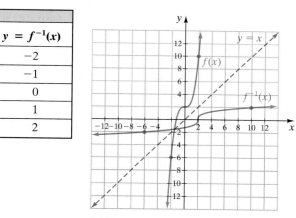

FIGURE 9.10

Notice for each point (a, b) on the graph of $f(x)$, the point (b, a) appears on the graph of $f^{-1}(x)$. For example, the points $(2, 10)$ and $(-2, -6)$, indicated in blue, appear on the graph of $f(x)$, and the points $(10, 2)$ and $(-6, -2)$, indicated in red, appear on the graph of $f^{-1}(x)$. The graphs of $f(x)$ and $f^{-1}(x)$ are shown in Figure 9.10.

NOW TRY EXERCISE 61

✳

Using Your Graphing Calculator

In Example 5, we were given $f(x) = x^3 + 2$ and found that $f^{-1}(x) = \sqrt[3]{x - 2}$. The graphs of these functions are symmetric about the line $y = x$, although it may not appear that way on a graphing calculator. Figure 9.11 on page 612 shows the two graphs using a standard calculator window.

Since the horizontal axis is longer than the vertical axis, and both axes have 10 tick marks, the graphs appear distorted. Many calculators have a feature to "square the axes." When this feature is used, the window is still rectangular, but the distance between the tick marks is equalized. To equalize the spacing between the tick marks on the vertical and horizontal axes on the TI-83 Plus calculator, press ZOOM then select

(continued on the next page)

option 5, ZSquare. Figure 9.12 shows the graphs after this option is selected. A third illustration of the graphs can be obtained using ZOOM, option 4, ZDecimal. This option resets the x-axis to go from -4.7 to 4.7 and the y-axis to go from -3.1 to 3.1, as shown in Figure 9.13.

Standard window	ZSquared window	ZDecimal window
$-10, 10, 1, -10, 10, 1$	$\approx -15.2, \approx 15.2, 1, -10, 10, 1$	$-4.7, 4.7, 1, -3.1, 3.1, 1$
FIGURE 9.11	FIGURE 9.12	FIGURE 9.13

4 Find the Composition of a Function and Its Inverse

If two functions $f(x)$ and $f^{-1}(x)$ are inverses of each other, $(f \circ f^{-1})(x) = x$ and $(f^{-1} \circ f)(x) = x$.

EXAMPLE 6 In Example 4, we determined that for $f(x) = 4x + 2$, $f^{-1}(x) = \dfrac{x-2}{4}$. Show that

a) $(f \circ f^{-1})(x) = x$ **b)** $(f^{-1} \circ f)(x) = x$

Solution **a)** To determine $(f \circ f^{-1})(x)$, substitute $f^{-1}(x)$ for each x in $f(x)$.

$$f(x) = 4\boxed{x} + 2$$
$$(f \circ f^{-1})(x) = 4\left(\boxed{\dfrac{x-2}{4}}\right) + 2$$
$$= x - 2 + 2 = x$$

b) To determine $(f^{-1} \circ f)(x)$, substitute $f(x)$ for each x in $f^{-1}(x)$.

$$f^{-1}(x) = \dfrac{\boxed{x} - 2}{4}$$
$$(f^{-1} \circ f)(x) = \dfrac{\boxed{4x + 2} - 2}{4}$$
$$= \dfrac{4x}{4} = x$$

Thus, $(f \circ f^{-1})(x) = (f^{-1} \circ f)(x) = x$. ✳

EXAMPLE 7 In Example 5, we determined that $f(x) = x^3 + 2$ and $f^{-1}(x) = \sqrt[3]{x-2}$ are inverse functions. Show that

a) $(f \circ f^{-1})(x) = x$ **b)** $(f^{-1} \circ f)(x) = x$

Solution **a)** To determine $(f \circ f^{-1})(x)$, substitute $f^{-1}(x)$ for each x in $f(x)$.

$$f(x) = \boxed{x}^3 + 2$$
$$(f \circ f^{-1})(x) = (\boxed{\sqrt[3]{x-2}})^3 + 2$$
$$= x - 2 + 2 = x$$

b) To determine $(f^{-1} \circ f)(x)$, substitute $f(x)$ for each x in $f^{-1}(x)$.

$$f^{-1}(x) = \sqrt[3]{x - 2}$$
$$(f^{-1} \circ f)(x) = \sqrt[3]{(x^3 + 2) - 2}$$
$$= \sqrt[3]{x^3} = x$$

Thus, $(f \circ f^{-1})(x) = (f^{-1} \circ f)(x) = x$.

Because a function and its inverse "undo" each other, the composite of a function with its inverse results in the given value from the domain. For example, for any function $f(x)$ and its inverse $f^{-1}(x)$, $(f^{-1} \circ f)(3) = 3$, and $(f \circ f^{-1})\left(-\frac{1}{2}\right) = -\frac{1}{2}$. ✳

NOW TRY EXERCISE 77

Exercise Set 9.1

Concept/Writing Exercises

1. Explain how to find $(f \circ g)(x)$ when you are given $f(x)$ and $g(x)$.

2. Explain how to find $(g \circ f)(x)$ when you are given $f(x)$ and $g(x)$.

3. **a)** What are one-to-one functions?

   b) Explain how you may determine whether a function is a one-to-one function.

4. Do all functions have inverse functions? If not, which functions do?

5. Consider the set of ordered pairs $\{(3, 5), (4, 2), (-1, 3), (0, -2)\}$.

   a) Is this set of ordered pairs a function? Explain.

   b) Does this function have an inverse? Explain.

   c) If this function has an inverse, give the inverse function. Explain how you determined your answer.

6. Suppose $f(x)$ and $g(x)$ are inverse functions.

   a) What is $(f \circ g)(x)$ equal to?

   b) What is $(g \circ f)(x)$ equal to?

7. What is the relationship between the domain and range of a function and the domain and range of its inverse function?

8. What is the value of $(f \circ f^{-1})(6)$? Explain.

Practice the Skills

*For each pair of functions, find **a)** $(f \circ g)(x)$, **b)** $(f \circ g)(4)$, **c)** $(g \circ f)(x)$, and **d)** $(g \circ f)(4)$.*

9. $f(x) = x^2 + 1, g(x) = x + 5$

10. $f(x) = x^2 - 3, g(x) = x + 2$

11. $f(x) = x + 3, g(x) = x^2 + x - 4$

12. $f(x) = x + 2, g(x) = x^2 + 4x - 2$

13. $f(x) = \dfrac{1}{x}, g(x) = 2x + 3$

14. $f(x) = 3x + 1, g(x) = \dfrac{2}{x}$

15. $f(x) = \dfrac{2}{x}, g(x) = x^2 + 1$

16. $f(x) = x^2 - 2, g(x) = \dfrac{4}{x}$

17. $f(x) = x^2 + 1, g(x) = x^2 + 5$

18. $f(x) = x^2 - 4, g(x) = x^2 + 3$

19. $f(x) = x - 4, g(x) = \sqrt{x + 5}, x \geq -5$

20. $f(x) = \sqrt{x + 6}, x \geq -6, g(x) = x + 5$

In Exercises 21–42, determine whether each function is a one-to-one function.

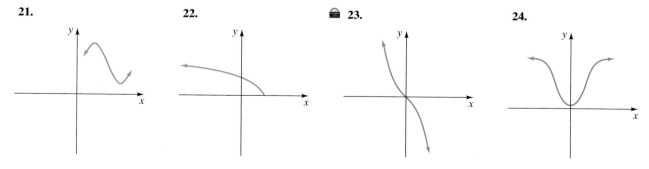

21.

22.

23.

24.

25. $\{(2, 4), (3, -7), (5, 3), (-6, 0)\}$ **26.** $\{(-4, 2), (2, 3), (4, 1), (0, 4)\}$ **27.** $\{(-4, 2), (5, 3), (0, 2), (3, 7)\}$

28. $\{(0, 5), (1, 4), (-3, 5), (4, 2)\}$ **29.** $y = 2x + 4$ **30.** $y = 3x - 6$

31. $y = x^2 - 4$ **32.** $y = -x^2 + 1$ **33.** $y = x^2 - 2x + 6$

34. $y = x^2 - 4, x \le 0$ **35.** $y = x^2 - 4, x \ge 0$ **36.** $y = x^2 - 2x + 6, x \ge 1$

37. $y = |x|$ **38.** $y = \sqrt{x}$ **39.** $y = -\sqrt{x}$

40. $y = \sqrt[3]{x}$ **41.** $y = x^3$ **42.** $y = (x - 2)^3$

For the given function, find the domain and range of both $f(x)$ and $f^{-1}(x)$.

43. $\{(4, 0), (9, 3), (2, 7), (-1, 6), (-2, 4)\}$

44. $\left\{(-2, -3), (-4, 0), (5, 3), (6, 2), \left(2, \dfrac{1}{2}\right)\right\}$

45.   **46.**   **47.**   **48.**

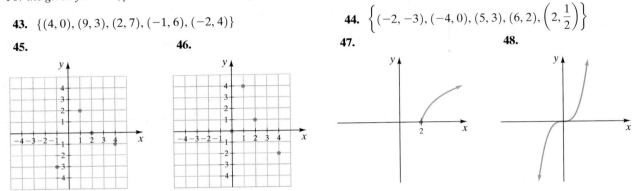

*For each function, **a)** determine whether it is one-to-one; **b)** if it is one-to-one, find its inverse function.*

49. $f(x) = x - 3$ **50.** $g(x) = x + 4$

51. $h(x) = 5x$ **52.** $k(x) = 2x - 7$

53. $p(x) = 2x^2$ **54.** $r(x) = |x|$

55. $t(x) = x^2 + 3$ **56.** $m(x) = -x^2 + x + 6$

57. $g(x) = \dfrac{1}{x}$ **58.** $h(x) = \dfrac{3}{x}$

59. $f(x) = x^2 + 4$ **60.** $g(x) = x^3 + 6$

61. $g(x) = x^3 - 5$ **62.** $f(x) = \sqrt{x}, x \ge 0$

63. $g(x) = \sqrt{x + 2}, x \ge -2$ **64.** $f(x) = x^2 - 1, x \ge 0$

65. $h(x) = x^2 - 4, x \ge 0$ **66.** $h(x) = |x|$

*For each one-to-one function, **a)** find $f^{-1}(x)$ and **b)** graph $f(x)$ and $f^{-1}(x)$ on the same axes.*

67. $f(x) = 2x + 8$ **68.** $f(x) = -3x + 6$

69. $f(x) = \sqrt{x}, x \ge 0$ **70.** $f(x) = -\sqrt{x}, x \ge 0$

71. $f(x) = \sqrt{x + 4}, x \ge -4$ **72.** $f(x) = \sqrt{x - 1}, x \ge 1$

73. $f(x) = \sqrt[3]{x}$ **74.** $f(x) = \sqrt[3]{x + 3}$

75. $f(x) = \dfrac{1}{x}, x > 0$ **76.** $f(x) = \dfrac{1}{x}$

For each pair of inverse functions, show that $(f \circ f^{-1})(x) = x$ and $(f^{-1} \circ f)(x) = x$.

77. $f(x) = x - 4, f^{-1}(x) = x + 4$

78. $f(x) = 2x + 3, f^{-1}(x) = \dfrac{x - 3}{2}$

79. $f(x) = \dfrac{1}{2}x + 5, f^{-1}(x) = 2x - 10$

80. $f(x) = -\dfrac{1}{3}x + 4, f^{-1}(x) = -3x + 12$

81. $f(x) = \sqrt[3]{x - 2}$, $f^{-1}(x) = x^3 + 2$

82. $f(x) = \sqrt[3]{x + 8}$, $f^{-1}(x) = x^3 - 8$

83. $f(x) = \dfrac{2}{x}$, $f^{-1}(x) = \dfrac{2}{x}$

84. $f(x) = \sqrt{x + 1}$, $f^{-1}(x) = x^2 - 1$, $x \geq 0$

Problem Solving

85. Is $(f \circ g)(x) = (g \circ f)(x)$ for all values of x? Explain and give an example to support your answer.

86. Consider the functions $f(x) = \sqrt{x + 5}$, $x \geq -5$, and $g(x) = x^2 - 5$, $x \geq 0$.

   a) Show that $(f \circ g)(x) = (g \circ f)(x)$ for $x \geq 0$.

   b) Explain why we need to stipulate that $x \geq 0$ for part **a)** to be true.

87. Consider the functions $f(x) = x^3 + 2$ and $g(x) = \sqrt[3]{x - 2}$.

   a) Show that $(f \circ g)(x) = (g \circ f)(x)$.

   b) What are the domains of $f(x)$, $g(x)$, $(f \circ g)(x)$, and $(g \circ f)(x)$? Explain.

88. For the function $f(x) = x^3$, $f(2) = 2^3 = 8$. Explain why $f^{-1}(8) = 2$.

89. For the function $f(x) = x^4$, $x > 0$, $f(2) = 16$. Explain why $f^{-1}(16) = 2$.

90. The function $f(x) = 12x$ converts feet, x, into inches. Find the inverse function that converts inches into feet. In the inverse function, what do x and $f^{-1}(x)$ represent?

91. The function $f(x) = 3x$ converts yards, x, into feet. Find the inverse function that converts feet into yards. In the inverse function, what do x and $f^{-1}(x)$ represent?

92. The function $f(x) = \frac{22}{15}x$ converts miles per hour, x, into feet per second. Find the inverse function that converts feet per second into miles per hour.

93. The function $f(x) = \frac{5}{9}(x - 32)$ converts degrees Fahrenheit, x, to degrees Celsius. Find the inverse function that changes degrees Celsius into degrees Fahrenheit.

94. **a)** Does the function $f(x) = |x|$ have an inverse? Explain.

   b) If the domain is limited to $x \geq 0$, does the function have an inverse? Explain.

   c) Find the inverse function of $f(x) = |x|$, $x \geq 0$.

Use your graphing calculator to determine whether the following functions are inverses.

95. $f(x) = 3x - 4$, $g(x) = \dfrac{x}{3} + \dfrac{4}{3}$

96. $f(x) = \sqrt{4 - x^2}$, $g(x) = \sqrt{4 - 2x}$

97. $f(x) = x^3 - 12$, $g(x) = \sqrt[3]{x + 12}$

98. $f(x) = x^5 + 5$, $g(x) = \sqrt[5]{x - 5}$

Challenge Problems

99. *Area* When a pebble is thrown into a pond, the circle formed by the pebble hitting the water expands with time. The area of the expanding circle may be found by the formula $A = \pi r^2$. The radius, r, of the circle, in feet, is a function of time, t, in seconds. Suppose that the function is $r(t) = 2t$.

   a) Find the radius of the circle at 3 seconds.

   b) Find the area of the circle at 3 seconds.

   c) Express the area as a function of time by finding $A \circ r$.

   d) Using the function found in part **c)**, find the area of the circle at 3 seconds.

   e) Do your answers in parts **b)** and **d)** agree? If not, explain.

100. Surface Area The surface area, S, of a spherical balloon of radius r, in inches, is found by $S(r) = 4\pi r^2$. If the balloon is being blown up at a constant rate by a machine, then the radius of the balloon is a function of time. Suppose that this function is $r(t) = 1.2t$, where t is in seconds.

a) Find the radius of the balloon at 2 seconds.

b) Find the surface area at 2 seconds.

c) Express the surface area as a function of time by finding $S \circ r$.

d) Using the function found in part b), find the surface area after 2 seconds.

e) Do your answers in parts b) and d) agree? If not, explain why not.

Group Activity

Discuss and answer Exercise 101 as a group.

101. Consider the function $f(x) = 2^x$. This is an example of an *exponential function*, which we will discuss in the next section.

a) Graph this function by substituting values for x and finding the corresponding values of $f(x)$.

b) Do you think this function has an inverse? Explain your answer.

c) Using the graph in part a), draw the inverse function, $f^{-1}(x)$ on the same axes.

d) Explain how you obtained the graph of $f^{-1}(x)$.

Cumulative Review Exercises

[1.3] 102. Divide $\left|\dfrac{-9}{4}\right| \div \left|\dfrac{-4}{9}\right|$.

[3.5] 103. Find the equation of a line in standard form that passes through $\left(\frac{1}{2}, 3\right)$ and is parallel to the graph of $2x + 3y - 9 = 0$.

[6.3] 104. Simplify $\dfrac{\dfrac{3}{x^2} - \dfrac{2}{x}}{\dfrac{x}{4}}$.

[6.4] 105. Solve the formula $\dfrac{1}{f} = \dfrac{1}{p} + \dfrac{1}{q}$ for p.

[8.1] 106. Solve the equation $x^2 + 2x - 6 = 0$ by completing the square.

9.2 EXPONENTIAL FUNCTIONS

1 Graph exponential functions.

2 Solve applications of exponential functions.

1 Graph Exponential Functions

We often read about things that are growing exponentially. For example, you may read that the world population is growing exponentially or that the use of e-mail is growing exponentially. What does this indicate? The graph in Figure 9.14 on page 617 shows population growth worldwide. The graph in Figure 9.15, also on page 617, shows the shipment of smart handheld devices. Both graphs have the same general shape, as indicated by the curves. Both graphs are exponential functions and rise rapidly.

In the quadratic function $f(x) = x^2$, the variable is the base and the exponent is a constant. In the function $f(x) = 2^x$, the constant is the base and the variable is the exponent. The function $f(x) = 2^x$ is an example of an *exponential function*, which we define on page 617.

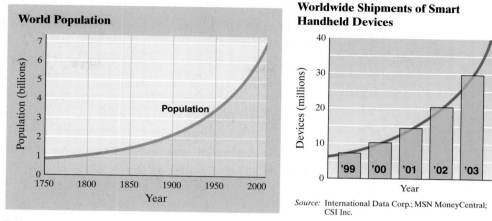

FIGURE 9.14

FIGURE 9.15

DEFINITION

Exponential Function
For any real number $a > 0$ and $a \neq 1$,

$$f(x) = a^x$$

is an **exponential function**.

An exponential function is a function of the form $f(x) = a^x$, where a is a positive real number not equal to 1. Notice the variable is in the exponent.

Examples of Exponential Functions

$$f(x) = 2^x, \qquad g(x) = 5^x, \qquad h(x) = \left(\frac{1}{2}\right)^x$$

Since $y = f(x)$, functions of the form $y = a^x$ are also exponential functions. Exponential functions can be graphed by selecting values for x, finding the corresponding values of y [or $f(x)$], and plotting the points.

Before we graph exponential functions, let's discuss some characteristics of the graphs of exponential functions.

For all exponential functions of the form $y = a^x$ or $f(x) = a^x$, where $a > 0$ and $a \neq 1$,

1. The domain of the function is $(-\infty, \infty)$.

2. The range of the function is $(0, \infty)$.

3. The graph passes through the points $\left(-1, \frac{1}{a}\right)$, $(0, 1)$, and $(1, a)$.

In most cases, a reasonably good exponential graph can be drawn from only the three points listed in item 3. When $a > 1$, the graph becomes almost horizontal to the left of $\left(-1, \frac{1}{a}\right)$ and somewhat vertical to the right of $(1, a)$, see Example 1. When $0 < a < 1$, the graph becomes almost horizontal to the right of $(1, a)$ and somewhat vertical to the left of $\left(-1, \frac{1}{a}\right)$, see Example 2.

EXAMPLE 1   Graph the exponential function $y = 2^x$. State the domain and range of the function.

Solution   The function is of the form $y = a^x$, where $a = 2$. First, construct a table of values. In the table, the three points listed in item 3 of the box are shown in red.

x	-4	-3	-2	-1	0	1	2	3	4
y	$\frac{1}{16}$	$\frac{1}{8}$	$\frac{1}{4}$	$\frac{1}{2}$	1	2	4	8	16

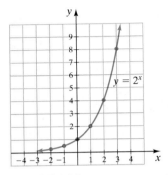

FIGURE 9.16

Now plot these points and connect them with a smooth curve (Fig. 9.16). The three ordered pairs in red in the table are marked in red on the graph.

Domain: $\mathbb{R}$

Range: $\{y | y > 0\}$

The domain of this function is the set of real numbers, $\mathbb{R}$. The range is the set of values greater than 0. If you study the equation $y = 2^x$, you should realize that y must always be positive because 2 is positive.

EXAMPLE 2   Graph $y = \left(\frac{1}{2}\right)^x$. State the domain and range of the function.

Solution   This function is of the form $y = a^x$, where $a = \frac{1}{2}$. Construct a table of values and plot the curve (Fig. 9.17).

x	-4	-3	-2	-1	0	1	2	3	4
y	16	8	4	2	1	$\frac{1}{2}$	$\frac{1}{4}$	$\frac{1}{8}$	$\frac{1}{16}$

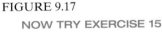

FIGURE 9.17

NOW TRY EXERCISE 15

The domain is the set of real numbers, $\mathbb{R}$. The range is $\{y | y > 0\}$.

Note that the graphs in Figures 9.16 and 9.17 are both graphs of one-to-one functions. *The graphs of exponential functions of the form $y = a^x$ are similar to Figure 9.16 when $a > 1$ and similar to Figure 9.17 when $0 < a < 1$.* Note that $y = 1^x$ is not a one-to-one function, so we exclude it from our discussion of exponential functions.

What will the graph of $y = 2^{-x}$ look like? Remember that 2^{-x} means $\frac{1}{2^x}$ or $\left(\frac{1}{2}\right)^x$. Thus, the graph of $y = 2^{-x}$ will be identical to the graph in Figure 9.17. Now consider the equation $y = \left(\frac{1}{2}\right)^{-x}$. This equation may be rewritten as $y = 2^x$ since $\left(\frac{1}{2}\right)^{-x} = 2^x$. Thus, the graph of $y = \left(\frac{1}{2}\right)^{-x}$ will be identical to the graph in Figure 9.16.

Using Your Graphing Calculator

In Figure 9.18, we show the graph of the function $y = 2^x$ on the standard window of a graphing calculator. In this chapter, we will sometimes use equations like $y = 2000(1.08)^x$. If you were to graph this function on a standard calculator window, you would not see any of the graph. Can you explain why? By observing the function, can you determine the y-intercept of the graph? To determine the y-intercept, substitute 0 for x. When you do so, you find that the y-intercept is at $2000(1.08)^0 = 2000(1) = 2000$. In Figure 9.19, we show the graph of $y = 2000(1.08)^x$.

FIGURE 9.18

$-10, 25, 5, 0, 15000, 1000$

FIGURE 9.19

2 Solve Applications of Exponential Functions

Exponential functions are often used to describe the growth and decay of certain quantities. Example 3 illustrates an exponential equation used in genetics.

EXAMPLE 3 **Plant Reproduction** The number of gametes (a special type of reproductive cell), g, in a certain species of plant is determined by the function $g(n) = 2^n$, where n is the total number of cells that an individual of the species commonly has. Determine total the number of gametes if the plant has 12 cells.

Solution By evaluating 2^{12} on a calculator, we can determine that an individual of this species, with 12 cells, has 4096 gametes. ✳

EXAMPLE 4 **Value of a Jeep** Ronald Yates just bought a new Jeep for $18,000. Assume the value of the Jeep depreciates at a rate of 20% per year. Therefore, the value of the Jeep is 80% of the previous year's value. One year from now, its value will be $18,000(0.80). Two years from now, its value will be $18,000(0.80)(0.80) = $18,000(0.80)^2$ and so on. Therefore, the formula for the value of the Jeep is

$$v(t) = 18,000(0.80)^t$$

where t is time in years. Find the value of the Jeep **a)** 1 year from now and **b)** 5 years from now.

Solution **a)** To find the value 1 year from now, substitute 1 for t.

$$v(t) = 18,000(0.80)^t$$
$$v(1) = 18,000(0.80)^1 \qquad \text{Substitute 1 for } t.$$
$$= 14,400$$

One year from now, the value of the Jeep will be $14,400.

b) To find the value 5 years from now, substitute 5 for t.

$$v(t) = 18,000(0.80)^t$$
$$v(5) = 18,000(0.80)^5 \qquad \text{Substitute 5 for } t.$$
$$= 18,000(0.32768)$$
$$= \$5898.24$$

NOW TRY EXERCISE 47   Five years from now, the value of the Jeep will be $5898.24 ✳

EXAMPLE 5 **Compound Interest** We have seen the *compound interest formula* $A = p\left(1 + \dfrac{r}{n}\right)^{nt}$ in earlier chapters. When interest is compounded periodically (yearly, monthly, quarterly), this formula can be used to find the amount, A.

    In the formula, r is the interest rate, p is the principal, n is the number of compounding periods per year, and t is the number of years. Suppose that $10,000 is invested at 5% interest compounded quarterly for 6 years. Find the amount in the account after 6 years.

Solution Understand   We are given that the principal, p, is $10,000. We are also given that the interest rate, r, is 5%. Because the interest is compounded quarterly, the number of compounding periods, n, is 4. The money is invested for 6 years. Therefore, t is 6.

Translate   Now we substitute these values into the formula

$$A = p\left(1 + \frac{r}{n}\right)^{nt}$$

$$= 10{,}000\left(1 + \frac{0.05}{4}\right)^{4(6)}$$

Carry Out

$$= 10{,}000(1 + 0.0125)^{24}$$
$$= 10{,}000(1.0125)^{24}$$
$$\approx 10{,}000(1.347351) \qquad \textit{From a calculator}$$
$$\approx 13{,}435.11$$

NOW TRY EXERCISE 39   Answer   The original \$10,000 has grown to about \$13,435.11 after 6 years.   ✳

EXAMPLE 6   **Carbon 14 Dating**   Carbon 14 dating is used by scientists to find the age of fossils and artifacts. The formula used in carbon dating is

$$A = A_0 \cdot 2^{-t/5600}$$

where A_0 represents the amount of carbon 14 present when the fossil was formed and A represents the amount of carbon 14 present after t years. If 500 grams of carbon 14 were present when an organism died, how many grams will be found in the fossil 2000 years later?

Solution   Understand   When the fossil died, 500 grams of carbon 14 were present. Therefore, $A_0 = 500$. To find out how many grams of carbon 14 will be present 2000 years later, we substitute 2000 for t in the formula.

Translate $$A = A_0 \cdot 2^{-t/5600}$$
$$= 500(2)^{-2000/5600}$$

Carry Out $$\approx 500(0.7807092) \qquad \textit{From a calculator}$$
$$\approx 390.35 \text{ grams}$$

Answer   After 2000 years, about 390.35 of the original 500 grams of carbon 14 are
NOW TRY EXERCISE 41   still present.   ✳

EXAMPLE 7   **Consumer Credit**   The amount of outstanding consumer credit has been on the rise since 1996. The graph in Figure 9.20 that appeared in *The Washington Post* shows the outstanding consumer credit in the United States from 1996 through 2001.

Consumer Credit Outstanding (seasonally adjusted)

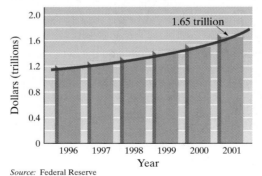

FIGURE 9.20   *Source:* Federal Reserve

An exponential function that closely approximates this curve is $f(t) = 1.156(1.074)^t$. In this function, $f(t)$ represents the outstanding credit, in trillions of dollars, and t is years since 1996. Assume that the outstanding consumer credit continues rising as in the past. Use this function to estimate the outstanding consumer credit in **a)** 2003 and **b)** 2006.

Solution **a)** Understand   In this function, t is years since 1996. The year 2003 would represent $t = 7$ since 2003 is 7 years after 1996. To find the outstanding consumer credit in 2003, we need to evaluate the function for $t = 7$.

Translate and Carry Out

$$f(t) = 1.156(1.074)^t$$
$$f(7) = 1.156(1.074)^7 \approx 1.905407 \qquad \textit{From a calculator}$$

Answer   Therefore, the outstanding consumer credit in 2003 is about $1.905407 trillion.

b) Since 2006 is 10 years after 1996, to find the outstanding consumer credit, we need to evaluate the function for $t = 10$.

$$f(t) = 1.156(1.074)^t$$
$$f(10) = 1.156(1.074)^{10} \approx 2.360482 \qquad \textit{From a calculator}$$

Answer   Therefore, the outstanding consumer credit in 2006 will be about

NOW TRY EXERCISE 51 $2.360482 trillion.

Mathemathics in Action

Linear and Exponential Models
The following data illustrate the growth in the number of people getting transplants and the number of people waiting for transplants from 1990 to 1999.

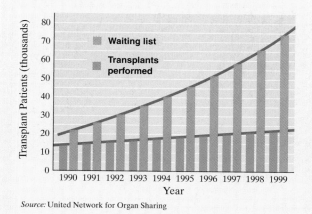

Organ Transplants

Source: United Network for Organ Sharing

Quite often data are reported in newspapers and magazines using two groups of bars so that trends and patterns can be analyzed. For example, this graph, taken from the February 23, 2001, issue of the *St. Petersburg Times*, shows that the number of people getting transplants (green bars) appears to be growing linearly (straight line) while the number of people waiting for transplants (red bars) appears to be growing exponentially. People studying graphs can often model the graphs as linear or exponential functions. For instance, to study this material, scientists, mathematicians, or statisticians might let $g(t)$ represent the straight line (green line) and $w(t)$ represent the exponential curve (in red).

The number of people waiting for transplants, $w(t)$, in thousands, can then be approximated by the function

$$w(t) = 21.303(1.147)^t$$

where t is years after 1990 ($t = 0$ represents 1990).

The number of people getting transplants, $g(t)$, in thousands, can then be approximated by the function

$$g(t) = 0.702t + 15.302$$

where t is years after 1990.

To estimate the numbers of people who will get transplants or who will be waiting for transplants in the year 2005, we compute $w(t)$ and $g(t)$ for $t = 15$ (since $2005 - 1990 = 15$).

$w(15) = 21.303(1.147)^{15} \approx 166.683$ thousand
or 166,683 people

$g(15) = 0.702(15) + 15.302 \approx 25.832$ thousand
or 25,832 people

This means that in 2005, about six times as many people will be waiting for transplants as will actually get transplants.

Exercise Set 9.2

Concept/Writing Exercises

1. What are exponential functions?

2. Consider the exponential function $y = 2^x$.
   a) As x increases, what happens to y?
   b) Can y ever be 0? Explain.
   c) Can y ever be negative? Explain.

3. Consider the exponential function $y = \left(\frac{1}{2}\right)^x$.
   a) As x increases, what happens to y?
   b) Can y ever be 0? Explain.
   c) Can y ever be negative? Explain.

4. Consider the exponential function $y = 2^{-x}$. Write an equivalent exponential function that does not contain a negative sign in the exponent. Explain how you obtained your answer.

5. Consider the equations $y = 2^x$ and $y = 3^x$.
   a) Will both graphs have the same or different y-intercepts? Determine their y-intercepts.
   b) How will the graphs of the two functions compare?

6. Consider the equations $y = \left(\frac{1}{2}\right)^x$ and $y = \left(\frac{1}{3}\right)^x$.
   a) Will both graphs have the same or different y-intercepts? Determine their y-intercepts.
   b) How will the graphs of the two functions compare?

Practice the Skills

Graph each exponential function.

7. $y = 2^x$

8. $y = 3^x$

9. $y = \left(\frac{1}{2}\right)^x$

10. $y = \left(\frac{1}{3}\right)^x$

11. $y = 4^x$

12. $y = 5^x$

13. $y = \left(\frac{1}{4}\right)^x$

14. $y = \left(\frac{1}{5}\right)^x$

15. $y = 3^{-x}$

16. $y = 4^{-x}$

17. $y = \left(\frac{1}{3}\right)^{-x}$

18. $y = \left(\frac{1}{4}\right)^{-x}$

19. $y = 2^{x-1}$

20. $y = 2^{x+1}$

21. $y = \left(\frac{1}{3}\right)^{x+1}$

22. $y = \left(\frac{1}{3}\right)^{x-1}$

23. $y = 2^x - 1$

24. $y = 2^x + 1$

25. $y = 3^x - 1$

26. $y = 3^x + 2$

Problem Solving

27. We stated earlier that, for exponential functions $f(x) = a^x$, the value of a cannot equal 1.
    a) What does the graph of $f(x) = a^x$ look like when $a = 1$?
    b) Is $f(x) = a^x$ a function when $a = 1$?
    c) Does $f(x) = a^x$ have an inverse function when $a = 1$? Explain your answer.

28. How will the graphs of $y = a^x$ and $y = a^x + k, k > 0$, compare?

29. How will the graphs of $y = a^x$ and $y = a^x - k, k > 0$, compare?

30. For $a > 1$, how will the graphs of $y = a^x$ and $y = a^{x+1}$ compare?

31. For $a > 1$, how will the graphs of $y = a^x$ and $y = a^{x+2}$ compare?

32. a) Is $y = x^{\pi}$ an exponential function? Explain.

b) Is $y = \pi^x$ an exponential function? Explain.

33. *Population Growth* The following graph shows world population growth through history. The world population has been growing exponentially since the beginning of the industrial revolution. The following table indicates the population in 1650 (about 0.5 billion people) and each later year for which the world population doubled.

World Population Growth through History

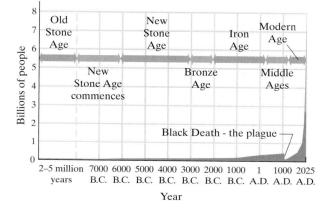

a) Draw a graph of the data provided in the table below. Place the year on the horizontal axis.

Year	Population (billions)
1650	0.5
1850	1
1930	2
1976	4
2016	8

b) How long did it take for the population to double (called the *doubling time*) from 0.5 billion to 1 billion people?

c) What is the doubling time for the population to increase from 4 billion to the projected 8 billion people?

34. *Simple and Compound Interest* The following graph indicates linear growth of $100 invested at 7% simple interest and exponential growth at 7% interest compounded annually. In the formulas, A represents the amount in dollars and t represents the time in years.

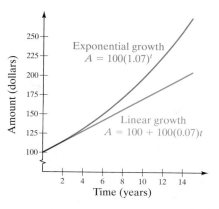

a) Use the graph to estimate the doubling time for $100 invested at 7% simple interest.

b) Estimate the doubling time for $100 invested at 7% interest compounded annually.

c) Estimate the difference in amounts after 10 years for $100 invested by each method.

d) Most banks compound interest daily instead of annually. What effect does this have on the total amount? Explain.

35. *Plant Reproduction* Use the formula $g = 2^n$ to determine the number of gametes a plant has if it has 8 cells (see Example 3).

36. *Plant Reproduction* Determine the number of gametes a plant has if it has 10 cells.

37. *Bacteria in a Petri Dish* Four bacteria are placed in a petri dish. The population will triple every day. The formula for the number of bacteria in the dish on day t is

$$N(t) = 4(3)^t$$

where t is the number of days after the four bacteria are placed in the dish. How many bacteria are in the dish 2 days after the four bacteria are placed in the dish?

38. *Bacteria in a Petri Dish* Refer to Exercise 37. How many bacteria are in the dish 6 days after the four bacteria are placed in the dish?

39. *Compound Interest* If Don Gecewicz invests $5000 at 6% interest compounded quarterly, find the amount after 4 years (see Example 5).

40. *Compound Interest* If Don Treadwell invests $8000 at 8% interest compounded quarterly, find the amount after 5 years.

41. *Carbon 14 Dating* If 12 grams of carbon 14 are originally present in a certain animal bone, how much will remain at the end of 1000 years? Use $A = A_0 \cdot 2^{-t/5600}$ (see Example 6).

42. *Carbon 14 Dating* If 60 grams of carbon 14 are originally present in the fossil Tim Jonas found at an archeological site, how much will remain after 10,000 years?

43. *Radioactive Substance* The amount of a radioactive substance present, in grams, at time t, in years, is given by the formula $y = 80(2)^{-0.4t}$. Find the number of grams present in **a)** 10 years. **b)** 100 years.

44. *Radioactive Substance* The amount of a radioactive substance present, in grams, at time t, in years, is given by the formula $y = 20(3)^{-0.6t}$. Find the number of grams present in 3 years.

45. *Population* The expected future population of Ackworth, which now has 2000 residents, can be approximated by the formula $y = 2000(1.2)^{0.1t}$, where t is the number of years in the future. Find the expected population of the town in **a)** 10 years. **b)** 100 years.

46. *Population* The expected future population of Antwerp, which currently has 6800 residents, can be approximated by the formula $y = 6800(1.4)^{-0.2t}$, where t is the number of years in the future. Find the expected population of the town 20 years in the future.

47. *Value of an SUV* The cost of a new SUV is $24,000. If it depreciates at a rate of 18% per year, the value of the SUV in t years can be approximated by the formula

$$V(t) = 24{,}000(0.82)^t$$

Find the value of the SUV in 4 years.

48. *Value of an ATV* The cost of a new all-terrain vehicle is $6200. If it depreciates at a rate of 15% per year, the value of the ATV in t years can be approximated by the formula

$$V(t) = 6200(0.85)^t$$

Find the value of the ATV in 10 years.

See Exercise 48.

49. *Water Use* The average U.S. resident used about 116,000 gallons of water in 2002. Suppose that each year after 2002 the average resident is able to reduce the amount of water used by 5%. The amount of water used by the average resident t years after 2002 could then be found by the formula $A = 116{,}000(0.95)^t$.

a) Explain why this formula may be used to find the amount of water used.

b) What would be the average amount of water used in the year 2007?

50. *Recycling Aluminum* Currently, about $\frac{2}{3}$ of all aluminum cans are recycled each year, while about $\frac{1}{3}$ are disposed of in landfills. The recycled aluminum is used to make new cans. Americans used about 190,000,000 aluminum cans in 2002. The number of new cans made each year from recycled 2002 aluminum cans n years later can be estimated by the formula $A = 190{,}000{,}000 \left(\frac{2}{3}\right)^n$.

a) Explain why the formula may be used to estimate the number of cans made from recycled aluminum cans n years after 2002.

b) How many cans will be made from 2002 recycled aluminum cans in 2009?

51. *Atmospheric Pressure* Atmospheric pressure varies with altitude. The greater the altitude, the lower the pressure, as shown in the following graph.

The equation $A = 41.97(0.996)^x$ can be used to estimate the altitude, A, in kilometers, for a given pressure, x, in millibars (mb). If the atmospheric pressure on top of Mt. Everest is about 389 mb, estimate the altitude of the top of Mt. Everest.

52. *Centenarians* Based on projections of the U.S. Census Bureau, the number of centenarians (people age 100 or older) will grow exponentially beyond 1995 (see the graph below). The function

$$f(t) = 71.24(1.045)^t$$

can be used to approximate the number of centenarians, in thousands, in the United States where t is time in years since 1995. Use this function to estimate the number of centenarians in **a)** 2060. **b)** 2070.

Number of Centenarians in the United States

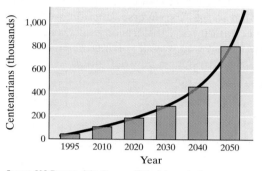

Source: U.S. Bureau of the Census; middle series projections

53. In Exercise 34, we graphed the amount for various years when $100 is invested at 7% simple interest and at 7% interest compounded annually.

a) Use the compound interest formula given in Example 5 to determine the amount if $100 is compounded daily at 7% for 10 years (assume 365 days per year).

b) Estimate the difference in the amount in 10 years for the $100 invested at 7% simple interest versus the 7% interest compounded daily.

54. Graph $y = 2^x$ and $y = 3^x$ on the same window.

55. a) Graph $y = 3^{x-5}$.

b) Use your grapher to solve the equation $4 = 3^{x-5}$. Round your answer to the nearest hundredth.

56. a) Graph $y = \left(\frac{1}{2}\right)^{2x+3}$.

b) Use your grapher to solve the equation $-3 = \left(\frac{1}{2}\right)^{2x+3}$.

Challenge Problem

57. Suppose Bob Jenkins gives Carol Dantuma $1 on day 1, $2 on day 2, $4 on day 3, $8 on day 4, and continues this doubling process for 30 days.

a) Determine how much Bob will give Carol on day 15.

b) Determine how much Bob will give Carol on day 20.

c) Express the amount, using exponential form, that Bob gives Carol on day n.

d) How much, in dollars, will Bob give Carol on day 30? Write the amount in exponential form. Then use a calculator to evaluate.

e) Express the total amount Bob gives Carol over the 30 days as a sum of exponential terms. (Do not find the actual value.)

Group Activity

58. Functions that are exponential or are approximately exponential are commonly seen.

a) Have each member of the group individually determine a function not given in this section that may approximate an exponential function. You may use newspapers, books, or other sources.

b) As a group, discuss one another's functions. Determine whether each function presented is an exponential function.

c) As a group, write a paper that discusses each of the exponential functions and state why you believe each function is exponential.

Cumulative Review Exercises

[5.1] **59.** Consider the polynomial

$$2.3x^4y - 6.2x^6y^2 + 9.2x^5y^2$$

a) Write the polynomial in descending order of the variable x.

b) What is the degree of the polynomial?

c) What is the leading coefficient?

[5.2] **60.** If $f(x) = x + 3$ and $g(x) = x^2 - 2x + 4$, find $(f \cdot g)(x)$.

[7.1] **61.** Write $\sqrt{a^2 - 8a + 16}$ as an absolute value.

[7.4] **62.** Simplify $\sqrt[4]{\dfrac{32x^5y^9}{2y^3z}}$.

9.3 LOGARITHMIC FUNCTIONS

1 Convert from exponential form to logarithmic form.
2 Graph logarithmic functions.
3 Compare the graphs of exponential and logarithmic functions.
4 Solve applications of logarithmic functions.

SSM   Study Guide   CD/Video

MathPro 4/5   PH Math Tutor Center   prenhall.com/Angel

1 Convert from Exponential Form to Logarithmic Form

Now we are ready to introduce **logarithms**. Consider the exponential function $y = 2^x$. Recall from Section 9.1 that to find the inverse function, we interchange x and y and solve the resulting equation for y. Interchanging x and y gives the equation $x = 2^y$. But at this time we have no way of solving the equation $x = 2^y$ for y. To solve this equation for y, we introduce a new definition.

DEFINITION | **Logarithm**
For all positive numbers a, where $a \neq 1$,
$$y = \log_a x \quad \text{means} \quad x = a^y$$

By the definition of logarithm, $x = 2^y$ means $y = \log_2 x$. We can therefore reason that $y = 2^x$ and $y = \log_2 x$ are inverse functions. In general, $y = a^x$ and $y = \log_a x$ are inverse functions.

In the equation $y = \log_a x$, the word *log* is an abbreviation for the word *logarithm*; $y = \log_a x$ is read "y is the logarithm of x to the base a." The letter y represents the logarithm, the letter a represents the base, and the letter x represents the number.

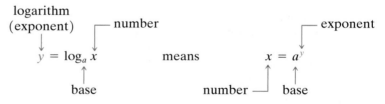

In words, the logarithm of the number x to the base a is the *exponent* to which the base a must be raised to equal the number x. In short, *a logarithm is an exponent*. For example,
$$2 = \log_{10} 100 \quad \text{means} \quad 100 = 10^2$$
In $\log_{10} 100 = 2$, the logarithm is 2, the base is 10, and the number is 100. The logarithm, 2, is the *exponent* to which the base, 10, must be raised to equal the number, 100. Note $10^2 = 100$.

Following are some examples of how an exponential expression can be converted to a logarithmic expression.

Exponential Form	Logarithmic Form
$10^0 = 1$	$\log_{10} 1 = 0$
$4^2 = 16$	$\log_4 16 = 2$
$\left(\dfrac{1}{2}\right)^5 = \dfrac{1}{32}$	$\log_{1/2} \dfrac{1}{32} = 5$
$5^{-2} = \dfrac{1}{25}$	$\log_5 \dfrac{1}{25} = -2$

Now let's do a few examples involving conversion from exponential form into logarithmic form, and vice versa.

EXAMPLE 1   Write each equation in logarithmic form.

NOW TRY EXERCISE 31

a) $3^4 = 81$ **b)** $\left(\dfrac{1}{5}\right)^3 = \dfrac{1}{125}$ **c)** $2^{-4} = \dfrac{1}{16}$

Solution   **a)** $\log_3 81 = 4$ **b)** $\log_{1/5} \dfrac{1}{125} = 3$ **c)** $\log_2 \dfrac{1}{16} = -4$

EXAMPLE 2   Write each equation in exponential form.

a) $\log_6 36 = 2$ **b)** $\log_3 9 = 2$ **c)** $\log_{1/3} \dfrac{1}{81} = 4$

Solution   **a)** $6^2 = 36$ **b)** $3^2 = 9$ **c)** $\left(\dfrac{1}{3}\right)^4 = \dfrac{1}{81}$

EXAMPLE 3   Write each equation in exponential form; then find the unknown value.

a) $y = \log_5 25$ **b)** $2 = \log_a 16$ **c)** $3 = \log_{1/2} x$

Solution   **a)** $5^y = 25$. Since $5^2 = 25$, $y = 2$.

b) $a^2 = 16$. Since $4^2 = 16$, $a = 4$. Note that a must be greater than 0, so -4 is not a possible answer for a.

NOW TRY EXERCISE 65

c) $\left(\dfrac{1}{2}\right)^3 = x$. Since $\left(\dfrac{1}{2}\right)^3 = \dfrac{1}{8}$, $x = \dfrac{1}{8}$.

2   Graph Logarithmic Functions

Now that we know how to convert from exponential form into logarithmic form and vice versa, we can graph logarithmic functions. Equations of the form $y = \log_a x$, $a > 0$, $a \neq 1$, and $x > 0$, are called **logarithmic functions**. The graphs of logarithmic functions pass the vertical line test. To graph a logarithmic function, change it to exponential form and then plot points. This procedure is illustrated in Examples 4 and 5.

Before we graph logarithmic functions, let's discuss some characteristics of the graphs of logarithmic functions.

> For all logarithmic functions of the form $y = \log_a x$ or $f(x) = \log_a x$, where $a > 0$, $a \neq 1$, and $x > 0$
>
> **1.** The domain of the function is $(0, \infty)$.
> **2.** The range of the function is $(-\infty, \infty)$.
> **3.** The graph passes through the points $\left(\dfrac{1}{a}, -1\right)$, $(1, 0)$, and $(a, 1)$.

In most cases, a reasonably good logarithmic graph can be drawn from just the three points listed in item 3. When $a > 1$, the graph becomes almost vertical to the left of $\left(\dfrac{1}{a}, -1\right)$ and somewhat horizontal to the right of $(a, 1)$, see Example 1.

When $0 < a < 1$, the graph becomes almost vertical to the left of $(a, 1)$ and somewhat horizontal to the right of $\left(\dfrac{1}{a}, -1\right)$, see Example 2.

EXAMPLE 4 Graph $y = \log_2 x$. State the domain and range of the function.

Solution This is an equation of the form $y = \log_a x$, where $a = 2$. $y = \log_2 x$ means $x = 2^y$. Using $x = 2^y$, construct a table of values. The table will be easier to develop by selecting values for y and finding the corresponding values for x. In the table, the three points listed in item 3 in the box are shown in blue.

FIGURE 9.21

x	$\dfrac{1}{16}$	$\dfrac{1}{8}$	$\dfrac{1}{4}$	$\dfrac{1}{2}$	1	2	4	8	16
y	-4	-3	-2	-1	0	1	2	3	4

Now draw the graph (Fig. 9.21). The three ordered pairs in blue in the table are marked in blue on the graph. The domain, the set of x-values, is $\{x \mid x > 0\}$. The range, the set of y-values, is the set of all real numbers, $\mathbb{R}$.

EXAMPLE 5 Graph $y = \log_{1/2} x$. State the domain and range of the function.

Solution This is an equation of the form $y = \log_a x$, where $a = \frac{1}{2}$. $y = \log_{1/2} x$ means $x = \left(\frac{1}{2}\right)^y$. Construct a table of values by selecting values for y and finding the corresponding values of x.

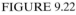

FIGURE 9.22

x	16	8	4	2	1	$\dfrac{1}{2}$	$\dfrac{1}{4}$	$\dfrac{1}{8}$	$\dfrac{1}{16}$
y	-4	-3	-2	-1	0	1	2	3	4

The graph is illustrated in Fig. 9.22. The domain is $\{x \mid x > 0\}$. The range is the set of real numbers, $\mathbb{R}$.

If we study the domains in Examples 4 and 5, we see that the domains of both $y = \log_2 x$ and $y = \log_{1/2} x$ are $\{x \mid x > 0\}$. In fact, **for any logarithmic function $y = \log_a x$, the domain is $\{x \mid x > 0\}$.** Also note that the graphs in Examples 4 and 5 are both graphs of one-to-one functions.

3 Compare the Graphs of Exponential and Logarithmic Functions

Recall that to find inverse functions we switch x and y and solve the resulting equation for y. Consider $y = a^x$. If we switch x and y, we get $x = a^y$. By our definition of logarithm, this function may be rewritten as $y = \log_a x$, which is an equation solved for y. Therefore, $y = a^x$ and $y = \log_a x$ are *inverse functions*. We may therefore write: if $f(x) = a^x$, then $f^{-1}(x) = \log_a x$.

In Figure 9.23, we show general graphs of $y = a^x$ and $y = \log_a x$, $a > 1$, on the same axes. Notice they are symmetric about the line $y = x$. Also, notice the following boxed information.

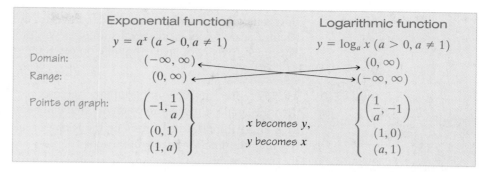

	Exponential function $y = a^x \, (a > 0, a \neq 1)$	Logarithmic function $y = \log_a x \, (a > 0, a \neq 1)$
Domain:	$(-\infty, \infty)$	$(0, \infty)$
Range:	$(0, \infty)$	$(-\infty, \infty)$
Points on graph:	$\left(-1, \dfrac{1}{a}\right)$ $(0, 1)$ $(1, a)$	*x* becomes *y*, *y* becomes *x*	$\left(\dfrac{1}{a}, -1\right)$ $(1, 0)$ $(a, 1)$

From the information in the box, we can see that the range of the exponential function is the domain of the logarithmic function, and vice versa. We can also see that the *x*- and *y*-values in the ordered pairs are switched in the exponential and logarithmic functions.

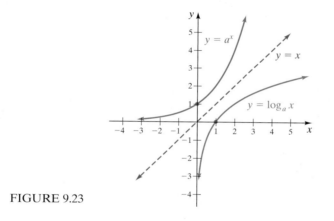

FIGURE 9.23

The graphs of $y = 2^x$ and $y = \log_2 x$ are illustrated in Figure 9.24. The graphs of $y = \left(\frac{1}{2}\right)^x$ and $y = \log_{1/2} x$ are illustrated in Figure 9.25. In each figure, the graphs are inverses of each other and are symmetric with respect to the line $y = x$.

NOW TRY EXERCISE 15   FIGURE 9.24   FIGURE 9.25

4 Solve Applications of Logarithmic Functions

We will see many applications of logarithms later, but let's look at one important application now.

EXAMPLE 6

Earthquakes Logarithms are used to measure the magnitude of earthquakes. The Richter scale for measuring earthquakes was developed by Charles R. Richter. The magnitude, R, of an earthquake on the Richter scale is given by the formula

$$R = \log_{10} I$$

where I represents the number of times greater (or more intense) the earthquake is than the smallest measurable activity that can be measured on a seismograph.

a) If an earthquake measures 4 on the Richter scale, how many times more intense is it than the smallest measurable activity?

b) How many times more intense is an earthquake that measures 5 than an earthquake that measures 4?

Solution **a)** Understand The Richter number, R, is 4. To find how many times more intense the earthquake is than the smallest measurable activity, I, we substitute $R = 4$ into the formula and solve for I.

Translate
$$R = \log_{10} I$$
$$4 = \log_{10} I$$

Carry Out
$$10^4 = I \qquad \textit{Change to exponential form.}$$
$$10{,}000 = I$$

Answer Therefore, an earthquake that measures 4 is 10,000 times more intense than the smallest measurable activity.

b)
$$5 = \log_{10} I$$
$$10^5 = I \qquad \textit{Change to exponential form.}$$
$$100{,}000 = I$$

NOW TRY EXERCISE 113

Since $(10{,}000)(10) = 100{,}000$, an earthquake that measures 5 is 10 times more intense than an earthquake that measures 4. ✳

Exercise Set 9.3

Concept/Writing Exercises

1. Consider the logarithmic function $y = \log_a x$.
  a) What are the restrictions on a?
  b) What is the domain of the function?
  c) What is the range of the function?

2. Write $y = \log_a x$ in exponential form.

3. If some points on the graph of the exponential function $f(x) = a^x$ are $\left(-3, \frac{1}{27}\right)$, $\left(-2, \frac{1}{9}\right)$, $\left(-1, \frac{1}{3}\right)$, $(0, 1)$, $(1, 3)$, $(2, 9)$, and $(3, 27)$, list some points on the graph

of the logarithmic function $g(x) = \log_a x$. Explain how you determined your answer.

4. For the logarithmic function $y = \log_a(x - 3)$, what must be true about x? Explain.

5. Discuss the relation between the graphs of $y = a^x$ and $y = \log_a x$ for $a > 0$ and $a \neq 1$.

6. What is the x-intercept of the graph of an equation of the form $y = \log_a x$?

Practice the Skills

Graph the logarithmic function.

7. $y = \log_2 x$

8. $y = \log_3 x$

9. $y = \log_{1/2} x$

10. $y = \log_{1/3} x$

11. $y = \log_5 x$

12. $y = \log_4 x$

13. $y = \log_{1/5} x$

14. $y = \log_{1/4} x$

Graph each pair of functions on the same axes.

15. $y = 2^x$, $y = \log_{1/2} x$ **16.** $y = \left(\frac{1}{2}\right)^x$, $y = \log_2 x$ 🔒 **17.** $y = 2^x$, $y = \log_2 x$ **18.** $y = \left(\frac{1}{2}\right)^x$, $y = \log_{1/2} x$

Write each equation in logarithmic form.

19. $2^3 = 8$

22. $3^5 = 243$

25. $8^{1/3} = 2$

28. $\left(\frac{1}{3}\right)^4 = \frac{1}{81}$

31. $4^{-3} = \frac{1}{64}$

34. $81^{1/2} = 9$

37. $32^{-1/5} = \frac{1}{2}$

40. $10^{1.0792} = 12$

43. $c^b = w$

20. $2^6 = 64$

23. $16^{1/2} = 4$

26. $16^{1/4} = 2$

🔒 **29.** $2^{-3} = \frac{1}{8}$

32. $5^{-4} = \frac{1}{625}$

35. $16^{-1/2} = \frac{1}{4}$

38. $81^{-1/4} = \frac{1}{3}$

41. $e^2 = 7.3891$

44. $a^n = b$

21. $3^2 = 9$

24. $49^{1/2} = 7$

27. $\left(\frac{1}{2}\right)^5 = \frac{1}{32}$

30. $6^{-2} = \frac{1}{36}$

33. $64^{1/3} = 4$

36. $8^{-1/3} = \frac{1}{2}$

39. $10^{0.6990} = 5$

42. $e^{-1/2} = 0.6065$

Write each equation in exponential form.

45. $\log_2 8 = 3$

48. $\log_{1/2} \frac{1}{64} = 6$

51. $\log_{81} 9 = \frac{1}{2}$

54. $\log_{10} \frac{1}{100} = -2$

57. $\log_6 216 = 3$

60. $\log_{10} 0.62 = -0.2076$

63. $\log_r c = -a$

46. $\log_5 125 = 3$

49. $\log_5 \frac{1}{625} = -4$

52. $\log_{64} 4 = \frac{1}{3}$

55. $\log_{27} \frac{1}{3} = -\frac{1}{3}$

58. $\log_4 1024 = 5$

61. $\log_e 6.52 = 1.8749$

64. $\log_w s = -p$

47. $\log_{1/3} \frac{1}{27} = 3$

50. $\log_4 \frac{1}{16} = -2$

53. $\log_8 \frac{1}{64} = -2$

56. $\log_{10} 1000 = 3$

59. $\log_{10} 8 = 0.9031$

62. $\log_e 30 = 3.4012$

Write each equation in exponential form; then find the unknown value.

65. $\log_4 16 = y$

69. $\log_2 x = 5$

73. $\log_{1/2} x = 2$

66. $\log_5 125 = y$

70. $\log_3 x = 3$

74. $\log_{1/3} x = 4$

67. $\log_a 81 = 4$

71. $\log_2 \frac{1}{8} = y$

🔒 **75.** $\log_a \frac{1}{27} = -3$

68. $\log_a 25 = 2$

72. $\log_7 \frac{1}{49} = y$

76. $\log_9 \frac{1}{81} = y$

Evaluate the following.

77. $\log_{10} 100$

81. $\log_{10} \frac{1}{10}$

85. $\log_4 64$

89. $\log_8 \frac{1}{64}$

🔒 **93.** $\log_9 9$

78. $\log_{10} 1000$

82. $\log_{10} \frac{1}{1000}$

86. $\log_{11} 121$

90. $\log_{12} \frac{1}{12}$

94. $\log_8 8$

79. $\log_{10} 10$

🔒 **83.** $\log_{10} 10{,}000$

87. $\log_3 \frac{1}{81}$

91. $\log_7 1$

95. $\log_4 1024$

80. $\log_{10} 1$

84. $\log_{10} 100{,}000$

88. $\log_5 \frac{1}{125}$

92. $\log_{13} 1$

96. $\log_2 128$

Problem Solving

97. If $f(x) = 4^x$, what is $f^{-1}(x)$?

98. If $f(x) = \log_5 x$, what is $f^{-1}(x)$?

99. Between which two integers must $\log_{10} 425$ lie? Explain.

100. Between which two integers must $\log_{10} 0.672$ lie? Explain.

101. Between which two integers must $\log_3 62$ lie? Explain.

102. Between which two integers must $\log_5 0.3256$ lie? Explain.

103. For $x > 1$, which will grow faster as x increases, 2^x or $\log_{10} x$? Explain.

104. For $x > 1$, which will grow faster as x increases, x or $\log_{10} x$? Explain.

Change to exponential form, then solve for x. We will discuss rules for solving problems like this in Section 9.4.

105. $x = \log_{10} 10^5$

106. $x = \log_7 7^9$

107. $x = \log_b b^3$

108. $x = \log_e e^5$

Change to logarithmic form, then solve for x. We will discuss rules for solving problems like this in Section 9.4.

109. $x = 10^{\log_{10} 8}$

110. $x = 6^{\log_6 4}$

111. $x = b^{\log_b 9}$

112. $x = c^{\log_c 2}$

113. *Earthquake* If the magnitude of an earthquake is 7 on the Richter scale, how many times more intense is the earthquake than the smallest measurable activity? Use $R = \log_{10} I$ (see Example 6).

114. *Earthquake* If the magnitude of an earthquake is 5 on the Richter scale, how many times more intense is the earthquake than the smallest measurable activity? Use $R = \log_{10} I$.

115. *Earthquake* How many times more intense is an earthquake that measures 6 on the Richter scale than an earthquake that measures 2?

116. *Earthquake* How many times more intense is an earthquake that measures 3 on the Richter scale than an earthquake that measures 1?

117. Graph $y = \log_2(x - 1)$.

118. Graph $y = \log_3(x - 2)$.

Cumulative Review Exercises

[5.4–5.7] Factor.

119. $3x^3 - 12x^2 - 36x$

120. $x^4 - 16$

121. $40x^2 + 52x - 12$

122. $6r^2s^2 + rs - 1$

9.4 PROPERTIES OF LOGARITHMS

1 Use the product rule for logarithms.

2 Use the quotient rule for logarithms.

3 Use the power rule for logarithms.

4 Use additional properties of logarithms.

1 Use the Product Rule for Logarithms

When finding the logarithm of an expression, the expression is called the **argument** of the logarithm. For example, in $\log_{10} 3$, the 3 is the argument, and in $\log_{10}(2x + 4)$, the $(2x + 4)$ is the argument. When the argument contains a variable, we assume that the argument represents a positive value. *Remember, only logarithms of positive numbers exist.*

To be able to do calculations using logarithms, you must understand their properties. The first property we discuss is the product rule for logarithms.

Product Rule for Logarithms

For positive real numbers x, y, and a, $a \neq 1$,

$$\log_a xy = \log_a x + \log_a y \qquad \textbf{Property 1}$$

This rule tells us that the logarithm of a product of two factors equals the sum of the logarithms of the factors.

To prove this property, we let $\log_a x = m$ and $\log_a y = n$. Remember, logarithms are exponents. Now we write each logarithm in exponential form.

$$\log_a x = m \quad \text{means} \quad a^m = x$$
$$\log_a y = n \quad \text{means} \quad a^n = y$$

By substitution and using the rules of exponents, we see that

$$xy = a^m \cdot a^n = a^{m+n}$$

We can now convert $xy = a^{m+n}$ into logarithmic form.

$$xy = a^{m+n} \quad \text{means} \quad \log_a xy = m + n$$

Finally, substituting $\log_a x$ for m and $\log_a y$ for n, we obtain

$$\log_a xy = \log_a x + \log_a y$$

which is property 1.

Examples of Property 1

$$\log_3 (5 \cdot 7) = \log_3 5 + \log_3 7$$
$$\log_4 3x = \log_4 3 + \log_4 x$$
$$\log_8 x^2 = \log_8 (x \cdot x) = \log_8 x + \log_8 x \text{ or } 2 \log_8 x$$

Property 1, the product rule, can be expanded to three or more factors, for example, $\log_a xyz = \log_a x + \log_a y + \log_a z$.

2 Use the Quotient Rule for Logarithms

Now we give the quotient rule for logarithms, which we refer to as property 2.

Quotient Rule for Logarithms

For positive real numbers x, y, and a, $a \neq 1$,

$$\log_a \frac{x}{y} = \log_a x - \log_a y \qquad \textbf{Property 2}$$

This rule tells us that the logarithm of a quotient equals the difference between the logarithm of the numerator and the logarithm of the denominator.

Examples of Property 2

$$\log_3 \frac{17}{4} = \log_3 17 - \log_3 4$$
$$\log_6 \frac{x}{3} = \log_6 x - \log_6 3$$
$$\log_5 \frac{x}{x+2} = \log_5 x - \log_5 (x+2)$$

3 Use the Power Rule for Logarithms

The next property we discuss is the power rule for logarithms.

> **Power Rule for Logarithms**
>
> If x and a are positive real numbers, $a \neq 1$, and n is any real number, then
>
> $$\log_a x^n = n \log_a x \qquad \textbf{Property 3}$$

This rule tells us that the logarithm of a number raised to a power equals the exponent times the logarithm of the number.

Examples of Property 3

$$\log_2 4^3 = 3 \log_2 4$$
$$\log_{10} x^2 = 2 \log_{10} x$$
$$\log_5 \sqrt{12} = \log_5 (12)^{1/2} = \frac{1}{2} \log_5 12$$
$$\log_8 \sqrt[5]{x+3} = \log_8 (x+3)^{1/5} = \frac{1}{5} \log_8 (x+3)$$

Properties 2 and 3 can be proved in a manner similar to that given for property 1 (see Exercises 79 and 80).

EXAMPLE 1   Use properties 1 through 3 to expand.

a) $\log_8 \dfrac{27}{43}$ **b)** $\log_4 (64 \cdot 180)$ **c)** $\log_{10} (32)^{1/5}$

Solution   **a)** $\log_8 \dfrac{27}{43} = \log_8 27 - \log_8 43$ *Quotient rule*

b) $\log_4 (64 \cdot 180) = \log_4 64 + \log_4 180$ *Product rule*

c) $\log_{10} (32)^{1/5} = \dfrac{1}{5} \log_{10} 32$ *Power rule*

NOW TRY EXERCISE 11

Often we will have to use two or more of these properties in the same problem.

EXAMPLE 2   Expand.

a) $\log_{10} 4(x+2)^3$

b) $\log_5 \dfrac{(4-x)^2}{3}$

c) $\log_5 \left(\dfrac{4-x}{3}\right)^2$

d) $\log_5 \dfrac{[x(x+4)]^3}{2}$

Solution   **a)** $\log_{10} 4(x+2)^3 = \log_{10} 4 + \log_{10} (x+2)^3$ *Product rule*

$= \log_{10} 4 + 3 \log_{10} (x+2)$ *Power rule*

b) $\log_5 \dfrac{(4-x)^2}{3} = \log_5 (4-x)^2 - \log_5 3$ *Quotient rule*

$= 2 \log_5 (4-x) - \log_5 3$ *Power rule*

c) $\log_5\left(\dfrac{4-x}{3}\right)^2 = 2\log_5\left(\dfrac{4-x}{3}\right)$ *Power rule*

$= 2[\log_5(4-x) - \log_5 3]$ *Quotient rule*

$= 2\log_5(4-x) - 2\log_5 3$ *Distributive property*

d) $\log_5\dfrac{[x(x+4)]^3}{2} = \log_5[x(x+4)]^3 - \log_5 2$ *Quotient rule*

$= 3\log_5 x(x+4) - \log_5 2$ *Power rule*

$= 3[\log_5 x + \log_5(x+4)] - \log_5 2$ *Product rule*

NOW TRY EXERCISE 19 $= 3\log_5 x + 3\log_5(x+4) - \log_5 2$ *Distributive property* ✳

HELPFUL HINT

In Example 2**b)**, when we expanded $\log_5\dfrac{(4-x)^2}{3}$, we first used the quotient rule. In Example 2**c)**, when we expanded $\log_5\left(\dfrac{4-x}{3}\right)^2$, we first used the power rule. Do you see the difference in the two problems? In $\log_5\dfrac{(4-x)^2}{3}$, just the numerator of the argument is squared; therefore, we use the quotient rule first. In $\log_5\left(\dfrac{4-x}{3}\right)^2$, the entire argument is squared, so we use the power rule first.

EXAMPLE 3   Write each of the following as the logarithm of a single expression.

a) $3\log_8(x+2) - \log_8 x$

b) $\log_7(x+1) + 2\log_7(x+4) - 3\log_7(x-5)$

Solution   **a)** $3\log_8(x+2) - \log_8 x = \log_8(x+2)^3 - \log_8 x$ *Power rule*

$= \log_8\dfrac{(x+2)^3}{x}$ *Quotient rule*

b) $\log_7(x+1) + 2\log_7(x+4) - 3\log_7(x-5)$

$= \log_7(x+1) + \log_7(x+4)^2 - \log_7(x-5)^3$ *Power rule*

$= \log_7(x+1)(x+4)^2 - \log_7(x-5)^3$ *Product rule*

$= \log_7\dfrac{(x+1)(x+4)^2}{(x-5)^3}$ *Quotient rule*

NOW TRY EXERCISE 39   ✳

AVOIDING COMMON ERRORS

The Correct Rules Are

$$\log_a xy = \log_a x + \log_a y$$

$$\log_a \frac{x}{y} = \log_a x - \log_a y$$

Note that:

$$\log_a(x+y) \neq \log_a x + \log_a y \qquad \log_a(xy) \neq (\log_a x)(\log_a y)$$

$$\log_a(x-y) \neq \log_a x - \log_a y \qquad \log_a(x/y) \neq \frac{\log_a x}{\log_a y}$$

4 Use Additional Properties of Logarithms

The last properties we discuss in this section will be used to solve equations in Section 9.6.

> ### Additional Properties of Logarithms
>
> If $a > 0$, and $a \neq 1$, then
>
> $$\log_a a^x = x \qquad \textbf{Property 4}$$
> and $\qquad a^{\log_a x} = x \, (x > 0) \qquad \textbf{Property 5}$

Examples of Property 4	Examples of Property 5
$\log_6 6^5 = 5$	$3^{\log_3 7} = 7$
$\log_6 6^x = x$	$5^{\log_5 x} = x \, (x > 0)$

EXAMPLE 4 Evaluate. **a)** $\log_5 25$ **b)** $\sqrt{16}^{\,\log_4 9}$

Solution **a)** $\log_5 25$ may be written as $\log_5 5^2$. By property 4,

$$\log_5 25 = \log_5 5^2 = 2$$

b) $\sqrt{16}^{\,\log_4 9}$ may be written $4^{\log_4 9}$. By property 5,

$$\sqrt{16}^{\,\log_4 9} = 4^{\log_4 9} = 9$$

NOW TRY EXERCISE 55

Exercise Set 9.4

Concept/Writing Exercises

1. In your own words, explain the product rule for logarithms.

2. In your own words, explain the quotient rule for logarithms.

3. In your own words, explain the power rule for logarithms.

4. Explain why we need to stipulate that x and y are positive real numbers when discussing the product and quotient rules.

5. Is $\log_a (xyz) = \log_a x + \log_a y + \log_a z$ a true statement? Explain.

6. Is $\log_b (x + y + z) = \log_b x + \log_b y + \log_b z$ a true statement? Explain.

Practice the Skills

Use properties 1–3 to expand.

7. $\log_4 (3 \cdot 10)$

8. $\log_5 (2 \cdot 7)$

9. $\log_8 7(x + 3)$

10. $\log_9 x(x + 2)$

11. $\log_2 \dfrac{27}{5}$

12. $\log_8 \dfrac{\sqrt{x}}{13}$

13. $\log_{10} \dfrac{\sqrt{x}}{x - 9}$

14. $\log_5 3^7$

15. $\log_6 x^7$

16. $\log_4 (r + 7)^5$

17. $\log_9 10(4)^6$

18. $\log_8 x^3(x - 2)$

19. $\log_4 \sqrt{\dfrac{a^3}{a + 2}}$

20. $\log_9 (x - 6)^3 x^2$

21. $\log_3 \dfrac{d^6}{(a - 5)^4}$

22. $\log_7 x^2(x - 12)$

23. $\log_8 \dfrac{y(y + 2)}{y^3}$

24. $\log_{10} \left(\dfrac{x}{6}\right)^2$

25. $\log_{10} \dfrac{2m}{3n}$

26. $\log_5 \dfrac{\sqrt{a} \, \sqrt[3]{b}}{\sqrt[4]{c}}$

Write as a logarithm of a single expression.

27. $\log_5 2 + \log_5 3$

29. $\log_2 9 - \log_2 5$

31. $5 \log_4 2$

33. $\log_{10} x + \log_{10}(x + 3)$

35. $2 \log_9 z - \log_9(z - 2)$

37. $2(\log_5 p - \log_5 3)$

39. $\log_2 n + \log_2(n + 4) - \log_2(n - 3)$

41. $\dfrac{1}{2}[\log_5(x - 4) - \log_5 x]$

43. $2 \log_9 5 + \dfrac{1}{3} \log_9(r - 6) - \dfrac{1}{2} \log_9 r$

45. $4 \log_6 3 - [2 \log_6(x + 3) + 4 \log_6 x]$

46. $2 \log_7(m - 4) + 3 \log_7(m + 3) - [5 \log_7 2 + 3 \log_7(m - 2)]$

28. $\log_3 4 + \log_3 11$

30. $\log_7 10 - \log_7 3$

32. $\dfrac{1}{3} \log_8 6$

34. $\log_5(x + 1) - \log_5(x + 2)$

36. $3 \log_8 x + 2 \log_8(x + 9)$

38. $\dfrac{1}{2}[\log_6(r - 1) - \log_6 r]$

40. $2 \log_5 t + 5 \log_5(t - 6) + \log_5(t - 7)$

42. $5 \log_7(a + 3) + 2 \log_7(a - 1) - \dfrac{1}{2} \log_7 a$

44. $5 \log_6(x + 3) - [2 \log_6(x - 4) + 3 \log_6 x]$

Find the value by writing each argument using the numbers 2 and/or 5 and using the values $\log_a 2 = 0.3010$ and $\log_a 5 = 0.6990$.

47. $\log_a 10$

50. $\log_a 4$

48. $\log_a 0.4$

51. $\log_a 25$

49. $\log_a 2.5$

52. $\log_a \sqrt[3]{5}$

Evaluate (see Example 4).

53. $5^{\log_5 10}$

57. $\log_3 27$

54. $\log_5 5$

58. $2 \log_9 \sqrt{9}$

55. $(2^3)^{\log_8 5}$

59. $5(\sqrt[3]{27})^{\log_3 5}$

56. $\log_8 64$

60. $\dfrac{1}{2} \log_6 \sqrt[3]{6}$

Problem Solving

61. For $x > 0$ and $y > 0$, is $\log_a \dfrac{x}{y} = \log_a xy^{-1} = \log_a x + \log_a y^{-1} = \log_a x + \log_a \dfrac{1}{y}$?

62. Read Exercise 61. By the quotient rule, $\log_a \dfrac{x}{y} = \log_a x - \log_a y$. Can we therefore conclude that $\log_a x - \log_a y = \log_a x + \log_a \dfrac{1}{y}$?

63. Use the product rule to show that
$$\log_a \dfrac{x}{y} = \log_a x + \log_a \dfrac{1}{y}$$

64. a) Explain why
$$\log_a \dfrac{3}{xy} \neq \log_a 3 - \log_a x + \log_a y$$
b) Expand $\log_a \dfrac{3}{xy}$ correctly.

65. Express $\log_a(x^2 - 4) - \log_a(x + 2)$ as a single logarithm and simplify.

66. Express $\log_a(x - 3) - \log_a(x^2 + 4x - 21)$ as a single logarithm and simplify.

67. Is $\log_a(x^2 + 8x + 16) = 2 \log_a(x + 4)$? Explain.

68. Is $\log_a(4x^2 - 20x + 25) = 2 \log_a(2x - 5)$? Explain.

If $\log_{10} x = 0.4320$, find the following.

69. $\log_{10} x^2$

71. $\log_{10} \sqrt[4]{x}$

70. $\log_{10} \sqrt{x}$

72. $\log_{10} x^{10}$

If $\log_{10} x = 0.5000$ and $\log_{10} y = 0.2000$, find the following.

73. $\log_{10} xy$

74. $\log_{10}\left(\dfrac{x}{y}\right)$

75. Using the information given in the instructions for Exercises 73 and 74, is it possible to find $\log_{10}(x + y)$? Explain.

76. Are the graphs of $y = \log_b x^2$ and $y = 2 \log_b x$ the same? Explain your answer by discussing the domains of each equation.

Use properties 1–3 to expand.

78. $\log_3 \left[\dfrac{(a^2 + b^2)(c^2)}{(a - b)(b + c)(c + d)} \right]^2$

77. $\log_2 \dfrac{\sqrt[4]{xy} \ \sqrt[3]{a}}{\sqrt[5]{a - b}}$

79. Prove the quotient rule for logarithms.

80. Prove the power rule for logarithms.

Group Activity

Discuss and answer Exercise 81 as a group.

81. Consider $\log_a \dfrac{\sqrt{x^4 y}}{\sqrt{xy^3}}$, where $x > 0$ and $y > 0$.

  a) Group member 1: Expand the expression using the quotient rule.

  b) Group member 2: Expand the expression using the product rule.

  c) Group member 3: First simplify $\dfrac{\sqrt{x^4 y}}{\sqrt{xy^3}}$, then expand the resulting logarithm.

  d) Check each other's work and make sure all answers are correct. Can this expression be simplified by all three methods?

Cumulative Review Exercises

[2.5] **82.** Solve the inequality $\dfrac{x - 4}{2} - \dfrac{2x - 5}{5} > 3$ and indicate the solution in
  a) set builder notation.
  b) interval notation.

[5.7] **83.** **a)** Write an expression for the shaded area of the figure.

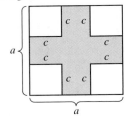

  b) Write the expression in part **a)** in factored form.

[6.4] **84.** Solve $\dfrac{15}{x} + \dfrac{9x - 7}{x + 2} = 9$ for x.

[7.7] **85.** Multiply $(3i + 4)(2i - 3)$.

[8.4] **86.** Solve $2a - 7\sqrt{a} = 30$ for a.

9.5 COMMON LOGARITHMS

1   Find common logarithms of powers of 10.
2   Find common logarithms.
3   Find antilogarithms.

1   Find Common Logarithms of Powers of 10

The properties discussed in Section 9.4 can be used with any valid base (a real number greater than 0 and not equal to 1). However, since we are used to working in base 10, we will often use the base 10 when computing with logarithms. **Base 10 logarithms** are called **common logarithms**. When we are working with common logarithms, it is not necessary to list the base. Thus, $\log x$ means $\log_{10} x$.

The properties of logarithms written as common logarithms follow. For positive real numbers x and y, and any real number n,

1. $\log xy = \log x + \log y$

2. $\log \dfrac{x}{y} = \log x - \log y$

3. $\log x^n = n \log x$

The logarithms of most numbers are irrational numbers. Even the values given by calculators are usually only approximations of the actual values. *Even though we are working with approximations when evaluating most logarithms, we generally write the logarithm with an equal sign.* Thus, rather than writing $\log 6 \approx 0.77815$, we will write $\log 6 = 0.77815$. The values given for logarithms are accurate to at least four decimal places.

In Chapter 1 we learned that 1 can be expressed as 10^0 and 10 can be expressed as 10^1. Since, for example, 5 is between 1 and 10, it must also be between 10^0 and 10^1.

$$1 < 5 < 10$$
$$10^0 < 5 < 10^1$$

The number 5 can be expressed as the base 10 raised to an exponent between 0 and 1. The number 5 is approximately equal to $10^{0.69897}$. As is done with logarithms, when writing exponential expressions, we often use the equal sign, even when the values are only approximations. Thus, for example, we will generally write $10^{0.69897} = 5$ rather than $10^{0.69897} \approx 5$.

If you look up log 5 on a calculator, as will be explained shortly, it will display a value about 0.69897. Notice that

$$\log 5 = \boxed{0.69897} \quad \text{and} \quad 5 = 10^{0.69897}$$

We can see that *the common logarithm*, 0.69897, *is the exponent* on the base 10. Now we are ready to define common logarithms.

DEFINITION

> **Common Logarithm**
> The **common logarithm** of a positive real number is the *exponent* to which the base 10 is raised to obtain the number.
>
> $$\text{If } \log N = L, \quad \text{then} \quad 10^L = N.$$

For example, if $\log 5 = 0.69897$, then $10^{0.69897} = 5$.

Now consider the number 50.

$$10 < 50 < 100$$
$$10^1 < 50 < 10^2$$

The number 50 can be expressed as the base 10 raised to an exponent between 1 and 2. The number $50 = 10^{1.69897}$; thus $\log 50 = 1.69897$.

2 Find Common Logarithms

To find common logarithms of numbers, we can use a calculator that has a logarithm key, $\boxed{\text{LOG}}$.

Using Your Calculator

Finding Common Logarithms

Scientific Calculator

To find common logarithms, enter the number, then press the logarithm key. The answer will then be displayed.

EXAMPLE	KEYS TO PRESS	ANSWER DISPLAYED
Find log 400.	400 LOG	2.60206
Find log 0.0538.	0.0538 LOG	−1.2692177

Graphing Calculator

On some graphing calculators, you first press the LOG key and then you enter the number. For example, on the TI-83 Plus, you would do the following:

EXAMPLE	KEYS TO PRESS	ANSWER DISPLAYED
Find log 400.	LOG (400) ENTER	2.602059991

↑
Generated by calculator

EXAMPLE 1 Find the exponent to which the base 10 must be raised to obtain the number 43,600.

Solution *We are asked to find the exponent, which is a logarithm.* We need to determine log 43,600. Using a calculator, we find

$$\log 43{,}600 = 4.6394865$$

NOW TRY EXERCISE 7 Thus, the exponent is 4.6394865. Note that $10^{4.6394865} = 43{,}600$. ✳

3 Find Antilogarithms

The question that should now be asked is, "If we know the common logarithm of a number, how do we find the number?" For example, if log $N = 3.406$, what is N? To find N, the number, we need to determine the value of $10^{3.406}$. Since

$$10^{3.406} = 2546.830253$$

$N = 2546.830253$. The number 2546.830253 is the *antilogarithm* of 3.406.

When we find the value of the number from the logarithm, we say we are finding the **antilogarithm** or **inverse logarithm**. If the logarithm of N is L, then N is the antilogarithm or inverse logarithm of L.

DEFINITION | **Antilogarithm**
If log $N = L$, then $N = $ antilog L.

When we are given the common logarithm, which is the exponent on the base 10, the *antilog is the number* obtained when the base 10 is raised to that exponent.

Examples

Number *Exponent*
↓ ↓
$\log 962 = 2.9831751$

Exponent *Number*
↓ ↓
antilog $2.9831751 = 962$

Number *Exponent*
↓ ↓
$\log 0.00046 = -3.3372422$

Exponent *Number*
↓ ↓
antilog $-3.3372422 = 0.00046$

When finding an antilog, we start with the logarithm, or the exponent, and end with the number equal to 10 raised to that logarithm or exponent. If antilog $-3.3372422 = 0.00046$ then $10^{-3.3372422} = 0.00046$.

Using Your Calculator

Finding Antilogarithms
Scientific Calculator

To find antilogarithms on a scientific calculator, enter the logarithm, and press the $\boxed{2^{nd}}$, $\boxed{INV}$, or $\boxed{Shift}$ key, depending upon which of these keys your calculator has. Then press the $\boxed{LOG}$ key. After the $\boxed{LOG}$ key is pressed, the antilog will be displayed.

EXAMPLE	KEYS TO PRESS	ANSWER DISPLAYED
Find antilog 2.9831751.	2.9831751 $\boxed{INV}$ $\boxed{LOG}$	962.00006*
Find antilog -3.3372422.	3.3372422 $\boxed{+/-}$ $\boxed{INV}$ $\boxed{LOG}$	0.00046**

When you are finding the antilog of a negative value, enter the value and then press the $\boxed{+/-}$ key before pressing the inverse and logarithm keys.

* Some calculators give slightly different answers, depending on their electronics.
** Some calculators may display answers in scientific notation form.

Graphing Calculator

On most graphing calculators, you press the $\boxed{2^{nd}}$ then the $\boxed{LOG}$ key before you enter the logarithm.

On the TI-83 Plus and on certain other calculators, 10^x is printed directly above the $\boxed{LOG}$ key. The antilog is actually the value of 10^x, where x is the logarithm. When you press $\boxed{2^{nd}}$ $\boxed{LOG}$, the TI-83 Plus displays 10^ followed by a left parentheses. You then enter the logarithm followed by the $\boxed{)}$ key. After you press $\boxed{ENTER}$, the antilog is displayed.

EXAMPLE	KEYS TO PRESS	ANSWER DISPLAYED
Find antilog 2.9831751.	$\boxed{2^{nd}}$ $\boxed{LOG}$ * (2.9831751 $\boxed{)}$ $\boxed{ENTER}$	962.0000619
Find antilog -3.3372422.	$\boxed{2^{nd}}$ $\boxed{LOG}$ ($\boxed{(-)}$ 3.3372422 $\boxed{)}$ $\boxed{ENTER}$	4.599999664E⁻4**

* Left parenthesis is generated by the TI-83 Plus.
** Recall from scientific notation that this number is 0.0004599999664.

Since we generally do not need the accuracy given by most calculators, in the exercise set that follows we will round logarithms to four decimal places and antilogarithms to three **significant digits**. In a number written in decimal form, any zeros preceding the first nonzero digit are not significant digits. The first nonzero digit in a number, moving from left to right, is the first significant digit.

Examples

0.0063402	First significant digit is shaded.
3.0424080	First three significant digits are shaded.
0.0000138483	First three significant digits are shaded.
206,435.05	First four significant digits are shaded.

Although most antilogarithms will be irrational numbers, when writing antilogarithms we will use an equal sign rather than an approximately equal to sign, just as we did when evaluating logarithms. All antilogarithms will be accurate to at least three significant digits.

EXAMPLE 2 Find the value obtained when the base 10 is raised to the -1.052 power.

Solution We are asked to find the value of $10^{-1.052}$. Since we are given the exponent, or logarithm, we can find the value by taking the antilog of -1.052.

$$\text{antilog } -1.052 = 0.0887156$$

NOW TRY EXERCISE 55   Thus, $10^{-1.052} = 0.0887$ rounded to three significant digits.

EXAMPLE 3 Find N if $\log N = 3.742$.

Solution We are given the logarithm and asked to find the antilog, or the number N.

$$\text{antilog } 3.742 = 5520.7744$$

NOW TRY EXERCISE 33   Thus, $N = 5520.7744$.

EXAMPLE 4 Find the following antilogs and round to three significant digits.

a) antilog 6.827   **b)** antilog -2.35

Solution **a)** Using a calculator, we find antilog $6.827 = 6,714,288.5$. Rounding to three significant digits, we get antilog $6.827 = 6,710,000$.

b) Using a calculator, we find antilog $-2.35 = 0.0044668$. Rounding to three significant digits, we get antilog $-2.35 = 0.00447$.

EXAMPLE 5 **Earthquake** The magnitude of an earthquake on the Richter scale is given by the formula $R = \log I$, where I is the number of times more intense the quake is than the smallest measurable activity. How many times more intense is an earthquake measuring 7.8 on the Richter scale than the smallest measurable activity?

Solution We want to find the value for I. We are given that $R = 7.8$. Substitute 7.8 for R in the formula $R = \log I$, and then solve for I.

$$R = \log I$$
$$7.8 = \log I \qquad \textit{Substitute 7.8 for R.}$$

To find I, we need to take the antilog of both sides of the equation.

$$\text{antilog } 7.8 = I$$
$$63,100,000 = I$$

Thus, this earthquake is about 63,100,000 times more intense than the smallest
NOW TRY EXERCISE 87   measurable activity.

Mathematics in Action

Radioactive Waste in Nevada

The disposal of waste radioactive material is one of the great challenges facing societies whose use of nuclear materials continues to grow. In 2002, after more than 20 years and $4 billion in scientific study, the President of the United States signed a bill authorizing the creation of a nuclear waste depository at Yucca Mountain, Nevada. Spent nuclear fuel and high-level radioactive waste make up most of the material to be disposed of at Yucca Mountain. About 90 percent of this waste is from commercial nuclear power plants; the remaining is from defense programs. This waste is currently stored at facilities in 43 states.

(continued on the next page)

Spent nuclear fuel and high-level radioactive waste contain short- and long-lived radiation-emitting radionuclides. Most of the radionuclides in this waste will decay to insignificant levels within several hundred years. However, some radionuclides will take many thousands of years—300,000 to 800,000 years for some—to decay to non-threatening levels. The decay of radioactivity can be described with exponential functions.

Yucca Mountain is a 1,200-foot high, flat-topped, volcanic ridge extending six miles from north to south. Initially thought to be a good candidate because it seemed arid, Yucca has turned out to be quite wet underground. The mountain has a deep water table. The repository is being built approximately 1,000 feet below the land surface and 1,000 feet above the water table.

Yucca's location is in an active earthquake region. There are about 33 earthquake faults in and around the site. Department of Energy scientists have said a repository in Yucca Mountain could be built to withstand earthquakes registering up to 6.5 on the logarithmic Richter scale for faults close to the mountain.

Exercise Set 9.5

Concept/Writing Exercises

1. What are common logarithms?
2. Write log $N = L$ in exponential form.
3. What are antilogarithms?
4. If log 652 = 2.8142, what is antilog 2.8142?

Practice the Skills

Find the common logarithm of each number. Round the answer to four decimal places.

5. 45
6. 352
7. 19,200
8. 1000
9. 0.0613
10. 27,700
11. 100
12. 0.000835
13. 3.75
14. 0.375
15. 0.000472
16. 0.00872

Find the antilog of each logarithm, Round the answer to three significant digits.

17. 0.4193
18. 1.3845
19. 4.6283
20. 5.8149
21. −1.7086
22. −3.0192
23. 0.0000
24. 5.5922
25. 2.7625
26. −4.1390
27. −0.1543
28. −1.2549

Find each number N. Round N to three significant digits.

29. log $N = 2.0000$
30. log $N = 1.2948$
31. log $N = -2.103$
32. log $N = 1.9330$
33. log $N = 4.1409$
34. log $N = 3.3817$
35. log $N = -1.06$
36. log $N = -1.1469$
37. log $N = -0.6218$
38. log $N = 1.9036$
39. log $N = -0.3936$
40. log $N = -1.3206$

To what exponent must the base 10 be raised to obtain each value? Round your answer to four decimal places.

41. 3560
42. 817,000
43. 0.0727
44. 0.00612
45. 102
46. 8.92
47. 0.00128
48. 73,700,000

Find the value obtained when 10 is raised to the following exponents. Round your answer to three significant digits.

49. 2.9153
50. 3.2473
51. −0.158
52. −2.2351
53. −1.4802
54. 4.5619
55. 1.3503
56. −2.1918

By changing the logarithm to exponential form, evaluate the common logarithm without the use of a calculator.

57. log 1
58. log 100
59. log 0.1
60. log 1000
61. log 0.01
62. log 10
63. log 0.001
64. log 10,000

In Section 9.4, we stated that for $a > 0$, and $a \neq 1$, $\log_a a^x = x$ and $a^{\log_a x} = x\,(x > 0)$. Rewriting these properties using common logarithms $(a = 10)$, we obtain $\log 10^x = x$ and $10^{\log x} = x\,(x > 0)$, respectively. Use these properties to evaluate the following.

65. $\log 10^7$

66. $\log 10^{8.3}$

67. $10^{\log 7}$

68. $10^{\log 8.3}$

69. $6 \log 10^{5.2}$

70. $8 \log 10^{1.2}$

71. $5(10^{\log 9.4})$

72. $2.3(10^{\log 5.2})$

Problem Solving

73. On your calculator, you find log 462 and obtain the value 1.6646. Can this value be correct? Explain.

74. On your calculator, you find log 6250 and obtain the value 2.7589. Can this value be correct? Explain.

75. On your calculator, you find log 0.163 and obtain the value −2.7878. Can this value be correct? Explain.

76. On your calculator, you find log−1.23 and obtain the value 0.08991. Can this value be correct? Explain.

77. Is $\log \dfrac{y}{3x} = \log y - \log 3 + \log x$? Explain.

78. Is $\log \dfrac{5x^2}{2} = 2\,(\log 5 + \log x) - \log 2$? Explain.

If log 25 = 1.3979 and log 5 = 0.6990, find the answer if possible. If it is not possible to find the answer, indicate so. Do not find the logarithms on your calculator except to check answers.

79. $\log 125$

80. $\log \dfrac{1}{5}$

81. $\log 30$

82. $\log 625$

83. $\log \dfrac{1}{25}$

84. $\log \sqrt{5}$

Solve Exercises 85–88 using $R = \log I$ (see Example 5). Round your answer to three significant digits.

85. Find I if $R = 3.41$.

86. Find I if $R = 4.96$.

87. Find I if $R = 6.37$.

88. Find I if $R = 8.15$.

89. *Astronomy* In astronomy, a formula used to find the diameter, in kilometers, of minor planets (also called asteroids) is $\log d = 3.7 - 0.2g$, where g is a quantity called the absolute magnitude of the minor planet. Find the diameter of a minor planet if its absolute magnitude is **a)** 11 and **b)** 20. **c)** Find the absolute magnitude of the minor planet whose diameter is 5.8 kilometers.

90. *Standardized Test* The average score on a standardized test is a function of the number of hours studied for the test. The average score, $f(x)$, in points, can be approximated by $f(x) = \log 0.3x + 1.8$, where x is the number of hours studied for the test. The maximum possible score on the test is 4.0. Find the score received by the average person who studied for **a)** 15 hours. **b)** 55 hours.

91. *Learning Retention* Sammy Barcia just finished a course in biology. The percent of the course he will re-member t months later can be approximated by the function

$$R(t) = 94 - 46.8 \log(t + 1)$$

for $0 \le t \le 48$. Find the percent of the course Sammy will remember **a)** 2 months later and **b)** 48 months later.

92. *Learning Retention* Karen Frye just finished a course in psychology. The percent of the course she will remember t months later can be approximated by the function

$$R(t) = 85 - 41.9 \log(t + 1)$$

for $0 \le t \le 48$. Find the percent of the course she will remember **a)** 10 months later. **b)** 36 months later.

93. *Earthquake* How many times more intense is an earthquake having a Richter scale number of 4.6 than the smallest measurable activity? See Example 5.

94. Sport-Utility Vehicles Since 1992, Sport-utility vehicles (SUVs) sales in the United States have been on the rise. The number of sales each year, $f(t)$, in millions, can be approximated by the function $f(t) = 0.98 + 1.97 \log(t + 1)$ where $t = 0$ represents 1992, $t = 1$ represents 1993, and so on. If this trend continues, estimate the number of SUVs sold in **a)** 2003. **b)** 2005.

95. Energy of an Earthquake A formula sometimes used to estimate the seismic energy released by an earthquake is $\log E = 11.8 + 1.5 m_s$, where E is the seismic energy and m_s is the surface wave magnitude.

a) Find the energy released in an earthquake whose surface wave magnitude is 6.

b) If the energy released during an earthquake is 1.2×10^{15}, what is the magnitude of the surface wave?

96. Sound Pressure The sound pressure level, s_p, is given by the formula $s_p = 20 \log \dfrac{p_r}{0.0002}$, where p_r is the sound pressure in dynes/cm^2.

a) Find the sound pressure level if the sound pressure is 0.0036 dynes/cm^2

b) If the sound pressure level is 10.0, find the sound pressure.

97. Earthquake The Richter scale, used to measure the strength of earthquakes, relates the magnitude, M, of the earthquake to the release of energy, E, in ergs, by the formula

$$M = \frac{\log E - 11.8}{1.5}$$

An earthquake releases 1.259×10^{21} ergs of energy. What is the magnitude of such an earthquake on the Richter scale?

98. pH of a Solution The pH is a measure of the acidity or alkalinity of a solution. The pH of water, for example, is 7. In general, acids have pH numbers less than 7 and alkaline solutions have pH numbers greater than 7. The pH of a solution is defined as pH $= -\log[H_3O^+]$, where H_3O^+ represents the hydronium ion concentration of the solution. Find the pH of a solution whose hydronium ion concentration is 2.8×10^{-3}.

Challenge Problems

99. Solve the formula $R = \log I$ for I.

100. Solve the formula $\log E = 11.8 + 1.5 m$ for E.

101. Solve the formula $R = 85 - 41.9 \log(t + 1)$ for t.

102. Solve the formula $f = 30 - \log x$ for x.

Group Activity

103. In section 9.7, we introduce the *change of base formula*, $\log_a x = \dfrac{\log_b x}{\log_b a}$, where a and b are bases and x is a positive number.

a) Group member 1: Use the change of base formula to evaluate $\log_3 45$. (*Hint*: Let $b = 10$.)

b) Group member 2: Repeat part **a)** for $\log_5 30$.

c) Group member 3: Repeat part **a)** for $\log_6 40$.

d) As a group, use the fact that $\log_a x = \dfrac{\log_b x}{\log_b a}$, where $b = 10$, to graph the equation $y = \log_2 x$ for $x > 0$. Use a grapher if available.

Cumulative Review Exercises

[4.3] **104. Cars** Two cars start at the same point in Alexandria, Virginia, and travel in opposite directions. One car travels 5 miles per hour faster than the other car. After 4 hours, the two cars are 420 miles apart. Find the speed of each car.

[4.5] **105.** Solve the system of equations.
$$3r = -4s - 6$$
$$3s = -5r + 1$$

[5.8] **106.** Solve $4x^3 + 4x^2 - 48x = 0$ for x.

[7.1] **107.** Write $\sqrt{(3x^2 - y)^2}$ as an absolute value.

[8.6] **108.** Solve $(x - 3)(x + 4)(x - 2) \le 0$ and give the solution in interval notation.

9.6 EXPONENTIAL AND LOGARITHMIC EQUATIONS

1 Solve exponential and logarithmic equations.

2 Solve applications.

1 Solve Exponential and Logarithmic Equations

In Sections 9.2 and 9.3 we introduced **exponential** and **logarithmic equations**. In this section we give more examples of their use and discuss further procedures for solving such equations.

To solve exponential and logarithmic equations, we often use the following properties.

> ### Properties for Solving Exponential and Logarithmic Equations
>
> **a.** If $x = y$, then $a^x = a^y$.
>
> **b.** If $a^x = a^y$, then $x = y$.
>
> **c.** If $x = y$, then $\log_b x = \log_b y$ $(x > 0, y > 0)$.
>
> **d.** If $\log_b x = \log_b y$, then $x = y$ $(x > 0, y > 0)$.   **Properties 6a–6d**

We will be referring to these properties when explaining the solutions to the examples in this section.

EXAMPLE 1 Solve the equation $8^x = \dfrac{1}{2}$.

Solution To solve this equation, we will write both sides of the equation with the same base, 2, and then use property 6b.

$$8^x = \frac{1}{2}$$

$$(2^3)^x = \frac{1}{2} \quad \textit{Write 8 as } 2^3.$$

$$2^{3x} = 2^{-1} \quad \textit{Write 1/2 as } 2^{-1}.$$

Using property 6b, we can write

$$3x = -1$$

$$x = -\frac{1}{3}$$

NOW TRY EXERCISE 7

When both sides of the exponential equation cannot be written as a power of the same base, we often begin by taking the logarithm of both sides of the equation, as in Example 2. In the following examples, we will round logarithms to the nearest ten-thousandth.

EXAMPLE 2 Solve the equation $5^n = 20$.

Solution Take the logarithm of both sides of the equation and solve for n.

$$\log 5^n = \log 20$$
$$n \log 5 = \log 20 \qquad \text{Power rule}$$
$$n = \frac{\log 20}{\log 5} \qquad \text{Divide both sides by } \log 5.$$
$$\approx \frac{1.3010}{0.6990} \approx 1.8612$$

NOW TRY EXERCISE 23

Some logarithmic equations can be solved by expressing the equation in exponential form. **It is necessary to check logarithmic equations for extraneous solutions.** When checking a solution, if you obtain the logarithm of a nonpositive number, the solution is extraneous.

EXAMPLE 3 Solve the equation $\log_2 (x + 1)^3 = 4$.

Solution Write the equation in exponential form.

$$(x + 1)^3 = 2^4 \qquad \text{Write in exponential form.}$$
$$(x + 1)^3 = 16$$
$$x + 1 = \sqrt[3]{16} \qquad \text{Take the cube root of both sides.}$$
$$x = -1 + \sqrt[3]{16} \quad \text{Solve for } x.$$

Check
$$\log_2 (x + 1)^3 = 4$$
$$\log_2 \left[(-1 + \sqrt[3]{16}) + 1\right]^3 \overset{?}{=} 4$$
$$\log_2 (\sqrt[3]{16})^3 \overset{?}{=} 4$$
$$\log_2 16 \overset{?}{=} 4 \qquad \left(\sqrt[3]{16}\right)^3 = 16$$
$$2^4 \overset{?}{=} 16 \qquad \text{Write in exponential form.}$$
$$16 = 16 \qquad \text{True}$$

NOW TRY EXERCISE 43

Other logarithmic equations can be solved using the properties of logarithms given in earlier sections.

EXAMPLE 4 Solve the equation $\log (3x + 2) + \log 9 = \log (x + 5)$.

Solution
$$\log (3x + 2) + \log 9 = \log (x + 5)$$
$$\log [(3x + 2)(9)] = \log (x + 5) \quad \text{Product rule}$$
$$(3x + 2)(9) = (x + 5) \qquad \text{Property 6d}$$
$$27x + 18 = x + 5$$
$$26x + 18 = 5$$
$$26x = -13$$
$$x = -\tfrac{1}{2}$$

Check for yourself that the solution is $-\tfrac{1}{2}$.

EXAMPLE 5 Solve the equation $\log x + \log (x + 1) = \log 12$.

Solution
$$\log x + \log (x + 1) = \log 12$$
$$\log x(x + 1) = \log 12 \qquad \text{Product rule}$$

$$x(x + 1) = 12 \qquad \text{Property 6d}$$
$$x^2 + x = 12$$
$$x^2 + x - 12 = 0$$
$$(x + 4)(x - 3) = 0$$
$$x + 4 = 0 \quad \text{or} \quad x - 3 = 0$$
$$x = -4 \qquad\qquad x = 3$$

Check $\quad x = -4 \qquad\qquad\qquad x = 3$

$\log x + \log (x + 1) = \log 12 \qquad \log x + \log (x + 1) = \log 12$

$\log (-4) + \log (-4 + 1) \overset{?}{=} \log 12 \qquad \log 3 + \log (3 + 1) \overset{?}{=} \log 12$

$\log (-4) + \log (-3) \overset{?}{=} \log 12 \qquad \log 3 + \log 4 \overset{?}{=} \log 12$

Stop. ↑   ↑ $\qquad\qquad\qquad \log [(3)(4)] \overset{?}{=} \log 12$

Logarithms of negative numbers are not $\qquad\qquad \log 12 = \log 12 \quad$ *True*
real numbers.

Thus, −4 is an extraneous solution. The only solution is 3.

Using Your Graphing Calculator

We have shown how equations in one variable may be solved graphically. Logarithmic and exponential equations may also be solved graphically by graphing each side of the equation and finding the x-coordinate of the point of intersection of the two graphs. In Example 5, we found that the solution to the equation $\log x + \log (x + 1) = \log 12$ was $x = 3$. Figure 9.26 shows the graphical solution to this equation. The horizontal line is the graph of $y = \log 12$ since $\log 12$ is a constant. Notice that the x-coordinate of the point of intersection of the two graphs, 3, is the solution to the equation.

−2, 10, 1, −1, 2, 1

FIGURE 9.26

EXAMPLE 6 Solve the equation $\log (3x - 5) - \log 5x = 1.23$.

Solution
$$\log (3x - 5) - \log 5x = 1.23$$
$$\log \frac{3x - 5}{5x} = 1.23 \qquad \text{Quotient rule}$$
$$\frac{3x - 5}{5x} = \text{antilog } 1.23 \qquad \begin{array}{l}\text{Take the antilog of}\\\text{both sides.}\end{array}$$
$$\frac{3x - 5}{5x} = 17.0 \qquad \begin{array}{l}\text{Rounded to three}\\\text{significant digits.}\end{array}$$
$$3x - 5 = 5x(17.0) \qquad \text{Multiply both sides by 5x.}$$
$$3x - 5 = 85x$$
$$-5 = 82x$$
$$x = -\frac{5}{82} \approx -0.061$$

Check $\qquad\qquad \log (3x - 5) - \log 5x = 1.23$

$$\log [3(-0.061) - 5] - \log [(5)(-0.061)] \overset{?}{=} 1.23$$
$$\log (-5.183) - \log (-0.305) \overset{?}{=} 1.23 \quad \text{Stop.}$$

NOW TRY EXERCISE 57

Since we have the logarithms of negative numbers, -0.061 is an extraneous solution. Thus, this equation has no solution. Its solution is the empty set, $\varnothing$. ✳

HELPFUL HINT

Below we show some of the steps used in the solutions of Example 3 and Example 6 from this section.

Example 3

$$\log_2 (x + 1)^3 = 4$$

$$(x + 1)^3 = 2^4 \quad \text{\textit{Write in exponential form.}}$$

Example 6

$$\log \frac{3x - 5}{5x} = 1.23$$

$$\frac{3x - 5}{5x} = \text{antilog } 1.23 \quad \text{\textit{Take the antilog of both sides.}}$$

Notice the steps we used were different in Examples 3 and 6. In Example 3, we write the equation in exponential form, while in Example 6, we take the antilog of both sides of the equation. In Example 6, we could have also written the second step (line) as $10^{1.23} = \dfrac{3x - 5}{5x}$, and then evaluated $10^{1.23}$ on a calculator to obtain 17.0 (rounded to three significant digits). Then we could finish the problem to find the solution. However, since Example 6 is given in base 10, we decided to just take the antilog of both sides. Antilogs of base 10 numbers are easily evaluated on a calculator. You may solve problems similar to Example 6 using either method.

2 Solve Applications

Now we will look at an application that involves an exponetial equation.

EXAMPLE 7   **Bacteria**   If there are initially 1000 bacteria in a culture, and the number of bacteria doubles each hour, the number of bacteria after t hours can be found by the formula

$$N = 1000(2)^t$$

How long will it take for the culture to grow to 30,000 bacteria?

Solution

$$N = 1000(2)^t$$

$$30{,}000 = 1000(2)^t \quad \text{\textit{Substitute 30,000 for N.}}$$

$$30 = (2)^t \quad \text{\textit{Divide both sides by 1000.}}$$

We want to find the value for t. To accomplish this we will use logarithms. Begin by taking the logarithm of both sides of the equation.

$$\log 30 = \log (2)^t$$

$$\log 30 = t \log 2 \quad \text{\textit{Power rule}}$$

$$\frac{\log 30}{\log 2} = t \quad \text{\textit{Divide both sides by log 2.}}$$

$$\frac{1.4771}{0.3010} \approx t$$

$$4.91 \approx t$$

NOW TRY EXERCISE 69   It will take about 4.91 hours for the culture to grow to 30,000 bacteria. ✳

Exercise Set 9.6

Concept/Writing Exercises

1. If $\log c = \log d$, then what is the relationship between c and d?

2. If $c^r = c^s$, then what is the relationship between r and s?

3. After solving a logarithmic equation, what must you do?

4. In properties 6c and 6d, we specify that both x and y must be positive. Explain why.

5. How can you tell quickly that $\log (x + 4) = \log (-2)$ has no real solution?

6. Can $x = -1$ be a solution of the equation $\log_3 x + \log_3(x - 8) = 2$? Explain.

Practice the Skills

Solve each exponential equation without using a calculator.

7. $5^x = 125$

8. $2^x = 128$

9. $3^x = 243$

10. $4^x = 256$

11. $49^x = 7$

12. $81^x = 3$

13. $5^{-x} = \dfrac{1}{25}$

14. $6^{-x} = \dfrac{1}{216}$

15. $27^x = \dfrac{1}{3}$

16. $16^x = \dfrac{1}{4}$

17. $2^{x+1} = 64$

18. $3^{x-5} = 81$

19. $2^{3x-2} = 16$

20. $64^x = 4^{4x+1}$

21. $27^x = 3^{2x+3}$

22. $\left(\dfrac{1}{2}\right)^x = 8$

Use a calculator to solve each equation. Round your answers to the nearest hundredth.

23. $7^x = 50$

24. $1.05^x = 15$

25. $4^{x-1} = 20$

26. $2.3^{x-1} = 5.6$

27. $1.63^{x+1} = 25$

28. $4^x = 9^{x-2}$

29. $3^{x+4} = 6^x$

30. $5^x = 2^{x+5}$

Solve each logarithmic equation. Use a calculator where appropriate. If the answer is irrational, round the answer to the nearest hundredth.

31. $\log_{16} x = \dfrac{1}{2}$

32. $\log_{49} x = \dfrac{1}{2}$

33. $\log_{125} x = \dfrac{1}{3}$

34. $\log_{81} x = \dfrac{1}{4}$

35. $\log_2 x = -3$

36. $\log_6 x = -2$

37. $\log x = 1$

38. $\log x = 3$

39. $\log_2 (5 - 3x) = 3$

40. $\log_4 (3x + 1) = 3$

41. $\log_5 (x + 2)^2 = 2$

42. $\log_3 (a - 2)^2 = 2$

43. $\log_2 (r + 4)^2 = 4$

44. $\log_2 (p - 3)^2 = 4$

45. $\log (x + 3) = 2$

46. $\log (2x - 1) = 1$

47. $\log_2 x + \log_2 5 = 2$

48. $\log_3 2x + \log_3 x = 3$

49. $\log (r + 2) = \log (3r - 1)$

50. $\log 2a = \log (1 - a)$

51. $\log (2x + 1) + \log 4 = \log (7x + 8)$

52. $\log (x + 3) + \log x = \log 4$

53. $\log n + \log (3n - 5) = \log 2$

54. $\log (x + 4) - \log x = \log (x + 1)$

55. $\log 5 + \log y = 0.72$

56. $\log (x + 4) - \log x = 1.22$

57. $2 \log x - \log 4 = 2$

58. $\log 6000 - \log (x + 2) = 3.15$

59. $\log x + \log (x - 3) = 1$

60. $2 \log_2 x = 2$

61. $\log x = \dfrac{1}{3} \log 27$

62. $\log_7 x = \dfrac{3}{2} \log_7 64$

63. $\log_8 x = 3 \log_8 2 - \log_8 4$

64. $\log_4 x + \log_4 (6x - 7) = \log_4 5$

65. $\log_5 (x + 3) + \log_5 (x - 2) = \log_5 6$

66. $\log_7 (x + 6) - \log_7 (x - 3) = \log_7 4$

67. $\log_2 (x + 3) - \log_2 (x - 6) = \log_2 4$

68. $\log (x - 7) - \log (x + 3) = \log 6$

Problem Solving

Solve each problem. Round your answers to the nearest hundredth.

69. **Bacteria** If the initial number of bacteria in the culture in Example 7 is 4500, when will the number of bacteria in the culture reach 50,000? Use $N = 4500(2)^t$.

70. **Bacteria** If after 4 hours the culture in Example 7 contains 2224 bacteria, how many bacteria were present initially?

71. Radioactive Decay The amount, A, of 200 grams of a certain radioactive material remaining after t years can be found by the equation $A = 200(0.800)^t$. When will 40 grams remain?

72. Radioactive Decay The amount, A, of 70 grams of a certain radioactive material remaining after t years can be found by the equation $A = 70(0.620)^t$. When will 10 grams remain?

73. Savings Account Carmen Strawn invests $2000 in a savings account earning interest at a rate of 5% compounded annually. How long will it take for the $2000 to grow to $4600? Use the compound interest formula, $A = p\left(1 + \dfrac{r}{n}\right)^{nt}$, which was introduced on page 619.

74. Savings Account If Tekar Werner invests $800 in a savings account earning interest at a rate of 6% compounded semiannually, how long will it take for the $800 to grow to $2400?

🔒 **75. Infant Mortality Rate** The infant mortality rate (deaths per 1000 live births) in the United States has been decreasing since before 1959. Although it has fallen significantly, it is still higher than in many other nations. The U.S. infant mortality rate can be approximated by the function

$$f(t) = 26 - 12.1 \log(t + 1)$$

where t is the number of years since 1960 and $0 \le t \le 45$. Use this function to estimate the U.S. infant mortality rate in **a)** 1990. **b)** 2005.

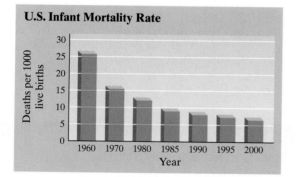

U.S. Infant Mortality Rate

76. Homicides Since 1993, the number of homicides in New York City has been on the decline. The number of homicides can be approximated by the function

$$f(t) = 1997 - 1576 \log(t + 1)$$

where t is the number of years since 1993. Use this function to estimate the number of homicides in New York City in 2004.

77. Depreciation A machine purchased for business use can be depreciated to reduce income tax. The value of the machine at the end of its useful life is called its *scrap value*. When the machine depreciates by a constant percentage annually, its scrap value, S, is $S = c(1 - r)^n$,

where c is the original cost, r is the annual rate of depreciation as a decimal, and n is the useful life in years. Find the scrap value of a machine that costs $50,000, has a useful life of 12 years, and has an annual depreciation rate of 15%.

78. Depreciation If the machine in Exercise 77 costs $100,000, has a useful life of 15 years, and has an annual depreciation rate of 8%, find its scrap value.

79. Power Gain of an Amplifier The power gain, P, of an amplifier is defined as

$$P = 10 \log\left(\frac{P_{\text{out}}}{P_{\text{in}}}\right)$$

where P_{out} is the output power in watts and P_{in} is the input power in watts. If an amplifier has an output power of 12.6 watts and an input power of 0.146 watts, find the power gain.

80. Earthquake Measured on the Richter scale, the magnitude, R, of an earthquake of intensity I is defined by $R = \log I$, where I is the number of times more intense the earthquake is than the minimum level for comparison.

  a) How many times more intense was the 1906 San Francisco earthquake, which measured 8.25 on the Richter scale, than the minimum level for comparison?

  b) How many times more intense is an earthquake that measures 6.4 on the Richter scale than one that measures 4.7?

81. Magnitude of Sound The decibel scale is used to measure the magnitude of sound. The magnitude d, in decibels, of a sound is defined to be $d = 10 \log I$, where I is the number of times greater (or more intense) the sound is than the minimum intensity of audible sound.

  a) An airplane engine (nearby) measures 120 decibels. How many times greater than the minimum level of audible sound is the airplane engine?

  b) The intensity of the noise in a busy city street is 70 decibels. How many times greater is the intensity of the sound of the airplane engine than the sound of the city street?

82. In the following procedure, we begin with a true statement and end with a false statement. Can you find the error?

$$2 < 3 \qquad \textit{True}$$
$$2 \log (0.1) < 3 \log (0.1) \qquad \textit{Multiply both sides by } \log(0.1).$$
$$\log (0.1)^2 < \log (0.1)^3 \qquad \textit{Property 3}$$
$$(0.1)^2 < (0.1)^3 \qquad \textit{Property 6d}$$
$$0.01 < 0.001 \qquad \textit{False}$$

83. Solve $8^x = 16^{x-2}$.

84. Solve $27^x = 81^{x-3}$.

85. Use equations that are quadratic in form to solve the equation $2^{2x} - 6(2^x) + 8 = 0$.

86. Use equations that are quadratic in form to solve the equation $2^{2x} - 18(2^x) + 32 = 0$.

Change the exponential or logarithmic equation to the form $ax + by = c$, and then solve the system of equations.

87. $2^x = 8^y$
$x + y = 4$

88. $3^{2x} = 9^{y+1}$
$x - 2y = -3$

89. $\log (x + y) = 2$
$x - y = 8$

90. $\log (x + y) = 3$
$2x - y = 5$

Use your calculator to estimate the solutions to the nearest tenth. If a real solution does not exist, so state.

91. $\log (x + 3) + \log x = \log 16$

92. $\log (3x + 5) = 2.3x - 6.4$

93. $5.6 \log (5x - 12) = 2.3 \log (x - 5.4)$

94. $5.6 \log (x + 12.2) - 1.6 \log (x - 4) = 20.3 \log (2x - 6)$

Cumulative Review Exercises

[2.2] **95.** Consider the following two figures. Which has a greater volume, and by how much?

4 ft
3 ft

4 ft
3 ft
3 ft

[3.6] **96.** Let $f(x) = x^2 - x$ and $g(x) = x - 6$.
Find $(g - f)(3)$.

[4.6] **97.** Determine the solution set to the system of inequalities.

$$3x - 4y \le 6$$
$$y > -x + 4$$

[7.4] **98.** Simplify $\dfrac{2\sqrt{xy} - \sqrt{xy}}{\sqrt{x} + \sqrt{y}}$.

[8.3] **99.** Solve $E = mc^2$ for c.

[8.5] **100.** Write the function for the parabola that has the shape of $f(x) = 2x^2$ and has its vertex at $(3, -2)$.

9.7 NATURAL EXPONENTIAL AND NATURAL LOGARITHMIC FUNCTIONS

SSM Study Guide CD/Video

MathPro 4/5 PH Math Tutor Center prenhall.com/Angel

1 Identify the natural exponential function.

2 Identify the natural logarithmic function.

3 Find values on a calculator.

4 Find natural logarithms using the change of base formula.

5 Solve natural logarithmic and natural exponential equations.

6 Solve applications.

The natural exponential function and *its inverse*, the **natural logarithmic** function, are exponential functions and logarithmic functions of the type presented in the previous sections. They share all the properties of exponential functions and logarithmic functions discussed earlier. The importance of these special functions lies in the many varied applications in real life of a unique irrational number designated by the letter e.

1   Identify the Natural Exponential Function

In Section 9.2 we indicated that exponential functions were of the form $f(x) = a^x$, $a > 0$ and $a \neq 1$. Now we introduce a very special exponential function. It is called the **natural exponential function**, and it uses the number e. Like the irrational number π, the number e is an irrational number whose value can only be approximated by a decimal number. The number e plays a very important role in higher-level mathematics courses. The value of e is approximately 2.7183. Now we define the natural exponential function.

DEFINITION	**The Natural Exponential Function**
	The natural exponential function is
	$$f(x) = e^x$$
	where $e \approx 2.7183$.

2   Identify the Natural Logarithmic Function

We discussed common logarithms in Section 9.5. Now we will discuss natural logarithms.

DEFINITION	**Natural Logarithms**
	Natural logarithms are logarithms to the base e. Natural logarithms are indicated by the letters ln.
	$$\log_e x = \ln x$$

The notation $\ln x$ is read the "natural logarithm of x." The function $f(x) = \ln x$ is called the **natural logarithmic function**.

You must remember that the base of the natural logarithm is e. Thus, when you change a natural logarithm to exponential form, the base of the exponential expression will be e.

> For $x > 0$ if $y = \ln x$, then $e^y = x$.

EXAMPLE 1   Find the value of the expression by changing the natural logarithm to exponential form.

a) $\ln 1$ **b)** $\ln e$

Solution   **a)** Let $y = \ln 1$; then $e^y = 1$. Since any nonzero value to the 0th power equals 1, y must equal 0. Thus, $\ln 1 = 0$.

b) Let $y = \ln e$; then $e^y = e$. For e^y to equal e, y must equal 1. Thus, $\ln e = 1$.   ✳

The functions $y = a^x$ and $y = \log_a x$ are inverse functions. Similarly, the functions $y = e^x$ and $y = \ln x$ are inverse functions. (Remember, $y = \ln x$ means $y = \log_e x$.) That is, if $f(x) = e^x$, then $f^{-1}(x) = \ln x$.

The graphs of $y = e^x$ and $y = \ln x$ are illustrated in Figure 9.27. Notice that the graphs are symmetric about the line $y = x$, which is what we expect of inverse functions.

Note that the graph of $y = e^x$ is similar to graphs of the form $y = a^x$, $a > 1$, and that the graph of $y = \ln x$ is similar to graphs of the form $y = \log_a x$, $a > 1$.

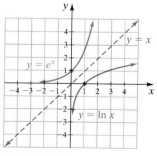

FIGURE 9.27

3 Find Values on a Calculator

Now we will learn how to find natural logarithms on a calculator.

Using Your Calculator
Find Natural Logarithms

Natural logarithms can be found using a calculator that has a $\boxed{\text{LN}}$ key. Natural logarithms are found in the same manner that we found common logarithms on a calculator, except we use the natural log key, $\boxed{\text{LN}}$, instead of the common log key, $\boxed{\text{LOG}}$.

Scientific Calculator

EXAMPLE	KEYS TO PRESS	ANSWER DISPLAYED
Find ln 242.	242 $\boxed{\text{LN}}$	5.4889377
Find ln 0.85.	.85 $\boxed{\text{LN}}$	−0.1625189

Graphing Calculator*

On the TI-83 Plus, after the $\boxed{\text{LN}}$ key is pressed, the calculator displays ln (on the screen.

EXAMPLE	KEYS TO PRESS	ANSWER DISPLAYED
Find ln 242.	$\boxed{\text{LN}}$ (242 $\boxed{)}$ $\boxed{\text{ENTER}}$	5.488937726
Find ln 0.85.	$\boxed{\text{LN}}$ (.85 $\boxed{)}$ $\boxed{\text{ENTER}}$	−.1625189295

*The keys are for a TI-83 Plus. Read your manual for instructions on how to find natural logarithms on your calculator.

When finding the natural logarithm of a number, we are finding an exponent. The natural logarithm of a number is the exponent to which the base e must be raised to obtain that number. For example,

if ln 242 = 5.4889377, then $e^{5.4889377} = 242$
if ln 0.85 = −0.1625189, then $e^{-0.1625189} = 0.85$

Since $y = \ln x$ and $y = e^x$ are inverse functions, we can use the inverse key $\boxed{\text{INV}}$, in combination with the natural log key, $\boxed{\text{LN}}$, to obtain values for e^x.

Using Your Calculator
Finding Values of e^x
Scientific Calculator

To find values of e^x, first enter the exponent on e. Then press either $\boxed{\text{shift}}$, $\boxed{\text{2}^{\text{nd}}}$, or $\boxed{\text{INV}}$, depending on your calculator. Then press the natural log key, $\boxed{\text{LN}}$. After the $\boxed{\text{LN}}$ key is pressed, the value of e^x will be displayed.

EXAMPLE	KEYS TO PRESS	ANSWER DISPLAYED
Find $e^{5.24}$.	5.24 $\boxed{\text{INV}}$ $\boxed{\text{LN}}$	188.6701
Find $e^{-1.639}$.	1.639 $\boxed{+/-}$ $\boxed{\text{INV}}$ $\boxed{\text{LN}}$	0.1941741

(continued on the next page)

Graphing Calculator*

On the TI-83 Plus, after $\boxed{2^{nd}}$ $\boxed{LN}$ is pressed, the calculator displays $e\wedge($ on the screen.

EXAMPLE	KEYS TO PRESS	ANSWER DISPLAYED
Find $e^{5.24}$.	$\boxed{2^{nd}}$ $\boxed{LN}$ $(5.24\boxed{)}$ $\boxed{ENTER}$	188.6701024
Find $e^{-1.639}$.	$\boxed{2^{nd}}$ $\boxed{LN}$ $(\boxed{(-)} 1.639 \boxed{)}$ $\boxed{ENTER}$	.1941741194

* Keys are for a TI-83 Plus. Read your manual for instructions on how to evaluate natural exponential expressions on your calculator.

Remember e is about 2.7183. When we evaluated $(2.7183)^{5.24}$ in the previous calculator box we obtained a value close to 188.6701. If we found ln 188.6701 on a calculator, we would obtain a value close to 5.24. What do you think we would get if we evaluated ln 0.1941741 on a calculator? If you answered, "A value close to -1.639," you answered correctly.

EXAMPLE 2 Find N if **a)** ln $N = 4.92$ and **b)** ln $N = -0.0253$.

Solution **a)** If we write ln $N = 4.92$ in exponential form, we get $e^{4.92} = N$. Thus, we simply need to evaluate $e^{4.92}$ to determine N.

$$e^{4.92} = 137.00261 \quad \textit{From a calculator}$$

Thus, $N = 137.00261$.

b) If we write ln $N = -0.0253$ in exponential form, we get $e^{-0.0253} = N$.

$$e^{-0.0253} = 0.9750174 \quad \textit{From a calculator}$$

NOW TRY EXERCISE 19 Thus, $N = 0.9750174$.

4 Find Natural Logarithms Using the Change of Base Formula

If you are given a logarithm in a base other than 10 or e, you will not be able to evaluate it on your calculator directly. When this occurs, you can use the **change of base formula**.

Change of Base Formula

For any logarithm bases a and b, and positive number x,

$$\log_a x = \frac{\log_b x}{\log_b a}$$

We can prove the change of base formula by beginning with $\log_a x = m$.

$$\log_a x = m$$
$$a^m = x \qquad \textit{Change to exponential form.}$$
$$\log_b a^m = \log_b x \qquad \textit{By property 6c on page 646}$$
$$m \log_b a = \log_b x \qquad \textit{Power rule}$$
$$(\log_a x)(\log_b a) = \log_b x \qquad \textit{Substitution for } m$$
$$\log_a x = \frac{\log_b x}{\log_b a} \qquad \textit{Divide both sides by } \log_b a.$$

In the change of base formula, 10 is often used in place of base b because we can find common logarithms on a calculator. Replacing base b with 10, we get

$$\log_a x = \frac{\log_{10} x}{\log_{10} a} \quad \text{or} \quad \log_a x = \frac{\log x}{\log a}$$

EXAMPLE 3 Use the change of base formula to find $\log_3 24$.

Solution If we substitute 3 for a and 24 for x in $\log_a x = \dfrac{\log x}{\log a}$, we obtain

$$\log_3 24 = \frac{\log 24}{\log 3} \approx \frac{1.3802}{0.4771} \approx 2.8929$$

NOW TRY EXERCISE 23 Note that $3^{2.8929} \approx 24$. ✳

We can use the same procedure as in Example 3 to find natural logarithms using the change of base formula. For example, to evaluate $\ln 20$ (or $\log_e 20$), we can substitute e for a and 20 for x in the formula $\log_a x = \dfrac{\log x}{\log a}$.

$$\log_e 20 = \frac{\log 20}{\log e} \approx \frac{1.3010}{0.4343} \approx 2.9956$$

Thus, $\ln 20 \approx 2.9956$. If you find $\ln 20$ on a calculator, you will obtain a very close value.

Since $\log e \approx 0.4343$, to evaluate natural logarithms using common logarithms, we use the formula

$$\ln x = \frac{\log x}{\log e} \approx \frac{\log x}{0.4343}$$

EXAMPLE 4 Use the change of base formula to find $\ln 95$.

Solution

$$\ln 95 = \frac{\log 95}{\log e} \approx \frac{1.9777}{0.4343} \approx 4.5538$$

✳

5 Solve Natural Logarithmic and Natural Exponential Equations

The properties of logarithms discussed in Section 9.4 still hold true for natural logarithms. Following is a summary of these properties in the notation of natural logarithms.

Properties for Natural Logarithms		
$\ln xy = \ln x + \ln y$	($x > 0$ and $y > 0$)	*Product rule*
$\ln \dfrac{x}{y} = \ln x - \ln y$	($x > 0$ and $y > 0$)	*Quotient rule*
$\ln x^n = n \ln x$	($x > 0$)	*Power rule*

Consider the expression $\ln e^x$, which means $\log_e e^x$. From property 4 on page 636, $\log_e e^x = x$. Thus, $\ln e^x = x$. Similarly, $e^{\ln x} = e^{\log_e x} = x$ by property 5. Although $\ln e^x = x$ and $e^{\ln x} = x$ are just special cases of properties 4 and 5, respectively, we will call these properties 7 and 8 so that we can make reference to them.

Additional Properties for Natural Logarithms and Natural Exponential Expressions	
$\ln e^x = x$	**Property 7**
$e^{\ln x} = x, \quad x > 0$	**Property 8**

Using property 7, $\ln e^x = x$, we can state, for example, that $\ln e^{kt} = kt$, and $\ln e^{-2.06t} = -2.06t$. Using property 8, $e^{\ln x} = x$, we can state, for example, that $e^{\ln (t+2)} = t + 2$ and $e^{\ln kt} = kt$.

EXAMPLE 5   Solve the equation $\ln y - \ln (x + 6) = t$ for y.

Solution

$$\ln y - \ln (x + 6) = t$$

$$\ln \frac{y}{x + 6} = t \qquad \text{\textit{Quotient rule}}$$

$$\frac{y}{x + 6} = e^t. \qquad \text{\textit{Change to exponential form.}}$$

NOW TRY EXERCISE 63

$$y = e^t(x + 6). \qquad \text{\textit{Solve for y.}} \qquad \text{✳}$$

EXAMPLE 6   Solve the equation $225 = 450e^{-0.4t}$ for t.

Solution   Begin by dividing both sides of the equation by 450 to isolate $e^{-0.4t}$.

$$\frac{225}{450} = \frac{\cancel{450}e^{-0.4t}}{\cancel{450}}$$

$$0.5 = e^{-0.4t}$$

Now take the natural logarithm of both sides of the equation to eliminate the exponential expression on the right side of the equation.

$$\ln 0.5 = \ln e^{-0.4t}$$

$$\ln 0.5 = -0.4t \qquad \text{\textit{Property 7}}$$

$$-0.6931472 = -0.4t$$

$$\frac{-0.6931472}{-0.4} = t$$

$$1.732868 = t \qquad \text{✳}$$

EXAMPLE 7   Solve the equation $P = P_0 e^{kt}$ for t.

Solution   We can follow the same procedure as used in Example 6.

$$P = P_0 e^{kt}$$

$$\frac{P}{P_0} = \frac{\cancel{P_0}e^{kt}}{\cancel{P_0}} \qquad \text{\textit{Divide both sides by P_0.}}$$

$$\frac{P}{P_0} = e^{kt}$$

$$\ln \frac{P}{P_0} = \ln e^{kt} \qquad \text{\textit{Take natural log of both sides.}}$$

$$\ln P - \ln P_0 = \ln e^{kt} \qquad \text{\textit{Quotient rule}}$$

$$\ln P - \ln P_0 = kt \qquad \text{\textit{Property 7}}$$

NOW TRY EXERCISE 59

$$\frac{\ln P - \ln P_0}{k} = t \qquad \text{\textit{Solve for t.}} \qquad \text{✳}$$

6 Solve Applications

Now let's look at some applications that involve the natural exponential function and natural logarithms.

When a quantity increases or decreases at an *exponential rate*, a formula often used to find the value of P after time t is

$$P = P_0 e^{kt}$$

where P_0 is the initial or starting value and k is the constant growth rate or decay rate compounded continuously. We will refer to this formula as the **exponential growth (or decay) formula**. In the formula, other letters may be used in place of P. When $k > 0$, P increases as t increases. When $k < 0$, P decreases and gets closer to 0 as t increases.

EXAMPLE 8

Interest Compounded Continuously   Banks often credit compound interest continuously. When interest is compounded continuously, the balance, P, in the account at any time, t, can be calculated by the exponential growth formula $P = P_0 e^{kt}$, where P_0 is the principal initially invested and k is the interest rate.

a) Suppose the interest rate is 6% compounded continuously and $1000 is initially invested. Determine the balance in the account after 1 year.

b) How long will it take the account to double in value?

Solution   **a)** Understand and Translate   We are told that the principal initially invested, P_0, is $1000. We are also given that the time, t, is 1 year and that the interest rate, k, is 6% or 0.06. We substitute these values into the given formula and solve for P.

$$P = P_0 e^{kt}$$
$$P = 1000 e^{(0.06)(1)}$$

Carry Out
$$= 1000 e^{0.06} = 1000(1.0618365) \quad \textit{From a calculator}$$
$$= 1061.8365$$

Answer   After 1 year, the balance in the account is $\approx$\$1061.84.

b) Understand and Translate   For the value of the account to double, the balance in the account would have to reach $2000. Therefore, we substitute 2000 for P and solve for t.

$$P = P_0 e^{kt}$$
$$2000 = 1000 e^{0.06t}$$
$$2 = e^{0.06t} \quad \textit{Divide both sides by 1000.}$$

Carry Out
$$\ln 2 = \ln e^{0.06t} \quad \textit{Take natural log of both sides.}$$
$$\ln 2 = 0.06t \quad \textit{Property 7}$$
$$\frac{\ln 2}{0.06} = t$$
$$\frac{0.6931472}{0.06} = t$$
$$11.552453 \approx t$$

Answer   Thus, with an interest rate of 6% compounded continuously, the account will double in about 11.6 years.   ✳

EXAMPLE 9   **Radioactive Decay**   Strontium 90 is a radioactive isotope that decays exponentially at 2.8% per year. Suppose there are initially 1000 grams of strontium 90 in a substance.

a) Find the number of grams of strontium 90 left after 50 years.

b) Find the half-life of strontium 90.

Solution **a)** Understand Since the strontium 90 is decaying over time, the value of k in the formula $P = P_0 e^{kt}$ is negative. Since the rate of decay is 2.8% per year, we use $k = -0.028$. Therefore, the formula we use is $P = P_0 e^{-0.028t}$.

Translate $\qquad P = P_0 e^{-0.028t}$

$\qquad\qquad = 1000e^{-0.028(50)}$

Carry Out $\qquad = 1000e^{-1.4} = 1000(0.246597) = 246.597$

Answer Thus, after 50 years, 246.597 grams of strontium 90 remain.

b) To find the half-life, we need to determine when 500 grams of strontium 90 are left.

$$P = P_0 e^{-0.028t}$$
$$500 = 1000e^{-0.028t}$$
$$0.5 = e^{-0.028t} \qquad \textit{Divide both sides by 1000.}$$
$$\ln 0.5 = \ln e^{-0.028t} \qquad \textit{Take natural log of both sides.}$$
$$-0.6931472 = -0.028t \qquad \textit{Property 7}$$
$$\frac{-0.6931472}{-0.028} = t$$
$$24.755257 \approx t$$

NOW TRY EXERCISE 71 Thus, the half-life of strontium 90 is about 24.8 years.

EXAMPLE 10 **Selling Toys** The formula for estimating the amount of money, A, spent on advertising a certain toy is $A = 350 + 650 \ln n$ where n is the expected number of toys to be sold.

a) If the company wishes to sell 2500 toys, how much money should the company expect to spend on advertising?

b) How many toys can be expected to be sold if $6000 is spent on advertising?

Solution **a)** $A = 350 + 650 \ln n$

$\qquad = 350 + 650 \ln 2500 \qquad \textit{Substitute 2500 for n.}$

$\qquad = 350 + 650(7.824046)$

$\qquad = 5435.63$

Thus, $5435.63 should be expected to be spent on advertising.

b) Understand and Translate We are asked to find the number of toys expected to be sold, n, if $6000 is spent on advertising. We substitute the given values into the equation and solve for n.

$$A = 350 + 650 \ln n$$

Carry Out
$$6000 = 350 + 650 \ln n \qquad \textit{Substitute 6000 for A.}$$
$$5650 = 650 \ln n \qquad \textit{Subtract 350 from both sides.}$$
$$\frac{5650}{650} = \ln n \qquad \textit{Divide both sides by 650.}$$
$$8.69231 \approx \ln n$$
$$e^{8.69231} \approx n \qquad \textit{Change to exponential form.}$$
$$5957 \approx n \qquad \textit{Obtain answer from a calculator.}$$

NOW TRY EXERCISE 75 Answer   Thus, about 5957 toys can be expected to be sold if $6000 is spent on advertising.

Using Your Graphing Calculator

Equations containing natural logarithms and natural exponential functions can be solved on your graphing calculator. For example, to solve the equation $\ln x + \ln (x + 3) = \ln 8$, we set

$$Y_1 = \ln x + \ln (x + 3)$$
$$Y_2 = \ln 8$$

and find the intersection of the graphs, as shown in Figure 9.28.

In Figure 9.28, we used the CALC, INTERSECT option to find the intersection of the graphs. The solution is the x-coordinate of the intersection. The solution to the equation is $x = 1.7016$, to the nearest ten-thousandth.

To solve the equation $4e^{0.3x} - 5 = x + 3$, we set

$$Y_1 = 4e^{0.3x} - 5$$
$$Y_2 = x + 3$$

and find the intersection of the graphs, as shown in Figure 9.29. This equation has two solutions since there are two intersections.

The solutions to the equation are approximately $x = -7.5896$ and $x = 3.5284$. In Exercises 89 through 93, we use the grapher to check or solve equations.

FIGURE 9.28

FIGURE 9.29

Exercise Set 9.7

Concept/Writing Exercises

1. **a)** What is the base in the natural exponential function?

   b) What is the approximate value of e?

2. What is another way of writing $\log_e x$?

3. What is the domain of $\ln x$?

4. Under what conditions will $\ln x < 0$?

5. Give the change of base formula.

6. Is $n \log_e x = \ln x^n$? Explain.

7. To what is $\ln e^x$ equal?

8. To what is $e^{\ln x}$ equal?

9. What is the inverse of $\ln x$?

10. Under what circumstances will P in the formula $P = P_0 e^{kt}$ increase when t increases?

11. Under what conditions will P in the formula $P = P_0 e^{kt}$ decrease when t increases?

12. Is it possible to find the value of $\ln(-3.52)$? Explain.

Practice the Skills

Find the following values. Round values to four decimal places.

13. $\ln 62$

14. $\ln 451$

15. $\ln 0.813$

16. $\ln 0.000568$

Find the value of N. Round values to three significant digits.

17. $\ln N = 1.6$

18. $\ln N = 4.1$

19. $\ln N = -2.41$

20. $\ln N = -0.674$

21. $\ln N = -0.0287$

22. $\ln N = 0.543$

Use the change of base formula to find the value of the following logarithms. Do not round logarithms in the change of base formula. Write the answer rounded to the nearest ten-thousandth.

23. $\log_3 56$

24. $\log_3 198$

25. $\log_2 21$

26. $\log_2 89$

27. $\log_4 11$

28. $\log_4 158$

29. $\log_5 63$

30. $\log_5 719$

31. $\log_6 123$

32. $\log_6 806$

33. $\log_7 51$

34. $\log_9 3294$

35. $\log_5 0.463$

36. $\log_3 0.0365$

Solve the following logarithmic equations.

37. $\ln x + \ln (x - 1) = \ln 12$

38. $\ln (x + 3) + \ln (x - 2) = \ln 14$

39. $\ln x + \ln (x + 4) = \ln 5$

40. $\ln (x + 3) + \ln (x - 3) = \ln 40$

41. $\ln x = 5 \ln 2 - \ln 8$

42. $\ln x = \dfrac{5}{2} \ln 9$

43. $\ln (x^2 - 4) - \ln (x + 2) = \ln 1$

44. $\ln (x + 12) - \ln (x + 4) = \ln 2$

Each of the following equations is in the form $P = P_0 e^{kt}$. Solve each equation for the remaining variable. Remember, e is a constant. Write the answer rounded to the nearest ten-thousandth.

45. $P = 120e^{2.3(1.6)}$

46. $900 = P_0 e^{(0.4)(3)}$

47. $50 = P_0 e^{-0.5(3)}$

48. $120 = 60e^{2t}$

49. $90 = 30e^{1.4t}$

50. $20 = 40e^{-0.5t}$

51. $80 = 40e^{k(3)}$

52. $10 = 50e^{k(4)}$

53. $20 = 40e^{k(2.4)}$

54. $100 = A_0 e^{-0.02(3)}$

55. $A = 6000e^{-0.08(3)}$

56. $75 = 100e^{-0.04t}$

Solve for the indicated variable.

57. $V = V_0 e^{kt}$, for V_0

58. $P = P_0 e^{kt}$, for P_0

59. $P = 150e^{4t}$, for t

60. $200 = P_0 e^{kt}$, for t

61. $A = A_0 e^{kt}$, for k

62. $140 = R_0 e^{kt}$, for k

63. $\ln y - \ln x = 2.3$, for y

64. $\ln y + 5 \ln x = \ln 2$, for y

65. $\ln y - \ln (x + 3) = 6$, for y

66. $\ln(x + 2) - \ln(y - 1) = \ln 5$, for y

Problem Solving

Use a calculator to solve.

67. If $e^x = 12.183$, find the value of x. Explain how you obtained your answer.

68. To what exponent must the base e be raised to obtain the value 184.93? Explain how you obtained your answer.

69. *Interest Compounded Continuously* If $5000 is invested at 8% compounded continuously,

  a) determine the balance in the account after 2 years.

  b) How long would it take the value of the account to double? (See Example 8.)

70. *Interest Compounded Continuously* If $3000 is invested at 5% compounded continuously,

  a) determine the balance in the account after 30 years.

  b) How long would it take the value of the account to double?

71. *Radioactive Decay* Refer to Example 9. Determine the amount of strontium 90 remaining after 20 years if there were originally 70 grams.

72. *Strontium 90* Refer to Example 9. Determine the amount of strontium 90 remaining after 40 years if there were originally 200 grams.

73. *Soft Drinks* For a certain soft drink, the percent of a target market, $f(t)$, that buys the soft drink is a function of the number of days, t, that the soft drink is advertised. The function that describes this relationship is $f(t) = 1 - e^{-0.04t}$.

  a) What percent of the target market buys the soft drink after 50 days of advertising?

  b) How many days of advertising are needed if 75% of the target market is to buy the soft drink?

74. *Trout in a Lake* In 2000, a lake had 300 trout. The growth in the number of trout is estimated by the function $g(t) = 300e^{0.07t}$ where t is the number of years after 2000. How many trout will be in the lake in **a)** 2003? **b)** 2010?

75. *Walking Speed* It was found in a psychological study that the average walking speed, $f(P)$, of a person living in a city is a function of the population of the city. For a city of population P, the average walking speed in feet per second is given by $f(P) = 0.37 \ln P + 0.05$. The population of Nashville, Tennessee, is 972,000.

a) What is the average walking speed of a person living in Nashville?

b) What is the average walking speed of a person living in New York City, population 8,567,000?

c) If the average walking speed of the people in a certain city is 5.0 feet per second, what is the population of the city?

76. *Advertising* For a certain type of tie, the number of ties sold, $N(a)$, is a function of the dollar amount spent on advertising, a (in thousands of dollars). The function that describes this relationship is $N(a) = 800 + 300 \ln a$.

a) How many ties were sold after $1500 (or $1.5 thousand) was spent on advertising?

b) How much money must be spent on advertising to sell 1000 ties?

77. Assume that the value of the island of Manhattan has grown at an exponential rate of 8% per year since 1626 when Peter Minuet of the Dutch West India Company purchased it for $24. Then the value of Manhattan can be determined by the equation $V = 24e^{0.08t}$, where t is the number of years since 1626. Determine the value of the island of Manhattan in 2003, that is when $t = 377$ years.

78. *Prescribing a Drug* The percent of doctors who accept and prescribe a new drug is given by the function $P(t) = 1 - e^{-0.22t}$, where t is the time in months since the drug was placed on the market. What percent of doctors accept a new drug 2 months after it is placed on the market?

79. *World Population* The world population in January 2003 was estimated to be about 6.30 billion people. Let's assume that the world population will continue to grow exponentially at the current growth rate of about 1.3% per year. Then, the expected world population, in billions of people, in t years, is given by the function

$$P(t) = 6.30e^{0.013t}$$

where t is the number of years since 2003.

a) Find the world population in 2010.

b) In how many years will the world population double?

80. *Girls Weights* The shaded area of the following graph shows the normal range (from the 5th to the 95th percentile) of weights for girls from birth up to 36 months. The median, or 50th percentile, of weights is indicated by the orange line. The function $y = 3.17 + 7.32 \ln x$ can be used to estimate the median weight of girls from 3 months to 36 months. Use this function to estimate the median weight for girls at **a)** 18 months. **b)** 30 months.

Source: *Newsweek* Special 2000 Edition, Your Child

81. *Boys Heights* The shaded area of the graph on the top of the next page shows the normal range (from the 5th to the 95th percentile) of heights for boys from birth up to 36 months. The median, or 50th percentile, of heights is indicated by the blue line. The function

$y = 15.29 + 5.93 \ln x$ can be used to estimate the median height of boys from 3 months to 36 months. Use this function to estimate the median height for boys at age **a)** 18 months. **b)** 30 months.

Boys

95th
50th
5th

Source: Newsweek Special 2000 Edition, Your Child

82. **Radioactive Decay** Plutonium, which is commonly used in nuclear reactors, decays exponentially at a rate of 0.003% per year. The formula $A = A_0 e^{kt}$ can be used to find the amount of plutonium remaining from an initial amount, A_0, after t years. In the formula, the k is replaced with -0.00003.

   a) If 1000 grams of plutonium are present in 2003, how many grams of plutonium will remain in the year 2103, 100 years later?

   b) Find the half-life of plutonium.

83. **Carbon Dating** Carbon dating is used to estimate the age of ancient plants and objects. The radioactive element, carbon 14, is most often used for this purpose. Carbon 14 decays exponentially at a rate of 0.01205% per year. The amount of carbon 14 remaining in an object after t years can be found by the function $f(t) = v_0 e^{-0.0001205t}$, where v_0 is the initial amount present.

   a) If an ancient animal bone originally had 20 grams of carbon 14, and when found it had 9 grams of carbon 14, how old is the bone?

   b) How old is an item that has 50% of its original carbon 14 remaining?

84. **Compound Interest** At what rate, compounded continuously, must a sum of money be invested if it is to double in 6 years?

85. **Compound Interest** How much money must be deposited today to become $20,000 in 18 years if invested at 6% compounded continuously?

86. **Radioisotope** The power supply of a satellite is a radioisotope. The power P, in watts, remaining in the power supply is a function of the time the satellite is in space.

   a) If there are 50 grams of the isotope originally, the power remaining after t days is $P = 50e^{-0.002t}$. Find the power remaining after 100 days.

   b) When will the power remaining drop to 10 watts?

87. **Radioactive Decay** During the nuclear accident at Chernobyl in Ukraine in 1986, two of the radioactive materials that escaped into the atmosphere were cesium 137, with a decay rate of 2.3% and strontium 90, with a decay rate of 2.8%.

   a) Which material will decompose more quickly? Explain.

   b) What percentage of the cesium will remain in 2036, 50 years after the accident?

88. **Radiometric Dating** In the study of radiometric dating (using radioactive isotopes to determine the age of items), the formula

$$t = \frac{t_h}{0.693} \ln\left(\frac{N_0}{N}\right)$$

is often used. In the formula, t is the age of the item, t_h is the half-life of the radioactive isotope used, N_0 is the original number of radioactive atoms present, and N is the number remaining at time t. Suppose a rock originally contained 5×10^{12} atoms of uranium 238. Uranium 238 has a half-life of 4.5×10^9 years. If at present there are 4×10^{12} atoms, how old is the rock?

 In Exercises 89–93, use your graphing calculator. In Exercises 91–93, round your answers to the nearest thousandth.

89. Check your answer to Exercise 37.

90. Check your answer to Exercise 39.

91. Solve the equation $e^{x-4} = 12 \ln (x + 2)$.

92. Solve the equation $\ln (4 - x) = 2 \ln x + \ln 2.4$.

93. Solve the equation $3x - 6 = 2e^{0.2x} - 12$.

Challenge Exercises

In Exercises 94–97, when you solve for the given variable, write the answer without using the natural logarithm.

94. **Intensity of Light** The intensity of light as it passes through a certain medium is found by the formula $x = k(\ln I_0 - \ln I)$. Solve this equation for I_0.

95. **Velocity** The distance traveled by a train initially moving at velocity v_0 after the engine is shut off can be calculated by the formula $x = \frac{1}{k} \ln (kv_0t + 1)$. Solve this equation for v_0.

96. **Molecule** A formula used in studying the action of a protein molecule is $\ln M = \ln Q - \ln(1 - Q)$. Solve this equation for Q.

97. **Electric Circuit** An equation relating the current and time in an electric circuit is $\ln i - \ln I = \frac{-t}{RC}$. Solve this equation for i.

Cumulative Review Exercises

[3.3] **98.** Let $h(x) = \dfrac{x^2 + 4x}{x + 6}$. Find **a)** $h(-3)$. **b)** $h\left(\dfrac{2}{5}\right)$.

[4.3] **99.** *Tickets* The admission at an ice hockey game is $15 for adults and $11 for children. A total of 650 tickets were sold. Determine how many children's tickets and how many adult's tickets were sold if a total of $8790 was collected.

[5.2] **100.** Multiply $(3xy^2 + y)(4x - 3xy)$.

[5.6] **101.** Find two values of b that will make $4x^2 + bx + 9$ a perfect square trinomial.

[7.3] **102.** Multiply $\sqrt[3]{x}\left(\sqrt[3]{x^2} + \sqrt[3]{x^5}\right)$.

CHAPTER SUMMARY
Key Words and Phrases

9.1
Composite function
Composition of functions
Horizontal line test
Inverse function
One-to-one function
Vertical line test

9.2
Exponential function

9.3
Logarithm
Logarithmic function

9.4
Argument of a logarithm
Power rule for logarithms
Product rule for logarithms

Quotient rule for logarithms

9.5
Antilogarithms
Base 10 logarithm
Common logarithm
Inverse logarithm
Significant digits

9.6
Exponential equation

Logarithmic equation

9.7
Change of base formula
e
Exponential growth (or decay) formula
Natural exponential function
Natural logarithm
Natural logarithmic function

IMPORTANT FACTS

Composite function of f with g: $(f \circ g)(x) = f[g(x)]$
Composite function of g with f: $(g \circ f)(x) = g[f(x)]$
If $f(x)$ and $g(x)$ are inverse functions, then $(f \circ g)(x) = (g \circ f)(x) = x$.
$y = \log_a x$ means $x = a^y$
$y = a^x$ and $y = \log_a x$ are inverse functions. Therefore, if $f(x) = a^x$, then $f^{-1}(x) = \log_a x$.
$y = e^x$ and $y = \ln x$ are inverse functions. Therefore, if $f(x) = e^x$, then $f^{-1}(x) = \ln x$.
The domain of a logarithmic function of the form $y = \log_a x$ is $x > 0$.

(continued on the next page)

Exponential Function

$y = a^x, a > 0$ and $a \neq 1$

Logarithmic Function

$y = \log_a x, a > 0, a \neq 1,$ and $x > 0$

Domain: $(-\infty, \infty)$ → $(0, \infty)$

Range: $(0, \infty)$ → $(-\infty, \infty)$

Points:
$\left(-1, \dfrac{1}{a}\right)$
$(0, 1)$
$(1, a)$

x becomes y
y becomes x

$\left(\dfrac{1}{a}, -1\right)$
$(1, 0)$
$(a, 1)$

Properties of Logarithms

1. $\log_a xy = \log_a x + \log_a y$

2. $\log_a \dfrac{x}{y} = \log_a x - \log_a y$

3. $\log_a x^n = n \log_a x$

4. $\log_a a^x = x$

5. $a^{\log_a x} = x \, (x > 0)$

Change of Base Formula

$\log_a x = \dfrac{\log_b x}{\log_b a}$

To solve exponential and logarithmic equations, we also use these properties:

6. a. If $x = y$, then $a^x = a^y$.

b. If $a^x = a^y$, then $x = y$.

c. If $x = y$, then $\log_b x = \log_b y$ $(x > 0, y > 0)$.

d. If $\log_b x = \log_b y$, then $x = y$ $(x > 0, y > 0)$.

Natural Logarithms

7. $\ln e^x = x$

8. $e^{\ln x} = x, x > 0$

Chapter Review Exercises

[9.1] Given $f(x) = x^2 - 3x + 4$ and $g(x) = 2x - 5$, find the following.

1. $(f \circ g)(x)$ **2.** $(f \circ g)(2)$ **3.** $(g \circ f)(x)$ **4.** $(g \circ f)(-3)$

Given $f(x) = 6x + 1$ and $g(x) = \sqrt{x - 3}, x \geq 3$, find the following.

5. $(f \circ g)(x)$ **6.** $(g \circ f)(x)$

Determine whether each function is a one-to-one function.

7.

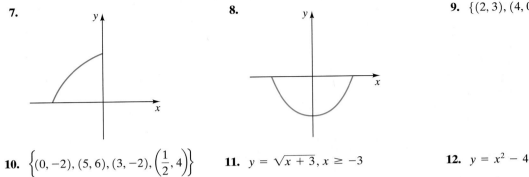

8.

9. $\{(2, 3), (4, 0), (-5, 7), (3, 8)\}$

10. $\left\{(0, -2), (5, 6), (3, -2), \left(\dfrac{1}{2}, 4\right)\right\}$ **11.** $y = \sqrt{x + 3}, x \geq -3$ **12.** $y = x^2 - 4$

For each function, find the domain and range of both $f(x)$ and $f^{-1}(x)$.

13. $\{(5, 3), (6, 2), (-4, -3), (0, 7)\}$ **14.**

Find $f^{-1}(x)$ and graph $f(x)$ and $f^{-1}(x)$ on the same axes.

15. $y = f(x) = 4x - 2$ **16.** $y = f(x) = \sqrt[3]{x - 1}$

17. *Yards to Feet* The function $f(x) = 36x$ converts yards, x, into inches. Find the inverse function that converts inches into yards. In the inverse function, what do x and $f^{-1}(x)$ represent?

18. *Gallons to Quartz* The function $f(x) = 4x$ converts gallons, x, into quarts. Find the inverse function that converts quarts into gallons. In the inverse function, what do x and $f^{-1}(x)$ represent?

[9.2] Graph the following functions.

19. $y = 2^x$

20. $y = \left(\dfrac{1}{2}\right)^x$

21. *Smart Handheld Devices* Since 1999, the number of worldwide shipments of smart handheld devices has been growing exponentially (see the graph on the right). The number of shipments, $f(t)$, in millions, can be approximated by the function $f(t) = 7.02e^{0.365t}$ where t is the number of years since 1999. Use this function to estimate the number of worldwide shipments of these devices in **a)** 2003, **b)** 2005, **c)** 2007.

Worldwide Shipments of Smart Handheld Devices

Source: International Data Corp.; MSN MoneyCentral; CSI Inc.

[9.3] *Write each equation in logarithmic form.*

22. $7^2 = 49$

23. $81^{1/4} = 3$

24. $5^{-2} = \dfrac{1}{25}$

Write each equation in exponential form.

25. $\log_2 16 = 4$

26. $\log_{1/3} \dfrac{1}{9} = 2$

27. $\log_6 \dfrac{1}{36} = -2$

Write each equation in exponential form and find the missing value.

28. $3 = \log_4 x$

29. $3 = \log_a 8$

30. $-3 = \log_{1/4} x$

Graph the following functions.

31. $y = \log_3 x$

32. $y = \log_{1/2} x$

[9.4] *Use the properties of logarithms to expand each expression.*

33. $\log_5 17^3$

34. $\log_3 \sqrt{x-5}$

35. $\log \dfrac{6(a+1)}{b}$

36. $\log \dfrac{x^4}{9(2x+3)^5}$

Write the following as the logarithm of a single expression.

37. $2 \log x - 3 \log (x+1)$

38. $3(\log 2 + \log x) - \log y$

39. $\dfrac{1}{2}[\ln x - \ln (x+2)] - \ln 2$

40. $3 \ln x + \dfrac{1}{2} \ln (x+1) - 3 \ln (x+4)$

Evaluate.

41. $8^{\log_8 9}$

42. $\log_4 4^5$

43. $7 \log_9 81$

44. $4^{\log_8 \sqrt{8}}$

[9.5, 9.7] *Use a calculator to find each logarithm. Round your answers to the nearest ten-thousandth.*

45. $\log 763$

46. $\ln 0.0281$

Use a calculator to find the antilog of each number. Give the antilog to three significant digits.

47. 3.159

48. -2.645

Use a calculator to find N. Round your answer to three significant digits.

49. $\log N = 2.3304$

50. $\log N = -1.2262$

Evaluate.

51. $\log 10^5$

52. $10^{\log 4}$

53. $9 \log 10^{3.2}$

54. $2\left(10^{\log 4.7}\right)$

[9.6] *Solve without using a calculator.*

55. $125 = 5^x$

56. $81^x = \dfrac{1}{9}$

57. $2^{3x-1} = 32$

58. $27^x = 3^{2x+5}$

Solve using a calculator. Round your answers to the nearest thousandth.

59. $7^x = 89$

60. $2.6^x = 714$

61. $12.5^{x+1} = 381$

62. $3^{x+2} = 8^x$

Solve the logarithmic equation.

63. $\log_3 (5x + 1) = 4$

64. $\log x + \log (4x - 7) = \log 2$

65. $\log_3 x + \log_3 (2x + 1) = 1$

66. $\ln (x + 1) - \ln (x - 2) = \ln 4$

[9.7] Solve each exponential equation for the remaining variable. Round your answer to the nearest thousandth.

67. $40 = 20e^{0.6t}$

68. $100 = A_0 e^{-0.42(3)}$

Solve for the indicated variable.

69. $A = A_0 e^{kt}$, for t

70. $150 = 600e^{kt}$, for k

71. $\ln y - \ln x = 4$, for y

72. $\ln (y + 1) - \ln (x + 5) = \ln 3$, for y

Use the change of base formula to evaluate. Write the answer rounded to the nearest ten-thousandth.

73. $\log_2 196$

74. $\log_3 74$

[9.2–9.7]

75. ***Compound Interest*** Find the amount of money accumulated if Justine Elwood puts \$12,000 in a savings account yielding 10% interest per year for 8 years. Use $A = p(1 + r)^n$.

76. ***Interest Compounded Continuously*** If \$10,000 is placed in a savings account paying 4% interest compounded continuously, find the time needed for the account to double in value.

77. ***Bacteria*** The bacteria *Escherichia coli* are commonly found in the bladders of humans. Suppose that 2000 bacteria are present at time 0. Then the number of bacteria present t minutes later may be found by the function $N(t) = 2000(2)^{0.05t}$.

a) When will 50,000 bacteria be present?

b) Suppose that a human bladder infection is classified as a condition with 120,000 bacteria. When would a person develop a bladder infection if he or she started with 2000 bacteria?

78. ***Atmospheric Pressure*** The atmospheric pressure, P, in pounds per square inch at an elevation of x feet above sea level can be found by the formula $P = 14.7e^{-0.00004x}$.

Find the atmospheric pressure at the top of the half dome in Yosemite National Park, an elevation of 8842 feet.

79. ***Remembering*** A class of history students is given a final exam at the end of the course. As part of a research project, the students are also given equivalent forms of the exam each month for n months. The average grade of the class after n months may be found by the function $A(n) = 72 - 18 \log (n + 1)$, $n \geq 0$.

a) What was the class average when the students took the original exam $(n = 0)$?

b) What was the class average for the exam given 2 months later?

c) After how many months was the class average 59.4?

Chapter Practice Test

1. a) Determine whether the following function is a one-to-one function.

$$\{(4, 2), (-3, 8), (-1, 3), (5, 7)\}$$

b) List the set of ordered pairs in the inverse function.

2. Given $f(x) = x^2 - 3$ and $g(x) = x + 2$, find **a)** $(f \circ g)(x)$. **b)** $(f \circ g)(5)$.

3. Given $f(x) = x^2 + 7$ and $g(x) = \sqrt{x - 5}$, $x \geq 5$, find **a)** $(g \circ f)(x)$. **b)** $(g \circ f)(4)$.

In Exercises 4 and 5, **a)** *find $f^{-1}(x)$ and* **b)** *graph $f(x)$ and $f^{-1}(x)$ on the same axes.*

4. $y = f(x) = -3x - 5$
5. $y = f(x) = \sqrt{x - 1}, x \geq 1$
6. What is the domain of $y = \log_4 x$?
7. Evaluate $\log_3 \dfrac{1}{81}$.
8. Graph $y = 3^x$.
9. Graph $y = \log_2 x$.
10. Write $2^{-5} = \frac{1}{32}$ in logarithmic form.
11. Write $\log_5 625 = 4$ in exponential form.

Write Exercises 12 and 13 in exponential form and find the missing value.

12. $4 = \log_2 (x + 1)$
13. $y = \log_{64} 16$
14. Expand $\log_2 \dfrac{x^3(x - 4)}{x + 2}$.
15. Write as the logarithm of a single expression.

$$5 \log_6 (x - 4) + 2 \log_6 (x + 3) - \frac{1}{2} \log_6 x.$$

16. Evaluate $8 \log_9 \sqrt{9}$.
17. **a)** Find $\log 4620$ rounded to 4 decimal places.
    b) Find $\ln 0.0692$ rounded to 4 decimal places.
18. Solve $3^x = 519$ for x.
19. Solve $\log 4x = \log (x + 3) + \log 2$ for x.
20. Solve $\log (x + 5) - \log (x - 2) = \log 6$ for x.
21. Find N, rounded to 4 decimal places, if $\ln N = 3.52$.
22. Evaluate $\log_6 40$, rounded to 4 decimal places, using the change of base formula.
23. Solve $200 = 500e^{-0.03t}$ for t, rounded to 4 decimal places.
24. **Saving Account** What amount of money accumulates if Kim Lee puts $3500 in a savings account yielding 6% interest compounded quarterly for 10 years?
25. **Carbon 14** The amount of carbon 14 remaining after t years is found by the formula $v = v_0 e^{-0.0001205t}$, where v_0 is the original amount of carbon 14. If a fossil originally contained 60 grams of carbon 14, and now contains 40 grams of carbon 14, how old is the fossil?

Cumulative Review Test

Take the following test and check your answers with those that appear at the end of the test. Review any questions that you answered incorrectly. The section and objective where the material was covered are indicated after the answer.

1. Simplify $\dfrac{(2xy^2z^{-3})^2}{(3x^{-1}yz^2)^{-1}}$.
2. Evaluate $4^2 - (2 - 3^2)^2 + 4^3$.
3. **Perfume** Thomas Furgeson bought a bottle of perfume as a gift for his wife. The perfume cost $92 before tax. If the total price, including tax, was $98.90, find the tax rate.
4. Solve the inequality $-3 \leq 2x - 7 < 8$ and write the answer as a solution set and in interval rotation.
5. Solve $2x - 3y = 5$ for y.
6. Let $h(x) = \dfrac{x^2 + 4x}{x + 6}$. Find $h(-3)$.
7. Find the slope of the line in the figure below. Then write the equation of the given line.

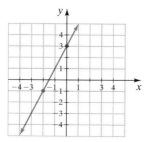

8. Graph $4x = 3y - 3$.
9. Graph $y \leq \dfrac{1}{3}x + 6$.
10. Solve the system of equations

$$\frac{1}{2}x + \frac{1}{3}y = 13$$

$$\frac{1}{5}x + \frac{1}{8}y = 5$$

11. Divide $\dfrac{x^3 + 3x^2 + 5x + 4}{x + 1}$.
12. Factor $x^2 - 2xy + y^2 - 25$.
13. Solve $(2x + 5)^2 - 9 = 0$.
14. Solve $\dfrac{2x + 3}{x + 1} = \dfrac{3}{2}$.
15. Solve $a_n = a_1 + nd - d$, for d.
16. L varies inversely as the square of P. Find L when $P = 4$ and $K = 100$.
17. Simplify $3\sqrt{45x^3} + \sqrt{5x}$.

18. Solve $\sqrt{2a + 9} - a + 3 = 0$.

19. Solve $(x^2 - 5)^2 + 3(x^2 - 5) - 10 = 0$.

20. Let $g(x) = x^2 - 4x - 5$.

  a) Express $g(x)$ in the form $g(x) = a(x - h)^2 + k$.

  b) Draw the graph and label the vertex.

Answers to Cumulative Review Test

1. $\dfrac{12xy^5}{z^4}$; [Sec. 1.5, Obj. 7]   **2.** 31; [Sec. 1.4, Obj. 3]   **3.** 7.5%; [Sec. 2.3, Obj. 2]   **4.** $\left\{ x \middle| 2 \le x < \dfrac{15}{2} \right\}$, $\left[2, \dfrac{15}{2} \right)$; [Sec. 2.5, Obj. 3]

5. $y = \dfrac{2x - 5}{3}$; [Sec. 2.2, Obj. 2]   **6.** -1; [Sec. 3.2, Obj. 4]   **7.** $m = 2$, $y = 2x + 3$; [Sec. 3.4, Obj. 2]

8.   ; [Sec. 3.4, Obj. 5]   **9.**   ; [Sec. 3.7, Obj. 1]

$4x = 3y - 3$

10. $(10, 24)$; [Sec. 4.1, Obj. 3]   **11.** $x^2 + 2x + 3 + \dfrac{1}{x + 1}$; [Sec. 5.3, Obj. 3]   **12.** $(x - y + 5)(x - y - 5)$;

[Sec. 5.6, Obj. 2]   **13.** $-1, -4$; [Sec. 8.1, Obj. 1]   **14.** -3; [Sec. 6.4, Obj. 3]   **15.** $d = \dfrac{a_n - a_1}{n - 1}$; [Sec. 6.4, Obj. 6]

16. 6.25; [Sec. 6.6, Obj. 2]   **17.** $(9x + 1)\sqrt{5x}$; [Sec. 7.4, Obj. 1]   **18.** 8; [Sec. 7.6, Obj. 1]   **19.** $0, \pm\sqrt{7}$; [Sec. 8.4, Obj. 1]

20. a) $g(x) = (x - 2)^2 - 9$ **b)**   ; [Sec. 8.5, Obj. 6]

$(2, -9)$

$g(x) = x^2 - 4x - 5$

Chapter 10

Characteristics of Functions and Their Graphs

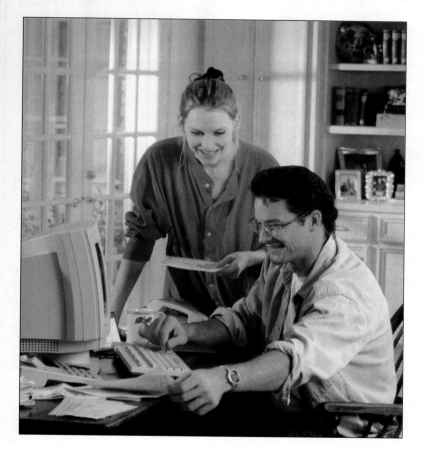

10.1 Continuous and Discontinuous Functions

10.2 Characteristics of Graphs of Polynomial Functions

10.3 Finding the Rational Zeros of a Polynomial Function

10.4 Finding the Real Zeros of a Polynomial Function

10.5 Characteristics of Graphs of Rational Functions

Chapter Summary
Chapter Review Exercises
Chapter Practice Test
Cumulative Review Test

To operate a business at a profit, its revenues must be greater than its costs. If a company produces x items and $R(x)$ represents the company's revenue from these items and $C(x)$ represents the company's cost of producing these items, this profit can be represented by the function $P(x) = R(x) - C(x)$. By analyzing the profit function, a business manager can determine how production influences a company's profit. On page 698 we write a polynomial function to represent the profit of a company selling computers. Then, by finding the zeros of this function, we determine how many computers must be produced for the business to make a profit.

SSM   Study Guide   CD/Video   MathPro 4/5   PH Math Tutor Center   prenhall.com/Angel

A Look Ahead

In this chapter we look at the general characteristics of functions. We begin with a discussion of continuous and discontinuous functions. Then we discuss the specific characteristics of polynomial functions and rational functions.

In Section 3.1, we saw that we could graph a function by plotting a sufficient number of points. In this chapter, we graph functions using a minimum number of points. We choose the points to graph based on the function's key features. In Sections 10.2–10.4, we develop strategies for finding the number and location of the real zeros of polynomial functions. In Section 10.5, we find the asymptotes of rational functions and use them to graph the function.

10.1 CONTINUOUS AND DISCONTINUOUS FUNCTIONS

SSM   Study Guide   CD/Video

MathPro 4/5   PH Math Tutor Center   prenhall.com/Angel

1 Learn new terminology.
2 Identify local maxima and minima.
3 Identify continuous and discontinuous functions.
4 Graph functions defined piecewise.
5 Use the greatest integer function.

1   Learn New Terminology

In this chapter we will discuss functions in more depth. We will expand our knowledge of some functions that were introduced earlier, such as polynomial functions and rational functions, and we will introduce some new functions such as piecewise functions and the greatest integer function. We will also discuss what it means for a graph to be continuous or discontinuous.

Let us start by examining the graph in Figure 10.1.

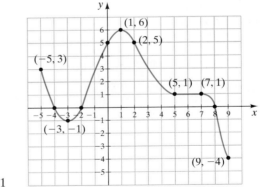

FIGURE 10.1

First notice that this graph represents a function since it passes the vertical line test. Starting at the left side of the x-axis moving to the right, the values of x increase. If the function increases as the values of x increase, we say the **function is increasing**. If the function decreases as the values of x increase, we say the **function is decreasing**. On the graph we have indicated 10 points. We can use the x-values of the ordered pairs to designate intervals. For example, the interval from $x = -5$ to $x = -3$ may be indicated by $[-5, -3]$. Below we illustrate some intervals and indicate whether the function is increasing, decreasing, or staying constant in the interval.

Interval	Function
$[-5, -3]$	decreasing
$[-3, 1]$	increasing
$[1, 5]$	decreasing
$[5, 7]$	constant
$[7, 9]$	decreasing

Do not confuse the terms increasing function and decreasing function with positive values of the function and negative values of the function. Consider the interval $[-5, -3]$. Within this interval the function is *positive* from $[-5, -4)$ since it is above the x-axis, and negative from $(-4, -3]$ since the function is below the x-axis. The function is 0 at $x = -4$.

2 Identify Local Maxima and Minima

In Figure 10.1 the point at $(-3, -1)$ is the lowest point in a small interval centered around the point $x = -3$. The point $(-3, -1)$ is called a **local minimum**. The point $(1, 6)$ is the highest point in a small interval centered around the point $x = 1$. The point $(1, 6)$ is called a **local maximum**. Now look at the x-intercepts $(-4, 0)$, $(-2, 0)$, $(8, 0)$. The x-coordinates of the x-intercepts are called the **real zeros of the function**, for these are the real-number values of x where the function has a value of zero. The real zeros of the function given in Figure 10.1 are -4, -2, and 8. A function may have non-real zeros. At non-real zeros the values of x are complex numbers.

The graph does not cross the x-axis at non-real zeros. We will discuss non-real zeros later in the chapter.

3 Identify Continuous and Discontinuous Functions

Now consider Figures 10.2 and 10.3.

The graph in Figure 10.2 is an example of a **continuous curve** in the interval $[-5, 6]$. It is also a **smooth curve** since there are no sharp corners on the graph. Notice that you can trace the entire graph without removing the pencil from the paper. Now look at Figure 10.3. The graph is continuous in the interval $[-2, 7)$. At $x = 7$ there is a **gap** in the graph, which makes it discontinuous. We say that there is a **point of discontinuity** at $x = 7$, or that the graph is **discontinuous** at $x = 7$. Now look at the graph where $x = 9$. There is a *hole* in the graph at $x = 9$, which indicates that

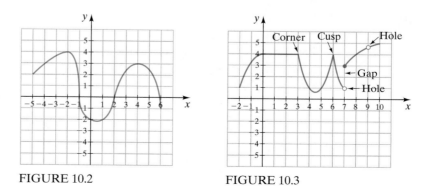

FIGURE 10.2   FIGURE 10.3

the function is not defined at $x = 9$. Therefore, there is also a point of discontinuity at $x = 9$. A continuous curve can have no gaps or holes. A smooth continuous curve can have no corners or cusps or sharp points.

4 Graph Functions Defined Piecewise

Most of the functions we have discussed so far have been continuous functions. Linear, quadratic, cubic, logarithmic, and exponential functions are all continuous functions. It is important to recognize and graph discontinuous functions. We will now introduce and graph two special types of functions: piecewise functions and greatest integer functions. We begin with piecewise functions.

While most functions have a single equation that defines the behavior over its entire domain, **functions defined piecewise**, or **piecewise functions**, are defined using two or more equations. One familiar piecewise defined function is the function used to determine how much income tax we pay. The income tax owed is different for different taxable incomes. Each tax bracket has its own equation to determine the amount of tax owed as a function of taxable income. We present a function defined piecewise in Example 1.

EXAMPLE 1   Graph the function

$$f(x) = \begin{cases} x \text{ if } x < 3 \\ -x \text{ if } x \geq 3 \end{cases}$$

and determine if the function is continuous.

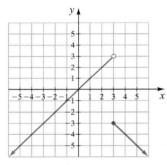

FIGURE 10.4

Solution   This function defined piecewise has two equations. To graph the function in the interval $(-\infty, 3)$ we use $f(x) = x$, and to graph the function in the interval $[3, \infty)$ we use $f(x) = -x$. The function $f(x) = x$ is defined for $x < 3$. If this function were defined at $x = 3$, the value of the function would be 3. Since this part of the function is not defined at 3, we place an open circle, or hole, at $(3, 3)$. We then continue to graph the equation for values of x less than 3, see the red line in Figure 10.4. We now graph $f(x) = -x$ for $x \geq 3$. At $x = 3$, $f(x)$ has a value of -3. The point $(3, -3)$ is included on the graph of $f(x) = -x$, so we place a solid dot at $(3, -3)$. Now continue to graph $f(x) = -x$ for values of $x > 3$; see the green line in Figure 10.4.

NOW TRY EXERCISE 31

Notice the function has a point of discontinuity at $x = 3$.   ✳

EXAMPLE 2   Graph the function

$$f(x) = \begin{cases} x - 3 \text{ if } x \leq 3 \\ -x + 3 \text{ if } x > 3 \end{cases}$$

and determine if the graph is continuous or discontinuous.

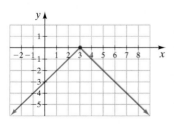

FIGURE 10.5

Solution   First graph $f(x) = x - 3$ for values of x less than or equal to 3. When x is 3, $f(x) = 0$; see the red line in Figure 10.5.

Now graph $f(x) = -x + 3$ for values of x greater than 3. If x were allowed to be 3, $f(x)$ would be 0. The point $(3, 0)$ is already included on the graph. Now graph $f(x) = -x + 3$ for values greater than 3; see the green line in Figure 10.5.

By observing the graph, we see it is continuous for all values of x. However,

NOW TRY EXERCISE 35

the graph is not smooth since there is a corner at $x = 3$.   ✳

Notice that the piecewise function in Example 1 is not continuous, but the piecewise function in Example 2 is continuous.

5   Use the Greatest Integer Function

A function that will always have a discontinuity is the greatest integer function. The **greatest integer function** is a function that truncates, or rounds down, a value to the greatest integer less than or equal to the value. A symbol used to indicate the greatest integer less than or equal to x is $[x]$. For example,

$$[8.7] = 8, \quad \left[\frac{19}{5}\right] = 3, \quad [-6] = -6 \quad \text{and} \quad [-4.03] = -5.$$

EXAMPLE 3   Graph $f(x) = [x]$.

Solution   For any value of x in the interval $[0, 1)$, the value of $[x] = 0$. For any value of x in the interval $[1, 2)$, $[x] = 1$, and so on. The values in each interval are constant, but the function jumps at integer values of x. The graph of the function is shown in Figure 10.6. Notice it consists of a series of line segments. The left endpoint of each line segment is included, and the right endpoint is excluded. The domain of the function is $(-\infty, \infty)$. The range is the set of integers

$$\{\dots, -3, -2, -1, 0, 1, 2, 3, \dots\}$$

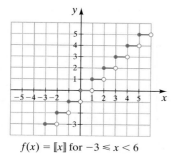

$f(x) = [x]$ for $-3 \leqslant x < 6$

FIGURE 10.6

NOW TRY EXERCISE 39

Notice the graph in Figure 10.6 is made up of a series of horizontal lines that resemble steps. Such graphs are called **step functions**.

EXAMPLE 4   Graph $f(x) = \left[\frac{1}{2}x + 3\right]$.

Solution   We need to determine the intervals for the steps of this greatest integer function. To find the closed endpoint for $f(x) = 0$, set the expression $\frac{1}{2}x + 3$ equal to 0 and solve for x.

$$\frac{1}{2}x + 3 = 0$$

$$\frac{1}{2}x = -3$$

$$x = -6$$

To find the closed endpoint for $f(x) = 1$, set the expression equal to 1 and solve for x.

$$\frac{1}{2}x + 3 = 1$$

$$\frac{1}{2}x = -2$$

$$x = -4$$

Since the closed endpoint for $f(x) = 0$ is -6 and the closed endpoint for $f(x) = 1$ is -4, $f(x)$ must equal 0 in the interval $[-6, -4)$.
To find the closed endpoint for $f(x) = 2$, set the expression equal to 2 and solve for x.

$$\frac{1}{2}x + 3 = 2$$

$$\frac{1}{2}x = -1$$

$$x = -2$$

Since the closed endpoint for $f(x) = 1$ is -4 and the closed endpoint for $f(x) = 2$ is -2, $f(x)$ must equal 1 in the interval $[-4, -2)$.

We continue this process to obtain the following table of values.

$f(x) = \left[\!\left[\frac{1}{2}x + 3\right]\!\right]$ for $-6 \leqslant x < 6$

FIGURE 10.7

Interval	Value of $f(x)$
$[-6, -4)$	0
$[-4, -2)$	1
$[-2, 0)$	2
$[0, 2)$	3
$[2, 4)$	4
$[4, 6)$	5

NOW TRY EXERCISE 43   Using this table, we get the graph in Figure 10.7.   ✳

The greatest integer function has many practical applications. Example 5 shows a practical application of the greatest integer function.

EXAMPLE 5   **Purchasing CDs**

a) Write a greatest integer function that can be used to determine the maximum number of CD's that a person can purchase for x dollars if each CD costs \$14.

b) Graph the function.

Solution   **a)** A person could purchase 1 CD with \$14, 2 CD's with \$28, 3 CD's with \$42, and so on. To determine the maximum number of CD's, divide the given dollar amount by 14, and round down. Thus, the greatest integer function is $f(x) = \left[\!\left[\dfrac{x}{14}\right]\!\right]$.

b) By selecting values for x, we can determine the following table.

FIGURE 10.8

Interval	Value of $f(x)$
$[0, 14)$	0
$[14, 28)$	1
$[28, 42)$	2
$[42, 56)$	3
$[56, 70)$	4

NOW TRY EXERCISE 47   Figure 10.8 shows the graph for values from \$0 up to but not including \$70.   ✳

In the Exercise Set you will be asked to determine whether the graphs of certain functions are continuous or discontinuous. On pages 386 and 387 we discussed finding the domains of rational functions, and in the Using Your Graphing Calculator box we graphed a rational function. Every rational function will have a point of discontinuity at any value of x for which the function is not defined. For example, in $f(x) = \dfrac{x}{x^2 - 4}$ the function is not defined at $x = 2$ and $x = -2$, and so the graph of $f(x)$ is discontinuous at those points. We will discuss the graphs of rational functions further in Section 10.5.

Using Your Graphing Calculator

Graphing calculators can graph the greatest integer function. On the TI-83 Plus calculator, the greatest integer is given as [int (]. It is found under the *CATALOG* menu. Scroll down until you reach [int (]. It is also found using [NUM] of [MATH] menu. Other calculators may use a different representation. If you are using a different calculator, consult the manual to learn how to use the greatest integer function.

Exercise Set 10.1

Concept/Writing Exercises

1. Does every function have a local minimum or maximum? Explain.

2. If a function is defined for all real numbers, is it always continuous? Explain.

3. Is $f(x) = x^2 - 4$ a continuous function? Explain.

4. Is $f(x) = \begin{cases} 3x - 4 \text{ if } x < \dfrac{4}{3} \\ 4 - 3x \text{ if } x \geq \dfrac{4}{3} \end{cases}$ a continuous function? Explain.

5. Is $g(x) = \begin{cases} 2x - 1 \text{ if } x < -5 \\ 2x + 6 \text{ if } x \geq -5 \end{cases}$ a continuous function? Explain.

6. Describe the graph of a greatest integer function.

Practice the Skills

Classify each function as continuous or discontinuous.

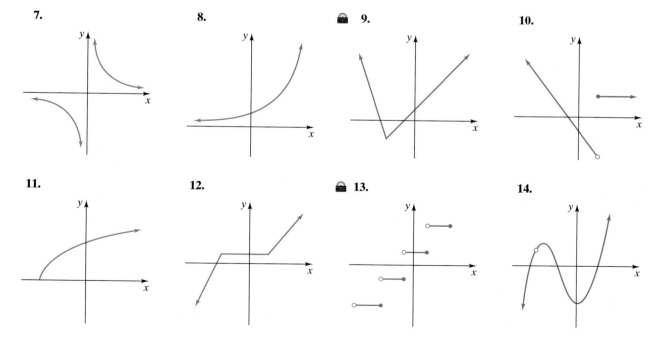

7.
8.
9.
10.

11.
12.
13.
14.

Classify each function as continuous or discontinuous. If the function is not continuous, identify the point(s) of discontinuity.

15. $f(x) = 5x^2 + 3x + 1$

16. $f(x) = x^3 + x - 4$

17. $g(x) = 7^x$

18. $g(x) = 2^x - 1$

19. $h(x) = \log_7 x$

20. $h(x) = \log_2(x + 3)$

21. $m(x) = \dfrac{1}{x + 5}$

22. $m(x) = \dfrac{1}{x^2 + 8}$

23. $j(x) = [\![2x]\!]$

24. $j(x) = \left[\!\left[\dfrac{x}{2}\right]\!\right]$

25. $f(x) = -x + 7$

26. $f(x) = 2(x - 11)^2$

27. $g(x) = \dfrac{3}{x - 3}$

28. $g(x) = \dfrac{5}{x + 6}$

29. $p(x) = \dfrac{1}{x^2 + 4}$

30. $p(x) = \dfrac{7}{x + 10}$

Graph each piecewise function.

31. $f(x) = \begin{cases} x + 5 \text{ if } x < -2 \\ -x + 5 \text{ if } x \geq -2 \end{cases}$

32. $h(x) = \begin{cases} \dfrac{1}{2}x \text{ if } x \leq 1 \\ 2x \text{ if } x > 1 \end{cases}$

33. $g(x) = \begin{cases} -x + 4 \text{ if } x \leq 0 \\ 4 \qquad \text{ if } x > 0 \end{cases}$

34. $r(x) = \begin{cases} 3 \quad \text{ if } x < -1 \\ -3 \text{ if } x \geq -1 \end{cases}$

35. $h(x) = \begin{cases} -2x + 1 \text{ if } x < \dfrac{1}{2} \\ 2x - 1 \quad \text{ if } x \geq \dfrac{1}{2} \end{cases}$

36. $f(x) = \begin{cases} -x + 4 \text{ if } x < 4 \\ -2x + 8 \text{ if } x > 4 \end{cases}$

37. $p(x) = \begin{cases} \dfrac{3}{2}x - 4 \quad \text{ if } x \leq 2 \\ -\dfrac{2}{3}x + \dfrac{19}{3} \text{ if } x > 2 \end{cases}$

38. $g(x) = \begin{cases} 3x \qquad \text{ if } x < 5 \\ -\dfrac{1}{5}x + 16 \text{ if } x \geq 5 \end{cases}$

Graph each greatest integer function.

39. $f(x) = [\![3x]\!]$

40. $g(x) = [\![x - 5]\!]$

41. $h(x) = [\![-2x]\!]$

42. $f(x) = \left[\!\left[\dfrac{x + 2}{3}\right]\!\right]$

43. $g(x) = \left[\!\left[-\dfrac{1}{2}x + 1\right]\!\right]$

44. $h(x) = \left[\!\left[\dfrac{3}{4}x - 2\right]\!\right]$

45. $f(x) = \left[\!\left[\dfrac{2x + 7}{10}\right]\!\right]$

46. $g(x) = \left[\!\left[\dfrac{8 - x}{3}\right]\!\right]$

Problem Solving

47. *Greeting Cards* Kelly Chen wants to stock up on greeting cards which are selling for $1.25 a piece. Write a greatest integer function that represents the number of cards she can purchase for x dollars.

48. *Cat Food* Don Jacobs wants to stock up on cans of cat food which are selling for $0.89 per can. Write a greatest integer function that represents the number of cans he can purchase for x dollars.

49. *Calendars* The office supply manager wants to purchase calendars which are selling for $3.99 each. Write a greatest integer function that represents the number of calendars she can purchase for x dollars.

50. *Flyers* When Teri Lovelace lost her cat Kundera, she made a flyer to distribute around the neighborhood.

She went to Copy Cats to make copies of her flyer. Copies cost $0.12 each. Write a greatest integer function that represents the number of copies she can make for x dollars.

51. *Bikes* A bike rental company rents bikes by the hour. They charge $5.00 per hour for the first 3 rental hours and $3.00 per hour for each additional hour. Write a piecewise function to represent the cost of renting a bike as a function of hours it is rented. *Hint:* The cost, in dollars, for $x > 3$ hours will be the cost of the first 3 hours, $15, plus $3(x - 3)$, where x is the number of hours.

52. *Electricity* An electric company's monthly charge is $9.27 plus an economy rate of 11.589 cents per kilowatt-hour for the first 345 kWhr and premium rate of

13.321 cents per kilowatt-hour for kilowatt-hours over 345 kWhr. Write a piecewise function to represent the monthly electric cost as a function of kilowatt-hours. *Hint*: The cost for 345 kWhr is $0.11589(345) + 9.27$, or

$49.25205. The cost for any kilowatt-hours over $345(x > 345)$ will be the cost for the first 345 kWhr, $49.25205, plus $0.13321(x - 345)$, where x is the number of kilowatt-hours.

A function $f(x)$ is an even function if $f(-x) = f(x)$ for all x in the domain of f. A function $f(x)$ is an odd function if $f(-x) = -f(x)$ for all x in the domain of f. For example $f(x) = x^4 + 2$ is an even function because $f(-x) = (-x)^4 + 2 = x^4 + 2$. Thus $f(-x) = f(x)$. The function $f(x) = x^3 + x$ is an odd function because $f(-x) = (-x)^3 + (-x) = -x^3 - x$ and $-f(x) = -(x^3 + x) = -x^3 - x$. Therefore $f(-x) = -f(x)$. For Exercises 53 through 62 determine whether the function is an even function, an odd function, or neither.

53. $f(x) = x^2 - 15$

54. $f(x) = 3x^3 - x$

55. $f(x) = 3x^3 + x^2 - 10$

56. $g(x) = 2x^4 - 6x^2 + \frac{1}{3}$

57. $g(x) = 3x^6 - 15x^2 + 4$

58. $g(x) = e^{2x} + 8$

59. $g(x) = e^{x^2} + 5$

60. $h(x) = 5x^5 + 3x^3 - 2x$

61. $h(x) = x^4 - x^3 + x^2$

62. $h(x) = 6x^5 - 3x^4 + \frac{5}{2}$

Challenge Problems

Graph each piecewise function.

63. $f(x) = \begin{cases} 2x & \text{if } x < -1 \\ -x + 5 & \text{if } -1 \le x < 2 \\ x + 1 & \text{if } x \ge 2 \end{cases}$

64. $h(x) = \begin{cases} -x - 4 & \text{if } x \le -1 \\ 2x - 1 & \text{if } -1 < x < 2 \\ x + 1 & \text{if } x \ge 2 \end{cases}$

65. $p(x) = \begin{cases} 2x + 3 & \text{if } x < -1 \\ x + 1 & \text{if } -1 \le x < 1 \\ 2x & \text{if } 1 \le x < 3 \\ x - 1 & \text{if } x \ge 3 \end{cases}$

66. $g(x) = \begin{cases} -x + 3 & \text{if } x < 1 \\ 5 & \text{if } x = 1 \\ \sqrt{x} & \text{if } x > 1 \end{cases}$

Group Activity

We discussed even and odd functions before Exercises 53–62. Review that material now. Then solve Exercises 67 and 68.

67. a) Explain why for every even function if the ordered pair (x, y) is a point on the graph of the function, the ordered pair $(-x, y)$ will also be on the graph.

b) Explain why the graph of an even function will always be symmetric with respect to the y-axis.

68. a) Explain why for every odd function if the ordered pair (x, y) is a point on the graph of the function, the ordered pair $(-x, -y)$ will also be on the graph.

b) Explain why the graph of an odd function will always be symmetric with respect to the origin.

Cumulative Review Exercises

[3.1] **69.** Graph $y = -x^3 - 1$.

[3.1] **70.** Graph $y = \dfrac{1}{x - 2}$.

[5.1] **71.** Consider the polynomial $4x^5 - 7x^4 - 6x^3 + x^2 - 9$.

a) Give the degree of the polynomial.

b) Give its leading coefficient.

[6.1] **72.** Determine the domain of the rational function $g(x) = \dfrac{x - 7}{x^2 - 4x - 21}$.

10.2 CHARACTERISTICS OF GRAPHS OF POLYNOMIAL FUNCTIONS

1   Define a polynomial function.
2   Define power functions.
3   Determine the *x*-intercepts of a graph and the zeros of a polynomial function.
4   Determine the maximum number of turning points of the graph of a polynomial function.
5   Graph polynomial functions.

1   Define a Polynomial Function

We introduced polynomial functions in Section 5.1. We will now give a definition of a polynomial function.

> A **polynomial function** is a function of the form
>
> $$P(x) = a_n x^n + a_{n-1} x^{n-1} + a_{n-2} x^{n-2} + \ldots + a_1 x + a_0$$
>
> where the coefficients $a_n, a_{n-1}, a_{n-2}, \ldots, a_0$ are real numbers,* and all exponents on the variable are nonnegative integers. The domain of all polynomial functions is all real numbers.

Examples of polynomial functions are:

$$P(x) = x^5 + 5x^3 - 2x \quad \text{and} \quad f(x) = -4x^3 + 2x^2 + \frac{1}{2}x - 5$$

An important part of this chapter involves graphing polynomial functions. We will discuss a number of items to help us graph polynomial functions. In this section we first discuss the end behavior of polynomial functions. By **end behavior** we mean how the graph behaves for large positive values of *x* and for large negative values of *x*. That is, whether the graph will be above or below the *x*-axis for large positive values of *x* and for large negative values of *x*.

2   Define Power Functions

Functions of the form $f(x) = ax^n$ are called **power functions**. Power functions are also polynomial functions. A power function of degree *n* can be used to determine the end behavior of a polynomial function of the form $f(x) = a_n x^n + a_{n-1} x^{n-1} + \ldots + a_0$. For example, the graph of $f(x) = x^5$ can be used to determine the end behavior of the graph of $f(x) = 6x^5 + 5x^4 + 2x^3 - 6$ because the leading term of the polynomial, $6x^5$, has degree 5. For your convenience, in Figure 10.9, on the next page, we give the graphs of power functions of the form $f(x) = x^n$ for $n = 1$, 2, 3, 4, 5, and 6.

Consider the graph of $f(x) = x^4$. Suppose *x* is a large positive value, then $f(x)$ will be a large positive value, and the graph will be above the *x*-axis. Suppose *x* is a large negative value, then $f(x)$ will still be a large positive value, since a negative number raised to an even exponent is positive. For a large negative value of *x* the function will be positive and the graph will be above the *x*-axis. Using this example, we can reason that the graph of a function of the form $f(x) = x^n$, where *x*

---

*A polynomial function can be defined using complex numbers for $a_n, a_{n-1}, a_{n-2}, \ldots, a_0$. We will not discuss such polynomials in this book.

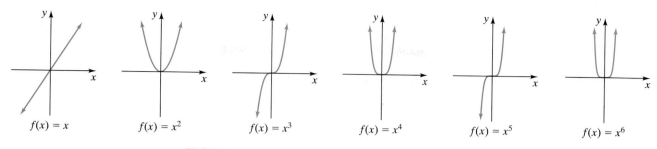

FIGURE 10.9

is an even number $(2, 4, 6, \ldots)$ will be above the x-axis for both large positive and large negative values of x.

Consider the graph of $f(x) = x^5$. Suppose that x is a large positive value, then $f(x)$ will be a large positive value, and the graph will be above the x-axis. Suppose x is a large negative value, then $f(x)$ will be a large negative value since a negative number raised to an odd exponent is negative. For a large negative value of x, the function will be negative, and its graph will be below the x-axis. Using this example, we can reason that the graph of a function of the form $f(x) = x^n$, where x is an odd number $(1, 3, 5, \ldots)$, will be above the x-axis when x is a large positive value and below the x-axis when x is a large negative value. The graphs of $y = -x^n$ for $n = 1, 2, 3, 4, 5$, and 6 are shown in Figure 10.10.

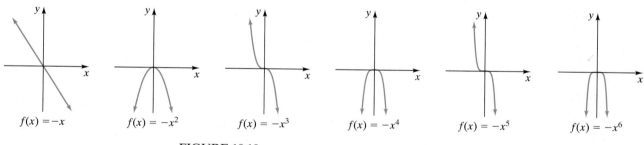

FIGURE 10.10

When the leading coefficient in a polynomial function is negative, the end behavior of the graph of the function can be determined from the graphs in Figure 10.10.

EXAMPLE 1 By considering only the end behavior of the graphs of the given polynomial function, determine which, if any, of graphs 1, 2, or 3 could possibly be the graph of the polynomial function.

a) $P(x) = 3x^4 + 3x^3 - 6$

b) $P(x) = -2x^5 + 6x^4 - 3x + 2$

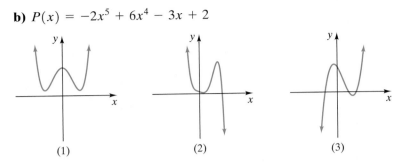

(1)　　　　　　　　(2)　　　　　　　　(3)

Solution　**a)** Since the leading term is $3x^4$, the graph has the same end behavior as the graph of $f(x) = x^4$. The graph must be above the x-axis for both large positive and large negative values of x. Therefore the only possible answer is 3.

b) Since the leading term is $-2x^5$, the graph has the same end behavior as the graph of $f(x) = -x^5$. The graph must be below the x-axis for large positive values of x and above the x-axis for large negative values of x. Therefore the only

NOW TRY EXERCISE 7　possible answer is 2.　✳

3　Determine the *x*-intercepts of a Graph and the Zeros of a Polynomial Function

Now we are ready to discuss how to find the x-intercepts of the graph of a polynomial function and the zeros of a polynomial function. A **zero of a polynomial function** is a value of x where the function is equal to 0.

> If $P(x)$ is a polynomial function and x is a real number for which $P(x) = 0$, then x is called a **zero** of $P(x)$.

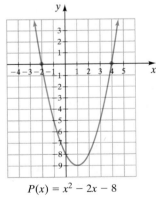

$P(x) = x^2 - 2x - 8$

FIGURE 10.11

Consider the quadratic function $P(x) = x^2 - 2x - 8$ which we graphed on page 371. Quadratic functions are one type of polynomial function. The graph is shown in Figure 10.11 for your convenience.

The x-intercepts are $(-2, 0)$ and $(4, 0)$. The zeros of the function are therefore -2 and 4. We can obtain the zeros by setting the function equal to 0, and solving the resulting equation for x.

$$P(x) = x^2 - 2x - 8$$
$$0 = x^2 - 2x - 8$$
$$0 = (x + 2)(x - 4)$$
$$x + 2 = 0 \quad \text{or} \quad x - 4 = 0$$
$$x = -2 \qquad\qquad x = 4$$

As we just illustrated, if we can find the linear factors of a polynomial function, we can find the real zeros of the function by setting the product of the factors equal to zero and then using the zero-factor property. Also, we can work backward, if we know the real zeros of a function, then we can determine the x-intercepts of the graph of the function.

> If a real number r is a zero of a polynomial function, that is $P(r) = 0$,
> a) $(r, 0)$ is an x-intercept of the graph, and
> b) $(x - r)$ is a factor of $P(x)$.

For example, consider $P(x) = x^3 + 6x^2 - x - 30$. Notice $P(2) = 0$.

$$P(x) = x^3 + 6x^2 - x - 30$$
$$P(2) = 8 + 24 - 2 - 30 = 0$$

Since $P(2) = 0$, then 2 is a zero of the polynomial function, $(2, 0)$ is an x-intercept of the graph of $P(x)$, and $(x - 2)$ is a factor of $P(x)$.

EXAMPLE 2 Consider the function $P(x) = x^3 - 2x^2 - 8x$.

a) Determine the zeros of the function.

b) Determine the x-intercepts of the graph of $P(x)$.

Solution **a)** First factor the expression, as follows.

$$P(x) = x(x^2 - 2x - 8)$$
$$P(x) = x(x - 4)(x + 2)$$

Next set $P(x) = 0$ and solve for x.

$$0 = x(x - 4)(x + 2)$$

$x = 0$ or $x - 4 = 0$ or $x + 2 = 0$

$x = 0$ $x = 4$ $x = -2$

Thus, the zeros are $0, 4, -2$.

NOW TRY EXERCISE 21 **b)** The x-intercepts of the graph are $(0, 0)$, $(4, 0)$, and $(-2, 0)$. ✳

Consider the function $P(x) = (x + 5)(x - 4)^2$. This function can be written $P(x) = (x + 5)(x - 4)(x - 4)$. Since 4 makes two factors of $(x - 4)$ equal 0, we say that 4 is a zero of **multiplicity** 2.

EXAMPLE 3 Determine the zeros of $P(x) = (x - 3)^2(x + 4)^3(x - 5)^6$ and give their multiplicities.

Solution The factor $(x - 3)^2$ is zero when $x = 3$. Since there are two factors of $(x - 3)$, its multiplicity is 2.

The factor $(x + 4)^3$ is zero when $x = -4$. Thus, -4 is a zero of multiplicity 3.

NOW TRY EXERCISE 25 The factor $(x - 5)^6$ is zero when $x = 5$. Thus, 5 is a zero of multiplicity 6. ✳

Let's now discuss x-intercepts a little further. Consider Figure 10.12.

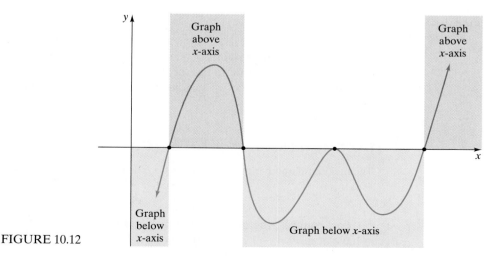

FIGURE 10.12

The graph shows four x-intercepts. That is, four values of x where $P(x) = 0$. Notice at three of the four intercepts the graph crosses the x-axis, thus there is a change in the sign of $P(x)$. At one x-intercept, the graph touches, but does not cross, the x-axis. Notice that the function has the same sign, negative, on both sides of that x-intercept.

If we know the multiplicity of a zero we can determine whether the function crosses or just touches the x-axis at that zero.

> For a real zero of an even multiplicity, the graph of $P(x)$ touches but does not cross the x-axis at that zero.
>   For a real zero of an odd multiplicity, the graph of $P(x)$ crosses the x-axis at that zero.

For a zero of an even multiplicity, the sign of $P(x)$ does not change from one side of the zero to the other side of the zero. For a zero of an odd multiplicity, the sign of $P(x)$ changes from one side of the zero to the other side of the zero.

4 Determine the Maximum Number of Turning Points of the Graph of a Polynomial Function

In Figure 10.1 we introduced local maxima and local minima. Local maxima and local minima are points at which the graph changes direction (from increasing to decreasing or vice versa). Therefore such points are called **turning points**, see Figure 10.13.

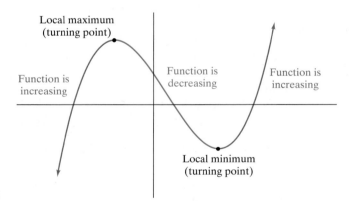

FIGURE 10.13

To find the turning points of a graph requires a knowledge of calculus, so we will not ask you to determine the location of turning points on the graphs. However, the maximum number of turning points on a graph can be found by the following theorem.

Maximum Number of Turning Points

The maximum number of turning points on the graph of a polynomial function of degree n is $n - 1$.

For example, the graph of the polynomial function $P(x) = 3x^4 + 2x - 6$ can have at most $4 - 1$ or 3 turning points.

5 Graph Polynomial Functions

In Examples 4 and 5 we will use our knowledge of end behavior, zeros of polynomial functions, and turning points to graph the functions.

EXAMPLE 4 For the polynomial function $P(x) = (x + 5)(x + 2)(x - 2)$

a) Determine the x-intercepts and whether the graph crosses or touches the x-axis at each x-intercept.

b) Explain the end behavior of the graph of the function for large positive values of x and for large negative values of x.

c) Determine the maximum number of turning points on the graph of the function.

d) Use the x-intercepts and test values to find the intervals on which the graph is above the x-axis, and the intervals on which the graph is below the x-axis.

e) Using all the information, connect the x-intercepts and the test points with a smooth continuous curve to sketch the graph of the function.

Solution **a)** To determine the x-intercepts, set the function equal to 0.

$$P(x) = (x + 5)(x + 2)(x - 2)$$
$$0 = (x + 5)(x + 2)(x - 2)$$

$$x + 5 = 0 \quad \text{or} \quad x + 2 = 0 \quad \text{or} \quad x - 2 = 0$$
$$x = -5 \qquad\qquad x = -2 \qquad\qquad x = 2$$

The x-intercepts are $(-5, 0)$, $(-2, 0)$ and $(2, 0)$. Since each of the zeros is of multiplicity one, the graph will cross the axis at each x-intercept.

b) If we were to multiply the factors, the leading term would be x^3. Therefore, the end behavior of the given function is the same as the end behavior of $f(x) = x^3$. The function will be positive and above the x-axis for large positive values of x and negative and below the x-axis for large negative values of x.

c) Since the polynomial is of degree 3, the maximum number of turning points is $3 - 1$ or 2.

d) The x-intercepts break the graph into 4 regions as illustrated below. We will select a test value in each region. The test values we arbitrarily selected are $-6, -4,$ 0 and 3, as shown in Figure 10.14.

FIGURE 10.14

To determine the value of $P(x)$ at a test value, substitute the test value for each x in the function. For example, for the test value -6, we get

$$P(x) = (x + 5)(x + 2)(x - 2)$$
$$P(-6) = (-6 + 5)(-6 + 2)(-6 - 2)$$
$$= (-1)(-4)(-8) = -32$$

Using the intervals and the test values, we complete the following table.

Interval	Test Value	Value of $P(x)$	Location of $P(x)$ with regard to the x-axis
$(-\infty, -5)$	-6	-32	below
$(-5, -2)$	-4	12	above
$(-2, 2)$	0	-20	below
$(2, \infty)$	3	40	above

e) Plot the x-intercepts and the test points and draw a smooth continuous curve through the points. Notice that when $x = 0$ and $x = 1$, $P(x) = -20$, see Figure 10.15.

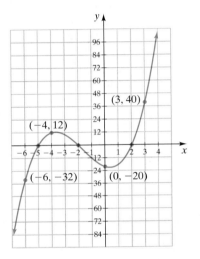

FIGURE 10.15

Notice that the end behavior is as we expected, and the function crosses the x-axis at each x-intercept. The graph has 2 turning points.

NOW TRY EXERCISE 49

EXAMPLE 5 Sketch the graph of $P(x) = 3x^4 - x^3 - 4x^2$.

Solution Begin by factoring the expression on the right of the equal sign.

$$P(x) = 3x^4 - x^3 - 4x^2$$
$$= x^2(3x^2 - x - 4)$$
$$= x^2(3x - 4)(x + 1)$$

Now set $P(x) = 0$ and solve for x.

$$0 = x^2(3x - 4)(x + 1)$$

$$x^2 = 0 \quad \text{or} \quad 3x - 4 = 0 \quad \text{or} \quad x + 1 = 0$$

$$x = 0 \qquad\qquad x = \frac{4}{3} \qquad\qquad x = -1$$

Therefore, the x-intercepts are $(0, 0)$, $\left(\frac{4}{3}, 0\right)$ and $(-1, 0)$. Notice the zero at $x = 0$ is of multiplicity 2. Now use the x-values obtained above to set up intervals. When setting up the intervals, start with the smallest value of x and work from the smallest toward the largest value of x. The test values we selected are shown in Figure 10.16.

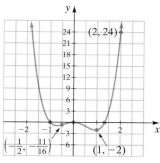

FIGURE 10.16

FIGURE 10.17

NOW TRY EXERCISE 65

Interval	Test Value	Value of $P(x)$	Location of $P(x)$ with regard to the x-axis
$(-\infty, -1)$	-2	40	above
$(-1, 0)$	$-\frac{1}{2}$	$-\frac{11}{16}$	below
$\left(0, \frac{4}{3}\right)$	1	-2	below
$\left(\frac{4}{3}, \infty\right)$	2	24	above

Now we plot the points and draw the graph, see Figure 10.17.

In Figure 10.17 we did not illustrate the ordered pair $(-2, 40)$ since it is off the graph. Notice the end behavior of the function is the same as the end behavior of $f(x) = x^4$. Also notice that at $x = 0$ the graph touches, but does not cross the x-axis. We expect this behavior because this zero has multiplicity 2, an even number. The graph has 3 turning points which is the maximum number it can have.   ✳

Exercise Set 10.2

Concept/Writing Exercises

1. Consider the polynomial function graphed below.

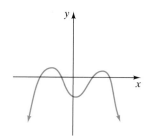

a) Is the constant term positive or negative? Explain.

b) Is the leading coefficient positive or negative? Explain.

c) Is the degree of the polynomial function even or odd? Explain.

d) Can the degree of the polynomial function be 6? Explain.

e) Can the degree of the polynomial function be 7? Explain.

f) How many real zeros does the polynomial function have? Explain.

2. Consider the polynomial function graphed below.

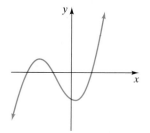

a) Is the constant term positive or negative? Explain.

b) Is the leading coefficient positive or negative? Explain.

c) Is the degree of the polynomial function even or odd? Explain.

d) Can the degree of the polynomial function be 1? Explain.

e) Can the degree of the polynomial function be 4? Explain.

f) How many real zeros does the polynomial function have? Explain.

3. a) Describe the end behavior of the graph of $f(x) = x^{10}$

  b) Describe the end behavior of the graph of $g(x) = x^{11}$.

4. a) If the zeros of a polynomial function are $x = 2$, $x = 1$, and $x = 5$, can the function be $P(x) = a(x + 2)(x + 1)(x + 5)$? Explain.

  b) If the zeros of a polynomial function are $x = 3$, $x = -1$, and $x = -5$, can the function be $P(x) = a(x - 3)(x + 1)(x + 5)$? Explain.

Practice the Skills

For Exercises 5–8, determine which of the given graphs could be the graph of the given polynomial function. Explain your answer.

5. $P(x) = -2x^3 + 5x - 8$

6. $P(x) = -\dfrac{1}{3}x^4 + 5x + 4$

7. $P(x) = 3x^5 + 5x^3 - 25x^2 + 20$

8. $P(x) = x^6 - 2x^5 + 4x^4 - 8x^3 - 16x^2 + 32x - 64$

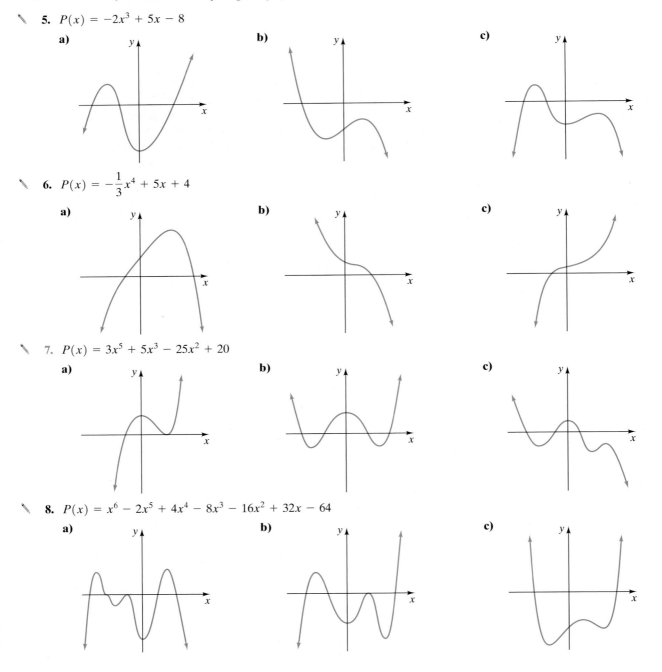

Describe the location, either above the x-axis or below the x-axis, of the graph of each function for a) large positive values of x, and b) large negative values of x.

9. $P(x) = x^7$

10. $f(x) = 2x^7$

11. $f(x) = -3x^7$

12. $P(x) = x^7 + x^6 + x^5 + x^4 + x + 10$

13. $f(x) = -\dfrac{1}{4}x^2 + 2x$

14. $f(x) = 4x^2 + 2x$

15. $f(x) = -3x^4 + x^3 - 2x + 17$

16. $P(x) = 7x^4 - 3x + 5$

17. $f(x) = -5x^7 + 2x^5 - x^4 + x^2 - x + 1$

18. $f(x) = -x^6 + 7x^4 - 2x^3 + x^2 - x + 3$

Find the real zeros of each function and state their multiplicities if it is a number other than 1.

19. $f(x) = x^2 - 2x - 15$

20. $g(x) = x^2 - 8x + 16$

21. $P(x) = x^3 + 6x^2 + 5x$

22. $h(x) = 2x^3 + 9x^2 - 5x$

23. $f(x) = x^2 - 25$

24. $g(x) = x^2 - 3$

25. $P(x) = (x + 7)(x - 2)^3$

26. $f(x) = (x - 9)^3$

27. $g(x) = (x + 3)(2x + 1)(3x - 1)$

28. $h(x) = (x + 7)(x + 1)(x - 3)^2$

29. $h(x) = (x + 2)^3(x - 4)^2$

30. $f(x) = (x + 6)(x - 5)(x^2 - 49)$

31. $P(x) = x(x + 1)^2(x - 1)^3(x - 6)^4$

32. $g(x) = x^4(x^2 - 25)(x^2 - 9)$

Determine the maximum possible number of turning points for the graph of each function.

33. $f(x) = 4x^5 + x^3 - 8x^2 + 64$

34. $g(x) = 3x^2 + 2x - 9$

35. $h(x) = \frac{1}{2}x - 7$

36. $P(x) = 5x^5 - x^7$

37. $g(x) = 3x^6 + x^5 - x^4 + 6x^2 - 5x + 12$

38. $h(x) = x^3 - 4x^2 + 9x - 24$

39. $f(x) = x(x - 5)(3x + 4)$

40. $P(x) = x^2(x^2 - 16)(x + 3)$

41. $h(x) = (x - 1)(x + 2)$

42. $f(x) = (x - 6)^2(4x - 1)$

43. $g(x) = (2x - 3)^3(x + 8)(x - 1)$

44. $P(x) = (x + 8)(x - 5)(x + 3)(x - 2)(x + 1)$

Graph each function. See Example 4.

45. $f(x) = x(x + 3)(x - 1)$

46. $f(x) = x^2(x + 3)(x + 1)$

47. $P(x) = (x + 3)^2(x - 1)$

48. $g(x) = (x + 2)(x + 1)(x - 3)$

49. $P(x) = (x + 3)(x - 3)(x - 5)$

50. $g(x) = x(x + 4)^2$

51. $g(x) = (x + 1)(x^2 - 36)$

52. $h(x) = (x + 3)(x + 1)(x - 2)(x - 4)$

Problem Solving

Determine the minimum degree of the polynomial function graphed.

53. **54.** **55.** **56.**

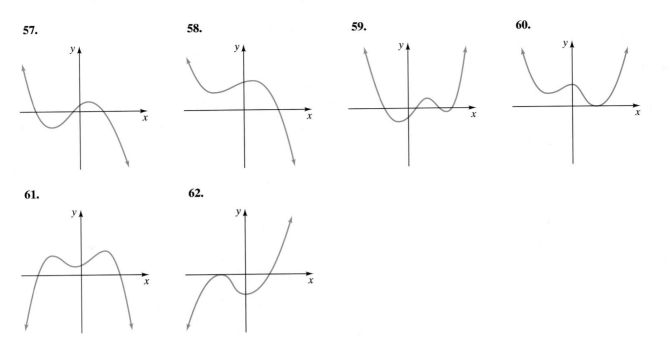

57. 58. 59. 60.

61. 62.

Graph each function. See Example 5.

63. $f(x) = x^3 - 2x^2 - 3x$

64. $g(x) = 2x^3 + 5x^2 - 3x$

65. $f(x) = 5x^3 + 4x^2 - x$

66. $g(x) = 6x^4 - 10x^3 - 4x^2$

67. $f(x) = x^3 - 4x^2 + 4x$

68. $g(x) = x^4 - 4x^3 + 4x^2$

In the Using Your Graphing Calculator box on page 183 we explained how to find the zeros of a function using a graphing calculator. A similar procedure can be used to find the turning points (the local maxima or local minima) of a polynomial function. On the TI-83 Plus you use the CALC menu then select item 3:minimum or item 4:maximum. Read page 183 now. Then read your graphing calculator manual if further explanation is needed on how to find the local maxima and local minima of a polynomial function. Then for each of the following exercises:

a) On your graphing calculator, graph the polynomial function over the given interval.

b) Determine if the function has a local maximum or local minimum in the given interval.

c) Use your graphing calculator to find the turning point of the polynomial function in the given interval. Round the coordinates to the nearest hundredth when appropriate.

69. $y = 3x^2 - 2x + 3, (-1, 1)$

70. $y = -\frac{1}{2}x^2 - 5x + 4, (-6, -4)$

71. $y = -x^3 + 2x - 3, (-2, 0)$

72. $y = x^3 + 3x^2 - 5, (-3, -1)$

Group Activity

73. In Exercise 47, we graphed $P(x) = (x + 3)^2(x - 1)$ where the linear factor $(x + 3)$ is squared. Now we will look at graphs of functions having two linear factors, each of which is squared.

    a) Each group member should graph the appropriate function given below.
Group Member 1: Graph $f(x) = (x + 1)^2(x - 1)^2$
Group Member 2: Graph $f(x) = (x + 2)^2(x - 1)^2$
Group Member 3: Graph $f(x) = (x + 2)^2(x + 1)^2$

    b) Is $f(x) \geq 0$ for every value of x for your graph?

    c) Share your graph with the other members. As a group, what can you conclude from the three graphs?

    d) As a group, graph $g(x) = (x + 2)^2(x + 1)^2(x - 1)^2$.

    e) Discuss why the graph of $g(x)$ is similar to the three graphs of $f(x)$ from part **a)**.

Cumulative Review Exercises

[5.3] **74.** Use synthetic division to divide
$(4x^4 - 3x^3 + 2x^2 - x - 250) \div (x - 3)$.

[5.3] **75.** Use the Remainder Theorem to determine the remainder for the division
$(5x^3 - 7x^2 + 9x - 11) \div (x + 4)$.

[6.2] **76.** Add $\dfrac{3}{5x} + \dfrac{x}{4}$.

[6.2] **77.** Subtract $\dfrac{3x - 1}{y - 4} - \dfrac{2x + 5}{4 - y}$.

10.3 FINDING THE RATIONAL ZEROS OF A POLYNOMIAL FUNCTION

SSM   Study Guide   CD/Video

MathPro 4/5   PH Math Tutor Center   prenhall.com/Angel

1 Determine the maximum number of real zeros of a polynomial function.

2 Use Descartes' rule of signs.

3 Determine the possible rational zeros of a polynomial function.

4 Use synthetic division to determine the rational zeros of a polynomial function.

5 Find complex zeros of a polynomial function.

---

1 Determine the Maximum Number of Real Zeros of a Polynomial Function

In the previous section we learned that the maximum number of turning points of the graph of a polynomial of degree n is $n - 1$. Consider Figure 10.18. Figure 10.18 shows the graph of a cubic or third-degree polynomial function with 2 turning points, the maximum number possible. Figure 10.19 shows the graph of a fourth-degree polynomial with 3 turning points, the maximum number possible.

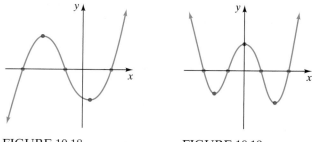

FIGURE 10.18   FIGURE 10.19

Notice that in Figure 10.18, the number of x-intercepts is 3, one more than the maximum number of turning points. In Figure 10.19 the number of x-intercepts is 4, one more than the maximum number of turning points. By examining other graphs, we can determine that the maximum number of x-intercepts is one more than the maximum number of turning points. Since the maximum number of turning points for a polynomial of degree n is $n - 1$, the maximum number of x-intercepts is $(n - 1) + 1$ or n.

For a polynomial function $P(x)$ of degree n

a) the maximum number of x-intercepts of its graph is n, and

b) the maximum number of real zeros of $P(x)$ is n.

An **odd-degree polynomial function** will always have at least one real zero because one end of the graph will be above the x-axis and the other end will be below the x-axis, see Figure 10.20.

Since an **even-degree polynomial function** will have both ends of its graph either above the x-axis or below the x-axis, it is possible for the polynomial function to have no real zeros, see Figure 10.21.

FIGURE 10.20   FIGURE 10.21

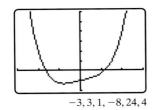

$-3, 3, 1, -8, 24, 4$

FIGURE 10.22

An even-degree polynomial function with a leading coefficient that has the opposite sign as its constant term has at least 2 real zeros. For example, $P(x) = x^4 + 2x - 4$ has a minimum of 2 real zeros since the leading coefficient, 1, is positive and the constant, -4, is negative. Note that $P(-2) = 8$, $P(0) = -4$ and $P(2) = 16$. Since there is a change of sign of $P(x)$ from $x = -2$ to $x = 0$, there must be a zero between these values. Similarly, since there is a change of sign of $P(x)$ between $x = 0$ and $x = 2$, there must be another zero between these values. See Figure 10.22.

EXAMPLE 1 Determine the maximum and minimum number of x-intercepts of the graph of each polynomial function.

a) $P(x) = 4x^3 - 5x^2 + 9x + 7$

b) $P(x) = -x^4 + 2x^3 + 13x^2 - 5x + 20$

Solution **a)** Because $P(x) = 4x^3 - 5x^2 + 9x + 7$ is a third-degree polynomial function, its graph has a maximum of three x-intercepts. Since its degree is an odd number, its graph has a minimum of one x-intercept.

b) Because $P(x) = -x^4 + 2x^3 + 13x^2 - 5x + 20$ is a fourth-degree polynomial function, its graph has a maximum of four x-intercepts. Since its degree is an even number, and its leading coefficient and constant term have opposite signs, its graph has at least two x-intercepts.   ✳

NOW TRY EXERCISE 11

2 Use Descartes' Rule of Signs

We have recently shown that the maximum number of real zeros of an nth degree polynomial function is n. Therefore we know that the maximum number of positive zeros is n and the maximum number of negative zeros is n.

Using **Descartes' Rule of Signs**, we can give an even better estimate of the maximum number of positive zeros and the maximum number of negative zeros.

Descartes' Rule of Signs

For any polynomial function $P(x)$ written in descending order,

1. the number of positive real zeros of $P(x)$ either equals the number of variations in signs of the nonzero coefficients of $P(x)$ or equals that number less an even integer.

2. the number of negative real zeros of $P(x)$ either equals the number of variations in sign of the nonzero coefficients of $P(-x)$ or equals that number less an even integer.

Although the Descartes' Rule of Signs is intended for higher degree polynomial functions, we will look at a linear function to help us develop our intuitive understanding of the rule.

The function $P(x) = 3x + 4$ has 1 zero. Notice that because both coefficients are positive there are no variations in sign in the coefficients of $P(x)$. Thus, the zero cannot be positive. Now let's look at the variations in $P(-x)$.

$$P(x) = 3x + 4$$
$$P(-x) = 3(-x) + 4$$
$$P(-x) = -3x + 4$$

FIGURE 10.23

Notice that there is 1 variation in sign in the coefficients, from minus to plus. By Descartes' Rule of Signs, the number of negative zeros is 1. Thus, the 1 real zero in $P(x)$ occurs at a negative value of x.

We will now look at a quadratic function. Quadratic functions have a maximum of 2 real zeros. The quadratic function $P(x) = (x - 2)(x - 4)$ has zeros at $x = 2$ and $x = 4$. From the graph of the function, shown in Figure 10.23, we see that the function has 2 positive real zeros and no negative real zeros.

Now let's see how Descartes' Rule of Signs could have been used to determine the possible number of real zeros of the function. When multiplied out we obtain $P(x) = x^2 - 6x + 8$. $P(x)$ has 2 variations in sign. By Descartes' Rule of Signs, we know that the function either has 2 or 0 positive zeros. Now let's look at the sign changes in the coefficients of $P(-x)$.

$$P(x) = x^2 - 6x + 8$$
$$P(-x) = (-x)^2 - 6(-x) + 8$$
$$P(-x) = x^2 + 6x + 8$$

Because there are no variations in the sign of the coefficients of $P(-x)$, we know that the function does not have any negative zeros.

Consider $P(x) = x^2 + x - 6$. Since this is an even-degree polynomial function whose leading coefficient, 1, is opposite in sign from the constant term, -6, we know the polynomial has two real zeros. Since there is one change in sign in $P(x)$ there is one positive real zero. $P(-x) = x^2 - x - 6$. Since there is one change of sign in $P(-x)$ there is one negative real zero. Thus, this polynomial has one positive real zero and one negative real zero. Use factoring now to verify the real zeros are -3 and 2.

Now we will use Descartes' Rule of Signs to determine the possible number of positive and negative real zeros in a sixth-degree polynomial function.

EXAMPLE 2   Determine the possible number of positive and negative real zeros in $P(x) = 3x^6 - 5x^5 + 3x^4 + x^2 - 10$.

Solution   The function, $P(x) = 3x^6 - 5x^5 + 3x^4 + x^2 - 10$, is written in descending order. Since the nonzero coefficients of $P(x)$, which are $3, -5, 3, 1, -10$, have 3 sign changes, the graph of the function has either 3 or 1 positive zeros. Now find $P(-x)$.

$$P(x) = 3x^6 - 5x^5 + 3x^4 + x^2 - 10$$
$$P(-x) = 3(-x)^6 - 5(-x)^5 + 3(-x)^4 + (-x)^2 - 10$$
$$= 3x^6 + 5x^5 + 3x^4 + x^2 - 10$$

NOW TRY EXERCISE 21

Since the nonzero coefficients of $P(-x)$, which are $3, 5, 3, 1, -10$, have 1 sign change, the function has 1 negative real zero.   ✳

3   Determine the Possible Rational Zeros of a Polynomial Function

Now that we know the possible number of positive and negative real zeros of a polynomial function, we will turn our attention to finding the zeros. A real zero of a polynomial function will either be a rational number or an irrational number. The set of possible rational zeros for a polynomial function can be determined from the leading coefficient and the constant term of the polynomial.

> ### Rational Zeros Theorem
>
> For any polynomial function $P(x) = a_n x^n + a_{n-1}x^{n-1} + \ldots + a_1 x + a_0$ where $a_n \neq 0$, $a_0 \neq 0$, and all coefficients are integers, the possible rational zeros of $P(x)$ are of the form $\dfrac{p}{q}$ where p is a factor of a_0 and q is a factor of a_n.

To develop our intuitive understanding of the **Rational Zeros Theorem**, we will use it to determine the possible rational zeros of the function $P(x) = 2x^2 + x - 3$. By the Rational Zeros Theorem, we know that the possible rational zeros are of the form $\dfrac{p}{q}$ where p is a factor of the constant term, -3, and q is a factor of the leading coefficient, 2.

Factors of -3: $\pm 1, \pm 3$

Factors of 2: $\pm 1, \pm 2$

Now use the factors of -3 in the numerator and the factors of 2 in the denominator. For $p = \pm 1$ and $q = \pm 1$, we get ± 1. For $p = \pm 3$ and $q = \pm 1$, we get ± 3. For $p = \pm 1$ and $q = \pm 2$, we get $\pm\frac{1}{2}$. For $p = \pm 3$ and $q = \pm 2$, we get $\pm\frac{3}{2}$. Therefore the list of possible rational zeros is as follows:

Possible rational zeros: $\pm 1, \pm 3, \pm\dfrac{1}{2}, \pm\dfrac{3}{2}$

If we find the actual zeros of $P(x) = 2x^2 + x - 3$ using factoring and the zero factor property, we obtain the real zeros $-\frac{3}{2}$ and 1.

$$P(x) = 2x^2 + x - 3$$
$$P(x) = (2x + 3)(x - 1)$$

$$0 = (2x + 3)(x - 1)$$

$$2x + 3 = 0 \qquad \text{or} \qquad x - 1 = 0$$
$$2x = -3 \qquad\qquad\qquad x = 1$$
$$x = -\frac{3}{2}$$

Notice the real zeros are among the list of possible rational zeros.

The Rational Zeros Theorem is used to find the zeros of higher degree polynomial functions. In Example 3 we will use it to find the possible zeros of a fifth-degree polynomial function.

EXAMPLE 3 Find the possible rational zeros of
$$P(x) = 3x^5 - 4x^4 - 6x^3 + 2x^2 - x + 8$$

Solution The possible rational zeros of $P(x) = 3x^5 - 4x^4 - 6x^3 + 2x^2 - x + 8$ are of the form $\frac{p}{q}$ where p is a factor of 8 and q is a factor of 3.

Factors of 8: ±1, ±2, ±4, ±8

Factors of 3: ±1, ±3

NOW TRY EXERCISE 35 Possible rational zeros: $\pm 1, \pm 2, \pm 4, \pm 8, \pm\frac{1}{3}, \pm\frac{2}{3}, \pm\frac{4}{3}, \pm\frac{8}{3}$.

4 Use Synthetic Division to Determine the Rational Zeros of a Polynomial Function

Now that we know how to find the possible rational zeros of a polynomial function, we will discuss procedures to find the actual zeros. We will discuss two methods that can be used to find zeros: the **remainder theorem** and **synthetic division**.

In Section 5.3 we gave a brief introduction to the remainder theorem. We will repeat it below.

Remainder Theorem

If the polynomial $P(x)$ is divided by $x - a$, the remainder is equal to $P(a)$.

We can use the remainder theorem with the factor theorem that follows to identify zeros of a polynomial function.

Factor Theorem

Let $P(x)$ be a polynomial function, then $x - a$ is a factor of $P(x)$ if and only if $P(a) = 0$.

If we use the remainder theorem for a specific value a, and obtain a remainder of 0, then that specific value, a, is a zero of the polynomial and $(x - a)$ is a factor of the polynomial.

EXAMPLE 4 Determine which of the following rational numbers are zeros of the polynomial $P(x) = 3x^3 + 2x^2 - 19x + 6$.
a) 2   **b)** −5   **c)** $\frac{1}{2}$ **d)** $\frac{1}{3}$

Solution To determine if any of the values is a zero of the polynomial, we will evaluate the polynomial at the value. If the value of the polynomial is 0, then the given value is a zero of the polynomial. If it is not 0, then it is not a zero of the polynomial.

a) $P(x) = 3x^3 + 2x^2 - 19x + 6$
$$P(2) = 3(2)^3 + 2(2)^2 - 19(2) + 6$$
$$= 3(8) + 2(4) - 38 + 6$$
$$= 24 + 8 - 38 + 6 = 0$$

Since $P(2) = 0$, the number 2 is a zero of the polynomial function and $(x - 2)$ is a factor of the polynomial.

b) $P(x) = 3x^3 + 2x^2 - 19x + 6$
$$P(-5) = 3(-5)^3 + 2(-5)^2 - 19(-5) + 6 = -224$$

Since $P(-5) \neq 0$, -5 is not a zero of the polynomial function.

c) $P(x) = 3x^3 + 2x^2 - 19x + 6$
$$P\left(\tfrac{1}{2}\right) = 3\left(\tfrac{1}{2}\right)^3 + 2\left(\tfrac{1}{2}\right)^2 - 19\left(\tfrac{1}{2}\right) + 6 = -\tfrac{21}{8}$$

Since $P\left(\tfrac{1}{2}\right) \neq 0, \tfrac{1}{2}$ is not a zero.

d) $P(x) = 3x^3 + 2x^2 - 19x + 6$
$$P\left(\tfrac{1}{3}\right) = 3\left(\tfrac{1}{3}\right)^3 + 2\left(\tfrac{1}{3}\right)^2 - 19\left(\tfrac{1}{3}\right) + 6$$
$$= 3\left(\tfrac{1}{27}\right) + 2\left(\tfrac{1}{9}\right) - \tfrac{19}{3} + 6$$
$$= \tfrac{1}{9} + \tfrac{2}{9} - \tfrac{57}{9} + \tfrac{54}{9} = 0$$

Since $P\left(\tfrac{1}{3}\right) = 0, \tfrac{1}{3}$ is a zero of the polynomial function and $\left(x - \tfrac{1}{3}\right)$ is a factor. ✳

Another method to determine if a specific value is a zero of a polynomial uses synthetic division. Synthetic division was discussed in Section 5.3. Please review that material before continuing with this section.

We will do one synthetic division problem now.

EXAMPLE 5 Divide $x^3 - 4x^2 - 11x + 30$ by $x + 3$ and determine the remainder.

Solution From Section 5.3 we know we write $x + 3$ as $x - (-3)$ and set up and work the synthetic division as follows.

$$
\begin{array}{r|rrrr}
-3 & 1 & -4 & -11 & 30 \\
   &   & -3 & 21 & -30 \\
\hline
   & 1 & -7 & 10 & 0 \\
\end{array}
$$
$\leftarrow$ remainder

Since the number below the line to the far right is 0, the remainder when $x^3 - 4x^2 - 11x + 30$ is divided by $x + 3$ is 0, and $x + 3$ is a factor of $x^3 - 4x^2 - 11x + 30$.
✳

In Example 5, suppose we were given the polynomial function $P(x) = x^3 - 4x^2 - 11x + 30$, rather than the polynomial $x^3 - 4x^2 - 11x + 30$. Because the polynomial divided by $x + 3$ has a remainder of 0, $x + 3$ must be a factor of the polynomial function, $P(x)$, and -3 must be a zero of the polynomial function. If we evaluate the polynomial function at -3 we obtain

$$P(x) = x^3 - 4x^2 - 11x + 30$$
$$P(-3) = (-3)^3 - 4(-3)^2 - 11(-3) + 30$$
$$= -27 - 36 + 33 + 30 = 0$$

Thus, -3 is a zero and $(x + 3)$ is a factor of $P(x)$.

EXAMPLE 6 Determine the rational zeros of $P(x) = x^4 + 4x^3 - 7x^2 - 34x - 24$.

Solution The possible rational zeros of $P(x) = x^4 + 4x^3 - 7x^2 - 34x - 24$ are of the form $\dfrac{p}{q}$ where p is a factor of -24 and q is a factor of 1.

Factors of -24: $\pm 1, \pm 2, \pm 3, \pm 4, \pm 6, \pm 8, \pm 12, \pm 24$

Factors of 1: ± 1

Possible rational zeros: $\pm 1, \pm 2, \pm 3, \pm 4, \pm 6, \pm 8, \pm 12, \pm 24$

We will use Descartes' Rule of Signs to determine the possible number of positive and negative zeros. Since $P(x)$ has only 1 variation in sign, it has 1 positive real zero.

$$P(x) = x^4 + 4x^3 - 7x^2 - 34x - 24$$
$$P(-x) = x^4 - 4x^3 - 7x^2 + 34x - 24$$

Since $P(-x)$ has 3 variations in sign, it can have 3 or 1 negative real zeros.

Let us find the 1 positive zero first. We can use either synthetic division or the remainder theorem to find the zeros. In this example we will use synthetic division. We will try the possible rational zeros in increasing order, starting with 1, until we find the positive zero.

Using synthetic division

$$
\begin{array}{r|rrrrr}
1 & 1 & 4 & -7 & -34 & -24 \\
  &   & 1 & 5 & -2 & -36 \\
\hline
  & 1 & 5 & -2 & -36 & -60 \\
\end{array}
$$

Since the remainder is -60, 1 is not a zero.

$$
\begin{array}{r|rrrrr}
2 & 1 & 4 & -7 & -34 & -24 \\
  &   & 2 & 12 & 10 & -48 \\
\hline
  & 1 & 6 & 5 & -24 & -72 \\
\end{array}
$$

Since the remainder is -72, 2 is not a zero.

$$
\begin{array}{r|rrrrr}
3 & 1 & 4 & -7 & -34 & -24 \\
  &   & 3 & 21 & 42 & 24 \\
\hline
  & 1 & 7 & 14 & 8 & 0 \\
\end{array}
$$

Since the remainder is 0, 3 is a zero of the polynomial function. From Section 5.3 we know that the polynomial can now be written in factored form as follows.

$$P(x) = x^4 + 4x^3 - 7x^2 - 34x - 24$$
$$= (x - 3)(x^3 + 7x^2 + 14x + 8)$$

Now we will find the negative zeros of $P(x) = x^4 + 4x^3 - 7x^2 - 34x - 24$. To do so we use the fact that the remaining zeros of $P(x)$ must also be zeros of the remaining reduced polynomial $Q(x) = x^3 + 7x^2 + 14x + 8$. Therefore, we work with $x^3 + 7x^2 + 14x + 8$ to find the remaining zeros of $P(x)$.

Now we find the possible rational zeros of $Q(x) = x^3 + 7x^2 + 14x + 8$.

Factors of 8: $\pm 1, \pm 2, \pm 4, \pm 8$

Factors of 1: ± 1

Possible rational zeros: $\pm 1, \pm 2, \pm 4, \pm 8$

We can eliminate 1 and 2 because they were not zeros of $P(x) = x^4 + 4x^3 - 7x^2 - 34x - 24$. In fact, we can eliminate all of the possible positive zeros because by Descartes' rule there is only 1 positive zero, and we have already found it. This leaves the following.

Possible rational zeros: $-1, -2, -4, -8$

Now use synthetic division with $x^3 + 7x^2 + 14x + 8$.

$$
\begin{array}{r|rrrr}
-1 & 1 & 7 & 14 & 8 \\
   &   & -1 & -6 & -8 \\
\hline
   & 1 & 6 & 8 & 0
\end{array}
$$

Since the remainder is 0, -1 is a zero.
We can now factor the polynomial further as

$$
\begin{aligned}
P(x) &= (x - 3)(x^3 + 7x^2 + 14x + 8) \\
&= (x - 3)(x + 1)(x^2 + 6x + 8) \\
&= (x - 3)(x + 1)(x + 2)(x + 4)
\end{aligned}
$$

The function has 4 rational zeros $x = 3$, $x = -1$, $x = -2$, and $x = -4$. Notice there is 1 positive zero and 3 negative zeros, which agrees with what we expected by Descartes' Rule of Signs.

NOW TRY EXERCISE 47

❋

When analyzing a situation that can be modeled with a polynomial function, it can be useful to identify the zeros of the polynomial function. In Example 7, we use the zeros to find the production requirement for a business to make a profit.

Businesses often deal with costs, $C(x)$, profits, $P(x)$, and revenue, $R(x)$. In each case, x deals with the number of items manufactured and $C(x)$, $P(x)$, and $R(x)$ are represented in terms of money. Revenue is the money taken in from the selling of the items, cost is the money required to operate the business, and profit is the difference between these two. Thus, $P(x) = R(x) - C(x)$. The company makes a profit when more money is earned than is spent.

EXAMPLE 7 **Manufacturing Computers** The Big Screen Computer Company manufactures computers. The costs to operate the plant on a daily basis are given by $C(x) = x^2 + 5x + 9$ and the revenue is given by $R(x) = x^3 + 2x^2 - 4x$. Both C and R are expressed in thousands of dollars and x is the number (in hundreds) of computers manufactured each day. Due to constraints at the plant, no more than 900 computers can be manufactured in one day.

a) Find $P(x)$.

b) Find $C(0)$, $R(0)$, and $P(0)$, and interpret each of these.

c) Find the rational zero(s) of $P(x)$.

d) How many computers must be manufactured each day to make a profit?

Solution **a)** Since profit is the difference between revenue and cost,

$$
\begin{aligned}
P(x) &= R(x) - C(x) \\
&= (x^3 + 2x^2 - 4x) - (x^2 + 5x + 9) \\
&= x^3 + x^2 - 9x - 9
\end{aligned}
$$

b) $C(x) = x^2 + 5x + 9$
$$
\begin{aligned}
C(0) &= 0^2 + 5 \cdot 0 + 9 \\
&= 9
\end{aligned}
$$

It costs $9000 to operate the plant daily even though no computers are being manufactured.

$$
\begin{aligned}
R(x) &= x^3 + 2x^2 - 4x \\
R(0) &= 0^3 + 2(0)^2 - 4 \cdot 0 \\
&= 0
\end{aligned}
$$

There is no revenue when no computers are being manufactured.

$$P(x) = x^3 + x^2 - 9x - 9$$
$$P(0) = 0^3 + 0^2 - 9 \cdot 0 - 9$$
$$= -9$$

The company shows a loss of $9000 (since $P(0) = -9$) each day when no computers are manufactured. This amount is the money needed to operate the plant when no computers are being made.

c) The possible rational zeros of $P(x) = x^3 + x^2 - 9x - 9$ are of the form $\dfrac{p}{q}$ where p is a factor of -9 and q is a factor of 1.

Factors of -9: $\pm 1, \pm 3, \pm 9$

Factors of 1: ± 1

Possible rational zeros: $\pm 1, \pm 3, \pm 9$

By Descartes' Rule of Signs, we know there is one positive real zero. We will try the possible positive rational zeros in increasing order, starting with 1.

$$
\begin{array}{r|rrrr}
1 & 1 & 1 & -9 & -9 \\
  &   & 1 & 2 & -7 \\
\hline
  & 1 & 2 & -7 & -16
\end{array}
$$

Since the remainder is -16, 1 is not a zero.

$$
\begin{array}{r|rrrr}
3 & 1 & 1 & -9 & -9 \\
  &   & 3 & 12 & 9 \\
\hline
  & 1 & 4 & 3 & 0
\end{array}
$$

Since the remainder is 0, 3 is a zero of the polynomial function. The polynomial can now be written in factored form as follows.

$$
\begin{aligned}
P(x) &= x^3 + x^2 - 9x - 9 \\
&= (x - 3)(x^2 + 4x + 3) \\
&= (x - 3)(x + 3)(x + 1)
\end{aligned}
$$

The rational zeros are $x = 3$, $x = -3$, and $x = -1$.

d) In this real-life problem, the number of computers, x, cannot be negative. Therefore we can discard -3 and -1 as x-intercepts of the graph of $P(x)$ for this problem. At $x = 3$, which represents 300 computers, the profit is $0. This is called the *break-even point*. At the break-even point the revenue and cost are the same, and their difference is $0.

To decide the number of computers to be made to make a profit let's look at the graph of $P(x)$ for $0 \le x \le 9$ in Figure 10.24. (Remember the company cannot manufacture more than 900 computers each day.)

FIGURE 10.24

From the graph we see that profit is negative (representing a loss) when fewer than 300 computers are manufactured. On the other hand, a profit is shown when more than 300 computers are made. The manager should plan to manufacture more than 300 computers each day to make a profit.

NOW TRY EXERCISE 69

5 Find Complex Zeros of a Polynomial Function

Thus far we have discussed only real zeros of polynomial functions. Polynomial functions can also have nonreal zeros, that is, complex numbers that are zeros of the function. There are no x-intercepts of the graph of the polynomial function at nonreal zeros. In Example 8 the polynomial function has nonreal zeros.

EXAMPLE 8 Find all zeros of $P(x) = x^2 + 4$.

Solution We set $P(x) = 0$ and solve for x.

$$P(x) = x^2 + 4$$
$$0 = x^2 + 4$$
$$\text{or} \quad x^2 + 4 = 0$$
$$x^2 = -4$$
$$x = \pm\sqrt{-4} \quad \text{\textit{Square root property}}$$
$$x = \pm 2i$$

Thus the zeros are $2i$ and $-2i$. We can check the zeros in Example 8 using the remainder theorem as follows.

$$P(x) = x^2 + 4 \qquad\qquad P(x) = x^2 + 4$$
$$P(2i) = (2i)^2 + 4 \qquad P(-2i) = (-2i)^2 + 4$$
$$= 4i^2 + 4 \qquad\qquad = 4i^2 + 4$$
$$= 4(-1) + 4 \qquad\qquad = 4(-1) + 4$$
$$= 0 \qquad\qquad\qquad = 0$$

Since $2i$ and $-2i$ are zeros of $P(x) = x^2 + 4$, $(x - 2i)$ and $(x + 2i)$ are factors of $x^2 + 4$. Thus we can factor $P(x)$ as $P(x) = (x - 2i)(x + 2i)$.

EXAMPLE 9 Determine all zeros of $P(x) = x^5 + 5x^4 + 12x^3 + 24x^2 + 32x + 16$.

Solution The possible rational zeros of $P(x) = x^5 + 5x^4 + 12x^3 + 24x^2 + 32x + 16$ are of the form $\frac{p}{q}$ where p is a factor of 16 and q is a factor of 1.

Factors of 16: $\pm 1, \pm 2, \pm 4, \pm 8, \pm 16$

Factors of 1: ± 1

Possible rational zeros: $\pm 1, \pm 2, \pm 4, \pm 8, \pm 16$

Since the coefficients of $P(x) = x^5 + 5x^4 + 12x^3 + 24x^2 + 32x + 16$ are all positive numbers, the coefficients have no variation in sign. By Descartes' Rule of Signs, $P(x)$ has no positive real zeros. As a result the possible rational zeros can be reduced to $-1, -2, -4, -8, -16$. Let's now find $P(-x)$.

$$P(x) = x^5 + 5x^4 + 12x^3 + 24x^2 + 32x + 16$$
$$P(-x) = (-x)^5 + 5(-x)^4 + 12(-x)^3 + 24(-x)^2 + 32(-x) + 16$$
$$\text{or} \quad P(-x) = -x^5 + 5x^4 - 12x^3 + 24x^2 - 32x + 16$$

There are 5 variations in sign in $P(-x)$. Thus there can be $5, 3,$ or 1 negative real zeros. Let's now try using synthetic division with $x = -1$.

$$
\begin{array}{r|rrrrrr}
-1 & 1 & 5 & 12 & 24 & 32 & 16 \\
   &   & -1 & -4 & -8 & -16 & -16 \\
\hline
   & 1 & 4 & 8 & 16 & 16 & 0
\end{array}
$$

The remainder is $0,$ therefore -1 is a zero. Now factor the polynomial.

$$
\begin{aligned}
P(x) &= x^5 + 5x^4 + 12x^3 + 24x^2 + 32x + 16 \\
     &= (x + 1)(x^4 + 4x^3 + 8x^2 + 16x + 16)
\end{aligned}
$$

We now use synthetic division to try to obtain a zero in the new nonlinear factor $x^4 + 4x^3 + 8x^2 + 16x + 16.$ Notice that its leading coefficient, $1,$ and its constant term, $16,$ are the same as the original polynomial, so the possible rational zeros are the same. Let's now try $x = -2.$

$$
\begin{array}{r|rrrrr}
-2 & 1 & 4 & 8 & 16 & 16 \\
   &   & -2 & -4 & -8 & -16 \\
\hline
   & 1 & 2 & 4 & 8 & 0
\end{array}
$$

Therefore, -2 is a zero. Now factor the polynomial further.

$$
\begin{aligned}
P(x) &= (x + 1)(x^4 + 4x^3 + 8x^2 + 16x + 16) \\
     &= (x + 1)(x + 2)(x^3 + 2x^2 + 4x + 8)
\end{aligned}
$$

We continue to use synthetic division to try to obtain a zero in the new nonlinear factor $x^3 + 2x^2 + 4x + 8.$ Notice that its leading coefficient is $1,$ and its constant term is $8,$ so -16 has been eliminated as a possible rational zero. Thus the possible negative real zeros are $-1, -2, -4,$ and $-8.$ We found -1 to be one real factor. However, it could have a multiplicity of $2,$ so we will try it again.

$$
\begin{array}{r|rrrr}
-1 & 1 & 2 & 4 & 8 \\
   &   & -1 & -1 & -3 \\
\hline
   & 1 & 1 & 3 & 5
\end{array}
$$

Since the remainder is not $0, -1$ is not a zero of multiplicity of $2.$ Let's try -2 again.

$$
\begin{array}{r|rrrr}
-2 & 1 & 2 & 4 & 8 \\
   &   & -2 & 0 & -8 \\
\hline
   & 1 & 0 & 4 & 0
\end{array}
$$

Since the remainder is $0, -2$ is a zero a second time. Thus, -2 is a zero of multiplicity $2.$ Let's now factor the polynomial further.

$$
\begin{aligned}
P(x) &= (x + 1)(x + 2)(x^3 + 2x^2 + 4x + 8) \\
     &= (x + 1)(x + 2)(x + 2)(x^2 + 4)
\end{aligned}
$$

In Example 8 we found that $x^2 + 4 = (x + 2i)(x - 2i).$ Therefore we can factor the polynomial as follows.

$$
P(x) = (x + 1)(x + 2)^2(x + 2i)(x - 2i)
$$

The function has 3 real zeros: -2 of multiplicity 2 and -1, and two complex zeros: $2i$ and $-2i$. Since -2 is of multiplicity 2, the x-intercept of the graph touches the x-axis at -2 but does not pass through the x-axis. The graph of $P(x) = x^5 + 5x^4 + 12x^3 + 24x^2 + 32x + 16$ is shown in Figure 10.25. The y-intercept of the graph is at $(0, 16)$. ✳

NOW TRY EXERCISE 59

FIGURE 10.25

Using Your Graphing Calculator

In this section we explained how to find zeros of polynomial functions. On page 183 we explain how the real zeros of a function can be found using a graphing calculator. You should read that material now. In this section, the emphasis is on finding the zeros algebraically. If you wish, you can use your graphing calculator to check your answers when finding zeros, or when graphing polynomial functions.

Exercise Set 10.3

Concept/Writing Exercises

1. What is the maximum number of x-intercepts of the graph of a fourth-degree polynomial function? Explain.

2. What is the maximum number of x-intercepts of the graph of a seventh-degree polynomial function? Explain.

3. What is the minimum number of x-intercepts of the graph of a sixth-degree polynomial function? Explain.

4. What is the minimum number of x-intercepts of the graph of a fifth-degree polynomial function? Explain.

5. If the coefficients of a polynomial function are all positive, how many positive x-intercepts does its graph have? Explain.

6. If the coefficients of a polynomial are positive for any term which is of even degree and negative for any term which is of odd degree, how many negative x-intercepts does the polynomial have? (Note that a constant term has degree 0.) Explain.

Practice the Skills

Determine the maximum and minimum number of x-intercepts for the graph of each polynomial function. Do not graph the function.

7. $f(x) = x^2 + 8x + 5$

8. $f(x) = 3x^2 - 2x - 5$

9. $g(x) = 2x^3 + x^2 - 5x - 3$

10. $g(x) = -2x^3 - x^2 + 7x - 15$

11. $r(x) = 5x^4 - x^3 + 4x^2 + 2x + 17$

12. $r(x) = 5x^4 - x^3 - 7x^2 + 21x - 17$

13. $m(x) = 7x^5 - 2x^4 - x^3 + 9x - 10$

14. $m(x) = -3x^5 + x^4 - \frac{1}{2}x^3 + x - 11$

15. $r(x) = 8x^6 + 7x^5 - x^4 + 3x^3 - 2x^2 + 11x + 5$

16. $r(x) = -2x^6 + x^5 - 3x^4 + 13x^3 - 2x^2 - 7x + 8$

Use Descartes' Rule of Signs to determine the possible number of positive and negative zeros of each polynomial function.

17. $f(x) = x^3 + 2x^2 + 4x + 7$

18. $f(x) = x^3 - 3x^2 + 5x + 11$

19. $f(x) = -6x^3 + 2x^2 + 3x - 1$

20. $g(x) = 2x^4 + 7x^3 - 2x^2 - 6x + 5$

21. $g(x) = 5x^4 - 2x^3 + x^2 - 9x - 14$

22. $g(x) = 7x^4 - 3x^3 + x^2 - x + 17$

23. $g(x) = -3x^4 - x^3 - 6x^2 - 2x - 3$

24. $g(x) = 17x^4 + 7x^3 - 2x^2 - x - 12$

25. $r(x) = x^5 - x^4 + x^3 + 2x^2 + 7x - 3$

26. $r(x) = x^5 + x^4 - 6x^3 - 15x^2 + 12x + 16$

27. $r(x) = 7x^5 + 2x^4 + x^3 + 5x^2 + 9x + 20$

28. $r(x) = -7x^5 - 8x^4 - 4x^3 + 6x^2 + x + 19$

List the possible rational zeros of each function.

29. $f(x) = x^2 + 3x + 2$

30. $f(x) = x^2 - 5x + 6$

31. $f(x) = 3x^2 + 4x + 3$

32. $g(x) = x^3 - x^2 + 3x - 5$

33. $g(x) = x^3 - 6x^2 + 3x + 12$

34. $g(x) = 4x^3 - x^2 + 9x + 5$

35. $g(x) = 3x^3 - 7x^2 + 17x - 4$

36. $h(x) = 5x^4 - x^3 + 7x^2 - 19x + 6$

37. $h(x) = 6x^4 + 5x^3 - x^2 + 9x + 4$

38. $h(x) = 2x^4 - 4x^3 + x + 5$

39. $r(x) = 6x^5 + x^4 - x^2 + 11x + 3$

40. $r(x) = 8x^5 - 7x^4 + 3x^2 - x + 1$

41. $r(x) = 7x^5 - x^4 + 15x^3 - 18x^2 + 4$

42. $t(x) = 9x^5 + 23x^4 - 7x^2 + 3$

Find all the rational zeros of each function.

43. $f(x) = x^2 - 6x + 8$

44. $f(x) = x^2 + 3x - 18$

45. $f(x) = 3x^2 + 14x - 5$

46. $f(x) = 6x^2 + 5x + 1$

47. $g(x) = x^3 + 7x^2 - x - 7$

48. $g(x) = x^3 + 2x^2 - 5x - 6$

49. $g(x) = 2x^3 + x^2 - 8x - 4$

50. $g(x) = x^3 - 5x^2 + 9x - 45$

51. $r(x) = x^4 - 12x^2 - 64$

52. $r(x) = x^4 - 13x^2 + 64$

53. $r(x) = x^4 - 8x^2 + 16$

54. $r(x) = 6x^4 - x^3 + 5x^2 - x - 1$

55. $r(x) = x^4 + x^3 - 7x^2 - 13x - 6$

56. $r(x) = 4x^4 - 8x^3 - 7x^2 + 17x - 6$

57. $r(x) = x^4 - 8x^3 + 22x^2 - 24x + 9$

58. $t(x) = x^5 + 7x^4 - 5x^3 - 35x^2 + 4x + 28$

Problem Solving

Find all the zeros (real and complex) of each function.

59. $f(x) = x^4 - 3x^2 - 4$

60. $g(x) = x^4 + 24x - 25$

61. $h(x) = x^4 + x^3 - x - 1$

62. $f(x) = x^4 - x^3 + x - 1$

63. $g(x) = x^4 + 3x^3 - x^2 - 12x - 12$

64. $g(x) = x^4 - 3x^3 - 5x^2 + 27x - 36$

65. $f(x) = x^5 - x^4 + 2x^3 - 2x^2 + x - 1$

66. $g(x) = x^5 + x^4 + x^3 + x^2 - 2x - 2$

For Exercises 67–71 you are given $R(x)$ and $C(x)$ where x represents the number of items produced in hundreds. $R(x)$ and $C(x)$ represent revenue and cost in thousands of dollars. You may want to review Example 7 at this time.

a) *Find $P(x)$.*

b) *Find $P(0)$, $R(0)$, and $C(0)$ and interpret each of these.*

c) *Find the break-even point (value where $P(x) = 0$) in the given domain.*

d) *For what values of x in the given domain is $P(x) > 0$?*

e) *For what values of x in the given domain is $P(x) < 0$?*

f) *Interpret the results from parts **d)** and **e)**.*

67. $R(x) = x^3 + 5x^2 + x, C(x) = x^2 + 6$ $0 \le x \le 3$

68. $R(x) = x^3 + 3x^2 - 5x, C(x) = x^2 + 4x + 18$ $0 \le x \le 7$

69. $R(x) = x^3 + x^2 + x, C(x) = 3x^2 + 14x + 10$ $0 \le x \le 8$

70. $R(x) = x^3 + 3x, C(x) = x^2 + 13x + 8$ $0 \le x \le 5$

71. $R(x) = 2x^3 + x^2 + 2x, C(x) = 2x^2 + 7x + 2$ $0 \le x \le 5$

Group Activity

72. In this section we have been finding the rational and complex zeros of polynomial functions. Now you will work with your group to create polynomial functions with certain characteristics. For each description, do the following:

a) Write your polynomial function in factored form.

b) Multiply the factors of your polynomial function.

c) Determine if this is the only polynomial function having these characteristics. If it is not the only one with

these characteristics, have every group member find another polynomial with the given characteristics.

1. Create a polynomial function with the rational zeros: $-4, -2, 0, 3$.

2. Create a polynomial function with the zeros: -5 multiplicity 2, 6.

3. Create a polynomial function with the zeros: -2 multiplicity 3, 1, 3.

Cumulative Review Exercises

[7.2] **73.** Simplify $\left(\dfrac{y^{-2/3}}{y^{-6}}\right)^{1/2}$.

[7.3] **74.** Simplify $3\sqrt{72x^3 y}$.

[7.5] **75.** Simplify $\dfrac{5}{\sqrt{5} - 10}$.

[7.6] **76.** Solve $\sqrt{5x + 6} = x$.

10.4 FINDING THE REAL ZEROS OF A POLYNOMIAL FUNCTION

1 Locate and determine the real zeros of a polynomial function.

2 Find upper and lower bounds for the real zeros of a polynomial function.

3 Graph a polynomial function using its key features.

4 Estimates irrational zeros.

1 Locate and Determine the Real Zeros of a Polynomial Function

Since polynomial functions are continuous, they have no gaps or holes in their graphs. If we evaluate a function at two points a and b, we know that the function has a value at every number between $f(a)$ and $f(b)$. For instance, if $a < b$, and $f(a) = 1$, and $f(b) = 2$, we know that there exists c between a and b such that $f(c) = \sqrt{2}$ because $\sqrt{2}$ is between 1 and 2 (see Figure 10.26).

Using this concept, we know that a polynomial function f has a real zero between any two numbers a and b for which the sign of $f(a)$ is opposite the sign of $f(b)$, see Figure 10.27.

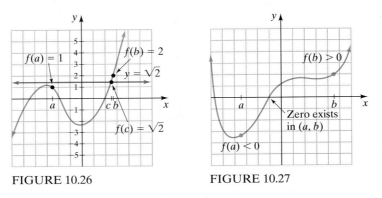

FIGURE 10.26

FIGURE 10.27

This fact is known as the Intermediate Value Theorem.

Intermediate Value Theorem

For a continuous function $f(x)$ defined over the interval $[a, b]$, if the sign of $f(a)$ is the opposite of the sign of $f(b)$, then a real zero exists in (a, b).

The Intermediate Value Theorem is used to locate intervals where real zeros occur.

Suppose we check the possible rational zeros by evaluating the function at each possible rational zero, in order, from smallest possible rational zero to largest possible rational zero. If we find that the sign of the value of the function changes from one possible rational zero to the next, we know that an irrational zero exists between those 2 possible rational zeros.

EXAMPLE 1 In which of the following intervals does the polynomial function, $f(x) = \frac{1}{2}x^4 - x^3 - 5$, have a real zero?

a) $(-2, -1)$ **b)** $(-1, 2)$

Solution A polynomial function will have a real zero in an interval if the sign of the function at one endpoint of the interval is opposite the sign of the function at the other endpoint of the interval.

a) We will substitute each endpoint in the function, and then evaluate the function to determine if there is a change of sign of $f(x)$.

$$f(x) = \frac{1}{2}x^4 - x^3 - 5 \qquad\qquad f(x) = \frac{1}{2}x^4 - x^3 - 5$$

$$f(-2) = \frac{1}{2}(-2)^4 - (-2)^3 - 5 \qquad f(-1) = \frac{1}{2}(-1)^4 - (-1)^3 - 5$$

$$= \frac{1}{2}(16) - (-8) - 5 \qquad\qquad = \frac{1}{2}(1) - (-1) - 5$$

$$= 8 + 8 - 5 \qquad\qquad\qquad = 0.5 + 1 - 5$$

$$= 11 \qquad\qquad\qquad\qquad = -3.5$$

Since the function changes sign in the interval $(-2, -1)$, it has at least one real zero in this interval.

b) From part **a)** we know that $f(-1) = -3.5$. We will now evaluate $f(2)$.

$$f(x) = \frac{1}{2}x^4 - x^3 - 5$$

$$f(2) = \frac{1}{2}(2)^4 - (2)^3 - 5$$

$$= \frac{1}{2}(16) - 8 - 5$$

$$= 8 - 8 - 5$$

$$= -5$$

The sign of the function at -1 is the same as the sign of the function at 2. From this information, we cannot determine whether a real zero exists in this interval. ✳

NOW TRY EXERCISE 9

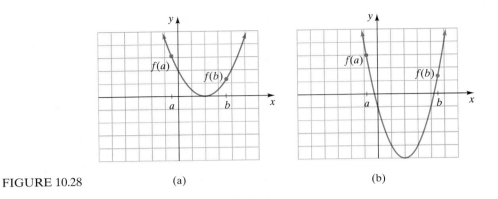

FIGURE 10.28   (a)   (b)

It is important to recognize that the Intermediate Value Theorem does not imply that there is no zero in (a, b), when $f(a)$ is the same sign as $f(b)$. As illustrated in Figure 10.28a, the function will have the same sign at both endpoints of the interval when the interval contains a zero of even multiplicity. The function may also have the same sign at both endpoints of the interval when the function has more than one zero in the interval (see Figure 10.28b).

2 Find Upper and Lower Bounds for the Real Zeros of a Polynomial Function

We have strategies for determining the possible number of positive and negative real zeros, for identifying all possible rational zeros, and for identifying intervals where a real zero occurs. We will now learn a strategy for recognizing a region over which all of the real zeros occur. We can identify such a region by finding an upper and lower bound for the zeros using the Boundedness Theorem.

Boundedness Theorem

Let $P(x)$ be a polynomial function with real coefficients and a positive leading coefficient.

a) If $P(x)$ is synthetically divided by $x - b$, where $b > 0$ and all the numbers in the bottom row of the synthetic division are nonnegative, then $P(x)$ has no real zero greater than b. We call b an **upper bound** for the real zeros of $P(x)$.

b) If $P(x)$ is synthetically divided by $x - c$ where $c < 0$ and all of the numbers in the bottom row of the synthetic division alternate in sign (with 0 considered positive or negative as explained below), then $P(x)$ has no real zero less than c. We call c a **lower bound** for the real zeros of $P(x)$.

In part **b)** where a 0 occurs in the bottom row of numbers, treat it as a negative number when it follows a positive number, and as a positive number when it follows a negative number. For example, if the bottom row of numbers is $2 -4\ 0\ 6 -2$ we treat the 0 as a positive number since it follows -4. Thus we get $+ - + + -$. Since there are two pluses in a row we do not consider all the numbers alternating in sign. Now consider $4 -3\ 0 -2\ 1$. Since the 0 follows -3 we treat it as a positive number. Thus we get $+ - + - +$. Since these signs alternate, the numbers $4 -3\ 0 -2\ 1$ are considered alternating in sign.

If a is a lower bound and b is an upper bound for the zeros of $P(x)$, we say that the zeros of $P(x)$ are *bounded over* the interval $[a, b]$.

We use the Boundedness Theorem after we have completed a synthetic division to reduce the number of rational zeros which need to be checked.

EXAMPLE 2 Determine if the real zeros of $f(x) = x^4 - 5x^3 + 7x^2 + 3x - 10$ are all included in the interval, or are bounded over the interval $[-2, 6]$.

Solution Use synthetic division with $x = -2$ for the possible lower bound and then use synthetic division with $x = 6$, the possible upper bound.

$$
\begin{array}{r|rrrrr}
-2 & 1 & -5 & 7 & 3 & -10 \\
   &   & -2 & 14 & -42 & 78 \\
\hline
   & 1 & -7 & 21 & -39 & 68 \\
\end{array}
$$

Since the bottom row of numbers in the synthetic division has alternating signs, -2 is a lower bound for the real zeros of the function. Therefore, there can be no real zero that is less than -2. Now use synthetic division with $x = 6$.

$$
\begin{array}{r|rrrrr}
6 & 1 & -5 & 7 & 3 & -10 \\
  &   & 6 & 6 & 78 & 486 \\
\hline
  & 1 & 1 & 13 & 81 & 476 \\
\end{array}
$$

Since the bottom row of numbers in the synthetic division are all greater than or equal to zero, 6 is an upper bound for the real zeros of the function. Therefore, there can be no real zero that is greater than 6. Thus, all real zeros of this function are in the interval $[-2, 6]$. ✳

NOW TRY EXERCISE 25

The Boundedness Theorem applies when there is a positive leading coefficient. The real zeros of $f(x)$ and $-f(x)$ will be the same since the graphs of $f(x)$ and $-f(x)$ are symmetric with regard to the x-axis. Therefore, if the leading coefficient of a polynomial function is negative, multiply both sides of the equation by -1, and proceed as in Example 2.

3 Graph a Polynomial Function Using Its Key Features

With graphing calculators widely available, you may question the need to be skilled at graphing polynomial functions. The process of graphing polynomial functions using the theorems presented in this chapter gives you the opportunity to develop your reasoning skills to understand the basic characteristics of polynomial functions.

For the graph of a polynomial function to be complete, it needs an accurate y-intercept, accurate rational x-intercepts, reasonable estimates for irrational x-intercepts, an accurate data point in each interval defined by the x-intercepts, and a smooth curve connecting these points. Although the graph can be made more accurate by plotting more points in each region, the basic sketch of the graph will be complete.

EXAMPLE 3 Consider $f(x) = x^3 + 3x^2 - x - 3$.

a) Determine the y-intercept of the graph.

b) Determine the possible number of positive zeros and the possible number of negative zeros of $f(x)$.

c) Determine the possible rational zeros of $f(x)$.

d) Determine all real zeros of $f(x)$.

e) Graph the function.

Solution   **a)** To find the y-intercept, substitute 0 for x.

$$f(x) = x^3 + 3x^2 - x - 3$$
$$f(0) = 0 + 0 - 0 - 3 = -3$$

The y-intercept is $(0, -3)$.

b) We find the possible number of positive and negative zeros using Descartes' Rule of Signs.

$$f(x) = x^3 + 3x^2 - x - 3 \quad \text{\textit{Has one variation in signs.}}$$
$$f(-x) = -x^3 + 3x^2 + x - 3 \quad \text{\textit{Has two variations in signs.}}$$

Since the number of variations in sign in $f(x)$ is 1, there is 1 positive real zero. Since the number of variations in sign in $f(-x)$ is 2, the number of real negative zeros is either 2 or zero.

c) The possible rational zeros are of the form $\dfrac{p}{q}$ where p is a factor of -3 and q is a factor of 1.

Factors of -3: $\pm 1, \pm 3$

Factors of 1: ± 1

Possible rational zeros: $\pm 1, \pm 3$

d) Since we definitely have 1 positive real zero we will look for that one first. We will use synthetic division with $x = 1$.

$$
\begin{array}{r|rrrr}
1 & 1 & 3 & -1 & -3 \\
  &   & 1 & 4 & 3 \\
\hline
  & 1 & 4 & 3 & 0
\end{array}
$$

Since the remainder is 0, 1 is a zero of $f(x)$ and $(x - 1)$ is a factor of $f(x)$. We have found the one positive zero. Notice the bottom row of the synthetic division contains all positive numbers.

This tells us that 1 is an upper bound. We can now factor $f(x)$ as follows.

$$f(x) = x^3 + 3x^2 - x - 3$$
$$f(x) = (x - 1)(x^2 + 4x + 3)$$
$$= (x - 1)(x + 3)(x + 1)$$

The other zeros, -3 and -1 can now be obtained using the zero factor property. The 3 zeros, from smallest to largest, are $-3, -1$, and 1.

e) We will set up a table to help in graphing the polynomial. From the 3 real zeros we get 3 x-intercepts on the graph, which break the graph into 4 intervals.

Interval	Test value	Value of $f(x)$
$(-\infty, -3)$	-4	-15
$(-3, -1)$	-2	3
$(-1, 1)$	0	-3
$(1, \infty)$	2	15

Now draw the graph, see Figure 10.29 on the next page.

FIGURE 10.29 $f(x) = x^3 + 3x^2 - x - 3$

NOW TRY EXERCISE 37

Notice the end behavior is the same end behavior as $f(x) = x^3$, which is what we expected.   ✳

EXAMPLE 4 Graph $f(x) = x^4 - x^3 - 5x^2 - x + 4$.

Solution We will follow the same basic procedure we used in Example 3. By substituting 0 for x we find that $(0, 4)$ is the y-intercept.

$$f(x) = x^4 - x^3 - 5x^2 - x + 4 \quad \textit{Has 2 variations in sign.}$$
$$f(-x) = x^4 + x^3 - 5x^2 + x + 4 \quad \textit{Has 2 variations in sign.}$$

Therefore the number of positive real zeros is either 2 or 0 and the number of negative real zeros is either 2 or 0.

Possible factors of 4: $\pm 1, \pm 2, \pm 4$

Possible factors of 1: ± 1

Possible rational zeros: $\pm 1, \pm 2, \pm 4$

Let us find the positive zeros first. We will start with 1.

$$
\begin{array}{r|rrrrr}
1 & 1 & -1 & -5 & -1 & 4 \\
  &   & 1 & 0 & -5 & -6 \\
\hline
  & 1 & 0 & -5 & -6 & -2 \\
\end{array}
$$

Therefore 1 is not a zero. Notice $f(1) = -2$. Also note that $f(0) = 4$. Therefore there must be a real positive irrational zero between $x = 0$ and $x = 1$. We now know that there is 1 real positive zero. By Descartes' Rule, we know that there must be a second real positive zero. Let's continue to find it. Let's try synthetic division with $x = 2$.

$$
\begin{array}{r|rrrrr}
2 & 1 & -1 & -5 & -1 & 4 \\
  &   & 2 & 2 & -6 & -14 \\
\hline
  & 1 & 1 & -3 & -7 & -10 \\
\end{array}
$$

Therefore, 2 is not a zero. Let's try synthetic division with $x = 4$.

$$
\begin{array}{r|rrrrr}
4 & 1 & -1 & -5 & -1 & 4 \\
  &   & 4 & 12 & 28 & 108 \\
\hline
  & 1 & 3 & 7 & 27 & 112 \\
\end{array}
$$

Therefore 4 is not a zero. Since $f(2) = -10$ and $f(4) = 112$, there must be a real irrational zero between $x = 2$ and $x = 4$. If we evaluate the function at 3, we find $f(3) = 10$. Therefore the positive irrational zero is in the interval $(2, 3)$. We have determined there are 2 positive irrational zeros. Since we know we have only 2 positive real zeros, we will now look for the negative zeros. Let's use synthetic division with $x = -1$.

$$
\begin{array}{r|rrrrr}
-1 & 1 & -1 & -5 & -1 & 4 \\
   &   & -1 & 2 & 3 & -2 \\
\hline
   & 1 & -2 & -3 & 2 & 2 \\
\end{array}
$$

Therefore, -1 is not a zero. Let's try $x = -2$.

$$
\begin{array}{r|rrrrr}
-2 & 1 & -1 & -5 & -1 & 4 \\
   &   & -2 & 6 & -2 & 6 \\
\hline
   & 1 & -3 & 1 & -3 & 10 \\
\end{array}
$$

Therefore, -2 is not a zero. Since the signs in the bottom row of the synthetic division alternate in sign, there can be no real zero less than -2. Therefore we can eliminate -4 as a possibility. If the function has negative zeros, it has 2 which occur in the interval $(-2, 0)$. From the y-intercept and the synthetic divisions we have already determined that the following points are on $f(x)$:

$$(0, 4), (1, -2), (2, -10), (4, 112), (-1, 2), (-2, 10).$$

Let us use these points and the point $(3, 10)$ to sketch the graph. We know the end behavior of $f(x) = x^4$. Therefore the graph must be positive both for large positive values and large negative values of x. The graph is sketched in Figure 10.30. Note that there is an irrational zero between $x = 0$ and $x = 1$ and an irrational zero between $x = 2$ and $x = 3$, and that there are no negative real zeros. Next, we will learn how to estimate the irrational zeros of a function.

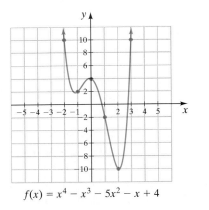

$$f(x) = x^4 - x^3 - 5x^2 - x + 4$$

NOW TRY EXERCISE 49   FIGURE 10.30

4 Estimate Irrational Zeros

In Example 4, we determined that one real zero was between $x = 0$ and $x = 1$. We can get close approximations to the actual irrational zero by evaluating the function for various values in the interval. If there is a change of sign in $f(x)$ between any 2 values of x, then there is a real zero between those values of x; we found in Example 4 that $f(0) = 4$ and $f(1) = -2$. Let's consider the graph between these 2 points, see Figure 10.31 on the top of the next page.

FIGURE 10.31

If the curve was straight, it appears that the x-intercept is about 0.7. Let's evaluate $f(x)$ for selected values of x and determine the values of $f(x)$. The table below shows some values.

x	$f(x) = x^4 - x^3 - 5x^2 - x + 4$
0.6	$f(0.6) = 0.6^4 - 0.6^3 - 5(0.6)^2 - 0.6 + 4 = 1.5136$
0.7	$f(0.7) = 0.7^4 - 0.7^3 - 5(0.7)^2 - 0.7 + 4 = 0.7471$
0.8	$f(0.8) = 0.8^4 - 0.8^3 - 5(0.8)^2 - 0.8 + 4 = -0.1024$

There is a change of sign in $f(x)$ between $x = 0.7$ and $x = 0.8$. Therefore the zero is between 0.7 and 0.8. If we desire more accuracy we could evaluate $f(x)$ for values between 0.7 and 0.8 and find where there is a change of sign in $f(x)$. If you evaluate additional values you will find that

$$f(0.77) \approx 0.1605$$
$$f(0.78) \approx 0.0736$$
$$f(0.79) \approx -0.0140$$

Thus the irrational zero is between 0.78 and 0.79. Since $f(0.785) \approx 0.0299$ the irrational root is between 0.785 and 0.790. Thus the irrational root, estimated to the nearest hundredth, is 0.79. If you were to determine the irrational zero of $f(x) = x^4 - x^3 - 5x^2 - x + 4$ between $x = 2$ and $x = 3$, you would find that

$$f(2.75) \approx -0.1680$$
$$f(2.76) \approx 0.1553$$
$$f(2.755) \approx -0.00716.$$

Thus, the irrational zero between $x = 2$ and $x = 3$ is 2.76 to the nearest hundredth.

Exercise Set 10.4

Concept/Writing Exercises

1. Let $f(x) = x^3 - 2x^2 - 5x + 2$.

a) Is $x = 5$ an upper bound for the real zeros?

b) Is $x = 4$ an upper bound for the real zeros?

c) Is $x = 3$ an upper bound for the real zeros? Explain your answers.

2. Let $f(x) = x^3 - 2x^2 - 5x + 2$.

a) Is $x = -3$ a lower bound for the real zeros?

b) Is $x = -2$ a lower bound for the real zeros?

c) Is $x = -1$ a lower bound for the real zeros?

3. Let $f(x) = x^3 - 2x^2 - 5x + 2$.

a) Using the Intermediate Value Theorem, can you determine whether $f(x)$ has a zero in $(1, 2)$? Explain.

b) Using the Intermediate Value Theorem, can you determine whether $f(x)$ has a zero in $(-2, -1)$? Explain.

4. Let $g(x) = x^4 - x^3 + 7x - 17$.

a) Using the Intermediate Value Theorem, can you determine whether $g(x)$ has a zero in $(0, 1)$? Explain.

b) Using the Intermediate Value Theorem, can you determine whether $g(x)$ has a zero in $(1, 2)$? Explain.

c) Is $x = 3$ an upper bound for the real zeros of $g(x)$?

Practice the Skills

Use the Intermediate Value Theorem to determine if a real zero of $f(x)$ occurs in the given interval. If this cannot be determined, indicate so.

5. $f(x) = x^2 - 5x + 1$ in $(2, 3)$
6. $f(x) = x^2 - 5x + 2$ in $(4, 5)$
7. $f(x) = 3x^2 - 2x - 4$ in $(1, 2)$
8. $f(x) = x^3 - 7x - 4$ in $(-4, -3)$
9. $f(x) = x^3 - 7x - 2$ in $(-3, -2)$
10. $f(x) = x^3 - 2x^2 - 5x + 2$ in $(-3, -2)$
11. $f(x) = x^3 - 2x^2 - 5x + 2$ in $(-2, -1)$
12. $f(x) = x^4 - x^3 + 7x - 17$ in $(2, 3)$
13. $f(x) = x^4 - x^3 + 7x - 15$ in $(1, 3)$
14. $f(x) = x^4 - x^3 + 7x - 17$ in $(-3, -2)$
15. $f(x) = x^4 - x^3 + 7x - 17$ in $(-2, -1)$
16. $f(x) = 2x^4 - 3x^3 + 8x^2 - 9x - 7$ in $(-2, -1)$
17. $f(x) = 2x^4 - 3x^3 + 8x^2 - 9x - 12$ in $(-1, 0)$
18. $f(x) = x^4 + 5x^3 - 12x^2 - 13x + 5$ in $(2, 3)$
19. $f(x) = x^5 - 2x^4 + 6x^3 - x^2 - 17$ in $(0, 1)$
20. $f(x) = x^5 - 2x^4 + 7x^3 - x^2 - 12$ in $(1, 2)$

Use the Boundedness Theorem to determine if the zeros of $f(x)$ are bounded over the given interval. If this cannot be determined, indicate so.

21. $f(x) = x^2 - 6x + 3$ in $[-1, 6]$
22. $f(x) = 3x^2 - x - 5$ in $[-1, 6]$
23. $f(x) = 5x^2 - 2x + 7$ in $[-2, 2]$
24. $g(x) = x^3 - 2x^2 - 5x + 11$ in $[-3, 2]$
25. $g(x) = x^3 + 2x^2 - 5x - 6$ in $[-2, 4]$
26. $g(x) = 2x^3 + x^2 - 8x + 7$ in $[-3, 5]$
27. $g(x) = x^3 - 5x^2 + 9x - 53$ in $[-5, 6]$
28. $g(x) = 2x^3 - 7x^2 + 6x + 8$ in $[-1, 1]$
29. $h(x) = x^4 + x^3 - 7x^2 - 13x - 6$ in $[-2, 2]$
30. $h(x) = x^4 + x^3 - 7x^2 - 13x - 6$ in $[-2, 5]$
31. $h(x) = 6x^4 - x^3 + 5x^2 - x + 10$ in $[-1, 2]$
32. $h(x) = 2x^4 - 3x^3 + 6x^2 - 7x - 15$ in $[-3, 1]$
33. $h(x) = x^4 + 5x^3 - 12x^2 - 13x + 2$ in $[-5, 3]$
34. $t(x) = x^5 - 2x^4 + 7x^3 - x^2 + 2x + 1$ in $[-2, 3]$

Graph each polynomial function using the procedure in Examples 3 and 4.

35. $f(x) = x^2 - x - 20$
36. $h(x) = x^2 + 2x - 8$
37. $g(x) = x^3 + 2x^2 - 5x - 6$
38. $g(x) = x^3 + 7x^2 - x - 7$
39. $g(x) = 2x^3 + x^2 - 8x - 4$
40. $g(x) = x^3 - 5x^2 + 9x - 45$
41. $g(x) = x^3 - x^2 - x - 2$
42. $g(x) = x^3 + x^2 + 5x + 5$
43. $g(x) = x^3 + x^2 - 8x - 12$
44. $g(x) = x^3 - 4x^2 - 3x + 18$
45. $g(x) = x^3 - 10x^2 + 17x + 28$
46. $g(x) = x^3 - 4x^2 - 11x + 30$
47. $h(x) = x^4 + 2x^3 - 16x^2 - 2x + 15$
48. $h(x) = x^4 - 20x^2 + 64$
49. $h(x) = x^4 + x^3 - 7x^2 - 13x - 6$
50. $h(x) = x^4 + 2x^3 - 16x^2 - 2x + 15$
51. $h(x) = 6x^4 - x^3 + 5x^2 - x - 1$
52. $h(x) = x^4 - 12x^2 - 64$
53. $h(x) = x^4 - x^3 - 5x^2 - x - 6$
54. $h(x) = x^5 + 7x^4 - 5x^3 - 35x^2 + 4x + 28$

Problem Solving

Find all real zeros (estimate the irrational zeros to the nearest hundredth).

55. $f(x) = x^3 + x^2 + x - 4$
56. $g(x) = 2x^4 + x^2 - 1$
57. $h(x) = 2x^4 - 3x^3 + x^2 - 12$
58. $f(x) = 3x^3 - 2x^2 - 20$
59. $g(x) = 8x^4 - 2x^2 + 5x - 1$
60. $h(x) = x^4 + 8x^3 - x^2 + 2$
61. $f(x) = 2x^3 + 6x^2 - 8x + 2$
62. $g(x) = 3x^3 - 10x + 9$

Cumulative Review Exercises

[6.1] **63.** Simplify $\dfrac{3x^2 + 9x^3 - 12x^4 y}{6x^4 y}$.

[6.2] **65.** Subtract $\dfrac{3}{4x} - \dfrac{5}{6y}$.

[6.2] **64.** Add $\dfrac{4}{x+2} + \dfrac{5}{x+3}$.

[6.4] **66.** Solve $\dfrac{3}{y} + \dfrac{1}{3} = \dfrac{2}{3y}$.

10.5 CHARACTERISTICS OF GRAPHS OF RATIONAL FUNCTIONS

1 Recognize the key features of graphs of rational functions.
2 Find vertical asymptotes.
3 Find holes in graphs of rational functions.
4 Find horizontal asymptotes.
5 Find oblique asymptotes.
6 Graph rational functions.

1 Recognize the Key Features of Graphs of Rational Functions

In Chapter 6 we briefly discussed rational functions. We will repeat the definition here for your convenience.

> A **rational function** is a function of the form $f(x) = \dfrac{p}{q}$ where p and q are polynomials and $q \neq 0$.

A rational function is **proper** if the degree of its numerator is less than the degree of its denominator. If the degree of its numerator is greater than or equal to the degree of its denominator, the function is an **improper** rational function.

One example of a proper rational function is the reciprocal function, $f(x) = \dfrac{1}{x}$. In Example 6 of Section 3.1 we graphed this function by plotting points.

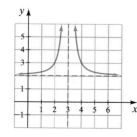

FIGURE 10.32

Although we can always graph functions by plotting points, it is often easier to graph them using their key features.

We will now revisit the graph of the reciprocal function so that we can identify its key features, see Figure 10.32. Shortly we will discuss how to find the key features of rational functions and use them as tools for graphing.

Notice that it appears that the graph has no x-intercept and no y-intercept. Instead, the graph appears to approach the x-axis and the y-axis. A line that a portion of a graph approaches but never reaches is called an **asymptote**. For instance the graph of $f(x) = \dfrac{1}{x}$ has a **horizontal asymptote** at $y = 0$ and a **vertical asymptote** at $x = 0$. The vertical and horizontal asymptotes of a rational function are two key features of its graph. The graph in Figure 10.33 has a horizontal asymptote at $y = 2$ and a vertical asymptote at $x = 3$.

Before we formally define horizontal and vertical asymptotes, we need to introduce a new symbol. We often use an arrow pointing to the right to represent the word "approaches." For example, $x \to 4$ is read "as x approaches 4," $p(x) \to \infty$ is read "as $p(x)$ approaches infinity," and $|f(x)| \to \infty$ is read "as the absolute value of $f(x)$ approaches infinity." Now let's define vertical and horizontal asymptotes.

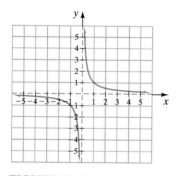

FIGURE 10.33

> For the rational function $f(x) = \dfrac{p(x)}{q(x)}$, written in the lowest terms,
>
> a) if $f(x) \to a$ as $|x| \to \infty$ then the line $y = a$ is a **horizontal asymptote**; and
>
> b) if $|f(x)| \to \infty$ as $x \to a$, then the line $x = a$ is a **vertical asymptote**.

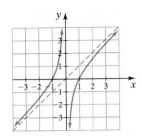

FIGURE 10.34

A horizontal asymptote is the *y*-value that a function approaches as $|x| \to \infty$. A vertical asymptote is the *x*-value where $|f(x)|$ approaches ∞.

Some asymptotes are neither vertical nor horizontal. These are called **oblique asymptotes**. Figure 10.34 illustrates an oblique asymptote. We will discuss oblique asymptotes in more depth shortly.

Consider the rational function $f(x) = \dfrac{x^2 - 4}{x + 2}$. Its graph is shown in Figure 10.35.

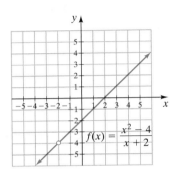

FIGURE 10.35

This graph appears linear with a **hole** at $x = -2$. Holes are another feature that may be present in the graph of a rational function. We will discuss holes in more detail shortly. Now let's learn how to find asymptotes.

2 Find Vertical Asymptotes

Rational functions are not defined for values of *x* for which the denominator is equal to 0. Rational functions will have a point of discontinuity, where the function is not defined. At values of *x* where the denominator is 0, the graph will either have a vertical asymptote or a hole. We will first discuss finding vertical asymptotes.

> **Locating vertical asymptotes**
>
> A rational function $f(x) = \dfrac{p(x)}{q(x)}$, in lowest terms, will have a vertical asymptote at *r* if *r* is a real zero of the denominator $q(x)$.

By the boxed information, if *after simplifying the rational function*, we find a value that results in the remaining denominator having a value of 0, then the graph of the rational function will have a vertical asymptote at that value.

EXAMPLE 1 Find the vertical asymptotes, if any, of the graph of each rational function.

a) $f(x) = \dfrac{x}{x^2 - 4}$ **b)** $g(x) = \dfrac{x - 2}{x + 5}$ **c)** $R(x) = \dfrac{x^2}{x^2 + 3}$

d) $f(x) = \dfrac{x^2 - 4}{x + 2}$ **e)** $H(x) = \dfrac{x^2 - 9}{x^2 + 3x - 18}$

Solution **a)** The function cannot be simplified further. Since both 2 and −2 make the denominator 0, the graph will have vertical asymptotes at $x = 2$ and $x = -2$.

b) The function cannot be simplified further. Since −5 makes the denominator 0, the graph has a vertical asymptote at $x = -5$.

c) The function cannot be simplified further. Since there is no value of x that makes the denominator 0, the graph of this function contains no vertical asymptotes.

d) The function can be simplified by factoring as follows

$$f(x) = \frac{x^2 - 4}{x + 2} = \frac{\cancel{(x + 2)}(x - 2)}{\cancel{x + 2}} = x - 2$$

Since there are no values that make the denominator of the simplified expression 0, there are no vertical asymptotes.

e) The function can be simplified by factoring as follows

$$H(x) = \frac{x^2 - 9}{x^2 + 3x - 18} = \frac{(x + 3)\cancel{(x - 3)}}{(x + 6)\cancel{(x - 3)}} = \frac{x + 3}{x + 6}$$

Because -6 makes the denominator of the simplified expression 0, there is a vertical asymptote at $x = -6$.

NOW TRY EXERCISE 11

HELPFUL HINT

A rational function can have no vertical asymptotes, one vertical asymptote, or more than one vertical asymptote. The graph of a rational function will never intersect any of its vertical asymptotes.

---

3 Find Holes in Graphs of Rational Functions

If a value makes both the numerator and denominator of a rational function zero, then there will be either a hole or a vertical asymptote at that value. If, when simplified, the value makes the denominator of the simplified rational function zero, then there will be a vertical asymptote at that value. If the value does not make the denominator of the simplified rational function zero, then there will be a hole at that value.

Notice that in the function in Example 1 **d)**, $f(x) = \dfrac{x^2 - 4}{x + 2} = \dfrac{(x + 2)(x - 2)}{x + 2}$, the function is not defined at $x = -2$. When $x = -2$ both the numerator and denominator of the function are 0. When this occurs there may be a *hole* in the graph of the function. Therefore, the graph of the function may have a hole at $x = -2$. To determine if there is a hole, simplify the function.

$$f(x) = \frac{\cancel{(x + 2)}(x - 2)}{\cancel{x + 2}}$$
$$f(x) = x - 2, x \neq -2$$

Since the denominator of the simplified function is 1, the value -2 does not make the denominator of the simplified function 0. Therefore, there is a hole, and not a vertical asymptote, in the graph of the function at $x = -2$

Now look at the function in Example 1 **e)** $H(x) = \dfrac{x^2 - 9}{x^2 + 3x - 18} = \dfrac{(x + 3)(x - 3)}{(x + 6)(x - 3)}$. This function is not defined at $x = 3$. When $x = 3$ both the numerator and denominator are 0. Therefore, the graph of this function may have a hole at $x = 3$. To determine if there is a hole in the function, simplify the function

$$f(x) = \frac{(x + 3)\cancel{(x - 3)}}{(x + 6)\cancel{(x - 3)}}$$
$$f(x) = \frac{x + 3}{x + 6}, x \neq 3$$

Since the denominator of the simplified function, $x + 6$, is not 0 when $x = 3$, the graph of the function has a hole at $x = 3$.

Now consider the function $G(x) = \dfrac{x - 4}{(x - 4)^2}$. Since the value 4 makes both the numerator and denominator 0, there may be a hole in the function at $x = 4$. To determine if there is a hole, simplify the function

$$G(x) = \frac{x - 4}{(x - 4)^2} = \frac{\cancel{x - 4}}{(x - 4)\cancel{(x - 4)}}$$

$$G(x) = \frac{1}{x - 4}, x \neq 4$$

Since 4 makes the denominator of the simplified function $\dfrac{1}{x - 4}$ equal to 0, there is a vertical asymptote, and not a hole, in the graph of $G(x)$ at $x = 4$.

4 Find Horizontal Asymptotes

To determine horizontal asymptotes we will look at two categories of rational functions: proper rational functions, and improper rational functions where the numerator and denominator contain polynomials of the same degree. Consider the **proper rational function** $f(x) = \dfrac{x + 1}{4x^2 + 9}$. We need to determine how the values of $f(x)$ behave as $x \to \infty$ or $x \to -\infty$. Suppose x gets larger and larger. Then the numerator of $f(x)$, which is $x + 1$, can be approximated by the power function $y = x$. The denominator, $4x^2 + 2$, can be approximated by the power function $y = 4x^2$. Therefore, $f(x) = \dfrac{x + 1}{4x^2 + 9} \approx \dfrac{x}{4x^2} = \dfrac{1}{4x}$.

The rational function $f(x) = \dfrac{1}{4x}$ will approach 0 as $x \to \infty$ or $x \to -\infty$. For example, if $x = 100$, $f(x) = \dfrac{1}{400} = 0.0025$ and if $x = -100$, $f(x) = \dfrac{1}{-400} = -0.0025$. Since the function approaches 0 as $x \to \infty$ or $x \to -\infty$ there is a horizontal asymptote at $y = 0$. Using the same reasoning we can determine the following theorem.

> If a rational function is proper, the line $y = 0$ is a horizontal asymptote of its graph.

EXAMPLE 2 Determine if the graph of the functions has a horizontal asymptote of $y = 0$.

a) $f(x) = \dfrac{x^2 - 6}{2x^3 + 9}$

b) $g(x) = \dfrac{x^3 + 5x - 2}{3x^6 - 6x}$

Solution **a)** Since the function is a proper rational function, its graph will have a horizontal asymptote at $y = 0$.

b) Since the function is a proper rational function, its graph will have a horizontal asymptote at $y = 0$.

NOW TRY EXERCISE 33

Now we look at rational functions whose polynomials in the numerator and denominator have the same degree. Consider $f(x) = \dfrac{3x^2 + 4x}{5x^2 - 2x + 6}$. Both the numerator and denominator are of degree 2. Let us divide each term in the numerator and each term in the denominator by x^2, the variable and exponent in the leading terms.

$$f(x) = \dfrac{\dfrac{3x^2}{x^2} + \dfrac{4x}{x^2}}{\dfrac{5x^2}{x^2} - \dfrac{2x}{x^2} + \dfrac{6}{x^2}} = \dfrac{3 + \dfrac{4}{x}}{5 - \dfrac{2}{x} + \dfrac{6}{x^2}}$$

As $x \to \infty$ or $x \to -\infty$, $\dfrac{4}{x}$, $-\dfrac{2}{x}$, and $\dfrac{6}{x^2}$ will all approach 0, and therefore the value of the function will approach $\frac{3}{5}$. This graph will have horizontal asymptote at $y = \frac{3}{5}$. Using the same reasoning we can conclude the following.

> For a rational function where the numerator and denominator are of the same degree, the ratio of the leading coefficients of the simplified function will give the horizontal asymptote of the graph. That is, the graph of a rational function with simplified form
> $$f(x) = \dfrac{a_n x^n + a_{n-1} x^{n-1} + \ldots + a_0}{b_n x^n + b_{n-1} x^{n-1} + \ldots + b_0}$$ will have a horizontal asymptote at $y = \dfrac{a_n}{b_n}$.

EXAMPLE 3   Determine the horizontal asymptotes of the graphs of the following functions.

a) $T(x) = \dfrac{3x^6 + 5x^2 - x}{2x^6 - 9x^3 + 5}$

b) $R(x) = \dfrac{15x^4 - 6x^3 + 3x - 2}{4x^4 - 6x^2 + 3}$

Solution   **a)** Since the numerator and denominator are of the same degree, the graph of the function will have a horizontal asymptote at $y = \frac{3}{2}$.

b) The graph of $R(x)$ will have a horizontal asymptote at $y = \frac{15}{4}$.   ✳

5   Find Oblique Asymptotes

Now we consider a rational function where the polynomial in the numerator is exactly 1 degree higher than the polynomial in the denominator. For example, consider $R(x) = \dfrac{3x^2 + 2x - 6}{x + 2}$. The numerator is of degree 2 and the denominator is of degree 1. If we divide the denominator by the numerator we obtain

$$
\require{enclose}
\begin{array}{r}
3x - 4 \\
x + 2 \enclose{longdiv}{3x^2 + 2x - 6} \\
\underline{3x^2 + 6x} \\
-4x - 6 \\
\underline{-4x - 8} \\
2
\end{array}
$$

Therefore, $R(x) = \dfrac{3x^2 + 2x - 6}{x + 2} = 3x - 4 + \dfrac{2}{x + 2}$.

For large values of x the fraction $\dfrac{2}{x+2}$ approaches 0 and $f(x)$ approaches $3x - 4$. Therefore there is an asymptote at $y = 3x - 4$. The asymptote at $y = 3x - 4$ is an oblique asymptote. The graph of $R(x)$ is shown in Figure 10.36.

$$R(x) = \frac{3x^2 + 2x - 6}{x + 2}$$

FIGURE 10.36

Using the same procedure and reasoning we can conclude the following.

> If the numerator of a rational function is of degree exactly 1 more than the denominator, to obtain the oblique asymptote (if it exists), divide the numerator by the denominator and disregard the remainder. The new quotient will be of the form $ax + b$. Set the quotient equal to y to obtain the equation of an oblique asymptote of the form $y = ax + b$. The oblique asymptote will be a line.

EXAMPLE 4 Determine if the function $g(x) = \dfrac{6x^3 + 4x^2 + 18 - 7}{2x^2 + 3}$ has an oblique asymptote.

Solution Divide the numerator by the denominator

$$
\begin{array}{r}
3x + 2 \\
2x^2 + 0x + 3 \overline{)6x^3 + 4x^2 + 18x - 7} \\
\underline{6x^3 + 0x^2 + 9x } \\
4x^2 + 9x - 7 \\
\underline{4x^2 + 0x + 6} \\
9x - 13
\end{array}
$$

Thus $g(x) = \dfrac{6x^3 + 4x^2 + 18x - 7}{2x^2 + 3} = 3x + 2 + \dfrac{9x - 13}{2x^2 + 3}$. If we drop the remainder and set the quotient equal to y, we get $y = 3x + 2$. The graph of $g(x)$ will have an oblique asymptote at $y = 3x + 2$. Note that there is no vertical asymptote since the denominator can never equal 0.

NOW TRY EXERCISE 39

If the degree of the numerator of a rational function is 2 or more greater than the degree of the denominator, then the graph of the function has no horizontal or oblique asymptotes.

The graph of a rational function may have only 1 nonvertical asymptote (either horizontal or oblique). It is possible for the graph of a rational function to intersect the nonvertical asymptote. However, the graph of a rational function cannot intersect a vertical asymptote.

Let Us Summarize What We Have Learned

1. **Vertical Asymptotes** occur at values that make the denominator of a rational function, in its lowest terms, equal to 0.

2. **Holes** may occur at values that result in both the numerator and denominator having a value of 0.

3. **Horizontal Asymptotes**

   a) The graph of a proper rational function has a horizontal asymptote at $y = 0$.

   b) If the degree of the numerator of a rational function is the same as the degree of the denominator, the graph of the rational function has a horizontal asymptote equal to the ratio of the leading coefficients.

4. **Oblique Asymptotes**
   If the degree of the numerator of a rational function is exactly 1 more than the degree of the denominator, the graph of the rational function may have an oblique asymptote. The asymptote can be found by dividing the numerator by the denominator, disregarding the remainder, and setting the quotient equal to y.

6 Graph Rational Functions

Now we will use the information we have learned to graph rational functions. It may be helpful to plot some points from time to time to get a more accurate graph.

EXAMPLE 5 Graph $f(x) = \dfrac{1}{x - 5}$

Solution When $x = 0$, $f(x) = -\frac{1}{5}$. Thus the y-intercept is $\left(0, -\frac{1}{5}\right)$. There is a vertical asymptote at $x = 5$. Since this is a proper rational function, there is a horizontal asymptote at $y = 0$, see Figure 10.37. We use dashed lines to indicate the asymptotes because the asymptotes are not a part of the graph. They are only an aid in drawing the graph. To determine the behavior of the graph as it approaches the vertical asymptote, select values for x approaching 5 *from the right*, that is, select values for x getting closer and closer to 5, but always greater than 5. Selecting values for x to the right of 5, we get the following table.

x	$f(x)$
6	1
5.5	2
5.1	10
5.01	100

FIGURE 10.37

We see that as we approach 5 from the right, $f(x)$ gets larger and larger. Therefore $f(x) \to \infty$ as $x \to 5$ from the right. Now select values for x approaching $x = 5$ *from the left*. That is, select values for x closer and closer to 5, but always less than 5. Selecting value for x to the left of 5, we get the following table.

x	$f(x)$
4	-1
4.5	-2
4.9	-10
4.99	-100

NOW TRY EXERCISE 47

We see that as we approach 5 from the left, $f(x)$ gets smaller and smaller. Therefore, $f(x) \to -\infty$ as $x \to 5$ from the left. We use the values in the table, the fact that a graph never crosses a vertical asymptote, and the fact that the horizontal asymptote is at $y = 0$ to draw the graph; see Figure 10.37. ✳

Using Your Graphing Calculator

In some of the following examples the calculations may be time consuming. If you have a graphing calculator, you should use it to evaluate functions for various values of the variable.

On page 36 we explained how to evaluate an expression for various values of the variable. If you have forgotten how to do so, please read that material now.

Consider the function $f(x) = \dfrac{4x - 1}{2x + 6}$, which we will graph in Example 6. To evaluate the function at $-4, -3.5$, and -3.1, we enter the following.

```
-4 -> x: (4x - 1)/(2x + 6)
                       8.5
-3.5 -> x: (4x - 1)/(2x + 6)
                        15
-3.1 -> x: (4x - 1)/(2x + 6)
                        67
```

Notice that when $x = -4$, the function equals 8.5; when $x = -3.5$, the function equals 15; and when $x = -3.1$, the function equals 67.

EXAMPLE 6 Graph $f(x) = \dfrac{4x - 1}{2x + 6}$.

Solution By substituting $x = 0$, we determine the y-intercept is $\left(0, -\frac{1}{6}\right)$. There is a vertical asymptote at $x = -3$, since -3 makes the denominator 0. Since the numerator and denominator are of the same degree, the horizontal asymptote is at $y = \frac{4}{2}$ or $y = 2$; see Figure 10.38a on the next page.

Now plot a few points on each side of the vertical asymptote. Remember to work on only one side of the vertical asymptote at a time.

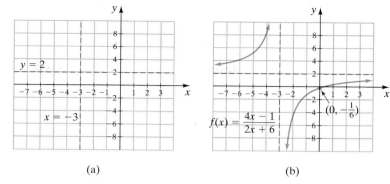

FIGURE 10.38   (a)   (b)

In $(-\infty, -3)$, approaching -3 from the left,

x	$f(x)$
-4	8.5
-3.5	15.0

In $(-3, \infty)$, approaching -3 from the right,

x	$f(x)$
-2	-4.5
-2.5	-11

We can see from the few points plotted and the vertical asymptote that as x approaches -3 from the left, the graph approaches ∞, and as x approaches -3 from the right, the graph approaches $-\infty$. Remember **a graph never crosses a vertical asymptote**. The graph is shown in Figure 10.38b.   ✳

NOW TRY EXERCISE 53

EXAMPLE 7   Graph $f(x) = \dfrac{x + 1}{(2x - 1)(x + 3)}$.

Solution   When $x = 0, f(x) = -\frac{1}{3}$. Thus the y-intercept is $\left(0, -\frac{1}{3}\right)$. There are vertical asymptotes at $x = \frac{1}{2}$ and $x = -3$. Since the numerator is of degree 1 and the denominator is of degree 2, this is a proper rational function. Therefore, the horizontal asymptote is $y = 0$; see Figure 10.39a.

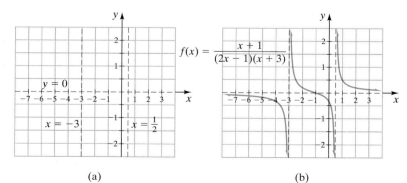

FIGURE 10.39   (a)   (b)

Now let's select some values in each of the three intervals $(\infty, -3)$, $\left(-3, \frac{1}{2}\right)$, and $\left(\frac{1}{2}, \infty\right)$. We work with each interval individually.

In $(-\infty, -3)$ approaching -3 from the left

x	$f(x)$
-4	-0.333
-3.5	-0.625
-3.1	-2.9
-3.01	-28.6

As x approaches -3 from the left, the function approaches $-\infty$.
In the interval $\left(-3, \frac{1}{2}\right)$ approaching -3 from the right

x	$f(x)$
-2	0.2
-2.5	0.5
-2.9	2.79
-2.99	28.5

As x approaches -3 from the right, the function approaches ∞.
In the interval $\left(-3, \frac{1}{2}\right)$ approaching $\frac{1}{2}$ from the left

x	$f(x)$
0	-0.333
0.4	-2.06
0.49	-21.3
0.499	-214

As x approaches $\frac{1}{2}$ from the left, the function approaches $-\infty$.
In the interval $\left(\frac{1}{2}, \infty\right)$ approaching $\frac{1}{2}$ from the right

x	$f(x)$
1.5	0.28
0.6	2.22
0.51	21.5
0.501	214

NOW TRY EXERCISE 49

As x approaches $\frac{1}{2}$ from the right, the function approaches ∞. The graph is illustrated in Figure 10.39b on page 653. Notice that the function approaches 0 (the horizontal asymptote) as x approaches ∞ and as x approaches $-\infty$. Also notice that the graph crosses the horizontal asymptote between $x = -3$ and $x = \frac{1}{2}$. ✳

EXAMPLE 8 Graph $f(x) = \dfrac{x^2 + 4}{x + 2}$.

Solution When $x = 0$, $f(x) = 2$. Thus the y-intercept of the graph is $(0, 2)$. This function is reduced to its lowest terms since the numerator cannot be factored over the set of real numbers. There is a vertical asymptote at $x = -2$. Since the numerator is exactly 1 degree greater than the denominator, we can obtain the oblique asymptote by dividing

$$\begin{array}{r} x - 2 \\ x + 2 \overline{\smash{\big)}\, x^2 + 0x + 4} \\ \underline{x^2 + 2x} \\ -2x + 4 \\ \underline{-2x - 4} \\ 8 \end{array}$$

The oblique asymptote is obtained from the quotient, $x - 2$. The oblique asymptote is $y = x - 2$, see Figure 10.40a.

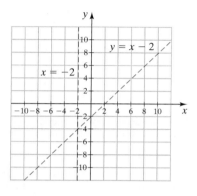

FIGURE 10.40a

Now select some values for x in the intervals $(-\infty\ -2)$ and $(-2, \infty)$ and evaluate $f(x)$ at those values.

In the interval $(-\infty, -2)$ approaching -2 from the left

x	$f(x)$
-5	-9.67
-3	-13
-2.5	-20.5
-2.1	-84.1
-2.01	-804

As x approaches -2 from the left, the function approaches $-\infty$.

In the interval $(-2, \infty)$ approaching -2 from the right

x	$f(x)$
5	4.14
-1	5
-1.5	12.5
-1.9	76.1
-1.99	796

As x approaches -2 from the right, the function approaches ∞.

The graph is drawn in Figure 10.40b. Notice the graph approaches $x = -2$, but is not defined at -2. Also the graph approaches the oblique asymptote $y = x - 2$ as $x \to \infty$ and as $x \to -\infty$.

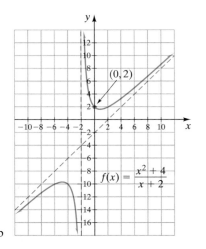

$$f(x) = \frac{x^2 + 4}{x + 2}$$

NOW TRY EXERCISE 59   FIGURE 10.40b

Using Your Graphing Calculator

If you have a graphing calculator, you should graph each function we graph on your graphing calculator after you have graphed the function using the procedure discussed in this section. On page 388 we discussed using the dot mode to graph functions. You should reread that material now.

In Figure 10.41 we illustrate the graph of $f(x) = \dfrac{x^2 + 4}{x + 2}$ which we graphed in Figure 10.40b.

$$-10, 10, 1, -30, 30, 3 \qquad\qquad -10, 10, 1, -30, 30, 3$$
$$\text{(a)} \qquad\qquad\qquad\qquad \text{(b)}$$

FIGURE 10.41

In Figure 10.41a we use the connected mode, and in Figure 10.41b we use the dot mode.

EXAMPLE 9   Graph $f(x) = \dfrac{x^2 - 4}{x - 2}$.

Solution   When $x = 0$, $f(x) = 2$. Thus, the y-intercept is $(0, 2)$. The function can be factored as follows.

$$f(x) = \frac{x^2 - 4}{x - 2} = \frac{(x + 2)(x - 2)}{(x - 2)} = x + 2, (x \neq 2)$$

Since $x + 2$ is defined for all values of x, there is no vertical asymptote. From the factored form of $f(x)$ we see that $f(x)$ is not defined at $x = 2$. However at $x = 2$, both the numerator and denominator have a value of 0. Therefore, there is a hole at 2 in the function $f(x) = x + 2, x \neq 2$. Following is a table of values. The graph is illustrated by Figure 10.42.

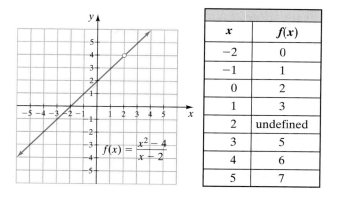

x	$f(x)$
-2	0
-1	1
0	2
1	3
2	undefined
3	5
4	6
5	7

$$f(x) = \frac{x^2 - 4}{x - 2}$$

NOW TRY EXERCISE 57   FIGURE 10.42

Exercise Set 10.5

Concept/Writing Exercises

1. Do the graphs of all rational functions have at least 1 asymptote? Explain.

2. Can the graph of a rational function have more than 1 vertical asymptote? Explain.

3. Can the graph of a rational function have more than 1 horizontal asymptote? Explain.

4. Explain how to find the y-intercept of the graph of a rational function.

5. Explain how to determine the points of discontinuity of the graph of a rational function.

6. Explain why the graph of a proper rational function has a horizontal asymptote at $y = 0$.

7. Under what conditions could the graph of a rational function have an oblique asymptote?

8. Under what conditions will the graph of a rational function have neither a horizontal nor an oblique asymptote?

Practice the Skills

For each rational function, find all points of discontinuity and determine whether there is a hole or a vertical asymptote at each point.

9. $f(x) = \dfrac{3 - x}{x}$

10. $g(x) = \dfrac{4 - 3x}{x - 9}$

11. $h(x) = \dfrac{x + 2}{x + 5}$

12. $f(x) = \dfrac{x - 6}{x - 7}$

13. $g(x) = \dfrac{x^2 - 25}{x + 5}$

14. $h(x) = \dfrac{x^2 - 7x + 10}{x - 5}$

15. $f(x) = \dfrac{x^2 + 8x - 9}{x - 1}$

16. $g(x) = \dfrac{6x^2 + x - 2}{3x + 2}$

17. $g(x) = \dfrac{4x - 7}{x^2 - 16}$

18. $h(x) = \dfrac{3x}{4x^2 - 2x}$

19. $f(x) = \dfrac{x - 1}{x^2 - 7x + 6}$

20. $g(x) = \dfrac{x + 3}{x^2 - 7x + 12}$

21. $h(x) = \dfrac{x + 7}{x^2 + 8x + 15}$

22. $f(x) = \dfrac{x - 4}{x^2 - x - 12}$

23. $g(x) = \dfrac{x + 5}{x^2 - 25}$

24. $h(x) = \dfrac{x^2 - 3x - 10}{x^2 - 8x + 15}$

25. $f(x) = \dfrac{x^2 + 4x + 3}{x^2 - 4x + 3}$

26. $g(x) = \dfrac{x^2 + 10x + 24}{x^2 - 36}$

27. $h(x) = \dfrac{x + 5}{x^2 + 5}$

28. $f(x) = \dfrac{5x - 10}{x^2 - x - 2}$

Determine whether the graph of the rational function has a horizontal asymptote, an oblique asymptote, or neither. Give the equation for the asymptote if it exists.

29. $f(x) = \dfrac{7}{x}$

30. $g(x) = \dfrac{5}{x - 6}$

🔒 **31.** $h(x) = \dfrac{6x - 1}{2x + 3}$

32. $f(x) = \dfrac{8x - 3}{4x + 11}$

33. $g(x) = \dfrac{12x + 5}{3x^2}$

34. $h(x) = \dfrac{16x - 5}{7x^2 + 3}$

35. $f(x) = \dfrac{4x^2 - 1}{x^2 - 9}$

36. $g(x) = \dfrac{3x^2 - 5x + 9}{2x^2 - x - 3}$

37. $f(x) = \dfrac{x^2 - 8x + 15}{x + 2}$

38. $f(x) = \dfrac{x^2 - 5x - 8}{x - 3}$

🔒 **39.** $f(x) = \dfrac{2x^2 + 8x + 12}{x - 2}$

40. $f(x) = \dfrac{3x^2 - 9x + 13}{x - 1}$

41. $f(x) = \dfrac{4x^2 - 6}{8x^3 - 1}$

42. $f(x) = \dfrac{8x^3 - 12x^2 + 3x - 1}{4x^2 - 1}$

43. $f(x) = \dfrac{x^3 + 2}{x^2 + 2x}$

44. $f(x) = \dfrac{x^2 + 2x}{x^3 + 5x}$

Graph the rational function.

45. $f(x) = \dfrac{1}{x + 2}$

46. $g(x) = \dfrac{1}{x - 3}$

47. $h(x) = \dfrac{5}{x - 1}$

48. $f(x) = -\dfrac{2}{x + 6}$

🔒 **49.** $h(x) = \dfrac{x + 1}{x(x + 4)}$

50. $f(x) = \dfrac{x}{(x + 1)(x - 2)}$

51. $g(x) = \dfrac{x + 1}{x - 1}$

52. $h(x) = \dfrac{6x - 5}{2x + 8}$

🔒 **53.** $f(x) = \dfrac{3x + 3}{2x + 4}$

54. $g(x) = \dfrac{2x + 4}{x - 1}$

55. $h(x) = \dfrac{15x^2 + 10}{3x^2}$

56. $f(x) = \dfrac{3 - 8x^2}{2x^2 + 1}$

57. $g(x) = \dfrac{x^2 - 3x - 4}{x - 2}$

58. $h(x) = \dfrac{x^2 - x - 12}{x - 2}$

🔒 **59.** $g(x) = \dfrac{x^2 + 2x - 8}{x + 4}$

60. $f(x) = \dfrac{x^2 + 9}{x - 5}$

61. $r(x) = \dfrac{x^2 - 5x + 6}{x - 2}$

62. $h(x) = \dfrac{2x^2 - x - 3}{x + 1}$

63. $g(x) = \dfrac{2x^3 - x^2 + x - 8}{x^2 - 2x + 1}$

64. $r(x) = \dfrac{x^3 - x^2 - x + 1}{x - 1}$

Problem Solving

65. Write a rational function that has a horizontal asymptote at $y = 0$ and a vertical asymptote at $x = -4$. Explain how you determined your answer.

66. Write a rational function that has a horizontal asymptote at $y = 5$ and a vertical asymptote at $x = 3$. Explain how you determined your answer.

🔒 **67.** Write a rational function that is the graph of $f(x) = 2x - 5$ with a hole at the point $(4, 3)$. Explain how you determined your answer.

68. Write a rational function that is the graph of $f(x) = x^2 - 3$ with a hole at the point $(-3, 6)$. Explain how you determined your answer.

69. Write a rational function that has a horizontal asymptote at $y = 0$ and no vertical asymptote. Explain how you determined your answer.

70. Write a rational function that has an oblique asymptote at $y = 2x + 1$ and a vertical asymptote at $x = 3$. Explain how you determined your answer.

Group Activity

71. In this section we have been graphing rational functions in which the degree of the numerator is at most 1 greater than the degree of the denominator. Now we will look at the graphs of rational functions in which the degree of the numerator is 2 greater than the degree of the denominator.

a) Graph the function $f(x) = \dfrac{x^4 + 1}{x^2}$.

b) Find the quotient of $x^4 + 1$ divided by x^2.

c) Graph the quotient (disregard the remainder) on the same coordinate axes.

d) What do you notice?

e) Explain why the graph of $f(x) = \dfrac{x^4 + 1}{x^2}$ is very similar to the graph of the quotient, $g(x) = x^2$, when $|x| \to \infty$.

Cumulative Review Exercises

[9.1] **72.** Find the inverse of $f(x) = \dfrac{1}{2}x + 3$.

[9.1] **73.** Find the inverse of $f(x) = \dfrac{5}{x}$.

[9.3] **74.** Solve $\log_a \dfrac{1}{144} = -2$ for a.

[9.3] **75.** Solve $\log_2 x = -5$ for x.

CHAPTER SUMMARY

Key Words and Phrases

10.1
Continuous
Decreasing
Discontinuous
Greatest integer function
Increasing
Local maxima
Local minima
Piecewise function
Point of discontinuity
Real zeros
Smooth
Step function

10.2
End behavior
Multiplicity
Polynomial function
Power function
Turning points
Zero of a polynomial

10.3
Descartes' Rule of Signs
Even-degree polynomial function

Factor Theorem
Odd-degree polynomial function
Rational Zeros Theorem
Remainder Theorem
Synthetic division

10.4
Intermediate Value Theorem
Irrational zeros
Lower bounds
Upper bounds

10.5
Asymptote
Approaches ($\longrightarrow$)
Hole
Horizontal asymptote
Improper rational function
Oblique asymptote
Proper rational function
Vertical asymptote

IMPORTANT FACTS

Greatest Integer Function, symbolized $f(x) = [\![x]\!]$, truncates, or rounds down, a value to the greatest integer less than or equal to the value.

Polynomial Functions

The end behavior of a polynomial function will match the end behavior of a power function having the same degree and the same sign for the leading coefficient.

A zero of a polynomial function, $P(x)$, is a value of x where $P(x) = 0$.

For a real zero of even multiplicity, the graph of $P(x)$ touches the x-axis but does not cross the x-axis at that zero.

For a real zero of odd multiplicity, the graph of $P(x)$ crosses the x-axis at that zero.

> The maximum number of turning points in the graph of a polynomial function of degree n is $n - 1$.
>
> The maximum number of real zeros of a polynomial function of degree n is n.

Descartes' Rule of Signs

For any polynomial function $P(x)$ written in descending order,

1. the number of positive real zeros of $P(x)$ either equals the number of variations in sign of the non-zero coefficients of $P(x)$ or equals that number less an even integer.

2. the number of negative real zeros of $P(x)$ either equals the number of variations in sign of the non-zero coefficients of $P(-x)$ or equals that number less an even integer.

Rational Zeros Theorem

For any polynomial function $P(x)$, the possible rational zeros of $P(x)$ are of the form $\dfrac{p}{q}$ where p is a factor of the constant term and q is a factor of the leading coefficient.

Remainder Theorem

If the polynomial $P(x)$ is divided by $x - a$, the remainder is equal to $P(a)$.

Intermediate Value Theorem

For a continuous function $f(x)$ defined over the interval (a, b), if the sign of $f(a)$ is the opposite that of the sign of $f(b)$, then a real zero exists in (a, b).

Boundedness Theorem is used to determine the upper and lower bounds for the zeros of a polynomial function.

Rational Functions

Vertical asymptotes occur at values that make the denominator of a rational function in lowest terms equal to 0. Holes may occur at values that result in both the numerator and denominator having a value of 0.

Horizontal Asymptotes

Proper rational functions have a horizontal asymptote at $y = 0$.
If the degree of the numerator is the same as the degree of the denominator, the function will have a horizontal asymptote at $y = \dfrac{a_n}{b_n}$ where a_n is the leading coefficient of the numerator and b_n is the leading coefficient of the denominator.

Oblique Asymptotes may occur when the degree of the numerator is exactly 1 greater than the degree of the denominator.

Chapter Review Exercises

[10.1] Classify each function as continuous or discontinuous.

1.

2.

Classify each function as continuous or discontinuous. If the function is not continuous, name the point(s) of discontinuity.

3. $f(x) = -4x^3 + 8x^2 - 12x + 19$

4. $g(x) = \dfrac{x + 8}{x^2 + 2}$

5. $h(x) = \left[\dfrac{x + 5}{4}\right]$

6. $f(x) = \begin{cases} x - 2 & \text{if } x < 2 \\ 2 - x & \text{if } x \geq 2 \end{cases}$

Graph each piecewise function.

7. $f(x) = \begin{cases} 2x & \text{if } x < 1 \\ 3 & \text{if } x \geq 1 \end{cases}$

8. $g(x) = \begin{cases} 4 - x & \text{if } x \leq -2 \\ \dfrac{1}{4}x & \text{if } x > -2 \end{cases}$

9. $h(x) = \begin{cases} -4 & \text{if } x < 0 \\ 4 & \text{if } x \geq 0 \end{cases}$

10. $f(x) = \begin{cases} 3 & \text{if } x \leq 3 \\ x & \text{if } x > 3 \end{cases}$

Graph each greatest integer function.

11. $f(x) = [2x - 3]$

12. $g(x) = \left[\dfrac{1}{2}x + 1\right]$

13. $h(x) = \left[\dfrac{x}{12}\right]$

14. $f(x) = [\sqrt{x}]$

15. *Postage Stamps* Sharon Bell wants to stock up on postage stamps, which cost $0.37 a piece. Write a greatest integer function that represents the number of stamps she can purchase for x dollars.

16. *Videos* Guillermo Lopez wants to rent some videos, which rent for $2.25 per video. Write a greatest integer function that represents the number of videos he can rent for x dollars.

[10.2] *Describe the location, either above the x-axis or below the x-axis, of the graph of $f(x)$ for **a)** large positive values of x, and **b)** large negative values of x.*

17. $f(x) = -2x^3 - 6x^2 + 7x - 3$

18. $f(x) = 2x^4 - x^3 + 7x^2 + 26x$

19. $f(x) = -x^6 + 3x^4 - 5x^2 + 11$

20. $f(x) = 7x^5 - 2x^4 - 5x^3 + 7x - 1$

Find the real zeros of each function and state their multiplicities if it is a number other than 1.

21. $f(x) = x^2 + 11x + 24$

22. $g(x) = x^3 + 5x^2 - 36x$

23. $h(x) = (x + 2)^3(2x - 1)$

24. $f(x) = x^2 - 7$

Determine the maximum number of turning points for the graph of each function. Do not graph the function.

25. $f(x) = 3x^2 + 6x - 16$

26. $g(x) = -2x^4 - 5x^3 + 6x + 11$

27. $h(x) = 2x^3 + 7x^2 - x + 9$

28. $f(x) = -5x^6 + 6x^5 + 13x - 2$

Determine the minimum degree of the polynomial function graphed.

29.

30.

Graph each function.

31. $f(x) = (x + 4)(x + 1)(x - 5)$

32. $g(x) = (x + 4)(x + 1)^2(x - 5)$

33. $h(x) = x^4 - 7x^3 + 6x^2$

34. $f(x) = x^5 - 7x^4 + 6x^3$

[10.3] Determine the maximum and minimum number of x-intercepts for the graph of each polynomial function.

35. $f(x) = -x^3 + 5x^2 - 7x - 9$

36. $g(x) = x^4 - x^3 - x^2 + x - 9$

37. $h(x) = x^5 - x^4 - 2x^3 + 20$

38. $f(x) = -x^6 + 5x^4 + 3x^2 - 18$

Use Descartes' Rule of Signs to determine

a) *the possible number of positive real zeros of each polynomial function.*

b) *the possible number of negative real zeros of each polynomial function.*

39. $f(x) = x^2 - 12x + 36$

40. $g(x) = x^2 + 3x - 21$

41. $h(x) = x^3 + 5x^2 + 3x + 9$

42. $f(x) = x^4 - x^3 - 6x^2 + 2x + 3$

List the possible rational zeros of each function.

43. $f(x) = x^3 + 3x^2 - 9x + 12$

44. $g(x) = 2x^5 - 9x^2 - 16x - 5$

45. $h(x) = 3x^4 - 12x^3 - 8x^2 + 4x - 9$

46. $f(x) = x^6 - 5x^4 + 6x - 64$

Determine the rational zeros of each function.

47. $f(x) = x^3 + 10x^2 - x - 10$

48. $g(x) = x^4 - 4x^3 + 5x^2 - 4x + 4$

49. $h(x) = 3x^4 + 10x^3 - 5x^2 + 10x - 8$

50. $f(x) = x^4 + x^3 - 35x^2 - 57x + 90$

Find all the zeros of each function.

51. $f(x) = x^4 - 4x^3 + 11x^2 - 64x - 80$

52. $g(x) = x^4 + x^3 + 19x^2 + 25x - 150$

[10.4] Use the Intermediate Value Theorem to determine if a real zero of $g(x) = x^3 - 4x^2 - 5x + 13$ occurs in the given interval. If this cannot be determined, indicate so.

53. $(-3, -1)$

54. $(-1, 1)$

55. $(1, 5)$

56. $(4, 5)$

Use the Boundedness Theorem to determine if the zeros of f(x) are bounded over the given interval. If this cannot be determined, indicate so.

57. $f(x) = x^3 - x^2 - 8$ in $[-1, 2]$

58. $g(x) = x^3 - x^2 - 11x - 6$ in $[-3, 3]$

59. $h(x) = x^4 - 5x^3 + 7x^2 - 12x - 96$ in $[-3, 6]$

60. $f(x) = x^5 - 3x^4 - 5x^2 + 18x + 3$ in $[-1, 4]$

By determining the y-intercept, all x-intercepts, and using a test value in each interval determined by the x-intercepts, graph each polynomial function.

61. $f(x) = 2x^3 + x^2 + 2x + 1$

62. $g(x) = x^3 + 2x^2 - 19x - 20$

63. $h(x) = x^4 - x^3 - 6x^2 + 4x + 8$

64. $h(x) = 4x^5 + 12x^4 - x - 3$

[10.5] Find all points of discontinuity and determine whether there is a hole or a vertical asymptote at each point.

65. $f(x) = \dfrac{4x}{x^2 - 25}$

66. $g(x) = \dfrac{x + 5}{x^2 - 25}$

67. $h(x) = \dfrac{x^2 - 16}{x + 4}$

68. $f(x) = \dfrac{x + 4}{x^2 + 16}$

Determine if the graph of the rational function has a horizontal asymptote, an oblique asymptote, or neither. Give the equation of the horizontal or oblique asymptote if it exists.

69. $f(x) = \dfrac{16x^5 + 5}{4x^7 - 12}$

70. $g(x) = \dfrac{24x^2 + x - 10}{8x^2 - 12x + 3}$

71. $h(x) = \dfrac{6x^2 - 5x + 7}{2x + 3}$

72. $f(x) = \dfrac{x^2 + 9x + 14}{x + 2}$

73. $g(x) = \dfrac{8x^2 + 6x - 25}{4x + 9}$

74. $h(x) = \dfrac{18x^3 - 12x^2 + 7x}{12x^3 - 9x^2 + 4x}$

Graph the rational function.

75. $f(x) = \dfrac{2}{x + 2}$

76. $g(x) = \dfrac{7x - 5}{2x + 1}$

77. $h(x) = \dfrac{x^2 + x - 20}{x - 4}$

78. $f(x) = \dfrac{x - 3}{x^2 + x - 2}$

79. $g(x) = \dfrac{x^2 + 5x - 12}{x - 3}$

80. $h(x) = \dfrac{2x^2 - 9x + 15}{x - 6}$

Chapter Practice Test

1. Describe the location, either above the x-axis or below the x-axis, of the graph of
$f(x) = -4x^6 + 3x^5 - 3x^4 - 5x^3 + 6x^2 - 7$ for

  a) large positive values of x.

  b) large negative values of x.

2. Find the zeros and their multiplicities for the polynomial function $f(x) = x(x - 4)^3(2x + 3)^2$.

3. Determine the maximum number of turning points for the function $f(x) = 3x^7 - 12x^5 + x^3 - x^2 - 11$.

4. Determine the minimum degree of the polynomial function graphed below.

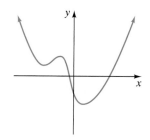

5. Determine the maximum and minimum number of x-intercepts for the graph of the polynomial function $f(x) = 3x^5 + 2x^4 - 20x^3 - 12x^2 + x + 15$.

6. Use Descartes' Rule of Signs to determine

  a) the possible number of positive real zeros for $f(x) = x^5 - 3x^4 - 7x^3 - x^2 + 6$

  b) the possible number of negative real zeros for $f(x) = x^5 - 3x^4 - 7x^3 - x^2 + 6$

7. List the possible rational zeros of
$g(x) = 4x^6 + 5x^4 - 12x^3 - 18x - 6$.

8. Find all the zeros of
$f(x) = x^4 + x^3 - 3x^2 + 9x - 108$

9. Use the Intermediate Value Theorem to determine if the function $f(x) = x^3 - 5x^2 - 4x + 8$ has a real zero in each of the following intervals. If this cannot be determined, indicate so.

  a) $(2, 5)$

  b) $(-2, 0)$

10. Use the Boundedness Theorem to determine if the real zeros of $f(x) = x^4 - 5x^3 - 4x^2 + 6x - 10$ are bounded over $[-2, 5]$.

Find all points of discontinuity and determine whether there is a hole or vertical asymptote at each point.

11. $f(x) = \dfrac{x^2 + 2x - 8}{x + 4}$

12. $g(x) = \dfrac{x + 5}{x - 5}$

13. $h(x) = \dfrac{x^2 + 3x - 10}{2x^2 + 3x}$

Determine if the graph of the rational function has a horizontal asymptote, an oblique asymptote, or neither. Give the equation of the horizontal or oblique asymptote if it exists.

14. $f(x) = \dfrac{15x^2 - 7x + 3}{3x^2 + 4}$

15. $g(x) = \dfrac{x^2 - 6x + 9}{x - 3}$

16. $h(x) = \dfrac{4x^2 + 9x - 7}{x + 3}$

17. $f(x) = \dfrac{4x^3 - 5x^2 + 64}{x^5 - 4x^4 + 16x + 32}$

Graph each function.

18. $f(x) = \begin{cases} 2x - 3 & \text{if } x \le -2 \\ -2x + 3 & \text{if } x > -2 \end{cases}$

19. $g(x) = \left\lceil \dfrac{x - 1}{4} \right\rceil$

20. $h(x) = x^4 + 3x^3 - 4x^2$

21. $f(x) = x^5 + 4x^4 - 3x^3 - 15x^2 + 18x + 27$

22. $g(x) = x^3 - 9x^2 + 6x + 16$

23. $h(x) = \dfrac{3x}{x^2 - 2x - 15}$

24. $f(x) = \dfrac{12x + 9}{4x + 3}$

25. $g(x) = \dfrac{x^2 - 2x - 24}{x - 1}$

Cumulative Review Test

Take the following test and check your answers with those that appear at the end of the test. Review any questions that you answered incorrectly. The section and objective where the material was covered is indicated after the answer.

1. Evaluate $-6^2 + 12 \div 3 - 4 \cdot 6$.

2. Simplify $\dfrac{5.2 \times 10^7}{1.3 \times 10^{-3}}$ and write the answer without exponents.

3. Solve $-\dfrac{1}{3}(9x - 4) = \dfrac{1}{2}(2 - 4x)$.

4. Find the solution set to $|3x - 5| > 7$.

5. Graph $y = \frac{1}{4}x - 3$.

6. Find the equation of the line through $(-4, 3)$ that is perpendicular to the graph of $4x - 6y = 7$.

7. Solve the following system of equations.
$$3x + y = 10$$
$$y = 3x - 2$$

8. Determine the solution to the system of inequalities.
$$y \le -x + 6$$
$$y > \dfrac{2}{3}x - 1$$

9. Multiply $(2x + 7)(2x - 7)$.

10. Divide $(x^4 - 11) \div (x - 3)$.

11. Factor $6x^2 + 9xy + 4xy + 6y^2$.

12. Add $\dfrac{7}{10x} + \dfrac{2}{3x^2}$.

13. Solve $\dfrac{15}{x} + \dfrac{9x - 7}{x + 2} = 9$.

14. **a)** State the domain of $f(x) = \sqrt{x + 3}$.
b) Graph $f(x)$.

15. Simplify $\dfrac{4}{3 + \sqrt{5}}$.

16. Solve $x^2 - 14x = -9$.

17. Evaluate $4\log_3 81$.

18. Solve $\log_4(x + 16) = 3$.

19. Graph $P(x) = 2x^3 - 3x^2 - 59x + 30$.

20. Graph $f(x) = \dfrac{x + 3}{x^2 + 5x + 6}$.

Answers to Cumulative Review Test

1. -56; [Sec. 1.4, obj. 3]   **2.** 40,000,000,000; [Sec. 1.6, obj. 3]   **3.** $\frac{1}{3}$; [Sec. 2.1, obj. 3]

4. $\left\{x \mid x < -\frac{2}{3} \text{ or } x > 4\right\}$; [Sec. 2.6, obj. 4]   **5.** 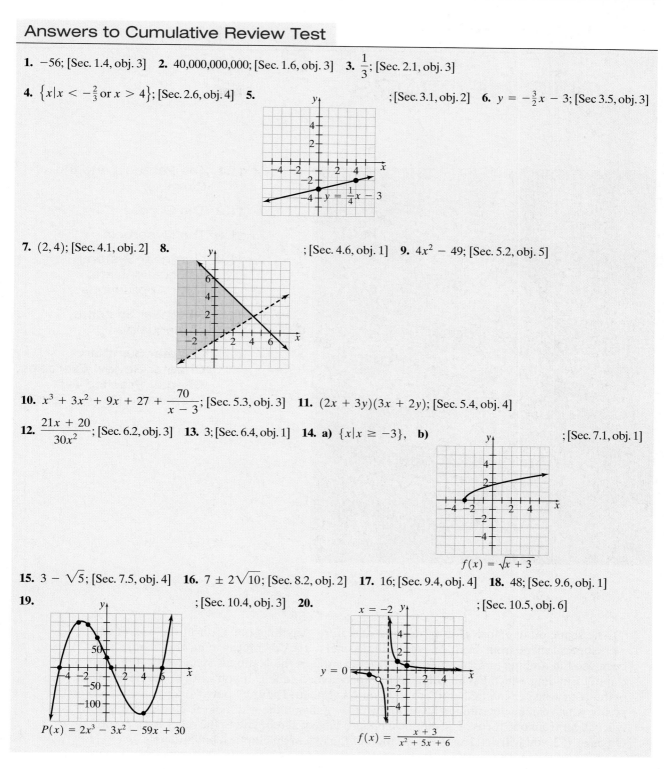 ; [Sec. 3.1, obj. 2]   **6.** $y = -\frac{3}{2}x - 3$; [Sec 3.5, obj. 3]

7. $(2, 4)$; [Sec. 4.1, obj. 2]   **8.** ; [Sec. 4.6, obj. 1]   **9.** $4x^2 - 49$; [Sec. 5.2, obj. 5]

10. $x^3 + 3x^2 + 9x + 27 + \dfrac{70}{x - 3}$; [Sec. 5.3, obj. 3]   **11.** $(2x + 3y)(3x + 2y)$; [Sec. 5.4, obj. 4]

12. $\dfrac{21x + 20}{30x^2}$; [Sec. 6.2, obj. 3]   **13.** 3; [Sec. 6.4, obj. 1]   **14. a)** $\{x \mid x \geq -3\}$,   **b)** ; [Sec. 7.1, obj. 1]

$f(x) = \sqrt{x + 3}$

15. $3 - \sqrt{5}$; [Sec. 7.5, obj. 4]   **16.** $7 \pm 2\sqrt{10}$; [Sec. 8.2, obj. 2]   **17.** 16; [Sec. 9.4, obj. 4]   **18.** 48; [Sec. 9.6, obj. 1]

19. ; [Sec. 10.4, obj. 3]   **20.** ; [Sec. 10.5, obj. 6]

$P(x) = 2x^3 - 3x^2 - 59x + 30$

$x = -2$

$y = 0$

$f(x) = \dfrac{x + 3}{x^2 + 5x + 6}$

Chapter 11

Conic Sections

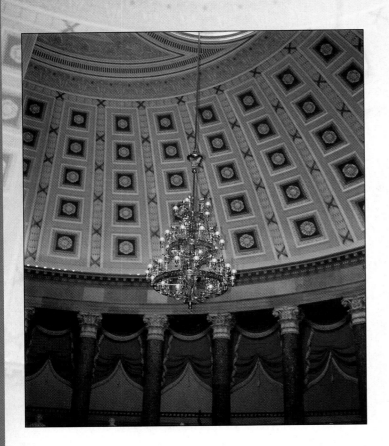

11.1 The Parabola and the Circle

11.2 The Ellipse

11.3 The Hyperbola

11.4 Nonlinear Systems of Equations and Their Applications

11.5 Nonlinear Systems of Inequalities

Chapter Summary
Chapter Review Exercises
Chapter Practice Test
Cumulative Review Test

The shape of an ellipse gives it an unusual feature. Anything bounced off the wall of an elliptical shape from one focal point will ricochet to the other focal point. This feature has been used in architecture and medicine. One example is the National Statuary Hall in the Capitol Building, which has an elliptically shaped domed ceiling. If you whisper at one focal point, your whisper can be heard at the other focal point. The trick only works at the focal points, which can be determined if you know the length and the width of the ellipse. Similarly, a ball hit from one focus of an elliptical billiard table will rebound to the other focal point. On page 752, you will determine the location of the foci of an elliptical billiard table.

SSM   Study Guide   CD/Video   MathPro 4/5   PH Math Tutor Center   prenhall.com/Angel

A Look Ahead   The focus of this chapter is on graphing conic sections. These include the circle, ellipse, parabola, and hyperbola. We have already discussed parabolas. In this chapter we will learn more about them.

We solved linear systems of equations graphically in Section 4.1. In Section 11.4, we solve nonlinear systems of equations. The systems we solve will contain the equation of at least one conic section. We will provide both graphical and algebraic solutions to the systems. In Section 11.5, we solve nonlinear systems of inequalities using the same method we used in Section 4.6 to solve linear systems of inequalities.

Conic sections are important in higher mathematics courses and in certain science and engineering courses. If you take additional mathematics courses, you will probably cover this material again, but in more depth.

11.1 THE PARABOLA AND THE CIRCLE

SSM   Study Guide   CD/Video

MathPro 4/5   PH Math Tutor Center   prenhall.com/Angel

1 Identify and describe the conic sections.
2 Review parabolas.
3 Graph parabolas of the form $x = a(y - k)^2 + h$.
4 Learn the distance and midpoint formulas.
5 Graph circles with centers at the origin.
6 Graph circles with centers at (h, k).

1 Identify and Describe the Conic Sections

In previous chapters, we discussed parabolas. A **parabola** is one type of conic section. Parabolas will be discussed further in this section. Other conic sections are circles, ellipses, and hyperbolas. Each of these shapes is called a conic section because each can be made by slicing a cone and observing the shape of the slice. The methods used to slice the cone to obtain each conic section are illustrated in Figure 11.1.

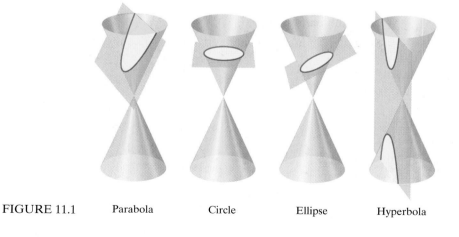

FIGURE 11.1   Parabola   Circle   Ellipse   Hyperbola

2 Review Parabolas

We discussed parabolas in Sections 8.5. Example 1 will refresh your memory on how to graph parabolas in the forms $y = ax^2 + bx + c$ and $y = a(x - h)^2 + k$.

EXAMPLE 1 Consider $y = 2x^2 + 4x - 6$.

a) Write the equation in $y = a(x - h)^2 + k$ form.

b) Determine whether the parabola opens upward or downward.

c) Determine the vertex of the parabola.

d) Determine the y-intercept of the parabola.

e) Determine the x-intercepts of the parabola.

f) Graph the parabola.

Solution **a)** First, factor 2 from the two terms containing the variable to make the coefficient of the squared term 1. (Do not factor 2 from the constant, -6.) Then complete the square.

$$
\begin{aligned}
y &= 2x^2 + 4x - 6 \\
&= 2(x^2 + 2x) - 6 \\
&= 2\left(x^2 + 2x \boxed{+ 1}\right)\boxed{- 2} - 6 \qquad \textit{Complete the square.} \\
&= 2(x + 1)^2 - 8
\end{aligned}
$$

b) The parabola opens upward because $a = 2$, which is greater than 0.

c) The vertex of the graph of an equation in the form $y = a(x - h)^2 + k$ is (h, k). Therefore, the vertex of the graph of $y = 2(x + 1)^2 - 8$ is $(-1, -8)$. The vertex of a parabola can also be found using

$$
\left(-\frac{b}{2a}, \frac{4ac - b^2}{4a}\right) \quad \text{or} \quad \left(-\frac{b}{2a}, f\left(-\frac{b}{2a}\right)\right)
$$

Show that both of these procedures give $(-1, -8)$ as the vertex of the parabola now.

d) To determine the y-intercept, let $x = 0$ and solve for y.

$$
\begin{aligned}
y &= 2(x + 1)^2 - 8 \\
&= 2(0 + 1)^2 - 8 \\
&= -6
\end{aligned}
$$

The y-intercept is $(0, -6)$.

e) To determine the x-intercepts, let $y = 0$ and solve for x.

$$
\begin{aligned}
y &= 2(x + 1)^2 - 8 \\
0 &= 2(x + 1)^2 - 8 &&\textit{Substitute 0 for y.} \\
8 &= 2(x + 1)^2 &&\textit{Add 8 to both sides.} \\
4 &= (x + 1)^2 &&\textit{Divide both sides by 2.} \\
\pm 2 &= x + 1 &&\textit{Square root property} \\
-1 \pm 2 &= x &&\textit{Subtract 1 from both sides.}
\end{aligned}
$$

$$
\begin{array}{lll}
x = -1 - 2 & \text{or} & x = -1 + 2 \\
x = -3 & & x = 1
\end{array}
$$

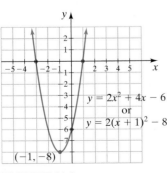

$y = 2x^2 + 4x - 6$
or
$y = 2(x + 1)^2 - 8$

$(-1, -8)$

FIGURE 11.2

NOW TRY EXERCISE 17

The x-intercepts are $(-3, 0)$ and $(1, 0)$. The x-intercepts could also be found by substituting 0 for y in $y = 2x^2 + 4x - 6$ and solving for x using factoring or the quadratic formula. Do this now and show that you get the same x-intercepts.

f) We use the vertex and the x- and y-intercepts to draw the graph, which is shown in Figure 11.2.

3 Graph Parabolas of the Form $x = a(y - k)^2 + h$

Parabolas can also open to the right or left. The graph of an equation of the form $x = a(y - k)^2 + h$ will be a parabola whose vertex is at the point (h, k). If a is a positive number, the parabola will open to the right, and if a is a negative number, the parabola will open to the left. The four different forms of a parabola are shown in Figure 11.3.

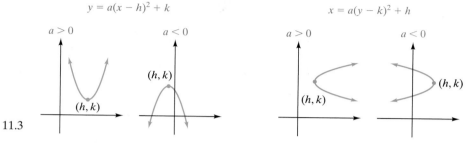

FIGURE 11.3

Parabola with Vertex at (h, k)

1. $y = a(x - h)^2 + k, a > 0$ (opens upward)
2. $y = a(x - h)^2 + k, a < 0$ (opens downward)
3. $x = a(y - k)^2 + h, a > 0$ (opens to the right)
4. $x = a(y - k)^2 + h, a < 0$ (opens to the left)

Note that equations of the form $y = a(x - h)^2 + k$ are functions since their graphs pass the vertical line test. However, equations of the form $x = a(y - k)^2 + h$ are not functions since their graphs do not pass the vertical line test.

EXAMPLE 2 Sketch the graph of $x = -2(y + 4)^2 - 1$.

Solution The graph opens to the left since the equation is of the form $x = a(y - k)^2 + h$ and $a = -2$, which is less than 0. The equation can be expressed as $x = -2[y - (-4)]^2 - 1$. Thus, $h = -1$ and $k = -4$. The vertex of the graph is $(-1, -4)$. See Figure 11.4. If we set $y = 0$, we see that the x-intercept is at $-2(0 + 4)^2 - 1 = -2(16) - 1$ or -33. By substituting values for y you can find the corresponding values of x. When $y = -2, x = -9$, and when $y = -6, x = -9$. These points are marked on the graph. Notice that this graph has no y-intercept.

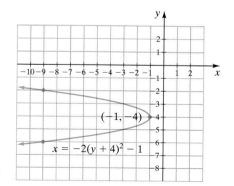

FIGURE 11.4

EXAMPLE 3 **a)** Write the equation $x = 2y^2 + 12y + 13$ in the form $x = a(y - k)^2 + h$.
b) Graph $x = 2y^2 + 12y + 13$.

Solution **a)** First factor 2 from the first two terms. Then, complete the square on the expression within the parentheses.

$$x = 2y^2 + 12y + 13$$
$$= \boxed{2}\,(y^2 + 6y) + 13$$
$$= \boxed{2}\,(y^2 + 6y \boxed{+ 9}) \boxed{- 18} + 13$$
$$= 2(y + 3)^2 - 5$$

b) Since $a > 0$, the parabola opens to the right. Note that when $y = 0$, $x = 2(0)^2 + 12(0) + 13 = 13$. Thus, the x-intercept is $(13, 0)$. The vertex of the parabola is $(-5, -3)$. When $y = -6$, we find that $x = 13$. Thus, another point on the graph is $(13, -6)$. Using the quadratic formula, we can determine that the y-intercepts are about $(0, -4.6)$ and $(0, -1.4)$. The graph is shown in Figure 11.5.

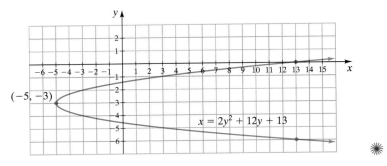

$x = 2y^2 + 12y + 13$

$(-5, -3)$

NOW TRY EXERCISE 45   FIGURE 11.5

4 Learn the Distance and Midpoint Formulas

Now we will derive a formula to find the **distance** between two points on a line. We will use this formula shortly to develop the formula for a circle. Consider Figure 11.6.

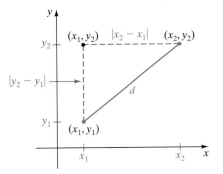

FIGURE 11.6

The horizontal distance between the two points (x_1, y_2) and (x_2, y_2), indicated by the red dashed line, is $|x_2 - x_1|$. We use the absolute value because we want the distance to be positive. If x_1 was larger than x_2, then $x_2 - x_1$ would be negative. The vertical distance between the points (x_1, y_1) and (x_1, y_2), indicated by the green dashed line, is $|y_2 - y_1|$. Using the Pythagorean Theorem where d is the distance between the two points, we get

$$d^2 = |x_2 - x_1|^2 + |y_2 - y_1|^2$$

Since any nonzero number squared is positive, we do not need absolute value signs. We can therefore write

$$d^2 = (x_2 - x_1)^2 + (y_2 - y_1)^2$$

Using the square root property, with the principal square root, we get the distance between the points (x_1, y_1) and (x_2, y_2), which is $d = \sqrt{(x_2 - x_1)^2 + (y_2 - y_1)^2}$.

Distance Formula

The distance, d, between any two points (x_1, y_1) and (x_2, y_2) can be found by the distance formula

$$d = \sqrt{(x_2 - x_1)^2 + (y_2 - y_1)^2}$$

The distance between any two points will always be a positive number. Can you explain why? When finding the distance, it makes no difference which point we designate as point 1, (x_1, y_1), or point 2, (x_2, y_2). Note that the square of any real number will always be greater than or equal to 0. For example, $(5 - 2)^2 = (2 - 5)^2 = 9$.

EXAMPLE 4 Determine the distance between the points $(2, 3)$ and $(-4, 1)$.

Solution As an aid, we plot the points (Fig. 11.7). Label $(2, 3)$ point 1 and $(-4, 1)$ point 2. Thus, (x_2, y_2) represents $(-4, 1)$ and (x_1, y_1) represents $(2, 3)$. Now use the distance formula to find the distance, d.

$$\begin{aligned} d &= \sqrt{(x_2 - x_1)^2 + (y_2 - y_1)^2} \\ &= \sqrt{(-4 - 2)^2 + (1 - 3)^2} \\ &= \sqrt{(-6)^2 + (-2)^2} \\ &= \sqrt{36 + 4} \\ &= \sqrt{40} \quad \text{or} \quad \approx 6.32 \end{aligned}$$

FIGURE 11.7

NOW TRY EXERCISE 57

Thus, the distance between the points $(2, 3)$ and $(-4, 1)$ is $\sqrt{40}$ or about 6.32 units.

AVOIDING COMMON ERRORS

Students will sometimes begin finding the distance correctly using the distance formula but will forget to take the square root of the sum $(x_2 - x_1)^2 + (y_2 - y_1)^2$ to obtain the correct answer. When taking the square root, remember that $\sqrt{a^2 + b^2} \neq a + b$.

It is often necessary to find the **midpoint** of a line segment between two given endpoints. To do this, we use the midpoint formula.

Midpoint Formula

Given any two points (x_1, y_1) and (x_2, y_2), the point halfway between the given points can be found by the midpoint formula:

$$\text{midpoint} = \left(\frac{x_1 + x_2}{2}, \frac{y_1 + y_2}{2} \right)$$

To find the midpoint, we take the average (the mean) of the x-coordinates and of the y-coordinates.

EXAMPLE 5 A line segment through the center of a circle intersects the circle at the points $(-3, 6)$ and $(4, 1)$. Find the center of the circle.

Solution To find the center of the circle, we find the midpoint of the line segment between $(-3, 6)$ and $(4, 1)$. It makes no difference which points we label (x_1, y_1) and (x_2, y_2). We will let $(-3, 6)$ be (x_1, y_1) and $(4, 1)$ be (x_2, y_2). See Figure 11.8.

$$\text{midpoint} = \left(\frac{x_1 + x_2}{2}, \frac{y_1 + y_2}{2} \right)$$

$$= \left(\frac{-3 + 4}{2}, \frac{6 + 1}{2} \right) = \left(\frac{1}{2}, \frac{7}{2} \right)$$

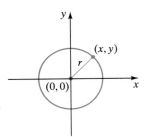

FIGURE 11.8

The point $\left(\frac{1}{2}, \frac{7}{2} \right)$ is halfway between the points $(-3, 6)$ and $(4, 1)$. It is also the center of the circle.

5 Graph Circles with Centers at the Origin

A **circle** may be defined as the set of points in a plane that are the same distance from a fixed point called its **center**.

The formula for the *standard form* of a circle whose center is at the origin may be derived using the distance formula. Let (x, y) be a point on a circle of radius r with center at $(0, 0)$. See Figure 11.9. Using the distance formula, we have

FIGURE 11.9

$$d = \sqrt{(x_2 - x_1)^2 + (y_2 - y_1)^2} \qquad \textit{Distance formula}$$

$$\text{or} \quad r = \sqrt{(x - 0)^2 + (y - 0)^2} \qquad \textit{Substitute r for d, (x, y) for}$$
$$\textit{(x}_2\textit{, y}_2\textit{), and (0, 0) for (x}_1\textit{, y}_1\textit{).}$$

$$r = \sqrt{x^2 + y^2} \qquad \textit{Simplify the radicand.}$$

$$r^2 = x^2 + y^2 \qquad \textit{Square both sides.}$$

Circle with Its Center at the Origin and Radius r

$$x^2 + y^2 = r^2$$

For example, $x^2 + y^2 = 16$ is a circle with its center at the origin and radius 4, and $x^2 + y^2 = 7$ is a circle with its center at the origin and radius $\sqrt{7}$. Note that $4^2 = 16$ and $(\sqrt{7})^2 = 7$.

EXAMPLE 6 Graph the following equations.

a) $x^2 + y^2 = 64$ **b)** $y = \sqrt{64 - x^2}$ **c)** $y = -\sqrt{64 - x^2}$

Solution **a)** If we rewrite the equation as

$$x^2 + y^2 = 8^2$$

we see that the radius of the circle is 8. The graph is illustrated in Figure 11.10 on page 741.

b) If we solve the equation $x^2 + y^2 = 64$ for y, we obtain

$$y^2 = 64 - x^2$$

$$y = \pm\sqrt{64 - x^2}$$

In the equation $y = \pm\sqrt{64 - x^2}$, the equation $y = +\sqrt{64 - x^2}$ or, simply, $y = \sqrt{64 - x^2}$, represents the top half of the circle while the equation $y = -\sqrt{64 - x^2}$ represents the bottom half of the circle. Thus, the graph of $y = \sqrt{64 - x^2}$, where y is the principal square root, lies above and on the x-axis. For any value of x in the domain of the function, the value of y must be greater than or equal to 0. Why? The graph is the semicircle shown in Figure 11.11.

c) The graph of $y = -\sqrt{64 - x^2}$ is also a semicircle. However, this graph lies below and on the x-axis. For any value of x in the domain of the function, the value of y must be less than or equal to 0. Why? The graph is shown in Figure 11.12.

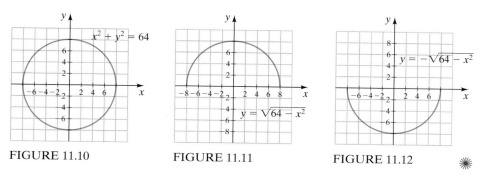

NOW TRY EXERCISE 101   FIGURE 11.10   FIGURE 11.11   FIGURE 11.12

Consider the equations $y = \sqrt{64 - x^2}$ and $y = -\sqrt{64 - x^2}$ in Example **6 b)** and **6 c)**. If you square both sides of the equations and rearrange the terms, you will obtain $x^2 + y^2 = 64$. Try this now and see.

Using Your Graphing Calculator

When using your calculator, you insert the function you wish to graph to the right of $y =$. Circles are not functions since they do not pass the vertical line test. To graph the equation $x^2 + y^2 = 64$, which is a circle of radius 8, we solve the equation for y to obtain $y = \pm\sqrt{64 - x^2}$. We then graph the two functions $Y_1 = \sqrt{64 - x^2}$ and $Y_2 = -\sqrt{64 - x^2}$ on the same axes to obtain the circle. These graphs are illustrated in Figure 11.13. Because of the distortion (described in the Using Your Graphing Calculator box in Section 9.1), the graph does not appear to be a circle. When you use the SQUARE *feature* of the calculator, the figure appears as a circle (see Fig. 11.14).

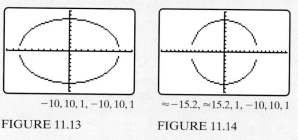

$-10, 10, 1, -10, 10, 1$ $\approx -15.2, \approx 15.2, 1, -10, 10, 1$

FIGURE 11.13   FIGURE 11.14

6   Graph Circles with Centers at (h, k)

The standard form of a circle with center at (h, k) and radius r can be derived using the distance formula. Let (h, k) be the center of the circle and let (x, y) be any point on the circle (see Fig. 11.15). If the radius, r, represents the distance

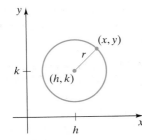

FIGURE 11.15

between a point, (x, y), on the circle and its center, (h, k), then by the distance formula

$$r = \sqrt{(x - h)^2 + (y - k)^2}$$

We now square both sides of the equation to obtain the standard form of a circle with center at (h, k) and radius r.

$$r^2 = (x - h)^2 + (y - k)^2$$

Circle with Its Center at (h, k) and Radius r

$$(x - h)^2 + (y - k)^2 = r^2$$

EXAMPLE 7   Determine the equation of the circle shown in Figure 11.16.

Solution   The center is $(-3, 2)$ and the radius is 3.

$$(x - h)^2 + (y - k)^2 = r^2$$
$$[x - (-3)]^2 + (y - 2)^2 = 3^2$$
$$(x + 3)^2 + (y - 2)^2 = 9$$

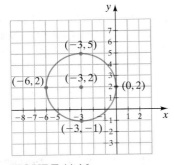

FIGURE 11.16

EXAMPLE 8

a) Show that the graph of the equation $x^2 + y^2 + 6x - 2y + 6 = 0$ is a circle.

b) Determine the center and radius of the circle and then draw the circle.

c) Find the area of the circle.

Solution   **a)** We will write this equation in standard form by completing the square. First we rewrite the equation, placing all the terms containing like variables together.

$$x^2 + 6x + y^2 - 2y + 6 = 0$$

Then we move the constant to the right side of the equation.

$$x^2 + 6x + y^2 - 2y = -6$$

Now we complete the square twice, once for each variable. We will first work with the variable x.

$$x^2 + 6x + 9 + y^2 - 2y = -6 + 9$$

Now we work with the variable y.

$$x^2 + 6x + 9 + y^2 - 2y + 1 = -6 + 9 + 1$$

or

$$\underbrace{x^2 + 6x + 9}_{} + \underbrace{y^2 - 2y + 1}_{} = 4$$
$$(x + 3)^2 + (y - 1)^2 = 4$$
$$(x + 3)^2 + (y - 1)^2 = 2^2$$

b) The center of the circle is at $(-3, 1)$ and the radius is 2. The circle is sketched in Figure 11.17.

c) The area is

$$A = \pi r^2 = \pi(2)^2 = 4\pi \approx 12.6 \text{ square units}$$

$$x^2 + y^2 + 6x - 2y + 6 = 0$$

FIGURE 11.17

NOW TRY EXERCISE 111

Exercise Set 11.1

Concept/Writing Exercises

1. Name the four conic sections. Draw a picture showing how each is formed.

2. Explain how to determine the direction a parabola will open by examining the equation.

3. Will all parabolas of the form $y = a(x - h)^2 + k, a > 0$ be functions? Explain. What will be the domain and range of $y = a(x - h)^2 + k, a > 0$?

4. Will all parabolas of the form $x = a(y - k)^2 + h, a > 0$ be functions? Explain. What will be the domain and range of $x = a(y - k)^2 + h, a > 0$?

5. How will the graphs of $y = 2(x - 3)^2 + 4$ and $y = -2(x - 3)^2 + 4$ compare?

6. Give the distance formula.

7. When the distance between two different points is found by using the distance formula, why must the distance always be a positive number?

8. Give the midpoint formula.

9. What is the definition of a circle?

10. What is the equation of a circle with center at (h, k)?

11. Is $x^2 - y^2 = 25$ an equation for a circle? Explain.

12. Is $-x^2 + y^2 = 36$ an equation for a circle? Explain.

13. Is $2x^2 + 3y^2 = 6$ an equation for a circle? Explain.

14. Is $x = y^2 - 6y + 2$ an equation for a parabola? Explain.

15. Is $x^2 = y^2 - 6y + 2$ an equation for a parabola? Explain.

16. Is $x = -6y + 2$ an equation for a parabola? Explain.

Practice the Skills

Graph each equation.

17. $y = (x - 2)^2 + 3$
18. $y = (x - 2)^2 - 3$
19. $y = (x + 3)^2 + 2$
20. $y = (x + 3)^2 - 2$
21. $y = (x - 2)^2 - 1$
22. $y = (x + 2)^2 + 1$
23. $y = -(x - 1)^2 + 1$
24. $y = -(x + 4)^2 - 5$
25. $y = -(x + 3)^2 + 4$
26. $y = 2(x + 1)^2 - 3$
27. $y = -3(x - 5)^2 + 3$
28. $x = (y - 1)^2 + 1$
29. $x = (y - 4)^2 - 3$
30. $x = -(y - 2)^2 + 1$
31. $x = -(y - 5)^2 + 4$
32. $x = -2(y - 4)^2 + 4$
33. $x = -5(y + 3)^2 - 6$
34. $x = 3(y + 1)^2 + 5$
35. $y = -2\left(x + \dfrac{1}{2}\right)^2 + 6$
36. $y = -\left(x - \dfrac{5}{2}\right)^2 + \dfrac{1}{2}$

In Exercises 37–50, **a)** *Write the equation in the form $y = a(x - h)^2 + k$ or $x = a(y - k)^2 + h$.* **b)** *Graph the equation.*

37. $y = x^2 + 2x$
38. $y = x^2 - 2x$
39. $y = x^2 + 6x$
40. $y = x^2 - 4x$
41. $x = y^2 + 4y$
42. $x = y^2 - 6y$
43. $y = x^2 + 7x + 10$
44. $y = x^2 + 2x - 7$
45. $x = -y^2 + 6y - 9$
46. $x = -y^2 - 5y - 4$
47. $y = -x^2 + 4x - 4$
48. $y = 2x^2 - 4x - 4$
49. $x = -y^2 + 3y - 4$
50. $x = 3y^2 - 12y - 36$

Determine the distance between each pair of points. Use a calculator where appropriate and round your answers to the nearest hundredth.

51. $(3, -1)$ and $(3, -6)$
52. $(-6, 2)$ and $(-2, 2)$
53. $(-4, 3)$ and $(5, 3)$
54. $(1, 2)$ and $(4, 6)$
55. $(-1, -3)$ and $(4, 9)$
56. $(-3, -5)$ and $(3, 3)$
57. $(-4, -5)$ and $(5, -2)$
58. $(1, 7)$ and $(6, 0)$
59. $(3, -1)$ and $\left(\dfrac{1}{2}, 4\right)$
60. $\left(\dfrac{3}{4}, 2\right)$ and $\left(-\dfrac{1}{2}, 6\right)$
61. $(-1.6, 3.5)$ and $(-4.3, -1.7)$
62. $(5.2, -3.6)$ and $(-1.6, 2.3)$
63. $(\sqrt{7}, \sqrt{3})$ and $(0, 0)$
64. $(-\sqrt{5}, -\sqrt{2})$ and $(0, 0)$

Determine the midpoint of the line segment between each pair of points.

65. $(1, 3)$ and $(5, 9)$
66. $(0, 8)$ and $(4, -6)$
67. $(-7, 2)$ and $(7, -2)$
68. $(4, 7)$ and $(1, -3)$
69. $(-1, 4)$ and $(4, 6)$
70. $(-2, -9)$ and $(-6, -3)$
71. $\left(3, \dfrac{1}{2}\right)$ and $(2, -4)$
72. $\left(\dfrac{5}{2}, 3\right)$ and $\left(2, \dfrac{9}{2}\right)$
73. $(\sqrt{3}, 2)$ and $(\sqrt{2}, 7)$
74. $(-\sqrt{7}, 8)$ and $(\sqrt{5}, \sqrt{3})$

Write the equation of each circle with the given center and radius.

75. Center $(0, 0)$, radius 6

77. Center $(2, 0)$, radius 5

79. Center $(0, 5)$, radius 2

81. Center $(3, 4)$, radius 9

83. Center $(2, -6)$, radius 10

85. Center $(1, 2)$, radius $\sqrt{7}$

76. Center $(0, 0)$, radius 7

78. Center $(-3, 0)$, radius 8

80. Center $(0, -6)$, radius 4

82. Center $(-5, 2)$, radius 1

84. Center $(-6, -1)$, radius 7

86. Center $(-7, -2)$, radius $\sqrt{11}$

Write the equation of each circle. Assume the radius is a whole number.

87.

88.

89.

90.

Graph each equation

91. $x^2 + y^2 = 16$

92. $x^2 + y^2 = 5$

93. $x^2 + y^2 = 10$

94. $(x - 1)^2 + y^2 = 5$

95. $(x + 4)^2 + y^2 = 25$

96. $x^2 + (y - 3)^2 = 4$

97. $x^2 + (y + 1)^2 = 9$

98. $(x - 2)^2 + (y + 3)^2 = 16$

99. $(x + 8)^2 + (y + 2)^2 = 9$

100. $(x + 3)^2 + (y - 4)^2 = 36$

101. $y = \sqrt{25 - x^2}$

102. $y = \sqrt{16 - x^2}$

103. $y = -\sqrt{4 - x^2}$

104. $y = -\sqrt{49 - x^2}$

*In Exercises 105–112, **a)** use the method of completing the square to write each equation in standard form. **b)** Draw the graph.*

105. $x^2 + y^2 + 8x + 15 = 0$

106. $x^2 + y^2 + 4y = 0$

107. $x^2 + y^2 + 6x - 4y + 9 = 0$

108. $x^2 + y^2 + 2x - 4y - 4 = 0$

109. $x^2 + y^2 + 6x - 2y + 6 = 0$

110. $x^2 + y^2 + 4x - 6y - 3 = 0$

111. $x^2 + y^2 - 8x + 2y + 13 = 0$

112. $x^2 + y^2 - x + 3y - \dfrac{3}{2} = 0$

Problem Solving

113. Find the area of the circle in Exercise 91.

114. Find the area of the circle in Exercise 95.

In Exercises 115–118, find the x- and y-intercepts, if they exist, of the graph of each equation.

115. $x = y^2 - 6y - 7$

116. $x = -y^2 + 8y - 12$

117. $x = 2(y - 5)^2 + 6$

118. $x = -(y + 2)^2 - 8$

119. If you know the midpoint of a line segment, is it possible to determine the length of the line segment? Explain.

120. If you know one endpoint of a line segment and the length of the line segment, can you determine the other endpoint? Explain.

121. Find the length of the line segment whose midpoint is $(4, -6)$ with one endpoint at $(7, -2)$.

122. Find the length of the line segment whose midpoint is $(-2, 4)$ with one endpoint at $(3, 6)$.

123. Find the equation of a circle with center at $(-5, 2)$ that is tangent to the x-axis (that is, the circle touches the x-axis at only one point).

124. Find the equation of a circle with center at $(-3, -4)$ that is tangent to the y-axis.

In Exercises 125 and 126, find **a)** *the radius of the circle whose diameter is along the line shown,* **b)** *the center of the circle, and* **c)** *the equation of the circle.*

125.

126.

127. ***Points of Intersection*** What is the maximum number and the minimum number of points of intersection possible for the graphs of $y = a(x - h_1)^2 + k_1$ and $x = a(y - k_2)^2 + h_2$? Explain.

128. ***Inscribed Triangle*** Consider the figure below.

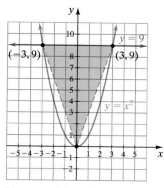

a) Find the area of the triangle outlined in green.
b) When a triangle is inscribed within a parabola, as in the figure, the area within the parabola from the base of the triangle is $\frac{4}{3}$ the area of the triangle. Find the area within the parabola from $x = -3$ to $x = 3$.

129. ***Ferris Wheel*** The Ferris wheel at Navy Pier in Chicago is 150 feet tall. The radius of the wheel itself is 68.2 feet.

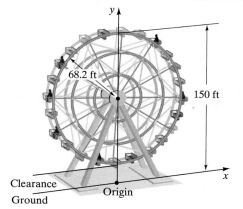

a) What is the clearance below the wheel?
b) How high is the center of the wheel from the ground?
c) Find the equation of the wheel. Assume the origin is on the ground directly below the center of the wheel.

130. ***Shaded Area*** Find the shaded area of the square in the figure. The equation of the circle is $x^2 + y^2 = 9$.

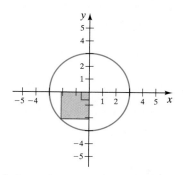

131. ***Shaded Area*** Consider the figure below. Write an equation for
a) the blue circle,
b) the red circle, and
c) the green circle.
d) Find the shaded area.

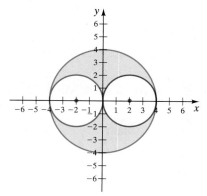

132. *Points of Intersection* Consider the equations $x^2 + y^2 = 16$ and $(x - 2)^2 + (y - 2)^2 = 16$. By considering the center and radius of each circle, determine the number of points of intersection of the two circles.

133. *Concentric Circles* Find the area between the two concentric circles whose equations are $(x - 2)^2 + (y + 4)^2 = 16$ and $(x - 2)^2 + (y + 4)^2 = 64$. *Concentric circles* are circles that have the same center.

134. *Tunnel* A highway department is planning to construct a semicircular one-way tunnel through a mountain. The tunnel is to be large enough so that a truck 8 feet wide and 10 feet tall will pass through the center of the tunnel with 1 foot to spare directly above the corner of the

truck when it is driving down the center of the tunnel (as shown in the figure below.) Determine the minimum radius of the tunnel.

Group Activity

Discuss and answer Exercise 135 as a group.

135. *Equation of a Parabola* The equation of a parabola can be found if three points on the parabola are known. To do so, start with $y = ax^2 + bx + c$. Then substitute the x- and y-coordinates of the first point into the equation. This will result in an equation in a, b, and c. Repeat the procedure for the other two points. This process yields a system of three equations in three variables. Next solve the system for a, b, and c. To find the equation of the parabola, substitute the values found for a, b, and c into the equation $y = ax^2 + bx + c$.

Three points on a parabola are $(0, 12)$, $(3, -3)$, and $(-2, 32)$.

a) Individually, find a system of equations in three variables that can be used to find the equation of the parabola. Then compare your answers. If each member

of the group does not have the same system, determine why.

b) Individually, solve the system and determine the values of a, b, and c. Then compare your answers.

c) Individually, write the equation of the parabola passing through $(0, 12)$, $(3, -3)$, and $(-2, 32)$. Then compare your answers.

d) Individually, write the equation in

$$y = a(x - h)^2 + k$$

form. Then compare your answers.

e) Individually, graph the equation in part **d)**. Then compare your answers.

Cumulative Review Exercises

[1.5] 136. Simplify $\dfrac{4x^{-3}y^4}{12x^{-2}y^3}$.

[2.5] 137. Solve the inequality $-4 < 3x - 4 < 8$. Write the solution in interval notation.

[4.5] 138. Evaluate the determinant.

$$\begin{vmatrix} 4 & 0 & 3 \\ 5 & 2 & -1 \\ 3 & 6 & 4 \end{vmatrix}$$

[5.2] 139. a) Write expressions to represent each of the four areas shown in the figure.

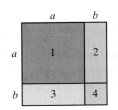

b) Express the total area shown as the square of a binomial.

[8.6] 140. Graph $y = (x - 4)^2 + 1$.

11.2 THE ELLIPSE

1 Graph ellipses.

2 Graph ellipses with centers at (h, k).

SSM   Study Guide   CD/Video

MathPro 4/5   PH Math Tutor Center   prenhall.com/Angel

1   Graph Ellipses

An **ellipse** may be defined as a set of points in a plane, the sum of whose distances from two fixed points is a constant. The two fixed points are called the **foci** (each is a focus) of the ellipse (see Fig. 11.18). In Figure 11.18, F_1 and F_2 represent the two foci.

FIGURE 11.18

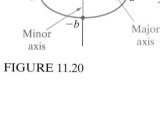

FIGURE 11.19

We can construct an ellipse using a length of string and two thumbtacks. Place the two thumbtacks fairly close together (Fig. 11.19). Then tie the ends of the string to the thumbtacks. With a pencil or pen pull the string taut, and, while keeping the string taut, draw the ellipse by moving the pencil around the thumbtacks.

In Figure 11.20, the line segment from $-a$ to a on the x-axis is the *longer* or **major axis** and the line segment from $-b$ to b is the *shorter* or **minor axis** of the ellipse. The major axis of an ellipse may also be on the y-axis. Figure 11.20 also shows the *center* of the ellipse and two vertices (the red dots). The vertices are the endpoints of the major axis.

The standard form of an ellipse with its center at the origin follows.

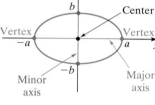

FIGURE 11.20

Ellipse with Its Center at the Origin

$$\frac{x^2}{a^2} + \frac{y^2}{b^2} = 1$$

where $(a, 0)$ and $(-a, 0)$ are the x-intercepts and $(0, b)$ and $(0, -b)$ are the y-intercepts.

Notice the x-intercepts are found using the constant in the denominator of the x^2-term, and the y-intercepts are found using the constant in the denominator of the y^2-term. If $a^2 > b^2$, the major axis of the ellipse will be along the x-axis. If $b^2 > a^2$, the major axis of the ellipse will be along the y-axis.

In Example 1, the major axis of the ellipse is along the x-axis.

FIGURE 11.21

NOW TRY EXERCISE 15

EXAMPLE 1   Graph $\dfrac{x^2}{9} + \dfrac{y^2}{4} = 1$.

Solution   We can rewrite the equation as

$$\frac{x^2}{3^2} + \frac{y^2}{2^2} = 1$$

Thus, $a = 3$ and the x-intercepts are $(3, 0)$ and $(-3, 0)$. Since $b = 2$, the y-intercepts are $(0, 2)$ and $(0, -2)$. The ellipse is illustrated in Figure 11.21.   ✳

An equation may be written so that it may not be obvious that its graph is an ellipse. This is illustrated in Example 2.

EXAMPLE 2 Graph $20x^2 + 9y^2 = 180$.

Solution To make the right side of the equation equal to 1, we divide both sides of the equation by 180. We then obtain an equation that we can recognize as an ellipse.

$$\frac{20x^2 + 9y^2}{180} = \frac{180}{180}$$

$$\frac{20x^2}{180} + \frac{9y^2}{180} = 1$$

$$\frac{x^2}{9} + \frac{y^2}{20} = 1$$

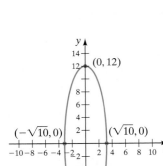

The equation can now be recognized as an ellipse in standard form.

$$\frac{x^2}{a^2} + \frac{y^2}{b^2} = 1$$

Since $a^2 = 9$, $a = 3$. We know that $b^2 = 20$; thus $b = \sqrt{20}$ (or approximately 4.47).

$$\frac{x^2}{3^2} + \frac{y^2}{(\sqrt{20})^2} = 1$$

The *x*-intercepts are $(3, 0)$ and $(-3, 0)$. The *y*-intercepts are $(0, -\sqrt{20})$ and $(0, \sqrt{20})$. The graph is illustrated in Figure 11.22. Note that the major axis lies along the *y*-axis.   ✳

$20x^2 + 9y^2 = 180$

FIGURE 11.22

NOW TRY EXERCISE 19

In Example 1, since $a^2 = 9$ and $b^2 = 4$ and $a^2 > b^2$, the major axis is along the *x*-axis. In Example 2, since $a^2 = 9$ and $b^2 = 20$ and $b^2 > a^2$, the major axis is along the *y*-axis. In the specific case where $a^2 = b^2$, the figure is a circle. Thus, the circle is a special case of an ellipse.

EXAMPLE 3 Write the equation of the ellipse illustrated in Figure 11.23.

Solution The *x*-intercepts are $(-\sqrt{10}, 0)$ and $(\sqrt{10}, 0)$; thus, $a = \sqrt{10}$ and $a^2 = 10$. The *y*-intercepts are $(0, -12)$ and $(0, 12)$; thus, $b = 12$ and $b^2 = 144$.

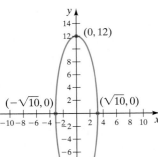

$$\frac{x^2}{a^2} + \frac{y^2}{b^2} = 1$$

$$\frac{x^2}{10} + \frac{y^2}{144} = 1$$ ✳

The formula for the **area of an ellipse** is $A = \pi ab$. In Example 1, where $a = 3$ and $b = 2$, the area is $A = \pi(3)(2) = 6\pi \approx 18.8$ square units.

In Example 2, where $a = 3$ and $b = \sqrt{20}$, the area is $A = \pi(3)(\sqrt{20}) = \pi(3)(2\sqrt{5}) = 6\pi\sqrt{5} \approx 42.1$ square units.

FIGURE 11.23

NOW TRY EXERCISE 41

2 Graph Ellipses with Centers at (*h, k*)

Horizontal and vertical translations, similar to those used in Chapter 8, may be used to obtain the equation of an ellipse with center at (h, k).

Ellipse with Center at (*h, k*)

$$\frac{(x - h)^2}{a^2} + \frac{(y - k)^2}{b^2} = 1$$

In the formula, the *h* shifts the graph left or right from the origin and *k* shifts the graph up or down from the origin, as shown in Figure 11.24.

FIGURE 11.24

EXAMPLE 4 Graph $\dfrac{(x-2)^2}{25} + \dfrac{(y+3)^2}{16} = 1$.

Solution This is the graph of $\dfrac{x^2}{25} + \dfrac{y^2}{16} = 1$ or $\dfrac{x^2}{5^2} + \dfrac{y^2}{4^2} = 1$ translated so that its center is at $(2, -3)$. Note that $a = 5$ and $b = 4$. The graph is shown in Figure 11.25.

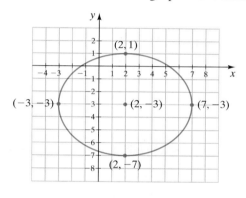

NOW TRY EXERCISE 33   FIGURE 11.25

An understanding of ellipses is useful in many areas. Astronomers know that planets revolve in elliptical orbits around the Sun. Communications satellites move in elliptical orbits around Earth (see Fig. 11.26).

Ellipses are used in medicine to smash kidney stones. When a signal emerges from one focus of an ellipse, the signal is reflected to the other focus. In kidney stone machines, the person is situated so that the stone to be smashed is at one focus of an elliptically shaped chamber called a lithotripter (see Fig. 11.27 and Exercises 57 and 58).

In certain buildings with ellipsoidal ceilings, a person standing at one focus can whisper something and a person standing at the other focus can clearly hear what the person whispered. There are many other uses for ellipses, including lamps that are made to concentrate light at a specific point.

FIGURE 11.26   FIGURE 11.27

Using Your Graphing Calculator

Ellipses are not functions. To graph ellipses on a graphing calculator, we solve the equation for y. This will give the two equations that we use to graph the ellipse.

In Example 1, we graphed $\dfrac{x^2}{9} + \dfrac{y^2}{4} = 1$. Solving this equation for y, we get

$$\frac{x^2}{9} + \frac{y^2}{4} = 1$$

$$36 \cdot \frac{x^2}{9} + 36 \cdot \frac{y^2}{4} = 1 \cdot 36 \qquad \textit{Multiply by the LCD.}$$

$$4x^2 + 9y^2 = 36$$

$$9y^2 = 36 - 4x^2$$

$$y^2 = \frac{36 - 4x^2}{9}$$

$$y^2 = \frac{4(9 - x^2)}{9} \qquad \textit{Factor 4 from the numerator.}$$

$$y = \pm\frac{2}{3}\sqrt{9 - x^2} \qquad \textit{Square root property}$$

To graph the ellipse, we let $Y_1 = \frac{2}{3}\sqrt{9 - x^2}$ and $Y_2 = -\frac{2}{3}\sqrt{9 - x^2}$ and graph both equations. The graphs of Y_1 and Y_2 are illustrated in Figure 11.28.

FIGURE 11.28

Mathematics in Action

Ellipses in the Gym

In the 1990s, a new breed of exercise device started appearing in health clubs and on television infomercials—the elliptical trainer. It is an exercise machine on which the leg motion of the upright user involves elliptical-shaped strides. Normally when you run or walk, the path of each foot through the air traces the top half of an ellipse, but the motion is stopped when your foot hits the ground or treadmill belt. An elliptical trainer allows your feet to complete their elliptical path without impact. The most popular elliptical trainer models feature a no-impact, forward or reverse stride with adjustable incline and resistance that yield impressive cross-training benefits. The device is widely recommended as part of the exercise routine in preparation for running a marathon.

A study showed that in the year 2000, 6.2 million Americans exercised with an elliptical motion trainer.

The non-impact feature of the machine allows for a less fatiguing workout compared to other popular aerobic equipment, while providing comparable cardiovascular benefits. A cross between a Nordic ski trainer, stair climber, and treadmill, the device is especially friendly to older fitness enthusiasts—particularly those with knee problems. Elliptical trainers have also been warmly welcomed by physical and rehabilitative therapists as a way to provide carefully calibrated natural motion using all of the major muscles in the legs.

Exercise Set 11.2

Concept/Writing Exercises

1. What is the definition of an ellipse?
2. What is the equation of an ellipse with its center at the origin?
3. What is the equation of an ellipse with its center at (h, k)?
4. In your own words, discuss the graphs of $\dfrac{x^2}{a^2} + \dfrac{y^2}{b^2} = 1$ when $a > b$, $a < b$, and $a = b$.
5. Explain why the circle is a special case of the ellipse.
6. In the formula $\dfrac{x^2}{a^2} + \dfrac{y^2}{b^2} = 1$, what do the a and b represent?
7. What is the first step in graphing the ellipse whose equation is $10x^2 + 36y^2 = 360$?
8. What is the first step in graphing the ellipse whose equation is $25x^2 + 9y^2 = 225$?
9. Is $\dfrac{x^2}{16} - \dfrac{y^2}{49} = 1$ an equation for an ellipse? Explain.
10. Is $-\dfrac{x^2}{36} + \dfrac{y^2}{81} = 1$ an equation for an ellipse? Explain.

Practice the Skills

Graph each equation.

11. $\dfrac{x^2}{4} + \dfrac{y^2}{1} = 1$
12. $\dfrac{x^2}{1} + \dfrac{y^2}{4} = 1$
13. $\dfrac{x^2}{4} + \dfrac{y^2}{9} = 1$
14. $\dfrac{x^2}{9} + \dfrac{y^2}{4} = 1$
15. $\dfrac{x^2}{25} + \dfrac{y^2}{9} = 1$
16. $\dfrac{x^2}{81} + \dfrac{y^2}{16} = 1$
17. $\dfrac{x^2}{16} + \dfrac{y^2}{25} = 1$
18. $\dfrac{x^2}{100} + \dfrac{y^2}{49} = 1$
19. $x^2 + 36y^2 = 36$
20. $x^2 + 16y^2 = 16$
21. $9x^2 + 25y^2 = 225$
22. $49x^2 + y^2 = 49$
23. $9x^2 + 4y^2 = 36$
24. $25x^2 + 4y^2 = 100$
25. $100x^2 + 25y^2 = 400$
26. $25x^2 + 100y^2 = 400$
27. $x^2 + 2y^2 = 8$
28. $x^2 + 36y^2 = 36$
29. $\dfrac{x^2}{16} + \dfrac{(y-2)^2}{9} = 1$
30. $\dfrac{(x-1)^2}{25} + \dfrac{y^2}{1} = 1$
31. $\dfrac{(x-4)^2}{9} + \dfrac{(y+3)^2}{25} = 1$
32. $\dfrac{(x-3)^2}{16} + \dfrac{(y+2)^2}{49} = 1$
33. $\dfrac{(x+1)^2}{9} + \dfrac{(y-2)^2}{4} = 1$
34. $\dfrac{(x-3)^2}{16} + \dfrac{(y-4)^2}{25} = 1$
35. $(x+3)^2 + 9(y+1)^2 = 81$
36. $18(x-1)^2 + 2(y+3)^2 = 72$
37. $4(x-2)^2 + 9(y+2)^2 = 36$
38. $(x-5)^2 + 4(y+4)^2 = 4$
39. $12(x+4)^2 + 3(y-1)^2 = 48$
40. $16(x-2)^2 + 4(y+3)^2 = 16$

Problem Solving

41. Find the area of the ellipse in Exercise 11.
42. Find the area of the ellipse in Exercise 15.
43. How many points are on the graph of $9x^2 + 36y^2 = 0$? Explain.
44. Consider the graph of the equation $\dfrac{x^2}{a^2} + \dfrac{y^2}{b^2} = 1$. Explain what will happen to the shape of the graph as the value of b gets closer to the value of a. What is the shape of the graph when $a = b$?

Find the equation of the ellipse that has the four points as endpoints of the major and minor axes.

45. $(3, 0), (-3, 0), (0, 5), (0, -5)$
46. $(7, 0), (-7, 0), (0, 4), (0, -4)$
47. $(2, 0), (-2, 0), (0, 3), (0, -3)$
48. $(1, 0), (-1, 0), (0, 6), (0, -6)$
49. How many points of intersection will the graphs of the equations $x^2 + y^2 = 16$ and $\dfrac{x^2}{4} + \dfrac{y^2}{9} = 1$ have? Explain.
50. How many points of intersection will the graphs of the equations $y = 2(x-2)^2 - 3$ and $\dfrac{(x-2)^2}{4} + \dfrac{(y+3)^2}{9} = 1$ have? Explain.

In Exercises 53 and 54, write the following equation in standard form. Determine the center of each ellipse.

51. $x^2 + 4y^2 - 4x - 8y - 92 = 0$

52. $x^2 + 4y^2 + 6x + 16y - 11 = 0$

53. *Art Gallery* An art gallery has an elliptical hall. The maximum distance from one focus to the wall is 90.2 feet and the minimum distance is 20.7 feet. Find the distance between the foci.

54. *Communications Satellite* A space shuttle transported a communications satellite to space. The satellite travels in an elliptical orbit around Earth. The maximum distance of the satellite from Earth is 23,200 miles and the minimum distance is 22,800 miles. Earth is at one focus of the ellipse. Find the distance from Earth to the other focus.

55. *Tunnel through a Mountain* The tunnel in the photo is the top half of an ellipse. The width of the tunnel is 20 feet and the height is 24 feet.

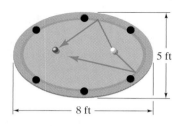

a) If you pictured a completed ellipse with the center of the ellipse being at the center of the road, determine the equation of the ellipse.

b) Determine the area of the ellipse found in part a).

c) Determine the area of the opening of the tunnel.

56. *Billiard Table* An elliptical billiard table is 8 feet long by 5 feet wide. Determine the location of the foci. On such a table, if a ball is put at each focus and one ball is hit with enough force, it would hit the ball at the other focus no matter where it banks on the table.

57. *Lithotripter Machine* Suppose the lithotripter machine described on page 749 is 6 feet long and 4 feet wide. Describe the location of the foci.

58. *Lithotripter* On page 749 we gave a brief introduction to the lithotripter, which uses ultrasound waves to shatter kidney stones. Do research and write a detailed report describing the procedure used to shatter kidney stones. Make sure that you explain how the waves are directed on the stone.

59. *Whispering Gallery* The National Statuary Hall in the Capitol Building in Washington, D.C., is a "whispering gallery." Do research and explain why one person standing at a certain point can whisper something and someone standing a considerable distance away can hear it.

60. Check your answer to Exercise 11 on your grapher.

61. Check your answer to Exercise 17 on your grapher.

Challenge Problems

Determine the equation of the ellipse that has the following four points as endpoints of the major and minor axes.

62. $(-4, 3), (2, 3), (-1, 5), (-1, 1)$

63. $(2, 2), (6, 2), (4, 5), (4, -1)$

Group Activity

Work Exercise 64 individually. Then compare your answers.

64. *Tunnel* The photo shows an elliptical tunnel (with the bottom part of the ellipse not shown) near Rockefeller Center in New York City. The maximum width of the tunnel is 18 feet and the maximum height *from the ground to the top* is 11.5 feet.

a) If the *completed ellipse* would have a maximum height of 15 feet, how high from the ground is the center of the elliptical tunnel?

b) Consider the following graph, which could be used to represent the tunnel.

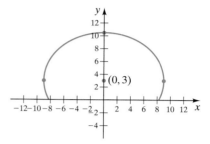

If the ellipse were continued, what would be the other *y*-intercept of the graph?

c) Write the equation of the ellipse, if completed, in part **b)**.

Cumulative Review Exercises

[2.2] **65.** Solve the formula $S = \frac{n}{2}(f + l)$ for l.

[5.4] **66.** Divide $\frac{2x^2 + 2x - 2}{2x - 3}$.

[7.6] **67.** Solve $\sqrt{8b - 15} = 10 - b$.

[8.6] **68.** Solve $\frac{3x + 5}{x - 2} \leq 0$, and give the solution in interval notation.

[9.7] **69.** Find $\log_6 4000$.

11.3 THE HYPERBOLA

SSM   Study Guide   CD/Video

MathPro 4/5   PH Math Tutor Center   prenhall.com/Angel

1   Graph hyperbolas.
2   Review conic sections.

1 Graph Hyperbolas

A **hyperbola** is the set of points in a plane, the difference of whose distances from two fixed points (called foci) is a constant. A hyperbola is illustrated in Figure 11.29a on the next page. In the figure, for every point on the hyperbola, the difference $M - N$, is the same constant. A hyperbola may look like a pair of parabolas. However, the shapes are actually quite different. A hyperbola has two **vertices.** The point halfway between the two vertices is the **center** of the hyperbola. The line through the vertices is called the **transverse axis.** In Figure 11.29b the transverse axis lies along the *x*-axis, and in Figure 11.29c, the transverse axis lies along the *y*-axis.

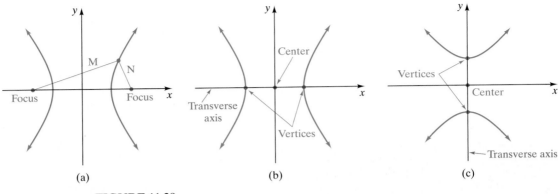

FIGURE 11.29

The dashed lines in Figure 11.30 are called **asymptotes.** The asymptotes are not a part of the hyperbola but are used as an aid in graphing hyperbolas. (We will discuss asymptotes shortly.) Also given in Figure 11.30 is the standard form of the equation for each hyperbola. In Figure 11.30a, both vertices are a units from the origin. In Figure 11.30b, both vertices are b units from the origin. Note that in the standard form of the equation, the denominator of the x^2 term is always a^2 and the denominator of the y^2 term is always b^2.

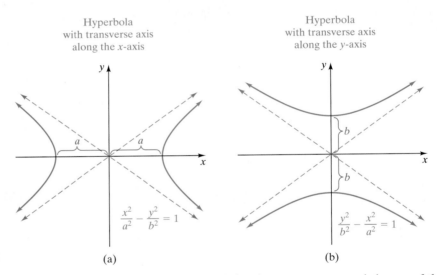

FIGURE 11.30

           (a)   (b)

A hyperbola centered at the origin whose transverse axis is one of the coordinate axes has either x-intercepts (Fig. 11.30a) or y-intercepts (Fig. 11.30b), but not both. When a hyperbola is centered at the origin, the intercepts are the vertices of the hyperbola. When written in standard form, the intercepts will be on the axis indicated by the variable with the positive coefficient. The intercepts will be the positive and the negative square root of the denominator of the positive term.

Examples	Intercepts on	Intercepts
$\dfrac{x^2}{25} - \dfrac{y^2}{9} = 1$	x-axis	$(-5, 0)$ and $(5, 0)$
$\dfrac{y^2}{9} - \dfrac{x^2}{25} = 1$	y-axis	$(0, -3)$ and $(0, 3)$

Asymptotes can help you graph hyperbolas. The asymptotes are two straight lines that go through the center of the hyperbola (see Fig. 11.30). As the values of x and y get larger, the graph of the hyperbola approaches the asymptotes. The equations of the asymptotes of a hyperbola whose center is the origin are

$$y = \frac{b}{a}x \quad \text{and} \quad y = -\frac{b}{a}x$$

The asymptotes can be drawn quickly by plotting the four points (a, b), $(-a, b)$, $(a, -b)$, and $(-a, -b)$, and then connecting these points with dashed lines to form a rectangle. Next, draw dashed lines through the opposite corners of the rectangle to obtain the asymptotes.

Hyperbola with Its Center at the Origin

Transverse Axis Along x-Axis (opens to the right and left)	Transverse Axis Along y-Axis (opens upward and downward)
$\dfrac{x^2}{a^2} - \dfrac{y^2}{b^2} = 1$	$\dfrac{y^2}{b^2} - \dfrac{x^2}{a^2} = 1$

Asymptotes

$$y = \frac{b}{a}x \quad \text{and} \quad y = -\frac{b}{a}x$$

EXAMPLE 1   **a)** Determine the equations of the asymptotes of the hyperbola with equation

$$\frac{x^2}{9} - \frac{y^2}{16} = 1$$

b) Draw the hyperbola using the asymptotes.

Solution   **a)** The value of a^2 is 9; the positive square root of 9 is 3. The value of b^2 is 16; the positive square root of 16 is 4. The asymptotes are

$$y = \frac{b}{a}x \quad \text{and} \quad y = -\frac{b}{a}x$$

or

$$y = \frac{4}{3}x \quad \text{and} \quad y = -\frac{4}{3}x$$

b) To graph the hyperbola, we first graph the asymptotes. To graph the asymptotes, we can plot the points $(3, 4)$, $(-3, 4)$, $(3, -4)$, and $(-3, -4)$ and draw the rectangle as illustrated in Figure 11.31 on the next page. The asymptotes are the dashed lines through the opposite corners of the rectangle.

Since the x-term in the original equation is positive, the graph intersects the x-axis. Since the denominator of the positive term is 9, the vertices are at $(3, 0)$ and $(-3, 0)$. Now draw the hyperbola by letting the hyperbola approach its asymptotes (Fig. 11.32). Note that the asymptotes are drawn using dashed lines since they are not part of the hyperbola. They are used merely to help draw the graph.

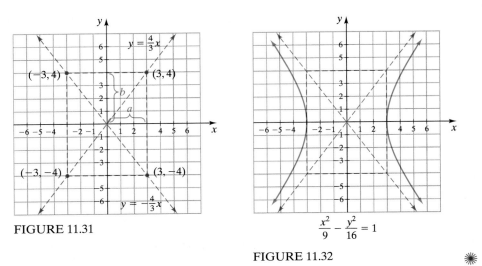

FIGURE 11.31

$$\frac{x^2}{9} - \frac{y^2}{16} = 1$$

NOW TRY EXERCISE 21

FIGURE 11.32

EXAMPLE 2 **a)** Show that the equation $-25x^2 + 4y^2 = 100$ is a hyperbola by expressing the equation in standard form.

b) Determine the equations of the asymptotes of the graph.

c) Draw the graph.

Solution **a)** We divide both sides of the equation by 100 to obtain 1 on the right side of the equation.

$$\frac{-25x^2 + 4y^2}{100} = \frac{100}{100}$$

$$\frac{-25x^2}{100} + \frac{4y^2}{100} = 1$$

$$\frac{-x^2}{4} + \frac{y^2}{25} = 1$$

Rewriting the equation in standard form (positive term first), we get

$$\frac{y^2}{25} - \frac{x^2}{4} = 1$$

b) Since $a = 2$ and $b = 5$, the equations of the asymptotes are

$$y = \frac{5}{2}x \quad \text{and} \quad y = -\frac{5}{2}x$$

c) The graph intersects the y-axis at $(0, 5)$ and $(0, -5)$. Figure 11.33a on the next page illustrates the asymptotes, and Figure 11.33b illustrates the hyperbola.

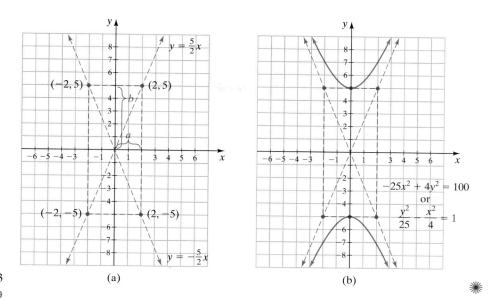

FIGURE 11.33   (a)   (b)
NOW TRY EXERCISE 29

We have discussed hyperbolas with their centers at the origin. Hyperbolas do not have to be centered at the origin. In this book, we will not discuss such hyperbolas.

Using Your Graphing Calculator

We can graph hyperbolas just as we did circles and ellipses. To graph hyperbolas on the graphing calculator, solve the equation for y and graph each part. Consider Example 1,

$$\frac{x^2}{9} - \frac{y^2}{16} = 1.$$

Show that if you solve this equation for y you get $y = \pm \frac{4}{3}\sqrt{x^2 - 9}$. Let $Y_1 = \frac{4}{3}\sqrt{x^2 - 9}$ and $Y_2 = -\frac{4}{3}\sqrt{x^2 - 9}$. Figures 11.34a, 11.34b, 11.34c, and 11.34d give the graphs of Y_1 and Y_2 for different window settings. The window settings used are indicated above each graph.

Standard setting	ZOOM: option 5 ZSquare setting	ZOOM: option 4, ZDecimal setting	Set window as shown below figure (called "friendly window settings")
$-10, 10, 1, -10, 10, 1$	$\approx -15.2, \approx 15.2, 1, -10, 10, 1$	$-4.7, 4.7, 1, -3.1, 3.1, 1$	$-14.1, 14.1, 1, -9.3, 9.3, 1$
(a)	(b)	(c)	(d)

FIGURE 11.34

In part (d), the "friendly window setting," the ratio of the length of the x-axis (28.2 units) to the length of the y-axis (18.6 units) is about 1.516. This is the same ratio as the length to the width of the display window of the calculator on the TI-83 Plus.

2 Review Conic Sections

The following chart summarizes conic sections.

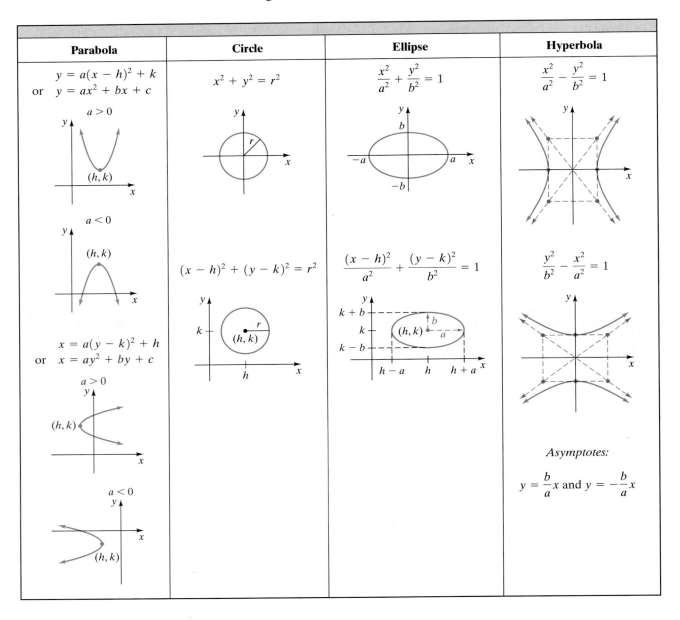

Parabola	Circle	Ellipse	Hyperbola
$y = a(x - h)^2 + k$ or $y = ax^2 + bx + c$	$x^2 + y^2 = r^2$	$\dfrac{x^2}{a^2} + \dfrac{y^2}{b^2} = 1$	$\dfrac{x^2}{a^2} - \dfrac{y^2}{b^2} = 1$
$x = a(y - k)^2 + h$ or $x = ay^2 + by + c$	$(x - h)^2 + (y - k)^2 = r^2$	$\dfrac{(x - h)^2}{a^2} + \dfrac{(y - k)^2}{b^2} = 1$	$\dfrac{y^2}{b^2} - \dfrac{x^2}{a^2} = 1$

Asymptotes:

$$y = \frac{b}{a}x \text{ and } y = -\frac{b}{a}x$$

EXAMPLE 3 Indicate whether each equation represents a parabola, a circle, an ellipse, or a hyperbola.

a) $6x^2 = -6y^2 + 48$ **b)** $x - y^2 = 7y + 3$ **c)** $2x^2 = 8y^2 + 72$

Solution **a)** This equation has an x-squared term and a y-squared term. Let's place all the squared terms on the left side of the equation.

$$6x^2 = -6y^2 + 48$$
$$6x^2 + 6y^2 = 48 \qquad \text{Add } 6y^2 \text{ to both sides.}$$

Since the coefficients of both squared terms are the same number, we divide both sides of the equation by this number. Divide both sides by 6.

$$\frac{6x^2 + 6y^2}{6} = \frac{48}{6}$$

$$x^2 + y^2 = 8$$

This equation is of the form $x^2 + y^2 = r^2$ where $r^2 = 8$.
The equation $6x^2 = -6y^2 + 48$ represents a circle.

b) This equation has a y-squared term, but no x-squared term. Let's solve the equation for x.

$$x - y^2 = 7y + 3$$

$$x = y^2 + 7y + 3 \quad \text{\small Add } y^2 \text{ \small to both sides.}$$

This equation is of the form $x = ay^2 + by + c$ where $a = 1$, $b = 7$, and $c = 3$.
The equation $x - y^2 = 7y + 3$ represents a parabola.

c) This equation has an x-squared term and a y-squared term. Let's place all the squared terms on the left side of the equation.

$$2x^2 = 8y^2 + 72$$

$$2x^2 - 8y^2 = 72 \quad \text{\small Subtract } 8y^2 \text{ \small from both sides.}$$

Since the coefficients of both squared terms are different numbers, we want to divide the equation by the constant on the right side. Divide both sides by 72.

$$\frac{2x^2 - 8y^2}{72} = \frac{72}{72}$$

$$\frac{2x^2}{72} - \frac{8y^2}{72} = 1$$

$$\frac{x^2}{36} - \frac{y^2}{9} = 1$$

This equation is of the form $\dfrac{x^2}{a^2} - \dfrac{y^2}{b^2} = 1$ where $a^2 = 36$ (or $a = 6$) and $b^2 = 9$ (or $b = 3$).

NOW TRY EXERCISE 53   The equation $2x^2 = 8y^2 + 72$ represents a hyperbola.   ✳

Exercise Set 11.3

Concept/Writing Exercises

1. What is the definition of a hyperbola?

2. What are asymptotes? How do you find the equations of the asymptotes of a hyperbola?

3. Discuss the graph of $\dfrac{x^2}{a^2} - \dfrac{y^2}{b^2} = 1$ for nonzero real numbers a and b. Include the transverse axis, vertices, and asymptotes.

4. Discuss the graph of $\dfrac{y^2}{b^2} - \dfrac{x^2}{a^2} = 1$ for nonzero real numbers a and b. Include the transverse axis, vertices, and asymptotes.

5. Is $\dfrac{x^2}{25} + \dfrac{y^2}{64} = 1$ an equation for a hyperbola? Explain.

6. Is $-\dfrac{x^2}{25} - \dfrac{y^2}{64} = 1$ an equation for a hyperbola? Explain.

7. Is $4x^2 - 25y^2 = 100$ an equation for a hyperbola? Explain.

8. Is $36x^2 - 9y^2 = -324$ an equation for a hyperbola? Explain.

9. What is the first step in graphing the hyperbola whose equation is $x^2 - 9y^2 = 81$? Explain.

10. What is the first step in graphing the hyperbola whose equation is $4x^2 - y^2 = -64$? Explain.

Practice the Skills

*a) Determine the equations of the asymptotes for each equation. **b)** Graph the equation.*

11. $\dfrac{x^2}{9} - \dfrac{y^2}{4} = 1$

12. $\dfrac{y^2}{4} - \dfrac{x^2}{9} = 1$

13. $\dfrac{x^2}{4} - \dfrac{y^2}{1} = 1$

14. $\dfrac{y^2}{1} - \dfrac{x^2}{4} = 1$

15. $\dfrac{x^2}{9} - \dfrac{y^2}{25} = 1$

16. $\dfrac{y^2}{25} - \dfrac{x^2}{9} = 1$

17. $\dfrac{x^2}{25} - \dfrac{y^2}{16} = 1$

18. $\dfrac{y^2}{16} - \dfrac{x^2}{25} = 1$

19. $\dfrac{y^2}{9} - \dfrac{x^2}{16} = 1$

20. $\dfrac{x^2}{16} - \dfrac{y^2}{9} = 1$

21. $\dfrac{y^2}{25} - \dfrac{x^2}{36} = 1$

22. $\dfrac{x^2}{36} - \dfrac{y^2}{25} = 1$

23. $\dfrac{y^2}{25} - \dfrac{x^2}{4} = 1$

24. $\dfrac{x^2}{4} - \dfrac{y^2}{25} = 1$

25. $\dfrac{y^2}{16} - \dfrac{x^2}{81} = 1$

26. $\dfrac{x^2}{81} - \dfrac{y^2}{16} = 1$

*In Exercises 27–36, **a)** write each equation in standard form and determine the equations of the asymptotes. **b)** Draw the graph.*

27. $x^2 - 25y^2 = 25$

28. $25y^2 - x^2 = 25$

29. $16x^2 - 4y^2 = 64$

30. $4y^2 - 16x^2 = 64$

31. $9y^2 - x^2 = 9$

32. $9y^2 - 25x^2 = 225$

33. $25x^2 - 9y^2 = 225$

34. $x^2 - 9y^2 = 9$

35. $4y^2 - 36x^2 = 144$

36. $64y^2 - 25x^2 = 1600$

In Exercises 37–60, indicate whether the equation represents a parabola, a circle, an ellipse, or a hyperbola. See Example 3.

37. $5x^2 + 5y^2 = 125$

38. $25x^2 - 5y^2 = 125$

39. $x^2 + 5y^2 = 125$

40. $x = 5y^2 + 5y + 1$

41. $4x^2 - 4y^2 = 29$

42. $10x^2 + 10y^2 = 90$

43. $2y = 12x^2 - 8x + 20$

44. $4y^2 - 6x^2 = 48$

45. $6x^2 + 7y^2 = 42$

46. $9.2x^2 + 9.2y^2 = 46$

47. $3x = -2y^2 + 9y - 15$

48. $8x^2 - 3y^2 = 48$

49. $6x^2 + 6y^2 = 36$

50. $9x^2 = -9y^2 + 54$

51. $5y^2 = 7x^2 + 35$

52. $9x^2 = -18y^2 + 36$

53. $x + y = 2y^2 + 5$

54. $2x^2 = -2y^2 + 72$

55. $3x^2 = 12y^2 + 48$

56. $-8x^2 = -9y^2 - 72$

57. $y - x + 2 = x^2$

58. $19x^2 = -2y^2 + 38$

59. $-3x^2 - 3y^2 = -243$

60. $x - y^2 = 9x$

Problem Solving

61. Find the equation of the hyperbola whose vertices are $(0, 2)$ and $(0, -2)$ and whose asymptotes are $y = \dfrac{1}{2}x$ and $y = -\dfrac{1}{2}x$.

62. Find an equation of a hyperbola whose vertices are $(0, 6)$ and $(0, -6)$ and whose asymptotes are $y = \dfrac{3}{2}x$ and $y = -\dfrac{3}{2}x$.

63. Find the equation of the hyperbola whose vertices are $(-3, 0)$ and $(3, 0)$ and whose asymptotes are $y = 2x$ and $y = -2x$.

64. Find an equation of a hyperbola whose vertices are $(7, 0)$ and $(-7, 0)$ and whose asymptotes are $y = \dfrac{2}{7}x$ and $y = -\dfrac{2}{7}x$.

65. Find an equation of a hyperbola whose transverse axis is along the x-axis and whose equations of the asymptotes are $y = \dfrac{5}{3}x$ and $y = -\dfrac{5}{3}x$. Is this the only possible answer? Explain.

66. Find an equation of a hyperbola whose transverse axis is along the y-axis and whose equations of the asymptotes are $y = \dfrac{2}{3}x$ and $y = -\dfrac{2}{3}x$. Is this the only possible answer? Explain.

67. Are any hyperbolas of the form $\dfrac{x^2}{a^2} - \dfrac{y^2}{b^2} = 1$ functions? Explain.

68. Are any hyperbolas of the form $\dfrac{y^2}{b^2} - \dfrac{x^2}{a^2} = 1$ functions? Explain.

69. Considering the graph of $\dfrac{x^2}{9} - \dfrac{y^2}{4} = 1$, determine the domain and range of the relation.

70. Considering the graph of $\dfrac{y^2}{16} - \dfrac{x^2}{9} = 1$, determine the domain and range of the relation.

71. If the equation $\dfrac{x^2}{a^2} - \dfrac{y^2}{b^2} = 1$, where $a > b$, is graphed, and then the values of a and b are interchanged, and the new equation is graphed, how will the two graphs compare? Explain your answer.

72. If the equation $\dfrac{x^2}{a^2} - \dfrac{y^2}{b^2} = 1$, where $a > b$, is graphed, and then the signs of each term on the left side of the equation are changed, and the new equation is graphed, how will the two graphs compare? Explain your answer.

73. Check your answer to Exercise 15 on your grapher.

74. Check your answer to Exercise 21 on your grapher.

Cumulative Review Exercises

[3.4] **75.** Write the equation, in slope-intercept form, of the line that passes through the points $(-6, 4)$ and $(-2, 2)$.

[3.6] **76.** Let $f(x) = 3x^2 - x + 4$ and $g(x) = 6 - 4x^2$. Find $(f + g)(x)$.

[4.4] **77.** Solve the system of equations.

$$-4x + 9y = 7$$
$$5x + 6y = -3$$

[6.2] **78.** Add $\dfrac{3x}{2x - 3} + \dfrac{3x + 6}{2x^2 + x - 6}$.

[8.3] **79.** Solve the formula $E = \dfrac{1}{2}mv^2$ for v.

[9.6] **80.** Solve the equation $\log(x + 3) = \log 4 - \log x$.

11.4 NONLINEAR SYSTEMS OF EQUATIONS AND THEIR APPLICATIONS

1 Solve nonlinear systems using substitution.
2 Solve nonlinear systems using addition.
3 Solve applications.

SSM   Study Guide   CD/Video

MathPro 4/5   PH Math Tutor Center   prenhall.com/Angel

1 Solve Nonlinear Systems Using Substitution

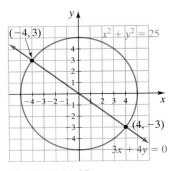

FIGURE 11.35

In Chapter 4, we discussed systems of linear equations. Here we discuss nonlinear systems of equations. A **nonlinear system of equations** is a system of equations in which at least one equation is not linear (that is, one whose graph is not a straight line).

The solution to a system of equations is the point or points that satisfy all equations in the system. Consider the system of equations

$$x^2 + y^2 = 25$$
$$3x + 4y = 0$$

Both equations are graphed on the same axes in Figure 11.35. Note that the graphs appear to intersect at the points $(-4, 3)$ and $(4, -3)$. The check shows that these points satisfy both equations in the system and are therefore solutions to the system.

Check $(-4, 3)$

$$x^2 + y^2 = 25 \qquad\qquad 3x + 4y = 0$$
$$(-4)^2 + 3^2 \overset{?}{=} 25 \qquad\qquad 3(-4) + 4(3) \overset{?}{=} 0$$
$$16 + 9 \overset{?}{=} 25 \qquad\qquad -12 + 12 \overset{?}{=} 0$$
$$25 = 25 \quad \textit{True} \qquad\qquad 0 = 0 \quad \textit{True}$$

Check $(4, -3)$

$$4^2 + (-3)^2 = 25 \qquad\qquad 3(4) + 4(-3) = 0$$
$$16 + 9 \overset{?}{=} 25 \qquad\qquad 12 - 12 \overset{?}{=} 0$$
$$25 = 25 \quad \textit{True} \qquad\qquad 0 = 0 \quad \textit{True}$$

The graphical procedure for solving a system of equations may be inaccurate since we have to estimate the point or points of intersection. An exact answer may be obtained algebraically.

To solve a system of equations algebraically, we often solve one or more of the equations for one of the variables and then use substitution. This procedure is illustrated in Examples 1 and 2.

EXAMPLE 1 Solve the previous system of equations algebraically using the substitution method.

$$x^2 + y^2 = 25$$
$$3x + 4y = 0$$

Solution We first solve the linear equation $3x + 4y = 0$ for either x or y. We will solve for y.

$$3x + 4y = 0$$
$$4y = -3x$$
$$y = -\frac{3x}{4}$$

Now we substitute $-\dfrac{3x}{4}$ for y in the equation $x^2 + y^2 = 25$ and solve for the remaining variable, x.

$$x^2 + y^2 = 25$$
$$x^2 + \left(-\frac{3x}{4}\right)^2 = 25$$
$$x^2 + \frac{9x^2}{16} = 25$$

$$16\left(x^2 + \frac{9x^2}{16}\right) = 16(25)$$

$$16x^2 + 9x^2 = 400$$

$$25x^2 = 400$$

$$x^2 = \frac{400}{25} = 16$$

$$x = \pm\sqrt{16} = \pm 4$$

Next, we find the corresponding value of y for each value of x by substituting each value of x (one at a time) into the equation solved for y.

$$x = 4 \qquad\qquad x = -4$$

$$y = -\frac{3x}{4} \qquad\qquad y = -\frac{3x}{4}$$

$$= -\frac{3(4)}{4} \qquad\qquad = -\frac{3(-4)}{4}$$

$$= -3 \qquad\qquad = 3$$

NOW TRY EXERCISE 9

The solutions are $(4, -3)$ and $(-4, 3)$. This checks with the solution obtained graphically in Figure 11.35.   ✳

Our objective in using substitution is to obtain a single equation containing only one variable.

HELPFUL HINT

STUDY TIP

In this section, we will be using the substitution method and addition method to solve non-linear systems of equations. Both methods were introduced in Chapter 4 to solve linear systems of equations. If you do not remember how to use both methods to solve linear systems of equations, now is a good time to review Chapter 4.

In Examples 1 and 2, we solve systems using the substitution method, while in Examples 3 and 4, we solve systems using the addition method.

You may choose to solve a system by the substitution method if addition of the two equations will not lead to an equation that can be easily solved, as is the case with the systems in Examples 1 and 2.

EXAMPLE 2 Solve the following system of equations using the substitution method.

$$y = x^2 - 3$$
$$x^2 + y^2 = 9$$

Solution Since both equations contain x^2, we will solve one of the equations for x^2. We will choose to solve $y = x^2 - 3$ for x^2.

$$y = x^2 - 3$$
$$y + 3 = x^2$$

Now substitute $y + 3$ for x^2 in the equation $x^2 + y^2 = 9$.

$$x^2 + y^2 = 9$$
$$(y + 3) + y^2 = 9$$
$$y^2 + y + 3 = 9$$
$$y^2 + y - 6 = 0$$

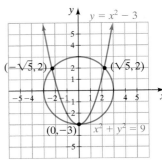

FIGURE 11.36

$$(y + 3)(y - 2) = 0$$
$$y + 3 = 0 \quad \text{or} \quad y - 2 = 0$$
$$y = -3 \qquad y = 2$$

Now find the corresponding values of x by substituting the values found for y.

$$
\begin{array}{ll}
\quad\; y = -3 & \quad\; y = 2 \\
y = x^2 - 3 & y = x^2 - 3 \\
-3 = x^2 - 3 & 2 = x^2 - 3 \\
0 = x^2 & 5 = x^2 \\
0 = x & \pm\sqrt{5} = x
\end{array}
$$

NOW TRY EXERCISE 19 This system has three solutions: $(0, -3)$, $(\sqrt{5}, 2)$, and $(-\sqrt{5}, 2)$. ✳

Note that in Example 2, the graph of the equation $y = x^2 - 3$ is a parabola and the graph of the equation $x^2 + y^2 = 9$ is a circle. The graphs of both equations are illustrated in Figure 11.36.

HELPFUL HINT Students will sometimes solve for one variable and assume that they have the solution. Remember that the solution, if one exists, to a system in two variables consists of one or more ordered pairs.

---

2 Solve Nonlinear Systems Using Addition

We can often solve systems of equations more easily using the addition method that was discussed in Section 4.1. As with the substitution method, our objective is to obtain a single equation containing only one variable.

EXAMPLE 3 Solve the system of equations using the addition method.

$$x^2 + y^2 = 9$$
$$2x^2 - y^2 = -6$$

Solution If we add the two equations, we will obtain one equation containing only one variable.

$$
\begin{array}{rl}
x^2 + y^2 = & 9 \\
\underline{2x^2 - y^2 = -6} & \\
3x^2 \quad\;\; = & 3 \\
x^2 = & 1 \\
x = & \pm 1
\end{array}
$$

Now solve for the variable y by substituting $x = \pm 1$ into *either* of the original equations.

$$
\begin{array}{ll}
\qquad x = 1 & \qquad x = -1 \\
x^2 + y^2 = 9 & x^2 + y^2 = 9 \\
1^2 + y^2 = 9 & (-1)^2 + y^2 = 9 \\
1 + y^2 = 9 & 1 + y^2 = 9 \\
y^2 = 8 & y^2 = 8 \\
y = \pm\sqrt{8} & y = \pm\sqrt{8} \\
\;\; = \pm 2\sqrt{2} & \;\; = \pm 2\sqrt{2}
\end{array}
$$

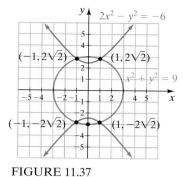

FIGURE 11.37

There are four solutions to this system of equations:

$$(1, 2\sqrt{2}), (1, -2\sqrt{2}), (-1, 2\sqrt{2}), \text{ and } (-1, -2\sqrt{2})$$

The graphs of the equations in the system solved in Example 3 are given in Figure 11.37. Notice the four points of intersection of the two graphs.

It is possible that a system of equations has no real solution (therefore, the graphs do not intersect). Example 4 illustrates such a case.

EXAMPLE 4 Solve the system of equations using the addition method.

$$x^2 + 4y^2 = 16 \quad (eq. 1)$$

$$x^2 + y^2 = 1 \quad (eq. 2)$$

Solution Multiply (eq. 2) by -1 and add the resulting equation to (eq. 1).

$$
\begin{aligned}
x^2 + 4y^2 &= 16 \\
-x^2 - y^2 &= -1 \quad (eq. 2) \text{ multiplied by } -1. \\
\hline
3y^2 &= 15 \\
y^2 &= 5 \\
y &= \pm\sqrt{5}
\end{aligned}
$$

Now solve for x.

$y = \sqrt{5}$	$y = -\sqrt{5}$
$x^2 + y^2 = 1$	$x^2 + y^2 = 1$
$x^2 + (\sqrt{5})^2 = 1$	$x^2 + (-\sqrt{5})^2 = 1$
$x^2 + 5 = 1$	$x^2 + 5 = 1$
$x^2 = -4$	$x^2 = -4$
$x = \pm\sqrt{-4}$	$x = \pm\sqrt{-4}$
$x = \pm 2i$	$x = \pm 2i$

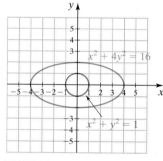

FIGURE 11.38

NOW TRY EXERCISE 29

Since x is an imaginary number for both values of y, this system of equations has no real solution. In solving nonlinear systems of equations, we are interested in finding all real number solutions.

The graphs of the equations in Example 4 are shown in Figure 11.38. Notice that the two graphs do not intersect; therefore, there is no real solution. This agrees with the answer we obtained in Example 4.

3 Solve Applications

Now we will study some applications of nonlinear systems.

EXAMPLE 5 **Flower Garden** Fred and Judy Vespucci want to build a rectangular flower garden behind their house. Fred went to a landscape shop and bought enough topsoil to cover 150 square meters of land. Judy went to the local hardware store and purchased 50 meters of fence for the perimeter of the garden. How should they build the garden to use all the topsoil he bought and all the fence she purchased?

Solution   Understand and Translate   We begin by drawing a sketch (see Fig 11.39).

$$\text{let } x = \text{length of garden}$$
$$y = \text{width of garden}$$

Since $A = xy$ and Fred bought topsoil to cover 150 square meters, we have

$$xy = 150$$

Since $P = 2x + 2y$ and Judy purchased 50 meters of fence for the perimeter of the garden, we have

$$2x + 2y = 50$$

The system of equations is

$$xy = 150$$
$$2x + 2y = 50$$

FIGURE 11.39

Carry Out   We will solve the system using substitution. The equation $2x + 2y = 50$ is a linear equation. We will solve this equation for y. (We could also solve for x.)

$$2x + 2y = 50$$
$$2y = 50 - 2x$$
$$y = \frac{50 - 2x}{2} = \frac{50}{2} - \frac{2x}{2} = 25 - x$$

Now substitute $25 - x$ for y in the equation $xy = 150$.

$$xy = 150$$
$$x(25 - x) = 150$$
$$25x - x^2 = 150$$
$$0 = x^2 - 25x + 150$$
$$0 = (x - 10)(x - 15)$$
$$x - 10 = 0 \quad \text{or} \quad x - 15 = 0$$
$$x = 10 \quad\quad\quad x = 15$$

Answer   If $x = 10$, then $y = 25 - 10 = 15$. And, if $x = 15$, then $y = 25 - 15 = 10$. Thus, in either case, the dimensions of the flower garden are 10 meters by 15 meters.

NOW TRY EXERCISE 41

EXAMPLE 6   **Bicycles**   Hike 'n' Bike Company produces and sells bicycles. Its weekly cost equation is $C = 50x + 400, 0 \le x \le 160$, and its weekly revenue equation is $R = 100x - 0.3x^2, 0 \le x \le 160$, where x is the number of bicycles produced and sold each week. Find the number of bicycles that must be produced and sold for Hike 'n' Bike to break even.

Solution   Understand and Translate   A company breaks even when its cost equals its revenue. When its cost is greater than its revenue, the company has a loss. When its revenue exceeds its cost, the company makes a profit.

The system of equations is

$$C = 50x + 400$$
$$R = 100x - 0.3x^2$$

For Hike 'n' Bike to break even, its cost must equal its revenue. Thus, we write

$$C = R$$
$$50x + 400 = 100x - 0.3x^2$$

Carry Out   Writing this quadratic equation in standard form, we obtain

$$0.3x^2 - 50x + 400 = 0, \quad 0 \le x \le 160$$

We will solve this equation using the quadratic formula.

$$a = 0.3, \quad b = -50, \quad c = 400$$

$$x = \frac{-b \pm \sqrt{b^2 - 4ac}}{2a}$$

$$= \frac{-(-50) \pm \sqrt{(-50)^2 - 4(0.3)(400)}}{2(0.3)}$$

$$= \frac{50 \pm \sqrt{2020}}{0.6}$$

$$x = \frac{50 + \sqrt{2020}}{0.6} \approx 158.2 \quad \text{or} \quad x = \frac{50 - \sqrt{2020}}{0.6} \approx 8.4$$

Profit region
(when revenue exceeds cost)

Cost:
$C = 50x + 400$

Revenue:
$R = 100x - 0.3x^2$

Cost or revenue (dollars)

Number of bicycles

FIGURE 11.40

Answer   The cost will equal the revenue and the company will break even when approximately 8 bicycles are sold. The cost will also equal the revenue when approximately 158 bicycles are sold. The company will make a profit when between 9 and 158 bicycles are sold. When fewer than 9 or more than 158 bicycles are sold, the company will have a loss (see Fig. 11.40).   ✳

Using Your Graphing Calculator

To solve nonlinear systems of equations graphically, graph the equations and find the intersections of the graphs. Consider the system in Example 1, $x^2 + y^2 = 25$ and $3x + 4y = 0$. To graph $x^2 + y^2 = 25$, we use $Y_1 = \sqrt{25 - x^2}$ and $Y_2 = -\sqrt{25 - x^2}$. Therefore, to solve this system we find the intersection of

$$Y_1 = \sqrt{25 - x^2}$$
$$Y_2 = -\sqrt{25 - x^2}$$
$$Y_3 = -\frac{3}{4}x$$

The system is graphed, using the ZOOM: 5 (ZSquared) feature in Figure 11.41a.* In Figure 11.41b, we graph the same three equations using the "friendly numbers" shown below the figure. Using the calculator with either the TRACE and ZOOM features, the TABLE feature, or the INTERSECT feature, you will find that the solutions are $(4, -3)$ and $(-4, 3)$.

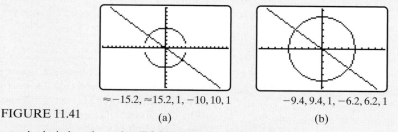

$\approx -15.2, \approx 15.2, 1, -10, 10, 1$ $-9.4, 9.4, 1, -6.2, 6.2, 1$

FIGURE 11.41   (a)   (b)

* Start with the standard window, then select ZOOM: 5 to get this graph.

Exercise Set 11.4

Concept/Writing Exercises

1. What is a nonlinear system of equations?

2. Explain how nonlinear systems of equations may be solved graphically.

3. Can a nonlinear system of equations have exactly one real solution? If so, give an example. Explain.

4. Can a nonlinear system of equations have exactly two real solutions: If so, give an example. Explain.

5. Can a nonlinear system of equations have exactly three real solutions? If so, give an example. Explain.

6. Can a nonlinear system of equations have no real solutions? If so, give an example. Explain.

Practice the Skills

Find all real solutions to each system of equations using the substitution method.

7. $x^2 + y^2 = 8$
   $x + y = 0$

8. $x^2 + y^2 = 8$
   $x - y = 0$

9. $x^2 + y^2 = 9$
   $x + 2y = 3$

10. $x^2 + y^2 = 4$
    $x - 2y = 4$

11. $y = x^2 - 5$
    $3x + 2y = 10$

12. $x^2 + y = 4$
    $y = x^2 + 2$

13. $x + y = 4$
    $x^2 - y^2 = 4$

14. $y - x = 1$
    $x^2 - y^2 = 1$

15. $x + y^2 = 4$
    $x^2 + y^2 = 6$

16. $2x^2 + y^2 = 16$
    $x^2 - y^2 = -4$

17. $x^2 + y^2 = 4$
    $y = x^2 - 12$

18. $x^2 - 4y^2 = 36$
    $x^2 + y^2 = 9$

19. $x^2 + y^2 = 9$
    $y = x^2 - 3$

20. $x^2 + y^2 = 4$
    $y - x = 5$

21. $2x^2 - y^2 = -8$
    $x - y = 6$

22. $x^2 + y^2 = 25$
    $x - 3y = -5$

Find all real solutions to each system of equations using the addition method.

23. $x^2 - y^2 = 4$
    $2x^2 + y^2 = 8$

24. $x^2 + y^2 = 16$
    $x^2 - y^2 = 16$

25. $x^2 + y^2 = 13$
    $2x^2 + 3y^2 = 30$

26. $x^2 + y^2 = 16$
    $2x^2 - 5y^2 = 25$

27. $x^2 + y^2 = 25$
    $x^2 - 2y^2 = 7$

28. $3x^2 - y^2 = 4$
    $x^2 + 4y^2 = 10$

29. $4x^2 + 9y^2 = 36$
    $2x^2 - 9y^2 = 18$

30. $x^2 + 4y^2 = 16$
    $-9x^2 + y^2 = 4$

31. $5x^2 - 2y^2 = -13$
    $3x^2 + 4y^2 = 39$

32. $2x^2 - y^2 = 7$
    $x^2 + 2y^2 = 6$

33. $x^2 - 2y^2 = 7$
    $x^2 + y^2 = 34$

34. $x^2 + y^2 = 25$
    $2x^2 - 3y^2 = -30$

35. $x^2 + y^2 = 9$
    $16x^2 - 4y^2 = 64$

36. $3x^2 + 4y^2 = 35$
    $2x^2 + 5y^2 = 42$

Problem Solving

37. Make up your own nonlinear system of equations whose solution is the empty set. Explain how you know the system has no solution.

38. If a system of equations consists of an ellipse and a hyperbola, what is the maximum number of points of intersection? Make a sketch to illustrate this.

39. *Dance Floor* Myra Majewsky wants to build a dance floor at her gym. The dance floor is to have a perimeter of 90 meters and an area of 500 square meters. Find the dimensions of the dance floor.

40. *Vegetable Garden* Roger Seeman is planning to build a rectangular vegetable garden behind his house. The

garden is to have a perimeter of 66 yards and an area of 270 square yards. Find the dimensions of the vegetable garden.

41. **Rectangular Region** Ellen Dupree fences in a rectangular area along a riverbank as illustrated. If 20 feet of fencing encloses an area of 48 square feet, find the dimensions of the enclosed area.

42. **Rectangular Region** A rectangular area is to be fenced along a river as illustrated in Exercise 41. If 40 feet of fencing encloses an area of 200 square feet, find the dimensions of the enclosed area.

🔒 43. **Currency** A country's currency includes a bill that has an area of 112 square centimeters with a diagonal of $\sqrt{260}$ centimeters. Find the length and width of the bill.

44. **Ice Rink** A rectangular ice rink has an area of 3000 square feet. If the diagonal across the rink is 85 feet, find the dimensions of the rink.

45. **Piece of Wood** Frank Samuelson, a carpenter, has a rectangular piece of wood. When he measures the diagonal it measures 17 inches. When he cuts the wood along the diagonal, the perimeter of each triangle formed is 40 inches. Find the dimensions of the original piece of wood.

46. **Sailboat** A sail on a sailboat is shaped like a right triangle with a perimeter of 36 meters and a hypotenuse of 15 meters. Find the length of the legs of the triangle.

47. **Baseball and Football** Paul Martin throws a football upward from the ground. Its height above the ground at any time, t, is given by the formula $d = -16t^2 + 64t$. At the same time that the football is thrown, Shannon Ryan throws a baseball upward from the top of an 80-foot-tall building. Its height above the ground at any time, t, is given by the formula $d = -16t^2 + 16t + 80$. Find the time at which the two balls will be the same height above the ground. (Neglect air resistance.)

48. **Tennis Ball and Snowball** Robert Snell throws a tennis ball downward from a helicopter flying at a height of 1000 feet. The height of the ball above the ground at any time t is found by the formula $d = -16t^2 - 10t + 1000$. At the instant the ball is thrown from the helicopter, Ramon Sanchez throws a snowball upward from the top of an 800-foot-tall building. The height above the ground of the snowball at any time, t, is found by the formula $d = -16t^2 + 80t + 800$. At what time will the ball and snowball pass each other? (Neglect air resistance.)

49. **Simple Interest** Simple interest is calculated using the simple interest formula, interest = principal · rate · time or $i = prt$. If Seana Hayden invests a certain principal at a specific interest rate for 1 year, the interest she obtains is $7.50. If she increases the principal by $25 and the interest rate is decreased by 1%, the interest remains the same. Find the principal and the interest rate.

50. **Simple Interest** If Claire Brooke invests a certain principal at a specific interest rate for 1 year, the interest she obtains is $72. If she increases the principal by $120 and the interest rate is decreased by 2%, the interest remains the same. Find the principal and the interest rate. Use $i = prt$.

For the given cost and revenue equations, find the break-even point(s).

51. $C = 10x + 300, R = 30x - 0.1x^2$

52. $C = 12.6x + 150, R = 42.8x - 0.3x^2$

53. $C = 80x + 900, R = 120x - 0.2x^2$

54. $C = 0.6x^2 + 9, R = 12x - 0.2x^2$

Solve the following systems using your graphing calculator. Round your answers to the nearest hundredth.

55. $3x - 5y = 12$
$x^2 + y^2 = 10$

56. $y = 2x^2 - x + 2$
$4x^2 + y^2 = 36$

Challenge Problems

57. *Intersecting Roads* The intersection of three roads forms a right triangle, as shown in the figure.

If the hypotenuse is 26 yards and the area is 120 square yards, find the length of the two legs of the triangle.

58. In the figure shown, R represents the radius of the larger orange circle and r represents the radius of the smaller orange circles. If $R = 2r$ and if the shaded area is 122.5π, find r and R.

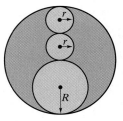

Cumulative Review Exercises

[1.4] **59.** List the order of operations we follow when evaluating an expression.

[5.8] **60.** Factor $(x + 1)^3 + 1$.

[6.6] **61.** x varies inversely as the square of P. If $x = 10$ when P is 6, find x when $P = 20$.

[7.4] **62.** Simplify $\dfrac{2}{\sqrt{x + 2} - 3}$.

[9.7] **63.** Solve $A = A_0 e^{kt}$ for k.

11.5 NONLINEAR SYSTEMS OF INEQUALITIES

1 Graph second-degree inequalities.
2 Solve a system of inequalities graphically.

SSM Study Guide CD/Video

MathPro 4/5 PH Math Tutor Center prenhall.com/Angel

1 Graph Second-Degree Inequalities

In Sections 11.1 through 11.3 we graphed second-degree equations such as $x^2 + y^2 = 25$. Now we will graph second-degree inequalities, such as $x^2 + y^2 \leq 25$. To graph second-degree inequalities, we use a procedure similar to the one we used to graph linear inequalities. That is, graph the equation using a solid line if the inequality is $\leq$ or $\geq$, and a dashed line if the inequality is $<$ or $>$, and then shade in the region that satisfies the inequality. Examples 1 and 2 illustrate this procedure.

EXAMPLE 1 Graph the inequality $\dfrac{x^2}{4} + \dfrac{y^2}{9} \geq 1$.

Solution First, graph the equation $\dfrac{x^2}{4} + \dfrac{y^2}{9} = 1$. Use a solid line when drawing the ellipse since the inequality contains $\geq$ (see Fig. 11.42a). Next select a point not on the graph and determine if this point satisfies the inequality. Often the easiest point to use is the point $(0, 0)$. In this example we will select the point $(0, 0)$.

Check Point $(0, 0)$

$$\frac{x^2}{4} + \frac{y^2}{9} \geq 1$$

$$\frac{0^2}{4} + \frac{0^2}{9} \overset{?}{\geq} 1$$

$$0 + 0 \overset{?}{\geq} 1$$

$$0 \geq 1 \qquad \textit{False}$$

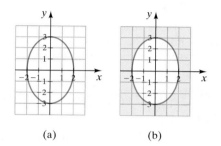

FIGURE 11.42   (a)   (b)

The point $(0, 0)$, which is within the ellipse, is not a solution. Thus all the points outside the ellipse will satisfy the inequality. Shade this outer region (see Fig. 11.42b). The shaded region, and the ellipse itself, form the solution to the inequality. ✳

NOW TRY EXERCISE 15

EXAMPLE 2 Graph the inequality $\dfrac{y^2}{25} - \dfrac{x^2}{9} < 1$.

Solution Graph the equation $\dfrac{y^2}{25} - \dfrac{x^2}{9} = 1$. Use a dashed line when drawing the hyperbola since the inequality contains $<$ (see Fig. 11.43a). Next select a point not on the graph to use as a check point. We will use $(0, 0)$.

Check Point $(0, 0)$

$$\frac{y^2}{25} - \frac{x^2}{9} < 1$$

$$\frac{0^2}{25} - \frac{0^2}{9} \overset{?}{<} 1$$

$$0 - 0 \overset{?}{<} 1$$

$$0 < 1 \qquad \textit{True}$$

Since $(0, 0)$ is a solution, we shade in the region containing the point $(0, 0)$. The solution is indicated in Figure 11.43b.

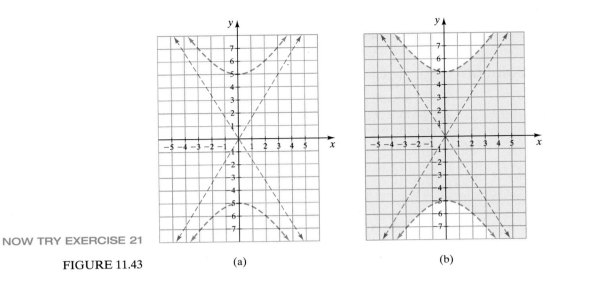

NOW TRY EXERCISE 21

FIGURE 11.43 (a) (b)

2 Solve a System of Inequalities Graphically

Now we will find the graphical solution to a system of inequalities in which at least one inequality is not linear. The solution to a system of inequalities is the set of ordered pairs that satisfy all inequalities in the system.

To Solve a System of Two Inequalities Graphically

1. Graph one inequality.
2. On the same axes, graph the second inequality. Use a different type of shading than was used in the first inequality.
3. The solution is the shaded region from both inequalities.

EXAMPLE 3 Solve the system of inequalities graphically.

$$x^2 + y^2 < 25$$
$$2x + y \geq 4$$

Solution First, graph the inequality, $x^2 + y^2 < 25$. The inner region of the circle satisfies the inequality (see Fig. 11.44). On the same axes, graph the second inequality, $2x + y \geq 4$. The solution to the system is the region containing both types of shading and that part of the straight line within the boundaries of the circle (see Fig. 11.45).

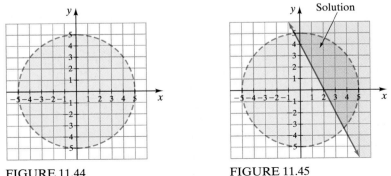

NOW TRY EXERCISE 25 **FIGURE 11.44**   **FIGURE 11.45**

STUDY TIP

In this section we are using a process covered in Section 4.6. If you do not remember this process, review Section 4.6 now.

EXAMPLE 4 Solve the system of inequalities graphically.

$$\frac{x^2}{4} - \frac{y^2}{9} \le 1$$

$$y > (x + 2)^2 - 4$$

Solution Graph each inequality on the same axes (see Fig. 11.46). The solution is the region containing both types of shading and the solid line within the parabola.

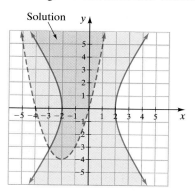

NOW TRY EXERCISE 29 FIGURE 11.46

Exercise Set 11.5

Concept/Writing Exercises

1. When graphing an inequality containing $>$ or $<$, why is the boundary graphed with a dashed line?

2. Can $(0, 0)$ always be used as a check point? Explain.

3. Describe the regions into which the graph of a circle divides the xy-plane.

4. Describe the regions into which the graph of a hyperbola divides the xy-plane.

Practice the Skills

Graph each inequality.

5. $y < (x - 3)^2 + 2$

6. $y > -(x + 4)^2 + 1$

7. $x \ge (y - 5)^2 - 3$

8. $x \le 2(y + 3)^2 + 4$

9. $y < -x^2 + 4x - 4$

10. $x^2 + y^2 \le 16$

11. $x^2 + y^2 > 9$

12. $(x - 1)^2 + (y + 4)^2 < 25$

13. $(x + 2)^2 + (y - 1)^2 \ge 25$

14. $(x - 4)^2 + y^2 \le 1$

15. $\frac{x^2}{1} + \frac{y^2}{4} \le 1$

16. $\frac{(x - 2)^2}{16} + \frac{(y + 1)^2}{25} > 1$

17. $\frac{x^2}{64} + \frac{y^2}{49} < 1$

18. $\frac{(x + 5)^2}{25} + \frac{(y + 3)^2}{9} \ge 1$

19. $16x^2 + 4y^2 \ge 64$

20. $\frac{x^2}{9} - \frac{y^2}{4} \ge 1$

21. $\frac{y^2}{9} - \frac{x^2}{25} > 1$

22. $\frac{y^2}{64} - \frac{x^2}{36} < 1$

23. $\frac{x^2}{4} - \frac{y^2}{36} \le 1$

24. $16y^2 - 25x^2 \le 400$

Determine the solution to each system of inequalities.

25. $x^2 + y^2 \ge 16$
$x + y < 3$

26. $\frac{x^2}{25} - \frac{y^2}{9} < 1$
$2x + 3y > 6$

27. $4x^2 + y^2 > 16$
$y \ge 2x + 2$

28. $x^2 - y^2 > 1$
$\dfrac{x^2}{9} + \dfrac{y^2}{4} \le 1$

29. $x^2 + y^2 \le 36$
$y < (x + 1)^2 - 5$

30. $y \ge x^2 - 2x + 1$
$x^2 + y^2 > 4$

31. $\dfrac{y^2}{16} - \dfrac{x^2}{4} \ge 1$
$\dfrac{x^2}{4} - \dfrac{y^2}{1} < 1$

32. $4x^2 + 9y^2 \le 36$
$x > (y - 2)^2 + 1$

33. $xy \le 6$
$2x - y \le 8$

34. $25x^2 - 4y^2 > 100$
$5x + 3y \le 15$

35. $(x - 3)^2 + (y + 2)^2 \ge 16$
$y \le 4x - 2$

36. $x^2 + (y - 3)^2 \le 25$
$y > (x - 2)^2 + 3$

Problem Solving

37. Is it possible for the solution to a system of inequalities containing 2 linear inequalities to contain

  a) no points?

  b) exactly 1 point?

  c) exactly 2 points?

  d) all points on the xy-plane?

  Explain your answers.

38. Is it possible for the solution to a system of inequalities containing 1 linear inequality and one second-degree inequality to contain

  a) no points?

  b) exactly 1 point?

  c) exactly 2 points?

  d) all points on the xy-plane?

  Explain your answers.

39. Is it possible for the solution to a system of inequalities containing 2 second-degree inequalities to contain

  a) no points?

  b) exactly 1 point?

  c) exactly 2 points?

  d) all points on the xy-plane?

  Explain your answers.

40. Sketch a system of inequalities containing, a parabola and line that has

  a) no solution

  b) 1 solution

41. Sketch a system of inequalities containing a circle and an ellipse that has

  a) no solution

  b) 1 solution

  c) 2 solutions

42. Sketch a system of inequalities containing an ellipse and a hyperbola that has

  a) no solution

  b) 1 solution

  c) 2 solutions

Group Activity

For Exercises 43 and 44, a) each group member should graph all of the inequalities on the same coordinate axes, b) as a group, determine the solution to the system.

43. $y > 4x - 6$
$x^2 + y^2 \ge 36$
$2x + y \le 8$

44. $x^2 + y^2 > 25$
$\dfrac{x^2}{25} + \dfrac{y^2}{9} > 1$
$2x - 3y < 12$

Cumulative Review Exercises

[6.1] **45.** Simplify $\left(\dfrac{4x^{-2}y^3}{2xy^{-4}}\right)^2 \left(\dfrac{3xy^{-1}}{6x^4y^{-3}}\right)^{-2}$

Solve the equation $x^2 - 2x - 4 = 0$:

[8.1] **46.** By completing the square.

[8.2] **47.** Using the quadratic formula

[8.6] **48.** Graph the solution to the inequality $(x + 4)(x - 2)(x - 4) \le 0$ on a number line.

CHAPTER SUMMARY

Key Words and Phrases

11.1
Center of a circle
Circle
Distance
Midpoint
Parabola

11.2
Area of an ellipse
Ellipse
Foci
Major axis
Minor axis

11.3
Asymptotes
Center of a hyperbola
Hyperbola
Transverse axis
Vertices of a hyperbola

11.4
Nonlinear system of
equations

11.5
Second-degree inequality
Nonlinear system of
inequalities

IMPORTANT FACTS

Distance Formula: $d = \sqrt{(x_2 - x_1)^2 + (y_2 - y_1)^2}$

Midpoint Formula: $\left(\dfrac{x_1 + x_2}{2}, \dfrac{y_1 + y_2}{2} \right)$

Circle	Ellipse
$(x - h)^2 + (y - k)^2 = r^2$	$\dfrac{(x - h)^2}{a^2} + \dfrac{(y - k)^2}{b^2} = 1$
center at (h, k)	
radius r	center at (h, k)

Parabola	Hyperbola
$y = a(x - h)^2 + k$	x-axis transverse axis
vertex at (h, k)	$\dfrac{x^2}{a^2} - \dfrac{y^2}{b^2} = 1$
opens upward when $a > 0$	
opens downward when $a < 0$	y-axis transverse axis
	$\dfrac{y^2}{b^2} - \dfrac{x^2}{a^2} = 1$
$x = a(y - k)^2 + h$	asymptotes:
vertex at (h, k)	
opens right when $a > 0$	$y = \dfrac{b}{a}x \quad \text{or} \quad y = -\dfrac{b}{a}x$
opens left when $a < 0$	

Chapter Review Exercises

[11.1] Find the length and the midpoint of the line segment between each pair of points.

1. $(0, 0), (3, -4)$ **2.** $(-4, 1), (-1, 5)$ **3.** $(-6, -2), (-1, 10)$ **4.** $(-4, 3), (-2, 5)$

Graph each equation.

5. $y = (x - 2)^2 + 1$ **6.** $y = (x + 3)^2 - 4$ **7.** $x = (y - 1)^2 + 4$ **8.** $x = -2(y + 4)^2 - 3$

*In Exercises 9–14, **a)** write each equation in the form $y = a(x - h)^2 + k$ or $x = a(y - k)^2 + h$. **b)** Graph the equation.*

9. $y = x^2 - 8x + 22$ **10.** $x = -y^2 - 2y + 5$ **11.** $x = y^2 + 5y + 4$ **12.** $y = 2x^2 - 8x - 24$

*In Exercises 13–18, **a)** write the equation of each circle in standard form. **b)** Draw the graph.*

13. Center $(0, 0)$, radius 4   **14.** Center $(-3, 4)$, radius 1   **15.** $x^2 + y^2 - 4y = 0$
16. $x^2 + y^2 - 2x + 6y + 1 = 0$ **17.** $x^2 - 8x + y^2 - 10y + 40 = 0$ **18.** $x^2 + y^2 - 4x + 10y + 17 = 0$

Graph each equation.

19. $y = \sqrt{9 - x^2}$ **20.** $y = -\sqrt{25 - x^2}$

Determine the equation of each circle.

21.

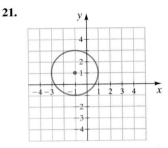

22.

[11.2] *Graph each equation.*

23. $\dfrac{x^2}{4} + \dfrac{y^2}{9} = 1$ **24.** $\dfrac{x^2}{36} + \dfrac{y^2}{64} = 1$ **25.** $4x^2 + 9y^2 = 36$ **26.** $9x^2 + 16y^2 = 144$

27. $\dfrac{(x - 3)^2}{16} + \dfrac{(y + 2)^2}{4} = 1$ **28.** $\dfrac{(x + 3)^2}{9} + \dfrac{y^2}{25} = 1$ **29.** $25(x - 2)^2 + 9(y - 1)^2 = 225$

30. For the ellipse in Exercise 23, find the area.

[11.3] *In Exercises 31–34, **a)** determine the equations of the asymptotes for each equation. **b)** Draw the graph.*

31. $\dfrac{x^2}{4} - \dfrac{y^2}{16} = 1$ **32.** $\dfrac{x^2}{4} - \dfrac{y^2}{4} = 1$ **33.** $\dfrac{y^2}{4} - \dfrac{x^2}{36} = 1$ **34.** $\dfrac{y^2}{25} - \dfrac{x^2}{16} = 1$

*In Exercises 35–38, **a)** write each equation in standard form. **b)** Determine the equations of the asymptotes.*
c) Draw the graph.

35. $x^2 - 9y^2 = 9$ **36.** $25x^2 - 16y^2 = 400$

37. $4y^2 - 25x^2 = 100$ **38.** $49y^2 - 9x^2 = 441$

[11.1–11.3] *Identify the graph of each equation as a circle, ellipse, parabola, or hyperbola.*

39. $\dfrac{x^2}{64} - \dfrac{y^2}{25} = 1$ **40.** $3x^2 + 7y^2 = 21$ **41.** $4x^2 + 4y^2 = 100$ **42.** $4x^2 - 25y^2 = 25$

43. $\dfrac{x^2}{7} + \dfrac{y^2}{9} = 1$ **44.** $y = (x - 3)^2 + 1$ **45.** $4x^2 + 9y^2 = 36$ **46.** $x = -y^2 + 6y - 7$

[11.4] *Find all real solutions to each system of equations using the substitution method.*

47. $x^2 + y^2 = 16$
     $x^2 - y^2 = 16$

48. $x^2 = y^2 + 4$
     $x + y = 4$

49. $x^2 + y^2 = 9$
     $y = 3x + 9$

50. $x^2 + 2y^2 = 7$
     $x^2 - 6y^2 = 29$

Find all real solutions to each system of equations using the addition method.

51. $x^2 + y^2 = 49$
     $x^2 - y^2 = 49$

52. $x^2 + y^2 = 25$
     $x^2 - 2y^2 = -2$

53. $-4x^2 + y^2 = -12$
     $8x^2 + 2y^2 = -8$

54. $-2x^2 - 3y^2 = -6$
     $5x^2 + 4y^2 = 15$

55. *Pool Table* Jerry and Denise Angel have a pool table in their house. It has an area of 36 square feet and a perimeter of 26 feet. Find the dimensions of the pool table.

56. *Bottles of Glue* The Dip and Dap Company has a cost equation of $C = 20.3x + 120$ and a revenue equation of $R = 50.2x - 0.2x^2$, where x is the number of bottles of glue sold. Find the number of bottles of glue the company must sell to break even.

57. *Savings Account* If Kien Kempter invests a certain principal at a specific interest rate for 1 year, the interest is $250. If he increases the principal by $1250 and the interest rate is decreased by 1%, the interest remains the same. Find the principal and interest rate. Use $i = prt$.

58. *Persian Carpet* Tanya Richardson just purchased a rectangular Persian carpet. If the carpet has an area of 300 square feet and the diagonal of the carpet is 25 feet, find the length and width of the carpet.

[11.5] *Graph each inequality.*

59. $y > -x^2 + 6x - 8$

60. $(x + 3)^2 + (y - 2)^2 \geq 16$

61. $\dfrac{(x - 1)^2}{36} + \dfrac{(y + 3)^2}{16} \leq 1$

62. $\dfrac{y^2}{25} - \dfrac{x^2}{49} \geq 1$

Graph each system of nonlinear inequalities.

63. $2x + y \geq 6$
     $x^2 + y^2 < 9$

64. $xy > 5$
     $y < 3x + 4$

65. $4x^2 + 9y^2 \leq 36$
     $x^2 + y^2 > 25$

66. $\dfrac{x^2}{4} - \dfrac{y^2}{9} > 1$
     $y \geq (x - 3)^2 - 4$

Chapter Practice Test

1. Why are parabolas, circles, ellipses, and hyperbolas called conic sections?

2. Determine the length of the line segment whose endpoints are $(-4, 5)$ and $(3, 4)$.

3. Determine the midpoint of the line segment whose endpoints are $(-3, 4)$ and $(7, -1)$.

4. Determine the vertex of the graph of $y = -2(x + 3)^2 + 1$, and then graph the equation.

5. Graph $x = y^2 - 2y + 4$.

6. Write the equation $x = -y^2 - 4y - 5$ in the form $x = a(y - k)^2 + h$, and then draw the graph.

7. Write the equation of a circle with center at $(2, 4)$ and radius 3, and then draw the graph of the circle.

8. Find the area of the circle whose equation is $(x + 5)^2 + (y - 5)^2 = 9$.

9. Write the equation of the circle shown.

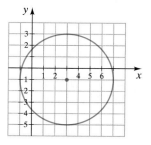

10. Graph $y = -\sqrt{16 - x^2}$.

11. Write the equation $x^2 + y^2 + 2x - 6y + 1 = 0$ in standard form, and then draw the graph.

12. Graph $4x^2 + 25y^2 = 100$.

13. Is the following graph the graph of
$$\frac{(x + 2)^2}{4} + \frac{(y + 1)^2}{16} = 1?$$ Explain your answer.

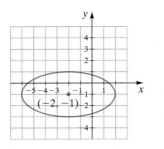

14. Graph $4(x - 4)^2 + 36(y + 2)^2 = 36$.

15. Find the center of the ellipse given by the equation $3(x - 5)^2 + 6(y + 3)^2 = 18$.

16. Explain how to determine whether the transverse axis of a hyperbola lies on the x- or y-axis.

17. What are the equations of the asymptotes of the graph of $\dfrac{x^2}{16} - \dfrac{y^2}{36} = 1$?

18. Graph $\dfrac{y^2}{25} - \dfrac{x^2}{1} = 1$.

19. Graph $\dfrac{x^2}{4} - \dfrac{y^2}{9} = 1$.

In Exercises 20 and 21, determine whether the graph of the equation is a parabola, circle, ellipse, or hyperbola.

20. $4x^2 - 16y^2 = 48$

21. $16x^2 + 4y^2 = 64$

Solve each system of equations.

22. $x^2 + y^2 = 7$
$2x^2 - 3y^2 = -1$

23. $x + y = 8$
$x^2 + y^2 = 9$

24. **Vegetable Garden** Frank Wong has a rectangular vegetable garden on his farm that has an area of 6000 square meters. Find the dimensions of the garden if the perimeter is 320 meters.

25. Graph the system of nonlinear inequalities.
$$\frac{x^2}{9} - \frac{y^2}{25} < 1$$
$$x^2 + y^2 \le 25$$

Cumulative Review Test

Take the following test and check your answers with those that appear at the end of the test. Review any questions that you answered incorrectly. The section and objective where the material was covered are indicated after the answer.

1. Simplify $(7x^2y^5)(-3xy^4)$.

2. Solve $4x - 2(3x - 7) = 2x - 5$.

3. Find the solution set $2(x - 3) + 2x = 4x - 5$.

4. Find the solution set $|3x + 1| > 4$

5. Graph $y = -2x + 2$.

6. If $f(x) = x^2 + 3x + 1$, find $f(10)$.

7. Solve the system of equations.
$$\frac{1}{2}x - \frac{1}{3}y = 2$$
$$\frac{1}{4}x + \frac{2}{3}y = 6$$

8. Factor $x^4 - x^2 - 6$.

9. A large triangular sign has a height that is 6 feet less than its base. If the area of the sign is 56 square feet, find the length of the base and the height of the sign.

10. Multiply $\dfrac{3x^2 - x - 4}{4x^2 + 5x + 1} \cdot \dfrac{2x^2 - 5x - 12}{6x^2 + x - 12}$.

11. Subtract $\dfrac{x}{x + 3} - \dfrac{x + 1}{2x^2 - 2x - 24}$.

12. Solve $\dfrac{3}{x + 3} + \dfrac{5}{x + 4} = \dfrac{12x + 19}{x^2 + 7x + 12}$.

13. Simplify $\left(\dfrac{81x^{1/2}y^3}{9x^{3/2}}\right)^{1/2}$.

14. Simplify $\dfrac{4\sqrt{x}}{\sqrt{x} - y}$.

15. Solve $3\sqrt[3]{2x + 2} = \sqrt[3]{80x - 24}$.

16. Solve $3x^2 - 4x + 5 = 0$ by the quadratic formula.

17. Solve $\log(3x - 4) + \log 4 = \log(x + 6)$.

18. Find the rational zeros of $f(x) = 4x^4 - 8x^3 - 7x^2 + 17x - 6$.

19. Graph $9x^2 + 4y^2 = 36$.

20. Graph $\dfrac{y^2}{25} - \dfrac{x^2}{16} = 1$.

Answers to Cumulative Review Test

1. $-21x^3y^9$; [Sec. 1.5, Obj. 1]   **2.** $\dfrac{19}{4}$; [Sec. 2.1, Obj. 3]   **3.** $\varnothing$; [Sec. 2.1, Obj. 5]   **4.** $\left\{x\,\middle|\,x < -\dfrac{5}{3} \text{ or } x > 1\right\}$; [Sec. 2.6, Obj. 4]

5. ; [Sec. 3.1, Obj. 2]   **6.** 131; [Sec. 3.2, Obj. 4]   **7.** $(8, 6)$; [Sec. 4.1, Obj. 3]

8. $(x^2 + 2)(x^2 - 3)$; [Sec. 5.5, Obj. 5]   **9.** base: 14 ft, height: 8 feet; [Sec. 5.8, Obj. 3]

10. $\dfrac{x - 4}{4x + 1}$; [Sec. 6.1, Obj. 3]   **11.** $\dfrac{2x^2 - 9x - 1}{2(x + 3)(x - 4)}$; [Sec. 6.2, Obj. 3]   **12.** 2; [Sec. 6.4, Obj. 2]

13. $\dfrac{3y^{3/2}}{x^{1/2}}$; [Sec. 7.2, Obj. 3]   **14.** $\dfrac{4x + 4y\sqrt{x}}{x - y^2}$; [Sec. 7.5; Obj. 2]   **15.** 3; [Sec. 7.6, Obj. 2]

16. $\dfrac{2 \pm i\sqrt{11}}{3}$; [Sec. 8.2, Obj. 2]   **17.** 2; [Sec. 9.6, Obj. 1]   **18.** $-\dfrac{3}{2}, \dfrac{1}{2}, 1, 2$; [Sec. 10.3, Obj. 4]

19. $9x^2 + 4y^2 = 36$; [Sec. 11.2, Obj. 1]   **20.** $\dfrac{y^2}{25} - \dfrac{x^2}{16} = 1$; [Sec. 11.3, Obj. 1]

Chapter 12

Sequences, Series, and Probability

12.1 Sequences and Series

12.2 Arithmetic Sequences and Series

12.3 Geometric Sequences and Series

12.4 Mathematical Induction

12.5 The Binomial Theorem

12.6 The Counting Principle, Permutations, and Combinations

12.7 Probability

Chapter Summary
Chapter Review Exercises
Chapter Practice Test
Cumulative Review Test

When a ball bounces, its rebound height is often a percentage of the height from which it was dropped. If a ball rebounds 4 feet when dropped from 6 feet, it has rebounded $66\frac{2}{3}\%$ of its original height. Theoretically, every rebound will have a rebound and the ball will never stop bouncing. Would it be possible to calculate the total distance traveled by a ball that never stops bouncing? In this chapter you will learn how to make that calculation, that is, you will learn how to sum an infinite geometric sequence. On page 806 you estimate the total vertical distance a dropped ball travels.

SSM

Study Guide

CD/Video

MathPro 4/5

PH Math
Tutor Center

prenhall.com/Angel

A Look Ahead **S**equences and series are discussed in this chapter. A sequence is a list of numbers in a specific order and a series is the sum of the numbers in a sequence. In this book we discuss two types of sequences and series: arithmetic and geometric. Series can be used to solve many real-life problems as illustrated in this chapter. If you take higher-level mathematics courses, you may use sequences to approximate some irrational numbers. Other types of sequences not discussed in this book, including the Fibonacci sequence, may be discussed in other mathematics courses.

The formulas we develop while discussing arithmetic and geometric series can be proved using mathematical induction. We discuss mathematical induction in Section 12.4.

In Section 12.5, we introduce the binomial theorem which is used to expand expressions of the form $(a + b)^n$. We learn to calculate binomial coefficients by finding "the number of combinations of n items taken r at a time." In Section 12.6, we develop this concept further when discussing permutations and combinations. We learn how to calculate the number of permutations and number of combinations of a given situation. These numbers are used for determining probabilities. In Section 12.7, we discuss probability and the related topic, odds.

12.1 SEQUENCES AND SERIES

1 Find the terms of a sequence.
2 Write a series.
3 Find partial sums.
4 Use summation notation, Σ.

1 Find the Terms of a Sequence

Many times we see patterns in numbers. For example, suppose you are given a job offer with a starting salary of $30,000. You are given two options for your annual salary increases. One option is an annual salary increase of $2000 per year. The salary you would receive under this option is shown below.

Year	1	2	3	4	...
	↓	↓	↓	↓	
Salary	$30,000	$32,000	$34,000	$36,000	...

Each year the salary is $2000 greater than the previous year. The three dots on the right of the lists of numbers indicate that the list continues in the same manner.

The second option is a 5% salary increase each year. The salary you would receive under this option is shown below.

Year	1	2	3	4	...
	↓	↓	↓	↓	
Salary	$30,000	$31,500	$33,075	$34,728.75	...

With this option, the salary in a given year after year 1 is 5% greater than the previous year's salary.

The two lists of numbers that illustrate the salaries are examples of sequences. A **sequence** of numbers is a list of numbers arranged in a specific order. Consider the list of numbers given below, which is a sequence.

$$5, 10, 15, 20, 25, 30, \ldots$$

The first term is 5. We indicate this by writing $a_1 = 5$. Since the second term is 10, $a_2 = 10$, and so on. The three dots, called an ellipsis, indicate that the sequence continues indefinitely and is an **infinite sequence**.

DEFINITION	An **infinite sequence** is a function whose domain is the set of natural numbers.

Consider the infinite sequence $5, 10, 15, 20, 25, 30, 35, \ldots$

Domain: $\{1, \quad 2, \quad 3, \quad 4, \quad 5, \quad 6, \quad 7, \quad \ldots, \quad n, \quad \ldots\}$

$\downarrow \quad \downarrow \quad \downarrow \quad \downarrow \quad \downarrow \quad \downarrow \quad \downarrow \qquad \downarrow$

Range: $\{5, \quad 10, \quad 15, \quad 20, \quad 25, \quad 30, \quad 35, \quad \ldots, \quad 5n, \quad \ldots\}$

Note that the terms of the sequence $5, 10, 15, 20, \ldots$ are found by multiplying each natural number by 5. For any natural number, n, the corresponding term in the sequence is $5 \cdot n$ or $5n$. The **general term of the sequence**, a_n, which defines the sequence, is $a_n = 5n$.

$$a_n = f(n) = 5n$$

To find the twelfth term of the sequence, substitute 12 for n in the general term of the sequence, $a_{12} = 5 \cdot 12 = 60$. Thus, the twelfth term of the sequence is 60. Note that the terms in the sequence are the function values, or the numbers in the range of the function. When writing the sequence, we do not use set braces. The general form of a sequence is

$$a_1, a_2, a_3, a_4, \ldots, a_n, \ldots$$

For the infinite sequence $2, 4, 8, 16, 32, \ldots, 2^n, \ldots$ we can write

$$a_n = f(n) = 2^n$$

Notice that when $n = 1, a_1 = 2^1 = 2$; when $n = 2, a_2 = 2^2 = 4$; when $n = 3$, $a_3 = 2^3 = 8$; when $n = 4, a_4 = 2^4 = 16$; and so on. What is the seventh term of this sequence? The answer is $a_7 = 2^7 = 128$.

A sequence may also be **finite**.

DEFINITION	A **finite sequence** is a function whose domain includes only the first n natural numbers.

A finite sequence has only a finite number of terms.

Examples of Finite Sequences

$5, 10, 15, 20$ domain is $\{1, 2, 3, 4\}$

$2, 4, 8, 16, 32$ domain is $\{1, 2, 3, 4, 5\}$

EXAMPLE 1   Write the finite sequence defined by $a_n = 2n + 1$, for $n = 1, 2, 3, 4$.

Solution

$$a_n = 2n + 1$$
$$a_1 = 2(1) + 1 = 3$$
$$a_2 = 2(2) + 1 = 5$$

$$a_3 = 2(3) + 1 = 7$$
$$a_4 = 2(4) + 1 = 9$$

Thus, the sequence is $3, 5, 7, 9$.

Since each term of the sequence in Example 1 is larger than the preceding term, it is called an **increasing sequence**.

EXAMPLE 2 Given $a_n = (2n + 3)/n^2$,
a) find the first term in the sequence.
b) find the third term in the sequence.
c) find the fifth term in the sequence.
d) find the tenth term in the sequence.

Solution **a)** When $n = 1$, $a_1 = \dfrac{2(1) + 3}{1^2} = \dfrac{5}{1} = 5$.

b) When $n = 3$, $a_3 = \dfrac{2(3) + 3}{3^2} = \dfrac{9}{9} = 1$.

c) When $n = 5$, $a_5 = \dfrac{2(5) + 3}{5^2} = \dfrac{13}{25} = 0.52$.

d) When $n = 10$, $a_{10} = \dfrac{2(10) + 3}{10^2} = \dfrac{23}{100} = 0.23$.

Note in Example 2 that since there is no restriction on n, a_n is the general term of an infinite sequence.

In Example 2, the first four terms of the sequence are $5, \dfrac{7}{4} = 1.75, 1,$

$\dfrac{11}{16} = 0.6875$. Since each term of the sequence generated by $a_n = (2n + 3)/n^2$ will be smaller than the preceding term, the sequence is called a **decreasing sequence**.

EXAMPLE 3 Find the first four terms of the sequence whose general term is $a_n = (-1)^n(n)$.

Solution
$$a_n = (-1)^n(n)$$
$$a_1 = (-1)^1(1) = -1$$
$$a_2 = (-1)^2(2) = 2$$
$$a_3 = (-1)^3(3) = -3$$
$$a_4 = (-1)^4(4) = 4$$

NOW TRY EXERCISE 25

If we write the sequence in Example 3, we get $-1, 2, -3, 4, \ldots, (-1)^n(n)$. Notice that each term alternates in sign. We call this an **alternating sequence**.

2 Write a Series

A **series** is the expressed sum of the terms of a sequence. A series may be finite or infinite, depending on whether the sequence it is based on is finite or infinite.

Examples

Finite Sequence

a_1, a_2, a_3, a_4, a_5

Finite Series
$$a_1 + a_2 + a_3 + a_4 + a_5$$
Infinite Sequence
$$a_1, a_2, a_3, a_4, a_5, \ldots, a_n, \ldots$$
Infinite Series
$$a_1 + a_2 + a_3 + a_4 + a_5 + \cdots + a_n + \cdots$$

EXAMPLE 4 Write the first eight terms of the sequence; then write the series that represents the sum of that sequence if

a) $a_n = \left(\dfrac{1}{2}\right)^n$ **b)** $a_n = (-2)^n$

Solution **a)** We begin with $n = 1$; thus, the first eight terms of the sequence whose general term is $a_n = \left(\frac{1}{2}\right)^n$ are

$$\left(\frac{1}{2}\right)^1, \left(\frac{1}{2}\right)^2, \left(\frac{1}{2}\right)^3, \left(\frac{1}{2}\right)^4, \left(\frac{1}{2}\right)^5, \left(\frac{1}{2}\right)^6, \left(\frac{1}{2}\right)^7, \left(\frac{1}{2}\right)^8$$

or

$$\frac{1}{2}, \frac{1}{4}, \frac{1}{8}, \frac{1}{16}, \frac{1}{32}, \frac{1}{64}, \frac{1}{128}, \frac{1}{256}$$

The series that represents the sum of the sequence is

$$\frac{1}{2} + \frac{1}{4} + \frac{1}{8} + \frac{1}{16} + \frac{1}{32} + \frac{1}{64} + \frac{1}{128} + \frac{1}{256} = \frac{255}{256}$$

b) We again begin with $n = 1$; thus, the first eight terms of the sequence whose general term is $a_n = (-2)^n$ are

$$(-2)^1, (-2)^2, (-2)^3, (-2)^4, (-2)^5, (-2)^6, (-2)^7, (-2)^8$$

or

$$-2, 4, -8, 16, -32, 64, -128, 256$$

The series that represents the sum of this sequence is

$$-2 + 4 + (-8) + 16 + (-32) + 64 + (-128) + 256 = 170 \qquad \ast$$

3 Find Partial Sums

For an infinite sequence with the terms $a_1, a_2, a_3, \ldots, a_n, \ldots$, a **partial sum** is the sum of a finite number of consecutive terms of the sequence, beginning with the first term.

$$s_1 = a_1 \qquad \text{\textit{First partial sum}}$$
$$s_2 = a_1 + a_2 \qquad \text{\textit{Second partial sum}}$$
$$s_3 = a_1 + a_2 + a_3 \qquad \text{\textit{Third partial sum}}$$
$$\vdots$$
$$s_n = a_1 + a_2 + a_3 + \cdots + a_n \qquad \text{\textit{nth partial sum}}$$

The sum of all the terms of the infinite sequence is called an **infinite series**, and is given by the following

$$s = a_1 + a_2 + a_3 + \cdots + a_n + \cdots$$

EXAMPLE 5 Given the infinite sequence defined by $a_n = (1 + n^2)/n$, find the indicated partial sums.

a) s_1 and **b)** s_4

Solution   **a)** $s_1 = a_1 = \dfrac{1 + 1^2}{1} = \dfrac{1 + 1}{1} = 2$

   b) $s_4 = a_1 + a_2 + a_3 + a_4$

$$= \frac{1 + 1^2}{1} + \frac{1 + 2^2}{2} + \frac{1 + 3^2}{3} + \frac{1 + 4^2}{4}$$

$$= 2 + \frac{5}{2} + \frac{10}{3} + \frac{17}{4}$$

$$= \frac{24}{12} + \frac{30}{12} + \frac{40}{12} + \frac{51}{12}$$

$$= \frac{145}{12} \quad \text{or} \quad 12\frac{1}{12}$$

NOW TRY EXERCISE 39

4 Use Summation Notation, Σ

When the general term of a sequence is known, the Greek letter **sigma**, Σ, can be used to write a series. The sum of the first n terms of the sequence whose nth term is a_n is represented by

$$\sum_{i=1}^{n} a_i = a_1 + a_2 + a_3 + \cdots + a_n$$

where i is called the **index of summation** or simply the **index**, n is the **upper limit of summation**, and 1 is the **lower limit of summation**. In this illustration, we used i for the index; however, any letter can be used for the index.

Consider the sequence $5, 7, 9, 11, 13, \ldots, 2n + 3, \ldots$. The sum of the first five terms can be represented using **summation notation**.

$$\sum_{i=1}^{5} (2i + 3)$$

This notation is read "the sum as i goes from 1 to 5 of $2i + 3$."

To evaluate the series represented by $\displaystyle\sum_{i=1}^{5} (2i + 3)$, we first substitute 1 for i in $2i + 3$ and list the value obtained. Then we substitute 2 for i in $2i + 3$ and list the value. We follow this procedure for the values 1 through 5. We then sum these values to obtain the series value.

$$\sum_{i=1}^{5} (2i + 3) = (2 \cdot 1 + 3) + (2 \cdot 2 + 3) + (2 \cdot 3 + 3) + (2 \cdot 4 + 3) + (2 \cdot 5 + 3)$$

$$= 5 + 7 + 9 + 11 + 13$$

$$= 45$$

EXAMPLE 6   Write out the series $\displaystyle\sum_{k=1}^{6} (k^2 + 1)$ and evaluate it.

Solution $\displaystyle\sum_{k=1}^{6} (k^2 + 1) = (1^2 + 1) + (2^2 + 1) + (3^2 + 1) + (4^2 + 1) + (5^2 + 1) + (6^2 + 1)$

$$= 2 + 5 + 10 + 17 + 26 + 37$$

NOW TRY EXERCISE 61 $= 97$

EXAMPLE 7   Consider the general term of a sequence $a_n = 2n^2 - 4$. Represent the third partial sum, s_3, in summation notation.

Solution The third partial sum will be the sum of the first three terms, $a_1 + a_2 + a_3$. We can represent the third partial sum as $\sum_{i=1}^{3} (2i^2 - 4)$.

EXAMPLE 8 For the following set of values $x_1 = 3$, $x_2 = 4$, $x_3 = 5$, $x_4 = 6$, and $x_5 = 7$, does

$$\sum_{i=1}^{5} (x_i)^2 = \left(\sum_{i=1}^{5} x_i\right)^2 ?$$

Solution

$$\sum_{i=1}^{5} (x_i)^2 = (x_1)^2 + (x_2)^2 + (x_3)^2 + (x_4)^2 + (x_5)^2$$

$$= 3^2 + 4^2 + 5^2 + 6^2 + 7^2$$

$$= 9 + 16 + 25 + 36 + 49 = 135$$

$$\left(\sum_{i=1}^{5} x_i\right)^2 = (x_1 + x_2 + x_3 + x_4 + x_5)^2$$

$$= (3 + 4 + 5 + 6 + 7)^2 = (25)^2 = 625$$

Since $135 \neq 625$, $\sum_{i=1}^{5} (x_i)^2 \neq \left(\sum_{i=1}^{5} x_i\right)^2$.

NOW TRY EXERCISE 75

When a summation symbol is written without any upper and lower limits, it means that all the given data are to be summed.

EXAMPLE 9 A formula used to find the arithmetic mean, $\bar{x}$ (read x bar), of a set of data is $\bar{x} = \dfrac{\Sigma x}{n}$, where n is the number of pieces of data.

Joan Sally's five test scores are 70, 90, 83, 74, and 92. Find the arithmetic mean of her scores.

Solution

$$\bar{x} = \frac{\Sigma x}{n} = \frac{70 + 90 + 83 + 74 + 92}{5} = \frac{409}{5} = 81.8$$

Exercise Set 12.1

Concept/Writing Exercises

1. What is a sequence?

2. What is an infinite sequence?

3. What is a finite sequence?

4. What is an increasing sequence?

5. What is a decreasing sequence?

6. What is an alternating sequence?

7. What is a series?

8. What is the nth partial sum of a series?

9. Write the following notation in words: $\sum_{i=1}^{5} (i + 2)$.

10. Consider the summation $\sum_{k=1}^{5} (k + 2)$.
   a) What is the 1 called?
   b) What is the 5 called?
   c) What is the k called?

11. Let $a_n = 2n - 1$. Is this an increasing sequence or a decreasing sequence? Explain.

12. Let $a_n = -3n + 7$. Is this an increasing sequence or a decreasing sequence? Explain.

13. Let $a_n = 1 + (-2)^n$. Is this an alternating sequence? Explain.

14. Let $a_n = (-1)^{2n}$. Is this an alternating sequence? Explain.

Practice the Skills

Write the first five terms of the sequence whose nth term is shown.

15. $a_n = 3n$

16. $a_n = -3n$

17. $a_n = 4n - 1$

18. $a_n = 2n + 3$

19. $a_n = \dfrac{2}{n}$

20. $a_n = \dfrac{3}{n^2}$

21. $a_n = \dfrac{n + 2}{n + 1}$

22. $a_n = \dfrac{n - 5}{n + 3}$

23. $a_n = (-1)^n$

24. $a_n = (-1)^{2n}$

25. $a_n = (-2)^{n+1}$

26. $a_n = 3^{n-1}$

Find the indicated term of the sequence whose nth term is shown.

27. $a_n = 2n + 7$, twelfth term

28. $a_n = 3n + 1$, sixth term

29. $a_n = \dfrac{n}{4} + 5$, sixteenth term

30. $a_n = \dfrac{n}{2} - 7$, fourteenth term

31. $a_n = (-1)^n$, eighth term

32. $a_n = (-2)^n$, fourth term

33. $a_n = n(n + 2)$, ninth term

34. $a_n = (n - 1)(n + 3)$, fifth term

35. $a_n = \dfrac{n^2}{2n + 1}$, ninth term

36. $a_n = \dfrac{n(n + 1)}{n^2}$, tenth term

Find the first and third partial sums, s_1 and s_3, for each sequence.

37. $a_n = 3n - 1$

38. $a_n = 2n + 5$

39. $a_n = 2^n + 1$

40. $a_n = 3^n - 2$

41. $a_n = \dfrac{n - 1}{n + 1}$

42. $a_n = \dfrac{n}{n + 3}$

43. $a_n = (-1)^n$

44. $a_n = (-2)^n$

45. $a_n = \dfrac{n^2}{2}$

46. $a_n = \dfrac{n^2}{n + 4}$

Write the next three terms of each sequence.

47. $2, 4, 8, 16, 32, \ldots$

48. $\dfrac{1}{2}, \dfrac{1}{3}, \dfrac{1}{4}, \dfrac{1}{5}, \ldots$

49. $5, 7, 9, 11, 13, \ldots$

50. $5, 10, 15, 20, 25, \ldots$

51. $1, \dfrac{1}{2}, \dfrac{1}{3}, \dfrac{1}{4}, \dfrac{1}{5}, \ldots$

52. $\dfrac{2}{3}, \dfrac{3}{4}, \dfrac{4}{5}, \dfrac{5}{6}, \dfrac{6}{7}, \ldots$

53. $-1, 1, -1, 1, -1, \ldots$

54. $-10, -20, -30, -40, \ldots$

55. $1, \dfrac{1}{3}, \dfrac{1}{9}, \dfrac{1}{27}, \ldots$

56. $\dfrac{1}{3}, \dfrac{2}{3}, \dfrac{3}{3}, \dfrac{4}{3}, \ldots$

57. $1, -\dfrac{1}{2}, \dfrac{1}{4}, -\dfrac{1}{8}, \ldots$

58. $\dfrac{2}{3}, \dfrac{1}{3}, \dfrac{1}{6}, \dfrac{1}{12}, \ldots$

59. $7, -1, -9, -17, \ldots$

60. $37, 32, 27, 22, \ldots$

Write out each series, then evaluate it.

61. $\displaystyle\sum_{i=1}^{5} (3i - 1)$

62. $\displaystyle\sum_{i=1}^{4} (4i + 7)$

63. $\displaystyle\sum_{k=1}^{6} (k^2 - 1)$

64. $\displaystyle\sum_{k=1}^{5} (2k^2 - 3)$

65. $\displaystyle\sum_{i=1}^{4} \dfrac{i^2}{2}$

66. $\displaystyle\sum_{i=1}^{3} \dfrac{i^2}{4}$

67. $\displaystyle\sum_{j=2}^{4} \dfrac{j^2 + j}{j + 1}$

68. $\displaystyle\sum_{j=2}^{5} \dfrac{j^3}{j + 1}$

For the given general term a_n, write an expression using Σ to represent the indicated partial sum.

69. $a_n = n + 3$, fifth partial sum

70. $a_n = n^2 + 1$, fourth partial sum

71. $a_n = \dfrac{n^2}{4}$, third partial sum

72. $a_n = \dfrac{n^2 + 1}{n + 1}$, third partial sum

For the set of values $x_1 = 2$, $x_2 = 3$, $x_3 = 5$, $x_4 = -1$, and $x_5 = 4$, find each of the following.

73. $\displaystyle\sum_{i=1}^{5} x_i$

74. $\displaystyle\sum_{i=1}^{5} (x_i + 5)$

75. $\left(\displaystyle\sum_{i=1}^{5} x_i\right)^2$

76. $\displaystyle\sum_{i=1}^{5} 3x_i$

77. $\displaystyle\sum_{i=1}^{5} (x_i)^2$

78. $\displaystyle\sum_{i=1}^{4} (x_i^2 + 3)$

Find the arithmetic mean, $\bar{x}$, of the following sets of data.

79. $15, 20, 25, 30, 35$

80. $16, 20, 96, 18, 25$

81. $72, 83, 4, 60, 18, 20$

82. $5, 12, 9, 12, 17, 36, 70$

Problem Solving

In Exercises 83 and 84, consider the following rectangles. For the nth rectangle, the length is 2n and the width is n.

83. Perimeter

a) Find the perimeters for the first four rectangles, and then list the perimeters in a sequence.

b) Find the general term for the perimeter of the nth rectangle in the sequence. Use p_n for perimeter.

84. Area

a) Find the areas for the four rectangles, and then list the areas in a sequence.

b) Find the general term for the area of the nth rectangle in the sequence. Use a_n for area.

85. Create your own sequence that is an increasing sequence and list the first five terms.

86. Create your own sequence that is a decreasing sequence and list the first five terms.

87. Create your own sequence that is an alternating sequence and list the first five terms.

88. Write **a)** $\displaystyle\sum_{i=1}^{n} x_i$ as a sum of terms and **b)** $\displaystyle\sum_{j=1}^{n} x_j$ as a sum of terms. **c)** For a given set of values of x, from x_1 to x_n, will $\displaystyle\sum_{i=1}^{n} x_i = \sum_{j=1}^{n} x_j$? Explain.

89. Solve $\bar{x} = \dfrac{\Sigma x}{n}$ for Σx.

90. Solve $\bar{x} = \dfrac{\Sigma x}{n}$ for n.

91. Is $\displaystyle\sum_{i=1}^{n} 2x_i = 2\sum_{i=1}^{n} x_i$? Illustrate your answer with an example.

92. Is $\displaystyle\sum_{i=1}^{n} \dfrac{x_i}{2} = \dfrac{1}{2}\sum_{i=1}^{n} x_i$? Illustrate your answer with an example.

93. Let $x_1 = 3$, $x_2 = 5$, $x_3 = 2$, and $y_1 = 4$, $y_2 = 1$, $y_3 = 6$. Find the following. Note that $\Sigma x = x_1 + x_2 + x_3$, $\Sigma y = y_1 + y_2 + y_3$, and $\Sigma xy = x_1 y_1 + x_2 y_2 + x_3 y_3$. **a)** Σx, **b)** Σy, **c)** $\Sigma x \cdot \Sigma y$, **d)** Σxy, **e)** Is $\Sigma x \cdot \Sigma y = \Sigma xy$?

Cumulative Review Exercises

[2.6] **94.** Solve $\left|\dfrac{1}{2}x + \dfrac{3}{5}\right| = \left|\dfrac{1}{2}x - 1\right|$.

[5.6] **95.** Factor $8y^3 - 64x^6$.

[7.6] **96.** Solve $\sqrt{x + 2} - 1 = \sqrt{x - 3}$.

[8.3] **97.** Solve $V = \pi r^2 h$ for r.

12.2 ARITHMETIC SEQUENCES AND SERIES

SSM Study Guide CD/Video

MathPro 4/5 PH Math Tutor Center prenhall.com/Angel

1 Find the common difference in an arithmetic sequence.
2 Find the *n*th term of an arithmetic sequence.
3 Find the *n*th partial sum of an arithmetic sequence.

1 Find the Common Difference in an Arithmetic Sequence

In the previous section, we started our discussion by assuming you got a job with a starting salary of $30,000. One option for salary increases was an increase of $2000 each year. This would result in the sequence

$$\$30,000, \$32,000, \$34,000, \$36,000, \dots$$

This is an example of an arithmetic sequence.

DEFINITION

An **arithmetic sequence** is a sequence in which each term after the first differs from the preceding term by a constant amount.

The constant amount by which each pair of successive terms differs is called the **common difference**, d. The common difference can be found by subtracting any term from the term that directly follows it.

Arithmetic Sequence	Common Difference
$1, 4, 7, 10, 13, 16, \dots$	$d = 4 - 1 = 3$
$5, 1, -3, -7, -11, -15, \dots$	$d = 1 - 5 = -4$
$\dfrac{7}{2}, \dfrac{2}{2}, -\dfrac{3}{2}, -\dfrac{8}{2}, -\dfrac{13}{2}, -\dfrac{18}{2}, \dots$	$d = \dfrac{2}{2} - \dfrac{7}{2} = -\dfrac{5}{2}$

Notice that the common difference can be a positive number or a negative number. If the sequence is increasing, then d is a positive number. If the sequence is decreasing, then d is a negative number.

EXAMPLE 1 Write the first five terms of the arithmetic sequence with
a) first term 6 and common difference 3.
b) first term 3 and common difference -2.
c) first term 1 and common difference $\frac{1}{3}$.

Solution **a)** Start with 6 and keep adding 3. The sequence is $6, 9, 12, 15, 18$.
b) $3, 1, -1, -3, -5$
c) $1, \dfrac{4}{3}, \dfrac{5}{3}, 2, \dfrac{7}{3}$

2 Find the *n*th Term of an Arithmetic Sequence

In general, an arithmetic sequence with first term, a_1, and common difference, d, has the following terms:

$$a_1 = a_1, \quad a_2 = a_1 + d, \quad a_3 = a_1 + 2d, \quad a_4 = a_1 + 3d, \quad \text{and so on.}$$

If we continue this process, we can see that the nth term, a_n, can be found by the following formula:

---

nth Term of an Arithmetic Sequence

$$a_n = a_1 + (n - 1)d$$

EXAMPLE 2 **a)** Write an expression for the general (or nth) term, a_n, of the arithmetic sequence whose first term is -3 and whose common difference is 4.

b) Find the twelfth term of the sequence.

Solution **a)** The nth term of the sequence is $a_n = a_1 + (n - 1)d$. Substituting $a_1 = -3$ and $d = 4$, we obtain

$$
\begin{aligned}
a_n &= a_1 + (n - 1)d \\
&= -3 + (n - 1)4 \\
&= -3 + 4(n - 1) \\
&= -3 + 4n - 4 \\
&= 4n - 7
\end{aligned}
$$

Thus, $a_n = 4n - 7$.

b) $a_n = 4n - 7$

$a_{12} = 4(12) - 7 = 48 - 7 = 41$

NOW TRY EXERCISE 11 The twelfth term in the sequence is 41. ✳

EXAMPLE 3 Find the number of terms in the arithmetic sequence $5, 9, 13, 17, \ldots, 41$.

Solution The first term, a_1, is 5; the nth term is 41, and the common difference, d, is 4. Substitute the appropriate values into the formula for the nth term and solve for n.

$$
\begin{aligned}
a_n &= a_1 + (n - 1)d \\
41 &= 5 + (n - 1)4 \\
41 &= 5 + 4n - 4 \\
41 &= 4n + 1 \\
40 &= 4n \\
10 &= n
\end{aligned}
$$

NOW TRY EXERCISE 23 The sequence has 10 terms. ✳

3 Find the nth Partial Sum of an Arithmetic Sequence

An **arithmetic series** is the sum of the terms of an arithmetic sequence. A finite arithmetic series can be written

$$s_n = a_1 + (a_1 + d) + (a_1 + 2d) + (a_1 + 3d) + \cdots + (a_n - 2d) + (a_n - d) + a_n$$

If we consider the last term as a_n, the term before the last term will be $a_n - d$, the second before the last term will be $a_n - 2d$, and so on.

A formula for the nth partial sum, s_n, can be obtained by adding the reverse of s_n to itself.

$$
\begin{array}{rcl}
s_n = & a_1 + (a_1 + d) + (a_1 + 2d) + \cdots + (a_n - 2d) + (a_n - d) + & a_n \\
s_n = & a_n + (a_n - d) + (a_n - 2d) + \cdots + (a_1 + 2d) + (a_1 + d) + & a_1 \\
\hline
2s_n = & (a_1 + a_n) + (a_1 + a_n) + (a_1 + a_n) + \cdots + (a_1 + a_n) + (a_1 + a_n) + & (a_1 + a_n)
\end{array}
$$

Since the right side of the equation contains n terms of $(a_1 + a_n)$, we can write

$$2s_n = n(a_1 + a_n)$$

Now divide both sides of the equation by 2 to obtain the following formula.

nth Partial Sum of an Arithmetic Sequence

$$s_n = \frac{n(a_1 + a_n)}{2}$$

EXAMPLE 4 Find the sum of the first 25 natural numbers.

Solution The arithmetic sequence is 1, 2, 3, 4, 5, 6, ..., 25. The first term, a_1, is 1; the last term, a_n, is 25. There are 25 terms; thus, $n = 25$. Using the formula for the nth partial sum, we have

$$s_n = \frac{n(a_1 + a_n)}{2} = \frac{25(1 + 25)}{2} = \frac{25(26)}{2} = 25(13) = 325$$

The sum of the first 25 natural numbers is 325. Thus, $s_{25} = 325$. ✳

EXAMPLE 5 The first term of an arithmetic sequence is 4, and the last term is 31. If $s_n = 175$, find the number of terms in the sequence and the common difference.

Solution We substitute the appropriate values, $a_1 = 4$, $a_n = 31$, and $s_n = 175$, into the formula for the nth partial sum and solve for n.

$$s_n = \frac{n(a_1 + a_n)}{2}$$

$$175 = \frac{n(4 + 31)}{2}$$

$$175 = \frac{35n}{2}$$

$$350 = 35n$$

$$10 = n$$

There are 10 terms in the sequence. We can now find the common difference by using the formula for the nth term of an arithmetic sequence.

$$a_n = a_1 + (n - 1)d$$
$$31 = 4 + (10 - 1)d$$
$$31 = 4 + 9d$$
$$27 = 9d$$
$$3 = d$$

NOW TRY EXERCISE 45 The common difference is 3. The sequence is 4, 7, 10, 13, 16, 19, 22, 25, 28, 31. ✳

Examples 6–7 illustrate some applications of arithmetic sequences and series.

EXAMPLE 6 **Salary** Mary Tufts is given a starting salary of $25,000 and is promised a $1200 raise after each of the next 8 years. Find her salary during her eighth year of work.

Solution Understand Her salaries during the first few years would be

$$\$25{,}000, \$26{,}200, \$27{,}400, \$28{,}600, \ldots$$

Since we are adding a constant amount each year, this is an arithmetic sequence. The general term of an arithmetic sequence is $a_n = a_1 + (n-1)d$.

Translate In this example, $a_1 = 25,000$ and $d = 1200$. Thus, for $n = 8$, Mary's salary would be

$$a_8 = 25,000 + (8-1)1200$$

Carry Out
$$= 25,000 + 7(1200)$$
$$= 25,000 + 8400$$
$$= 33,400$$

Answer During her eighth year of work, Mary's salary would be $33,400. If we listed all the salaries for the 8-year period, they would be $25,000, $26,200, $27,400, $28,600, $29,800, $31,000, $32,200, $33,400. ✳

EXAMPLE 7 **Pendulum** Each swing of a pendulum (left to right or right to left) is 3 inches shorter than the preceding swing. The first swing is 8 feet.

a) Find the length of the twelfth swing.

b) Determine the distance traveled by the pendulum during the first 12 swings.

Solution

a) **Understand** Since each swing is decreasing by a constant amount, this problem can be represented as an arithmetic series. Since the first swing is given in feet and the decrease in swing in inches, we will change 3 inches to 0.25 feet ($3 \div 12 = 0.25$). The twelfth swing can be considered a_{12}. The difference, d, is negative since the distance is decreasing with each swing.

Translate
$$a_n = a_1 + (n-1)d$$
$$a_{12} = 8 + (12-1)(-0.25)$$

Carry Out
$$= 8 + 11(-0.25)$$
$$= 8 - 2.75$$
$$= 5.25 \text{ feet}$$

Answer The twelfth swing is 5.25 feet.

b) **Understand and Translate** The distance traveled during the first 12 swings can be found using the formula for the nth partial sum. The first swing, a_1, is 8 feet and the twelfth swing, a_{12}, is 5.25 feet.

$$s_n = \frac{n(a_1 + a_n)}{2}$$

$$s_{12} = \frac{12(a_1 + a_{12})}{2}$$

Carry Out
$$= \frac{12(8 + 5.25)}{2} = \frac{12(13.25)}{2} = 6(13.25) = 79.5 \text{ feet}$$

NOW TRY EXERCISE 75 **Answer** The pendulum travels 79.5 feet during its first 12 swings. ✳

Exercise Set 12.2

Concept/Writing Exercises

1. What is an arithmetic sequence?

2. What is an arithmetic series?

3. What do we call the constant amount by which each pair of successive terms in an arithmetic sequence differs?

4. How can the common difference in an arithmetic sequence be found?

5. If an arithmetic sequence is increasing, is the value for d a positive number or a negative number?

6. If an arithmetic sequence is decreasing, is the value for d a positive number or a negative number?

7. Can an arithmetic sequence consist of only negative numbers? Explain.

8. Can an arithmetic sequence consist of only odd numbers? Explain.

9. Can an arithmetic sequence consist of only even numbers? Explain.

10. Can an alternating sequence be an arithmetic sequence? Explain.

Practice the Skills

Write the first five terms of the arithmetic sequence with the given first term and common difference. Write the expression for the general (or nth) term, a_n, of the arithmetic sequence.

11. $a_1 = 4, d = 3$

12. $a_1 = -15, d = 4$

13. $a_1 = 7, d = -2$

14. $a_1 = 3, d = -5$

15. $a_1 = \dfrac{1}{2}, d = \dfrac{3}{2}$

16. $a_1 = -\dfrac{5}{3}, d = -\dfrac{1}{3}$

17. $a_1 = 100, d = -5$

18. $a_1 = \dfrac{5}{4}, d = -\dfrac{3}{4}$

Find the indicated quantity of the arithmetic sequence.

19. $a_1 = 8, d = 3$; find a_4

20. $a_1 = 8, d = -3$; find a_5

21. $a_1 = -9, d = 4$; find a_{10}

22. $a_1 = -1, d = -2$; find a_{12}

23. $a_1 = -2, d = \dfrac{5}{3}$; find a_{13}

24. $a_1 = 5, a_8 = -21$; find d

25. $a_1 = 5, a_9 = 21$; find d

26. $a_1 = -\dfrac{1}{2}, a_7 = \dfrac{17}{2}$; find d

27. $a_1 = 4, a_n = 28, d = 3$; find n

28. $a_1 = -2, a_n = -20, d = -3$; find n

29. $a_1 = 100, a_n = 60, d = -8$; find n

30. $a_1 = -\dfrac{7}{3}, a_n = -\dfrac{17}{3}, d = -\dfrac{2}{3}$; find n

Find the sum, s_n, and common difference, d, of each sequence.

31. $a_1 = 1, a_{10} = 19, n = 10$

32. $a_1 = -5, a_7 = 13, n = 7$

33. $a_1 = \dfrac{3}{5}, a_8 = 2, n = 8$

34. $a_1 = 12, a_8 = -23, n = 8$

35. $a_1 = -3, a_6 = 15.5, n = 6$

36. $a_1 = \dfrac{12}{5}, a_5 = \dfrac{28}{5}, n = 5$

37. $a_1 = 7, a_{11} = 67, n = 11$

38. $a_1 = 14.25, a_{31} = 18.75, n = 31$

Write the first four terms of each sequence; then find a_{10} and s_{10}.

39. $a_1 = 4, d = 3$

40. $a_1 = -5, d = 4$

41. $a_1 = 5, d = -3$

42. $a_1 = -7, d = -4$

43. $a_1 = -8, d = -5$

44. $a_1 = \dfrac{7}{2}, d = \dfrac{5}{2}$

45. $a_1 = 100, d = -7$

46. $a_1 = -15, d = 4$

47. $a_1 = \dfrac{9}{5}, d = \dfrac{3}{5}$

48. $a_1 = 35, d = 3$

Find the number of terms in each sequence and find s_n.

49. $1, 4, 7, 10, \ldots, 43$

50. $-8, -6, -4, -2, \ldots, 42$

51. $-9, -5, -1, 3, \ldots, 27$

52. $7, 14, 21, 28, \ldots, 63$

53. $\dfrac{1}{2}, \dfrac{2}{2}, \dfrac{3}{2}, \dfrac{4}{2}, \dfrac{5}{2}, \ldots, \dfrac{17}{2}$

54. $-\dfrac{5}{6}, -\dfrac{7}{6}, -\dfrac{9}{6}, -\dfrac{11}{6}, \ldots, -\dfrac{21}{6}$

55. $9, 12, 15, 18, \ldots, 93$

56. $-12, -16, -20, \ldots, -52$

Problem Solving

57. Find the sum of the first 50 natural numbers.

58. Find the sum of the first 50 even numbers.

59. Find the sum of the first 50 odd numbers.

60. Find the sum of the first 30 multiples of 5.

61. Find the sum of the first 20 multiples of 3.

62. Find the sum of the numbers between 50 and 200, inclusive.

63. Determine how many numbers between 7 and 1610 are divisible by 6.

64. Determine how many numbers between 12 and 1470 are divisible by 8.

Pyramids occur everywhere. At athletic events, cheerleaders may form a pyramid where the people above stand on the shoulders of the people below. The illustration on the right shows a pyramid with 1 cheerleader on the top row, 2 cheerleaders in the middle row, and 3 cheerleaders in the bottom row. Notice that $a_1 = 1$, $a_2 = 2$, and $a_3 = 3$. Also, observe that $d = 1$, $n = 3$, and $s_3 = 6$.

At a bowling alley, the pins at the end of the bowling lane form a pyramid. The first row has 1 pin, the second row has 2 pins, the third row has 3 pins, and the fourth row has 4 pins. Thus, $a_1 = 1$, $d = 1$, $n = 4$, and $s_4 = 10$.

Use the idea of a pyramid to solve Exercises 65–70.

65. *Auditorium* An auditorium has 18 seats in the first row. Each successive row has two more seats than the previous row. How many seats are in the twelfth row? How many seats are in the first 12 rows?

66. *Auditorium* An auditorium has 22 seats in the first row. Each successive row has four more seats than the previous row. How many seats are in the ninth row? How many seats are in the first nine rows?

67. *Logs* Yamil Bermudez stacks logs so that there are 20 logs in the bottom layer, and each layer contains one log less than the layer below it. How many logs are in the pile?

68. *Logs* Suppose Yamil, in Exercise 67, stopped stacking the logs after completing the layer containing eight logs. How many logs are in the pile?

69. *Glasses in a Stack* At their fiftieth wedding anniversary, Mr. and Mrs. Carlson are about to pour champagne into the top glass in the photo shown on the next page. The top row has 1 glass, the second row has 3 glasses, the third row has 5 glasses, and so on. Each row has 2 more glasses than the row above it. This pyramid has 13 rows.

a) How many glasses are in the twelfth row (bottom row)?

b) How many glasses are there total?

70. *Candies in a Stack* Individually wrapped candies are stacked in rows such that the top row has 1 candy, the second row has 3 candies, the third row has 5 candies, and so on. Each row has 2 more candies than the row above it. There are 7 rows of candies.

a) How many candies are in the seventh row (bottom row)?

b) How many candies are there total?

71. *Sum of Numbers* Karl Friedrich Gauss (1777–1855), a famous mathematician, as a child found the sum of the first 100 natural numbers quickly in his head $(1 + 2 + 3 + \cdots + 100)$. Explain how he might have done this and find the sum of the first 100 natural numbers as you think Gauss might have. (*Hint:* $1 + 100 = 101, 2 + 99 = 101$, etc.)

72. *Sum of Numbers* Use the same process from Exercise 71 to find the sum of the numbers from 101 to 150.

73. *Sum of Numbers* Find a formula for the sum of the first *n* consecutive odd numbers starting with 1.
$$1 + 3 + 5 + \cdots + (2n - 1)$$

74. *Sum of Even Numbers* Find a formula for the sum of the first *n* consecutive even numbers starting with 2.
$$2 + 4 + 6 + 8 + \cdots + 2n$$

75. *Swinging on a Vine* A long vine is attached to the branch of a tree. Sally Wynn swings from the vine, and each swing (left to right or right to left) is $\frac{1}{2}$ foot less than the previous swing. If her first swing is 22 feet, find

a) the length of the seventh swing and

b) the distance traveled during her seven swings.

76. *Pendulum* Each swing of a pendulum is 2 inches shorter than the previous swing (left to right or right to left). The first swing is 6 feet. Find

a) the length of the eighth swing and

b) the distance traveled by the pendulum during the eight swings.

77. *Bouncing Ball* Frank Holyton drops a ball from a second-story window. Each time the ball bounces, the height reached is 6 inches less than on the previous bounce. If the first bounce reaches a height of 6 feet, find the height attained on the eleventh bounce.

78. *Ping-Pong Ball* A Ping-Pong ball falls from the table and bounces to a height of 3 feet. If each successive bounce is 3 inches less than the previous bounce, find the height attained on the twelfth bounce.

79. *Packages* On Monday, March 17, Brian Nguyen started a new job at a packing company. On that day, he was able to prepare 105 packages for shipment. His boss expects Brian to be more productive with experience. Each day for the first week, Brian is expected to prepare 10 more packages than the previous day's total.

a) How many packages is Brian expected to prepare on March 21?

b) How many packages is Brian expected to prepare during his first week of employment?

80. *Salary* Marion Nickelson is making an annual salary of $37,500 at the Thompson Frozen Food Factory. Her boss has promised her an increase of $1500 in her salary each year over the next 10 years.

a) What will be Marion's salary 10 years from now?

b) What will be her total salary for these 11 years?

81. *Money* If Craig Campanella saves $1 on day 1, $2 on day 2, $3 on day 3, and so on, how much money, in total, will he have saved on day 31?

82. *Money* If Dan Currier saves 50¢ on day 1, $1.00 on day 2, $1.50 on day 3, and so on, how much, in total, will he have saved by the end of 1 year (365 days)?

83. *Money* Carrie Dereshi recently retired and met with her financial planner. She arranged to receive $32,000 the first year. Because of inflation, each year she will get $400 more than she received the previous year.

a) What income will she receive in her tenth year of retirement?

b) How much money will she have received in total during her first 10 years of retirement?

84. *Salary* Fran Summers is given a starting salary of $20,000 and is told she will receive a $1000 raise at the end of each year.

a) Find her salary during year 12.

b) How much will she receive in total during her first 12 years?

85. *Angles* The sum of the interior angles of a triangle, a quadrilateral, a pentagon, and a hexagon are 180°, 360°, 540°, and 720°, respectively. Use the pattern here to find the formula for the sum of the interior angles of a polygon with n sides.

86. Another formula that may be used to find the nth partial sum of an arithmetic series is

$$s_n = \frac{n}{2}[2a_1 + (n-1)d]$$

Derive this formula using the two formulas presented in this section.

Group Activity

In calculus, a topic of importance is limits. Consider $a_n = \frac{1}{n}$. The first five terms of this sequence are $\frac{1}{1}, \frac{1}{2}, \frac{1}{3}, \frac{1}{4}, \frac{1}{5}$. Since the value of $\frac{1}{n}$ gets closer and closer to 0 as n gets larger and larger, we say that the limit of $\frac{1}{n}$ as n approaches infinity is 0. We write this as $\lim\limits_{n \to +\infty} \frac{1}{n} = 0$ or $\lim\limits_{n \to +\infty} a_n = 0$. Notice that $\frac{1}{n}$ can never equal 0, but its value approaches 0 as n gets larger and larger.

a) *Group member 1: Find $\lim\limits_{n \to +\infty} a_n$ for Exercises 87 and 88.*

b) *Group member 2: Find $\lim\limits_{n \to +\infty} a_n$ for Exercises 89 and 90.*

c) *Group member 3: Find $\lim\limits_{n \to +\infty} a_n$ for Exercises 91 and 92.*

d) *Exchange work and check each other's answers.*

87. $a_n = \dfrac{1}{n-2}$

88. $a_n = \dfrac{n}{n+1}$

89. $a_n = \dfrac{1}{n^2+2}$

90. $a_n = \dfrac{2n+1}{n}$

91. $a_n = \dfrac{4n-3}{3n+1}$

92. $a_n = \dfrac{n^2}{n+1}$

Cumulative Review Exercises

[2.2] 93. Solve $A = P + Prt$ for r.

[4.1] 94. Solve the system of equations.

$$y = -2x - 1$$
$$x + 2y = 4$$

[5.4] 95. Factor $8n^2 - 4n - 20n + 10$.

[10.1] 96. Graph $(x+4)^2 + y^2 = 25$.

12.3 GEOMETRIC SEQUENCES AND SERIES

SSM Study Guide CD/Video

MathPro 4/5 PH Math Tutor Center prenhall.com/Angel

1 Find the common ratio in a geometric sequence.

2 Find the nth term of a geometric sequence.

3 Find the nth partial sum of a geometric sequence.

4 Identify infinite geometric series.

5 Find the sum of an infinite geometric series.

6 Study applications of geometric series.

1 Find the Common Ratio in a Geometric Sequence

In Section 12.1, we assumed you got a job with a starting salary of $30,000. We also mentioned that an option for salary increases was a 5% salary increase each year. This would result in the following sequence.

$$\$30,000, \quad \$31,500, \quad \$33,075, \quad \$34,728.75, \quad \ldots$$

This is an example of a geometric sequence.

DEFINITION	A **geometric sequence** is a sequence in which each term after the first is a multiple of the preceding term.

The common multiple is called the **common ratio**.

The common ratio, r, in any geometric sequence can be found by dividing any term, except the first, by the preceding term. The common ratio of the previous geometric sequence is $\dfrac{31,500}{30,000} = 1.05$ (or 105%).

Consider the geometric sequence

$$1, 3, 9, 27, 81, \ldots, 3^{n-1}, \ldots$$

The common ratio is 3 since $3 \div 1 = 3$ (or $9 \div 3 = 3$, and so on).

Geometric Sequence	Common Ratio
$4, 8, 16, 32, 64, \ldots, 4(2^{n-1}), \ldots$	2
$3, 12, 48, 192, 768, \ldots, 3(4^{n-1}), \ldots$	4
$7, \dfrac{7}{2}, \dfrac{7}{4}, \dfrac{7}{8}, \dfrac{7}{16}, \ldots, 7\left(\dfrac{1}{2}\right)^{n-1}, \ldots$	$\dfrac{1}{2}$
$5, -\dfrac{5}{3}, \dfrac{5}{9}, -\dfrac{5}{27}, \dfrac{5}{81}, \ldots, 5\left(-\dfrac{1}{3}\right)^{n-1}, \ldots$	$-\dfrac{1}{3}$

EXAMPLE 1 Determine the first five terms of the geometric sequence if $a_1 = 4$ and $r = \frac{1}{2}$.

Solution $a_1 = 4, \quad a_2 = 4 \cdot \dfrac{1}{2} = 2, \quad a_3 = 2 \cdot \dfrac{1}{2} = 1, \quad a_4 = 1 \cdot \dfrac{1}{2} = \dfrac{1}{2}, \quad a_5 = \dfrac{1}{2} \cdot \dfrac{1}{2} = \dfrac{1}{4}$

Thus, the first five terms of the geometric sequence are

$$4, 2, 1, \dfrac{1}{2}, \dfrac{1}{4}$$

NOW TRY EXERCISE 15

2 Find the *n*th Term of a Geometric Sequence

In general, a geometric sequence with first term, a_1, and common ratio, r, has the following terms:

$$a_1, \qquad a_1 r, \qquad a_1 r^2, \qquad a_1 r^3, \qquad a_1 r^4, \ldots, \qquad a_1 r^{n-1}, \ldots$$

↑	↑	↑	↑	↑	↑
1st	2nd	3rd	4th	5th	*n*th
term, a_1	term, a_2	term, a_3	term, a_4	term, a_5	term, a_n

Thus, we can see that the *n*th term of a geometric sequence is given by the following formula:

***n*th Term of a Geometric Sequence**
$$a_n = a_1 r^{n-1}$$

EXAMPLE 2 **a)** Write an expression for the general (or *n*th) term, a_n, of the geometric sequence with $a_1 = 3$ and $r = -2$.

b) Find the twelfth term of this sequence.

Solution   **a)** The nth term of the sequence is $a_n = a_1 r^{n-1}$. Substituting $a_1 = 3$ and $r = -2$, we obtain

$$a_n = a_1 r^{n-1} = 3(-2)^{n-1}$$

Thus, $a_n = 3(-2)^{n-1}$.

b)
$$a_n = 3(-2)^{n-1}$$
$$a_{12} = 3(-2)^{12-1} = 3(-2)^{11} = 3(-2048) = -6144$$

The twelfth term of the sequence is -6144. The first 12 terms of the sequence are

NOW TRY EXERCISE 35 $3, -6, 12, -24, 48, -96, 192, -384, 768, -1536, 3072, -6144$.   ✷

HELPFUL HINT
STUDY TIP

In this chapter, you will be working with exponents and using rules for exponents. The rules for exponents were discussed in Section 1.5 and again in Chapter 6. If you do not remember the rules for exponents, now is a good time to review Section 1.5.

EXAMPLE 3   Find r and a_1 for the geometric sequence with $a_2 = 24$ and $a_5 = 648$.

Solution   The sequence can be represented with blanks for the missing terms.

$$\underline{}, 24, \underline{}, \underline{}, 648$$
$$\quad\;\; \uparrow \qquad\qquad\;\; \uparrow$$
$$\quad\;\; a_2 \qquad\qquad\;\; a_5$$

If we assume that a_2 is the first term of a sequence with the same common ratio, we obtain

$$24, \underline{}, \underline{}, 648$$
$$\;\uparrow \qquad\qquad \uparrow$$
$$\text{1st} \qquad\quad \text{4th}$$
$$\text{term} \qquad\quad \text{term}$$

Now we use the formula for the nth term of a geometric sequence to find r. We let the first term, a_1, be 24 and the number of terms, n, be 4.

$$a_n = a_1 r^{n-1}$$
$$648 = 24 r^{4-1}$$
$$648 = 24 r^3$$
$$\frac{648}{24} = r^3$$
$$27 = r^3$$
$$3 = r$$

Thus, the common ratio is 3.

The first term of the original sequence must be $24 \div 3$ or 8. Thus, $a_1 = 8$. The first term, a_1, could also be found using the formula with $a_n = 648, r = 3$, and

NOW TRY EXERCISE 85 $n = 5$. Find a_1 by the formula now.   ✷

---

3   Find the nth Partial Sum of a Geometric Sequence

A **geometric series** is the sum of the terms of a geometric sequence. The sum of the first n terms, s_n, of a geometric sequence can be expressed as

$$s_n = a_1 + a_1 r + a_1 r^2 + a_1 r^3 + \cdots + a_1 r^{n-2} + a_1 r^{n-1} \qquad (eq.\,1)$$

If we multiply both sides of the equation by r, we obtain

$$rs_n = a_1r + a_1r^2 + a_1r^3 + \cdots + a_1r^{n-1} + a_1r^n \qquad (eq.\,2)$$

Now we subtract the corresponding sides of $(eq.\,2)$ from $(eq.\,1)$. The red-colored terms drop out, leaving

$$s_n - rs_n = a_1 - a_1r^n$$

Now we solve the equation for s_n.

$$s_n(1 - r) = a_1(1 - r^n) \qquad \textit{Factor.}$$

$$s_n = \frac{a_1(1 - r^n)}{1 - r} \qquad \textit{Divide both sides by } 1 - r.$$

Thus, we have the following formula for the nth partial sum of a geometric sequence.

---

nth Partial Sum of a Geometric Sequence

$$s_n = \frac{a_1(1 - r^n)}{1 - r}, \qquad r \neq 1$$

EXAMPLE 4 Find the seventh partial sum of a geometric sequence whose first term is 8 and whose common ratio is $-\frac{1}{2}$.

Solution Substitute the appropriate value for a, r, and n.

$$s_n = \frac{a_1(1 - r^n)}{1 - r}$$

$$s_7 = \frac{8\left[1 - \left(-\frac{1}{2}\right)^7\right]}{1 - \left(-\frac{1}{2}\right)} = \frac{8\left(1 + \frac{1}{128}\right)}{\frac{3}{2}} = \frac{8\left(\frac{129}{128}\right)}{\frac{3}{2}} = \frac{129}{16} \cdot \frac{2}{3} = \frac{43}{8}$$

Thus, $s_7 = \dfrac{43}{8}$.

✳

EXAMPLE 5 Given $s_n = 93$, $a_1 = 3$, and $r = 2$, find n.

Solution

$$s_n = \frac{a_1(1 - r^n)}{1 - r}$$

$$93 = \frac{3(1 - 2^n)}{1 - 2} \qquad \textit{Substitute values for } s_n, a_1, \textit{ and } r.$$

$$93 = \frac{3(1 - 2^n)}{-1}$$

$$-93 = 3(1 - 2^n) \qquad \textit{Both sides were multiplied by } -1.$$

$$-31 = 1 - 2^n \qquad \textit{Both sides were divided by 3.}$$

$$-32 = -2^n \qquad \textit{1 was subtracted from both sides.}$$

$$32 = 2^n \qquad \textit{Both sides were divided by } -1.$$

$$2^5 = 2^n \qquad \textit{Write 32 as } 2^5.$$

Therefore, $n = 5$.

✳

When working with a geometric series, r can be a positive number as we saw in Example 5 or a negative number as we saw in Example 4.

4 Identify Infinite Geometric Series

All the geometric sequences that we have examined thus far have been finite since they have had a last term. The following sequence is an example of an infinite geometric sequence.

$$1, \frac{1}{2}, \frac{1}{4}, \frac{1}{8}, \frac{1}{16}, \dots, \left(\frac{1}{2}\right)^{n-1}, \dots$$

Note that the three dots at the end of the sequence indicate that the sequence continues indefinitely. The sum of the terms in an infinite geometric sequence forms an **infinite geometric series**. For example,

$$1 + \frac{1}{2} + \frac{1}{4} + \frac{1}{8} + \frac{1}{16} + \dots + \left(\frac{1}{2}\right)^{n-1} + \dots$$

is an infinite geometric series. Let's find some partial sums.

Partial sum	Series	Sum
Second	$1 + \frac{1}{2}$	1.5
Third	$1 + \frac{1}{2} + \frac{1}{4}$	1.75
Fourth	$1 + \frac{1}{2} + \frac{1}{4} + \frac{1}{8}$	1.875
Fifth	$1 + \frac{1}{2} + \frac{1}{4} + \frac{1}{8} + \frac{1}{16}$	1.9375
Sixth	$1 + \frac{1}{2} + \frac{1}{4} + \frac{1}{8} + \frac{1}{16} + \frac{1}{32}$	1.96875

With each successive partial sum, the amount being added is less than with the previous partial sum. Also, the sum seems to be getting closer and closer to 2. In Example 6, we will show that the sum of this infinite geometric series is indeed 2.

5 Find the Sum of an Infinite Geometric Series

Consider the formula for the sum of the first n terms of an infinite geometric series:

$$s_n = \frac{a_1(1 - r^n)}{1 - r}, \qquad r \neq 1$$

What happens to r^n if $|r| < 1$ and n gets larger and larger? Suppose that $r = \frac{1}{2}$; then

$$\left(\frac{1}{2}\right)^1 = 0.5, \quad \left(\frac{1}{2}\right)^2 = 0.25, \quad \left(\frac{1}{2}\right)^3 = 0.125, \quad \left(\frac{1}{2}\right)^{20} \approx 0.000001$$

We can see that when $|r| < 1$, the value of r^n gets exceedingly close to 0 as n gets larger and larger. Thus, when considering the sum of an infinite geometric series, symbolized s_∞, the expression r^n approaches 0 when $|r| < 1$. Therefore, replacing r^n with 0 in the formula $s_n = \dfrac{a_1(1 - r^n)}{1 - r}$ leads to the following formula.

> **Sum of an Infinite Geometric Series**
>
> $$s_\infty = \frac{a_1}{1 - r} \quad \text{where} \quad |r| < 1$$

EXAMPLE 6 Find the sum of the infinite geometric series $1 + \frac{1}{2} + \frac{1}{4} + \frac{1}{8} + \cdots + \left(\frac{1}{2}\right)^{n-1} + \cdots$.

Solution $a_1 = 1$ and $r = \frac{1}{2}$. Note that $\left|\frac{1}{2}\right| < 1$.

$$s_\infty = \frac{a_1}{1 - r} = \frac{1}{1 - \frac{1}{2}} = \frac{1}{\frac{1}{2}} = 2$$

Thus, $1 + \frac{1}{2} + \frac{1}{4} + \frac{1}{8} + \frac{1}{16} + \cdots + \left(\frac{1}{2}\right)^{n-1} + \cdots = 2$. ✳

EXAMPLE 7 Find the sum of the infinite geometric series

$$5 - 2 + \frac{4}{5} - \frac{8}{25} + \frac{16}{125} - \frac{32}{625} + \cdots$$

Solution The terms of the corresponding sequence are $5, -2, \frac{4}{5}, -\frac{8}{25}, \ldots$. Note that $a_1 = 5$. To find the common ratio, r, we can divide the second term, -2, by the first term, 5.

$$r = -2 \div 5 = -\frac{2}{5}$$

Since $\left|-\frac{2}{5}\right| < 1$,

$$s_\infty = \frac{a_1}{1 - r}$$

$$= \frac{5}{1 - \left(-\frac{2}{5}\right)} = \frac{5}{1 + \frac{2}{5}} = \frac{5}{\frac{7}{5}} = \frac{25}{7}$$

NOW TRY EXERCISE 71 ✳

EXAMPLE 8 Write $0.343434\ldots$ as a ratio of integers.

Solution We can write this decimal as

$$0.34 + 0.0034 + 0.000034 + \cdots + (0.34)(0.01)^{n-1} + \cdots$$

This is an infinite geometric series with $r = 0.01$. Since $|r| < 1$,

$$s_\infty = \frac{a_1}{1 - r} = \frac{0.34}{1 - 0.01} = \frac{0.34}{0.99} = \frac{34}{99}$$

NOW TRY EXERCISE 81 If you divide 34 by 99 on a calculator, you will see $.34343434$ displayed. ✳

What is the sum of a geometric series when $|r| > 1$? Consider the geometric sequence in which $a_1 = 1$ and $r = 2$.

$$1, 2, 4, 8, 16, 32, \ldots, 2^{n-1}, \ldots$$

The sum of its terms is

$$1 + 2 + 4 + 8 + 16 + 32 + \cdots + 2^{n-1} + \cdots$$

What is the sum of this series? As n gets larger and larger, the sum gets larger and larger. We therefore say that the sum "does not exist." For $|r| > 1$, the sum of an infinite geometric series does not exist.

6 Study Applications of Geometric Series

Now let's look at some applications of geometric sequences and series.

EXAMPLE 9   **Savings Account**   Jean Simmons invests $1000 at 5% interest compounded annually in a savings account. Determine the amount in his account and the amount of interest earned at the end of 10 years.

Solution   Understand   Suppose we let P represent any principal invested. At the beginning of the second year, the amount grows to $P + 0.05P$ or $1.05P$. This amount will be the principal invested for year 2. At the beginning of the third year the second year's principal will grow by 5% to $(1.05P)(1.05)$, or $(1.05)^2P$. The amount in Jean's account at the beginning of successive years is

Year 1	Year 2	Year 3	Year 4
P	$1.05P$	$(1.05)^2P$	$(1.05)^3P$

and so on. This is a geometric series with $r = 1.05$. The amount in his account at the end of 10 years will be the same as the amount in his account at the beginning of year 11. We will therefore use the formula

$$a_n = a_1 r^{n-1}, \quad \text{with} \quad r = 1.05 \quad \text{and} \quad n = 11$$

Translate   We have a geometric sequence with $a_1 = 1000$, $r = 1.05$, and $n = 11$. Substituting these values into the formula, we obtain the following.

$$a_n = a_1 r^{n-1}$$
$$a_n = 1000(1.05)^{n-1}$$
Carry Out $$a_{11} = 1000(1.05)^{11-1}$$
$$= 1000(1.05)^{10}$$
$$\approx 1000(1.62889)$$
$$\approx 1628.89$$

Answer   After 10 years, the amount in the account is about $1628.89. The amount of interest is $1628.89 − $1000 = $628.89.   ✳

EXAMPLE 10   **Money**   Suppose someone offered you $1000 a day for each day of a 30-day month. Or, you could elect to take a penny on day 1, 2¢ on day 2, 4¢ on day 3, 8¢ on day 4, and so on. The amount would continue to double each day for 30 days.

a) Without doing any calculations, take a guess at which of the two offerings would provide the greater total return for 30 days.

b) Calculate the total amount you would receive by selecting $1000 a day for 30 days.

c) Calculate the amount you would receive on day 30 by selecting 1¢ on day 1 and doubling the amount each day for 30 days.

d) Calculate the total amount you would receive for 30 days by selecting 1¢ on day 1 and doubling the amount each day for 30 days.

Solution   **a)** Each of you will have your own answer to part **a)**.

b) If you received $1000 a day for 30 days, you would receive 30($1000) = $30,000.

c) Understand   Since the amount is doubled each day, this represents a geometric sequence with $r = 2$. The chart that follows shows the amount you would receive in each of the first 7 days. We also show the amounts written with base 2, the common ratio.

Day	1	2	3	4	5	6	7
Amount (cents)	1	2	4	8	16	32	64
Amount (cents)	2^0	2^1	2^2	2^3	2^4	2^5	2^6

Notice that for any given day, the exponent on 2 is 1 less than the given day. For example, on day 7, the amount is 2^6. In general, the amount on day n is 2^{n-1}.

Translate   To find the amount received on day 30, we evaluate $a_n = a_1 r^{n-1}$ for $n = 30$.

$$a_n = a_1 r^{n-1}$$
$$a_{30} = 1(2)^{30-1}$$

Carry Out

$$a_{30} = 1(2)^{29}$$
$$= 1(536{,}870{,}912)$$
$$= 536{,}870{,}912$$

Answer   On day 30, the amount that you would receive is 536,870,912 cents or $5,368,709.12.

d) Understand and Translate   To find the total amount received over the 30 days, we find the thirtieth partial sum.

$$s_n = \frac{a_1(1 - r^n)}{1 - r}$$
$$s_{30} = \frac{1(1 - 2^{30})}{1 - 2}$$

Carry Out

$$= \frac{1(1 - 1{,}073{,}741{,}824)}{-1}$$
$$= 1{,}073{,}741{,}823$$

Answer   Therefore, over 30 days the total amount you would receive by this method would be 1,073,741,823 cents or $10,737,418.23. The amount received by this method greatly surpasses the $30,000 received by selecting $1000 a day for 30 days.   ✳

EXAMPLE 11   **Pendulum**   On each swing (left to right or right to left), a certain pendulum travels 90% as far as on its previous swing. For example, if the swing to the right is 10 feet, the swing back to the left is $0.9 \times 10 = 9$ feet (see Fig. 12.1). If the first swing is 10 feet, determine the total distance traveled by the pendulum by the time it comes to rest.

Solution   Understand   This problem may be considered an infinite geometric series with $a_1 = 10$ and $r = 0.9$. We can therefore use the formula $s_\infty = \dfrac{a_1}{1 - r}$ to find the total distance traveled by the pendulum.

Translate and Carry Out

$$s_\infty = \frac{a_1}{1 - r} = \frac{10}{1 - 0.9} = \frac{10}{0.1} = 100 \text{ feet}$$

FIGURE 12.1

NOW TRY EXERCISE 99   Answer   By the time the pendulum comes to rest, it has traveled 100 feet.   ✳

Mathematics in Action

Chain Letters

A chain letter is "a letter directing the recipient to send out multiple copies so that its circulation increases in a geometric progression as long as the instructions are carried out." *Webster's II, New Riverside University Dictionary*.

Today, using the Internet, it takes just the click of a button to forward a message to hundreds of people. Who hasn't received one of these e-mails promising wealth, or promoting the sale of some item. These e-mail messages, commonly known as junk e-mail or spam, are helping to clog the network and are interfering with the people receiving messages they actually want and, sometimes, urgently need to receive.

The geometric growth of the message traffic from a chain letter or e-mail is astounding. If ten people receive one and each person complies by sending it on to ten more people, the ninth resending results in over one billion e-mails. Even if only a portion of the recipients respond, as is almost always the case, growth of message volume to the thousands and tens of thousands is not hard to imagine.

Federal laws exist that make it a crime to induce anyone to send money through the mail for fraudulent purposes or to advance a game of chance for money or prizes. And the senders of such e-mails, chain or otherwise, can be prosecuted *even if the (re)sender is unaware of the nature of the crime*. So, in addition to helping keep the Internet healthy when you simply delete a chain letter, you may also be helping to keep yourself out of court, or worse.

Exercise Set 12.3

Concept/Writing Exercises

1. What is a geometric sequence?
2. What is a geometric series?
3. Explain how to find the common ratio in a geometric sequence.
4. What is an infinite geometric series?
5. In a geometric series, if $|r| < 1$, what does r^n approach as n gets larger and larger?
6. Does the sum of an infinite geometric series when $|r| > 1$ exist?

7. In a geometric sequence, can the value of r be a negative number?
8. In a geometric sequence, can the value of r be a positive number?
9. In a geometric series, if $a_1 = 6$ and $r = 1/4$, does s_∞ exist? If so, what is its value? Explain.
10. In a geometric series, if $a_1 = 6$ and $r = -2$, does s_∞ exist? If so, what is its value? Explain.

Practice the Skills

Determine the first five terms of each geometric sequence.

11. $a_1 = 2, r = 3$

12. $a_1 = 2, r = -3$

13. $a_1 = 6, r = \dfrac{1}{2}$

14. $a_1 = 6, r = -\dfrac{1}{2}$

15. $a_1 = 80, r = \dfrac{1}{4}$

16. $a_1 = \dfrac{1}{2}, r = 4$

17. $a_1 = 90, r = -\dfrac{1}{3}$

18. $a_1 = -5, r = -2$

19. $a_1 = -1, r = 3$

20. $a_1 = -1, r = -3$

21. $a_1 = 6, r = -2$

22. $a_1 = -12, r = -1$

23. $a_1 = \dfrac{1}{3}, r = \dfrac{1}{2}$

24. $a_1 = \dfrac{1}{2}, r = -\dfrac{1}{3}$

25. $a_1 = 3, r = \dfrac{3}{2}$

26. $a_1 = 15, r = -\dfrac{2}{5}$

Find the indicated term of each geometric sequence.

27. $a_1 = 4, r = 2$; find a_6

28. $a_1 = 4, r = -2$; find a_6

29. $a_1 = \dfrac{1}{8}, r = 2$; find a_{10}

30. $a_1 = 9, r = \dfrac{1}{3}$; find a_7

31. $a_1 = -12, r = \dfrac{1}{2}$; find a_9

32. $a_1 = 3, r = 3$; find a_7

🔒 **33.** $a_1 = -3, r = -2$; find a_{12}

34. $a_1 = -10, r = -2$; find a_{10}

35. $a_1 = 2, r = \dfrac{1}{2}$; find a_8

36. $a_1 = 5, r = \dfrac{2}{3}$; find a_9

37. $a_1 = 50, r = \dfrac{1}{3}$; find a_{10}

38. $a_1 = -7, r = -\dfrac{3}{4}$; find a_7

Find the indicated sum.

39. $a_1 = 3, r = 2$; find s_5

40. $a_1 = 5, r = -3$; find s_5

41. $a_1 = 4, r = 5$; find s_6

42. $a_1 = 9, r = \dfrac{1}{2}$; find s_6

🔒 **43.** $a_1 = 80, r = 2$; find s_7

44. $a_1 = 1, r = -2$; find s_{12}

45. $a_1 = -30, r = -\dfrac{1}{2}$; find s_9

46. $a_1 = \dfrac{3}{5}, r = 3$; find s_7

47. $a_1 = -9, r = \dfrac{2}{5}$; find s_5

48. $a_1 = 35, r = \dfrac{1}{5}$; find s_{12}

For each geometric sequence, find the common ratio, r, and then write an expression for the general (or nth) term, a_n.

49. $3, \dfrac{3}{2}, \dfrac{3}{4}, \dfrac{3}{8}, \ldots$

50. $3, -\dfrac{3}{2}, \dfrac{3}{4}, -\dfrac{3}{8}, \ldots$

51. $7, 14, 28, 56, \ldots$

52. $3, 9, 27, 81, \ldots$

53. $2, -6, 18, -54, \ldots$

54. $\dfrac{3}{4}, \dfrac{1}{2}, \dfrac{1}{3}, \dfrac{2}{9}$

55. $-1, -3, -9, -18, \ldots$

56. $\dfrac{4}{3}, \dfrac{8}{3}, \dfrac{16}{3}, \dfrac{32}{3}, \ldots$

Find the sum of the terms in each geometric sequence.

57. $1, \dfrac{1}{2}, \dfrac{1}{4}, \dfrac{1}{8}, \dfrac{1}{16}, \ldots$

58. $1, -\dfrac{1}{2}, \dfrac{1}{4}, -\dfrac{1}{8}, \dfrac{1}{16}, \ldots$

59. $1, \dfrac{1}{5}, \dfrac{1}{25}, \dfrac{1}{125}, \dfrac{1}{625}, \ldots$

60. $1, -\dfrac{1}{5}, \dfrac{1}{25}, -\dfrac{1}{125}, \dfrac{1}{625}, \ldots$

61. $6, 3, \dfrac{3}{2}, \dfrac{3}{4}, \dfrac{3}{8}, \ldots$

62. $\dfrac{1}{3}, \dfrac{1}{9}, \dfrac{1}{27}, \dfrac{1}{81}, \ldots$

🔒 **63.** $5, 2, \dfrac{4}{5}, \dfrac{8}{25}, \ldots$

64. $-\dfrac{4}{3}, -\dfrac{4}{9}, -\dfrac{4}{27}, -\dfrac{4}{81}, \ldots$

65. $\dfrac{1}{3}, \dfrac{4}{15}, \dfrac{16}{75}, \dfrac{64}{375}, \ldots$

66. $6, -2, \dfrac{2}{3}, -\dfrac{2}{9}, \dfrac{2}{27}, \ldots$

67. $9, -1, \dfrac{1}{9}, -\dfrac{1}{81}, \ldots$

68. $\dfrac{5}{3}, 1, \dfrac{3}{5}, \dfrac{9}{25}, \ldots$

Find the sum of each infinite geometric series.

69. $2 + 1 + \dfrac{1}{2} + \dfrac{1}{4} + \dfrac{1}{8} + \cdots$

70. $4 + 2 + 1 + \dfrac{1}{2} + \cdots$

71. $8 + \dfrac{16}{3} + \dfrac{32}{9} + \dfrac{64}{27} + \cdots$

72. $10 - 5 + \dfrac{5}{2} - \dfrac{5}{4} + \cdots$

73. $-60 + 20 - \dfrac{20}{3} + \dfrac{20}{9} - \cdots$

74. $4 + \dfrac{8}{3} + \dfrac{16}{9} + \dfrac{32}{27} + \cdots$

🔒 **75.** $-12 - \dfrac{12}{5} - \dfrac{12}{25} - \dfrac{12}{125} - \cdots$

76. $5 - 1 + \dfrac{1}{5} - \dfrac{1}{25} + \cdots$

Write each repeating decimal as a ratio of integers.

77. $0.2727\ldots$

78. $0.454545\ldots$

79. $0.7777\ldots$

80. $0.375375\ldots$

81. $0.515151\ldots$

82. $0.742742\ldots$

Problem Solving

83. In a geometric sequence, $a_3 = 28$ and $a_5 = 112$; find r and a_1.

84. In a geometric sequence, $a_2 = 27$ and $a_5 = 1$; find r and a_1.

85. In a geometric sequence, $a_2 = 15$ and $a_5 = 405$; find r and a_1.

86. In a geometric sequence, $a_2 = 12$ and $a_5 = -324$; find r and a_1.

87. *Loaf of Bread* A loaf of bread currently costs $1.40. Determine the cost of a loaf of bread after 8 years (the start of the 9th year) if inflation were to grow at a constant rate of 3% per year. *Hint:* After year 1 (the start of year 2), the cost of a loaf of bread is $1.40(1.03). After year 2 (the start of year 3), the cost would be $1.40(1.03)^2$, and so on.

88. *Bicycle* A specific bicycle currently costs $400. Determine the cost of the bicycle after 12 years if inflation were to grow at a constant rate of 5% per year.

89. *Mass* A substance loses half its mass each day. If there are initially 300 grams of the substance, find
 a) the number of days after which only 37.5 grams of the substance remain.
 b) the amount of the substance remaining after 8 days.

90. *Bacteria* The number of a certain type of bacteria doubles every hour. If there are initially 1000 bacteria, after how many hours will the number of bacteria reach 64,000?

91. *Population* In March 2002, the population of the United States was about 281 million people. If the population grows at a rate of 2.2% per year, find
 a) the population after 10 years.
 b) the number of years for the population to double.

92. *Farm Equipment* A piece of farm equipment that costs $75,000 decreases in value by 15% per year. Find the value of the equipment after 4 years.

93. *Filtered Light* The amount of light filtering through a lake diminishes by one-half for each meter of depth.
 a) Write a sequence indicating the amount of light remaining at depths of 1, 2, 3, 4, and 5 meters.
 b) What is the general term for this sequence?
 c) What is the remaining light at a depth of 7 meters?

94. *Investment* You invest $10,000 in a savings account paying 6% interest annually. Find the amount in your account at the end of 8 years.

95. *Injected Dye* A tracer dye is injected into Mark Damion for medical reasons. After each hour, two-thirds of the previous hour's dye remains. How much dye remains in Mark's system after 10 hours?

96. *Pendulum* On each swing (left to right or right to left), a pendulum travels 80% as far as on its previous swing. If the first swing is 8 feet, determine the total distance traveled by the pendulum by the time it comes to rest.

97. *Bungee Jumping* Shawna Kelly goes bungee jumping off a bridge above water. On the initial jump, the bungee cord stretches to 220 feet. Assume the first bounce reaches a height of 60% of the original jump and that each additional bounce reaches a height of 60% of the previous bounce.
 a) What will be the height of the third bounce?
 b) Theoretically, Shawna would never stop bouncing, but realistically, she will. Use the infinite geometric series to estimate the total distance Shawna travels in a *downward* direction.

98. *Bungee Jumping* Repeat Exercise 97 **b)**, but this time find the total distance traveled in an *upward* direction.

99. *Ping-Pong Ball* A Ping-Pong ball falls off a table 30 inches high. Assume that the first bounce reaches a height of 70% of the distance the ball fell and each additional bounce reaches a height of 70% of the previous bounce.
 a) How high will the ball bounce on the fourth bounce?
 b) Theoretically, the ball would never stop bouncing, but realistically, it will. Estimate the total distance the ball travels in the *downward* direction.

100. *Ping-Pong Ball* Repeat Exercise 99 **b)**, but this time find the total distance traveled in the *upward* direction.

101. Stack of Chips Suppose that you form stacks of blue chips such that there is one blue chip in the first stack and in each successive stack you double the number of chips. Thus, you have stacks of 1, 2, 4, 8, and so on, of blue chips. You also form stacks of red chips, starting with one red chip and then tripling the number in each successive stack. Thus the stacks will contain 1, 3, 9, 27 and so on, red chips. How many more would the sixth stack of red chips have than the sixth stack of blue chips?

102. Stack of Money If you start with $1 and double your money each day, how many days will it take to surpass $1,000,000?

103. Depreciation One method of depreciating an item on an income tax return is the declining balance method. With this method, a given percent of the cost of the item is depreciated each year. Suppose that an item has a 5-year life and is depreciated using the declining balance method. Then, at the end of its first year, it loses $\frac{1}{5}$ of its value and $\frac{4}{5}$ of its value remains. At the end of the second year it loses $\frac{1}{5}$ of the remaining $\frac{4}{5}$ of its value, and so on. A car has a 5-year life expectancy and costs $15,000.

a) Write a sequence showing the value of the car remaining for each of the first 3 years.

b) What is the general term of this sequence?

c) Find the value of the car at the end of 5 years.

104. Scrap Value On page 651, Exercise Set 9.6, Exercise 77, a formula for scrap value was given. The scrap value,

S, is found by $S = c(1 - r)^n$ where c is the original cost, r is the annual depreciation rate and n is the number of years the object is depreciated.

a) If you have not already done so, do Exercise 103 above to find the value of the car remaining at the end of 5 years.

b) Use the formula given to find the scrap value of the car at the end of 5 years and compare this answer with the answer found in part **a)**.

105. Bouncing Ball A ball is dropped from a height of 10 feet. The ball bounces to a height of 9 feet. On each successive bounce, the ball rises to 90% of its previous height. Find the *total vertical distance* traveled by the ball when it comes to rest.

106. Wave Action A particle follows the path indicated by the wave shown. Find the *total vertical distance* traveled by the particle.

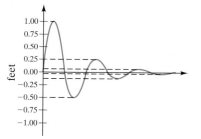

107. The formula for the nth term of a geometric sequence is $a_n = a_1 r^{n-1}$. If $a_1 = 1$, $a_n = r^{n-1}$.

a) How do you think the graphs of $y_1 = 2^{n-1}$ and $y_2 = 3^{n-1}$ will compare?

b) Graph both y_1 and y_2 and determine whether your answer to part **a)** was correct.

108. Use your grapher to decide the value of n to the nearest hundredth, where $100 = 3 \cdot 2^{n-1}$.

Challenge Problem

109. Find the sum of the sequence 1, 2, 4, 8, ..., 1,048,576 and the number of terms in the sequence.

Cumulative Review Exercises

[3.6] **110.** Let $f(x) = x^2 - 4$ and $g(x) = x - 2$. Find $(f \cdot g)(4)$.

[5.2] **111.** Multiply $(2x - 3y)(3x^2 + 4xy - 2y^2)$.

[6.4] **112.** Solve $S = \dfrac{a}{1 - r}$ for r.

[9.1] **113.** Let $g(x) = x^3 + 6$. Find $g^{-1}(x)$.

[9.6] **114.** Solve $\log x + \log(x + 1) = \log 12$.

[10.4] **115. Sail on a Sailboat** A sail on a sailboat is shaped like a right triangle with a perimeter of 36 meters and a hypotenuse of 15 meters. Find the length of each leg of the triangle.

12.4 MATHEMATICAL INDUCTION

1 Prove statements using mathematical induction.

SSM   Study Guide   CD/Video

MathPro 4/5   PH Math Tutor Center   prenhall.com/Angel

1 Prove Statements Using Mathematical Induction

There are times when we need formulas or rules to aid in determining sums of large quantities of numbers. Consider the case of adding up the first few natural numbers.

The sum of the first two is

$$1 + 2 = 3$$

The sum of the first three is

$$1 + 2 + 3 = 6$$

The sum of the first five is

$$1 + 2 + 3 + 4 + 5 = 15$$

These sums are easy to compute when only a few natural numbers are being added. To add up the first 100 natural numbers would take some time, and if no mistakes are made, you would find the sum is 5050. You would probably agree that it is somewhat cumbersome to find sums this way.

Fortunately, there is a formula which gives the sum of the first n natural numbers. It is

$$1 + 2 + 3 + 4 + \ldots + n = \frac{n(n + 1)}{2}$$

We can see that it works for some of the examples we just covered. When $n = 2$, the left side is $1 + 2 = 3$ and the right side is $\frac{2(2 + 1)}{2} = \frac{2(3)}{2} = \frac{6}{2} = 3$. Here we see the two sides produce the same value.

When $n = 3$, the left side is $1 + 2 + 3 = 6$ and the right side is $\frac{3(3 + 1)}{2} = \frac{3(4)}{2} = \frac{12}{2} = 6$. Here both values are the same. When $n = 100$, the left side is $1 + 2 + 3 + \ldots + 99 + 100$ which adds up to 5050 (after considerable work). The right side is $\frac{100(101)}{2} = \frac{10,100}{2} = 5050$. Again the values are the same.

As n becomes larger, it should be clear that it is easier to substitute a number into the right side to obtain the sum than to add up the numbers on the left side.

If we were to substitute values from 1 to 100 for n in the formula, and each case were true, it would suggest that the formula is true. However, we cannot assume that a formula is true just because it is true for 100 or even 1000 cases. We know that a formula is true for *all values of n* only when it has been proven true. Fortunately, we can do this using a process called mathematical induction. (In Example 4 we prove that this formula is true.)

Mathematical induction is a powerful method for proving that a mathematical statement involving natural numbers is true for all natural numbers. To prove a statement using mathematical induction, we must show that the statement is true for $n = 1$ and that if it is true for some natural number k, then it is also true for the next natural number, $k + 1$. Recall that the natural numbers are also called the positive integers.

The Principle of Mathematical Induction

Let S_n be a mathematical statement concerning the positive integer n. If the following two conditions are true, then S_n is true for every positive integer n.

    Condition 1. S_1 is true and

    Condition 2. for any positive integer $k, k \leq n$, if S_k is true, then S_{k+1} is true.

We will not prove this theorem, but we will try to give you some insight into why this theorem works. Suppose you have infinitely many dominos lined up in a row. Now suppose we know the following two facts are true:

1. The first domino is pushed over.

2. If one of the dominos falls over, say the kth domino, then so will the next one, the $k + 1$ domino.

Is it safe to conclude that *all* the dominos will fall over? The answer is yes, because if the first one falls (Condition 1 in the Principle of Mathematical Induction), then the second one falls (by Condition 2); and if the second one falls, then so does the third (by Condition 2), and so on.

Mathematically, proving that this statement is true for S_1 and for S_{k+1} when S_k is true, we know that it is true for S_2. Similarly, by knowing S_2 is true and that S_{k+1} is true when S_k is true, we know that S_3 is true. Continuing this way, we know that S_n is true for every positive integer n.

Condition 1 is shown to be true by substituting 1 for n in the expression and then evaluating the expression. If the result is a true statement, then Condition 1 holds. If you are asked to prove a formula is true by induction, Condition 2 may sometimes be accomplished by beginning with the formula using the variable k, then adding the $k + 1$ term to both sides of the formula. Then, using algebraic techniques, if you can show that the formula with the $k + 1$ term added to both sides simplifies to the formula obtained by replacing each k with $k + 1$, you have illustrated Condition 2. Example 1 illustrates this process. In this example we use the formula

$$1 + 3 + 5 + \ldots + (2n - 1) = n^2$$

For example, when $n = 1$ (only 1 term), we evaluate $(2n - 1)$ and n^2 to get

$$2(1) - 1 = 1^2$$
$$1 = 1$$

When $n = 2$, (the sum of 2 terms), we get

$$1 + [2(2) - 1] = 2^2$$
$$1 + 3 = 2^2$$

When $n = 3$, we get

$$1 + 3 + [2(3) - 1] = 3^2$$
$$1 + 3 + 5 = 3^2 \text{ and so on.}$$

In Example 1 we prove that the sum of the first n odd integers is n^2.

EXAMPLE 1 Prove $1 + 3 + 5 + \ldots + (2n - 1) = n^2$ for all positive integers n.

Solution Condition 1: Show that $1 + 3 + 5 + \ldots + (2n - 1) = n^2$ for $n = 1$. For $n = 1$, we get $1 = 1^2$ (see previous page). This is a true statement and Condition 1 is satisfied.

Condition 2: Assume that $1 + 3 + 5 + \ldots + (2k - 1) = k^2$ is true for some positive integer k. We need to show that the statement holds for $k + 1$. That is, we wish to show that

$$1 + 3 + 5 + \ldots + (2k - 1) + [2(k + 1) - 1] = (k + 1)^2.$$

Notice the $k + 1$ term $[2(k + 1) - 1]$, which was added to the left side of the equal sign, is obtained by substituting $k + 1$ for k in the term $(2k - 1)$. On the right side we substitute $k + 1$ for k. To continue our proof by induction, we begin with $1 + 3 + 5 + \ldots + (2k - 1) = k^2$, which we assume to be true, and add the $k + 1$ term, $[2(k + 1) - 1]$ to both sides of the formula. Then we simplify the right side of the formula to show that

$$1 + 3 + 5 + \ldots + (2k - 1) + [2(k + 1) - 1] = (k + 1)^2$$

We will now show the process.

$$1 + 3 + 5 + \ldots + (2k - 1) = k^2$$

$1 + 3 + 5 + \ldots + (2k - 1) + [2(k + 1) - 1] = k^2 + [2(k + 1) - 1]$ *Add $2(k + 1) - 1$ to both sides.*

$1 + 3 + 5 + \ldots + (2k - 1) + [2(k + 1) - 1] = k^2 + 2k + 2 - 1$ *Simplify the expression on the right.*

$1 + 3 + 5 + \ldots + (2k - 1) + [2(k + 1) - 1] = k^2 + 2k + 1$

$1 + 3 + 5 + \ldots + (2k - 1) + [2(k + 1) - 1] = (k + 1)(k + 1)$ *Factor.*

$1 + 3 + 5 + \ldots + (2k - 1) + [2(k + 1) - 1] = (k + 1)^2$ *Write factored expression as $(k + 1)^2$.*

NOW TRY EXERCISE 7 We have now shown Condition 2 holds. Since Conditions 1 and 2 hold, we have proven this formula true for all positive integers by mathematical induction. ✳

As we show in Example 2, mathematical induction can also be used to prove statements about inequalities.

EXAMPLE 2 Prove $3^n > n$ for all positive integers n.

Solution Condition 1: Show that $3^n > n$ for $n = 1$.

$$3^n > n$$
$$3^1 \overset{?}{>} 1$$
$$3 > 1$$

Since this is a true statement, Condition 1 holds.

Condition 2: Assume $3^k > k$ for some positive integer k. We wish to show that the statement holds for $k + 1$; that is, we wish to show that $3^{k+1} > k + 1$. In our proof we use the fact that $3 \cdot 3^k = 3^{k+1}$.

$3^k > k$

$3 \cdot 3^k > 3 \cdot k$ *Multiply both sides of the inequality by 3.*

$3^{k+1} > k + k + k$ *Write $3k$ as $k + k + k$.*

$3^{k+1} > k + 1$ *since $k \geq 1$, this inequality must also be true*

NOW TRY EXERCISE 37

Since we have shown that Conditions 1 and 2 hold, we have proven, using induction, that the inequality statement is true for all positive integers. ✳

EXAMPLE 3 Prove $6n$ is an even number for all positive integers.

Solution If $6n$ is an even number, then it is divisible by 2. If a number, a, is divisible by 2 then $a \div 2$ is an integer.

Condition 1: To show that $6n$ is an even number for $n = 1$, we need to show that $6(1) \div 2$ is an integer:

$$6(1) \div 2 = \frac{6}{2} = 3$$

Since 3 is an integer, Condition 1 holds.

Condition 2: Assume $6k \div 2$ is an integer for some integer k. We need to show that this statement holds for $k + 1$, that is, that $6(k + 1) \div 2$ is an integer.

Assume $\quad \dfrac{6k}{2} = m \qquad$ *where m is an integer. Now consider $k + 1$.*

$$\frac{6(k + 1)}{2} = \frac{6k + 6}{2} \qquad \text{\textit{Write the expression using $k + 1$ and then simplify.}}$$

$$= \frac{6k}{2} + \frac{6}{2}$$

$$= m + 3 \qquad \text{\textit{since $6k \div 2 = m$}}$$

NOW TRY EXERCISE 39

Since m is an integer, $m + 3$ is also an integer, so $6(k + 1) \div 2$ is an integer. Thus, Condition 2 is satisfied. As a result $6n$ is an even number for all positive integers, n. ✳

EXAMPLE 4 Prove $1 + 2 + 3 + \ldots + n = \dfrac{n(n + 1)}{2}$ for all positive integers n.

Solution Condition 1: We show the formula is true for $n = 1$ as follows.

$$1 = \frac{1(1 + 1)}{2}$$

$$1 = 1$$

Condition 2: We assume the formula holds for some k, and show that the formula holds for $k + 1$. Thus, we assume that for some k

$$1 + 2 + 3 + \ldots + k = \frac{k(k + 1)}{2}$$

We need to show the statement is true for $k + 1$. That is, we wish to show

$$1 + 2 + 3 + \ldots + k + (k + 1) = \frac{(k + 1)[(k + 1) + 1]}{2}$$

We begin with our original formula in k.

$$1 + 2 + 3 + \ldots + k = \frac{k(k + 1)}{2}$$

and add the $k + 1$ term to each side.

$$1 + 2 + 3 + \ldots + k + (k + 1) = \frac{k(k + 1)}{2} + (k + 1) \qquad \text{\textit{Add the $k + 1$ term to both sides.}}$$

Now factor out the common factor $k + 1$ on the right to get

$$= (k + 1)\left(\frac{k}{2} + 1\right)$$

$$= (k + 1)\left(\frac{k + 2}{2}\right)$$

$$1 + 2 + 3 + \ldots + k + (k + 1) = \frac{(k + 1)[(k + 1) + 1]}{2} \qquad \text{Write } k + 2 \text{ as } (k + 1) + 1.$$

This is the statement we wanted to show. Since we have shown that the statement we assumed true for k is also true for $k + 1$, we have shown Condition 2 true. We have therefore proven the given formula by induction. ✳

NOW TRY EXERCISE 9

EXAMPLE 5 **Pennies in a Piggy Bank** For a third grade project, Bob Henderson has decided to deposit pennies into a piggy bank by dropping in one penny on the first day, two pennies on the second day, three pennies on the third day, and so on, for 30 days.

a) How many pennies are in the piggy bank at the end of 30 days?

b) If Bob continues this plan for an additional 30 days, how many pennies are in the piggy bank at the end of 60 days?

Solution **a)** Use the formula from Example 4 with $n = 30$.

$$1 + 2 + 3 + \ldots + n = \frac{n(n + 1)}{2} \qquad \textit{Formula}$$

$$1 + 2 + 3 + \ldots + 29 + 30 = \frac{30(30 + 1)}{2} \qquad \textit{Substitute 30 for } n.$$

$$= \frac{30(31)}{2} \qquad \textit{Simplify.}$$

$$= 465$$

At the end of 30 days, Bob has 465 pennies ($4.65) in the piggy bank.

b) For 60 days, use the same formula with $n = 60$.

$$1 + 2 + 3 + \ldots + n = \frac{n(n + 1)}{2} \qquad \textit{Formula}$$

$$1 + 2 + 3 + \ldots + 59 + 60 = \frac{60(60 + 1)}{2} \qquad \textit{Substitute 60 for } n.$$

$$= \frac{60(61)}{2} \qquad \textit{Simplify.}$$

$$= 1830$$

At the end of 60 days, Bob has 1830 pennies ($18.30) in the piggy bank. ✳

Exercise Set 12.4

Concept/Writing Exercises

1. Can we prove a statement S_n is true for all n by testing a finite number of cases? Explain.

2. What is mathematical induction?

3. What two conditions must be proven to be true in order to prove S_n is true for all n?

4. If S_1 is true and, if S_k is true then S_{k+1} is true, how do we know that S_3 is true?

Practice the Skills

Use mathematical induction to prove the following statements for all positive intergers n.

5. $1 + 5 + 9 + \ldots + (4n - 3) = n(2n - 1)$

6. $1 + 4 + 7 + \ldots + (3n - 2) = \dfrac{n}{2}(3n - 1)$

7. $2 + 4 + 6 + \ldots + 2n = n(n + 1)$

8. $2 + 5 + 8 + \ldots + (3n - 1) = \dfrac{n}{2}(3n + 1)$

9. $3 + 4 + 5 + \ldots + (n + 2) = \dfrac{n}{2}(n + 5)$

10. $3 + 5 + 7 + \ldots + (2n + 1) = n(n + 2)$

🔒 **11.** $3 + 6 + 9 + \ldots + 3n = \dfrac{3n(n + 1)}{2}$

12. $4 + 8 + 12 + \ldots + 4n = \dfrac{4n(n + 1)}{2}$

13. $5 + 10 + 15 + \ldots + 5n = \dfrac{5n(n + 1)}{2}$

14. $7 + 14 + 21 + \ldots + 7n = \dfrac{7n(n + 1)}{2}$

15. $1 + 2 + 2^2 + \ldots + 2^{n-1} = 2^n - 1$

16. $1 + 3 + 3^2 + \ldots + 3^{n-1} = \dfrac{3^n - 1}{2}$

17. $1 + 4 + 4^2 + \ldots + 4^{n-1} = \dfrac{4^n - 1}{3}$

18. $1 + 6 + 6^2 + \ldots + 6^{n-1} = \dfrac{6^n - 1}{4}$

19. $2 + 2^2 + 2^3 + \ldots + 2^n = 2^{n+1} - 2$

20. $3 + 3^2 + 3^3 + \ldots + 3^n = 3\left(\dfrac{3^n - 1}{2}\right)$

🔒 **21.** $4 + 4^2 + 4^3 + \ldots + 4^n = 4\left(\dfrac{4^n - 1}{3}\right)$

22. $1^2 + 2^2 + 3^2 + \ldots + n^2 = \dfrac{n(n + 1)(2n + 1)}{6}$

23. $1^3 + 2^3 + 3^3 + \ldots + n^3 = \left[\dfrac{n(n + 1)}{2}\right]^2$

24. $1^3 + 3^3 + 5^3 + \ldots + (2n - 1)^3 = n^2(2n^2 - 1)$

25. $\dfrac{1}{1 \cdot 2} + \dfrac{1}{2 \cdot 3} + \dfrac{1}{3 \cdot 4} + \ldots + \dfrac{1}{n(n + 1)} = \dfrac{n}{n + 1}$

26. $\dfrac{1}{1 \cdot 3} + \dfrac{1}{3 \cdot 5} + \dfrac{1}{5 \cdot 7} + \ldots + \dfrac{1}{(2n - 1)(2n + 1)} = \dfrac{n}{2n + 1}$

27. $1 + \dfrac{1}{2} + \dfrac{1}{2^2} + \ldots + \dfrac{1}{2^{n-1}} = 2\left(1 - \dfrac{1}{2^n}\right)$

28. $1 + \dfrac{1}{3} + \dfrac{1}{3^2} + \ldots + \dfrac{1}{3^{n-1}} = \dfrac{3}{2}\left(1 - \dfrac{1}{3^n}\right)$

29. $1 + \dfrac{1}{4} + \dfrac{1}{4^2} + \ldots + \dfrac{1}{4^{n-1}} = \dfrac{4}{3}\left(1 - \dfrac{1}{4^n}\right)$

30. $1 + \dfrac{1}{5} + \dfrac{1}{5^2} + \ldots + \dfrac{1}{5^{n-1}} = \dfrac{5}{4}\left(1 - \dfrac{1}{5^n}\right)$

31. $1 + \dfrac{3}{4} + \left(\dfrac{3}{4}\right)^2 + \ldots + \left(\dfrac{3}{4}\right)^{n-1} = 4\left[1 - \left(\dfrac{3}{4}\right)^n\right]$

32. $1 + \dfrac{2}{5} + \left(\dfrac{2}{5}\right)^2 + \ldots + \left(\dfrac{2}{5}\right)^{n-1} = \dfrac{5}{3}\left[1 - \left(\dfrac{2}{5}\right)^n\right]$

33. $\left(\dfrac{3}{4}\right)^{n+1} < \dfrac{3}{4}$

34. $\left(\dfrac{2}{3}\right)^{n+1} < \dfrac{2}{3}$

🔒 **35.** $\left(\dfrac{1}{4}\right)^{n+1} < \dfrac{1}{4}$

36. $2^{n+4} > 4(n + 4)$

37. $3^n \geq 2n + 1$

38. $3^{n+3} > 9(n + 3)$

39. $n^2 + n$ is even.

40. $n^2 - n + 4$ is even.

🔒 **41.** $n^3 + 2n$ is divisible by 3.

42. $n(n + 1)(n + 2)$ is divisible by 6.

Problem Solving

🔒 **43.** Use mathematical induction to prove that when $r \neq 1$, $a + ar + ar^2 + \ldots + ar^{n-1} = \dfrac{a(1 - r^n)}{1 - r}$ for all positive integers n.

44. Use mathematical induction to prove $a + (a + d) + (a + 2d) + \ldots + [a + (n - 1)d] = \dfrac{2na + d(n^2 - n)}{2}$ for all positive integers n.

Group Activity

45. Chords A chord is a line segment that connects two points on a circle. The figure below is a circle with two chords. Notice that the two chords divide the circle into 4 regions.

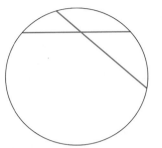

a) Draw a circle with no chords. How many regions does the circle have?

b) Draw a circle with 1 chord. How many regions does the circle have?

c) Draw your own diagram of a circle with 2 chords. How many regions does the circle have? Will all circles with 2 chords have the same number of regions?

d) Draw a circle with 3 chords. How many regions does the circle have? Will all circles with 3 chords have the same number of regions? How do you construct a diagram with the maximum number of regions?

e) Complete the table.

Number of Chords on a Circle	Maximum Number of Regions in the Circle
0	
1	
2	
3	
4	

f) What is the pattern in the sequence of numbers in the Maximum Number of Regions in the Circle column?

g) Write a formula for the maximum number of regions, r, in a circle with n chords.

h) Use your formula to find the maximum number of regions in a circle with 5 chords. Use a diagram to find the maximum number of regions in a circle with 5 chords.

i) Do you think your formula is true for all positive integers?

j) Explain how this activity relates to mathematical induction.

46. Nodes A node is a designated point in a diagram. The figure below is a circle with 3 nodes connected with line segments. Notice that the line segments connect each node to every other node. The line segments divide the circle into 4 regions.

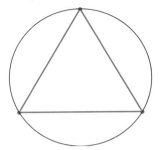

a) Draw a circle with one node. How many regions does the circle have?

b) Draw a circle with two nodes and connect them with a line segment. How many regions does the circle have?

c) Draw your own circle with three nodes and connect each node to every other node. How many regions does the circle have?

d) Draw a circle with four nodes and connect each node to every other node. How many regions does the circle have?

e) Draw a circle with five nodes and connect each node to every other node. How many regions does the circle have?

f) Complete the table.

Number of Nodes	Number of Regions
1	
2	
3	
4	
5	

g) Consider the formula $R = 2^{n-1}$ where R is the number of regions and n is the number of nodes. Show that this formula works for $n = 1, 2, 3, 4, 5$.

h) Use the formula in part **g)** to predict the number of regions in a circle with 6 nodes.

i) Draw a circle with six nodes and connect each node to every other node. What is the maximum number of regions in the circle? Is this what you expected? Explain. Hint: Don't let 3 line segments intersect inside the circle.

Cumulative Review Exercises

[5.2] Expand.

47. $(2x + 3y)^2$

48. $(5x^2 - 2y)^2$

49. $[4x + (y - 3)]^2$

50. $(2x + 3y)^3$

12.5 THE BINOMIAL THEOREM

SSM

Study Guide

CD/Video

1 Evaluate factorials.

2 Use Pascal's triangle.

3 Use the binomial theorem.

MathPro 4/5

PH Math Tutor Center

prenhall.com/Angel

1 Evaluate Factorials

To understand the binomial theorem, you must have an understanding of what **factorials** are. The notation $n!$ is read "n factorial." Its definition follows.

DEFINITION	**n Factorial**
	$$n! = n(n - 1)(n - 2)(n - 3)\ldots(1)$$ for any positive integer n.

Examples

$$6! = 6 \cdot 5 \cdot 4 \cdot 3 \cdot 2 \cdot 1 = 720$$
$$8! = 8 \cdot 7 \cdot 6 \cdot 5 \cdot 4 \cdot 3 \cdot 2 \cdot 1 = 40{,}320$$

Note that by definition **0! is 1**.

Below, we explain how to find factorials on a calculator.

Using Your Calculator

Scientific Calculator

Factorials can be found on calculators that contain an $\boxed{n!}$ or $\boxed{x!}$ key. Often, the factorial key is a second function key. In the following examples, the answers appear after $\boxed{n!}$.

Evaluate 6!   6 $\boxed{2^{nd}}$ $\boxed{n!}$ 720

Evaluate 9!   9 $\boxed{2^{nd}}$ $\boxed{n!}$ 362880

Graphing Calculator

Graphing calculators do not have a factorial key. On some graphing calculators, factorials are found under the $\boxed{\text{MATH}}$, Probability function menu.

On the TI-83 Plus calculator, to get the probability function, PRB, menu, you press $\boxed{\text{MATH}}$, and then scroll to the right using the right arrow key, $\boxed{\blacktriangleright}$, three times, until you get to PRB. The $n!$ (or !) is the fourth item down the menu.

(continued on the next page)

To find 5! or 6!, the keystrokes are as follows.

KEYSTROKES	ANSWER
5 MATH ▶ ▶ ▶ 4 ENTER	120
6 MATH ▶ ▶ ▶ 4 ENTER	720

2 Use Pascal's Triangle

Using polynomial multiplication, we can obtain the following expansions of the binomial $a + b$:

$$(a + b)^0 = 1$$
$$(a + b)^1 = a + b$$
$$(a + b)^2 = a^2 + 2ab + b^2$$
$$(a + b)^3 = a^3 + 3a^2b + 3ab^2 + b^3$$
$$(a + b)^4 = a^4 + 4a^3b + 6a^2b^2 + 4ab^3 + b^4$$
$$(a + b)^5 = a^5 + 5a^4b + 10a^3b^2 + 10a^2b^3 + 5ab^4 + b^5$$
$$(a + b)^6 = a^6 + 6a^5b + 15a^4b^2 + 20a^3b^3 + 15a^2b^4 + 6ab^5 + b^6$$

Note that when expanding a binomial of the form $(a + b)^n$:

1. There are $n + 1$ terms in the expansion.
2. The first term is a^n and the last term is b^n.
3. Reading from left to right, the exponents on a decrease by 1 from term to term, while the exponents on b increase by 1 from term to term.
4. The sum of the exponents on the variables in each term is n.
5. The coefficients of the terms equidistant from the ends are the same.

If we examine just the variables in $(a + b)^5$, we have a^5, a^4b, a^3b^2, a^2b^3, ab^4, and b^5.

The numerical coefficients of each term in the expansion of $(a + b)^n$ can be found by using **Pascal's triangle**, named after Blaise Pascal, a seventeenth-century French mathematician. For example, if $n = 5$, we can determine the numerical coefficients of $(a + b)^5$ as follows.

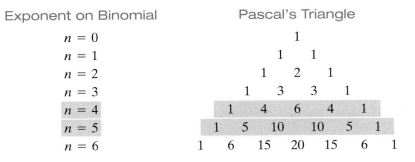

Exponent on Binomial	Pascal's Triangle
$n = 0$	1
$n = 1$	1 1
$n = 2$	1 2 1
$n = 3$	1 3 3 1
$n = 4$	1 4 6 4 1
$n = 5$	1 5 10 10 5 1
$n = 6$	1 6 15 20 15 6 1

Examine row 5 ($n = 4$) and row 6 ($n = 5$).

Notice that the first and last numbers in each row are 1, and the inner numbers are obtained by adding the two numbers in the row above (to the right and left). The numerical coefficients of $(a + b)^5$ are $1, 5, 10, 10, 5,$ and 1. Thus, we can write the expansion of $(a + b)^5$ by using the information in 1–5 on page 816 for the variables and their exponents, and by using Pascal's triangle for the coefficients.

$$(a + b)^5 = a^5 + 5a^4b + 10a^3b^2 + 10a^2b^3 + 5ab^4 + b^5$$

This method of expanding a binomial is not practical when n is large.

3 Use the Binomial Theorem

We will shortly introduce a more practical method, called the binomial theorem, to expand expressions of the form $(a + b)^n$. However, before we introduce this formula, we need to explain how to find *binomial coefficients* of the form $\binom{n}{r}$.

Binomial Coefficients

For n and r nonnegative integers, $n \geq r$,

$$\binom{n}{r} = \frac{n!}{r! \cdot (n - r)!}$$

The binomial coefficient $\binom{n}{r}$ is read "the number of *combinations* of n items taken r at a time." Combinations are used in many areas of mathematics, including the study of probability.

EXAMPLE 1 Evaluate $\binom{5}{2}$.

Solution Using the definition, if we substitute 5 for n and 2 for r, we obtain

$$\binom{5}{2} = \frac{5!}{2! \cdot (5 - 2)!} = \frac{5!}{2! \cdot 3!} = \frac{5 \cdot 4 \cdot 3 \cdot 2 \cdot 1}{(2 \cdot 1) \cdot (3 \cdot 2 \cdot 1)} = 10$$

Thus, $\binom{5}{2}$ equals 10.

EXAMPLE 2 Evaluate.

a) $\binom{7}{4}$ **b)** $\binom{4}{4}$ **c)** $\binom{5}{0}$

Solution **a)** $\binom{7}{4} = \frac{7!}{4! \cdot (7 - 4)!} = \frac{7!}{4! \cdot 3!} = \frac{7 \cdot 6 \cdot 5 \cdot 4 \cdot 3 \cdot 2 \cdot 1}{(4 \cdot 3 \cdot 2 \cdot 1)(3 \cdot 2 \cdot 1)} = 35$

b) $\binom{4}{4} = \frac{4!}{4! \cdot (4 - 4)!} = \frac{4!}{4! \cdot 0!} = \frac{1}{1} = 1$ *Remember that 0! = 1.*

c) $\binom{5}{0} = \frac{5!}{0! \cdot (5 - 0)!} = \frac{5!}{0! \cdot 5!} = \frac{1}{1} = 1$

NOW TRY EXERCISE 17

By studying Examples 2 **b)** and **c)**, you can reason that, for any positive integer n,

$$\binom{n}{n} = 1 \quad \text{and} \quad \binom{n}{0} = 1$$

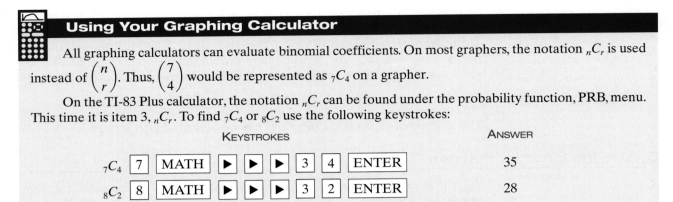

Using Your Graphing Calculator

All graphing calculators can evaluate binomial coefficients. On most graphers, the notation $_nC_r$ is used instead of $\binom{n}{r}$. Thus, $\binom{7}{4}$ would be represented as $_7C_4$ on a grapher.

On the TI-83 Plus calculator, the notation $_nC_r$ can be found under the probability function, PRB, menu. This time it is item 3, $_nC_r$. To find $_7C_4$ or $_8C_2$ use the following keystrokes:

	KEYSTROKES	ANSWER
$_7C_4$	7 MATH ▶ ▶ ▶ 3 4 ENTER	35
$_8C_2$	8 MATH ▶ ▶ ▶ 3 2 ENTER	28

If you are using a different graphing calculator, consult the manual to learn to evaluate combinations.

Now we introduce the binomial theorem.

Binomial Theorem

For any positive integer n,

$$(a + b)^n = \binom{n}{0}a^nb^0 + \binom{n}{1}a^{n-1}b^1 + \binom{n}{2}a^{n-2}b^2 + \binom{n}{3}a^{n-3}b^3 + \cdots + \binom{n}{n}a^0b^n$$

Notice in the binomial theorem that the sum of the exponents on the variables in each term is n. In the combination, the top number is always n and the bottom number is always the same as the exponent on the second variable in the term. For example, if we consider the term $\binom{n}{3}a^{n-3}b^3$, the sum of the exponents on the variables is $(n - 3) + 3 = n$. Also, the exponent on the variable b is 3, and the bottom number in the combination is also 3.

If the variables and exponents on one term of the binomial theorem are a^7b^5, then n must be $7 + 5$ or 12. Also, the combination preceding a^7b^5 must be $\binom{12}{5}$. Thus, the term would be $\binom{12}{5}a^7b^5$.

Now we will expand $(a + b)^5$ using the binomial theorem and see if we get the same expression as we did when we used polynomial multiplication and Pascal's triangle to obtain the expansion.

$$(a + b)^5 = \binom{5}{0}a^5b^0 + \binom{5}{1}a^{5-1}b^1 + \binom{5}{2}a^{5-2}b^2 + \binom{5}{3}a^{5-3}b^3 + \binom{5}{4}a^{5-4}b^4 + \binom{5}{5}a^{5-5}b^5$$

$$= \binom{5}{0}a^5b^0 + \binom{5}{1}a^4b^1 + \binom{5}{2}a^3b^2 + \binom{5}{3}a^2b^3 + \binom{5}{4}a^1b^4 + \binom{5}{5}a^0b^5$$

$$= \frac{5!}{0! \cdot 5!}a^5 + \frac{5!}{1! \cdot 4!}a^4b + \frac{5!}{2! \cdot 3!}a^3b^2 + \frac{5!}{3! \cdot 2!}a^2b^3 + \frac{5!}{4! \cdot 1!}ab^4 + \frac{5!}{5! \cdot 0!}b^5$$

$$= a^5 + 5a^4b + 10a^3b^2 + 10a^2b^3 + 5ab^4 + b^5$$

This is the same expression as we obtained earlier.

In the binomial theorem, the first and last terms of the expansion contain a factor raised to the zero power. Since any nonzero number raised to the 0th power equals 1, we could have omitted those factors. These factors were included so that you could see the pattern better.

EXAMPLE 3   Use the binomial theorem to expand $(2x + 3)^6$.

Solution   If we use $2x$ for a and 3 for b, we obtain

$(2x+3)^6 = \binom{6}{0}(2x)^6(3)^0 + \binom{6}{1}(2x)^5(3)^1 + \binom{6}{2}(2x)^4(3)^2 + \binom{6}{3}(2x)^3(3)^3 + \binom{6}{4}(2x)^2(3)^4 + \binom{6}{5}(2x)^1(3)^5 + \binom{6}{6}(2x)^0(3)^6$

$= 1(2x)^6 + 6(2x)^5(3) + 15(2x)^4(9) + 20(2x)^3(27) + 15(2x)^2(81) + 6(2x)(243) + 1(729)$

$= 64x^6 + 576x^5 + 2160x^4 + 4320x^3 + 4860x^2 + 2916x + 729$

NOW TRY EXERCISE 19

EXAMPLE 4   Use the binomial theorem to expand $(5x - 2y)^4$.

Solution   Write $(5x - 2y)^4$ as $[5x + (-2y)]^4$. Use $5x$ in place of a and $-2y$ in place of b in the binomial theorem.

$[5x + (-2y)]^4 = \binom{4}{0}(5x)^4(-2y)^0 + \binom{4}{1}(5x)^3(-2y)^1 + \binom{4}{2}(5x)^2(-2y)^2 + \binom{4}{3}(5x)^1(-2y)^3 + \binom{4}{4}(5x)^0(-2y)^4$

$= 1(5x)^4 + 4(5x)^3(-2y) + 6(5x)^2(-2y)^2 + 4(5x)(-2y)^3 + 1(-2y)^4$

$= 625x^4 - 1000x^3y + 600x^2y^2 - 160xy^3 + 16y^4$

NOW TRY EXERCISE 25

Exercise Set 12.5

Concept/Writing Exercises

1. Explain how to construct Pascal's triangle. Construct the first five rows of Pascal's triangle.
2. Explain in your own words how to find $n!$ for any whole number n.
3. Give the value of $1!$.
4. Give the value of $0!$.
5. Can you evaluate $(-4)!$? Explain.
6. Can you evaluate $(-6)!$? Explain.
7. How many terms are there in the expansion of $(a + b)^{12}$? Explain.
8. How many terms are there in the expansion of $(x + y)^{17}$? Explain.

Practice the Skills

Evaluate each combination.

9. $\binom{5}{2}$
10. $\binom{6}{3}$
11. $\binom{5}{5}$
12. $\binom{9}{3}$
13. $\binom{7}{0}$
14. $\binom{10}{8}$
15. $\binom{8}{4}$
16. $\binom{12}{8}$
17. $\binom{8}{2}$
18. $\binom{7}{5}$

Use the binomial theorem to expand each expression.

19. $(x + 4)^3$
20. $(x - 4)^3$
21. $(2x + 3)^3$
22. $(2x - 3)^3$
23. $(a - b)^4$
24. $(2r + s^2)^4$
25. $(3a - b)^5$
26. $(x + 2y)^5$
27. $\left(2x + \dfrac{1}{2}\right)^4$
28. $\left(\dfrac{2}{3}x + \dfrac{3}{2}\right)^4$
29. $\left(\dfrac{x}{2} - 3\right)^4$
30. $(3x^2 + y)^5$

Write the first four terms of each expansion.

31. $(x + 10)^{10}$
32. $(2x + 3)^8$
33. $(3x - y)^7$
34. $(3p + 2q)^{11}$
35. $(x^2 - 3y)^8$
36. $\left(2x + \dfrac{y}{5}\right)^9$

Problem Solving

37. Is $n!$ equal to $n \cdot (n-1)!$? Explain and give an example to support your answer.

38. Is $(n+1)!$ equal to $(n+1) \cdot n!$? Explain and give an example to support your answer.

39. Is $(n-3)!$ equal to $(n-3)(n-4)(n-5)!$ for $n \geq 5$? Explain and give an example to support your answer.

40. Is $(n+2)!$ equal to $(n+2)(n+1)(n)(n-1)!$ for $n \geq 1$? Explain and give an example to support your answer.

41. Under what conditions will $\binom{n}{m}$, where n and m are nonnegative integers, have a value of 1?

42. Can $\binom{n}{m}$ ever have a value of 0? Explain.

43. What are the first, second, next to last, and last terms of the expansion $(x+3)^8$?

44. What are the first, second, next to last, and last terms of the expansion $(2x+5)^6$?

45. Write the binomial theorem using summation notation.

46. Prove that $\binom{n}{r} = \binom{n}{n-r}$ for any whole numbers n and r, and $r \leq n$.

Cumulative Review Exercises

[3.4] **47.** Find the y-intercept for the line $2x + y = 6$.

[4.1] **48.** Solve the system of equations.

$$\frac{1}{5}x + \frac{1}{2}y = 4$$
$$\frac{2}{3}x - y = \frac{8}{3}$$

[5.8] **49.** Solve $x(x - 12) = -20$.

[7.3] **50.** Simplify $\sqrt{20xy^4}\,\sqrt{6x^5y^7}$.

[9.1] **51.** Find $f^{-1}(x)$ if $f(x) = 2x + 8$.

12.6 THE COUNTING PRINCIPLE, PERMUTATIONS, AND COMBINATIONS

SSM Study Guide CD/Video

MathPro 4/5 PH Math Tutor Center prenhall.com/Angel

1 Use the counting principle to determine the number of possible outcomes of an experiment.

2 Determine the number of permutations.

3 Deterimine the number of combinations.

In the last section we calculated binomial coefficients by finding the number of combinations of n items taken r at a time. In this section we develop this idea further and develop other counting methods.

1 Use the Counting Principle to Determine the Number of Possible Outcomes of an Experiment

Before we discuss the counting principle we need to learn the meanings of certain terms. An **experiment** is any process that leads to a set of results. If an experiment consists of more than one part, each part is called a **trial**. For example, an experiment may consist of tossing a coin 3 times. Each individual toss is considered a trial. The possible results of each trial are its **outcomes**. When a coin is tossed, the 2 possible outcomes are a head and a tail. When a die is rolled, its possible outcomes are 1, 2, 3, 4, 5, and 6. The set of all possible outcomes of an experiment is called its **sample space**. The sample space when a coin is tossed is $\{H, T\}$. The sample space when a die is rolled is $\{1, 2, 3, 4, 5, 6\}$. Any subset of a sample space is called an **event**. For example, if we roll a die, obtaining the number 2, it is one event. Rolling an even number may be considered another event. Rolling an even number can be satisfied by rolling either a 2, 4, or 6. When

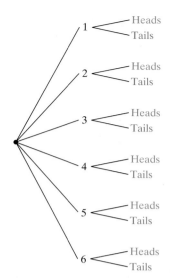

FIGURE 12.2

the outcome of one event has no effect on the outcome of any other event, we say the events are **independent events**. For example, if we roll a die three times, the outcome of each roll does not affect the outcomes of the other rolls. Therefore, the events are independent. In the experiment of rolling a die and tossing a coin, the events are independent events. What are the possible outcomes when a die is rolled and a coin is tossed? The possible outcomes are 1 with heads, 1 with tails, 2 with heads, 2 with tails, and so on. We can illustrate the possible outcomes with the *tree diagram* shown in Figure 12.2.

From the tree diagram, notice that every possible die outcome has 2 possible coin outcomes. Since we have 6 possible die outcomes, our total number of possible outcomes is $6 \cdot 2 = 12$. When we determine the total number of possible outcomes by multiplying the number of possible outcomes of each independent event, we are using the **counting principle**.

Counting Principle

If one event can occur in m distinct ways and a second event can occur in n distinct ways, then the two events, in that specific order, can occur in $m \cdot n$ distinct ways.

We will now work some problems using the counting principle.

EXAMPLE 1 **License Plates**   A license plate is to consist of 3 letters followed by 4 digits. Determine the number of distinct license plates possible if the letters and digits can be repeated.

Solution   Since the letter or digit used in any position has no effect on the letter or number used in any other position, the 7 events are independent events. Each event involving a letter has 26 possible outcomes. Each event involving a digit has 10 possible outcomes. The total number of possible outcomes is

$$26 \cdot 26 \cdot 26 \cdot 10 \cdot 10 \cdot 10 \cdot 10 = 175,760,000$$
$$\text{L} \quad \text{L} \quad \text{L} \quad \text{D} \quad \text{D} \quad \text{D} \quad \text{D}$$

NOW TRY EXERCISE 43

EXAMPLE 2 **License Plates**   A license plate is to consist of 3 letters followed by 4 digits. Determine the number of distinct license plates possible if the letters and digits cannot be repeated.

Solution   There are 26 possible outcomes for the first letter. Since repetition is not permitted, for the second letter only 25 letters are possible. For the third letter, only 24 letters are possible. For the first digit there are 10 possibilities. Since repetition of the digits is not permitted, for the second digit, 9 digits are possible. There are 8 possibilities for the third digit and 7 possibilities for the fourth digit. The total number of outcomes is therefore

$$26 \cdot 25 \cdot 24 \cdot 10 \cdot 9 \cdot 8 \cdot 7 = 78,624,000$$
$$\text{L} \quad \text{L} \quad \text{L} \quad \text{D} \quad \text{D} \quad \text{D} \quad \text{D}$$

EXAMPLE 3 **Baseball Cards**   At a baseball card show, Sharon Becker wishes to display 5 different Mark McGwire cards.

a) In how many different ways can she place the 5 cards in a straight row?

b) If she wants to place Mark McGwire's rookie card in the middle, in how many ways can she arrange the cards?

Solution   **a)** There are 5 positions to fill, using the 5 cards. In the first position, on the left, she can use any one of the 5 cards. In the second position she can use any of the 4 remaining cards. In the third position she can use any of the 3 remaining cards, and so on. The number of distinct possible arrangements is

$$\underline{5} \cdot \underline{4} \cdot \underline{3} \cdot \underline{2} \cdot \underline{1} \cdot = 120$$

b) We begin by satisfying the specified requirement. In this case the rookie card must be placed in the middle. Therefore there is only 1 possibility for the middle position.

$$\underline{} \ \underline{} \ \underline{1} \ \underline{} \ \underline{}$$

For the first position on the left, there are now 4 possibilities. For the second position there will be 3 possibilities. For the fourth position there will be 2 possibilities. Finally, in the last position there is only 1 possibility.

$$4 \cdot 3 \cdot 1 \cdot 2 \cdot 1 = 24$$

Thus there are 24 ways the cards can be arranged.   ✳

2 Determine the Number of Permutations

As we saw in Example 3a, the five baseball cards can be ordered in 120 different ways. Each different arrangement is one permutation of the baseball cards.

> A **permutation** is any ordered arrangement of a given set of objects.

When we find the number of possible arrangements of a set of objects, we are finding the **number of permutations**. In Example 3a, we found that the number of permutations of the 5 baseball cards is $5 \cdot 4 \cdot 3 \cdot 2 \cdot 1$. Recall from Section 12.5 that the product of all positive integers less than or equal to n is n factorial, symbolized $n!$ Thus we write $5 \cdot 4 \cdot 3 \cdot 2 \cdot 1 = 5!$.

We now explain how to determine the number of permutations of n distinct items.

Number of Permutations

The number of permutations of n distinct items is $n!$.

EXAMPLE 4   **Compact Disks**   In how many ways can 7 different CDs be stacked in a CD holder with 7 slots?

Solution   Since there are 7 different CDs, the same as the number of slots available, the number of permutations is $7!$.

NOW TRY EXERCISE 57

$$7! = 7 \cdot 6 \cdot 5 \cdot 4 \cdot 3 \cdot 2 \cdot 1 = 5040$$ ✳

In Example 4 the number of CDs is the same as the number of slots in the CD holder. In Example 5 we find the number of permutations when we have more CDs than our CD holder can hold.

EXAMPLE 5   **Compact Disks**   In how many ways can a CD holder with 7 slots be filled from a collection of 12 different CDs?

Solution Since there are 12 CDs, the first slot has 12 options. After 1 CD has been used in the first slot, any of the remaining 11 CDs can be used in the second slot. Then any of the remaining 10 CDs can be used in the third slot, and so on. Following, we use dashes for the 7 slots and show the number of possible outcomes for each slot. The number of permutations is

NOW TRY EXERCISE 49

$$\underline{12}\cdot\underline{11}\cdot\underline{10}\cdot\underline{9}\cdot\underline{8}\cdot\underline{7}\cdot\underline{6} = 3,991,680$$

We use the notation $_nP_r$ to indicate the number of permutations of n items taken r at a time. In Example 5 we found the number of permutations of 12 items taken 7 at a time, $_{12}P_7$.

We can use the counting principle to evaluate the number of permutations represented by $_nP_r$. The n represents the number of distinct possible outcomes for the first event. The r represents the number of events. The number of distinct possible outcomes for each event after the first one will be 1 less than the number of outcomes for the previous event. To evaluate $_nP_r$, multiply n by each consecutive integer less than n, up to and including $n - r + 1$. For example, when we evaluate $_{12}P_7$

$$_{12}P_7 = 12\cdot11\cdot10\cdot9\cdot8\cdot7\cdot6$$

$$12 - 7 + 1 = 6$$

In general, the number of permutations of n items taken r at a time, $_nP_r$, may be found by the formula $_nP_r = n(n - 1)(n - 2)\ldots(n - r + 1)$. Now let's look at some other examples.

$$_7P_4 = 7\cdot6\cdot5\cdot4$$
$$7 - 4 + 1 = 4$$
$$_9P_2 = 9\cdot8$$
$$9 - 2 + 1 = 8$$
$$_{10}P_3 = 10\cdot9\cdot8$$
$$10 - 3 + 1 = 8$$

Notice that the last number in the multiplication is always $n - r + 1$. For example, in $_{10}P_3$, the last number is $10 - 3 + 1 = 8$. Now let's develop an alternative formula for determining $_nP_r$.

$$_nP_r = n(n - 1)(n - 2)\ldots(n - r + 1)$$
$$= n(n - 1)(n - 2)\ldots(n - r + 1) \times \frac{(n - r)!}{(n - r)!}$$
$$= \frac{n(n - 1)(n - 2)\ldots(n - r + 1)(n - r)(n - r - 1)\ldots(3)(2)(1)}{(n - r)!}$$
$$= \frac{n!}{(n - r)!}$$

Multiply by $\dfrac{(n - r)!}{(n - r)!}$ which is equivalent to multiplying by 1.

Permutations Formula

The number of permutations of n items taken r at a time is

$$_nP_r = \frac{n!}{(n - r)!}$$

EXAMPLE 6   **Club Officers**   A club has 20 members who are eligible to hold office. How many different arrangements of president, vice-president, and secretary are possible?

Solution   Since there are 20 people and 3 people are to be selected, $n = 20$ and $r = 3$.

$$_{20}P_3 = \frac{20!}{(20 - 3)!}$$

$$= \frac{20!}{17!}$$

$$= \frac{20 \cdot 19 \cdot 18 \cdot \cancel{17!}}{\cancel{17!}}$$

$$= 20 \cdot 19 \cdot 18$$

$$= 6840$$

Therefore, there are 6840 different arrangements, or permutations, of the three officers possible.

NOW TRY EXERCISE 51

3   Determine the Number of Combinations

Sometimes the arrangement of items is not important to the problem. For instance, if you are selecting 7 CDs to be placed in your CD holder, you may be concerned only with the group of CDs selected, not the order in which they are placed in the holder. When we find the number of groups of items that are possible, we are finding the **number of combinations**.

> A **combination** is a distinct grouping of objects without regard to their arrangement.

It is important to distinguish a permutation from a combination. The letters a, b, c, and b, c, a are 2 different permutations because the ordering of the 3 letters is different, but they represent the same combination of letters because the same letters are used in each case.

EXAMPLE 7   **Permutation or Combination?**   Determine whether the situation represents a permutation or combination problem.

a) Two students from a class of 28 are selected to serve on a committee. In how many ways can this be done?

b) A group of 8 friends are throwing a surprise party for Shelia Pierce. They draw a name to determine who brings a cake to the surprise party and then draw a name to determine who brings Shelia. How many different assignments are possible?

Solution   **a)** Since the order in which the students are selected is not important, the problem is a combination problem.

b) Since the two friends have different tasks, one bringing a cake and the other bringing Shelia, the order that names are selected is important. Therefore, this problem is a permutation problem.

The notation we use for the number of combinations of n items taken r at a time is $_nC_r$. It can also be written as $\binom{n}{r}$ as we did in Section 12.5 when we found binomial coefficients.

Consider the set of elements $\{a, b, c, d, e\}$. The number of permutations of 2 letters from this set of 5 letters, $_5P_2$, is 20:

$$ab, ac, ad, ae, ba, bc, bd, be, ca, cb, cd, ce, da, db, dc, de, ea, eb, ec, ed$$

The number of combinations of 2 letters from this set of 5 letters, $_5C_2$, is 10:

$$ab, ac, ad, ae, bc, bd, be, cd, ce, de$$

Since every group of 2 elements can be written in 2! or 2 ways, the number of permutations is twice the number of combinations. We could calculate the number of combinations by dividing the number of permutations by 2!.

$$_5C_2 = \frac{_5P_2}{2!} = \frac{5 \cdot 4}{2 \cdot 1} = 10$$

From the counting principle we know that a group with r distinct elements can be arranged in $r!$ different ways. Thus a group containing 3 distinct elements can be arranged in 3! or 6 different ways. When determining the number of combinations, we only want to count the same groups of items once. We can calculate the number of combinations by dividing the number of permutations by $r!$, the number of permutations in one group of r distinct elements. Since $_nP_r = \dfrac{n!}{(n-r)!}$ we conclude the following.

Combination Formula

The number of combinations of n items taken r at a time is

$$_nC_r = \frac{n!}{(n-r)!r!}$$

EXAMPLE 8   **Nine Drivers Needed**   Twelve people have volunteered to drive the swim team to their out-of-town swim meet. Only 9 drivers are needed. How many different groups of drivers can be selected?

Solution   Since the order in which the drivers are selected is unimportant, this problem is a combination problem. We want to find the number of different groups of 9 that can be formed from 12 people, so $n = 12$ and $r = 9$.

$$_{12}C_9 = \frac{12!}{(12-9)!9!}$$
$$= \frac{12!}{3!9!}$$
$$= \frac{\overset{4}{12} \cdot 11 \cdot \overset{5}{10} \cdot 9!}{3 \cdot 2 \cdot 1 \cdot 9!}$$
$$= 4 \cdot 11 \cdot 5$$
$$= 220$$

NOW TRY EXERCISE 65   Thus there are 220 different combinations of 9 of the 12 drivers.

EXAMPLE 9 **Restaurants** On a 4-day vacation you want to eat dinner at a different local restaurant every evening. The beach town has 10 restaurants. How many different combinations of restaurants could you select?

Solution We want to find the number of different groups of 4 that can be formed from 10 restaurants, so $n = 10$ and $r = 4$.

$$_{10}C_4 = \frac{10!}{(10-4)!4!}$$

$$= \frac{10!}{6!4!}$$

$$= \frac{10 \cdot \overset{3}{\cancel{9}} \cdot 8 \cdot 7 \cdot \cancel{6!}}{\cancel{6!} \cdot \cancel{4} \cdot \cancel{3} \cdot \cancel{2} \cdot 1}$$

$$= 10 \cdot 3 \cdot 7$$

$$= 210$$

NOW TRY EXERCISE 67   Thus there are 210 different combinations of 4 of the 10 restaurants.   ✳

Up until now we have used only factorials of positive integers. In Example 10 we use 0!, which is defined to be 1.

EXAMPLE 10 **Pizzas** A pizza restaurant offers a cheese pizza with 5 possible toppings: pepperoni, sausage, green peppers, onions, and mushrooms. How many different kinds of pizza can be ordered?

Solution A pizza can have 0, 1, 2, 3, 4, or 5 toppings. We must find the sum of the number of combinations for each number of toppings. There is only 1 type of pizza that can be made with no toppings, the plain cheese pizza. We can calculate the number of combinations possible using 0 toppings, $_5C_0$, as follows.

$$_5C_0 = \frac{5!}{(5-0)!0!}$$

$$= \frac{5!}{5!0!}$$

$$= \frac{\cancel{5!}}{\cancel{5!} \cdot 1}$$

$$= 1$$

Thus, there is only 1 pizza possible that uses no toppings.

For 1 topping, we calculate the number of combinations possible using $_5C_1$.

$$_5C_1 = \frac{5!}{(5-1)!1!}$$

$$= \frac{5!}{4!1!}$$

$$= \frac{5 \cdot \cancel{4!}}{\cancel{4!} \cdot 1}$$

$$= 5$$

There are 5 types of pizza that can be made using exactly one of the 5 toppings listed. For 2 toppings, we calculate the number of combinations possible using $_5C_2$.

$$_5C_2 = \frac{5!}{(5-2)!2!}$$

$$= \frac{5!}{3!2!}$$

$$= \frac{5 \cdot \overset{2}{\cancel{4}} \cdot \cancel{3!}}{\cancel{3!} \cdot \cancel{2} \cdot 1}$$

$$= 5 \cdot 2$$

$$= 10$$

There are 10 types of pizza that can be made using exactly 2 of the 5 toppings listed. For 3 toppings, we calculate the number of combinations possible using $_5C_3$.

$$_5C_3 = \frac{5!}{(5-3)!3!}$$

$$= \frac{5!}{2!3!}$$

$$= \frac{5 \cdot \overset{2}{\cancel{4}} \cdot \cancel{3!}}{\cancel{2} \cdot 1 \cdot \cancel{3!}}$$

$$= 5 \cdot 2$$

$$= 10$$

There are 10 types of pizza that can be made using exactly 3 of the 5 toppings listed. For 4 toppings, we calculate the number of combinations possible using $_5C_4$.

$$_5C_4 = \frac{5!}{(5-4)!4!}$$

$$= \frac{5!}{1!4!}$$

$$= \frac{5 \cdot \cancel{4!}}{1 \cdot \cancel{4!}}$$

$$= 5$$

There are 5 types of pizza that can be made using exactly 4 of the 5 toppings listed. For 5 toppings, we calculate the number of combinations possible using $_5C_5$.

$$_5C_5 = \frac{5!}{(5-5)!5!}$$

$$= \frac{5!}{0!5!}$$

$$= \frac{\cancel{5!}}{1 \cdot \cancel{5!}}$$

$$= 1$$

There is 1 type of pizza that can be made using all 5 of the 5 toppings listed. To find the total number of types of pizzas we add the combinations.

$$_5C_0 + {_5C_1} + {_5C_2} + {_5C_3} + {_5C_4} + {_5C_5} = 1 + 5 + 10 + 10 + 5 + 1 = 32$$

Thus, there are 32 types of pizzas that can be ordered. ✳

Now we will do an example using combinations and the counting principle.

EXAMPLE 11 **Manuscripts** An editor has 8 manuscripts for mathematics books and 5 manuscripts for computer science books. If he is to select 5 mathematics and 3 computer science manuscripts for publication, how many different choices does he have?

Solution For the mathematics books, 5 of the 8 manuscripts will be selected. The number of combinations of mathematics manuscripts is $_8C_5 = 56$. For computer science books, 3 of the 5 manuscripts will be selected. The number of combinations of computer science manuscripts is $_5C_3 = 10$. Since each of the combinations of the mathematics manuscripts can be paired with each of the combinations of the computer science manuscripts, we use the counting principle to determine the total number of choices. That is, we multiply the number of combinations of mathematics manuscripts by the number of combinations of computer science manuscripts.

$$_8C_5 \cdot {_5C_3} = 56 \cdot 10 = 560$$

NOW TRY EXERCISE 83 Therefore the editor has 560 different choices.

Using Your Graphing Calculator

In Section 12.5, we learned how to compute $n!$ and $_nC_r$ on a TI-83 Plus calculator. We can also compute $_nP_r$. To find $_nP_r$ on the TI-83 Plus calculator, use item 2 under the PRB menu. To find $_7P_4$, use the following keystrokes:

KEYSTROKES   ANSWER

$_7P_4$ [7] [MATH] [▶] [▶] [▶] [2] [4] [ENTER]   840

If you are using a different graphing calculator, consult the manual to learn how to evaluate permutations.

Exercise Set 12.6

Concept/Writing Exercises

1. What is permutation?
2. What is a combination?
3. What does $_nP_r$ represent?
4. What does $_nC_r$ represent?
5. Explain how to evaluate $n!$
6. What is 0! equal to?
7. State the counting principle in your own words.
8. Explain the difference between a combination and a permutation. Give an example of a permutation problem and an example of a combination problem.

Practice the Skills

Evaluate each expression.

9. $7!$
10. $5!$
11. $\dfrac{12!}{8!}$
12. $\dfrac{18!}{13!}$
13. $\dfrac{13!}{(13-4)!}$
14. $\dfrac{25!}{(25-5)!}$
15. $\dfrac{9!}{2!4!}$
16. $\dfrac{12!}{3!4!5!}$
17. $_5P_0$
18. $_5P_1$
19. $_5P_2$
20. $_5P_3$
21. $_5P_4$
22. $_5P_5$
23. $_6P_4$
24. $_8P_5$
25. $_9P_4$
26. $_8P_7$
27. $_{10}P_0$
28. $_7P_7$
29. $_5C_0$
30. $_5C_1$
31. $_5C_2$
32. $_5C_3$

33. $_5C_4$

34. $_5C_5$

🔒 **35.** $_7C_3$

36. $_8C_4$

37. $_9C_6$

38. $_6C_0$

39. $_7C_7$

40. $_{10}C_9$

Problem Solving

Solve Exercises 41–48 using the counting principle.

41. *Race Tracks* The daily double at most race tracks consists of selecting the winning horse in both the first and second races. If the first race has 7 entries and the second race has 8 entries, how many daily double tickets must you purchase to guarantee a win?

42. *Teller Machines* To use an automated teller machine, you generally must enter a four-digit code, using the digits 0–9. How many four-digit codes are possible if repetition of digits is permitted?

43. *Security* A secured elevator allows access by pressing the correct sequence of buttons. A display of the 10 buttons by the elevator door is shown below.

   The correct sequence of 6 buttons must be pressed to unlock the door. If the same button may be pressed consecutively, how many possible ways can the 6 buttons be pressed?

44. *Elevators* A secured elevator allows access by pressing the correct sequence of letters on a control panel. If the control panel contains the 26 letters of the alphabet and a 4-letter code must be entered (repetition is permitted), how many different codes are possible?

45. *Social Security Numbers* A social security number consists of 9 digits. How many different social security numbers are possible if repetition of digits is permitted?

46. *Identification Numbers* A college identification number consists of a letter followed by 4 digits. How many different identification numbers are possible if repetition is not permitted?

🔒 **47. *Sound Systems*** The operator of the Sounds Great Audio Store is planning a grand opening. He wishes to advertise that the store has many different sound systems available. They stock 10 different tape decks, 12 different receivers, and 9 different speakers. Assuming that a sound system will consist of 1 of each, and that all pieces are compatible, how many different sound systems can they advertise?

48. *Race Tracks* The trifecta at most race tracks consists of selecting the first-, second-, and third-place finishers in a particular race in their proper order. If there are 6 entries in the trifecta race, how many tickets must you purchase to guarantee a win?

Solve Exercises 49–58 using permutations.

49. *Door Prizes* Three door prizes—a poster, a can of hair spray, and a bag of jelly beans—are to be awarded (gag gifts) at a get-together of 7 people. If 3 people are to be selected at random and awarded the gifts, how many different ways can the gifts be awarded?

50. *Club Officers* If a club consists of 8 members, how many different arrangements of president and vice-president are possible?

51. *Prizes* A teacher decides to give 6 different prizes to 6 of 30 students in her class. In how many ways can she do so?

52. *Paintings in a Museum* Nine Van Gogh paintings are to be displayed in a museum.
   a) In how many different ways can they be arranged if they must be next to one another?
   b) In how many different ways can they be displayed if *Starry, Starry Night* is to be in the middle?

🔒 **53. *Office Security*** A night guard at NSB Bank visits 8 different offices every hour. The pattern is varied each night so that the guard will not follow a specific routine. In how many different ways can this pattern be varied?

54. **Permutations of Words** Find the number of permutations of the letters in the word DOUBLE.

55. **Permutations of Words** Find the number of permutations of the letters in the word CURVE.

56. **Permutations of Words** Find the number of permutations of the letters in the word PYRAMID.

57. **History Tests** In one question of a history test the student is asked to match 10 dates with 10 events; each

date can only be matched with 1 event. In how many different ways can this question be answered?

58. **Using Flags for Messages** Six different colored flags will be placed on a pole, one beneath another. The arrangement of the colors indicates the message. How many messages are possible if 6 flags are to be selected from 8 different colored flags?

Solve Exercises 59–70 using combinations.

59. **Bookcases** A bookcase contains 12 different books. Four books will be selected at random and given away as prizes to 4 individuals. In how many ways can the selection be made?

60. **Banana Splits** John's ice cream parlor has 15 different flavors. Willie orders a banana split and has to select 3 different flavors. How many different selections are possible?

61. **Prizes** Mrs. Suarez, a teacher, decides to give identical prizes to 6 of the 18 students in her class. In how many ways can she do so?

62. **Essay Questions** Amy Drake must select and answer 4 of 5 essay questions on a test. In how many ways can she do so?

63. **Textbook Search Committee** A textbook search committee is considering 6 books for possible adoption. The

committee has decided to select 3 of the 6 for further consideration. In how many ways can it do so?

64. **Microwaves** Five of a sample of 18 microwaves will be selected and tested for defects. In how many ways can the quality control engineer do so?

65. **Scholarships** Three of 8 finalists will be selected and awarded a $10,000 scholarship. In how many ways can this selection be made?

66. **Records** Carter Fenton, a disc jockey has 20 records that he wants to play during the next hour, but he has time to play only 16 of them. How many different ways can he select a group of 16 to play?

67. **Grades in a Class** At the beginning of the semester your instructor informs the class that he marks on a curve and that exactly 6 of the 30 students in the class will receive a final grade of A. In how many ways can this result occur?

68. **Quiniela Bets** A quiniela bet consists of selecting the first- and second-place winners, in any order, in a particular event. For example, suppose you select a 2–5 quiniela. If 2 wins and 5 finishes second, or if 5 wins and 2 finishes second, you win. How many quiniela tickets are necessary to guarantee a win?

69. **Limousines** Zwick's Limousine Service is going to select and purchase 3 different limousines of 6 different limousines on display. How many different selections are possible?

70. **Lotteries** In the California State lottery you must select the 6 winning numbers (in any order) out of 51 possible numbers to win the grand prize. How many combinations of 6 numbers out of the 51 are possible?

Solve Exercises 71–86 using the counting principle, permutations, combinations, or a combination of these methods.

71. **ISBN Numbers** Each book registered in the Library of Congress must have an ISBN code number. For an ISBN number of the form D-DD-DDDDDD-D, where D represents a digit from 0 through 9, how many different ISBN numbers are possible if repetition of digits is allowed?

72. **Outfits** Mr. Johnston has 10 pairs of pants, 9 shirts, 12 ties, and 8 sport coats. Assuming he must wear 1 of each, how many different outfits can he wear?

73. **Permutations of Words** Find the number of permutations of the letters in the word NUMBER.

74. **Permutations of Words** Find the number of permutations of the letters in the word EQUATION.

75. **Combination Locks**

a) To open a combination lock, you must know the lock's 3-number sequence in its proper order. Repetition of numbers is permitted. Why is this lock more like a permutation lock than a combination lock? Why is it not a true permutation problem?

b) Assuming that a combination lock has 40 numbers, determine how many different 3-number arrangements are possible if repetition of numbers is allowed.

c) Answer the question in part **b)** if repetition is not allowed

76. *License Plates* A license plate is to consist of 4 digits followed by 2 letters. Determine the number of different license plates possible if

a) repetition of numbers and letters is permitted.

b) repetition of numbers and letters is not permitted.

c) the first and second digits must be odd, and repetition is not permitted.

d) the first digit cannot be a zero and repetition is not permitted.

77. *Telephone Numbers* A telephone number consists of 7 digits with the restriction that the first digit cannot be 0 or 1.

a) How many distinct telephone numbers are possible?

b) How many distinct telephone numbers are possible with 3-digit area codes preceding the 7-digit number, where the first digit of the area code is not 0 or 1?

c) With the increasing use of cellular phones and paging systems our society is beginning to run out of usable phone numbers. Various phone companies are developing phone numbers that use 11 digits instead of 7. How many distinct phone numbers can be made with 11 digits assuming that the area code remains 3 digits and the first digit of the area code and the phone number cannot be 0 or 1?

78. *Arrangements of Pictures* The 6 pictures shown are to be placed side by side along a wall. In how many ways can they be arranged from left to right if

a) they can be arranged in any order?

b) the bird must be on the far left?

c) the bird must be on the far left and the giraffe must be next to the bird?

d) a four-legged animal must be on the far right?

79. *Wedding Reception* At the reception line of Mr. and Mrs. Doan's wedding, Mrs. Doan, Mr. Doan, the best man, the maid of honor, the 3 ushers, and the 3 bridesmaids must line up to receive the guests.

a) If these individuals can line up in any order, how many arrangements are possible?

b) If Mr. Doan must be the last in line and Mrs. Doan must be next to him, and the others can line up in any order, how many arrangements are possible?

c) If Mr. Doan is to be last in line, Mrs. Doan next to him, and males and females are to alternate, how many arrangements are possible?

80. *Batting Orders* In how many ways can the manager of a National League baseball team arrange his batting order of 9 players if the pitcher must bat last?

81. *History Tests* On a history test, Frank Marsh must write an essay for 3 of the 4 questions in Part 1, and 3 of the 5 questions in Part 2. How many different combinations of questions can he answer?

82. *Food* Miguel is sent to the store to get 4 different boxes of cereal and 3 different containers of ice cream. If there are 8 different types of cereal and 7 different ice creams to choose from, how many different choices does Miguel have?

83. *Committees* How many different committees can be formed from 5 teachers and 30 students if the committee is to consist of 3 teachers and 3 students?

84. *Mathematics Tests* A teacher is constructing a mathematics test consisting of 10 questions. She has a pool of 28 questions, which are classified by level of difficulty as follows: 6 difficult questions, 10 average questions, and 12 easy questions. How many different 10-question tests can she construct from the pool of 28 questions if her test is to have 3 difficult, 4 average, and 3 easy questions?

85. *Cereals* A cereal company is testing 6 oat cereals, 5 wheat cereals, and 4 rice cereals. If it plans to market 3 of the oat cereals, 2 of the wheat cereals, and 2 of the rice cereals, how many different combinations are possible?

86. *Trays of Hors D'oeuvres* Derrick Williams' catering service is making up trays of hors d'oeuvres. The hors d'oeuvres are categorized as inexpensive, average, and expensive. If the client must select 3 of the 7 inexpensive, 5 of the 8 average, and 2 of the 4 expensive hors d'oeuvres, how many different choices are possible?

Group Activity

87. *Tapping Glasses*

   a) At Drew Archibald's restaurant 4 people at dinner make a toast. If each person is to tap glasses with each other person, how many taps will take place?

   b) Repeat part **a)** with 5 people.

   c) How many taps will there be if there are n people at the dinner table?

88. *Lotteries* Determine the number of combinations possible in a state lottery where you must select

   a) 6 of 46 numbers.

   b) 6 of 47 numbers.

   c) 6 of 48 numbers.

   d) 6 of 49 numbers.

   e) Does the number of combinations increase by the same amount going from part **a)** to part **b)** as from part **b)** to part **c)**?

   f) Guess the number of combinations possible for 6 of 50 numbers.

   g) Determine the number of combinations possible for 6 of 50 numbers. Compare your answer to your answer in part **f)**.

89. *License Plates* Some states in the United States are experiencing tremendous growth and must redesign license plates to accommodate the demand. Each group member will determine the number of distinct license plates possible using digits (0 through 9) or letters from the English alphabet (a through z). Each license plate must fill 6 positions.

   a) Group Member 1: Determine the number of license plates possible using 6 digits (repetition is permitted).

      Group Member 2: Determine the number of license plates possible using a letter in the first position followed by 5 digits (repetition is permitted).

      Group Member 3: Determine the number of license plates possible using 3 letters followed by 3 digits (repetition is permitted).

   b) Share your result with the other members.

   c) Suppose a state needs 1.5 million license plates. Which of the 3 plans described in part **a)** can be used. Explain.

   d) Suppose a state needs 25 million license plates. Can any of the plans described in part **a)** be used? If not, as a group, design a license plate containing a combination of 7 letters and digits which will satisfy this need.

Cumulative Review Exercises

[2.1] **90.** Solve $\dfrac{3}{4}x - 2 = 4$.

[5.8] **91.** Solve $3x^2 = 18x$.

[6.4] **92.** Solve $\dfrac{x^2}{x + 5} = \dfrac{25}{x + 5}$.

[7.6] **93.** Solve $\sqrt{x - 8} + 12 = 19$.

12.7 PROBABILITY

1   Determine the probability of an event.

2   Determine the probability an event will not occur.

3   Determine the probability of a compound event.

4   Determine the odds against an event.

1 Determine the Probability of an Event

In Section 12.6 we discussed outcomes. Recall that the possible results of an experiment are its **outcomes**. For instance the outcomes of rolling a die are 1, 2, 3, 4, 5, and 6.

An **event** is a subset of the possible outcomes of an experiment. For example, when a die is rolled, the event of rolling a number 2 can be satisfied by only one outcome, 2. The event of rolling a number greater than 2 can be satisfied by any of the four outcomes 3, 4, 5, or 6. The event of rolling an even number can be satisfied by any of three outcomes 2, 4, or 6.

When we determine the probability of an event, we are finding the likelihood of the event occurring. We use the notation $P(E)$ to represent the probability of an event E.

If each outcome of an experiment has the same chance of occurring as any other outcome, they are said to be **equally likely outcomes**. For instance, when rolling a die, the outcomes $1, 2, 3, 4, 5,$ and 6 are equally likely. An odd number is as likely to be rolled as an even number; therefore, odd and even are another set of equally likely outcomes. If the outcomes of an experiment are equally likely, the probability can be calculated using the probability formula.

Probability

$$P(E) = \frac{\text{number of outcomes favorable to } E}{\text{total number of possible outcomes}}$$

EXAMPLE 1

Rolling a Die   A die is rolled. Find the probability of rolling

a) a 4.

b) an odd number.

c) a number less than 3.

d) a 7.

e) a number less than 7.

Solution   **a)** There are 6 possible equally likely outcomes: $1, 2, 3, 4, 5,$ and 6. The event of rolling a 4 can occur in one way.

$$P(4) = \frac{\text{number of outcomes that result in rolling a 4}}{\text{total number of possible outcomes}}$$

$$= \frac{1}{6}$$

b) The event of rolling an odd number can occur in three ways (when a 1, 3, or 5 is rolled).

$$P(\text{odd number}) = \frac{\text{number of outcomes that result an odd number}}{\text{total number of possible outcomes}}$$

$$= \frac{3}{6} = \frac{1}{2}$$

c) The event of rolling a number less than 3 can occur in two ways (when a 1 or 2 is rolled).

$$P(\text{number less than 3}) = \frac{\text{number of outcomes that result in a number less than 3}}{\text{total number of possible outcomes}}$$

$$= \frac{2}{6} = \frac{1}{3}$$

d) A 7 cannot be rolled on a die. Therefore, the number outcomes that result in a 7 is 0.

$$P(7) = \frac{\text{number of outcomes that result in a 7}}{\text{total number of possible outcomes}}$$

$$= \frac{0}{6} = 0$$

e) The event of rolling a number less than 7 can occur in six ways (when a 1, 2, 3, 4, 5, or 6 is rolled).

$$P(\text{number less than 7}) = \frac{\text{number of outcomes that result in a number less than 7}}{\text{total number of possible outcomes}}$$

$$= \frac{6}{6} = 1$$

NOW TRY EXERCISE 11

Example 1 illustrates two important facts about probabilities. In Part **d)** we found that the probability of rolling a 7 on a die is 0. It is impossible to roll a 7 on a die since a die does not contain a 7. The probability of an event that cannot occur is 0. In Part **e)** we found that the probability of rolling a number less than 7 is 1. When a die is rolled, a number less than 7 must be rolled because all of the numbers on a die are less than 7. The probability of an event that must occur is 1.

We now present some important facts about probabilities.

Important Facts

1. The probability of an event that cannot occur is 0.

2. The probability of an event that must occur is 1.

3. Every probability will be a number between 0 and 1, inclusive; that is, $0 \le P(E) \le 1$.

4. The sum of the probabilities of all possible outcomes of an experiment is 1.

To illustrate statement 4, consider a die. There are 6 possible outcomes. Each individual outcome has a probability of $\frac{1}{6}$. The sum of the probabilities of all the possible outcomes is 1.

Roll a die						
Possible outcome	1	2	3	4	5	6
Probability	$\frac{1}{6}$	$\frac{1}{6}$	$\frac{1}{6}$	$\frac{1}{6}$	$\frac{1}{6}$	$\frac{1}{6}$

Sum of Probabilities = 1

2 Determine the Probability an Event Will Not Occur

In any experiment an event must either occur or not occur. The sum of the probability that an event will occur and the probability that the event will not occur is 1. Thus, for any event A, we conclude

$$P(A) + P(\text{not } A) = 1$$

or

$$P(\text{not } A) = 1 - P(A)$$

To find the probability of an event not happening, subtract the probability of the event happening from 1.

For instance, in Example 1 **c)** we determined the probability of obtaining a number less than 3 when rolling a die. When rolling a die, we will either get a number less than 3 or we will not get a number less than 3. Since we found that the probability of getting a number less than 3 is $\frac{1}{3}$, then the probability that we do not get a number less than 3 must be $1 - \frac{1}{3} = \frac{2}{3}$.

EXAMPLE 2   **Deck of Cards**   A deck of 52 playing cards (see photo) consists of 4 suits: spades, diamonds, hearts, and clubs. Each suit has 13 cards, including numbered cards ace (1) through 10 and 3 face (or picture) cards, the jack, the queen, and the king. Hearts and diamonds are red suits; clubs and spades are black suits. One card is selected at random from the deck of cards. Find the probability that the card selected is

a) a 7.

b) not a 7.

c) a spade.

d) a heart *and* a club.

e) a card greater than 6 *and* less than 9.

Solution   **a)** There are four 7s in a deck of 52 cards.

$$P(7) = \frac{4}{52} = \frac{1}{13}$$

b) $P(\text{not a } 7) = 1 - P(7) = 1 - \frac{1}{13} = \frac{12}{13}$

c) There are 13 spades in the deck.

$$P(\text{spade}) = \frac{13}{52} = \frac{1}{4}$$

d) The word *and* means that both events must occur. There is no card that is both a heart and a club.

$$P(\text{heart and club}) = \frac{0}{52} = 0$$

e) The cards that are both greater than 6 and less than 9 are the 7s and 8s. There are four 7s and four 8s, or a total of 8 cards.

NOW TRY EXERCISE 19

$$P(\text{greater than 6 and less than 9}) = \frac{8}{52} = \frac{2}{13}$$

3   Determine the Probability of a Compound Event

A **compound event** is one that provides an alternative such as A or B, where A and B are events. For example, suppose a die is rolled where A is the event of "rolling a 4" and B is the event of "rolling an odd number." If we are asked to find $P(\text{rolling a 4 or rolling an odd number})$, or $P(A \text{ or } B)$, then we are finding the probability of a compound event.

Let's discuss how to find the probability of a compound event using events A and B as given above.

Possible outcomes for   Possible outcomes for
rolling a 4   rolling an odd number

$\{4\}$ $\{1, 3, 5\}$

There is one possible outcome for rolling a 4 and three different possible outcomes for rolling an odd number. Thus, there are four outcomes $\{4, 1, 3, 5\}$ that satisfy either one of the events out of six possible outcomes. Thus,

$$P(\text{rolling a 4 or rolling an odd number}) = \frac{4}{6} = \frac{2}{3}$$

Suppose that event A is rolling a 3 rather than rolling a 4. Then $P(A$ or $B)$ be-comes P(rolling a 3 or rolling an odd number). Let's determine the probability.

Possible outcomes for   Possible outcomes for
rolling a 3   rolling an odd number

$\{3\}$ $\{1, 3, 5\}$

There is one possible outcome for rolling a 3 and three possible outcomes for rolling an odd number. However, the outcome for rolling a 3 is included in the outcomes for rolling an odd number. Thus, there are only three outcomes $\{1, 3, 5\}$ that satisfy either one of the events out of six possible outcomes.

$$P(\text{rolling a 3 or rolling an odd number}) = \frac{3}{6} = \frac{1}{2}$$

Notice in the last example we had to be careful not to count an outcome that satisfies both events more than once. To find $P(A$ or $B)$, we use the addition rule.

Addition Rule.

For events A and B,

$$P(A \text{ or } B) = P(A) + P(B) - P(A \text{ and } B)$$

Sometimes events A and B cannot happen simultaneously. When this happens, we say that events A and B are **mutually exclusive**. The events "rolling a 4" and "rolling an odd number" are mutually exclusive events because it is not possible for both events to occur simultaneously. The events "rolling a 3" and "rolling an odd number" are not mutually exclusive events because it is possible for both events to occur simultaneously. If a 3 were rolled, both events would be satisfied.

If events A and B are mutually exclusive, the $P(A$ and $B)$ in the addition formula equals 0. Now let's work an example.

EXAMPLE 3   **Deck of Cards**   One card is selected at random from a deck of cards. Find the probability of selecting

a) a queen *or* king.

b) a heart *or* a 6.

c) a red card *or* a diamond.

Solution   **a)** There are 4 queens and 4 kings. Since a card can never be both a queen and a king, the events are mutually exclusive.

$$P(\text{queen or king}) = P(\text{queen}) + P(\text{king}) - P(\text{queen and king})$$
$$= \frac{4}{52} + \frac{4}{52} - \frac{0}{52} = \frac{8}{52} = \frac{2}{13}$$

b) There are 13 hearts in the deck and four 6s. One of the 6s is a heart, the 6 of hearts. Since one of the 52 cards is both a heart and a 6, $P(\text{heart and 6}) = \frac{1}{52}$.

$$P(\text{heart or 6}) = P(\text{heart}) + P(6) - P(\text{heart and 6})$$
$$= \frac{13}{52} + \frac{4}{52} - \frac{1}{52} = \frac{16}{52} = \frac{4}{13}$$

c) There are 26 red cards and 13 diamond cards. All of the 13 diamonds are red cards.

$$P(\text{red or diamond}) = P(\text{red}) + P(\text{diamond}) - P(\text{red and diamond})$$

$$= \frac{26}{52} + \frac{13}{52} - \frac{13}{52} = \frac{26}{52} = \frac{1}{2}$$

NOW TRY EXERCISE 23

4 Determine the Odds Against an Event

We often hear the word "odds." Odds are generally quoted as "odds against an event." The **odds against an event** is a ratio of the probability that the event does not occur to the probability that the event occurs. Thus, to find the odds against an event we first find the probability of the event occurring and the probability of the event not occurring. Recall that the probability that an event does not occur can be calculated from the probability of the event occuring.

$$\text{Odds against an event} = \frac{P(\text{Event does not occur})}{P(\text{Event occurs})}$$

EXAMPLE 4 **Rolling a Die**   Find the odds against rolling a 3 on one roll of a die.

Solution First determine the probability of rolling a 3 and the probability of not rolling a 3.

$$P(3) = \frac{1}{6} \qquad P(\text{not a } 3) = 1 - P(3)$$

$$= 1 - \frac{1}{6} = \frac{5}{6}$$

Now that we know the probability of rolling a 3 and the probability of not rolling a 3, we can determine the odds against rolling a 3.

$$\text{Odds against rolling a } 3 = \frac{P(\text{not a } 3)}{P(3)}$$

$$= \frac{\frac{5}{6}}{\frac{1}{6}}$$

$$= \frac{5}{6} \cdot \frac{6}{1}$$

$$= \frac{5}{1}$$

When giving the odds, it is common to write the ratio with a colon. The ratio $\frac{5}{1}$ can be written as $5 : 1$ and is read "5 to 1." The odds against rolling 3 are $5 : 1$.

NOW TRY EXERCISE 55

EXAMPLE 5 **Businesses Going Bankrupt**   In a town, an estimated 2 of every 5 businesses go bankrupt within their first year of operation. If you open a new business in this town, what are the odds against it going bankrupt?

Solution The probability that your business will go bankrupt is $\frac{2}{5}$. The probability that your business will not go bankrupt is $1 - \frac{2}{5} = \frac{3}{5}$.

$$\text{Odds against business going bankrupt} = \frac{P(\text{business does not go bankrupt})}{P(\text{business goes bankrupt})}$$

$$= \frac{\dfrac{3}{5}}{\dfrac{2}{5}}$$

$$= \frac{3}{5} \cdot \frac{5}{2}$$

$$= \frac{3}{2}$$

NOW TRY EXERCISE 61   The odds against the business going bankrupt are $3:2$.   ✳

Although odds are generally given as odds against an event, they may also be given as odds in favor of an event. The **odds in favor of an event** is a ratio of the probability that the event occurs to the probability that the event does not occur.

$$\text{Odds in favor of an event} = \frac{P(\text{Event occurs})}{P(\text{Event does not occur})}$$

If the odds against an event are $a:b$, odds in favor of the event are $b:a$.

Exercise Set 12.7

Concept/Writing Exercises

1. What are equally likely outcomes?

2. How do you determine the probability of an event?

3. What is the sum of $P(A)$ and $P(\text{not } A)$?

4. **a)** If $P(A) = 1$, what is true about A?
   b) If $P(A) = 0$, what is true about A?

5. What does it mean for two events to be mutually exclusive?

6. How do you determine the odds against an event?

7. Between what two numbers, inclusive, must all probabilities be?

8. If the odds against an event are $5:3$, what are the odds in favor of the event?

Practice the Skills

For Exercises 9–16, a marble is taken out of a bag containing 75 marbles—15 red, 20 blue, 10 green, and 30 orange. Find each given probability.

9. P(a red marble is chosen)

10. P(a blue marble is chosen)

11. P(a green marble is chosen)

12. P(an orange marble is chosen)

13. P(a red marble is *not* chosen)

14. P(a red *and* green marble is chosen)

15. P(a blue *or* a green marble is chosen)

16. P(a red *or* an orange marble is chosen)

For Exercises 17–28, a card is selected at random from a deck of cards. Find each probability.

17. P(selecting a 9)

18. P(selecting a 10)

19. P(selecting the ace of spades)

20. P(selecting a black card)

21. P(selecting a red card)

22. P(selecting a king)

23. *P*(selecting a card that is not a king)

24. *P*(selecting a card that is a 9 and a black card)

25. *P*(selecting a card that is a 9 or a 10)

26. *P*(selecting a card that is a red card and a king)

27. *P*(selecting a card that is a red card or a king)

28. *P*(selecting a card that is a king or the ace spades)

For Exercises 29–34, a song is selected at random from a juke box containing 40 rock/pop songs, 27 country songs, and 3 classical pieces. Find each probability.

29. *P*(selecting a rock/pop song)

30. *P*(selecting a country song)

31. *P*(selecting a classical piece)

32. *P*(selecting a song that is *not* a rock/pop song)

33. *P*(selecting a country song *or* a classical piece)

34. *P*(selecting a song that is a rock/pop song and a classical piece)

For Exercises 35–40, a spinner is spun. Assuming that the spinner cannot land on a line, find the probability of landing on
a) *yellow,* **b)** *purple,* **c)** *green,* **d)** *yellow or purple,* **e)** *a color that is not purple.*

35.

36.

37.

38.

39.

40.

For Exercises 41–46, a traffic light is red for 40 seconds, yellow for 10 seconds, and green for 60 seconds. Find each probability.

41. P(the light is red)

42. P(the light is yellow)

43. P(the light is green)

44. P(the light is *not* red)

45. P(the light is yellow *or* green)

46. P(the light is *not* green)

47. If $P(A) = 0.4$, $P(B) = 0.5$, and $P(A \text{ and } B) = 0.2$, find $P(A \text{ or } B)$.

48. If $P(A) = 0.1$, $P(B) = 0.3$, and $P(A \text{ and } B) = 0.1$, find $P(A \text{ or } B)$.

49. If $P(A) = 0.8$, $P(B) = 0.2$, and $P(A \text{ and } B) = 0.1$, find $P(A \text{ or } B)$.

50. If $P(A) = 0.6$, $P(B) = 0.2$, and $P(A \text{ and } B) = 0.0$, find $P(A \text{ or } B)$.

51. *Blouses* Aimee's closet contains 12 blouses, 4 of which are white. Aimee selects 1 blouse at random.

  a) Find the probability that it is white.

  b) Find the probability that it is not white.

  c) Find the odds against it being white.

  d) Find the odds in favor of it being white.

52. *Dogs* There are 5 samoyeds, 3 boxers, and 7 poodles in the animal shelter. Paulo selects 1 dog at random.

  a) Find the probability that it is a boxer.

  b) Find the probability that it is not a boxer.

  c) Find the odds in favor of it being a boxer.

  d) Find the odds against it being a boxer.

In Exercises 53–56, a die is rolled.

53. Find the odds against rolling a 4.

54. Find the odds against rolling an even number.

55. Find the odds against rolling a number greater than 5.

56. Find the odds against rolling a number less than 4.

In Exercises 57–60, a card is selected from a deck of cards.

57. Find the odds against selecting a 9.

58. Find the odds against selecting a Jack, Queen, or King.

59. Find the odds in favor of selecting a red card.

60. Find the odds against and the odds in favor of selecting a card greater than 7 (ace is low).

Problem Solving

61. *Lotteries* One million tickets are sold for a lottery.

  a) If you purchase a ticket, find your odds against winning.

  b) If you purchase 10 tickets, find your odds against winning.

62. *Class of Students* One person is selected at random from a class of 16 males and 12 females.

  a) Find the odds against selecting a female.

  b) Find the odds against selecting a male.

63. *Odds of an Event* The odds in favor of an event are $7:3$.

  a) Find the probability that the event occurs.

  b) Find the probability that the event does not occur.

64. *Tennis Tournament* The odds in favor of Erin winning a tennis tournament are $2:7$.

  a) Find the probability that Erin will win the tournament.

  b) Find the probability that Erin will not win the tournament.

65. *Promotion* The odds against Jamal getting promoted to vice-president are $3:7$. Find the probability that Jamal gets promoted to vice-president.

66. *Horse Race* The odds against the horse Speedy winning a race are $5:2$.

  a) Find the probability that Speedy wins.

  b) Find the probability that Speedy loses.

67. *Buying a House* Suppose that the probability that you buy a house this week is 0.6. Find the odds against buying a house this week.

68. *Probability of Rain* Suppose that the probability that it will rain today is $\frac{3}{8}$. Find the odds against it raining today.

69. *Water Heater* Suppose that the probability that Triangle Pools fixes your pool's water heater correctly the first time is 0.7. Find the odds against your pool's water heater being repaired correctly by Triangle Pools on the first attempt.

70. *Horse Race* Find the probability of each horse winning the race if the odds against the horse winning are as given in the chart below. (Do not be concerned that the sum of the probabilities is not 1.)

Horse	Odds
Old Boy	$8:3$
Daylight	$3:1$
One For All	$12:1$
Sunday	$7:6$
Lightning	$1:1$

Group Activity

In this section we have calculated theoretical probabilities using possible outcomes of an experiment. We will now determine empirical (or experimental) probabilities using the results of actual data. Empirical probabilities are determined by dividing the number of times a specific outcome occurs by the total number of trials. For this activity, each group member will need 3 coins.

71. *Tossing Coins* a) Each member should toss the 3 coins 40 times and record the number of times each toss of the 3 coins results in exactly

    3 heads
    2 heads
    1 head
    0 heads.

b) Now, each group member should determine the experimental probabilities of obtaining each of the 4 events given in part **a)** by dividing the number of observations of 3 heads, 2 heads, 1 head, and 0 heads by the total number of trials, 40.

c) Compare your result with the results of the other members. Are the fractions (probabilities) close? Can you explain why or why not?

d) Construct a tree diagram for tossing 3 coins. Use this to determine the theoretical probabilities of getting 3 heads, 2 heads, 1 head, and 0 heads. Write these probabilities with a denominator of 120.

e) Combine the results from part **a)** for all 3 group members to find the total number of times each toss of the 3 coins resulted in exactly 3 heads, 2 heads, 1 head, and 0 heads. Divide these numbers by 120 to obtain the experimental probabilities of each of these outcomes.

f) Compare the results from parts **d)** and **e)**. Are the experimental probabilities close to the theoretical probabilities? Explain.

Cumulative Review Exercises

[3.3] 72. Graph $12x - 16y = 48$ using the x- and y-intercepts.

[10.4] 73. Graph $P(x) = x^4 - 2x^3 + 4x^2 - 14x - 21$.

[10.5] 74. Graph $f(x) = \dfrac{3x^2 - 9x - 12}{2x^2 - 10x + 8}$.

[11.2] 75. Graph $\dfrac{(x+3)^2}{25} + \dfrac{(y-5)^2}{9} = 1$.

CHAPTER SUMMARY
Key Words and Phrases

12.1
Alternating sequence
Decreasing sequence
Finite sequence
General term of a sequence
Increasing sequence
Index of summation
Infinite sequence
Lower limit of a summation
Partial sum of a series
Sequence
Series

Sigma
Summation notation
Upper limit of a summation

12.2
Arithmetic sequence
Arithmetic series
Common difference

12.3
Common ratio
Geometric sequence

Geometric series
Infinite geometric series

12.4
Mathematical induction

12.5
Binomial coefficient
Binomial theorem
Factorial
Pascal's triangle

12.6
Combination

Counting principle
Independent events
Outcomes
Permutation
Tree diagram

12.7
Compound event
Equally likely outcomes
Mutually exclusive events
Odds against an event
Odds in favor of an event
Probability

(continued on the next page)

IMPORTANT FACTS

$$\sum_{i=1}^{n} x_i = x_1 + x_2 + x_3 + \cdots + x_n$$

nth Term of a Geometric Sequence

$$a_n = a_1 r^{n-1}$$

Pascal's Triangle

$$
\begin{array}{ccccccccc}
 & & & & 1 & & & & \\
 & & & 1 & & 1 & & & \\
 & & 1 & & 2 & & 1 & & \\
 & 1 & & 3 & & 3 & & 1 & \\
1 & & 4 & & 6 & & 4 & & 1 \\
\end{array}
$$

nth Partial Sum of a Geometric Series

$$s_n = \frac{a_1(1 - r^n)}{1 - r}, r \neq 1$$

Sum of an Infinite Geometric Series

$$s_\infty = \frac{a_1}{1 - r}, |r| < 1$$

n Factorial

$$n! = n(n - 1)(n - 2) \cdots (2)(1)$$

nth Term of an Arithmetic Sequence

$$a_n = a_1 + (n - 1)d$$

The Principle of Mathematical Induction

S_n is true for every positive integer n if the following two conditions are true.

Condition 1: S_1 is true and

Condition 2: for any positive integer $k, k \leq n$, if S_k is true, then S_{k+1} is true.

nth Partial Sum of an Arithmetic Series

$$s_n = \frac{n(a_1 + a_n)}{2}$$

Binomial Coefficients

$$\binom{n}{r} = \frac{n!}{r!(n - r)!}$$

Binomial Theorem

$$(a + b)^n = \binom{n}{0}a^n b^0 + \binom{n}{1}a^{n-1}b^1 + \binom{n}{2}a^{n-2}b^2 + \binom{n}{3}a^{n-3}b^3 + \cdots + \binom{n}{n}a^0 b^n$$

Permutation Formula

$$_nP_r = \frac{n!}{(n - r)!}$$

Probability

$$P(E) = \frac{\text{Number of outcomes favorable to } E}{\text{Total number of possible outcomes}}$$

Combination Formula

$$_nC_r = \frac{n!}{(n - r)!r!}$$

Addition Rule

$$P(A \text{ or } B) = P(A) + P(B) - P(A \text{ and } B)$$

Odds

$$\text{Odds against an event} = \frac{P(\text{Event does not occur})}{P(\text{Event does occur})}$$

$$\text{Odds in favor of an event} = \frac{P(\text{Event does occur})}{P(\text{Event does not occur})}$$

Chapter Review Exercises

[12.1] Write the first five terms of each sequence.

1. $a_n = n + 2$

2. $a_n = n^2 + n - 3$

3. $a_n = \dfrac{4}{n}$

4. $a_n = \dfrac{n^2}{n + 4}$

Find the indicated term of each sequence.

5. $a_n = 3n - 11$, seventh term

6. $a_n = (-1)^n + 3$, seventh term

7. $a_n = \dfrac{n + 7}{n^2}$, ninth term

8. $a_n = (n)(n - 3)$, eleventh term

For each sequence, find the first and third partial sums, s_1 and s_3.

9. $a_n = 2n + 7$

10. $a_n = n^2 + 5$

11. $a_n = \dfrac{n + 3}{n + 2}$

12. $a_n = (-1)^n(n + 8)$

Write the next three terms of each sequence. Then write an expression for the general term, a_n.

13. $2, 4, 8, 16, \ldots$

14. $-27, 9, -3, 1, \ldots$

15. $\dfrac{1}{5}, \dfrac{2}{5}, \dfrac{4}{5}, \dfrac{8}{5}, \ldots$

16. $12, 8, 4, 0, \ldots$

Write out each series. Then find the sum of the series.

17. $\displaystyle\sum_{j=1}^{3} (j^2 + 4)$

18. $\displaystyle\sum_{k=1}^{4} k(k + 2)$

19. $\displaystyle\sum_{k=1}^{5} \dfrac{k^2}{6}$

20. $\displaystyle\sum_{i=1}^{4} \dfrac{i}{i + 1}$

For the set of values $x_1 = 3, x_2 = 9, x_3 = 5, x_4 = 10$, evaluate the indicated sum.

21. $\displaystyle\sum_{i=1}^{4} x_i$

22. $\displaystyle\sum_{i=1}^{4} (x_i)^2$

23. $\displaystyle\sum_{i=2}^{3} (x_i^2 + 1)$

24. $\left(\displaystyle\sum_{i=1}^{4} x_i \right)^2$

In Exercises 25 and 26, consider the following rectangles. For the nth rectangle, the length is $n + 3$ and the width is n.

25. *Perimeter*
   a) Find the perimeters for the four rectangles, and then list the perimeters in a sequence.
   b) Find the general term for the perimeter of the *n*th rectangle in the sequence. Use p_n for perimeter.

26. *Area*
   a) Find the areas for the four rectangles, and then list the areas in a sequence.
   b) Find the general term for the area of the *n*th rectangle in the sequence. Use a_n for area.

[12.2] *Write the first five terms of the arithmetic sequence with the indicated first term and common difference.*

27. $a_1 = 5, d = 2$

28. $a_1 = -12, d = -\dfrac{1}{2}$

29. $a_1 = \dfrac{1}{2}, d = -2$

30. $a_1 = -100, d = \dfrac{1}{5}$

For each arithmetic sequence, find the indicated value.

31. $a_1 = 6, d = 3$; find a_8

32. $a_1 = 10, a_7 = -14$; find d

33. $a_1 = -3, a_9 = 1$; find d

34. $a_1 = 22, a_n = -3, d = -5$; find n

Find s_n and d for each arithmetic sequence.

35. $a_1 = 7, a_8 = 21, n = 8$

36. $a_1 = -12, a_7 = -48, n = 7$

37. $a_1 = \dfrac{3}{5}, a_7 = 3, n = 7$

38. $a_1 = -\dfrac{10}{3}, a_9 = -6, n = 9$

Write the first four terms of each arithmetic sequence. Then find a_{10} and s_{10}.

39. $a_1 = -8, d = 4$

40. $a_1 = 5, d = -3$

41. $a_1 = \dfrac{5}{6}, d = \dfrac{2}{3}$

42. $a_1 = -60, d = 5$

Find the number of terms in each arithmetic sequence. Then find s_n.

43. $4, 9, 14, \ldots, 59$

44. $-7, -4, -1, \ldots, 14$

45. $\dfrac{6}{10}, \dfrac{9}{10}, \dfrac{12}{10}, \ldots, \dfrac{36}{10}$

46. $-9, -3, 3, 9, \ldots, 69$

47. *Barrels of Oil* Barrels of oil are stacked with 20 barrels in the bottom row, 19 barrels in the second row, 18 barrels in the third row, and so on, to the top row, which has only 1 barrel. How many barrels are there?

48. *Salary* Ahmed Mocanda just started a new job with an annual salary of $30,000. He has been told that his salary will increase by $1000 per year for the next 10 years.

a) Write a sequence showing his salary for the first 4 years.

b) Write a general term of this sequence.

c) What will his salary be 6 years from now?

d) How much money will he make in the first 11 years?

[12.3] *Determine the first five terms of each geometric sequence.*

49. $a_1 = 5, r = 2$

50. $a_1 = -12, r = \dfrac{1}{2}$

51. $a_1 = 20, r = -\dfrac{2}{3}$

52. $a_1 = -100, r = \dfrac{1}{5}$

Find the indicated term of each geometric sequence.

53. $a_1 = 6, r = \dfrac{1}{3}$; find a_5

54. $a_1 = 15, r = 2$; find a_7

55. $a_1 = -8, r = -3$; find a_6

56. $a_1 = \dfrac{1}{12}, r = \dfrac{2}{3}$; find a_5

Find each sum.

57. $a_1 = 7, r = 2$; find s_7

58. $a_1 = -84, r = -\frac{1}{4}$; find s_5

59. $a_1 = 9, r = \frac{3}{2}$; find s_4

60. $a_1 = \frac{3}{5}, r = \frac{5}{3}$; find s_7

For each geometric sequence, find the common ratio, r, and then write an expression for the general term, a_n.

61. $6, 12, 24, \ldots$

62. $-4, -20, -100, \ldots$

63. $7, \frac{7}{3}, \frac{7}{9}, \ldots$

64. $\frac{9}{5}, \frac{18}{15}, \frac{36}{45}, \ldots$

Find the sum of the terms in each infinite geometric sequence.

65. $3, \frac{3}{2}, \frac{3}{4}, \frac{3}{8}, \ldots$

66. $\frac{5}{2}, 1, \frac{2}{5}, \frac{4}{25}, \ldots$

67. $-8, \frac{8}{3}, -\frac{8}{9}, \frac{8}{27}, \ldots$

68. $-6, -4, -\frac{8}{3}, -\frac{16}{9}, \ldots$

Find the sum of each infinite series.

69. $8 + 4 + 2 + 1 + \cdots$

70. $7 + \frac{7}{3} + \frac{7}{9} + \frac{7}{27} + \cdots$

71. $5 - 1 + \frac{1}{5} - \frac{1}{25} + \cdots$

72. $-4, -\frac{8}{3}, -\frac{16}{9}, -\frac{32}{27}, \ldots$

Write the repeating decimal as a ratio of integers.

73. $0.262626\ldots$

74. $0.621621\ldots$

75. *Money* You begin with $100, double that to get $200, double that again to get $400, and so on. How much will you have after you perform this process 10 times?

76. *Salary* Gertude Dibble started a new job on January 1, 2003 with a monthly salary of $1600. Her boss has agreed to give her a 4% raise each month for the remainder of the year.
a) What is Gertude's salary in July?
b) What is Gertude's salary in December?
c) How much money does Gertude make in 2003?

77. *Inflation* If the inflation rate was a constant 8% per year (each year the cost of living is 8% greater than the previous year), how much would a product that costs $200 now cost after 12 years?

78. *Pendulum* On each swing (left to right or right to left), a pendulum travels 92% as far as on its previous swing. If the first swing is 8 feet, find the distance traveled by the pendulum by the time it comes to rest.

[12.4] *Use mathematical induction to prove the following statements for all positive integers n.*

79. $2 + 5 + 8 + \ldots + (3n - 1) = \dfrac{n(3n + 1)}{2}$

80. $2^2 + 4^2 + 6^2 + \ldots + (2n)^2 = \dfrac{2n(2n + 1)(n + 1)}{3}$

81. $1 \cdot 2 + 2 \cdot 3 + 3 \cdot 4 + \ldots + n(n + 1) = \dfrac{n(n + 1)(n + 2)}{3}$

82. $2 + 6 + 10 + \ldots + (4n - 2) = 2n^2$

[12.5] *Use the binomial theorem to expand the expression.*

83. $(3x + y)^4$

84. $(2x - 3y^2)^3$

Write the first four terms of the expansion.

85. $(x - 2y)^9$

86. $(2a^2 + 3b)^8$

[12.6] *Compute each permutation or combination.*

87. *Envelopes* Jeff has paper in 7 different colors and envelopes in 5 different colors. How many different paper and envelope color sets can he create?

88. *Seats in a Row* Eight people are to be seated in a row. In how many ways can they be arranged?

89. *Gifts* At a party 12 unique gifts are randomly distributed to the 12 guests. How many ways can the gifts be distributed?

90. *Dog Sled* Seven of 15 huskies are to be selected to pull a dogsled. They will be arranged in single file. How many arrangements are possible?

91. *Finalists in a Contest* Five finalists remain in a contest. The winner receives $1000 and the runner up

receives $500. How many different arrangements of prizes are possible?

92. *Employees* In a company with 37 employees, a group of 4 will be selected to attend a training session. How many different groups are possible?

93. *Clubs* A club consists of 6 males and 8 females. Three males and 3 females are to be selected to represent the club at dinner. How many different combinations are possible?

94. *Fitness Books* Ten fitness books are available to be checked out. A maximum of 3 can be checked out at a time. If at least 1 fitness book is checked out, how many different groups of fitness books can be checked out?

[12.7] *In Exercises 95–100 each of the digits 0–9 is written on a piece of paper, and all of the pieces of paper are placed in a hat. One number is selected at random.*

95. Find the probability that the number selected is odd.

96. Find the probability that the number selected is even and greater than 4.

97. Find the probability that the number selected is even or greater than 4.

98. Find the probability that the number selected is greater than 6 and less than 4.

99. Find the probability that the number selected is a multiple of 3.

100. Find the probability that the number selected is less than 10.

101. If the probability of rain is 0.75, what are the odds against rain?

102. Assume that the odds against winning money at a certain slot machine are 5:4. What is the probability of winning money at the slot machine?

Chapter Practice Test

1. Are the following sequences arithmetic, geometric, or neither? Explain your answer.
  a) $1, 1, 2, 3, 5, 8, 13, \ldots$
  b) $4, 1, -2, -5, -8, \ldots$
  c) $4, -2, 1, -\frac{1}{2}, \frac{1}{4}, -\frac{1}{8}, \ldots$

2. Write the first five terms of the sequence with $a_n = \dfrac{n + 2}{n^2}$.

3. Find the first and third partial sums of $a_n = \dfrac{2n + 3}{n}$.

4. Write out the following series and find the sum of the series.
$$\sum_{n=1}^{5} (2n^2 + 3)$$

5. For $x_1 = 4, x_2 = 2, x_3 = 8,$ and $x_4 = 12$, find $\sum_{i=1}^{4} x_i^2$.

6. Write the general term for the following arithmetic sequence.
$$\frac{1}{3}, \frac{2}{3}, \frac{3}{3}, \frac{4}{3}, \ldots$$

7. Write the general term for the following geometric sequence.
$$5, 10, 20, 40, \ldots$$

Write the first four terms of each sequence.

8. $a_1 = 12, d = -3$

9. $a_1 = \dfrac{5}{8}, r = \dfrac{2}{3}$

10. Find a_8 when $a_1 = 100$ and $d = -12$.

11. Find s_8 for the arithmetic series with $a_1 = 3$ and $a_8 = -11$.

12. Find the number of terms in the arithmetic sequence $-4, -16, -28, \ldots, -148$.

13. Find a_7 when $a_1 = 8$ and $r = \dfrac{2}{3}$.

14. Find s_7 when $a_1 = \dfrac{3}{5}$ and $r = -5$.

15. Find the sum of the following infinite geometric series.

$$3 + 2 + \frac{4}{3} + \frac{8}{9} + \cdots$$

16. Use mathematical induction to prove

$$5 + 10 + 15 + \ldots + 5n = \frac{5n(n+1)}{2}$$

for all positive integers n.

17. Use the binomial theorem to expand $(x + 2y)^4$.

18. **Sandwiches** If a deli has 3 choices for bread, 4 choices for meat, and 2 choices for cheese, how many different sandwiches can be prepared if 1 type of bread, 1 type of meat, and 1 type of cheese are to be used?

19. **Letters** If 7 letters arrive in the mail, in how many different ways can they be opened?

20. **Chores** If 8 chores need to be done today and you get to choose the 2 that you will do, how many options do you have?

21. **Photos** If 4 of 11 vacation photos are to be selected to be placed in a display using 4 identical frames, how many ways can this be done?

22. **Rolling a Die** If a die is rolled, what is the probability of rolling a number that is a multiple of 2 *or* a multiple of 3?

23. **Winning a Contest** If the probability of winning a contest is $\frac{1}{500}$, what are the odds against winning the contest?

24. **Retirement** To save for retirement, Jamie Monroe plans to save $1000 the first year, $2000 the second year, $3000 the third year, and to increase the amount saved by $1000 in each successive year. How much will she have saved by the end of her twentieth year of savings?

25. **Bacteria** The number of bacteria in a culture is tripling every hour. If there are initially 500 bacteria in the culture, how many bacteria will be in the culture by the end of the eighth hour?

Cumulative Review Test

Take the following test and check your answers with those that appear at the end of the test. Review any questions that you answered incorrectly. The section and objective where the material was covered are indicated after the answer.

1. Solve $A = \dfrac{1}{2}bh$, for b.

2. Find an equation of the line through $(4, -2)$ and $(1, 9)$. Write the equation in slope-intercept form.

3. Solve the system of equations.

$$\begin{aligned} x + y + z &= 1 \\ 2x + 2y + 2z &= 2 \\ 3x + 3y + 3z &= 3 \end{aligned}$$

4. Multiply $(5x^3 + 4x^2 - 6x + 2)(x + 5)$.

5. Factor $x^3 + 2x - 5x^2 - 10$.

6. Factor $(a + b)^2 + 6(a + b) + 9$.

7. Subtract $5 - \dfrac{x-1}{x^2 + 3x - 10}$.

8. y varies directly as the square of z. If y is 80 when z is 20, find y when z is 90.

9. If $f(x) = 3\sqrt[3]{x-2}$ and $g(x) = \sqrt[3]{7x-14}$ find all values of x for which $f(x) = g(x)$.

10. Solve $\sqrt{6x-5} - \sqrt{2x+6} - 1 = 0$.

11. Solve by completing the square.

$$x^2 + 2x + 12 = 0$$

12. Solve by the quadratic formula.

$$x^2 - \frac{x}{5} - \frac{1}{3} = 0$$

13. **Numbers** Twice the square of a positive number decreased by three times the number is 20. Find the number.

14. Graph $y = x^2 - 4x$ and label the vertex.

15. Solve $\log_a \dfrac{1}{64} = 3$ for a.

16. Graph $y = 2^x - 1$.

17. Find an equation of a circle with center at $(-6, 2)$ and radius 4.

18. Graph $(x + 3)^2 + (y + 1)^2 = 16$.

19. Find the sum of the infinite geometric series.

$$8 + \frac{16}{3} + \frac{32}{9} + \frac{64}{27} + \cdots$$

20. *Marbles* A bag has 15 red, 20 blue, 10 green, and 30 orange marbles. Find the probability that a blue marble is selected from the bag.

Answers to Cumulative Review Test

1. $b = \dfrac{2A}{h}$; [Sec. 2.2, Obj. 2] **2.** $y = -\dfrac{11}{3}x + \dfrac{38}{3}$; [Sec. 3.5, Obj. 1] **3.** Infinite number of solutions; [Sec. 4.2, Obj. 3]

4. $5x^4 + 29x^3 + 14x^2 - 28x + 10$; [Sec. 5.2, Obj. 3] **5.** $(x^2 + 2)(x - 5)$; [Sec. 5.4, Obj. 4] **6.** $(a + b + 3)^2$;

[Sec. 5.5, Obj. 5] **7.** $\dfrac{5x^2 + 14x - 49}{(x + 5)(x - 2)}$; [Sec. 6.2, Obj. 3] **8.** 1620; [Sec. 6.6, Obj. 1] **9.** 2; [Sec. 7.6, Obj. 2]

10. 5; [Sec. 7.6, Obj. 3] **11.** $-1 \pm i\sqrt{11}$; [Sec. 8.1, Obj. 3] **12.** $\dfrac{3 \pm \sqrt{309}}{30}$; [Sec. 8.2, Obj. 2] **13.** 4; [Sec. 5.8, Obj. 3]

14. $y = x^2 - 4x$; [Sec. 8.5, Obj. 3] **15.** $\dfrac{1}{4}$; [Sec. 9.3, Obj. 1] **16.** $y = 2^x - 1$; [Sec. 9.2, Obj. 1]

17. $(x + 6)^2 + (y - 2)^2 = 16$; [Sec. 11.1, Obj. 6] **18.** $(x + 3)^2 + (y + 1)^2 = 16$; [Sec. 11.1, Obj. 6]

19. 24; [Sec. 12.3, Obj. 5] **20.** $\dfrac{4}{15}$; [Sec. 12.7, Obj. 1]

Appendix

Areas and Perimeters			
Figure	**Sketch**	**Area**	**Perimeter**
Square	s	$A = s^2$	$P = 4s$
Rectangle	w l	$A = lw$	$P = 2l + 2w$
Parallelogram	h w l	$A = lh$	$P = 2l + 2w$
Trapezoid	b_1 s_1 h s_2 b_2	$A = \dfrac{1}{2}h(b_1 + b_2)$	$P = s_1 + s_2 + b_1 + b_2$
Triangle	s_1 h s_2 b	$A = \dfrac{1}{2}bh$	$P = s_1 + s_2 + b$

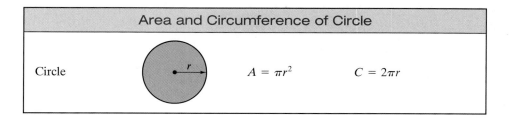

Area and Circumference of Circle		
Circle	r	$A = \pi r^2 \qquad C = 2\pi r$

Volumes and Surface Areas of Three-Dimensional Figures			
Figure	**Sketch**	**Volume**	**Surface Area**
Rectangular solid		$V = lwh$	$s = 2lh + 2wh + 2wl$
Right circular cylinder		$V = \pi r^2 h$	$s = 2\pi rh + 2\pi r^2$
Sphere		$V = \dfrac{4}{3}\pi r^3$	$s = 4\pi r^2$
Right circular cone		$V = \dfrac{1}{3}\pi r^2 h$	$s = \pi r \sqrt{r^2 + h^2}$
Square or rectangular pyramid		$V = \dfrac{1}{3}lwh$	

Answers

CHAPTER 1

Exercise Set 1.1
1–11. Answers will vary. **13.** Do all the homework and preview the new material to be covered in class. **15.** See the steps on page 5 of your text. **17.** The more you put into the course, the more you will get out of it. **19.** Answers will vary.

Exercise Set 1.2
1. A variable is a letter used to represent various numbers. **3.** A set is a collection of objects. **5.** A set that contains no elements **7.** $>$, is greater than; $\geq$, is greater than or equal to; $<$, is less than; $\leq$, is less than or equal to; $\neq$, is not equal to **9.** $\{3, 4, 5, 6\}$ **11.** An integer can be written with a denominator of 1. **13.** True **15.** False **17.** True **19.** True **21.** True **23.** $<$ **25.** $<$ **27.** $>$ **29.** $<$ **31.** $>$ **33.** $<$ **35.** $>$ **37.** $>$ **39.** $A = \{0\}$ **41.** $C = \{18, 20\}$ **43.** $E = \{0, 1, 2\}$ **45.** $H = \{0, 7, 14, 21, \ldots\}$ **47.** $J = \{1, 2, 3, 4, \ldots\}$ or $J = N$ **49. a)** 4 **b)** 4, 0 **c)** $-3, 4, 0$ **d)** $-3, 4, \frac{1}{2}, \frac{5}{9}, 0, -1.23, \frac{99}{100}$ **e)** $\sqrt{2}, \sqrt{8}$ **f)** $-3, 4, \frac{1}{2}, \frac{5}{9}, 0, \sqrt{2}, \sqrt{8}, -1.23, \frac{99}{100}$ **51.** $A \cup B = \{1, 2, 3, 4, 5, 6, 8\}$; $A \cap B = \{2, 4\}$ **53.** $A \cup B = \{-3, -2, -1, 0, 1, 2\}$; $A \cap B = \{-1, 0\}$ **55.** $A \cup B = \{2, 4, 6, 8, 10\}$; $A \cap B = \{ \ \}$ **57.** $A \cup B = \{0, 5, 10, 15, 20, 25, 30\}$; $A \cap B = \{ \ \}$ **59.** $A \cup B = \{-1, 0, 1, e, i, \pi\}$; $A \cap B = \{-1, 0, 1\}$ **61.** The set of natural numbers **63.** The set of whole number multiples of 3 **65.** The set of odd integers **67. a)** Set A is the set of all x such that x is a natural number less than 8. **b)** $A = \{1, 2, 3, 4, 5, 6, 7\}$ **69.** **71.** **73.** **75.** **77.** **79.** $\{x | x \geq 2\}$ **81.** $\{x | x < 5 \text{ and } x \in I\}$ or $\{x | x \leq 4 \text{ and } x \in I\}$ **83.** $\{x | -3 < x \leq 5\}$ **85.** $\{x | -2.5 \leq x < 4.2\}$ **87.** $\{x | -3 \leq x \leq 1 \text{ and } x \in I\}$ **89.** Yes **91.** Yes **93.** No **95.** Yes **97.** One example is $\left\{\frac{3}{2}, \frac{4}{3}, \frac{5}{4}, \frac{6}{5}, \frac{7}{6}\right\}$ **99.** One example is $A = \{2, 4, 5, 8, 9\}$, $B = \{4, 5, 6, 9\}$ **101. a)** {Old Navy, The Gap, Tommy Hilfiger, Abercrombie & Fitch, Express/The Limited, Nike, Adidas, Polo/Ralph Lauren, FUBU} **b)** Union, the union of two sets is the set of elements that belong to either or both sets. **c)** {Tommy Hilfiger} **d)** Intersection, the intersection of two sets is the set of elements common to both sets. **103. a)** $A = \{\text{Alex, James}\}$, $B = \{\text{Alex, James, George, Connor}\}$, $C = \{\text{Alex, Stephen}\}$, $D = \{\text{Alex, George, Connor}\}$ **b)** {Alex} **c)** Only Alex **105. a)** $\{1, 3, 4, 5, 6, 7\}$ **b)** $\{2, 3, 4, 6, 8\}$ **c)** $\{1, 2, 3, 4, 5, 6, 7, 8\}$ **d)** $\{3, 4, 6\}$ **107. a)** $\{x | x > 1\}$ includes fractions and decimal numbers which the other set does not contain. **b)** $\{2, 3, 4, 5, \ldots\}$ **c)** No, since it is not possible to list all real numbers greater than 1 in roster form. **109. a)** $0.\overline{1}; 0.\overline{2}; 0.\overline{3}$ **b)** $\frac{4}{9}; \frac{5}{9}; \frac{6}{9}$ or $\frac{2}{3}$ **c)** Based on a) and b), we deduce that $0.\overline{9} = \frac{9}{9} = 1$.

Exercise Set 1.3
1. Two numbers whose sum is zero **3.** No; $|0|$ is not positive **5.** Since a and $-a$ are the same distance from 0 on a number line, $|a| = |-a|$ for all real numbers, $\mathbb{R}$. **7.** Since $|5| = 5$ and $|-5| = 5$, the desired values for a are 5 and -5. **9.** { }, the absolute value of any real number must be greater than or equal to 0. **11.** Answers will vary. **13.** Answers will vary. **15.** $-\frac{a}{b}$ or $\frac{-a}{b}$ **17. a)** $a + b = b + a$ **b)** Answers will vary. **19.** Answers will vary. One example is $2 + (3 \cdot 4) \neq (2 + 3) \cdot (2 + 4), 14 \neq 30$ **21.** 5 **23.** 7 **25.** $\frac{7}{8}$ **27.** 0 **29.** -7 **31.** $-\frac{5}{9}$ **33.** $=$ **35.** $>$ **37.** $>$ **39.** $>$ **41.** $>$ **43.** $<$ **45.** $-|5|, -2, -1, |-3|, 4$ **47.** $-32, -|4|, 4, |-7|, 15$ **49.** $-|2.9|, -2.4, -2.1, -2, |-2.8|$ **51.** $-2, \frac{1}{3}, \left|-\frac{1}{2}\right|, \left|\frac{3}{5}\right|, \left|-\frac{3}{4}\right|$ **53.** 1 **55.** -1 **57.** -4 **59.** -0.99 **61.** $\frac{11}{24}$ **63.** 6.92 **65.** -17.2 **67.** 4 **69.** -2 **71.** $\frac{17}{20}$ **73.** -35 **75.** $\frac{5}{4}$ **77.** 12 **79.** 235.9192 **81.** 5 **83.** 1 **85.** $-\frac{3}{32}$ **87.** $\frac{5}{3}$ **89.** -4 **91.** -8 **93.** 5 **95.** 42.8064 **97.** -21.6 **99.** -4 **101.** $\frac{81}{16}$ **103.** -1 **105.** $-\frac{17}{45}$ **107.** 77 **109.** Commutative property of addition **111.** Multiplicative property of zero **113.** Associative property of addition **115.** Identity property of multiplication **117.** Associative property of multiplication **119.** Distributive property **121.** Identity property of addition **123.** Inverse property of addition **125.** Inverse property

of multiplication   **127.** Double negative property   **129.** $-6, \dfrac{1}{6}$ **131.** $\dfrac{22}{7}, -\dfrac{7}{22}$ **133.** 100°F   **135.** 148.2 feet below the
starting point, or -148.2 feet   **137. a)** Owe $12,400   **b)** Receive $27,500   **139. a)** She is entitled to a refund because her
payments totaled $12,000 which was greater than the tax she owed.   **b)** $1875 refund   **141.** Answers will vary.
143. a) $6784   **b)** $2908   **c)** $5411   **d)** $496   **145.** -50 **147.** -1 **149.** True   **150.** $\{1, 2, 3, 4, \ldots\}$ **151. a)** $3, 4, -2, 0$
b) $3, 4, -2, \dfrac{5}{6}, 0$ **c)** $\sqrt{3}$ **d)** $3, 4, -2, \dfrac{5}{6}, \sqrt{3}, 0$ **152. a)** $\{1, 4, 7, 9, 12, 15\}$ **b)** $\{4, 7\}$ **153.** ←—○————●—→
                                                                                                                                            -4 6

Exercise Set 1.4   **1. a)** Base   **b)** Exponent   **3. a)** Index   **b)** Radicand   **5.** The positive number whose square
equals the radicand   **7.** A negative number raised to an odd power is a negative number.   **9.** Parentheses, exponents and
roots, multiplication and division from left to right, addition and subtraction from left to right   **11. a)** Answers will vary.
b) 24   **13.** 9   **15.** -9 **17.** 9   **19.** $-\dfrac{81}{625}$ **21.** $-\dfrac{25}{36}$ **23.** -6 **25.** -3 **27.** 0.1   **29.** 0.015   **31.** 0.095   **33.** 76,183.335   **35.** 1.710
37. 3.250   **39.** -0.723 **41. a)** 9   **b)** -9 **43. a)** 1   **b)** -1 **45. a)** 1   **b)** -1 **47. a)** $\dfrac{1}{9}$ **b)** $-\dfrac{1}{9}$ **49. a)** 27   **b)** -27
51. a) -27 **b)** 27   **53. a)** -8 **b)** 8   **55. a)** $\dfrac{8}{27}$ **b)** $-\dfrac{8}{27}$ **57.** -14 **59.** -19 **61.** -22.221 **63.** $-\dfrac{5}{16}$ **65.** 42   **67.** 22
69. 0   **71.** $\dfrac{1}{2}$ **73.** -10 **75.** 5   **77.** 64   **79.** 16   **81.** $\dfrac{27}{5}$ **83.** Undefined   **85.** -4 **87.** 0   **89.** $-\dfrac{10}{3}$ **91.** $\dfrac{242}{5}$ **93.** $\dfrac{1}{4}$ **95.** 26
97. -41 **99.** -9 **101.** -80 **103.** 33   **105.** -3 **107.** $\dfrac{3}{2}$ **109.** $\dfrac{7y - 14}{2}, 14$ **111.** $6(3x + 6) - 9, 81$ **113.** $\left(\dfrac{x + 3}{2y}\right)^2 - 3, 1$
115. a) 9.51 billion trips   **b)** 22.51 billion trips   **117. a)** 7.62%   **b)** 21.78%   **119. a)** $1.262 billion   **b)** $8.902 billion
121. a) 2.3 million people   **b)** 105 million people   **122. a)** $A \cap B = \{b, c, f\}$ **b)** $A \cup B = \{a, b, c, d, f, g, h\}$
123. All real numbers, $\mathbb{R}$ **124.** $a \geq 0$ **125.** $4, -4$ **126.** $-|6|, -4, -|-2|, 0, |-5|$ **127.** Associative property of addition

Exercise Set 1.5   **1. a)** $a^m \cdot a^n = a^{m+n}$ **b)** Answers will vary.   **3. a)** $a^0 = 1, a \neq 0$ **b)** Answers will vary.
5. a) $(ab)^m = a^m b^m$ **b)** Answers will vary.   **7. a)** $\left(\dfrac{a}{b}\right)^m = \dfrac{a^m}{b^m}, b \neq 0$ **b)** Answers will vary.   **9.** $x = \dfrac{1}{5}$, because if $\dfrac{1}{x} = 5$,
then $x = \dfrac{1}{5}$ **11. a)** The opposite of x is $-x$; the reciprocal of x is $\dfrac{1}{x}$ **b)** $x^{-1}; \dfrac{1}{x}$ **c)** $-x$ **13.** 32   **15.** 9   **17.** $\dfrac{1}{36}$ **19.** 8
21. 1   **23.** 64   **25.** 36   **27.** $\dfrac{16}{81}$ **29. a)** $\dfrac{1}{9}$ **b)** $\dfrac{1}{9}$ **c)** $-\dfrac{1}{9}$ **d)** $-\dfrac{1}{9}$ **31. a)** 2   **b)** -2 **c)** -2 **d)** 2   **33. a)** 5   **b)** -5
c) 1   **d)** -1 **35. a)** $3xy$ **b)** 1   **c)** $3x$ **d)** 3   **37.** $\dfrac{5}{y^3}$ **39.** x^4 **41.** $2ab^3$ **43.** $\dfrac{5}{2m^2 n^3}$ **45.** $\dfrac{5z^4}{x^2 y^3}$ **47.** $\dfrac{1}{6xy}$ **49.** $\dfrac{1}{2}$ **51.** x^4
53. 36   **55.** $\dfrac{1}{49}$ **57.** $\dfrac{1}{m^{10}}$ **59.** $5w^5$ **61.** $\dfrac{12}{a^7}$ **63.** $3p$ **65.** $-10r^7$ **67.** $8x^7 y^2$ **69.** $\dfrac{3x^2}{y^6}$ **71.** $-\dfrac{3x^3 z^2}{y^5}$ **73. a)** 3   **b)** 6   **c)** 1
d) 0   **75. a)** $-\dfrac{1}{12}$ **b)** $\dfrac{7}{12}$ **c)** $\dfrac{11}{10}$ **d)** $\dfrac{23}{120}$ **77.** 81   **79.** $\dfrac{1}{81}$ **81.** b^6 **83.** $-c^3$ **85.** $\dfrac{9}{x^6}$ **87.** $\dfrac{7}{10}$ **89.** $\dfrac{13}{16}$ **91.** $\dfrac{9}{16b^2}$ **93.** $\dfrac{16x^4}{y^4}$
95. $\dfrac{q^{12}}{27p^6}$ **97.** $-\dfrac{g^{12}}{8h^9}$ **99.** $\dfrac{9j^2}{16k^4}$ **101.** $8r^6 s^{15}$ **103.** $\dfrac{y^6}{64x^3}$ **105.** $64x^9 y^3$ **107.** $\dfrac{z^3}{8x^3 y^3}$ **109.** $\dfrac{x^{16}}{y^{10}}$ **111.** $\dfrac{x^4 y^8}{4z^{12}}$ **113.** $-\dfrac{8b^{12}}{a^6 c^3}$
115. $\dfrac{27}{8x^{21} y^9}$ **117.** x^{7a+3} **119.** w^{5a-7} **121.** x^{w+7} **123.** x^{5p+2} **125.** x^{2m+2} **127.** $\dfrac{5m^{2b}}{n^{2a}}$ **129. a)** $x < 0$ or $x > 1$
b) $0 < x < 1$ **c)** $x = 0$ or $x = 1$ **d)** Not true for $0 \leq x \leq 1$ **131. a)** The product of an even number of negative
factors is positive.   **b)** The product of an odd number of negative factors is negative.   **133. a)** Yes   **b)** Yes, because
$x^{-2} = \dfrac{1}{x^2}$ and $(-x)^{-2} = \dfrac{1}{(-x)^2} = \dfrac{1}{x^2}$ **135.** -3, because $(y^{-2}/y^{-3})^2 = y^2$ **137.** $-1, 3$, because $(x^{-1}/x^4)^{-1} = x^5$, and
$(y^5/y^3)^{-1} = 1/y^2$ **139.** $x^{9/8}$ **141.** $\dfrac{1}{x^{9/2} y^{19/6}}$ **144. a)** $A \cup B = \{1, 2, 3, 4, 5, 6, 8\}$ **b)** $A \cap B = \{ \ \}$ **145.** ←—●————○—→
                                                                                                                                                                            -3 2
146. -6 **147.** -5

Exercise Set 1.6   **1.** A number greater than or equal to 1 and less than 10 multiplied by a power of 10
3. 1×10^{-2}, because $1 \times 10^{-2} = 0.01$ and $1 \times 10^{-3} = 0.001$.   **5.** 4.7×10^3 **7.** 3.1×10^{-2} **9.** 3.6×10^5 **11.** 1.86×10^{-6}
13. 5.78×10^6 **15.** 1.01×10^{-4} **17.** 31,000   **19.** 0.0000213   **21.** 0.917   **23.** 9,000,000   **25.** 207,000   **27.** 1,000,000
29. 240,000,000   **31.** 0.021   **33.** 0.000027   **35.** 10,660   **37.** 0.0003   **39.** 0.0000005733   **41.** 1.5×10^{-5} **43.** 5.0×10^3
45. 2.0×10^{-8} **47.** 1.645×10^{12} **49.** 3.2×10^5 **51.** 3.0×10^0 **53.** 9.369×10^{14} **55.** 9.905×10^2 **57.** 5.337×10^2

59. 3.868×10^{-25} **61.** 7.047×10^{-27} **63.** 3.333×10^{60} **65. a)** Subtract 1 from the exponent **b)** Subtract 2 from the exponent **c)** Subtract 6 from the exponent **d)** 6.58×10^{-10} **67. a)** 1.0×10^4 or 10,000 **b)** 4.725×10^5 or 472,500 **c)** The error in part **b)** because the answer is off by more **69.** 30,000 hours **71. a)** $\approx 5.859 \times 10^9$ people **b)** $\approx 4.58\%$ **73. a)** ≈ 478.26 **b)** 87,816,000 pounds **75. a)** $\approx 3.092 \times 10^7$ passengers **b)** $\approx 3.689 \times 10^7$ passengers **c)** ≈ 1.193 **77. a)** 988 million **b)** $\approx 20.74\%$ **c)** ≈ 344.1 people/square mile **d)** ≈ 78.7 people/square mile **79. a)** $\approx 5.87 \times 10^{12}$ miles **b)** 500 seconds or $8\frac{1}{3}$ minutes **c)** 6.72×10^{11} seconds or 21,309 years

Chapter Review Exercises

1. $\{4, 5, 6, 7\}$ **2.** $\{0, 3, 6, 9, \ldots\}$ **3.** Yes **4.** Yes **5.** Yes **6.** No **7.** $4, 6$ **8.** $4, 6, 0$ **9.** $-2, 4, 6, 0$ **10.** $-2, 4, 6, \frac{1}{2}, 0, \frac{15}{27}, -\frac{1}{5}, 1.47$ **11.** $\sqrt{7}, \sqrt{3}$ **12.** $-2, 4, 6, \frac{1}{2}, \sqrt{7}, \sqrt{3}, 0, \frac{15}{27}, -\frac{1}{5}, 1.47$ **13.** False **14.** True **15.** True **16.** True **17.** $A \cup B = \{1, 2, 3, 4, 5, 6, 8, 10\}; A \cap B = \{2, 4, 6\}$ **18.** $A \cup B = \{2, 3, 4, 5, 6, 7, 8, 9\}$; $A \cap B = \{ \ \}$ **19.** $A \cup B = \{1, 2, 3, 4, \ldots\}; A \cap B = \{ \ \}$ **20.** $A \cup B = \{3, 4, 5, 6, 9, 10, 11, 12\}; A \cap B = \{9, 10\}$ **21.** [number line at 5] **22.** [number line at -2] **23.** [number line from -1.3 to 2.4] **24.** [number line $-1\,0\,1\,2\,3\,4$] **25.** $<$ **26.** $<$ **27.** $<$ **28.** $=$ **29.** $<$ **30.** $>$ **31.** $>$ **32.** $>$ **33.** $-\pi, -3, 3, \pi$ **34.** $0, \frac{3}{5}, 2.3, |-3|$ **35.** $-2, 3, |-5|, |-7|$ **36.** $-7, -3, |-3|, |-7|$ **37.** $-4, -|-3|, 5, 6$ **38.** $-3, 0, |1.6|, |-2.3|$ **39.** Distributive property **40.** Commutative property of multiplication **41.** Associative property of addition **42.** Identity property of addition **43.** Associative property of multiplication **44.** Double negative property **45.** Multiplicative property of zero **46.** Inverse property of addition **47.** Inverse property of multiplication **48.** Identity property of multiplication **49.** 13 **50.** 11 **51.** 8 **52.** -7 **53.** 0 **54.** 21 **55.** 9 **56.** -47 **57.** 15 **58.** 34 **59.** 6 **60.** 512 **61.** Undefined **62.** $\frac{8}{3}$ **63.** 15 **64.** -67 **65. a)** \$816.37 million **b)** \$6,075.54 million **66. a)** 944.53 ton-miles **b)** 1626.29 ton-miles **67.** 32 **68.** x^5 **69.** a^8 **70.** y^9 **71.** b^9 **72.** $\frac{1}{c^2}$ **73.** $\frac{1}{125}$ **74.** 3 **75.** $4m^6$ **76.** $\frac{7}{5}$ **77.** $\frac{27}{8}$ **78.** $\frac{y^2}{x}$ **79.** $-15x^3y^4$ **80.** $\frac{6}{v^3w^3}$ **81.** $\frac{3y^7}{x^5}$ **82.** $\frac{3}{xy^9}$ **83.** $\frac{g^5}{h^5j^{14}}$ **84.** $\frac{3m}{n^4}$ **85.** $125a^3b^3$ **86.** $\frac{x^{10}}{9y^2}$ **87.** $\frac{p^{14}}{q^{12}}$ **88.** $-\frac{8a^3}{b^9c^6}$ **89.** $\frac{z^4}{36x^2y^6}$ **90.** $\frac{m^9}{27}$ **91.** $\frac{n^6}{4m^4}$ **92.** $\frac{625x^4y^4}{z^{20}}$ **93.** $\frac{9x^{10}}{4y^{14}z^{12}}$ **94.** $-\frac{x^6z^2}{8y^2}$ **95.** 7.42×10^{-5} **96.** 2.6×10^5 **97.** 1.83×10^5 **98.** 1.0×10^{-6} **99.** 30,000 **100.** 0.02 **101.** 200,000,000 **102.** 2000 **103. a)** $\$1.13 \times 10^8$ **b)** $\$5.9 \times 10^7$ **c)** ≈ 2.87 **104. a)** 10,400,000,000 **b)** 10.4 billion kilometers **c)** 5.2×10^8 kilometers or 520,000,000 kilometers **d)** 6.24×10^9 miles or 6,240,000,000 miles

Chapter Practice Test

1. $A = \{6, 7, 8, 9, \ldots\}$ **2.** False **3.** True **4.** $-\frac{3}{5}, 2, -4, 0, \frac{19}{12}, 2.57, -1.92$ **5.** $-\frac{3}{5}, 2, -4, 0, \frac{19}{12}, 2.57, \sqrt{8}, \sqrt{2}, -1.92$ **6.** $A \cup B = \{5, 7, 8, 9, 10, 11, 14\}; A \cap B = \{8, 10\}$ **7.** $A \cup B = \{1, 3, 5, 7, \ldots\}$; $A \cap B = \{3, 5, 7, 9, 11\}$ **8.** [number line from -2.3 to 5.2] **9.** [number line $-3\,-2\,-1\,0\,1\,2$] **10.** $-|4|, -2, |3|, 6$ **11.** Associative property of addition **12.** Commutative property of addition **13.** 0 **14.** 33 **15.** Undefined **16.** $-\frac{37}{22}$ **17.** 17 **18. a)** \$1200 **b)** \$2061.28 **19.** $\frac{1}{9}$ **20.** $\frac{16}{m^6n^4}$ **21.** $\frac{4c^2}{5ab^5}$ **22.** $-\frac{y^{21}}{27x^{12}}$ **23.** 2.42×10^8 **24.** 260,000,000 **25. a)** $\approx 7.642 \times 10^7$ **b)** $\approx 6.458 \times 10^7$ **c)** $\approx 1.184 \times 10^7$

CHAPTER 2

Using Your Calculator, 2.1 **1.** No **2.** Yes

Exercise Set 2.1 **1.** The terms of an expression are the parts added. **3. a)** $\frac{1}{4}$ **b)** -1 **c)** $-\frac{3}{5}$ **5. a)** Like terms have the same variables and exponents. **b)** No; the exponent on x is different for each term. **7.** No, 4 does not make the equation true **9.** If $a = b$, then $a + c = b + c$. **11. a)** An infinite number of solutions. **b)** $\mathbb{R}$ **13. a)** Answers will vary. **b)** -12 **15.** Symmetric property **17.** Transitive property **19.** Reflexive property **21.** Addition property of equality

23. Multiplication property of equality   **25.** Multiplication property of equality   **27.** Three   **29.** Two   **31.** Zero   **33.** One
35. Seven   **37.** Twelve   **39.** Cannot be simplified   **41.** $5x^2 - x - 5$ **43.** $8.7c^2 + 3.6c$ **45.** Cannot be simplified
47. $-pq + p + q$ **49.** $2d + 2$ **51.** $\frac{8}{3}x + \frac{13}{2}$ **53.** $-17x - 4$ **55.** $6x - 3y$ **57.** $-9b + 93$ **59.** $4r^2 - 2rs + 3r + 4s$
61. 3   **63.** $\frac{3}{2}$ **65.** 2   **67.** 16   **69.** 5   **71.** 1   **73.** 1   **75.** 0   **77.** 3   **79.** -1 **81.** 5   **83.** 5   **85.** -1 **87.** $-\frac{1}{2}$ **89.** 6   **91.** 2
93. 68   **95.** -64 **97.** -4 **99.** 24   **101.** -4 **103.** $\frac{15}{16}$ **105.** 1.00   **107.** 1.18   **109.** 0.43   **111.** 1701.39   **113.** -1.85
115. ∅; contradiction   **117.** {0}; conditional   **119.** ℝ; identity   **121.** ℝ; identity   **123.** ∅; contradiction   **125. a)** 10 minutes
b) 25 minutes   **c)** the fifth night   **d)** the seventh night   **127. a)** 62.74%   **b)** 2006   **129. a)** 2219   **b)** 2006
131. Answers will vary. One possible answer: $x = \frac{5}{2}, 2x - 4 = 1, 4x = 10$ **133.** Answers will vary. One possible answer:
$2x - 4 = 5x - 3(1 + x)$ **135.** Answers will vary. One possible answer: $3x + 3 = \frac{3}{2}x + x + 6$ **137.** -22, substitute -2
for x and solve for n.   **139.** $\triangle = \dfrac{\odot + \square}{\ast}$ **141.** $\odot = \dfrac{\otimes - \triangle}{\square}$ **143. a)** Answers will vary.   **b)** $|a| = \begin{cases} a \text{ if } a \geq 0 \\ -a \text{ if } a < 0 \end{cases}$
144. a) -9 **b)** 9   **145.** -4 **146.** $-\dfrac{27}{64}$

Exercise Set 2.2

1. An equation that is a mathematical model of a real-life situation   **3.** Understand, translate,
carry out, check, answer   **5. a)** $l = 5$ **b)** $l = \dfrac{P - 2w}{2}$ **c)** no   **d)** you should obtain the same answer.   **7.** 120   **9.** 300
11. 201.06   **13.** 80   **15.** 220   **17.** $\dfrac{7}{4}$ **19.** 66.67   **21.** 4   **23.** 119.10   **25.** $y = -3x + 5$ **27.** $y = \dfrac{1}{4}x - \dfrac{13}{4}$ **29.** $y = 3x - 8$
31. $y = \dfrac{3}{4}x - 1$ **33.** $y = x + 2$ **35.** $y = -\dfrac{4}{3}x + 11$ **37.** $t = \dfrac{d}{r}$ **39.** $t = \dfrac{i}{pr}$ **41.** $l = \dfrac{P - 2w}{2}$ **43.** $h = \dfrac{V}{lw}$ **45.** $h = \dfrac{V}{\pi r^2}$
47. $l = \dfrac{3V}{wh}$ **49.** $m = \dfrac{y - b}{x}$ **51.** $m = \dfrac{y - y_1}{x - x_1}$ **53.** $\mu = x - z\sigma$ **55.** $T_2 = \dfrac{T_1 P_2}{P_1}$ **57.** $h = \dfrac{2A}{b_1 + b_2}$ **59.** $n = \dfrac{2S}{f + l}$
61. $F = \dfrac{9}{5}C + 32$ **63.** $m_1 = \dfrac{Fd^2}{km_2}$ **65. a)** $p = 9.15d$ **b)** $d = \dfrac{p}{9.15}$ **c)** Answers will vary.   **67.** $154   **69.** 6.5 years
71. a) 3.14 square inches   **b)** 78.54 square inches   **73. a)** 75 cubic feet   **b)** 2.78 cubic yard   **c)** $105   **75.** The cylinder,
difference is 0.22 cubic inch   **77.** $11,264.93   **79.** $4958.41   **81.** 4.12%   **83. a)** 7.49%   **b)** 6.30%   **85. a)** 4 pounds per week
b) 2500 calories   **87. a)** $S = 100 - a$ **b)** 40%   **89. a)** $s = \dfrac{rt^2}{u}$ **b)** $u = \dfrac{rt^2}{s}$ **90.** -29 **91.** 1   **92.** -125 **93.** $\dfrac{4}{3}$

Exercise Set 2.3

1. $A = 72°, B = 18°$ **3.** $A = 36°, B = 144°$ **5.** $40°, 60°, 80°$,   **7.** $32   **9.** 25 rides
11. 225 miles   **13.** $623.65   **15.** 13 times   **17.** WMC: $\approx$\$464.4 million, SUI: $\approx$\$172 million   **19.** SAT: $7.19 billion,
HSF: $7.29 billion   **21.** Steel: 6000 pounds per square inch; bone: 24,000 pounds per square inch   **23.** $16.25   **25. a)** $2.55
b) $1.74   **27.** U.S.: 97, Russia: 88, China: 59, Australia: 58, Germany: 57   **29.** $1058.25 per month   **31. a)** $\approx$28 months or 2.33
years   **b)** Yes   **33.** 9 inches, 12 inches, 15 inches   **35.** $39°, 51°, 90°$ **37.** $l = 7$ feet, $w = 4$ feet   **39.** 3 feet by 6 feet   **41.** $60
43. a) $135.14   **b)** $85.14   **45.** $16   **47.** Animals: 250,000; plants: 350,000; non-beetle insects: 540,000; beetles: 360,000
49. a) 76   **b)** No; he needs a score of 111   **51. a)** and **b)** Answers will vary.   **53.** 220 miles   **55.** $\dfrac{13}{5}$ **56.** -2.7 **57.** $\dfrac{5}{16}$
58. -5 **59.** $\dfrac{y^{18}}{8x^{12}}$

Exercise Set 2.4

1. 11.4 miles   **3.** 2 hours   **5.** 6 hours   **7. a)** 6 miles per hour   **b)** 12 miles per hour
9. a) $\approx$3.13 hours   **b)** $\approx$21.3 miles   **11.** 13.8 hours   **13.** $\approx$0.58 hour or 35 minutes   **15.** $12,570 at 3%, $17,430 at 4.1%
17. 54 pounds   **19. a)** 2200 shares of Johnson & Johnson and 4400 shares of AOL.   **b)** $4480   **21.** 30 ounces   **23.** 2.8 tea-
spoons 30%, 1.2 teaspoons 80%   **25.** 35%   **27.** 5 pounds striped, 15 pounds black oil   **29.** $\approx$25.77 hours   **31.** 500 minutes
or $8\dfrac{1}{3}$ hours   **33.** 6 quarts   **35. a)** $\approx$3.71 hours   **b)** $\approx$2971.43 miles   **37.** 8 small, 4 large paintings   **39.** 9.6 ounces 80%
solution, 118.4 ounces water   **41.** $\approx$35.6 ounces sirloin, $\approx$28.4 ounces veal   **43.** 3 miles   **45.** $\approx$11.4 ounces   **47. a), b)** and **c)**
Answers will vary.   **49.** $\approx$149 miles   **51.** 6 quarts   **52.** 6.0×10^{10} **53.** -5.7 **54.** $\dfrac{18}{5}$ **55.** $y = \dfrac{x - 42}{30}$ **56.** 140 miles

Exercise Set 2.5

1. It is necessary to change the direction of the inequality symbol when multiplying or dividing both sides of the inequality by a negative number **3. a)** When the endpoints are not included **b)** When the endpoints are included **c)** Answers will vary. One example is $x > 4$ **d)** Answers will vary. One example is $x \geq 4$. **5.** $a < x$ and $x < b$ **7. a)** **b)** $(-2, \infty)$ **c)** $\{x | x > -2\}$ **9. a)** **b)** $(-\infty, \pi]$

c) $\{w | w \leq \pi\}$ **11. a)** **b)** $\left(-3, \dfrac{3}{4}\right]$ **c)** $\left\{q \middle| -3 < q \leq \dfrac{3}{4}\right\}$

13. a) **b)** $(-7, -4]$ **c)** $\{x | -7 < x \leq -4\}$ **15.**

17. **19.** **21.** **23.** **25.**

27. $\left(-\infty, \dfrac{3}{2}\right)$ **29.** $[2, \infty)$ **31.** $\left(-\infty, \dfrac{3}{2}\right]$ **33.** $(-\infty, \infty)$ **35.** $[-5, 1)$ **37.** $[-4, 5]$ **39.** $\left[4, \dfrac{11}{2}\right)$ **41.** $(-6, -4]$

43. $\{x | 3 \leq x < 7\}$ **45.** $\{x | 0 < x \leq 1\}$ **47.** $\left\{u \middle| 4 \leq u \leq \dfrac{19}{3}\right\}$ **49.** $\{c | -3 < c \leq 2\}$ **51.** $\varnothing$ **53.** $\{x | -5 < x < 2\}$

55. $(-\infty, 2) \cup [7, \infty)$ **57.** $[0, 2]$ **59.** $(-\infty, 0) \cup (6, \infty)$ **61.** $[0, \infty)$ **63. a)** $l + g \leq 130$ **b)** $l + 2w + 2d \leq 130$
c) 24.5 inches **65.** 11 boxes **67.** 77 minutes **69.** 1881 books **71.** 42 ounces **73.** For sales over \$5,000 per week **75.** 24
77. $76 \leq x \leq 100$ **79.** $6.97 < x < 8.77$ **81. a)** \$30,258.60 **b)** \$44,960.54 **83. a)** 1995, 1996, 1997, 1998, and 1999;
During these years, *both* events is occuring. **b)** 2000, 2001, 2002, 2003, 2004, and 2005; During these years, *at least one* of
the events is occuring. **85.** Answers will vary. **87.** $84 \leq x \leq 100$ **89. a)** Answers will vary. **b)** $(-3, \infty)$

91. a) $A \cup B = \{1, 2, 3, 4, 5, 6, 8, 9\}$ **b)** $A \cap B = \{1, 8\}$ **92. a)** 4 **b)** 0, 4 **c)** $-3, 4, \dfrac{5}{2}, 0, -\dfrac{29}{80}$ **d)** $-3, 4, \dfrac{5}{2}, \sqrt{7}, 0, -\dfrac{29}{80}$

93. Associative property of addition **94.** Commutative property of addition **95.** $V = \dfrac{R - L + Dr}{r}$

Exercise Set 2.6

1. Set $x = a$ or $x = -a$ **3.** $-a < x < a$ **5.** $x < -a$ or $x > a$ **7.** All real numbers except
for 0; the absolute value of every real number except 0 is greater than 0. **9.** Set $x = y$ or $x = -y$ **11. a)** Two **b)** Infinite
number **c)** Infinite number **13. a)** D **b)** B **c)** E **d)** C **e)** A **15.** $\{-2, 2\}$ **17.** $\left\{-\dfrac{1}{2}, \dfrac{1}{2}\right\}$ **19.** $\varnothing$ **21.** $\{-12, 2\}$

23. $\{-5\}$ **25.** $\left\{\dfrac{3}{2}, \dfrac{11}{6}\right\}$ **27.** $\{-17, 23\}$ **29.** $\{3\}$ **31.** $\{w | -11 < w < 11\}$ **33.** $\{q | -13 \leq q \leq 3\}$ **35.** $\{b | 1 < b < 5\}$

37. $\{x | -9 \leq x \leq 6\}$ **39.** $\left\{x \middle| \dfrac{1}{3} < x < \dfrac{13}{3}\right\}$ **41.** $\varnothing$ **43.** $\{j | -18 < j < 6\}$ **45.** $\left\{k \middle| -\dfrac{1}{4} < k < \dfrac{13}{4}\right\}$

47. $\{y | y < -7 \text{ or } y > 7\}$ **49.** $\{x | x < -9 \text{ or } x > 1\}$ **51.** $\left\{b \middle| b < \dfrac{2}{3} \text{ or } b > 4\right\}$ **53.** $\{h | h < 1 \text{ or } h > 4\}$

55. $\{x | x < 2 \text{ or } x > 6\}$ **57.** $\{x | x \leq -18 \text{ or } x \geq 2\}$ **59.** $\mathbb{R}$ **61.** $\{x | x < 2 \text{ or } x > 2\}$ **63.** $\{-1, 15\}$ **65.** $\{-3, 1\}$

67. $\left\{-23, \dfrac{13}{7}\right\}$ **69.** $\{10\}$ **71.** $\{-1, 1\}$ **73.** $\{q | q < -8 \text{ or } q > -4\}$ **75.** $\{w | -1 \leq w \leq 8\}$ **77.** $\left\{-\dfrac{8}{5}, 2\right\}$

79. $\left\{x \middle| x < -\dfrac{5}{2} \text{ or } x > -\dfrac{5}{2}\right\}$ **81.** $\left\{x \middle| -\dfrac{13}{3} \leq x \leq \dfrac{5}{3}\right\}$ **83.** $\varnothing$ **85.** $\{w | -16 < w < 8\}$ **87.** $\mathbb{R}$ **89.** $\left\{2, \dfrac{22}{3}\right\}$

91. $\left\{-\dfrac{3}{2}, \dfrac{9}{7}\right\}$ **93. a)** $[0.085, 0.093]$ **b)** 0.085 inch **c)** 0.093 inch **95. a)** $[132, 188]$ **b)** 132 to 188 feet below sea level,

inclusive **97.** $|x| = 5$ **99.** $|x| \geq 5$ **101.** $x = -\dfrac{b}{a}$; $|ax + b|$ is never less than 0, so set $|ax + b| = 0$ and solve for x.

103. a) Set $ax + b = -c$ or $ax + b = c$ and solve each equation for x. **b)** $x = \dfrac{-c - b}{a}$ or $x = \dfrac{c - b}{a}$

105. a) Write $ax + b < -c$ or $ax + b > c$ and solve each inequality for x. **b)** $x < \dfrac{-c - b}{a}$ or $x > \dfrac{c - b}{a}$ **107.** $\mathbb{R}$;

Since $3 - x = -(x - 3)$ **109.** $\{x | x \geq 0\}$; by definition of absolute value **111.** $\{2\}$; set $x + 1 = 2x - 1$ or

$x + 1 = -(2x - 1)$ **113.** $\{x | x \leq 2\}$; by definition $|x - 2| = -(x - 2)$ if $x \leq 2$ **115.** $\{3\}$ **117.** $\varnothing$ **119.** $\dfrac{29}{72}$ **120.** 25

121. 1.33 miles **122.** $\{x | x < 4\}$

Chapter Review Exercises **1.** Eight **2.** One **3.** Seven **4.** $a^2 - a + 4$ **5.** $7x^2 + 2xy - 4$ **6.** Cannot be simplified **7.** $4x - 3y + 6$ **8.** -4 **9.** 20 **10.** $-\dfrac{13}{3}$ **11.** -10 **12.** $-\dfrac{9}{2}$ **13.** No solution **14.** $\mathbb{R}$ **15.** $-\dfrac{1}{2}$ **16.** $\dfrac{1}{4}$ **17.** 64

18. -4 **19.** $R = \dfrac{E}{I}$ **20.** $h = \dfrac{A}{\pi r^2}$ **21.** $w = \dfrac{P - 2l}{2}$ **22.** $h = \dfrac{2A}{b}$ **23.** $m = \dfrac{y - b}{x}$ **24.** $y = \dfrac{2x - 5}{3}$

25. $R_2 = R_T - R_1 - R_3$ **26.** $a = \dfrac{2S - b}{3}$ **27.** $l = \dfrac{K - 2d}{2}$ **28.** $22 **29.** 3 years **30.** $5833.33 **31.** 150 miles **32.** $233.33

33. $2570 at 3.5%, $2430 at 4% **34.** 187.5 gallons of 20%, 62.5 gallons of 60% **35.** $2\dfrac{2}{3}$ hours **36. a)** 3000 miles per hour

b) 16,500 miles **37.** 15 pounds of $6.00 coffee; 25 pounds of $6.80 coffee **38.** $30 **39. a)** 1 hour **b)** 14.4 miles
40. 40°; 65°; 75° **41.** 300 gallons per hour; 450 gallons per hour **42.** 35°, 55° **43.** 7.5 ounces **44.** $4500 at 10%; $7500 at 6%
45. More than 5 **46.** 40 miles per hour, 50 miles per hour **47.** **48.**

49. $\dfrac{5}{2}$ **50.** $\dfrac{21}{4}$ **51.** $-\dfrac{9}{2}$ **52.** -10 **53.** 0

54. 0 **55.** 8 boxes **56.** 7 minutes **57.** 17 weeks **58.** $\{x | 81 \le x \le 100\}$ **59.** $(5, 11)$ **60.** $(-3, 5]$

61. $\left(\dfrac{7}{2}, 6\right)$ **62.** $\left(\dfrac{8}{3}, 6\right)$ **63.** $(-3, 1]$ **64.** $(2, 14)$ **65.** $\{h | -3 < h \le 2\}$ **66.** $\mathbb{R}$ **67.** $\{x | x \le -4\}$ **68.** $\{g | g < -5 \text{ or } g \ge 11\}$

69. $\{-2, 2\}$ **70.** $\{x | -3 < x < 3\}$ **71.** $\{x | x \le -4 \text{ or } x \ge 4\}$ **72.** $\{-16, 6\}$ **73.** $\{x | x \le -3 \text{ or } x \ge 7\}$ **74.** $\left\{-\dfrac{1}{2}, \dfrac{9}{2}\right\}$

75. $\{q | 1 < q < 8\}$ **76.** $\{-1, 4\}$ **77.** $\{x | -14 < x < 22\}$ **78.** $\left\{-5, -\dfrac{4}{5}\right\}$ **79.** $\mathbb{R}$ **80.** $\left[-5, -\dfrac{1}{3}\right]$ **81.** $(4, 7]$

82. $\left(-\dfrac{17}{2}, \dfrac{27}{2}\right]$ **83.** $[-1, 6)$ **84.** $(-\infty, \infty)$ **85.** $\left(\dfrac{2}{3}, 8\right]$

Chapter Practice Test **1.** Seven **2.** $16p - 3q - 4pq$ **3.** $10q + 44$ **4.** -26 **5.** $\dfrac{4}{3}$ **6.** $-\dfrac{35}{11}$ **7.** $\varnothing$ **8.** $\mathbb{R}$
9. $\dfrac{13}{3}$ **10.** $b = \dfrac{a - 2c}{3}$ **11.** $b_2 = \dfrac{2A - hb_1}{h}$ **12.** $625 **13.** 80 visits **14.** 4.2 hours **15.** 6.25 liters **16.** $7000 at 8%; $5000 at 7%
17. -10 **18.** 33 **19.** $\left(\dfrac{9}{2}, 7\right]$ **20.** $[4, 13)$ **21.** $\{-7, 2\}$ **22.** $\left\{-\dfrac{14}{3}, \dfrac{26}{5}\right\}$ **23.** $\{-3\}$
24. $\{x | x < -1 \text{ or } x > 4\}$ **25.** $\left\{x \left| \dfrac{1}{2} \le x \le \dfrac{5}{2}\right.\right\}$

CHAPTER 3

Exercise Set 3.1 **1. a)** A straight line **b)** Two; Two points uniquely determine a straight line. **3.** They are in a straight line. **5.** $A(3, 1), B(-6, 0), C(2, -4), D(-2, -4), E(0, 3), F(-8, 1), G\left(\dfrac{3}{2}, -1\right)$ **7.**
9. I **11.** IV **13.** II **15.** III **17.** No **19.** No **21.** Yes **23.** Yes **25.** No
27. **29.** **31.**

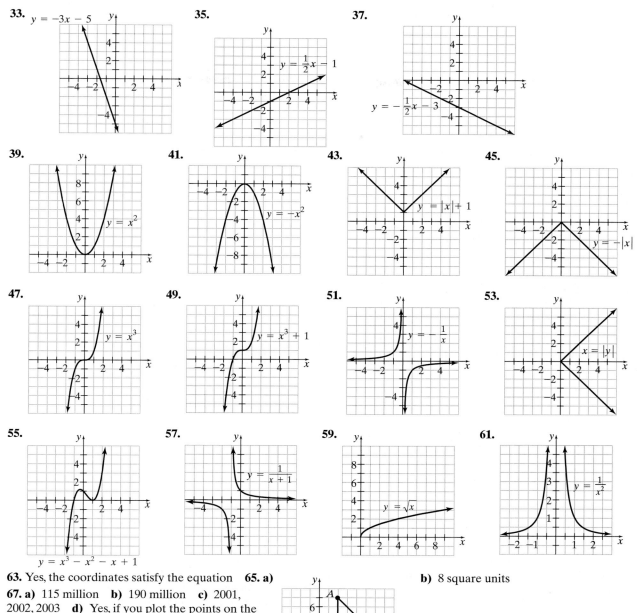

33. $y = -3x - 5$

35. $y = \frac{1}{2}x - 1$

37. $y = -\frac{1}{2}x - 3$

39. $y = x^2$

41. $y = -x^2$

43. $y = |x| + 1$

45. $y = -|x|$

47. $y = x^3$

49. $y = x^3 + 1$

51. $y = -\frac{1}{x}$

53. $x = |y|$

55. $y = x^3 - x^2 - x + 1$

57. $y = \frac{1}{x + 1}$

59. $y = \sqrt{x}$

61. $y = \frac{1}{x^2}$

63. Yes, the coordinates satisfy the equation   **65. a)**

b) 8 square units

67. a) 115 million   **b)** 190 million   **c)** 2001, 2002, 2003   **d)** Yes, if you plot the points on the top of the bars in the middle and connect them, they would approximate a straight line

69.

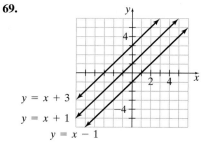

$y = x + 3$

$y = x + 1$

$y = x - 1$

a) Each graph crosses the y-axis at the point corresponding to the constant term in the graph's equation.   **b)** Yes

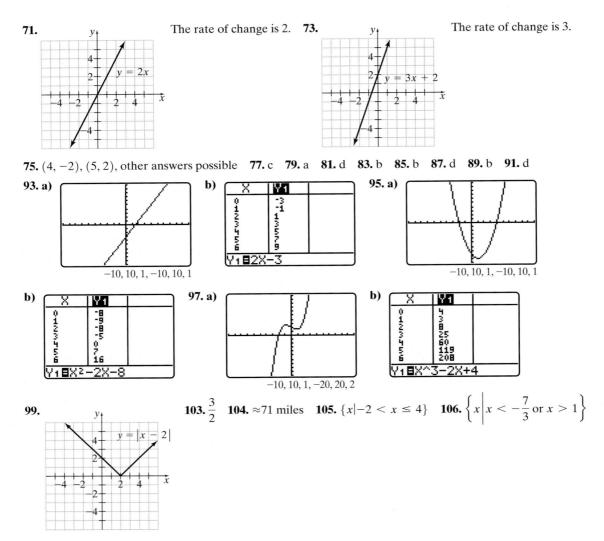

71. The rate of change is 2. **73.** The rate of change is 3.

75. $(4, -2), (5, 2)$, other answers possible **77.** c **79.** a **81.** d **83.** b **85.** b **87.** d **89.** b **91.** d

93. a) **b)** **95. a)**

$-10, 10, 1, -10, 10, 1$ $Y_1 \boxminus 2X - 3$ $-10, 10, 1, -10, 10, 1$

b) **97. a)** **b)**

$Y_1 \boxminus X^2 - 2X - 8$ $-10, 10, 1, -20, 20, 2$ $Y_1 \boxminus X^3 - 2X + 4$

99. **103.** $\dfrac{3}{2}$ **104.** ≈ 71 miles **105.** $\{x | -2 < x \le 4\}$ **106.** $\left\{ x \middle| x < -\dfrac{7}{3} \text{ or } x > 1 \right\}$

Exercise Set 3.2 **1.** A correspondence where each member in the domain corresponds to exactly one member of the range **3.** Yes, a relation is any set of ordered pairs. **5.** If a vertical line drawn through any part of the graph intersects the graph at more than one point, the graph is not a function. **7.** The set of values for the dependent variable **9.** Domain: $\{x | x \ne 0\}$, Range: $\{y | y \ne 0\}$; x cannot be 0 because you cannot divide by 0. y cannot be 0 because the numerator is 1. **11.** Domain: $\mathbb{R}$, Range: $\{y | y \ge 0\}$; x can be any real number, $|x|$ can never be negative **13.** If y depends on x, then x is the independent variable. **15. a)** Function **b)** Domain: $\{3, 5, 10\}$, Range: $\{6, 10, 20\}$ **17. a)** Function **b)** Domain: {Cameron, Tyrone, Vishnu}, Range: $\{1, 2\}$ **19. a)** Not a function **b)** Domain: $\{1990, 2001, 2002\}$; Range: $\{20, 34, 37\}$ **21. a)** Function **b)** Domain: $\{1, 2, 3, 4, 5\}$; Range: $\{1, 2, 3, 4, 5\}$ **23. a)** Function **b)** Domain: $\{1, 2, 3, 4, 5, 7\}$; Range: $\{-1, 0, 2, 4, 5\}$ **25. a)** Not a function **b)** Domain: $\{1, 2, 3\}$ **c)** $\{1, 2, 4, 5, 6\}$ **27. a)** Not a function **b)** Domain: $\{0, 1, 2\}$; Range: $\{-7, -1, 2, 3\}$ **29. a)** Function **b)** Domain: $\mathbb{R}$; Range: $\mathbb{R}$ **c)** 2 **31. a)** Not a function **b)** Domain: $\{x | 0 \le x \le 2\}$, Range: $\{y | -3 \le y \le 3\}$ **c)** ≈ 1.5 **33. a)** Function **b)** Domain: $\mathbb{R}$, Range: $\{y | y \ge 0\}$ **c)** 1, 3 **35. a)** Function **b)** Domain: $\{-1, 0, 1, 2, 3\}$, Range: $\{-1, 0, 1, 2, 3\}$ **c)** 2 **37. a)** Not a function **b)** Domain: $\{x | x \ge 2\}$, Range: $\mathbb{R}$ **c)** 3 **39. a)** Function **b)** Domain: $\{x | -2 \le x \le 2\}$, Range: $\{y | -1 \le y \le 2\}$ **c)** $-2, 2$ **41. a)** 1 **b)** 11 **43. a)** -6 **b)** -4 **45. a)** 2 **b)** 2 **47. a)** 7 **b)** 0 **49. a)** 0 **b)** 3 **51. a)** 1 **b)** Undefined **53. a)** 12 square feet **b)** 27 square feet **55. a)** $A(r) = \pi r^2$ **b)** ≈ 314.2 square yards **57. a)** $C(F) = \dfrac{5}{9}(F - 32)$ **b)** $-40°C$ **59. a)** $18.23°C$ **b)** $27.68°C$ **61. a)** $78.32°$ **b)** $73.04°$ **63. a)** 91 oranges **b)** 204 oranges **65.** Answers will vary. One possible interpretation: The person warms up slowly, possibly by walking, for 5 minutes. Then the person begins jogging slowly over a period of 5 minutes. For the next 15 minutes the person jogs. For the next 5 minutes the person walks slowly and his heart rate decreases to his normal heart rate. The rate stays the same for the next 5 minutes.   **67.** Answers will vary. One possible interpretation: The man walks on level ground, about 30 feet above sea level, for 5 minutes. For the next 5 minutes he walks uphill to 45 feet above sea level. For 5 minutes he walks on level ground, then walks quickly downhill for 3 minutes to an elevation of 20 feet above sea level. For 7 minutes he walks on level ground. Then he walks quickly uphill for 5 minutes.   **69.** Answers will vary. One possible interpretation: Driver is in stop-and-go traffic, then gets on highway for about 15 minutes, then stops car for a couple of minutes, then stop-and-go traffic.   **71. a)** Yes, it passes the vertical line test.   **b)** Year   **c)** ≈$115 billion   **d)** ≈43.75%   **73. a)** Yes, it passes the vertical line test.   **b)** Yes, it passes the vertical line test.   **c)** No, it does not appear to be linear   **d)** Yes, it appears to be a straight line   **e)** 1999   **f)** 2000   **75. a)**

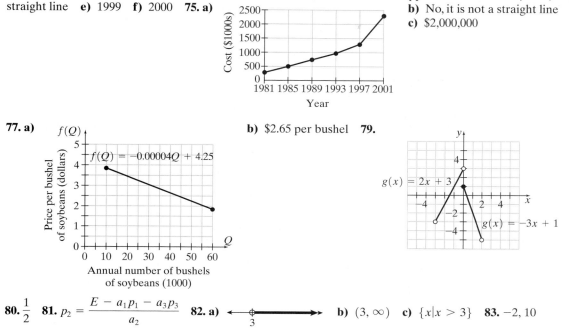

b) No, it is not a straight line   **c)** $2,000,000

77. a)   **b)** $2.65 per bushel   **79.**

80. $\frac{1}{2}$ **81.** $p_2 = \dfrac{E - a_1 p_1 - a_3 p_3}{a_2}$ **82. a)**   **b)** $(3, \infty)$ **c)** $\{x \mid x > 3\}$ **83.** $-2, 10$

Exercise Set 3.3

1. $ax + by = c$ **3.** To find the x-intercept, set $y = 0$ and solve for x. To find the y-intercept, set $x = 0$ and solve for y.   **5.** Vertical line   **7.** Horizontal line   **9.** Graph both sides of the equation. The solution is the x-coordinate of the point of intersection.   **11.** $2x + y = 4$ **13.** $3x - 4y = -14$

15.   **17.**   **19.**   **21.**

23.   **25.**   **27.**

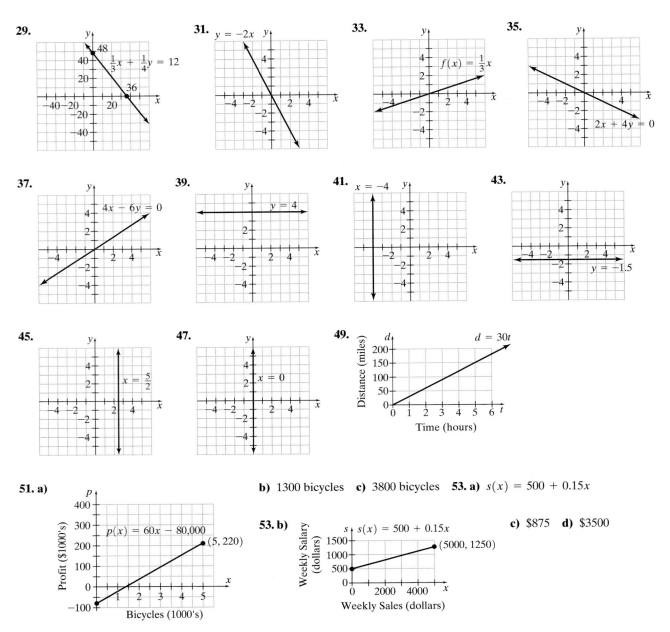

29. $\frac{1}{3}x + \frac{1}{4}y = 12$

31. $y = -2x$

33. $f(x) = \frac{1}{3}x$

35. $2x + 4y = 0$

37. $4x - 6y = 0$

39. $y = 4$

41. $x = -4$

43. $y = -1.5$

45. $x = \frac{5}{2}$

47. $x = 0$

49. $d = 30t$

51. a) $p(x) = 60x - 80,000$ — (5, 220)   **b)** 1300 bicycles   **c)** 3800 bicycles   **53. a)** $s(x) = 500 + 0.15x$

53. b) $s(x) = 500 + 0.15x$ — (5000, 1250)   **c)** \$875   **d)** \$3500

55. a) There is only one y-value for each x-value.   **b)** Independent: length; Dependent: weight   **c)** Yes
d) 11.5 kilograms   **e)** 65 centimeters   **f)** 12.0–15.5 kilograms   **g)** Increases; yes, as babies get older their weights vary
more.   **57.** When the graph goes through the origin, because at the origin both x and y are zero.   **59.** Answers will vary.
One possible answer is $f(x) = 4$.   **61.** Both intercepts will be at 0.   **63. a)**   **b)** 2 (or -2) units

c) 4 (or −4) units   **d)** $\frac{1}{2}$; slope   **65.** 1   **67.** −3   **69.** (−3.2, 0), (0, 6.4)   **71.** (−2, 0), (0, −2.5)   **73.** 102   **74.** −$\frac{18}{13}$

75. a) Answers will vary.   **b)** $x = a + b$ or $x = a − b$ **76. a)** Answers will vary.   **b)** $a − b < x < a + b$

77. a) Answers will vary.   **b)** $x < a − b$ or $x > a + b$ **78.** $\{−2, 2\}$

Exercise Set 3.4   **1.** Select two points on the line; find $\frac{\Delta y}{\Delta x}$.   **3.** The line rises going from left to right.   **5.** The

change in x is 0, and we cannot divide by zero.   **7.** Solve for y.   **9. a)** Moved down 3 units   **b)** (0, −6)   **11.** The change in

y for a unit change in x **13.** −2   **15.** −$\frac{1}{2}$ **17.** −1   **19.** Undefined   **21.** 0   **23.** −$\frac{2}{3}$ **25.** $b = 3$ **27.** $k = 1$ **29.** $x = 6$

31. $r = 0$ **33.** $m = −3$, $y = −3x$ **35.** $m = −\frac{1}{3}$, $y = −\frac{1}{3}x + 2$ **37.** m is undefined, $x = −2$ **39.** $m = 0$, $y = 3$

41. $m = −\frac{3}{2}$, $y = −\frac{3}{2}x + 15$ **43.**

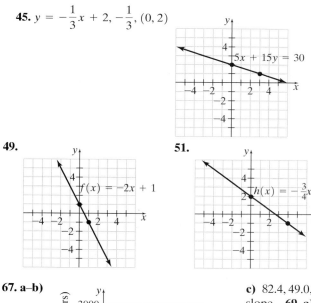

$y = −x + 2, −1, (0, 2)$

45. $y = −\frac{1}{3}x + 2, −\frac{1}{3}, (0, 2)$

47. $y = \frac{5}{2}x + 2, \frac{5}{2}, (0, 2)$

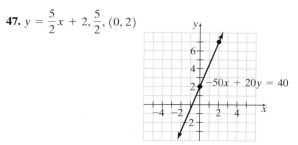

49.

51.

53. a) 2   **b)** 4   **c)** 1   **d)** 3   **55.** If the slopes are the same and the y-intercepts are different, the lines are parallel.   **57.** (0, −5)   **59. a)** $y = 3x + 1$ **b)** $y = 3x − 5$ **61. a)** 1   **b)** (0, 2)   **c)** $y = x + 2$ **63.** $y = \frac{3}{2}x − 7$ **65.** 0.2

67. a–b)

c) 82.4, 49.0, 123.8   **d)** 1995–2000, because its line segment has the greatest slope   **69. a)** $h(x) = −x + 200$ **b)** 186 beats per minute

71. a) $N(t) = −1.35t + 36.5$ **b)** 33.8 million people   **c)** 24.35 million people   **d)** 2004   **73. a)** $n(t) = −50.64t + 7156$ **b)** 6143 hospitals   **c)** 5637 hospitals   **d)** 2017   **75. a)** $P(t) = 5700t + 110,500$ **b)** $121,900   **c)** $196,000   **d)** 2010   **77.** The y-intercept is wrong.

79. The slope is wrong.   **81.** Height: 14.2 inches, width: 6.4 inches   **84.** 19

85. −$\frac{92}{5}$ **86.** 2.4   **87.** First: 75 miles per hour; second: 60 miles per hour

88. a) $x < −2$ or $x > 1$ **b)** $−2 < x < 1$

Exercise Set 3.5   **1.** $y − y_1 = m(x − x_1)$ **3.** Two lines are perpendicular if their slopes are negative reciprocals,

or if one line is vertical and the other is horizontal.   **5.** $y = 2x − 1$ **7.** $y = −\frac{1}{2}x + 1$ **9.** $y = \frac{1}{2}x − \frac{9}{2}$ **11.** $y = −\frac{3}{2}x$

13. $y = \frac{1}{2}x - 5$ **15.** Parallel **17.** Neither **19.** Perpendicular **21.** Perpendicular **23.** Parallel **25.** Neither

27. Perpendicular **29.** Parallel **31.** Neither **33.** $y = 2x + 1$ **35.** $2x - 5y = 19$ **37.** $y = -\frac{5}{3}x + 5$

39. $f(x) = -3x + 14$ **41.** $y = -\frac{2}{3}x + 6$ **43. a)** $C(s) = 45.7s + 95.8$ **b)** 324.3 calories **45. a)** $d(p) = -0.20p + 90$

b) 38 DVD players **c)** \$225 **47. a)** $s(p) = 95p - 60$ **b)** 206 kites **c)** \$3.00 **49. a)** $m(s) = -\frac{1}{3}s + 55$ **b)** 35 miles

per gallon **c)** 75 miles per hour **51. a)** $p(s) = 69.1s + 2786$ **b)** \$4029.80 **c)** ≈ 18 years **53. a)** $y(a) = -0.865a + 79.25$

b) 47.2 years **c)** 62.7 years old **55. a)** $w(a) = 0.189a + 10.6$ **b)** 14.758 kilograms **58.** $\left(-\infty, \frac{2}{5}\right)$ **59.** Reverse the

direction of the inequality symbol. **60. a)** Any set of ordered pairs **b)** A correspondence where each member of the domain
corresponds to a unique member in the range **c)** Answers will vary. **61.** Domain: $\{3, 4, 5, 6\}$; Range: $\{-2, -1, 2, 3\}$

Exercise Set 3.6 **1.** Yes, this is how addition of functions is defined. **3.** $g(x) \neq 0$ since division by zero is
undefined. **5.** No, subtraction is not commutative. One example is $5 - 3 = 2$ but $3 - 5 = -2$ **7. a)** 2 **b)** -8 **c)** -15

d) $-\frac{3}{5}$ **9. a)** $x^2 + 2x + 1$ **b)** $a^2 + 2a + 1$ **c)** 9 **11. a)** $x^3 + x - 4$ **b)** $a^3 + a - 4$ **c)** 6 **13. a)** $4x^3 - x + 4$

b) $4a^3 - a + 4$ **c)** 34 **15.** -7 **17.** -13 **19.** -60 **21.** Undefined **23.** 13 **25.** $-\frac{3}{4}$ **27.** $2x^2 - 6$ **29.** -6 **31.** 30

33. 0 **35.** $-\frac{3}{7}$ **37.** $-\frac{1}{45}$ **39.** $-2x^2 + 2x - 6$ **41.** 3 **43.** -4 **45.** 1 **47.** Undefined **49.** 4 **51.** 0 **53.** -3 **55.** -2

57. a) Total = public + private. **b)** 1970–1980 **c)** 1990–2000 **59. a)** 6,500,000 people **b)** 700,000 people
c) 7,200,000 people **61.** $f(a)$ and $g(a)$ must either be opposites or both be equal to 0. **63.** $f(a) = g(a)$ **65.** $f(a)$ and

$g(a)$ must have opposite signs. **67.**   **69.**   **72.** $-\frac{1}{27}$ **73.** 1.63×10^6

$-10, 10, 1, -10, 10, 1$ $-10, 10, 1, -10, 10, 1$

74. $h = \frac{2A}{b}$ **75.** \$450 **76.**   **77.**

$y = |x| - 2$

$3x - 4y = 12$

Exercise Set 3.7 **1.** Points on the line are solutions for the corresponding equation, and are not solutions if the
symbol used is $<$ or $>$. **3.** $(0, 0)$ cannot be used as a check point if the line passes through the origin.

5.   **7.**   **9.**   **11.**

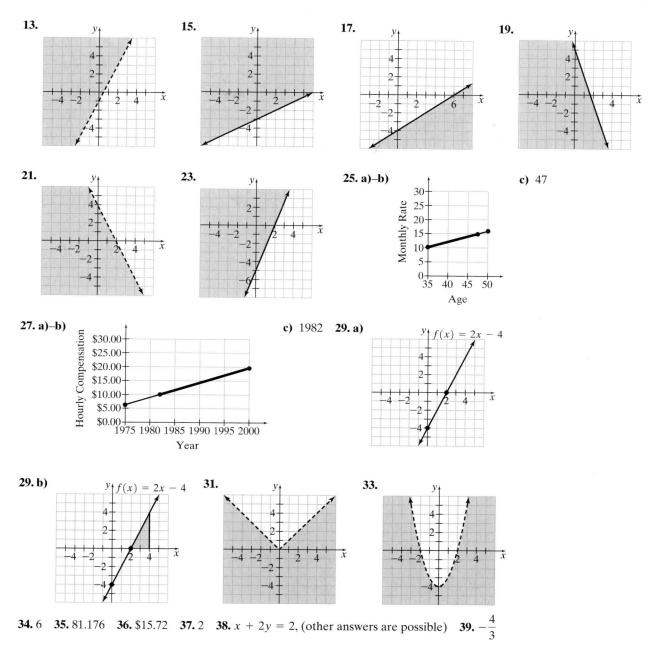

34. 6   **35.** 81.176   **36.** $15.72   **37.** 2   **38.** $x + 2y = 2$, (other answers are possible)   **39.** $-\dfrac{4}{3}$

Chapter Review Exercises   1.

2.

3.

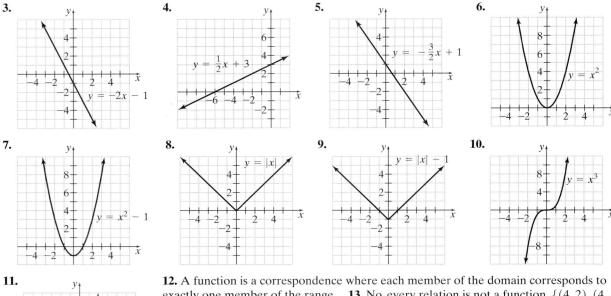

Graph of $y = -2x - 1$

4. Graph of $y = \frac{1}{2}x + 3$

5. Graph of $y = -\frac{3}{2}x + 1$

6. Graph of $y = x^2$

7. Graph of $y = x^2 - 1$

8. Graph of $y = |x|$

9. Graph of $y = |x| - 1$

10. Graph of $y = x^3$

11.

Graph of $y = x^3 + 4$

12. A function is a correspondence where each member of the domain corresponds to exactly one member of the range.   **13.** No, every relation is not a function. $\{(4, 2), (4, -2)\}$ is a relation but not a function. Yes, every function is a relation because it is a set of ordered pairs.   **14.** Yes, each member of the domain corresponds to exactly one member of the range.   **15.** No, the domain element 2 corresponds to more than one member of the range (5 and -5).   **16. a)** No, the relation is not a function   **b)** Domain: $\{x | -1 \le x \le 1\}$; Range: $\{y | -1 \le y \le 1\}$ **17. a)** No, the relation is not a function.   **b)** Domain: $\{x | -2 \le x \le 2\}$; Range: $\{y | -1 \le y \le 1\}$ **18. a)** Yes, the relation is a function.   **b)** Domain: $\mathbb{R}$; Range: $\{y | y \le 0\}$ **19. a)** Yes, the relation is a function.   **b)** Domain: $\mathbb{R}$; Range: $\mathbb{R}$ **20. a)** -3 **b)** $-h^2 + 3h - 5$ **21. a)** -4 **b)** $2a^3 - 3a^2 + 1$

22. Answers will vary. Here is one possible interpretation: The car speeds up to 50 mph. Stays at this speed for about 11 minutes. Speeds up to about 68 mph. Stays at that speed for about 5 minutes, then stops quickly. Stopped for about 5 minutes. Then in stop and go traffic for about 5 minutes.   **23. a)** 720 baskets   **b)** 1500 baskets   **24. a)** 84 feet   **b)** 36 feet

25. Graph of $3x - 4y = 6$

26. Graph of $\frac{2}{3}x = \frac{1}{4}y + 20$

27. Graph of $f(x) = 4$

28. Graph of $x = -2$

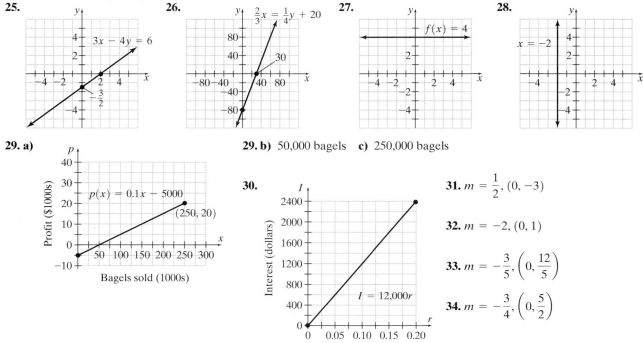

29. a) Graph of $p(x) = 0.1x - 5000$, point $(250, 20)$

29. b) 50,000 bagels   **c)** 250,000 bagels

30. Graph of $I = 12{,}000r$

31. $m = \dfrac{1}{2}$, $(0, -3)$

32. $m = -2$, $(0, 1)$

33. $m = -\dfrac{3}{5}$, $\left(0, \dfrac{12}{5}\right)$

34. $m = -\dfrac{3}{4}$, $\left(0, \dfrac{5}{2}\right)$

35. m is undefined, no y-intercept   **36.** $m = 0, (0, 6)$ **37.** $-\dfrac{1}{2}$ **38.** $-\dfrac{1}{3}$ **39.** $m = 0; y = 3$ **40.** m is undefined; $x = 2$

41. $m = -\dfrac{1}{2}; y = -\dfrac{1}{2}x + 2$ **42. a)** -2 **b)** $(0, -1)$ **c)** $y = -2x - 1$ **43.** $(0, 0)$

44. a)   **b)** 1970–1980: 16.4; 1980–1990: 4.2; 1990–2000: -23.5 **c)** 1970–1980

45. $n(t) = 0.7t + 35.6$ **46.** Parallel   **47.** Perpendicular   **48.** Neither   **49.** $y = \dfrac{1}{2}x + 3$ **50.** $y = -x - 2$ **51.** $y = -\dfrac{2}{3}x + 4$

52. $y = \dfrac{5}{2}x - 2$ **53.** $y = -\dfrac{5}{3}x - 4$ **54.** $y = -\dfrac{1}{2}x + 4$ **55.** Neither   **56.** Parallel   **57.** Perpendicular   **58.** Neither

59. a) $r(a) = 0.61a - 10.59$ **b)** \$15.03   **60. a)** $C(r) = 1.8r + 435$ **b)** 507 calories   **c)** 91.7 yards per minute

61. $x^2 - x - 1$ **62.** 5   **63.** $-x^2 + 5x - 9$ **64.** -15 **65.** -56 **66.** 70   **67.** $-\dfrac{2}{3}$ **68.** -2 **69. a)** 300   **b)** 750   **c)** 1450   **d)** 750

e) 1750   **f)** 1500   **70. a)** 490 million   **b)** 190 million   **c)** 680 million

71.   **72.**   **73.**   **74.**

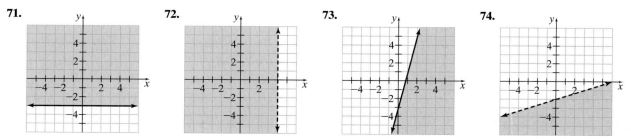

Chapter Practice Test

1.   **2.**   **3.**   **4.**

5. A function is a correspondence where each member in the domain corresponds with exactly one member in the range.
6. Yes, because each member in the domain corresponds to exactly one member in the range.   **7.** Yes; Domain: $\mathbb{R}$; Range:
$\{y | y \le 4\}$ **8.** No; Domain: $\{x | -3 \le x \le 3\}$; Range: $\{y | -2 \le y \le 2\}$ **9.** 26   **10.**
11.   **12.**

13.

14. a)

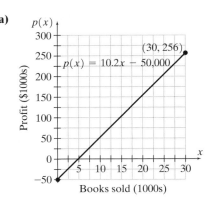

Books sold (1000s)

b) 4900 books   **c)** 14,700 books

15. $m = \dfrac{4}{3}, (0, -3)$ **16.** $y = 3x - 5$

17. $y = -2x + 7$

18. $p(t) = 2.386t + 274.634$ **19.** Parallel,

the slope of both lines is the same, $\dfrac{2}{3}$.

20. a) $r(t) = -3.2t + 362$ **b)** 282 deaths/
100,000   **c)** 234 deaths/100,000   **21.** 13

22. $-\dfrac{1}{2}$ **23.** $2a^2 - a$

24. a) 44 million tons
b) 18 million tons   **c)** 26 million tons

25.

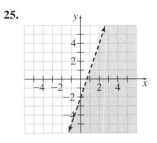

CHAPTER 4

Using Your Graphing Calculator, 4.1   **1.** $(2.76, 0.82)$ **2.** $(13.29, 9.57)$ **3.** $(-4.67, -4.66)$
4. $(-2.25, 10.52)$

Exercise Set 4.1   **1.** The solution to a system of linear equations is the point(s) that satisfy all equations in the
system.   **3.** A dependent system is a system that has an infinite number of solutions.   **5.** A consistent system of equations
has a solution.   **7.** Compare the slopes and y-intercepts of the equations. If the slopes are different, the system is consistent. If
the slopes and y-intercepts are the same, the system is dependent. If the slopes are the same and the y-intercepts are
different, the system is inconsistent.   **9.** You will get a true statement, like $0 = 0$.   **11.** None   **13. b)**   **15. b)**   **17.** Consistent;
one solution   **19.** Dependent; infinite number of solutions   **21.** Inconsistent; no solution   **23.** Inconsistent; no solution

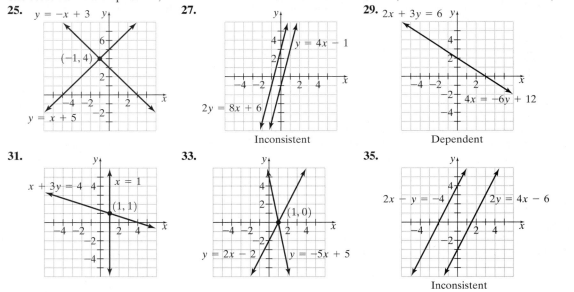

25. $y = -x + 3$ $(-1, 4)$ $y = x + 5$

27. $y = 4x - 1$ $2y = 8x + 6$ Inconsistent

29. $2x + 3y = 6$ $4x = -6y + 12$ Dependent

31. $x + 3y = 4$ $x = 1$ $(1, 1)$

33. $y = 2x - 2$ $(1, 0)$ $y = -5x + 5$

35. $2x - y = -4$ $2y = 4x - 6$ Inconsistent

37. $(-1, 0)$ **39.** $(-3, -3)$ **41.** $(2, 1)$ **43.** $\left(-1, \dfrac{1}{2}\right)$ **45.** Infinite number of solutions   **47.** No solution   **49.** $\left(-\dfrac{19}{5}, -3\right)$

51. $(8, 6)$ **53.** $(2, 5)$ **55.** $\left(-1, -\dfrac{5}{3}\right)$ **57.** Infinite number of solutions   **59.** No solution   **61.** $(1, 1)$ **63.** $(5, 6)$

65. $\left(\dfrac{14}{5}, -\dfrac{12}{5}\right)$ **67.** $\left(\dfrac{37}{7}, \dfrac{19}{7}\right)$ **69.** $(3, 2)$ **71.** $(4, 0)$ **73.** $(4, 3)$ **75.** $(10, 4)$ **77. a), b),** and **c)** Answers will vary.   **79.** 2085

81. Multiply the first equation by 2 and notice that the new equation is identical to the second equation.   **83. a)** Infinite number, because a system of equations can have no solution, one solution, or an infinite number of solutions **b)** $m = -4, y = -4x - 13, (0, -13)$ **c)** Yes   **85.** One example is: $x + y = 1, 2x + 2y = 2$, create one equation and then multiply it by a constant to get the second equation   **87. a)** One example is: $x + y = 7, x - y = -3$.   **b)** Choose coefficients for x and y, then use the given coordinates to find the constants.   **89.** $A = 2$ and $B = 5$ **91.** $m = 4, b = -2$

93. The system is dependent or one graph is not in the viewing window.   **95.** $(8, -1)$ **97.** $(-1, 2)$ **99.** $\left(\dfrac{1}{a}, 5\right)$

102. Rational numbers can be expressed as quotients of two integers, denominator not 0. Irrational numbers cannot.   **103. a)** Yes, the set of real numbers includes the set of rational numbers.   **b)** Yes, the set of real numbers includes the set of irrational numbers.   **104.** $-\dfrac{17}{4}$ **105.** $\mathbb{R}$ **106.** 520   **107.** No, the points $(-3, 4)$ and $(-3, 2)$ have the same first coordinate but different second coordinates.   **108.** Undefined

Exercise Set 4.2

1. The graph will be a plane.   **3.** $(1, -2, -4)$ **5.** $\left(-7, -\dfrac{35}{4}, -3\right)$ **7.** $(0, 3, 6)$ **9.** $(1, 2, 0)$

11. $(-3, 17, -7)$ **13.** $(3, 1, -2)$ **15.** $(2, -1, 3)$ **17.** $\left(\dfrac{2}{3}, -\dfrac{1}{3}, 1\right)$ **19.** $(0, -1, 0)$ **21.** $\left(-\dfrac{11}{17}, \dfrac{7}{34}, -\dfrac{49}{17}\right)$ **23.** $(0, 0, 0)$

25. $(4, 6, 8)$ **27.** $\left(\dfrac{2}{3}, \dfrac{23}{15}, \dfrac{37}{15}\right)$ **29.** $(1, 1, 2)$ **31.** Inconsistent   **33.** Dependent   **35.** Inconsistent   **37.** No point is common to all three planes. Therefore, the system is inconsistent.   **39.** One point is common to all three planes; therefore, the system is consistent.   **41. a)** Yes, the 3 planes can be parallel   **b)** Yes, the 3 planes can intersect at one point   **c)** no, the 3 planes cannot intersect at exactly two points   **43.** $A = 9, B = 6, C = 2; 9x + 6y + 2z = 1$ **45.** Answers will vary. One example is $x + y + z = 10, x + 2y + z = 11, x + y + 2z = 16$ **47. a)** $a = 1, b = 2, c = -4$ **b)** $y = x^2 + 2x - 4$, Substitute 1 for a, 2 for b, and -4 for c into $y = ax^2 + bx + c$ **49.** $(1, 2, 3, 4)$ **51. a)** $\dfrac{1}{4}$ hour or 15 minutes   **b)** 1.25 miles

52. $\left\{x \middle| x < -\dfrac{3}{2} \text{ or } x > \dfrac{27}{2}\right\}$ **53.** $\left\{x \middle| -\dfrac{8}{3} < x < \dfrac{16}{3}\right\}$ **54.** $\varnothing$

Exercise Set 4.3

1. Magic Kingdom: 14.8 million; Disneyland: 12.3 million   **3.** Hamburger, 21 grams; fries, 67 grams   **5.** $25°, 65°$ **7.** $52°, 128°$ **9.** 12.2 miles per hour, 3.4 miles per hour   **11.** $500, 4\%$ **13.** 1.2 ounces of 5%, 1.8 ounces of 30%   **15.** 10 gallons concentrate, 190 gallons water   **17.** $17\dfrac{1}{3}$ pounds birdseed, $22\dfrac{2}{3}$ pounds sunflower seeds   **19.** 65 adult's, 160 children's   **21.** $6000 at 5%, $4000 at 6%   **23.** 160 gallons whole, 100 gallons skim   **25.** 7 pounds Season's Choice, 13 pounds Garden Mix   **27.** 50 miles per hour, 55 miles per hour   **29.** Cabrina: 8 hours, Dabney: 3.4 hours   **31.** 80 grams A, 60 grams B **33.** 200 grams first alloy, 100 grams second alloy   **35.** Tom: 60 miles per hour, Melissa: 75 miles per hour   **37.** ≈ 2006 **39. a)** $C = 0.05t + 8.95, C = 0.07t + 5.95$ **39. b)**

c) 150 minutes   **d)** 150 minutes   **41. a)** $x + y + z = 24; x = 2y - 2; z = 5y + 2$ **b)** Personal mail, 3; bills and statements, 4; advertisements, 17   **43. a)** $x + y + z = 41; z = 3x - 14; y = 2x - 5$ **b)** Iraq: 10 million; Angola: 15 million; Iran: 16 million   **45. a)** $x + y + z = 180; \quad x = \dfrac{2}{3}y; z = 3y - 30$ **b)** $30°, 45°, 105°$ **47. a)** $x + y + z = 10,000;$ $y = 2x; 0.03x + 0.05y + 0.06z = 525$ **b)** $1500 at 3%, $3000 at 5%, $5500 at 6%   **49. a)** $x + y + z = 8; 0.10x + 0.12y + 0.20z = 0.13(8); z = x - 2$ **b)** 4 liters of the 10% solution; 2 liters of the 12% solution; 2 liters of the 20% solution   **51. a)** $5x + 4y + 7z = 154; 3x + 2y + 5z = 94; 2x + 2y + 4z = 76$ **b)** 10 children's chairs; 12 standard chairs; 8 executive chairs

53. $I_A = \dfrac{27}{38}; I_B = -\dfrac{15}{38}; I_C = -\dfrac{6}{19}$ **56.** $-\dfrac{35}{8}$ **57.** 4   **58.** Use the vertical line test.   **59.** $y = x - 10$

Exercise Set 4.4

1. It has the same number of rows and columns   **3.** Change the -1 in the second row to 1 by multiplying row 2 by -1, or $-1R_2$ **5.** Switch R_2 and R_3 to get a 1 in the second row, second column.   **7.** Dependent

9. $\begin{bmatrix} 1 & -2 & | & -3 \\ 3 & 1 & | & -4 \end{bmatrix}$ **11.** $\begin{bmatrix} 1 & 1 & 3 & | & -8 \\ 3 & 2 & 1 & | & -5 \\ 4 & 7 & 2 & | & -1 \end{bmatrix}$ **13.** $\begin{bmatrix} 1 & 3 & | & 12 \\ 0 & 17 & | & 30 \end{bmatrix}$ **15.** $\begin{bmatrix} 1 & 0 & 8 & | & \frac{1}{4} \\ 0 & 2 & -38 & | & -\frac{13}{4} \\ 6 & -3 & 1 & | & 0 \end{bmatrix}$ **17.** $(3, 0)$ **19.** $(-2, 2)$

21. $(0, 1)$ **23.** Dependent system   **25.** $(-2, 1)$ **27.** Inconsistent system   **29.** $\left(\frac{2}{3}, \frac{1}{4}\right)$ **31.** $\left(\frac{4}{5}, -\frac{7}{8}\right)$ **33.** $(2, 1, 3)$

35. $(3, 1, 2)$ **37.** $\left(1, -1, \frac{1}{2}\right)$ **39.** Dependent system   **41.** $\left(\frac{1}{2}, 2, 4\right)$ **43.** Inconsistent system   **45.** $\left(5, \frac{1}{3}, -\frac{1}{2}\right)$ **47.** No, this is the same as switching the order of the equations.   **49.** $\angle x = 30°, \angle y = 65°, \angle z = 85°$ **51.** 26% by Chiquita, 25% by Dole, 14% by Del Monte, 35% by other   **53. a)** $\{1, 2, 3, 4, 5, 6, 9, 10\}$ **b)** $\{4, 6\}$ **54. a)**

b) $\{x| -2 < x \le 4\}$ **c)** $(-2, 4]$ **55.** A graph is an illustration of the set of points whose coordinates satisfy an equation.
56. -76

Exercise Set 4.5

1. Answers will vary   **3.** If $D = 0$ and $D_x, D_y,$ or $D_z \ne 0$, the system is inconsistent.

5. $\left(-2, -\frac{1}{2}\right)$ **7.** 7   **9.** -8 **11.** -12 **13.** 44   **15.** $(-5, 2)$ **17.** $(3, 2)$ **19.** $(-7, -2)$ **21.** $(-3, 2)$ **23.** Infinite number of

solutions   **25.** $(2, -3)$ **27.** No solution   **29.** $(2, 5)$ **31.** $(1, -1, 2)$ **33.** $\left(\frac{1}{2}, -\frac{1}{2}, 2\right)$ **35.** $\left(\frac{1}{2}, -\frac{1}{8}, 2\right)$ **37.** $(-1, 0, 2)$

39. Infinite number of solutions   **41.** $(1, -1, 2)$ **43.** No solution   **45.** $(3, 4, 1)$ **47.** $(-1, 5, -2)$ **49.** It will have the opposite sign. This can be seen by comparing $a_1 b_2 - a_2 b_1$ to $a_2 b_1 - a_1 b_2$ **51.** 0   **53.** 0   **55.** Yes, it will have the opposite sign.

57. 5   **59.** 6   **61. a)** $x = \dfrac{c_1 b_2 - c_2 b_1}{a_1 b_2 - a_2 b_1}$ **b)** $y = \dfrac{a_1 c_2 - a_2 c_1}{a_1 b_2 - a_2 b_1}$ **62.** $\left(-\infty, \dfrac{14}{11}\right)$

63.   **64.**   **65.**

Exercise Set 4.6

1. Answers will vary.   **3.** Yes, since the point of intersection satisfies both inequalities, it also satisfies the system of inequalities.

5.   **7.**   **9.**   **11.**

13.   **15.**   **17.**   **19.**

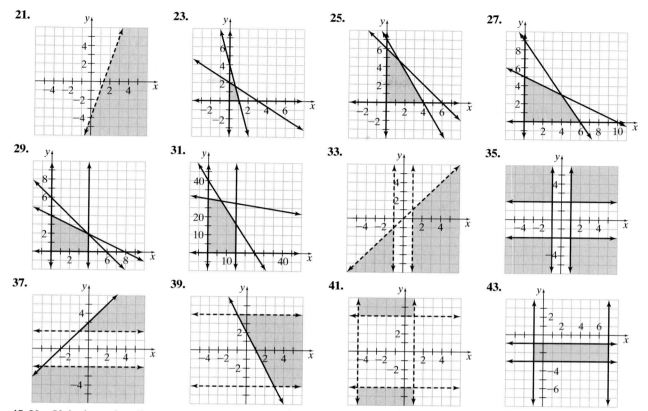

21. **23.** **25.** **27.**

29. **31.** **33.** **35.**

37. **39.** **41.** **43.**

45. Yes. If the boundary lines are parallel, there may be no solution. One example is $y > 3x + 1$; $y < 3x - 2$
47. There is no solution. Opposite sides of the same line are being shaded, but not the line itself.   **49.** There are an infinite number of solutions. Both inequalities include the line $5x - 2y = 3$.   **51.** There are an infinite number of solutions. The lines are parallel, but the same side of each line is being shaded.   **53.**   **55.**

57. $f_2 = \dfrac{f_3 d_3 - f_1 d_1}{d_2}$

58. Domain: $\{-1, 0, 4, 5\}$; Range: $\{-5, -2, 2, 3\}$
59. Domain: $\mathbb{R}$; Range: $\mathbb{R}$
60. Domain: $\mathbb{R}$; Range: $\{y | y \geq -1\}$

Chapter Review Exercises   **1.** Inconsistent; no solution   **2.** Consistent; one solution   **3.** Consistent; one solution   **4.** Consistent; one solution

5. **6.** **7.** **8.**

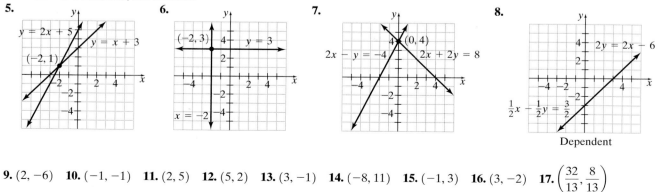

9. $(2, -6)$ **10.** $(-1, -1)$ **11.** $(2, 5)$ **12.** $(5, 2)$ **13.** $(3, -1)$ **14.** $(-8, 11)$ **15.** $(-1, 3)$ **16.** $(3, -2)$ **17.** $\left(\dfrac{32}{13}, \dfrac{8}{13}\right)$

18. $\left(-1, \dfrac{13}{3}\right)$ **19.** $(1, 2)$ **20.** $\left(\dfrac{7}{5}, \dfrac{13}{5}\right)$ **21.** $(6, -2)$ **22.** $\left(-\dfrac{78}{7}, -\dfrac{48}{7}\right)$ **23.** Infinite number of solutions   **24.** No solution

25. $(1, 2, -4)$ **26.** $(-1, 3, -2)$ **27.** $(-5, 1, 2)$ **28.** $(3, -2, -2)$ **29.** $\left(\dfrac{8}{3}, \dfrac{2}{3}, 3\right)$ **30.** $(0, 2, -3)$ **31.** No solution

32. Infinite number of solutions   **33.** Dennis: 38, Jennifer: 28   **34.** Airplane: 520 mph, wind: 40 mph   **35.** James should combine 2 liters of the 20% acid solution with 4 liters of the 50% acid solution.   **36.** 410 adult tickets and 240 children tickets were sold.   **37.** His ages were 41 years and 77 years.   **38.** $20,000 invested at 7%, $15,000 invested at 5%, and $5000 was invested at 3%.   **39.** $(11, -2)$ **40.** $(3, 1)$ **41.** Infinite number of solutions   **42.** $(2, 1, -2)$
43. No solution   **44.** $(1, -1, 3)$ **45.** $(2, 3)$ **46.** $(-3, 2)$ **47.** $(-1, 2)$ **48.** $(-2, 3, 4)$ **49.** $(1, 1, 2)$ **50.** No solution
51.   **52.**   **53.**   **54.** No solution

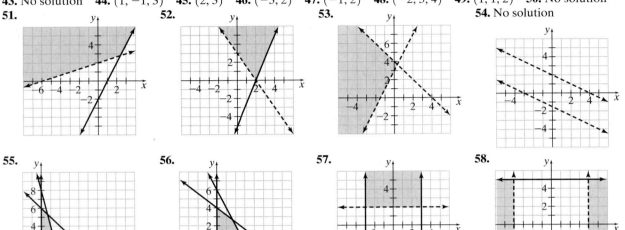

55.   **56.**   **57.**   **58.**

Practice Test   **1.** Answers will vary.   **2.** Consistent; one solution   **3.** Dependent; infinite number of solutions

4. Inconsistent; no solution   **5.**   **6.**   **7.** $(1, 1)$ **8.** $(-3, 2)$

9. $\left(-\dfrac{1}{2}, 4\right)$

10. Infinite number of solutions   **11.** $\left(\dfrac{44}{19}, \dfrac{48}{19}\right)$ **12.** $(1, -1, 2)$ **13.** $\begin{bmatrix} -2 & 3 & 7 & \bigm| & 5 \\ 3 & -2 & 1 & \bigm| & -2 \\ 1 & -6 & 5 & \bigm| & -13 \end{bmatrix}$ **14.** $\begin{bmatrix} 6 & -2 & 4 & \bigm| & 4 \\ 0 & 5 & -3 & \bigm| & 12 \\ 2 & -1 & 4 & \bigm| & -3 \end{bmatrix}$

15. $(4, -1)$ **16.** $(3, -1, 2)$ **17.** -2 **18.** 165   **19.** $(-3, 2)$ **20.** $(3, 1, -1)$ **21.** $6\dfrac{2}{3}$ pounds cashews; $13\dfrac{1}{3}$ pounds peanuts

22. $6\dfrac{2}{3}$ liters 6% solution; $3\dfrac{1}{3}$ liters 15% solution   **23.** 4, 9, and 12   **24.**   **25.**

CHAPTER 5

Using Your Graphing Calculator, 5.1

1.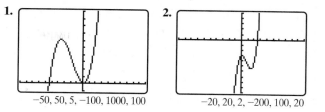

$-50, 50, 5, -100, 1000, 100$ $-20, 20, 2, -200, 100, 20$

Exercise Set 5.1

1. The terms are the parts that are added. **3.** A polynomial is a finite sum of terms in which all variables have whole number exponents and no variable appears in a denominator. **5.** The leading coefficient is the coefficient of the leading term. **7. a)** It is the same as that of the highest-degree term. **b)** 7 **9. a)** A polynomial is linear if its degree is 0 or 1. **b)** Answers will vary. One example is $x + 4$. **11. a)** A polynomial is cubic if it has degree 3 and is in one variable. **b)** Answers will vary. One example is $x^3 + x - 4$. **13.** Answers will vary. One example is $x^5 + x + 1$. **15.** Monomial **17.** Monomial **19.** Not a polynomial; -3 exponent **21.** Not a polynomial; $\frac{1}{2}$ exponent **23.** $-x^2 + 4x - 5, 2$ **25.** $10x^2 + 3xy + 9y^2, 2$ **27.** In descending order, 4 **29. a)** 6 **b)** 3

31. a) 6 **b)** 9 **33. a)** 17 **b)** $-\frac{1}{3}$ **35.** -7 **37.** -7 **39.** -2.0312 **41.** $x^2 + 9x - 12$ **43.** $x^2 - 13x - 7$

45. $2y^2 + 9y - 11$ **47.** $-\frac{2}{3}a^2 - \frac{29}{36}a + 7$ **49.** $-3.5x^2 - 2.1x - 19.6$ **51.** $-\frac{4}{3}x^3 - \frac{1}{4}x^2y + 9xy^2$ **53.** $5a - 10b + 13c$

55. $8a^2b - 10ab + 11b^2$ **57.** $7r^2 - 4rt - 3t^2$ **59.** $10x^2 - 5x - 9$ **61.** $-3w^2 + 6w$ **63.** $3x + 14$ **65.** $-3x^2 + 2x - 12$

67. $-4.4a^2 - 5.7a - 26.4$ **69.** $-\frac{11}{2}x^2y + xy^2 + \frac{2}{45}$ **71.** $5x^{2r} - 10x^r + 3$ **73.** $-x^{2s} - 4x^s + 15$ **75.** $7b^{4n} - 3b^{3n} - 4b^{2n} + 1$

77. $4x^2 + 8x + 20$ **79.** $3x^2 + 4x + 19$ **81.** $2x^2 + 12x + 9$ **83.** No, for example $(x^2 + x + 1) + (x^3 - 2x^2 + x) = x^3 - x^2 + 2x + 1$ **85.** No, for example $(x^2 + 3x - 5) + (-x^2 - 4x + 2) = -x - 3$ **87.** $A \approx 113.10$ square inches **89.** 674 feet **91. a)** $P(x) = 2x^2 + 360x - 8050$ **b)** $\$47,950$ **93. c)** The y-intercept is $(0, -4)$ and the leading coefficient is positive **95. c)** The y-intercept is $(0, -6)$ and the leading coefficient is negative **97. a)** 36,610 **b)** Yes. **99.** $\$88,210$ **101. a)** **b)** Increase **c)** Answers will vary. **d)** Decrease **e)** Answers will vary. **103. b)** The y-intercept is $(0, -5)$ and the leading coefficient is negative

105. c) cubic equation and the leading coefficient is positive **107.** 3 **108.** $\frac{15}{16}$

109. 6 hours. **110.** $-\frac{2}{9}$ **111.** $(-4, 0, -1)$

$-10, 10, 1, -10, 10, 1$

Exercise Set 5.2

1. a)–d) Answers will vary. **3. a)** Answers will vary. **b)** $x^3 - 2x^2 - 21x + 12$
5. a) Answers will vary. **b)** Answers will vary. One possible answer is $(x + 4)(x - 4)$. **c)** Answers will vary. **d)** Answers will vary. One possible answer is $x^2 - 16$. **7.** Yes, for example $(x + 2)(x - 1) = x^2 + x - 2$ **9.** $24x^2y^5$ **11.** $\frac{1}{9}x^7y^8z^2$

13. $6x^6y^3 - 9x^3y^4 - 12x^2y$ **15.** $2xyz + \frac{8}{3}y^2z - 6y^3z$ **17.** $0.6x^2 - 1.5x + 2.1y$ **19.** $2.85a^{11}b^5 - 1.38a^9b^7 + 0.36a^6b^9$

21. $12x^2 - 38x + 30$ **23.** $-2x^3 + 8x^2 - 3x + 12$ **25.** $x^2 + \frac{23}{6}xy - \frac{2}{3}y^2$ **27.** $0.6a^2 + 9.79ab - 3.5b^2$ **29.** $x^3 + x^2 - 5x - 2$

31. $2a^3 - 7a^2b + 5ab^2 - 6b^3$ **33.** $x^4 + 2x^2 + 10x + 7$ **35.** $5x^4 + 29x^3 + 14x^2 - 28x + 10$ **37.** $3m^4 - 11m^3 - 5m^2 - 2m - 20$
39. $8x^3 - 12x^2 + 6x - 1$ **41.** $10r^4 - 2r^3s - r^2s^2 + rs^3 - 2s^4$ **43.** $x^2 + 4x + 4$ **45.** $4x^2 - 12x + 9$ **47.** $16x^2 - 24xy + 9y^2$
49. $25m^4 - 4n^2$ **51.** $y^2 + 8y - 4xy + 16 - 16x + 4x^2$ **53.** $25x^2 + 20xy + 30x + 4y^2 + 12y + 9$ **55.** $a^2 - b^2 - 4b - 4$

57. $2x^3y + 2x^2y^2 + 6xy^3$ **59.** $2x^3y^2 + \frac{3}{2}x^2y^3 - \frac{7}{2}xy^4$ **61.** $\frac{3}{5}x^2y^5z^7 + 3x^2y^4z^2 - \frac{1}{10}x^2y^3z^9$ **63.** $21a^2 + 10a - 24$

65. $64x^2 - \frac{1}{16}$ **67.** $4x^2 - 36xy + 81y^2$ **69.** $2x^3 + 10x^2 + 9x - 9$ **71.** $6p^3 - p^2q - 16pq^2 + 6q^3$ **73.** $9x^2 + 12x + 4 - y^2$
75. $a^4 - 2a^2b^2 + b^4$ **77.** $2x^3 - 4x^2 - 64x + 192$ **79. a)** $x^2 - x - 20$ **b)** -8 **81. a)** $10x^3 + 36x^2 - 2x - 12$ **b)** 1196
83. a) $-x^4 + 3x^3 - 2x^2 + 6x$ **b)** -72 **85.** $x^2 + 5x$ **87.** $x^2 + y^2$ **89. a) and b)** $x^2 + 8x + 15$ **91.** $36 - x^2$

93. a) $11x + 12$ **b)** 117 square inches, 50 square inches. **95.** $(x + 5)(x - 5)$, product of the sum and difference of the same two terms **97.** $(x + 6)(x + 6)$, square of a binomial formula **99.** $a(x - n)(x - n)(x - n)$
101. a) Answers will vary. **b)** $a^2 + 2ab + b^2$ **c)** $a^2 + 2ab + b^2$ **d)** Same. **103. a)** $A = P(1 + r)^t$ **b)** \$1123.60
105. $a^2 + 2ab + b^2 - 3a - 3b + 5$ **107.** $15x^{3t-1} + 18x^{4t}$ **109.** $12x^{3m} - 18x^m - 10x^{2m} + 15$ **111.** $y^{a^2-b^2}$
113. $x^4 - 12x^3y + 54x^2y^2 - 108xy^3 + 81y^4$ **115. a)** Answers will vary. **b)** It is correct. **117.** $y^2 - 2y - 2xy + 2x + x^2 + 1$
119. $\dfrac{43}{60}$ **120.** $8r^6s^{15}$ **121.** $\left(-\dfrac{7}{3}, \dfrac{1}{3}\right]$ **122.** $\dfrac{7}{4}$

Exercise Set 5.3 **1. a)** Answers will vary. **b)** $\dfrac{5}{3}x^3 - 2x^2 - \dfrac{4}{3}x - 4 + \dfrac{7}{3x}$ **3.** Yes; answers will vary.

5. Place them in descending order of the variable. **7. a)** Answers will vary. **b)** $x + 8 + \dfrac{36}{x - 5}$ **9.** No because the

remainder is not equal to 0. **11.** $2x + 9$ **13.** $2x + 1$ **15.** $\dfrac{5}{3}y^2 + 2y - 3$ **17.** $x^3 - \dfrac{3}{2}x^2 + 3x - 2$ **19.** $4x^2 - 5xy - \dfrac{5}{2y}$

21. $\dfrac{9x}{2y} - 6x^2 + \dfrac{7y}{2x}$ **23.** $\dfrac{z}{2} + z^2 - \dfrac{3}{2}x^2y^4z^7$ **25.** $x + 2$ **27.** $2x + 3$ **29.** $3x + 2$ **31.** $x + 5 + \dfrac{2}{x + 1}$ **33.** $2b + 5 + \dfrac{1}{b - 2}$

35. $4x + 9 + \dfrac{2}{2x - 3}$ **37.** $2x + 5$ **39.** $x^2 + 2x + 3 + \dfrac{1}{x + 1}$ **41.** $3b^2 - 3b + 1 + \dfrac{2}{3b + 2}$ **43.** $2x^2 - 8x + 38 - \dfrac{156}{x + 4}$

45. $3x^3 + 6x + 4$ **47.** $x + 4$ **49.** $2c^2 - 6c + 3$ **51.** $x + 6$ **53.** $x + 3$ **55.** $x - 7$ **57.** $x + 8 + \dfrac{12}{x - 3}$ **59.** $3x + 5 + \dfrac{10}{x - 4}$

61. $4x^2 + x + 3 + \dfrac{3}{x - 1}$ **63.** $3c^2 - 2c + 2 + \dfrac{12}{c + 3}$ **65.** $y^3 + y^2 + y + 1$ **67.** $x^3 - 4x^2 + 16x - 64 + \dfrac{272}{x + 4}$

69. $x^4 - \dfrac{10}{x + 1}$ **71.** $b^4 + 3b^3 - 3b^2 + 3b - 3 - \dfrac{9}{b + 1}$ **73.** $3x^2 + 3x - 3$ **75.** $2x^3 + 2x - 2$ **77.** 10 **79.** 0; factor

81. $-\dfrac{19}{4}$ or -4.75 **83.** $3x + 2$ **85.** 3 times greater, find the areas by multiplying the polynomials, then compare
87. No, the dividend is a binomial **89.** If the remainder is 0, $x - a$ is a factor. **91.** $x^2 - 2x - 8$ **93.** $x^2 + 9x + 24$
95. $2x^2 + 3xy - y^2$ **97.** $x + \dfrac{5}{2} + \dfrac{11}{2(2x - 3)}$ **99.** $w = r + 1$ **101.** $x^3 - 6x^2 + 13x - 10$, multiply $(x - 3)(x^2 - 3x + 4)$

and then add 2. **103.** $2x + 1 - \dfrac{3}{2x} - \dfrac{1}{2x^2}$ **105.** Not a factor; compute $P(1)$. $P(1) = 101$, which is not 0, so $x - 1$ is not a

factor **107.** Factor; compute $P(-1)$. $P(-1) = 0$, so $x + 1$ is a factor. **109. a)** $3x^2 - 2x + 5 - \dfrac{13}{3x + 5}$ **b)** Because we

are expressing the remainder in terms of $3x + 5$ **110.** 2.0×10^{10} **111.** $30°, 60°, 90°$ **112.** $\left\{-1, \dfrac{11}{5}\right\}$ **113.** -864

114. $3r + 3s - 6t$

Exercise Set 5.4 **1.** Determine if all the terms contain a greatest common factor and, if so, factor it out.
3. a) Answers will vary. **b)** $2x^2y$ **c)** $2x^2y(3y^4 - x + 6x^7y^2)$ **5. a)** Answers will vary. **b)** $(3x^2 - y^3)(2x + y^2)$
7. xy^4; the largest power of x common to all the terms is x, and the largest power of y common to all the terms is y^4.
9. $7(n + 1)$ **11.** $2(x^2 - 2x + 4)$ **13.** $4(3y^2 - 4y + 6)$ **15.** $x^2(9x^2 - 3x + 11)$ **17.** $-3a^2(8a^5 - 3a^4 + 1)$
19. $3xy(x + 2xy + 1)$ **21.** $8a^2c(10a^3b^4 - 2a^2b^2c + 1)$ **23.** $3pq^2r(3p^3q^3 - pr + 2q^3r^2)$ **25.** $-2(26p^2q^2 + 8pq^3 - 13r)$
27. $-4(2x - 1)$ **29.** $-(x^2 - 4x + 12)$ **31.** $-3(r^2 + 2r - 3)$ **33.** $-2rs^3(3r^3 - 2rs - s^2)$ **35.** $-a^2b(a^2bc - 5ac^2 - 1)$
37. $(a + 3)(x + 1)$ **39.** $(x - 4)(3c + 2)$ **41.** $-(x - 2)(2x - 9)$ **43.** $-2(a + 2)(a + 2)$ or $-2(a + 2)^2$
45. $(x + 3)(x - 5)$ **47.** $2(2y - 1)(2y - 5)$ **49.** $(x + y)(a + b)$ **51.** $(x - 3)(x^2 + 4)$ **53.** $(5m - 6n)(2m - 5n)$
55. $5(a + 3)(a^2 - 2)$ **57.** $c^2(c - 1)(c^2 + 1)$ **59.** $(2x + 1)(6x - 5)$ **61.** $(x + 4)(3x - 2)$ **63.** $(3x + 2)(9x - 5)$
65. a) 240 feet **b)** $h(t) = -16t(t - 8)$ **c)** 240 feet **67. a)** ≈ 2856.64 square feet **b)** $A = r(\pi r + 2l)$
c) ≈ 2856.64 square feet **69. a)** $(1 - 0.06)(x + 0.06x) = 0.94(1.06x)$ **b)** $0.9964x$; slightly lower than the price of the 2002
model (99.64% of the original cost) **71. a)** $(x + 0.15x) - 0.20(x + 0.15x) = 0.80(x + 0.15x)$ **b)** $0.80(1.15x) = 0.92x$;
92% of the regular price **73.** $(3x - 2)^4(15ax - 10a + 4)$ **75.** $2(x - 3)(2x^4 - 12x^3 + 15x^2 + 9x + 2)$
77. $(x^2 + 2x - 3)(a + b)$ **79.** $x^{4m}(x^{2m} - 2)$ **81.** $x^{2m}(3x^{2m} - 2x^m + 1)$ **83.** $(a^r + c^r)(b^r - d^r)$
85. a) Yes **b)** 0; subtracting the same quantity from itself **c)** Answers will vary.
87. a) They should be the same graph; they represent the same function

87. b)

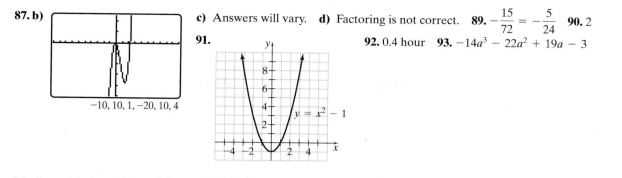

$-10, 10, 1, -20, 10, 4$

c) Answers will vary. **d)** Factoring is not correct. **89.** $-\dfrac{15}{72} = -\dfrac{5}{24}$ **90.** 2

91.

$y = x^2 - 1$

92. 0.4 hour **93.** $-14a^3 - 22a^2 + 19a - 3$

Using Your Graphing Calculator, 5.5   **1.** Yes   **2.** No

Exercise Set 5.5   **1.** Factor out the greatest common factor if it is present.   **3. a)** Answers will vary.

b) $(2x - 3)(3x + 4)$ **5.** No; $2(x + 3)(x + 1)$; $(2x + 2)$ has a GCF of 2 **7.** No, $3x(x + 4)(x - 2)$; $(3x - 6)$ has a GCF of 3
9. Both are $+$ **11.** One is $+$, one is $-$ **13.** $(x + 3)(x + 4)$ **15.** $(b - 1)(b - 9)$ **17.** $(c - 6)^2$ **19.** $(y - 9)^2$
21. $(x - 32)(x - 2)$ **23.** $(x + 2)(x - 15)$ **25.** $-(a - 15)(a - 3)$ **27.** Prime **29.** $(m - 2n)(m - 5n)$ **31.** $4(r + 4)(r - 1)$
33. $x(x - 6)(x + 3)$ **35.** $(a - 1)(5a - 3)$ **37.** $3(x - 2)(x + 1)$ **39.** $(3c + 7)(2c - 9)$ **41.** $(2b + 1)(4b - 3)$
43. $(3c - 2)(2c + 5)$ **45.** $4(2p - 3q)(2p + q)$ **47.** Prime **49.** $2(3a + 4b)(3a - b)$ **51.** $(9y + 4)(y - 12)$
53. $10(5b - 2)(2b - 1)$ **55.** $ab^5(a - 4)(a + 3)$ **57.** $3b^2c(b - 3c)^2$ **59.** $4m^6n^3(m + 2n)(2m - 3n)$ **61.** $(6x - 5)(5x + 4)$
63. $8x^2y^4(x + 4)(x - 1)$ **65.** $(x^2 + 3)(x^2 - 2)$ **67.** $(b^2 + 5)(b^2 + 4)$ **69.** $(2a^2 + 5)(3a^2 - 5)$ **71.** $(2x + 5)(2x + 3)$
73. $(3a + 1)(2a + 5)$ **75.** $(xy + 7)(xy + 2)$ **77.** $(2xy - 11)(xy + 1)$ **79.** $(5 - y)(y - 1)(2y - 5)$
81. $(p - 3)(2p + 3)(p + 2)$ **83.** $(a^3 - 10)(a^3 + 3)$ **85.** $(x + 3)(x + 2)(x + 1)$ **87.** $a^3b^2(5a - 3b)(a - b)$
89. $(x + 6)(x + 1)$ **91.** $(x + 6)(x + 3)$ **93.** $2x^2 - 5xy - 12y^2$; multiply $(2x + 3y)(x - 4y)$ **95.** Divide; $x + 6$
97. a) Answers will vary. **b)** $(6x - 5)(5x + 8)$; $(7x - 1)(7x - 13)$ **99.** $\pm 3, \pm 9$ **101.** 6 or -6; b is the sum of the factors
of 5 **103. a)** 4 **b)** $(x - 3)(x - 5)$ **105. a)** -8 **b)** Not factorable **107.** Answers will vary. One example is $x^2 + 2x + 1$.
109. $(2a^n + 3)(2a^n - 5)$ **111.** $(x + y)^2(x - 4y)(x - 3y)$ **113.** $(x^n - 2)(x^n + 5)$ **115. a)** Answers will vary. **b)** Correct

117. $C = \dfrac{5}{9}(F - 32)$ **118.**

$y = -3x + 4$

119. 4 **120.** $x^2 + 2xy + y^2 + 10x + 10y + 25$

121. $(2x^2 - 5)(x + 2)$

Exercise Set 5.6   **1. a)** Answers will vary. **b)** $(x + 4)(x - 4)$ **3.** Answers will vary.
5. $a^3 + b^3 = (a + b)(a^2 - ab + b^2)$ **7.** No, $(x + 7)(x - 7) = x^2 - 49$ **9.** No, $(x - 9)^2 = x^2 - 18x + 81$

11. $(x + 9)(x - 9)$ **13.** $(a + 10)(a - 10)$ **15.** $(1 + 6b)(1 - 6b)$ **17.** $(5 + 4y^2)(5 - 4y^2)$ **19.** $\left(\dfrac{1}{5} + y\right)\left(\dfrac{1}{5} - y\right)$

21. $(xy + 11c)(xy - 11c)$ **23.** $(0.2x + 0.3)(0.2x - 0.3)$ **25.** $x(12 - x)$ **27.** $(a + 3b + 2)(a - 3b - 2)$
29. $(x + 5)^2$ **31.** $(2 + a)^2$ **33.** $(2x - 5y)^2$ **35.** $(0.9x - 0.2)^2$ **37.** $(y^2 + 2)^2$ **39.** $(x + y + 1)^2$ **41.** $(y + 1)^2$
43. $(x + 3 + y)(x + 3 - y)$ **45.** $(x + 7)(3 - x)$ **47.** $(c - 1)(7 - c)$ **49.** $(y^2 - 3)^2$ **51.** $(x - 3)(x^2 + 3x + 9)$
53. $(4 - a)(16 + 4a + a^2)$ **55.** $(p - 3a)(p^2 + 3ap + 9a^2)$ **57.** $(3y - 2x)(9y^2 + 6xy + 4x^2)$
59. $4(2a - 3b)(4a^2 + 6ab + 9b^2)$ **61.** $2(b - 5c)(b^2 + 5bc + 25c^2)$ **63.** $(x + 2)(x^2 + x + 1)$
65. $(a - b - 3)(a^2 - 2ab + b^2 + 3a - 3b + 9)$ **67.** $-9(b^2 + 3b + 3)$ **69.** $(11y^2 + 7x)(11y^2 - 7x)$
71. $(4y + 9x)(4y - 9x)$ **73.** $(5x^2 + 9y^3)(5x^2 - 9y^3)$ **75.** $(x - 4)(x^2 + 4x + 16)$ **77.** $(3xy + 4)^2$ **79.** $(a^2 + b^2)^2$
81. $(x - 1 + y)(x - 1 - y)$ **83.** $(x + y + 1)(x^2 + 2xy + y^2 - x - y + 1)$ **85.** $3m(-m + 2n)$
87. $(3x - 2)(9x^2 + 6x + 4)$ **89.** $(6a - b)(36a^2 + 6ab + b^2)$ **91.** $(4x + 3a)(16x^2 - 12ax + 9a^2)$ **93. a)** $a^2 - b^2$

b) $(a + b)(a - b)$ **95. a)** $6a^3 - 6ab^2$ **b)** $6a(a + b)(a - b)$ **97. a)** $\dfrac{4}{3}\pi R^3 - \dfrac{4}{3}\pi r^3$ **b)** $\dfrac{4}{3}\pi(R - r)(R^2 + Rr + r^2)$

99. $12; -12;$ write $4x^2 + bx + 9$ as $(2x)^2 + bx + (3)^2$; $bx = 2(2x)(3)$ or $bx = -2(2x)(3)$ **101.** $c = 4;$ write
$25x^2 + 20x + c$ as $(5x)^2 + 20x + (a)^2$ then $20x = 2(5x)(a)$. Since $a = 2, c = 4$ **103. a)** Find an expression whose
square is $25x^2 - 30x + 9$ **b)** $s(x) = 5x - 3$ **c)** 7   **105.** $(x^2 + 4x + 8)(x^2 - 4x + 8)$ **107.** $h(2a + h)$ **109. a)** 16
b) $x^2 + 8x + 16$ **c)** $(x + 4)^2$ **111.** $(8x^{2a} + 3y^{3a})(8x^{2a} - 3y^{3a})$ **113.** $(a^n - 8)^2$ **115.** $(x^n - 2)(x^{2n} + 2x^n + 4)$
117. Correct   **119. a)** $(x^3 + 1)(x^3 - 1)$ **b)** $(x^2 - 1)(x^4 + x^2 + 1)$ **121.** $4x + 5y - 2$ **122.** -15
123. $20°, 30°, 40°$ **124.** $15y^{10}(3y^2 + 2)$ **125.** $(4x - 3y)(3x + y)$

Exercise Set 5.7   **1.** Answers will vary.   **3.** $3(x + 5)(x - 5)$ **5.** $(5s - 3)(2s + 5)$ **7.** $-(4r - 3)(2r - 5)$
9. $0.4(x + 0.3)(x - 0.3)$ **11.** $6x(x^2 + 3)(x^2 - 3)$ **13.** $3x^4(x - 1)(x + 4)$ **15.** $5x^2y^2(x + 4)(x + 3)$
17. $x^2(x + y)(x - y)$ **19.** $x^4y^2(x - 1)(x^2 + x + 1)$ **21.** $x(x^2 + 4)(x + 2)(x - 2)$ **23.** $2(x^2 + 2y)(x^4 - 2x^2y + 4y^2)$
25. $2(a + b + 5)(a + b - 5)$ **27.** $6(x + 3y)^2$ **29.** $x(x + 4)$ **31.** $(2a + b)(a - 2b)$ **33.** $(y + 7)^2$ **35.** $5a^2(3a - 1)^2$
37. $\left(x + \dfrac{1}{4}\right)\left(x^2 - \dfrac{1}{4}x + \dfrac{1}{16}\right)$ **39.** $(3x + 2)(x + 3)(x - 3)$ **41.** $ab(a + 8b)(a - 8b)$ **43.** $(9 + x + y)(9 - x - y)$
45. $2(3x - 2)(4x - 3)$ **47.** $(8x - 3)(2x + 5)$ **49.** $(x^2 + 4)(x + 2)(x - 2)$ **51.** $(b - 2x)(5c - 7y)$
53. $(3x^2 - 4)(x^2 + 1)$ **55.** $(y + x - 6)(y - x + 6)$ **57.** $(2y + 5)(y + 8)$ **59.** $(a + 6b + 4c)(a + 6b - 4c)$
61. $3x^2y(x + 3)(2x - 1)$ **63.** $(x + y)^2(x - y)^2$ **65. e)**   **67. a)**   **69. f)**   **71. c)**   **73.** $2(x + 3)(x + 2)$
75. $(x + 6)(x + 2)$ **77.** $(x + 8)(x + 1)$ **79.** $(5x - 3)(25x^2 + 15x + 9)$ **81. a)** $a(a + b) - b(a + b) = a^2 - b^2$
b) $(a + b)(a - b)$ **83. a)** $a^2 + 2ab + b^2$ **b)** $(a + b)^2$ **85. a)** $a(a - b) + a(a - b) + b(a - b) + b(a - b)$ or
$2a(a - b) + 2b(a - b)$ **b)** $2(a + b)(a - b)$ **87. a)** Answers will vary.   **b)** Answers will vary.
89. a) $x^{-5}(x^2 - 2x - 3)$ **b)** $x^{-5}(x - 3)(x + 1)$ **91. a)** $x^{-3/2}(5x^2 + 2x - 3)$ **b)** $x^{-3/2}(5x - 3)(x + 1)$ **92.** 1
93. $\{z | z < -6 \text{ or } z > 0\}$ **94.** ≈ 17.3 pounds at \$5.20; ≈ 12.7 pounds at \$6.30   **95.** $5x^3 - x^2 + 16x + 16$
96. $(x + 2)(2x^2 - 5)$

Using Your Graphing Calculator, 5.8   **1.** $y = x^2 - 6x + 5$ **2.** $y = x^2 - x - 6$ **3.** $y = x^2 + 4x$

Exercise Set 5.8   **1.** The degree of a polynomial function is the same as the degree of the leading term.
3. $ax^2 + bx + c = 0$ **5. a)** The zero-factor property only holds when one side of the equation is 0.   **b)** -2 and -5
7. a) Answers will vary.   **b)** $\dfrac{4}{3}, -\dfrac{5}{4}$ **9. a)** Legs   **b)** Hypotenuse   **11.** -8 and -2; at the x-intercepts $y = 0$ **13.** Yes, if
the graph does not cross the x-axis   **15.** Yes, if the graph crosses the x-axis twice   **17.** 0, 4   **19.** 0, 1   **21.** -1, 7   **23.** 0, -4, 9
25. $-\dfrac{3}{2}, -\dfrac{5}{4}$ **27.** 0, 3   **29.** 0, -5 **31.** 0, 6   **33.** 0, 5   **35.** 1, 5   **37.** -4, 3   **39.** -4 **41.** 2, 10   **43.** -6, -1 **45.** -4, 6   **47.** 0, 6, -3
49. $-\dfrac{1}{6}, \dfrac{3}{2}$ **51.** 0, -4, 3   **53.** 6, -6 **55.** $-\dfrac{3}{2}, \dfrac{3}{2}$ **57.** 0, $-\dfrac{4}{5}, \dfrac{4}{5}$ **59.** -11, 9   **61.** 0, -8 **63.** -1, -4 **65.** $\dfrac{5}{2}, \dfrac{4}{3}$ **67.** 0, -3, -5
69. $-2, -\dfrac{1}{3}$ **71.** $\dfrac{3}{5}, \dfrac{5}{2}$ **73.** 6, -5 **75.** $(4, 0), (-6, 0)$ **77.** $(-8, 0)$ **79.** $(0, 0), \left(\dfrac{4}{3}, 0\right), \left(\dfrac{5}{2}, 0\right)$ **81.** $x = 1$ **83.** $x = 2$
85. $x = 8$ **87. d)**   **89. b)**   **91.** $y = x^2 - 7x + 10$ **93.** $y = x^2 - 3x - 4$ **95.** $y = 6x^2 - 7x - 10$ **97.** Width = 2 feet,
length = 5 feet   **99.** Base = 10 feet, height = 16 feet   **101.** 2.5 centimeters   **103.** 3 feet   **105.** 2 seconds   **107.** Tim: 5 miles;
Bob: 12 miles   **109.** 13 feet   **111.** 50 bicycles   **113.** 13 inches by 13 inches   **115. a)** $V = a^3 - ab^2$ **b)** $V = a(a + b)(a - b)$
c) 3 inches   **117. a)** $f(x) = x^2 + 7x + 10$ **b)** $x^2 + 7x + 10 = 0$ **c)** An infinite number; any function of the form
$f(x) = a(x^2 + 7x + 10), a \neq 0$ **d)** An infinite number; any equation of the form $a(x^2 + 7x + 10) = 0, a \neq 0$

119. a) Answers will vary. Examples are:

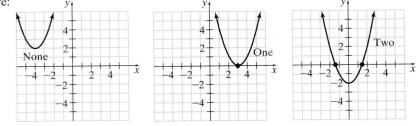

b) None (no x-intercepts), one (one x-intercept), or two (two x-intercepts)   **121.** 73.721949 mph   **123.** $\pm 2, \pm 3$

125. $y' = \dfrac{1}{x + y}$ **127.** $y' = \dfrac{x + xy}{2xy + 3}$ **128.** $\dfrac{x^4}{16y^6}$ **129.** [number line from $\frac{5}{12}$ to $\frac{23}{12}$]   **130.** $(2, -1)$ **131.** -72 **132.** $(x + 3)(x - 2)$

Chapter Review Exercises

1. a) Binomial, **b)** $3x^2 + 2$, **c)** 2 **2. a)** Trinomial, **b)** $4x^3 + 5x - 9$, **c)** 3 **3.** Not a polynomial **4. a)** Polynomial, **b)** $2x^4 - 10x^2y + 6xy^3 - 3$, **c)** 4 **5.** $x^2 - 3x + 9$ **6.** $5x^2 + 11x - 4$ **7.** $3a - 8b + 6$ **8.** $6x^3 - 9x + 10$ **9.** $3x^2y + 3xy - 9y^2$ **10.** $-3ab + b^2 - 2a$ **11.** $5x^2 + 7x - 4$ **12.** $-10a^2b + 4ab$ **13.** 15 **14.** -75 **15.** $3x^2 + 27$ **16.** $2x^2 + 24x + 23$ **17. a)** \$780.46 billion **b)** Yes **18. a)** \$773.13 billion **b)** Yes **19.** $6x^3 - 14x^2 + 10x$ **20.** $-3x^4y^2 - 3x^2y^6 + 18xy^7$ **21.** $6x^2 - 7x - 5$ **22.** $50a^2 + 75a - 27$ **23.** $x^2 + 16xy + 64y^2$ **24.** $a^2 - 22ab + 121b^2$ **25.** $10x^2y + 8xy^2 - 5x - 4y$ **26.** $6p^2q^2 + 11pqr - 7r^2$ **27.** $4a^2 + 36ab + 81b^2$ **28.** $9x^2 - 12xy + 4y^2$ **29.** $49x^2 - 25y^2$ **30.** $4a^2 - 25b^4$ **31.** $25x^2y^2 - 36$ **32.** $81a^4 - 4b^4$ **33.** $x^2 + 6xy + 9y^2 + 4x + 12y + 4$ **34.** $4p^2 - 4pq + q^2 - 20p + 10q + 25$ **35.** $6x^3 - x^2 - 24x + 18$ **36.** $4x^4 + 12x^3 + 6x^2 + 13x - 15$ **37.** $x^2 + 8x + 10$ **38.** $x^2 + xy + 4y + xz$ **39. a)** $x^2 - x - 6$ **b)** 0 **40. a)** $2x^3 - 4x^2 - 6x + 12$ **b)** 12 **41. a)** $x^3 - x^2 - 5x + 6$ **b)** 9 **42. a)** $x^4 - 4$ **b)** 77 **43.** $2x^2 + 3x + 4$ **44.** $\frac{7}{4}a^2 - 4a + 5$ **45.** $7y^2 + 2 + \frac{2}{3y}$ **46.** $9p - 5q - 2$ **47.** $\frac{x^2}{2y} + x + \frac{3y}{2}$ **48.** $4x - 3$ **49.** $x^3 - 2x^2 + 3x + 7$ **50.** $2a^3 + a^2 - 3a - 4$ **51.** $x + 4 - \frac{6}{x - 3}$ **52.** $2x^2 + 3x - 4 + \frac{2}{2x + 3}$ **53.** $3x^2 + 7x + 21 + \frac{73}{x - 3}$ **54.** $2y^4 - 2y^3 - 8y^2 + 8y - 7 + \frac{6}{y + 1}$ **55.** $x^4 + 2x^3 + 4x^2 + 8x + 16 + \frac{12}{x - 2}$ **56.** $2x^2 + 2x + 6$ **57.** 8 **58.** -236 **59.** $-\frac{53}{9}$ or $-5.\overline{8}$ **60.** 0; factor **61.** $4(x^2 + 2x + 6)$ **62.** $3x^3(5x + 2 - 4xy^3)$ **63.** $2a^2b^3(5a - 6b^3)$ **64.** $3xy^2z^2(4y^2z + 2xy - 5x^2z)$ **65.** $(5x - y)(x + 4y)$ **66.** $(3a - 2b)(4a + 5b)$ **67.** $(2x - 3)(x + 9)$ **68.** $(3x - 5)(16x - 15)$ **69.** $(5x + 2)(13x - 7)$ **70.** $(7x + 9)(2x - 1)$ **71.** $(17x + 3)(9x - 7)$ **72.** $(4x + 5)(5x - 2)$ **73.** $(x + 6)(x + 2)$ **74.** $(x - 3)(x + 5)$ **75.** $(x - 7)(x + 3)$ **76.** $(x - 8)(x - 2)$ **77.** $-(x - 15)(x + 3)$ **78.** $-(x - 12)(x - 1)$ **79.** $x(2x + 1)(x + 6)$ **80.** $x(4x - 5)(2x + 5)$ **81.** $a(4a - 5)(a - 1)$ **82.** $y^2(12y + 1)(y + 5)$ **83.** $(x - 18y)(x + 3y)$ **84.** $(2p - 5q)(3p - 2q)$ **85.** $(x^2 + 3)(x^2 + 5)$ **86.** $(x^2 - 7)(x^2 + 9)$ **87.** $(x + 9)(x + 11)$ **88.** $(x - 7)(x + 2)$ **89.** $(x + 10)(x + 1)$ **90.** $(x + 10)(x + 2)$ **91.** $(x + 7)(x - 7)$ **92.** $(x + 10)(x - 10)$ **93.** $(x^2 + 4)(x + 2)(x - 2)$ **94.** $(x^2 + 9)(x + 3)(x - 3)$ **95.** $(2a + 1)^2$ **96.** $(2y - 3)^2$ **97.** $(x - 1)(x + 5)$ **98.** $3(y - 2)(3y + 4)$ **99.** $(p^2 + 8)^2$ **100.** $(b^2 - 7)^2$ **101.** $(x + 4 + y)(x + 4 - y)$ **102.** $(a + 3b + 2c)(a + 3b - 2c)$ **103.** $(3x + y)^2$ **104.** $(6b - 5c)^2$ **105.** $(x - 3)(x^2 + 3x + 9)$ **106.** $(y + 4)(y^2 - 4y + 16)$ **107.** $(5x - 1)(25x^2 + 5x + 1)$ **108.** $(2a - 3b)(4a^2 + 6ab + 9b^2)$ **109.** $(x - 1)(x^2 + 4x + 7)$ **110.** $(x - 5)(x^2 - x + 7)$ **111.** $(y - 4z)(y^2 + 4yz + 16z^2)$ **112.** $(a + 4)(a^2 + 5a + 7)$ **113.** $(x + 3)(x - 3)$ **114.** $(a + 2b)(a - 2b)$ **115.** $(2x - y)(4x^2 + 2xy + y^2)$ **116.** $4a(a + c)(a - c)$ **117.** $y^2(x + 3)(x - 5)$ **118.** $3x(x - 4)(x - 2)$ **119.** $3xy^4(x - 2)(x + 6)$ **120.** $3y(y^2 + 3)(y^2 - 3)$ **121.** $2y(x + 2)(x^2 - 2x + 4)$ **122.** $5x^2y(x + 2)^2$ **123.** $3x(2x + 1)(x - 4)$ **124.** $(x + 5 + y)(x + 5 - y)$ **125.** $3(x + 2y)(x^2 - 2xy + 4y^2)$ **126.** $(x + 4)^2(x - 1)$ **127.** $(4x + 1)(4x + 5)$ **128.** $(2x^2 - 1)(2x^2 + 3)$ **129.** $(x - 1)(x - 2)(x + 1)$ **130.** $(3a - b)(3x + 4y)$ **131.** $(2pq - 3)(3pq + 2)$ **132.** $(3x^2 - 2)^2$ **133.** $(2y + x + 2)(2y - x - 2)$ **134.** $2(3a + 5)(4a + 3)$ **135.** $3x^2y^4(x + 3)(2x - 3)$ **136.** $\left(x - \frac{2}{3}y^2\right)\left(x^2 + \frac{2}{3}xy^2 + \frac{4}{9}y^4\right)$ **137.** $(x + 9)(x + 2)$ **138.** $(y + 10)(y + 5)$ **139.** $(a + 2b)(a - 2b)$ **140.** $2b(a + b)$ **141.** $(2a + b)(a + 3b)$ **142.** $(a + b)^2$ **143.** $3, -\frac{1}{4}$ **144.** $-\frac{5}{2}, -\frac{7}{3}$ **145.** $0, 2$ **146.** $0, -\frac{4}{3}$ **147.** $-4, -3$ **148.** $5, -6$ **149.** $7, 1$ **150.** $0, 2, 4$ **151.** $\frac{4}{3}, -\frac{1}{4}$ **152.** $\frac{3}{4}, -\frac{2}{5}$ **153.** $2, -2$ **154.** $2, -3$ **155.** $(-5, 0), (6, 0)$ **156.** $\left(\frac{6}{5}, 0\right), \left(\frac{5}{4}, 0\right)$ **157.** $y = x^2 - 3x - 18$ **158.** $y = 12x^2 + 28x + 15$ **159.** Width = 9 feet, length = 11 feet **160.** Height = 4 feet, base = 11 feet **161.** 3 inches, 7 inches **162.** 9 seconds **163.** 9

Chapter Practice Test

1. a) Trinomial **b)** $-6x^4 - 4x^2 + 2x$ **c)** 4 **d)** -6 **2.** $4x^2y - 14y^2 + 2x + 6y$ **3.** $-8x^8y^3 + 20x^6y^4 - 12x^4y^2$ **4.** $10a^2 - 13ab - 3b^2$ **5.** $4x^3 + 8x^2y - 9xy^2 - 6y^3$ **6.** $4x^4 - 5y + \frac{7}{x^2}$ **7.** $x - 5 + \frac{25}{2x + 3}$ **8.** $3x^3 + 3x^2 + 15x + 15 + \frac{79}{x - 5}$ **9.** -89 **10.** $2xy(6x^2 + 5xy^3 - 4y^2)$ **11.** $x(x - 3)(x + 1)$ **12.** $(a + 2b)(2a + 3b)$ **13.** $(2b^2 + 9)(b^2 - 2)$ **14.** $4(x - 3)(x + 2)$ **15.** $(x + 8)(x + 4)$ **16.** $q^6(3p - 2)(9p^2 + 6p + 4)$ **17. a)** $3x^2 - 19x + 20$ **b)** -6 **18.** $4(x + y)(x - y)$ **19.** $(x + 11)(x + 4)$ **20.** $\frac{3}{7}, -4$ **21.** $0, -4, 1$ **22.** $\left(\frac{1}{4}, 0\right), \left(-\frac{3}{2}, 0\right)$ **23.** $y = x^2 - 8x + 12$ **24.** Height = 4 meters, base = 14 meters **25.** 7 seconds

CHAPTER 6

Exercise Set 6.1 **1. a)** A rational expression is an expression of the form $\dfrac{p}{q}$, p and q polynomials, $q \neq 0$.

b) Answers will vary. **3. a)** A rational function is a function of the form $f(x) = \dfrac{p}{q}$, p and q polynomials, $q \neq 0$.

b) Answers will vary. **5. a)** The domain of a rational function is the set of values that can replace the variable
b) $\{x | x \neq -4 \text{ and } x \neq 4\}$ **7. a)** Factor -1 from numerator or denominator and reduce. **b)** -1 **9. a)** Invert second

fraction, factor all expressions, simplify, then multiply numerators and multiply denominators. **b)** $\dfrac{1}{r + 6}$ **11.** 4 **13.** $5, \dfrac{5}{2}$

15. None **17.** $6, -6$ **19.** $\{p | p \neq 2\}$ **21.** $\{x | x \neq -3 \text{ and } x \neq 2\}$ **23.** $\left\{ a \Big| a \neq \dfrac{1}{2} \text{ and } a \neq -2 \right\}$ **25.** $\{x | x \text{ is a real number}\}$

27. $\{a | a \neq -7 \text{ and } a \neq 7\}$ **29.** $1 - y$ **31.** $\dfrac{x - 2y}{3}$ **33.** x **35.** -1 **37.** $-(p + 4)$ **39.** $\dfrac{a - 5}{a + 3}$ **41.** $4x^2 + 10xy + 25y^2$

43. $\dfrac{2x - 5}{2}$ **45.** $\dfrac{a + 3}{a + 5}$ **47.** $\dfrac{a^2 + ab + b^2}{a + b}$ **49.** $\dfrac{xy^2}{9}$ **51.** $12x^3y^2$ **53.** 1 **55.** 1 **57.** $\dfrac{1}{r - 8}$ **59.** $\dfrac{1}{x - 4}$ **61.** $\dfrac{(a + 1)^2}{9(a + b)^2}$

63. $\dfrac{x - 4}{4x + 1}$ **65.** $\dfrac{(x + 2)(x - 2)}{(x^2 + 2x + 4)(x^2 + 4)}$ **67.** x^4 **69.** $\dfrac{x^3}{x + 2}$ **71.** $\dfrac{(a - b)(a + b)}{a^2 + ab + b^2}$ **73.** 1 **75.** $\dfrac{p - q}{p + q}$ **77.** $\dfrac{x + 1}{x(x - 1)}$

79. One possible answer is $\dfrac{1}{(x - 2)(x + 3)}$; denominator 0 at $x = 2$ and $x = -3$ **81.** Numerator is never 0.

83. a) 4, makes numerator 0 **b)** 2 and -2, makes denominator 0 **85.** One possible answer is $f(x) = \dfrac{x - 2}{(x - 3)(x + 1)}$;

numerator 0 at $x = 2$, denominator 0 at $x = 3$ and $x = -1$ **87.** $x + 5$; numerator must be $x + 5$ **89.** $y^2 - 4y - 5$,
factors must be $(y - 5)(y + 1)$ **91.** $x^2 + x - 2$; factors must be $(x - 1)(x + 2)$ **93.** $2x^2 + x - 6$; factors must

be $(x + 2)(2x - 3)$. **95.** $\dfrac{3a + b}{2}$ **97.** $2(a + b)$ **99.** $\dfrac{(x + 2)(3x + 1)}{(2x - 3)(x + 1)}$ **101.** $\dfrac{x - 1}{x + 3}$ **103.** $\dfrac{1}{x^2(x - p)^n}$ **105.** x^y

107. a) $\{x | x \neq 2\}$ **b)**

c) Decreasing

d) Increasing

$-10, 10, 1, -10, 10, 1$

109. a) $\{x | x \neq 2\}$ **b)**

c) Decreasing

d) Increasing

$-10, 10, 1, -10, 10, 1$

111. a) $\{x | x \neq 0\}$ **b)** $-0.1, -1, -2, -10, -100, 100, 10, 2, 1, 0.1$ **c)**
d) No; numerator can never be 0

113. $y = x + 2$ **114.** $\left(-\infty, \dfrac{3}{2} \right)$ **115.** $-28, 32$

116. 0.1 **117.** $(2, -1)$ **118.** $(3x - y + 2)(3x - y - 2)$

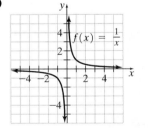

$f(x) = \dfrac{1}{x}$

Exercise Set 6.2 **1. a), b)** Answers will vary. **c)** $(8x + 11)(8x - 11)(x - 2)$ **3. a)** The entire numerator

was not subtracted. **b)** $\dfrac{x^2 - 4x - x^2 - x + 2}{(x + 3)(x - 2)}$ **5.** $\dfrac{3x + 5}{x + 2}$ **7.** $\dfrac{7x - 2}{x - 5}$ **9.** $\dfrac{x + 7}{x + 3}$ **11.** $\dfrac{7x - 11}{x - 2}$ **13.** $\dfrac{x - 3}{x - 1}$ **15.** $\dfrac{5}{r - 2}$

17. $\dfrac{x + 5}{x + 3}$ **19.** $6a^3$ **21.** $40x^4y^5$ **23.** $6a^4b^5$ **25.** $(x + 3)(x + 2)$ **27.** $z - 4$ **29.** $x^4(x - 2)^3$ **31.** $(a - 8)(a + 3)(a + 8)$

33. $(x - 1)(x + 4)(4x + 9)$ **35.** $\dfrac{17}{3r}$ **37.** $\dfrac{7x - 3}{12x^2}$ **39.** $\dfrac{25y^2 + 8x^2}{40x^4y^3}$ **41.** $\dfrac{8 - 3b}{b - 2}$ **43.** $\dfrac{2a}{a - b}$ **45.** $\dfrac{5x^2 + 3x - 12}{(x - 4)(x + 1)}$

47. $\dfrac{4a+7}{(a+2)^2}$ **49.** $\dfrac{2x^2+6x+12}{(x-1)(x+4)(x-2)}$ **51.** $\dfrac{2x+3}{(x-8)(x-1)}$ **53.** $\dfrac{5x^2+14x-49}{(x+5)(x-2)}$ **55.** $\dfrac{3a-1}{4a+1}$ **57.** $\dfrac{2x^2-4xy+4y^2}{(x-2y)^2(x+2y)}$

59. $\dfrac{16}{r-4}$ **61.** 0 **63.** $\dfrac{15x^2-70x+30}{(3x-2)(x-4)}$ **65.** $\dfrac{26r^2+13r-40}{(4r-5)(2r+3)}$ **67.** $\dfrac{x^2-18x-30}{(5x+6)(x-2)}$ **69.** $\dfrac{8m^2+5mn}{(2m+3n)(3m+2n)(2m+n)}$

71. 0 **73.** $\dfrac{1}{2x+3y}$ **75.** No **77.** Yes, if you multiply either fraction by $\dfrac{-1}{-1}$ you get the other fraction. **79. a)** $\{x|x\neq 3\}$

b) $\{x|x\neq -4\}$ **c)** $\dfrac{2x^2+3x+8}{(x-3)(x+4)}$ **d)** $\{x|x\neq 3 \text{ and } x\neq -4\}$ **81.** $\dfrac{2x^2+8x-3}{(x+1)(x+2)}$ **83.** $\dfrac{3x^2+19x+7}{(x+2)(x+3)}$

85. D: $\{x|x\neq 2\}$, R: $\{y|y\neq 1\}$ **87.** $\dfrac{x^2+5x+4}{(x+2)(x-2)(x+3)}$ **89.** $\dfrac{2x}{x^4+x^3-10x^2-4x+24}$ **91.** $\dfrac{ad}{bd}+\dfrac{bc}{bd}=\dfrac{ad+bc}{bd}$

93. a) 4 **b)** $\dfrac{a^2-b^2}{a^2}$ **95.** $7x^2-6x+6$; $5x^2-(7x^2)=-2x^2$, $-(-6x)=6x$, $-6-(6)=-12$ **97.** $\dfrac{3x+10}{x-2}$

99. $\dfrac{1}{(a-5)(a+3)}$ **101.** $-x^2+4x+5$ **103. a)** $\dfrac{ax+bn-bx}{n}$ **b)** 79.2 **105.** $\dfrac{a-b+1}{(a-b)^2}$ **107.** No **109. a)** $\dfrac{x+1}{x}$

b) $\dfrac{x^2+x+1}{x^2}$ **c)** $\dfrac{x^4+x^3+x^2+x+1}{x^4}$ **d)** $\dfrac{x^n+x^{n-1}+x^{n-2}+\cdots+1}{x^n}$ **111.** $\dfrac{-h}{(a+1)(a+h+1)}$

112. a) 6 minutes **b)** 960 bottles **113.** $\{x|-2<x<8\}$ **114.** $-\dfrac{2}{3}$ **115.** -21 **116.** $3x-7+\dfrac{27}{2x+3}$ **117.** $\dfrac{1}{3}$, 7

Exercise Set 6.3 **1.** One that has a fractional expression in its numerator or denominator or both. **3.** $\dfrac{75a}{b^5}$

5. $\dfrac{y-x}{3xy}$ **7.** $\dfrac{12x^3}{y^6}$ **9.** $\dfrac{x(y-1)}{3+x}$ **11.** $\dfrac{5}{3a^2}$ **13.** $\dfrac{x-y}{y}$ **15.** $\dfrac{m+2}{2m^2+1}$ **17.** $\dfrac{2(x+2)}{x^5}$ **19.** $\dfrac{-a+1}{(a+1)(2a+1)}$ **21.** $\dfrac{x-1}{x+1}$

23. $\dfrac{a^2+1}{2a}$ **25.** $\dfrac{x}{5(x-3)}$ **27.** $\dfrac{x^2+5x-6}{x(x-2)}$ **29.** $\dfrac{(x+6)(5x+11)}{5x^2+13x-78}$ **31.** $\dfrac{3+a^2b}{a^2}$ **33.** $\dfrac{ab}{b+a}$ **35.** $\dfrac{b(1+a)}{a(1-b)}$

37. $\dfrac{y-x}{x+y}$ **39.** $\dfrac{(a+b)^2}{ab}$ **41.** $\dfrac{6y-x}{3xy}$ **43.** $\dfrac{2y-3xy+5y^2}{3y^2-4x}$ **45.** $\dfrac{x^2+9x+14}{x+1}$ **47.** $\dfrac{x^2+3x-4}{x+1}$ **49. a)** $\dfrac{2}{7}$ **b)** $\dfrac{4}{13}$

51. $R_T=\dfrac{R_1R_2R_3}{R_2R_3+R_1R_3+R_1R_2}$ **53.** a **55.** $\dfrac{-1}{a(a+h)}$ **57.** $\dfrac{-1}{(a+1)(a+h+1)}$ **59.** $\dfrac{-2a-h}{a^2(a+h)^2}$ **61.** $\dfrac{4a^2+1}{4a(2a^2+1)}$

63. $\dfrac{5}{12}$ **64.** $\dfrac{13}{48}$ **65.** $\left(-23,-\dfrac{34}{5}\right)$ **66.** $\left\{3,\dfrac{5}{3}\right\}$ **67.** Neither

Exercise Set 6.4 **1.** A number obtained when solving an equation that is not a true solution **3. a)** Multiply both sides by 12 to remove fractions **b)** -24 **c)** Write each term with LCD of 12 so you can add and subtract.

d) $\dfrac{-x+24}{12}$ **5.** Similar figures are figures whose corresponding angles are the same and whose corresponding sides are in proportion. **7.** No, $x=3$ makes $\dfrac{7}{x-3}$ undefined **9.** 7 **11.** 5 **13.** 1 **15.** 3 **17.** All real numbers. **19.** $\dfrac{1}{4}$ **21.** $\dfrac{14}{3}$

23. No solution **25.** $\dfrac{6}{5}$ **27.** ≈ 1.63 **29.** 8 **31.** $-\dfrac{4}{3}$, 1 **33.** 3.76 **35.** $-2,-3$ **37.** -5 **39.** $-\dfrac{5}{2}$ **41.** 5 **43.** No solution

45. No solution **47.** $\dfrac{17}{4}$ **49.** 12, 2 **51.** 12, 4 **53.** 1, -2 **55.** $-5, \dfrac{1}{3}$ **57.** 2 **59.** $P_2=\dfrac{V_1P_1}{V_2}$ **61.** $V_2=\dfrac{V_1P_1}{P_2}$

63. $y=y_1+m(x-x_1)$ **65.** $x=zs+\bar{x}$ **67.** $w=\dfrac{fl-df}{d}$ **69.** $q=\dfrac{pf}{p-f}$ **71.** $a=\dfrac{v_2-v_1}{t_2-t_1}$ **73.** $d=\dfrac{a_n-a_1}{n-1}$

75. $G=\dfrac{Fd^2}{m_1m_2}$ **77.** $T_1=\dfrac{T_2P_1V_1}{P_2V_2}$ **79.** $V_0=\dfrac{S-S_0-gt^2}{t}$ **81. a)** $\dfrac{2x+7}{(x-2)(x+2)}$ **b)** $-\dfrac{7}{2}$ **83. a)** $\dfrac{4}{b+5}$ **b)** No solution

85. $c\neq 0$; cannot divide by 0 **87.** $f(x)$: graph **b)**; $g(x)$: graph **a)**; $f(x)$ is undefined when $x=3$ **89. a)** \$6250 **b)** $R=\dfrac{AC}{0.80I}$

91. a) 20 feet per minute squared **b)** $t_1=t_2+\dfrac{v_1-v_2}{a}$ **93.** 35 units **95.** ≈ 176.47 ohms **97.** ≈ 0.101 meter **99. a)** $\approx 9.71\%$

b) Tax-Free Money Market Portfolio since $9.71\% > 7.68\%$. **101.** One answer is $\dfrac{1}{x-4}+\dfrac{1}{x+2}=0$; 4 and -2 make

the fraction undefined **103.** One answer is $\dfrac{1}{x} + \dfrac{1}{x} = \dfrac{2}{x}$. **105.** $-1 < x \le 3$ **106.** $m = -\dfrac{1}{3}$; y-intercept, $\left(0, \dfrac{14}{3}\right)$.
107. $3x^2y - 7xy - 4y^2 - 2x$ **108.** 2 feet

Exercise Set 6.5
1. Equal to $\dfrac{1}{2}$ since it takes exactly the same time for each of them.

3. a)

Worker	Rate of work	Time completed	Part of Task completed
Bill	$\dfrac{1}{7}$	x	$\dfrac{x}{7}$
Bob	$\dfrac{1}{8}$	x	$\dfrac{x}{8}$

b) $\dfrac{x}{7} + \dfrac{x}{8} = 1$ **c)** Less; it should take less time than the fastest person because they are working together.
5. 1.5 months **7.** 2 hours **9.** 18.75 minutes
11. 2.4 hours **13.** ≈ 3.08 hours **15.** 100 hours
17. 7.8 months **19.** 75 minutes **21.** 15.27 minutes
23. 1.62 hours **25.** 12 hours **27.** 5 **29.** 2, 4 **31.** 4, 6

33. 20 **35.** $\dfrac{2}{3}$, 1 **37.** 0.36 miles per hour **39.** 1.53 feet per second **41.** 7.5 miles **43.** 36 miles per hour **45.** 30.59 yards

47. Local: 10.93 miles per hour, express: 16.13 miles per hour. **49.** 350 miles **51.** 60 miles per hour **53.** 120 kilometers per hour **55.** 18 feet per minute **57.** 108,000 miles **59.** Answers will vary. **61. a)** 10 minutes **b)** 15 miles

c) 690 miles per hour **62.** $\dfrac{x^5}{72}$ **63.** 5.26×10^9 **64.** $\$2500$ **65.**

$y = |x| - 2$

66. $a(2a^2 - 5)(a - 1)$

Exercise Set 6.6
1. a) As one quantity increases, the other increases. **b), c)** Answers will vary. **3.** One quantity varies as a product of two or more quantities. **5. a)** Decrease **b)** Inverse variation; by the definition of inverse variation **7.** Direct **9.** Inverse **11.** Direct **13.** Direct **15.** Direct **17.** Inverse **19.** Direct **21.** Inverse **23.** Inverse

25. a) $x = ky$ **b)** 72 **27. a)** $y = kR$ **b)** 306 **29. a)** $R = \dfrac{k}{W}$ **b)** $\dfrac{1}{20}$ **31. a)** $A = \dfrac{kB}{C}$ **b)** 9 **33. a)** $x = ky$ **b)** 20

35. a) $y = kR^2$ **b)** 20 **37. a)** $C = \dfrac{k}{J}$ **b)** ≈ 0.41 **39. a)** $F = \dfrac{kM_1M_2}{d}$ **b)** 40 **41.** Doubled **43.** Halved **45.** Doubled

47. Unchanged **49.** Doubled **51.** $y = \dfrac{k}{x}$; $k = 5$ **53.** $\$8814$ **55.** 3096 milligrams **57.** 10 hours **59.** 6400 cubic centimeters

61. 3.12 hours **63.** ≈ 0.32 seconds **65.** 117.6 feet **67.** $\$438.90$ **69.** 4600 tapes **71.** 25 ohms **73.** $\approx \$124.92$

75. a) $F = \dfrac{km_1m_2}{d^2}$ **b)** Force is 24 times the original force **77.** $h = \dfrac{3V}{4\pi r^2}$ **78.** 276 **79.** $-14x^3 - 22x^2 + 19x - 3$
80. $(x + 3)(x - 2)$

Chapter Review Exercises
1. 5 **2.** -2 **3.** None **4.** $\{x|x \ne -1\}$ **5.** $\{x|x \ne 0\}$ **6.** $\{x|x \ne 2 \text{ and } x \ne -5\}$

7. x **8.** $x - 7$ **9.** -1 **10.** $\dfrac{x - 1}{x - 2}$ **11.** $\dfrac{x - 3}{x + 1}$ **12.** $\dfrac{a^2 + 2ab + 4b^2}{a + 2b}$ **13.** $\dfrac{9x^2 + 3xy + y^2}{3x + y}$ **14.** $\dfrac{2x - 3}{x^2 - 2x + 4}$ **15.** $x(x + 1)$

16. $(x + 2y)(x - 2y)$ **17.** $(x + 7)(x - 5)(x + 2)$ **18.** $(x + 2)^2(x - 2)(x + 1)$ **19.** $6xz^2$ **20.** $-\dfrac{x}{4}$ **21.** $8x^3z^5$

22. $\dfrac{7x + 6}{3x^2}$ **23.** $\dfrac{(x - y)y^2}{2x^3}$ **24.** $3x + 2$ **25.** $\dfrac{10x + 3y^2}{5x^2y}$ **26.** 1 **27.** $\dfrac{2x + 1}{3x + 1}$ **28.** $\dfrac{4a + 5}{a + 1}$ **29.** $\dfrac{6b - 8}{b - 1}$ **30.** $\dfrac{a^2 - b^2}{a^2}$

31. $\dfrac{1}{3(a + 3)}$ **32.** $\dfrac{a^2 + c^2}{ac}$ **33.** $\dfrac{x + 1}{2x - 1}$ **34.** 1 **35.** $2x(x - 5y)$ **36.** $\dfrac{2a^2 + 9a + 4}{4a(a + 2)}$ **37.** $\dfrac{x^2 - 2x - 5}{(x + 5)(x - 5)}$ **38.** $-\dfrac{2(x + 1)}{x^2 - 4}$

39. $\dfrac{x + 5}{x + 6}$ **40.** $\dfrac{1}{(x + 2)(x - 2)(x - 3)}$ **41.** $\dfrac{16(x - 2y)}{3(x + 2y)}$ **42.** $\dfrac{3}{a^3}$ **43.** $\dfrac{22x + 5}{(x - 5)(x - 10)(x + 5)}$ **44.** $\dfrac{2(x - 4)}{(x - 3)(x - 5)}$

45. $-\dfrac{1}{x-3}$ **46.** $\dfrac{a+3}{a+5}$ **47.** $\dfrac{x+6}{x-4}$ **48.** $\dfrac{a+2b^2}{3}$ **49.** $\dfrac{x^2+6x-24}{(x-1)(x+9)}$ **50.** $\dfrac{x-4}{x-6}$ **51. a)** $\{x|x\neq-2\}$ **b)** $\{x|x\neq-4\}$

c) $\dfrac{2x^2+7x+4}{(x+2)(x+4)}$ **d)** $\{x|x\neq-2 \text{ and } x\neq-4\}$ **52. a)** $\{x|x\neq3 \text{ and } x\neq-3\}$ **b)** $\{x|x\neq3\}$ **c)** $\dfrac{x^2+8x+12}{(x+3)(x-3)}$

d) $\{x|x\neq3 \text{ and } x\neq-3\}$ **53.** $\dfrac{3ac^2}{b^3}$ **54.** $\dfrac{x+y}{x^2+xy^3}$ **55.** $\dfrac{3y-1}{5y^2+1}$ **56.** $\dfrac{2a+1}{2}$ **57.** $\dfrac{x+1}{-x+1}$ **58.** $\dfrac{3x^2-29x+68}{4x^2-6x-54}$

59. $\dfrac{x^2+3x+2}{x+5}$ **60.** $\dfrac{x^2+6x+8}{x+3}$ **61.** $\dfrac{18}{5}$ or $3\dfrac{3}{5}$ **62.** -2 **63.** 52 **64.** 2.4 **65.** 5 **66.** -9 **67.** -18 **68.** -28 **69.** -6

70. -10 **71.** $b=\dfrac{ac}{a-c}$ **72.** $\bar{x}=x-sz$ **73.** 60 ohms **74.** 1.5 centimeters **75.** $10,2$ **76.** $21,3$ **77.** 17.14 minutes

78. 14 hours **79.** 3 **80.** $\dfrac{5}{6}$ **81.** 5 miles per hour **82.** car: 50 miles per hour, plane: 150 miles per hour **83.** 20 **84.** $\dfrac{25}{2}$

85. ≈ 426.7 **86.** $\$8.40$ **87.** 1600 feet **88.** 200.96 square units **89.** 2.38 minutes

Chapter Practice Test

1. -7 and 4 **2.** $\left\{x\Big|x\neq-4 \text{ and } x\neq\dfrac{1}{2}\right\}$ **3.** $4x^5y+8+9xy^2$ **4.** $\dfrac{x-5y}{x+y}$ **5.** $\dfrac{1}{x^4y^2}$

6. $\dfrac{1}{x+2}$ **7.** $\dfrac{5}{a(a+b)}$ **8.** x^2+y^2 **9.** $\dfrac{3x^2+2x+2}{x^2(x+1)}$ **10.** $\dfrac{-3x-1}{(x-3)(x+3)(x+1)}$ **11.** $\dfrac{m(6m+n)}{(6m+5n)(2m-n)(2m+3n)}$

12. $\dfrac{x(x+10)}{(2x-1)^2(x+3)}$ **13.** $x+3$ **14. a)** $\dfrac{3x^2+2x-9}{(x+5)(2x+3)}$ **b)** $\left\{x\Big|x\neq-5 \text{ and } x\neq-\dfrac{3}{2}\right\}$ **15.** $\dfrac{x+5}{x+2}$ **16.** $\dfrac{y+x}{y-x}$

17. $\dfrac{b(a-b)}{a}$ **18.** $\dfrac{3x-4}{5x^2-x}$ **19.** 30 **20.** 12 **21.** $C=\dfrac{2b+Ad}{A}$ **22.** 0.75 watt **23.** 6 **24.** 4.12 hours **25.** $6\dfrac{2}{3}$ miles

CHAPTER 7

Exercise Set 7.1

1. a) Two, positive and negative. **b)** $7, -7$ **c)** Principal square root **d)** 7 **3.** There is no real number which, when squared, gives -81. **5.** No; if the radicand is negative, the answer is not a real number. **7. a)** 1.3 **b)** 1.3
9. a) 3 **b)** -3 **c)** -3 **11.** 8 **13.** -4 **15.** -5 **17.** 1 **19.** 1 **21.** Not a real number **23.** -7 **25.** Not a real number

27. Not a real number **29.** $\dfrac{1}{5}$ **31.** $\dfrac{1}{2}$ **33.** $\dfrac{2}{3}$ **35.** $-\dfrac{2}{3}$ **37.** ≈ -2.07 **39.** 9 **41.** 19 **43.** 152 **45.** 235.23 **47.** 0.06 **49.** $\dfrac{11}{13}$

51. $|x-9|$ **53.** $|x-3|$ **55.** $|3x^2-2|$ **57.** $|6a^3-5b^4|$ **59.** $|a^5|$ **61.** $|z^{15}|$ **63.** $|a-4|$ **65.** $|3a+2b|$ **67.** 2 **69.** ≈ 9.381

71. ≈ 5.290 **73.** -3 **75.** 95 **77.** 14 **79.** 55 **81.** Select a value less than $-\dfrac{1}{2}$ **83.** $x-1\geq0; x\geq1$ **85.** $2x-6\geq0; x\geq3$

87. a) All real numbers **b)** $a\geq0$ **c)** All real numbers **89.** If n is even, finding even root of positive number. If n is odd, expression is real. **91.** $x+5>0; x>-5$ **93.** d **95.** a **97.** One answer is $f(x)=\sqrt{x-6}$. **99. a)** No **b)** Yes, when $x=0$ **c)** Yes **101. a)** $\sqrt{1288}\approx 35.89$ feet per second **b)** $\sqrt{2576}\approx 50.75$ feet per second.

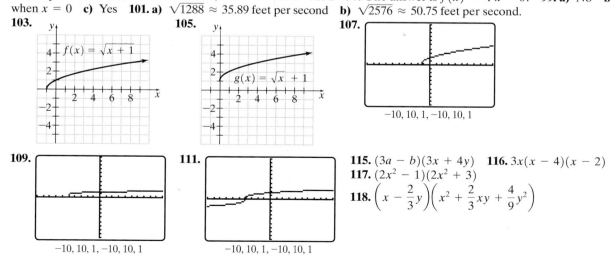

103.

105.

107.
$-10, 10, 1, -10, 10, 1$

109.
$-10, 10, 1, -10, 10, 1$

111.
$-10, 10, 1, -10, 10, 1$

115. $(3a-b)(3x+4y)$ **116.** $3x(x-4)(x-2)$
117. $(2x^2-1)(2x^2+3)$
118. $\left(x-\dfrac{2}{3}y\right)\left(x^2+\dfrac{2}{3}xy+\dfrac{4}{9}y^2\right)$

Exercise Set 7.2

1. a) When n is even and $a \geq 0$, or n is odd **b)** $a^{1/n}$ **3. a)** Always real **b)** a **c)** a **d)** $|a|$

5. a) No; $(xy)^{1/2} = x^{1/2}y^{1/2}$ **b)** No; $(xy)^{-1/2} = x^{-1/2}y^{-1/2} = \dfrac{1}{x^{1/2}y^{1/2}}$ **7.** $a^{3/2}$ **9.** $9^{5/2}$ **11.** $z^{5/3}$ **13.** $7^{10/3}$ **15.** $9^{7/4}$ **17.** $y^{13/3}$

19. $(a^3b)^{1/4}$ **21.** $(x^9z^5)^{1/4}$ **23.** $(5x^6y^7)^{1/3}$ **25.** $(3a + 5b)^{1/6}$ **27.** $\sqrt{a}$ **29.** $\sqrt{c^5}$ **31.** $\sqrt[3]{17^5}$ **33.** $\sqrt{24x^3}$ **35.** $\left(\sqrt[5]{7b^2c}\right)^3$

37. $\sqrt[5]{6a + 5b}$ **39.** $\dfrac{1}{\sqrt[3]{b^3 - c}}$ **41.** a^2 **43.** x^3 **45.** $\sqrt[3]{y}$ **47.** $\sqrt{b}$ **49.** 18.1 **51.** x^3y^6 **53.** $\sqrt{xyz}$ **55.** $\sqrt[4]{x}$ **57.** $\sqrt[8]{y}$

59. $\sqrt[9]{x^2y}$ **61.** $\sqrt[10]{a^7}$ **63.** 3 **65.** 4 **67.** 64 **69.** Not a real number **71.** $\dfrac{4}{3}$ **73.** $\dfrac{1}{2}$ **75.** -6 **77.** -4 **79.** $\dfrac{1}{4}$ **81.** $\dfrac{1}{16}$

83. $\dfrac{3}{4}$ **85.** Not a real number **87.** 22 **89.** $\dfrac{9}{14}$ **91.** $x^{9/2}$ **93.** $x^{1/6}$ **95.** $\dfrac{1}{x}$ **97.** 1 **99.** $\dfrac{y^{5/3}}{12}$ **101.** $\dfrac{8}{x^{11/6}}$ **103.** $\dfrac{1}{2x^{1/3}}$ **105.** $\dfrac{121}{x^{1/7}}$

107. $\dfrac{125}{a^{66/5}}$ **109.** $\dfrac{x}{y^{16}}$ **111.** $6z^{7/2} - 3$ **113.** $\dfrac{5}{x^5} + \dfrac{20}{x^{3/2}}$ **115.** $8x^{13/6} - 28x^2$ **117.** ≈ 11.83 **119.** ≈ 3.32 **121.** ≈ 12.65 **123.** ≈ 0.03

125. n is odd, or n is even and $a \geq 0$. **127.** $(4^{1/2} + 9^{1/2})^2 \neq 4 + 9$; $25 \neq 13$ **129.** $(1^{1/3} + 1^{1/3})^3 \neq 1 + 1$; $8 \neq 2$

131. $x^{1/2}(x^2 + 1)$ **133.** $y^{1/3}(1 - y)(1 + y)$ **135.** $\dfrac{1 + y^2}{y^{2/5}}$ **137. a)** $2^{10} = 1024$ bacteria **b)** $2^{10}\sqrt{2} \approx 1448$ bacteria

139. a) $2.69\sqrt{7^3} \approx \$49.82$ billion **b)** $2.69\sqrt{14^3} \approx \$140.91$ billion **141.** 9 **143.** $\{x|x \geq 5\}$ **145. a)** $(x - 4)^2$ **b)** $(x - 4)^2$

147. 2; $z^{\frac{1}{3}} \cdot z^{\frac{1}{a} \cdot \frac{1}{5} \cdot \frac{1}{4}} = z^{\frac{1}{60a}}$; $z^{\frac{1}{60a}} = z^{\frac{1}{120}}$, $60a = 120$; $a = 2$ **149. c)** is a function **150.** $\dfrac{b^2 + a^3b}{a^3 - b}$ **151.** $0, 3$

152. ≈ 441.67 miles per hour.

Exercise Set 7.3

1. a) Square the natural numbers. **b)** $1, 4, 9, 16, 25, 36$ **3. a)** Raise the natural numbers to the fifth power. **b)** $1, 32, 243, 1024, 3125$ **5.** If n is even and a or b are negative, the numbers are not real numbers.
7. If n is even and a or b is negative; the numbers are not real numbers; **9.** $2\sqrt{2}$ **11.** $2\sqrt{7}$ **13.** $2\sqrt{3}$ **15.** $5\sqrt{2}$ **17.** $5\sqrt{3}$
19. $2\sqrt{10}$ **21.** $2\sqrt[3]{2}$ **23.** $3\sqrt[3]{2}$ **25.** $2\sqrt[3]{4}$ **27.** $3\sqrt[3]{4}$ **29.** $3\sqrt[4]{2}$ **31.** $-2\sqrt[5]{2}$ **33.** $x\sqrt{x}$ **35.** $a^5\sqrt{a}$ **37.** $-z^{10}\sqrt{z}$ **39.** $a^2\sqrt[3]{a}$
41. $b^4\sqrt[3]{b}$ **43.** $x\sqrt[4]{x}$ **45.** $z\sqrt[5]{z^2}$ **47.** $3y^4\sqrt[5]{y^3}$ **49.** $2x\sqrt{6x}$ **51.** $2x^2y^3\sqrt{2y}$ **53.** $-2x^3y^3z^6\sqrt{5y}$ **55.** $xy^2\sqrt[3]{y}$ **57.** $3a^2b^2\sqrt[3]{3b^2}$
59. $2x^2y^2z^4\sqrt[4]{2yz^3}$ **61.** $3a^2b^2\sqrt[4]{b}$ **63.** $2a^2b^2\sqrt[5]{b^2}$ **65.** 3 **67.** $\dfrac{2}{5}$ **69.** 3 **71.** $\dfrac{1}{4}$ **73.** $\dfrac{1}{2}$ **75.** $\dfrac{1}{3}$ **77.** $\dfrac{1}{2}$ **79.** 2 **81.** $\dfrac{r^2}{5}$ **83.** $\dfrac{6x^2}{5y^5}$

85. $\dfrac{c^2}{3}$ **87.** a^2b^4 **89.** $2\sqrt{2}$ **91.** $3x^2$ **93.** $2x^2y^2\sqrt{2y}$ **95.** $\dfrac{\sqrt[3]{7y}}{2x^4}$ **97.** $\dfrac{y^2\sqrt[3]{5y}}{x^2}$ **99.** $\dfrac{x^3\sqrt[4]{20y}}{3}$ **101.** $(a \cdot b)^{1/2} = a^{1/2}b^{1/2} = \sqrt{a}\,\sqrt{b}$

103. No; One example is $\sqrt{18}/\sqrt{2} = 3$. **105. a)** No **b)** When $\sqrt[n]{x}$ is a real number and not equal to 0.

106. $C = \dfrac{5}{9}(F - 32)$ **107.** $\{-28, 32\}$ **108.** $3x^6 - x^3 + 6$ **109.** $(x - 1)(x^2 - 8x + 19)$

Exercise Set 7.4

1. Radicals with the same radicand and index **3.** ≈ 5.97
5. No: one example is $\sqrt{9} + \sqrt{16} \neq \sqrt{9 + 16}$, $3 + 4 \neq 5$, $7 \neq 5$. **7.** 0 **9.** $4\sqrt{3}$ **11.** $-4\sqrt{3} + 5$ **13.** $-6\sqrt[4]{y}$ **15.** $2\sqrt[3]{x} + 7\sqrt{5}$
17. $7\sqrt{x} - 2\sqrt{y}$ **19.** $2(\sqrt{2} - \sqrt{3})$ **21.** $-30\sqrt{3} + 20\sqrt{5}$ **23.** $-4\sqrt{10}$ **25.** $18y\sqrt{5x}$ **27.** $-16\sqrt{5x}$ **29.** $-27a\sqrt{2}$
31. $7\sqrt[3]{4}$ **33.** -7 **35.** $5a\sqrt[3]{ab^2}$ **37.** $3r^3s^2\sqrt{rs}$ **39.** 0 **41.** 10 **43.** $2\sqrt[3]{7}$ **45.** $3m^2n^5\sqrt{3n}$ **47.** $3x^3y^4\sqrt[3]{2x^2y}$
49. $3x^3y^4\sqrt[4]{2xy^3}$ **51.** $2xyz^2\sqrt[4]{x^2z^2}$ **53.** $5 - \sqrt{15}$ **55.** $2\sqrt[3]{y^2} - y^3$ **57.** $4x^5y^3\sqrt[3]{x} + 4xy^4\sqrt[3]{2x^2y^2}$ **59.** 20 **61.** $6 - x^2$
63. $5 - z$ **65.** $23 + 9\sqrt{3}$ **67.** $16 - 10\sqrt{2}$ **69.** $10 - 3\sqrt{6}$ **71.** $29 - 12\sqrt{5}$ **73.** $18x - \sqrt{3xy} - y$
75. $8 - 2\sqrt[3]{18} - \sqrt[3]{12}$ **77.** $4x - 8\sqrt{x}$ **79.** $x^2 + x\sqrt[3]{x^2}$ **81.** $x\sqrt[4]{27x^2} - x^2\sqrt[4]{3x}$ **83.** $2\sqrt{6}$ **85.** $7\sqrt{5}$ **87.** $-14 + 11\sqrt{2}$
89. $4\sqrt{6} - 2\sqrt{3}$ **91.** $15\sqrt{2}$ **93.** $2x^3\sqrt[3]{10x^2}$ **95.** $2b^2c\sqrt[4]{2ab^5c^3}$ **97.** $4ab\sqrt[4]{b}$ **99.** $x - 2\sqrt[3]{x^2y^2} - \sqrt[3]{xy} + 2y$
101. $ab\sqrt[3]{12a^2b^2} - 2a^2b^2\sqrt[3]{3}$ **103.** $2x + 5$ **105.** $2|r - 4|$ **107.** $P = 14\sqrt{5}, A = 60$ **109.** $P = 17\sqrt{5}, A = 52.5$
111. No, $-\sqrt{2} + \sqrt{2} = 0$ **113. a)** ≈ 45.17 miles per hour **b)** ≈ 35.33 miles per hour **115. a)** 37 inches **b)** ≈ 37.97 inches
117. a)　　　　　　　**119. a)**　　　　　　　**121.**

b) Raises the graph 2 units　　　**b)** $\{x|x \geq 0\}$

123. A quotient of two integers, denominator not 0. **124.** A number that can be represented on a real number line.
125. A real number that cannot be expressed as a quotient of two integers. **126.** $|a| = \begin{cases} a, & a \geq 0 \\ -a, & a < 0 \end{cases}$ **127.** $m = \dfrac{2E}{v^2}$

128. a) 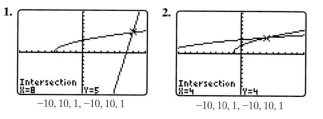 **b)** $\left(-\dfrac{1}{2}, 4\right]$ **c)** $\left\{x \,\middle|\, -\dfrac{1}{2} < x \leq 4\right\}$

Exercise Set 7.5

1. a) Same two terms with the sign of the second term changed. **b)** $x + \sqrt{3}$
3. a) Answers will vary. **b)** $\dfrac{4\sqrt{3y}}{3y}$ **5.** (1) No perfect powers are factors of any radicand. (2) No radicand contains

fractions. (3) There are no radicals in any denominator. **7.** $\dfrac{\sqrt{3}}{3}$ **9.** $\dfrac{3\sqrt{7}}{7}$ **11.** $\dfrac{\sqrt{17}}{17}$ **13.** $\sqrt{7}$ **15.** $\dfrac{p\sqrt{2}}{2}$ **17.** $\dfrac{\sqrt{5y}}{5}$

19. $\sqrt{15}$ **21.** $\dfrac{\sqrt{xy}}{y}$ **23.** $\dfrac{\sqrt{10m}}{4}$ **25.** $\dfrac{\sqrt{2n}}{3}$ **27.** $\dfrac{3x^2y\sqrt{yz}}{z^2}$ **29.** $\dfrac{2y^3z\sqrt{15xz}}{3x}$ **31.** $\dfrac{5x^3y^2\sqrt{yz}}{z^2}$ **33.** $\dfrac{\sqrt[3]{4}}{2}$ **35.** $\dfrac{3\sqrt[3]{x^2}}{x}$ **37.** $\dfrac{\sqrt[4]{8}}{2}$

39. $\dfrac{z\sqrt[4]{2}}{2}$ **41.** $\dfrac{x\sqrt[4]{z^2}}{z}$ **43.** $\dfrac{13\sqrt[5]{y^3}}{y}$ **45.** $\dfrac{3\sqrt[7]{a^4}}{a}$ **47.** $\dfrac{\sqrt[3]{2x^2}}{2x}$ **49.** $\dfrac{5m\sqrt[4]{8}}{2}$ **51.** $\dfrac{\sqrt[4]{135x}}{3x}$ **53.** $\dfrac{\sqrt[3]{12x^2y}}{2y}$ **55.** $\dfrac{\sqrt[3]{4xy^2z}}{z}$

57. $\dfrac{\sqrt{3}-1}{2}$ **59.** $2 - \sqrt{3}$ **61.** $\dfrac{-4\sqrt{2}-28}{47}$ **63.** $\dfrac{10+\sqrt{30}}{14}$ **65.** $\dfrac{12-2\sqrt{x}}{36-x}$ **67.** $\dfrac{4x+4y\sqrt{x}}{x-y^2}$ **69.** $\dfrac{-13+3\sqrt{6}}{23}$

71. $a + a^3$ **73.** $\dfrac{2\sqrt{x+2}+6}{x-7}$ **75.** $\dfrac{\sqrt{x}}{3}$ **77.** $\dfrac{\sqrt{10}}{5}$ **79.** -1 **81.** $\dfrac{2xy^3\sqrt{30xz}}{5z}$ **83.** $\dfrac{\sqrt{6}}{x}$ **85.** $\dfrac{\sqrt{a}-3}{a-9}$ **87.** $-\dfrac{\sqrt{2x}}{2}$

89. $\dfrac{\sqrt[4]{24x^3y^2}}{2x}$ **91.** $\dfrac{2y^4z^3\sqrt[3]{2x^2z}}{x}$ **93.** $\dfrac{a\sqrt{r}+2r\sqrt{a}}{a-4r}$ **95.** $\dfrac{\sqrt[3]{150y^2}}{5y}$ **97.** $\dfrac{y^3z\sqrt[4]{54x^2}}{3x}$ **99.** $\sqrt{2}$ **101.** $\dfrac{4\sqrt{5}}{5}$ **103.** $\dfrac{13\sqrt{6}}{6}$

105. $\dfrac{15\sqrt{2}}{2}$ **107.** $\dfrac{13\sqrt{2}}{2}$ **109.** $-\dfrac{301\sqrt{2}}{20}$ **111.** $\dfrac{3\sqrt{6}}{4}$ **113.** $\left(-\dfrac{2}{y}+\dfrac{3}{x}\right)\sqrt{xy}$ **115.** $\sqrt{a}$ **117.** $\sqrt[3]{(a+b)^5}$ **119.** $\sqrt[15]{(a+2b)^2}$

121. $\sqrt[6]{rs^5}$ **123.** $\sqrt[15]{x^2y^8}$ **125.** ≈ 3.69 meters **127.** ≈ 12 inches **129. a)** 6.21 million **b)** 2.36 million **131.** $\dfrac{3}{\sqrt{3}}$;

$\dfrac{2}{\sqrt{2}} = \sqrt{2}, \dfrac{3}{\sqrt{3}} = \sqrt{3}$ **133.** $2 + \sqrt{3}$; rationalize the denominator and compare **135. a)** $4, 8, 12$ **b)** $9, 18, 27$

c) $x^{(3a+2b)/6}$ **d)** $x^{(3a-2b)/6}$ **137.** $\dfrac{3\sqrt{2a}-3b}{2a-3b}$ **139.** $\dfrac{10}{15+3\sqrt{5}}$ **141.** $\dfrac{1}{\sqrt{x+h}+\sqrt{x}}$ **144.** $b_2 = \dfrac{2A}{h} - b_1$
145. 40 mph, 50 mph. **146.** $4x^3 + x^2 - 20x + 4$ **147.** $-8, 1$

Using Your Graphing Calculator, 7.6 1.

![Intersection X=8 Y=5 graph; $-10, 10, 1, -10, 10, 1$] 2. ![Intersection X=4 Y=4 graph; $-10, 10, 1, -10, 10, 1$]

Exercise Set 7.6

1. a) Answers will vary. **b)** 5 **3.** 0 **5.** Answers will vary. **7.** 1; Answers will vary.
9. 16 **11.** No real solution **13.** -64 **15.** 16 **17.** 9 **19.** -1 **21.** 16 **23.** 76 **25.** No real solution **27.** No real solution

29. 1 **31.** 8 **33.** 5 **35.** $\dfrac{2}{3}$ **37.** 20 **39.** 1 **41.** 10 **43.** 6 **45.** 8 **47.** 0 **49.** -3 **51.** $\dfrac{3}{2}$ **53.** No real Solution **55.** 2

57. 9 **59.** No real solution **61.** 9 **63.** 3, 7 **65.** -1 **67.** 7 **69.** 4 **71.** 5 **73.** $v = \dfrac{p^2}{2}$ **75.** $g = \dfrac{v^2}{2h}$ **77.** $F = \dfrac{Mv^2}{R}$ **79.** $m = \dfrac{x^2k}{V_0^2}$
81. $A = \pi r^2$ **83.** $\sqrt{87}$ **85.** $4\sqrt{2}$ **87.** 4 **89.** No Solution **91.** 7 **93.** 5 **95.** 1 **97.** 3 **99.** $\sqrt{2000} \approx 44.7$ feet **101.** 12 feet

103. a) ≈ 3.51 seconds **b)** $\sqrt{2} \cdot T$; compare $\sqrt{\dfrac{l}{32}}$ with $\sqrt{\dfrac{l}{16}}$ **c)** $\sqrt{24} \approx 4.90$ seconds **105.** $R = \dfrac{8\mu l}{\pi r^4}$
107. $0.2(\sqrt{149.4})^3 \approx 365.2$ days. **109.** $\sqrt{10,000} = 100$ pounds **111.** $\sqrt{320} \approx 17.89$ feet per second

113. $\sqrt{1649} \approx 40.61$ meters **115.** $2, -2$ **117.** $-4, \dfrac{3}{2}$ **119.** 21 **121.** $5, -5$ **123. a)** $3, 7$; points of intersection **b)** Yes
c) $3, 7$; yes **125.** At $x = 4$, $g(x)$ or $y = 0$. Therefore the graph must have an x-intercept at 4. **127.** $L_1 \approx 0.44, L_2 \approx 0.76$

129. All real numbers   **131.** 1.5   **133.** ≈ -3.7; ≈ 3.7 **135.** No real solution   **137.** $n = \dfrac{z^2\sigma^2}{(\overline{x} - \mu)^2}$ **140.** $P_2 = \dfrac{P_1 P_3}{P_1 - P_3}$ **141.** x

142. $\dfrac{3a}{2b(2a + 3b)}$ **143.** $t(t - 4)$ **144.** $\dfrac{3}{x + 3}$ **145.** 2

Exercise Set 7.7

1. a) $\sqrt{-1}$ **b)** -1 **3.** Yes   **5.** Yes   **7.** $a - bi$ **9. a)** $\sqrt{2}$ **b)** 1   **c)** $\sqrt{-3}$ or $2i$ **d)** 6
e) Every number we have studied is a complex number.   **11.** $5 + 0i$ **13.** $7 + 0i$ **15.** $21 - 6i$ **17.** $0 + 2i\sqrt{6}$ **19.** $8 - 2i\sqrt{3}$
21. $1 + 7i\sqrt{2}$ **23.** $9 - 5i$ **25.** $0 + (2 - 3\sqrt{5})i$ **27.** $21 + 5i$ **29.** 0   **31.** $-12 - 12i$ **33.** $(4\sqrt{2} + \sqrt{3}) - 2i\sqrt{2}$
35. $9 - 4i\sqrt{2}$ **37.** $-7 - 2i\sqrt{5}$ **39.** $10 - 2i$ **41.** $-9 + 2i$ **43.** $-33 + 21i$ **45.** $20 + 4i\sqrt{3}$ **47.** $-16 + 8i$ **49.** $1 + 5i$
51. 409   **53.** $39 - 9i\sqrt{2}$ **55.** $\dfrac{25}{72} + \dfrac{1}{4}i$ **57.** $-\dfrac{2}{3}i$ **59.** $\dfrac{3 - 2i}{2}$ **61.** $\dfrac{8 + 4i}{5}$ **63.** $\dfrac{9 + 18i}{5}$ **65.** $\dfrac{9 - 12i}{10}$ **67.** $\dfrac{3 + i}{5}$
69. $\dfrac{5\sqrt{2} - 2i\sqrt{6}}{37}$ **71.** $\dfrac{(5\sqrt{10} - 2\sqrt{15}) + (10\sqrt{2} + 5\sqrt{3})i}{45}$ **73.** 5   **75.** $\dfrac{2\sqrt{2}}{3}$ **77.** $7 - 7i$ **79.** $\sqrt{2} + 2i\sqrt{3}$
81. $20.8 - 16.64i$ **83.** $37 - 39i$ **85.** $\dfrac{4 - 5i}{2}$ **87.** $\dfrac{5\sqrt{3} + 10i}{7}$ **89.** $1 + \dfrac{2}{45}i$ **91.** $\dfrac{1}{4} - \dfrac{31}{50}i$ **93.** 2   **95.** $-4.33 - 10.91i$ **97.** -1
99. 1   **101.** i **103.** $-i$ **105. a)** $-2 - 3i$ **b)** $\dfrac{2 - 3i}{13}$ **107.** True; $(2i)(2i) = -4$ **109.** False; $(1 + i)(1 + 2i) = -1 + 3i$
111. Even values; i^n where n is even will either be -1 or 1   **113.** -1 **115.** $16 - 4i$ **117.** $11 + 7i$ **119.** 0   **121.** 1   **123.** Yes
125. No   **127.** $\approx 0.83 - 3i$ **129.** $\approx 1.5 - 0.33i$ **131.** $-i$ **133.** $2 + i\sqrt{2}, 2 - i\sqrt{2}$ **135.** $6 + 3i\sqrt{3}$ **137.** $-1 + 7i\sqrt{3}$
139. 15 pounds at \$5.50, 25 pounds at \$6.30   **140.** $2c - 3 + \dfrac{2}{4c + 9}$ **141.** $\dfrac{a^2}{b(a - b)}$ **142.** 4

Chapter Review Exercises

1. 7   **2.** -3 **3.** 4   **4.** -5 **5.** 7   **6.** 93.4   **7.** $|x|$ **8.** $|x - 2|$ **9.** $|x - y|$
10. $|x^2 - 4x + 12|$ **11.** 7   **12.** 57   **13.** ≈ 2.2 **14.** 11 meters   **15.** $x^{5/2}$ **16.** $x^{5/3}$ **17.** $y^{15/4}$ **18.** $5^{2/7}$ **19.** $\sqrt{x}$ **20.** $\sqrt[5]{a^4}$
21. $\left(\sqrt[4]{3m^2n}\right)^7$ **22.** $\left(\dfrac{1}{\sqrt[3]{x + 3y}}\right)^5$ **23.** 9   **24.** x^5 **25.** 49   **26.** $\sqrt[4]{a}$ **27.** -5 **28.** Not a real number   **29.** $\dfrac{3}{4}$ **30.** $\dfrac{3}{8}$
31. $x^{4/15}$ **32.** $\dfrac{4}{y^2}$ **33.** $\dfrac{1}{a^{14/15}}$ **34.** $\dfrac{49x^{10}}{y^7}$ **35.** $5a^2 - 2a^{5/2}$ **36.** $\dfrac{4}{x^{7/6}} + 9$ **37.** $x^{2/5}(1 + x)$ **38.** $\dfrac{1 + a^3}{a^{1/2}}$ **39.** 5   **40.** ≈ 2.668

41.

42.

43. $5\sqrt{3}$ **44.** $4\sqrt[3]{2}$ **45.** $\dfrac{7}{2}$ **46.** $\dfrac{3}{5}$ **47.** $-\dfrac{9}{8}$ **48.** $-\dfrac{2}{5}$ **49.** 10
50. 4   **51.** $2xyz^2\sqrt{2y}$ **52.** $5xy^3\sqrt{2xy}$ **53.** $3a^2b^3\sqrt[3]{2ab}$
54. $5x^2y^3z^5\sqrt[3]{x^2z}$ **55.** $x^{14}y^{21}z^{35}$ **56.** $8a^3b^{12}c^{18}$ **57.** $2x^3\sqrt{10}$
58. $2x^3y\sqrt[3]{x^2y^2}$ **59.** $2x^2y^3\sqrt[3]{4x^2}$ **60.** $2x^2y^4\sqrt[4]{x}$
61. $6x - 2\sqrt{15x}$ **62.** $2x^2y^2\sqrt[3]{y^2} + x\sqrt[3]{18y}$ **63.** $\sqrt[4]{a^3b^2}$
64. $\sqrt[6]{x^5y^2}$ **65.** $\dfrac{27r^{9/2}}{p^3}$ **66.** $\dfrac{y^{1/5}}{4xz^{1/3}}$ **67.** $\dfrac{\sqrt{15}}{5}$ **68.** $\dfrac{\sqrt[3]{14}}{2}$
69. $\dfrac{\sqrt[4]{54}}{3}$ **70.** $\dfrac{x\sqrt{11}}{11}$ **71.** $\dfrac{6\sqrt{x}}{x}$ **72.** $\dfrac{a\sqrt[3]{5}}{5}$ **73.** $\dfrac{9\sqrt[3]{y}}{y}$ **74.** $\dfrac{7\sqrt[4]{z^3}}{z}$ **75.** $\dfrac{x}{2}$ **76.** $\dfrac{x}{2}$ **77.** $\dfrac{4y^2}{x}$ **78.** $2x^2y^3$ **79.** $\dfrac{x^2\sqrt{5y}}{y}$
80. $\dfrac{2\sqrt{21ab}}{7b}$ **81.** $\dfrac{x^2y^2\sqrt{6yz}}{z}$ **82.** $\dfrac{5xy^2\sqrt{15yz}}{3z}$ **83.** $3xy\sqrt[3]{2y}$ **84.** $\dfrac{\sqrt[3]{75xy^2}}{5y}$ **85.** $\dfrac{y\sqrt[3]{4x^2}}{x}$ **86.** $\dfrac{y^2\sqrt[3]{4x}}{2x}$ **87.** $\dfrac{b^2\sqrt[4]{3ab^2}}{a}$
88. $\dfrac{y\sqrt[4]{4x^3y^2}}{2x}$ **89.** 7   **90.** $x - y^2$ **91.** $x^2 - y$ **92.** $28 + 10\sqrt{3}$ **93.** $x + \sqrt{5xy} - \sqrt{3xy} - y\sqrt{15}$
94. $\sqrt[3]{6x^2} - \sqrt[3]{4xy} - \sqrt[3]{9xy} + \sqrt[3]{6y^2}$ **95.** $-10 + 5\sqrt{5}$ **96.** $\dfrac{3x - x\sqrt{x}}{9 - x}$ **97.** $\dfrac{7a + a\sqrt{b}}{49 - b}$ **98.** $\dfrac{x\sqrt{y} + 8x}{y - 64}$ **99.** $\dfrac{x - \sqrt{xy}}{x - y}$
100. $\dfrac{x - \sqrt{xy} - 2y}{x - y}$ **101.** $\dfrac{3\sqrt{a - 1} + 6}{a - 5}$ **102.** $\dfrac{4\sqrt{y + 2} + 12}{y - 7}$ **103.** $2\sqrt[3]{x}$ **104.** $-4\sqrt{3}$ **105.** $8 - 13\sqrt[3]{2}$ **106.** $\dfrac{69\sqrt{2}}{8}$
107. $(3x^2y^3 - 4x^3y^4)\sqrt{x}$ **108.** $(2x^2y^2 - x + 3x^3)\sqrt[3]{xy}$ **109.** $3x\sqrt{2} - \sqrt{30x}$ **110.** $2x^2 + 2x^2\sqrt[3]{2x}$ **111.** $2x + 4$
112. $\sqrt{5}|2a + 5|$ **113.** $\sqrt[6]{x + 5}$ **114.** $\sqrt[12]{b^5}$ **115. a)** $12\sqrt{3}$ **b)** 24   **116. a)** $8\sqrt{5} + \sqrt{130}$ **b)** $10\sqrt{13}$

117. a) **b)** $x \geq 0$ **118. a)**   **b)** $x \geq 0$ **119.** 64 **120.** No solution
121. 125 **122.** -125 **123.** 49
124. 64 **125.** No solution **126.** 2
127. -3 **128.** 2 **129.** 0, 9 **130.** 5
131. 4 **132.** 6 **133.** $L = \dfrac{V^2 w}{2}$
134. $A = \pi r^2$ **135.** $2\sqrt{14}$ **136.** $5\sqrt{3}$
137. $\sqrt{29} \approx 5.39$ meters.
138. $\sqrt{1280} \approx 35.78$ feet per second

139. $2\pi\sqrt{2} \approx 2.83\pi \approx 8.89$ seconds **140.** $\sqrt{\dfrac{90}{0.145}} \approx 25$ meters per second **141.** $m \approx 5m_0$. Thus, it is ≈ 5 times its original

mass. **142.** $5 + 0i$ **143.** $-6 + 0i$ **144.** $2 - 16i$ **145.** $3 + 4i$ **146.** $7 + i$ **147.** $1 - 2i$ **148.** $3\sqrt{3} + (\sqrt{5} - \sqrt{7})i$

149. $-6 + 6i$ **150.** $17 - 6i$ **151.** $(24 + 3\sqrt{5}) + (4\sqrt{3} - 6\sqrt{15})i$ **152.** $-\dfrac{2i}{3}$ **153.** $\dfrac{(-2 - \sqrt{3})i}{2}$ **154.** $\dfrac{15 - 10i}{13}$

155. $\dfrac{5\sqrt{3} + 3i\sqrt{2}}{31}$ **156.** 0 **157.** 7 **158.** i **159.** $-i$ **160.** 1 **161.** -1

Chapter Practice Test **1.** $|5x - 1|$ **2.** $\dfrac{1}{x^{12/5}}$ **3.** $\dfrac{1 + x^2}{x^{2/3}}$ **4.**   **5.** $4x^3 y^5 \sqrt{3x}$

6. $2x^3 y^3 \sqrt[3]{5x^2 y}$ **7.** $\dfrac{x^3 y \sqrt{6yz}}{4z}$ **8.** $\dfrac{4\sqrt[3]{x^2}}{x}$ **9.** $\dfrac{3 - \sqrt{3}}{6}$ **10.** $8\sqrt{6}$

11. $(2xy + 2x^2 y^2)\sqrt[3]{y^2}$ **12.** $6\sqrt{3} - 2\sqrt{6} - 30 + 10\sqrt{2}$ **13.** $\sqrt[8]{x^5 y^3}$

14. $\sqrt[12]{(7x + 2)^7}$ **15.** 7 **16.** -3 **17.** 9 **18.** 3 **19.** $g = \dfrac{8w^2}{h}$

20. $\sqrt{12,880} \approx 113.49$ feet per second **21.** 13 feet

22. $2\pi\sqrt{\dfrac{1400}{65,000}} \approx 0.92$ second **23.** $18 + 14i$ **24.** $\dfrac{33 - 17i}{53}$ **25.** 2

CHAPTER 8

Exercise Set 8.1 **1.** ± 6 **3.** If $x^2 = a$, then $x = \pm\sqrt{a}$. **5.** $\left(\dfrac{b}{2}\right)^2$ must equal c. **7. a)** Yes **b)** No, ± 3

9. Multiply by $\dfrac{1}{2}$ to make $a = 1$. **11.** $\left(-\dfrac{6}{2}\right)^2 = 9$ **13.** ± 7 **15.** $\pm 7i$ **17.** $\pm 4i\sqrt{3}$ **19.** $\pm i\sqrt{61}$ **21.** 8, 0 **23.** $-3 \pm 7i$

25. $4 \pm 3i\sqrt{5}$ **27.** $-1, \dfrac{1}{3}$ **29.** $\dfrac{1 \pm 2i}{3}$ **31.** $-0.9, -2.7$ **33.** $\dfrac{5 \pm 2\sqrt{3}}{2}$ **35.** $-\dfrac{1}{20}, -\dfrac{9}{20}$ **37.** $1, -4$ **39.** $3, -5$ **41.** 2, 4

43. 1, 5 **45.** $-1, \dfrac{1}{2}$ **47.** $-\dfrac{1}{2}, 4$ **49.** 4, 7 **51.** $-1, 4$ **53.** 4, 5 **55.** 7, -4 **57.** 1, -10 **59.** $2 \pm \sqrt{14}$ **61.** $-4 \pm \sqrt{11}$

63. $\dfrac{1 \pm \sqrt{13}}{2}$ **65.** $\dfrac{-3 \pm i\sqrt{15}}{2}$ **67.** 0, 1 **69.** 0, -2 **71.** 0, $\dfrac{1}{3}$ **73.** 1, -3 **75.** 9, -5 **77.** $-8, -3$ **79.** $\dfrac{1}{3}, -1$ **81.** $\dfrac{1 \pm i\sqrt{39}}{4}$

83. $\dfrac{-3 \pm \sqrt{59}}{10}$ **85. a)** $21 = (x + 2)(x - 2)$ **b)** S **87. a)** $18 = (x + 4)(x + 2)$ **b)** $-3 + \sqrt{19}$ **89.** 12 mph

91. 7, 9 **93.** 5 feet by 12 feet **95.** $\dfrac{12 + \sqrt{288}}{2} \approx 14.49$ feet by 14.49 feet **97.** $\sqrt{200} \approx 14.14$ inches

99. $\sqrt{24} \approx 4.90$ feet **101.** 4% **103.** $\approx 6\%$ **105. a)** $S = 32 + 80\sqrt{\pi} \approx 173.80$ square inches **b)** $r = \dfrac{4\sqrt{\pi}}{\pi} \approx 2.26$ inches

c) $r = -5 + \sqrt{\dfrac{80 + 25\pi}{\pi}} \approx 2.1$ inches **107.** 2 **108.** \$4200 at 7%, \$5800 at $6\dfrac{1}{4}\%$ **109.** $\left\{10, \dfrac{4}{3}\right\}$

110. 0 **111.** $4x^3 + x^2 - 20x + 4$

Exercise Set 8.2   **1.** $x = \dfrac{-b \pm \sqrt{b^2 - 4ac}}{2a}$ **3.** $a = -3, b = 6, c = 5$ **5.** Yes; if you multiply both sides of one equation by -1 you get the other equation   **7. a)** $b^2 - 4ac$ **b)** -204 **c)** Answers will vary.   **9.** Two real solutions **11.** No real solution   **13.** Two real solutions   **15.** No real solution   **17.** One real solution   **19.** One real solution   **21.** $3, 6$ **23.** $2, 4$ **25.** $1, -3$ **27.** $-2 \pm 2\sqrt{6}$ **29.** ± 7 **31.** $\dfrac{2 \pm i\sqrt{11}}{3}$ **33.** $0, 3$ **35.** $\dfrac{2 \pm i\sqrt{2}}{2}$ **37.** -1 **39.** 5 **41.** $1 \pm \sqrt{2}$ **43.** $\dfrac{2 \pm \sqrt{10}}{3}$ **45.** $-3, \dfrac{1}{2}$ **47.** $\dfrac{1}{2}, -\dfrac{5}{3}$ **49.** $4, -6$ **51.** $\dfrac{1}{3}, \dfrac{2}{3}$ **53.** $\dfrac{-6 \pm 2\sqrt{6}}{3}$ **55.** $\dfrac{3 \pm \sqrt{309}}{30}$ **57.** $\dfrac{3 \pm \sqrt{33}}{2}$ **59.** $\dfrac{2 \pm i\sqrt{2}}{2}$ **61.** $\dfrac{-1 \pm i\sqrt{23}}{4}$ **63.** $\dfrac{-0.6 \pm \sqrt{0.84}}{0.2}$ or $-3 \pm \sqrt{21}$ **65.** $0, 2$ **67.** $-5, 6$ **69.** $\dfrac{7 \pm \sqrt{17}}{4}$ **71.** No real number **73.** $f(x) = x^2 - 7x + 10$ **75.** $f(x) = x^2 + 2x - 24$ **77.** $f(x) = 15x^2 - x - 6$ **79.** $f(x) = x^2 - 3$ **81.** $f(x) = x^2 + 4$ **83.** $f(x) = x^2 - 6x + 7$ **85.** $f(x) = x^2 - 4x + 13$ **87. a)** $n(10 - 0.02n) = 450$;   **b)** 50;   **89. a)** $n(50 - 0.2n) = 1680$; **b)** 40 **91.** Answers will vary.   **93.** Yes   **95.** 2 **97.** $w = 3$ feet, $l = 7$ feet   **99.** 2 inches   **101. a)** ≈ 4.57 seconds **b)** ≈ 4.79 seconds.   **103.** $2\sqrt{5}, -\sqrt{5}$ **105.** $(-0.12 + \sqrt{14.3952})/1.2 \approx 3.0618$ millimeters   **107. a)** ≈ 1.94 seconds **b)** ≈ 2.74 seconds   **c)** Courtney's   **d)** Yes, at 1.5 seconds   **108.** 3.0×10^2 or 300 **109.** 10 **110.** $(2, -1)$ **111.** $\dfrac{6y - x}{3xy}$ **112.** No real solution

Exercise Set 8.3   **1.** Answers will vary.   **3.** $S = \sqrt{A}$ **5.** $S = \sqrt{A + s^2}$ **7.** $i = \sqrt{\dfrac{E}{r}}$ **9.** $t = \dfrac{\sqrt{d}}{4}$ **11.** $c = \sqrt{\dfrac{E}{m}}$ **13.** $r = \sqrt{\dfrac{3V}{\pi h}}$ **15.** $W = \sqrt{d^2 - L^2}$ **17.** $b = \sqrt{c^2 - a^2}$ **19.** $H = \sqrt{d^2 - L^2 - W^2}$ **21.** $t = \sqrt{\dfrac{h - s_0}{-16}}$ or $t = \dfrac{\sqrt{s_0 - h}}{4}$ **23.** $v = \sqrt{\dfrac{2E}{m}}$ **25.** $v_1 = \sqrt{v_2^2 - 2ad}$ **27.** $c = \sqrt{(v')^2 + v^2}$ **29. a)** $\$13,740$ **b)** ≈ 8 **31. a)** $32°$ F   **b)** $50.2°$ F   **c)** ≈ 2.92 minutes   **33. a)** $\$117,500$ **b)** ≈ 2268 square feet   **35. a)** 4 seconds **b)** 29.26 feet per second   **37. a)** $\approx \$18.5$ billion   **b)** 2006 **39.** $l = 25$ meters, $w = 20$ meters   **41.** 4 feet per hour **43.** Going 6 mph, returning 8 mph   **45.** Bonita ≈ 11.52 hours; Pamela ≈ 12.52 hours   **47.** 130 mph   **49.** Chris ≈ 11.76 hours; John ≈ 12.26 hours   **51.** 60 mph   **53.** Answers will vary.   **55.** 6 meters by 3 meters or 2 meters by 9 meters   **57.** -24 **58.** $R = \dfrac{E - Ir}{I}$ **59.** $\dfrac{16}{r - 4}$ **60.** $\dfrac{x}{y^{16}}$ **61.** No solution.

Exercise Set 8.4   **1.** Can be written in the form $au^2 + bu + c = 0$.   **3.** $u = x^2$; gives equation $3u^2 - 5u + 1 = 0$. **5.** $u = z^{-1}$; gives equation $u^2 - u = 56$.   **7.** $\pm 1, \pm 3$ **9.** $\pm 1, \pm 5$ **11.** $\pm 2, \pm 3$ **13.** $\pm 2, \pm \sqrt{3}$ **15.** $\pm \dfrac{1}{2}, \pm 1$ **17.** $\pm \sqrt{3}, \pm \sqrt{5}$ **19.** $\pm 3, \pm i\sqrt{2}$ **21.** $\pm 1, \pm i\sqrt{5}$ **23.** 4 **25.** 4 **27.** $\dfrac{1}{9}$ **29.** $1, -9$ **31.** $\dfrac{4}{3}, -\dfrac{1}{2}$ **33.** $\pm 2, \pm i$ **35.** $-5, -\dfrac{3}{2}$ **37.** $\pm \dfrac{5\sqrt{6}}{6}, \pm \dfrac{\sqrt{39}}{3}$ **39.** $-\dfrac{1}{5}$ **41.** $2, 3$ **43.** $2, \dfrac{1}{3}$ **45.** $1, -\dfrac{1}{10}$ **47.** $-\dfrac{1}{2}, \dfrac{1}{6}$ **49.** $1, 8$ **51.** $-64, -343$ **53.** $\dfrac{1}{4}$ **55.** $1, 32$ **57.** $(1, 0), (16, 0)$ **59.** None   **61.** $(-4, 0), \left(\dfrac{1}{5}, 0\right)$ **63.** $(-27, 0), (8, 0)$ **65.** $(-1, 0), (4, 0)$ **67.** $(\pm 2, 0), (\pm 4, 0)$ **69.** Let $u = x^2$ **71.** Let $u = x^{-1}$ **73.** $x^4 - 20x^2 + 64 = 0$; start with $(x - 2)(x + 2)(x - 4)(x + 4) = 0$ **75.** $x^4 - 5x^2 + 6 = 0$; start with $(x + \sqrt{2})(x - \sqrt{2})(x + \sqrt{3})(x - \sqrt{3}) = 0$ **77.** No; imaginary solution always occur in pairs.   **79. a) and b)** $\dfrac{1}{5}, -\dfrac{1}{4}$ **81.** $-\dfrac{14}{5}, -\dfrac{8}{3}$ **83.** $3, \dfrac{5}{4}$ **85.** $2, 1$ **87.** $-3, 1, 2, -4$ **89.** $\pm \sqrt{\dfrac{3 \pm \sqrt{15}}{2}}$ **91.** $\dfrac{43}{60}$ **92.** 1 **93.** D: $\mathbb{R}$, R: $\{y | y \geq 0\}$ **94.** $2xy^2\sqrt[3]{2}$ **95.** $11\sqrt{3}$

Exercise Set 8.5   **1.** The graph of a quadratic equation is called a parabola.   **3.** The axis of symmetry of a parabola is the line where, if the graph is folded, the two sides overlap.   **5.** $\left(-\dfrac{b}{2a}, \dfrac{4ac - b^2}{4a}\right)$ **7. a)** When $a > 0, f(x)$ will have a minimum since the graph opens upward.   **b)** When $a < 0, f(x)$ will have a maximum since the graph opens downward. **9.** Set $x = 0$ and solve for y

11. a) 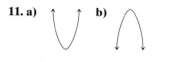 **b)**

13. Minimum value; the graph opens upward

15. a) Upward **b)** $(0, 15)$
c) $(-4, -1)$ **d)** $(-5, 0), (-3, 0)$
e)

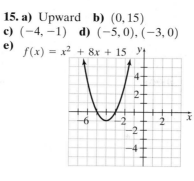

$f(x) = x^2 + 8x + 15$

17. a) Upward **b)** $(0, -3)$
c) $(-1, -4)$ **d)** $(-3, 0), (1, 0)$
e)

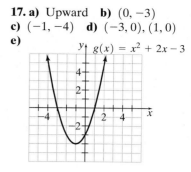

$g(x) = x^2 + 2x - 3$

19. a) Downward **b)** $(0, 8)$
c) $(-1, 9)$ **d)** $(-4, 0), (2, 0)$
e)

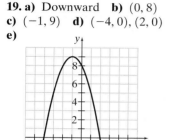

$f(x) = -x^2 - 2x + 8$

21. a) Downward **b)** $(0, -15)$
c) $(4, 1)$ **d)** $(3, 0), (5, 0)$
e)

$p(x) = -x^2 + 8x - 15$

23. a) Downward **b)** $(0, -5)$
c) $(2, -1)$ **d)** No x-intercepts
e)

$t(x) = -x^2 + 4x - 5$

25. a) Upward **b)** $(0, 4)$
c) $(2, 0)$ **d)** $(2, 0)$
e)

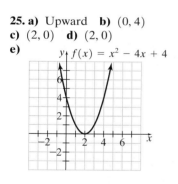

$f(x) = x^2 - 4x + 4$

27. a) Upward **b)** $(0, 2)$
c) $(0, 2)$ **d)** No x-intercepts
e)

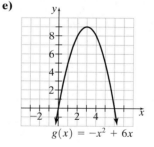

$r(x) = x^2 + 2$

29. a) Downward **b)** $(0, 0)$
c) $(3, 9)$ **d)** $(0, 0), (6, 0)$
e)

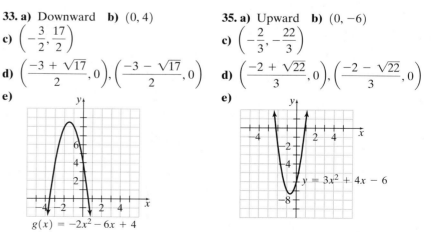

$g(x) = -x^2 + 6x$

31. a) Downward **b)** $(0, -8)$
c) $(1, -6)$ **d)** No x-intercepts
e)

$f(x) = -2x^2 + 4x - 8$

33. a) Downward **b)** $(0, 4)$
c) $\left(-\dfrac{3}{2}, \dfrac{17}{2} \right)$
d) $\left(\dfrac{-3 + \sqrt{17}}{2}, 0 \right), \left(\dfrac{-3 - \sqrt{17}}{2}, 0 \right)$
e)

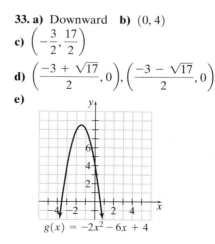

$g(x) = -2x^2 - 6x + 4$

35. a) Upward **b)** $(0, -6)$
c) $\left(-\dfrac{2}{3}, -\dfrac{22}{3} \right)$
d) $\left(\dfrac{-2 + \sqrt{22}}{3}, 0 \right), \left(\dfrac{-2 - \sqrt{22}}{3}, 0 \right)$
e)

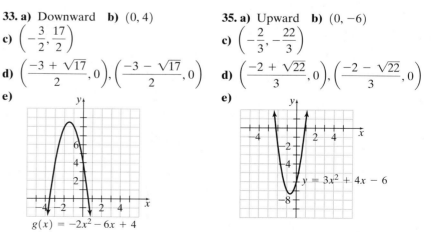

$y = 3x^2 + 4x - 6$

37. a) Upward **b)** $(0, -6)$
c) $\left(\dfrac{1}{4}, -\dfrac{49}{8}\right)$ **d)** $\left(-\dfrac{3}{2}, 0\right), (2, 0)$
e)

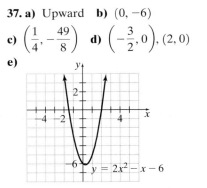

$y = 2x^2 - x - 6$

39. a) Downward **b)** $(0, -5)$
c) $\left(\dfrac{3}{2}, -\dfrac{11}{4}\right)$ **d)** No x-intercepts
e)

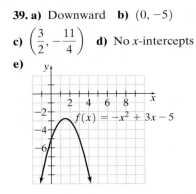

$f(x) = -x^2 + 3x - 5$

41.

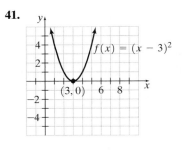

$f(x) = (x - 3)^2$
$(3, 0)$

43. $f(x) = (x + 1)^2$

$(-1, 0)$

45.

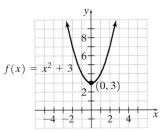

$f(x) = x^2 + 3$
$(0, 3)$

47.

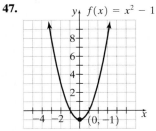

$f(x) = x^2 - 1$
$(0, -1)$

49.

$f(x) = (x - 2)^2 + 3$
$(2, 3)$

51. $f(x) = (x + 4)^2 + 4$

$(-4, 4)$

53.

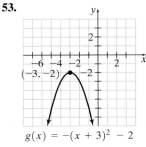

$(-3, -2)$
$g(x) = -(x + 3)^2 - 2$

55.

$(2, 2)$
$y = -2(x - 2)^2 + 2$

57.

$(-1, -3)$
$h(x) = -2(x + 1)^2 - 3$

59. a) $f(x) = (x - 3)^2 - 1$
b)

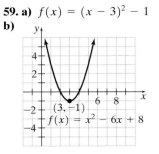

$(3, -1)$
$f(x) = x^2 - 6x + 8$

61. a) $f(x) = \left(x - \dfrac{1}{2}\right)^2 + \dfrac{3}{4}$
b)

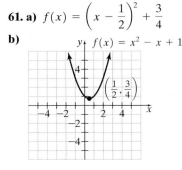

$f(x) = x^2 - x + 1$
$\left(\dfrac{1}{2}, \dfrac{3}{4}\right)$

63. a) $f(x) = -(x + 2)^2 - 2$
b)

$(-2, -2)$
$f(x) = -x^2 - 4x - 6$

65. a) $g(x) = (x - 2)^2 - 9$
b)

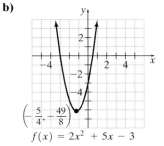

$(2, -9)$
$g(x) = x^2 - 4x - 5$

67. a) $f(x) = 2\left(x + \dfrac{5}{4}\right)^2 - \dfrac{49}{8}$
b)

$\left(-\dfrac{5}{4}, -\dfrac{49}{8}\right)$
$f(x) = 2x^2 + 5x - 3$

69. d) **71.** b) **73. a)** $x = 7$ **b)** $A = 121$ **75. a)** $x = 10.5$ **b)** $A = 240.25$

77. a) $n = 250$ **b)** $R = \$1250$ **79.** 2003 **81.** 4 units **83.** 3 units **85.** $f(x) = 2(x - 3)^2 - 2$ **87.** $f(x) = -4\left(x + \dfrac{3}{5}\right)^2 - \sqrt{2}$

89. a) The graphs will have the same x-intercepts but $f(x) = x^2 - 8x + 12$ will open upward and $g(x) = -x^2 + 8x - 12$ will open downward. **b)** Yes, both at $(6, 0)$ and $(2, 0)$ **c)** No; vertex of $f(x)$ at $(4, -4)$, vertex of $g(x)$ at $(4, 4)$

d)

91. a)

b) \$2 **c)** \$22 **d)** \$12 **e)** \$10,000

93. a) 100 **b)** \$3800

95. a) 40.425 meters **b)** 2.5 seconds **c)** $\approx$5.37 seconds

97. a) $\approx$\$40.17 **b)** 1992 **c)** $\approx$\$16

99. 225 square feet

101. -16; 4 and -4

103. 400; 20 and 20

105. a) \$142,400 **b)** 380

$I = -x^2 + 24x - 44, 0 \le x \le 24$

107. a) $f(t) = -16(t - 1.625)^2 + 45.25$ **b)** 45.25 feet, 1.625 seconds **c)** Same **109.** 200π square feet

110.
111. $(-2, 3, 2)$ **112.** -8 **113.** $\dfrac{x}{x + 6}$

Exercise Set 8.6

1. a) $x < 2$ or $x > 5$ **b)** $2 < x < 5$ **3.** Yes; $\ge$

5. Yes, -2 and 1 make the fraction 0; No, -1 makes the fraction undefined **7.**   **9.**

11.   **13.**   **15.**   **17.**

19.   **21.** $[-3, -1] \cup [2, \infty)$ **23.** $(-\infty, -4) \cup (-2, 3)$ **25.** $\left(-6, -\dfrac{5}{2}\right) \cup (2, \infty)$

27. $\left(-4, -\dfrac{5}{3}\right) \cup (3, \infty)$ **29.** $\left[\dfrac{8}{3}, \infty\right)$ **31.** $(-\infty, 0)$ **33.**

35.   **37.**   **39.**   **41.**

43. $\{x | x < -2 \text{ or } x > 4\}$ **45.** $\{x | -5 < x < 3\}$ **47.** $\{x | x \le -1 \text{ or } x > 1\}$ **49.** $\{a | -5 < a < 4\}$ **51.** $\{c | c < 4 \text{ or } c > 8\}$

53. $\{y | -6 < y \le -2\}$ **55.** $\left\{a \middle| a \le -2 \text{ or } a > \dfrac{1}{2}\right\}$ **57.** $\left\{x \middle| -\dfrac{4}{3} < x < \dfrac{1}{2}\right\}$ **59.** $\left\{x \middle| -\dfrac{5}{3} \le x < 2\right\}$ **61.** $(-\infty, -3) \cup (-1, 5)$

63. $(-3, 2) \cup (4, \infty)$ **65.** $(-2, 1] \cup [6, \infty)$ **67.** $(-\infty, -7) \cup [0, 3)$ **69.** $(-\infty, -4) \cup (1, 6]$ **71.** $\left[-\dfrac{5}{2}, 3\right] \cup (6, \infty)$

73.   **75.**   **77.**   **79.**   **81.**

83.   **85. a)** $(4, \infty)$; $y > 0$ in this interval **b)** $(-\infty, 2) \cup (2, 4)$; $y < 0$ in this interval

87. $x^2 + 2x - 8 > 0$ **89.** $\dfrac{x + 3}{x - 4} \ge 0$ **91.** All real numbers; for any value of x, the expression is ≥ 0. **93.** All real numbers except -1; for any value of x except -1, the expression is ≥ 0. **95.** No solution; the graph opens upward and has no x-intercepts, so it is always above the x-axis. **97.**

99. $x^2 - 3x > 0$; multiply factors containing boundary values. **101.** $x^2 < 0$; x^2 is always ≥ 0. **103.** $(-\infty, -3) \cup (-1, 1) \cup (3, \infty)$

105. $[-2, -1] \cup [2, \infty)$ **109.** 6 quarts **110.** -1 **111.** $3r + 3s - 6t$ **112.** $\dfrac{x - 1}{x + 1}$ **113.** $38 - 9i$

Chapter Review Exercises

1. $3 \pm 2\sqrt{6}$ **2.** $\dfrac{-5 \pm 2\sqrt{15}}{2}$ **3.** $-\dfrac{1}{3}, 1$ **4.** $\dfrac{5}{4}, -\dfrac{3}{4}$ **5.** $3, 4$ **6.** $7, -3$

7. $-1 \pm \sqrt{6}$ **8.** $-3 \pm \sqrt{19}$ **9.** $1 \pm 3i$ **10.** $2 \pm 2i\sqrt{7}$ **11. a)** $32 = (x+1)(x+5)$ **b)** 3 **12. a)** $63 = (x+2)(x+4)$
b) 5 **13.** $8, 9$ **14.** ≈ 16.90 ft by ≈ 16.90 ft **15.** Two real solutions **16.** No real solution **17.** One real solution

18. No real solution **19.** One real solution **20.** Two real solutions **21.** $0, -\dfrac{7}{3}$ **22.** $5, 6$ **23.** $8, -1$ **24.** $0, \dfrac{9}{4}$ **25.** $\dfrac{3}{2}, -\dfrac{5}{3}$

26. $\dfrac{1}{4}, -3$ **27.** $3 \pm \sqrt{2}$ **28.** $-2 \pm 2\sqrt{3}$ **29.** $\dfrac{-2 \pm \sqrt{10}}{2}$ **30.** $\dfrac{3 \pm \sqrt{33}}{3}$ **31.** $\dfrac{1 \pm i\sqrt{119}}{2}$ **32.** $\dfrac{-5.7 \pm \sqrt{43.53}}{2.4}$ **33.** $\dfrac{5}{2}, -\dfrac{5}{3}$

34. $\dfrac{1}{4}, -\dfrac{3}{2}$ **35.** $10, -6$ **36.** $\dfrac{2}{3}, -\dfrac{3}{2}$ **37.** $\dfrac{7 \pm \sqrt{89}}{10}$ **38.** $\dfrac{3 \pm 3\sqrt{3}}{2}$ **39.** $f(x) = x^2 - 3x - 4$ **40.** $f(x) = 3x^2 + 4x - 4$

41. $f(x) = x^2 - 7$ **42.** $f(x) = x^2 - 6x + 13$ **43.** 8 feet by 11 feet **44.** $\sqrt{128} \approx 11.31$ **45.** 5% **46.** $7, 9$ **47.** 8 inches by

12 inches **48.** \$475 **49. a)** 640 feet **b)** 7 seconds **50. a)** 68 feet **b)** $\dfrac{4 + \sqrt{416}}{8} \approx 3.05$ seconds **51. a)** 40 milliliters

b) $150°$ C **52.** larger ≈ 23.51 hours; smaller ≈ 24.51 hours **53.** 50 mph **54.** 1.6 mph **55.** $l \approx 9.6$ units, $w \approx 7.6$ units

56. 30 **57.** $a = \sqrt{c^2 - b^2}$ **58.** $t = \sqrt{\dfrac{c - h}{4.9}}$ **59.** $v_y = \sqrt{v^2 - v_x^2}$ **60.** $v_2 = \sqrt{v_1^2 + 2ad}$ **61.** $\pm 2, \pm 3$ **62.** $\pm 4, \pm \sqrt{3}$

63. $\pm 2\sqrt{2}, \pm i\sqrt{3}$ **64.** $\dfrac{3}{2}, -\dfrac{2}{5}$ **65.** $\dfrac{1}{16}$ **66.** $\dfrac{27}{8}, 8$ **67.** $4, \dfrac{13}{8}$ **68.** $-\dfrac{6}{5}, -\dfrac{7}{2}$ **69.** $(\pm 2, 0), (\pm 5, 0)$ **70.** $\left(\dfrac{4}{25}, 0\right)$ **71.** None

72. $(3 \pm \sqrt{17}, 0), (3 \pm \sqrt{6}, 0)$

73. a) Upward **b)** $(0, 0)$
c) $\left(-\dfrac{5}{2}, -\dfrac{25}{4}\right)$ **d)** $(0, 0), (-5, 0)$
e)

$\left(-\dfrac{5}{2}, -\dfrac{25}{4}\right)$ $y = x^2 + 5x$

74. a) Upward **b)** $(0, -8)$
c) $(1, -9)$ **d)** $(-2, 0), (4, 0)$
e)

$y = x^2 - 2x - 8$ $(1, -9)$

75. a) Downward **b)** $(0, -9)$
c) $(0, -9)$ **d)** No x-intercepts
e)

$(0, -9)$ $y = -x^2 - 9$

76. a) Downward **b)** $(0, 15)$
c) $\left(-\dfrac{1}{4}, \dfrac{121}{8}\right)$ **d)** $(-3, 0), \left(\dfrac{5}{2}, 0\right)$
e)

$\left(-\dfrac{1}{4}, \dfrac{121}{8}\right)$ $y = -2x^2 - x + 15$

77. a) \$11 **b)** \$9100
78. a) 2.5 seconds **b)** 160 feet
79.

$f(x) = (x - 3)^2$ $(3, 0)$

80.

$(-2, -3)$ $f(x) = -(x + 2)^2 - 3$

81.

$(-4, -1)$ $g(x) = -2(x + 4)^2 - 1$

82.

$(1, 3)$ $h(x) = \dfrac{1}{2}(x - 1)^2 + 3$

83.

-6 -1

84.

-5 2

85.

$\dfrac{11 - \sqrt{41}}{2}$ $\dfrac{11 + \sqrt{41}}{2}$

86.

-4 $\dfrac{4}{3}$

87.

$-\dfrac{3}{2}$ $\dfrac{3}{2}$

88. $-\sqrt{5}$ $\sqrt{5}$ **89.** $\{x|x < -1 \text{ or } x > 3\}$ **90.** $\{x|-2 < x \le 5\}$ **91.** $\{x|x < -3 \text{ or } x \ge 2\}$

92. $\left\{x\left|-\dfrac{5}{3} < x < 6\right.\right\}$ **93.** $\{x|-3 < x < -1 \text{ or } x > 2\}$ **94.** $\{x|x \le 0 \text{ or } 3 \le x \le 5\}$ **95.** $\left[-\dfrac{4}{3}, 1\right] \cup [3, \infty)$

96. $(-\infty, -5) \cup (-2, 0)$ **97.** $(-2, 0) \cup (4, \infty)$ **98.** $(-\infty, -3) \cup (2, 5)$ **99.** $(-2, 3] \cup (5, \infty)$ **100.** $(-\infty, -3) \cup [0, 5]$

101. -7 -4 **102.** -2 2 **103.** $\dfrac{5}{3}$ $\dfrac{23}{10}$

Chapter Practice Test **1.** $3, -5$ **2.** $2 \pm i\sqrt{13}$ **3.** $7, -2$ **4.** $-4 \pm \sqrt{11}$ **5.** $\dfrac{2}{3}, -1$ **6.** $\dfrac{-7 \pm \sqrt{29}}{2}$

7. $f(x) = 5x^2 - 18x - 8$ **8.** $v = \sqrt{\dfrac{2K}{m}}$ **9. a)** \$138,000 **b)** ≈ 2712.57 square feet **10.** 50 mph **11.** $\pm \dfrac{\sqrt{10}}{2}, \pm i\sqrt{10}$

12. $\dfrac{343}{27}, -216$ **13.** $\left(\dfrac{25}{16}, 0\right)$ **14.** **15.** **16.** Two real solutions

$y \; f(x) = (x - 3)^2 + 2$

$h(x) = -\dfrac{1}{2}(x - 2)^2 - 2$

17. a) Upward **b)** $(0, -8)$ **c)** $(-1, -9)$ **d)** $(-4, 0), (2, 0)$ **e)** <graph $y = x^2 + 2x - 8$>

18. $f(x) = 2x^2 + 11x - 6$ **19.** <number line> -6 7

20. <number line> -3 -1 4

21. a) $\left[-\dfrac{5}{2}, -2\right)$ **b)** $\left\{x\left|-\dfrac{5}{2} \le x < -2\right.\right\}$

22. $w = 4$ feet, $l = 12$ feet **23.** 5 seconds

24. a) 20 **b)** \$500 **25.** 20

CHAPTER 9

Exercise Set 9.1 **1.** To find $(f \circ g)(x)$, substitute $g(x)$ for x in $f(x)$. **3. a)** Each y has a unique x. **b)** Use the horizontal line test. **5. a)** Yes; each first coordinate is paired with only one second coordinate. **b)** Yes; each second coordinate is paired with only one first coordinate. **c)** $\{(5, 3), (2, 4), (3, -1), (-2, 0)\}$; reverse each ordered pair.
7. The domain of f is the range of f^{-1} and the range of f is the domain of f^{-1}. **9. a)** $x^2 + 10x + 26$ **b)** 82 **c)** $x^2 + 6$
d) 22 **11. a)** $x^2 + x - 1$ **b)** 19 **c)** $x^2 + 7x + 8$ **d)** 52 **13. a)** $\dfrac{1}{2x + 3}$ **b)** $\dfrac{1}{11}$ **c)** $\dfrac{2}{x} + 3$ **d)** $3\dfrac{1}{2}$ **15. a)** $\dfrac{2}{x^2 + 1}$
b) $\dfrac{2}{17}$ **c)** $\dfrac{4}{x^2} + 1$ **d)** $1\dfrac{1}{4}$ **17. a)** $x^4 + 10x^2 + 26$ **b)** 442 **c)** $x^4 + 2x^2 + 6$ **d)** 294 **19. a)** $\sqrt{x + 5} - 4$ **b)** -1
c) $\sqrt{x + 1}$ **d)** $\sqrt{5}$ **21.** No **23.** Yes **25.** Yes **27.** No **29.** Yes **31.** No **33.** No **35.** Yes **37.** No **39.** Yes
41. Yes **43.** $f(x)$: Domain: $\{-2, -1, 2, 4, 9\}$; Range: $\{0, 3, 4, 6, 7\}$; $f^{-1}(x)$: Domain: $\{0, 3, 4, 6, 7\}$; Range: $\{-2, -1, 2, 4, 9\}$
45. $f(x)$: Domain: $\{-1, 1, 2, 4\}$; Range: $\{-3, -1, 0, 2\}$; $f^{-1}(x)$: Domain: $\{-3, -1, 0, 2\}$; Range: $\{-1, 1, 2, 4\}$ **47.** $f(x)$:
Domain: $\{x|x \ge 2\}$; Range: $\{y|y \ge 0\}$; $f^{-1}(x)$: Domain: $\{x|x \ge 0\}$; Range: $\{y|y \ge 2\}$ **49. a)** Yes **b)** $f^{-1}(x) = x + 3$
51. a) Yes **b)** $h^{-1}(x) = \dfrac{x}{5}$ **53. a)** No **55. a)** No **57. a)** Yes **b)** $g^{-1}(x) = \dfrac{1}{x}$ **59. a)** No **61. a)** Yes
b) $g^{-1}(x) = \sqrt[3]{x + 5}$ **63. a)** Yes **b)** $g^{-1}(x) = x^2 - 2, x \ge 0$ **65. a)** Yes **b)** $h^{-1}(x) = \sqrt{x + 4}, x \ge -4$

67. a) $f^{-1}(x) = \dfrac{x-8}{2}$

b)

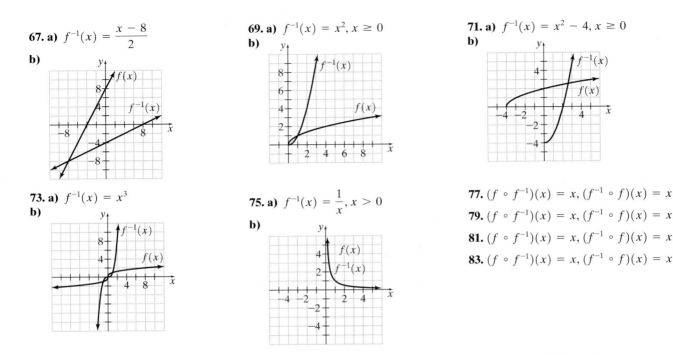

69. a) $f^{-1}(x) = x^2, x \geq 0$

b)

71. a) $f^{-1}(x) = x^2 - 4, x \geq 0$

b)

73. a) $f^{-1}(x) = x^3$

b)

75. a) $f^{-1}(x) = \dfrac{1}{x}, x > 0$

b)

77. $(f \circ f^{-1})(x) = x, (f^{-1} \circ f)(x) = x$

79. $(f \circ f^{-1})(x) = x, (f^{-1} \circ f)(x) = x$

81. $(f \circ f^{-1})(x) = x, (f^{-1} \circ f)(x) = x$

83. $(f \circ f^{-1})(x) = x, (f^{-1} \circ f)(x) = x$

85. No, composition of functions is not commutative. Let $f(x) = x^2$ and $g(x) = x + 1$. Then $(f \circ g)(x) = x^2 + 2x + 1$, while $(g \circ f)(x) = x^2 + 1$. **87. a)** $(f \circ g)(x) = x; (g \circ f)(x) = x$ **b)** The Domain is $\mathbb{R}$ for all of them. **89.** The range of $f^{-1}(x)$ is the domain of $f(x)$. **91.** $f^{-1}(x) = \dfrac{x}{3}; x$ is feet and $f^{-1}(x)$ is yards **93.** $f^{-1}(x) = \dfrac{9}{5}x + 32$. **95.** Yes

97. Yes **99. a)** 6 feet **b)** $36\pi \approx 113.10$ square feet **c)** $A(t) = 4\pi t^2$ **d)** $36\pi \approx 113.10$ square feet **e)** Answers should agree **102.** $\dfrac{81}{16}$ **103.** $2x + 3y = 10$ **104.** $\dfrac{12 - 8x}{x^3}$ **105.** $p = \dfrac{fq}{q - f}$ **106.** $-1 \pm \sqrt{7}$

Exercise Set 9.2 **1.** Exponential functions are functions of the form $f(x) = a^x, a > 0, a \neq 1$.

3. a) As x increases, y decreases. **b)** No, $\left(\dfrac{1}{2}\right)^x$ can never be 0. **c)** No, $\left(\dfrac{1}{2}\right)^x$ can never be negative. **5. a)** Same; $(0, 1)$.

b) $y = 3^x$ will be steeper than $y = 2^x$ for $x > 0$.

7.   **9.**   **11.**   **13.**

15.   **17.**   **19.**   **21.**

23. $y = 2^x - 1$

25. $y = 3^x - 1$

27. a) It is a horizontal line through $y = 1$.   **b)** Yes
c) No, not a one-to-one function
29. $y = a^x - k$ is $y = a^x$ lowered k units.
31. The graph of $y = a^{x+2}$ is the graph of $y = a^x$ shifted 2 units to the left.

33. a)

b) 200 years   **c)** 40 years
35. 256   **37.** 36
39. $\approx\$6344.93$
41. ≈10.6 grams
43. a) 5 grams.   **b)** $\approx7.28 \times 10^{-11}$ grams
45. a) 2400   **b)** $\approx12{,}383$
47. $\approx\$10{,}850.92$
49. a) Answers will vary.   **b)** $\approx89{,}758.6$ gallons
51. ≈8.83 kilometers
53. a) $\approx\$201.36$ **b)** $\approx\$31.36$

55. a)

b) ≈6.26
57. a) \$16,384   **b)** \$524,288   **c)** 2^{n-1} **d)** $\$2^{29} = \$536{,}870{,}912$
e) $2^0 + 2^1 + 2^2 + \cdots + 2^{29}$
59. a) $-6.2x^6y^2 + 9.2x^5y^2 + 2.3x^4y$ **b)** 8   **c)** -6.2

60. a) $x^3 + x^2 - 2x + 12$ **61.** $|a - 4|$ **62.** $\dfrac{2xy\sqrt[4]{xy^2z^3}}{z}$

$-10, 10, 1, -10, 10, 1$

Exercise Set 9.3

1. a) $a > 0$ and $a \neq 1$ **b)** $\{x \mid x > 0\}$ **c)** $\mathbb{R}$

3. $\left(\dfrac{1}{27}, -3\right)\left(\dfrac{1}{9}, -2\right), \left(\dfrac{1}{3}, -1\right)$ $(1, 0), (3, 1), (9, 2),$ and $(27, 3)$; the functions $f(x) = a^x$ and $g(x) = \log_a x$ are inverses.

5. The functions $y = a^x$ and $y = \log_a x$ for $a \neq 1$ are inverses of each other, thus the graphs are symmetric with respect to the line $y = x$. For each ordered pair (x, y) on the graph of $y = a^x$, the ordered pair (y, x) is on the graph of $y = \log_a x$.

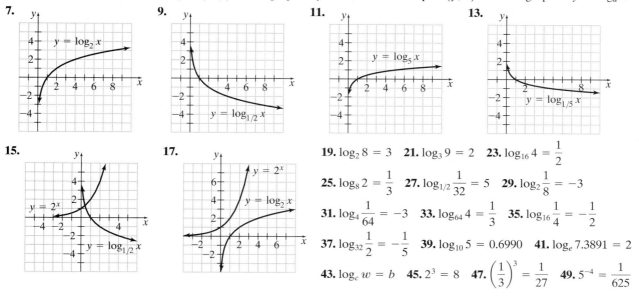

7. $y = \log_2 x$

9. $y = \log_{1/2} x$

11. $y = \log_5 x$

13. $y = \log_{1/5} x$

15. $y = 2^x$ $y = \log_{1/2} x$

17. $y = 2^x$ $y = \log_2 x$

19. $\log_2 8 = 3$ **21.** $\log_3 9 = 2$ **23.** $\log_{16} 4 = \dfrac{1}{2}$

25. $\log_8 2 = \dfrac{1}{3}$ **27.** $\log_{1/2} \dfrac{1}{32} = 5$ **29.** $\log_2 \dfrac{1}{8} = -3$

31. $\log_4 \dfrac{1}{64} = -3$ **33.** $\log_{64} 4 = \dfrac{1}{3}$ **35.** $\log_{16} \dfrac{1}{4} = -\dfrac{1}{2}$

37. $\log_{32} \dfrac{1}{2} = -\dfrac{1}{5}$ **39.** $\log_{10} 5 = 0.6990$ **41.** $\log_e 7.3891 = 2$

43. $\log_c w = b$ **45.** $2^3 = 8$ **47.** $\left(\dfrac{1}{3}\right)^3 = \dfrac{1}{27}$ **49.** $5^{-4} = \dfrac{1}{625}$

51. $81^{1/2} = 9$ **53.** $8^{-2} = \dfrac{1}{64}$ **55.** $27^{-1/3} = \dfrac{1}{3}$ **57.** $6^3 = 216$ **59.** $10^{0.9031} = 8$ **61.** $e^{1.8749} = 6.52$ **63.** $r^{-a} = c$ **65.** 2   **67.** 3

69. 32   **71.** -3 **73.** $\dfrac{1}{4}$ **75.** 3   **77.** 2   **79.** 1   **81.** -1 **83.** 4   **85.** 3   **87.** -4 **89.** -2 **91.** 0   **93.** 1   **95.** 5   **97.** $f^{-1}(x) = \log_4 x$

99. 2 and 3, since 425 lies between $10^2 = 100$ and $10^3 = 1000$.   **101.** 3 and 4, since 62 lies between $3^3 = 27$ and $3^4 = 81$.
103. 2^x; Note that for $x = 10, 2^x = 1024$ while $\log_{10} x = 1$.   **105.** 5   **107.** 3   **109.** 8   **111.** 9   **113.** 10,000,000   **115.** 10,000
117.

119. $3x(x + 2)(x - 6)$
120. $(x - 2)(x + 2)(x^2 + 4)$
121. $4(2x + 3)(5x - 1)$
122. $(3rs - 1)(2rs + 1)$

Exercise Set 9.4

1. Answers will vary.   **3.** Answers will vary.   **5.** Yes, it is an expansion of property 1.

7. $\log_4 3 + \log_4 10$ **9.** $\log_8 7 + \log_8 (x + 3)$ **11.** $\log_2 27 - \log_2 5$ **13.** $\dfrac{1}{2}\log_{10} x - \log_{10}(x - 9)$

15. $7 \log_6 x$ **17.** $\log_9 10 + 6 \log_9 4$ **19.** $\dfrac{3}{2}\log_4 a - \dfrac{1}{2}\log_4 (a + 2)$ **21.** $6 \log_3 d - 4 \log_3 (a - 5)$ **23.** $\log_8 (y + 2) - 2 \log_8 y$

25. $\log_{10} 2 + \log_{10} m - \log_{10} 3 - \log_{10} n$ **27.** $\log_5 6$ **29.** $\log_2 \dfrac{9}{5}$ **31.** $\log_4 32$ **33.** $\log_{10} x(x + 3)$ **35.** $\log_9 \dfrac{z^2}{z - 2}$

37. $\log_5 \left(\dfrac{p}{3}\right)^2$ **39.** $\log_2 \dfrac{n(n + 4)}{n - 3}$ **41.** $\log_5 \sqrt{\dfrac{x - 4}{x}}$ **43.** $\log_9 \dfrac{25\sqrt[3]{r - 6}}{\sqrt{r}}$ **45.** $\log_6 \dfrac{81}{(x + 3)^2 x^4}$ **47.** 1   **49.** 0.3980

51. 1.3980   **53.** 10   **55.** 5   **57.** 3   **59.** 25   **61.** Yes   **63.** $\log_a \dfrac{x}{y} = \log_a \left(x \cdot \dfrac{1}{y}\right) = \log_a x + \log_a \dfrac{1}{y}$ **65.** $\log_a (x - 2)$

67. Yes, $\log_a (x^2 + 8x + 16) = \log_a (x + 4)^2 = 2 \log_a (x + 4)$ **69.** 0.8640   **71.** 0.1080   **73.** 0.7000   **75.** No, there is no

relationship between $\log_{10} (x + y)$ and $\log_{10} xy$ or $\log_{10} \left(\dfrac{x}{y}\right)$ **77.** $\dfrac{1}{4}\log_2 x + \dfrac{1}{4}\log_2 y + \dfrac{1}{3}\log_2 a - \dfrac{1}{5}\log_2 (a - b)$

79. Answers will vary.   **82. a)** $\{x | x > 40\}$ **b)** $(40, \infty)$ **83. a)** $a^2 - 4c^2$ **b)** $(a + 2c)(a - 2c)$ **84.** 3   **85.** $-18 - i$ **86.** 36

Exercise Set 9.5

1. Common logarithms are logarithms with base 10.   **3.** Antilogarithms are numbers
obtained by taking 10 to the power of the logarithm.   **5.** 1.6532   **7.** 4.2833   **9.** -1.2125 **11.** 2.0000   **13.** 0.5740
15. -3.3261 **17.** 2.63   **19.** 42,500   **21.** 0.0196   **23.** 1.00   **25.** 579   **27.** 0.701   **29.** 100   **31.** 0.00789   **33.** 13,800   **35.** 0.0871
37. 0.239   **39.** 0.404   **41.** 3.5514   **43.** -1.1385 **45.** 2.0086   **47.** -2.8928 **49.** 823   **51.** 0.695   **53.** 0.0331   **55.** 22.4
57. 0   **59.** -1 **61.** -2 **63.** -3 **65.** 7   **67.** 7   **69.** 31.2   **71.** 47   **73.** No; $10^2 = 100$ and since $462 > 100$, log 462 must be
greater than 2.   **75.** No; $10^0 = 1$ and $10^{-1} = 0.1$ and since $1 > 0.163 > 0.1$, log 0.163 must be between 0 and -1.

77. No; $\log \dfrac{y}{3x} = \log y - \log 3 - \log x$ **79.** 2.0970   **81.** Not possible   **83.** -1.3979 **85.** 2570   **87.** 2,340,000

89. a) ≈ 31.62 kilometers   **b)** ≈ 0.50 kilometer   **c)** ≈ 14.68 **91. a)** $\approx 72\%$ **b)** $\approx 15\%$ **93.** $\approx 39,800$ times more intense

95. a) $\approx 6.31 \times 10^{20}$ **b)** ≈ 2.19 **97.** ≈ 6.2 **99.** $I = $ antilog R **101.** $t = $ antilog $\left(\dfrac{85 - R}{41.9}\right) - 1$ **104.** 50 mph, 55 mph

105. $(2, -3)$ **106.** $0, -4, 3$ **107.** $|3x^2 - y|$ **108.** $(-\infty, -4] \cup [2, 3]$

Exercise Set 9.6

1. $c = d$ **3.** Check for extraneous solutions.   **5.** $\log (-2)$ is not a real number.   **7.** 3   **9.** 5
11. $\dfrac{1}{2}$ **13.** 2   **15.** $-\dfrac{1}{3}$ **17.** 5   **19.** 2   **21.** 3   **23.** 2.01   **25.** 3.16   **27.** 5.59   **29.** 6.34   **31.** 4   **33.** 5   **35.** $\dfrac{1}{8}$ **37.** 10   **39.** -1

41. $-7, 3$ **43.** $0, -8$ **45.** 97   **47.** $\dfrac{4}{5}$ **49.** $\dfrac{3}{2}$ **51.** 4   **53.** 2   **55.** 1.05   **57.** 20   **59.** 5   **61.** 3   **63.** 2   **65.** 3   **67.** 9

69. ≈ 3.47 hours   **71.** ≈ 7.21 years   **73.** ≈ 17.07 years   **75. a)** ≈ 7.95 **b)** ≈ 5.88 **77.** $\approx \$7112.09$ **79.** ≈ 19.36
81. a) 1,000,000,000,000   **b)** 100,000 times greater   **83.** 8   **85.** $x = 1$ and $x = 2$ **87.** $(3, 1)$ **89.** $(54, 46)$ **91.** 2.8
93. No solution   **95.** The box is greater by ≈ 7.73 cubic feet   **96.** -9

97.

98. $\dfrac{x\sqrt{y}-y\sqrt{x}}{x-y}$ **99.** $c=\sqrt{\dfrac{E}{m}}$ **100.** $f(x)=2(x-3)^2-2$

Exercise Set 9.7

1. a) e **b)** ≈2.7183 **3.** $\{x|x>0\}$ **5.** $\log_a x=\dfrac{\log_b x}{\log_b a}$ **7.** x **9.** e^x **11.** $k<0$ **13.** 4.1271

15. -0.2070 **17.** 4.95 **19.** 0.0898 **21.** 0.972 **23.** 3.6640 **25.** 4.3923 **27.** 1.7297 **29.** 2.5743 **31.** 2.6857 **33.** 2.0206

35. -0.4784 **37.** 4 **39.** 1 **41.** 4 **43.** 3 **45.** $P=4757.5673$ **47.** $P_0=224.0845$ **49.** $t=0.7847$ **51.** $k=0.2310$

53. $k=-0.2888$ **55.** $A=4719.7672$ **57.** $V_0=\dfrac{V}{e^{kt}}$ **59.** $t=\dfrac{\ln P-\ln 150}{4}$ **61.** $k=\dfrac{\ln A-\ln A_0}{t}$ **63.** $y=xe^{2.3}$

65. $y=(x+3)e^6$ **67.** ≈2.5000; find $\ln 12.183$ **69. a)** $\approx\$5867.55$ **b)** ≈8.66 years **71.** ≈39.98 grams **73. a)** $\approx86.47\%$

b) ≈34.66 days **75. a)** ≈5.15 feet per second **b)** ≈5.96 feet per second **c)** $\approx646{,}000$ **77.** $\approx\$300{,}977{,}000{,}000{,}000$

79. a) ≈6.9 billion **b)** ≈53 years **81. a)** ≈32.43 inches **b)** ≈35.46 inches **83. a)** ≈6626.62 years **b)** ≈5752.26 years

85. $\approx\$6791.91$ **87. a)** Strontium 90, since it has a higher decay rate **b)** $\approx31.66\%$ of original amount **89.** Answers will vary.

91. 7.286 **93.** $-1.507, 16.659$ **95.** $v_0=\dfrac{e^{xk}-1}{kt}$ **97.** $i=Ie^{-t/RC}$ **98. a)** -1 **b)** $\dfrac{11}{40}$ or 0.275 **99.** 240 children, 410 adults

100. $-9x^2y^3+12x^2y^2-3xy^2+4xy$ **101.** $-12, 12$ **102.** $x+x^2$

Review Exercises

1. $4x^2-26x+44$ **2.** 8 **3.** $2x^2-6x+3$ **4.** 39 **5.** $6\sqrt{x-3}+1, x\ge3$

6. $\sqrt{6x-2}, x\ge\dfrac{1}{3}$ **7.** One-to-one **8.** Not one-to-one **9.** One-to-one **10.** Not one-to-one **11.** One-to-one

12. Not one-to-one **13.** $f(x)$: Domain: $\{-4, 0, 5, 6\}$; Range: $\{-3, 2, 3, 7\}$; $f^{-1}(x)$: Domain: $\{-3, 2, 3, 7\}$; Range: $\{-4, 0, 5, 6\}$

14. $f(x)$: Domain: $\{x|x\ge0\}$; Range: $\{y|y\ge2\}$; $f^{-1}(x)$: Domain: $\{x|x\ge2\}$; Range: $\{y|y\ge0\}$

15. $f^{-1}(x)=\dfrac{x+2}{4}$,

16. $f^{-1}(x)=x^3+1$;

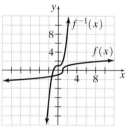

17. $f^{-1}(x)=\dfrac{x}{36}$, x is inches, $f^{-1}(x)$ is yards. **18.** $f^{-1}(x)=\dfrac{x}{4}$, x is quarts, $f^{-1}(x)$ is gallons

19.   **20.**

21. a) 30.23 million **b)** 62.73 million **c)** 130.16 million

22. $\log_7 49=2$ **23.** $\log_{81} 3=\dfrac{1}{4}$ **24.** $\log_5 \dfrac{1}{25}=-2$ **25.** $2^4=16$ **26.** $\left(\dfrac{1}{3}\right)^2=\dfrac{1}{9}$ **27.** $6^{-2}=\dfrac{1}{36}$ **28.** $4^3=x; 64$

29. $a^3=8; 2$ **30.** $\left(\dfrac{1}{4}\right)^{-3}=x; 64$

31.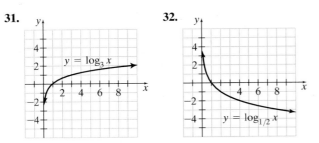

32.

33. $3 \log_5 17$ **34.** $\dfrac{1}{2} \log_3 (x - 5)$

35. $\log 6 + \log (a + 1) - \log b$

36. $4 \log x - \log 9 - 5 \log (2x + 3)$ **37.** $\log \dfrac{x^2}{(x + 1)^3}$

38. $\log \dfrac{(2x)^3}{y}$ **39.** $\ln \dfrac{\sqrt{\dfrac{x}{x + 2}}}{2}$ **40.** $\ln \dfrac{x^3 \sqrt{x + 1}}{(x + 4)^3}$ **41.** 9

42. 5   **43.** 14   **44.** 2   **45.** 2.8825   **46.** -3.5720 **47.** 1440   **48.** 0.00226   **49.** 214   **50.** 0.0594   **51.** 5   **52.** 4   **53.** 28.8   **54.** 9.4

55. 3   **56.** $-\dfrac{1}{2}$ **57.** 2   **58.** 5   **59.** 2.307   **60.** 6.877   **61.** 1.353   **62.** 2.240   **63.** 16   **64.** 2   **65.** 1   **66.** 3   **67.** $t \approx 1.155$

68. $A_0 \approx 352.542$ **69.** $t = \dfrac{\ln A - \ln A_0}{k}$ **70.** $k = \dfrac{\ln 0.25}{t}$ **71.** $y = xe^4$ **72.** $y = 3x + 14$ **73.** 7.6147   **74.** 3.9177

75. $\approx\$25,723.07$ **76.** ≈ 17.3 years   **77. a)** ≈ 92.88 minutes   **b)** ≈ 118.14 minutes   **78.** ≈ 10.32 pounds per square inch
79. a) 72   **b)** ≈ 63.4 **c)** ≈ 4 months

Chapter Practice Test   **1. a)** Yes   **b)** $\{(2, 4), (8, -3), (3, -1), (7, 5)\}$ **2. a)** $x^2 + 4x + 1$ **b)** 46

3. a) $\sqrt{x^2 + 2}$ **b)** $3\sqrt{2}$ **4. a)** $f^{-1}(x) = -\dfrac{1}{3}(x + 5)$ **b)**

5. a) $f^{-1}(x) = x^2 + 1, x \geq 0$ **b)**

6. $\{x \mid x > 0\}$ **7.** -4 **8.**

9.

10. $\log_2 \dfrac{1}{32} = -5$ **11.** $5^4 = 625$ **12.** $2^4 = x + 1, 15$ **13.** $64^y = 16, \dfrac{2}{3}$

14. $3 \log_2 x + \log_2 (x - 4) - \log_2 (x + 2)$ **15.** $\log_6 \dfrac{(x - 4)^5 (x + 3)^2}{\sqrt{x}}$

16. 4   **17. a)** 3.6646   **a)** -2.6708 **18.** ≈ 5.69 **19.** 3   **20.** $\dfrac{17}{5}$ **21.** 33.7844   **22.** 2.0588

23. 30.5430   **24.** $\approx\$6349.06$ **25.** ≈ 3364.86 years old

CHAPTER 10

Exercise Set 10.1   **1.** No, a function will only have a local maximum or local minimum if the function has an interval where it changes from increasing to decreasing. If a function never increases or if it never decreases, it would never have an interval where it changes from increasing to decreasing. A linear function is an example of a function that would not have a local maximum or local minimum.   **3.** Yes, since $f(x) = x^2 - 4$ is a polynomial function, it is continuous.

5. No, since the two pieces give different values at $x = -5$, the function is not continuous. **7.** Discontinuous
9. Continuous **11.** Continuous **13.** Discontinuous **15.** Continuous **17.** Continuous **19.** Continuous **21.** Discontinuous;
Point of discontinuity at $x = -5$. **23.** Discontinuous; Points of discontinuity occur at $\ldots, -1, -\frac{1}{2}, 0, \frac{1}{2}, 1, \ldots$ **25.** Continuous
27. Discontinuous; Point of discontinuity at $x = 3$ **29.** Continuous **31.**

33.

35.

37.

39.

41.

43.

45.

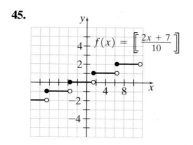

47. $f(x) = \left[\!\left[\dfrac{x}{1.25}\right]\!\right]$ **49.** $f(x) = \left[\!\left[\dfrac{x}{3.99}\right]\!\right]$ **51.** $f(x) = \begin{cases} 5x & \text{if } x \le 3 \\ 15 + 3(x-3) & \text{if } x > 3 \end{cases}$.
53. Since $f(-x) = f(x)$, $f(x) = x^2 - 15$ is an even function. **55.** Since $f(-x) \ne f(x)$
and $f(-x) \ne -f(x)$, $f(x) = 3x^3 + x^2 - 10$ is neither even nor odd. **57.** Since
$g(-x) = g(x)$, $g(x) = 3x^6 - 15x^2 + 4$ is an even function. **59.** Since $g(-x) = g(x)$,
$g(x) = e^{x^2} + 5$ is an even function. **61.** Since $h(-x) \ne h(x)$ and $h(-x) \ne -h(x)$,
$h(x) = x^4 - x^3 + x^2$ is neither even nor odd.

63.

65.

69.

70.

71. a) The degree of the polynomial is 5.   **b)** The leading coefficient is 4.
72. $\{x|x \neq 7 \text{ and } x \neq -3\}$

Exercise Set 10.2   1. a) The constant term is negative because the function crosses the y-axis below the origin.   b) The leading coefficient is negative because the value of the function is negative for large positive values of x.   c) The degree of the polynomial function is even because the end behavior for large negative values of x is below the x-axis and the end behavior for large positive values of x is also below the x-axis.   d) Yes, the degree of the polynomial function can be 6 because the function has even degree and has fewer than 5 turning points.   e) No, the degree of the polynomial function cannot be 7 because the degree of the polynomial is even.   f) The polynomial has 4 real zeros because it crosses the x-axis 4 times.   3. a) $f(x) = x^{10}$ is an even-degree polynomial function with a positive leading coefficient thus it is above the x-axis for large positive values of x and above the x-axis for large negative values of x.   b) $f(x) = x^{11}$ is an odd-degree polynomial function with a positive leading coefficient thus it is above the x-axis for large positive values of x and below the x-axis for large negative values of x.   5. b; $P(x) = -2x^3 + 5x - 8$ is a third-degree polynomial function with a negative leading coefficient. Thus, large positive values of x will be below the x-axis and large negative values of x will be above the x-axis.   7. a; $P(x) = 3x^5 + 5x^3 - 25x^2 + 20$ is a fifth-degree polynomial function with a positive leading coefficient. Thus, large positive values of x will be above the x-axis and large negative values of x will be below the x-axis.   9. a) Above the x-axis   b) Below the x-axis   11. a) Below the x-axis   b) Above the x-axis   13. a) Below the x-axis   b) Below the x-axis   15. a) Below the x-axis   b) Below the x-axis   17. a) Below the x-axis   b) Above the x-axis   19. 5, -3 21. 0, -5, -1 23. $-5, 5$ 25. $-7, 2$ of multiplicity 3   27. $-3, -\dfrac{1}{2}, \dfrac{1}{3}$ 29. -2 of multiplicity 3, 4 of multiplicity 2   31. 0, -1 of multiplicity 2, 1 of multiplicity 3, 6 of multiplicity 4   33. 4   35. 0   37. 5   39. 2   41. 1   43. 4

45.   **47.**   **49.**   **51.**

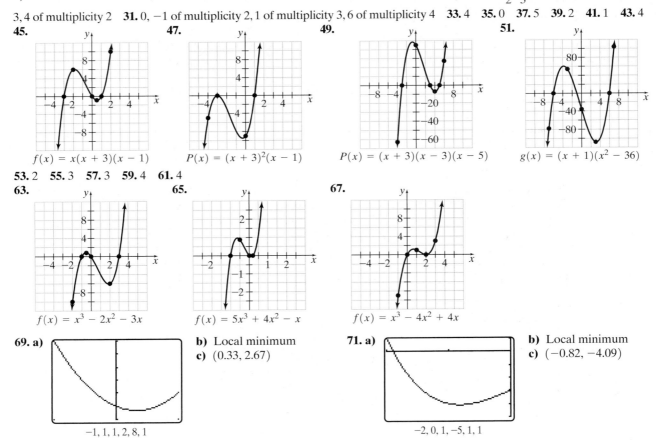

$f(x) = x(x + 3)(x - 1)$ $P(x) = (x + 3)^2(x - 1)$ $P(x) = (x + 3)(x - 3)(x - 5)$ $g(x) = (x + 1)(x^2 - 36)$

53. 2   **55.** 3   **57.** 3   **59.** 4   **61.** 4

63.   **65.**   **67.**

$f(x) = x^3 - 2x^2 - 3x$ $f(x) = 5x^3 + 4x^2 - x$ $f(x) = x^3 - 4x^2 + 4x$

69. a)   **b)** Local minimum
c) (0.33, 2.67)

$-1, 1, 1, 2, 8, 1$

71. a)   **b)** Local minimum
c) $(-0.82, -4.09)$

$-2, 0, 1, -5, 1, 1$

74. $4x^3 + 9x^2 + 29x + 86 + \dfrac{8}{x-3}$ **75.** -479 **76.** $\dfrac{5x^2 + 12}{20x}$ **77.** $\dfrac{5x+4}{y-4}$

Exercise Set 10.3

1. The graph of a fourth-degree polynomial function has a maximum of 4 x-intercepts because the maximum number of x-intercepts of the graph of a polynomial function is the degree of the polynomial function.   **3.** The minimum number of x-intercepts of the graph of a sixth-degree polynomial is 0. Because the graph of an even-degree polynomial is either above the x-axis on both ends or below the x-axis on both ends, it is possible for the graph of the polynomial to be completely above the x-axis or completely below the x-axis.   **5.** If the coefficients of a polynomial are all positive, the coefficients never change sign. By Descartes' Rule of Signs, the graph of the polynomial has no positive x-intercepts.   **7.** Maximum of 2 x-intercepts; Minimum of 0 x-intercepts   **9.** Maximum of 3 x-intercepts; Minimum of 1 x-intercept   **11.** Maximum of 2 x-intercepts; Minimum of 0 x-intercepts   **13.** Maximum of 5 x-intercepts; Minimum of 1 x-intercept   **15.** Maximum of 6 x-intercepts; Minimum of 0 x-intercepts   **17.** 0 positive zeros; either 3 or 1 negative zeros   **19.** Either 2 or 0 positive zeros; 1 negative zero   **21.** Either 3 or 1 positive zeros; 1 negative zero   **23.** 0 positive zeros; either 4, 2, or 0 negative zeros   **25.** Either 3 or 1 positive zeros; either 2 or 0 negative zeros   **27.** 0 positive zeros; either 5, 3, or 1 negative zeros   **29.** $\pm 1, \pm 2$ **31.** $\pm 1, \pm 3, \pm \dfrac{1}{3}$ **33.** $\pm 1, \pm 2, \pm 3, \pm 4, \pm 6, \pm 12$ **35.** $\pm 1, \pm 2, \pm 4, \pm \dfrac{1}{3}, \pm \dfrac{2}{3}, \pm \dfrac{4}{3}$

37. $\pm 1, \pm 2, \pm 4, \pm \dfrac{1}{2}, \pm \dfrac{1}{3}, \pm \dfrac{2}{3}, \pm \dfrac{4}{3}, \pm \dfrac{1}{6}$ **39.** $\pm 1, \pm 3, \pm \dfrac{1}{2}, \pm \dfrac{3}{2}, \pm \dfrac{1}{3}, \pm \dfrac{1}{6}$ **41.** $\pm 1, \pm 2, \pm 4, \pm \dfrac{1}{7}, \pm \dfrac{2}{7}, \pm \dfrac{4}{7}$ **43.** $2, 4$ **45.** $-5, \dfrac{1}{3}$

47. $-7, -1, 1$ **49.** $-2, -\dfrac{1}{2}, 2$ **51.** $-4, 4$ **53.** -2 of multiplicity 2, 2 of multiplicity 2   **55.** $-2, -1$ of multiplicity 2, 3

57. 1 of multiplicity 2, 3 of multiplicity 2   **59.** $2, -2, i, -i$ **61.** $1, -1, \dfrac{-1 + i\sqrt{3}}{2}, \dfrac{-1 - i\sqrt{3}}{2}$ **63.** $2, -2, \dfrac{-3 + i\sqrt{3}}{2}, \dfrac{-3 - i\sqrt{3}}{2}$

65. $1, i$ of multiplicity 2, $-i$ of multiplicity 2   **67. a)** $P(x) = x^3 + 4x^2 + x - 6$ **b)** $P(0) = -6$, The company has a loss of $6000 when no items are produced.; $R(0) = 0^3 + 5(0)^2 + 0 = 0$, The company has no revenue when no items are produced.; $C(0) = 6$, The company has a cost of $6000 when no items are produced.   **c)** The break-even point is $x = 1$ (when 100 items are produced).   **d)** $1 < x \le 3$;   **e)** $0 \le x < 1$ **f)** The company operates at a profit when $x > 1$ (when more than 100 items are produced). It operates at a loss when $x < 1$ (when less than 100 items are produced).
69. a) $P(x) = x^3 - 2x^2 - 13x - 10$ **b)** $P(0) = -10$, The company has a loss of $10,000 when no items are produced.; $R(0) = 0$, The company has no revenue when no items are produced.; $C(0) = 10$, The company has a cost of $10,000 when no items are produced.   **c)** The break-even point is $x = 5$ (when 500 items are produced).   **d)** $5 < x \le 8$ **e)** $0 \le x < 5$ **f)** The company operates at a profit when $x > 5$ (when more than 500 items are produced). It operates at a loss when $x < 5$ (when less than 500 items are produced).   **71. a)** $P(x) = 2x^3 - x^2 - 5x - 2$ **b)** $P(0) = -2$, The company has a loss of $2000 when no items are produced.; $R(0) = 0$; The company has no revenue when no items are produced.; $C(0) = 2$; The company has a cost of $2000 when no items are produced.   **c)** The break-even point is $x = 2$ (when 200 items are produced).   **d)** $2 < x \le 5$ **e)** $0 \le x < 2$ **f)** The company operates at a profit when $x > 2$ (when more than 200 items are produced). It operates at a loss when $x < 2$ (when less than 200 items are produced).   **73.** $y^{8/3}$ **74.** $18x\sqrt{2xy}$
75. $-\dfrac{10 + \sqrt{5}}{19}$ **76.** 6

Exercise Set 10.4

1. a) Yes, we know that $x = 5$ is an upper bound for the real zeros of $f(x) = x^3 - 2x^2 - 5x + 2$ because the numbers in the bottom row of the synthetic division are nonnegative.   **b)** Yes, we know that $x = 4$ is an upper bound for the real zeros of $f(x) = x^3 - 2x^2 - 5x + 2$ because the numbers in the bottom row of the synthetic division are nonnegative.   **c)** No, $x = 3$ is not an upper bound for the real zeros of $f(x) = x^3 - 2x^2 - 5x + 2$. Since the remainder of the synthetic division is -4, we know that $f(3) = -4$. Thus the graph of the function is below the x-axis at $x = 3$. Since the leading coefficient is positive, we know that the graph of the function is eventually above the x-axis as x gets larger. Because the graph is below the x-axis at $x = 3$ and it remains above the x-axis for all x greater than some x-value, there exist a real zero greater than $x = 3$.   **3. a)** No. Since the value of the function has the same sign at $x = 1$ that it has at $x = 2$, you cannot tell if it has a zero in the interval $(1, 2)$.   **b)** Yes. Since the value of the continuous function is negative at $x = -2$ and positive at $x = -1$, it must have a zero in the interval $(-2, -1)$.   **5.** Cannot determine   **7.** Yes   **9.** Yes   **11.** Yes   **13.** Yes   **15.** Cannot determine   **17.** Yes   **19.** Cannot determine   **21.** Yes   **23.** Yes   **25.** Cannot determine   **27.** Yes   **29.** Cannot determine   **31.** Yes   **33.** Cannot determine

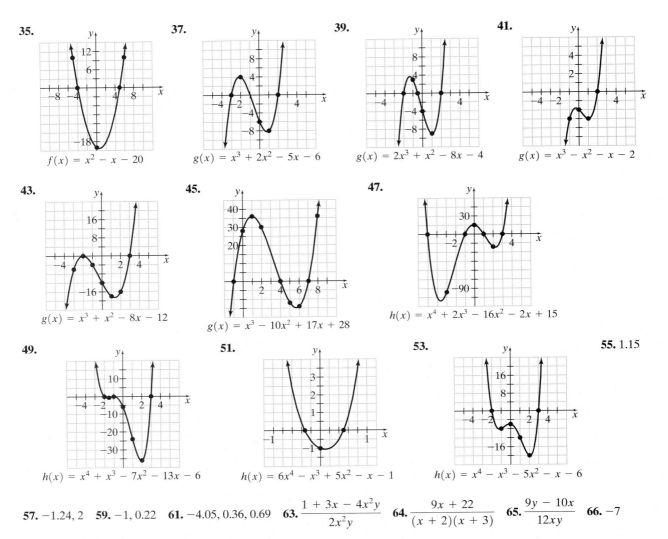

35. $f(x) = x^2 - x - 20$

37. $g(x) = x^3 + 2x^2 - 5x - 6$

39. $g(x) = 2x^3 + x^2 - 8x - 4$

41. $g(x) = x^3 - x^2 - x - 2$

43. $g(x) = x^3 + x^2 - 8x - 12$

45. $g(x) = x^3 - 10x^2 + 17x + 28$

47. $h(x) = x^4 + 2x^3 - 16x^2 - 2x + 15$

49. $h(x) = x^4 + x^3 - 7x^2 - 13x - 6$

51. $h(x) = 6x^4 - x^3 + 5x^2 - x - 1$

53. $h(x) = x^4 - x^3 - 5x^2 - x - 6$

55. 1.15

57. $-1.24, 2$ **59.** $-1, 0.22$ **61.** $-4.05, 0.36, 0.69$ **63.** $\dfrac{1 + 3x - 4x^2y}{2x^2y}$ **64.** $\dfrac{9x + 22}{(x + 2)(x + 3)}$ **65.** $\dfrac{9y - 10x}{12xy}$ **66.** -7

Exercise Set 10.5

1. No. If a rational function has no variables in the denominator when it is written in lowest terms, it will not have asymptotes.   **3.** No. We have 3 cases to consider. 1) If $f(x)$ is a proper rational function, the values of $f(x)$ approach 0 as $x \to \infty$ and as $x \to -\infty$. 2) If the degree of the numerator of $f(x)$ is the same as the degree of the denominator of $f(x)$, the values of $f(x)$ approach the ratio of the leading coefficients of the numerator and the denominator as $x \to \infty$ and as $x \to -\infty$. 3) If the degree of the numerator is greater than the degree of the denominator, the graph of $f(x)$ does not have a horizontal asymptote.   **5.** The point of discontinuity can be found by finding the values of x that make the denominator of the function equal to 0.   **7.** A graph of a rational function can have an oblique asymptote when the degree of the numerator is one greater than the degree of the denominator.   **9.** $x = 0$: Vertical asymptote   **11.** $x = -5$: Vertical asymptote   **13.** $x = -5$: Hole   **15.** $x = 1$: Hole   **17.** $x = 4$: Vertical asymptote; $x = -4$: Vertical asymptote   **19.** $x = 6$: Vertical asymptote; $x = 1$: Hole   **21.** $x = -3$: Vertical asymptote; $x = -5$: Vertical asymptote   **23.** $x = 5$: Vertical asymptote; $x = -5$: Hole   **25.** $x = 3$: Vertical asymptote; $x = 1$: Vertical asymptote   **27.** No points of discontinuity   **29.** Horizontal asymptote; $y = 0$ **31.** Horizontal asymptote; $y = 3$ **33.** Horizontal asymptote; $y = 0$ **35.** Horizontal asymptote; $y = 4$ **37.** Oblique asymptote; $y = x - 10$ **39.** Oblique asymptote; $y = 2x + 12$ **41.** Horizontal asymptote; $y = 0$ **43.** Oblique asymptote; $y = x - 2$

45. $f(x) = \dfrac{1}{x + 2}$

47. $h(x) = \dfrac{5}{x - 1}$

49. $h(x) = \dfrac{x + 1}{x(x + 4)}$

51. $g(x) = \dfrac{x + 1}{x - 1}$

53. $f(x) = \dfrac{3x + 3}{2x + 4}$

55. $h(x) = \dfrac{15x^2 + 10}{3x^2}$

57. $g(x) = \dfrac{x^2 - 3x - 4}{x - 2}$

59. $g(x) = \dfrac{x^2 + 2x - 8}{x + 4}$

61. $r(x) = \dfrac{x^2 - 5x + 6}{x - 2}$

63. $g(x) = \dfrac{2x^3 - x^2 + x - 8}{x^2 - 2x + 1}$

65. Answers may vary. Since the function has a vertical asymptote at $x = -4$, the denominator contains the factor $x + 4$. Since the function has a horizontal asymptote at $y = 0$, the function is a proper rational function. A function that meets these requirements is $f(x) = \dfrac{1}{x + 4}$. **67.** Answers may vary. Since the function has a hole at $(4, 3)$, it is undefined at $x = 4$ and has the factor $x - 4$ in its denominator and numerator. A function that meets these requirements is $f(x) = \dfrac{(2x - 5)(x - 4)}{x - 4}$.

69. Answers may vary. Since the function has a horizontal asymptote at $y = 0$, it is a proper rational function. Since it has no vertical asymptotes, its denominator is never 0. A function that meets these requirements is $f(x) = \dfrac{1}{x^2 + 1}$.

72. $f^{-1}(x) = 2x - 6$ **73.** $f^{-1}(x) = \dfrac{5}{x}$ **74.** $a = 12$ **75.** $x = \dfrac{1}{32}$

Review Exercises   **1.** Continuous   **2.** Discontinuous   **3.** Continuous   **4.** Continuous
5. Discontinuous; ..., $-5, -1, 3, 7, ...$ **6.** Continuous

7.   **8.**   **9.**   **10.**

11.   **12.**   **13.**   **14.**

$f(x) = \left[\!\left[2x - 3 \right]\!\right]$ $g(x) = \left[\!\left[\frac{1}{2}x + 1 \right]\!\right]$ $h(x) = \left[\!\left[\frac{x}{12} \right]\!\right]$

15. $f(x) = \left[\!\left[\frac{x}{0.37} \right]\!\right]$ **16.** $f(x) = \left[\!\left[\frac{x}{2.25} \right]\!\right]$ **17. a)** Below the x-axis   **b)** Above the x-axis   **18. a)** Above the x-axis
b) Above the x-axis   **19. a)** Below the x-axis   **b)** Below the x-axis   **20. a)** Above the x-axis   **b)** Below the x-axis
21. $-8, -3$ **22.** $0, -9, 4$ **23.** -2 of multiplicity $3, \frac{1}{3}$ **24.** $\sqrt{7}, -\sqrt{7}$ **25.** 1   **26.** 3   **27.** 2   **28.** 5   **29.** 5   **30.** 6

31.   **32.**   **33.**   **34.**

$f(x) = (x+4)(x+1)(x-5)$ $g(x) = (x+4)(x+1)^2(x-5)$ $h(x) = x^4 - 7x^3 + 6x^2$ $f(x) = x^5 - 7x^4 + 6x^3$

35. Maximum: 3; minimum: 1   **36.** Maximum: 4; minimum: 2   **37.** Maximum: 5; minimum: 1   **38.** Maximum: 6; minimum: 0
39. a) 2 or 0   **b)** 0   **40. a)** 1   **b)** 1   **41. a)** 0   **b)** 3 or 1   **42. a)** 2 or 0   **b)** 2 or 0   **43.** $\pm1, \pm2, \pm3, \pm4, \pm6, \pm12$
44. $\pm1, \pm5, \pm\frac{1}{2}, \pm\frac{5}{2}$ **45.** $\pm1, \pm3, \pm9, \pm\frac{1}{3}$ **46.** $\pm1, \pm2, \pm4, \pm8, \pm16, \pm32, \pm64$ **47.** $1, -1, -10$ **48.** 2 of multiplicity 2
49. $\frac{2}{3}, -4$ **50.** $1, 6, -3, -5$ **51.** $5, -1, 4i, -4i$ **52.** $2, -3, 5i, -5i$ **53.** Yes   **54.** Cannot be determined
55. Cannot be determined   **56.** Yes   **57.** Cannot be determined   **58.** Cannot be determined   **59.** Yes   **60.** Yes
61.   **62.**   **63.**   **64.**

$f(x) = 2x^3 + x^2 + 2x + 1$ $g(x) = x^3 + 2x^2 - 19x - 20$ $h(x) = x^4 - x^3 - 6x^2 + 4x + 8$ $h(x) = 4x^5 + 12x^4 - x - 3$

65. 5: Vertical asymptote; −5: Vertical asymptote   **66.** 5: Vertical asymptote; −5: Hole   **67.** −4: Hole   **68.** No points of discontinuity   **69.** Horizontal asymptote; $y = 0$ **70.** Horizontal asymptote; $y = 3$ **71.** Oblique asymptote; $y = 3x - 7$

72. Neither   **73.** Oblique asymptote; $y = 2x - 3$ **74.** Horizontal asymptote; $y = \dfrac{3}{2}$

75.   **76.**   **77.**

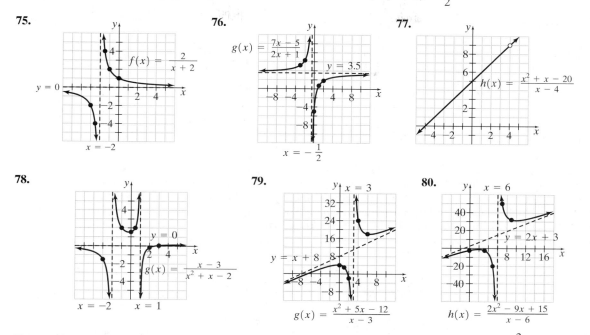

78.   **79.**   **80.**

Practice Test   **1. a)** Below the x-axis   **b)** Below the x-axis   **2.** $0, 4$ of multiplicity 3, $-\dfrac{3}{2}$ of multiplicity 2 **3.** 6

4. 4 **5.** Maximum: 5; minimum: 1   **6. a)** 2 or 0   **b)** 3 or 1   **7.** $\pm 1, \pm 2, \pm 3, \pm 6, \pm \dfrac{1}{2}, \pm \dfrac{3}{2}, \pm \dfrac{1}{4}, \pm \dfrac{3}{4}$ **8.** $3, -4, 3i, -3i$

9. a) Cannot be determined   **b)** Yes   **10.** Cannot be determined   **11.** −4: Hole   **12.** 5: Vertical asymptote

13. 0: Vertical asymptote; $-\dfrac{3}{2}$: Vertical asymptote   **14.** Horizontal asymptote; $y = 5$ **15.** Neither   **16.** Oblique asymptote;

$y = 4x - 3$ **17.** Horizontal asymptote; $y = 0$ **18.**   **19.**

20.   **21.**   **22.**

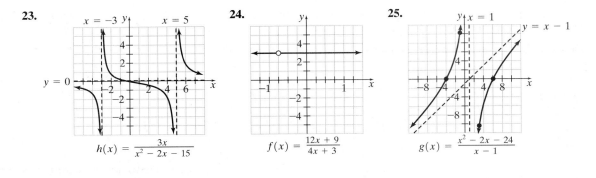

23. $h(x) = \dfrac{3x}{x^2 - 2x - 15}$

24. $f(x) = \dfrac{12x + 9}{4x + 3}$

25. $g(x) = \dfrac{x^2 - 2x - 24}{x - 1}$

CHAPTER 11

Exercise Set 11.1 **1.** Parabola, circle, ellipse, and hyperbola; see page 735 for picture. **3.** Yes, because each value of x corresponds to only one value for y. The domain is $\mathbb{R}$, and the range is $\{y|y \geq k\}$ **5.** The graphs have the same vertex, $(3, 4)$. The first graph opens upward, and the second graph opens downward. **7.** The distance is always a positive number because both differences are squared and we use the principal square root. **9.** A circle is the set of all points in a plane that are the same distance from a fixed point. **11.** No, $x^2 + y^2 = 25$ would be an equation of a circle **13.** No, the coefficients of x^2 and y^2 term would need to be the same **15.** No, if x^2 were replaced by x it would be an equation of a parabola

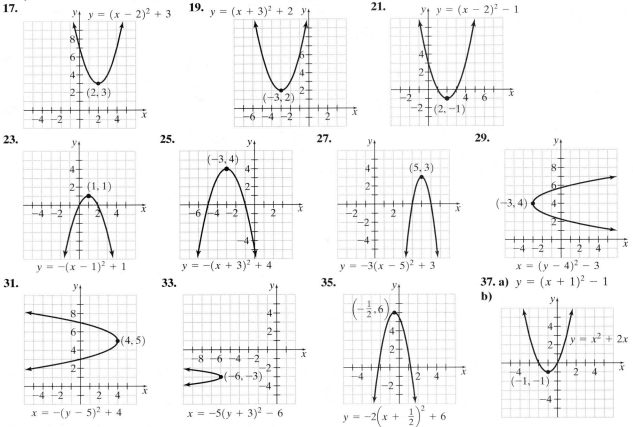

17. $y = (x - 2)^2 + 3$ **19.** $y = (x + 3)^2 + 2$ **21.** $y = (x - 2)^2 - 1$

23. $y = -(x - 1)^2 + 1$ **25.** $y = -(x + 3)^2 + 4$ **27.** $y = -3(x - 5)^2 + 3$ **29.** $x = (y - 4)^2 - 3$

31. $x = -(y - 5)^2 + 4$ **33.** $x = -5(y + 3)^2 - 6$ **35.** $y = -2\left(x + \dfrac{1}{2}\right)^2 + 6$ **37. a)** $y = (x + 1)^2 - 1$ **b)** $y = x^2 + 2x$

39. a) $y = (x + 3)^2 - 9$
b) $y = x^2 + 6x$

41. a) $x = (y + 2)^2 - 4$
b)

43. a) $y = \left(x + \dfrac{7}{2}\right)^2 - \dfrac{9}{4}$
b) $y = x^2 + 7x + 10$

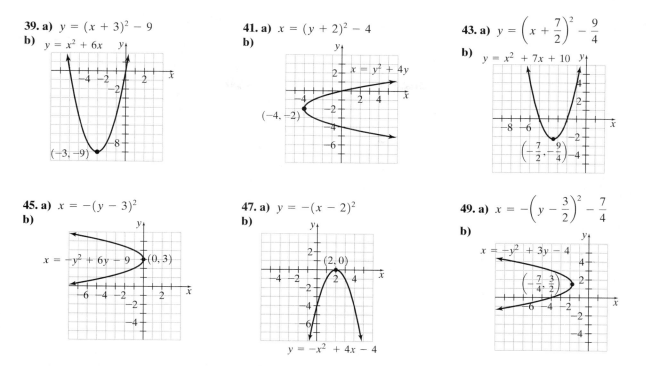

45. a) $x = -(y - 3)^2$
b)

47. a) $y = -(x - 2)^2$
b)

49. a) $x = -\left(y - \dfrac{3}{2}\right)^2 - \dfrac{7}{4}$
b)

51. 5   **53.** 9   **55.** 13   **57.** $\sqrt{90} \approx 9.49$ **59.** $\sqrt{\dfrac{125}{4}} \approx 5.59$ **61.** $\sqrt{34.33} \approx 5.86$ **63.** $\sqrt{10} \approx 3.16$ **65.** $(3, 6)$ **67.** $(0, 0)$

69. $\left(\dfrac{3}{2}, 5\right)$ **71.** $\left(\dfrac{5}{2}, -\dfrac{7}{4}\right)$ **73.** $\left(\dfrac{\sqrt{3} + \sqrt{2}}{2}, \dfrac{9}{2}\right)$ **75.** $x^2 + y^2 = 36$ **77.** $(x - 2)^2 + y^2 = 25$ **79.** $x^2 + (y - 5)^2 = 4$

81. $(x - 3)^2 + (y - 4)^2 = 81$ **83.** $(x - 2)^2 + (y + 6)^2 = 100$ **85.** $(x - 1)^2 + (y - 2)^2 = 7$ **87.** $x^2 + y^2 = 16$

89. $(x - 3)^2 + (y + 2)^2 = 9$ **91.**   **93.**   **95.**

97.   **99.**   **101.**

103.

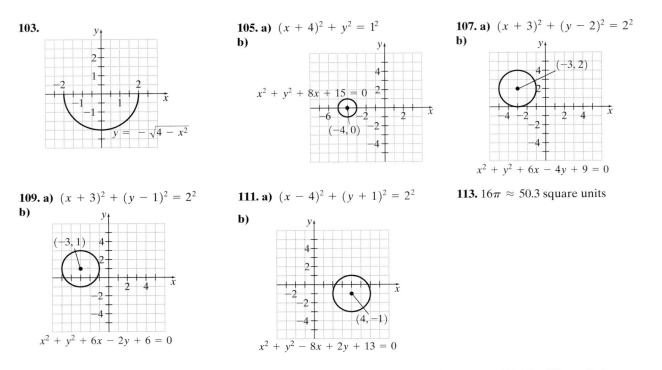

105. a) $(x + 4)^2 + y^2 = 1^2$
b) $x^2 + y^2 + 8x + 15 = 0$

107. a) $(x + 3)^2 + (y - 2)^2 = 2^2$
b) $x^2 + y^2 + 6x - 4y + 9 = 0$

109. a) $(x + 3)^2 + (y - 1)^2 = 2^2$
b) $x^2 + y^2 + 6x - 2y + 6 = 0$

111. a) $(x - 4)^2 + (y + 1)^2 = 2^2$
b) $x^2 + y^2 - 8x + 2y + 13 = 0$

113. $16\pi \approx 50.3$ square units

115. x-intercept: $(-7, 0)$; y-intercepts: $(0, -1), (0, 7)$ **117.** x-intercept: $(56, 0)$; no y-intercepts   **119.** No, different line segments can have the same midpoint   **121.** 10   **123.** $(x + 5)^2 + (y - 2)^2 = 4$ **125. a)** $2\sqrt{2}$ **b)** $(7, 6)$ **c)** $(x - 7)^2 + (y - 6)^2 = 8$
127. 4, 0, a parabola opening up or down and a parabola opening right or left can be drawn to have a maximum of 4 intersections, or a minimum of 0 intersections.   **129. a)** 13.6 feet **b)** 81.8 feet **c)** $x^2 + (y - 81.8)^2 = 4651.24$
131. a) $x^2 + y^2 = 16$ **b)** $(x - 2)^2 + y^2 = 4$ **c)** $(x + 2)^2 + y^2 = 4$ **d)** 8π square units   **133.** 48π square units

136. $\dfrac{y}{3x}$ **137.** $(0, 4)$ **138.** 128   **139. a)** $1 : a^2, 2 : ab, 3 : ab, 4 : b^2$ **b)** $(a + b)^2$ **140.**

Exercise Set 11.2
1. An ellipse is a set of points in a plane, the sum of whose distances from two fixed points is constant.   **3.** $\dfrac{(x - h)^2}{a^2} + \dfrac{(y - k)^2}{b^2} = 1$ **5.** If $a = b$, the formula for a circle is obtained.   **7.** Divide both sides by 360.

9. No, equation for an ellipse is $\dfrac{x^2}{a^2} + \dfrac{y^2}{b^2} = 1$.   **11.**   **13.**

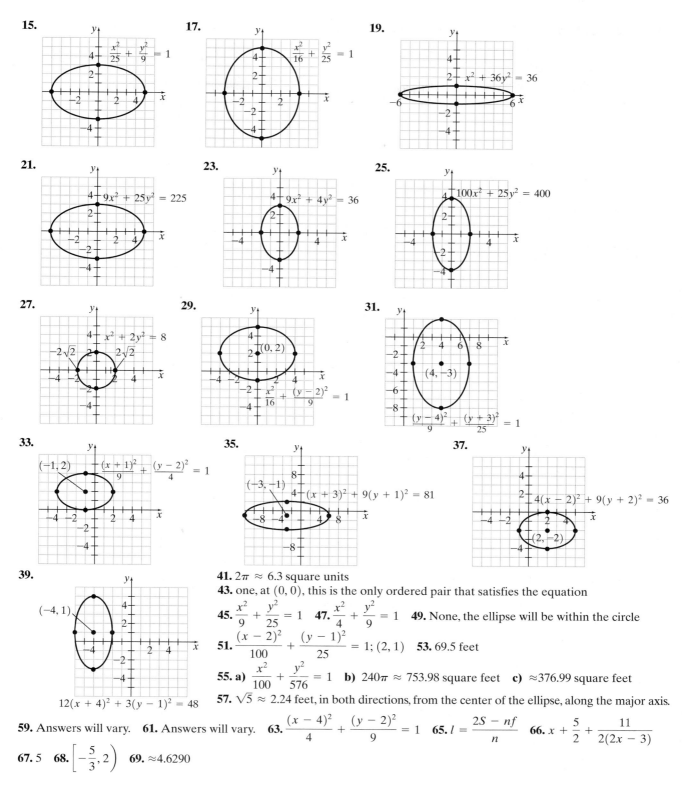

15.

$\dfrac{x^2}{25} + \dfrac{y^2}{9} = 1$

17.

$\dfrac{x^2}{16} + \dfrac{y^2}{25} = 1$

19.

$x^2 + 36y^2 = 36$

21.

$9x^2 + 25y^2 = 225$

23.

$9x^2 + 4y^2 = 36$

25.

$100x^2 + 25y^2 = 400$

27.

$x^2 + 2y^2 = 8$

$-2\sqrt{2}$ $2\sqrt{2}$

29.

$(0, 2)$

$\dfrac{x^2}{16} + \dfrac{(y - 2)^2}{9} = 1$

31.

$(4, -3)$

$\dfrac{(y - 4)^2}{9} + \dfrac{(y + 3)^2}{25} = 1$

33.

$(-1, 2)$

$\dfrac{(x + 1)^2}{9} + \dfrac{(y - 2)^2}{4} = 1$

35.

$(-3, -1)$

$(x + 3)^2 + 9(y + 1)^2 = 81$

37.

$4(x - 2)^2 + 9(y + 2)^2 = 36$

$(2, -2)$

39.

$(-4, 1)$

$12(x + 4)^2 + 3(y - 1)^2 = 48$

41. $2\pi \approx 6.3$ square units

43. one, at $(0, 0)$, this is the only ordered pair that satisfies the equation

45. $\dfrac{x^2}{9} + \dfrac{y^2}{25} = 1$ **47.** $\dfrac{x^2}{4} + \dfrac{y^2}{9} = 1$ **49.** None, the ellipse will be within the circle

51. $\dfrac{(x - 2)^2}{100} + \dfrac{(y - 1)^2}{25} = 1; (2, 1)$ **53.** 69.5 feet

55. a) $\dfrac{x^2}{100} + \dfrac{y^2}{576} = 1$ **b)** $240\pi \approx 753.98$ square feet **c)** ≈ 376.99 square feet

57. $\sqrt{5} \approx 2.24$ feet, in both directions, from the center of the ellipse, along the major axis.

59. Answers will vary. **61.** Answers will vary. **63.** $\dfrac{(x - 4)^2}{4} + \dfrac{(y - 2)^2}{9} = 1$ **65.** $l = \dfrac{2S - nf}{n}$ **66.** $x + \dfrac{5}{2} + \dfrac{11}{2(2x - 3)}$

67. 5 **68.** $\left[-\dfrac{5}{3}, 2 \right)$ **69.** ≈ 4.6290

Exercise Set 11.3 **1.** A hyperbola is the set of points in a plane the differences of whose distances from two fixed points is a constant. **3.** The graph of $\dfrac{x^2}{a^2} - \dfrac{y^2}{b^2} = 1$ is a hyperbola with vertices at $(a, 0)$ and $(-a, 0)$. Its transverse axis lies along the x-axis. The asymptotes are $y = \pm\dfrac{b}{a}x$. **5.** No, the signs of the x and y terms must differ. **7.** Yes, divide both sides of the equation by 100 and you will see the equation is that of a hyperbola. **9.** Divide both sides of the equation by 81.

11. a) $y = \pm\dfrac{2}{3}x$

b)

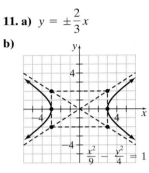

13. a) $y = \pm\dfrac{1}{2}x$

b)

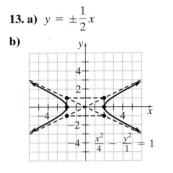

15. a) $y = \pm\dfrac{5}{3}x$

b)

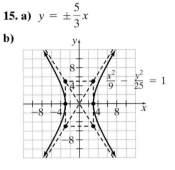

17. a) $y = \pm\dfrac{4}{5}x$

b)

19. a) $y = \pm\dfrac{3}{4}x$

b)

21. a) $y = \pm\dfrac{5}{6}x$

b)

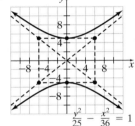

23. a) $y = \pm\dfrac{5}{2}x$

b)

25. a) $y = \pm\dfrac{4}{9}x$

b)

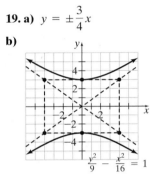

27. a) $\dfrac{x^2}{25} - \dfrac{y^2}{1} = 1$, $y = \pm\dfrac{1}{5}x$

b)

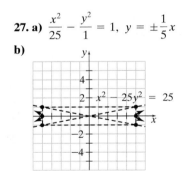

29. a) $\dfrac{x^2}{4} - \dfrac{y^2}{16} = 1$, $y = \pm 2x$

b)

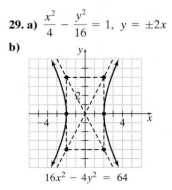

$16x^2 - 4y^2 = 64$

31. a) $\dfrac{y^2}{1} - \dfrac{x^2}{9} = 1$, $y = \pm\dfrac{1}{3}x$

b)

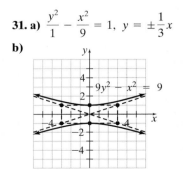

33. a) $\dfrac{x^2}{9} - \dfrac{y^2}{25} = 1$, $y = \pm\dfrac{5}{3}x$

b)

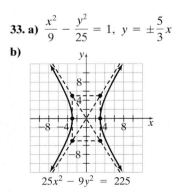

$25x^2 - 9y^2 = 225$

35. a) $\dfrac{y^2}{36} - \dfrac{x^2}{4} = 1$, $y = \pm 3x$

b)

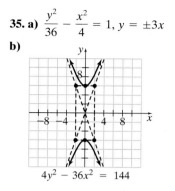

$4y^2 - 36x^2 = 144$

37. Circle **39.** Ellipse **41.** Hyperbola **43.** Parabola **45.** Ellipse **47.** Parabola **49.** Circle **51.** Hyperbola **53.** Parabola **55.** Hyperbola **57.** Parabola

59. Circle **61.** $\dfrac{y^2}{4} - \dfrac{x^2}{16} = 1$ **63.** $\dfrac{x^2}{9} - \dfrac{y^2}{36} = 1$

65. $\dfrac{x^2}{9} - \dfrac{y^2}{25} = 1$, no, $\dfrac{x^2}{18} - \dfrac{y^2}{50} = 1$ and others will work. The ratio of $\dfrac{b}{a}$ must be $\dfrac{5}{3}$.

67. No, graphs of hyperbolas of this form do not pass the vertical line test
69. D: $(-\infty, -3] \cup [3, \infty)$; R: $\mathbb{R}$
71. The transverse axes of both graphs are along the x-axis. The vertices of the second graph will be closer to the origin, and the second graph will open wider.

73. Answers will vary. **75.** $y = -\dfrac{1}{2}x + 1$ **76.** $-x^2 - x + 10$ **77.** $\left(-1, \dfrac{1}{3}\right)$ **78.** $\dfrac{3(x+1)}{2x-3}$ **79.** $v = \sqrt{\dfrac{2E}{m}}$ **80.** 1

Exercise Set 11.4
1. A nonlinear system of equations is a system in which at least one equation is nonlinear.
3. Yes, for example ⌀ **5.** Yes, for example ⬥ **7.** $(2, -2), (-2, 2)$ **9.** $(3, 0), \left(-\dfrac{9}{5}, \dfrac{12}{5}\right)$ **11.** $(-4, 11), \left(\dfrac{5}{2}, \dfrac{5}{4}\right)$ **13.** $\left(\dfrac{5}{2}, \dfrac{3}{2}\right)$
15. $(2, \sqrt{2}), (2, -\sqrt{2}), (-1, \sqrt{5}), (-1, -\sqrt{5})$ **17.** No real solution **19.** $(0, -3), (\sqrt{5}, 2)(-\sqrt{5}, 2)$ **21.** $(2, -4), (-14, -20)$
23. $(2, 0), (-2, 0)$ **25.** $(3, 2), (3, -2), (-3, 2), (-3, -2)$ **27.** $(\sqrt{19}, \sqrt{6}), (\sqrt{19}, -\sqrt{6}), (-\sqrt{19}, \sqrt{6}), (-\sqrt{19}, -\sqrt{6})$
29. $(3, 0), (-3, 0)$ **31.** $(1, 3), (1, -3), (-1, 3), (-1, -3)$ **33.** $(5, 3), (5, -3), (-5, 3), (-5, -3)$
35. $(\sqrt{5}, 2), (\sqrt{5}, -2), (-\sqrt{5}, 2), (-\sqrt{5}, -2)$ **37.** Answers will vary. **39.** 20 meters by 25 meters
41. 6 feet by 8 feet or 4 feet by 12 feet **43.** length: 14 centimeters, width: 8 centimeters **45.** 8 inches by 15 inches
47. ≈ 1.67 seconds **49.** $r = 6\%$, $p = \$125$ **51.** ≈ 16 and ≈ 184 **53.** ≈ 26 and ≈ 174 **55.** $(-1, -3), (3.12, -0.53)$
57. 10 yards, 24 yards **59.** Parentheses, exponents, multiplication or division, addition or subtraction.

60. $(x + 2)(x^2 + x + 1)$ **61.** 0.9 **62.** $\dfrac{2\sqrt{x+2} + 6}{x - 7}$ **63.** $k = \dfrac{\ln A - \ln A_0}{t}$

Exercise Set 11.5
1. When graphing an inequality containing $>$ or $<$, the boundary is graphed with a dashed line because it is not part of the solution. **3.** The graph of a circle divides the xy-plane into 2 regions: a region inside the circle and a region outside the circle.

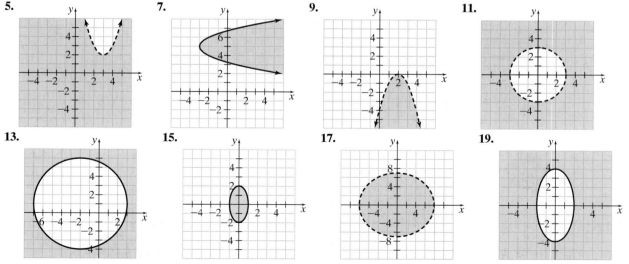

21. **23.** **25.** **27.**

29. **31.** **33.** **35.**

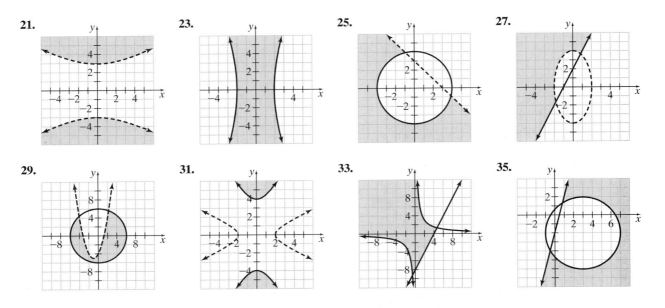

37. a) Yes, the solution to a system of linear inequalities containing two linear inequalities can contain no points. For instance, if the lines are parallel and the top line was shaded above and the bottom line was shaded below, the shaded regions would not overlap. Thus, the solution would contain no points.   **b)** No, the solution to a system of linear inequalities containing two linear inequalities cannot contain exactly 1 point. If there is a solution, either the lines overlap completely or the shaded regions overlap. In either case, more than 1 solution exists.   **c)** No, the solution to a system of linear inequalities containing two linear inequalities cannot contain exactly 2 points. If there is a solution, either the lines overlap completely or the shaded regions overlap. In either case, more than 2 solutions exists.   **d)** No, the solution to a system of linear inequalities containing two linear inequalities cannot contain all points on the xy-plane. The solution to a system of inequalities is the intersection of the solutions to each inequality. Since the solution to an inequality cannot contain all points on the xy-plane, the intersection of two inequalities cannot contain all points on the xy-plane.

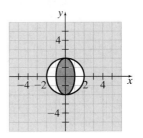

39. a) Yes, the solution to a system of inequalities containing two second-degree inequalities can contain no points. For instance, two circles that do not overlap and are shaded within have no overlapping regions. Thus, the solution would contain no points.

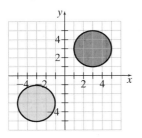

b) Yes, the solution to a system of inequalities containing two second-degree inequalities can contain exactly 1 point. For instance two circles which are shaded within and intersect at exactly one point would have exactly one point in common. Thus, the solution would contain exactly 1 point.

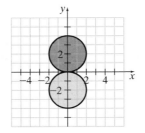

c) Yes, the solution to a system of inequalities containing two second-degree inequalities can contain exactly 2 points. For instance an ellipse and a circle can intersect in exactly 2 points. If the ellipse is within the circle and its interior is shaded and the circle's exterior is shaded, they intersect in exactly two points. Thus, the solution would contain exactly 2 points.

d) No, the solution to a system of inequalities containing two second-degree inequalities cannot contain all points on the *xy*-plane. The solution to a system of inequalities is the intersection of the solutions to each inequality. Since the solution to an inequality cannot contain all points on the *xy*-plane, the intersection of two inequalities cannot contain all points on the *xy*-plane.

41. a) **b)** **c)**

45. $16y^{10}$
46. $1 \pm \sqrt{5}$
47. $1 \pm \sqrt{5}$
48. a)

Review Exercises

1. $5; \left(\dfrac{3}{2}, -2\right)$ **2.** $5; \left(-\dfrac{5}{2}, 3\right)$ **3.** $13; \left(-\dfrac{7}{2}, 4\right)$ **4.** $\sqrt{8} \approx 2.83; (-3, 4)$

5. $y = (x - 2)^2 + 1$

6. $y = (x + 3)^2 - 4$

7. $x = (y - 1)^2 + 4$

8. $x = -2(y + 4)^2 - 3$

9. a) $y = (x - 4)^2 + 6$
b) $y = x^2 - 8x + 22$

10. a) $x = -(y + 1)^2 + 6$
b) $x = -y^2 - 2y + 5$

11. a) $x = \left(y + \dfrac{5}{2}\right)^2 - \dfrac{9}{4}$
b) $x = y^2 + 5y + 4$

12. a) $y = 2(x - 2)^2 - 32$
b) $y = 2x^2 - 8x - 24$

13. a) $x^2 + y^2 = 4^2$
b)

14. a) $(x + 3)^2 + (y - 4)^2 = 1^2$
b) $(x + 3)^2 + (y - 4)^2 = 1^2$

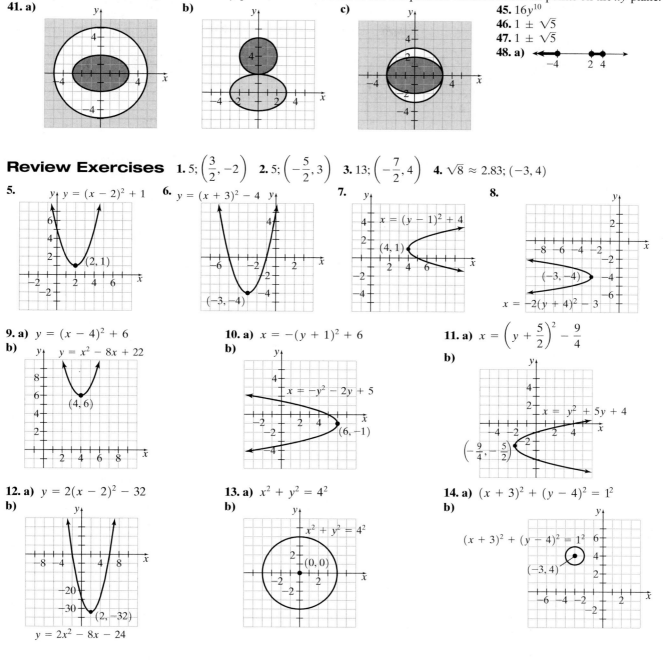

15. a) $x^2 + (y - 2)^2 = 2^2$
b)

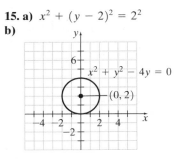

$x^2 + y^2 - 4y = 0$
$(0, 2)$

16. a) $(x - 1)^2 + (y + 3)^2 = 3^2$
b)

$(1, -3)$
$x^2 + y^2 - 2x + 6y + 1 = 0$

17. a) $(x - 4)^2 + (y - 5)^2 = 1^2$
b)

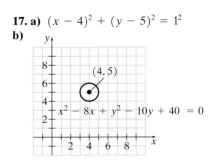

$(4, 5)$
$x^2 - 8x + y^2 - 10y + 40 = 0$

18. a) $(x - 2)^2 + (y + 5)^2 = (\sqrt{12})^2$
b)

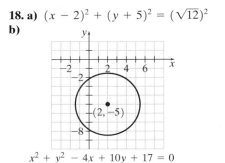

$(2, -5)$
$x^2 + y^2 - 4x + 10y + 17 = 0$

19.

$y = \sqrt{9 - x^2}$

20.

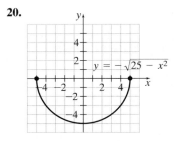

$y = -\sqrt{25 - x^2}$

21. $(x + 1)^2 + (y - 1)^2 = 4$ **22.** $(x - 5)^2 + (y + 3)^2 = 9$

23.

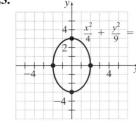

$\dfrac{x^2}{4} + \dfrac{y^2}{9} = 1$

24.

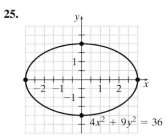

$\dfrac{x^2}{36} + \dfrac{y^2}{64} = 1$

25.

$4x^2 + 9y^2 = 36$

26.

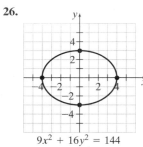

$9x^2 + 16y^2 = 144$

27.

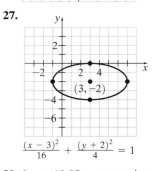

$(3, -2)$
$\dfrac{(x - 3)^2}{16} + \dfrac{(y + 2)^2}{4} = 1$

28.

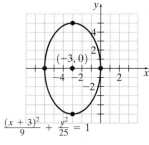

$(-3, 0)$
$\dfrac{(x + 3)^2}{9} + \dfrac{y^2}{25} = 1$

29.

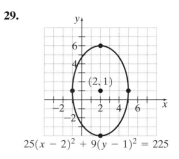

$(2, 1)$
$25(x - 2)^2 + 9(y - 1)^2 = 225$

30. $6\pi \approx 18.85$ square units
31. a) $y = \pm 2x$
b)

$\dfrac{x^2}{4} - \dfrac{y^2}{16} = 1$

32. a) $y = \pm x$
b)

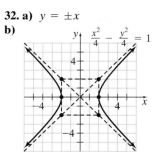

$\dfrac{x^2}{4} - \dfrac{y^2}{4} = 1$

33. a) $y = \pm\dfrac{1}{3}x$

34. a) $y = \pm\dfrac{5}{4}x$

35. a) $\dfrac{x^2}{9} - \dfrac{y^2}{1} = 1$ **b)** $y = \pm\dfrac{1}{3}x$

b)

b)

c)

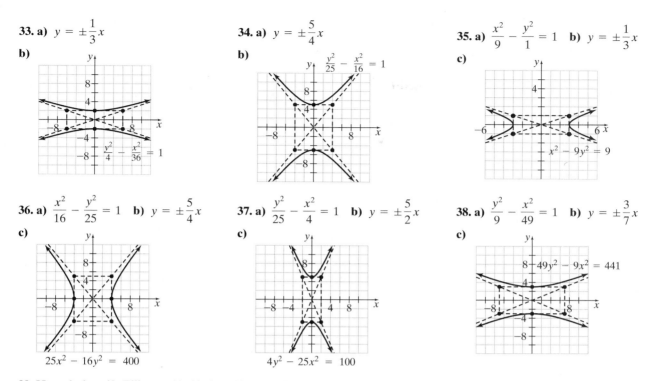

36. a) $\dfrac{x^2}{16} - \dfrac{y^2}{25} = 1$ **b)** $y = \pm\dfrac{5}{4}x$

37. a) $\dfrac{y^2}{25} - \dfrac{x^2}{4} = 1$ **b)** $y = \pm\dfrac{5}{2}x$

38. a) $\dfrac{y^2}{9} - \dfrac{x^2}{49} = 1$ **b)** $y = \pm\dfrac{3}{7}x$

c)

c)

c)

39. Hyperbola **40.** Ellipse **41.** Circle **42.** Hyperbola **43.** Ellipse **44.** Parabola **45.** Ellipse **46.** Parabola

47. $(4, 0), (-4, 0)$ **48.** $\left(\dfrac{5}{2}, \dfrac{3}{2}\right)$ **49.** $(-3, 0), \left(-\dfrac{12}{5}, \dfrac{9}{5}\right)$ **50.** No real solution **51.** $(7, 0), (-7, 0)$

52. $(4, 3), (4, -3), (-4, 3), (-4, -3)$ **53.** No real solution **54.** $(\sqrt{3}, 0), (-\sqrt{3}, 0)$ **55.** 4 feet by 9 feet **56.** ≈ 4 and ≈ 145

57. $r = 5\%$, $p = \$5000$ **58.** 15 feet by 20 feet

59.

60.

61.

62.

63.

64.

65.

66.

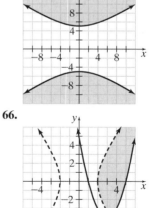

Chapter Practice Test
1. They are formed by cutting a cone or a pair of cones.   **2.** $\sqrt{50} \approx 7.07$ **3.** $\left(2, \dfrac{3}{2}\right)$

4. $(-3, 1)$,

$y = -2(x + 3)^2 + 1$

5.

$x = y^2 - 2y + 4$

6. $x = -(y + 2)^2 - 1$

$x = -y^2 - 4y - 5$

7.

$(x - 2)^2 + (y - 4)^2 = 9$

8. $9\pi \approx 28.27$ square units
9. $(x - 3)^2 + (y + 1)^2 = 16$
10.

$y = -\sqrt{16 - x^2}$

11.

$(x + 1)^2 + (y - 3)^2 = 9$

$(-1, 3)$

$x^2 + y^2 + 2x - 6y + 1 = 0$

12.

$4x^2 + 25y^2 = 100$

13. No, the major axis should be along the y-axis.   **14.**

$4(x - 4)^2 + 36(y + 2)^2 = 36$

$(4, -2)$

15. $(5, -3)$

16. The transverse axis lies along the axis corresponding to the positive term of the equation in standard form.

17. $y = \pm\dfrac{3}{2}x$

18.

$\dfrac{y^2}{25} - \dfrac{x^2}{1} = 1$

19.

$\dfrac{x^2}{4} - \dfrac{y^2}{9} = 1$

20. Hyperbola, divide both sides of the equation by 48.
21. Ellipse, divide both sides of the equation by 64.
22. $(2, \sqrt{3})$, $(2, -\sqrt{3})$, $(-2, \sqrt{3})$, $(-2, -\sqrt{3})$
23. No real solution
24. 60 meters by 100 meters

25.

CHAPTER 12

Exercise Set 12.1
1. A sequence is a list of numbers arranged in a specific order. **3.** A finite sequence is a function whose domain includes only the first n natural numbers. **5.** In a decreasing sequence, the terms decrease.

7. A series is the sum of the terms of a sequence. **9.** $\sum_{i=1}^{5} (i + 2)$ is the sum as i goes from 1 to 5 of $i + 2$.

11. It is an increasing sequence. Each number in the sequence is greater than the preceeding number **13.** Yes, the signs of the terms alternate. **15.** $3, 6, 9, 12, 15$ **17.** $3, 7, 11, 15, 19$ **19.** $2, 1, \frac{2}{3}, \frac{1}{2}, \frac{2}{5}$ **21.** $\frac{3}{2}, \frac{4}{3}, \frac{5}{4}, \frac{6}{5}, \frac{7}{6}$ **23.** $-1, 1, -1, 1, -1$

25. $4, -8, 16, -32, 64$ **27.** 31 **29.** 9 **31.** 1 **33.** 99 **35.** $\frac{81}{19}$ **37.** $2, 15$ **39.** $3, 17$ **41.** $0, \frac{5}{6}$ **43.** $-1, -1$ **45.** $\frac{1}{2}, 7$

47. $64, 128, 256$ **49.** $15, 17, 19$ **51.** $\frac{1}{6}, \frac{1}{7}, \frac{1}{8}$ **53.** $1, -1, 1$ **55.** $\frac{1}{81}, \frac{1}{243}, \frac{1}{729}$ **57.** $\frac{1}{16}, -\frac{1}{32}, \frac{1}{64}$ **59.** $-25, -33, -41$

61. $2 + 5 + 8 + 11 + 14 = 40$ **63.** $0 + 3 + 8 + 15 + 24 + 35 = 85$ **65.** $\frac{1}{2} + 2 + \frac{9}{2} + 8 = 15$ **67.** $2 + 3 + 4 = 9$

69. $\sum_{i=1}^{5} (i + 3)$ **71.** $\sum_{k=1}^{3} \frac{k^2}{4}$ **73.** 13 **75.** 169 **77.** 55 **79.** 25 **81.** ≈ 42.83 **83. a)** $6, 12, 18, 24$ **b)** $p_n = 6n$
85. Answers will vary. **87.** Answers will vary. **89.** $\Sigma x = n\overline{x}$ **91.** Yes, for example if $n = 3$, you obtain $2x_1 + 2x_2 + 2x_3 = 2(x_1 + x_2 + x_3)$ **93. a)** 10 **b)** 11 **c)** 110 **d)** 29 **e)** No **94.** $\frac{2}{5}$ **95.** $8(y - 2x^2)(y^2 + 2x^2y + 4x^4)$ **96.** 7 **97.** $r = \sqrt{\frac{V}{\pi h}}$

Exercise Set 12.2
1. In an arithmetic sequence, each term differs by a constant amount. **3.** It is called the common difference. **5.** Positive number **7.** Yes, for example $-1, -2, -3, \ldots$ **9.** Yes, for example $2, 4, 6, \ldots$

11. $4, 7, 10, 13, 16; a_n = 3n + 1$ **13.** $7, 5, 3, 1, -1; a_n = -2n + 9$ **15.** $\frac{1}{2}, 2, \frac{7}{2}, 5, \frac{13}{2}; a_n = \frac{3}{2}n - 1$ **17.** $100, 95, 90, 85, 80;$

$a_n = -5n + 105$ **19.** 17 **21.** 27 **23.** 18 **25.** 2 **27.** 9 **29.** 6 **31.** $s_{10} = 100; d = 2$ **33.** $s_8 = \frac{52}{5}; d = \frac{1}{5}$

35. $s_6 = 37.5; d = 3.7$ **37.** $s_{11} = 407; d = 6$ **39.** $4, 7, 10, 13; a_{10} = 31; s_{10} = 175$ **41.** $5, 2, -1, -4; a_{10} = -22, s_{10} = -85$

43. $-8, -13, -18, -23; a_{10} = -53; s_{10} = -305$ **45.** $100, 93, 86, 79; a_{10} = 37; s_{10} = 685$ **47.** $\frac{9}{5}, \frac{12}{5}, 3, \frac{18}{5}; a_{10} = \frac{36}{5}; s_{10} = 45$

49. $n = 15, s_{15} = 330$ **51.** $n = 10; s_{10} = 90$ **53.** $n = 17; s_{17} = \frac{153}{2}$ **55.** $n = 29; s_{29} = 1479$ **57.** 1275 **59.** 2500 **61.** 630

63. 267 **65.** $40, 348$ **67.** 210 **69. a)** 25 **b)** 169 **71.** $101 \cdot 50 = 5050$ **73.** $s_n = n^2$ **75. a)** 19 feet **b)** 143.5 feet **77.** 1 foot

79. a) 145 **b)** 625 **81.** $\$496$ **83. a)** $\$35,600$ **b)** $\$338,000$ **85.** $a_n = 180°(n - 2)$ **93.** $r = \frac{A - P}{Pt}$ **94.** $(-2, 3)$

95. $2(2n - 5)(2n - 1)$ **96.**

Exercise Set 12.3
1. A geometric sequence is a sequence in which each term after the first is the same multiple of the preceding term. **3.** To find the common ratio; take any term except the first and divide by the term that precedes it.

5. 0 **7.** Yes **9.** Yes, s_∞ exists because $|r| < 1$, $s_\infty = 8$ **11.** $2, 6, 18, 54, 162$ **13.** $6, 3, \frac{3}{2}, \frac{3}{4}, \frac{3}{8}$ **15.** $80, 20, 5, \frac{5}{4}, \frac{5}{16}$

17. $90, -30, 10, -\dfrac{10}{3}, \dfrac{10}{9}$ **19.** $-1, -3, -9, -27, -81$ **21.** $6, -12, 24, -48, 96$ **23.** $\dfrac{1}{3}, \dfrac{1}{6}, \dfrac{1}{12}, \dfrac{1}{24}, \dfrac{1}{48}$ **25.** $3, \dfrac{9}{2}, \dfrac{27}{4}, \dfrac{81}{8}, \dfrac{243}{16}$

27. 128 **29.** 64 **31.** $-\dfrac{3}{64}$ **33.** 6144 **35.** $\dfrac{1}{64}$ **37.** $\dfrac{50}{19{,}683}$ **39.** 93 **41.** 15,624 **43.** 10,160 **45.** $-\dfrac{2565}{128}$ **47.** $-\dfrac{9279}{625}$

49. $r = \dfrac{1}{2}; \, a_n = 3\left(\dfrac{1}{2}\right)^{n-1}$ **51.** $r = 2; \, a_n = 7(2)^{n-1}$ **53.** $r = -3; \, a_n = 2(-3)^{n-1}$ **55.** $r = 3; \, a_n = -(3)^{n-1}$ **57.** 2 **59.** $\dfrac{5}{4}$

61. 12 **63.** $\dfrac{25}{3}$ **65.** $\dfrac{5}{3}$ **67.** $\dfrac{81}{10}$ **69.** 4 **71.** 24 **73.** -45 **75.** -15 **77.** $\dfrac{3}{11}$ **79.** $\dfrac{7}{9}$ **81.** $\dfrac{17}{33}$ **83.** $r = 2$ or $r = -2; \, a_1 = 7$

85. $r = 3; \, a_1 = 5$ **87.** $\approx \$1.77$ **89. a)** 3 days **b)** ≈ 1.172 grams **91. a)** ≈ 349.31 million **b)** ≈ 31.9 years

93. a) $\dfrac{1}{2}, \dfrac{1}{4}, \dfrac{1}{8}, \dfrac{1}{16}, \dfrac{1}{32}$ **b)** $a_n = \dfrac{1}{2}\left(\dfrac{1}{2}\right)^{n-1} = \left(\dfrac{1}{2}\right)^{n}$ **c)** $\dfrac{1}{128} \approx 0.78\%$ **95.** $\left(\dfrac{2}{3}\right)^{10}$ or $\approx 1.7\%$ **97. a)** 47.52 feet **b)** 550 feet

99. a) 7.203 inches **b)** 100 inches **101.** 211 **103. a)** $\$12{,}000, \$9000, \$7680, \$6144$ **b)** $a_n = 12{,}000\left(\dfrac{4}{5}\right)^{n-1}$ **c)** $\approx \$4915.20$

105. 190 feet **107. a)** y_2 goes up more steeply. **b)** **109.** $n = 21; \, s_n = 2{,}097{,}151$ **110.** 24

111. $6x^3 - x^2y - 16xy^2 + 6y^3$ **112.** $r = \dfrac{S - a}{S}$

$-10, 10, 1, -1, 19, 1$

113. $g^{-1}(x) = \sqrt[3]{x - 6}$ **114.** 3
115. 9 meters, 12 meters

Exercise Set 12.4

1. No. When we test a case, we are only able to prove that particular case. If there are an infinite number of cases to test, we will never be able to prove all are true by testing a finite number of cases. **3.** S_n is proven to be true for all n, if it is proved that S_1 is true and, if S_k is true then S_{k+1} is true.

5. Prove $1 + 5 + 9 + \ldots + (4n - 3) = n(2n - 1)$ for all positive integers n.

Condition 1: For $n = 1$,
$1 \overset{?}{=} 1(2(1) - 1)$
$1 \overset{?}{=} 1(2 - 1)$
$1 \overset{?}{=} 1(1)$
$1 = 1$ True

Condition 2:
Assume $1 + 5 + 9 + \ldots + (4k - 3) = k(2k - 1)$ for some positive integer k.
Then
$1 + 5 + 9 + \ldots + (4k - 3) + [4(k + 1) - 3] = k(2k - 1) + [4(k + 1) - 3]$
$1 + 5 + 9 + \ldots + (4k - 3) + [4(k + 1) - 3] = 2k^2 - k + 4k + 4 - 3$
$1 + 5 + 9 + \ldots + (4k - 3) + [4(k + 1) - 3] = 2k^2 + 3k + 1$
$1 + 5 + 9 + \ldots + (4k - 3) + [4(k + 1) - 3] = (k + 1)(2k + 1)$
$1 + 5 + 9 + \ldots + (4k - 3) + [4(k + 1) - 3] = (k + 1)(2k + 2 - 1)$
$1 + 5 + 9 + \ldots + (4k - 3) + [4(k + 1) - 3] = (k + 1)[2(k + 1) - 1]$

7. Prove $2 + 4 + 6 + \ldots + 2n = n(n + 1)$ for all positive integers n.

Condition 1: For $n = 1$,
$2 \overset{?}{=} 1(1 + 1)$
$2 \overset{?}{=} 1(2)$
$2 = 2$ True

Condition 2:
Assume $2 + 4 + 6 + \ldots + 2k = k(k + 1)$ for some positive integer k.
Then
$2 + 4 + 6 + \ldots + 2k + 2(k + 1) = k(k + 1) + 2(k + 1)$
$2 + 4 + 6 + \ldots + 2k + 2(k + 1) = k^2 + k + 2k + 2$
$2 + 4 + 6 + \ldots + 2k + 2(k + 1) = k^2 + 3k + 2$
$2 + 4 + 6 + \ldots + 2k + 2(k + 1) = (k + 1)(k + 2)$
$2 + 4 + 6 + \ldots + 2k + 2(k + 1) = (k + 1)[(k + 1) + 1]$

9. Prove $3 + 4 + 5 + \ldots + (n + 2) = \dfrac{n}{2}(n + 5)$ for all positive integers n.

Condition 1: For $n = 1$
$3 \overset{?}{=} \dfrac{1}{2}(1 + 5)$

$3 \overset{?}{=} \dfrac{1}{2}(6)$

$3 = 3$ True

Condition 2:
Assume $3 + 4 + 5 + \ldots + (k + 2) = \dfrac{k}{2}(k + 5)$ for some positive integer k.
Then
$3 + 4 + 5 + \ldots + (k + 2) + [(k + 1) + 2] = \dfrac{k}{2}(k + 5) + [(k + 1) + 2]$

$3 + 4 + 5 + \ldots + (k + 2) + [(k + 1) + 2] = \dfrac{k^2}{2} + \dfrac{5k}{2} + k + 3$

$3 + 4 + 5 + \ldots + (k + 2) + [(k + 1) + 2] = \dfrac{k^2 + 7k + 6}{2}$

$$3 + 4 + 5 + \ldots + (k + 2) + [(k + 1) + 2] = \frac{(k + 1)(k + 6)}{2}$$

$$3 + 4 + 5 + \ldots + (k + 2) + [(k + 1) + 2] = \frac{k + 1}{2}[(k + 1) + 5]$$

11. Prove $3 + 6 + 9 + \ldots + 3n = \frac{3n(n + 1)}{2}$ for all positive integers n.

Condition: For $n = 1$

$3 \overset{?}{=} \frac{3(1)(1 + 1)}{2}$

$3 \overset{?}{=} \frac{3(2)}{2}$

$3 = 3$ True

Condition 2:

Assume $3 + 6 + 9 + \ldots + 3k = \frac{3k(k + 1)}{2}$ for some positive integer k.

Then

$$3 + 6 + 9 + \ldots + 3k + 3(k + 1) = \frac{3k(k + 1)}{2} + 3(k + 1)$$

$$3 + 6 + 9 + \ldots + 3k + 3(k + 1) = \frac{3k^2}{2} + \frac{3k}{2} + 3k + 3$$

$$3 + 6 + 9 + \ldots + 3k + 3(k + 1) = \frac{3k^2 + 9k + 6}{2}$$

$$3 + 6 + 9 + \ldots + 3k + 3(k + 1) = \frac{3(k^2 + 3k + 2)}{2}$$

$$3 + 6 + 9 + \ldots + 3k + 3(k + 1) = \frac{3(k + 1)(k + 2)}{2}$$

$$3 + 6 + 9 + \ldots + 3k + 3(k + 1) = \frac{3(k + 1)[(k + 1) + 1]}{2}$$

13. Prove $5 + 10 + 15 + \ldots + 5n = \frac{5n(n + 1)}{2}$ for all positive integers n.

Condition 1:

$5 \overset{?}{=} \frac{5(1)(1 + 1)}{2}$

$5 \overset{?}{=} \frac{5(2)}{2}$

$5 = 5$ True

Condition 2:

Assume $5 + 10 + 15 + \ldots + 5k = \frac{5k(k + 1)}{2}$ for some positive integer k.

Then

$$5 + 10 + 15 + \ldots + 5k + 5(k + 1) = \frac{5k(k + 1)}{2} + 5(k + 1)$$

$$5 + 10 + 15 + \ldots + 5k + 5(k + 1) = \frac{5k^2}{2} + \frac{5k}{2} + 5k + 5$$

$$5 + 10 + 15 + \ldots + 5k + 5(k + 1) = \frac{5k^2 + 15k + 10}{2}$$

$$5 + 10 + 15 + \ldots + 5k + 5(k + 1) = \frac{5(k^2 + 3k + 2)}{2}$$

$$5 + 10 + 15 + \ldots + 5k + 5(k + 1) = \frac{5(k + 1)(k + 2)}{2}$$

$$5 + 10 + 15 + \ldots + 5k + 5(k + 1) = \frac{5(k + 1)[(k + 1) + 1]}{2}$$

15. Prove $1 + 2 + 2^2 + \ldots + 2^{n-1} = 2^n - 1$ for all positive integers n.

Condition 1:

$1 \overset{?}{=} 2^1 - 1$

$1 \overset{?}{=} 2 - 1$

$1 = 1$ True

Condition 2:

Assume $1 + 2 + 2^2 + \ldots + 2^{k-1} = 2^k - 1$ for some positive integer k.

Then

$1 + 2 + 2^2 + \ldots + 2^{k-1} + 2^{(k+1)-1} = 2^k - 1 + 2^{(k+1)-1}$

$1 + 2 + 2^2 + \ldots + 2^{k-1} + 2^{(k+1)-1} = 2^k - 1 + 2^k$

$1 + 2 + 2^2 + \ldots + 2^{k-1} + 2^{(k+1)-1} = 2^k + 2^k - 1$

$1 + 2 + 2^2 + \ldots + 2^{k-1} + 2^{(k+1)-1} = 2(2^k) - 1$

$1 + 2 + 2^2 + \ldots + 2^{k-1} + 2^{(k+1)-1} = 2^{k+1} - 1$

17. Prove $1 + 4 + 4^2 + \ldots + 4^{n-1} = \dfrac{4^n - 1}{3}$ for all positive integers n.

Condition 1:

$1 \stackrel{?}{=} \dfrac{4^1 - 1}{3}$

$1 \stackrel{?}{=} \dfrac{4 - 1}{3}$

$1 \stackrel{?}{=} \dfrac{3}{3}$

$1 = 1$ True

Condition 2:

Assume $1 + 4 + 4^2 + \ldots + 4^{k-1} = \dfrac{4^k - 1}{3}$ for some positive integer k.

Then

$1 + 4 + 4^2 + \ldots + 4^{k-1} + 4^{(k+1)-1} = \dfrac{4^k - 1}{3} + 4^{(k+1)-1}$

$1 + 4 + 4^2 + \ldots + 4^{k-1} + 4^{(k+1)-1} = \dfrac{4^k - 1}{3} + \dfrac{3 \cdot 4^k}{3}$

$1 + 4 + 4^2 + \ldots + 4^{k-1} + 4^{(k+1)-1} = \dfrac{4^k + 3 \cdot 4^k - 1}{3}$

$1 + 4 + 4^2 + \ldots + 4^{k-1} + 4^{(k+1)-1} = \dfrac{4^k(1 + 3) - 1}{3}$

$1 + 4 + 4^2 + \ldots + 4^{k-1} + 4^{(k+1)-1} = \dfrac{4^k(4) - 1}{3}$

$1 + 4 + 4^2 + \ldots + 4^{k-1} + 4^{(k+1)-1} = \dfrac{4^{k+1} - 1}{3}$

19. Prove $2 + 2^2 + 2^3 + \ldots + 2^n = 2^{n+1} - 2$ for all positive integers n.

Condition 1:

$2 \stackrel{?}{=} 2^{1+1} - 2$

$2 \stackrel{?}{=} 2^2 - 2$

$2 \stackrel{?}{=} 4 - 2$

$2 = 2$ True

Condition 2:

Assume $2 + 2^2 + 2^3 + \ldots + 2^k = 2^{k+1} - 2$ for some positive integer k.

Then

$2 + 2^2 + 2^3 + \ldots + 2^k + 2^{k+1} = 2^{k+1} - 2 + 2^{k+1}$

$2 + 2^2 + 2^3 + \ldots + 2^k + 2^{k+1} = 2^{k+1} + 2^{k+1} - 2$

$2 + 2^2 + 2^3 + \ldots + 2^k + 2^{k+1} = 2(2^{k+1}) - 2$

$2 + 2^2 + 2^3 + \ldots + 2^k + 2^{k+1} = 2^{(k+1)+1} - 2$

21. Prove $4 + 4^2 + 4^3 + \ldots + 4^n = 4\left(\dfrac{4^n - 1}{3}\right)$ for all positive integers n.

Condition 1:

$4 \stackrel{?}{=} 4\left(\dfrac{4^1 - 1}{3}\right)$

$4 \stackrel{?}{=} 4\left(\dfrac{4 - 1}{3}\right)$

$4 \stackrel{?}{=} 4\left(\dfrac{3}{3}\right)$

$4 = 4$ True

Condition 2:

Assume $4 + 4^2 + 4^3 + \ldots + 4^k = 4\left(\dfrac{4^k - 1}{3}\right)$ for some positive integer k.

Then

$4 + 4^2 + 4^3 + \ldots + 4^k + 4^{k+1} = 4\left(\dfrac{4^k - 1}{3}\right) + 4^{k+1}$

$4 + 4^2 + 4^3 + \ldots + 4^k + 4^{k+1} = \dfrac{4(4^k - 1)}{3} + \dfrac{3 \cdot 4 \cdot 4^k}{3}$

$4 + 4^2 + 4^3 + \ldots + 4^k + 4^{k+1} = \dfrac{4[(4^k - 1) + 3 \cdot 4^k]}{3}$

$4 + 4^2 + 4^3 + \ldots + 4^k + 4^{k+1} = \dfrac{4(4^k + 3 \cdot 4^k - 1)}{3}$

$4 + 4^2 + 4^3 + \ldots + 4^k + 4^{k+1} = \dfrac{4[4^k(1 + 3) - 1]}{3}$

$4 + 4^2 + 4^3 + \ldots + 4^k + 4^{k+1} = \dfrac{4(4^k \cdot 4 - 1)}{3}$

$4 + 4^2 + 4^3 + \ldots + 4^k + 4^{k+1} = 4\left(\dfrac{4^{k+1} - 1}{3}\right)$

23. Prove $1^3 + 2^3 + 3^3 + \ldots + n^3 = \left[\dfrac{n(n+1)}{2}\right]^2$ for all positive integers n.

Condition 1:

$1^3 \stackrel{?}{=} \left[\dfrac{1(1+1)}{2}\right]^2$

$1 \stackrel{?}{=} \left[\dfrac{1(2)}{2}\right]^2$

$1 \stackrel{?}{=} (1)^2$

$1 = 1$ True

Condition 2:

Assume $1^3 + 2^3 + 3^3 + \ldots + k^3 = \left[\dfrac{k(k+1)}{2}\right]^2$ for some positive integer k.

Then

$1^3 + 2^3 + 3^3 + \ldots + k^3 + (k+1)^3 = \left[\dfrac{k(k+1)}{2}\right]^2 + (k+1)^3$

$1^3 + 2^3 + 3^3 + \ldots + k^3 + (k+1)^3 = \dfrac{k^2(k+1)^2}{4} + \dfrac{4(k+1)^2(k+1)}{4}$

$1^3 + 2^3 + 3^3 + \ldots + k^3 + (k+1)^3 = \dfrac{(k+1)^2[k^2 + 4(k+1)]}{4}$

$1^3 + 2^3 + 3^3 + \ldots + k^3 + (k+1)^3 = \dfrac{(k+1)^2(k^2 + 4k + 4)}{2^2}$

$1^3 + 2^3 + 3^3 + \ldots + k^3 + (k+1)^3 = \dfrac{(k+1)^2(k+2)^2}{2^2}$

$1^3 + 2^3 + 3^3 + \ldots + k^3 + (k+1)^3 = \left[\dfrac{(k+1)[(k+1)+1]}{2}\right]^2$

25. Prove $\dfrac{1}{1\cdot 2} + \dfrac{1}{2\cdot 3} + \dfrac{1}{3\cdot 4} + \ldots + \dfrac{1}{n(n+1)} = \dfrac{n}{n+1}$ for all positive integers n.

Condition 1:

$\dfrac{1}{1\cdot 2} \stackrel{?}{=} \dfrac{1}{1+1}$

$\dfrac{1}{2} = \dfrac{1}{2}$ True

Condition 2:

Assume $\dfrac{1}{1\cdot 2} + \dfrac{1}{2\cdot 3} + \dfrac{1}{3\cdot 4} + \ldots + \dfrac{1}{k(k+1)} = \dfrac{k}{k+1}$ for some positive integer k.

Then

$\dfrac{1}{1\cdot 2} + \dfrac{1}{2\cdot 3} + \dfrac{1}{3\cdot 4} + \ldots + \dfrac{1}{k(k+1)} + \dfrac{1}{(k+1)[(k+1)+1]} = \dfrac{k}{k+1} + \dfrac{1}{(k+1)[(k+1)+1]}$

$\dfrac{1}{1\cdot 2} + \dfrac{1}{2\cdot 3} + \dfrac{1}{3\cdot 4} + \ldots + \dfrac{1}{k(k+1)} + \dfrac{1}{(k+1)[(k+1)+1]} = \dfrac{k[(k+1)+1]+1}{(k+1)[(k+1)+1]}$

$\dfrac{1}{1\cdot 2} + \dfrac{1}{2\cdot 3} + \dfrac{1}{3\cdot 4} + \ldots + \dfrac{1}{k(k+1)} + \dfrac{1}{(k+1)[(k+1)+1]} = \dfrac{k(k+2)+1}{(k+1)(k+2)}$

$\dfrac{1}{1\cdot 2} + \dfrac{1}{2\cdot 3} + \dfrac{1}{3\cdot 4} + \ldots + \dfrac{1}{k(k+1)} + \dfrac{1}{(k+1)[(k+1)+1]} = \dfrac{k^2 + 2k + 1}{(k+1)(k+2)}$

$\dfrac{1}{1\cdot 2} + \dfrac{1}{2\cdot 3} + \dfrac{1}{3\cdot 4} + \ldots + \dfrac{1}{k(k+1)} + \dfrac{1}{(k+1)[(k+1)+1]} = \dfrac{(k+1)^2}{(k+1)(k+2)}$

$\dfrac{1}{1\cdot 2} + \dfrac{1}{2\cdot 3} + \dfrac{1}{3\cdot 4} + \ldots + \dfrac{1}{k(k+1)} + \dfrac{1}{(k+1)[(k+1)+1]} = \dfrac{k+1}{k+2}$

$\dfrac{1}{1\cdot 2} + \dfrac{1}{2\cdot 3} + \dfrac{1}{3\cdot 4} + \ldots + \dfrac{1}{k(k+1)} + \dfrac{1}{(k+1)[(k+1)+1]} = \dfrac{k+1}{(k+1)+1}$

27. Prove $1 + \dfrac{1}{2} + \dfrac{1}{2^2} + \ldots + \dfrac{1}{2^{n-1}} = 2\left(1 - \dfrac{1}{2^n}\right)$ for all positive integers n.

Condition 1:

$1 \stackrel{?}{=} 2\left(1 - \dfrac{1}{2^1}\right)$

$1 \stackrel{?}{=} 2\left(1 - \dfrac{1}{2}\right)$

Condition 2:

Assume $1 + \dfrac{1}{2} + \dfrac{1}{2^2} + \ldots + \dfrac{1}{2^{k-1}} = 2\left(1 - \dfrac{1}{2^k}\right)$ for some positive integer k.

Then

$1 + \dfrac{1}{2} + \dfrac{1}{2^2} + \ldots + \dfrac{1}{2^{k-1}} + \dfrac{1}{2^{(k+1)-1}} = 2\left(1 - \dfrac{1}{2^k}\right) + \dfrac{1}{2^{(k+1)-1}}$

$$1 \stackrel{?}{=} 2\left(\frac{1}{2}\right)$$

$$1 = 1 \quad \text{True}$$

$$1 + \frac{1}{2} + \frac{1}{2^2} + \ldots + \frac{1}{2^{k-1}} + \frac{1}{2^{(k+1)-1}} = 2 - \frac{2}{2^k} + \frac{1}{2^k}$$

$$1 + \frac{1}{2} + \frac{1}{2^2} + \ldots + \frac{1}{2^{k-1}} + \frac{1}{2^{(k+1)-1}} = 2 - \frac{1}{2^k}$$

$$1 + \frac{1}{2} + \frac{1}{2^2} + \ldots + \frac{1}{2^{k-1}} + \frac{1}{2^{(k+1)-1}} = 2 - \frac{2}{2 \cdot 2^k}$$

$$1 + \frac{1}{2} + \frac{1}{2^2} + \ldots + \frac{1}{2^{k-1}} + \frac{1}{2^{(k+1)-1}} = 2 - \frac{2}{2^{k+1}}$$

$$1 + \frac{1}{2} + \frac{1}{2^2} + \ldots + \frac{1}{2^{k-1}} + \frac{1}{2^{(k+1)-1}} = 2\left(1 - \frac{1}{2^{k+1}}\right)$$

29. Prove $1 + \frac{1}{4} + \frac{1}{4^2} + \ldots + \frac{1}{4^{n-1}} = \frac{4}{3}\left(1 - \frac{1}{4^n}\right)$ for all positive integers n.

Condition 1:

$$1 \stackrel{?}{=} \frac{4}{3}\left(1 - \frac{1}{4^1}\right)$$

$$1 \stackrel{?}{=} \frac{4}{3}\left(1 - \frac{1}{4}\right)$$

$$1 \stackrel{?}{=} \frac{4}{3}\left(\frac{3}{4}\right)$$

$$1 = 1 \quad \text{True}$$

Condition 2:

Assume $1 + \frac{1}{4} + \frac{1}{4^2} + \ldots + \frac{1}{4^{k-1}} = \frac{4}{3}\left(1 - \frac{1}{4^k}\right)$ for some positive integer k.

Then

$$1 + \frac{1}{4} + \frac{1}{4^2} + \ldots + \frac{1}{4^{k-1}} + \frac{1}{4^{(k+1)-1}} = \frac{4}{3}\left(1 - \frac{1}{4^k}\right) + \frac{1}{4^{(k+1)-1}}$$

$$1 + \frac{1}{4} + \frac{1}{4^2} + \ldots + \frac{1}{4^{k-1}} + \frac{1}{4^{(k+1)-1}} = \frac{4}{3} - \frac{4}{3 \cdot 4^k} + \frac{1}{4^k}$$

$$1 + \frac{1}{4} + \frac{1}{4^2} + \ldots + \frac{1}{4^{k-1}} + \frac{1}{4^{(k+1)-1}} = \frac{4}{3} - \frac{4}{3 \cdot 4^k} + \frac{3}{3 \cdot 4^k}$$

$$1 + \frac{1}{4} + \frac{1}{4^2} + \ldots + \frac{1}{4^{k-1}} + \frac{1}{4^{(k+1)-1}} = \frac{4}{3} - \frac{1}{3 \cdot 4^k}$$

$$1 + \frac{1}{4} + \frac{1}{4^2} + \ldots + \frac{1}{4^{k-1}} + \frac{1}{4^{(k+1)-1}} = \frac{4}{3} - \frac{4}{3 \cdot 4 \cdot 4^k}$$

$$1 + \frac{1}{4} + \frac{1}{4^2} + \ldots + \frac{1}{4^{k-1}} + \frac{1}{4^{(k+1)-1}} = \frac{4}{3} - \frac{4}{3 \cdot 4^{k+1}}$$

$$1 + \frac{1}{4} + \frac{1}{4^2} + \ldots + \frac{1}{4^{k-1}} + \frac{1}{4^{(k+1)-1}} = \frac{4}{3}\left(1 - \frac{1}{4^{k+1}}\right)$$

31. Prove $1 + \frac{3}{4} + \frac{9}{16} + \ldots + \left(\frac{3}{4}\right)^{n-1} = 4\left[1 - \left(\frac{3}{4}\right)^n\right]$ for all positive integers n.

Condition:

$$1 \stackrel{?}{=} 4\left[1 - \left(\frac{3}{4}\right)^1\right]$$

$$1 \stackrel{?}{=} 4\left[1 - \frac{3}{4}\right]$$

$$1 \stackrel{?}{=} 4\left(\frac{1}{4}\right)$$

$$1 = 1 \quad \text{True}$$

Condition 2:

Assume $1 + \frac{3}{4} + \frac{9}{16} + \ldots + \left(\frac{3}{4}\right)^{k-1} = 4\left[1 - \left(\frac{3}{4}\right)^k\right]$ for some positive integer k.

Then

$$1 + \frac{3}{4} + \frac{9}{16} + \ldots + \left(\frac{3}{4}\right)^{k-1} + \left(\frac{3}{4}\right)^{(k+1)-1} = 4\left[1 - \left(\frac{3}{4}\right)^k\right] + \left(\frac{3}{4}\right)^{(k+1)-1}$$

$$1 + \frac{3}{4} + \frac{9}{16} + \ldots + \left(\frac{3}{4}\right)^{k-1} + \left(\frac{3}{4}\right)^{(k+1)-1} = 4 - \frac{4 \cdot 3^k}{4^k} + \left(\frac{3}{4}\right)^k$$

$$1 + \frac{3}{4} + \frac{9}{16} + \ldots + \left(\frac{3}{4}\right)^{k-1} + \left(\frac{3}{4}\right)^{(k+1)-1} = 4 - \frac{4 \cdot 3^k}{4^k} + \frac{3^k}{4^k}$$

$$1 + \frac{3}{4} + \frac{9}{16} + \ldots + \left(\frac{3}{4}\right)^{k-1} + \left(\frac{3}{4}\right)^{(k+1)-1} = 4 + \frac{-4 \cdot 3^k + 3^k}{4^k}$$

$$1 + \frac{3}{4} + \frac{9}{16} + \ldots + \left(\frac{3}{4}\right)^{k-1} + \left(\frac{3}{4}\right)^{(k+1)-1} = 4 + \frac{3^k(-4 + 1)}{4^k}$$

$$1 + \frac{3}{4} + \frac{9}{16} + \ldots + \left(\frac{3}{4}\right)^{k-1} + \left(\frac{3}{4}\right)^{(k+1)-1} = 4 + \frac{3^k(-3)}{4^k}$$

$$1 + \frac{3}{4} + \frac{9}{16} + \ldots + \left(\frac{3}{4}\right)^{k-1} + \left(\frac{3}{4}\right)^{(k+1)-1} = 4 - \frac{3^{k+1}}{4^k}$$

$$1 + \frac{3}{4} + \frac{9}{16} + \ldots + \left(\frac{3}{4}\right)^{k-1} + \left(\frac{3}{4}\right)^{(k+1)-1} = 4 - \frac{4 \cdot 3^{k+1}}{4 \cdot 4^k}$$

$$1 + \frac{3}{4} + \frac{9}{16} + \ldots + \left(\frac{3}{4}\right)^{k-1} + \left(\frac{3}{4}\right)^{(k+1)-1} = 4\left(1 - \frac{3^{k+1}}{4^{k+1}}\right)$$

$$1 + \frac{3}{4} + \frac{9}{16} + \ldots + \left(\frac{3}{4}\right)^{k-1} + \left(\frac{3}{4}\right)^{(k+1)-1} = 4\left[1 - \left(\frac{3}{4}\right)^{k+1}\right]$$

33. Prove $\left(\frac{3}{4}\right)^{n+1} < \frac{3}{4}$ for all positive integers n.

Condition 1:

$\left(\frac{3}{4}\right)^{n+1} \overset{?}{<} \frac{3}{4}$

$\left(\frac{3}{4}\right)^{1+1} \overset{?}{<} \frac{3}{4}$

$\left(\frac{3}{4}\right)^{2} \overset{?}{<} \frac{3}{4}$

$\frac{9}{16} < \frac{3}{4}$ True

Condition 2:

Assume $\left(\frac{3}{4}\right)^{k+1} < \frac{3}{4}$ for some positive integer k.

Then $\left(\frac{3}{4}\right)\left(\frac{3}{4}\right)^{k+1} < \left(\frac{3}{4}\right)\left(\frac{3}{4}\right)$

$\left(\frac{3}{4}\right)^{(k+1)+1} < \frac{9}{16} < \frac{3}{4}$

35. Prove $\left(\frac{1}{4}\right)^{n+1} < \frac{1}{4}$ for all positive integers n.

Condition 1:

$\left(\frac{1}{4}\right)^{n+1} \overset{?}{<} \frac{1}{4}$

$\left(\frac{1}{4}\right)^{1+1} \overset{?}{<} \frac{1}{4}$

$\left(\frac{1}{4}\right)^{2} \overset{?}{<} \frac{1}{4}$

$\frac{1}{16} \overset{?}{<} \frac{1}{4}$ True

Condition 2:

Assume $\left(\frac{1}{4}\right)^{k+1} < \frac{1}{4}$ for some positive integer k.

Then $\left(\frac{1}{4}\right) \cdot \left(\frac{1}{4}\right)^{k+1} \overset{?}{<} \frac{1}{4} \cdot \frac{1}{4}$

$\left(\frac{1}{4}\right)^{(k+1)+1} < \frac{1}{16} < \frac{1}{4}$

37. Prove $3^n \geq 2n + 1$ for all positive integers n.

Condition 1:

$3^n \overset{?}{\geq} 2n + 1$

$3^1 \overset{?}{\geq} 2(1) + 1$

$3 \overset{?}{\geq} 2 + 1$

$3 \geq 3$ True

Condition 2:

Assume $3^k \geq 2k + 1$ for some positive integer k.

$3 \cdot 3^k \geq 3(2k + 1)$

$3^{k+1} \geq 6k + 3$

$3^{k+1} \geq 4k + 2k + 2 + 1$

$3^{k+1} \geq 4k + 2(k + 1) + 1 \geq 2(k + 1) + 1$

39. Prove $n^2 + n$ is even for all positive integers n.

Condition 1:

Is $(n^2 + n) \div 2$ an integer when $n = 1$?

$(1^2 + 1) \div 2$

$(1 + 1) \div 2$

$2 \div 2$

1

Condition 2:

Assume $(k^2 + k) \div 2 = m$ for some integers k and m.

Then

$[(k + 1)^2 + (k + 1)] \div 2$

$(k^2 + 2k + 1 + k + 1) \div 2$

$(k^2 + 3k + 2) \div 2$

$(k^2 + k + 2k + 2) \div 2$

$(k^2 + k) \div 2 + (2k + 2) \div 2$

$m + 2(k + 1) \div 2$

$m + k + 1$

Since m is an integer and k is an integer, $m + k + 1$ is an integer.

Thus, $[(k + 1)^2 + (k + 1)] \div 2$ is an integer.

41. Prove $n^3 + 2n$ is divisible by 3 for all positive integers n.

Condition 1:

Is $(n^3 + 2n) \div 3$ an integer when $n = 1$?
$$[1^3 + 2(1)] \div 3$$
$$(1 + 2) \div 3$$
$$3 \div 3$$
$$1$$

Condition 2:

Assume $(k^3 + 2k) \div 3 = m$ for some integers k and m.

Then
$$[(k + 1)^3 + 2(k + 1)] \div 3$$
$$[(k + 1)^2(k + 1) + 2k + 2] \div 3$$
$$[(k^2 + 2k + 1)(k + 1) + 2k + 2] \div 3$$
$$(k^3 + k^2 + 2k^2 + 2k + k + 1 + 2k + 2) \div 3$$
$$(k^3 + 3k^2 + 5k + 3) \div 3$$
$$(k^3 + 3k^2 + 2k + 3k + 3) \div 3$$
$$(k^3 + 2k) \div 3 + 3k^2 \div 3 + 3k \div 3 + 3 \div 3$$
$$m + k^2 + k + 1$$

Since m is an integer and k is an integer, $m + k^2 + k + 1$ is an integer. Thus, $[(k + 1)^3 + 2(k + 1)] \div 3$ is an integer.

43. Prove $a + ar + ar^2 + \ldots + ar^{n-1} = \dfrac{a(1 - r^n)}{1 - r}$ for all positive integers n.

Condition 1:
$$a \stackrel{?}{=} \frac{a(1 - r^1)}{1 - r}$$
$$a \stackrel{?}{=} \frac{a(1 - r)}{1 - r}$$
$$a = a \quad \text{True}$$

Condition 2:

Assume $a + ar + ar^2 + \ldots + ar^{k-1} = \dfrac{a(1 - r^k)}{1 - r}$ for some positive integer k.

Then
$$a + ar + ar^2 + \ldots + ar^{k-1} + ar^{(k+1)-1} = \frac{a(1 - r^k)}{1 - r} + ar^{(k+1)-1}$$
$$a + ar + ar^2 + \ldots + ar^{k-1} + ar^{(k+1)-1} = \frac{a - ar^k}{1 - r} + \frac{ar^k(1 - r)}{1 - r}$$
$$a + ar + ar^2 + \ldots + ar^{k-1} + ar^{(k+1)-1} = \frac{a - ar^k + ar^k - ar^{k+1}}{1 - r}$$
$$a + ar + ar^2 + \ldots + ar^{k-1} + ar^{(k+1)-1} = \frac{a - ar^{k+1}}{1 - r}$$
$$a + ar + ar^2 + \ldots + ar^{k-1} + ar^{(k+1)-1} = \frac{a(1 - r^{k+1})}{1 - r}$$

47. $4x^2 + 12xy + 9y^2$ **48.** $25x^4 - 20x^2y + 4y^2$ **49.** $16x^2 + 8xy - 24x + y^2 - 6y + 9$ **50.** $8x^3 + 36x^2y + 54xy^2 + 27y^3$

Exercise Set 12.5

1. Answers will vary. **3.** 1 **5.** No, only factorials of nonnegative numbers can be found.
7. 13, the number of terms is one more than the exponent **9.** 10 **11.** 1 **13.** 1 **15.** 70 **17.** 28 **19.** $x^3 + 12x^2 + 48x + 64$
21. $8x^3 + 36x^2 + 54x + 27$ **23.** $a^4 - 4a^3b + 6a^2b^2 - 4ab^3 + b^4$ **25.** $243a^5 - 405a^4b + 270a^3b^2 - 90a^2b^3 + 15ab^4 - b^5$
27. $16x^4 + 16x^3 + 6x^2 + x + \dfrac{1}{16}$ **29.** $\dfrac{1}{16}x^4 - \dfrac{3}{2}x^3 + \dfrac{27}{2}x^2 - 54x + 81$ **31.** $x^{10} + 100x^9 + 4500x^8 + 120,000x^7$
33. $2187x^7 - 5103x^6y + 5103x^5y^2 - 2835x^4y^3$ **35.** $x^{16} - 24x^{14}y + 252x^{12}y^2 - 1512x^{10}y^3$ **37.** Yes, $4! = 4 \cdot 3!$
39. Yes, $(7 - 3)! = (7 - 3)(7 - 4)(7 - 5)! = 4 \cdot 3 \cdot 2!$ **41.** $m = n$ or $m = 0$ **43.** $x^8, 24x^7, 17,496x, 6561$
45. $(a + b)^n = \displaystyle\sum_{i=0}^{n} \binom{n}{i} a^{n-i}b^i$ **47.** $(0, 6)$ **48.** $(10, 4)$ **49.** 2, 10 **50.** $2x^3y^5\sqrt{30y}$ **51.** $f^{-1}(x) = \dfrac{x - 8}{2}$

Exercise Set 12.6

1. A permutation is any ordered arrangement of a given set of objects. **3.** $_nP_r$ is a symbol used to represent the number of arrangements of r objects that can be made from a set of n objects. **5.** Answers will vary. Possible answer: Find the product of all natural numbers less than or equal to n. **7.** Answers will vary. **9.** 5040 **11.** 11,880
13. 17,160 **15.** 7560 **17.** 1 **19.** 20 **21.** 120 **23.** 360 **25.** 3024 **27.** 1 **29.** 1 **31.** 10 **33.** 5 **35.** 35 **37.** 84 **39.** 1
41. 56 **43.** 1,000,000 **45.** 1,000,000,000 **47.** 1080 **49.** 210 **51.** 427,518,000 **53.** 40,320 **55.** 120 **57.** 3,628,800
59. 495 **61.** 18,564 **63.** 20 **65.** 56 **67.** 593,775 **69.** 20 **71.** 10,000,000,000 **73.** 720 **75. a)** Since the 3-number sequence is ordered this is more like a permutation problem. The fact that repetition is allowed prevents it from being a true permutation problem. **b)** 64,000 **c)** 59,280 **77. a)** 8,000,000 **b)** 6,400,000,000 **c)** 64,000,000,000,000
79. a) 3,628,800 **b)** 40,320 **c)** 576 **81.** 40 **83.** 40,600 **85.** 1200 **90.** 8 **91.** 0, 6 **92.** 5 **93.** 57

Exercise Set 12.7

1. Equally likely outcomes are outcomes that have the same chance of occurring.
3. Since either A will occur or A will not occur, $P(A) + P(\text{not } A) = 1$. **5.** If two events are mutually exclusive, they cannot occur simultaneously. **7.** All probabilities are between 0 and 1 inclusive. **9.** $\frac{1}{5}$ **11.** $\frac{2}{15}$ **13.** $\frac{4}{5}$ **15.** $\frac{2}{5}$ **17.** $\frac{1}{13}$

19. $\frac{1}{52}$ **21.** $\frac{1}{2}$ **23.** $\frac{12}{13}$ **25.** $\frac{2}{13}$ **27.** $\frac{7}{13}$ **29.** $\frac{4}{7}$ **31.** $\frac{3}{70}$ **33.** $\frac{3}{7}$ **35. a)** $\frac{1}{2}$ **b)** $\frac{1}{4}$ **c)** $\frac{1}{4}$ **d)** $\frac{3}{4}$ **e)** $\frac{3}{4}$

37. a) $\frac{1}{4}$ **b)** $\frac{1}{2}$ **c)** $\frac{1}{4}$ **d)** $\frac{3}{4}$ **e)** $\frac{1}{2}$ **39. a)** $\frac{1}{3}$ **b)** $\frac{1}{3}$ **c)** $\frac{1}{3}$ **d)** $\frac{2}{3}$ **e)** $\frac{2}{3}$ **41.** $\frac{4}{11}$ **43.** $\frac{6}{11}$ **45.** $\frac{7}{11}$ **47.** 0.7 **49.** 0.9

51. a) $\frac{1}{3}$ **b)** $\frac{2}{3}$ **c)** 2:1 **d)** 1:2 **53.** 5:1 **55.** 5:1 **57.** 12:1 **59.** 1:1 **61. a)** 999,999:1 **b)** 99,999:1

63. a) $\frac{7}{10}$ **b)** $\frac{3}{10}$ **65.** $\frac{7}{10}$ **67.** 2:3 **69.** 3:7 **72.**

$12x - 16y = 48$

73.

$P(x) = x^4 - 2x^3 + 4x^2 - 14x - 21$

74.

$f(x) = \dfrac{3x^2 - 9x - 12}{2x^2 - 10x + 8}$

75.

$\dfrac{(x+3)^2}{25} + \dfrac{(y-5)^2}{9} = 1$

Chapter Review Exercises

1. 3, 4, 5, 6, 7 **2.** $-1, 3, 9, 17, 27$ **3.** $4, 2, \frac{4}{3}, 1, \frac{4}{5}$ **4.** $\frac{1}{5}, \frac{2}{3}, \frac{9}{7}, 2, \frac{25}{9}$ **5.** 10 **6.** 2
7. $\frac{16}{81}$ **8.** 88 **9.** $s_1 = 9, s_3 = 33$ **10.** $s_1 = 6, s_3 = 29$ **11.** $s_1 = \frac{4}{3}, s_3 = \frac{227}{60}$ **12.** $s_1 = -9, s_3 = -10$ **13.** 32, 64, 128; $a_n = 2^n$
14. $-\frac{1}{3}, \frac{1}{9}, -\frac{1}{27}; a_n = (-1)^n(3^{4-n})$ **15.** $\frac{16}{5}, \frac{32}{5}, \frac{64}{5}; a_n = \frac{2^{n-1}}{5}$ **16.** $-4, -8, -12; a_n = 16 - 4n$ **17.** $5 + 8 + 13 = 26$
18. $3 + 8 + 15 + 24 = 50$ **19.** $\frac{1}{6} + \frac{4}{6} + \frac{9}{6} + \frac{16}{6} + \frac{25}{6} = \frac{55}{6}$ **20.** $\frac{1}{2} + \frac{2}{3} + \frac{3}{4} + \frac{4}{5} = \frac{163}{60}$ **21.** 27 **22.** 215 **23.** 108
24. 729 **25. a)** 10, 14, 18, 22 **b)** $p_n = 4n + 6$ **26. a)** 4, 10, 18, 28 **b)** $a_n = n(n+3) = n^2 + 3n$ **27.** 5, 7, 9, 11, 13.
28. $-12, -\frac{25}{2}, -13, -\frac{27}{2}, -14$ **29.** $\frac{1}{2}, -\frac{3}{2}, -\frac{7}{2}, -\frac{11}{2}, -\frac{15}{2}$ **30.** $-100, -\frac{499}{5}, -\frac{498}{5}, -\frac{497}{5}, -\frac{496}{5}$ **31.** 27 **32.** -4
33. $\frac{1}{2}$ **34.** 6 **35.** $s_8 = 112; d = 2$ **36.** $s_7 = -210; d = -6$ **37.** $s_7 = \frac{63}{5}; d = \frac{2}{5}$ **38.** $s_9 = -42; d = -\frac{1}{3}$ **39.** $-8, -4, 0, 4$;
$a_{10} = 28; s_{10} = 100$ **40.** $5, 2, -1, -4, a_{10} = -22; s_{10} = -85$ **41.** $\frac{5}{6}, \frac{3}{2}, \frac{13}{6}, \frac{17}{6}; a_{10} = \frac{41}{6}; s_{10} = \frac{115}{3}$ **42.** $-60, -55, -50, -45$;
$a_{10} = -15, s_{10} = -375$ **43.** $n = 12, s_{12} = 378$ **44.** $n = 8, s_8 = 28$ **45.** $n = 11; s_{11} = \frac{231}{10}$ **46.** $n = 14; s_{14} = 420$ **47.** 210
48. a) $30,000, $31,000, $32,000, $33,000 **b)** $a_n = 29,000 + 1000n$ **c)** $36,000 **d)** $385,000 **49.** 5, 10, 20, 40, 80
50. $-12, -6, -3, -\frac{3}{2}, -\frac{3}{4}$ **51.** $20, -\frac{40}{3}, \frac{80}{9}, -\frac{160}{27}, \frac{320}{81}$ **52.** $-100, -20, -4, -\frac{4}{5}, -\frac{4}{25}$ **53.** $\frac{2}{27}$ **54.** 960 **55.** 1944

56. $\dfrac{4}{243}$ **57.** 889 **58.** $-\dfrac{4305}{64}$ **59.** $\dfrac{585}{8}$ **60.** $\dfrac{37{,}969}{1215}$ **61.** $r = 2; a_n = 6(2)^{n-1}$ **62.** $r = 5; a_n = -4(5)^{n-1}$

63. $r = \dfrac{1}{3}; a_n = 7\left(\dfrac{1}{3}\right)^{n-1}$ **64.** $r = \dfrac{2}{3}; a_n = \dfrac{9}{5}\left(\dfrac{2}{3}\right)^{n-1}$ **65.** 6 **66.** $\dfrac{25}{6}$ **67.** -6 **68.** -18 **69.** 16 **70.** $\dfrac{21}{2}$ **71.** $\dfrac{25}{6}$ **72.** -12

73. $\dfrac{26}{99}$ **74.** $\dfrac{23}{37}$ **75.** $\$102{,}400$ **76. a)** $\approx\$2024.51$ **b)** $\approx\$2463.13$ **c)** $\approx\$24{,}041.29$ **77.** $\approx\$503.63$ **78.** 100 feet

79. Prove $2 + 5 + 8 + \ldots + (3n - 1) = \dfrac{n(3n + 1)}{2}$ for all positive integers n.

Condition 1: For $n = 1$

$2 \overset{?}{=} \dfrac{1[3(1) + 1]}{2}$

$2 \overset{?}{=} \dfrac{1(3 + 1)}{2}$

$2 \overset{?}{=} \dfrac{1(4)}{2}$

$2 \overset{?}{=} \dfrac{4}{2}$

$2 = 2$ True

Condition 2:

Assume $2 + 5 + 8 + \ldots + (3k - 1) = \dfrac{k(3k + 1)}{2}$ for some positive integer k.

Then

$2 + 5 + 8 + \ldots + (3k - 1) + [3(k + 1) - 1] = \dfrac{k(3k + 1)}{2} + [3(k + 1) - 1]$

$2 + 5 + 8 + \ldots + (3k - 1) + [3(k + 1) - 1] = \dfrac{3k^2 + k}{2} + 3k + 3 - 1$

$2 + 5 + 8 + \ldots + (3k - 1) + [3(k + 1) - 1] = \dfrac{3k^2 + k + 6k + 6 - 2}{2}$

$2 + 5 + 8 + \ldots + (3k - 1) + [3(k + 1) - 1] = \dfrac{3k^2 + 7k + 4}{2}$

$2 + 5 + 8 + \ldots + (3k - 1) + [3(k + 1) - 1] = \dfrac{(k + 1)(3k + 4)}{2}$

$2 + 5 + 8 + \ldots + (3k - 1) + [3(k + 1) - 1] = \dfrac{(k + 1)[3(k + 1) + 1]}{2}$

80. Prove $2^2 + 4^2 + 6^2 + \ldots + (2n)^2 = \dfrac{2n(2n + 1)(n + 1)}{3}$ for all positive integers n.

Condition 1: For $n = 1$

$2^2 \overset{?}{=} \dfrac{2(1)[2(1) + 1](1 + 1)}{3}$

$4 \overset{?}{=} \dfrac{2(2 + 1)(2)}{3}$

$4 \overset{?}{=} \dfrac{2(3)(2)}{3}$

$4 = 4$ True

Condition 2:

Assume $2^2 + 4^2 + 6^2 + \ldots + (2k)^2 = \dfrac{2k(2k + 1)(k + 1)}{3}$ for some positive integer k.

Then

$2^2 + 4 + 6^2 + \ldots + (2k)^2 + [2(k + 1)]^2 = \dfrac{2k(2k + 1)(k + 1)}{3} + [2(k + 1)]^2$

$2^2 + 4 + 6^2 + \ldots + (2k)^2 + [2(k + 1)]^2 = \dfrac{2k(2k^2 + 3k + 1)}{3} + (2k + 2)^2$

$2^2 + 4 + 6^2 + \ldots + (2k)^2 + [2(k + 1)]^2 = \dfrac{4k^3 + 6k^2 + 2k}{3} + \dfrac{3(4k^2 + 8k + 4)}{3}$

$2^2 + 4 + 6^2 + \ldots + (2k)^2 + [2(k + 1)]^2 = \dfrac{4k^3 + 6k^2 + 2k}{3} + \dfrac{12k^2 + 24k + 12}{3}$

$2^2 + 4 + 6^2 + \ldots + (2k)^2 + [2(k + 1)]^2 = \dfrac{4k^3 + 18k^2 + 26k + 12}{3}$

$2^2 + 4 + 6^2 + \ldots + (2k)^2 + [2(k + 1)]^2 = \dfrac{(4k^2 + 10k + 6)(k + 2)}{3}$

$2^2 + 4 + 6^2 + \ldots + (2k)^2 + [2(k + 1)]^2 = \dfrac{(2k + 2)(2k + 3)(k + 2)}{3}$

$2^2 + 4 + 6^2 + \ldots + (2k)^2 + [2(k + 1)]^2 = \dfrac{2(k + 1)[2(k + 1) + 1][(k + 1) + 1]}{3}$

81. Prove $1 \cdot 2 + 2 \cdot 3 + 3 \cdot 4 + \ldots + n(n + 1) = \dfrac{n(n + 1)(n + 2)}{3}$ for all positive integers n.

Condition 1: For $n = 1$,

$1 \cdot 2 \overset{?}{=} \dfrac{1(1 + 1)(1 + 2)}{3}$

$2 \overset{?}{=} \dfrac{1(2)(3)}{3}$

$2 = 2$ True

Condition 2:

Assume $1 \cdot 2 + 2 \cdot 3 + 3 \cdot 4 + \ldots + k(k + 1) = \dfrac{k(k + 1)(k + 2)}{3}$ for some positive integer k.

Then

$1 \cdot 2 + 2 \cdot 3 + 3 \cdot 4 + \ldots + k(k + 1) + (k + 1)[(k + 1) + 1] = \dfrac{k(k + 1)(k + 2)}{3} + (k + 1)[(k + 1) + 1]$

$1 \cdot 2 + 2 \cdot 3 + 3 \cdot 4 + \ldots + k(k + 1) + (k + 1)[(k + 1) + 1] = \dfrac{k(k^2 + 3k + 2)}{3} + (k + 1)(k + 2)$

$1 \cdot 2 + 2 \cdot 3 + 3 \cdot 4 + \ldots + k(k + 1) + (k + 1)[(k + 1) + 1] = \dfrac{k^3 + 3k^2 + 2k}{3} + \dfrac{3(k^2 + 3k + 2)}{3}$

$1 \cdot 2 + 2 \cdot 3 + 3 \cdot 4 + \ldots + k(k + 1) + (k + 1)[(k + 1) + 1] = \dfrac{k^3 + 3k^2 + 2k + 3k^2 + 9k + 6}{3}$

$1 \cdot 2 + 2 \cdot 3 + 3 \cdot 4 + \ldots + k(k + 1) + (k + 1)[(k + 1) + 1] = \dfrac{k^3 + 6k^2 + 11k + 6}{3}$

$1 \cdot 2 + 2 \cdot 3 + 3 \cdot 4 + \ldots + k(k + 1) + (k + 1)[(k + 1) + 1] = \dfrac{(k^2 + 3k + 2)(k + 3)}{3}$

$1 \cdot 2 + 2 \cdot 3 + 3 \cdot 4 + \ldots + k(k + 1) + (k + 1)[(k + 1) + 1] = \dfrac{(k + 1)(k + 2)(k + 3)}{3}$

$1 \cdot 2 + 2 \cdot 3 + 3 \cdot 4 + \ldots + k(k + 1) + (k + 1)[(k + 1) + 1] = \dfrac{(k + 1)[(k + 1) + 1][(k + 1) + 2]}{3}$

82. Prove $2 + 6 + 10 + \ldots + (4n - 2) = 2n^2$ for all positive integers n.

Condition 1: For $n = 1$

$2 \overset{?}{=} 2 \cdot 1^2$

$2 = 2$ True

Condition 2:

Assume $2 + 6 + 10 + \ldots + (4k - 2) = 2k^2$ for some positive integer k.

Then

$2 + 6 + 10 + \ldots + (4k - 2) + [4(k + 1) - 2] = 2k^2 + [4(k + 1) - 2]$

$2 + 6 + 10 + \ldots + (4k - 2) + [4(k + 1) - 2] = 2k^2 + 4k + 4 - 2$

$2 + 6 + 10 + \ldots + (4k - 2) + [4(k + 1) - 2] = 2k^2 + 4k + 2$

$2 + 6 + 10 + \ldots + (4k - 2) + [4(k + 1) - 2] = 2(k^2 + 2k + 1)$

$2 + 6 + 10 + \ldots + (4k - 2) + [4(k + 1) - 2] = 2(k + 1)(k + 1)$

$2 + 6 + 10 + \ldots + (4k - 2) + [4(k + 1) - 2] = 2(k + 1)^2$

83. $81x^4 + 108x^3y + 54x^2y^2 + 12xy^3 + y^4$ **84.** $8x^3 - 36x^2y^2 + 54xy^4 - 27y^6$ **85.** $x^9 - 18x^8y + 144x^7y^2 - 672x^6y^3$
86. $256a^{16} + 3072a^{14}b + 16{,}128a^{12}b^2 + 48{,}384a^{10}b^3$ **87.** 35   **88.** 40,320   **89.** 479,001,600   **90.** 32,432,400   **91.** 20
92. 66,045   **93.** 1120   **94.** 175   **95.** $\dfrac{1}{2}$ **96.** $\dfrac{1}{5}$ **97.** $\dfrac{4}{5}$ **98.** 0   **99.** $\dfrac{3}{10}$ **100.** 1   **101.** 1: 3   **102.** $\dfrac{4}{9}$

Practice Test **1. a)** This sequence is neither arithmetic or geometric because the terms do not differ by a constant amount nor by a common multiple. **b)** This sequence is arithmetic because the terms differ by -3.

c) This sequence is geometric because the terms differ by the multiple $-\frac{1}{2}$. **2.** $3, 1, \frac{5}{9}, \frac{3}{8}, \frac{7}{25}$ **3.** $s_1 = 5; s_3 = \frac{23}{2}$

4. $5 + 11 + 21 + 35 + 53 = 125$ **5.** 228 **6.** $a_n = \frac{1}{3} + \left(\frac{1}{3}\right)(n-1) = \frac{1}{3}n$ **7.** $a_n = 5(2)^{n-1}$ **8.** $12, 9, 6, 3$

9. $\frac{5}{8}, \frac{5}{12}, \frac{5}{18}, \frac{5}{27}$ **10.** 16 **11.** -32 **12.** 13 **13.** $\frac{512}{729}$ **14.** $\frac{39,063}{5}$ **15.** 9

16. Prove $5 + 10 + 15 + \ldots + 5n = \dfrac{5n(n+1)}{2}$ for all positive integers n.

Condition 1: For $n = 1$

$5 \overset{?}{=} \dfrac{5(1)(1+1)}{2}$

$5 \overset{?}{=} \dfrac{5(2)}{2}$

$5 = 5$ True

Condition 2:

Assume $5 + 10 + 15 + \ldots + 5k = \dfrac{5k(k+1)}{2}$ for some positive integer k.

Then

$5 + 10 + 15 + \ldots + 5k + 5(k+1) = \dfrac{5k(k+1)}{2} + 5(k+1)$

$5 + 10 + 15 + \ldots + 5k + 5(k+1) = \dfrac{5k^2 + 5k}{2} + 5k + 5$

$5 + 10 + 15 + \ldots + 5k + 5(k+1) = \dfrac{5k^2 + 5k + 10k + 10}{2}$

$5 + 10 + 15 + \ldots + 5k + 5(k+1) = \dfrac{5k^2 + 15k + 10}{2}$

$5 + 10 + 15 + \ldots + 5k + 5(k+1) = \dfrac{5(k^2 + 3k + 2)}{2}$

$5 + 10 + 15 + \ldots + 5k + 5(k+1) = \dfrac{5(k+1)(k+2)}{2}$

$5 + 10 + 15 + \ldots + 5k + 5(k+1) = \dfrac{5(k+1)[(k+1)+1]}{2}$

17. $x^4 + 8x^3y + 24x^2y^2 + 32xy^3 + 16y^4$ **18.** 24 **19.** 5040 **20.** 28 **21.** 330 **22.** $\dfrac{2}{3}$ **23.** $499{:}1$ **24.** $\$210{,}000$ **25.** $3{,}280{,}500$

Applications Index

Advertising and Marketing
TV advertisers, 16
Car manufacturers TV ads, 64
Cost of commercials during Super Bowl, 178
Newspaper preferences, 17
Ties sold, 662

Agriculture and Home Gardening
California farmland, 227
Cornfield, 110
Dairy farms, 434
Farm division, 103
Farm equipment, 806
Farms in U.S., 496
Fence dimension, 102
Fertilizer solutions, 142, 266
Flower garden, 765
Genetically modified crops, 159
Grass seed mixture, 112
Greatest area, 584
Irrigating crops, 435
Landscaping, 427
Lawn fertilizer, 266
Milk tank, 434
Mowing lawns, 113, 434
Picking string beans, 450
Planting a garden, 450
Planting flowers, 435
Plowing a field, 434
Price of mower, 339
Rectangle garden, 374
Rectangular corral, 573-574
Rectangular garden, 535, 548, 597, 769
Side of a garden, 506
Small orchard, 601
Square garden, 374
Sunflower seed mixture, 112
Tractors in a field, 433
Tree wires, 369
Triangular garden, 102
Vegetable garden, 374, 768-769, 778
Weed killer, 266
Wheat crop, 598

Animals
Animal diet, 267
Birdseed, 266
Cat food purchase, 678
Feeding manatees, 435
Snail races, 109
Snooty the manatee, 110
Trout in a lake, 662

Astronomy
Communications satellite orbit, 752
Diameter of minor planets, 644
Distance to Proxima Centauri, 59
Distance to the sun, 58
Earth days, 507
Earth orbit, 58
Gravity on the moon, 548
Launching rockets, 437-438
Light-year, 60
Milky Way, 58
Periods of planets, 427
Return to space, 295
Space shuttles, 143
Two rockets, 113-114
Voyager, 64

Automobile
Antifreeze, 595
Auto jack, 413
Car antitheft system, 101
Car rental comparison, 142
Depreciation rate, 204, 807

Gas mileage of Buick Park Avenue, 213
Gas mileage of Honda Civic, 213
Limousine purchase possibilities, 830
Motor vehicle registrations, 232
Porsche sales, 128
Price of cars, 338
Sport-utility vehicle sales, 645
Stopping distance on dry pavement, 535
Stopping distance on snow, 535
Stopping distance, 175, 446, 376
Truck bed area, 710
Truck rental, 104, 114, 163, 265
Value of a jeep, 619
Value of an SUV, 624

Aviation
Air speed with headwind and tailwind, 265, 295, 431, 558
Airline flight, 487
Airports, 59
Carry-on luggage dimensions, 127
Flying a plane, 450
Helicopter ride, 437
Helipad area, 88
Refueling a jet, 113
Trip to Hawaii, 113

Banking and Finance. *See also* Economics
Compound interest, 619-620, 663, 668
Currency exchange, 87-88
Interest compounded continuously, 658, 661, 668
Investments, 426-427, 806
Money conversion, 445
Mortgage payment, 446
Mutual fund accounts, 218-219
Mutual funds, 100
Retirement plans, 472
Simple and compound interest, 623
Simple interest, 174, 421, 769
Tax-free investment, 425

Biology
Bacteria, 472, 649-650, 806, 847
in a Petri dish, 624
Plant reproduction, 619, 624

Business
Avon conference, 267
Bicycle manufacturing, 269, 766-767
Bicycle cost 806
Bicycle profit, 189
Bicycle shop, 375
Book contract, 137
Book manuscripts, 828
Bottles of glue, 777
Calendar sale/purchase, 142, 678
Cell phones, 544-545
Chair manufacturing, 267
Computer manufacturing, 698-700
Computer exports, 177
Computer sales, 177
Demand for DVD players, 212
Demand for new sandwiches, 212
Dinner seminars, 102
Dry cleaning profit, 127
Electronics sale, 143
Employees sent to training session, 846
First-class mail, 127
Furniture manufacturing, 269
Going bankrupt, 837
Hot dog stand sales, 108
Hourly compensation, 227

Inflatable boat sizes, 262-263
Insurance rates, 231
Internet sales, 472
Janitorial service, 558
Labor union membership, 41
Life insurance, 227
Microwave quality control, 830
Milk carton machine, 113
Multiple-bank loans, 261-262
Natural gas use, 219-220
Office security, 829
Online travel revenue, 302
Online travel, 177-178
Packages, 795
Packing spaghetti, 111
Painting sales, 113
Parking garage, 127
Personal computer sales, 159
Photocopy costs, 268, 678
Planting flowers, 428-429
Plywood guarantee, 139
Presorted first-class mail, 127
Pretzel shop, 443
Price of dress, 339
Price reductions, 103
Printing checks, 434
Profit, 308, 445, 556, 557, 583, 601
Promotion to vice president, 840
Sale, 142
Self-publishing a book, 127
Selling products
batteries, 547, 582
brooms, 601
chairs, 547
lamps, 547
paintings, 103
tables, 598
toys, 659-660
watches, 547, 582
Shopping in Tampa Bay Area, 95-96
Silk plants sold, 375
Small business, 29
Smart handheld devices, 666
Soda pop production, 106-107
Soft drinks marketing, 661
Sound systems, 829
Store sale, 227
Supply of baby strollers, 213
Supply of kites, 212-213
Taxi operating costs, 189
Tire store profit, 184-185
Toy store sales, 185
UPS packages, 126-127
Videotape rental, 446
Working at a vineyard, 430

Chemistry/General Science
Acetic acid solution, 111
Acid solutions, 112, 295
Alcohol solution, 113
Carbon 14 dating, 472, 620, 624, 663
Hydrogen peroxide solution, 112, 269
Mass, 806
Mixing solutions, 297
Molecule action, 663
pH of a solution, 645
Pressure and volume, 445
Radiometric dating, 663
Salt solution, 145
Science experiment, 58
Speed of computers, 202
Sulfuric acid solution, 269
Trisodium phosphate solution, 259-261
Water acidity in pool, 128

Construction and Home Improvement
Bedroom, 535
Blacktopping a driveway, 535

Bookcase dimensions, 102
Building a brick wall, 445
Cabinet makers, 434
Concrete driveway volume, 88
Dance floor at gym, 768
Drilling a well, 557-558
Installing windows, 435
Painting a room, 433
Painting a wall, 433
Piece of wood, 769
Planning a sandbox, 88
Roofing a house, 435
Room in a house, 584
Sandbox construction, 102
Wading pool, 535

Demographics
Aging newlyweds, 247
Cable viewers, 265
Centenarians, 40, 625
Employment by gender, 65
Immigration, 171-172
Infant mortality rate, 496-497, 651
Life expectancy, 213-214, 550-551
Most populous countries, 59
Number of people below poverty threshold, 202-203
Population, 624
density, 78
growth/increase, 142, 623
of Jamestown, 79
of U.S., 806
of U.S. and world, 58
of world, 59, 662

Economics
American automakers' market share, 79
Break-even analysis, 426
Comparing debts, 128-129
Consumer credit, 620-621
Consumer Price Index, 227
Goods and Services in Consumer Price Index, 17
Gross Domestic Product, 58
Inflation, 310, 749, 845
Public debt per person, 54-55
Purchasing power of the dollar, 203
State distributions of federal funds, 28-29
Stock prices, 28
Supply and demand, 178
Tax collections in U.S., 55-56
Trade deficit with China, 177
U.S. poverty threshold, 202
Web browsers market, 279

Education
Averaging grades, 128
Calculating grades, 123-124, 129
Class of students, 840
College employment, 127
College identification numbers, 829
Doctorate degrees, 268
Drug free-schools, 310, 557, 582-583
Earning an A, 103
Enrollment, 582
Essay questions, 830
Faculty-student committee, 831
Grade point average, 556-557
Grades in class, 830
History honor society, 99-100
History tests, 830, 831
Learning retention, 644
Mathematics tests, 831
Minimum grade, 127
Passing grade, 127
Photocopying syllabi, 110
Physics exam average, 103

Education *(cont.)*
Remembering, 668
SAT scores, 223
Scholarships, 830
School enrollment, 556
Standardized test, 644
Test average, 384
Textbook search committee, 830

Electronics
Current flow, 269
Electric circuit, 663
Electrical resistance, 446
Impedance, 518
Resistors, 413, 449
Total resistance, 420, 426
Wattage rating, 446, 451, 601

Environmental and Earth Sciences
Carbon dioxide in atmosphere, 40
Clean air, 128
Depth of ocean trenches, 20
Earthquake shock waves, 109-110
Earthquakes, 630, 632, 642-643, 644, 645, 651
Environment, 624
Pollen sources, 101
Recycling aluminum, 624-625
Recycling plastic, 59

Food and Beverage
Banana splits, 830
Buffalo wings, 266
Candies in a stack, 795
Cereal testing, 831
Food purchase, 831
Coffee mixture, 111
Coffee mixture, 143, 266, 518
Horseradish sauces, 112
Ice cream containers, 88-89
Juice mixture, 266-267
Loaf of bread, 806
Meatloaf, 113
Milk fat concentrations, 266
Milk mixture, 113
Mixed nuts, 111, 146
Organically grown food, 41
Peanuts and cashews, 296
Pizza toppings, 826-827
Quiche Lorraine recipe, 266
Restaurant combinations, 826
Sandwich choices, 847
Trays of hors d'oeuvres, 831
Vinegar solutions, 111
World's bananas, 278

Geometry
Angle measures, 143
in triangle, 99, 102, 268, 269, 297, 332, 509
complementary, 99, 265
of polygon, 796
supplementary, 99, 265
Area(s), 316, 330-331, 338
of rectangle, 598
sum of, 358
of a circle, 175, 451, 535
of multiple rectangles, 788, 843
of rectangle, 572-573
of shaded region, 346, 745
of triangle, 443, 509
Balloon volume, 308
Circle area, 175, 451, 535
Circular steel blade, 375
Concentric circles, 746
Chords, 814
Cubes, 331, 355, 375
Currency dimensions, 769
Cylinder volume, 175
Inscribed triangle, 535, 745
Intersecting roads, 770
Living room dimensions, 597
Nodes, 814
Perimeter
of quadrilateral, 305
of multiple rectangles, 788, 843
Rectangle, 598
area, 174, 572-573
region, 548

Sail on sailboat, 769
Similar triangles, 418-419
Squares, 175, 382
Stairway railing, 102
Surface area and volume, 363, 536, 625, 626
Triangle and circle, 597
Triangle
angle measurements, 99, 102, 268, 269, 297, 332, 509
area, 443, 509
perimeter, 102
similar, 418-419
Triangular sign, 382
Triangular tent entrance, 367-368
Volume of a spherical fish tank, 496
Volume
of a cylinder, 175
of balloon, 308
of cubes, 331, 355

Government Budget
CDC spending, 94-95
NASA budget, 101
Public debt, 195

Health, Nutrition, and Medicine
Administering a drug, 439-440
Alternative remedies, 35-36
Antibiotic dosage, 445
Blood flowing in an artery, 506-507
Body mass index, 89
Bone strength, 101
Boys' lengths from 18 to 36 months, 214
Boys' weights, 214
Burning calories, 206-208, 231-232
Exercise, 267
Fat content, 265
Fitness center weight loss, 144
Girls' growth from birth to 36 months, 190
Girls' head circumference, 215
Health care expenditures, 222
Heart disease, 233-234
Heart rate, 202
Inclined treadmill, 212
Injected dye, 806
Lithotripter, 752
Medicare beneficiaries, 222
Medicine mixture, 108-109
Mortality from firearms and motor vehicles, 248
Prescribing a drug, 662
Sleeping babies, 78
Stress test, 90
Tetanus decline, 203
Treadmill, 212
Typhoid fever, 231
U.S. hospitals, 203
Weight for girls/boys from birth to 36 months, 662-663
Weight loss, 89

Law Enforcement
Car theft, 309
Homicides, 651
Traffic light, 840
Skid marks, 487

Marine
Motion of a wave, 507
Ocean salinity, 113
Sailboats, 403-404
Ships at sea, 105
Speed of the *Titanic*, 88
Submarine depth, 139
Submarine dive, 27
Traveling by boat, 450
Triangular sail, 374
Wave action, 463, 807

Military
Chief Warrant Officer pay, 213
Land mines, 268
Submarine personnel, 268

Miscellaneous
Ages, 295

Antifreeze solution, 113, 114
Arrangement of pictures, 831
Art gallery, 752
Auditorium seats, 794
Baltimore revenue, 234
Blue dye, 143
Books in bookcases, 830
Brass alloy, 267
Bucket capacity, 89
Build an engine, 558
Carpet dimensions, 601, 777
Castle at Chichén Itzá, 204
Catching up, 111
Cellular phone use, 41
CD purchase, 676
Chores, 847
Club members, 836
Club officers, 824, 829
Combination locks, 830-831
Comparing state sizes, 146
Confidence interval, 508
Cub Scouts badges, 16
Daily newspapers, 232
Diagonal of a suitcase, 506
Door prizes, 829
Elevators, 829
Federal food assistance, 222-223
Filling
a pool, 434
a tank, 435
a tub, 429-430
bottles, 408
Finalists in a contest, 846
Fitness books, 846
Flags for messages, 830
Fractions, 450
Gifts, 846
Glass thickness, 139
Glasses in a stack, 794-795
Greeting card purchase, 678
Guarneri del Gesù violin, 214
Incandescent bulbs, 103
ISBN numbers, 830
Ladder against a house, 506, 523
Latchkey kids, 41
Lavender oil, 266
Letters, 847
License plates, 821, 831, 832
Log stack, 794
Mail volume, 268
Making a box, 375
Making copies, 384
Minimum wage increase, 101
Money, 795-796, 802-803, 845
Money market account, 81-82, 104
Moving sidewalk, 436
New River George bridge, 58
New suit, 101
Newspaper readers, 41
Newspapers, 247
Octane ratings, 112
Odds of an event, 840
Oil leak, 598
Paintings in a museum, 829
Permutations of words, 830
Photos, 548, 847
Picture frame, 374
Playground, 557
Political campaign costs, 63
Postage stamp purchase, 729
Price, 339
Prizes, 829, 830
Pumping water, 435
Pyramid of oil barrels, 844
Rectangular carpet, 382
Rectangular coffee table, 374
Rectangular shed, 374
Room speakers, 375
Shampooing a carpet, 434
Silver alloy, 267
Seats in a row, 846
Security system, 829
Social security numbers, 829
Sources of electricity, 222
Spelling bee, 308
Stack of chips, 807
Stack of money, 807
Stacking CDs, 822-823
Tangent line, 204
Tapping glasses, 832

Teenage clothing, 15-16
Telephone numbers, 831
Teller machines, 829
Tossing coins, 841
Totem pole, 434
Triangular tent, 374
Tunnel, 746, 752, 753
U.S. adults who read a newspaper, 197-198
Value of an ATV, 624
Value of Manhattan, 662
Veterinary bills, 557
Video rentals, 729
Water heater repair, 840
Water through a fire hose, 487
Wedding reception, 831
Weight limit on elevator, 127
Window washers, 434, 451
Winning a contest, 847

Number Problems
Consecutive integers, 597
Consecutive numbers, 496
Doubling a penny, 52
Integers, 535
Maximum product, 584
Sum of even numbers, 795
Sum of numbers, 297, 795

Personal Finances
Balancing a portfolio, 90
Certificate of deposit, 88
Checking account balance, 27
Comparing
accounts, 89
investments, 82-83, 89
mortgage rates, 97, 101, 102
salary plans, 127
toy sales, 103
Determine the loan length, 88
Determine the loan rate, 88
Dinner bill, 103
Father and son investing, 89
Health insurance policy, 129
Insurance, 425
Investments, 266, 269, 536
bonus investment, 142, 269
inheritance investment, 111
Laundry costs, 100
Money, 795-796, 802-803
Money market account, 81-82, 104
Monthly compounding, 89
Ordering lunch, 101
Phone booth call fee, 144
Piggy bank savings, 190, 812
Retirement, 847
Salary and increases, 791-792, 796, 844, 845
Salary plus commission, 142, 189, 257-258, 795
Savings account interest, 89, 266, 295, 535, 598, 651, 802
Savings interest, 777
Simple interest, 769
loan, 80-81, 88
Two investments, 107-108, 111, 112, 143, 145
Waitress pay, 100
Working two jobs, 113
Variable annuities, 79

Physics
Auto jack, 413
Average acceleration, 426
Average velocity, 425
Baseball, 383, 571-572, 584
Baseball and football, 769
Bouncing ball, 795, 807
Electrostatic force, 443
Falling object, 507, 524, 598
Falling rock, 446
Firing a cannon, 583
Flare, 338
Force of attraction, 447
Forces on a beam, 269
Forces on a car, 507
Free fall, 450
Heating a metal cube, 548
Height of a cannonball, 368-369

Inclined plane, 310
Kinetic and potential energy, 522
Levers, 421-422
Light filtering through lake, 806
Lighting/illumination, 441-442, 446, 447, 496, 663
Magnitude of sound, 651
Mass, 738
Optics, 413, 420-421
    focal length, 426, 427, 449
Pendulum
    swings, 792, 795, 803, 806, 845
    period of a pendulum, 503, 506, 522
Ping-pong ball, 795, 806
Power gain of an amplifier, 651
Pressure of an object, 447
Projectile, 374
Quadratic means, 508
Radioactive decay, 651, 659, 663
Radioactive substance, 624
Radioisotope, 663
Sound pressure, 645
Speed of light, 522
Speed of sound, 507
Springs, 445, 523-524
    bouncing spring, 139
Strength of a board, 496
Swinging on a vine, 795
Tennis ball and snowball, 769
Throwing a ball, 545-546, 583, 601
Throwing a horseshoe, 548
Throwing a rock, 548
Throwing an object, 583, 598
Velocity, 382, 522, 663
    of an object, 463
Water sculpture, 374
Weight of an object, 446
Whispering Gallery, 752

Real Estate
Apartment rental fees, 101
Average sale price of homes, 37-38
Cost of a house, 556, 600
House purchase odds, 840
Median home sale price, 203
Renting office space, 583

Recreation and Entertainment
Balloon flights, 110
Bicycle rentals, 678
Billiard table, 752
Bungee jumping, 806

Concert tickets, 268
Dartboard area, 88
Deck of cards, 835
Ferris wheel, 745
Horse race, 840
Lotteries, 830, 832, 840
Outing in a park, 436
Pool table, 777
Race tracks, 829
Riding a horse, 437
Rock concert, 297
Roller coaster, 557
Rolling a die, 833-834, 837. 847
Quiniela, 830
Selecting a card, 836
Selling tickets, 583, 599
Spinner experiment, 839
Theme parks, 265
Trip in a park, 600
Visiting a resort, 436
Walt Disney World visit, 96-97

Sports and Fitness
Baseball cards, 821-822
Baseball diamond, 503
Baseball equipment sale, 146
Baseball game cost, 265
Batting orders, 831
Bicycle travel, 145
Boating trip, 436
Bowling alley rates, 119
Canoe speed with and against current, 257
Canoe trip, 598
Canoe weight limit, 144
Canoeing, 435
Canyon hiking, 110
Charity race, 110
Cheerleader pyramids, 794
Cross-country skiing, 256
Dog sled, 846
Drivers to swim meet, 82
Fitness center membership, 143
Football game, 436
Golf club membership, 100
Hiking distance around Bear Lake, 111
Ice hockey, 295
Ice rink, 769
Jogging, 143, 267, 452
Jump shot, 338
Little League, 266
Masters golf tournament, 169-171

Olympic medals, 101
Paddling a water bike, 432
Pitching a ball, 446
Race cars, 114
Radius of a basketball hoop, 506
Riding a bike, 558
Riding horses, 258-259
Rowing speed, 265
Runner, 558
Running a race, 446
Running home, 105-106
Sailboat race, 437
Sailboats, 437
Skiing, 436
Sports apparel, 583-584
Swimming pool filling, 143
Tennis ball, 446
Tennis club court fees, 101
Tennis tournament, 840
Tickets to hockey game, 664
Walking and jogging, 297
Walking speed, 662

Taxes
Average tax refunds, 65
Depreciation deduction, 651, 807
Estimated taxes, 28
European gas taxes, 101
Hotel tax rate, 101
Income taxes, 128
Missouri sales tax, 100
Tax returns, 267
Tax strategy, 112
Tax tables, 119-120

Temperature and Weather
Air conditioning, 175
Atmospheric pressure, 625
Atmospheric pressure, 668
Converting between Celsius and Fahrenheit, 175
Electric heater, 558
Extreme temperatures, 21-22, 28
Film Gold location, 27
Melting ice, 441, 451
Most unusual change, 27
Probability of rain, 840
Sauna temperature, 175
Temperature in car radiator, 556

Time and Distance
Airport trip, 111
Car carrier, 558

Cars, 645
Car travel, 267
Car trip, 432-433
Catching up, 111
Distance to Calais, 114
Distance, 189, 445
Garden of the Gods to Grand Junction, 558
Hiking distance around Bear Lake, 111
Map distance, 450
Meeting at a restaurant, 112
Moving vehicles, 497
Road construction, 267
Route 66, 112
Skyline drive, 436
Sump pump rates 112
Suspension bridge, 437
Texas driving, 267
Trains in Alaska, 143
Travel time, 598
Trip to work, 113
Two cars, 146
Two trains, 142-143
Walkie-talkie range, 111

Transportation
Amtrak expenses, 201-202
Auto train, 436
Bus pass, 100
Comparing transportation, 113
George Washington Bridge tolls, 100
Public transportation usage, 40
Railroad traffic, 63
Trains sharing a track, 110
Transportation, 436-437
Travel/traveling, 437, 557, 558

Utilities
Cellular antennas, 370
Digging a trench, 434
Electricity costs, 678-679
Electricity surcharge, 450
Long distance phone plans/charges, 94, 127, 268
Telephone pole, 382, 522
Water bill, 446-447
Wire from telephone pole, 506

Work Problems
Molding machine, 598
Working together, 433

Index

A

Absolute value
  defined, 18, 130
  equations and inequalities
      containing, 130-136
  evaluating, 18-19
  evaluating radicals using, 460-461
  geometric interpretation of, 130
  graphs of, 154
  summary of procedures for solving
      equations and inequalities
      containing, 136
  systems of linear equalities
      containing, 289-290
Addition
  associative property of, 24-25
  commutative property of, 24-25
  distributive property of, 24-25
  identity property of, 24-25
  inverse property of, 24-25
  nonlinear systems of equations
      using, 764-765
  of complex numbers, 512
  of polynomials, 304-305
  of radicals, 481-482
  of rational expressions, 396-398, 399-403
  of real numbers, 19-20
Addition method of solving systems of
      linear equations, 241-244
Addition property of equality, 70
Additive identity element, 24-25
Additive inverse(s), 24-25
  definition, 18
  using, 21-22
Algebraic expression, 7
Alternating sequence, 783
"And," 121
Angles
  complementary, 97-98, 265
  supplementary, 98, 265
Antilogarithm, 640-643
  finding on graphing calculator, 642
Application(s). See also Problem solving
  of algebra. See also Problem solving
    mixture problems, 107-109
    motion problems, 104-107
    problem solving procedure for,
        93-99
    translating a verbal statement into
        an algebraic expression or
        equation, 90-93
  of exponential functions, 619-622
  of geometric series, 802-804
  of logarithmic functions, 630
  of nonlinear systems of equations,
      765-767
  involving radical equations, 503
  of rational expressions, 403-404
  solving
    using factoring, 367-370
    using linear systems in three
        variables, 261-264
    using rational expressions, 420
    using systems of equations, 256-261
Argument of the logarithm, 633
Arithmetic sequences
  defined, 789
  finding common difference in, 789
  finding nth partial sum of, 790-792
  finding nth term of, 789-790
Arithmetic series, 790-792
Associative property, 24-25
Asymptotes, 406
  in graphs of functions, 713-719
  of hyperbolas, 754
Augmented matrix, 271
  using row transformation to rewrite,
      271-272
  writing in triangular form, 271

B

Base 10 logarithms, 639
Base of expression, 30
Binomial theorem, 817-819
Binomial(s). See also Polynomials(s)
  evaluating coefficients on graphing
      calculator, 818
  described, 301
  dividing a polynomial by, 323-325
  expansion of, 816-817
  factoring common binomial factor,
      334-335
  finding the square of, 315
  multiplying by a binomial, 313-314
  Pascal's triangle and, 816-817
Boundary values, 586
Boundary, 224
Boundedness theorem, 706-707

C

Calculator. See also Graphing
      calculator; Scientific
      calculator
  evaluating exponential expressions
      on, 31
  evaluating roots on, 33
  importance of learning to use, 6
  checking linear equation solutions
      by substitution, 72
Cartesian coordinate system
  defined, 148
  plotting points in, 148-149
Change of base formula, 655-656
Circle(s)
  center, 740
  defined, 740
  graphing, 740-741
    with center at the origin, 740-741
    with centers at (h, k), 741-742
Coefficient, 68
Collinear points, 150
Combinations
  defined, 824
  determining number of, 824-828
  formula for, 825
  vs. permutations, 824
Combined variations, 443
Common logarithms, 639-643
  antilogarithms, 640-641
  using graphing calculator to find,
      640
Common ratio, 797
Commutative property, 24-25
Complementary angles, 97-98, 265
Complex fractions
  defined, 408
  recognizing, 408-409
  simplifying
    by multiplying a common
        denominator, 409-410
    by simplifying the numerator and
        denominator, 410-411
Complex numbers
  adding and subtracting, 512
  defined, 510
  dividing, 513-514
  multiplying, 512-513
  recognizing, 510-511
Complex zeros of polynomial
  functions, 700-702
Composite functions, 604-607
Compound events, 835-837
Conditional equations, 74-75
Conic sections summary, 758. See also
      Ellipse(s); Hyperbola(s);
      Parabola(s)
Conjugate
  defined, 490
  of complex number, 513

using to rationalize the
      denominator, 490-492
Constant, 68
Constant of proportionality, 439
Constant value, 7
Continuous functions, 673-674
Contradictions, 74-75
Coordinates of a point, 149
Counting numbers, 7, 11
Counting principle, 821-822
Cramer's rule
  defined, 280
  for linear equations in three
      variables, 283-284
  for linear equations in two variables,
      280-281
Cube root(s), 32, 457-458
Cubic equations, 367
Cubic polynomials, 301

D

Decimals, 54
Decreasing functions, 672-673
Decreasing sequence, 783
Degree of a term, 68, 300
Denominator
  rationalizing, 488-490
    in an addition problem, 493
    using the conjugate, 490-492
Dependent variable, 163
Descartes, René, 148
Descartes' Rule of Signs, 692-694
Determinant
  of a matrix, 279
  minor, 281
Difference of functions, 215-217
Difference of two squares, 316
  formula, 351-352
Direct variation, 438-440
Discontinuous functions, 673-674
Discriminant
  of trinomials, 350
  using to determine the number of
      real solutions to a quadratic
      equation, 542-543
Distance problems, 105
  formula, 105, 189, 738-739
Distributive property
  expanded form of, 312-313
  of multiplication over addition, 24-25
Division
  of polynomials, 321-328
  of radical expressions with different
      indices, 493
  of rational expressions, 391-392
  of real numbers, 23-24
Domain, 164
  of a rational function, 387

E

Elements
  of a matrix, 270
  of a set, 7
Ellipse(s)
  area of, 748
  defined, 747
  graphing
    with center at origin, 747-748
    with centers at (h, k), 748-750
    with calculator, 750
Empty set, 7
End behavior, 680
Equality
  addition property of, 70
  multiplication property of, 70
  properties of, 67
Equation(s)
  concepts involved in solving, 75-76
  containing fractions, 73-74

cubic, 367
defined, 69
equivalent, 69
exponential and logarithmic, 646-649
  applications involving, 649-650
  using graphing calculator to solve,
      648-649
of form $|x| = |y|$, 135-136
of form $|x| = a, a > 0$, 130-131
of form $ax + b = c, a \neq 0$. See
      Linear equations
linear. See Linear equations
natural logarithmic and natural
      exponential, 656-660
nonlinear systems of, 762-767
nonlinear. See Nonlinear equations
polynomial. See Polynomial
      equations
proportions, 417-419
quadratic. See Quadratic equation(s)
radical. See Radical equation(s)
rational. See Rational equations
solution sets, 69
solutions, 69
third-degree, 367
variations, 438-443
with rational exponents, 563-564
writing in quadratic form, 559-562
Equivalent equations, 69
Even degree polynomial functions, 692
Event(s)
  compound, 835-837
  defined, 820
  determining odds against, 837-838
  determining probability event will
      not occur, 834-835
  determining probability of, 832-834
  mutually exclusive, 836
Experiment
  defined, 820
  number of possible outcomes of,
      820-822
  outcomes, 834
Exponent(s). See also Exponential
      expressions; Exponential
      functions
  defined, 30
  negative exponent rule, 43-44, 466
  power rule for, 45-46, 466
  product rule for, 312
  product rule for, 42, 466
  quotient role for, 42-43, 466
  raising a power to a power, 45-46
  raising a product to a power, 46
  raising a quotient to a power, 46-48
  rules of, 45-48, 466-468
  zero exponent rule, 44-45, 466
Exponents
Exponential equations, 656-660
Exponential expressions. See also
      Exponents
  changing from/to a radical
      expression, 464
  defined, 30
  evaluating, 29-31
  evaluating, on a calculator, 31
Exponential functions
  applications of, 619-622
  defined, 616-617
  graphing, 616-618
Expression(s)
  defined, 7
  evaluating
    on a graphing calculator, 36-37
    using order of operations, 34-35
  exponential, 29-31
  index of, 455
  radical, 455-457
  rational. See Rational expressions

with rational exponents, 468-469
root of, 455
simplifying, 68-69
terms of, 68
undefined, 35

F
Factor(s)/factoring, 29
common binomial factor, 334-335
difference of two squares, 351-352
expressions with rational
    exponents, 469
greatest common factor, 332-333
by grouping, 335-336
monomial from a polynomial,
    333-334
polynomials using a combination of
    techniques, 359-361
solving quadratic equations by,
    365-367
trinomials
    by substitution, 347-348
    of form $ax^2 + bx + c, a \neq 1$
        using grouping, 346-347
        using trial and error, 342-346
    of form $x^2 + bx + c$, 340-342
    perfect square, 352-354
    special formulas for, 351-355
    sum and difference of two cubes,
        354-355
using graphing calculator to check
    problems on, 345
using to solve applications, 367-370
Factorials, 815-816
Finite sequence, 782-783
Finite series, 783-784
First-degree equations, 150. *See also*
    Linear equations
Foci of ellipse, 747
FOIL method, 313-314
Formula(s), 80
area of a circle, 175, 553
area of ellipse, 748
area of a trapezoid, 85
average acceleration, 426
average velocity, 425
break-even analysis, 426
carbon dating, 663
Celsius to Fahrenheit
    conversion, 175
combinations, 825
compound interest, 532, 619
containing rational expressions,
    420-422
diagonal of a box, 554
difference of two cubes, 354
difference of two squares, 351-352
distance, 105, 189, 738-739
Doppler effect, 422
earthquake magnitude, 630
energy of an earthquake, 645
equivalent taxable yield, 425
exponential growth/decay, 658
illuminance of light source, 441-442
interest compounded
    continuously, 658
levers, 421-422
magnitude of sound, 651
melting ice, 441
midpoint, 739-740
motion, 104
Newton's law, 553
optics, 420-421
perfect square trinomials, 352-354
perimeter of a square, 175
period of a pendulum, 503
permutations, 823-824
power gain of an amplifier, 651
radiometric dating, 663
simple interest, 81, 174, 189
stopping distance, 175, 535
solving for a variable in, 83-86
solving for the square of a variable
    in, 553-555
sum of two cubes, 354
surface area of a right circular cone,
    554-555
volume of a cylinder, 175
Fraction bar, 34

Fraction(s)
complex
    defined, 408
    recognizing, 408-409
    simplifying, 409-411
sign of, 24
inequalities with, 117-118
solving equations containing, 73-74
Function(s)
applications in daily life, 169-172
composite
    defined, 604, 605
    finding, 604
composition of, 605
decreasing, 672-673
defined, 163, 164, 165
domain, 164
exponential
    applications of, 619-622
    defined, 616-617
    identifying natural, 652-653
finding composition of a function
    and its inverse, 612-613
graphs of
    asymptotes in, 713-719
    continuous curve, 673-674
    continuous, 673-674
    defined piecewise, 674
    discontinuous, 673-674
    examples of, 719-725
    gaps in, 673-674
    greatest integer, 675-677
    holes in, 673-674, 714, 715-716, 719
    piecewise, 674
    rational
        improper, 713
        on graphing calculator, 719, 724
        proper, 713, 716-717
        recognizing key features of,
            713-714
    step, 675-677
increasing, 672-673
inverse, 609-611
linear
    defined, 179
    graphing, 179-180
        using intercepts, 180-183
        on graphing calculator, 183
local maxima and minima, 673
logarithmic
    applications of, 630
    defined, 626
    graphs of, 627-630
    identifying natural, 652-653
notation, 167-169
on graphing calculator, 611-612
one-to-one, 607-608
operations on, 215-220
polynomial, 301-304, 316-317, 680-
    687. *See also* Polynomial(s)
quadratic
    factoring
        to find the x-intercepts of,
            370-372
        to find the x-intercepts of, on
        graphing calculator, 371-372
    graphing, 566-580
        using the axis of symmetry,
            vertex,
        and intercepts, 569-571
range, 164
rational
    defined, 386
    domains of, 386-388
    on a graphing calculator, 388
    real zeros of, 673-674
solving problems involving, 419
square root, 456-457
study applications of, 184-185
vertical line test of, 165-167

G
General term of the sequence, 782
Geometric formulas, 849-850
Geometric sequences, 796-804
    defined, 797
    finding common ratio in, 796-797
    finding nth partial sum of, 798-800
    finding nth term of, 797-798

Geometric series
    applications of, 802-804
    defined, 798-799
    finding sum of, 800-801
    series, infinite, 800
Geometry
    similar figures, 417-419
    formulas for, 849-850
Graph(s)
    absolute values, 154
    circles, 740-742
        with centers at (h, k), 741-742
    comparing graphs of exponential
        and logarithmic functions,
        629-630
    defined, 150
    drawing by plotting points, 149-152
    ellipses
        with centers at (h, k), 748-750
        with center at origin, 747-748
    exponential functions, 616-618
    functions, 165-167
    interpreting, 156-157
    linear equations, 150
        using slope and y-intercept,
            196-197
    linear inequalities in two variables,
        224-226
    logarithmic functions, 627-629
    nonlinear equations, 152-155
    parabolas of form
        $x = a(y - k)^2 + h$, 737-738
    polynomial functions, 302-304,
        680-687
    quadratic functions, 566-580
        using axis of symmetry, vertex,
            and intercepts, 569-571
    relations, 165-167
    translated, 191-192
    using slope-intercept form to
        construct models from,
        197-198
    x-intercepts of, 370-372
Graphing calculator. *See also*
    Calculators
    adding and subtracting rational
        expressions, 402-403
    antilogarithms, 641-642
    binomial coefficients, 818
    checking factoring problems, 345
    circle, 741
    common logarithms, 640
    ellipses, 750
    equations, 155-156
        involving natural logarithms and
            natural
    exponential functions, 660
    evaluating expressions, 36-37
    evaluating roots, 33-34
    factorials, 815-816
    factoring to find the x-intercepts of
        quadratic functions, 371-372
    functions on, 611-612, 618
    graphing calculator window, 155
    greatest integer functions, 677
    inequalities, 226
    linear equations in one variable, 74
    linear functions, 183
    matrices, 276, 285
    natural logarithms, 654
    operations on functions, 220
    permutations, 826
    point of intersection of two
        functions, 187-188
    polynomial functions, 303-304
    quadratic equations that are not
        factorable, 369
    rational functions, 719, 724
    real zeros of polynomial functions, 702
    roots or expressions with rational
        exponents, 468-469
    scientific notation, 56
    solving linear equations in one
        variable, 187-188
    solving or checking radical
        equations, 500
    TABLE feature, 156
    TRACE feature, 156
    ZOOM feature, 187-188

Graphing equations, of form $x = a$ and
    $y = b$, 184
Graphing linear functions, using
    intercepts, 180-183
Greatest common factor, 332-333
Greatest integer functions, 675-677
Grouping
    factoring polynomials by, 335-336
    factoring trinomials by, 347-348

H
Half-planes, 224
Holes, in graphs of functions, 673-674,
    714, 715-716, 719
Horizontal asymptotes, 713-714, 716-
    717, 719
Horizontal line test, 608
Hyperbola(s)
    center of, 753
    defined, 753
    graphing
        with center at origin, 753-757
        on graphing calculator, 757
Hypotenuse, 369

I
Identities, 74-75
Identity property, 24-25
Imaginary numbers, 510
Imaginary unit, 510
Improper rational functions, 713
Increasing functions, 672-673
Increasing sequence, 783
Independent events, 821
Independent variable, 163
Index
    of cube root, 457
    odd, 459
    of radical expressions, 455
    of root, 32
    of summation, 785
Inequality(ies)
    defined, 115
    endpoints in, 8
    of form $|x| < a, a > 0$, 131-132
    of form $|x| > 0$ or $|x| < 0$, 135
    of form $|x| > a$ or $|x| < a, a < 0$,
        134-135
    of form $|x| > a, a > 0$, 133-134
    graphing, 770-773
        second-degree, 770-773
    identifying and using, 7-9
    indicated on the number line, 9, 115
    symbols for, 115
    linear. *See* Linear inequalities
    quadratic. *See* Quadratic inequalities
    rational, 590-592
    solving, 586-587
        polynomial, 588-590
        systems graphically, 772-773
    symbols, 115
Infinite geometric series, 800
    finding sum of, 800-801
Infinite sequence, 782
Infinite series, 783-784
Infinity, 116
Integers, 11
Intermediate value theorem, 705-706
Intersection
    of sets, 10
    of solution of sets, 121
Inverse functions, 609-611
    finding composition of a function
        and its inverse, 612-613
Inverse logarithm, 640-643
Inverse property, 24-25
Inverse variation, 440-442
Irrational numbers, 11
Irrational zeros, 710-711

J
Joint variations, 442

L
Leading coefficient, 300
Leading term, 300
Least common denominator, 73-74
    in rational expressions, 398-399

Least common multiple, 73-74
Like radicals, 481
Like terms, 68
Line(s)
  boundary, 224
  finding slope of, 192-194
  parallel, 192, 208-211
  perpendicular, 208-211
  point-slope form of, 205-206
  vertical, 194
Linear equation(s), 150. *See also* Equations
  checking solutions by substitution, 72
  with decimals, 71-73
  of form $x = a$ and $y = b$, 184
  graphing using slope and y-intercept, 196-197
  with nested parentheses, 72-73
  in one variable, 70
    solving graphically, 186-188
  point-slope form of, 205
    using to construct models from graphs, 206-208
  properties of equality related to, 67
  slope-intercept form of, 191-198
  solving, 69-72
  solving for a variable in, 83-86
  standard form of, 180
  systems of
    Cramer's rule for solving, 280-281, 283-284
    in three variables
      geometric interpretation of, 252
      inconsistent and dependent, 252-253
      solving, 249-252
      using matrices, 273-275
    in two variables
      solving by substitution, 240-241
      solving graphically, 237-240
      solving with addition method, 241-244
      using matrices to solve, 271-273
    using to solve applications, 256-261
  types of, 74-75
Linear functions. *See* Function(s)
Linear inequalities
  compound
    involving "and," 120-124
    involving "or," 24-125
  constraints, 268
  defined, 224
  with fractions, 117-118
  graphing on a number line, 116
  linear programming problems, 288-289
  properties used to solve, 115
  solving, 115-116
  symbols for, 115
  systems of
    containing absolute value, 289-290
    solving graphically, 287-288
  two variables, 224-226
Linear polynomials, 301
Linear programming problems, 288-289
Local maximum functions, 673
Logarithm(s) Logarithms. *See also* Functions
  additional properties of, 636
  base 10, 639
  common
    defined, 639-640
    finding, 640
  converting from exponential form, 626-627
  defined, 626
  logarithmic and exponential equations, 646-650
  natural, 653
    equations involving, 656-660
    finding on calculator, 654
    finding using change of base formula, 655-656
  power rule for, 634-636
  product rule for, 633
  quotient rule for, 634

Logarithmic functions. *See* Function(s)
Lower bound for real zeroes of $P(x)$, 706-707
Lower limit of summation, 785

M
Main fraction line, 408
Major axis of ellipse, 747
Mathematical induction
  defined, 809
  principles of, 809
  proving statements using, 808-812
Mathematical models, 80
Matrix/matrices, 270
  augmented, 271
  evaluating a determinant of a 2 x 2, 279-280
  evaluating a determinant of a 3 x 3, 281-282
  solving systems of equations with, 271-275
  using graphing calculators to work with, 276, 285
Maximum and minimum values of parabola, 571
Midpoint formula, 739-740
Minor axis of ellipse, 747
Minor determinant, 281
Mixture problems, 107-109
Monomial(s). *See also* Polynomials
  described, 301
  dividing a polynomial by, 321-323
  factoring from a polynomial, 333-334
  multiplying by polynomial, 312-313
Motion problems, 104, 431-433
  using quadratic formula to solve, 551-552
Multiplication
  associative property of, 24-25
  commutative property of, 24-25
  of complex numbers, 512-513
  distributive property of, 24-25
  identity property of, 24-25
  inverse property of, 24-25
  of polynomials, 312-317
  of radicals, 483-485
  of rational expressions, 390-391
  of real numbers, 22-23
Multiplication property of equality, 70
Multiplicative
  identity element, 24-25
  inverse, 24-25
  property of zero, 23
Mutually exclusive events, 836

N
Natural numbers
  defined, 7
  set of, 11
Negative exponent rule, 43-44, 466
Negative slope, 194
Nonlinear equation graphs, 152-155
  avoiding common errors in, 154-155
Nonlinear systems of equations
  applications of, 765-767
  defined, 762
  solving
    using addition, 764-765
    graphically on calculator, 767
    with substitution method, 762-764
Nonlinear systems of inequalities, 770-773
Notation, function, 167-169
nth term of arithmetic sequence, 789-790
Null set, 7
Number problems, 430-431
Number(s)
  constant, 68
  complex, 510. *See also* Complex numbers
  converting from scientific notation to decimal form, 54
  counting, 7, 11
  imaginary, 510
  integers, 11
  irrational, 11
  natural, 7, 11
  prime, 398

  rational, 11
  subscripts, 82
  whole, 11
  writing in scientific notation, 52-53
Numerical coefficient, 68

O
Oblique asymptotes, 714, 717-719
Odds against an event, 837-838
One-to-one functions, 607-608
Opposites. *See* Additive inverses
"Or," 124
Order
  of the inequality, 115
  of operations, 34-35
Ordered pair(s), 149, 163
Origin, 148
Outcomes, 820-822, 834

P
Parabola(s)
  axis of symmetry
    defined, 567
    finding, 567-568
  defined, 566, 735
  determining number of x-intercepts of, 569
  determining opening of, 566-567
  graphing form $x = a(y - k)^2 + h$, 735-736
  maximum and minimum problems, 571-574
  symmetry in, 567
  translating, 574-577
  vertex of
    at (h, k), 737
    defined, 567
    finding, 567-568
  writing functions in the form $f(x) = a(x - h)^2 + k$, 577-579
Parallel lines, 192
Partial sums, 784-785
Pascal, Blaise, 816
Pascal's triangle, 816-817
Perfect cube, 473
Perfect square, 350
  defined, 473
  trinomials, 528
    factoring, 352-354
Permutations
  defined, 822
  determining number of, 822-824
  formula for, 823-824
  on graphing calculator, 826
  vs. combinations, 824
Piecewise functions, 674
Point-slope form, 205-206
  to construct models from graphs, 206-208
Polya, George, 95
Polynomial(s)
  adding and subtracting, 304-305
  and expanded form of the distributive property, 312-313
  defined, 300
  division of, 321-328
    by synthetic division, 326-327
  equations. *See also* Quadratic equations
    defined, 364
    degree of, 364
  factoring, 336. *See also* Factor(s)/Factoring
    using a combination of techniques, 359-361
  functions, 316-317
    evaluating, 301-302
    graphs of
      boundedness theorem, 706-707
      defined, 680
      and Descartes' Rule of Signs, 692-694
      determining maximum number of turning points in, 684
      determining real zeros, 704-707
      determining the x-intercepts of, 682-684
      end behavior, 680

      estimating irrational zeros, 710-711
      even degree, 692
      examples, 685-687
      intermediate value theorem, 705-706
      key features, 707-710
      multiplicity in, 683
      understanding, 302-304
      zeros of, 682-684, 691-702, 704-707
  inequalities, 588-590
  multiplying, 312-317
    by a polynomial, 314-315
    product of the sum and difference of the same two terms, 315-316
    using FOIL method, 313-314
  prime, 347
  and product rule for exponents, 312
  types of, 301
Positive slope, 194
Positive square root, 455-456
Power functions, 680-682
Power
  raising to, 45-46
  rule
    for exponents, 45-46, 466
    for logarithms, 634-636
Powers of i, 514-515
Prime number, 398
Prime polynomials, 347
Principal square root, 455-456
Probability
  an event will not occur, 834-835
  an event will occur, 832-834
  defined, 833
  formula, 833
  of a compound event, 835-837
Problem solving. *See also* Applications
  distance problems, 105
  mixture problems, 107-109
  motion problems, 104-107
  procedure for, 79-83
  using scientific notation in, 54-56
  using rational equations in, 428-433
Product
  of functions, 215-217
  raising to a power, 46
  of the sum and difference of the same two terms, 315-316
Product rule
  for exponents, 42, 466
    and polynomial multiplication, 312
  for logarithms, 633
  for radicals
    defined, 474
    simplifying radicals using, 473-478
Proper rational functions, 713
Properties of equality, 67
Proportions, 417
Pythagorean theorem, 369

Q
Quadrants, 148
Quadratic equation(s)
  defined, 364
  determining number of real solutions, 542-543
  solving
    by completing the square, 529-533
    using factoring, 365-367
    using quadratic formula, 532, 537-541
    using square root property, 526-528
    using zero-factor property, 364-365
  study applications using, 544-546, 549-555
  writing given the solutions, 541-542
Quadratic formula
  deriving, 537
  solving
    motion problems with, 551-552
    quadratic equations with, 532, 537-541
    work problems with , 552-553

Quadratic functions. *See* Function(s)
Quadratic in form equations
  defined, 559
  solving, 559-562
Quadratic inequalities
  defined, 585
  solution to, 585
  solving, 585-588
Quadratic polynomials, 301
Quotient of functions, 215-217
Quotient rule
  for exponents, 42-43, 466
  for logarithms, 634
  for radicals, 473, 478-479
Quotient, raising to a power, 46-48

R
Radical equation(s)
  containing one radical, 498-500
  containing two radical terms and a
    nonradical term, 501-502
  containing two radicals, 501
  solving applications involving, 503
  solving for a variable in a
    radicand, 504
  solving or checking with graphing
    calculator, 500
Radical expressions
  addition and subtraction of,
    481-482
  changing from/to an exponential
    expression, 464
  defined, 455
  dividing, 488-490
    with different indices, 493
  like, 481
  multiplying, 483-485
  simplifying, 464-466
    using product rule for, 473-478
  square roots of, 455-457
  understanding when an expression is
    simplified, 492
  unlike, 481
Radical sign, 31, 455
Radicand, 31
  variable in, 504
Range, 164
Rate of change, 195
Rational equation(s)
  checking solutions to, 415-417
  defined, 414
  problem-solving with, 428-433
  proportions, 417-419
  solving, 414-415
Rational exponents
  factoring expressions with, 469
  solving equations with, 563-564
Rational expression(s)
  application of, 403-404
  defined, 386
  dividing, 391-392
  finding least common denominator
    in, 398-399
  formulas containing, 420-422
  multiplying, 390-391
  simplifying, 388-390
  adding and subtracting
    with a common denominator,
      396-398
    with unlike denominators, 399-403
Rational functions. *See* Functions
Rational inequalities, 590-592
Rational numbers, 11
Rational zeros
  of polynomial functions, 694-700
  theorem, 694-695
  using synthetic division to
    determine, 695-700
Rationalizing denominator, 488-490
  defined, 488

in an addition problem, 493
  using the conjugate, 490-492
Real numbers
  adding, 19-20
  dividing, 23-24
  multiplying, 22-23
  sets of, 11, 12
  subtracting of, 21-22
  using properties of, 24-25
Real zeros
  of a function, 673
  of polynomial functions, 691-692,
    702, 704-707
Reciprocal, 24
Rectangular coordinate system, 148
Reflexive property, 67
Relations, 163
Remainder theorem, 327-328
Root of function, 183
Roots
  cube, 457-458
  even, 458-459
  finding with graphing calculator,
    468-469
  odd, 458-459
  square, 455-457
Row transformation, 271-272
Rules of exponents, 45-48
  applying to rational and negative
    exponents, 466-468
  summary, 48, 466

S
Sample space, 820
Scientific calculator
  finding natural logarithms, 655
  scientific notation on, 56
Scientific notation
  converting to decimal form, 54
  using for problem solving, 54-56
  writing numbers in, 52-53
Sense of the inequality, 115
Sequences
  arithmetic, 789-792
  defined, 781
  geometric, 797
  terms of, 781-783
  types of, 782-783
Series
  defined, 783
  writing, 783-784
Set(s)
  crisp, 13
  defined, 7
  elements of, 7
  empty, 7
  fuzzy, 13
  identifying, 7
  in roster form 7
  intersection of, 10
  null, 7
  of numbers, 11-13
  of real numbers, 11
  set builder notation, 9-10
  subsets, 12
  union of, 10
Sigma, 785
Sign graph, 585
Similar figures, 417-419
Simple interest formula, 81
Simplifying
  an expression, 68-69
  rational expressions, 388-390
Slope
  of a line, 192-194
  as a rate of change, 195
  of vertical line, 194
Slope-intercept form, 196
  using to construct models from
    graphs, 197-198

Smooth curve functions, 673
Solution
  to an equation, 69. *See also*
    Equation(s)
  set, 69
Square matrix, 270
  determinant of, 279
Square root(s), 455-457
  defined, 31
  evaluating, 31-32
  functions of, 456-457
  positive, 31
  principal, 31
Square root property, 526
Standard form of quadratic
  equations, 364
Step functions, 675-677
Study skills
  attitude, 2
  finding help, 5-6
  homework, 3
  preparation for and attendance at
    class, 2-4
  preparation for and taking
    examinations, 4-5
  time management, 4
Subscripts, 82
Subsets, 12
Substitution method
  of factoring trinomials, 347-348
  of solving nonlinear systems of
    equations, 762-764
  of solving systems of linear
    equations, 240-241
Subtraction
  of complex numbers, 512
  of polynomials, 304-305
  of radicals, 481-482
  of rational expressions, 396-398,
    399-403
  of real numbers, 21-22
Sum and difference of two cubes,
  354-355
Sum of functions
  finding, 215-217
  graphing, 217-220
Summation notation, 785-786
Supplementary angles, 98, 265
Symbol(s)
  equal to, 93
  fraction bar, 34
  greater than or equal to, 8
  greater than, 8
  inequality, 8, 115
  infinity, 116
  is not equal to, 8
  less than or equal to, 8
  less than, 8
  radical sign, 31, 455
  sigma, 785
Symmetric property, 67
Symmetry in parabolas, 567
Synthetic division, 326-328
Systems of equations
  consistent, 237
  dependent, 238, 252-253, 276
  inconsistent, 238, 252-253, 276
  linear. *See* Linear equation(s)
  solving on graphing
    calculator, 239
Systems of linear inequalities. *See*
  Linear inequalities

T
Tangent line, 204
Terms
  degree of, 300
  of an expression, 68
  leading, 300
  of polynomials, 300

Third-degree equations, 367
Transitive property, 67
Translated graphs, 191-192
Translations of parabolas, 574-577
Transverse axis, of hyperbola, 753
Trapezoid, area of, 85
Trial and error factoring, 342-346
Trial, 820
Trinomial(s)
  described, 301
  factoring
    a common factor, 342
    form $ax^2 + bx + c, a \neq 1$
      using grouping, 346-347
      using trial and error,
        342-346
    form $x^2 + bx + c$, 340-342
    perfect square, 352-354
    special formulas for, 351-355
    sum and difference of two cubes,
      354-355
    using substitution, 347-348
  perfect square, 352-354, 528

U
Undefined expressions, 35
Union
  of sets, 10
  of solution of sets, 124
Unlike radicals, 481
Unlike terms, 68
Upper bound for real zeros of $P(x)$,
  706-707
Upper limit of summation, 785

V
Variable(s)
  defined, 6, 163
  in a radicand, 504
Variations
  combined, 443
  defined, 438
  direct, 438-440
  inverse, 440-442
  joint, 442
  summary, 442
Vertex/vertices
  of hyperbolas, 753
  of a parabola, 567
Vertical asymptotes, 713-715, 719
Vertical lines, 194
Vertical line test, 165-167, 608

W
Whole numbers, 11
Work problems, 428-430
  using quadratic formula to solve,
    552-553

X
x-axis, 148
x-intercept, 180

Y
y-axis, 148
y-intercept, 180

Z
Zero(s)
  complex, 700-702
  in division of polynomials, 321
  exponent rule, 44-45, 466
  of a function, 183
  irrational, 710-711
  multiplicative property of, 23
  of a polynomial function, 682-684
  rational, 694-700
  real, 691-692, 702, 704-707
  slope, 194
Zero-factor property, 364-365

Photo Credits

Chapter 7   Roots, Radicals, and Complex Numbers

If n is even and $a \geq 0$: $\sqrt[n]{a} = b$ if $b \geq 0$ and $b^n = a$

If n is odd: $\sqrt[n]{a} = b$ if $b^n = a$

Rules of radicals

$\sqrt{a^2} = |a|$

$\sqrt{a^2} = a, \ a \geq 0$

$\sqrt[n]{a^n} = a, \ a \geq 0$

$\sqrt[n]{a} = a^{1/n}, \ a \geq 0$

$\sqrt[n]{a^m} = \left(\sqrt[n]{a}\right)^m = a^{m/n}, a \geq 0$

$\sqrt[n]{a}\,\sqrt[n]{b} = \sqrt[n]{ab}, \ a \geq 0, b \geq 0$

$\dfrac{\sqrt[n]{a}}{\sqrt[n]{b}} = \sqrt[n]{\dfrac{a}{b}}, a \geq 0, b \geq 0$

A radical is simplified when the following are all true:

1. No perfect powers are factors of any radicand.
2. No radicand contains a fraction.
3. No denominator contains a radical.

Complex numbers: numbers of the form $a + bi$.

Powers of i: $i = \sqrt{-1}, i^2 = -1, i^3 = -i, i^4 = 1$

Chapter 8   Quadratic Functions

Square Root Property:

If $x^2 = a$, where a is a real number, then $x = \pm\sqrt{a}$.

A quadratic equation may be solved by factoring, completing the square, or the quadratic formula

Quadratic Formula: $x = \dfrac{-b \pm \sqrt{b^2 - 4ac}}{2a}$

Discriminant: $b^2 - 4ac$

If $b^2 - 4ac > 0$ then equation has two distinct real number solutions.

If $b^2 - 4ac = 0$ then equation has a single real number solution.

If $b^2 - 4ac < 0$ then equation has no real number solution.

Parabolas

For $f(x) = ax^2 + bx + c$, the vertex of the parabola is

$\left(-\dfrac{b}{2a}, \dfrac{4ac - b^2}{4a}\right)$ or $\left(-\dfrac{b}{2a}, f\left(-\dfrac{b}{2a}\right)\right)$.

For $f(x) = a(x - h)^2 + k$, the vertex of the parabola is (h, k).

If $f(x) = ax^2 + bx + c, a > 0$, the function will have a minimum

value of $\dfrac{4ac - b^2}{4a}$ at $x = -\dfrac{b}{2a}$.

If $f(x) = ax^2 + bx + c, a < 0$, the function will have a maximum

value of $\dfrac{4ac - b^2}{4a}$ at $x = -\dfrac{b}{2a}$.

Chapter 9   Exponential and Logarithmic Functions

Composite function of function f with function g: $(f \circ g)(x) = f[g(x)]$

To find the **inverse function**, $f^{-1}(x)$, interchange all x's and y's and solve the resulting equation for y.

If $f(x)$ and $g(x)$ are inverse functions, then $(f \circ g)(x) = (g \circ f)(x) = x$.

Exponential function: $f(x) = a^x, a > 0, a \neq 1$

Logarithm: $y = \log_a x$ means $x = a^y, a > 0, a \neq 1$

Properties of Logarithms:

$\log_a xy = \log_a x + \log_a y$

$\log_a (x/y) = \log_a x - \log_a y$

$\log_a x^n = n \log_a x$

$\log_a a^x = x$

$a^{\log_a x} = x, x > 0$

Common Logarithms are logarithms to the base 10.

Natural Logarithms are logarithms to the base e, where $e \approx 2.7183$.

Antilogarithm: If $\log N = L$ then $N = $ antilog L

Change of base formula: $\log_a x = \dfrac{\log_b x}{\log_b a}$

Natural exponential function: $f(x) = e^x$

To solve Exponential and Logarithmic equations we also use these properties:

If $x = y$, then $a^x = a^y$

If $a^x = a^y$, then $x = y$

If $x = y$, then $\log x = \log y$ $(x > 0, y > 0)$

If $\log x = \log y$, then $x = y$ $(x > 0, y > 0)$

$\ln e^x = x$

$e^{\ln x} = x, x > 0$

$f(x) = a^x,$ $a > 1$	$f(x) = a^x,$ $0 < a < 1$	$f(x) = \log_a x,$ $a > 1$	$f(x) = \log_a x,$ $0 < a < 1$